Modern Active Filter Design

OTHER IEEE PRESS BOOKS

Adjustable Speed AC Drive Systems, *Edited by B. K. Bose*
Optical Fiber Technology, II, *Edited by C. K. Kao*
Protective Relaying for Power Systems, *Edited by S. H. Horowitz*
Analog MOS Integrated Circuits, *Edited by P. R. Gray, D. A. Hodges, and R. W. Brodersen*
Interference Analysis of Communication Systems, *Edited by P. Stavroulakis*
Integrated Injection Logic, *Edited by J. E. Smith*
Sensory Aids for the Hearing Impaired, *Edited by H. Levitt, J. M. Pickett, and R. A. Houde*
Data Conversion Integrated Circuits, *Edited by Daniel J. Dooley*
Semiconductor Injection Lasers, *Edited by J. K. Butler*
Satellite Communications, *Edited by H. L. Van Trees*
Frequency-Response Methods in Control Systems, *Edited by A.G.J. MacFarlane*
Programs for Digital Signal Processing, *Edited by the Digital Signal Processing Committee*
Automatic Speech & Speaker Recognition, *Edited by N. R. Dixon and T. B. Martin*
Speech Analysis, *Edited by R. W. Schafer and J. D. Markel*
The Engineer in Transition to Management, *I. Gray*
Multidimensional Systems: Theory & Applications, *Edited by N. K. Bose*
Analog Integrated Circuits, *Edited by A. B. Grebene*
Integrated-Circuit Operational Amplifiers, *Edited by R. G. Meyer*
Modern Spectrum Analysis, *Edited by D. G. Childers*
Digital Image Processing for Remote Sensing, *Edited by R. Bernstein*
Reflector Antennas, *Edited by A. W. Love*
Phase-Locked Loops & Their Application, *Edited by W. C. Lindsey and M. K. Simon*
Digital Signal Computers and Processors, *Edited by A. C. Salazar*
Systems Engineering: Methodology and Applications, *Edited by A. P. Sage*
Modern Crystal and Mechanical Filters, *Edited by D. F. Sheahan and R. A. Johnson*
Electrical Noise: Fundamentals and Sources, *Edited by M. S. Gupta*
Computer Methods in Image Analysis, *Edited by J. K. Aggarwal, R. O. Duda, and A. Rosenfeld*
Microprocessors: Fundamentals and Applications, *Edited by W. C. Lin*
Machine Recognition of Patterns, *Edited by A. K. Agrawala*
Turning Points in American Electrical History, *Edited by J. E. Brittain*
Charge-Coupled Devices: Technology and Applications, *Edited by R. Melen and D. Buss*
Spread Spectrum Techniques, *Edited by R. C. Dixon*
Electronic Switching: Central Office Systems of the World, *Edited by A. E. Joel, Jr.*
Electromagnetic Horn Antennas, *Edited by A. W. Love*
Waveform Quantization and Coding, *Edited by N. S. Jayant*
Communication Satellite Systems: An Overview of the Technology, *Edited by R. G. Gould and Y. F. Lum*
Literature Survey of Communication Satellite Systems and Technology, *Edited by J. H. W. Unger*
Solar Cells, *Edited by C. E. Backus*
Computer Networking, *Edited by R. P. Blanc and I. W. Cotton*
Communications Channels: Characterization and Behavior, *Edited by B. Goldberg*
Large-Scale Networks: Theory and Design, *Edited by F. T. Boesch*
Optical Fiber Technology, *Edited by D. Gloge*
Selected Papers in Digital Signal Processing, II, *Edited by the Digital Signal Processing Committee*
A Guide for Better Technical Presentations, *Edited by R. M. Woelfle*
Career Management: A Guide to Combating Obsolescence, *Edited by H. G. Kaufman*
Energy and Man: Technical and Social Aspects of Energy, *Edited by M. G. Morgan*
Magnetic Bubble Technology: Integrated-Circuit Magnetics for Digital Storage and Processing, *Edited by H. Chang*
Frequency Synthesis: Techniques and Applications, *Edited by J. Gorski-Popiel*
Literature in Digital Processing: Author and Permuted Title Index (Revised and Expanded Edition), *Edited by H. D. Helms, J. F. Kaiser, and L. R. Rabiner*
Data Communications via Fading Channels, *Edited by K. Brayer*
Nonlinear Networks: Theory and Analysis, *Edited by A. N. Willson, Jr.*
Computer Communications, *Edited by P. E. Green, Jr. and R. W. Lucky*
Stability of Large Electric Power Systems, *Edited by R. T. Byerly and E. W. Kimbark*
Automatic Test Equipment: Hardware, Software, and Management, *Edited by F. Liguori*
Key Papers in the Development of Coding Theory, *Edited by E. R. Berkekamp*
Technology and Social Institutions, *Edited by K. Chen*
Key Papers in the Development of Information Theory, *Edited by D. Slepian*

Modern Active Filter Design

Edited by

Rolf Schaumann
Associate Professor of Electrical Engineering
University of Minnesota

Michael A. Soderstrand
Associate Professor of Electrical and Computer Engineering
University of California, Davis

Kenneth R. Laker
Member of Technical Staff
Bell Telephone Laboratories

A volume in the IEEE Press Selected Reprint Series, prepared under the sponsorship of the IEEE Circuits and Systems Society.

IEEE PRESS

The Institute of Electrical and Electronics Engineers, Inc. New York

PRINTED IN THE UNITED STATES OF AMERICA

Sole Worldwide Distributor (Exclusive of IEEE):

JOHN WILEY & SONS, INC.
605 Third Ave.
New York, NY 10016

Wiley Order Numbers: Clothbound: 0-471-09734-9
Paperbound: 0-471-09733-0

IEEE Order Numbers: Clothbound: PC01420
Paperbound: PP01438

Library of Congress Cataloging in Publication Data
Main entry under title:

Modern active filter design.

(IEEE Press selected reprint series)
Bibliography: p.
Includes indexes.
1. Electric filters, Active. I. Schaumann, Rolf, 1941- . II. Soderstrand, Michael A., 1946- III. Laker, Kenneth R., 1946-
TK7872.F5M62 621.3815'324 81-2368
ISBN 0-87942-147-9 AACR2
ISBN 0-87942-148-7 (pbk.)

Contents

Preface

This is the second volume published within the IEEE Press Selected Reprint series that is concerned with active filters, i.e., filters comprised of capacitors, resistors and active devices, usually operational amplifiers. It has been nine years since the appearance of the first volume, "Active Inductorless Filters" edited by S. K. Mitra. Since that time, considerable progress has been made in all aspects of active filter theory and design; we felt, therefore, that both students and practicing engineers would benefit from a new volume that collects in one place the more important and useful papers concerned with analog active filters. The reprints in this book, as well as the bibliography at the end, start in 1971, the year of publication of S. K. Mitra's "Active Inductorless Filters." For earlier work the reader is referred to that volume.

As a result of the huge number of papers that have appeared during the last decade we had, of course, to eliminate numerous valuable contributions to the field. Although aided by careful reviews from colleagues and from the IEEE Circuits and Systems Society, the final choice was often difficult and influenced not only by page limitations but naturally also by our preferences and biases. We sincerely hope that no workers in this field feel slighted because their area of research and their contributions are not represented as prominently as they feel necessary. For any serious omission or oversight we apologize in advance.

Several areas of importance in active filter design are intentionally not included in the present volume. One deals with the properties of operational amplifiers, whose performance critically affects the operation of active filters. Our justification is the availability of a separate IEEE Press book, dealing specifically with operational amplifiers ("Integrated Circuit Operational Amplifiers," edited by R. G. Meyer). However, the effect of the most critical amplifier parameter, the frequency-dependent and finite gain, is addressed in several of the papers in this book (e.g., papers #1-3, 1-5, 2-2, 2-4). The second area deals with computer-aided design and analysis, design centering, tolerancing and optimization. Nevertheless, some aspects of these questions are touched upon in paper 4-1. It is our belief that by including three or four papers we could not have done justice to this important area; rather, a separate volume seems to be called for. Also, some aspects of this field are treated in the IEEE Press book "Computer-Aided Filter Design", edited by G. Szentirmai. Finally, no papers about approximation have been included because many excellent textbooks discussing this subject are available, and also no papers dealing with n-port synthesis because, for this book, filters are considered to be single-input single-output systems.

This volume contains 42 reprints, divided into six parts. Part 1 is concerned with general considerations that are of importance in active filter design: sensitivity analysis, effects of operational amplifiers, noise and dynamic range. Part 2 contains eight papers discussing various aspects of second-order filter blocks, that in many respects can be regarded as the workhorses of active filter design. Part 3 deals with the design of high-order filters, built using the three approaches that have proven successful in practice: cascade, multiple feedback, and component or operational simulation of lossless ladders. Part 4 treats some theoretical and practical aspects of the fabrication of active filters, centered on hybrid technology. The problem of completely monolithic implementation of continuous analog active filters, to date solved only for a few special cases, is addressed in Part 5. Finally, Part 6 contains five representative papers about the switched-capacitor approach, a method currently under intense investigation for building precision integrated low-frequency (sampled data) active filters.

To guide the reader, short introductions are provided for each section, highlighting the papers' important contributions and giving, where appropriate, crossreferences to other papers in the book, or to papers which could not be included. References in the papers and the extensive bibliography at the end of the book should ease and encourage further study.

Although not specifically written as a textbook, we feel that the topics addressed and their order should make this volume useful for short courses, seminars, and as a supplementary text for senior or graduate courses on analog active filters. In addition, students, instructors and practicing engineers should find it a useful reference work.

The preparation of this book has been aided greatly by the advice and constructive criticism of a number of our colleagues in the area of active filters. Among them, we would like to acknowledge particularly Prof. S. K. Mitra of the University of California, Santa Barbara, for his help and advice in the early phases of this volume, Prof. M. S. Ghausi of Oakland University, Michigan, who provided us with continuous encouragement throughout the book's preparation, and Prof. A. S. Sedra of the University of Toronto, whose careful review and especially thoughtful comments helped make this a better book. We are grateful to Messrs. D. G. Hawkins and B. A. Stevens of Bell Laboratories for preparing the camera-ready bibliography. The contents of the bibliography were obtained from the INSPEC data base. INSPEC is an information service provided by the IEE in Great Britain, and we are grateful for permission to publish this material. Finally, we would like to thank IEEE Press and its managing editor, Mr. W. R. Crone, for the rapid production of this volume.

R. Schaumann
M. A. Soderstrand
K. R. Laker

Part I
Design Considerations

This first group of eight papers addresses three important items of the many considerations that must be born in mind in the design of practical active filters: sensitivity, frequency limitations and dynamic range.

The first four papers are concerned with different questions of sensitivity and its minimization. Soderstrand and Mitra in *paper 1-1* use the classical definition of sensitivity of a function F to a parameter k, $S^F_k = (dF/F)/(dk/k)$, to investigate the behavior of third-order filter functions. The analysis and results have practical applications in those cases where a second-order filter is obtained after, ideally, a pole-zero cancellation in a third-order function and, more importantly, where an ideally second-order filter becomes of third order due to the frequency-dependence of operational amplifiers (comp. paper 2-2). The pole-frequency sensitivity $S^{\omega_o}_k$ and pole-Q sensitivity S^Q_k which are of great importance in the evaluation of second-order sections are expressed in terms of the more easily calculated sensitivities to the coefficients of the third-order polynomial. In *paper 1-2,* Fleischer presents a lucid discussion of sensitivity minimization of a single-amplifier second-order filter, showing trade-offs between active and passive sensitivities, optimum choice of parameters and amplifier requirements. Although treating a specific biquad, the one underlying the STAR of paper 2-3, the methods are sufficiently general to be applicable to other topologies, even multi-amplifier ones. In *paper 1-3,* Moschytz and Horn discuss how the effects of finite and frequency-dependent amplifier gain, and of its variations, can be reduced by minimizing the so-called gain-sensitivity product (GSP) and a number of useful and generally valid results are derived. The GSP measure was introduced in an earlier paper [1] as a general figure of merit; it has been used in a number of studies, e.g. papers 1-4, 2-1, 2-4, and has as one of its main advantages that it retains a well-defined finite value that can be minimized whereas the simple sensitivity often approaches zero or infinity.

Although providing many useful insights and general results, the above papers have as their main weakness that they assume changes in only one parameter. In reality, one must consider the variations of all parameters in a circuit and, in addition, observe any statistical interdependence between the variations. One of the first studies addressing this question is *paper 1-4* by Rosenblum and Ghausi. In this paper, a statistical multiparameter sensitivity measure is introduced that is calculated relatively easily, permits quite arbitrary component statistics, and can readily be optimized so that filters of minimum sensitivity can be designed and comparisons between optimized filters can be made. The measure, possibly in some modified form, has been used in many studies (see e.g. paper 3-2 and the references therein); the results have been found to be quite realistic and do compare well with the more complete Monte-Carlo analysis (comp. paper 4-1), although considerably less computation time is used.

In practice, most serious difficulties in the performance of active filters can be traced to amplifier limitations. For an excellent review paper discussing these and their origins, the reader is referred to [2]. In general, at all but the lowest frequencies, the most troublesome source of errors is the frequency-dependence of the amplifier gain A; rather than being constant as is assumed for first-order design and analysis of active filters, the gain should be modeled more accurately as $A(s) = GB/(s + \sigma)$ where GB is the gain-bandwidth product and σ is the -3 dB frequency (for inexpensive 741-type operational amplifiers, $GB \cong 2\pi \cdot 1$ MHz and $\sigma \cong 2\pi \cdot 10$ Hz).

To eliminate the resulting errors, various methods have been proposed, ranging from predistortion, as in paper 2-2 and [3], to passive (see e.g. [4]) and active compensation. The first two methods work well for limited applications, but suffer from the disadvantage that they cannot account for any post-design variations of amplifier parameters. Active compensation, although in principle more expensive and resulting in higher power consumption, avoids this difficulty by largely cancelling the effects of the frequency-dependent gain of one amplifier by that of a second—identical—amplifier, fabricated on the same IC chip. Several practical circuits (e.g. paper 2-6) use this principle which is analyzed in detail in *paper 1-5* by Brackett and Sedra. The authors discuss active compensation and the resulting improvement for the case of integrators which, of course, form the essential basic building blocks of many active filters of second and higher order. The principle presented can be used in numerous active filter applications.

The last topic addressed in the paper selection in this first chapter is dynamic range. The dynamic range of active filters is limited at the low end by resistor and amplifier noise generated in the filter and at the upper end by nonlinearities in the operational amplifiers, due to slew rate and/or power supply limits.

In *paper 1-6,* Trofimenkoff, Treleaven and Bruton consider the noise analysis of second-order filter sections in general terms and derive results that are very convenient for practical calculations and for comparisons of different circuits. A number of integrals that occur frequently in noise calculations are evaluated; with their help, noise calculations for a wide variety of biquadratic sections can readily be completed. Lee and Mecrow in *paper 1-7* present a simple readily evaluated expression for the noise of second-order sections, that is related to the sensitivity of the section transfer function to the amplifier gain. The results can be used to calculated the noise of cascaded filters and, further, to determine the ordering of sections in the cascade to reduce the output noise and thereby increase dynamic range. This dynamic range problem in cascade filters is also addressed in *paper 1-8* by Luder. The author develops an algorithm that assigns the poles and zeros to second-order sections and determines their ordering such that nonlinear distortion is minimized. The same problem in multiple-loop feedback topologies is addressed in paper 3-2 and in [5].

[1] G. S. Moschytz, "The gain-sensitivity product—A figure of merit for hybrid integrated filters using single operational amplifiers", IEEE J. Solid-State Circuits, vol. SC-6, pp. 103-110, June 1971.

[2] J. E. Solomon, "The monolithic op amp: A tutorial study", IEEE J. Solid-State Circuits, vol. SC-9, pp. 314-332, December 1974.

[3] A. Budak and D. M. Petrela, "Frequency limitations of active filters using operational amplifiers", IEEE Trans. Circuit Theory, vol. CT-19, No. 5, pp. 322-328, July 1972.

[4] G. Wilson, "Compensation of some operational amplifier based RC-active networks", IEEE Trans. Circuits and Systems, vol. CAS-23, No. 7, pp. 443-446, July 1975.

[5] C. F. Chiou and R. Schaumann, "Comparison of dynamic range properties of high-order active bandpass filters", IEE Proc., vol. 127, Pt. G, Electronic Circuits and Systems, No. 3, pp. 101-108, June 1980.

Sensitivity analysis of third-order filters†

M. A. SODERSTRAND‡ and S. K. MITRA
University of California, Davis, California 95616

[Received 28 July 1970]

Expressions relating Q and pole-sensitivities of a third-order filter to the readily calculable denominator polynomial coefficient sensitivities are derived. Use of these expressions for the sensitivity analysis of two new third-order active RC filters are illustrated.

1. Introduction

Active RC filters are usually designed as a cascade of second-order filter sections to minimize pole-sensitivities and also to facilitate post-design corrective adjustments. Many such second-order filters are primarily reduced third-order systems obtained by pole-zero cancellation. Examples of these type of filters are the networks designed using the phantom-zero approach (Kerwin and Shaffer 1968, Kerwin 1970) and sensitivity-compensation approach (Dagget and Vlach 1969).

It appears that no analytic method exists for evaluation of the performance of these third-order systems due to incremental change in network parameters. Here we advance such a method. First, the commonly used definitions for pole-Q and pole-frequency sensitivities are extended for the third-order case. Then, expressions are derived relating these quantities to the readily calculable coefficient sensitivities. Finally, these relations are applied for sensitivity analysis of two new band-pass active RC third-order filters.

2. Basic considerations

A second-order filter section is characterized by a transfer function of the form

$$t(s)=\frac{\alpha_2 s^2+\alpha_1 s+\alpha_0}{s^2+\beta_1 s+\beta_0}, \tag{1}$$

where $\beta_1^2 < 4\beta_0$ for complex poles. If a second-order filter is obtained from a third-order system by cancellation of a real pole and real zero, then the transfer function is nominally of the form

$$\begin{aligned} t(s) &= \frac{(s+\gamma)(\alpha_2 s^2+\alpha_1 s+\alpha_0)}{(s+\gamma)(s^2+\beta_1 s+\beta_0)} \\ &= \frac{(s+\gamma)(\alpha_2 s^2+\alpha_1 s+\alpha_0)}{s^3+\delta_2 s^2+\delta_1 s+\delta_0}. \end{aligned} \tag{2}$$

† Communicated by the Authors. To be presented at the Mexico 1971 International I.E.E.E. Conference on Systems, Networks and Computer, January 1971.

‡ Now at Sandia Laboratories, Livermore, California, U.S.A.

Reprinted with permission from *Int. J. of Electron.*, vol. 30, pp. 265-272, Mar. 1971.

For second-order filters, the two system parameters of most interest are the pole-Q and pole-frequency ω_n defined as:

$$Q = \sqrt{\beta_0}/\beta_1, \tag{3}$$

$$\omega_n = \sqrt{\beta_0}. \tag{4}$$

This implies that the denominator polynomial $D(s)$ of the transfer function can also be written as

$$D(s) = s^2 + \frac{\omega_n}{Q} s + \omega_n^2, \tag{5}$$

which alternately can be expressed as

$$D(s) = (s - p_0)(s - p_0^*) = s^2 - 2\sigma_0 s + (\sigma_0^2 + \omega_0^2), \tag{6}$$

where

$$p_0 = \sigma_0 + j\omega_0.$$

Extending the above to the third-order case we can thus write the denominator polynomial in four different forms as follows :

$$D(s) = s^3 + \delta_2 s^2 + \delta_1 s + \delta_0 \tag{7}$$

$$= (s+\gamma)(s^2 + \beta_1 s + \beta_0) \tag{8}$$

$$= (s+\gamma)\left(s^2 + \frac{\omega_n}{Q} s + \omega_n^2\right) \tag{9}$$

$$= (s+\gamma)(s - p_0)(s - p_0^*). \tag{10}$$

3. Sensitivity definitions (Mitra 1969)

Commonly used sensitivity indices are the pole-Q sensitivity and pole-frequency sensitivity defined as :

$$S_x^Q = \frac{d(\ln Q)}{d(\ln x)} = \frac{x}{Q} \cdot \frac{dQ}{dx}, \tag{11}$$

$$S_x^{\omega_n} = \frac{d(\ln \omega_n)}{d(\ln x)} = \frac{x}{\omega_n} \cdot \frac{d\omega_n}{dx}. \tag{12}$$

An alternate sensitivity index is the pole-sensitivity :

$$S_x^{p_0} = \frac{dp_0}{dx/x} \tag{13}$$

$$= S_x^{\sigma_0} + j\, S_x^{\omega_0}, \tag{14}$$

where

$$S_x^{\sigma_0} = \frac{d\sigma_0}{dx/x} \quad \text{and} \quad S_x^{\omega_0} = \frac{d\omega_0}{dx/x}.$$

The above sensitivities can be related to the readily calculable denominator coefficient sensitivities,

$$S_x^{\beta_i} = \frac{d\beta_i/\beta_i}{dx/x}, \tag{15 a}$$

$$S_x^{\delta_i} = \frac{d\delta_i/\delta_i}{dx/x}, \tag{15 b}$$

as illustrated next.

4. Relation between sensitivities

We now derive the relations between various sensitivities. Consider first the second-order case. Using definitions (11), (12) and (15) in eqns. (3) and (4), we easily obtain the following :

$$S_x^{\omega_n} = \tfrac{1}{2} S_x^{\beta_0}, \tag{16}$$

$$S_x^{Q} = \tfrac{1}{2} S_x^{\beta_0} - S_x^{\beta_1}. \tag{17}$$

Likewise, from eqns. (1), (6), (13) and (14) we obtain

$$S_x^{\sigma_0} = \frac{-\beta_1}{2} S_x^{\beta_1}, \tag{18}$$

$$S_x^{\omega_0} = \frac{1}{\sqrt{(4\beta_0 - \beta_1^2)}} \left[\beta_0 S_x^{\beta_0} - \frac{\beta_1^2}{2} S_x^{\beta_1} \right]. \tag{19}$$

Similarly, the pole sensitivities $S_x^{\sigma_0}$ and $S_x^{\omega_0}$ can be expressed in terms of the Q sensitivity and ω_n sensitivity as illustrated in eqns. (20) and (21):

$$S_x^{\sigma_0} = \frac{-\omega_n}{2Q} (S_x^{\omega_n} - S_x^{Q}), \tag{20}$$

$$S_x^{\omega_0} = \frac{\omega_n}{2Q} \left[[\sqrt{(4Q^2 - 1)}] S_x^{\omega_n} + \frac{S_x^{Q}}{\sqrt{(4Q^2 - 1)}} \right]. \tag{21}$$

Similar expressions can be derived for the third-order case. From eqns. (7) and (9) we have

$$\left.\begin{aligned} \delta_2 &= \frac{\omega_n}{Q} + \gamma, \\ \delta_1 &= \omega_n^2 + \frac{\omega_n \gamma}{Q}, \\ \delta_0 &= \omega_n^2 \gamma, \end{aligned}\right\} \tag{22}$$

from which one obtains

$$\begin{bmatrix} \dfrac{\omega_n}{\delta_2 Q} & -\dfrac{\omega_n}{\delta_2 Q} & \dfrac{\gamma}{\delta_2} \\ \left[2\omega_n^2 + \dfrac{\gamma \omega_n}{Q} \right] \dfrac{1}{\delta_1} & -\dfrac{\gamma\omega_n}{Q\delta_1} & \dfrac{\gamma\omega_n}{Q\delta_1} \\ \dfrac{2\omega_n^2 \gamma}{\delta_0} & 0 & \dfrac{\omega_n^2 \gamma}{\delta_0} \end{bmatrix} \cdot \begin{bmatrix} S_x^{\omega_n} \\ S_x^{Q} \\ S_x^{\gamma} \end{bmatrix} = \begin{bmatrix} S_x^{\delta_2} \\ S_x^{\delta_1} \\ S_x^{\delta_0} \end{bmatrix} \tag{23}$$

Equation (23) can be solved to yield the desired results :

$$\left.\begin{aligned} S_x^{\omega_n} &= \frac{(Q\gamma-\omega_n)\delta_0 S_x^{\delta_0} + Q\omega_n^2\delta_1 S_x^{\delta_1} - Q\omega_n^2\gamma\delta_2 S_x^{\delta_2}}{2\omega_n^2[Q(\omega_n^2+\gamma^2)-\omega_n\gamma]}, \\ S_x^Q &= \frac{\begin{gathered}\delta_0(2\omega_n Q^2+\gamma Q-\omega_n)S_x^{\delta_0}+\delta_1(\omega_n^2 Q-2\omega_n\gamma Q^2)S_x^{\delta_1} \\ +\delta_2\omega_n^2(Q\gamma-2\omega_n Q^2)S_x^{\delta_2}\end{gathered}}{2\omega_n^2[(\omega_n^2+\gamma^2)Q-\omega_n\gamma]}, \\ S_x^\gamma &= \frac{Q\left[\frac{\delta_0}{\gamma}S_x^{\delta_0}-\delta_1 S_x^{\delta_1}+\delta_2\gamma S_x^{\delta_2}\right]}{(\omega_n^2+\gamma^2)Q-\omega_n\gamma}. \end{aligned}\right\} \quad (24)$$

Examples

To illustrate the application of above results, consider the circuit of fig. 1 (Soderstrand 1969) which is similar to a circuit proposed elsewhere (Kerwin and Shaffer 1968). Analysis yields

$$\frac{V_2}{V_1} = \frac{KG_1(sC_3+G_2+G_3)s/C_1C_3(1-K)}{s^3+\delta_2 s^2+\delta_1 s+\delta_0}, \quad (25)$$

Fig. 1

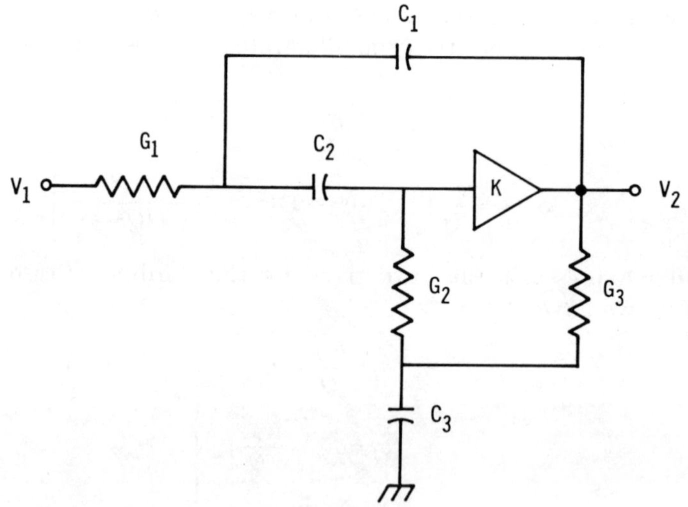

where

$$\left.\begin{aligned} \delta_2 &= \frac{C_1C_2(G_2+G_3)(1-K)+(C_1+C_2)G_2C_3+C_2G_1C_3}{C_1C_2C_3(1-K)}, \\ \delta_1 &= \frac{(C_1+C_2)G_2G_3(1-K)+C_2G_1(G_2+G_3)+C_3G_1G_2}{C_1C_2C_3(1-K)}, \\ \delta_0 &= \frac{G_1G_2G_3}{C_1C_2C_3}. \end{aligned}\right\} \quad (26)$$

If the normalized element values are chosen as

$$\left.\begin{aligned} &G_2=G_3=C_1=C_2=1,\\ &G_1=C_3=2,\\ &K=-(4Q-1), \end{aligned}\right\} \tag{27}$$

then the transfer function becomes

$$t_v=\frac{-\dfrac{4Q-1}{2Q}s(s+1)}{(s+1)\left(s^2+\dfrac{1}{Q}s+1\right)}=\frac{-\dfrac{4Q-1}{2Q}s}{s^2+\dfrac{1}{Q}s+1}. \tag{28}$$

Note that $\omega_n=\gamma=1$ and $\delta_2=\delta_1=1+1/Q$ and $\delta_0=1$. Substituting these in eqn. (24), we obtain

$$\left.\begin{aligned} S_x^{\omega_n}&=\frac{(Q-1)S_x^{\delta_0}+(Q+1)(S_x^{\delta_1}-S_x^{\delta_2})}{2(2Q-1)},\\ S_x^{Q}&=\frac{(Q+1)[S_x^{\delta_0}-S_x^{\delta_1}-S_x^{\delta_2}]}{2},\\ S_x^{\gamma}&=\frac{QS_x^{\delta_0}+(Q+1)(S_x^{\delta_2}-S_x^{\delta_1})}{(2Q-1)}. \end{aligned}\right\} \tag{29}$$

Consider now the calculation of these sensitivities with respect to the gain K of the amplifier. From eqn. (26) we calculate

$$\begin{aligned} &S_K^{\delta_0}=0,\\ &S_K^{\delta_2}=S_K^{\delta_1}=-\frac{(4Q-1)}{4Q(Q+1)}. \end{aligned} \tag{30}$$

Substitution of these values in eqn. (29) yields

$$\left.\begin{aligned} &S_K^{\omega_n}=S_K^{\gamma}=0,\\ &S_K^{Q}=1-\frac{1}{4Q}. \end{aligned}\right\} \tag{31}$$

Calculating the passive parameter sensitivities in the same way we obtain

$$\left.\begin{aligned} &S_{G_1}^{\omega_n}=\tfrac{1}{4}, \quad S_{G_2}^{\omega_n}=\tfrac{3}{8}, \quad S_{G_3}^{\omega_n}=\tfrac{3}{8},\\ &S_{C_1}^{\omega_n}=-\tfrac{3}{8}, \quad S_{C_2}^{\omega_n}=-\tfrac{3}{8}, \quad S_{C_3}^{\omega_n}=-\tfrac{1}{4},\\ &S_{G_1}^{Q}=\tfrac{1}{4}(2Q-1), \quad S_{G_2}^{Q}=-\tfrac{1}{8}(2Q+1), \quad S_{G_3}^{Q}=-\tfrac{1}{8}(2Q-3),\\ &S_{C_1}^{Q}=-\tfrac{1}{8}(2Q-3), \quad S_{C_2}^{Q}=-\tfrac{1}{8}(2Q+1), \quad S_{C_3}^{Q}=\tfrac{1}{4}(2Q-1),\\ &S_{G_1}^{\gamma}=\tfrac{1}{2}, \quad S_{G_2}^{\gamma}=\tfrac{1}{4}, \quad S_{G_3}^{\gamma}=\tfrac{1}{4},\\ &S_{C_1}^{\gamma}=-\tfrac{1}{4}, \quad S_{C_2}^{\gamma}=-\tfrac{1}{4}, \quad S_{C_3}^{\gamma}=-\tfrac{1}{2}. \end{aligned}\right\} \tag{32}$$

As a second example consider the circuit of fig. 2 which is a modified version of the previous circuit. Routine analysis yields for this circuit :

$$\frac{V_2}{V_1}=\frac{\dfrac{K_1K_2G_1}{(1-K_1K_2)C_1C_3}(sC_3+G_2+G_3)s}{s^3+\delta_2 s^2+\delta_1 s+\delta_0}, \tag{33}$$

Fig. 2

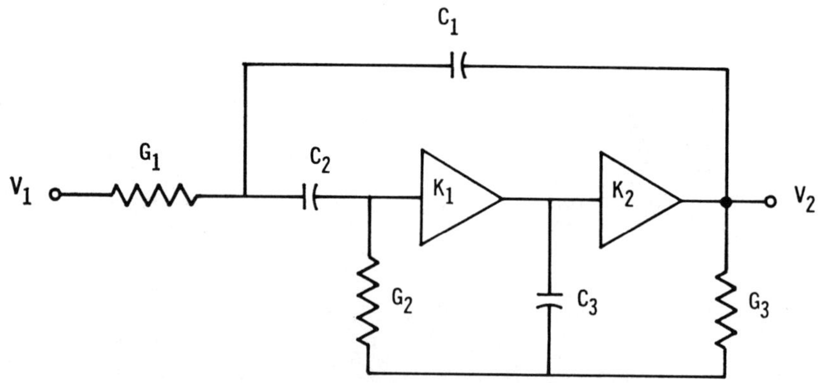

where

$$\left.\begin{aligned}
\delta_2&=\frac{G_2+G_3}{C_3}+\frac{G_2C_3(1-K_1)(C_1+C_2)+G_1C_2C_3}{(1-K_1K_2)C_1C_2C_3},\\
\delta_1&=\frac{(C_1+C_2)(1-K_1K_2)G_2G_3+C_2G_1(G_2+G_3)+G_1G_2C_3(1-K_1)}{(1-K_1K_2)C_1C_2C_3},\\
\delta_0&=\frac{G_1G_2G_3}{C_1C_2C_3}.
\end{aligned}\right\} \tag{34}$$

Normalized design values are :

$$\left.\begin{aligned}
&C_1=C_2=G_2=G_3=1,\\
&G_1=C_3=2,\\
&K_2=-[2Q(2-K_1)-1]/K_1,
\end{aligned}\right\} \tag{35}$$

for which the transfer function reduces to :

$$\begin{aligned}
\frac{V_2}{V_1}&=\frac{\dfrac{1-2Q(2-K_1)}{Q(2-K_1)}s(s+1)}{s^3+\left(1+\dfrac{1}{Q}\right)s^2+\left(1+\dfrac{1}{Q}\right)s+1}\\
&=\frac{\dfrac{1-2Q(2-K_1)}{Q(2-K_1)}s}{s^2+\dfrac{1}{Q}s+1}.
\end{aligned} \tag{36}$$

This implies that $\gamma = \omega_n = 1$. Thus the sensitivities may be calculated using eqn. (29), leading to :

$$\left.\begin{aligned}
&S_{K_1}{}^{\gamma}=0, && S_{K_2}{}^{\gamma}=0,\\
&S_{G_1}{}^{\gamma}=\frac{(2Q-1)-K_1(Q-1)}{(2Q-1)(2-K_1)}, && S_{C_1}{}^{\gamma}=-\frac{(2Q-1)-K_1(Q-1)}{2(2Q-1)(2-K_1)},\\
&S_{G_2}{}^{\gamma}=\frac{(2Q-1)-K_1Q}{2(2Q-1)(2-K_1)}, && S_{C_2}{}^{\gamma}=-\frac{(2Q-1)-K_1(Q-1)}{2(2Q-1)(2-K_1)},\\
&S_{G_3}{}^{\gamma}=\frac{(2Q-1)-K_1Q}{2(2Q-1)(2-K_1)}, && S_{C_3}{}^{\gamma}=-\frac{(2Q-1)-K_1Q}{(2Q-1)(2-K_1)},\\
&S_{K_1}{}^{\omega_n}=0, && S_{K_2}{}^{\omega_n}=0,\\
&S_{G_1}{}^{\omega_n}=\frac{(2Q-1)-K_1Q}{2(2-K_1)(2Q-1)}, && S_{C_1}{}^{\omega_n}=-\frac{3(2Q-1)-K_1(3Q-1)}{4(2Q-1)(2-K_1)},\\
&S_{G_2}{}^{\omega_n}=\frac{3(2Q-1)-K_1(3Q-2)}{4(2Q-1)(2-K_1)}, && S_{C_2}{}^{\omega_n}=-\frac{3(2Q-1)-K_1(3Q-1)}{4(2Q-1)(2-K_1)},\\
&S_{G_3}{}^{\omega_n}=\frac{3(2Q-1)-K_1(3Q-2)}{4(2Q-1)(2-K_1)}, && S_{C_3}{}^{\omega_n}=-\frac{(2Q-1)-K_1(Q-1)}{2(2Q-1)(2-K_1)},\\
&S_{K_1}{}^{Q}=\frac{4Q-1}{2Q(2-K_1)}, && S_{K_2}{}^{Q}=1-\frac{1}{2Q(2-K_1)},\\
&S_{G_1}{}^{Q}=\frac{(2Q-1)-K_1Q}{2(2-K_1)}, && S_{C_1}{}^{Q}=-\frac{(2Q-3)-K_1(Q-1)}{4(2-K_1)},\\
&S_{G_2}{}^{Q}=-\frac{(2Q+1)-K_1(Q+2)}{4(2-K_1)}, && S_{C_2}{}^{Q}=-\frac{(2Q+1)-K_1(Q-1)}{4(2-K_1)},\\
&S_{G_3}{}^{Q}=-\frac{(2Q-3)-K_1(Q-2)}{4(2-K_1)}, && S_{C_3}{}^{Q}=\frac{(2Q-1)-K_1(Q-1)}{2(2-K_1)}.
\end{aligned}\right\} \quad (37)$$

Notice that as $K_1 \to 0$ these sensitivities approach those of eqn. (32). It should be noted that the introduction of the amplifier K_1 has allowed us to reduce the required gain with a corresponding increase in sensitivity without moving the zeros of the feedback loop (phantom zeros) into the right-hand plane as was done by Kerwin and Shaffer. Thus the Q for the circuit of fig. 2 is determined by the gain product K_1K_2 while the resonant frequency is determined by the passive elements. It can be seen that the trade off between the gain product K_1K_2 and the sensitivity $S_{K_1}{}^{Q}$ is identical to that obtained by Kerwin and Shaffer. Furthermore, the sensitivity $S_{K_2}{}^{Q}$ can be kept less than 1. This is desirable, since K_2 is the larger gain and therefore inherently more difficult to hold to tight tolerances, while K_1 is low (2) thus being realizable to much tighter tolerance. Kerwin and Shaffer's gain K is the same magnitude as K_1K_2, and thus would be more difficult to stabilize.

5. Conclusion

The three sets of parameter triplets $(\delta_2, \delta_1, \delta_0)$, (Q, ω_n, γ), and $(\sigma_0, \omega_0, \gamma)$ have been shown to be useful parameters in describing a third-order system.

Sensitivities of these parameters were defined and expressions relating these sensitivities have been derived. Their use was demonstrated in the case of two new active RC band-pass filters reduced from third-order systems by pole-zero cancellations. These expressions can also be used for sensitivity analysis of non-reduced third-order systems.

It should be observed from eqns. (32) and (37) that $\sum_{\text{over all } i} S_{G_i}{}^{Q} = 0$ and $\sum_{\text{over all } i} S_{C_i}{}^{Q} = 0$ as expected (Mitra 1968).

Acknowledgments

This work was supported in part by a National Science Foundation Grant No. GK 14736 and by Sandia Corporation for the Atomic Energy Commission. The authors would like to acknowledge Shanta Pai for her comments during the preparation of this paper.

References

Daggett, K. E., and Vlach, J., 1969, *I.E.E.E. Trans. Circuit Theory*, **16,** 416.

Kerwin, W. J., 1970, *Active Filters*, edited by L. P. Huelsman (McGraw-Hill).

Kerwin, W. J., and Shaffer, C. V., 1968, Eleventh Midwest Symposium on Circuit Theory, University of Notre Dame, 13–14 May.

Mitra, S. K. 1969 a, WESCON Technical Papers, August ; 1969 b, *Analysis and Synthesis of Linear Active Network* (Wiley).

Soderstrand, M. A., 1969, Thesis M.S., University of California, Davis, California.

Paper 1-2

Sensitivity Minimization in a Single Amplifier Biquad Circuit

PAUL E. FLEISCHER, MEMBER, IEEE

***Abstract*—The single amplifier biquad (SAB) circuit of J. J. Friend is critically examined with a view to minimize its sensitivity properties. This is shown to involve the proper choice of the free design parameters to achieve a trade off between the contributions due to passive and active element variations. First the bandpass configuration is analyzed. A statistical measure of the overall pole variation due to both passive and active components is formulated. The analysis includes a novel but very simple approach for estimating the effects due to op amp variations which is applicable to many of the active filter topologies known.**

Using the figure of merit obtained in this fashion, it is shown that a near optimal design exists which permits ready physical realization, and which has the property that the transmission error terms are dominated by the irreducible fixed contribution due to the *RC* product which defines the resonant frequency. From the aspect of sensitivity, therefore, it is concluded that no other circuit topology can have superior performance. The frequency range over which these conclusions remain valid are determined by using reasonable estimates of circuit component tolerances. For a simple operational amplifier of the 741-type this range is shown to extend to approximately 2 kHz, while if a more sophisticated two-pole single zero compensation is used for the operational amplifier the range of validity is extended to about 20 kHz.

The above results are then shown to hold for all the other important

Manuscript received November 22, 1974; revised August 8, 1975.

The author is with Bell Laboratories, Holmdel, N.J. 07733.

Reprinted from *IEEE Trans. Circuits Syst.*, vol. CAS-23, pp. 45-55, Jan. 1976.

transfer functions which may be realized with the SAB. Since the SAB has previously been shown to be an economical and efficient building block for the realization of the different biquadratic transfer functions, the low sensitivity properties reported here demonstrate that it is a highly desirable topology.

I. Introduction

BASED ON THEIR MANY demonstrable advantages, active RC filters have largely supplanted passive filters in the audio frequency band. As a result of extensive research activity in this area, a very large number of different active filter topologies for realizing the basic biquadratic transfer functions have been developed. In particular, many alternative designs which use a single operational amplifier have been proposed. Although a comparison of these different circuits to determine which is "best" is quite impossible and fraught with considerable controversy, it is generally agreed that the two most important criteria for comparison are cost and performance. In the category of cost, attributes such as the ease of realizing the different biquadratic transfer functions, the number of components required, the resulting spread in element values, and the convenience of tuning are considered. In the area of performance, the sensitivity of the circuit, i.e., the variation of the transfer function due to changes in the passive and active components is considered.

Recently, Sedra [1] has presented a unified approach for classifying the available single amplifier biquadratic filters and for characterizing their sensitivities. The two main categories of filters he considers are "positive feedback" and "negative feedback" realizations. These two categories unfortunately omit the single amplifier biquad (SAB) section proposed by Friend [2] which is based on a bandpass design first published by Delyannis [3]. While the basic Delyannis circuit is of the negative feedback type, optional positive feedback is available to improve sensitivity performance and reduce element spreads. Numerical examples by Geffe [4], show the beneficial effects obtainable through proper choice of the amount of positive feedback by trading off the active and passive sensitivities. Friend, by introducing multiple inputs has extended the basic bandpass to a general-purpose biquadratic building block.

The main purpose of this paper is to present a detailed analysis of sensitivities in the SAB circuit and to show how the overall sensitivity may be minimized in high Q cases by proper adjustment of the amount of positive feedback used. More specifically, it will be shown that while the true minimum in sensitivity is achieved for an inconvenient choice of element values, a near optimum design exists which has very reasonable element spreads. Further, for the resulting realization it is shown that, in the audio frequency band, the total variation in transfer function is completely dominated by the component due to the variation in the resonant frequency which is only a function of the drift in the passive RC product. Since this term is irreducible, regardless of topology, it is concluded that the SAB, from a sensitivity point of view, is as good as any other circuit in the audio frequency band. This conclusion is consistent with Geffe's empirical observation [5], as well as the extensive practical experience of the author and his colleagues at Bell Laboratories.

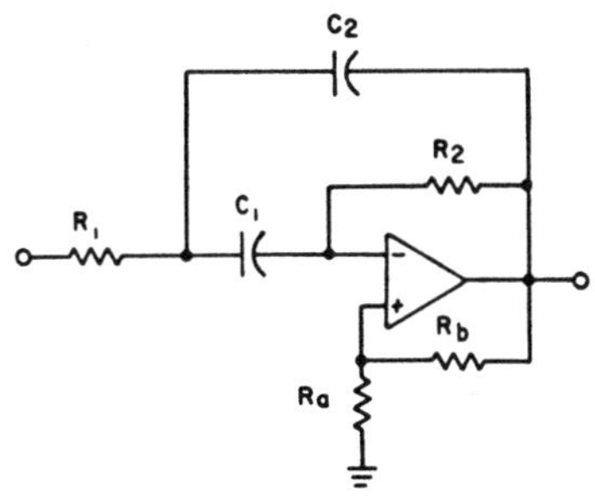

Fig. 1. SAB bandpass circuit.

A thorough demonstration of the flexibility and general usefulness of the SAB topology has been given by Friend, Harris, and Hilberman [6]. They show how this basic section may be used to realize any of the biquadratic transfer functions. This effort has culminated in the development of the STAR (standard tantalum active resonator) which provides a thin-film hybrid integrated circuit realization for all the SAB topologies. This circuit is now one of the topologies extensively used in the Bell System to realize active filter sections.

Based on its general applicability and flexibility, as well as its superior sensitivity properties, it is concluded that the SAB is the topology of choice in realizing active filter sections in the audio frequency band.

II. Analysis of Bandpass Configuration

In the first half of the paper the analysis will be restricted to the bandpass case as this is the only configuration for which the equations remain tractable. Once all the results are derived for the bandpass, it will be shown that they may be extended to the other circuit configurations.

The SAB bandpass topology is shown in Fig. 1. Note that the circuit shown has maximum gain. Gain reduction is normally achieved by splitting R_1 into a voltage divider. This aspect of the problem, however, will be ignored in the analysis as it represents only a trivial extension of results but would lead to additional notational complexity.

The transfer function of the bandpass circuit may readily be shown [6] to be

$$T(s) = \frac{-\frac{G_1}{C_2}(1+b)s}{D_1(s) + \frac{1+b}{A(s)} D_2(s)} \tag{1}$$

where

$$b = \frac{G_b}{G_a} \tag{2}$$

$$D_1(s) = s^2 + \left(\frac{C_1 + C_2}{C_1 C_2} G_2 - \frac{G_1}{C_2} b\right) s + \frac{G_1 G_2}{C_1 C_2} \tag{3}$$

$$D_2(s) = s^2 + \left(\frac{C_1 + C_2}{C_1 C_2} G_2 + \frac{G_1}{C_2}\right) s + \frac{G_1 G_2}{C_1 C_2} \tag{4}$$

and $A(s)$ represents the gain of the operational amplifier.

A. Circuit Design Parameters

The sensitivities to passive components will be derived by assuming infinite amplifier gain. This leads to the following design equations which characterize the nominal circuit:

$$\omega_0 = \left(\frac{G_1 G_2}{C_1 C_2}\right)^{1/2} \tag{5}$$

and

$$\frac{1}{Q} = \sqrt{\frac{G_2}{G_1}}\left(\sqrt{\frac{C_1}{C_2}} + \sqrt{\frac{C_2}{C_1}}\right) - b\sqrt{\frac{G_1}{G_2}}\sqrt{\frac{C_1}{C_2}}. \tag{6}$$

Note that there are six circuit elements available for achieving the desired circuit Q and resonant frequency. As already mentioned, the choice of flat gain level is not being considered, thus, there remain four degrees of freedom for the design. Two of these parameters are trivial and represent the impedance levels of the positive and negative feedback circuits. These choices would be based on practical considerations such as the availability of components, output impedance of the amplifier and effects of parasitics. The other two degrees of freedom, or design parameters, are the ones which are exploited here to minimize the overall sensitivity of the circuit. Although many possible alternative choices for the design parameters exist, the two which are chosen here lead to particularly simple equations and also provide a good deal of insight into the sensitivity behavior of the circuit.

Accordingly, the first parameter which is related to the capacitor ratios is defined as

$$\gamma \triangleq \frac{C_2}{C_1 + C_2} \tag{7}$$

while the second is related to the amount of positive feedback used. In examining (6) it is clear that even without positive feedback ($b = 0$) an arbitrary Q may be realized by adjusting the ratio of G_2 and G_1. The addition of positive feedback then enhances this basic Q factor which is being realized. Thus, for a given circuit, we define Q_0 as the Q which would be realized if the positive feedback were removed while all other circuit parameters remained intact:

$$\frac{1}{Q_0} \triangleq \sqrt{\frac{G_2}{G_1}}\left(\sqrt{\frac{C_1}{C_2}} + \sqrt{\frac{C_2}{C_1}}\right). \tag{8}$$

The amount of positive feedback is then measured by the amount of Q enhancement obtained due to the presence of positive feedback:

$$\eta \triangleq \frac{Q - Q_0}{Q_0} = \frac{Q}{Q_0} - 1. \tag{9}$$

Turning this equation around yields

$$Q = Q_0(1 + \eta) \tag{10}$$

which again clearly shows why η is referred to as the Q-enhancement factor.

It should be noted that physical realizability requires that

$$0 \leq \gamma \leq 1 \tag{11}$$

and that

$$\eta \geq 0. \tag{12}$$

Note that the equal capacitor case is represented by $\gamma = \frac{1}{2}$ and that $\eta = 0$ implies no positive feedback.

In later parts of the paper the relationship between η and the positive feedback conductance ratio (b) will be of interest. It is shown in Appendix A that

$$b = \gamma \frac{\eta(1 + \eta)}{Q^2} \tag{13a}$$

or

$$b = \gamma \frac{1}{Q_0}\left(\frac{1}{Q_0} - \frac{1}{Q}\right). \tag{13b}$$

Thus, if Q_0 is moderately large, say greater than 3 (which will turn out to be the desirable case), it will be reasonable to assume that $b \ll 1$. Note, incidentally, that circuit stability requires that

$$b \leq \frac{\gamma}{Q_0^2} \tag{14}$$

with equality corresponding to the case where the Q is being enhanced to infinity.

B. Sensitivities to Passive Components

The passive sensitivities of the circuit are most conveniently characterized by considering the ω_0 and Q sensitivities. These describe changes in the pole locations along an orthogonal set of axes and are the dominant components of the transfer function variation for moderate or high Q circuits.[1] The resonant frequency sensitivities are obtained by inspection of (5):

$$S_{G1}^{\omega_0} = S_{G_2}^{\omega_0} = \tfrac{1}{2} \tag{15}$$

$$S_{C_1}^{\omega_0} = S_{C_2}^{\omega_0} = -\tfrac{1}{2}. \tag{16}$$

Note that $S_{G_a}^{\omega_0}$ and $S_{G_b}^{\omega_0}$ are zero. Derivation of the Q sensitivities requires a fair amount of manipulations, details of which are shown in Appendix A. There, the following results are obtained:

$$S_{G_1}^{Q} = -S_{G_2}^{Q} = \eta + \tfrac{1}{2} \tag{17}$$

$$S_{G_b}^{Q} = -S_{G_a}^{Q} = \eta \tag{18}$$

and

$$S_{C_1}^{Q} = -S_{C_2}^{Q} = \gamma(1 + \eta) - \tfrac{1}{2}. \tag{19}$$

Since in an active *RC* filter the most efficient realization of the ω_0^2 term requires two resistors and two capacitors, it is clear that the fixed ω_0 sensitivities are the minimum which may be realized. Regarding the Q sensitivities the simplicity of the resulting expressions should be noted. In particular, if no positive feedback is used ($\eta = 0$) and the capacitors are equal ($\gamma = \frac{1}{2}$), four of the six sensitivity terms vanish. The remaining two, $S_{G_1}^{Q}$ and $S_{G_2}^{Q}$, become $+\frac{1}{2}$ and $-\frac{1}{2}$, respectively. This would represent an optimum design in

[1] In low Q circuits the variation of the gain constant is also relatively significant. Since low Q circuits are quite insensitive, they are of no interest in the present study.

the absence of active element variation. As will be seen in the next section, however, the sensitivities with respect to the active element decrease with increasing positive feedback.

C. Amplifier Sensitivity

The approach taken to characterize the behavior of the circuit due to finite gain in the operational amplifier involves a slight extension of the root sensitivity method due to Papoulis [7], [8]. While such techniques have been used in the past to analyze the movement of singularities due to passive components, it is somewhat surprising that this approach has not been applied to assess the effect of finite amplifier gain. Instead, the method one finds in the literature involves the calculation of the higher order polynomial which describes the system, followed by suitable approximations to estimate the movement of the dominant singularities. This approach, while valid, tends to obscure insight and also makes it difficult to generalize the results obtained.

In contrast, the method presented here is quite general, being applicable with trivial modifications to most one amplifier circuits and also to multiamplifier circuits if equal op amp gains are assumed. Further, the formulation is very simple and is obtained without specific reference to the op amp gain characteristics. Consequently, a good deal of insight is provided regarding the movement of the circuit singularities and, in particular, the influence op amp compensation is clearly discernible.

The poles of the SAB transfer function (1), are given by the zeros of the expression

$$D(s) = D_1(s) + \frac{1+b}{A(s)} D_2(s). \tag{20}$$

The poles of most other active filter building blocks are given by similar expressions.

Let s_0 and $s_0{}^*$ denote the location of the circuit poles when an ideal operational amplifier ($A = \infty$) is used,[2] i.e.,

$$D(s)|_{A=\infty} = D_1(s) = (s - s_0)(s - s_0{}^*). \tag{21}$$

Due to the presence of a real operational amplifier two effects take place. First, the zeros of $D(s)$ move from their nominal locations at s_0 and $s_0{}^*$ and, second, additional zeros are introduced due to the singularities of $A(s)$.

Consider this latter effect first. In most designs of practical interest the op amp generated singularities occur at frequencies substantially higher than the intended resonant frequency of the circuit. Their location is, therefore, approximately determined by replacing the $D_1(s)$ and $D_2(s)$ terms in (1) by their leading coefficients. This results in the following characteristic equation:

$$s^2\left(1 + \frac{1+b}{A(s)}\right) = 0. \tag{22}$$

Therefore, the high-frequency zeros of $D(s)$ are determined by the zeros of

$$A(s) + 1 + b = 0. \tag{23}$$

Since b is limited by (14)—the corresponding term in most other topologies is also small—it may be concluded that the high-frequency singularity or singularities lie in the vicinity of the unity gain crossover frequency of the compensated amplifier. Since this frequency is roughly 1 MHz for most currently used operational amplifiers, the effect of these extra singularities below say 50 kHz is quite negligible.

The dominant effect of finite amplifier gain, therefore, will be that due to the movement of the singularities at s_0 and $s_0{}^*$. If these singularities move to $s_0 + \Delta s_0$ and $s_0{}^* + \Delta s_0{}^*$, respectively, (20) may be rewritten as

$$\begin{aligned} D(s) &= D_1(s) + \frac{1+b}{A(s)} D_2(s) \\ &= (s - s_0 - \Delta s_0)(s - s_0{}^* - \Delta s_0{}^*)K(s) \end{aligned} \tag{24}$$

where $K(s)$ represents the additional high-frequency singularities just discussed.

Substituting $s = s_0 + \Delta s_0$ into this expression, yields, exactly:

$$D_1(s_0 + \Delta s_0) + \frac{1+b}{A(s_0 + \Delta s_0)} D_2(s_0 + \Delta s_0) = 0. \tag{25}$$

But, from (21) it follows that

$$D_1(s_0 + \Delta s_0) = \Delta s_0(s_0 + \Delta s_0 - s_0{}^*) \tag{26}$$

so that the following exact relat onship for Δs_0 is obtained from (25)

$$\Delta s_0 = -\frac{1+b}{A(s_0 + \Delta s_0)} \cdot \frac{D_2(s_0 + \Delta s_0)}{s_0 + \Delta s_0 - s_0{}^*}. \tag{27}$$

Since none of the terms on the right-hand side change rapidly with Δs_0 the following approximation is very good:

$$\Delta s_0 \approx -\frac{1+b}{A(s_0)} \frac{D_2(s_0)}{s_0 - s_0{}^*}. \tag{28}$$

This expression, which is quite general, leads to several interesting observations, the most important one being that the *fractional change in the complex pole location is inversely proportional to the magnitude of the amplifier gain at the pole frequency*. Further, the direction of this change depends on the phase angle of the amplifier gain at the pole frequency. The importance of operational amplifier compensation is thus clearly revealed. In Fig. 2 the commonly used single-pole compensation (741-type) is contrasted with the more sophisticated two-pole single zero compensation [9], [10]. Note that the latter compensation method provides at least an order of magnitude greater op amp gain in most of the audio frequency band. It is, therefore, highly preferable for the realization of precision filter building blocks. Regardless of the direction of the pole shift, its use leads to pole shift magnitudes at least an order smaller than that available when single pole compensated op amps are used.

Since (28) gives a good approximation of the change in

[2] Here it is assumed that s_0 and $s_0{}^*$ are a complex conjugate pair with s_0 denoting the singularity in the second quadrant.

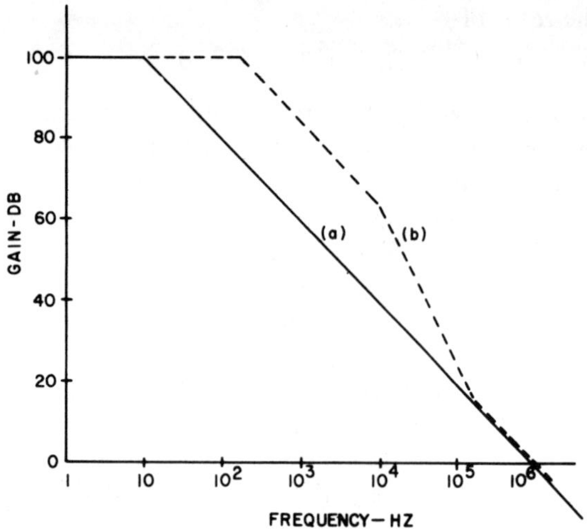

Fig. 2. Typical operational amplifier gains. (a) One-pole compensation. (b) Two-pole single zero compensation.

pole location, it becomes a valuable design tool for the predistortion of the nominal design to compensate for finite op amp gain. If $E\{\cdot\}$ represents the expected value, the expected pole shift due to finite op amp gain is given by

$$E\{\Delta s_0\} \approx -\frac{1+b}{s_0 - s_0^*} D_2(s_0) \cdot E\left\{\frac{1}{A(s_0)}\right\}. \tag{29}$$

Applying the negative of this shift to the nominal transfer function is a simple and effective method for predistortion which has been found to be quite accurate in practice. The simplicity of this method for predistorting for finite amplifier gain compares very favorably with the method proposed by Geffe [4].

Once the design is modified to compensate for the nominal expected amplifier gain, the effect of further amplifier changes are readily seen to be

$$\begin{aligned}\Delta s_0 &\approx -\frac{1+b}{s_0 - s_0^*} D_2(s_0)\Delta\left[\frac{1}{A(s_0)}\right] \\ &\approx +\frac{1+b}{s_0 - s_0^*} D_2(s_0)\frac{\Delta A(s_0)}{A^2(s_0)}.\end{aligned} \tag{30}$$

Upon rewriting this expression,

$$\frac{\Delta s_0}{s_0} \approx \frac{1+b}{s_0 - s_0^*} \cdot \frac{D_2(s_0)}{s_0} \cdot \frac{1}{A(s_0)} \cdot \frac{\Delta A(s_0)}{A(s_0)} \tag{31}$$

it becomes immediately apparent that a general expression for the gain sensitivity product [11] of the complex pole has been obtained:

$$GS \triangleq A(s_0) \cdot \frac{\Delta s_0}{s_0} \cdot \frac{A(s_0)}{\Delta A(s_0)} \approx \frac{1+b}{s_0 - s_0^*} \cdot \frac{D_2(s_0)}{s_0}. \tag{32}$$

Having demonstrated the general applicability of the method, let us now simplify (31) by substituting for $D_2(s_0)$ the expressions valid for the SAB circuit. Since, according to (21), $D_1(s_0) = 0$, it follows from (3) and (4) that

$$D_2(s_0) = D_2(s_0) - D_1(s_0) = (1+b)\frac{G_1}{C_2} s_0. \tag{33}$$

Also, using (3), as well as the definitions of Q and Q_0, results in

$$\frac{G_1}{C_2} b = \frac{\omega_0}{Q_0} - \frac{\omega_0}{Q}. \tag{34a}$$

This expression, together with (13b) and (10) yields

$$\frac{G_1}{C_2} = \frac{\omega_0 Q_0}{\gamma} = \frac{\omega_0 Q}{\gamma(1+\eta)} \tag{34b}$$

so that

$$D_2(s_0) = \frac{(1+b)\omega_0 Q}{\gamma(1+\eta)} s_0. \tag{35}$$

Substitution into (31) then yields

$$\frac{\Delta s_0}{s_0} \approx \frac{(1+b)^2 Q}{\gamma(1+\eta)} \cdot \frac{\omega_0}{s_0 - s_0^*} \cdot \frac{1}{A(s_0)} \cdot \frac{\Delta A(s_0)}{A(s_0)}. \tag{36}$$

Two further approximations may be made in this expression. First, as discussed before, we shall find that $b \ll 1$ so that b may be neglected. Second, for circuits of moderate or higher Q, $s_0 - s_0^*$ may be replaced by $j2\omega_0$. The final result is thus given by

$$\frac{\Delta s_0}{s_0} \approx \frac{-jQ}{2\gamma(1+\eta)A(s_0)} \cdot \frac{\Delta A(s_0)}{A(s_0)}. \tag{37}$$

As a final note we observe that, for a given Q, the pole shift is inversely proportional to the amount of positive feedback used in the circuit, so that, the variations due to the active component are minimized when η is large.

III. Optimum Bandpass Design

In the previous section the active and passive sensitivities of the bandpass section were derived. In order to minimize the overall circuit variability it is necessary to choose an overall figure of merit which equilibrates the effects due to active and passive element variations. The quantity chosen to be minimized is the variance of the magnitude of the fractional pole displacement, viz.,

$$\sigma^2 \triangleq E\left(\frac{\Delta s_0}{s_0}\right)^2. \tag{38}$$

To relate the ω_0 and Q sensitivities to the pole sensitivities the following result of Moschytz [12] is used

$$\frac{\Delta s_0}{s_0} = \frac{\Delta\omega_0}{\omega_0} - j\frac{\Delta Q}{Q}\left[\frac{1}{4Q^2 - 1}\right]^{1/2} \doteq \frac{\Delta\omega_0}{\omega_0} - j\frac{1}{2Q}\frac{\Delta Q}{Q}. \tag{39}$$

Before proceeding further it is worthwhile to relate the figure of merit in (38) with the resulting variation in transfer function. Consider first that we are interested in $T(s)$. Then, as shown by Moschytz [12],

$$\left.\frac{\Delta T(s)}{T(s)}\right|_{s=j\omega_0} = \frac{\Delta Q}{Q} + j2Q\frac{\Delta\omega_0}{\omega_0} \tag{40}$$

so that

$$\left|\frac{\Delta T(s)}{T(s)}\right|_{s=j\omega_0} = 2Q\left|\frac{\Delta s_0}{s_0}\right|. \tag{41}$$

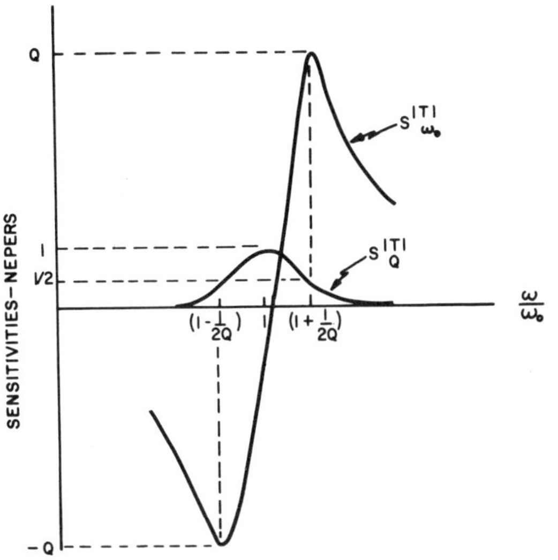

Fig. 3. Typical gain sensitivities for high Q case.

Thus the variation of the transfer function is directly related to the variation of the pole regardless of its direction.

For most applications, however, it is of greater interest to limit the variation of $|T(s)|$. The work of Hilberman [13] is relevant in examining this situation. As shown there, the direction of the pole change has a marked effect on the behavior of $\Delta|T(s)|$. Typical sensitivity curves of $|T(s)|$ with respect to pole Q and ω_0 variations are shown in Fig. 3. They show that the maximum deviation of $|T(s)|$ may occur anywhere in the band $\omega_0(1 - 1/2Q) \leq \omega \leq \omega_0(1 + 1/2Q)$. As a result the relative weighting of the variations due to Q and ω_0 cannot be determined in general. In practice, this dilemma is easily resolved by relying on experience with such circuits or their Monte Carlo simulation. It is almost universally found that the worst deviations in $|T(s)|$ occur at or near the band edge. Later results in this paper will also corroborate this observation, as it will be shown that, after optimization, the dominant remaining error term is that caused by the variation of ω_0 due to the passive elements. This portion of the variation, by the way, is not subject to minimization, as already observed in the discussion following (15) to (19).

A statistical error criterion which also incorporates the mean shift has recently been proposed by Shenoi [14]. If such a criterion is used, the above conclusions are further reinforced since the ω_0 shift caused by RC product drift will always be much larger than the Q shift which is a function of the drift in the ratios of like elements.

Looking at the band edge then we find from Hilberman's results that

$$S_{\omega_0}^{|T|} \doteq \pm Q \tag{42}$$

and

$$S_Q^{|T|} = \tfrac{1}{2} \tag{43}$$

so that a $2Q{:}1$ weighting ratio is proper. This in turn justifies the use of pole variability as a figure of merit. The influence of the angle of $\Delta s_0/s_0$ may readily be assessed from the following.

Let

$$\frac{\Delta s_0}{s_0} = \left|\frac{\Delta s_0}{s_0}\right| \varepsilon^{-j\alpha}. \tag{44}$$

Comparing this with (39) yields

$$\frac{\Delta\omega_0}{\omega_0} = \left|\frac{\Delta s_0}{s_0}\right| \cos\alpha$$

$$\frac{\Delta Q}{Q} = 2Q\left|\frac{\Delta s_0}{s_0}\right| \sin\alpha. \tag{45}$$

Thus at the band edges we get

$$\frac{\Delta|T|}{|T|} = S_{\omega_0}^{|T|}\frac{\Delta\omega_0}{\omega_0} + S_Q^{|T|}\frac{\Delta Q}{Q} = Q\left|\frac{\Delta s_0}{s_0}\right|(\pm\cos\alpha + \sin\alpha). \tag{46}$$

Since at one band edge or the other the trigonometric terms will add, it is easily shown that

$$Q\left|\frac{\Delta s_0}{s_0}\right| \leq \left.\frac{\Delta|T|}{|T|}\right|_{\max} \leq \sqrt{2}\,Q\left|\frac{\Delta s_0}{s_0}\right|. \tag{47}$$

This expression will aid later in relating the pole variance to the worst case transfer function deviation.

A. Formulation of Figure of Merit

As discussed previously the figure of merit is chosen to be the variance of the fractional pole shift. It is further assumed that like elements have the same statistics, thus $\Delta R/R$, $\Delta C/C$, and $\Delta A/A$ are described by their means μ_R, μ_C, and μ_A and their variances σ_R^2, σ_C^2, and σ_A^2. From the sensitivity relations, then, both the mean shift and the variance of the pole may be readily determined. The mean shift is not considered here since in a good design this can be removed through predistortion techniques. In particular, the effect of finite amplifier gain is assumed to be part of the original design. The only other contribution to the mean would result from a predictable RC product drift due to aging or temperature. This component can also be eliminated in the original design. If it is not, it will cause an additional error term but, as discussed previously, will not affect the optimum design.

The variance of the fractional pole change is readily obtained from the following summation over all the elements and should include random changes due to component value variations as well as tuning errors

$$\sigma^2 = \sum_X |S_X^{s_0}|^2\sigma_X^2. \tag{48}$$

By breaking down the passive contributions into their quadrature components, see (39), and adding all the components in a two-dimensional random walk fashion the following is obtained

$$\sigma^2 = \sum_{R,C} |S_{R,C}^{\omega_0}|^2\sigma_{R,C}^2 + \sum_{R,C}\frac{1}{4Q^2}|S_{R,C}^{Q}|^2\sigma_{R,C}^2 + |S_A^{s_0}|^2\sigma_A^2. \tag{49}$$

Substituting the various sensitivity relations, (15) through (19) and (37) yields, after some simplification

$$\sigma^2 = \frac{\sigma_R{}^2}{Q^2}\left[(\eta + \tfrac{1}{4})^2 + \tfrac{1}{16} + \frac{Q^2}{2}\right]$$

$$+ \frac{\sigma_C{}^2}{2Q^2}\{[\gamma(1+\eta) - \tfrac{1}{2}]^2 + Q^2\}$$

$$+ \frac{Q^2}{4}\frac{1}{\gamma^2(1+\eta)^2}\frac{1}{A^2}\sigma_A{}^2 \tag{50}$$

A^2 has been written for $|A(s_0)|^2$ in the last term.

B. Minimum Sensitivity, Design

The optimum design is obtained by minimizing (50) with respect to the two design parameters η and γ. Recall that η represents the amount of Q enhancement being realized due to the presence of positive feedback, and thus $\eta \geq 0$. The variable γ is related to the capacitor ratios as given in (7) and must satisfy the inequality $0 \leq \gamma \leq 1$.

First consider the problem of minimizing (50) without reference to physical realizability. Due to the manner in which the variables appear it can be concluded by inspection that the minimum occurs at $\eta = -\frac{1}{4}$ and some value of $\gamma(1+\eta)$. Therefore, no relative minimum lies inside the region corresponding to physical realizability and, consequently, the physically realizable minimum must lie on one of the boundaries. Inspection of (50) further shows that $\eta = \infty$, or $\gamma = 0$ correspond to a maximum. This leads to the following conclusion

The physically realizable minimum of (50) *is either on the line* $\eta = 0$ *or on the line* $\gamma = 1$.

Thus the search for the minimum is reduced to two one-dimensional problems. Before proceeding with the solution it should be observed that $\gamma = 1$ corresponds to $C_1 = 0$, and does not represent a practical solution. Therefore, as a practical matter γ will be limited by $\gamma = \gamma_0 < 1$. In particular $\gamma_0 = \frac{1}{2}$ would correspond to the equal capacitor case.

First consider the minimum which corresponds to the zero positive feedback case. Setting $\eta = 0$ in (50) and differentiating with respect to γ yields the following equation after some straightforward manipulation:

$$\gamma^3\left(\gamma - \frac{1}{2}\right) = \frac{1}{2}\frac{Q^4}{A^2}\frac{\sigma_A{}^2}{\sigma_C{}^2}. \tag{51}$$

Solution to this equation is readily obtained in an implicit fashion by assuming values of γ and computing the right-hand side. The result is given in graphical form in Fig. 4. It shows that for very low Q values $\gamma = \frac{1}{2}$ and then as Q increases γ increases. Since $\gamma \leq 1$, this solution is valid only while

$$Q \leq \left[\frac{A^2\sigma_C{}^2}{\sigma_A{}^2}\right]^{1/4} \tag{52}$$

and thus it represents the optimum design for the low Q case. It is instructive to substitute some typical numbers

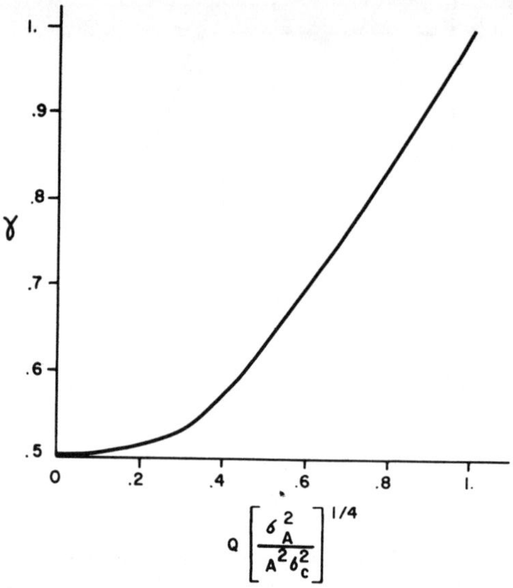

Fig. 4. Curve for determining optimum value of γ.

in this relation to find where approximately this Q limit lies. Choosing

$$\sigma_C = 0.002$$

$$\sigma_A = 0.15$$

$$A = |A(s_0)| = 10^3 \tag{53}$$

yields a Q limit of 3.65. Thus, the no positive feedback case will, for reasonable types of elements, correspond to low Q situations. Therefore, this solution is of no particular practical significance.

The important high Q case has an optimum solution when γ is fixed and the amount of positive feedback is varied. Although this optimum can readily be derived for any fixed γ_0, the relationships become somewhat simpler if γ is fixed at $\frac{1}{2}$ which corresponds to the practical and commonly used equal capacitor case. As will be shown later, the optimum is broad enough so that this choice results in a negligible increase in the figure of merit.

With $\gamma = \frac{1}{2}$ the relationship (50) becomes

$$\sigma^2 = \frac{\sigma_R{}^2}{Q^2}\left[(\eta + \tfrac{1}{4})^2 + \tfrac{1}{16} + \frac{Q^2}{2}\right]$$

$$+ \frac{\sigma_C{}^2}{8Q^2}[\eta^2 + 4Q^2] + \frac{Q^2}{(1+\eta)^2}\frac{1}{A^2}\sigma_A{}^2. \tag{54}$$

Differentiating this expression with respect to η yields, after some manipulation

$$\left(\eta + \frac{2\sigma_R{}^2}{8\sigma_R{}^2 + \sigma_C{}^2}\right)(\eta + 1)^3 = \frac{Q^4}{A^2}\frac{8}{8\sigma_R{}^2 + \sigma_C{}^2}\sigma_A{}^2. \tag{55}$$

Unless $\sigma_C{}^2 \gg \sigma_R{}^2$ the fraction on the left-hand side may be approximated. However, in any case it is less than $\frac{1}{4}$. This leads to the following relation

$$(\eta + \tfrac{1}{4})(\eta + 1)^3 = \frac{Q^4}{A^2}\frac{8\sigma_A{}^2}{8\sigma_R{}^2 + \sigma_C{}^2}. \tag{56}$$

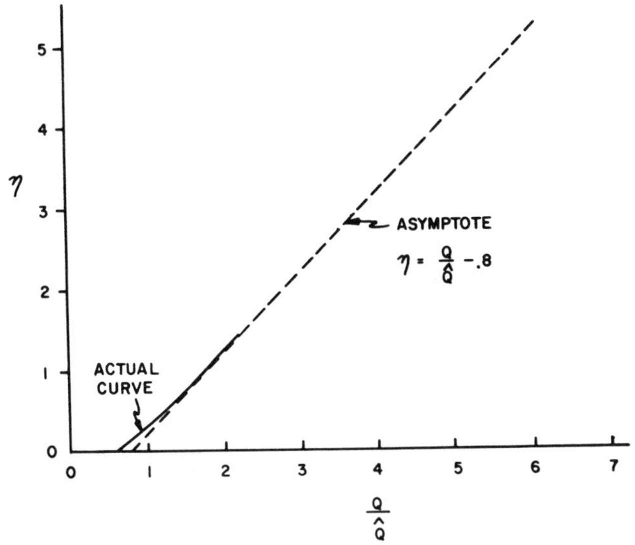

Fig. 5. Curve for determining optimum value of η.

Now let

$$\hat{Q} = \left[A^2 \frac{8\sigma_R^2 + \sigma_C^2}{8\sigma_A^2}\right]^{1/4}. \tag{57}$$

Then we get the simplified relationship

$$(\eta + \tfrac{1}{4})(\eta + 1)^3 = \left(\frac{Q}{\hat{Q}}\right)^4 \tag{58}$$

for determining the optimum choice of η. Again an implicit solution is easy to obtain. The plot (see Fig. 5) clearly shows that to a very close approximation

$$\eta \doteq -0.8 + \frac{Q}{\hat{Q}} \tag{59}$$

and thus, indeed, this represents the solution to the high Q case. If relation (9), which defined η, is substituted in the above the following simple equality is obtained:

$$\frac{Q}{Q_0} \doteq \frac{Q}{\hat{Q}} + 0.2. \tag{60}$$

Therefore, we have approximately

$$Q_0 = \hat{Q}. \tag{61}$$

This simple relationship leads to the following fascinating rule for choosing the optimum positive feedback ratio:

1) determine $\hat{Q}$ which is a function solely of the element tolerances and the available amplifier gain at the resonant frequency;
2) design the circuit without positive feedback to realize a quality factor of $\hat{Q}$;
3) Add sufficient positive feedback to enhance the quality factor to Q.

C. Properties of Optimum Design

In the previous section a near optimum design was obtained. The compromises involved both the choice to use equal capacitors and also some approximations in determining the optimal positive feedback. We shall now examine the results which obtain when these choices are made.

Substituting $\gamma = \frac{1}{2}$ and $\eta = Q/\hat{Q} - 1$ into (50) the variance of the pole position for the proposed suboptimal design is obtained:

$$\sigma^2 = \frac{\sigma_R^2}{Q^2}\left[\left(\frac{Q}{\hat{Q}} - \frac{3}{4}\right)^2 + \frac{1}{16} + \frac{Q^2}{2}\right] + \frac{\sigma_C^2}{2Q^2}\left\{\frac{1}{4}\left(\frac{Q}{\hat{Q}} - 1\right)^2 + Q^2\right\} + \frac{\hat{Q}^2}{A^2}\sigma_A^2. \tag{62}$$

Substituting (57) into the last term and neglecting the $\frac{1}{16}$ term in the first term yields

$$\sigma^2 \doteq \tfrac{1}{2}(\sigma_R^2 + \sigma_C^2) + \sigma_R^2\left[\frac{1}{\hat{Q}^2} + \left(\frac{1}{\hat{Q}} - \frac{3}{4Q}\right)^2\right] + \frac{\sigma_C^2}{8}\left[\frac{1}{\hat{Q}^2} + \left(\frac{1}{\hat{Q}} - \frac{1}{Q}\right)^2\right]. \tag{63}$$

Therefore

$$\sigma^2 < \tfrac{1}{2}(\sigma_R^2 + \sigma_C^2) + \left(2\sigma_R^2 + \frac{\sigma_C^2}{4}\right)\frac{1}{\hat{Q}^2}. \tag{64}$$

For the high Q case where a large amount of positive feedback is used $\hat{Q} \ll Q$ and the inequality approaches equality.

By tracing back through the derivations, or by simple observation, it is readily seen that the first term in this expression is the variance due to the irreducible resonant frequency variations caused by RC product shift. The second term incorporates the effects of Q variations as well as the changes due to amplifier variations which were the terms minimized.[3] This leads to the important conclusion that as long as $\hat{Q} > 2$ the irreducible resonant frequency term will dominate the overall error. A better perspective on when this condition holds may be obtained by substitution into (57)

$$\hat{Q} = \left[A^2 \frac{8\sigma_R^2 + \sigma_C^2}{8\sigma_A^2}\right]^{1/4} > 2 \tag{65}$$

which yields

$$A > 4\left[\frac{\sigma_A^2}{\sigma_R^2 + \frac{1}{8}\sigma_C^2}\right]^{1/2} \tag{66}$$

For a precision filter the following represent a reasonable set of element tolerances:

$$\sigma_A = 0.15$$

$$\sigma_R = 0.001$$

$$\sigma_C = 0.002.$$

Substituting these in the above relation yields approximately $A > 500$.

[3] It is also interesting to note that at the minimum the contributions due to amplifier gain variation and those due to Q variations are approximately equal. This is evident from (62) and (63).

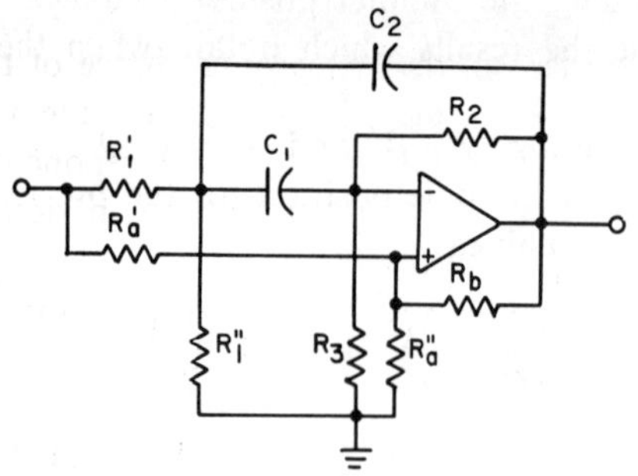

Fig. 6. SAB low-pass notch circuit.

For a single-pole compensated operational amplifier of the 741-type, this gain is realized up to about 2 kHz. Below this frequency, therefore, the fixed variance terms will dominate, but, at the higher end of the audio frequency band the Q and amplifier gain error terms will begin to take over.

On the other hand, for the two-pole single zero compensated operational amplifier the above gain is exceeded up to about 20 kHz. Thus, in the audio band the fixed irreducible error components will be highly dominating and, therefore, the extra expense of the more complex compensation scheme is highly indicated in critical designs. As an ancillary but very important benefit, note that the complete dominance of the fixed error terms implies that the minimum is very shallow, so that sizeable departures from the optimum design, for the sake of easier realization, may be made with relative impunity. A more careful analysis of this situation, details of which are omitted here, reveals that Q_0 need never be chosen to be greater than 4. This leads to R_2/R_1 and R_b/R_a ratios which are less than 60, permitting easy physical realization.

IV. Other Biquadratic Configurations

In this section it is shown that the results obtained for the bandpass circuit may be extended to other important biquadratic configurations: the low-pass notch (imaginary zeros larger than pole magnitude), the high-pass notch (imaginary zeros smaller than the pole magnitude), the high pass, and the all-pass (delay) section.

The low-pass (no finite zero) case is beyond the scope of this work as it is realized by a different circuit topology [2], [6].

A. Low-Pass Notch Configuration

A circuit diagram of the SAB realization of the low-pass notch characteristic is shown in Fig. 6. For this circuit the transfer function is given by[4]

$$T(s) = K \frac{s^2 + \omega_Z^2}{s^2 + \frac{\omega_0}{Q} s + \omega_0^2} = K \frac{s^2 + \omega_Z^2}{D_1(s) + \frac{1 + b}{A(s)} D_2(s)} \tag{67}$$

where

$$D_1(s) = s^2 + \left[\frac{C_1 + C_2}{C_1 C_2}(G_2 - bG_3) - \frac{G_1}{C_2} b\right] s + \frac{G_1(G_2 - bG_3)}{C_1 C_2} \tag{68}$$

$$D_2(s) = s^2 + \left[\frac{C_1 + C_2}{C_1 C_2}(G_2 + G_3) + \frac{G_1}{C_2}\right] s + \frac{G_1(G_2 + G_3)}{C_1 C_2} \tag{69}$$

$$b = \frac{G_b}{G_a' + G_a''} \tag{70}$$

$$K = \frac{G_a'}{G_a' + G_a''}. \tag{71}$$

These equations for determining the poles are quite similar to those given for the bandpass. In addition to the trivial splitting of the G_1 and G_a resistors, the only change in the nominal characteristics (described by $D_1(s)$) is the replacement of G_2 by $G_2 - bG_3$. The effect of this replacement is quite negligible since $b \ll 1$ as shown in (14). To show this more convincingly, consider the following relation which, it may be shown, G_3 must satisfy in order to produce the desired zero:

$$G_3 = G_2 \frac{\omega_Z^2 - \omega_0^2}{b\omega_Z^2 + \omega_0^2} \leq G_2 \left(\frac{\omega_Z^2}{\omega_0^2} - 1\right). \tag{72}$$

Unless the zero and pole frequencies are widely separated, G_3 compares to G_2 and therefore $bG_3 \ll G_2$. We conclude, therefore, that the passive sensitivity terms for this topology are approximately the same as for the bandpass.

Regarding the contribution of the active term the counterpart of (29) becomes

$$D_2(s_0) = D_2(s_0) - D_1(s_0) = (1 + b)\left\{\frac{G_1}{C_2} s_0 + \frac{C_1 + C_2}{C_1 C_2} G_3 s_0 + \frac{G_1 G_3}{C_1 C_2}\right\}. \tag{73}$$

In this expression the first term is the only one which appears in the bandpass case. To assess the influence of the other terms we assume $C_1 = C_2$ and use the following relation

$$\frac{G_1 G_3}{C_1 C_2} = \frac{1}{1 + b}(\omega_Z^2 - \omega_0^2) \doteq \omega_Z^2 - \omega_0^2. \tag{74}$$

Substituting this relation and (31) into the above and using $j\omega_0$ in place of s_0 yields

$$D_2(s_0) \doteq 2j\omega_0^2 Q_0 + j\frac{\omega_Z^2 - \omega_0^2}{Q_0} + (\omega_Z^2 - \omega_0^2). \tag{75}$$

It is seen that the first term, which is the one present in bandpass case, will tend to dominate. To the extent that it does not, the amplifier term in the variance relation can

[4] Here, only cases having zeros on the $j\omega$ axis are examined. The conclusions, however, remain equally valid for off axis zeros, as this degree of freedom is controlled exclusively by the R', R'' split.

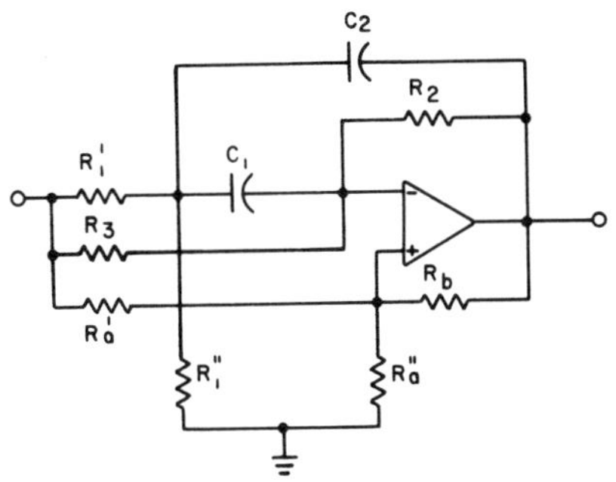

Fig. 7. SAB high-pass notch circuit.

easily be adjusted by using an artificially increased σ_A^2 to reflect the increase in $D_2(s_0)$.

The net result is that in most practical cases, i.e., where G_3 is not very large, the relations derived for the bandpass still hold, perhaps with a slightly more heavily weighted σ_A^2. Note also that the sensitivity of the numerator terms need not be considered as there is no design freedom to change them. In effect, the fixed error terms are even larger than in the bandpass case and the minimum shallower yet. Thus the various approximations made are further justified.

B. High-Pass Notch Configuration

The SAB circuit which realizes a high-pass notch is shown in Fig. 7. It is the same circuit as the low-pass notch except that G_3 is connected to the input. The poles of this topology are also given by relations (67) through (71), but the value of G_3, which leads to the correct zero placement may be shown to be given by

$$\frac{G_1G_3}{C_1C_2} = \frac{1}{1+b}\frac{K}{1-K}(\omega_0^2 - \omega_Z^2)$$
$$\doteq \frac{K}{1-K}(\omega_0^2 - \omega_Z^2). \qquad (76)$$

It is also easy to show that

$$G_3 \doteq G_2 \frac{K}{1-K}\left(1 - \frac{\omega_Z^2}{\omega_0^2}\right). \qquad (77)$$

An important difference, when compared to the previous topology, is the fact that there is a degree of freedom in choosing G_3. Thus a moderate value of G_3 is attainable if the gain constant K is not chosen too high. In particular, for $K = \frac{1}{2}$, $G_3 < G_2$ and all the arguments of the previous section hold. If, on the other hand, the gain is unduly increased, at most 6 dB is available, very large sensitivity penalties may result. Not only will all the pole variance terms increase, but, it may be shown that the zero location also becomes increasingly sensitive.

Thus it is very desirable to limit the gain constant which in turn limits G_3. When this is done, the optimal design rules of the bandpass case, as well as all the other conclusions remain valid for this case.

As a final note it is pointed out that the high-pass (zeros at $s = 0$) and the all-pass configurations are special cases of the high-pass notch design, and, therefore, all the above conclusions also hold.

V. Conclusion

It has been shown that by proper choice of the amount of positive feedback in the SAB circuit, the variation of transmission due to active and passive component variations may be minimized. When this minimum design is used the transmission variations are dominated by the fixed irreducible terms due to RC product variation. Therefore, from the sensitivity point of view, the SAB circuit represents a design which cannot be materially improved upon. These conclusions remain valid up to about 2 kHz if the simple one-pole compensation is used for the operational amplifier. However, the more complicated two-pole single zero compensation extends these results up to about 20 kHz.

The optimum design proposed here also leads to relatively convenient physical realization as the maximum resistance ratio is about 60:1.

Appendix

In this appendix the Q sensitivities of the bandpass configuration are derived. Consider relation (3) which defines the poles of the transfer function. It readily follows that

$$\frac{\omega_0}{Q} = \frac{C_1 + C_2}{C_1C_2}G_2 - \frac{G_1\,G_b}{C_2\,G_a} \qquad (A1)$$

and, therefore,

$$\frac{1}{Q} = \sqrt{\frac{G_2}{G_1}}\left(\sqrt{\frac{C_1}{C_2}} + \sqrt{\frac{C_2}{C_1}}\right) - \sqrt{\frac{G_1}{G_2}}\sqrt{\frac{C_1}{C_2}}\frac{G_b}{G_a}. \qquad (A2)$$

Now let

$$a = \sqrt{\frac{G_1}{G_2}} \qquad (A3)$$

$$b = \frac{G_b}{G_a} \qquad (A4)$$

$$c = \sqrt{\frac{C_1}{C_2}} \qquad (A5)$$

and note that according to (7)

$$\gamma = \frac{1}{1+c^2}. \qquad (A6)$$

Substituting these variables into (A2) yields

$$\frac{1}{Q} = \frac{1}{a}\left(c + \frac{1}{c}\right) - abc. \qquad (A7)$$

Also, rewriting (8) we have

$$\frac{1}{Q_0} = \frac{1}{a}\left(c + \frac{1}{c}\right) \qquad (A8)$$

and, therefore,

$$\frac{1}{Q_0} - \frac{1}{Q} = abc. \qquad (A9)$$

Now differentiate (A7) with respect to a

$$\frac{d\left(\frac{1}{Q}\right)}{da} = -\frac{1}{a^2}\left(c + \frac{1}{c}\right) - bc = \frac{-1}{a}\left[\frac{1}{a}\left(c + \frac{1}{c}\right) + abc\right]. \quad (A10)$$

Substituting (A8) and (A9) leads to

$$\frac{d\left(\frac{1}{Q}\right)}{da} = -\frac{1}{a}\left[\frac{2}{Q_0} - \frac{1}{Q}\right] = -\frac{1}{aQ}\left[2\frac{Q}{Q_0} - 1\right] \quad (A11)$$

and, therefore, using (9) which defines η yields

$$S_a^{1/Q} = Qa\frac{d\left(\frac{1}{Q}\right)}{da} = -1 - 2\eta. \quad (A12)$$

In a similar fashion we get

$$S_b^{1/Q} = Qb\frac{d\left(\frac{1}{Q}\right)}{db} = Qb(-ac)$$

$$= Q\left[\frac{1}{Q} - \frac{1}{Q_0}\right] = -\eta \quad (A13)$$

and also

$$S_c^{1/Q} = Qc\frac{d\left(\frac{1}{Q}\right)}{dc} = Qc\left[\frac{1}{a}\left(1 - \frac{1}{c^2}\right) - ab\right]$$

$$= Q\left[\frac{1}{a}\left(c - \frac{1}{c}\right) - abc\right]$$

$$= Q\left[\frac{1}{a}\left(c + \frac{1}{c}\right) - abc - \frac{2}{ac}\right]. \quad (A14)$$

But, it follows from (A8) and (A5) that

$$\frac{1}{Q_0} = \frac{c^2 + 1}{ac} = \frac{1}{\gamma ac}. \quad (A15)$$

Substituting this relation, as well as (A7), into (A14) yields

$$S_c^{1/Q} = Q\left[\frac{1}{Q} - \frac{2\gamma}{Q_0}\right] = 1 - 2\gamma(1 + \eta). \quad (A16)$$

From the relations (A12), (A13), (A16), and the well-known properties of the sensitivity function, viz.,

$$S_X^{1/Q} = -S_X^Q \quad (A17)$$

$$S_{1/X}^Q = -S_X^Q \quad (A18)$$

$$S_{\sqrt{X}}^Q = 2S_X^Q \quad (A19)$$

the sensitivity formulas of the main text follows directly.

One more equation relating b to the circuit design parameters needs to be derived. Using (A15) and (A9) we have

$$ac = \frac{Q_0}{\gamma} = \frac{1}{b}\left(\frac{1}{Q_0} - \frac{1}{Q}\right) \quad (A20)$$

and, therefore,

$$b = \frac{\gamma}{Q_0}\left(\frac{1}{Q_0} - \frac{1}{Q}\right) \quad (A21)$$

which is relation (13b) given in the text. The companion relation (13a) readily follows when (9), the defining equation for η, is substituted in the above:

$$b = \frac{\gamma}{Q^2}\cdot\frac{Q}{Q_0}\cdot\left(\frac{Q}{Q_0} - 1\right) = \frac{\gamma}{Q^2}\eta(1 + \eta). \quad (A22)$$

ACKNOWLEDGMENT

The author wishes to thank his colleagues J. J. Friend and D. Hilberman for many valuable and illuminating discussions and comments in the course of this work.

REFERENCES

[1] A. S. Sedra, "Generation and classification of single amplifier filters," *Int. J. Circuit Theory Appl.*, vol. 2, pp. 51–67, Mar. 1974.

[2] J. J. Friend, "A single operational-amplifier biquadratic filter section," in *1970 IEEE Int. Symp. Circuit Theory Digest Tech. Papers*, p. 189.

[3] T. Deliyannis, "High Q factor circuit with reduced sensitivity," *Electron. Lett.*, vol. 4, p. 577, Dec. 1968.

[4] P. R. Geffe, "Exact synthesis with real amplifiers," *IEEE Trans. Circuits Syst.*, vol. CAS-21. pp. 369–376, May 1974.

[5] P. R. Geffe, "Designers' guide to: active bandpass filters, Part 5," *EDN*, vol. 19, pp. 65–72, June 1974.

[6] J. J. Friend, C. A. Harris, and D. Hilberman, "STAR: An active biquadratic filter section," *IEEE Trans. Circuits Syst.*, vol. CAS-22, pp. 115–121, Feb. 1975.

[7] A. Papoulis, "Displacement of the zeros of the impedance $Z(p)$ due to incremental variations in the network elements," *Proc. IRE*, vol. 43, pp. 79–82, Jan. 1955.

[8] L. P. Huelsman, *Theory and Design of Active RC Circuits*. New York: McGraw-Hill, 1968, chapt. 2.

[9] National Semiconductor Corp., LM101A/LM102A Operational Amplifiers, data sheet, Aug. 1969.

[10] M. L. Embree and D. G. Marsh, "Family planning for Bell System op amps," *Bell Lab. Rec.*, pp. 120–129, Apr. 1974.

[11] G. S. Moshytz, "Gain sensitivity product: A figure of merit for hybrid integrated filters using single operational amplifiers," *IEEE J. Solid-State Circuits*, vol. SC-6, pp. 103–110, June 1971.

[12] G. S. Moschytz, "A note on pole, frequency and Q sensitivity," *IEEE J. Solid-State Circuits*, vol. SC-6, pp. 267–269, Aug. 1971.

[13] D. Hilberman, "An approach to the sensitivity and statistical variability of biquadratic filters," *IEEE Trans. Circuit Theory*, vol. CT-20, pp. 382–390, July 1973.

[14] B. A. Shenoi, "Optimum variability design and comparative evaluation of thin-film *RC* active filters," *IEEE Trans. Circuits and Systems*, vol. CAS-21, pp. 263–267, Mar. 1974.

Reducing Nonideal Op-Amp Effects in Active Filters by Minimizing the Gain-Sensitivity Product (GSP)

GEORGE S. MOSCHYTZ, SENIOR MEMBER, IEEE, AND PETER HORN, MEMBER, IEEE

***Abstract*—It is shown that the effects of finite amplifier gain and gain-bandwidth product, and of the variations of these quantities, on the transfer characteristics of an active filter can be expressed in terms of the gain-sensitivity product (GSP). To minimize these effects, it is sufficient, in most practical cases, to minimize the GSP with respect to the pole *Q*. Among other things, the expressions obtained also demonstrate the influence of the op-amp frequency compensation on the poles of an active filter.**

Experimental results comparing the pole variation of active filters with and without minimized GSP, coincide accurately with the theoretical results obtained in the paper. It is shown that considerable improvement in pole and transfer-characteristic stability of active filters can be expected by minimizing the GSP.

I. Introduction

THE gain-sensitivity product (GSP) was originally introduced as a figure of merit for hybrid-integrated filters comprising an operational amplifier and tracking RC components [1]. In this paper we show that the effects of nonideal op amp characteristics on the transfer characteristics of general active filters can be significantly reduced by minimizing the gain-sensitivity product. This conclusion is valid irrespective of the technology used, i.e., irrespective of the tracking properties of the RC components.

Consider an active network function F that is dependent on the open-loop gain A_i of m op amps. Then

$$\frac{\Delta F}{F}=\sum_{i=1}^{m} S_{A_i}^{F}\frac{\Delta A_i}{A_i}=\sum_{i=1}^{m}\Gamma_{A_i}^{F}\frac{\Delta A_i}{A_i^2} \tag{1}$$

where[1]

$$\Gamma_{A_i}^{F}=A_i S_{A_i}^{F}. \tag{(2)}$$

$\Gamma_{A_i}^{F}$ is the GSP of F with respect to the open-loop gain A_i. If the op amps are on one silicon chip (e.g., a quadamplifier where $m=4$), then it can be assumed that $\Delta A_i/A_i \equiv \Delta A/A$ and

$$\frac{\Delta F}{F}=\frac{\Delta A}{A^2}\cdot\sum_{i=1}^{m}\Gamma_{A_i}^{F}. \tag{3}$$

If only one amplifier is incorporated in the active

Manuscript received March 15, 1977; revised April 8, 1977.

The authors are with the Swiss Federal Institute of Technology, Zurich, Switzerland.

[1]To emphasize certain results, double parentheses (( )) are used.

Reprinted from *IEEE Trans. Circuits Syst.*, vol. CAS-24, pp. 437-445, Aug. 1977.

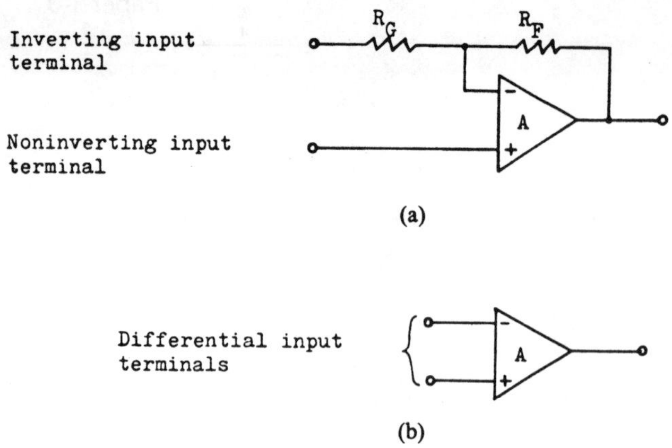

Fig. 1. Op amp. (a) Single-ended input mode. (b) Differential input mode.

network, then

$$\frac{\Delta F}{F}=\Gamma_A^F\cdot\frac{\Delta A}{A^2}. \tag{(4)}$$

For an ideal op amp, A will approach infinity and S_A^F zero; Γ_A^F *will however retain a well-defined value that can generally be minimized*. This is one reason why the definition of the GSP with respect to the *open*-loop gain is a useful one. Another becomes apparent when the op amp is used in the open-loop mode (as an inverting or differential amplifier).

In many active filters the op amp is used in the *closed*-loop mode as indicated in Fig. 1(a). For the inverting and the noninverting mode we obtain, respectively,

$$\beta_I=-\frac{A}{1+A/\beta_0}\cdot\left(\frac{\beta_0-1}{\beta_0}\right) \tag{5a}$$

and

$$\beta_N=\frac{A}{1+A/\beta_0} \tag{5b}$$

where

$$\beta_0=\frac{R_F+R_G}{R_G}. \tag{6}$$

For $A\gg 1$ the expressions in (5) simplify to

$$\beta_I\approx-(\beta_0-1) \tag{7a}$$

and

$$\beta_N\approx\beta_0. \tag{7b}$$

It follows that

$$S_A^{\beta_I}=S_A^{\beta_N}=\frac{1}{1+A/\beta_0}=\frac{\beta_N}{A}\approx\frac{|\beta_I|+1}{A} \tag{8}$$

and, since $\Gamma_A^{\beta_I}=\Gamma_A^{\beta_N}=\beta_N$

$$\frac{\Delta\beta_I}{\beta_I}=\frac{\Delta\beta_N}{\beta_N}=\left(\frac{\beta_N}{A}\right)\frac{\Delta A}{A}. \tag{9}$$

β_N is the closed-loop gain (CLG) or $1+|\mathrm{CLG}|$ for the noninverting and the inverting op amp, respectively. When $R_F/R_G\gg 1$, then $\beta_N\approx$CLG also for the inverting mode. Hence, β_N is often referred to simply as the closed-loop gain β. The GSP of F with respect to β can now readily be found since, with (8)

$$\underline{\underline{\Gamma_A^F}}=A\cdot S_\beta^F\cdot S_A^\beta=\beta S_\beta^F=\underline{\underline{\Gamma_\beta^F}}. \tag{(10)}$$

If the op amp is used in the open-loop mode, e.g., as a differential amplifier, then $\Delta F/F$ can readily be evaluated from (4). If, however, the variation of F with variations of CLG are of interest, then we have from (10)

$$\frac{\Delta F}{F}=\Gamma_\beta^F\cdot\frac{\Delta A}{A^2}. \tag{(11)}$$

Whether Γ_β^F or Γ_A^F is used is essentially a matter of convenience; the two are interchangeable, provided that a CLG quantity β is definable for the circuit in question. Since this is not always the case, we shall, in what follows, restrict ourselves to the use of the more general Γ_A^F with the understanding that wherever advantageous, β and A can be interchanged as in (10).

At this point the reader may well wonder what there is to be gained by using the general GSP, Γ_A^F rather than the sensitivity S_A^F. At first sight, the one does indeed seem to be a trivial expansion of the other, namely;

$$\frac{\Delta F}{F}=S_A^F\frac{\Delta A}{A}=\left[AS_A^F\right]\frac{\Delta A}{A^2}=\Gamma_A^F\frac{\Delta A}{A^2}.$$

Consider, however, the general active network function F introduced above, which is dependent on the op amp open-loop gain A. Assuming that F is continuous and differentiable when $A\to\infty$, then it can be expanded in a Taylor series of the form

$$F(x)=F(0)+x\frac{dF(0)}{dx}+\frac{x^2}{2!}\frac{d^2F(0)}{dx^2}+\cdots$$

where $x=A^{-1}$ and $x=0$ for $A=\infty$. Truncating the series after the first two terms, we obtain

$$F(A)\cong F_i+\frac{F_a}{A}$$

where F_i and F_a *are independent of A* and depend only on the network being characterized by F. Calculating the sensitivity of F with respect to A, we obtain

$$S_A^F=-\left.\frac{F_a/F}{A}\right|_{A\to\infty}=0.$$

Thus for large A the sensitivity goes to zero. Calculating the GSP of F with respect to A we obtain

$$A\cdot S_A^F=\Gamma_A^F=-\left.\frac{F_a}{F}\right|_{A\to\infty}=-\frac{F_a}{F_i}.$$

This is a well-defined analytical expression irrespective of whether A is infinite or not. Hence, a network can invariably be optimized by minimizing its GSP, whereas the same is not true of its sensitivity S_A^F. This holds true also if F is a function of m amplifiers with gains A_i as long as their variations are equal, i.e., $\Delta A_i/A_i\equiv\Delta A/A$. Furthermore[2], the multiplicand of the GSP, namely, $(\Delta A/A^2)$,

[2]This was pointed out by an unknown reviewer.

can be seen to be finite and normally small even though both its numerator and denominator may be very large, since

$$\frac{\Delta A}{A^2} \cong -\Delta\left(\frac{1}{A}\right)$$

where, of course, $1/A$ is very small.

Recently, various authors [2], [3] have dealt with the effect of finite gain, of gain variations, and of the gain-bandwidth product, on a dominant pole p_0. Invariably, some form or other of the GSP appeared in their results. It is the purpose of this contribution to show that these results can be generalized. Furthermore, it will be demonstrated that minimizing the GSP does indeed minimize not only the variation of the pole, but the variation of most other functions associated with the network that are caused by amplifier imperfections. As in [1], we shall see that in most practical cases it is sufficient to minimize only the GSP of q_p with respect to open- or closed-loop gain, in order to obtain a newtork that is as independent of op amp imperfections as is possible with this technique.

II. Pole Variation Due to Gain Variation

The pole variation due to a change in open-loop gain follows from (4) as

$$\frac{\Delta p_0}{p_0} = S_A^{p_0}\frac{\Delta A}{A} = \Gamma_A^{p_0}\frac{\Delta A}{A^2} \tag{12}$$

where, for convenience, A instead of $A(s)$ is used to denote the frequency dependent and finite gain of the op amp. $A(s)$, a function of the complex frequency $s=\sigma+j\omega$, is in general a complex function that is determined by the frequency compensation scheme used to stabilize the op amp.

From [4, p. 241], we car write the pole variation $\Delta p_0/p_0$ in terms of the frequency variation $\Delta\omega_p/\omega_p$ and the Q variation $\Delta q_p/q_p$ as follows:

$$\frac{\Delta p_0}{p_0} = \frac{\Delta\omega_p}{\omega_p} - j\frac{\Delta q_p/q_p}{\sqrt{4q_p^2-1}} \tag{13}$$

where

$$p_0 = -\frac{\omega_p}{2q_p} + j\omega_p\sqrt{1-\frac{1}{4q_p^2}}\,. \tag{14}$$

With

$$\frac{\Delta\omega_p}{\omega_p} = \Gamma_A^{\omega_p}\frac{\Delta A}{A^2} \tag{15a}$$

and

$$\frac{\Delta q_p}{q_p} = \Gamma_A^{q_p}\frac{\Delta A}{A^2} \tag{15b}$$

we can rewrite (13) as follows:

$$\frac{\Delta p_0}{p_0} = \left[\Gamma_A^{\omega_p} - j\frac{\Gamma_A^{q_p}}{\sqrt{4q_p^2-1}}\right]\frac{\Delta A}{A^2} = \Gamma_A^{p_0}\frac{\Delta A}{A^2}. \tag{(16)}$$

Comparing (13) with (16), we obtain

$$\frac{\Delta\omega_p}{\omega_p} = \Gamma_A^{\omega_p}\,\mathrm{Re}\left[\frac{\Delta A}{A^2}\right] + \frac{\Gamma_A^{q_p}}{\sqrt{4q_p^2-1}}\cdot\mathrm{Im}\left[\frac{\Delta A}{A^2}\right] \tag{(17)}$$

and

$$\frac{\Delta q_p}{q_p} = \Gamma_A^{\omega_p}\sqrt{4q_p^2-1}\cdot\mathrm{Im}\left[\frac{\Delta A}{A^2}\right] + \Gamma_A^{q_p}\cdot\mathrm{Re}\left[\frac{\Delta A}{A^2}\right]. \tag{(18)}$$

In (17) and (18) the open-loop gain $A(s)$ is evaluated for $s=p_0$, where p_0 is the nominal pole of interest. Whereas the quantity $A(p_0)$ will be complex, the change $\Delta A(s)/A(s)$ for $s=p_0$ will be real for the commonly used frequency range below the unity-gain cutoff frequency where the phase of the op amp is constant. For such cases, the real and imaginary parts of $1/A(p_0)$ will be decisive in establishing whether $\Delta\omega_p/\omega_p$ or $\Delta q_p/q_p$ dominate in (17) and (18). The factor $\Delta A/A$ will merely be a real coefficient common to all the terms on the right-hand side of (17) and (18); it may therefore be placed outside a pair of parentheses in both expressions.

In many practical circuits the pole frequency ω_p is *nominally* a function of only the passive components and not of the op amp gain.[3] For such circuits we have

$$\mathrm{Re}\left[\Gamma_A^{p_0}\right] = \Gamma_A^{\omega_p} = 0. \tag{19}$$

Then (16) simplifies to

$$\frac{\Delta p_0}{p_0} = -j\frac{\Gamma_A^{q_p}}{\sqrt{4q_p^2-1}}\frac{\Delta A}{A^2} \approx -j\frac{\Gamma_A^{q_p}}{2q_p}\frac{\Delta A}{A^2}. \tag{(20)}$$

Clearly, minimizing $\Gamma_A^{q_p}$ will minimize the variation of the pole p_0 due to variations in the open-loop gain.

III. Pole Variation Due to Finite Gain and Gain-Bandwidth Product

To find the effect of finite amplifier gain A on the pole p, we require p as a function of A. To obtain this, we can rewrite (12) in differential form in terms of the general pole p

$$\frac{dp}{p} = \frac{\Gamma_A^{p_0}}{A^2}dA \tag{21a}$$

where

$$p_0 = p(A=\infty). \tag{21b}$$

Thus p_0 is a pole of the network using an ideal op amp (i.e., infinite gain and bandwidth). Integrating (21a), whereby we assume that $\Gamma_A^{p_0}$ is, to a first approximation, independent of A (for $A \gg 1$ it is a constant), we obtain

$$\ln p + \ln C = -\frac{\Gamma_A^{p_0}}{A} \tag{22}$$

[3] In practice ω_p will, of course, depend on A (or β) due to the nonideal characteristics of the op amp. It is, among other things, precisely this dependence that will be derived and evaluated in the course of this paper.

and with (21b)

$$p = p_0 e^{-\Gamma_A^{p_0}/A}. \tag{(23)}$$

Equation (23) is the desired function $p(A)$. We obtain the relative pole shift $\Delta p_0/p_0$ resulting from finite amplifier gain $A(p_0)$ by expanding the exponential term as follows:

$$\frac{\Delta p_0}{p_0} = \frac{p - p_0}{p_0} = -\frac{\Gamma_A^{p_0}}{A(p_0)} + \frac{1}{2}\left(\frac{\Gamma_A^{p_0}}{A(p_0)}\right)^2 - \frac{1}{6}\left(\frac{\Gamma_A^{p_0}}{A(p_0)}\right)^3 + - \cdots. \tag{24}$$

For $|\Gamma_A^{p_0}/A(p_0)| \ll 1$, this simplifies to

$$\frac{\Delta p_0}{p_0} = -\frac{\Gamma_A^{p_0}}{A(p_0)}. \tag{(25)}$$

Note the change of sign between this expression for *finite* gain and (12) which pertains to a gain variation $\Delta A/A$. The reason for this polarity change can be explained as follows. The change of open-loop gain $\Delta A/A$ from infinite to finite gain A' is [6]

$$\frac{\Delta A}{A} = \left.\frac{A' - A}{A}\right|_{A\to\infty} = -1. \tag{26}$$

Consequently, the error caused by *finite* open-loop gain A of any function $F(A)$ is

$$\left(\frac{\Delta F}{F}\right)_f = \frac{F(A) - F(A=\infty)}{F(A=\infty)} = -\frac{\Gamma_A^F}{A}. \tag{(27)}$$

On the other hand, an error caused by a *change* in the finite open-loop gain will be as in (4)

$$\left(\frac{\Delta F}{F}\right)_c = \frac{F(A+\Delta A) - F(A)}{F(A)} = \left(\frac{\Gamma_A^F}{A}\right)\cdot\frac{\Delta A}{A}. \tag{(28)}$$

As pointed out earlier, either Γ_A^F or Γ_β^F can be used in (27) and (28) depending on which is more appropriate, or convenient, for the computation.

A comparison of (27) and (28) brings up the important but not generally appreciated fact, that the error caused by *finite* open-loop gain is always larger than the error caused by a *change* in the open-loop gain. This follows directly from the fact that $(\Delta A/A)_f = -1$, whereas $|\Delta A/A|_c < 1$. The error in finite open-loop gain can be accounted for during the design process of a network (e.g., tuning, predistortion), whereas the subsequent change in A due to ambient conditions (e.g., temperature, aging) is generally beyond the designer's control.

We can now compute the pole shift of p_0 due to finite op amp gain, namely, with (25)

$$\left(\frac{\Delta p_0}{p_0}\right)_f = -\frac{1}{A(p_0)}\left[\Gamma_A^{\omega_p} - j\frac{\Gamma_A^{q_p}}{\sqrt{4q_p^2 - 1}}\right]. \tag{(29)}$$

Similar to (17) and (18) we then obtain the frequency and q_p shift, respectively,

$$\left(\frac{\Delta\omega_p}{\omega_p}\right)_f = -\Gamma_A^{\omega_p}\mathrm{Re}\left[\frac{1}{A(p_0)}\right] - \frac{\Gamma_A^{q_p}}{\sqrt{4q_p^2 - 1}}\mathrm{Im}\left[\frac{1}{A(p_0)}\right] \tag{(30)}$$

$$\left(\frac{\Delta q_p}{q_p}\right)_f = -\Gamma_A^{\omega_p}\mathrm{Im}\left[\frac{1}{A(p_0)}\right]\sqrt{4q_p^2 - 1} - \Gamma_A^{q_p}\mathrm{Re}\left[\frac{1}{A(p_0)}\right]. \tag{(31)}$$

For the case that (19) holds, only the second terms on the right-hand side of (30) and (31) remain. To minimize the pole shift due to finite amplifier gain, only the GSP with respect to q_p, i.e., $\Gamma_A^{q_p}$ must then be minimized.

Whether calculating the pole shift due to gain variations as in (17) and (18), or to finite gain as in (30) and (31), it is clear that the quantity $1/A(p_0)$, and in particular its real and imaginary part, plays an important role. In other words, the frequency and q_p shifts, $\Delta\omega_p/\omega_p$ and $\Delta q_p/q_p$ will critically depend on the frequency compensation used for the op amp. This was first pointed out by Fleischer in [3], where results concurring with those above were obtained.

Consider, for example, two common methods of frequency compensation.

i. Single-Pole Compensation (See [5, p. 368])

Here the frequency-dependent open-loop amplifier gain can be written as

$$\frac{1}{A(s)} = \frac{s+\Omega}{A_0\Omega} = \frac{s+\Omega}{\omega_g} \tag{32}$$

where Ω is the 3-dB frequency of the open-loop gain and $\omega_g = A_0\,\Omega$ the gain bandwidth product. We require the inverse of the open-loop gain at the pole p_0 and obtain with (14)

$$\frac{1}{A(p_0)} = \frac{\left(\Omega - \dfrac{\omega_p}{2q_p}\right) + j\dfrac{\omega_p}{2q_p}\sqrt{4q_p 2 - 1}}{\omega_g}. \tag{33}$$

Hence with (19), (30), and (31) we obtain for the pole frequency variation due to finite amplifier gain

$$\left(\frac{\Delta\omega_p}{\omega_p}\right)_f = -\frac{\Gamma_A^{q_p}}{2q_p}\frac{\omega_p}{\omega_g} \tag{(34)}$$

and for the pole Q variation

$$\left(\frac{\Delta q_p}{q_p}\right)_f = -\Gamma_A^{q_p}\left(\frac{1}{A_0} - \frac{1}{2q_p}\cdot\frac{\omega_p}{\omega_g}\right). \tag{(35)}$$

Equations (34) and (35) give the pole shift due to finite gain-bandwidth product for the case of one-pole frequency compensation. The equivalent pole frequency and q_p variation due to variation in the op amp open-loop

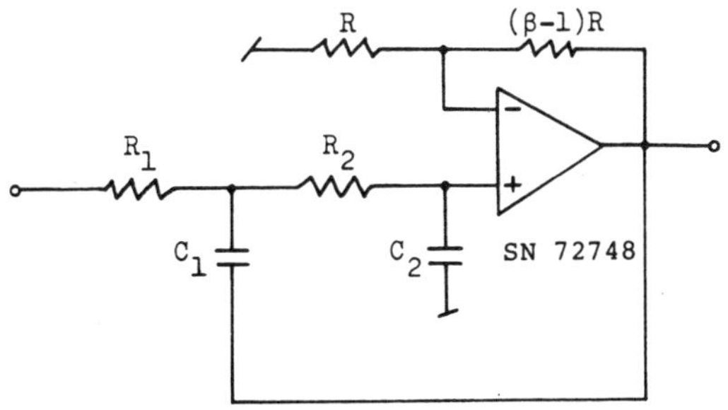

Fig. 2. Second-order low-pass filter.

gain is obtained from (17) and (18). Note that in the case of one-pole compensation both a frequency and a q_p shift from nominal must be expected, and that, in general

$$\left|\frac{\Delta q_p}{q_p}\right| \leqslant \left|\frac{\Delta \omega_p}{\omega_p}\right| \tag{36}$$

i.e., the frequency shift due to the finite op amp gain-bandwidth product will generally be larger than the change in q_p.

The results obtained above can best be verified experimentally by measuring the frequency shift $\Delta\omega_p/\omega_p$ of a second-order network resulting from the finite gain of an op amp and, in addition, on measuring the dependence of this frequency shift on the value of $\Gamma_A^{q_p}$. The measured results are then compared with those computed from (34).

The procedure for conducting these measurements was the following. The simple low-pass filter shown in Fig. 2 was designed using a Texas Instrument SN 72748 op amp with a 15 pF external frequency compensation capacitor. The approximate cutoff frequency and open-loop gain of the op amp were found to be $\omega_g \approx 2\pi \cdot 1.4$ MHz and $A_0 \approx 10^4$, respectively. The circuit of Fig. 2 was dimensioned three times, whereby each circuit provided a pole Q, $q_p = 5$ and a pole frequency $\omega_p = 2\pi \cdot 10$ kHz—but the circuits differed with respect to $S_\beta^{q_p}$ and $\Gamma_A^{q_p}$. The circuits were built using resistors and capacitors measured accurately to within 0.1 percent. The pole frequency was then measured for each circuit using the 90° phase shift as an indicator. Using a phase meter accurate to 0.1°, the resulting frequency measurements were accurate to within 0.02 percent (see [5, p. 406]). The measured deviations from the theoretical frequency were then compared with those computed from (34). The results obtained follow below in Table I.

TABLE I

	Circuit 1	Circuit 2	Circuit 3
C_1 [pF]	10,047	10,047	10,047
C_2 [pF]	4,704	10,047	10,047
R_1 [Ω]	1,676	7,921	693.7
R_2 [Ω]	3,195	316.8	3,617.5
β	2.171	2	6.758
$S_\beta^{q_p}$	11.5	50	14.75
$\Gamma_A^{q_p}$	25	100	100
$\frac{\Delta\omega_p}{\omega_p}$ [%] calculated	-1.78	-7.14	- 7.14
$\frac{\Delta\omega_p}{\omega_p}$ [%] measured	-1.88	-6.63	- 6.87

Note the close agreement between measurement and calculation. As expected, by reducing $\Gamma_A^{q_p}$ by a factor of 4, the resulting frequency shift due to finite gain-bandwidth product is reduced by approximately the same amount. This holds true irrespective of the value of the sensitivity $S_\beta^{q_p}$, as a comparison of the results obtained with circuits 2 and 3 clearly shows.

The experimental verification of (35) is less straightforward than that of (34). For one thing the term in brackets may be positive or negative depending on the values of A_0 and ω_g, neither of which is measurable very accurately. Moreover, this verification, which entails measuring the $\Delta q_p/q_p$ of a circuit built with passive components whose values ideally equal the calculated ones exactly, requires a precision in resistor and capacitor measurements that far exceeds the accuracy required for the measurement of $\Delta\omega_p/\omega_p$. Where 0.1-percent accuracy was quite sufficient for the measurement of $\Delta\omega_p/\omega_p$ because the sensitivities $S_{R_i,C_i}^{\omega_p}$ are low (i.e., less than unity), this same accuracy results in nonnegligible errors in the measurement of $\Delta q_p/q_p$ due to the significantly higher sensitivities $S_{R_i,C_i}^{q_p}$ and $S_\beta^{q_p}$. As a consequence, these errors, which are due to component inaccuracy (these are eliminated in practice by a functional "touch-up" tuning step), are indistinguishable from the errors due to finite op amp gain. Since no ambiguity of this kind exists with the measurement of $\Delta\omega_p/\omega_p$, we have limited ourselves here to frequency measurements, to demonstrate the efficacy of minimizing $\Gamma_A^{q_p}$ of networks incorporating op amps with finite gain-bandwidth product.

ii. Double-Pole Single-Zero Compensation (See [5, p. 547])

Here we have

$$\frac{1}{A(s)} = \frac{1}{A_0} \cdot \frac{s+\Omega_1}{\Omega_1} \cdot \frac{s+\Omega_2}{\Omega_2} \cdot \frac{\Omega_3}{s+\Omega_3}. \tag{37}$$

With (14) we have

$$p_0^2 \approx -\omega_p^2\left(1 + j\frac{1}{q_p}\right). \tag{38}$$

Furthermore, we can assume that

$$\Omega_1, \Omega_2 \ll \omega_p \ll \Omega_3 \tag{39}$$

in which case we obtain

$$\frac{1}{A(p_0)} = -\frac{\Omega_3}{A_0\Omega_1\Omega_2} \cdot \frac{\omega_p^2 + j\omega_p\left(\frac{\omega_p}{q_p} - \Omega_1 - \Omega_2\right)}{\Omega_3 + j\omega_p}. \tag{(40)}$$

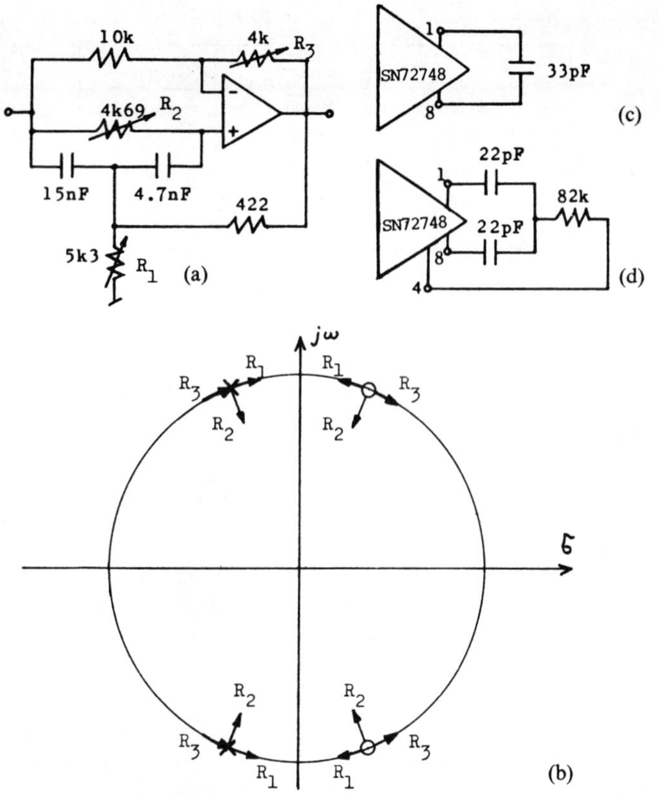

Fig. 3. Second-order all pass circuit. (a) Circuit diagram. (b) Tuning components. (c) Single-pole compensation. (d) Double-pole single-zero compensation.

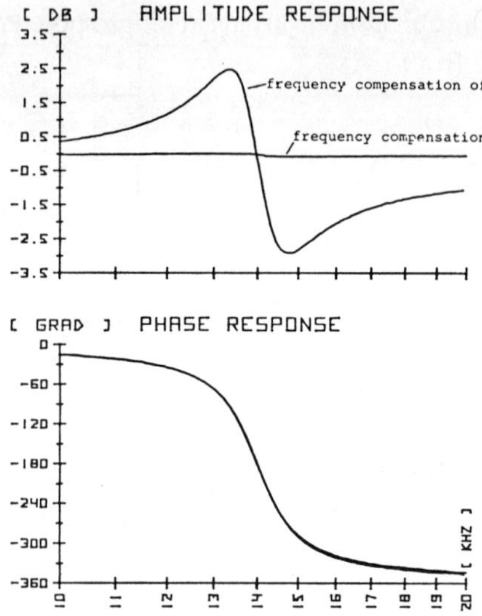

Fig. 4. Amplitude and phase response of second-order all pass circuit using frequency compensation of Fig. 3(c) and (d).

With (19) and (30) we then obtain

$$\left(\frac{\Delta\omega_p}{\omega_p}\right)_f = \left[\frac{\frac{1}{q_p} - \frac{(\Omega_1+\Omega_2)}{\omega_p} - \frac{\omega_p}{\Omega_3}}{\sqrt{4q_p^2-1}}\right]\left[\frac{\Gamma_A^{q_p}}{A_0}\cdot\frac{\omega_p^2}{\Omega_1\Omega_2}\right] \tag{41}$$

and with (31)

$$\left(\frac{\Delta q_p}{q_p}\right)_f = \left[1+\frac{\omega_p}{\Omega_3}\left\{\frac{1}{q_p} - \frac{(\Omega_1+\Omega_2)}{\omega_p}\right\}\right]\left[\frac{\Gamma_A^{q_p}}{A_0}\cdot\frac{\omega_p^2}{\Omega_1\Omega_2}\right]. \tag{42}$$

As in the previous cases, both $\Delta\omega_p/\omega_p$ and $\Delta q_p/q_p$ are reduced as $\Gamma_A^{q_p}$ is minimized.

If the second term in square brackets in (41) and (42) is designated C, then the first square bracket in (41) multiplies C by some small term $\epsilon/2q_p$, while the first square bracket in (42) multiplies C by a term $(1+\epsilon')$ where ϵ, $\epsilon' \ll 1$. It therefore follows that for the double-pole single-zero compensation:

$$\left|\frac{\Delta\omega_p}{\omega_p}\right| \ll \left|\frac{\Delta q_p}{q_p}\right|. \tag{43}$$

Since, in general, the transfer characteristics of active networks are much more sensitive to pole-frequency than to pole Q variations, a comparison of (43) with (36) clearly shows the advantage of double-pole single-zero frequency compensation over the commonly used single-pole form. This advantage has important practical consequences. In the all pass circuit shown in Fig. 3(a) (see [5, p. 532]), for example, the finite gain-bandwidth product of the amplifier must not cause the pole frequency to be shifted, because there is no available component in the circuit that can correct for such a shift without at the same time changing other network characteristics. This is indicated in Fig. 3(b). Thus a parasitic shift in the pole frequency will violate the requirement that the pole and zero frequencies be equal and will cause large amplitude errors that cannot be corrected for by tuning. However, using the double-pole single-frequency compensation, this error can be avoided almost completely thanks to the property given in (43). The measurements shown in Fig. 4 confirm this impressively. The all pass of Fig. 3(a) was designed for $f_p = 14$ kHz, $q_p = 10$ and tuned according to the procedure given in [5, p. 537]. Using an SN 72748 amplifier, the circuit was tuned first with single-pole compensation (Fig. 3(c)), and then with double-pole single-zero compensation (Fig. 3(d)). Using the single-pole compensation the maximum amplitude error was almost 3 dB; with the double-pole single-zero compensation it was less than 0.1 dB.

IV. Pole Variations Due to Variations of the Gain-Bandwidth Product

In the preceding section we considered the effect of a *finite* gain-bandwidth product on the pole location in the s-plane; in this section, we wish to examine the effect of a *change* in the gain-bandwidth product. Since

$$\omega_g = A_0\Omega \tag{44}$$

we have

$$\frac{\Delta\omega_g}{\omega_g} = \frac{\Delta A_0}{A_0} + \frac{\Delta\Omega}{\Omega}. \tag{45}$$

We must now find the effect of the variation $\Delta\omega_g/\omega_g$ on the gain $A(s)$, namely,

$$\frac{\Delta A}{A} = S_{A_0}^{A}\frac{\Delta A_0}{A_0} + S_{\Omega}^{A}\frac{\Delta\Omega}{\Omega}. \tag{46}$$

Considering the commonly used single-pole compensation we obtain from (32)

$$S_{A_0}^{A}=1 \tag{47a}$$

and

$$S_{\Omega}^{A}=1-\frac{\Omega}{s+\Omega}=\frac{s}{s+\Omega} \tag{47b}$$

hence

$$\frac{\Delta A}{A}=\frac{\Delta A_0}{A_0}+\frac{s}{s+\Omega}\cdot\frac{\Delta\Omega}{\Omega}. \tag{48}$$

Since Ω is very small in practice (typically 10 Hz) we can readily assume that our frequency of interest is very much higher (i.e., in the frequency range in which the finite gain-bandwidth product begins to make itself noticed). Hence for $s \gg \Omega$ a comparison of (45) and (48) shows that

$$\frac{\Delta\omega_g}{\omega_g}\approx\frac{\Delta A}{A}. \tag{(49)}$$

Thus the variation of finite gain-bandwidth will have the same effect on pole location as the variation of finite gain. The latter was treated under Section II and we may adopt the results obtained there directly. From (16) we find

$$\left(\frac{\Delta p_0}{p_0}\right)_c=\left[\Gamma_A^{\omega_p}-j\frac{\Gamma_A^{q_p}}{\sqrt{4q_p^2-1}}\right]\frac{1}{A(p_0)}\cdot\frac{\Delta\omega_q}{\omega_g}. \tag{(50)}$$

If (19) is valid, this simplifies to

$$\left(\frac{\Delta p_0}{p_0}\right)_c=-j\frac{\Gamma_A^{q_p}}{\sqrt{4q_p^2-1}}\cdot\frac{1}{A(p_0)}\frac{\Delta\omega_g}{\omega_g}. \tag{(51)}$$

With (33) (which applies only to single-pole compensation), (17), (18), and (19), we then obtain

$$\left(\frac{\Delta\omega_p}{\omega_p}\right)_c=\frac{\Gamma_A^{q_p}}{2q_p}\cdot\frac{\omega_p}{\omega_g}\cdot\frac{\Delta\omega_g}{\omega_g} \tag{(52)}$$

and

$$\left(\frac{\Delta q_p}{q_p}\right)_c=\Gamma_A^{q_p}\left(\frac{1}{A_0}-\frac{1}{2q_p}\frac{\omega_p}{\omega_g}\right)\frac{\Delta\omega_g}{\omega_g}. \tag{(53)}$$

Equation (52) was given by Sedra and Espinoza in [2]. Note that here again the only quantity that can be minimized, in order to minimize the frequency and q_p drift due to changes in the gain-bandwidth product, is the GSP, $\Gamma_A^{q_p}$.

V. Variations of Nth-Order Transfer Functions Due to Nonideal Amplifiers

With the results derived in the preceding sections we are now in a position to provide a basic approach according to which general nth-order active filter structures may be optimized. The approach has been applied successfully to arbitrary nth-order multiple amplifier structures, as will be reported elsewhere [7], [8].

We assume that an nth-order transfer function $T(s)$ is given by

$$T(s)=\frac{\sum_{i=0}^{n} b_i s^i}{\sum_{i=0}^{n} a_i s^i}=\frac{N(s)}{D(s)}. \tag{54}$$

The sensitivities of $T(s)$ to the coefficients a_i and b_i are

$$S_{b_i}^{T(s)}=S_{b_i}^{N(s)}=\frac{b_i s^i}{N(s)} \tag{55a}$$

and

$$S_{a_i}^{T(s)}=-S_{a_i}^{D(s)}=-\frac{a_i s^i}{D(s)}. \tag{55b}$$

These sensitivities are frequency-dependent functions, each of which is associated with its corresponding coefficient, and is determined by the function $T(s)$. The variation of $T(s)$ with changes in the coefficients is then

$$\frac{\Delta T}{T}(s)=\sum_{i=0}^{n}\frac{b_i s^i}{N(s)}\cdot\frac{\Delta b_i}{b_i}-\sum_{i=0}^{n}\frac{a_i s^i}{D(s)}\cdot\frac{\Delta a_i}{a_i}. \tag{56}$$

Hence, each coefficient variation is weighted by its corresponding frequency-dependent sensitivity function (see also [9]). Using (1) and assuming that there are m amplifiers in the network, we obtain the coefficient variations due to changes in amplifier gain as

$$\left(\frac{\Delta b_i}{b_i}\right)_c=\sum_{j=1}^{m}\Gamma_{A_j}^{b_i}\frac{\Delta A_j}{A_j^2} \tag{57a}$$

$$\left(\frac{\Delta a_i}{a_i}\right)_c=\sum_{j=1}^{m}\Gamma_{A_j}^{a_i}\frac{\Delta A_j}{A_j^2}. \tag{57b}$$

Assuming, for convenience, that the open-loop characteristics of all the amplifiers are the same, we readily obtain the variation $[\Delta T/T]_c$ due to a change in these amplifiers, namely, with (28), (56), and (57)

$$\left[\frac{\Delta T}{T}(s)\right]_c=\frac{1}{A(s)}\cdot\left[\sum_{i=0}^{n}\frac{b_i s^i}{N(s)}\cdot\sum_{j=1}^{m}\Gamma_{A_j}^{b_i}-\sum_{i=0}^{n}\frac{a_i s^i}{D(s)}\cdot\sum_{j=1}^{m}\Gamma_{A_j}^{a_i}\right]\cdot\frac{\Delta A}{A}. \tag{(58)}$$

To obtain the variation $[\Delta T/T]_f$ caused by finite amplifier gain we obtain with (27)

$$\left[\frac{\Delta T}{T}(s)\right]_f=-\frac{1}{A(s)}\cdot\left[\sum_{i=0}^{n}\frac{b_i s^i}{N(s)}\cdot\sum_{j=1}^{m}\Gamma_{A_j}^{b_i}-\sum_{i=0}^{n}\frac{a_i s^i}{D(s)}\cdot\sum_{j=1}^{m}\Gamma_{A_j}^{a_i}\right]. \tag{(59)}$$

Equations (58) and (59) indicate that *if a network is optimized with respect to the initial error caused by finite open-loop gain, then it is also optimized with respect to subsequent errors caused by changes in gain.* The right-hand expression in square brackets in the two equations corresponds to a form of GSP which we shall designate the *general GSP*. This is the GSP of the transfer function of a network with respect to its gain elements. Because of the interchangeability of β and A (see (10)) we can define it as follows:

$$\Gamma_A^{T(s)} = \sum_{i=0}^{n} \frac{b_i s^i}{N(s)} \cdot \sum_{j=1}^{m} \Gamma_{A_j}^{b_i} - \sum_{i=0}^{n} \frac{a_i s^i}{D(s)} \cdot \sum_{j=1}^{m} \Gamma_{A_j}^{a_i} \qquad ((60))$$

or

$$\Gamma_\beta^{T(s)} = \sum_{i=0}^{n} \frac{b_i s^i}{N(s)} \cdot \sum_{j=1}^{m} \Gamma_{\beta_j}^{b_i} - \sum_{i=0}^{n} \frac{a_i s^i}{D(s)} \cdot \sum_{j=1}^{m} \Gamma_{\beta_j}^{a_i}. \qquad ((61))$$

In these expressions, we use the *coefficient* GSP's where, for example,

$$\Gamma_{A_j}^{a_i} = A_j \cdot S_{A_j}^{a_i}. \qquad (62)$$

As in all other GSP expressions, A_j can also here be replaced by the closed-loop gain β_j.

The significance of (58) and (59) lies in the fact that the optimization of a circuit for minimum variation of $T(s)$ due to nonideal op amp characteristics now involves the minimization of *only* the coefficient GSP's, $\Gamma_{A_j}^{a_i}$ and $\Gamma_{A_j}^{b_i}$. The coefficient GSP's are weighted by the frequency functions (55a) and (55b), and can be assumed to be *frequency independent*. The optimization is only valid in a frequency range in which this is the case. The weighting functions (55a) and (55b) are, of course, directly available from the given transfer function.

By inspection of (58) and (59), it is clear that the weighting frequency functions $b_i s^i/N(s)$ and $a_i s^i/D(s)$ will have poles at the critical frequencies of $T(s)$ (i.e., at the roots of $N(s)=0$, and $D(s)=0$). In the case of notch networks the roots of $N(s)$ are on the $j\omega$-axis, in which case the optimization in the vicinity of the resulting attenuation poles cannot be carried out. This is generally not a serious limitation since the attenuation poles are in the stopband, whereas the optimization for minimum $\Gamma_A^{T(s)}$ will be carried out in the passband. When this is not the case, however, other optimization techniques must be used (e.g., [10]).

To illustrate the use of the expressions given above, we consider a simple second-order transfer function realized by an active network comprising a single amplifier A. The transfer function will have the form

$$T(s) = \frac{N(s)}{s^2 + \frac{\omega_p}{q_p} s + \omega_p^2}. \qquad (63)$$

To find the error $[\Delta T/T]$ due to finite amplifier gain, we obtain from (10) and (59)

$$\left[\frac{\Delta T}{T}(s)\right]_f = \frac{1}{A(s)} \left[\sum_{i=0}^{2} \frac{b_i s^i}{N(s)} \Gamma_A^{b_i} - \sum_{i=0}^{2} \frac{a_i s^i}{D(s)} \Gamma_A^{a_i}\right]. \qquad (64)$$

Assuming the network is type I (see [5, p. 47]), i.e., the zeros of $T(s)$ are independent of A, then $\Gamma_A^{b_i} \equiv 0$. With (63) we then obtain from (64)

$$\left[\frac{\Delta T}{T}(s)\right]_f = \frac{1}{A(s)} \left[\Gamma_A^{\omega_p}\left(\frac{\omega_p}{q_p} s + 2\omega_p^2\right) - \frac{\omega_p}{q_p} s \cdot \Gamma_A^{q_p}\right] \cdot \frac{1}{D(s)}. \qquad (65)$$

Assuming that ω_p is also nominally independent of A, i.e., $S_A^{\omega_p}=0$, we obtain the error due to finite amplifier gain as

$$\left[\frac{\Delta T}{T}(s)\right]_f = -\frac{\Gamma_A^{q_p}}{A(s)} \cdot \frac{\frac{\omega_p}{q_p} s}{s^2 + \frac{\omega_p}{q_p} s + \omega_p^2}. \qquad (66)$$

Similarly, the error due to changes in gain will be

$$\left[\frac{\Delta T}{T}(s)\right]_c = \frac{\Gamma_A^{q_p}}{A(s)} \cdot \frac{\frac{\omega_p}{q_p} s}{s^2 + \frac{\omega_p}{q_p} s + \omega_p^2} \cdot \frac{dA}{A}. \qquad (67)$$

The frequency-dependent function in (66) and (67) is a bandpass function with peak value $\Gamma_\beta^{q_p}$, which is modified by the function $1/A(s)$. Thus again, here, *minimizing the gain-sensitivity product implies minimizing the amplitude error with respect to nonideal op-amp characteristics*. Experimental results have verified this impressively, both for networks of second-order, as well as of higher order [7].

VI. Conclusions

It has been shown that the effects of finite gain, gain-bandwidth product, and the variations of these quantities on the pole frequency of an active filter using an op amp can be expressed in terms of the GSP. The resulting expressions are conceptually simple and completely general. They demonstrate that the minimization of the GSP will always minimize the errors of the pole frequency and Q caused by nonideal amplifier characteristics. Furthermore, in the case of second-order filters, it is generally sufficient to minimize the GSP with respect to the pole Q, i.e., $\Gamma_A^{q_p}$. Experimental results verify these findings.

The expressions derived also provide insight into the influence of the op amp frequency compensation on the location of the filter pole. An example is given of a single-amplifier all pass circuit first using single-pole, and then double-pole single-zero compensation. As anticipated by the theory, the experimental results obtained by the latter are far superior to those obtained with the former.

Finally, an expression is derived for the error of a general nth-order transfer function due to nonideal amplifier gain characteristics. It is shown that a general

GSP can be defined whose minimization involves only the minimization of the coefficient GSP's which are frequency independent. Such a minimization reduces the effects of the nonideal op amps in an nth-order multiple-amplifier active network. Experimental results coincide extremely well with these theoretical predictions; they will be published in a forthcoming paper.

References

[1] G. S. Moschytz, "Gain-sensitivity product—A figure of merit for hybrid-integrated filters using single operational amplifiers," *IEEE J. Solid-State Circuits*, vol. SC-6, no. 3, June 1971.

[2] A. S. Sedra and J. L. Espinoza, "Sensitivity and frequency limitations of biquadratic active filters," *IEEE Trans. Circuits and Systems*, vol. CAS-22, no. 2, Feb. 1975.

[3] P. E. Fleischer, "Sensitivity minimization in a single amplifier biquad circuit," *IEEE Trans. Circuits and Systems*, vol. CAS-23, no. 1, pp. 45–55, Jan. 1976.

[4] G. S. Moschytz, *Linear Integrated Networks: Fundamentals*. New York: Van Nostrand-Reinhold, 1974.

[5] ——, *Linear Integrated Networks: Design*. New York: Van Nostrand-Reinhold, 1975.

[6] J. K. Fidler, "Network sensitivity calculation," *IEEE Trans. Circuits and Systems*, vol. CAS-23, no. 9, pp. 567–571, Sept. 1976.

[7] J. Vlach, G. S. Moschytz, K. Singhal, and P. Horn, "Computer design of higher-order active networks: Part I—Analysis and experimental results," *1977 Proc. IEEE Int. Symp. on Circuits and Cystems*, Apr. 1977, pp. 207–214.

[8] ——, "Computer design of higher-order active networks: Part II—Optimization and study of various technologies," *1977 Proc. IEEE Int. Symp. on Circuits and Systems*, Apr. 1977, pp. 207–214.

[9] D. Hilberman, "An approach to the sensitivity and statistical variability of biquadratic filters," *IEEE Trans. Circuit Theory*, vol. CT-20, no. 4, pp. 382–390, July 1973.

[10] K. R. Laker and M. S. Ghausi, "A statistical multiparameter sensitivity measure for notch filters," *IEEE Trans.* vol. CAS-23, no. 3, pp. 173–176, Mar. 1976.

Paper 1-4

Multiparameter Sensitivity in Active *RC* Networks

ALAN L. ROSENBLUM AND MOHAMMED S. GHAUSI, MEMBER, IEEE

Abstract—A new statistical multiparameter measure of sensitivity is presented. The new measure realistically accounts for component tolerances and component tolerance correlations and is shown to be particularly useful in the design of integrated circuits. For illustration the measure is used to analyze several high-Q low-sensitivity networks. It is shown that for high-Q networks the new measure is readily interpreted in terms of the Q and center frequency sensitivities of the circuit elements. The measure is also generalized for high-Q analysis to account for the error in realization due to the ampliler pole as well as the element variations. This enables an evaluation of the circuit response versus center frequency to be performed.

I. Introduction

THE SENSITIVITY of the response of a network to component tolerances is an important practical consideration in the design of active *RC* networks. Deviations from the nominal element values can distort the response and result in an undesirable or unstable circuit. It is desirable, therefore, to employ design techniques that minimize this effect.

A measure must be defined in order to optimize the sensitivity of the response. In general, the choice depends upon the design criteria. Here the frequency response is assumed to be the design specification for which a given network function has been chosen. Since integrated circuit technology is an important aspect in the design of active *RC* networks, another significant factor is the applicability of the measure to an integrated circuit.

In order for the sensitivity measure to be practical, all parameter variations must be included; that is, a multiparameter measure is required. The component tolerances must also be accounted for in the definition of the measure. A general approach is to consider the element variations as random variables. In the production of integrated circuits, the actual nominal values are distributed about the design values and component tolerances are correlated due to the manufacturing process.

In the subsequent sections of this paper, a new multiparameter statistical sensitivity measure is defined. The gen-

Manuscript received July 9, 1970; revised February 10, 1971 and May 21, 1971. This research was supported by National Science Foundation under Grant GK18465.

A. L. Rosenblum was with the Department of Electrical Engineering, New York University, Bronx, N. Y. 10453. He is now with I. V. Computer Consultants, Inc., Greenvale, N. Y.

M. S. Ghausi is with the Department of Electrical Engineering, New York University, Bronx, N. Y. 10453.

Reprinted from *IEEE Trans. Circuits Theory*, vol. CT-18, pp. 592-599, Nov. 1971.

eral nature of the new measure and its utilization in integrated circuit design is described. A special application of the measure to high-Q analysis is also presented.

The measure is readily amenable to optimization and thus a minimum sensitivity network can be designed. It also provides a useful means of comparing sensitivities of transfer functions realized by a cascade of second-order sections with those realized by a section of higher order.

II. A General Statistical Sensitivity Measure

Let $T(j\omega, \mathbf{x})$ be a given network transfer function, where $\mathbf{x}$ is a column vector whose components are the elements of the network. The network function deviation due to parameter variations $\mathbf{\Delta x}$ is defined as

$$\Delta T = T(j\omega, \mathbf{x} + \mathbf{\Delta x}) - T(j\omega, \mathbf{x}). \tag{1}$$

The variation of the ith element of $\mathbf{\Delta x}$ (i.e., Δx_i) is assumed to be a random variable with zero mean and known statistics (distribution or variance). For a specified frequency band, the sensitivity measure is defined as

$$M(\mathbf{x}) = E\left[\int_{\omega_1}^{\omega_2} \left|\frac{\Delta T}{T}\right|^2 d\omega\right] \tag{2}$$

where E denotes the expected value.

The explicit functional dependency of ΔT on $\mathbf{\Delta x}$ is unknown. It can be approximated, however, under restrictions imposed upon the magnitudes of the components of $\mathbf{\Delta x}$. Here it is assumed that the element variations are small enough to utilize a linear approximation. In view of this, $\Delta T/T$ is expressed as follows:

$$\frac{\Delta T}{T} \doteq \frac{[\nabla_{\mathbf{x}} T]^t}{T} \mathbf{\Delta x} \tag{3}$$

where $\nabla_{\mathbf{x}} T = [\partial T/\partial x_1 \cdots \partial T/\partial x_k]^t$ and k is the number of elements in the network. It is convenient to consider a parameter variation normalized with respect to a nominal design value, i.e., $\Delta x_i/x_i$. Thus $\Delta T/T$ is written as

$$\frac{\Delta T}{T} \doteq \frac{[\nabla_{\mathbf{x}} T]^t}{T} D\mathbf{\hat{\Delta}x} = \boldsymbol{d}^t \mathbf{\hat{\Delta}x} \tag{4}$$

where

$$\mathbf{\hat{\Delta}x} = \begin{bmatrix} \Delta x_1/x_1 \\ \vdots \\ \Delta x_k/x_k \end{bmatrix} \quad \text{and} \quad D = \begin{bmatrix} x_1 & & & 0 \\ & x_2 & & \\ & & \ddots & \\ 0 & & & x_k \end{bmatrix}$$

and

$$\boldsymbol{d} = D\frac{\nabla_{\mathbf{x}} T}{T}, \quad \text{a } k \times 1 \text{ vector.}$$

Using (4) in (2) the measure becomes

$$M(\mathbf{x}) = E\left[\int_{\omega_1}^{\omega_2} (\boldsymbol{d}^t \mathbf{\Delta x})^* \mathbf{\hat{\Delta}x}^t \boldsymbol{d}\, d\omega\right] = \int_{\omega_1}^{\omega_2} \boldsymbol{d}^{t*} P \boldsymbol{d}\, d\omega \tag{5}$$

where $P = E[\mathbf{\hat{\Delta}x}\mathbf{\hat{\Delta}x}^t]$ is the covariance matrix ($k \times k$) of the component tolerances, * denotes the complex conjugate, and t the transpose. It is assumed that the covariance matrix P is known from either sample statistics or given distributions.

Note that the multiparameter measure presented by Schoeffler [1] is a special case of the measure defined in (5). In particular, consider the integrand of (5). The ith component of the vector $\boldsymbol{d}$ is

$$d_i = \frac{x_i}{T}\frac{\partial T}{\partial x_i} = S_{x_i}{}^T. \tag{6}$$

If the covariance matrix is assumed diagonal (i.e., zero cross-correlation terms), then the integrand in (5) can be expressed as

$$\boldsymbol{d}^{t*} P \boldsymbol{d} = \sum_{i=1}^{k} |S_{x_i}{}^T|^2 \sigma_{x_i}{}^2 \tag{7}$$

where

$$\sigma_{x_i}{}^2 = E[(\Delta x_i/x_i)^2], \qquad \text{for } i = 1, 2, \cdots, k.$$

If $\sigma_{x_i}{}^2$ is taken as unity for all x_i, then (7) is the multiparameter measure of sensitivity defined by Schoeffler. The sensitivity definition proposed by Kuo and Goldstein [2] may also be obtained as a special case of the measure in (5) [3].

In order for the measure in (5) to be useful, a general computing algorithm must be developed. First consider determination of the vector $\boldsymbol{d}$. The ith component of $\boldsymbol{d}$ is given in (6).

An effective representation of d_i is described from the structure of the transfer function. Consider the general transfer function of an active RC network which can be expressed as the ratio of two polynomials:

$$T(j\omega) = \left.\frac{N(s)}{D(s)}\right|_{s=j\omega} = \left.\frac{b_n s^n + b_{n-1}s^{n-1} + \cdots + b_1 s + b_0}{s^n + a_{n-1}s^{n-1} + \cdots + a_1 s + a_0}\right|_{s=j\omega} \tag{8}$$

(where some of the b_i may be zero). Examining (8), it is noted that $T(j\omega)$ can be considered as an explicit function of the coefficients of the polynomials $N(s)$ and $D(s)$ and an implicit function of the parameter vector $\mathbf{x}$. With this interpretation in mind, d_i in (6) is expressed as

$$d_i\Big|_{s=j\omega} = \frac{x_i}{T}\left[\frac{\partial T}{\partial b_n}\frac{\partial b_n}{\partial x_i} + \cdots + \frac{\partial T}{\partial b_0}\frac{\partial b_0}{\partial x_i} + \frac{\partial T}{\partial a_{n-1}}\frac{\partial a_{n-1}}{\partial x_i} + \cdots + \frac{\partial T}{\partial a_0}\frac{\partial a_0}{\partial x_i}\right]\Bigg|_{s=j\omega} \tag{9}$$

Equation (9) denoted in a compact form becomes

$$d_i\Big|_{s=j\omega} = \frac{x_i}{T}\frac{\partial \boldsymbol{b}^t}{\partial x_i}\nabla_b T\Big|_{s=j\omega} + \frac{x_i}{T}\frac{\partial \boldsymbol{a}^t}{\partial x_i}\nabla_a T\Big|_{s=j\omega} \tag{10}$$

where

$$\boldsymbol{a} = [a_{n-1}a_{n-2} \cdots a_1 a_0]^t \qquad \boldsymbol{b} = [b_n b_{n-1} \cdots b_1 b_0]^t$$

and

$$\nabla_a T = [\partial T/\partial a_{n-1} \cdots \partial T/\partial a_0]^t$$
$$\nabla_b T = [\partial T/\partial b_n \cdots \partial T/\partial b_0]^t.$$

Utilizing the relationship of (10) in (5), the measure is described by three terms [4], namely,

$$M(\mathbf{x}) = \int_{\omega_1}^{\omega_2} \left[\frac{\nabla_a T}{T}\right]^{*t} C_1{}^t P C_1 \left[\frac{\nabla_a T}{T}\right] d\omega + \int_{\omega_1}^{\omega_2} 2\,\mathrm{Re}\left(\left[\frac{\nabla_a T}{T}\right]^{*t} C_1{}^t P C_2 \left[\frac{\nabla_b T}{T}\right]\right) d\omega + \int_{\omega_1}^{\omega_2} \left[\frac{\nabla_b T}{T}\right]^{*t} C_2{}^t P C_2 \left[\frac{\nabla_b T}{T}\right] d\omega \quad (11)$$

where

$$C_1 = \underset{k \times n}{\begin{bmatrix} x_1 \dfrac{\partial \mathbf{a}^t}{\partial x_1} \\ \cdot \\ \cdot \\ \cdot \\ x_k \dfrac{\partial \mathbf{a}^t}{\partial x_k} \end{bmatrix}} \quad \text{and} \quad C_2 = \underset{k \times (n+1)}{\begin{bmatrix} x_1 \dfrac{\partial \mathbf{b}^t}{\partial x_1} \\ \cdot \\ \cdot \\ \cdot \\ x_k \dfrac{\partial \mathbf{b}^t}{\partial x_k} \end{bmatrix}}.$$

The matrices C_1 and C_2 are computed directly from the coefficients of the transfer function described symbolically in terms of the element values. For a given transfer function the terms $\nabla_a T$ and $\nabla_b T$ are fixed and independent of the parameter values.

A. Examples of the Measure Computation for Uncorrelated and Correlated Tolerances

In this section a simple active *RC* network configuration is analyzed on the basis of the sensitivity measure defined in (11) for uncorrelated and correlated component tolerances. In particular, a Sallen and Key (SK) circuit [5] (Fig. 1) that utilizes one operational amplifier is analyzed.

The transfer function to be realized is normalized with respect to ω_0 (center radian frequency) and represented as

$$T(s_n) = \frac{N(s_n)}{s_n{}^2 + a_1 s_n + a_0} = \frac{N(s_n)}{s_n{}^2 + (1/Q)s_n + 1} \quad (12)$$

where s_n is the complex frequency variable normalized with respect to ω_0. It is assumed that $N(s)$ has no zeros in the vicinity of the roots of the denominator.

The bandpass region is taken from $\omega_1 = 1 - 1/Q$ to $\omega_2 = 1 + 1/Q$ for $Q = 10$ and the normalized variations in resistors and capacitors (i.e., $\Delta R/R$) are taken to be uniform distributions with variances (i.e., $E[(\Delta R/R)^2]$) of 10^{-4}. Here for simplicity only the first term of the measure expression (11) is considered.

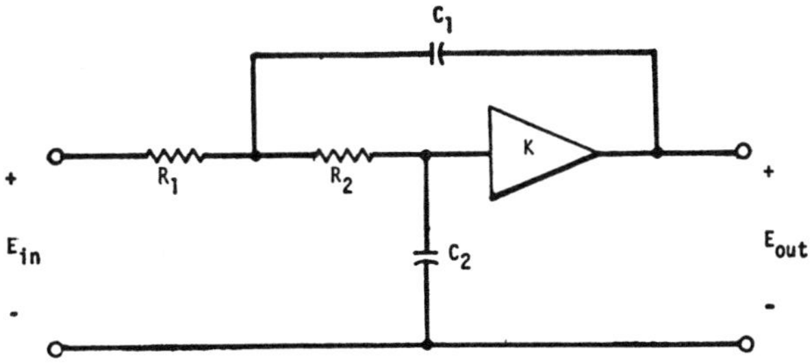

Fig. 1. Sallen and Key network (SK).

The actual transfer function for the network, assuming an ideal voltage-controlled voltage source with a gain K, is

$$T(s) = \frac{N(s)}{s^2 + s[G_1S_1 + G_2S_1 + G_2S_2(1-K)] + G_1S_1G_2S_2} \quad (13)$$

where

$$G_i = 1/R_i \qquad S_i = 1/C_i, \qquad i = 1, 2.$$

The design equations for the normalization used later in (17) are

$$G_1S_1 + G_2S_1 + G_2S_2(1 - K) = 1/Q \quad (14a)$$
$$G_1S_1G_2S_2 = 1. \quad (14b)$$

The matrix C_1 is determined using the definition in (11). Choosing the parameter vector as $\mathbf{x} = [G_1G_2S_1S_2K]^t$, the first row of the C_1 matrix becomes

$$G_1 \frac{\partial \mathbf{a}^t}{\partial G_1} = G_1 \frac{\partial}{\partial G_1} [G_1S_1 + G_2S_1 + G_2S_2(1 - K) \,\vdots\, G_1S_1G_2S_2] = [G_1S_1 \,\vdots\, G_1S_1G_2S_2]. \quad (15)$$

The other rows of the C_1 matrix are similarly determined and the matrix is expressed as

$$C_1 = \begin{bmatrix} G_1S_1 & G_1S_1G_2S_2 \\ G_2S_1 + G_2S_2(1 - K) & G_1S_1G_2S_2 \\ S_1(G_1 + G_2) & G_1S_1G_2S_2 \\ G_2S_2(1 - K) & G_1S_1G_2S_2 \\ -KG_2S_2 & 0 \end{bmatrix}. \quad (16)$$

Letting $x = G_1S_1$, the terms of the matrix C_1 in (16) can be expressed as functions of the variables K and x, namely,

$$C_1 = \begin{bmatrix} x & 1 \\ 1/Q - x & 1 \\ 1/Q + \dfrac{K-1}{x} & 1 \\ \dfrac{1-K}{x} & 1 \\ \dfrac{-K}{x} & 0 \end{bmatrix}. \quad (17)$$

In view of (17) and the measure expression in (11), the sensitivity measure is a function of the variables x and K. For parameter values $x=0.7$ and $K=1.5$ and a diagonal covariance matrix (uncorrelated component tolerances) the measure is 0.012.

B. Correlated Example

To illustrate the applicability of the measure to an integrated circuit, the correlation of the component tolerances must be taken into account. In most integrated circuits the temperature coefficients of capacitors and resistors are inversely proportional to each other, thus reducing the effects of component tolerances on the RC products. When the component tolerances are assumed to be random, this then implies that the correlation between the tolerances of the same type elements (e.g., two resistors) is positive and the correlation between the tolerances of different type elements (e.g., a resistor and a capacitor) is negative.

As an example, consider the SK network in Fig. 1 as an integrated circuit with a correlation coefficient of -0.7 for resistor and capacitor tolerances, namely,

$$E\left[\frac{\Delta R_i/R_i}{\sigma_{R_i}}\frac{\Delta C_j/C_j}{\sigma_{C_j}}\right] = -0.7$$

and a correlation coefficient of 0.7 for resistor tolerances and capacitor tolerances, i.e.,

$$E\left[\frac{\Delta R_i/R_i}{\sigma_{R_i}}\frac{\Delta R_j/R_j}{\sigma_{R_j}}\right] = 0.7$$

and

$$E\left[\frac{\Delta C_i/C_i}{\sigma_{C_i}}\frac{\Delta C_j/C_j}{\sigma_{C_j}}\right] = 0.7, \qquad \text{for } i \neq j.$$

Note

$$\sigma_{x_i} = \sqrt{E[(\Delta x_i/x_i)^2]}.$$

Assuming no correlation between the passive element tolerances and the gain tolerance, the covariance matrix P using component variances of 10^{-4} previously considered becomes

$$P = \begin{matrix} & \begin{matrix} R_1 & R_2 & C_1 & C_2 & K \end{matrix} \\ \begin{matrix} R_1 \\ R_2 \\ C_1 \\ C_2 \\ K \end{matrix} & \begin{bmatrix} 1 & 0.7 & -0.7 & -0.7 & 0 \\ 0.7 & 1 & -0.7 & -0.7 & 0 \\ -0.7 & -0.7 & 1 & 0.7 & 0 \\ -0.7 & -0.7 & 0.7 & 1 & 0 \\ 0 & 0 & 0 & 0 & 1 \end{bmatrix} \end{matrix} 10^{-4}. \tag{18}$$

Using the covariance matrix in (18) with the same parameter values as in the uncorrelated case the measure is computed as 0.0072.

Comparing this result with the uncorrelated case previously considered, it is noted that the measure for the correlated case is smaller. As the correlation for resistor and capacitor tolerances decreases (approaches -1) the measure becomes smaller as expected. For example, with correlation coefficients of ± 0.9 the measure is 0.0051. From the above, it is seen that the measure of sensitivity in (11) is practical for use in integrated circuits since it accounts for component tolerances and component tolerance correlations.

III. High-Q Sensitivity Considerations

For high-Q design, the measure in (11) can be simplified and interpreted in terms of Q and center frequency sensitivities when the covariance matrix is diagonal. The denominators for the first, second, and third terms of the measure in (11) are $|D|^2$, and $|D||N|$, and $|N|^2$, respectively, where $N(s)$ is the numerator and $D(s)$ is the denominator of the transfer function $T(s)$. For high-Q second-order networks $|D|$ is of the order of $1/Q$ in the bandpass region of the transfer function (normalized with respect to the center frequency). Since the denominator of the first term of the measure ($|D|^2$) is much smaller than the denominators of the other two terms ($|D||N|$, $|N|^2$) for high Q, the last terms of the measure in (11) are significantly smaller than the first and can be neglected.

Considering only the first term in (11), the measure becomes

$$M(x) \doteq \int_{\omega_1}^{\omega_2} \left[\frac{\nabla_a T}{T}\right]^{t*} C_1{}^t P C_1 \left[\frac{\nabla_a T}{T}\right] d\omega. \tag{19}$$

It is also noted that under a linear approximation

$$\left[\frac{\nabla_a T}{T}\right]^{*t} C_1{}^t P C_1 \left[\frac{\nabla_a T}{T}\right] \doteq E\left[\left|\frac{\Delta D}{D}\right|^2\right]. \tag{20}$$

It is particularly useful to develop an interpretation of (19) in terms of Q and center frequency sensitivities. If the covariance matrix P is taken as a diagonal matrix (i.e., zero cross-correlation terms), (19) is expressed as

$$M(x) = \int_{\omega_1}^{\omega_2} \sum_{i=1}^{k} |S_{x_i}{}^D|^2 \sigma_{x_i}{}^2 \, d\omega \tag{21}$$

where

$$S_{x_i}{}^D = \frac{x_i}{D}\frac{\partial D}{\partial x_i} \qquad \sigma_{x_i}{}^2 = E\left[\left(\frac{\Delta x_i}{x_i}\right)^2\right]$$

and k is the number of elements in the network. Consider the denominator of the transfer function, normalized with respect to ω_0 (center radian frequency), namely,

$$D(s) = s^2 + \frac{1}{Q}s + 1 = s^2 + a_1 s + a_0 \tag{22a}$$

where

$$a_1 = 1/Q \quad \text{and} \quad a_0 = 1 \tag{22b}$$

are the design equations and s is the complex frequency normalized with respect to ω_0. Utilizing (22a), $|S_{x_i}{}^D|^2_{\omega=\omega_0=1}$ is written as

$$|S_{x_i}{}^D|^2_{\omega=\omega_0=1} = Q^2\left[\left(\frac{x_i\partial a_1}{\partial x_i}\right)^2 + \left(\frac{x_i\partial a_0}{\partial x_i}\right)^2\right]\Bigg|_{\omega=\omega_0=1}. \quad (23)$$

From (6) we have

$$S_{x_i}{}^Q = \frac{x_i}{Q}\frac{\partial Q}{\partial x_i} \quad \text{and} \quad S_{x_i}{}^{\omega_0} = \frac{x_i}{\omega_0}\frac{\partial \omega_0}{\partial x_i}.$$

The terms

$$\frac{x_i\partial a_1}{\partial x_i} \quad \text{and} \quad \frac{x_i\partial a_0}{\partial x_i}$$

in (23) are expressed as

$$\frac{x_i\partial a_1}{\partial x_i} = \frac{-x_i}{Q^2}\frac{\partial Q}{\partial x_i} = -\frac{1}{Q}S_{x_i}{}^Q$$

$$\frac{x_i\partial a_0}{\partial x_i} = \frac{2x_i}{\omega_0}\frac{\partial \omega_0}{\partial x_i}\bigg|_{\omega_0=1} = 2S_{x_i}{}^{\omega_0}\bigg|_{\omega_0=1}. \quad (24)$$

Substituting (24) into (23) yields[1]

$$|S_{x_i}{}^D|^2_{\omega=\omega_0=1} = (S_{x_i}{}^Q)^2 + 4Q^2(S_{x_i}{}^{\omega_0}|_{\omega_0=1})^2. \quad (25)$$

The expression in (25) provides a means of obtaining a lower bound in the sensitivity measure in terms of Q and center frequency sensitivities (for the case of a diagonal covariance matrix). Examining (21), it is apparent that

$$M(\mathbf{x}) \geq \left(\sum_{i=1}^{k} |S_{x_i}{}^D|^2_{\omega=\omega_0=1}\sigma_{x_i}{}^2\Delta\omega\right) \quad (25a)$$

where $\Delta\omega$ is a weighting factor to account for the step size in the numerical integration.[2] Utilizing (25) and (25a), a lower bound on the measure is expressed as

$$M(\mathbf{x}) \geq \left[\sum_{i=1}^{k} (S_{x_i}{}^Q)^2\sigma_{x_i}{}^2\right]\Delta\omega + \left[4Q^2\sum_{i=1}^{k} (S_{x_i}{}^{\omega_0}|_{\omega_0=1})^2\sigma_{x_i}{}^2\right]\Delta\omega. \quad (26)$$

As demonstrated in the next section, the expression in (26) facilitates the comparison of networks on the basis of the sensitivity measure defined in (5).

IV. High-Q Low-Sensitivity Networks

In this section two high-Q low-sensitivity networks are analyzed and compared for Q's of 10 and 500 to illustrate the application of the measure to high-Q analysis.

[1] A somewhat similar expression has also been proposed and arrived at from a different approach in [6].

[2] The weighting factor ($\Delta\omega$) is chosen by examining (21) and the numerical integration process. It can be shown that when two or more intervals are chosen in the numerical integration (trapezoidal rule), (21) is greater than $(1/Q)\sum_{i=1}^{k}|S_{x_1}{}^D|^2_{\omega=\omega_0=1}\sigma_{x_1}{}^2$. Hence $1/Q$ is used as the weighting factor $\Delta\omega$.

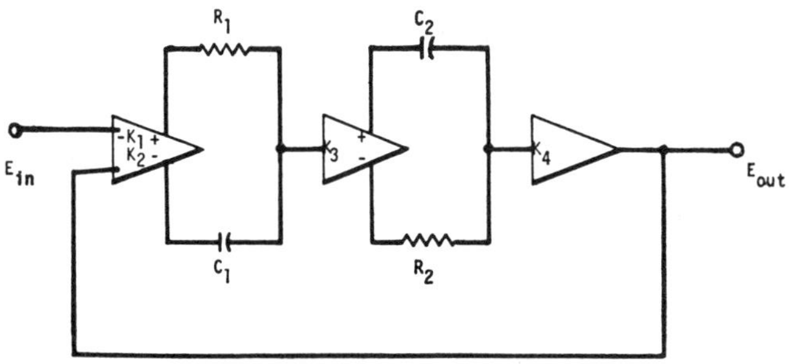

Fig. 2. Tarmy and Ghausi network (TG).

The transfer function to be realized is normalized with respect to ω_0 (center radian frequency) and represented as

$$T(s) = \frac{N(s)}{s^2 + a_1 s + a_0} = \frac{N(s_n)}{s_n{}^2 + \frac{1}{Q}s_n + 1} \quad (27)$$

where s_n is the complex frequency variable normalized with respect to ω_0. It is assumed that $N(s)$ has no zeros in the vicinity of the roots of the denominator.

The design equations are

$$a_1 = 1/Q \quad \text{and} \quad a_0 = 1. \quad (28)$$

The bandpass region is taken from $\omega_1 = 1 - 1/Q$ to $\omega_2 = 1 + 1/Q$, and the normalized variations in resistors and capacitors (i.e., $\Delta R/R$) are taken to be uniform distributions with variances (i.e., $E[(\Delta R/R)^2]$) of 10^{-4}.

A. *Tarmy and Ghausi Network* (TG)

A high-Q low-sensitivity configuration presented by Tarmy and Ghausi [6] is shown in Fig. 2. The transfer function of the network, assuming frequency independent voltage gains, is

$$T(s) = \frac{N(s)}{s^2 + (G_1S_1 + G_2S_2)\left[\frac{1 - K_2K_3K_4}{1 + K_2K_3K_4}\right]s + G_1G_2S_1S_2} \quad (29)$$

and the design equations are

$$(G_2S_2 + G_1S_1)\left[\frac{1 - K_2K_3K_4}{1 + K_2K_3K_4}\right] = \frac{1}{Q} \quad (30a)$$

$$G_1S_1G_2S_2 = 1. \quad (30b)$$

Defining $x = G_1S_1$ and with manipulation of (30a) and (30b), it can also be shown that each term of the matrix C_1 is only a function of x. Thus in view of (19), the measure is dependent only on the parameter x.

Examining (30), it is seen that the gain product of $y = K_2K_3K_4$ is always slightly less than unity, enabling K_2, K_3, and K_4 to be designed at approximately unity gain. In the TG realization of the gains (K_2, K_3, K_4) it can be shown that a 1-percent variation in resistor values will cause a change of gain of less than 0.01 percent. Accounting for the small gain variations (i.e., $E[(\Delta R/R)^2] = 10^{-4} => E[(\Delta K/K)^2] = 10^{-8}$), the measure becomes relatively independent of the parameter

TABLE I

SUMMARY OF SENSITIVITY ANALYSIS

Network Configuration	Lower Bound of Sensitivity[a] Measure [(3)–(8)]		Sensitivity Measure	
	$Q=10$	$Q=500$	$Q=10$	$Q=500$
Sallen and Key [3] (SK)	0.011	b	0.12	b
Two operational amplifiers [8], [9] G, SM	0.006	b	0.0066	b
Tarmy and Ghausi [6] (TG)	0.004	0.2	0.0045	0.22
Kerwin, Huelsman, and Newcomb [7] (KHN)	0.006	0.3	0.0067	0.33
Biquad [10], [11] (Tow and Thomas)	0.006	0.3	0.0066	0.33

[a] Weighting factor $1/Q$.
[b] Cannot be built for $Q=500$.

x (changes accuracy in fourth significant figure). The results are summarized in Table I for the following parameters:

$$G_1S_1 = G_2S_2 = 1 \quad \text{and} \quad K_2K_3K_4 = \frac{1 - 1/2Q}{1 + 1/2Q}.$$

These results are interpreted quite readily in view of the expression in (26). The center frequency sensitivity ($S_{x_i}^{\omega_0}$) is $\frac{1}{2}$ for $x_i = G_1, S_1, G_2, S_2$. Thus the second term in (26) is $4Q^2\sigma_{x_i}^2\Delta\omega$ and independent of the parameter values. The network has low-Q sensitivities ($S_{x_i}^Q = 1$) for the elements G_1, S_1, G_2, and S_2. For the gains, however, $S_{x_i}^Q$ is approximately Q. Since the variances of the gain variations are much smaller than the variances of resistor and capacitor tolerances, the product $(S_{x_i}^Q)^2\sigma_{x_i}^2\Delta\omega$ for each gain term (K_2, K_3, K_4), although dependent upon the nominal element values, is negligible compared to the second expression in (26). Thus the lower bound is relatively independent of the nominal element values and dependent upon the variations of the components of the center frequency term.

B. Kerwin, Huelsman, and Newcomb Network (KHN)

A high-Q low-sensitivity network presented by Kerwin, Huelsman, and Newcomb [7] is shown in Fig. 3. The transfer function, assuming ideal operational amplifiers, is

$$T(s) = \frac{N(s)}{s^2 + \dfrac{G_2}{G}\left[\dfrac{G_1 + G}{G_2 + G_5}\right]S_1G_8s + \dfrac{G_1}{G}S_1G_8S_2G_9} \tag{31}$$

and the design equations are

$$\frac{1}{Q} = \frac{G_2}{G}\left[\frac{G_1 + G}{G_2 + G_5}\right]S_1G_8$$

$$1 = \frac{G_1}{G}S_1G_8S_2G_9. \tag{32}$$

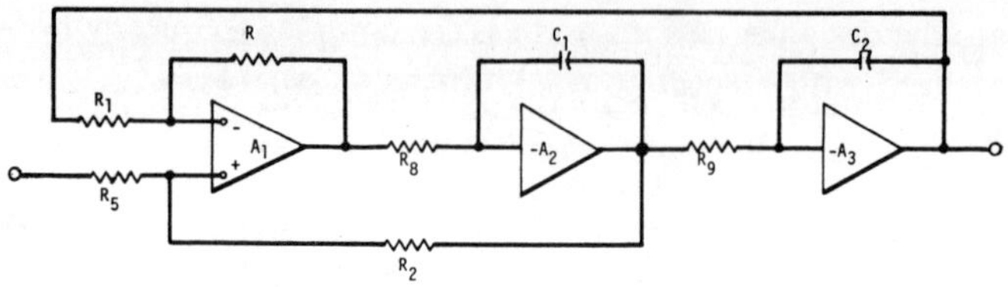

Fig. 3. Kerwin, Huelsman, and Newcomb network (KHN).

Defining $x = G_1/G$ and $y = G_8S_1$, it can be shown that the terms of the C_1 matrix in (19) are functions of x and y. The measure, however, is relatively independent of the parameters x and y (i.e., changes occur in the fourth significant figure). The results are summarized in Table I[3] for parameter values $y = x = 1$. They can readily be interpreted by examining the lower bound of the measure developed in (26). The second term of the expression [in (26)] is $6Q^2\sigma_{x_i}^2\Delta\omega$ and independent of the parameters x and y. Although the element Q sensitivities that comprise the first term in (26) are functions of x and y, they are small (i.e., $S_x^Q \leq 1$) and thus provide an insignificant contribution to the lower bound expression. It is noted from (32) that the center frequency of the KHN configuration depends on six elements as opposed to four in the TG network. Therefore, although both networks are Q-insensitive, the KHN circuit has a higher sensitivity measure than the TG network. The results of a sensitivity analysis for some other active RC configurations are also summarized in Table I.

V. NONIDEAL AMPLIFIER CONSIDERATIONS IN HIGH-Q DESIGN

In the previous section, the sensitivity analysis was performed assuming ideal network elements. In practical designs, however, the poles of the amplifiers cannot be neglected at high center frequencies because they affect not only the sensitivity of the network but also introduce errors in the design realization. In this section a general measure that accounts for the error in realization due to the poles of the amplifiers and the sensitivity of the network due to the variation of the components is developed.

Let the denominator of a transfer function be denoted as $D(\mathbf{a}, \mathbf{x})$, where $\mathbf{x}$ is the parameter vector and the components of $\mathbf{a}$ are the coefficients of the denominator polynomial. Defining $\mathbf{a}^0$ as the set of coefficients that gives the desired denominator, $\Delta D/D$ is written as

$$\frac{\Delta D}{D} = \frac{D(\mathbf{a}^0 + \Delta \mathbf{a}, \mathbf{x} + \Delta \mathbf{x}) - D(\mathbf{a}^0, \mathbf{x})}{D(\mathbf{a}^0, \mathbf{x})}. \tag{33}$$

Expanding (33) under a linear approximation yields

$$\frac{\Delta D}{D} \approx \frac{(\nabla_{\mathbf{a}} D)^t}{D(\mathbf{a}^0, \mathbf{x})}\Delta \mathbf{a} + \mathbf{d}^t\hat{\Delta}\mathbf{x} \tag{34}$$

[3] The biquad network [10], [11] has also been included. However, its performance is very similar to the KHN network and is therefore not treated separately here [4].

where $\boldsymbol{d}$ and $\hat{\Delta}x$ are defined in (4) and $\hat{\Delta}a$ is the change in denominator coefficients due to the nonideal effects of the amplifier. Utilizing the computational relationship for $\boldsymbol{d}$ in (10), (34) is written as

$$\frac{\Delta D}{D} = \frac{[\nabla_a D]^t}{D}\,\Delta a + \frac{[\nabla_a D]^t}{D}\,C_1\Delta x. \tag{35}$$

Noting the relationship in (20) and the fact that $\nabla_a D$ is independent of $\boldsymbol{a}$, the measure is expressed as

$$M(\mathbf{x}) = \int_{\omega_1}^{\omega_2} E\left[\left|\frac{\Delta D}{D}\right|^2\right] d\omega$$

$$= \int_{\omega_1}^{\omega_2} \frac{[\nabla_a D]^t}{D}\,(\Delta a \Delta a^t + C_1{}^t P C_1)\,\frac{\nabla_a D^*}{D^*}\,d\omega \tag{36a}$$

or

$$M(\mathbf{x}) = \int_{\omega_1}^{\omega_2} \frac{[\nabla_a D]^t}{D}\,\Delta a \Delta a^t\,\frac{\nabla_a D^*}{D^*}\,d\omega + \int_{\omega_1}^{\omega_2} \frac{[\nabla_a D]^t}{D}\,C_1{}^t P C_1\,\frac{\nabla_a D^*}{D^*}\,d\omega. \tag{36b}$$

Note that $E[\Delta\hat{x}]=0$. From (36b) it is seen that the general measure consists of two terms: the second expression is the sensitivity term previously considered and the first expression, denoted as the error term, accounts for the error in the desired realization due to the amplifier poles.

As an example (for simplicity, since only one amplifier is required) consider the SK network in Fig. 1. The form of the gain function is taken as

$$A(s) = \frac{K}{\dfrac{s}{p_0} + 1} \tag{37}$$

where K is the dc (ideal) gain, p_0 is the pole of the amplifier, and s is the complex frequency variable. (More accurate expressions for the gain function, such as one involving two poles, etc., can be used in which case the expressions will be more involved.) If Kp_0 is defined as the gain-bandwidth product (assumed constant) and s_n is the normalized complex frequency variable with respect to the center (radian) frequency ω_0, then the gain term in (37) is written as

$$\frac{K}{s_n \dfrac{K}{A_0} + 1} \tag{38}$$

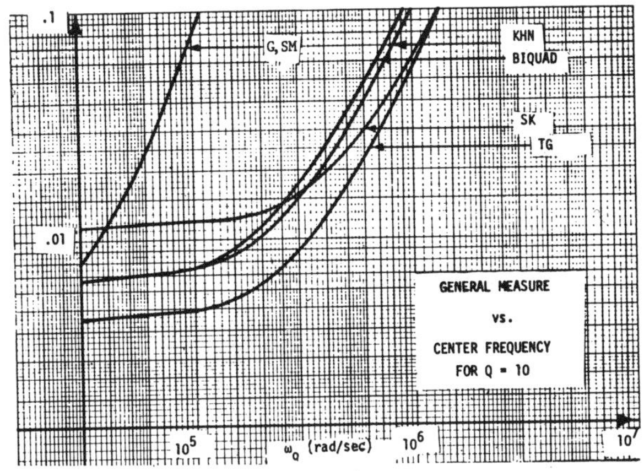

Fig. 4. General measure versus center frequency for $Q=10$.

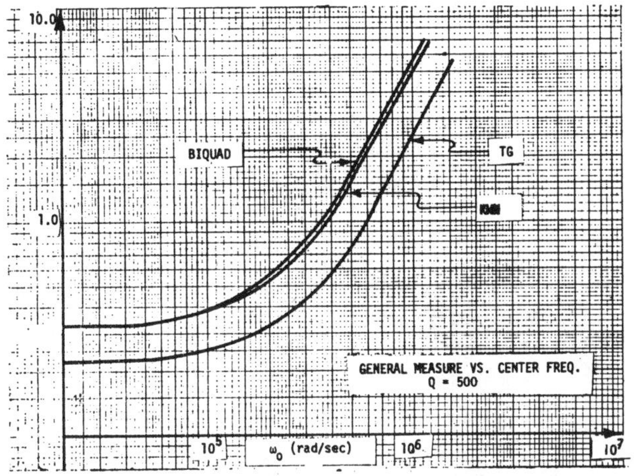

Fig. 5. General measure versus center frequency for $Q=500$.

where

$$A_0 = \frac{\text{gain bandwidth product}}{\omega_0} = \frac{Kp_0}{\omega_0}.$$

Employing the gain relationship in (38), the actual transfer function of the SK network (Fig. 1) is

$$T(s_n) = \frac{N(s_n)}{s_n{}^3 \dfrac{K}{A_0} + s_n{}^2\left[1 + \dfrac{K}{A_0}(G_1S_1 + G_2S_1 + G_2S_2)\right] + s_n\left[G_1S_1 + G_2S_1 + G_2S_2(1-K) + \dfrac{K}{A_0}G_1S_1G_2S_2\right] + G_1S_1G_2S_2}. \tag{39}$$

Note for the ideal case $A_0 \rightarrow \infty$ and (39) becomes (13).

The desired normalized denominator is $s_n{}^2 + (1/Q)s_n + 1$. If the general denominator is taken as a cubic, then

$$\mathbf{a}^0 = \begin{bmatrix} 0 \\ 1 \\ 1/Q \\ 1 \end{bmatrix}.$$

Comparing the desired and actual denominator in (39), it is readily seen that

$$\Delta a = \begin{bmatrix} K/A_0 \\ K/A_0(G_1S_1 + G_2S_1 + G_2S_2) \\ (K/A_0)G_1S_1G_2S_2 \\ 0 \end{bmatrix}.$$

Note that for the ideal case ($A_0 \rightarrow \infty$), $\Delta a = 0$.

The general measure, (36b), accounting for both the effects of element variations and nonideal amplifiers was used to compare the network performance as a function of center frequency for several active RC configurations at Q's of 10 and 500 with a gain-bandwidth product (Kp_0) of 50 MHz, which is typical for good operational amplifiers commercially available. The results are summarized in Figs. 4 and 5. It is noted that in these examples at low center frequencies ($f_0 < 1.6$ kHz), the sensitivity expression of the general measure (36b) is the predominant term, and for center frequencies greater than 169 kHz the error in the realization due to the nonidealness of the amplifiers has the dominate effect.

VI. Summary

A new multiparameter statistical sensitivity measure has been defined. It has been demonstrated that the new measure effectively accounts for component tolerances and component tolerance correlations. This makes it particularly applicable in the design of integrated circuits. Note that although one is usually concerned more with sensitivity in active RC networks, the measure and the methods of this paper are equally applicable to any linear network.

The measure was applied to the analysis of high-Q networks, where it was shown that it could be interpreted in terms of the Q and center frequency sensitivities of the element values and expanded to include the effects of the amplifier poles.

References

[1] J. D. Schoeffler, "The synthesis of minimum sensitivity networks," *IEEE Trans. Circuit Theory*, vol. CT-11, June 1964, pp. 271–276.

[2] A. J. Goldstein and F. F. Kuo, "Multiparameter sensitivity," *IRE Trans. Circuit Theory* (Corresp.), vol. CT-8, June 1961, pp. 177–178.

[3] A. Rosenblum, "Multiparameter sensitivity and sensitivity minimization in active *RC* networks," Ph.D. dissertation, New York University, New York, N. Y., June 1971.

[4] A. Rosenblum and M. S. Ghausi, "Multiparameter sensitivity in active *RC* networks," New York University, New York, N. Y., N.Y.U. Tech. Rep. 400-206, July 1, 1970.

[5] R. P. Sallen and E. L. Key, "A practical method of designing *RC* active networks," *IRE Trans. Circuit Theory*, vol. CT-2, Mar. 1955, pp. 74–85.

[6] R. Tarmy and M. S. Ghausi, "Very high-*Q* insensitive active *RC* networks," *IEEE Trans. Circuit Theory*, vol. CT-17, Aug. 1970, pp. 358–366.

[7] W. J. Kerwin, L. P. Huelsman, and R. W. Newcomb, "State-variable synthesis for insensitive integrated circuit transfer functions," *IEEE J. Solid-State Circuits*, vol. SC-2, Sept. 1967, pp. 87–92.

[8] P. R. Geffe, "A *Q*-invariant active resonator," *Proc. IEEE* (Lett.), vol. 57, Aug. 1969, p. 1442.

[9] M. A. Soderstrand and S. K. Mitra, "Extremely low sensitivity active *RC* filter," *Proc. IEEE* (Lett.), vol. 57, Dec. 1969, pp. 2175–2176.

[10] J. Tow, "A step-by-step active filter design," *IEEE Spectrum*, vol. 6, Dec. 1969, pp. 64–68.

[11] L. C. Thomas, "The Biquad: Pt. I—Some practical design considerations," *IEEE Trans. Circuit Theory*, vol. CT-18, May 1971, pp. 350–357.

[12] L. C. Thomas, "The Biquad: Pt. II—A multipurpose active filtering system," *IEEE Trans. Circuit Theory*, vol. CT-18, May 1971, pp. 358–361.

Active Compensation for High-Frequency Effects in Op-Amp Circuits with Applications to Active *RC* Filters

PETER O. BRACKETT AND ADEL S. SEDRA

Abstract—**The use of "active compensation" in precision electronic circuits employing modern IC op amps, whereby the op amps in a circuit provide "self-compensating" effects to mitigate their finite bandwidth limitations is discussed. In particular, several rather subtle applications in the field of *RC* active filters are presented. A number of novel integrator circuits having high performance are presented, and experimental results which support the theoretical derivations are given. Using these integrators considerably extends the useful frequency range of the "leap-frog" and "coupled-biquad" active filters, at no extra cost.**

I. INTRODUCTION

IN MANY APPLICATIONS of operational amplifiers (op amps) to high quality and precision electronic circuits, such as active-*RC* filters, the frequency response characteristic of the op amp becomes the primary limiting factor in circuit performance [1], [2]. There are several well-known passive compensation techniques which make use of additional passive components to introduce a controlled amount of phase lead that cancels the excess phase lag due to the amplifier's limited bandwidth [1], [3]. The main difficulty with passive compensation is that the additional compensating elements must be individually tailored to the given op amp and adjusted at specific ambient conditions of temperature and power supply voltage. Consequently, under changing ambient conditions the compensation will no longer be satisfactory.

With the introduction of low cost "dual" and "quad" op amps having closely matched characteristics which track with changes in temperature and voltage, it becomes feasible to consider "active compensation" methods whereby op amps in a circuit provide compensation for other op amps in the same circuit. As will be shown, the AcKerberg–Mossberg biquad [11], [12] employs, in effect, this concept. In many cases, however, the techniques for achieving this active compensation are quite subtle and it is the object of this paper to describe a number of such techniques as well as some of their applications.

In the following sections we make use of a single-pole model of the op amp response, that is the op amp gain is given by

$$A(s) = \frac{A_0\omega_p}{s+\omega_p} = \frac{\omega_t}{s+\omega_p}$$

where ω_t is the unity-gain bandwidth. In the frequency range of interest, $\omega \gg \omega_p$, we have

$$A(s) \simeq \frac{\omega_t}{s}$$

and

$$|A(j\omega)| \simeq \frac{\omega_t}{\omega}.$$

Manuscript received January 20, 1975; revised October 1, 1975. This work was supported in part by the National Research Council of Canada under Grant A-7394.

The authors are with the Department of Electrical Engineering, University of Toronto, Toronto, Ont., Canada M5S 1A4.

II. OP AMP INTEGRATORS

We shall direct our attention to the active compensation of op amp integrators which find wide use in many applications and in particular in certain types of active-*RC* filters [1]–[5]. The concept of the integrator Q-factor introduced in [1] will be extensively used. Specifically, if the op-amp frequency response is taken into account the transfer function of an integrator circuit may be expressed as

$$T(j\omega) = \frac{1}{R(\omega) + jX(\omega)}.$$

Then the integrator Q-factor is defined as

$$Q_I = \frac{X(\omega)}{R(\omega)}.$$

Fig. 1 is a compilation of seven different op-amp integrator circuits (both inverting and noninverting), a number of which are new. Also given on Fig. 1 are approximate expressions for the Q-factor of each integrator. These expressions are derived assuming operation at frequencies $\omega_p \ll \omega \ll \omega_t$ and in case of multiple op amps matching has been assumed. A number of observations follow.

1) Fig. 1(a) shows the familiar Miller integrator (inverting) whose Q-factor, derived in [1], is negative and approximately equal to the op amp gain, $|A|$.

2) Fig. 1(b) presents the most obvious way of realizing a noninverting integrator namely, by a Miller-inverter cascade. Note that due to the additional phase shift contributed by the inverter the Q-factor is considerably reduced in magnitude to $\frac{1}{3}|A|$.

3) The Deboo integrator [6] shown in Fig. 1(c) is a single amplifier noninverting integrator with a grounded capacitor. Although its Q-factor is higher than that of the Miller-inverter integrator the Deboo integrator is quite sensitive to resistor ratio matching [7] and therefore has limited practical use.

4) Fig. 1(d) shows a novel noninverting integrator found in the biquad circuits of Ackerberg and Mossberg [11], [12] and which we shall call the "phase-lead" integrator.

Reprinted from *IEEE Trans. Circuits Syst.*, vol. CAS-23, pp. 68-72, Feb. 1976.

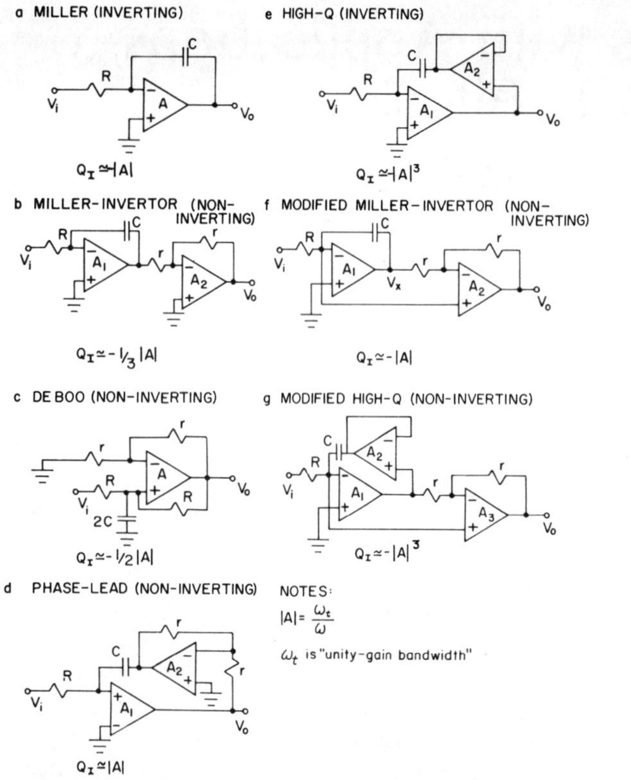

Fig. 1. Seven op-amp integrators and their Q-factors.

In this circuit active compensation, through a matched amplifier in the feedback path, is used to obtain a Q-factor equal in magnitude but opposite in sign to that of the inverting Miller integrator.

5) The feedback active compensation technique is employed in the circuit of Fig. 1(e) to realize an inverting integrator having an ultra-high Q-factor, namely, $Q \simeq -|A|^3$. This improvement of many orders of magnitude is obtained assuming amplifier matching. We have termed this integrator circuit "high-Q" integrator.

6) Another form of active compensation is the feedforward compensation technique employed in the noninverting integrator of Fig. 1(f). This is a modified Miller-inverter integrator which achieves a three-fold increase in the Q-factor magnitude by changing only one connection. Specifically, the summing junction potential of A_1 is fed to the noninverting input of A_2 causing the inverter to have a "transfer function" V_0/V_x given by

$$\frac{V_0}{V_x} = -\left(\frac{1 + 2/A_1}{1 + 2/A_2}\right)$$

which for matched amplifiers, of arbitrary frequency response, becomes $V_0/V_x = -1$. Note that the inverting input terminal of A_2 becomes a "true" virtual ground. It should be obvious that this compensation scheme is applicable, wherever a signal inversion is required, independent of the function the first amplifier is performing (integration in this case).

7) Both the feedforward and feedback compensation techniques are combined to obtain the "modified high-Q" integrator of Fig. 1(g). This is a noninverting integrator having an ultra-high Q-factor; $Q_I \simeq |A|^3$.

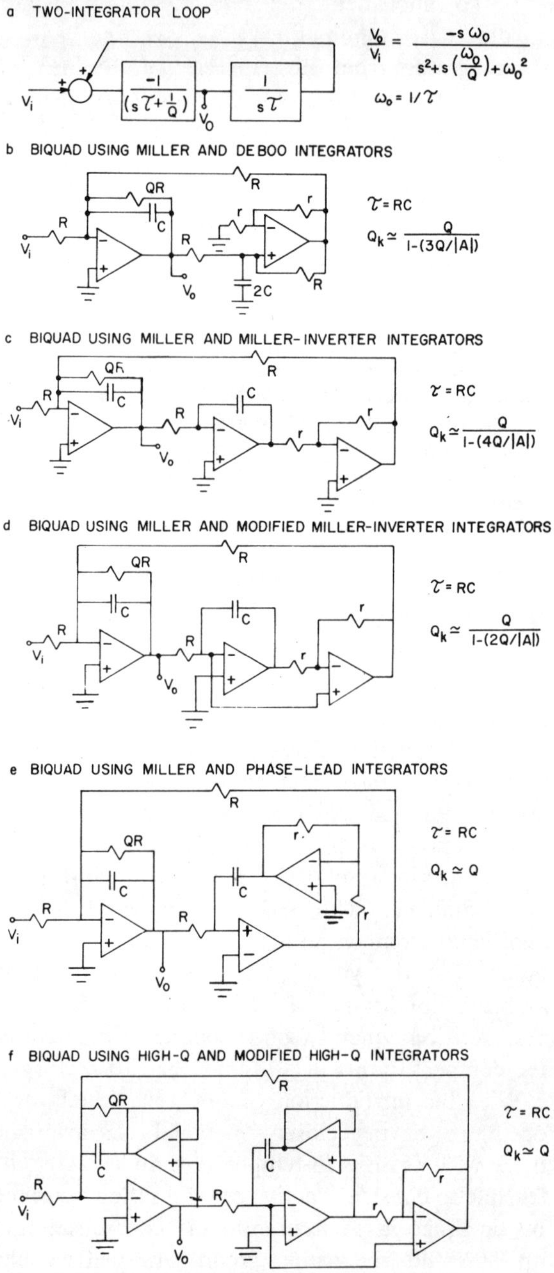

Fig. 2. Biquads using two-integrator loops and their nonideal Q-factors, Q_k.

III. Application to the Design of Biquadratic Filter Sections

As an example of the use of the tabulation of integrators given in Fig. 1 we briefly consider the design of a biquadratic filter section as a two-integrator loop [3], [5]. Fig. 2 shows a block diagram of the loop consisting of two integrators, one inverting and one noninverting. If the integrators are not ideal but have finite Q-factors, Q_1 and Q_2, the realized pole-Q is given by

$$Q_k = \frac{1}{\left[\frac{1}{Q} + \frac{1}{Q_1} + \frac{1}{Q_2}\right]} \tag{1}$$

where Q_1 and Q_2 are evaluated at ω_0.

To implement the two-integrator biquad, we select different combinations of inverting and noninverting integrators from those in Fig. 1 and use Fig. 1 to evaluate the actual realized pole-Q. A number of such realizations together with their Q_k values are given in Fig. 2. Note that the drastic Q-enhancement exhibited by the circuits b, c, and d in Fig. 2 results from choosing integrator implementations whose Q-factors have the same sign. In circuit e, which is the one proposed by Ackerberg and Mossberg [11], [12] the two integrators ideally have Q-factors of equal magnitude and opposite sign leading to cancellation of the error terms. Thus, theoretically the realized pole-Q is equal to its nominal value. However, detailed analysis of this biquad shows that both ω_0 and Q change to ω_{0k} and Q_k where

$$\delta = \frac{\omega_{0k} - \omega_0}{\omega_0} \simeq -\frac{3}{2}\frac{\omega_0}{\omega_t}$$

and

$$Q_k \simeq \frac{Q}{(1 + \delta)\left(1 - 4\delta Q \dfrac{\omega_0}{\omega_t}\right)}.$$

These deviations are quite small and of the same order as those obtained with state-of-the-art biquad sections [8].

The final biquad circuit in Fig. 2 uses the high-Q circuit ($Q \simeq -|A|^3$) for the inverting integrator and the modified high-Q circuit ($Q \simeq -|A|^3$) for the noninverting integrator. The result is a pole-Q that is closer to the design value than that obtained in any other circuit. The present circuit, however, is wasteful in its use of amplifiers and hence its practicality is doubtful. It is included to illustrate a point and for completeness.

IV. Application to the Active Operational Simulations of Ladder Filters

The second application of the results of Fig. 1 is to the realization of active operational simulations of ladder filters. Basically there are two such methods: the "leap-frog" technique [5] which is useful for low-pass filters and other filters derived from low-pass prototypes via reactance transformations, and the recently introduced [2], [14] direct simulation technique which applies to any arbitrary LC ladder structure. In both schemes a one-to-one correspondence is maintained between the reactive elements in the LC ladder and the integrators of the active operational simulation. Consequently, the sensitivity formulas of LC ladders [9], [10], [13] can be applied directly to the active analogs.

Specifically, for an LC ladder filter having a semi-uniform dissipation with Q_L denoting the Q-factor for the inductors and Q_C that for the capacitors, the deviation $\Delta\alpha$ in the attenuation function α is given by [9],

$$\Delta\alpha = 4.34 \left\{\left(\frac{1}{Q_L} + \frac{1}{Q_C}\right)\omega\tau(\omega) + \frac{1}{2}\left(\frac{1}{Q_L} - \frac{1}{Q_C}\right)\operatorname{Im}(\rho_1 + \rho_2)\right\} \text{ dB} \quad (2)$$

TABLE I

Deviation in Attenuation Function of Filters Realized as Operational Simulations of LC Ladder Filters. Different Combinations of Integrators (from Fig. 1) are Used to Simulate Capacitance and Inductance of LC Ladder.

	Capacitance	Inductance	$\Delta\alpha$ in dB
1	Miller	Miller-Inverter	$\Delta\alpha = -\frac{4.34}{\lvert A\rvert}[4\,\omega\tau(\omega) + \operatorname{Im}(\rho_1 + \rho_2)]$ In passband, $\Delta\alpha \approx -\frac{17.36}{\lvert A\rvert}[\omega\tau(\omega)]$
2	Miller	Modified Miller-Inverter	$\Delta\alpha = -\frac{8.68}{\lvert A\rvert}[\omega\tau(\omega)]$
3	Miller	Phase-Lead	$\Delta\alpha = \frac{4.34}{\lvert A\rvert}[\operatorname{Im}(\rho_1 + \rho_2]$ In passband, $\Delta\alpha \approx 0$
4	High-Q	Modified High-Q	$\Delta\alpha = -\frac{8.68}{\lvert A\rvert^3}[\omega\tau(\omega)]$

where $\tau(\omega)$ is the filter's group delay, Im denotes imaginary part, and ρ_1 and ρ_2 are the front and back reflection coefficients. In an active operational simulation one needs integrators of both signs and it has been common practice to use inverting integrators to simulate capacitors and noninverting integrators for the inductors. Equation (2) can thus be used, together with the integrator Q-factors given in Fig. 1, to predict the expected deviaton in the filter attenuation corresponding to different choices of integrator circuit implementations. It should be noted that in most classes of filter networks the first term on the right side of (2) dominates in the passband and in the transition band where $\tau(\omega)$ is large and we can use the approximation

$$\Delta\alpha \simeq 4.34\left(\frac{1}{Q_L} + \frac{1}{Q_C}\right)\omega\tau(\omega). \quad (3)$$

On the other hand, the second term dominates in the stopband where the reflection coefficients are large.[1]

Table I provides the results of evaluating the deviation $\Delta\alpha$ for four different combinations of integrator circuits. Case 1) is the usual implementation and the results agree with those in [1]. We note the large deviation in the passband attenuation. Case 2) which results by a simple modification of the Miller-inverter integrator results in reducing the passband distortion term by a factor of two while the stopband distortion term is ideally reduced to almost zero.

A dramatic improvement in performance is obtained in case 3) which uses the Ackerberg–Mossberg integrator. Here the passband deviation is reduced to almost zero (for perfect matching). Thus using the compensated integrator of Fig. 1(d) in a leap-frog or more general simulation [14] of an LC ladder tremendously extends the useful operating frequency range of this form of active-RC filter realization. Therefore, active filter realizations using integrators to simulate doubly terminated LC ladders do not have "inherent" high-frequency limitations as has recently been claimed [1].

Finally, case 4) makes use of the two high-Q integrators of Fig. 1. The resulting improvement in performance is

[1] In [1] the approximation (3) was used. However, as noted above, the second term cannot be neglected in the stopband specially when Q_L and Q_C are of opposite signs.

evident from the extremely small value for $\Delta\alpha$. This is, however, obtained at the expense of an additional amplifier per integrator. Furthermore, the high-Q values are obtained only if very close amplifier matching and tracking is maintained. Thus the practicability of this realization remains to be evaluated.

V. Experimental Results

To verify some of the results of the previous sections we present experimental data obtained for an active "leap-frog" ladder network. The network chosen for the tests is a 5th-order Chebyshev low-pass filter having a 1-dB passband ripple. This is the same filter considered in [1]. Fig. 3 illustrates the ideal attenuation and group delay response for this filter. The element values of the passive prototype, shown in Fig. 4(a), are $C_1 = C_5 = 2.135$ F, $C_3 = 3.001$ F, and $L_2 = L_4 = 1.091$ H, which correspond to a cutoff frequency of 5.333 kHz.

The standard "leap-frog" simulation of Fig. 4(b) using the Miller integrator of Fig. 1(a) for capacitor simulation and the Miller-inverter cascade of Fig: 1(b) for inductor simulation was constructed using 1458 op amps (which are dual 741's). These devices have a nominal gain-bandwidth product of $f_t = 1$ MHz. No attempt at matching the amplifier packages was made. Amplifiers A_3 and A_4 simulating L_2 were from the same package, and amplifiers A_6 and A_5 simulating L_4 were also in the same package. Amplifiers A_7 and A_8 simulating C_3 and C_5 were from the same package, but amplifier A_2 simulating C_1 was a single amplifier from an additional "dual" package. All integrating capacitors are of polystyrene dielectric, and were assigned a nominal value of 5.1 kpF, and subsequently measured to 0.1 percent accuracy on an impedance bridge. The corresponding resistors were metal film types, and were selected to provide the appropriate RC products to 0.2 percent accuracy. Thus the errors in the transfer function due to the passive component tolerances are quite negligible.

At this point, the transfer function of the filter across the passband was measured, and is shown in Fig. 5 as plot number 1. We note the severe Q-enhancement as predicted, and the resulting passband distortion is $\Delta\alpha = 0.69$ dB which confirms the results of [1], and is very close to that predicted from Table I.

Next, the noninverting inputs of A_4 and A_5 were unsoldered from ground and connected to the inverting inputs of A_3 and A_6, respectively, to form the modified Miller inverters of Fig. 1(f). As shown in Table I, this should reduce the passband distortion by a factor of 2. The plot labeled number 2 of Fig. 5 is the curve measured for this case. Note that the worst case distortion is now only $\Delta\alpha = 0.32$ dB, almost exactly one half that of the standard "leap-frog" network.

Finally, amplifiers A_4 and A_5 were wired in the feedback paths of A_3 and A_6, respectively, and the inputs to A_3 and A_6 were reversed, resulting in the phase-lead integrators of Fig. 1(d). This again does not involve any new components but is merely a wiring change. Plot number 3 of Fig. 5 is the resulting measured transfer function. Here the distortion is very nearly zero as predicted by Table I, and is in fact

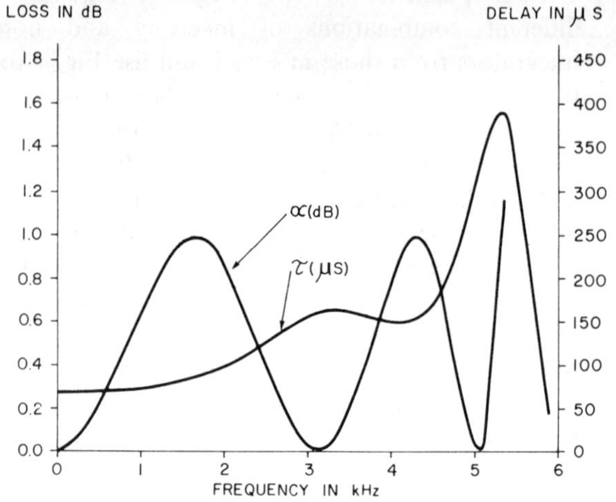

Fig. 3. 5th-order Chebyshev low-pass filter characteristics.

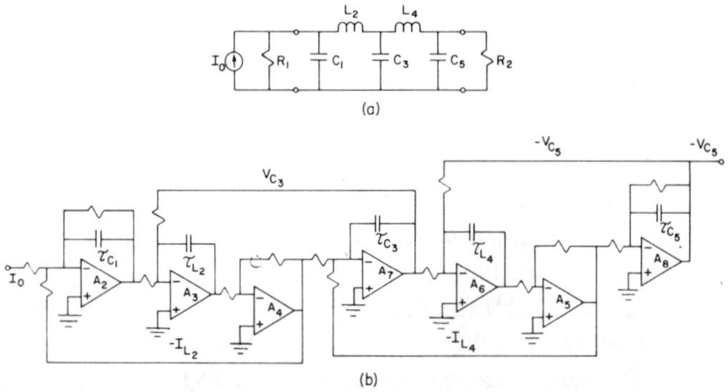

Fig. 4. (a) LC ladder implementing 5th-order Chebyshev and (b) its leap-frog active realization.

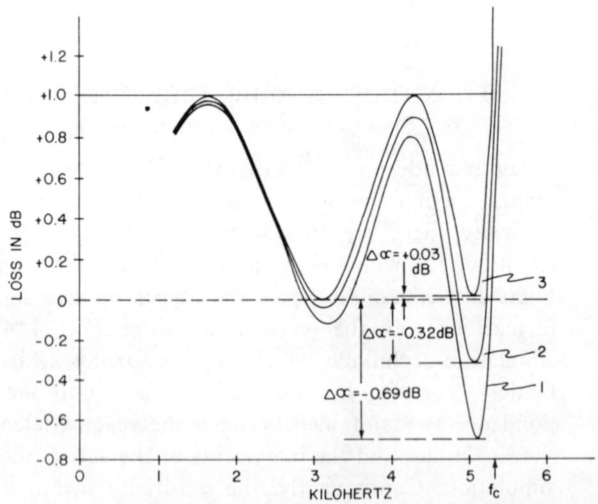

Fig. 5. Measured loss responses for 5th-order Chebyshev filter for various integrator combinations.

so small that it is completely masked by the effect of passive component tolerances!

As a further practical example, we next present experimental results obtained for a 12th-order 20-kHz SSB channel filter designed to pass the lower sideband (17 to 19.7 kHz) with a maximum passband ripple of 0.2 dB. This filter has two loss poles at each of zero and infinite frequency with four finite nonzero loss peaks, three of which are in the upper stopband. Fig. 6 is a plot of the ideal attenuation and

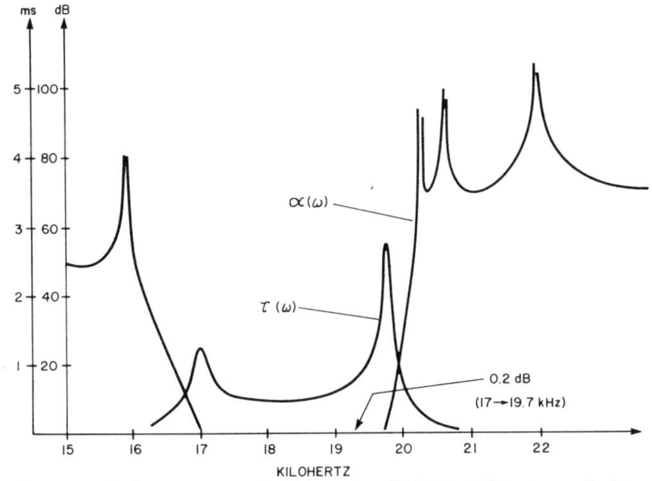

Fig. 6. Ideal response of 12th-order 20-kHz SSB channel filter.

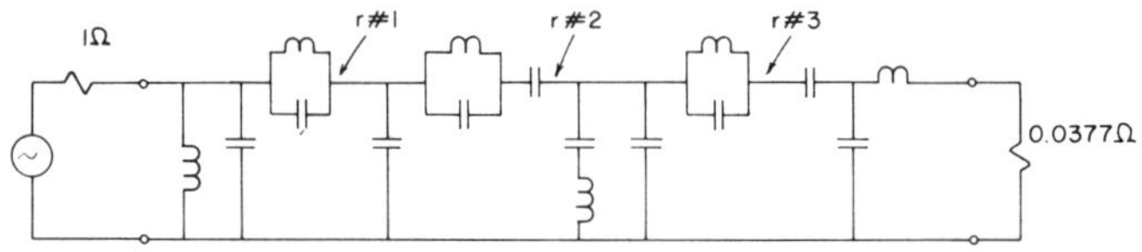

Fig. 7. *LC* ladder realization of the channel filter. Note that in active simulation three reciprocators [2] are required at locations shown.

group delay characteristics of this filter, and Fig. 7 is the schematic diagram for the prototype passive *LC* ladder which was simulated using the techniques of [2]–[14]. For those interested in further details, three "reciprocators"[2] were required in the design and placed in the locations denoted by "*r*" in Fig. 7. The maximum natural mode pole-*Q* is approximately 150 at a frequency of approximately 20 kHz. The op amps used throughout were 1458's (dual 741's) with no conscious attempt made at matching ω_t's. The complete filter was constructed using 1-percent resistors and 2-percent capacitors and was tuned by simply resonating the two integrator loops by trimming one resistor per loop. The integrators used were those of Fig. 1(a) and (d). Fig. 8 illustrates the experimental results obtained across the passband at an input (and output) level of approximately 0 dBm. As can be seen, the worst case loss deviation here is about 0.25 dB illustrating the excellent performance that can be obtained at high frequencies using these methods.

VI. Conclusions

In conclusion, we see that in precision applications of op amps, such as in the field of active *RC* filters, "active compensation" of op amps is not only possible but highly desirable. We have demonstrated theoretically the advantages of such techniques, and have confirmed the results by actual measurements. These techniques allow the limited bandwidth effects of the op amps to be self-compensating and extend the useful operating frequency range of active filters designed as operational simulations of doubly terminated *LC* ladders.

[2] For the definition and implementation of the "reciprocator" see [2], [14].

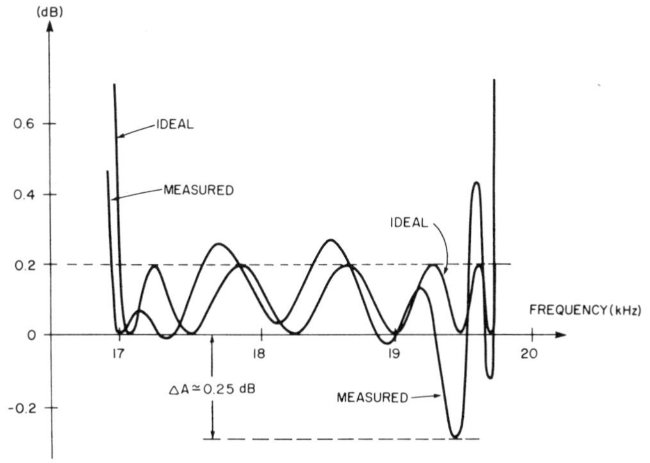

Fig. 8. Ideal and measured passband response of channel filter. Dual 741-type op amps were used and each two-integrator loop was resonant tuned by adjusting one resistor.

Acknowledgment

The authors wish to acknowledge the work of K. Martin at the University of Toronto in the design and construction of the 20-kHz channel filter.

References

[1] L. T. Bruton and A. I. A. Salama, "Frequency limitations of coupled-biquadratic active ladder structures," *IEEE Trans. J. Solid-State Circuits*, vol. SC-9, pp. 70–72, Apr. 1974.

[2] P. O. Brackett and A. S. Sedra, "Application of signal flow graphs to the synthesis of linear systems: Part I—A general theory of network simulation, Part II—Design of active ladder filters," in *Proc. 1975 IEEE Int. Symp. Circuits Syst.*, Apr. 1975, pp. 242–249.

[3] L. C. Thomas, "The biquad: Part I—Same practical design considerations, Part II—A multipurpose active filtering system," *IEEE Trans. Circuit Theory*, vol. CT-18, pp. 350–361, May 1971.

[4] J. Tow and Y. L. Kuo, "Coupled biquad active filters," *Proc. 1972 IEEE Int. Symp. Circuit Theory*, pp. 164–167, Apr. 1972.

[5] F. E. J. Girling and E. F. Good, "Active filters—7. The two integrator loop," *Wireless World*, pp. 76–80, Feb. 1970; "Active Filters—8. The two integrator loop, continued," pp. 134–139, Mar. 1970; "Active Filters—12. The leap frog or active ladder synthesis," pp. 341–345, July 1970.

[6] G. J. DeBoo, "A novel integrator results by grounding its capacitor," *Electron. Des*,. vol. 15, June 7, 1967.

[7] P. R. Geffe, "*RC*-amplifier resonators for active filters," *IEEE Trans. Circuit Theory*, vol. CT-15, pp. 415–419, Dec. 1968.

[8] A. S. Sedra and J. L. Espinoza, "Sensitivity and frequency limitations of biquadratic active filters," *IEEE Trans. Circuits Syst.*, vol. CAS-22, pp. 122–130, Feb. 1975.

[9] M. L. Blostein, "Sensitivity analysis of parasitic effects in resistance terminated *LC* two-ports," *IEEE Trans. Circuit Theory*, vol. CT-14, pp. 21–25, Mar. 1967.

[10] G. C. Temes and H. J. Orchard, "First-order sensitivity and worst case analysis of doubly terminated reactance two-ports," *IEEE Trans. Circuit Theory*, vol. CT-20, pp. 567–571, Sept. 1973.

[11] D. Akerberg and K. Mossberg, "Low sensitive easily trimmed standard building block for active *RC*-filters," *Electron. Lett.*, vol. 5, pp. 528–529, 1969.

[12] D. Akerberg and K. Mossberg, "A versatile active *RC* building block with inherent compensation for the finite bandwidth of the amplifier," *IEEE Trans. Circuits Syst.*, vol. CAS-21, pp. 75–78, Jan. 1974.

[13] G. C. Temes, "The effects of semi-uniform losses in reactance two-ports," *Electron. Lett.*, vol. 8, pp. 161–163, Mar. 1972.

[14] P. O. Brackett and A. S. Sedra, "Direct SFG simulation of *LC* ladder networks with application to active filter design," this issue, pp. 61–67.

Noise Performance of *RC*-Active Quadratic Filter Sections

F. N. TROFIMENKOFF, DAVID H. TRELEAVEN, AND L. T. BRUTON

***Abstract*—A method for calculating the noise voltage at the output of quadratic filter sections is developed. Multiple-feedback low-pass, bandpass, and high-pass quadratic filter sections realized using differential-input single-ended output operational amplifiers are analyzed. The amplifiers are assumed to have infinite input impedance, infinite gain, and zero output impedance. The noise sources associated with the amplifiers are assumed to be statistically independent, but can have both white and $1/f$ noise components. A noise analysis of a fourth-order maximally flat low-pass filter realized by cascading two quadratic filter sections is included.**

Manuscript received April 12, 1972; revised November 6, 1972 and January 8, 1973.

F. N. Trofimenkoff and L. T. Bruton are with the Department of Electrical Engineering, University of Calgary, Calgary, Alta., Canada.

D. H. Treleaven is with the Integrated Circuit Development Group, Microsystems International, Ottawa, Ont., Canada.

I. Introduction

Designers of active filters are concerned with noise performance because both the dynamic range of an active filter and the effectiveness with which a signal can be recovered from a noise background using an active filter are dependent on the output noise of the filter itself. Since $2n$th-order filters can be realized by cascading suitably designed *RC*-active quadratic filter sections, a study of the noise performance of such filter sections is a logical first step in the study of the noise performance of higher order filters of this type.

The output noise of *RC*-active quadratic filter sections is due to Johnson noise [1] in the circuit resistances and to the noise sources associated with the active device or devices used in the

Reprinted from *IEEE Trans. Circuits Theory,* vol. CT-20, pp. 524-532, Sept. 1973.

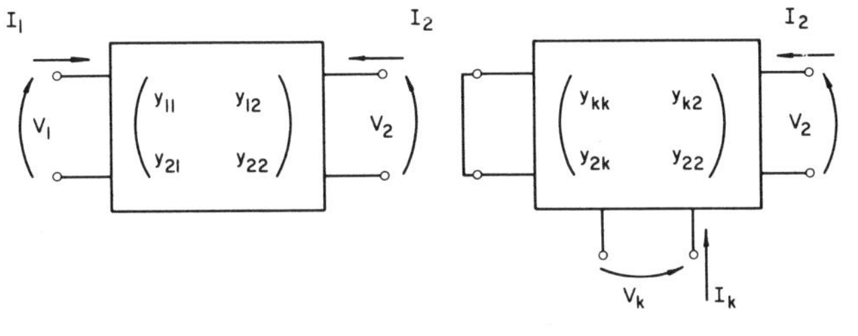

Fig. 1. (a) Conventional two-port network. (b) Technique for dealing with an internal voltage source in a two-port network.

realization of the section. In any particular case, the output noise level will depend on the voltage transfer function of the section, the quality of the active device or devices, the general impedance level of the circuit, and on the topology of the particular realization. Even single-amplifier realizations of quadratic filter sections employing two or three additional network resistors may require as many as five or six noise sources for a complete noise characterization. Thus because a complete noise analysis of an *RC*-active quadratic filter section can be complicated, considerable emphasis has been placed on the form of the noise analysis and on the method of presenting the results. The noise-analysis techniques are applicable to all two-capacitor quadratic section realizations and to some special types of realizations that employ more than two capacitors. In particular, subject to the above reservation, the techniques are directly applicable to quadratic sections that are realized using Shenoi-type gyrators [2], biquadratic generalized impedance converters [3], Sallen-Key structures [4], single-feedback operational amplifier structures [5], and to structures obtained using the Bach procedure [6].

II. Noise-Analysis Technique

A filter section can be viewed as a two-port network, as shown in Fig. 1(a), and can consequently be described by the equations

$$I_1 = y_{11} V_1 + y_{12} V_2$$
$$I_2 = y_{21} V_1 + y_{22} V_2 \tag{1}$$

where y_{11}, y_{12}, y_{21}, and y_{22} are the short-circuit admittance parameters. The open-circuit output voltage V_2 due to an input voltage V_1 will be given by

$$\frac{V_2}{V_1} = -\frac{y_{21}}{y_{22}}. \tag{2}$$

The total open-circuit output noise voltage of the filter section must be obtained by forming the appropriate sum of the contributions to the open-circuit output noise voltage arising from the various noise sources in the filter section. If the input is short-circuited, as shown in Fig. 1(b), a noise source V_k in a branch of the filter network can be regarded as an input to a new two-port network described by the short-circuit admittance parameters y_{kk}, y_{k2}, y_{2k}, and y_{22}. The open-circuit output voltage V_2 due to the noise voltage V_k will be given by

$$\frac{V_2}{V_k} = -\frac{y_{2k}}{y_{22}}. \tag{3}$$

If the discussion is limited to two-capacitor *RC*-active networks in which the electrical parameters of the active devices are independent of frequency, all the short-circuit admittance parameters must take the form

$$y_{mn} = \frac{a_{mn} + b_{mn} s + c_{mn} s^2}{a_0 + b_0 s + c_0 s^2} \tag{4}$$

where a_{mn}, b_{mn}, c_{mn}, a_0, b_0, and c_0 are constants and s is the usual Laplace transform variable. It therefore follows that all of the y_{2k}/y_{22} must take the form

$$\frac{a_{2k} + b_{2k} s + c_{2k} s^2}{a_{22} + b_{22} s + c_{22} s^2}. \tag{5}$$

The low-pass (LP), bandpass (BP), and high-pass (HP) quadratic filter sections that are considered in this work have input-output voltage transfer functions of the form

$$T_{\mathrm{LP}}(j\omega) = H_{O\mathrm{LP}} H_{\mathrm{LP}}(j\omega) = \frac{H_{O\mathrm{LP}}}{1 - \left(\frac{\omega}{\omega_0}\right)^2 + j\frac{\omega}{\omega_0}\frac{1}{Q}} \tag{6}$$

$$T_{\mathrm{BP}}(j\omega) = H_{O\mathrm{BP}} H_{\mathrm{BP}}(j\omega) = \frac{H_{O\mathrm{BP}}\, j\frac{\omega}{\omega_0}}{1 - \left(\frac{\omega}{\omega_0}\right)^2 + j\frac{\omega}{\omega_0}\frac{1}{Q}} \tag{7}$$

$$T_{\mathrm{HP}}(j\omega) = H_{O\mathrm{HP}} H_{\mathrm{HP}}(j\omega) = \frac{H_{O\mathrm{HP}}\left(j\frac{\omega}{\omega_0}\right)^2}{1 - \left(\frac{\omega}{\omega_0}\right)^2 + j\frac{\omega}{\omega_0}\frac{1}{Q}} \tag{8}$$

where $j = \sqrt{-1}$, ω is the angular frequency, ω_0 is the resonant angular frequency of the filter, Q is the quality factor of the filter, $H_{O\mathrm{LP}}$ is the LP filter gain when $\omega = 0$, $QH_{O\mathrm{BP}}$ is the BP filter gain when $\omega = \omega_0$, and $H_{O\mathrm{HP}}$ is the HP filter gain when $\omega \gg \omega_0$. If such quadratic filter sections are realized as two-capacitor *RC*-active networks in which the electrical parameters of the active devices are independent of frequency, it is possible to conclude that V_2/V_k must take the form

$$\frac{V_2}{V_k} = \xi_{\mathrm{HP}}^k T_{\mathrm{HP}}(j\omega) + \xi_{\mathrm{BP}}^k T_{\mathrm{BP}}(j\omega) + \xi_{\mathrm{LP}}^k T_{\mathrm{LP}}(j\omega) \tag{9}$$

where ξ_{HP}^k, ξ_{BP}^k, and ξ_{LP}^k are independent of frequency.

If all of the noise sources are uncorrelated, the total output noise voltage spectral density can be obtained by an rms addition of the contribution to the output noise voltage spectral density due to the various noise sources. Thus if e_{nk} is the noise voltage spectral density associated with the noise voltage source V_k, the total output noise voltage spectral density will be given by [7]

$$e_{no}^2 = \sum_k \left|\frac{V_2}{V_k}\right|^2 e_{nk}^2. \tag{10}$$

If use is made of the relationship [8]

$$|H_{\mathrm{HP}}(j\omega)|^2 = \left(2 - \frac{1}{Q^2}\right) |H_{\mathrm{BP}}(j\omega)|^2 - |H_{\mathrm{LP}}(j\omega)|^2 + 1 \tag{11}$$

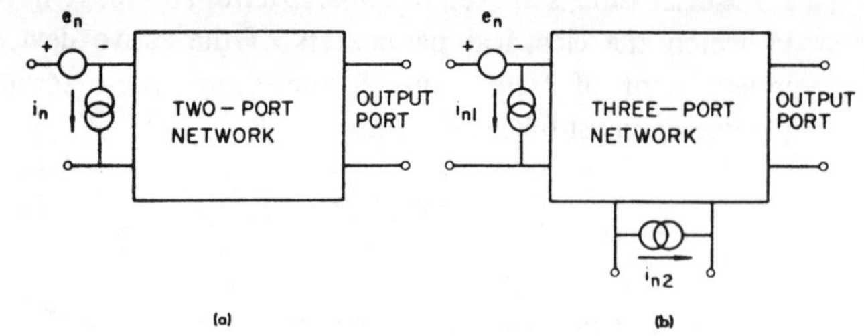

Fig. 2. (a) Representation of noise in a two-port network. (b) Representation of noise in a three-port network.

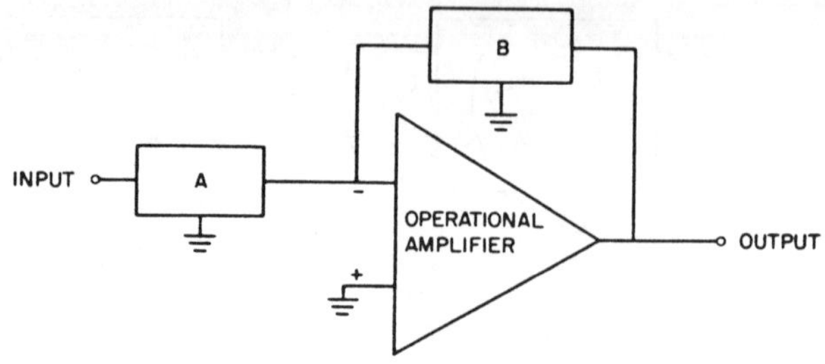

Fig. 3. Single-feedback operational amplifier circuit.

(9) and (10) can be combined to show that (10) will always take the form

$$e_{no}^2 = \phi_{BP}|H_{BP}(j\omega)|^2 + \phi_{LP}|H_{LP}(j\omega)|^2 + \phi_{AP}. \tag{12}$$

ϕ_{BP}, ϕ_{LP}, and ϕ_{AP} (the subscript AP refers to all-pass) are functions of the passive circuit elements, the frequency-independent electrical properties of the active devices, the noise sources associated with the passive elements, and the noise sources associated with the active elements.

An examination of (12) shows that $e_{no}^2 = \phi_{AP}$ for $\omega \to \infty$. The contribution to ϕ_{AP} by a noise source can therefore be calculated by short-circuiting the capacitors. Similarly, $e_{no}^2 = \phi_{LP} + \phi_{AP}$ when $\omega \to 0$. The sum of the contributions to ϕ_{LP} and ϕ_{AP} by any noise source can therefore be calculated by open-circuiting the capacitors. These observations provide a convenient method of checking the contributions to ϕ_{LP} and ϕ_{AP} when these are calculated with the aid of (9)–(11).

The total open-circuit output noise voltage in the frequency range f_l to f_h–$V_{no}(f_l,f_h)$–can be calculated by evaluating

$$V_{no}^2(f_l,f_h) = \int_{f_l}^{f_h} e_{no}^2\, df. \tag{13}$$

Thus when the e_{nk} are independent of frequency, ϕ_{BP}, ϕ_{LP}, and ϕ_{AP} will be independent of frequency and only the two integrals

$$\int_{f_l}^{f_h} |H_{BP}(j\omega)|^2\, df \tag{14}$$

and

$$\int_{f_l}^{f_h} |H_{LP}(j\omega)|^2\, df \tag{15}$$

are required to calculate $V_{no}(f_l,f_h)$.

In practical two-capacitor RC-active networks, resistive noise and active device noise must be considered. The noise voltage spectral density e_{nR} that can be associated with a resistance R at the thermodynamic temperature T is

$$e_{nR}^2 = 4kTR \tag{16}$$

where k is Boltzmann's constant. The noise properties of two-port [9] and three-port [10] active networks can be represented by noise voltage and noise current sources as shown in Fig. 2. These noise voltage and noise current sources often have noise voltage and noise current spectral densities that can be described analytically in the low frequency range by

$$e_{na}^2 = e_{nav}^2 + \left(\frac{f_e}{f}\right) e_{nav}^2 \tag{17}$$

$$i_{na}^2 = i_{nav}^2 + \left(\frac{f_i}{f}\right) i_{nav}^2. \tag{18}$$

The noise voltage and noise current spectral densities e_{nav} and i_{nav} and the characteristic $1/f$ noise corner frequencies f_e and f_i would normally be determined by fitting (17) and (18) to experimental data. There would, in the general case, be some correlation between the noise voltage source and the noise current source or sources, and between the noise current sources. The present work, however, is concerned with the case when the noise sources are statistically independent so that (10) may be used.

Since a $1/f$ dependence of the power spectrum at low frequencies implies an unbounded total noise power, it should be understood that (17) and (18) must be restricted to the frequency above some arbitrary frequency f_d. This frequency f_d would be the dividing point between what should be considered to be noise and what should be considered to be dc drift in a particular application.

The output noise voltage in the frequency range from f_d to some upper frequency f_x would generally be of interest in the case of all three types of sections. The output noise in the frequency range f_1 to f_2 where f_1 and f_2 are the usual upper and lower 3-dB cutoff frequencies would be of interest in the BP case. If only white (independent of frequency) noise sources and $1/f$ noise sources are considered, the evaluation of (13) will involve the first nine integrals listed in the Appendix [11]-[13]. The approximations involved in extending the integrals to range from $f = 0$ to $f = \infty$ in (48), (50), (52), and (54) are relatively good ones. For example, if $(f_d/f_0) = 0.2$ and $(f_x/f_0) = 5$ in (48) and (52), the exact values of the integrals would be $1.37 f_0 Q$ rather than $1.57 f_0 Q$.

Single-feedback RC-active circuits of the type shown in Fig. 3 are frequently used to realize quadratic transfer functions having a complex pole pair. If the networks A and B have no more than two capacitors each and if the operational amplifier input is a virtual earth so that A and B are effectively separated as far as the admittance parameter calculations for each network are concerned, the technique for noise calculations that has been described above will be applicable. Thus many practical RC-active filter noise analyses can be performed in this way despite the seemingly very severe restriction to two-capacitor circuits.

TABLE I
COMPONENTS OF ϕ_{BP}, ϕ_{LP}, AND ϕ_{AP}—LP QUADRATIC FILTER SECTION

	ϕ_{BP}	ϕ_{LP}	ϕ_{AP}
R_1	0	$4kTR_1H^2$	0
R_3	$4kTR_3Q^2\left(1+H+\frac{R_4}{R_3}\right)^2$	$4kTR_3(1+H)^2$	0
R_4	0	$4kTR_4$	0
R_6	$4kTR_6\left[Q^2\left(1+H+\frac{R_4}{R_3}\right)^2+2\left(1+\frac{R_4}{R_3}\right)\right]$	$4kTR_6(2+H)H$	$4kTR_6$
e_{na}	$e_{na}^2\left[Q^2\left(1+H+\frac{R_4}{R_3}\right)^2+2\left(1+\frac{R_4}{R_3}\right)\right]$	$e_{na}^2(2+H)H$	e_{na}^2
i_{na1}	$(i_{na1}R_3)^2Q^2\left(1+H+\frac{R_4}{R_3}\right)^2$	$(i_{na1}R_3)^2\left(1+H+\frac{R_4}{R_3}\right)^2$	0
i_{na2}	$(i_{na2}R_6)^2\left[Q^2\left(1+H+\frac{R_4}{R_3}\right)+2\left(1+\frac{R_4}{R_3}\right)\right]$	$(i_{na2}R_6)^2(2+H)H$	$(i_{na2}R_6)^2$

III. EXAMPLES INVOLVING OPERATIONAL AMPLIFIER REALIZATIONS OF QUADRATIC SECTIONS

Multiple-feedback LP, BP, and HP quadratic sections realized with the aid of differential-input single-ended output operational amplifiers serve as a convenient vehicle for illustrating the procedure for noise analysis that has been outlined in Section II. The amplifiers are assumed to have infinite input impedance and zero output impedance. The voltage gain is assumed to be very high and independent of frequency. The contributions to ϕ_{BP}, ϕ_{LP}, and ϕ_{AP} due to e_{na}, i_{na1}, i_{na2}, and the noise sources associated with the resistors would be obtained as outlined in Section II. Since the results are complicated even for the very simplest circuits, a systematic scheme for tabulating these contributions has been adopted.

A. Multiple-Feedback LP Section

The LP single-amplifier multiple-feedback quadratic section shown in Fig. 4 has the voltage transfer function [14]

$$T_{LP}(j\omega)=\frac{-H}{1-\left(\frac{\omega}{\omega_0}\right)^2+j\left(\frac{\omega}{\omega_0}\right)\frac{1}{Q}} \tag{19}$$

where

$$H=\frac{R_4}{R_1} \tag{20}$$

$$\omega_0^2=\frac{1}{C_2C_5R_3R_4} \tag{21}$$

and

$$Q=\frac{\sqrt{\frac{C_2}{C_5}\frac{R_4}{R_3}}}{\left(1+\frac{R_4}{R_1}+\frac{R_4}{R_3}\right)}. \tag{22}$$

The various contributions to ϕ_{BP}, ϕ_{LP}, and ϕ_{AP} are summarized in Table I. An examination of Table I shows that contributions due to e_{na} are independent of impedance level, contributions due to R_1, R_3, R_4, and R_6 are proportional to the impedance level, and contributions due to i_{na1} and i_{na2} are proportional to the square of the impedance level. If a choice of impedance level is available then for a given Q, ω_0, H, and a given amplifier, the output noise level will be a minimum when the contribution to the output noise due to e_{na} is dominant. A specification of Q, ω_0, H, and the general impedance level of the circuit via R_1 still leaves one degree of freedom at the designer's discretion. A large capacitance ratio is undesirable and R_4/R_3 is sometimes chosen so that C_2/C_5 has the smallest possible value for a given Q and H. This occurs when $(R_4/R_3)=1+H$. Moreover, the resistor R_6 is usually chosen to be equal to $R_3+R_1R_4/(R_1+R_4)$ to minimize dc drift due to amplifier input current variations. With the aid of these additional design constraints, Table I can be rewritten in terms of Q, H, R_1, e_{na}, i_{na1}, and i_{na2} only. The ratio of R_4/R_3 can, of course, be selected to minimize the output noise in some fashion, and Table I is thus a useful starting point for such minimization procedures.

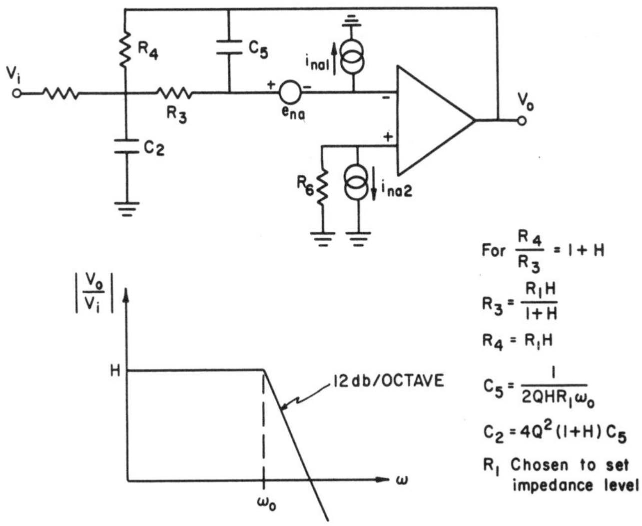

Fig. 4. LP multiple-feedback RC-active filter network.

It is useful to divide the output noise into an "inherent noise" component (that due to R_1, R_3, and R_4), and an "amplifier noise" component (that due to e_{na}, i_{na1}, i_{na2}, and

R_6) [8]. Since Johnson noise has a white spectrum, f_d can be taken to be zero in calculating the output noise due to R_1, R_3, and R_4. Thus for $f_x >> f_0$, (13), (12), and Table I can be used to show that the inherent output noise of the circuit of Fig. 4 is given by

$$V_{n0i}^2 = 4kTR_1H\left[\left\{Q^2\left(1 + H + \frac{R_4}{R_3}\right)^2 + (1 + H)^2\right\}\frac{R_3}{R_4} + (1 + H)\right]\frac{\pi}{2}f_0Q. \quad (23)$$

If the design with $(R_4/R_3) = 1 + H$ is used, the inherent noise will be given by

$$V_{n0i}^2 = 4kTR_1H(1 + H)(2Q^2 + 1)\,\pi f_0Q. \quad (24)$$

B. Multiple-Feedback BP Section

The BP single-amplifier multiple-feedback quadratic section shown in Fig. 5 has the voltage transfer function [14]

$$T_{BP}(j\omega) = \frac{-H\left(j\dfrac{\omega}{\omega_0}\right)}{1 - \left(\dfrac{\omega}{\omega_0}\right)^2 + j\left(\dfrac{\omega}{\omega_0}\,\dfrac{1}{Q}\right)} \quad (25)$$

where

$$H = \sqrt{\frac{\dfrac{R_5}{R_1}\dfrac{C_3}{C_4}}{1 + \dfrac{R_1}{R_2}}} \quad (26)$$

$$\omega_0^2 = \frac{1}{C_4R_5\left(\dfrac{C_3R_1R_2}{R_1 + R_2}\right)} \quad (27)$$

and

$$Q = \frac{\sqrt{\dfrac{R_5C_3}{R_1C_4}\left(1 + \dfrac{R_1}{R_2}\right)}}{\left(1 + \dfrac{C_3}{C_4}\right)}. \quad (28)$$

The various contributions to ϕ_{BP}, ϕ_{LP}, and ϕ_{AP} are summarized in Table II, where

$$\gamma = 1 + Q^2\left(1 + \frac{C_3}{C_4}\right). \quad (29)$$

It will be noted that the contributions to ϕ_{BP}, ϕ_{LP}, and ϕ_{AP} due to e_{na}, i_{na1}, i_{na2}, and the resistors have the same dependence on the impedance level as in the LP filter section example.

The inherent output noise of the circuit of Fig. 5 is given by

$$V_{n0i}^2 = 4kTR_1H^2\left[1 + \frac{R_1}{R_2}\right]\frac{\pi}{2}f_0Q. \quad (30)$$

The value of Q is a maximum for a given value of $[R_5/(R_1R_2/R_1 + R_2)]$ when the capacitors are selected so that $C_3/C_4 = 1$.

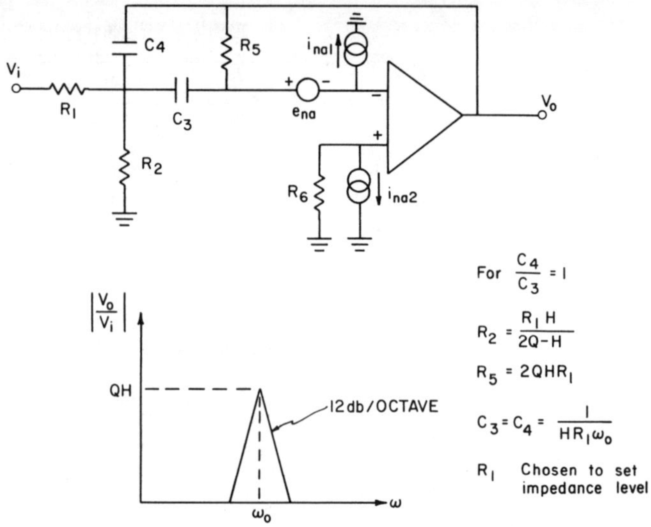

Fig. 5. BP multiple-feedback *RC*-active filter network.

TABLE II
COMPONENTS OF ϕ_{BP}, ϕ_{LP}, AND ϕ_{AP}—BP QUADRATIC FILTER SECTION

	ϕ_{BP}	ϕ_{LP}	ϕ_{AP}
R_1	$4kTR_1H^2$	0	0
R_2	$4kTR_2\left(\frac{R_1}{R_2}\right)^2H^2$	0	0
R_5	$\frac{4kTR_5}{Q^2}$	$4kTR_5$	0
R_6	$4kTR_6\frac{(\gamma^2-1)}{Q^2}$	0	$4kTR_6$
e_{na}	$e_{na}^2\frac{(\gamma^2-1)}{Q^2}$	0	e_{na}^2
i_{na1}	$\frac{(i_{na1}R_5)}{Q^2}$	$(i_{na1}R_5)^2$	0
i_{na2}	$(i_{na2}R_6)^2\frac{(\gamma^2-1)}{Q^2}$	0	$(i_{na2}R_6)^2$

Under these conditions

$$V_{n0i}^2 = 4kTR_1HQ^2\,\pi f_0. \quad (31)$$

If the resistor R_6 is made equal to R_5 in order to minimize dc drift due to amplifier input current variations, the entries in Table II can be rewritten in terms of Q, H, R_1, e_{na}, i_{na1}, and i_{na2} only.

C. Multiple-Feedback HP Section

The HP single-amplifier multiple-feedback quadratic section shown in Fig. 6 has the voltage transfer function [14]

$$T_{HP}(j\omega) = \frac{-H\left[j\left(\dfrac{\omega}{\omega_0}\right)\right]^2}{1 - \left(\dfrac{\omega}{\omega_0}\right)^2 + j\left(\dfrac{\omega}{\omega_0}\right)\dfrac{1}{Q}} \quad (32)$$

where

$$H = \frac{C_1}{C_4} \quad (33)$$

TABLE III
COMPONENTS OF ϕ_{BP}, ϕ_{LP}, AND ϕ_{AP}—HP QUADRATIC FILTER SECTION

	ϕ_{BP}	ϕ_{LP}	ϕ_{AP}
R_2	$4kTR_2Q^2\left(1+H+\frac{C_3}{C_4}\right)^2$	0	0
R_5	$\frac{4kTR_5}{Q^2}$	$4kTR_5$	0
R_6	$4kTR_6\left[\left(\frac{\gamma}{Q}\right)^2+(1+H)^2\left(2-\frac{1}{Q^2}\right)-2(1+H)\right]$	$4kTR_6[1-(1+H)^2]$	$4kTR_6(1+H)^2$
e_{na}	$e_{na}^2\left[\left(\frac{\gamma}{Q}\right)^2+(1+H)^2\left(2-\frac{1}{Q^2}\right)-2(1+H)\right]$	$e_{na}^2[1-(1+H)^2]$	$e_{na}^2(1+H)^2$
i_{na1}	$\frac{(i_{na1}R_5)^2}{Q^2}$	$(i_{na1}R_5)^2$	0
i_{na2}	$(i_{na2}R_6)^2\left[\left(\frac{\gamma}{Q}\right)^2+(1+H)^2\left(2-\frac{1}{Q^2}\right)-2(1+H)\right]$	$(i_{na2}R_6)^2[1-(1+H)^2]$	$(i_{na2}R_6)^2(1+H)^2$

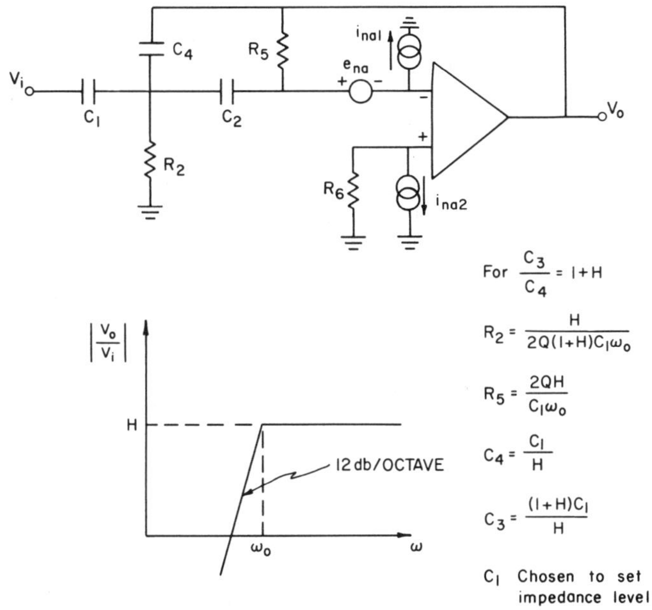

Fig. 6. HP multiple-feedback RC-active filter network.

$$\omega_0^2 = \frac{1}{C_4 R_2 C_3 R_5} \tag{34}$$

and

$$Q = \frac{\sqrt{\dfrac{C_3 R_5}{C_4 R_2}}}{\left(1+\dfrac{C_1}{C_4}+\dfrac{C_3}{C_4}\right)}. \tag{35}$$

The circuit of Fig. 6 has three capacitors in it. However, because the negative input to the amplifier is a virtual earth, C_2 will be in parallel with C_1 during the y_{22} calculation and in parallel with C_4 during the y_{21} calculation. The y_{2k} calculations for $k > 1$ will all be done with the input and output short-circuited so that C_4 will be in parallel with C_1. Thus the three-capacitor circuit is effectively a two-capacitor circuit as far as noise calculation is concerned. The various contributions to ϕ_{BP}, ϕ_{LP}, and ϕ_{AP} are summarized in Table III, where

$$\gamma = 1 + Q^2\left(1+\frac{C_1}{C_4}+\frac{C_3}{C_4}\right). \tag{36}$$

The contributions to ϕ_{BP}, ϕ_{LP}, and ϕ_{AP} due to e_{na}, i_{na1}, i_{na2}, and the resistors once again have the same dependence on the impedance level that has been noted for the case of the LP and BP filter section examples.

The inherent output noise of the circuit of Fig. 6 is given by

$$V_{nOi}^2 = 4kTR_2\left(1+\frac{C_1}{C_4}+\frac{C_3}{C_4}\right)^2\left[Q^2+\frac{C_4}{C_3}(1+Q^2)\right]\frac{\pi}{2}f_0Q. \tag{37}$$

The value of Q is a maximum for a given H and (R_5/R_2) when $(C_3/C_4) = 1 + (C_1/C_4)$. For this design

$$V_{nOi}^2 = 8kTR_2\,[Q^2(1+H)^2+(1+H)(1+Q^2)]\,\pi f_0 Q. \tag{38}$$

The resistor R_6 will normally be selected to be equal to R_5 in order to minimize dc drift due to amplifier input current variations. Thus the entries in Table III can be rewritten in terms of $Q, H, R_2, e_{na}, i_{na1}$, and i_{na2} only.

D. A Numerical Example–LP Case

Since Tables I, II, and III differ only in detail, the $(R_4/R_3) = (1+H)$ LP design of Fig. 4 with $f_0 = 100$ Hz, $H = 10$, $Q = (1/\sqrt{2})$ (maximally flat case), $R_1 = 1125\ \Omega$, $R_3 = 1020\ \Omega$, $R_4 = 11.25$ kΩ, $R_6 = 2040\ \Omega$, $C_5 = 0.1\ \mu$F, and $C_2 = 2.2\ \mu$F is sufficient to illustrate a complete noise calculation for all three types of filter sections. Measurements of e_{na}, i_{na1}, and i_{na2} for the RCA CA3741CT, a bipolar input operational amplifier, yield

$$e_{na}^2 = e_{nav}^2\left(1+\frac{f_e}{f_0}\cdot\frac{f_0}{f}\right)$$

$$\simeq 4.4\times 10^{-16}\left(1+1.25\frac{f_0}{f}\right)(\mathrm{V})^2/\mathrm{Hz} \tag{39}$$

and

$$i_{na1}^2 = i_{na2}^2 = i_{nav}^2\left(1+\frac{f_i}{f_0}\cdot\frac{f_0}{f}\right)$$

$$\simeq 1.5\times 10^{-24}\frac{f_0}{f}(\mathrm{A})^2/\mathrm{Hz}. \tag{40}$$

The entries in Table I can be evaluated to produce Table IV. If $f_d = 1$ Hz and $f_x = 10\,000$ Hz, (48)–(55) can be used to evaluate (13) to show that $V_{no}(f_x, f_d) = 8.32$-μV rms as compared to a measured value of 8.8-μV rms. The output noise due to the resistors R_1, R_3, R_4, and R_6 is 1.64-μV rms, that due to e_{na} is 8.10-μV rms, and that due to i_{na1} and i_{na2} is 0.927-μV rms. If the impedance level of the circuit is increased by a factor of ten by making all resistors larger by a factor of ten and all capacitors smaller by a factor of ten, the output noise due to the resistors R_1, R_3, R_4, and R_6 will be 5.19-μV rms, that due to e_{na} will remain at 8.10-μV rms, and that due to i_{na1} and i_{na2} will be 9.27-μV rms. The total output noise voltage would then be 13.4-μV rms.

IV. Tandem Combinations of Two Quadratic Sections

The foregoing analysis of single quadratic sections can be readily extended to include the case of filters produced by the tandem combination of two quadratic sections. The present discussion is limited to a consideration of a fourth-order maximally flat LP filter, but it should be evident that other types of fourth-order filters are also amenable to such analysis.

If an LP filter section labeled a is followed by an LP filter section labeled b to produce a fourth-order filter, the total noise voltage $V_{n0t}(f_d, f_x)$ at the output of section b in the frequency range f_d to f_x will be given by

$$V_{n0t}^2(f_d, f_x) = \int_{f_d}^{f_x} [\phi_{BPb} |H_{BPb}(j\omega)|^2 + \phi_{LPb} |H_{LPb}(j\omega)|^2 + \phi_{APb}]\, df + \int_{f_d}^{f_x} [\phi_{BPa} |H_{BPa}(j\omega)|^2 + \phi_{LPa} |H_{LPa}(j\omega)|^2 + \phi_{APa}] \cdot [H_{0LPb}^2 |H_{LPb}(j\omega)|^2]\, df. \quad (41)$$

A maximally flat response [15] will be obtained if $\omega_{0a} = \omega_{0b} = \omega_0$, $Q_a = 0.541$ and $Q_b = 1.307$, or $Q_a = 1.307$ and $Q_b = 0.541$. In that case,

$$|H_{LPa}(j\omega) H_{LPb}(j\omega)|^2 = \frac{1}{1 + \left(\frac{f}{f_0}\right)^8}. \quad (42)$$

If white noise and $1/f$ noise sources are considered, the list of integrals represented by (48)–(56) in the Appendix will have to be augmented by (57)–(60) in order to evaluate V_{n0t}^2. Once tables like Table IV are prepared for section a and section b, the analysis of the fourth-order maximally flat LP filter is not significantly different from that of any other quadratic filter section.

As in the case of single sections, (41) may be used as a starting point for noise minimization procedures. For example, suppose the filter discussed in Section III-D is required to be a maximally flat fourth-order one with $f_0 = 100$ Hz, $H_t = H_a H_b = 10$, and $R_{1a} \simeq 1$ kΩ. For the $(R_4/R_3) = 1 + (R_4/R_1)$ design, amplifiers of the quality discussed in Section III-D, and the impedance level in question, the noise at the output due to e_{naa} and e_{nab} can be assumed to be dominant. It is then easy to show that, provided $e_{naa}^2 = e_{nab}^2 \simeq e_{na}^2$ of (39),

$$V_{n0t}^2 / e_{nav}^2 f_0 = [4Q_b^2(1 + H_b)^2 + 2(2 + H_b)]\, \rho_1 + [2H_b(1 + H_b)]\, \rho_2 + [4Q_a^2(1 + H_a)^2 + 2(2 + H_a)]\, H_b^2 \rho_3 + [(2 + H_a) H_a H_b^2]\, \rho_4 + \frac{f_x}{f_0} + \frac{f_e}{f_0} \ln \frac{f_x}{f_d} \quad (43)$$

where

$$\rho_1 = \frac{\pi}{2} Q_b + \frac{\frac{f_e}{f_0} Q_b}{\sqrt{1 - \frac{1}{4Q_b^2}}} \tan^{-1} 2Q_b \sqrt{1 - \frac{1}{4Q_b^2}} \quad (44)$$

$$\rho_2 = \frac{\pi}{2} Q_b + \frac{f_e}{f_0} \frac{Q_b\left[1 - \frac{1}{2Q_b^2}\right]}{\sqrt{1 - \frac{1}{4Q_b^2}}} \tan^{-1} 2Q_b \sqrt{1 - \frac{1}{4Q_b^2}} + \frac{f_e}{f_0} \ln \frac{f_0}{f_d} \quad (45)$$

$$\rho_3 = \frac{\frac{\pi}{8}}{\sin \frac{3\pi}{8}} + \frac{\pi \frac{f_e}{f_0}}{4\sqrt{2}} \quad (46)$$

and

$$\rho_4 = \frac{\frac{\pi}{8}}{\sin \frac{\pi}{8}} + \frac{\frac{f_e}{f_0}}{8} \ln \frac{f_x}{f_d} \frac{\left(1 + \frac{f_d}{f_0}\right)}{\left(1 + \frac{f_x}{f_0}\right)}. \quad (47)$$

Since ρ_1, ρ_2, ρ_3, and ρ_4 are all greater than zero, V_{n0t}^2 does not have a maximum for positive H_a and H_b. The value of V_{n0t}^2 is least when H_a is large and $Q_a = 0.541$. If $(f_d/f_0) = 0.01$, $(f_x/f_0) = 100$, and e_{naa}^2 and e_{nab}^2 are as given by (39), V_{n0t} will take on the values 6.65, 5.47, and 4.96-μV rms for $H_a = 5$, 10, and 20, respectively. If Q_a is taken to be 1.307, V_{n0t} will take on the values 8.45, 7.60, and 7.20 for $H_a = 5$, 10, and 20, respectively.

V. Summary

The quadratic filter section noise-analysis procedure that has been outlined in this work is very convenient for carrying out practical calculations and should prove to be useful in carrying out comparative studies of various realizations of a given transfer function. In addition, the noise performance of tandem combinations of quadratic filter sections can be handled in a straightforward manner. Indeed, an example has been given to show that the analysis of a fourth-order maximally flat LP filter can be carried out without recourse to numerical integration if the noise analysis of each quadratic section in the tandem combination is available.

TABLE IV
COMPONENTS OF ϕ_{BP}, ϕ_{LP}, AND ϕ_{AP}—LP QUADRATIC FILTER SECTION

	ϕ_{BP}	ϕ_{LP}	ϕ_{AP}
R_1	0	1.86×10^{-15}	0
R_3	4.09×10^{-15}	2.04×10^{-15}	0
R_4	0	1.86×10^{-16}	0
R_6	1.30×10^{-14}	4.06×10^{-15}	3.38×10^{-17}
e_{na}	$1.17 \times 10^{-13}\left(1 + 1.25\frac{f_0}{f}\right)$	$5.28 \times 10^{-14}\left(1 + 1.25\frac{f_0}{f}\right)$	$4.4 \times 10^{-16}\left(1 + 1.25\frac{f_0}{f}\right)$
i_{na1}	$3.78 \times 10^{-16}\frac{f_0}{f}$	$7.55 \times 10^{-16}\frac{f_0}{f}$	0
i_{na2}	$1.66 \times 10^{-15}\frac{f_0}{f}$	$7.51 \times 10^{-16}\frac{f_0}{f}$	$6.25 \times 10^{-18}\frac{f_0}{f}$

Note: All entries are in square volts per hertz.

APPENDIX

$$\int_{f_d}^{f_x} |H_{BP}(j\omega)|^2\, df \simeq \int_0^{\infty} |H_{BP}(j\omega)|^2\, df, \quad \text{for } \frac{f_d}{f_0} < 0.2,\ \frac{f_x}{f_0} > 5$$

$$= \frac{\pi}{2} f_0 Q \tag{48}$$

$$\int_{f_1}^{f_2} |H_{BP}(j\omega)|^2\, df = \frac{\pi}{4} f_0 Q \tag{49}$$

$$\int_{f_d}^{f_x} \frac{f_0}{f} |H_{BP}(j\omega)|^2\, df \simeq \int_0^{\infty} \frac{f_0}{f} |H_{BP}(j\omega)|^2\, df, \quad \text{for } \frac{f_d}{f_0} < 0.2,\ \frac{f_x}{f_0} > 5$$

$$= \frac{Qf_0}{\sqrt{1 - \frac{1}{4Q^2}}} \tan^{-1}\left[2Q\sqrt{1 - \frac{1}{4Q^2}}\right], \quad Q > \frac{1}{2}$$

$$\simeq \frac{\pi}{2} f_0 Q, \quad Q > 4 \tag{50}$$

$$\int_{f_1}^{f_2} \frac{f_0}{f} |H_{BP}(j\omega)|^2\, df = \frac{Qf_0}{\sqrt{1 - \frac{1}{4Q^2}}} \tan^{-1} \frac{\sqrt{1 - \frac{1}{4Q^2}}}{\sqrt{1 + \frac{1}{4Q^2}}}, \quad Q > \frac{1}{2}$$

$$\int_{f_1}^{f_2} \frac{f_0}{f} H_{BP}(j\omega)|^2\, df \simeq \frac{\pi}{4} f_0 Q, \quad Q > 2 \tag{51}$$

$$\int_{f_d}^{f_x} |H_{LP}(j\omega)|^2\, df \simeq \int_0^{\infty} |H_{LP}(j\omega)|^2\, df, \quad \text{for } \frac{f_d}{f_0} < 0.2,\ \frac{f_x}{f_0} > 5$$

$$\int_{f_d}^{f_x} |H_{LP}(j\omega)|^2\, df = \frac{\pi}{2} f_0 Q \tag{52}$$

$$\int_{f_1}^{f_2} |H_{LP}(j\omega)|^2\, df = \frac{\pi}{4} f_0 Q, \quad f_1 \text{ and } f_2 \text{ defined for } H_{BP}(j\omega) \tag{53}$$

$$\int_{f_d}^{f_x} \frac{f_0}{f} |H_{LP}(j\omega)|^2\, df \simeq \frac{\left[1 - \frac{1}{2Q^2}\quad Qf_0\right]}{\sqrt{1 - \frac{1}{4Q^2}}} \tan^{-1}\left[2Q\sqrt{1 - \frac{1}{4Q^2}}\right] + f_0 \ln\left(\frac{f_0}{f_d}\right), \quad Q > \frac{1}{2},\ \frac{f_d}{f_0} < 0.2,\ \frac{f_x}{f_0} > 5$$

$$\simeq \frac{\pi}{2} f_0 Q + f_0 \ln\left(\frac{f_0}{f_d}\right), \quad Q > 4,\ \frac{f_d}{f_0} < 0.2,\ \frac{f_x}{f_0} > 5 \tag{54}$$

$$\int_{f_1}^{f_2} \frac{f_0}{f} |H_{LP}(j\omega)|^2\, df = \frac{\left[1 - \frac{1}{2Q^2}\right]}{\sqrt{1 - \frac{1}{4Q^2}}} Qf_0 \tan^{-1} \frac{\sqrt{1 - \frac{1}{4Q^2}}}{\sqrt{1 + \frac{1}{4Q^2}}} + \frac{f_0}{2} \ln\left(\frac{f_2}{f_1}\right), \quad Q > \frac{1}{2},$$

f_1 and f_2 defined for $H_{BP}(j\omega)\omega$

$$\simeq \frac{\pi}{4} f_0 Q + \frac{f_0}{2} \ln \frac{\sqrt{4Q^2 + 1} + 1}{\sqrt{4Q^2 + 1} - 1}, \quad Q > 2 \tag{55}$$

$$\int_{f_d}^{f_x} \frac{f_0}{f}\, df = f_0 \ln \frac{f_x}{f_d} \tag{56}$$

$$\int_{f_d}^{f_x} \frac{\left(\frac{f}{f_0}\right)^2 df}{1+\left(\frac{f}{f_0}\right)^8} \simeq \int_0^{\infty} \frac{\left(\frac{f}{f_0}\right)^2 df}{1+\left(\frac{f}{f_0}\right)^8}, \quad \text{for } \frac{f_d}{f_0} < 0.2 \text{ and } \frac{f_x}{f_0} > 5$$

$$= \frac{\frac{\pi}{8} f_0}{\sin\left(\frac{3\pi}{8}\right)} \tag{57}$$

$$\int_{f_d}^{f_x} \frac{\left(\frac{f}{f_0}\right) df}{1+\left(\frac{f}{f_0}\right)^8} \simeq \int_0^{\infty} \frac{\left(\frac{f}{f_0}\right) df}{1+\left(\frac{f}{f_0}\right)^8}, \quad \text{for } \frac{f_d}{f_0} < 0.2 \text{ and } \frac{f_x}{f_0} > 5$$

$$= \frac{\pi f_0}{4\sqrt{2}} \tag{58}$$

$$\int_{f_d}^{f_x} \frac{df}{1+\left(\frac{f}{f_0}\right)^8} \simeq \int_0^{\infty} \frac{df}{1+\left(\frac{f}{f_0}\right)^8}, \quad \text{for } \frac{f_d}{f_0} < 0.2 \text{ and } \frac{f_x}{f_0} > 5$$

$$= \frac{\frac{\pi}{8} f_0}{\sin \frac{\pi}{8}} \tag{59}$$

$$\int_{f_d}^{f_x} \frac{\left(\frac{f_0}{f}\right) df}{1+\left(\frac{f}{f_0}\right)^8} = \frac{f_0}{8} \ln \frac{f_x}{f_d} \frac{\left(1+\frac{f_d}{f_0}\right)}{\left(1+\frac{f_x}{f_0}\right)}$$

$$\simeq \frac{f_0}{8} \ln \frac{f_0}{f_d}, \quad \frac{f_d}{f_0} < 0.2 \text{ and } \frac{f_x}{f_0} > 5. \tag{60}$$

REFERENCES

[1] W. R. Bennett, *Electrical Noise*. New York: McGraw-Hill, 1960.

[2] B. A. Shenoi, "Practical realization of a gyrator circuit and *RC*-gyrator filters," *IEEE Trans. Circuit Theory*, vol. CT-12, pp. 374–380, Sept. 1965.

[3] L. T. Bruton, "Nonideal performance of two-amplifier positive-impedance converters," *IEEE Trans. Circuit Theory*, vol. CT-17, pp. 541–549, Nov. 1970.

[4] R. P. Sallen and E. L. Key, "A practical method of designing *RC* active filters," *IRE Trans. Circuit Theory*, vol. CT-2, pp. 74–85, Mar. 1955.

[5] S. K. Mitra, *Analysis and Synthesis of Linear Active Networks.* New York: Wiley, pp. 474–475.

[6] R. E. Bach, "Selecting *RC* values for active filters," *Electronics*, vol. 33, pp. 82–85, May 1960.

[7] H. C. Montgomery, "Transistor noise in circuit applications," *Proc. IRE*, vol. 40, pp. 1461–1471, Nov. 1952.

[8] L. T. Bruton, F. N. Trofimenkoff, and D. H. Treleaven, "Noise performance of low-sensitivity active filters," *IEEE J. Solid-State Circuits*, vol. SC-8, pp. 85–91, Feb. 1973.

[9] W. R. Huber, "Two-port equivalent noise generators," *Proc. IEEE* (Lett.), vol. 58, pp. 807–809, May 1970.

[10] D. H. Sheingold and L. Smith, "Operational-amplifier-circuit noise characteristics," *Electron. Inst. Digest*, pp. 50–57, Sept. 1969.

[11] F. N. Trofimenkoff, "Noise margins of band-pass filters," *IEEE Trans. Circuit Theory*, to be published.

[12] E. F. Vandivere, "Noise output of a multipole filter relative to that of the ideal square-response filter," *Proc. IEEE* (Corresp.), vol. 51, p. 1771, Dec. 1963.

[13] I. S. Gradshteyn and I. M. Ryzhik, *Table of Integrals, Series and Products.* New York: Academic, 1965.

[14] *Handbook of Operational Amplifier Active RC Networks*, Burr-Brown Research Corp., Tucson, Ariz., 1966.

[15] L. Weinberg, *Network Analysis and Synthesis*, New York: McGraw-Hill, 1962.

NOISE CHARACTERISTICS OF CASCADED 2ND-ORDER ACTIVE FILTERS

Indexing terms: Active filters, Noise

The absolute noise level at the output of single-amplifier, 2nd-order active filters, suitable for cascade connection, is derived in terms of the sensitivity of the filter response to the amplifier gain. A very simple expression is obtained, which is shown to be valid under practical conditions. A similar expression is obtained for cascaded 2nd-order sections, and a 4th-order example shows that the position of the sections in the cascade has a significant effect on the output noise level. The cascade positions giving lowest noise level are the opposite of those normally used to give maximum dynamic range.

Introduction: As advances in component technology and construction techniques have reduced the importance of the sensitivity problem, active inductorless filters have become more widely used. However, this wider use has revealed other disadvantageous properties; for example, noise levels in active filters compare unfavourably with those in passive LC filters. The noise characteristics of active filters have been the subject of a number of recent publications, which have considered 2nd-order sections[1-4] or active simulation of LC filters.[5,6] This letter uses an earlier result[7] to derive the noise characteristics of high-order sections.

Noise in 2nd-order sections: Reference 7 gives general relationships between the noise level at the output of a circuit and the noise arising in its constituent components. That part of the filter output noise attributable to an amplifier is given by:

$$\frac{\sqrt{\{\overline{e_0^2(\omega)}\}}}{E_0(\omega)} = |S_A^{F(j\omega)}| \frac{\sqrt{\{\overline{e^2(\omega)}\}}}{E_A(\omega)} \quad . \quad . \quad . \quad . \quad . \quad (1)$$

where A is the open-loop amplification of the amplifier, $e_0(\omega)$ is the filter output noise voltage in a narrow bandwidth centred on ω, $e(\omega)$ is the open-loop amplifier output noise voltage in the same bandwidth, $E_0(\omega)$ is the r.m.s. filter output signal voltage, $E_A(\omega)$ is the r.m.s. signal level at the amplifier output and $S_A^{F(j\omega)}$ is the differential sensitivity of the filter transfer function $F(j\omega)$ to the amplification A, defined as

$$S_A^{F(j\omega)} = \frac{\partial \ln F(j\omega)}{\partial \ln A} \quad . \quad . \quad . \quad . \quad . \quad . \quad . \quad . \quad (2)$$

In the special case of a 2nd-order section with a single amplifier, where the amplifier output is used as the filter output, it is obvious that $E_A = E_0$ and the expression for the output noise due to the amplifier reduces to

$$\sqrt{\{\overline{e_0^2(\omega)}\}} = |S_A^{F(j\omega)}|\sqrt{\{\overline{e^2(\omega)}\}} \quad . \quad . \quad . \quad . \quad . \quad (3)$$

This is a remarkably simple result and, since the sensitivity characteristics of 2nd-order sections are well known, requires very little analysis to reveal the form of this component of the output noise. However, it is necessary that the frequency spectrum of the open-loop amplifier output noise is known. In general, the output noise of the amplifier depends on the source impedance and, since this source impedance is presented by the passive RC network, it is frequency dependent. Thus, in general, the calculation of the amplifier output noise voltage is far from simple, but it is known that with some common operational-amplifier types the output noise voltage may be made effectively independent of source impedance if the source impedance is reduced below some value.[8] Reducing the impedance level of the passive RC network also reduces the thermal noise contribution to the filter output noise, allowing amplifier noise to predominate. Experimental results[9] indicate that this condition may be achieved with 741C type amplifiers when the source resistance is of the order of 1 kΩ.

The amplifier open-loop output noise spectrum under these conditions is readily obtained by measurement or from manufacturer's data sheets. The magnitude of the sensitivity function may then be applied as a simple multiplying factor to obtain the filter output noise. The validity of this approach has been demonstrated experimentally.[9]

Noise in high-order cascade schemes: When designing high-order filters as a cascade of 2nd-order sections, it is usual to arrange the Q-factors of the sections to increase along the cascade. This procedure ensures maximum available undistorted signal at the filter output. However, if the absolute noise level at the filter output is important, it is necessary to reassess the position of the sections in the cascade and to choose those positions resulting in minimum noise level.

Consider a 4th-order filter realised as a cascade of two 2nd-order sections, which obey the conditions described above. The filter output noise contributions from the two amplifiers can be obtained from equations 2 and 3 as:

Reprinted with permission from *Electron. Lett.*, vol. 12, pp. 301-303, 10 June 1976.

First amplifier:

$$\sqrt{\{\overline{e_{01}^2}(\omega)\}} = |S_{A_1}^{F_1(j\omega)}|\,|F_2(j\omega)|\sqrt{\{\overline{e_1^2}(\omega)\}} \quad . \quad . \quad . \quad (4)$$

Second amplifier:

$$\sqrt{\{\overline{e_{02}^2}(\omega)\}} = |S_{A_2}^{F_2(j\omega)}|\sqrt{\{\overline{e_2^2}(\omega)\}} \quad . \quad . \quad . \quad . \quad . \quad (5)$$

The total filter output noise is then given by

$$\sqrt{\{\overline{e_0^2}(\omega)\}} = \sqrt{\{\overline{e_{01}^2}(\omega) + \overline{e_{02}^2}(\omega)\}}$$

$$= \sqrt{\{|S_{A_1}^{F_1(j\omega)}|^2\,|F_2(j\omega)|^2\,\overline{e_1^2}(\omega) + |S_{A_2}^{F_2(j\omega)}|^2\,\overline{e_2^2}(\omega)\}} \quad (6)$$

where $e_1(\omega)$ and $e_2(\omega)$ are the average noise outputs from the first and second amplifiers, with amplifications A_1 and A_2, respectively, and $F_1(j\omega)$ and $F_2(j\omega)$ are the frequency responses of the first and second cascaded sections.

If it is assumed that the amplifications A_1 and A_2 are realised from identical operational amplifiers, with resistive negative feedback, the noise voltages $e_1(\omega)$ and $e_2(\omega)$ can be written as

$$e_1(\omega) = A_1\,e_u(\omega) \quad . \quad . \quad . \quad . \quad . \quad . \quad . \quad . \quad . \quad (7)$$

$$e_2(\omega) = A_2\,e_u(\omega) \quad . \quad . \quad . \quad . \quad . \quad . \quad . \quad . \quad . \quad (8)$$

where $e_u(\omega)$ is the operational-amplifier input noise voltage in a narrow bandwidth around ω. Note that these expressions do not exclude the case where A_1 and A_2 represent the open-loop amplifications of the operational amplifiers. Substituting eqns. 7 and 8 into eqn. 6 now gives

$$\sqrt{\{\overline{e_0^2}(\omega)\}}$$

$$= \{\sqrt{[A_1^2\,|S_{A_1}^{F_1(j\omega)}|^2\,|F_2(j\omega)|^2 + A_2^2\,|S_{A_2}^{F_2(j\omega)}|^2]}\}\sqrt{\{\overline{e_u^2}(\omega)\}} \quad (9)$$

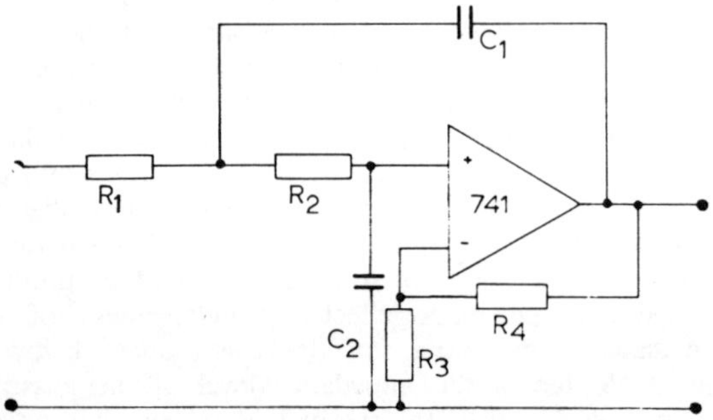

Fig. 1 *Second-order section used in cascade*

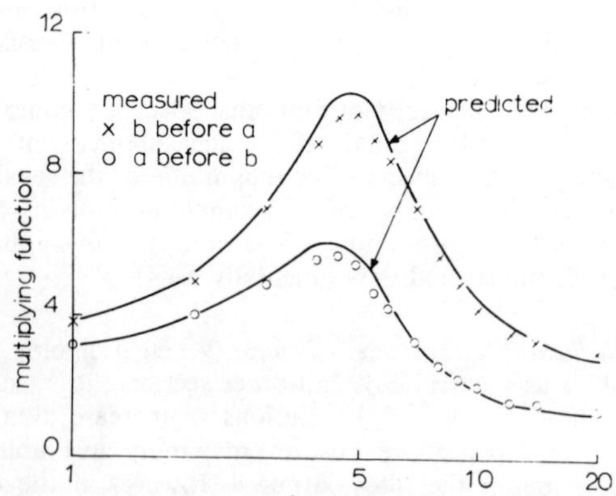

Fig. 2 *Measured and predicted multiplying functions for 4th-order, lowpass, Butterworth response realised as cascade of two 2nd-order sections*

Table 1 PARAMETERS AND COMPONENT VALUES FOR 2ND-ORDER SECTIONS

Position in cascade	ω_n	Q	$R_1 = R_2$	$C_1 = C_2$	$\frac{R_3+R_4}{R_3}$
	kHz		Ω	μF	
a	5·0	1·31	470	0·068	2·24
b	5·0	0·54	470	0·068	1·15

For any given operational-amplifier type the input noise level $e_u(\omega)$ will be fixed, so that minimisation of the filter output noise level requires the minimisation of the term in braces. If only the position of sections in the cascade is to be considered, this minimisation involves only the calculation of this function for each of the two possible arrangements. Since the constituents of this function will almost certainly have been calculated for other purposes, this involves only minor computational effort.

It is apparent from the above derivation that a similar expression may be obtained for any number of sections in cascade and used to find the positions giving optimum output noise.

Fourth-order example: It is difficult to draw general conclusions from the form of eqn. 9, because the sensitivities and amplifications depend very much on the form of the 2nd-order sections. However, it is interesting to consider the results of evaluating eqn. 9 for a particular case. The example used is a 4th-order, lowpass Butterworth response realised with a cascade of two 2nd-order Sallen-and-Key-type circuits.[10] Each 2nd-order section has the form shown in Fig. 1 with the parameters shown in Table 1.

The predicted and measured multiplying factors (the term in braces in eqn. 9) are shown in Fig. 2 as functions of frequency for both cascade configurations. It is interesting to note that the lower noise levels are obtained by placing the section with the highest Q-factor at the front of the cascade; that is, the opposite condition to that normally used.

Summary and conclusions: A simple expression has been obtained for the noise level at the output of a type of 2nd-order active filter commonly used in cascade, under certain conditions which may be readily obtained in practice. A similar expression was then derived for such 2nd-order sections in cascade, having the advantage that all the terms in the expression would normally be calculated for other purposes. Very little extra computational effort is then required to find the cascade arrangement giving minimum output noise level.

A 4th-order example has been analysed to reveal that the positions of the 2nd-order sections in the cascade significantly affect the output noise level. Further, the cascade positions giving lowest output noise are different to those normally used. The results have been confirmed experimentally. While this result is not shown to be general, it should be noted that the expression from which it is derived (eqn. 9) is heavily dependent upon the moduli of the complex sensitivity functions $S_A^{F(j\omega)}$. There is a marked similarity between these moduli for the various common forms of 2nd-order sections,[11] which suggests that this result may be general or, at least, common. The generality of this result is to be the subject of a further investigation.

Acknowledgment: S. Mecrow wishes to acknowledge financial support from the British Post Office.

M. R. LEE

S. MECROW

Department of Electrical & Electronic Engineering
Newcastle upon Tyne Polytechnic
Ellison Building,
Ellison Place, Newcastle upon Tyne NE1 8ST, England

29th April 1976

References

1 BOZIC, S. M., and EMMETT, S. G.: 'Noise in microelectronic active *RC* filters', *Microelectronics & Reliability*, 1971, **10**, pp. 31–36

2 BRUTON, L. T., TROFIMENKOFF, F. N., and TRELEAVEN, D. H.: 'Noise performance of low-sensitivity active filters', *IEEE J. Solid-State Circuits*, 1973, **SC–8**, pp. 85–91

3 TROFIMENKOFF, F. N., SMALLWOOD, R. E., and BRUTON, L. T.: 'Noise in positive gain Sallen-Key *RC*-active filters'. Proceedings of European conference on circuit theory, 1974, pp. 325–330

4 TROFIMENKOFF, F. N., TRELEAVEN, D. H., and BRUTON, L. T.: 'Noise performance of *RC*-active quadratic filter sections', *IEEE Trans.*, 1975, **CT–20**, pp. 524–532

5 FETTWEIS, A.: 'On noise performance of capacitor gyrator filters', *Int. J. Circuit Theory & Appl.*, 1974, **2**, pp. 181–186

6 ROBERTS, G. M., CREASEY, D. J., and BOZIC, S. M.: 'The performance of gyrator and leapfrog inductorless filters, using integrated circuits', *Microelectronics & Reliability*, 1973, **12**, pp. 159–161

7 HOLT, A. G. J., and LEE, M. R.: 'A relationship between sensitivity and noise', *Int. J. Electron.*, 1969, **26**, pp. 591–594

8 SHEINGOLD, D. H., and SMITH, L.: 'Operational amplifier circuit noise characteristics'. Electronic Instrument Digest, 1969, pp. 50–57

9 MECROW, S.: 'Noise characteristics of cascaded second-order active filters'. Undergraduate Project Report, 1976, Newcastle upon Tyne Polytechnic

10 SALLEN, R. P., and KEY, E. L.: 'A practical method of designing *RC* active filters', *IRE Trans.*, 1955, **CT–2**, pp. 74–85

11 LEE, M. R.: 'Sensitivity characteristics of high-order inductorless filters'. Ph.D. thesis, University of Newcastle upon Tyne, 1972

OPTIMIZATION OF THE DYNAMIC RANGE AND THE NOISE DISTANCE OF RC-ACTIVE FILTERS BY DYNAMIC PROGRAMMING

E. LUEDER

Institut für Netzwerk-und Systemtheorie, University of Stuttgart, Federal Republic of Germany

SUMMARY

RC-active filters are usually realized by cascading 2nd order stages. The degrees-of-freedom in this procedure are: the pole-zero assignments to form the second order transfer functions, the gain-factors for the stages and the sequence in the cascade. These freedoms will simultaneously be used to maximize the dynamic range and the noise distance of the filters. The exact solution to the problem will be obtained by dynamic programming. A rule of thumb avoiding a computer program and an example are given.

PROBLEM

A given rational transfer function $T(s)$ of an RC-active filter is usually factored into terms of at most 2nd degree in s which are realized by a cascade structure. We confine ourselves to the most important case of complex pairs of poles and zeros of $T(s)$. Thus $T(s)$ of degree $2m$ in s can be factored into

$$T(s) = \prod_{\nu=1}^{m} K_\nu \frac{\delta_\nu s^2 + c_{1\nu} s + c_{0\nu}}{s^2 + d_{1\nu} s + d_{0\nu}} = \prod_{\nu=1}^{m} K_\nu t_\nu(s) \tag{1a}$$

with

$$t_\nu(s) = \frac{\delta_\nu s^2 + c_{1\nu} s + c_{0\nu}}{s^2 + d_{1\nu} s + d_{0\nu}}, \qquad \delta_\nu = 0, 1 \tag{1b}$$

The cascade of the m two-ports in Figure 1 produces $T(s)$. The voltages $V_\nu(j\omega)$, $\nu = 0, 1, \ldots m-1, m$ are the input—and output voltages of the two-ports in Figure 1. As shown in Reference 1 and in greater detail in Reference 2 the dynamic range and the input-noise distance† of the filter are minimized if

$$d_\nu = \log \frac{\max |V_\nu(j\omega)|}{\min |V_\nu(j\omega)|} \to \min \qquad \text{for} \quad \nu = 1, 2, \ldots, m-1 \tag{2a}$$

with

$$\max |V_\nu(j\omega)| \text{calculated in the interval } \omega \in [0, \infty] \tag{2b}$$

and

$$\min |V_\nu(j\omega)| \text{ calculated in the interval } \omega \in [\omega_1, \omega_2] \tag{2c}$$

where $\omega \in [\omega_1, \omega_2]$ with $\omega_1 \geqslant 0$ and $\omega_2 > \omega_1$ is the pass band of the filter. The constraint (2a) insures that the shape of $|V_\nu(j\omega)|$ is as flat as possible. Thus high peaks which overdrive the amplifiers and therefore limit the dynamic range are avoided. We confine ourselves to stages with only one amplifier located at the output of the stage. Then the voltages $V_\nu(j\omega)$ coincide with the dynamic range V_{dyn} of the output of the amplifiers. In addition, low minimal values of $|V_\nu(j\omega)|$ which lead to a small noise distance are also eliminated by (2a). The

† Noise generated in the filter itself is not considered.

Received 16 September, 1974
Revised 20 January, 1975

Reprinted with permission from *Int. J. of Circuit Theory Appl.*, vol. 3, pp. 365-370, Dec. 1975.

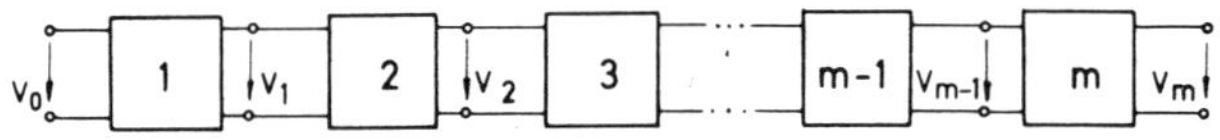

Figure 1. The cascade of m two-ports

voltages $V_\nu(j\omega)$ depend on the input voltage $V_0(j\omega)$. As filters are mainly used in telephone channels and as we are confronted with a great variety of speakers, we assume henceforth an average input voltage

$$|V_0(j\omega)| = 1 \tag{3}$$

Equation (3) yields:

$$V_\nu(j\omega) = \prod_{\mu=1}^{\nu} K_\mu t_\mu(j\omega)$$

and with (2a):

$$d_\nu = \log \frac{\max \prod_{\mu=1}^{\nu} t_\mu(j\omega)}{\min \prod_{\mu=1}^{\nu} t_\mu(j\omega)} \rightarrow \min \tag{4}$$

The minimum in (4) will be obtained by optimizing the cascade. The degrees-or-freedom for the selection of the cascade are

(a) The pole-zero assignments forming the 2nd order transfer functions $t_\nu(s)$ in equation (1b).
(b) The sequence in the cascade.
(c) The allocation of a gain factor K_ν to $t_\nu(s)$.

Obviously the criterion (4) is independent of K_ν. Therefore K_ν may be chosen according to one of the following three guidelines.

(a) To achieve the largest voltage swing at the output each K_ν should be chosen as high as possible without causing overdrive of the amplifier. For this choice of K_ν all maximum voltages $\max |V_\nu(j\omega)|$ in equation (2a) have the same value which equals the dynamic range V_{dyn} of the amplifiers. Inserting $\max |V_\nu(j\omega)| = V_{dyn}$ for all ν into equation (2a) yields

$$\min |V_\nu(j\omega)| = \frac{V_{dyn}}{10^{d_\nu}} \tag{5}$$

The smallest value $\min |V_\nu(j\omega)|$ or in other words the smallest noise distance occurs where d_ν is largest.

(b) If the noise distance is of prime importance the values for K_ν are selected so that the values of $\min |V_\nu(j\omega)|$ for all $\nu = 1, 2, \ldots m$ are of equal height and as large as possible.
(c) If both the dynamic range and the noise distance are important the K_ν's are selected so that the average values

$$\frac{\max |V_\nu(j\omega)| + \min |V_\nu(j\omega)|}{2}$$

are for all ν equal and are as large as possible as proposed in.[4]

The main problem is now to find the pole-zero-assignment and the sequence in the cascade which minimizes d_ν in (4). In previous solutions[1–3] this has been achieved in consecutive steps which is an approximation. An exact solution is found by the simultaneous optimization of the pole-zero assignment and the sequence with the help of dynamic programming[5] which is a known procedure providing an exact solution.

Step 1 of the procedure will provide an optimal solution for the stages 1 and 2 in the cascade. From the m given pole pairs and the m zero pairs in (1a) we select in all possible ways two pole pairs P_1 and P_2 and two zero pairs Z_1 and Z_2. There are $\binom{m}{2}^2$ possibilities to do this. The proof for this number will be omitted. For each possibility we have to allocate P_1, P_2, Z_1 and Z_2 to the stages 1 and 2 in Figure 1. To do this we have 2^2 possible allocations listed in Table I. The 1st possibility is to allocate pole pair P_1 and zero pair Z_1

Table I

Stage 1	Stage 2
Z_1 P_1	Z_2 P_2
Z_1 P_2	Z_2 P_1
Z_2 P_1	Z_1 P_2
Z_2 P_2	Z_1 P_1

to stage 1 and the remaining pairs P_2 and Z_2 to stage 2. Similarly we obtain the remaining allocations. For each of the 4 possible allocations in Table I we calculate the values d_1 and d_2 in equation (4). The best allocation is that with

$$\max(d_1, d_2) = \min$$

This allocation is retained, the others are disregarded. We apply the same procedure to all $\binom{m}{2}^2$ possibilities and obtain $\binom{m}{2}^2$, solutions in step 1 all of which we retain as candidates for the final solution. This method is called the optimal strategy of the dynamic programming procedure.

Step 2 will provide an optimum solution for the stages 1, 2 and 3. In step 2 we pick three pole pairs and three zero pairs in all possible ways. There are $\binom{m}{3}^2$ possibilities. Each possibility offers 3^2 ways to allocate a pole pair and a zero pair to stage 3. The remaining best assignments to the stages 2 and 1 are already known from step 1. For each allocation we calculate d_3 due to (4) and retain this allocation with

$$\max(d_1, d_2, d_3) = \min$$

Again we do this for all $\binom{m}{3}^2$ possibilities which yields $\binom{m}{3}^2$ solutions to be stored. The $\binom{m}{2}^2$ solutions from step 1 can now be deleted.

We continue in this manner until step $m-2$ which is the last step and provides the final solution for the cascade.

In this step we have $\binom{m}{m-1}^2$ possibilities to select $m-1$ pole pairs and $m-1$ zero pairs. Each possibility offers $(m-1)^2$ ways to allocate a pole pair and a zero pair to stage $m-1$. The remaining optimum allocations to the stages 1, 2, ... up to $m-2$ are already known from the previous steps. Only one pole pair and one zero pair are now left and have to be assigned to stage m. Now we calculate d_{m-1} for all $(m-1)^2$ cases and retain the assignment with

$$\max (d_1, d_2, \dots d_{m-2}, d_{m-1}) = \min$$

We do this for all $\binom{m}{m-1}^2$ possibilities. The result with the least value for $\max (d_1, \dots d_{m-1})$ is the exact optimal solution.

Before giving an example of this method we must cite three additional requirements often practical in actual applications especially in telephone channel filters:

1. The 1st stage of the cascade should be a low-pass, if possible. This prevents interference, usually located at higher frequencies, to enter the cascade.
2. The last stage should be a high-pass to suppress hum and low frequency noise originating in the amplifiers.
3. Sometimes a certain pole-zero assignment is required by the realization of the circuit.

In all these cases we make the required choice and then optimize the remaining stages.

EXAMPLE

As an example we investigate a bandpass with a transfer function of the 10th degree in s. The poles and zeros are listed in Table II and shown in the s-plane in Figure 2(a). The poles p_3 and zeros z_3 have to be paired because of the realization by an FEN-circuit.[6] The K_v are chosen by criterion a. The result is also plotted in Figure 2(a). The pertinent poles and zeros are looped. The number at each loop indicates the number in the sequence of the stages. The gain factors K_v and the values d_v are listed below Figure 2(a). The largest d_v-value is d_4. According to equation (5) this yields the lowest signal amplitude

$$\min |V_4(j\omega)| = \frac{V_{\mathrm{dyn}}}{10^{d_4}} = \frac{V_{\mathrm{dyn}}}{2 \cdot 15}$$

after stage 4.

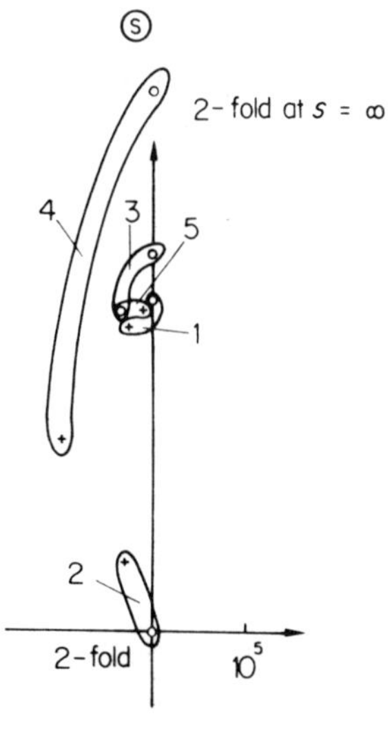

Figure 2(a)

$K_1 = 3{\cdot}15$ $\quad d_1 = 0{\cdot}17$
$K_2 = 0{\cdot}88$ $\quad d_2 = 0{\cdot}26$
$K_3 = 0{\cdot}58$ $\quad d_3 = 0{\cdot}331$
$K_4 = 1{\cdot}26$ $\quad d_4 = 0{\cdot}333$
$K_5 = 0{\cdot}84$

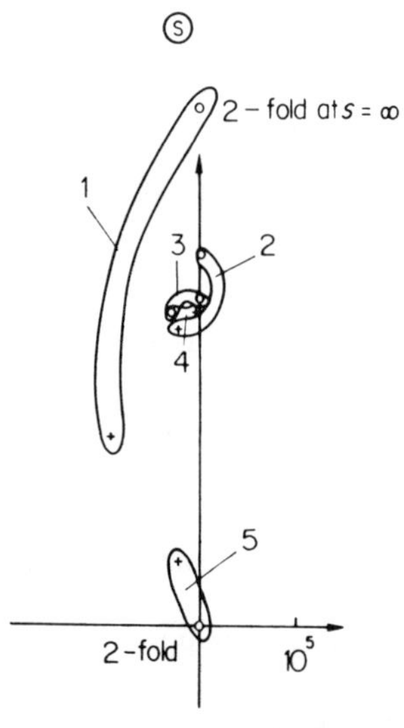

Figure 2(b)

$K_1 = 4{\cdot}75$ $\quad d_1 = 0{\cdot}569$
$K_2 = 0{\cdot}48$ $\quad d_2 = 0{\cdot}24$
$K_3 = 0{\cdot}96$ $\quad d_3 = 0{\cdot}28$
$K_4 = 0{\cdot}93$ $\quad d_4 = 0{\cdot}32$
$K_5 = 0{\cdot}83$

Table II

$z_1 = \pm j\,.\,0{\cdot}32233523\,.\,10^6$
$z_2 = \pm j\,.\,0{\cdot}36742346\,.\,10^6$
$z_3 = -0{\cdot}31480\,.\,10^5 + j\,.\,0{\cdot}3132295\,.\,10^6$
$z_4 = 0$ (twice)
$z_5 = \infty$ (twice)
$p_1 = -0{\cdot}276100\,.\,10^5 \pm j\,.\,0{\cdot}2961048\,.\,10^6$
$p_2 = -0{\cdot}31480\,.\,10^5 \pm j\,.\,0{\cdot}3132295\,.\,10^6$
$p_3 = -0{\cdot}8706\,.\,10^4 \pm j\,.\,0{\cdot}314697\,.\,10^6$
$p_4 = -0{\cdot}93340\,.\,10^5 \pm j\,.\,0{\cdot}18670202\,.\,10^6$
$p_5 = -0{\cdot}25280\,.\,10^5 \pm j\,.\,0{\cdot}62888167\,.\,10^5$

Now we optimize again, however with the constraints: low pass at the beginning, high pass at the end and pole-zero assignment p_3 to z_3. The result is plotted in Figure 2(b). The minimum voltage is now given by

$$\min |V_1(j\omega)| = \frac{V_{\text{dyn}}}{10^{d_1}} = \frac{V_{\text{dyn}}}{3{\cdot}63}$$

after stage 1. Thus the noise distance has been reduced by a factor of 1·7.

Finally a rule of thumb for the optimum pole-zero assignment and the optimal sequence is given. This rule avoids the use of a computer program, however, the solution is only an approximation.

The rule of thumb is:

(α) Pole-zero assignment: The pole with the highest pole-Q should be assigned to the zero nearest to it. Thus the peak created by the pole is reduced most effectively. Then the pole with the 2nd highest Q is assigned to the zero next to it and so on until all poles and zeros are paired. This procedure yields m second order functions.

(β) Determination of the sequence: Let the m 2nd order functions obtained from (α) be designated by $t_\nu(j\omega)$ as defined in equation (1b). To each transfer function t_ν belongs a D_ν-value defined by

$$D_\nu = \log \frac{\max |t_\nu(j\omega)|}{\min |t_\nu(j\omega)|}$$

where the maxima and minima are taken for the ω-intervals in equations (2b) and (2c). Now we order the sequence from input to output by increasing D_ν-values. Thus the stage with the 'flattest' magnitude of $t_\nu(j\omega)$ is the first one, with the '2nd flattest' magnitude the 2nd one etc.

With this rule of thumb one usually obtains a solution close to the optimum. If one, however, needs the optimum solution, the use of the computer program for the dynamic programming method is necessary.

For the example in Table II the rule of thumb yields the same solution as in Figure 2(a). If we meet the requirements, low pass at the beginning and high pass at the end, the rule of thumb yields the solution:

$$(p_4 z_5)(p_1 z_1)(p_2 z_2)(p_3 z_3)(p_5 z_4)$$

with the smallest noise distance

$$\min |V_2(j\omega)| = \frac{V_{\text{dyn}}}{10^{d_2}} = \frac{V_{\text{dyn}}}{3{\cdot}837}$$

which is by a factor of 1·057 (5·7 per cent) worse than the optimal solution. The gain factors for this solution are $K_1 = 4{\cdot}75$, $K_2 = 0{\cdot}48$, $K_3 = 0{\cdot}96$, $K_4 = 0{\cdot}93$ and $K_5 = 0{\cdot}83$.

ACKNOWLEDGEMENT

The author wishes to acknowledge the support of Mr. W. Schöller who designed the computer program and calculated the examples.

REFERENCES

1. E. Lueder, 'A decomposition of a transfer function minimizing distortion and inband losses', *BSTJ*, **49**, 455–469 (1970).
2. E. Lueder, Eine Kettenschaltung RC-aktiver Vierpole mit maximalem Dynamikbereich', *AEÜ*, **24**, 467–470 (1970).
3. S. Halfin, Optimization method for cascading filters', *BSTJ*, **49**, 185–190 (1970).
4. W. Schöller, Optimale Kettenschaltung RC-aktiver Vierpole', *Technical Report*, Institute of Network and Systems Theory, 1973, University of Stuttgart, Germany.
5. G. L. Nemhauser, *Dynamic Programming*, Wiley, New York, 1966.
6. G. S. Moschytz, 'Active RC-filter building blocks using frequency emphasizing networks', *IEEE J. Solid State Circuits*, SC-**2**, 59–62 (1967).

Part II
Second-Order Filter Sections

A large majority of all active filter designs is based on second-order sections that realize a function

$$T(s) = \frac{V_{out}}{V_{in}} = \frac{a_2 s^2 + a_1 s + a_o}{s^2 + b_1 s + b_o}$$

By an appropriate choice of the coefficients a_i, $i=0,1,2$, T(s) can represent all types of second-order filters, in particular those realizing lowpass, highpass, bandpass, band reject and all pass functions. Filters of higher order are obtained by interconnecting second-order sections in cascade (see paper 3-1) or, for reduced sensitivity, in some kind of feedback topology (see papers 3-2, 3-3, and 3-7). Thus, due to the importance of second-order "building blocks", this chapter contains eight papers out of the hundreds available in the literature which present general design and classification approaches and a few select circuits that have been proven practical and successful.

In *paper 2-1,* Sedra and Brown discuss a simple classification scheme for single-amplifier filters, placing circuits into groups that have, in principle, the same sensitivity performance. The gain-sensitivity product (compare paper 1-3) is used for the comparison and a principle is indicated by which a number of different filter functions can be obtained from a given topology by feeding the input signal into the appropriate node (see also the earlier companion paper [1] and the alternative method in [2]). The classification should help in the selection of a suitable circuit out of the numerous available ones. Furthermore, it is discussed how active and passive sensitivities can be traded off, allowing the user to, possibly, further improve the performance of a chosen circuit for a given application and technology (compare the discussion in paper 1-2).

It was pointed out in the introduction to chapter 1 that the frequency-dependence of amplifier gain, $A(s) = GB/(s + \sigma)$, if neglected, can cause serious deviations in the response of active filters. Thus, even at relatively low frequencies, the design may have to be based on the form of A(s) in order to avoid these problems. Geffe, in *paper 2-2,* in connection with a number of circuits presents a suitable method that can readily be applied to other filter structures. The reader should consult papers 1-1 and 2-1 in this context and paper 5-1 for the case where the operating frequency becomes a significant fraction of GB.

A very highly developed 'Single Amplifier Biquad' (SAB) is discussed in *paper 2-3* by Friend, Harris and Hilberman. This circuit is probably the most widely used and manufactured active filter section. The authors describe the work that leads to a well performing, optimized filter (compare also papers 1-2, 2-1 and, for fabrication details, paper 4-3), with emphasis on universal applicability and cost-effective manufacturing as a laser-trimmed hyrid circuit.

A number of performance limitations of single-amplifier sections can be overcome by incorporating additional amplifiers into the filter. This does not imply, though, that multi-amplifier filters necessarily show an improved behavior; thus, comparative evaluations of these filters, such as the one in *paper 2-4* by Sedra and Espinoza, are necessary in order to identify superior structures. The authors investigate 15 two-amplifier filters of which one, based on a number of criteria, emerges as a superior circuit. In *paper 2-5,* Fliege discusses this circuit that is based on a general impedance converter (GIC) and can realize any type of biquadratic transfer function. For further details about GIC-based sections, the reader should also consult [3] and [4]. A three-amplifier filter with good high-frequency performance is introduced by Åckerberg and Mossberg in *paper 2-6.* This circuit is a modification of a state-variable circuit [5] and achieves its improved high-frequency performance by using the active compensation method discussed in paper 1-5.

A problem requiring attention especially in the design of data transmission systems is that of gain and delay equalization. While, traditionally, the requirements were solved with passive RLC circuits, it was soon found that more economical solutions could be obtained with active circuits due to their greater flexibility and because they are easier to tune and adjust. One such approach is described by Fleischer in *paper 2-7;* the filter, based on a four-amplifier state-variable circuit, can be configured to provide both loss and delay equalization. A (loss) equalizer using fewer active elements is discussed by Brglez in *paper 2-8.* The circuit provides a full range of equalization (gain or loss) by replacing a simulated negative resistor by a switch, thereby saving one amplifier. For work closely related to paper 2-8, the reader is also referred to [6].

[1] A. S. Sedra, "Generation and classification of single amplifier filters", Int. J. of Circuit Theory and Applications, vol. 2, pp. 51-67, March 1974.

[2] M. A. Soderstrand and S. K. Mitra, "Gain and sensitivity limitations of active RC filters", IEEE Trans. Circuit Theory, vol. CT-18, No. 6, pp. 600-607, November 1971.

[3] B. B. Battacharyya, W. B. Mikhael and A. Antoniou, "Design of RC active networks using generalized immittance converters", J. Franklin Inst., vol. 297, No. 1, pp. 45-58, January 1974.

[4] C.F. Chiou and R. Schaumann, "Performance of GIC-derived active RC biquads with variable gain", Proc. IEE, vol. 128, Pt. G., Electronic Circuits and Systems, No. 1, pp. 46-52, Feb. 1981.

[5] W. J. Kerwin, L. P. Huelsman and R. W. Newcomb, "State variable synthesis for insensitive integrated circuit transfer functions", IEEE J. Solid-State Circuits, vol. SC-2, pp. 87-92, September 1967.

[6] W. Saraga, "Some new active RC variable equalizer circuits", Proc. IEEE Int. Symp. Circuits Systems, pp. 141-145, April 1980.

A REFINED CLASSIFICATION OF SINGLE AMPLIFIER FILTERS

ADEL S. SEDRA AND LES BROWN

Department of Electrical Engineering, University of Toronto, Toronto, Ontario, M5S 1A4, Canada

SUMMARY

It is shown that Single Amplifier Biquads (SAB's) may be grouped into two major categories, the Enhanced Negative Feedback (ENF) and the Enhanced Positive Feedback (EPF). These two major classes of circuit are related by the complementary transformation and hence exhibit identical sensitivity performance. In the extreme, the ENF class degenerates to the familiar Negative Feedback (NF) class while the EPF class reduces to the Positive Feedback (PF) case where the op amp is connected in a unity-gain configuration. Depending on the order of the RC feedback network, each class includes two subclasses with different sensitivity performance. The paper examines in detail the 'enhanced feedback' mechanism, and investigates the optimum design of the Deliyannis–Friend circuit and its complement the Sallen and Key positive feedback configuration.

1. INTRODUCTION

In a recent paper[1] a method for generating and classifying Single Amplifier Biquads (SAB's) was presented. The classification scheme was based on the network structure and enabled categorizing the large number of biquad circuits[2,3] into a finite number of families with the members of each family having identical sensitivity performance.

The method of Reference 1 starts by classifying SAB's into two major categories, namely, the Negative Feeback (NF) class whose general configuration is shown in Figure 1(a), and the Positive Feeback (PF) class whose general configuration is shown in Figure 1(b). Both circuits employ an operational amplifier

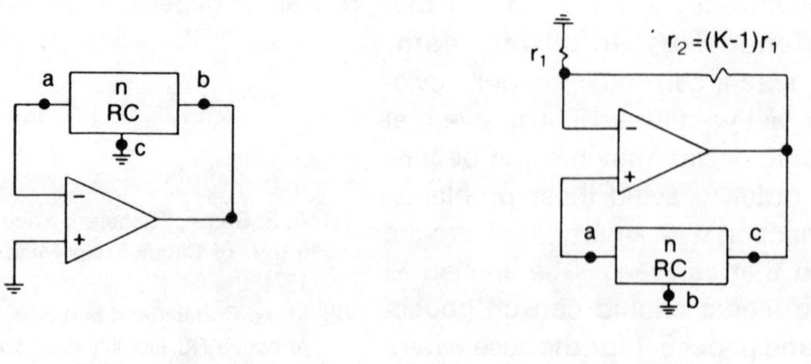

Figure 1. (a) General configuration of the Negative Feedback (NF) class of SAB's, (b) general configuration of the Positive Feedback class which under our new classification is termed Enhanced Positive Feedback (EPF)

(op amp) together with a two-port RC feedback network. For the NF class the RC network is placed in the negative feedback path of the op amp, while in the PF case the RC network is placed in the positive feedback path. Depending on the order of the RC network and depending on whether the value of the parameter K in the PF circuit is chosen equal to unity or greater than unity, six distinct subclasses or families of biquad circuits are obtained.

Reprinted with permission from *Int. J. of Circuit Theory Appl.*, vol. 7, pp. 127-137, Jan. 1979.

It should be noted that we are primarily concerned here with the mechanism of generating the filter poles and with the sensitivities of these poles. Therefore, we have not shown in Figure 1 the method of injecting the input signal. As explained in Reference 1, the input signal source may be connected to any terminal of the feedback network that was previously connected to earth. Clearly, this will not affect the natural modes (or poles) of the network. For the details of the method of injecting the input signal to form different transmission zeros the reader is referred to Reference 1.

The classification of Reference 1 has two major shortcomings. Firstly, it does not include those 'negative feedback' circuits to which some resistive positive feedback is applied. The general configuration for this class of circuits is shown in Figure 2. A notable example of circuits belonging to this class is the Deliyannis circuit[4] which has been subsequently generalized[5–7] and is currently widely employed.

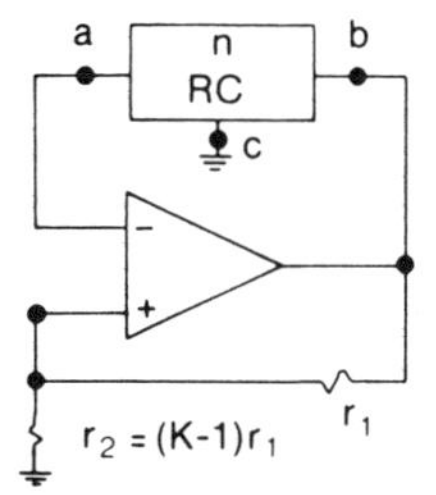

Figure 2. General configuration of the Enhanced Negative Feedback (ENF) class of SAB's

Secondly, the two major configurations of Figure 1 exhibit a disturbing lack of 'duality' or 'parallelism'. While the PF circuit includes some resistive negative feedback, the NF circuit does not contain any resistive positive feedback. Nevertheless, it has been observed[1] that certain subclasses of the PF configuration exhibit identical performance to subclasses of the NF configuration. Furthermore, it has been recently shown[8] that by applying the complementary transformation[9,10,18] to the circuit of Figure 1(b) one obtains the circuit of Figure 2.

The object of this paper is to present a refined and simplified classification of SAB's. In addition to the pedagogical value of this classification, its practical utility will be illustrated by an example. Specifically, we will study the 'optimum' design of the Deliyannis–Friend circuit and its complement, the Sallen and Key positive feedback circuit[11] (with $K>1$). It will be shown that traditional designs of the latter[1,11,12] are far from being optimum. Also, although an optimized design has been recently presented,[8] only the active sensitivities were considered in the optimization, resulting in circuits with high sensitivities to the passive components. An attempt to optimize both sensitivities was reported by Saraga.[13]

2. SENSITIVITY MEASURES

Before presenting the proposed classification scheme, we shall briefly review the sensitivity measures used to compare the different biquad circuits. It should be noted at the outset that all useful biquad circuits are designed such that the sensitivities of the pole frequency ω_0 are at their theoretical minima. This means that all $S_{x_i}^{\omega_0}$, where x_i denotes a passive component, are smaller than or equal to $(\frac{1}{2})$. Also, if we assume that the op amp has a real gain A_0 (which is valid only at very low frequencies), then $S_{A_0}^{\omega_0}$ is either zero or very close to zero. At higher frequencies the op amp frequency response should be taken into account, and one usually assumes a single-pole model, that is, the op amp gain is given by

$$A(p)=A_0\Big/\left[1+\left(\frac{p}{\omega_b}\right)\right]$$

where ω_b is the 3 dB frequency. For $\omega \gg \omega_b$, which is usually the case, $A(p)$ may be approximated by

$$A(p) \simeq (\omega_t/p) \tag{1}$$

where ω_t is the unity-gain bandwidth of the op amp ($\omega_t = A_0\omega_b$).

It has been shown[1,14,15] that the single-pole model of the op amp results in a fractional deviation $(\Delta\omega_0/\omega_0)$ in the pole frequency of the biquad, given by

$$\frac{\Delta\omega_0}{\omega_0} \simeq -\frac{1}{2Q} \cdot GS. \left(\frac{\omega_0}{\omega_t}\right) \tag{2}$$

where Q is the value of the realized Q-factor and GS is the gain–sensitivity product of Q, defined as

$$GS = A_0 . S_{A_0}^{Q}$$

Note that GS is evaluated assuming that the op amp has a constant gain A_0. Nevertheless, the value of GS can be used to determine the deviation in ω_0 due to the op amp's limited bandwidth. For this and other reasons,[14–16] the gain–sensitivity product of Q has become a figure-of-merit for comparing biquads *as far as their active sensitivities are concerned.*

Another way of expressing the result in (2) is to evaluate the sensitivity of ω_0 relative to ω_t. This is given by

$$S_{\omega_t}^{\omega_0} = \frac{1}{2Q} \cdot GS. \left(\frac{\omega_o}{\omega_t}\right) \tag{3}$$

Expressions for the effect of the op amp's frequency response on the value of the pole Q have been given elsewhere.[14] For the SAB circuits considered here the deviation in Q factor is usually small and ω_0 deviations are far more important in determining the variability in the biquad transfer function.

The only remaining sensitivity figures are the passive Q-sensitivities, $S_{x_i}^{Q}$. In a number of SAB circuits $S_{x_i}^{Q}$ are quite large, rendering such circuits of limited use especially if the filter is to be constructed using discrete passive components.

In conclusion, two factors are important in comparing the sensitivity performance of biquads, the gain–sensitivity product of Q, GS, and the passive Q-sensitivities, $S_{x_i}^{Q}$.

3. THE CLASSIFICATION SCHEME

We classify SAB's into the two major classes† whose general configurations are shown in Figures 2 and 1(b). The configuration shown in Figure 2 will be referred to as the Enhanced Negative Feedback (ENF) circuit. We will provide a justification for this name. In a parallel fashion, the configuration in Figure 1(b) will be referred to as the Enhanced Positive Feedback (EPF) circuit. It has been shown in Reference 8, and will be briefly proved here, that these two circuits are related to each other *via* the complementary transformation. That is, the two classes, though different in structure, have identical sensitivity performance. Thus, we need only concern ourselves with the study of either one of the two classes and obtain circuits belonging to the other class by applying the complementary transformation.

Each of these two major classes provides a degenerate class by selecting K equal to unity. Specifically, setting $K = 1$ in the ENF circuit of Figure 2 eliminates the resistive positive feedback and the circuit degenerates to the negative feedback (NF) configuration of Figure 1(a). In a parallel fashion, setting $K = 1$ in the EPF circuit of Figure 1(b) results in the circuit of Figure 3 in which the op amp is used in a unity-gain configuration. This class will be referred to as Positive Feedback (PF). Just as the ENF and

† We note that there is a class of SAB's which may be termed the Mixed Feedback class. Circuits belonging to this class have RC networks in both the positive and negative feedback paths. A special case has been advanced by Boctor.[17] As this class does not appear to be an extensive one and since it does not seem to have sensitivity advantages over some of the circuits considered here, it will not be included.

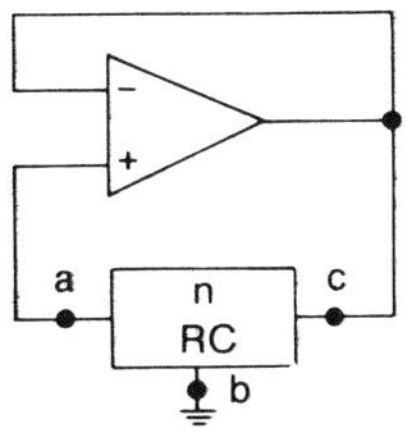

Figure 3. General configuration of the class of SAB's obtained by setting $K=1$ in the circuit of Figure 1(b). This class will be referred to as Positive Feedback (PF)

EPF circuits are complementary, the two degenerate configurations of Figure 1(a) and Figure 3 are also complementary.

Within each of the four classes considered, two subclasses exist based on the choice of the RC feedback network. Specifically, the RC feedback network can be either of a second order or a third order. In the latter case, the RC network embodies a pole–zero cancellation that reduces its apparent order to two.

Thus, SAB's may be classified into eight groups or families as shown in Figure 4. Four of these families (ENF1, ENF2, NF1, NF2) are complementary to the other four (EPF1, EPF2, PF2, PF4). That is, though different in structure, the sensitivity performance of each two corresponding groups is identical. Furthermore, two of the four major classes (NF and PF) are obtained as extreme degenerate cases of the other two (ENF and EPF). It should be noted that we are now using the name PF exclusively for the unity-gain case, reserving the name EPF for the more general case with $K>1$. Other than this, we have retained the family names used in Reference 1 to avoid confusion. For each of the eight families, we have shown in Figure 4 a circuit implementation as an example.

Let us now consider the reasons behind the name 'Enhanced'. We will do so for the ENF case and the results will apply directly to the complementary EPF configuration. To generate ENF circuits one requires a two-port RC network n having a pair of complex conjugate transmission zeros. Consider the circuit of Figure (2) and let n have the open-circuit voltage transfer function $T_{ab}(p)$ given by

$$T_{ab}(p)=\frac{p^2+p(\omega_1/q_1)+\omega_1^2}{p^2+p(\omega_1/q_2)+\omega_1^2} \tag{4}$$

where† (ω_1, q_1) characterize the pair of zeros and (ω_1, q_2) characterize the pair of real axis poles $(q_2<0{\cdot}5)$. Analysis of the circuit of Figure 2, assuming an ideal op amp with infinite gain, results in the characteristic equation

$$T_{ab}-1+\frac{1}{K}=0 \tag{5}$$

Substituting for T_{ab} from (4), and with some simple manipulations, we may express the characteristic equation in the form

$$p^2+p\frac{\omega_1}{q_1}\left[K-(K-1)\frac{q_1}{q_2}\right]+\omega_1^2=0$$

Comparing this with the standard form

$$p^2+p\left(\frac{\omega_0}{Q}\right)+\omega_0^2=0$$

† We have assumed that the poles and zeros of $T_{ab}(p)$ have equal frequencies ω_1. Though not general, this assumption simplifies the analysis to follow and focuses attention on the Q-enhancement mechanism.

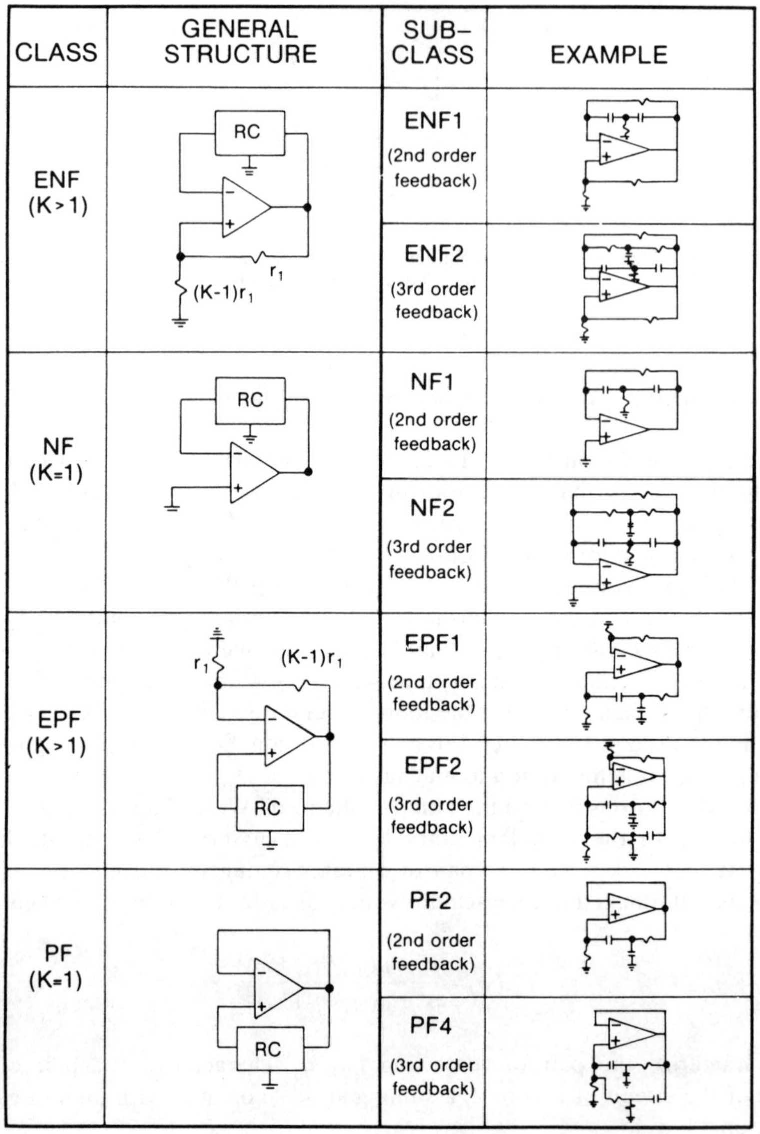

Figure 4. The new classification of SAB's into eight families

results in

$$\omega_0 = \omega_1 \tag{6}$$

$$Q = q_1 \Big/ \left[K - (K-1)\frac{q_1}{q_2} \right] \tag{7}$$

In the case $K = 1$, the positive feedback is destroyed and Q is given by

$$Q = q_1$$

Thus, while in the NF case ($K = 1$) the poles are coincident with the transmission zeros of the RC network, application of positive feedback ($K > 1$) moves the poles on a circle of radius ω_1 closer to the $j\omega$-axis by an amount determined by K. This implies that we design the RC network to have a value of q_1

smaller than the required Q and increase (enhance) this value to the required one through application of positive feedback whose amount is controlled by K. The denominator of (7) can be shown to be smaller than unity as follows. Since $K>1$ and $q_1>q_2$ then

$$(K-1)\left[1-\frac{q_1}{q_2}\right]<0$$

Thus,

$$K-1-(K-1)\frac{q_1}{q_2}<0$$

i.e.,

$$K-(K-1)\frac{q_1}{q_2}<1$$

From this we conclude that the selectivity of the ENF circuit is obtained by two processes. Part of the selectivity is realized by the negative feedback process through 'inverting' the zeros of the RC network by the infinite gain op amp, and part is realized by the positive feedback which moves the poles closer to the $j\omega$-axis on a circular locus. Obviously, a continuous tradeoff exists between the degree of dependence on the two mechanisms involved. This tradeoff manifests itself as a tradeoff between the values of the active and the passive sensitivities. As will be shown in Section 4, relying heavily on the negative feedback mechanism entails high active sensitivities and low passive sensitivities. On the other hand, increasing the dependence on the positive feedback (that is, increasing K) will reduce the active sensitivities at the expense of an increase in the passive sensitivities. This continuous tradeoff that is available to the designer of ENF circuits, and their complementary EPF circuits, is considered an extremely useful attribute which does not exist in the degenerate cases obtained by setting $K=1$.

We next consider the complementary transformation and show that the ENF and the EPF circuits are indeed related by such a transformation. Consider the ENF circuit of Figure 2 and exchange the op amp output terminal with earth. That is, terminals of the feedback networks that are connected to earth should be connected to the op amp output, while terminals that are connected to the op amp output should be earthed. In addition, the input terminals of the op amp should be interchanged. The reader can easily verify that this process will result in the EPF circuit of Figure 1(b).

Analysis of the ENF circuit of Figure 2 shows that its characteristic equation is given by

$$T_{\mathrm{ab}}-1+\frac{1}{K}+\frac{1}{A}=0 \tag{8}$$

while analysis of the EPF circuit of Figure 1(b) results in the characteristic equation

$$-T_{\mathrm{ac}}+\frac{1}{K}+\frac{1}{A}=0 \tag{9}$$

From the complementary transformation[9] we have

$$T_{\mathrm{ac}}=1-T_{\mathrm{ab}}$$

Substituting in (9) obtains equation (8). Thus, the two networks have identical characteristic functions and hence identical poles and pole sensitivities. The reader can easily verify that the circuits of Figure 1(a) (NF) and Figure 3 (PF) are complementary, as they should.

We conclude this section by presenting in Figure 5 the eight SAB classes together with their performance indices. The sensitivity tradeoff exhibited by the ENF1 and EPF1 classes is indicated in the table. Note that the two degenerate classes (NF1 and PF2) have very low passive sensitivities but unacceptably high values for GS and hence for $S^{\omega_0}_{\omega_t}$.

The information presented in Figure 5 raises two points. Firstly, the EPF1 circuit is nothing but the familiar Sallen and Key positive feedback configuration[11,12] with $K>1$. This is the case referred to as PF1

CLASS		FEEDBACK NETWORK	PASSIVE Q-SENSITIVITIES	GS	MAX. ELEMENT SPREAD
ENF & EPF (K > 1)	ENF1 & EPF1	2nd Order	Low ⟶ Q	Q^2 ⟶ Q	Q^2 ⟶ Low
	ENF2 & EPF2	3rd Order	Q	Q	Q
NF & PF (K = 1)	NF1 & PF2	2nd Order	Low	Q^2	Q^2
	NF2 & PF4	3rd Order	Q	Q	Q

Notes 1. For all SAB's: $S^{wo}_{wt} = \frac{1}{2Q} \cdot GS \cdot \left(\frac{wo}{wt}\right)$

2. Only order of magnitudes are indicated

Figure 5. The eight classes of SAB's and their properties: only order of magnitudes is indicated, and for all SAB's

$$S^{\omega_0}_{\omega_t} = \frac{1}{2Q} \cdot GS \cdot \left(\frac{\omega_0}{\omega_t}\right)$$

in Reference 1. Traditionally, such circuits exhibited high passive Q-sensitivities (of the order of Q). The fact that these circuits may be designed to feature sensitivity tradeoff that can be utilized to obtain optimum performance will be illustrated by an example in Section 4.

Secondly, we note that no significant sensitivity tradeoff exists in the ENF2 (EPF2) case. The reason is that these circuits use twin-T type networks which are known to exhibit high passive sensitivities. Thus, it should be clear that in the ENF and EPF circuits no significant sensitivity advantage is gained by using third-order feedback networks. One possible advantage, however, would be the availability of more terminals for signal injection and hence the versatility in realizing a variety of transmission zero configurations with the same network structure. Another possible advantage of circuits using twin-T type networks is that with hybrid-thin film technologies, systematic tuning methods exist.[16] On the other hand, if one is to construct the filter using discrete RC components, it may be recommended to use ENF1 and EPF1 type circuits.

4. STUDY OF A PARTICULAR ENF1(EPF1) CIRCUIT

As an illustrative example, we consider the design of a circuit belonging to the ENF1 class, and its EPF1 complement. We should mention at the outset that the 'optimum' design of these two circuits was the subject of a recent paper.[8] However, only the value of GS was optimized. As will be shown, designs with minimum GS exhibit large passive sensitivities, and in many cases one may prefer to reduce the latter at the expense of a slight increase in GS.

Figure 6(a) shows the Deliyannis–Friend circuit and Figure 6(b) shows its complement, the Sallen and Key positive feedback circuit. For $K = 1$ the circuit in Figure 6(a) degenerates to the familiar multiple-feedback configuration, while the circuit in Figure 6(b) reduces to the Sallen and Key unity-gain realization.

Both circuits in Figure 6 have the characteristic equation given in (8), which in the ideal case ($A = \infty$) reduces to that in (5), where $T_{ab}(p)$ can be easily derived by analyzing the RC network

$$T_{ab}(p) = \frac{p^2 + p(2/CR) + (a/C^2R^2)}{p^2 + p[(2+a)/CR] + (a/C^2R^2)}$$

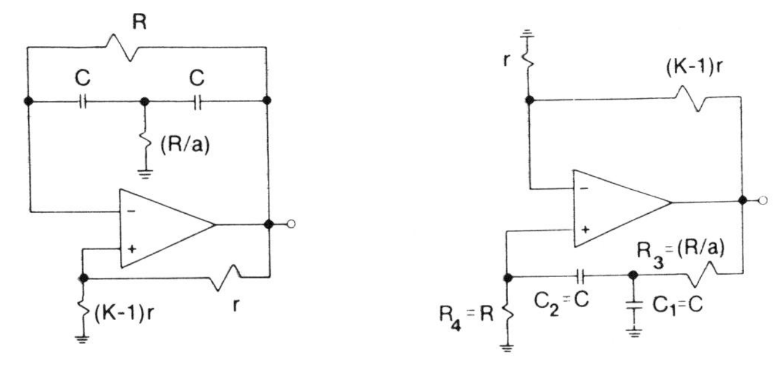

Figure 6. (a) The Deliyannis–Friend circuit, (b) its complement the Sallen and Key circuit

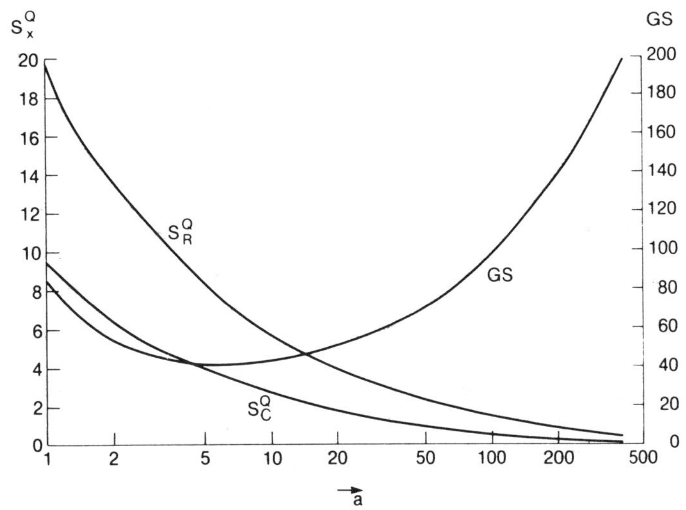

Figure 7. The gain–sensitivity product *GS* and passive *Q*-sensitivities versus the value of the parameter '*a*' for the circuits in Figure 6

Substituting in (5), and with some manipulations, we obtain

$$p^2 + p\frac{1}{CR}(2 + a - aK) + \frac{a}{C^2R^2} = 0$$

Comparing this with the standard form of the second-order characteristic equation results in

$$\omega_0 = \frac{\sqrt{a}}{CR} \tag{10}$$

$$Q = \frac{\sqrt{a}}{2 + a - aK} \tag{11}$$

These are two equations in the three unknowns, (CR), 'a' and K. Therefore, we may use the one available degree of freedom to optimize the sensitivity performance.

If we use 'a' as a design parameter, the design equations become

$$CR = \frac{\sqrt{a}}{\omega_0} \tag{12}$$

$$K = 1 + \frac{2}{a} - \frac{1}{\sqrt{(a)}Q} \tag{13}$$

From (13) we note that for $K \geqslant 1$ there is an upper limit on the value of 'a', given by

$$a \leqslant 4Q^2$$

This extreme value of 'a' corresponds to the NF1 and PF2 circuits for which $K = 1$ and $CR = 2Q/\omega_0$.

Let us now evaluate the active and passive sensitivities. It can be easily shown that

$$S_{A_0}^{\omega_0} = 0, \qquad \text{and} \qquad S_{A_0}^{Q} = \frac{\sqrt{(a)}K^2Q}{A_0}$$

Thus, the gain–sensitivity product is given by

$$GS = \sqrt{(a)}K^2Q \tag{14}$$

Substituting for K from (13) we obtain

$$GS = \sqrt{(a)}\left[1 + \frac{2}{a} - \frac{1}{\sqrt{(a)}Q}\right]^2 Q \tag{15}$$

Using (3) and (15) results in

$$S_{\omega_t}^{\omega_0} = \frac{\sqrt{(a)}}{2}\left[1 + \frac{2}{a} - \frac{1}{\sqrt{(a)}Q}\right]^2 \left(\frac{\omega_0}{\omega_t}\right) \tag{16}$$

Figure 7 shows a plot for GS versus the design parameter 'a' for $Q = 10$ and for 'a' ranging from 1 to the upper limit of 400. We observe that GS exhibits a 'broad' minimum. The value of 'a' for minimum GS can be obtained by differentiating (15) with respect to 'a' and equating the derivative to zero. The result for $Q > 5$ is

$$\check{a} \simeq 6$$

The corresponding minimum value of GS is given by

$$GS|_{\min} \simeq 3{\cdot}2Q \qquad \text{for } Q > 5$$

These values are identical to those derived in Reference 8.

Next, we consider the passive Q-sensitivities which are given by

$$S_{C_2}^{Q} = -S_{C_1}^{Q} = -\frac{1}{2} + \frac{Q}{\sqrt{a}}$$

$$S_{R_4}^{Q} = -S_{R_3}^{Q} = -\frac{1}{2} + \frac{2Q}{\sqrt{a}}$$

Obviously the passive Q-sensitivities decrease as 'a' is increased. At the upper limit of 'a', which corresponds to NF1 and PF2 designs, we have

$$S_{C_2}^{Q} = S_{C_1}^{Q} = 0, \qquad S_{R_4}^{Q} = -S_{R_3}^{Q} = \tfrac{1}{2}$$

which are the lowest possible, confirming earlier results. Note, however, the large value of GS, and correspondingly of $S_{\omega_t}^{\omega_0}$, obtained with such classes.

The passive sensitivities are plotted in Figure 7 for the case $Q = 10$. Note that at the point of minimum GS, the passive sensitivities can be unacceptably high. Specifically, their values are

$$|S_C^Q| \simeq -\tfrac{1}{2} + 0{\cdot}4Q, \qquad |S_R^Q| \simeq -\tfrac{1}{2} + 0{\cdot}8Q$$

An important observation can be made from the graphs of Figure 7. For values of $a < \check{a}$ ($\simeq 6$) both the active and passive sensitivities increase as 'a' is decreased. Thus, it is clearly inappropriate to select $a < \check{a}$. Unfortunately, traditional designs of the Sallen and Key positive feedback circuits were based (implicitly) on a value of 'a' in this region. In particular, many designs were based on $a = 1$, for which case

$$GS \simeq 9Q, \qquad S_{\omega_t}^{\omega_0} \simeq -4{\cdot}5\left(\frac{\omega_0}{\omega_t}\right)$$

$$|S_C^Q| = Q - \tfrac{1}{2}; \qquad |S_R^Q| = 2Q - \tfrac{1}{2}$$

Having decided to choose $a \geqslant 6$, the question still remains as to the exact value of 'a'. The answer depends on the specific application as well as on the technology to be employed in constructing the filter. If high passive sensitivities are of no concern, then 'a' should be selected equal or close to 6. On the other hand, if one desires smaller passive sensitivities and is willing to accept an increase in GS, then a higher value of 'a' is appropriate. Furthermore, the results in Figure 7 may be used to optimize a sensitivity measure that includes both the active and passive sensitivities with appropriate weighting functions, thus arriving at an optimum value for 'a'. The present authors have designed many circuits with 'a' in the range 20–100 with favourable results.

This example clearly shows why the ENF and EPF circuits (with second-order feedback) are superior to the extreme cases obtained by setting $K = 1$. The question may be asked as to which circuit to use, ENF or EPF? The answer to this will depend primarily on the transmission zeros desired. The complementary transformation does not, of course, preserve the biquad transmission zeros. Certain transmission zero configurations will be found to be easier to realize with one class than with the other. Furthermore, in

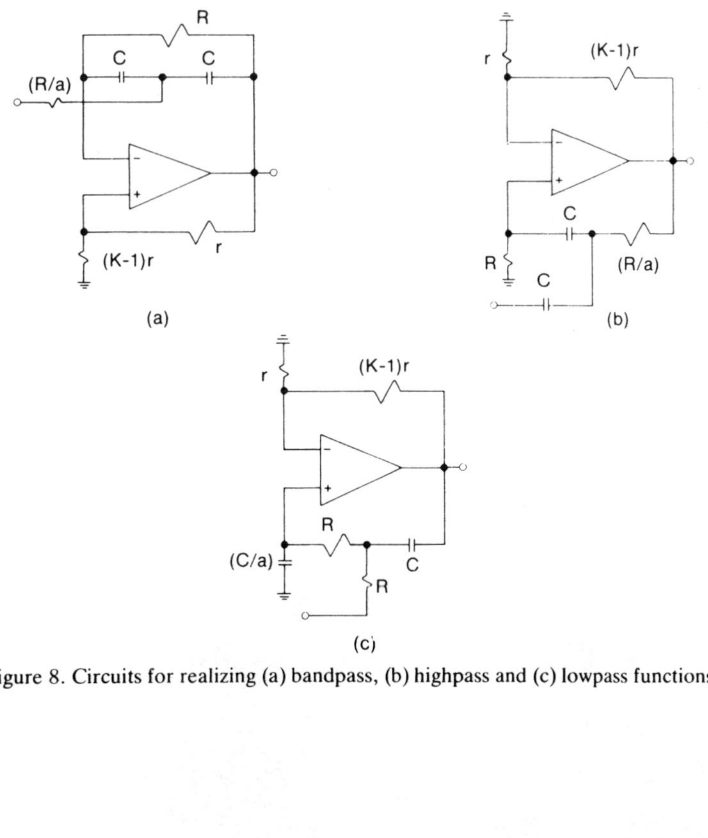

Figure 8. Circuits for realizing (a) bandpass, (b) highpass and (c) lowpass functions

some cases one requires an inverting biquad (ENF) while in other cases a positive sign might be specified (EPF).

Figure 8 provides three circuits for realizing bandpass, highpass and lowpass functions. In all three, the value of 'a' is used as a design parameter. All the analysis, graphs and conclusions presented in the foregoing apply to each of the three circuits.

5. CONCLUSIONS

It has been shown that Single Amplifier Biquads (SAB's) may be classified into two major groups, the Enhanced Negative Feedback (ENF) and the Enhanced Positive Feedback (EPF). These two classes are related by the complementary transformation and therefore exhibit identical sensitivity performance. Each of the two major classes provides in the extreme case a degenerate class, namely, the Negative Feedback (NF) and the Positive Feedback (PF). Since in each of these four cases, the RC feedback network can be either a second-order or a third-order one, a total of eight classes of SAB circuits is obtained.

The availability of a continuous tradeoff between the active and the passive sensitivities in the ENF and EPF circuits enables the designer to optimize the circuit to suit the particular application and the technology employed in constructing the active filter. The detailed example presented clearly illustrated this point.

It is hoped that the simple classification presented will enable workers in the area to quickly assess the relative merits of newly invented circuits. Also, designers of active filters might find it easier to answer the everlasting question: which circuit is best for my application?

REFERENCES

1. A. Sedra, 'Generation and classification of single amplifier filters', *Int. J. Cir. Theor. Appl.*, **2**, 51–67 (1974).
2. G. Daryanani, *Principles of Active Network Synthesis and Design*, Wiley, New York, 1976.
3. A Sedra and P. O. Brackett, *Filter Theory and Design: Active and Passive*, Matrix Publishers, Champaign, Il., 1978.
4. T. Deliyannis, 'High-Q factor circuit with reduced sensitivity', *Electron. Lett.*, **4**, 577 (1968).
5. J. J. Friend, 'A single operational amplifier biquadratic filter section', *1970 IEEE Int. Symp. on Circuit Theory, Atlanta, Ga.* (1970).
6. J. J. Friend, C. A. Harris and D. Hilberman, 'STAR: An active biquadratic filter section', *IEEE Trans. Cir. Syst.*, **CAS-22**, 115–121 (1975).
7. P. E. Fleischer, 'Sensitivity minimization in a single amplifier biquad circuit', *IEEE Trans. Cir. Syst.*, **CAS-23**, 45–55 (1976).
8. G. S. Moschytz and P. Horn, "Optimizing two commonly used active filter building blocks using the complementary transformation', *Proc. 1977 Int. Symp. on Circuits and Systems*, 332–335 (1977).
9. N. Fliege, 'Complementary transformation of feedback systems', *IEEE Trans. Cir. Theor.*, **CT-20**, 137–139 (1973).
10. D. Hilberman, 'Input and ground as complements in active filters', *IEEE Trans. Cir. Theor.*, **CT-20**, 540–547 (1973).
11. R. P. Sallen and E. L. Kay, 'A practical method of designing RC active filters', *IRE Trans. Cir. Theor.*, **CT-2**, 78–85 (1955).
12. W. J. Kerwin, 'Active RC network synthesis using voltage amplifiers', in *Active Filters*, Ed. L. P. Huelsman, Chapt. 2, McGraw-Hill, New York, 1970.
13. W. Saraga, 'Sensitivity of 2nd-order Salley–Key active RC filters', *Electron. Lett.*, **3**, 442–443 (1967).
14. A. Sedra and J. L. Espinoza, 'Sensitivity and frequency limitations of biquadratic active filters', *IEEE Trans. Cir. Syst.*, **CAS-22**, 122–130 (1975).
15. G. S. Moschytz and P. Horn, 'The gain–sensitivity product as a universal measure for active-filter pole variations due to nonideal amplifiers', *Proc. 1976 IEEE Int. Symp. on Circuits and Systems*, 122–125, Munich (1976).
16. G. S. Moschytz, *Linear Integrated Networks: Design*, Van Nostrand Reinhold, New York, 1975.
17. S. A. Boctor, 'A novel second-order canonical RC-active realization of high-pass notch filter', *IEEE Trans. Cir. Syst.*, **CAS-22**, 397–404 (1975).
18. M. N. S. Swamy, 'On the matrix parameters of a three-dimensional network', *Proc. IEEE*, **54**, 1081–1082 (1966).

Exact Synthesis with Real Amplifiers

PHILIP R. GEFFE, SENIOR MEMBER, IEEE

Abstract—**Second-degree active filter circuits are synthesized with operational amplifiers having a single pole or two poles and one zero. The method is applied to obtain designs and sensitivities for six useful filter sections. It is shown 1) that the two-pole amplifier is sometimes not the better choice, even though it has more gain at the operating frequency, and 2) that sensitivities are optimized by tradeoffs involving positive feedback.**

I. Introduction

IT HAS generally been the practice, in active synthesis, to assume that amplifier gain is infinite, or that it is a finite real number. Both assumptions are false, and they lead to severe difficulties in realizing active filters at moderately high frequencies [1], even below 1 kHz, with some micropower operational amplifiers. Nevertheless, excellent circuit performance can be obtained with real amplifiers having a surprisingly modest amount of gain if the problem is approached in a rational way.

In this paper we show how to obtain second-degree transfer functions with prescribed poles and zeros by a method of exact synthesis, which involves solving a system of nonlinear simultaneous equations. The active element is an operational amplifier with a specified open-loop frequency response, which may comprise a single dominant pole, or two poles and one zero. The mathematical method will readily yield solutions for amplifiers having more critical frequencies, but any attempt to synthesize higher degree networks will increase the degree of the resultant of the simultaneous set, with an attendant increase in numerical complications.

While this technique introduces parasitic poles and zeros into the transfer function, it will be seen that they are not cause for concern. Using a one-pole amplifier will result in a single parasitic pole whose magnitude is roughly equal to ω_0 multiplied by the amplifier gain at that frequency, and this may safely be neglected if the gain is 50 or more. In the more general case, the parasitics typically consist of one very remote pole and one moderately remote pole which is effectively canceled by a nearby parasitic zero.

It will be shown that the difficulties resulting from the use of real amplifiers lie in the sensitivity calculations and tradeoffs, rather than in the parasitic critical frequencies.

II. Amplifier Gain Functions

The magnitude response of a one-pole amplifier is shown in Fig. 1(a), and the gain function is

$$A(s) = \frac{A_0\omega_A}{s + \omega_A}. \tag{1}$$

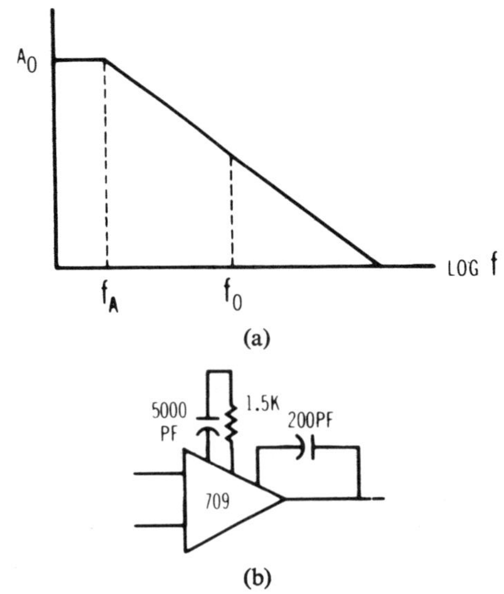

Fig. 1. (a) One-pole magnitude response. (b) Compensation for one-pole 709 amplifier.

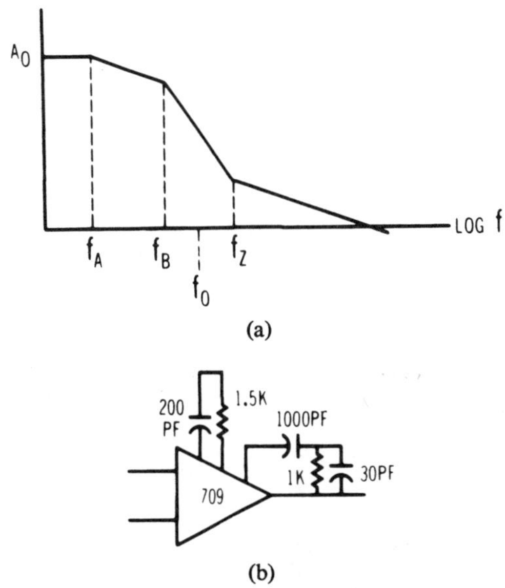

Fig. 2. (a) Two-pole one-zero response. (b) Compensation for 709 amplifier.

Fig. 1(b) shows the method, recommended by the manufacturer, of obtaining such a response from a 709 amplifier. Since the nominal parameter values are $A_0 = 45\,000$ and $f_A = 20$ Hz, the open-loop gain will fall to 60 dB at about 890 Hz.

Fig. 2(a) shows the response of an amplifier with two poles and one zero, and Fig. 2(b) shows how to obtain this from a 709. Here the frequency parameters are $f_A = 0.5$ kHz, $f_B = 4$ kHz, and $f_Z = 160$ kHz, so that the gain is

Manuscript received October 18, 1972; revised March 14, 1973 and July 17, 1973.

The author is with Westinghouse Electric Corporation, Baltimore, Md. 21203.

Reprinted from *IEEE Trans. Circuits Syst.*, vol. CAS-21, pp. 369-376, May 1974.

significantly higher than that of the one-pole 709, remaining above 60 dB until about 9 kHz. In this case the gain function may be written

$$A(s) = \frac{A_0(s/\omega_Z + 1)}{(s/\omega_A + 1)(s/\omega_B + 1)}. \tag{2}$$

If the operating frequency is f_0, the system can be frequency scaled to $\omega_0 = 1$ by dividing all frequencies by f_0:

$$A(s) = \frac{A_0(s/\sigma_0 + 1)}{(s/\sigma_1 + 1)(s/\sigma_2 + 1)} \tag{3}$$

where

$$\sigma_0 = f_Z/f_0$$
$$\sigma_1 = f_A/f_0$$
$$\sigma_2 = f_B/f_0. \tag{4}$$

While it is obvious that (3) includes (1), we would like the gain function to be still more general. In certain networks considered hereafter the noninverting input terminal of the amplifier receives a positive feedback fraction ρ of the output voltage. On analyzing the network we find that the gain function appears only in the form $1/A - \rho$, so we write

$$\frac{1}{A(s)} - \rho = \frac{g_2 s^2 + g_1 s + g_0}{g_3 s + A_0} \tag{5}$$

where

$$g_0 = 1 - \rho A_0$$
$$g_1 = \frac{1}{\sigma_1} + \frac{1}{\sigma_2} - \frac{\rho A_0}{\sigma_0}$$
$$g_2 = \frac{1}{\sigma_1 \sigma_2}$$
$$g_3 = \frac{A_0}{\sigma_0}. \tag{6}$$

We may, of course, set $\rho = 0$. For the one-pole amplifier case we have

$$\frac{1}{\sigma_0} = \frac{1}{\sigma_2} = 0. \tag{7}$$

The normalized gain of the one-pole amplifier is

$$A(s) = \frac{A_0}{g_1 s + g_0} = \frac{\sigma_1 A_0}{s + \sigma_1} = \frac{\gamma}{s + \sigma_1}. \tag{8}$$

It usually happens that σ_1 is small, in which case γ is an accurate approximation to the gain magnitude at the operating frequency and can be read off the amplifier manufacturer's data sheet. On this point it should be mentioned that the author has had no difficulty in applying the technique described with commercial self-compensated integrated-circuit operational amplifiers. Their transfer functions appear to approximate a one-pole response very well, except for some excess phase in the neighborhood of the 0-dB crossover frequency.

III. Synthesis Technique

We begin by analyzing a specific circuit to obtain the voltage gain function

$$\frac{e_2}{e_1} = \frac{N(s)}{D(s)}. \tag{9}$$

The passive elements R_1, C_1, etc., appear in the coefficients of this expression, but we no longer use the familiar symbols A or K to represent the amplifier gain. Instead, the gain of the amplifier is identified with the rational function (5) whose coefficients g_i will therefore be seen in the coefficients of (9).

Having reconciled ourselves to accepting two parasitic poles, we write the denominator as

$$D(s) = t_4 s^4 + t_3 s^3 + t_2 s^2 + t_1 s + t_0. \tag{10}$$

The desired gain function is

$$\frac{e_2}{e_1} = \frac{N(s)}{(\theta_2 s^2 + \theta_1 s + 1)(s^2 + d\omega_0 s + \omega_0^2)} \tag{11}$$

where the factor $(s^2 + d\omega_0 s + \omega_0^2)$ gives the frequency response we are trying to obtain, and $(\theta_2 s^2 + \theta_1 s + 1)$ is the parasitic factor that we are obliged to accept. Here d is the reciprocal of the pole Q, and ω_0 is the pole magnitude, which is the nominal tuning frequency of the section. From (11) we have

$$\begin{aligned} D(s) = {} & \theta_2 s^4 + (d\omega_0\theta_2 + \theta_1)s^3 \\ & + (d\omega_0\theta_1 + \omega_0^2\theta_2 + 1)s^2 \\ & + (\omega_0^2\theta_1 + d\omega_0)s + \omega_0^2. \end{aligned} \tag{12}$$

Our purpose is to force (10) and (12) to be the same polynomial, so we equate coefficients of like powers to obtain the simultaneous set

$$\frac{\theta_2}{\omega_0^2} = \frac{t_4}{t_0} \tag{13}$$

$$\frac{d\omega_0\theta_2 + \theta_1}{\omega_0^2} = \frac{t_3}{t_0} \tag{14}$$

$$\frac{1 + d\omega_0\theta_1 + \omega_0^2\theta_2}{\omega_0^2} = \frac{t_2}{t_0} \tag{15}$$

$$\frac{\omega_0^2\theta_1 + d\omega_0}{\omega_0^2} = \frac{t_1}{t_0} \tag{16}$$

and we note that use of a one-pole amplifier will automatically yield

$$t_4 = \theta_2 = 0.$$

After eliminating θ_1 and θ_2 from these equations, we are left with the simultaneous pair

$$(1 - d^2)\omega_0^4 t_4 + d\omega_0^3 t_3 - \omega_0^2 t_2 + t_0 = 0 \tag{17}$$

$$d\omega_0^4 t_4 - \omega_0^3 t_3 + \omega_0 t_1 - d t_0 = 0 \tag{18}$$

in which we shall later set $\omega_0 = 1$. These equations contain more than two unknowns, so that some may be assigned arbitrary values. By making a judicious choice of these (such as $R_1 = R_2 = 1$ or $C_1 = C_2 = 1$) the preceding equations can be written in the form

$$a_3xy + a_2x + a_1y + a_0 = 0 \quad (19)$$

$$b_3xy + b_2x + b_1y + b_0 = 0 \quad (20)$$

where x and y are the remaining unknown values of resistance or capacitance, and the a_i and b_i are known numbers. The solution of these equations is obtained by solving the resultant

$$\boldsymbol{R} = \begin{vmatrix} a_3x + a_1 & a_2x + a_0 \\ b_3x + b_1 & b_2x + b_0 \end{vmatrix} = 0. \quad (21)$$

After solving for x, y is given by

$$y = -\frac{a_2x + a_0}{a_3x + a_1} = -\frac{b_2x + b_0}{b_3x + b_1}. \quad (22)$$

θ_1 and θ_2 can now be calculated from (16) and (13), and we can use (14) and (15) as a check on numerical accuracy. It will then be found that the unknowns in $N(s)$, such as the constant multiplier, can be calculated without difficulty. It should be noted that (17) and (18) are solved without ever dividing by t_4 or θ_2 (which may be zero), so that the solution applies, without change, to both amplifier cases.

This procedure gives valid solutions for all the circuits considered in this paper, except the upper peaked section. Here the denominator quantities are constrained by the numerator conditions so that the simultaneous set must be augmented with another equation. This leads to a cubic resultant which must be solved for a real root.

IV. Design Formulas

The following symbols and formulas will be used:

$$T_1 = R_1C_1 \quad (23)$$

$$T_2 = R_2C_2 \quad (24)$$

$$T_{12} = R_1C_2 \quad (25)$$

$$R_1 = \frac{R_AR_B}{R_A + R_B} \quad (26)$$

$$K_0 = \frac{R_1}{R_A} \quad (27)$$

$$K_1 = 1 + \frac{R_2}{R_3} \quad (28)$$

$$R_3 = \frac{R_2}{K_1 - 1} \quad (29)$$

$$R_A = \frac{R_1}{K_0} \quad (30)$$

$$R_B = \frac{R_1}{1 - K_0} \quad (31)$$

$$R_4 = \frac{1 - (\alpha + \rho)}{\alpha} \quad (32)$$

$$R_5 = 1 \quad (33)$$

$$R_6 = \frac{1 - (\alpha + \rho)}{\rho}. \quad (34)$$

Then α is a feed-forward fraction, ρ a positive-feedback fraction, ω_1 an attenuation peak ($j\omega -$ axis zero), and H is the constant multiplier of the transfer function.

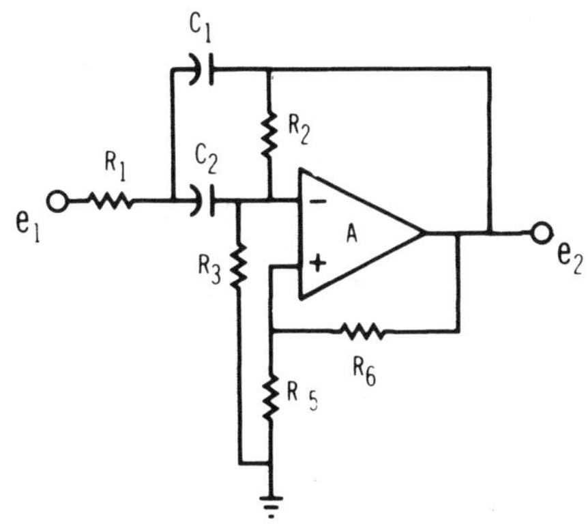

Fig. 3. Inverting bandpass section.

Inverting Bandpass Section (Fig. 3)

Analysis of the circuit gives

$$N(s) = -T_2s \quad (35)$$

$$D(s) = (T_1T_2)s^2 + (T_1 + T_{12})s + 1 + \left(\frac{1}{A} - \rho\right) \cdot \{(T_1T_2)s^2 + [K_1(T_1 + T_{12}) + T_2]s + K_1\}. \quad (36)$$

Substituting (5) in (35) and (36) and clearing fractions,

$$N(s) = -(g_3s + A_0)T_2s \quad (37)$$

$$\begin{aligned} D(s) = {} & (g_2T_1T_2)s^4 + [(g_1 + g_3)T_1T_2 + g_2K_1(T_1 + T_{12}) \\ & + g_2T_2]s^3 + [(g_0 + A_0)T_1T_2 \\ & + (g_3 + g_1K_1)(T_1 + T_{12}) + g_1T_2 + g_2K_1]s^2 \\ & + [(g_0K_1 + A_0)(T_1 + T_{12}) + g_0T_2 \\ & + g_1K_1 + g_3]s + (g_0K_1 + A_0). \end{aligned} \quad (38)$$

In this circuit it is convenient to choose

$$C_1 = C_2 = 1, \quad x = R_1, \quad y = R_2, \quad K_1 \geq 1. \quad (39)$$

The coefficients of (21) and (22) are given by

$$a_3 = -(A_0 + g_0) + d(g_1 + g_3) + 1 - d^2 \quad (40)$$

$$a_2 = 2\,dg_2K_1 - 2(g_3 + g_1K_1) \quad (41)$$

$$a_1 = g_2 - g_1 \quad (42)$$

$$a_0 = A_0 + (g_0 - g_2)K_1 \quad (43)$$

$$b_3 = -(g_1 + g_3) + dg_2 \quad (44)$$

$$b_2 = 2(A_0 + g_0K_1) - 2g_2K_1 \quad (45)$$

$$b_1 = g_0 - g_2 \quad (46)$$

$$b_0 = g_1K_1 + g_3 - d(A_0 + g_0K_1). \quad (47)$$

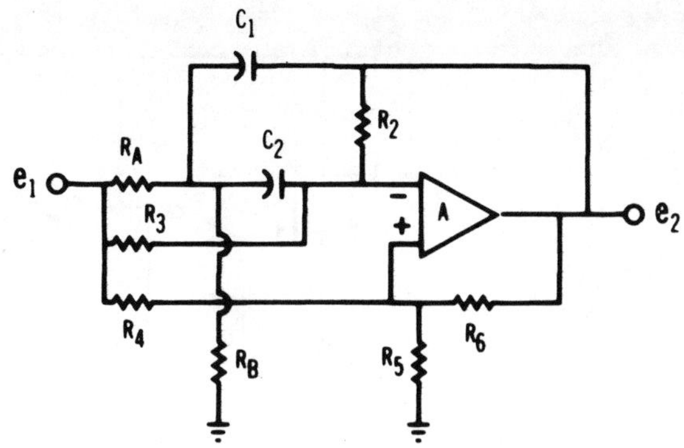

Fig. 4. Lower peaked section.

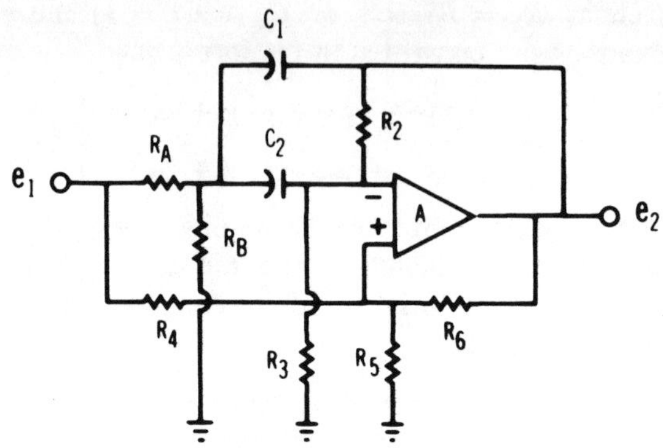

Fig. 5. Upper peaked section.

Then

$$\theta_2 = \frac{g_2 R_1 R_2}{A_0 + g_0 K_1} \tag{48}$$

$$\theta_1 = 2R_1 + \frac{g_0 R_2 + g_1 K_1 + g_3}{A_0 + g_0 K_1} - d \tag{49}$$

$$H = -\frac{A_0 R_2}{A_0 + g_0 K_1} \tag{50}$$

$$\sigma_0 = \frac{A_0}{g_3} \tag{51}$$

$$\frac{e_2}{e_1} = \frac{H(s/\sigma_0 + 1)s}{(\theta_2 s^2 + \theta_1 s + 1)(s^2 + ds + 1)}. \tag{52}$$

Lower Peaked Section (Fig. 4)

Equations (36)–(49) apply to this section and

$$N(s) = (\alpha T_1 T_2)s^2 + [(\alpha K_1 - K_1 + 1)(T_1 + T_{12}) + (\alpha - K_0)T_2]s + (\alpha K_1 - K_1 + 1). \tag{53}$$

Since (53) provides a $j\omega$-axis zero, we have, with $C_1 = C_2 = 1$,

$$\omega_1 \leq \omega_0 \tag{54}$$

$$2(\alpha K_1 - K_1 + 1)R_1 + (\alpha - K_0)R_2 = 0 \tag{55}$$

$$\omega_1^2 = \frac{(\alpha - 1)K_1 + 1}{\alpha R_1 R_2}. \tag{56}$$

Accordingly, we have the design formulas

$$\alpha = \frac{K_1 - 1}{K_1 - R_1 R_2 \omega_1^2} \tag{57}$$

$$K_0 = \frac{2R_1}{R_2}(\alpha K_1 - K_1 + 1) + \alpha \tag{58}$$

$$\frac{e_2}{e_1} = \frac{H(s/\sigma_0 + 1)(s^2 + \omega_1^2)}{(\theta_2 s^2 + \theta_1 s + 1)(s^2 + ds + 1)} \tag{59}$$

$$H = \frac{\alpha R_1 R_2 A_0}{A_0 + g_0 K_1} \tag{60}$$

$$\sigma_0 = \frac{A_0}{g_3}. \tag{61}$$

This circuit can, of course, be used as a high pass section by setting $\omega_1 = 0$.

Upper Peaked Section (Fig. 5)

Here the denominator of the transfer function is given by (38), and the numerator is

$$N(s) = (\alpha T_1 T_2)s^2 + [\alpha K_1(T_1 + T_{12}) + (\alpha - K_0)T_2]s + \alpha K_1. \tag{62}$$

Again, this provides a $j\omega$-axis zero; hence with $C_1 = C_2 = 1$ we have $x = R_1$, $y = R_2$, and

$$2\alpha K_1 R_1 + (\alpha - K_0)R_2 = 0 \tag{63}$$

$$\omega_1 \geq \omega_0 \tag{64}$$

$$K_1 = \omega_1^2 R_1 R_2. \tag{65}$$

The last equation must be solved simultaneously with (19) and (20), for which the a_i and b_i are given by (40)–(47). Substituting (65) in a_2, a_0, b_2, b_0, eliminates K_1 and gives the simultaneous pair

$$a_4 x^2 y + a_3 xy + a_2 x + a_1 y + a_0 = 0 \tag{66}$$

$$b_4 x^2 y + b_3 xy + b_2 x + b_1 y + b_0 = 0 \tag{67}$$

where

$$a_4 = 2\omega_1^2(dg_2 - g_1) \tag{68}$$

$$a_3 = \omega_1^2(g_0 - g_1) + d(g_1 + g_3) - (A_0 + g_0) + 1 - d^2 \tag{69}$$

$$a_2 = -2g_3 \tag{70}$$

$$a_1 = g_2 - g_1 \tag{71}$$

$$a_0 = A_0 \tag{72}$$

$$b_4 = 2\omega_1^2(g_0 - g_2) \tag{73}$$

$$b_3 = \omega_1^2(g_1 - dg_0) + dg_2 - (g_1 + g_3) \tag{74}$$

$$b_2 = 2A_0 \tag{75}$$

$$b_1 = g_0 - g_2 \tag{76}$$

$$b_0 = g_3 - dA_0. \tag{77}$$

The resultant of (66) and (67) is

$$R = \begin{vmatrix} a_4 x^2 + a_3 x + a_1 & a_2 x + a_0 \\ b_4 x^2 + b_3 x + b_1 & b_2 x + b_0 \end{vmatrix} = 0. \tag{78}$$

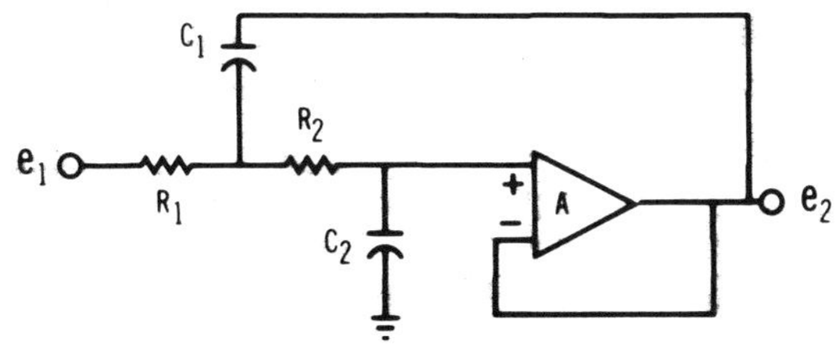

Fig. 6. Voltage-follower low-pass section.

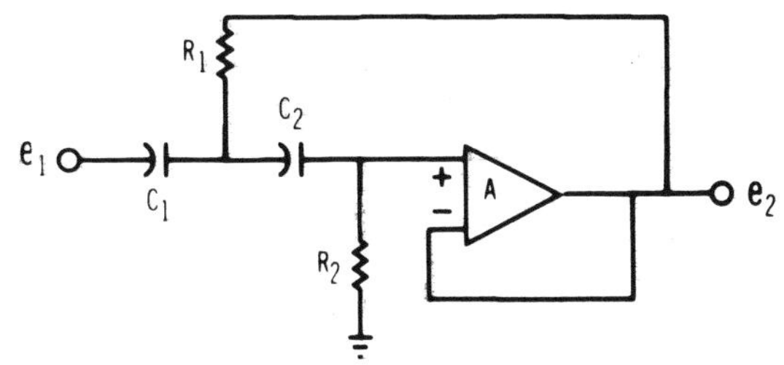

Fig. 7. Voltage-follower high-pass section.

Solving this for $x = R_1$, we follow with $y = R_2$:

$$y = \frac{-(a_2x + a_0)}{a_4x^2 + a_3x + a_1} = \frac{-(b_2x + b_0)}{b_4x^2 + b_3x + b_1}. \quad (79)$$

Then (65) gives K_1, and we choose $\alpha < 1$ and $K_0 < 1$ such that

$$\frac{\alpha}{K_0} = \frac{R_2}{2K_1R_1 + R_2}. \quad (80)$$

Finally, (59)–(61) apply to this circuit also.

Voltage-Follower Low-Pass Section (Fig. 6)

Here the amplifier closed-loop gain is

$$A_c = \frac{A(s)}{A(s) + 1} = \frac{g_3s + A_0}{g_2s^2 + (g_1 + g_3)s + (g_0 + A_0)} \quad (81)$$

$$N(s) = A_c \quad (82)$$

$$D(s) = (T_1T_2)s^2 + (T_2 + T_{12})s + 1 + (1 - A_c)T_1s. \quad (83)$$

Since $\rho = 0$ for this section, we have

$$g_0 = 1. \quad (84)$$

Substituting (81) in (82) and (83) and clearing fractions,

$$N(s) = g_3s + A_0 \quad (85)$$

$$\begin{aligned} D(s) = {} & (g_2T_1T_2)s^4 + [(g_1 + g_3)T_1T_2 + g_2(T_2 + T_{12}) \\ & + g_2T_1]s^3 + [g_2 + (A_0 + 1)T_1T_2 \\ & + (g_1 + g_3)(T_2 + T_{12}) + g_1T_1]s^2 + [g_1 + g_3 \\ & + (A_0 + 1)(T_2 + T_{12}) + T_1]s + (A_0 + 1). \end{aligned} \quad (86)$$

We may now write $A_0 + 1 \cong A_0$. Choosing $R_1 = R_2 = 1$, $x = C_1$, and $y = C_2$, we obtain the coefficients of (19) and (20):

$$a_3 = g_2 - A_0 + d(g_1 + g_3) - d^2g_2 \quad (87)$$

$$a_2 = dg_2 - g_1 \quad (88)$$

$$a_1 = 2dg_2 - 2(g_1 + g_3) \quad (89)$$

$$a_0 = A_0 - g_2 \quad (90)$$

$$b_3 = g_1 + g_3 - dg_2 \quad (91)$$

$$b_2 = g_2 - 1 \quad (92)$$

$$b_1 = 2g_2 - 2A_0 \quad (93)$$

$$b_0 = dA_0 - (g_1 + g_3) \quad (94)$$

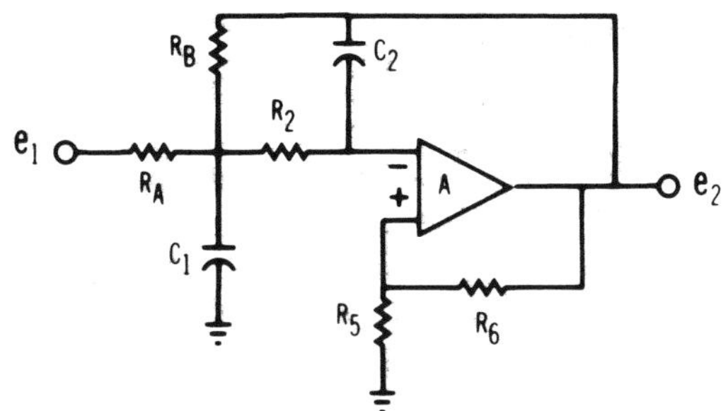

Fig. 8. Friend low-pass section.

and the constant multiplier is

$$H = \frac{A_0}{A_0 + 1} \cong 1. \quad (95)$$

Also

$$\frac{e_2}{e_1} = \frac{H(s/\sigma_0 + 1)}{(\theta_2s^2 + \theta_1s + 1)(s^2 + ds + 1)} \quad (96)$$

$$\sigma_0 = \frac{A_0}{g_3} \quad (97)$$

and θ_1 and θ_2 are obtained from (13) and (16).

Voltage-Follower High-Pass Section (Fig. 7)

First design the normalized low-pass section of Fig. 6 for the given value of Q. If the low-pass capacitance values are C_1' and C_2', then the high-pass elements are

$$C_1 = C_2 = 1 \quad (98)$$

$$R_1 = \frac{1}{C_1'} \quad (99)$$

$$R_2 = \frac{1}{C_2'}. \quad (100)$$

Friend Low-Pass Section (Fig. 8)

Analysis of the circuit, together with (5), gives

$$N(s) = -K_0(g_3s + A_0) \quad (101)$$

$$\begin{aligned} D(s) = {} & (g_2T_1T_2)s^4 + [(g_1 + g_3)T_1T_2 \\ & + g_2(T_1 + T_2 + T_{12})]s^3 + [(g_0 + A_0)T_1T_2 \\ & + g_1(T_1 + T_2 + T_{12}) + g_2 + g_3(T_2 + T_{12})]s^2 \\ & + [A_0(T_2 + T_{12}) + g_0(T_1 + T_2 + T_{12}) \\ & + g_1 + g_3(1 - K_0)]s + A_0(1 - K_0) + g_0. \end{aligned} \quad (102)$$

Choosing $R_1 = R_2 = 1$, $K_0 < 1$, $x = C_2$, $y = C_1$, we obtain the coefficients of (19) and (20):

$$a_3 = (1 - d^2)g_2 + d(g_1 + g_3) - (g_0 + A_0) \quad (103)$$

$$a_2 = 2dg_2 - 2(g_1 + g_3) \quad (104)$$

$$a_1 = dg_2 - g_1 \quad (105)$$

$$a_0 = g_0 + A_0(1 - K_0) - g_2 \quad (106)$$

$$b_3 = dg_2 - (g_1 + g_3) \quad (107)$$

$$b_2 = 2(g_0 + A_0) - 2g_2 \quad (108)$$

$$b_1 = g_0 - g_2 \quad (109)$$

$$b_0 = g_1 + g_3(1 - K_0) - dA_0(1 - K_0) - dg_0. \quad (110)$$

Then

$$H = \frac{-K_0A_0}{A_0(1 - K_0) + g_0} \quad (111)$$

$$\sigma_0 = \frac{A_0}{g_3} \quad (112)$$

$$\frac{e_2}{e_1} = \frac{H(s/\sigma_0 + 1)}{(\theta_2s^2 + \theta_1s + 1)(s^2 + ds + 1)}. \quad (113)$$

V. Sensitivities

Objective evaluation and comparison of active filter designs can only be made through sensitivity analysis, although, of course, other practical matters must also be considered. Analysis of circuits made with real amplifiers is less trivial than we are accustomed to, because they are described by higher degree polynomials. Formulas for cubic system sensitivities were published by Soderstrand and Mitra [2], as were formulas for still higher degree systems [3]. In our circuits, however, we can simplify the analytical work considerably by taking advantage of the fact that each circuit has only two dominant poles and zeros. All other critical frequencies are parasitics, and we do not contemplate designing for the case in which their effects on the transfer function are not negligible. But if they are negligible, then we need not take the trouble to calculate their sensitivities.

Accordingly, our sensitivity analysis derives from (17) and (18), from which the parasitic parameters have been eliminated. Derivation of the sensitivity formulas is made easy by use of the previously published identities [4]:

$$S_x^y \triangleq \frac{dy}{dx}\left(\frac{x}{y}\right) \quad (114)$$

$$S_x^K = 0 \quad (115)$$

$$S_x^{Ky^n} = nS_x^y \quad (116)$$

$$S_x^{\Pi y_i} = \Sigma S_x^{y_i} \quad (117)$$

$$S_x^{\Sigma y_i} = \frac{\Sigma(y_iS_x^{y_i})}{\Sigma y_i} \quad (118)$$

where y and y_i are differentiable functions of x, and K and n are constants.

Application of these identities to (17) gives

$$\omega_0{}^4t_4(4S_x^{\omega_0} + S_x^{t_4}) - d^2\omega_0{}^4t_4(2S_x^d + 4S_x^{\omega_0} + S_x^{t_4}) + d\omega_0{}^3t_3(S_x^d + 3S_x^{\omega_0} + S_x^{t_3}) - \omega_0{}^2t_2(2S_x^{\omega_0} + S_x^{t_2}) + t_0S_x^{t_0} = 0 \quad (119)$$

and (18) gives

$$d\omega_0{}^4t_4(S_x^d + 4S_x^{\omega_0} + S_x^{t_4}) - \omega_0{}^3t_3(3S_x^{\omega_0} + S_x^{t_3}) + \omega_0t_1(S_x^{\omega_0} + S_x^{t_1}) - dt_0(S_x^d + S_x^{t_0}) = 0. \quad (120)$$

These are linear equations in the two unknowns $S_x^{\omega_0}$ and S_x^d, and by letting $\omega_0 = 1$ and rearranging terms, we easily obtain

$$m_1S_x^{\omega_0} + m_2S_x^d = m_3 \quad (121)$$

$$m_4S_x^{\omega_0} + m_5S_x^d = m_6 \quad (122)$$

where

$$m_1 = 4(1 - d^2)t_4 + 3\,dt_3 - 2t_2 \quad (123)$$

$$m_2 = -d(2\,dt_4 - t_3) \quad (124)$$

$$m_3 = -(1 - d^2)t_4S_x^{t_4} - dt_3S_x^{t_3} + t_2S_x^{t_2} - t_0S_x^{t_0} \quad (125)$$

$$m_4 = 4\,dt_4 - 3t_3 + t_1 \quad (126)$$

$$m_5 = d(t_4 - t_0) \quad (127)$$

$$m_6 = -dt_4S_x^{t_4} + t_3S_x^{t_3} - t_1S_x^{t_1} + dt_0S_x^{t_0} \quad (128)$$

and the solution is

$$S_x^{\omega_0} = \frac{\begin{vmatrix} m_3 & m_2 \\ m_6 & m_5 \end{vmatrix}}{\begin{vmatrix} m_1 & m_2 \\ m_4 & m_5 \end{vmatrix}} \quad (129)$$

$$S_x^d = \frac{\begin{vmatrix} m_1 & m_3 \\ m_4 & m_6 \end{vmatrix}}{\begin{vmatrix} m_1 & m_2 \\ m_4 & m_5 \end{vmatrix}}. \quad (130)$$

From (116) we have

$$S_x^Q = -S_x^d. \quad (131)$$

A computer program, based on these equations, was written to investigate the performance of the bandpass circuit of Fig. 3, using the 709 amplifier as shown in Figs. 1 and 2. The results are displayed in Tables I, II, and III. From these numbers, some interesting and useful conclusions can be drawn.

First, a useful reduction in $S_{A_0}^{\omega_0}$ and $S_{A_0}^Q$ can be obtained by using positive feedback, $\rho > 0$. On the other hand, this leads to a rise in the passive sensitivities of Q. Since the latter necessitates tighter element tolerances, the design tradeoffs must be made with some care.

Second, one naturally draws the "obvious" conclusion that the two-pole amplifier should be preferred at higher frequencies because it has considerably more gain. The sensitivity numbers, however, do not confirm this. In general, the two-pole amplifier gives larger $S_{A_0}^Q$ and

TABLE I
$Q = 10, f_0 = 10$ kHz

AMP. POLES	ρ	$S^{W_o}_{A_o}$	$S^{W_o}_{f_A}$	$S^{Q}_{A_o}$	S^{Q}_{R1}	S^{Q}_{R2}	S^{Q}_{C1}	S^{Q}_{C2}
1	0	.0929	.0929	-.0517	-.552	.496	-.0639	.00843
2	0	.00404	.00358	-.181	-.863	.496	-.274	-.0932
1	.01	.0528	.0528	-.0282	-1.47	1.44	-.505	.474
2	.01	.00231	.00204	-.106	-1.68	1.47	-.645	.431
1	.05	.0314	.0314	-.0159	-3.20	3.18	-1.36	1.35
2	.05	.00135	.00120	-.0627	-3.38	3.25	-1.47	1.34
1	.10	.0255	.0255	-.0122	-4.63	4.62	-2.08	2.06
2	.10	.00109	.000963	-.507	-4.81	4.71	-2.18	2.08

TABLE II
$Q = 50, f_0 = 10$ kHz

AMP. POLES	ρ	$S^{W_o}_{A_o}$	$S^{W_o}_{f_A}$	$S^{Q}_{A_o}$	S^{Q}_{R1}	S^{Q}_{R2}	S^{Q}_{C1}	S^{Q}_{C2}
1	0	.331	.331	.227	-.273	.434	.0100	.151
2	0	.0101	.00921	-1.80	-4.15	.467	-2.74	-.943
1	.01	.0691	.0691	.0891	-6.57	6.64	-3.04	3.11
2	.01	.00357	.00322	-6.70	-8.14	6.78	-4.16	2.80
1	.02	.0517	.0517	.0698	-9.54	9.60	-4.52	4.58
2	.02	.00266	.00240	-.504	-10.9	9.84	-5.44	4.41
1	.04	.0388	.0388	.0552	-13.8	13.9	-6.67	6.72
2	.04	.00198	.00179	-3.78	-15.0	14.2	-7.44	6.67
1	.08	.0298	.0298	.0460	-20.2	20.2	-9.85	9.89
2	.08	.00152	.00137	-.290	-21.3	20.7	-10.5	9.94

TABLE III
$Q = 100, f_0 = 20$ Hz

AMP. POLES	ρ	$10^6 S^{W_o}_{A_o}$	$10^6 S^{W_o}_{f_A}$	$S^{Q}_{A_o}$	S^{Q}_{R1}	S^{Q}_{R2}	S^{Q}_{C1}	S^{Q}_{C2}
1	.001	449.	449.	.0894	-4.45	4.45	-1.98	1.98
2	.001	20.2	18.0	.0899	-4.45	4.45	-1.98	1.98
1	.002	327.	327.	.0651	-6.31	6.31	-2.91	2.91
2	.002	14.7	13.1	.0654	-6.31	6.31	-2.91	2.91
1	.004	237.	237.	.0472	-8.95	8.95	-4.22	4.22
2	.004	10.6	9.48	.0474	-8.95	8.95	-4.23	4.23

smaller $S_{A_0}^{\omega_0}$. But the latter result may be a disadvantage in some cases. This point is worth some discussion.

In some circuits ω_0 is independent of gain with ideal amplifiers. But ω_0 always acquires functional dependence on the active parameters when real amplifiers are used. In the silicon technology, the active thermal coefficients are approximately constant at -2000 ppm/°C in the range from about -10°C to $+60$°C. Accordingly this leads to a very useful negative contribution to the thermal coefficient of ω_0; and this can be optimized by making the right choices of f_A and ρ.

Third, in looking over the sensitivity numbers in the tables, the reader should note that they cannot be expected to satisfy the usual sensitivity invariants, because one or two RC products are implicit in the amplifier. That some of the number sets come close to showing invariant sums is merely a reflection of the fact that the active Q sensitivity decreases with increasing ρ.

VI. Discussion

In designing active filters with voltage amplifiers, the designer does not, of course, have the option of using real or ideal amplifiers. His only option in this respect lies in whether or not to accept misleading calculations based on simplistic modeling. How misleading this can be is revealed by considering the circuit of Fig. 6. Using an ideal amplifier with closed-loop gain, $A_c = 1$, we obtain

$$S_{A_c}^{\omega_0} = 0, \qquad S_{A_c}^{Q} = 2Q^2.$$

But an inspection of the results given by Budak and Petrela [1] shows that $S_{A_c}^{Q}$ is actually close to zero until the open-loop gain has fallen off drastically. On the other hand, $S_{A_c}^{\omega_0}$ is not only nonzero, but it is large enough to be a matter of serious concern in practical design.

Since the temperature compensation of ω_0 is a routine necessity in most active bandpass filters, it is highly desirable to understand the sensitivity performance in advance of laboratory work, since this permits circuit substitutions and tradeoffs that would be difficult or impossible to accomplish empirically without the sensitivity information.

Another advantage of the new technique is that it gives a prompt disclosure of unrealizability. Thus, in Table III the circuits are not realizable, with the choices in (39), without positive feedback, regardless of the type of phase compensation used. The author was notified of this fact by the computer program without having to go through any experimental cut and try.

Probably the most useful network discussed in this paper is the bandpass circuit of Fig. 3. Here R_3 is deleted if we choose $K_1 = 1$, and $R_3 = R_2$ if $K_1 = 2$. The reader may be interested to know that, in the latter instance, removing or inserting R_3 has very little effect in many cases. This indicates that the circuit is insensitive to the input impedance of the amplifier, which is a practical feature of some value. On the other hand, small values of R_3 result from choosing large values of K_1. This reduces negative feedback and leads to higher sensitivities, as noted by Friend [5].

Acknowledgment

The author wishes to acknowledge his indebtedness to Dr. M. Soderstrand of Sandia Corporation, Livermore, Calif., who checked the sensitivities displayed in Tables I, II, and III using a computer program which was independent of the work in this paper.

The circuits of Figs. 6 and 7 derive from Sallen and Key. The others were first developed by J. J. Friend [5], based on an idea of Deliyannis [6]. The compensation circuit of Fig. 2(b) was developed by C. C. Glover, of Westinghouse Electric Corporation, for use in active filters in 1968.

The author is grateful to the three anonymous reviewers of this paper whose criticisms led to substantial improvement.

References

[1] A. Budak and D. M. Petrela, "Frequency limitations of active filters using operational amplifiers," *IEEE Trans. Circuit Theory*, vol. CT-19, pp. 322–328, July 1972.

[2] M. A. Soderstrand and S. K. Mitra, "Sensitivity analysis of third-order filters," *Internat. J. Electron.*, vol. 30, pp. 265–272, 1971.

[3] M. A. Soderstrand and S. K. Mitra, "Computer-aided sensitivity analysis of higher order filters," in *Proc. 2nd Internal. Symp. on Network Theory*, Herceg Novi, Yugoslovia, July 1972.

[4] P. R. Geffe, "*RC*-amplifier resonators for active filters," *IEEE Trans. Circuit Theory*, vol. CT-15, pp. 415–419, Dec. 1968.

[5] J. J. Friend, "A single Op-amp biquadratic filter section," presented at the IEEE Internat. Symposium on Circuit Theory, Atlanta, Ga., Dec. 1970.

[6] T. Deliyannis, "High *Q*-factor circuit with reduced sensitivity," *Electron. Lett.*, vol. 4, p. 577, Dec. 1968.

STAR: An Active Biquadratic Filter Section

JOSEPH J. FRIEND, MEMBER, IEEE, CLIFF A. HARRIS, MEMBER, IEEE, AND DAN HILBERMAN, MEMBER, IEEE

Abstract—**This paper describes a thin-film hybrid integrated circuit (HIC) which is specifically manufactured to realize cascadable biquadratic active *RC* filter sections. The HIC consists of an *RC* network with one operational amplifier. This network is interconnected by a printed wiring board to realize Friend's Single Amplifier Biquad (SAB) and other filter sections. Design equations, components, predistortion method, and laser trimming are all discussed.**

I. Introduction

THIS PAPER describes a thin-film hybrid integrated circuit (HIC)—Standard Tantalum Active Resonator (STAR), which is specifically manufactured to realize cascadable biquadratic active filter sections. The active filter realization is based on a building block concept using the Single Amplifier Biquad (SAB) circuits developed by Friend [1].

The circuitry on the substrate is quite simple (Fig. 1), but was arrived at through extensive study to assure maximum utility and versatility in active filter realization while satisfying certain physical and economic constraints. The components on the STAR were intentionally configured as disconnected groups so that the one physical HIC could be used to realize many different electrical circuits.

There have been many commercial universal building blocks for active filters. The TRW,[1] Helipot,[2] and KTI[3] models provide most of the electronics in one package but require the user to provide a few critical external resistors to tune the filter. Fujitsu [2] and the Bell System [3] have made building blocks in thin-film form but these require mask changes with each filter for resistor, capacitor and conductor definition, i.e., the value and interconnection of components. The STAR avoids both the necessity for external components and the different masks. It does this by using one HIC for all filters, setting resistor values with a laser rather than changing masks, and interconnecting the components on the HIC via the printed wiring board. In this way we are able to use specific filter circuits which

Manuscript received February 13, 1974.
The authors are with the Bell Laboratories, Holmdel, N.J. 07733.

[1] TRW Semiconductors Inc., 14520 Aviation Blvd., Lawndale, Calif. 90260. Discontinued Model BP-200 Active Filter.
[2] Helipot Model 881, Universal Active Filter, Beckman Instruments, Helipot Division, Fullerton, Calif. 92634.
[3] KTI Type FS-20, "Universal Active Filter: Theory and Application," Kinetic Technology, Inc., 3393 De La Cruz Blvd., Santa Clara, Calif. 95051.

Reprinted from *IEEE Trans. Circuits Syst.*, vol. CAS-22, pp. 115-121, Feb. 1975.

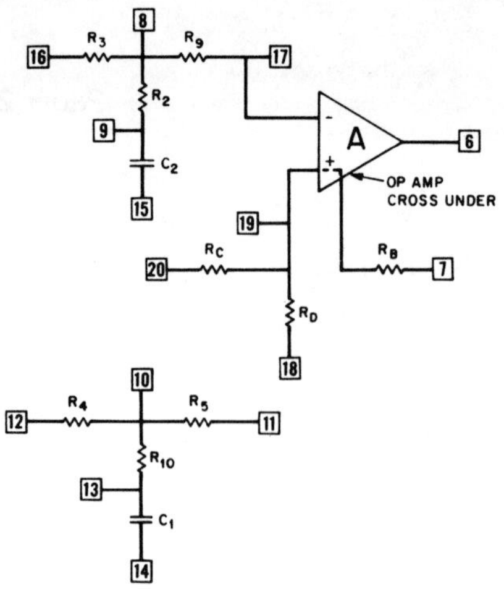

Fig. 1. Resistor–capacitor T's for the STAR (boxed numbers are pin numbers).

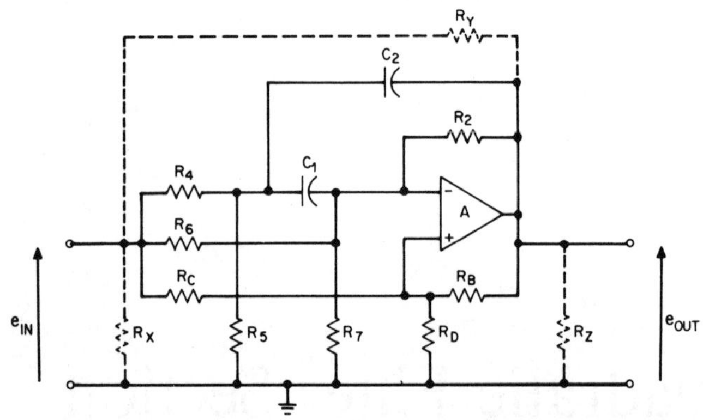

Fig. 2. Friend's SAB circuit.

are best suited for particular transfer functions and situations.

The circuits, with their design equations and STAR lead interconnections, are discussed first. Then the components are briefly described. The nonideal nature of the capacitors and amplifier requires a predistortion of the transfer function and the method used is outlined. The last section of the paper discusses the specific method used by the laser to set resistor values.

II. Active Filter Circuits Using The Star

A. SAB Circuits

The STAR is simply an RC network with an operational amplifier on a substrate. Experience has shown us that there is a family of circuits for realizing filters, each with its own advantages in specific applications. Here we will discuss those circuits and show how the STAR realizes all of them with one package.

Friend's SAB circuit [1] is shown in Fig. 2 (ignore the phantom resistors $\{R_X, R_Y, R_Z\}$ for the moment). It realizes an arbitrary biquadratic function with the exception of the low-pass case, which is discussed later.

For the circuit in Fig. 2 the transfer function is

$$t(s) \equiv \frac{e_{\text{out}}(s)}{e_{\text{in}}(s)} = \frac{n_2 s^2 + n_1 s + n_0}{d_2 s^2 + d_1 s + d_0}. \tag{1}$$

Assuming a finite op amp gain A, and not using $\{R_X, R_Y, R_Z\}$, the coefficients are given by

$$d_2 = C_1 C_2 (G_A + A^{-1}(G_A + G_B)) \tag{2}$$

$$d_1 = (C_1 + C_2)(G_A G_2 - G_B G_3) - C_1 G_1 G_B + A^{-1}(G_A + G_B)[C_1 G_1 + (C_1 + C_2) \cdot (G_2 + G_3)] \tag{3}$$

$$d_0 = G_1[G_A G_2 - G_B G_3 + A^{-1}(G_2 + G_3)(G_A + G_B)] \tag{4}$$

$$n_2 = C_1 C_2 G_C \tag{5}$$

$$n_1 = G_C[C_1 G_1 + (C_1 + C_2)(G_2 + G_3)] - C_1 G_4 (G_A + G_B) - G_6 (C_1 + C_2)(G_A + G_B) \tag{6}$$

$$n_0 = G_1[G_C(G_2 + G_3) - G_6(G_A + G_B)] \tag{7}$$

where

$$G_A \equiv G_C + G_D \tag{8}$$

$$G_1 \equiv G_4 + G_5 \tag{9}$$

$$G_3 \equiv G_6 + G_7. \tag{10}$$

Usually, (1) is normalized by dividing (3)–(7) by (2). Doing this and setting A to infinity the synthesis equations for (1), with $d_2 = 1$, are

$$G_1 = \frac{C_2 G_A}{2 G_B}\left[-d_1 + \left[d_1^2 + 4\left(1 + \frac{C_1}{C_2}\right) d_0 \frac{G_B}{G_A}\right]^{1/2}\right] \tag{11}$$

$$G_4 = \frac{G_1 n_2 + \left(1 + \frac{C_1}{C_2}\right) n_0 \frac{C_2^2}{G_1} - n_1 C_2}{1 + \frac{G_B}{G_A}} \tag{12}$$

$$G_3 = \frac{C_1 C_2 G_A (n_0 - d_0 n_2)}{G_1 (G_A + G_B)(n_2 - K_2)} \tag{13}$$

$$G_2 = \frac{C_1 C_2 d_0}{G_1} + \frac{G_B G_3}{G_A} \tag{14}$$

$$G_5 = G_1 - G_4 \tag{15}$$

$$G_C = n_2 G_A \tag{16}$$

$$G_D = G_A - G_C \tag{17}$$

$$\text{HPN}: G_6 = G_3, \quad (K_2 = 1)$$

$$\text{LPN}: G_7 = G_3, \quad (K_2 = 0) \tag{18}$$

where

$$K_2 \equiv \frac{G_6}{G_6 + G_7}. \tag{19}$$

The input constants for these equations are the transfer function coefficients and circuit element values for C_1, C_2, G_A, and G_B. Notice that the high-pass notch (HPN) and low-pass notch (LPN) create a situation where G_6 and G_7 need never be used simultaneously. While some situations may exist [1] in which it would be desirable to choose $0 < K_2 < 1$, for the STAR it was decided that this

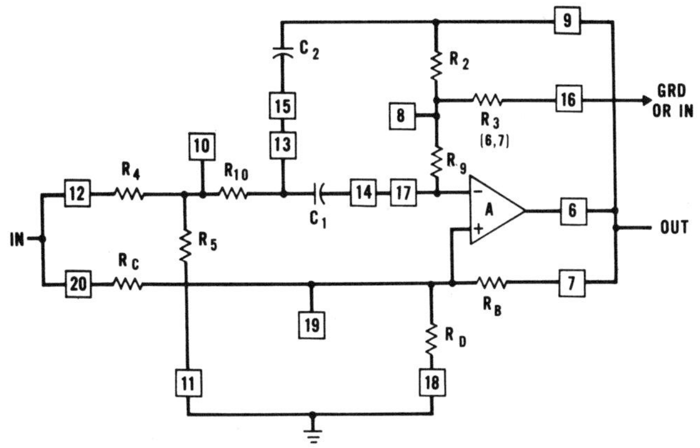

Fig. 3. STAR SAB circuit after transformations.

option would be dropped and one physical resistor R_3 would be used for both circuit elements.

The nominal capacitance for C_1 and C_2 is about 5 nF. We have found in many cases that when specifying this capacitance value the resulting values of some resistors, such as R_2, R_4, R_6, or R_7, exceed a few hundred kilohms. These become impractical to realize with material that has a 100-Ω/□ sheet resistance. To overcome this problem we have used three pi-to-tee transformations. The pi's are created by hypothetically placing three additional resistors $\{R_X,R_Y,R_Z\}$ in the circuit of Fig. 2. These resistors do not change the circuit's transfer function because they are placed between or across zero-impedance sources.

The three resistor pi's in Fig. 2 $\{R_X,R_4,R_5\}$, $\{R_Y,R_6,R_2\}$, and $\{R_Z,R_2,R_7\}$ can be transformed to three resistor tees. However, as we have already indicated, the resistors R_6 and R_7 never occur together so only two transformations need to be made $\{R_X,R_4,R_5\}$ and $\{R_Y,R_2,R_3\}$. These produce the tees $\{R_4,R_5,R_{10}\}$ and $\{R_2,R_3,R_9\}$, respectively. The transformed circuit, shown in Fig. 3, is the one the STAR realizes. Obviously, if the resistors are not large enough to require a transformation, it need not be done and R_{10} or R_9 can be short-circuited on the printed wiring board.

B. The Low-Pass Circuit

We have just seen that the STAR can realize the basic SAB circuit and therefore the common biquadratic filtering functions of a high-pass, high-pass notch, band-reject, all-pass, and low-pass notch. For a simple low-pass, we must use the STAR to realize the SAB low-pass filter [1], as in Fig. 4.

The coefficients of the biquadratic transfer function for this circuit are

$$d_2 = C_1C_2[G_D + A^{-1}(G_B + G_D)] \tag{20}$$

$$d_1 = C_2G_D(G_2 + G_3 + G_9) - C_1G_BG_9 + A^{-1}(G_B + G_D)(C_2(G_2 + G_3 + G_9) + C_1G_9) \tag{21}$$

$$d_0 = G_9(G_2G_D - G_3G_B) + A^{-1}G_9(G_B + G_D)(G_2 + G_3) \tag{22}$$

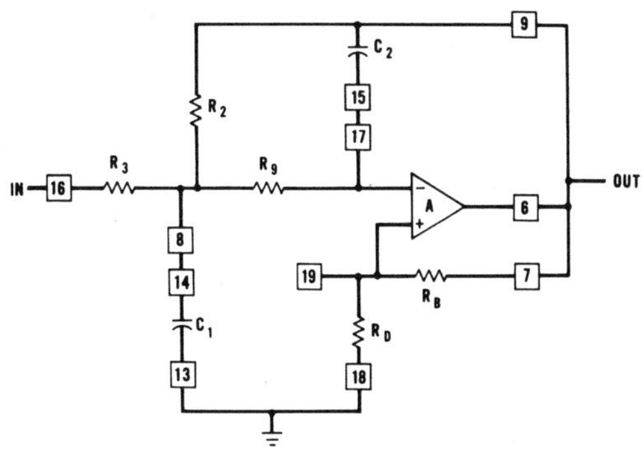

Fig. 4. STAR low-pass SAB filter.

$$n_2 = n_1 = 0 \tag{23}$$

$$n_0 = -G_3G_9(G_B + G_D). \tag{24}$$

The synthesis equations for the normalized transfer function are ($A = \infty$)

$$G_9 = \frac{d_1C_1 \pm \left[d_1^2C_1^2 - 4C_1C_2\left(1 - \dfrac{G_BC_1}{G_DC_2}\right)(d_0 + n_0)\right]^{1/2}}{2\left(1 - \dfrac{G_BC_1}{G_DC_2}\right)} \tag{25}$$

$$G_2 = \frac{C_1C_2}{G_9}\left[d_0 + \frac{n_0G_B}{G_D + G_B}\right] \tag{26}$$

$$G_3 = \frac{n_0C_1C_2G_D}{G_9(G_B + G_D)}. \tag{27}$$

The inputs to these equations are the transfer function coefficients, G_B, G_D and the capacitor values. In order to guarantee a real G_9 in (25) the following condition must be met

$$C_2 \leq C_1\left[\frac{G_B}{G_D} + \frac{d_1^2}{4(d_0 + n_0)}\right]. \tag{28}$$

There are two sets of values that emerge from the solutions of (25) through (28) because of the ± sign in (25). Both sets may be acceptable, but experience has shown that the choice of the negative sign yields smaller element value spreads.

If the values needed for resistors R_2 or R_3 exceed their allowable limits one can add a phantom resistor from input to output and perform a pi-to-tee transformation. This new circuit can be realized by the STAR in several ways.

C. The All-Pass Circuit

Although the SAB circuit of Fig. 3 can realize the all-pass function, a certain amount of flat loss is encountered. This loss is inversely proportional to Q and, therefore, presents a problem in realizing the common low-Q all-pass section. The alternative is the Steffen all-pass circuit [4], shown in Fig. 5, which provides the all-pass function with 0 dB loss for any Q. It is, however, more sensitive than the SAB circuit and should only be used for low Q's.

The biquadratic coefficients of this circuit are

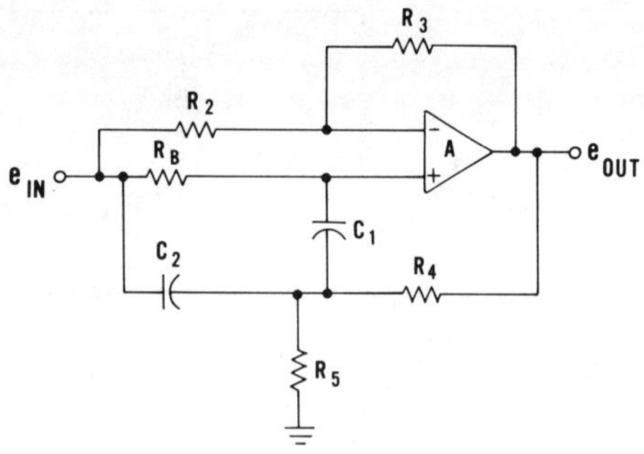

Fig. 5. Steffen all-pass circuit.

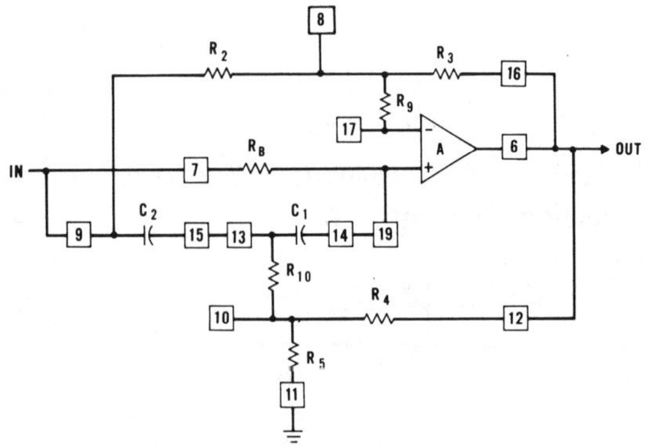

Fig. 6. Steffen all-pass using a STAR.

$$d_2 = C_1C_2(1 + A^{-1}\mu) \tag{29}$$

$$d_1 = C_2G_B(1 + A^{-1}\mu) - C_1G_4(G_2/G_3 - A^{-1}\mu) + C_1(G_B + G_5)(1 + A^{-1}\mu) \tag{30}$$

$$d_0 = G_B(G_4 + G_5)(1 + A^{-1}\mu) \tag{31}$$

$$n_2 = C_1C_2 \tag{32}$$

$$n_1 = G_B(C_1 + C_2) - C_1(G_4 + G_5)G_2/G_3 \tag{33}$$

$$n_0 = G_B(G_4 + G_5) \tag{34}$$

where

$$\mu \equiv 1 + G_2/G_3. \tag{35}$$

The ideal transfer function for this circuit is

$$t(s) = \frac{s^2 - d_1s + d_0}{s^2 + d_1s + d_0} \tag{36}$$

for which the synthesis equations are (with A infinite):

$$G_5 = 2Cd_1/\mu \tag{37}$$

$$G_4 = \frac{G_5(3\mu - 4) \pm [G_5{}^2(3\mu - 4)^2 - 8(1 - \mu)(4d_0C^2 + G_5{}^2(2 - \mu))]^{1/2}}{4(1 - \mu)} \tag{38}$$

and

$$G_B = \frac{d_0C^2}{G_4 + G_5}. \tag{39}$$

The input quantities to these equations are the transfer function coefficients, d_1 and d_0, the capacitor values ($C_1 = C_2 = C$), and the resistors R_2 and R_3.

Resistors R_9 and R_{10} may be used to reduce offset voltage and element values respectively as shown in Fig. 6.

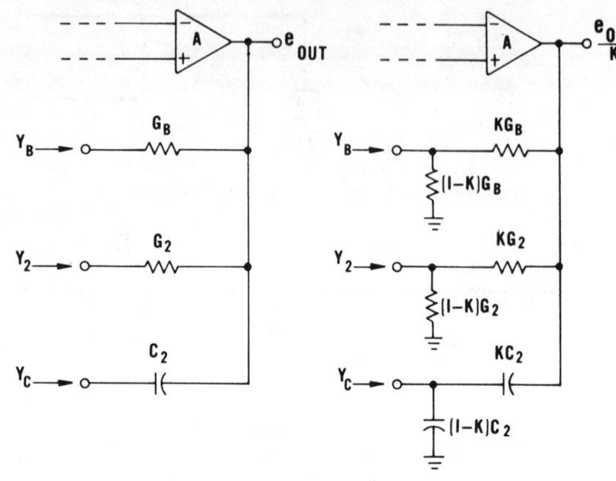

(a) (b)

Fig. 7. Capacitive gain enhancement (CGE $= 1/K$, $0 < K \leq 1$). (a) Original circuit. (b) Thevenin equivalent.

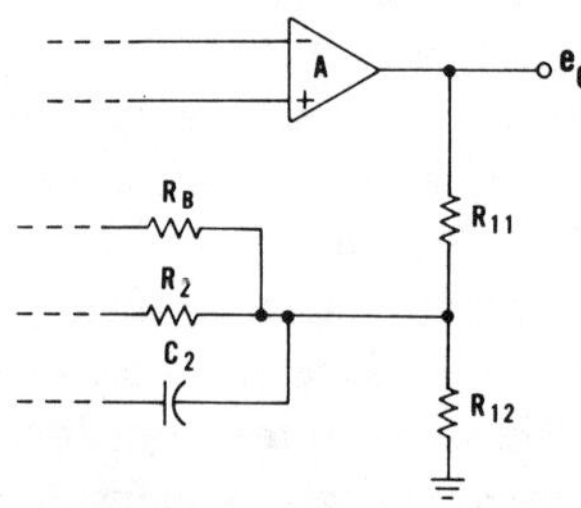

Fig. 8. Resistive gain enhancement (RGE $= 1 + R_{11}/R_{12}$).

III. Resistive Gain Enhancement

Occasionally, more gain is required in a biquad section than that which is available in the SAB realization. The section types which most often need it are the all-pass, high-pass, and high-pass notch. All have pass-band gains (at high frequencies) of less than unity. The low-pass notch may sometimes need additional gain. The low-pass circuit has no theoretical limit on the amount of flat gain it can realize.

Two methods can be used to gain enhance SAB sections and both involve the introduction of additional flat loss in the feedback networks surrounding the op amp. This reduces the effective open-loop gain of the op amp. The first method uses an ideal transformation wherein the series feedback elements G_B, G_2, and C_2 (see Fig. 2), and the common voltage source (the op-amp output) are replaced by three Thevenin equivalent circuits as shown in Fig. 7. The extra resistors generated, $(1 - K)G_2$ and $(1 - K)G_B$, appear across resistors that already exist in the STAR topology so that no new resistors are needed. However, the new (third) capacitor $(1 - K)C_2$ would have to be added external to the STAR, complicating the manufacturing process and adding considerable cost.

The second gain-enhancement method uses a resistive divider to attenuate the fed-back signals as shown in Fig. 8. This resistive gain enhancement (RGE) is not an ideal conversion for several reasons. First, the effective open-loop gain of the op amp is reduced so its affects on the transfer function are larger. Second, the impedances R_2, R_B, and C_2 now see a finite source resistance and this changes the transfer function coefficients too. Finally the RGE due to R_{11} and R_{12} is itself only approximate because the feedback circuitry loads the divider network somewhat.

The ratio of the RGE resistors determine the additional gain, $\text{RGE} = 1 + R_{11}/R_{12}$, but the impedance level remains to be chosen. The series value of $R_{11} + R_{12}$ must be high enough not to excessively load down the op amp. Excessive loading will either change the transfer function due to decreased effective op-amp gain or it will affect the signal drive range. On the other hand, if the parallel combination of R_{11} with R_{12} is too large, this source impedance on the feedback circuit will change the transfer function substantially. This latter effect can be predistorted for in the design stage.

IV. Components

Just as there are circuits that are best for certain filter applications, so too are there components that are best. In the Bell System, the least-expensive high-quality components, for volumes of over a million a year, are made with thin-film tantalum [5]. Tantalum-oxynitride resistors and tantalum-metal (TM) capacitors can be batch fabricated on the same ceramic substrate. This is done through a complicated sequence of processing steps but the batch nature of this processing makes the individual IC's inexpensive.

From a filter standpoint, the important component properties are the temperature coefficient and aging of the *RC* products and the dissipation factor of the capacitors. For thin-film resistors and capacitors these parameters are quite good for most applications, but there are some situations in which the TM capacitor aging is too large or in which the dissipation factor presents a problem. For these cases we must spend more money to get better performance. We designed a slightly more expensive STAR using a tantalum-nitride resistor network upon which one appliques NPO ceramic chip capacitors. This combination suffers from a temperature coefficient mismatch but the capacitor has better aging and dissipation characteristics. By placing some restrictions on the physical designs of both STAR versions, we were able to obtain a commonality in handling and testing fixtures that helps keep costs low for both HIC types.

The dc and ac voltages expected in the STAR circuit were analyzed and it was decided that the TM capacitor should be of a bipolar design. This involves putting two polar 10-nF capacitors in series with a common electrode but this was done in the same space that had been allocated to the chip capacitors. The op amp used in the STAR is a standard Western Electric silicon IC which has internal frequency compensation. This compensation produces a response which has two real poles and one real zero. The op amp gain is large enough to permit STAR filters to have complex poles up to 10 kHz and zeros beyond that.

V. Tuning and Predistortion

The method of tuning being used in manufacturing is to measure the two capacitors on the substrate and then compute resistor values based on these measurements and the desired transfer function using the appropriate equations from Section II. The resistors are then laser-trimmed to these values. The desired transfer function coefficients are predistorted to account for nominal nonidealities such as capacitor dissipation, finite amplifier gain, parasitic capacitance and resistive gain enhancement, if used.

If a filter application is sensitive to other small parasitics that are not modeled, such as stray capacitance on the printed wiring board, then one must build several models. By making appropriate measurements of gain or phase one can determine the extra predistortion that might be needed.

VI. Laser Trimming

The STAR is built with a tantalum technology. Trimming of tantalum resistors is normally anodic [5]. In the simplest form of anodic trimming an electrolyte is placed on the resistor, a current is passed through it and this thickens the oxide layer on the surface of the tantalum, reducing its effective cross-section and thereby increasing its resistance. There are two problems with this method for the STAR. First, anodic trimming of resistors over 50 kΩ to about ±0.1 percent is quite difficult. Second, and most important, resistor values can be increased by only 20 to 30 percent if the minimum resistor aging is to be obtained.

Laser trimming of the resistors can avoid both of these problems. Furthermore, by using the laser to generate whole new resistor meanders one can also eliminate the need for a special resistor-defining mask for each filter designed. This has its own advantages in production.

Laser trimming is simple in concept; one uses the energy of the laser beam to vaporize resistive material until the desired resistor value is reached. The resistor is clean, so measurements are limited in accuracy only by the resistance bridge. Not all of the heated resistive material is evaporated, however. The material that is left behind has degraded stability. Early experiments with tantalum nitride indicated that this heat-affected material changed value with time (aged) more than the unaffected material. Thus, if we had a large amount of heat-affected resistance in our STAR filters, we would have more aging of the filter performance than desired. To minimize the aging degradation in a laser-trimmed resistor, the resistor pattern of Fig. 9 was developed.

The laser-trimming requires accurate positioning. The method chosen for accomplishing this is by placing three dummy "edge-sensing" resistors on each circuit. When the STAR is first placed under the laser the beam searches for these edge-sensing resistors. When the value of each of these resistors begins to change note is made of the position of the laser. This location is then compared with the ideal position. The computer can then correct for horizontal, vertical, and angular position errors. After corrections are made the edge-sensing resistors are cut open completely and are not part of the circuit.

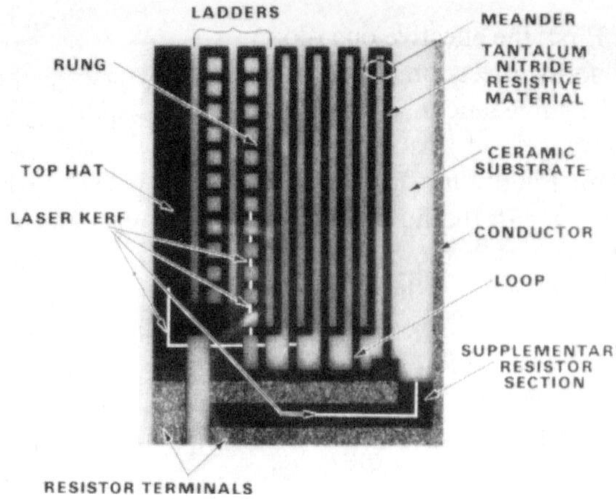

Fig. 9. Resistor pattern for laser trimming.

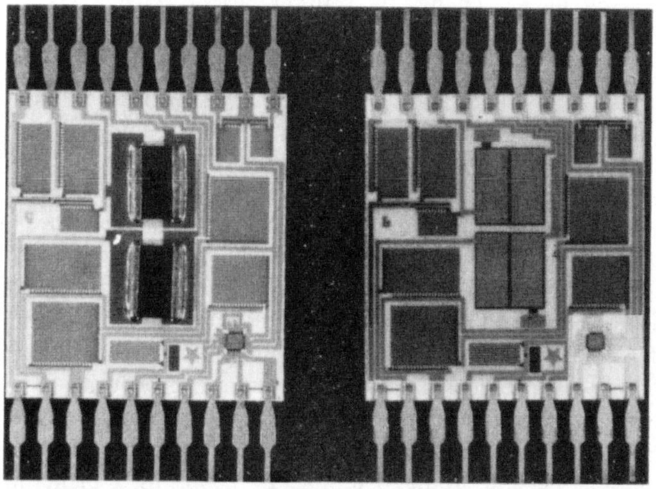

Fig. 10. STAR's with thin-film (left) and ceramic chip capacitors.

Laser-trimming has two more manufacturing advantages. First, because no code-to-code mask changes are involved, production decisions on product mix can be made at the laser-trimming station rather than the IC processing line. Second, for the same reason, engineering changes and new products can be introduced as soon as the laser's computer can be fed the new data.

Summary

The ceramic chip and TM capacitor versions of the STAR are shown in Fig. 10. These HIC's have the following features:

1) the STAR is expected to cost $\frac{2}{3}$ to $\frac{1}{2}$ that of an equivalent passive filter section;
2) it realizes all the filter circuits, SAB's and others, needed for biquadratic sections;
3) manufacturing fixtures are common to all codes, easing production problems;
4) both NPO ceramic chip and TM thin-film capacitors are available, as filter requirements dictate;
5) no filter-to-filter mask changes are required for a specific capacitor type;
6) a large range of resistor values can be obtained with the laser making the filter design more flexible;
7) resistors can be laser trimmed to an 0.1-percent accuracy;
8) the STAR takes a 1.10 × 1.00 in area on the PWB;
9) there are no crossovers on the HIC;
10) filters normally require only single-sided wiring on the printed wiring board; and
11) all resistors and capacitors can be measured directly from the lead frame.

Acknowledgment

The STAR was not the product of the authors' work alone. F. P. Cue and L. L. Taylor (formerly) of Bell Laboratories, Holmdel, N.J., made significant contributions to the early STAR designs. R. J. Moore, F. N. Fuss, and J. S. Fisher, of Bell Laboratories, Allentown, Pa., contributed to making the TM design and ceramic chip capacitor designs compatible. G. A. Bulger, W. H. D'Zio, and B. A. Unger contributed significantly to the resistor design. We should also acknowledge some 30 other Bell Laboratories and Western Electric employees, from managers to draftsmen, for contributing their efforts to making the STAR a reality.

References

[1] J. J. Friend, "A single operational-amplifier biquadratic filter section," in *1970 IEEE ISCT Dig. Tech. Papers,* p.189.
[2] N. Nakayama, T. Yao, M. Nakamura, and H. Sasaki, "Tantalum thin-film active filter," in *Proc. 1972 Electronic Components Conference*, pp. 207–212, Washington, D.C., May 1972.
[3] G. S. Moschytz, "FEN filter design using tantalum and silicon integrated circuits," *Proc. IEEE*, vol. 58, pp. 550–566, Apr. 1970.
[4] C. J. Steffen, Bell Labs. internal memo.
[5] Bell Laboratories, *Physical Design of Electronic Systems*, vol. 3, "Integrated device and connection technology," Englewood Cliffs, N.J.: Prentice-Hall, 1971, pt. 1.

Sensitivity and Frequency Limitations of Biquadratic Active Filters

ADEL S. SEDRA, MEMBER, IEEE, AND JUAN L. ESPINOZA

***Abstract*—A comparative study of single-amplifier filters, fifteen different two-amplifier circuits, and the state-variable three-amplifier configuration is given. The figures-of-merit used are the passive sensitivities, the ω_0 gain-sensitivity products, the Q gain-sensitivity products, and the ω_0 and Q deviations due to the amplifiers' limited bandwidth. In addition, practical matters such as input impedance and component spread are considered. Among the circuits considered a particular two-amplifier resonator emerges as the "best" biquadratic section. Some of the interesting results include a unified expression, applicable to a large class of low-sensitivity circuits, for the ω_0 deviation due to the amplifier's limited frequency response. In this expression, which applies for all active filters which have low ω_0 active sensitivities, the only configuration-dependent quantities are the Q gain-sensitivity products.**

I. INTRODUCTION

THE NUMBER of amplifiers in a biquadratic filter section varies from a minimum of one to a maximum of four. A recent study of single-amplifier filters [1] provides a systematic procedure to generate circuit configurations for implementing the various second-order functions. More importantly, the general procedure adopted allows the classification of single-amplifier circuits into categories having identical sensitivity characteristics.[1] Thus the limitations of single-amplifier circuits, both at low and high frequency, become evident and, hopefully, workers in the area would direct their efforts to more fruitful tasks than inventing "new" configurations.

Although the task of extending this theory to multi-amplifier circuits is understandably difficult, we will be content to critically evaluate a large number of such circuits. The objectives of this study are 1) to identify performance criteria for biquadratic active filters, 2) to assess existing configurations according to these criteria, and 3) to focus the attention on the circuit or circuits that best meet these criteria. The latter configurations represent the state-of-the-art and any new circuit should be measured against them.

In Section II we briefly summarize the results obtained for single-amplifier filters in order to outline the limitations of these minimal realizations. In Section III we will identify and justify the performance criteria or figures-of-merit used. Section IV provides the results of a study of fifteen different two-amplifier circuits. As will be shown, only 3 configurations will emerge as the "best" two-amplifier realizations. Consequently, in Section V the high-frequency limitations of these three circuits together with those for the 3-amplifier state-variable circuit are given. An interesting result will emerge regarding the significance of the Q gain-sensitivity product in determining the high-frequency limitations. Other practical matters and conclusions deemed useful to active filter designers will follow.

II. SINGLE-AMPLIFIER FILTERS

A broad study of single-amplifier filters [1] from a configuration point of view provides a method for generating circuits to realize a general biquadratic transfer function

$$T(p) = \frac{F(p)}{p^2 + p\left(\frac{\omega_0}{Q}\right) + \omega_0^2}$$

where p is the complex frequency variable, ω_0 and Q characterize the pair of complex poles, and $F(p)$ is a polynomial of order ≤ 2 determining the type of filter function. Most of the known single-amplifier filter realizations follow as special cases of the resulting biquads. Thus, in addition to providing a straightforward method for generating filter configurations, this unified approach allows cataloguing the large number of single-amplifier circuits available, as well as new ones, into distinct classes, where the members of each class have identical sensitivity and other characteristics.

The classification starts with the two major categories, namely, negative feedback (NF) and positive feedback (PF), shown schematically in Fig. 1. It has been shown that for the NF configuration to realize a pair of poles, characterized by ω_0 and Q, with ω_0 independent (or almost independent) of the amplifier gain at low frequencies,[2] network n_1, viewed as a two-port network between terminals 3 and 1 with 4 grounded, has to have a pair of zeros identical, in frequency and Q, to the desired poles. On the other hand, realization using the PF configuration requires an RC network n_2 having a bandpass transfer function, between terminals 3 and 2 with 4 grounded. Thus the first step in the generation of an NF filter realization is the selection of an RC two-port network capable of realizing a pair of complex transmission zeros while to generate a PF filter realization one starts with an RC bandpass network. The active filter transmission zeros are then controlled by injecting the signal V_i to the appropriate node, or nodes, 4,

Manuscript received October 22, 1973; revised April 10, 1974, and June 27, 1974. This work was supported in part by the National Research Council of Canada under Grant A-7394.

A. S. Sedra is with the Department of Electrical Engineering, University of Toronto, Toronto, Ont., Canada.

J. L. Espinoza was with the Department of Electrical Engineering, University of Toronto, Toronto, Ont., Canada. He is now with the Department of Electrical Engineering, Universite Tecnica del Estado, Santiago, Chile.

[1] Another classification scheme based on polynomial decomposition is given in [11].

[2] Although this is only true at very low frequencies, the reader should not be alarmed as we will consider the effect of the amplifier pole on ω_0 and Q at a later part of this paper.

Reprinted from *IEEE Trans. Circuits Syst.*, vol. CAS-22, pp. 122-130, Feb. 1975.

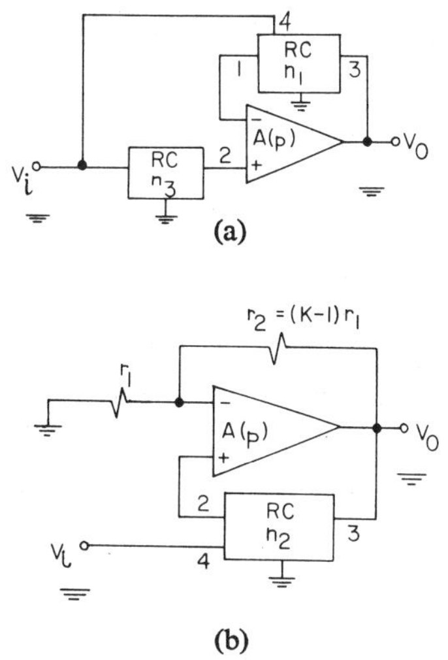

Fig. 1. (a) General single-amplifier NF configuration. (b) General single-amplifier PF configuration.

in the feedback network which are normally connected to ground. In the NF configuration, additional control on the zeros' locations may be obtained through network n_3.

As expected, the type of the *RC* feedback network determines the sensitivity properties of the resulting active filter. Distinct filter categories are obtained if the feedback network is a second-order one (such as the various forms of bridged-T networks for the NF realizations and the ladder networks for the PF realizations) or if it is a third-order network that provides a second-order transfer function through an internal pole-zero cancellation (such as the notch twin-T or parallel ladder networks for the NF realizations and their bandpass forms for the PF realizations). For the cataloguing method and design tables, the interested reader is referred to [1]. As an example of the results, it might be interesting to note that the common practice of treating the NF Sallen and Key circuits as different from the so-called multiple feedback infinite gain realization is unjustified. A redrawing of the Sallen and Key realizations, using operational amplifiers, reveals that the feedback networks, as in the multiple feedback case, are of the bridged-T type. Both circuits, therefore, belong to the same class and exhibit identical sensitivity performance.

It should be emphasized that the requirement stated above on the ω_0 sensitivity, namely,

$$S_{A_0}^{\omega_0} \simeq 0 \text{ for } \omega_0 \ll \omega_t$$

where A_0 is the low-frequency gain of the *OA* and ω_t is its unity-gain bandwidth, is a rather important one. Moschytz [2] has shown that at resonance the transfer function is $2Q$ times more sensitive to variations in ω_0 than it is to variations in Q. Hilberman [3], while confirming Moschytz's result, provides additional useful information about the relative contribution of ω_0 and Q sensitivities over the frequency range of interest. Again, he concludes that the designer can afford a much larger variation in Q than in ω_0 for a given allowable tolerance on the transfer function magnitude or phase.

Having decided on a configuration that allows the low ω_0 sensitivity to the active element, the resultant ω_0 passive sensitivities are usually the minimum possible, namely $\leq \frac{1}{2}$. The remaining sensitivity parameters are, therefore, the active and passive Q sensitivities. As far as the active Q sensitivities are concerned, a better measure is the gain-sensitivity product introduced by Msochytz [4]. Since we recognize that variabilities, not sensitivities, are what are ultimately important, and since we will be comparing circuits requiring widely different closed-loop gains, the gain-sensitivity product is a better index for assessing the effect of variations in amplifier gain. Moschytz, however, assumes tracking of passive component variations and hence does not consider the passive sensitivities. A better picture would be obtained by considering both the Q gain-sensitivity product, $GS_{A_0}{}^Q$, where[3]

$$GS_{A_0}{}^Q \equiv S_{A_0}{}^Q \cdot A_0$$

as well as the order of magnitude of the passive sensitivities. In our opinion, these two parameters provide a reasonable assessment of the utility of the particular configuration. Although we recognize the more elaborate studies of sensitivity carried out on statistical basis [3], [5], [6], these two parameters still provide a simple but reasonable measure of the circuit goodness at low frequencies.

Another important question is the effect of the amplifier limited bandwidth on the filter response. A convenient way for expressing this limitation is through the shift in ω_0 and Q. Specifically, the actual ω_0 and Q are denoted by ω_{0k} and Q_k and their values are evaluated using a first-order model for the *OA* frequency response, namely,

$$A = \frac{A_0}{1 + \dfrac{pA_0}{\omega_t}}.$$

[3] Note that we have added the superscript Q to the symbol *GS* in order to distinguish it from the ω_0 gain-sensitivity product defined later. Also, the subscript A_0 is added since we will consider multi-amplifier circuits.

TABLE I
COMPARISON OF THE SIX DIFFERENT CLASSES OF SINGLE AMPLIFIER FILTERS

CLASS		FEEDBACK NETWORK	PASSIVE Q SENSITIVITIES	$A_0 S_{A_0}^{Q}$ OR GS^Q	MAX. ELEMENT SPREAD
NF	NF1	2ND ORDER	LOW	Q^2	Q^2
	NF2	3RD ORDER	Q	Q	Q
PF K>1	PF1	2ND ORDER	Q	Q	INDEPENDENT OF Q
	PF3	3RD ORDER	Q	Q	Q
PF K=1	PF2	2ND ORDER	LOW	Q^2	Q^2
	PF4	3RD ORDER	Q	Q	Q

Only order of magnitudes are indicated.

A detailed analysis [1] shows that, for single-amplifier filters, the deviation $\Delta\omega_0 = \omega_{0k} - \omega_0$ due to the amplifier pole is given to a good approximation by

$$\delta \equiv \frac{\Delta\omega_0}{\omega_0} \simeq -\frac{1}{2}\left(\frac{\omega_0}{\omega_t}\right)\frac{1}{Q}(GS_{A_0}{}^{Q}). \quad (1)$$

It is interesting to note that, in this equation, the only configuration-dependent quantity is the gain-sensitivity product. Thus, although GS^Q is evaluated at low frequency, its significance in determining the filter nonideal response extends to high frequencies where the amplifier pole is taken into account. Similar results are also obtained for the deviation in the pole-Q, namely,

$$\frac{Q_k}{Q} \simeq 1\Big/\left[1 + \delta - 2\delta KQ\left(\frac{\omega_0}{\omega_t}\right)\right] \quad (2)$$

where $K = 1$ for NF circuits and ranges from 1 to 3 for PF circuits. Again we note that the Q deviations are a function of GS^Q.

Based on the possible choice of the RC networks n_1 and n_2 of Fig. 1,[4] six classes of single-amplifier filters are obtained:

1) the NF1 class, obtained from a second-order network implementation of n_1, such as the bridged-T;

2) the NF2 class, obtained from a third-order network implementation of n_1 which provides a second-order function through an internal pole-zero cancellation, such as the twin-T;

3) the PF1 class, obtained from a second-order network implementation of n_2, such as the ladder, with $K > 1$;

4) the PF3 class, obtained from a third-order network implementation of n_2 that provides a second-order function through an internal pole-zero cancellation, such as the bandpass [28] twin-T, with $K > 1$;

5) the PF2 class, generated as the PF1 class but with $K = 1$, that is, the OA is connected as a unity-gain follower; and

6) the PF4 class, generated as the PF3 class but with $K = 1$, that is, the OA is connected as a unity-gain follower.

A comparison of these six possible classes of single-amplifier filters is shown in Table I where only order of magnitude is indicated. Note that "low" refers to values around unity. An interesting observation is that the product of $GS_{A_0}{}^{Q}$ and the order of magnitude of the passive sensitivities is always of the order of Q^2. At this juncture we note the useful contribution of Friend [7] where he has applied some positive feedback to the NF1 class in an attempt to reduce the high active Q-sensitivities. In this way, he obtains a realization somewhere in between NF1 and PF1, depending on the amount of positive feedback used.

[4] Network n_3 in Fig. 1(a) is included to increase the configuration flexibility in realizing different transmission zeros. However, as n_3 is not in the feedback loop, it does not affect the selectivity performance of the filter.

III. PERFORMANCE CRITERIA FOR MULTIAMPLIFIER CIRCUITS

From active filter designers' experience, it seems that a $GS_{A_0}{}^{Q}$ of the order of Q (such as in NF2, PF1, PF3, and PF4 circuits) is quite acceptable. This is indeed the value encountered in the popular filter realizations, such as the state-variable [8]. The single-amplifier realizations that attain this goal suffer, however, from high passive sensitivities. The motivation, then, would be to keep GS^Q at its acceptable value of Q but substantially reduce the passive Q-sensitivities. The state-variable realization accomplishes this aim through the use of three or four amplifiers with the increased power dissipation and a rather drastic Q deviation at high frequencies (due to the limited amplifiers' bandwidth). It is natural, therefore, to inquire about the possibility of two-amplifier realizations. Many authors addressed themselves to this question, and the object of Section IV is to evaluate the utility of the resulting circuits. First, however, we summarize the performance criteria used in the comparisons.

As will become evident in Section IV, twelve of the fifteen two-amplifier configurations considered either obtain reduced active Q sensitivities at the very expensive price of increased active ω_0 sensitivities or they offer no advantage over their single-amplifier counterpart. Therefore, to provide a comparative analysis of two-amplifier circuits we had to introduce a new criterion, namely, the ω_0 gain-sensitivity product $GS_K{}^{\omega_0}$,

$$GS_K{}^{\omega_0} \equiv S_K{}^{\omega_0} \cdot K$$

where K denotes the nominal gain of a finite gain amplifier. At this juncture, we note that considering GS^{ω_0} rather that $S_K{}^{\omega_0}$ for comparison purposes is much more appropriate as it avoids the wide differences in the nominal value of K and, correspondingly, their different tolerances. Except for one circuit (circuit 12), the passive ω_0 sensitivities are at their theoretical minima ($\leq \frac{1}{2}$). Thus five performance criteria have to be evaluated for every realization. These

TABLE II

SENSITIVITIES AND Z_{in} OF THE TWO-AMPLIFIER CIRCUITS IN FIG. 2

No.	S_x^Q	$GS_{K_1}^{\omega_0}$	$GS_{K_2}^{\omega_0}$	$GS_{K_1}^Q$	$GS_{K_2}^Q$	$Z_{in}(j\omega_0)$
1a	LOW	0	0	Q^2	Q^2	R_1/jQ
1b	Q	0	0	$4Q/K_2$	$2K_2Q$	$R_1/Q(-1+j)$
2	0	$-\frac{1}{2}$	0	$\frac{1}{2}$	$2Q^2$	R_1/jQ
3	LOW	$\frac{1}{2}K_1$	LOW	$\frac{1}{2}K_1$	$Q^2/2K_1$	$1/\omega_0C_1Q$
4	LOW	LOW	$\frac{1}{2}K_2$	$2Q^2/K_2$	$\frac{1}{2}K_2$	R_3
5	LOW	$-9Q^2$	$9Q^2$	$9Q^2$	$-9Q^2$	R_1/Q
6	LOW	$-K_1/2$	$-2Q^2/K_1$	$K_1/2$	$2Q^2/K_1$	R_1/jQ
7	LOW	$K_1/2$	$4Q^2/K_1$	$K_1/2$	$4Q^2/K_1$	$4R_1/3$
8	LOW	$K_1/2$	$Q^2(4+K_1)^2/4K_1$	$K_1(4-K_1)/2(4+K_1)$	$Q^2(4+K_1)^2/4K_1$	$2/\omega_0CQ$
9	LOW	$K_1/2$	$8Q^2/(K_1-1)^2$	0	$8Q^2/(K_1-1)^2$	$1/\omega_0CQ$
10	LOW	LOW	$-4.5Q^2/(1+\beta)$	$\beta/3$	$4.5Q^2/(1+\beta)$	$R_1/\sqrt{2}$
11	LOW	$\frac{1}{2}$	$8Q^2$	0	0	$2R_1/Q^2$
12	Q^2	$-8Q^2$	$-16Q^2$	0	0	$R_1/2Q$
13	LOW *	LOW	LOW	$Q(1+\frac{R_3}{R_1})$	2Q	R_1
14	LOW	LOW	LOW	LOW	2Q	∞
15	LOW	LOW	LOW	2Q	2Q	∞

* Except for (R_2/R_4)

are $GS_{K_1}^{\omega_0}$, $GS_{K_2}^{\omega_0}$, $GS_{K_1}^{Q}$, $GS_{K_2}^{Q}$, and S_x^Q. The last figure denotes the order of magnitude of the passive Q-sensitivities. In the cases where K_1 and/or K_2 is ideally infinite, the gain sensitivity products are evaluated by first finding the sensitivity expression, multiplying by K, and substituting $K = \infty$. This process is similar to evaluating limits and results in a finite number.

Apart from the above-mentioned sensitivity figures, the deviations in ω_0 and Q due to the amplifiers' limited frequency response should be considered. This will be done for the three promising circuits, as well as for the state-variable realization, for comparison purposes. From a practical point of view, the filter designer is also interested in other matters such as the component spread and the input impedance of the filter section. The latter is an important consideration if high-order filters are to be constructed by cascading second-order sections. Information regarding these practical matters are therefore included.

IV. Comparative Evaluation of Fifteen Two-Amplifier Resonators

In this section we consider the 15 different two-amplifier circuits shown in Fig. 2. Each circuit is given an identifying number shown on the diagram, together with the reference(s) in which the circuit is reported. Circuit 15, based on the Antoniou gyrator [31] and included in the type A_1 positive impedance converter of Bruton's classification [9], is used here as a second-order section. The design equations given under the circuits diagrams are based on the component choices of the originator of each circuit. In case no such choices were reported, a reasonable compromise design is given. Also, for circuits reported by more than one author, each giving different design parameters, all the given designs are considered. In some cases, however, it has been decided to leave some parameter as a design choice with its effects on the different performance criteria clearly visible.

Using the design data of Fig. 2, the resulting performance criteria in Table II are obtained. On each of the fifteen circuits we make some observations and conclusions.

Observations and Conclusions

(The paragraph numbers correspond with the circuits' numbers.)

1) Design 1a has GS^Q of the order of Q^2 and thus offers no advantage over the single-amplifier NF1 or PF2 realizations. Also, the component spread, for equal capacitor, is of the order of Q^2. For design 1b, K_2 is left as a design parameter. However, since its value is around unity, the value of GS^Q is of the order of Q and the passive Q sensitivities are of the order of Q. This circuit, therefore, does not offer any advantage over the single-amplifier realizations of the NF2, PF1, PF3, or PF4 classes. Note, however, that the component spread is low.

2) The value of $GS_{K_2}^Q$ is of the order of Q^2 and no

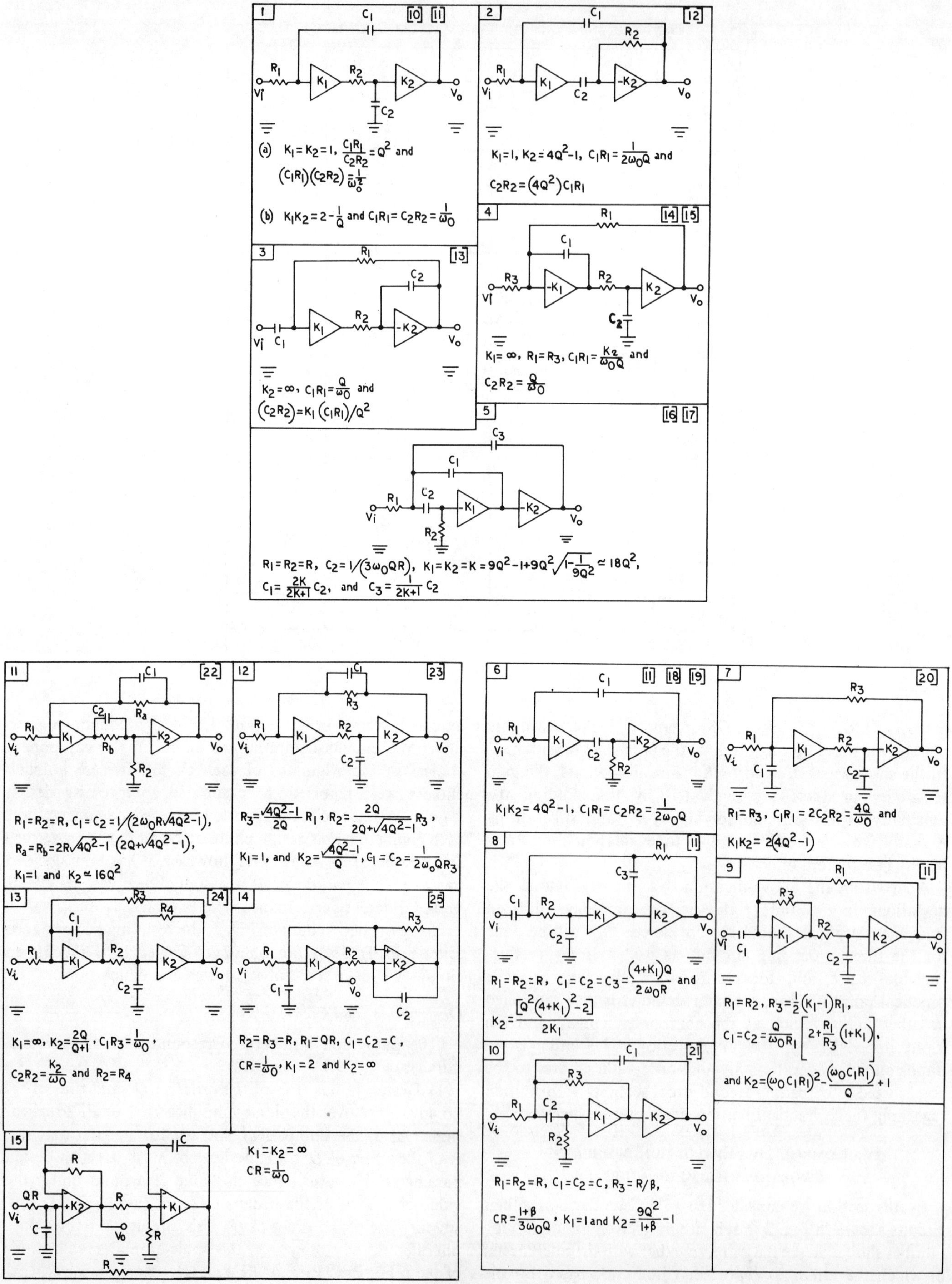

Fig. 2. Fifteen different two-amplifier resonators.

advantage is obtained over NF1 or PF2 circuits. The component spread is of the order of Q^2. Note also the need for an amplifier of large negative gain which is difficult to realize with a single OA.

3) The value of K_1 is left as a design choice. In order to keep $GS_{K_1}{}^{\omega_0}$ small, a low value for K_1 is desired with the resulting large $GS_{K_2}{}^Q$ (of the order of Q^2 as in NF1 and PF2). The component spread is also of the order of Q^2.

4) The value of K_2 is left as a design choice. In order to obtain a low value for $GS_{K_2}{}^{\omega_0}$, the value of K_2 should be low with the resulting large $GS_{K_1}{}^Q$. No significant advantage over NF1 or PF2 is obtained. Note that circuits 3 and 4 are similar from a "loop" standpoint.

5) All the values of GS are very large. It has been argued, however, that since $GS_{K_1}{}^{\omega_0} = -GS_{K_2}{}^{\omega_0}$ and $GS_{K_1}{}^Q = -GS_{K_2}{}^Q$ and then if the amplifiers' gain track, the resulting variations in ω_0 and Q will be small. In spite of this and due to the large element spread, the circuit cannot be recommended for high-Q realizations.

6) The value of K_1 is left as a free parameter in Table II. We note that it is impossible to choose a value for K_1 that yields low $GS_{K_1}{}^{\omega_0}$ and $GS_{K_2}{}^{\omega_0}$ simultaneously. It follows then that a compromise would be K_1 of the order of $2Q$ with the attendant large ω_0 sensitivities and GS^Q no better than many of the single-amplifier circuits.

7) As in circuit 6, it is impossible to obtain simultaneous low ω_0-sensitivities. Therefore, all GS values will be of the order of Q. Note also that the realization of K_2 poses a practical problem.

8) The free parameter K_1 should be chosen as small as possible to minimize $GS_{K_1}{}^{\omega_0}$. However, $GS_{K_2}{}^{\omega_0}$ remains unacceptably large. The value of $GS_{K_2}{}^Q$ is also of the order of Q^2. This circuit is indeed quite unattractive.

9) To equalize GS^{ω_0} one chooses $K_1 \simeq \sqrt[3]{16Q^2}$ with the resulting GS^{ω_0} as high as $2\sqrt[3]{2Q^2}$. Also, $GS_{K_2}{}^Q$ becomes of the order of $Q^{2/3}$ with a slight advantage over single-amplifier circuits. It is also practically difficult to obtain K_1 using a single OA.

10) As evident from Table II, to obtain low $GS_{K_2}{}^{\omega_0}$, β has to be chosen of the order of Q^2 leading to large $GS_{K_1}{}^Q$ and a realization almost identical to the NF1 and PF2 circuits. Other choices of β though might result in small GS^Q the obtained $GS_{K_2}{}^{\omega_0}$ becomes impractical.

11) This circuit has the interesting property of zero Q gain-sensitivity products. This, however, is obtained at the expensive price of $GS_{K_2}{}^{\omega_0} = 8Q^2$!

12) Similar to circuit 11, this configuration results in zero Q sensitivities. However, not only are *both* the ω_0 gain-sensitivity products of the order of Q^2 but the passive Q sensitivities are of the order of Q^2. Moreover, this configuration is unique in that the passive ω_0 sensitivities are also very large! ($S_{R_1}{}^{\omega_0} \simeq -4Q^2$ and $S_{R_3}{}^{\omega_0} \simeq 4Q^2$.) The result is an impractical circuit.

13) This circuit is the two-amplifier version of the state-variable realization replacing the integrator-inverter combination by the noninverting integrator formed by R_2, C_2, R_4, and K_2. Table II indicates that this circuit provides low ω_0 sensitivities and GS^Q of the order of Q. The only drawback is the high Q sensitivity to the resistive ratio (R_2/R_4). This is the first two-amplifier configuration in our list that provides an advantage over the single-amplifier circuits.

14) This circuit is based on the Riordan [26] inductance simulation circuit. It has been included in Bruton's classification of PIC's [9] as class $B2$ and used by Hamilton and Sedra [25] as a second-order filter section. As evident from Table II, circuit 14 offers low passive sensitivities, low ω_0 sensitivities and GS^Q of the order of Q. Thus this configuration meets all the performance criteria of interest.

15) This circuit is also derived from the inductance simulation circuits [31] belonging to class $A1$ of Bruton's classification and is used in [29]. The sensitivity figures in Table II indicate performance almost identical to that of circuit 14. The present circuit is, however, superior in performance at the higher frequency range where the amplifiers' poles have to be considered. This will be illustrated in Section V.

V. Frequency Limitations

Of the fifteen different two-amplifier circuits considered, and for the given design equations, only three configurations, circuits 13, 14, and 15, provide better performance criteria than single-amplifier circuits. In this section we consider the high-frequency limitations of these three circuits in an attempt to determine the "best" configuration. For comparison purposes, we also consider the three-amplifier state variable realization.

We assume that an amplifier of a finite gain K is obtained from an OA with a unity-gain bandwidth ω_t. Thus, in the characteristic polynomial of each circuit, a gain K is replaced by $K/(1 + p(K/\omega_t))$ where K now denotes a resistive ratio. In the case of an infinite gain amplifier, its gain A is replaced by $A_0/(1 + p(A_0/\omega_t)) \simeq \omega_t/p$. In the following, analysis is done in detail for circuit 15 while the results for the other circuits are given without proof.

The analysis of circuit 15 taking into account the finite amplifier gains $K_1 = A_1$ and $K_2 = A_2$ leads to the denominator polynomial

$$D(p) = p^2\left(1 + \frac{2}{A_2} + \frac{2}{A_1A_2}\right) + P\frac{1}{TQ} \cdot \left[1 + \frac{2}{A_2} + \frac{2}{A_1A_2} + 2Q\left(\frac{1}{A_1} + \frac{1}{A_2} + \frac{2}{A_1A_2}\right)\right] + \frac{1}{T^2}\left[1 + \frac{2}{A_1} + \frac{1}{Q}\frac{2}{A_1}\left(1 + \frac{1}{A_2}\right)\right].$$

Dropping the second-order terms (A_1A_2), replacing $1/T$ by its nominal value ω_0, and assuming $Q \gg 1$,

$$D(p) \simeq p^2\left(1 + \frac{2}{A_2}\right) + p\frac{\omega_0}{Q}\left(1 + \frac{2Q}{A_1} + \frac{2Q}{A_2}\right) + \omega_0{}^2\left(1 + \frac{2}{A_1}\right).$$

TABLE III
FREQUENCY LIMITATIONS OF PRACTICAL CONFIGURATIONS

Circuit	$\delta \equiv \frac{\Delta\omega_0}{\omega_0}$ and Q_k	δ and Q_k/Q for identical amplifiers
ALL CIRCUITS WITH LOW GS^{ω_0}	$\delta \simeq -\frac{1}{2Q}\sum_i GS^Q_{A_i}\left(\frac{\omega_0}{\omega_{t_i}}\right)$	$\delta \simeq -\frac{1}{2Q}\left(\frac{\omega_0}{\omega_t}\right)\sum_i GS^Q_{A_i}$
Single Amplifier Filters	$\delta \simeq -\frac{1}{2Q} GS^Q_A\left(\frac{\omega_0}{\omega_t}\right)$ $\frac{Q_k}{Q} \simeq 1\Big/\left[1+\delta-2K\delta Q\left(\frac{\omega_0}{\omega_t}\right)\right]$	Notes: (a) The values of GS^Q_A are given in Table (I). (b) K = 1 for NF, PF2 and PF4 circuits and K = 1 to 3 for PF1 and PF4 circuits.
Circuit (13)	$\delta \simeq -\left(\frac{\omega_0}{\omega_{t_1}}+\frac{\omega_0}{\omega_{t_2}}\right)$ $\frac{Q_k}{Q} \simeq 1\Big/\left[1-Q\left(\frac{\omega_0}{\omega_{t1}}+2\frac{\omega_0}{\omega_{t2}}\right)\right]$	$\delta \simeq -2\left(\frac{\omega_0}{\omega_t}\right)$ $\frac{Q_k}{Q} \simeq 1\Big/\left[1-3Q\left(\frac{\omega_0}{\omega_t}\right)\right]$
Circuit (14) ($R_3 = R_1$)	$\delta \simeq -\left(\frac{\omega_0}{\omega_{t_2}}\right)$ $\frac{Q_k}{Q} \simeq 1\Big/\left[1-4Q\left(\frac{\omega_0}{\omega_{t1}}\right)\right]$	$\delta \simeq -\left(\frac{\omega_0}{\omega_t}\right)$ $\frac{Q_k}{Q} \simeq 1\Big/\left[1-4Q\left(\frac{\omega_0}{\omega_t}\right)\right]$
Circuit (15)	$\delta \simeq -\left(\frac{\omega_0}{\omega_{t_1}}+\frac{\omega_0}{\omega_{t_2}}\right)$ $\frac{Q_k}{Q} \simeq 1\Big/(1+\delta)\left[1-2Q\left(\frac{\omega_0}{\omega_{t_2}}\right)\left(1-\frac{\omega_{t2}}{\omega_{t1}}+2\delta\right)\right]$	$\delta \simeq -2\left(\frac{\omega_0}{\omega_t}\right)$ $\frac{Q_k}{Q} \simeq 1\Big/\left[1+\delta-4\delta Q\left(\frac{\omega_0}{\omega_t}\right)\right]$
State Variable Circuit [Fig.(3)]	$\delta \simeq -\left(\frac{\omega_0}{\omega_{t1}}+\frac{1}{2}\frac{\omega_0}{\omega_{t2}}\right)$ $\frac{Q_k}{Q} \simeq 1\Big/\left[1-Q\left(\frac{\omega_0}{\omega_{t1}}+\frac{\omega_0}{\omega_{t2}}+2\frac{\omega_0}{\omega_{t3}}\right)\right]$	$\delta \simeq -\frac{3}{2}\left(\frac{\omega_0}{\omega_t}\right)$ $\frac{Q_k}{Q} \simeq 1\Big/\left[1-4Q\left(\frac{\omega_0}{\omega_t}\right)\right]$

Replacing $1/A_1$ and $1/A_2$ by p/ω_{t_1} and p/ω_{t_2}, respectively, leads to

$$D(p) = p^3\frac{2}{\omega_{t_2}} + p^2\left(1 + 2\frac{\omega_0}{\omega_{t_1}} + 2\frac{\omega_0}{\omega_{t_2}}\right) + p\frac{\omega_0}{Q}\left(1 + 2Q\frac{\omega_0}{\omega_{t_1}}\right) + \omega_0^2. \quad (3)$$

Thus, in addition to the dominant pair of poles having changed their parameters to ω_{0k} and Q_k, the filter acquires a third parasitic pole. The polynomial $D(p)$ may be written as

$$D(p) = \left(p^2 + p\frac{\omega_{0k}}{Q_k} + \omega_{0k}^2\right)\left(\frac{2p}{\omega_{t_2}} + \frac{\omega_0^2}{\omega_{0k}^2}\right). \quad (4)$$

Denoting $\omega_{0k} = \omega_0(1 + \delta)$ where $\delta \ll 1$ such that $\omega_{0k}^2 \simeq \omega_0^2(1 + 2\delta)$, then equating the p^2 coefficients in (3) and (4) obtains

$$\delta = -\frac{\frac{\omega_0}{\omega_{t_1}} + \frac{\omega_0}{\omega_{t_2}} - \frac{1}{Q_k}\frac{\omega_0}{\omega_{t_2}}}{1 - \frac{\omega_0}{\omega_{t_2}}\frac{1}{Q_k}}.$$

For $Q \gg 1$,

$$\delta \simeq -\left(\frac{\omega_0}{\omega_{t_1}} + \frac{\omega_0}{\omega_{t_2}}\right).$$

Note that this can be written in the form

$$\delta \simeq -\frac{1}{2Q}\left[GS_{A_1}{}^Q\left(\frac{\omega_0}{\omega_{t_1}}\right) + GS_{A_2}{}^Q\left(\frac{\omega_0}{\omega_{t_2}}\right)\right] \quad (5)$$

which is similar to that for single-amplifier filters given in (1). In fact for the single-amplifier filters, for the 3 two-amplifier circuits considered, as well as for the state-variable filter, it can be shown that the frequency deviation due to the amplifiers' limited frequency response is given by the general form

$$\delta \equiv \frac{\Delta\omega_0}{\omega_0} \simeq -\frac{1}{2Q}\sum_i GS_{A_i}{}^Q\left(\frac{\omega_0}{\omega_{t_i}}\right). \quad (6)$$

This result is very significant: *for these circuits in which the GS^{ω_0} values are low, the frequency deviation at higher frequencies is determined by GS^Q values which are evaluated at low frequencies.*

To obtain the Q deviation, the p-term coefficients in (3) and (4) are equated

$$\frac{Q_k}{Q} = 1\Big/(1+\delta)\left[1 - 2Q\left(\frac{\omega_0}{\omega_{t_2}}\right)\left(1 - \frac{\omega_{t_2}}{\omega_{t_1}} + 2\delta\right)\right]. \quad (7)$$

Substituting for δ from (5), it follows that the only configuration-dependent quantities in the Q_k expression are the GS^Q. An interesting observation follows from (7): if the cutoff frequencies of the two amplifiers are nearly equal, as is the case with dual OA's, the Q deviation is much smaller than that for the state-variable realization, as will be shown below. For $\omega_{t_1} = \omega_{t_2} = \omega_t$,

$$\frac{Q_k}{Q} \simeq 1\Big/\left[1 + \delta - 4\delta Q\left(\frac{\omega_0}{\omega_t}\right)\right].$$

This form is identical to that obtained for single-amplifier filters except for the multiplicative factor 4 in the third term (2 in the case of single-amplifier NF circuits and about 4 for PF circuits).

For comparison purposes, Table III provides the frequency limitations for single-amplifier filters, the three practical two-amplifier filters, and the state-variable three-amplifier filter, given in [8]. For the latter circuit,

$$GS_{A_1}{}^{Q} \simeq 2Q + 1 \qquad GS_{A_2}{}^{Q} \simeq Q + 1$$

while $GS_{A_3}{}^{Q}$ and all GS^{ω_0} are small.

A number of conclusions follow from Table III.

1) A general form for the frequency deviation is available and given in the first row. This form has been analytically and computationally tested and found to apply for all circuits having low values for GS^{ω_0}. These, of course, are the only circuits of interest. Unfortunately, no similar form is available for the Q deviations.

2) Single-amplifier circuits of the NF2, PF1, PF3, and PF4 classes exhibit ω_0 deviations of the same order as those for the other two- and three-amplifier circuits. Such deviations are independent of Q. On the other hand, single-amplifier circuits of the NF1 and PF2 classes (including the Sallen and Key NF and unity-gain circuits) suffer a larger frequency deviation, proportional to Q.

3) Although the ω_0 deviations of the two- and three-amplifier circuits considered are of the same order, only circuit 15 has a reasonably small Q deviation, of the same order as that for single-amplifier circuits. The Q deviations of the other circuits, including the state-variable realization, are quite large.

4) The Q deviations of the state-variable circuit has been pointed out before in [8] and [27]. We note, however, that our result for ω_0 deviation for this circuit is different from that reported in [27].

5) In spite of the large Q deviation of the state-variable realization, a compensation scheme has been reported in [8]. This, however, could be uneconomical.

6) Circuit 15 obtains its low Q deviation in the case ω_{t_1} and ω_{t_2} are close in value. Measurements on dual IC OA's [30] indicate excellent tracking of ω_{t_1} and ω_{t_2} over a wide temperature range.

7) In all respects, circuit 15 is the "optimum" configuration among the circuits considered.

VI. Other Practical Considerations

Other parameters are of some importance in selecting a biquadratic section such as the component spread, the possibility of obtaining the amplifiers from integrated circuit OA's, the cascadability of the section, the ability of the circuit to provide more than one filtering function simultaneously, and the configuration ability to realize all the different functions. Throughout this paper, comments are given regarding the first two matters. We note that circuits 13, 14, 15, and the state-variable circuit satisfy these requirements.

The cascadability of the filter section is an important matter in the design of high-order filters. The last column in Table II provides approximate values for the input impedances of circuits 1–15 evaluated at resonance. Again we note the superior performance of circuits 14 and 15. We also note that some circuits exhibit extremely low input impedance at resonance. This may lead to oscillations in the preceding filter section.

The state-variable realization is noted for its unique ability of providing two or three different filtering functions simultaneously. Circuit 15, however, is capable of providing all possible filtering functions with simple modifications that do not affect its superior performance. Some such configurations have been recently reported [29], while others, as well as their application to the design of high-order filters in a multiple-feedback structure, will be given in a further communication.

VII. Conclusions

Many important conclusions have already been given. In summary, the state-of-the-art biquadratic realizations display low ω_0 gain-sensitivity products, low passive sensitivities, and Q gain-sensitivity products of the order of Q. These requirements are obtained in three two-amplifier resonators (circuits 13, 14, 15), as well as in the three-amplifier state-variable realization. Among these four configurations, circuit 15 has the lowest Q deviation due to the limited amplifiers' bandwidth. It also features a high input impedance as well as a low spread in component value.

References

[1] A. S. Sedra, "Generation and classification of single amplifier filters," *Int. J. Circuit Theory Appl.*, vol. 2, pp. 51–67, Mar. 1974.

[2] G. S. Moschytz, "A note on pole, frequency and Q sensitivity," *IEEE J. Solid-State Circuits*, vol. SC-6, pp. 267–269, Aug. 1971.

[3] D. Hilberman, "An approach to the sensitivity and statistical variability of biquadratic filters," *IEEE Trans. Circuit Theory*, vol. CT-20, pp. 382–390, July 1973.

[4] G. S. Moschytz, "Gain-sensitivity product: a figure of merit for hybrid integrated filters using single op. amp.," *IEEE J. Solid State Circuits*, vol. SC-6, pp. 103–110, June 1971.

[5] A. L. Rosenblum and M. S. Ghausi, "Multiparameter sensitivity in active RC networks," *IEEE Trans. Circuit Theory*, vol. CT-18, pp. 592–599, Nov. 1971.

[6] B. A. Shenoi, "On a new statistical variability criterion for optimal design and comparative evaluation of RC-active filters," in *Proc. 1973 Int. Symp. Circuit Theory*, Toronto, Ont., Canada, pp. 122–125.

[7] J. J. Friend, "A single operational amplifier biquadratic filter section," in *Digest 1970 Int. Symp. Circuit Theory*, Atlanta, Ga., pp. 179–180

[8] L. C. Thomas, "The biquad: Part I—some practical design considerations," *IEEE Trans. Circuit Theory*, vol. CT-18, pp. 350–357, May 1971.

[9] L. T. Bruton, "Nonideal performance of two-amplifier positive impedance converters," *IEEE Trans. Circuit Theory*, vol. CT-17, pp. 541–549, Nov. 1970.

[10] R. Bach, Jr., "Selecting RC values for active filters," *Electron.*, pp. 82–85, May 13, 1960.

[11] M. Soderstrand and S. K. Mitra, "Gain and sensitivity limitations of active-RC filters," *IEEE Trans. Circuit Theory*, vol. CT-18, pp. 600–609, Nov. 1971.

[12] A. E. Sanderson, "A Q-invariant, ω_0-invariant active resonator," *Proc. IEEE* (Letters), vol. 60, pp. 908–909 [illegible] 1972.

[13] Y. Bedri and T. Deliyannis, "Active RC [illegible]ass and high-pass networks," *Electron Lett.*, vol. 5, pp. 331–3[illegible], July 24, 1969.

[14] M. Hills, "Active filters: part 13, narrowing the choice," *Electron.*, pp. 106–111, October 27, 1969.

[15] Y. Bedri and T. Deliyannis, "Selectivity improvement in a useful second-order active RC section," *Int. J. Electron.*, vol. 13, pp. 243–248, 1971.

[16] A. G. Holt and M. R. Lee, "General rule for the compensation of the sensitivity of a network to its active devices," *Electron. Lett.*, vol. 5, pp. 324–325, July 10, 1969.

[17] P. R. Geffe, "Passive sensitivities of gain compensated networks," *IEEE Trans. Circuit Theory*, vol. CT-18, pp. 302–304, Mar. 1971.
[18] ——, "A Q-invariant active resonator," *Proc. IEEE* (Letters), vol. 57, p. 1442, Aug. 1969.
[19] M. Soderstrand and S. K. Mitra, "Extremely low sensitivity active-RC filters," *Proc. IEEE* (Letters), vol. 57, pp. 2175–2176, Dec. 1969.
[20] T. Deliyannis, "A low pass filter with extremely low sensitivity," *Proc. IEEE* (Letters), vol. 58, pp. 1366–1367, Sept. 1970.
[21] B. K. Conant and Y. S. Wong, "Pitfalls in state-variable filter design."
[22] M. Soderstrand and S. K. Mitra, "Design of active RC filters with zero gain-sensitivity product," *IEEE Trans. Circuit Theory*, vol. CT-20, pp. 441–445, July 1973.
[23] ——, "Active-RC filters with zero gain-sensitivity product," in *Digest 1972 Int. Symp. Circuit Theory*, North Hollywood, Calif., pp. 340–344.
[24] P. R. Geffe, "RC-amplifier resonators for active filters," *IEEE Trans. Circuit Theory*, vol. CT-15, pp. 415–419, Dec. 1968.
[25] T. A. Hamilton and A. Sedra, "A novel application of a gyrator-type circuit," in *Fifth Asilomar Conf. Circuits and Systems, Conf. Rec.*, pp. 343–348, Nov. 1971.
[26] R. H. S. Riordan, "Simulated inductors using differential amplifiers," *Electron. Lett.*, vol. 3, pp. 50–51, Feb. 1967.
[27] A. Budak and D. M. Petrela, "Frequency limitations of active filters using operational amplifiers," *IEEE Trans. Circuit Theory*, vol. CT-19, pp. 322–328, July 1972.
[28] S. A. Boctor and A. S. Sedra, "On the synthesis of RC bandpass filters," *IEEE Trans. Circuit Theory*, vol. CT-19, pp. 374–376, July 1972.
[29] W. B. Mikhael, "Design of high performance RC-active filters," Ph.D. thesis, Sir George Williams Univ., Montreal, P.Q., Canada, May 1973.
[30] J. Valihora and J. T. Lim, "The feasibility of active filtering in frequency division multiplex systems," presented at the *1974 IEEE Int. Symp. Circuits and Systems*, San Francisco, Calif., Apr. 1974.
[31] A. Antoniou, "Realization of gyrators using operational amplifiers, and their use in RC-active-network synthesis," *Proc. Inst. Elec. Eng.*, vol. 116, pp. 1838–1850, 1969.

A New Class of Second-Order RC-Active Filters with two Operational Amplifiers

By Norbert Fliege, VDE

Communication from the Institut für Nachrichtensysteme, Universität Karlsruhe

1. Introduction

In recent years many second-order *RC*-active filters have been published. They have become popular because of their simple and transparent circuitry and because of their capability of realizing higher-order active filters by cascading technique.

A great deal of published one-operational amplifier second-order active filters are Sallen-Key-type filters [1] the sensitivities of which were investigated in great detail by Saraga [2]. These filters are restricted to low-Q or medium-Q applications because of their greater sensitivities to element variations compared with some other types. Better results have been achieved with active filters employing two operational amplifiers [3–8]. The relative pole sensitivities of these filters with respect to the passive elements are mostly not greater than unity. The relative Q-sensitivity with respect to the gain μ of the operational amplifiers is normally Q/μ. Thus, these filters are suitable for high-Q applications.

In this paper a new class of two-operational amplifier *RC*-active filters is presented including realizations for all types of commonly used second-order transfer functions.

2. Resonator and General Filter Structure

The principle of the resonator used in the following is shown in *Fig. 1.* The resonator consists of six admittances Y_1 to Y_6, two norators No 1 and No 2 and two nullators Nu 1 and Nu 2. The two nullators are to provide equal voltage U_0 at the nodes 1, 2 and 3 which can be done in two different ways as shown in Fig. 1a and 1b. Fig. 1a may be designated as symmetrical case (type S), Fig. 1b as unsymmetrical case (type U).

Since the currents flowing through the nullators are zero the voltage U_0 can be calculated from the norator voltages U_1 and U_2 to be

$$U_0 = \frac{Y_2}{Y_1 + Y_2} U_1 , \qquad (1)$$

$$U_0 = \frac{Y_3}{Y_3 + Y_4} U_1 + \frac{Y_4}{Y_3 + Y_4} U_2 , \qquad (2)$$

$$U_0 = \frac{Y_5}{Y_5 + Y_6} U_2 . \qquad (3)$$

Substituting Eqns. (1) and (3) into Eqn. (2) yields the characteristic equation of the resonators given in Fig. 1:

$$Y_1 Y_3 Y_5 + Y_2 Y_4 Y_6 = 0 . \qquad (4)$$

Combining one nullator and one norator we get a two-port element as shown in *Fig. 2* which is called a nullor. Such a nullor may be represented by an ideal operational amplifier having infinite gain. For each resonator in Fig. 1 there exist two different kinds of combining the nullators and norators: type 1: Nu 1 with No 2 and Nu 2 with No 1 and type 2: Nu 1 with No 1 and Nu 2 with No 2. As a result, *Fig. 3* shows the four equivalent resonators with two operational amplifiers the common characteristic equation of which is given in Eqn. (4).

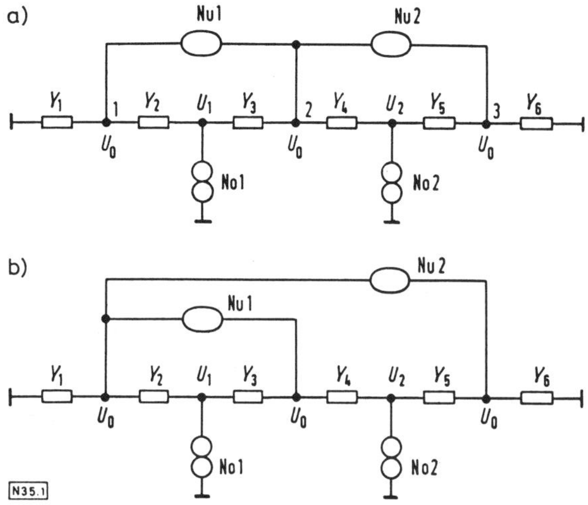

Fig. 1.
Two fundamental resonators.

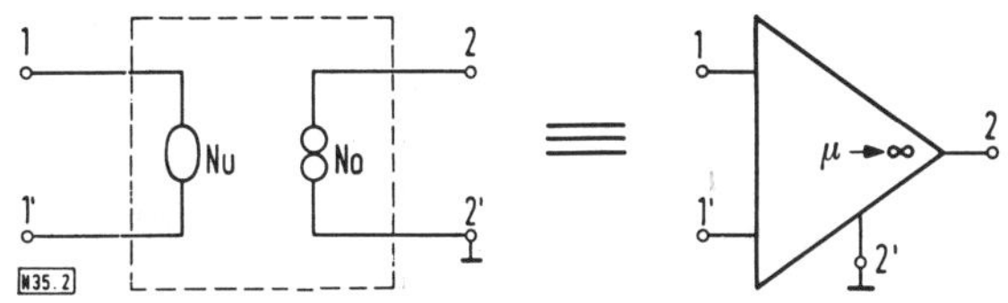

Fig. 2.
The nullor and its operational amplifier equivalent.

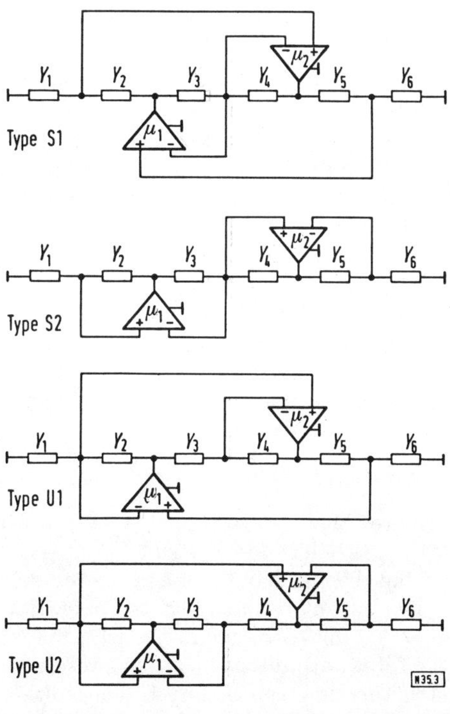

Fig. 3.
Four equivalent resonators with two operational amplifiers.

Reprinted with permission from *Nachrichtentech. Z.*, vol. 26, pp. 279-282, June 1973.

Table 1.
Specifications of the general filter structure elements for some second-order transfer function realizations.
PZC: Pole-zero configuration.

Fig. and PZC	$Y_{1a} =$	$Y_{1b} =$	$Y_2 =$	$Y_3 =$	$Y_4 =$	$Y_5 =$	$Y_{6a} =$	$Y_{6b} =$	condition
Fig. 6a, PZC1:	G_1	0	G_2	$s\,C_3$	G_4	$G_5 + s\,C_5$	0	G_6	—
Fig. 6b, PZC1:	0	G_1	$s\,C_2$	G_3	G_4	$G_5 + s\,C_5$	$s\,C_6$	0	$G_3\,C_5 = G_4\,C_6$
Fig. 6c, PZC2:	G_1	0	G_2	G_3	$s\,C_4$	G_5	$s\,C_6$	G_6	—
Fig. 6d, PZC2:	0	G_1	$s\,C_2$	G_3	G_4	G_5	$G_6 + s\,C_6$	0	$G_3\,G_5 = G_4\,G_6$
Fig. 6e, PZC3:	G_1	0	G_2	G_3	$s\,C_4$	G_5	G_6	$s\,C_6$	—
Fig. 6f, PZC3:	0	$s\,C_1$	G_2	G_3	G_4	$G_5 + s\,C_5$	G_6	0	$G_3\,G_5 = G_4\,G_6$
Fig. 6g, PZC4:	G_1	0	$s\,C_2$	G_3	G_4	G_5	$s\,C_6$	G_6	—
Fig. 6h, PZC5:	$s\,C_1$	G_1	G_2	G_3	G_4	$s\,C_5$	G_6	0	—
Fig. 6i, PZC7:	G_1	$s\,C_1$	G_2	G_3	G_4	$s\,C_5$	0	G_6	—
Fig. 6k, PZC8:	0	G_1	$s\,C_2$	G_3	G_4	G_5	G_6	$s\,C_6$	—
Fig. 6l, PZC6:	0	G_1	G_2	G_3	$s\,C_4$	G_5	G_6	$s\,C_6$	$G_1 = G_2$
Fig. 6m, PZC9:	0	G_1	G_2	G_3	$s\,C_4$	G_5	G_{6a}	$G_{6b} + s\,C_6$	$G_1\,G_{6a} = G_2\,G_{6b}$

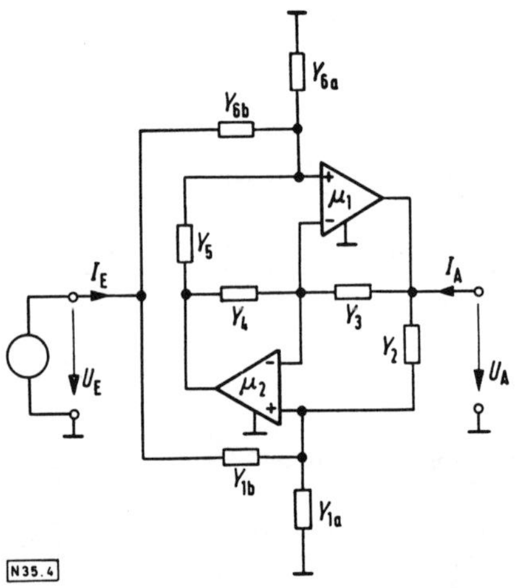

Fig. 4.
General filter structure.
Deutsche Patentanmeldung P 23 17 644.5.

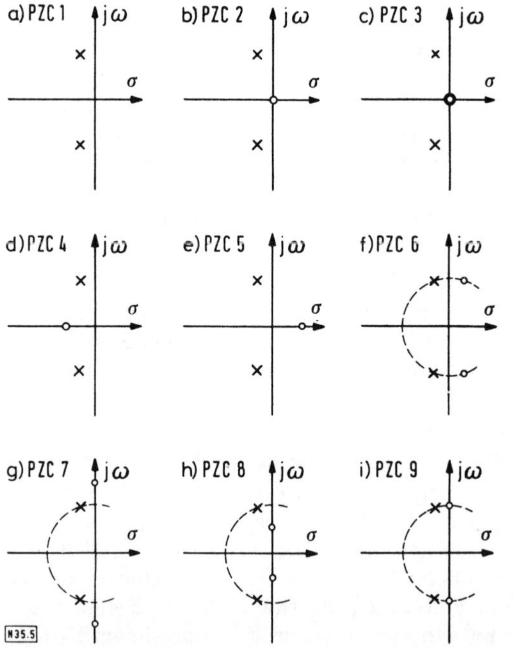

Fig. 5.
Pole-zero configurations (PZC) of second-order transfer functions.

The signs of the input terminals of the operational amplifiers in Fig. 3 have been chosen in such a way that the filter circuits described in the following section will be stable. The stability may be proved by assuming the operational amplifier gain μ to be a function of frequency $\mu = \mu(s)$ [9].

The resonators depicted in Fig. 3 have been utilized for realizing gyrators, generalized-immittance converters, frequency-dependent negative resistances and admittances [10—13]. In the following active filter stages with zero output impedance will be derived from the resonators given in Fig. 3. Representatively of the four resonators in Fig. 3 the type S1 may be considered.

The characteristic equation of the resonator will not be affected if the admittances Y_1 and Y_6 are not realized by single elements but by a shunt connection of two elements:

$$Y_1 = Y_{1a} + Y_{1b}\,, \qquad Y_6 = Y_{6a} + Y_{6b}\,. \tag{5}$$

Furthermore, the characteristic equation will remain unaffected if the elements Y_{1b} and Y_{6b} are not grounded directly but by an input voltage source as shown in *Fig. 4.* The output terminals of the operational amplifiers μ_1 or μ_2 (or both simultaneously) may be chosen to be the filter output terminals. As a result, the general filter structure is given in Fig. 4. The voltage transfer function of this network can be calculated to be:

$$T(s) = \frac{U_A}{U_E} =$$

$$= \frac{Y_{6b}(Y_2\,Y_4 + Y_{1a}\,Y_4) + Y_{1b}(Y_3\,Y_5 - Y_{6a}\,Y_4)}{Y_1\,Y_3\,Y_5 + Y_2\,Y_4\,Y_6}\,. \tag{6}$$

The same transfer function holds for the output terminals of the operational amplifier μ_2 if the admittance Y_1 will be substituted for Y_6, Y_2 for Y_5 and Y_3 for Y_4 and vice versa.

3. Filter Catalogue

Several different filter networks can be derived from the general structure given in Fig. 4 by specifying the admittances Y_{1a} to Y_{6b} to be either a conductance or a capacitance or a parallel connection of both. In *Table 1* the admittances are specified for some commonly used second-order transfer functions. The corresponding pole-zero configurations (PZC) can be seen from *Fig. 5,* the corresponding filter circuits from *Fig. 6.*

a) low-pass (PZC1)

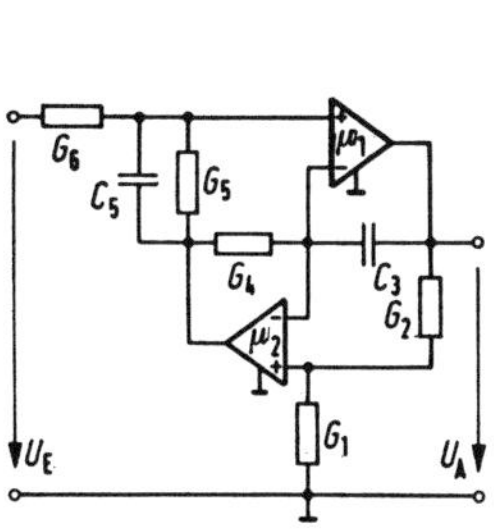

b) low-pass (PZC1)

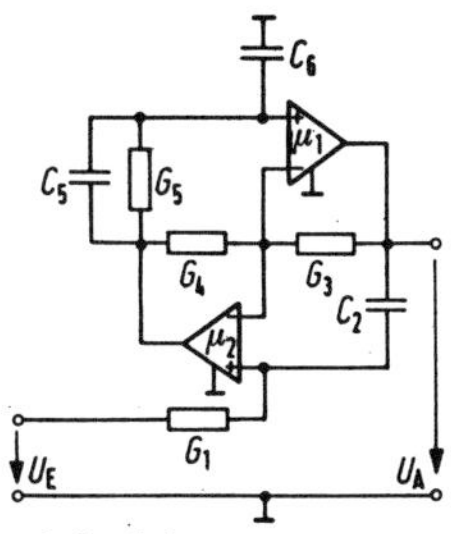

$G_3 C_5 = G_4 C_6$

c) band-pass (PZC 2)

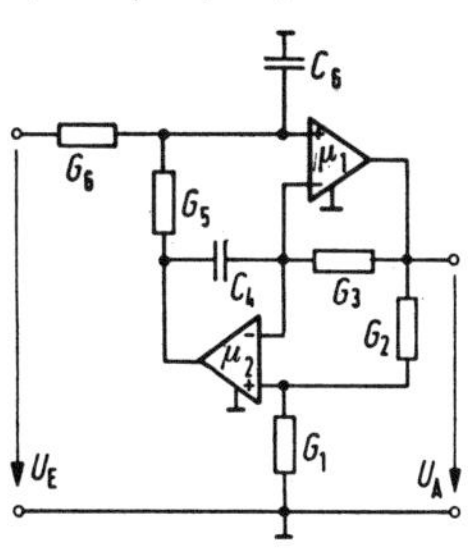

d) band-pass (PZC 2)

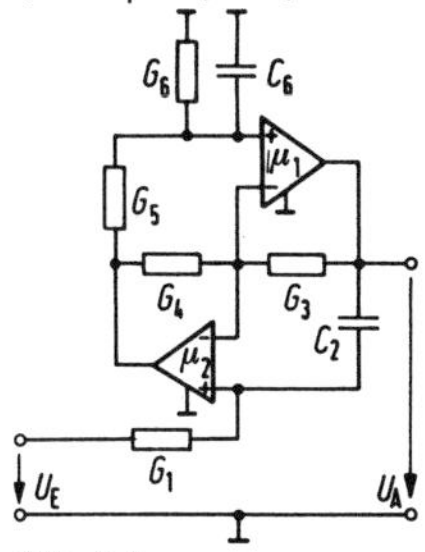

$G_3 G_5 = G_4 G_6$

a) $$T(s) = \frac{1+\alpha_{21}}{T_{34}\,T_{56}} \cdot \frac{1}{s^2 + \frac{1}{T_{55}}\,s + \frac{\alpha_{21}}{T_{34}\,T_{56}}}$$

b) $$T(s) = \frac{\alpha_{34}}{T_{21}\,T_{65}} \cdot \frac{1}{s^2 + \frac{1}{T_{21}}\,s + \frac{\alpha_{34}}{T_{21}\,T_{65}}}$$

c) $$T(s) = \frac{1+\alpha_{12}}{T_{66}} \cdot \frac{s}{s^2 + \frac{1}{T_{66}}\,s + \frac{\alpha_{12}}{T_{43}\,T_{65}}}$$

d) $$T(s) = -\frac{1}{T_{21}} \cdot \frac{s}{s^2 + \frac{1}{T_{66}}\,s + \frac{\alpha_{34}}{T_{21}\,T_{65}}}$$

e) high-pass (PZC 3)

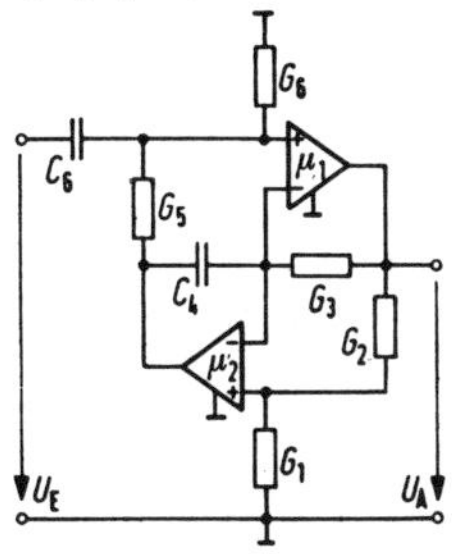

f) high-pass (PZC 3)

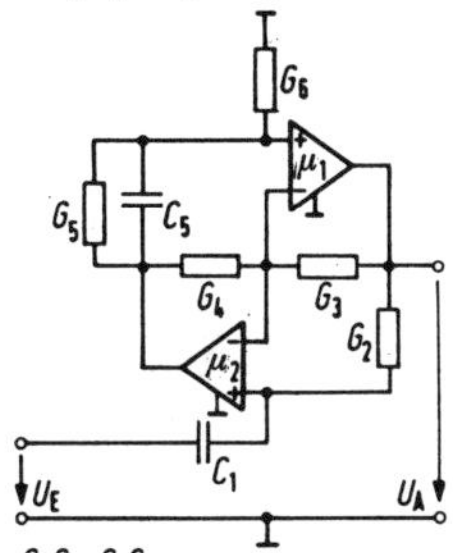

$G_3 G_5 = G_4 G_6$

g) one negative real zero (PZC 4)

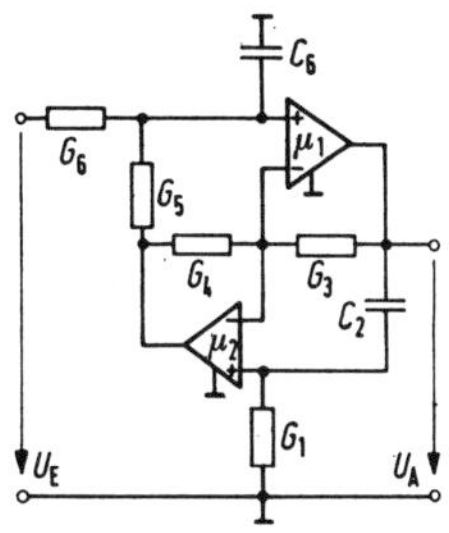

h) one positive real zero (PZC 5)

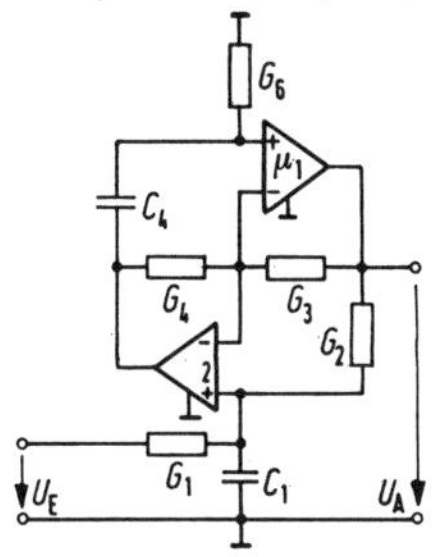

e) $$T(s) = (1+\alpha_{12})\,\frac{s^2}{s^2 + \frac{1}{T_{66}}\,s + \frac{\alpha_{12}}{T_{43}\,T_{65}}}$$

f) $$T(s) = \frac{s^2}{s^2 + \frac{1}{T_{55}}\,s + \frac{\alpha_{43}}{T_{12}\,T_{56}}}$$

g) $$T(s) = \frac{1}{T_{66}} \cdot \frac{s + \frac{1}{T_{21}}}{s^2 + \frac{1}{T_{66}}\,s + \frac{\alpha_{34}}{T_{21}\,T_{65}}}$$

h) $$T(s) = \frac{1}{T_{11}} \cdot \frac{s - \frac{\alpha_{43}}{T_{56}}}{s^2 + \frac{1}{T_{11}}\,s + \frac{\alpha_{43}}{T_{12}\,T_{56}}}$$

i) imaginary zeros (PZC 7)

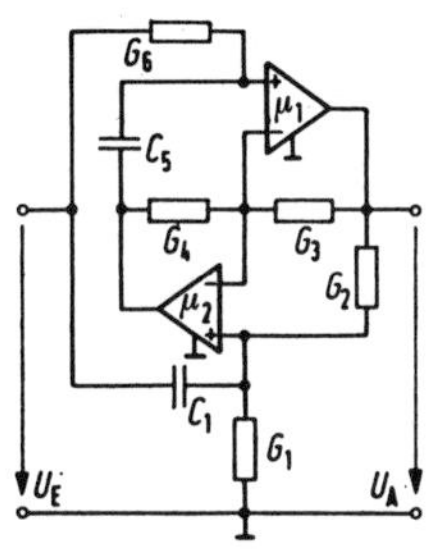

k) imaginary zeros (PZC 8)

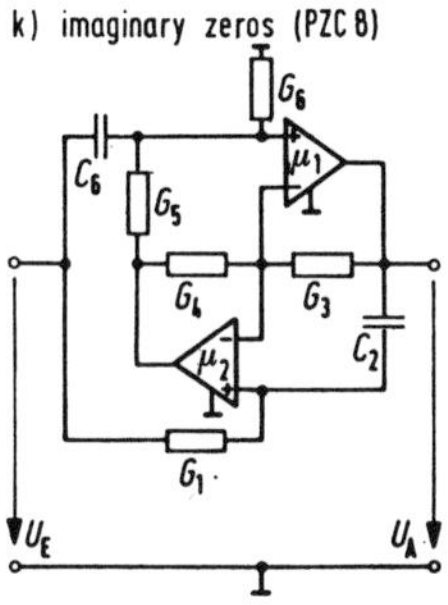

l) all-pass (PZC 6)

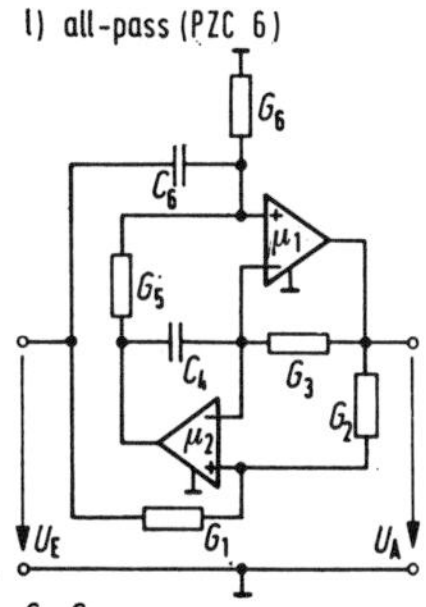

$G_1 = G_2$

m) band-rejection (PZC 9)

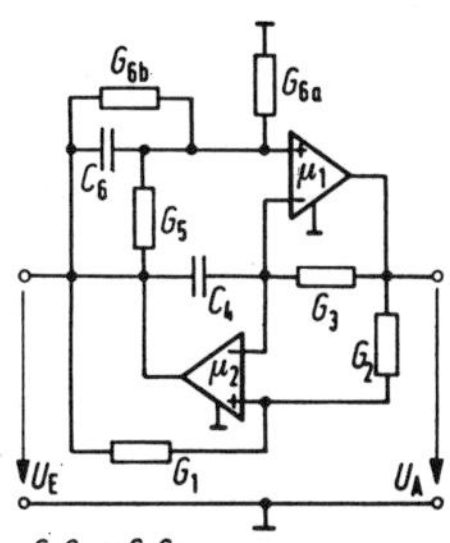

$G_1 G_{6a} = G_2 G_{6b}$

i) $$T(s) = \frac{s^2 + \frac{\alpha_{43}}{T_{12}\,T_{56}}\,(1+\alpha_{12})}{s^2 + \frac{1}{T_{11}}\,s + \frac{\alpha_{43}}{T_{12}\,T_{56}}}$$

k) $$T(s) = \frac{s^2 + \frac{1}{T_{21}\,T_{65}}\,(\alpha_{34} - \alpha_{65})}{s^2 + \frac{1}{T_{66}}\,s + \frac{\alpha_{34}}{T_{21}\,T_{65}}}$$

l) $$T(s) = \frac{s^2 - \frac{1}{T_{66}}\,s + \frac{1}{T_{43}\,T_{65}}}{s^2 + \frac{1}{T_{66}}\,s + \frac{1}{T_{43}\,T_{65}}}$$

m) $$T(s) = \frac{s^2 + \frac{\alpha_{12}}{T_{43}\,T_{65}}}{s^2 + \frac{1}{T_{66}}\,s + \frac{\alpha_{12}}{T_{43}\,T_{65}}}$$

Fig. 6.
Filter catalogue.
Deutsche Patentanmeldung P 23 17 644.5

For example, substituting the admittances specified in the first row of Table 1 into Eqn. (6) yields

$$T(s) = \frac{G_6 (G_2 G_4 + G_1 G_4)}{G_1 s C_3 (G_5 + s C_5) + G_2 G_4 G_6} =$$

$$= \frac{(G_1 + G_2) G_4 G_6}{G_1 C_3 C_5} \cdot \frac{1}{s^2 + s \frac{G_5}{C_5} + \frac{G_2 G_4 G_6}{G_1 C_3 C_5}},$$

$$T(s) = \frac{1 + \alpha_{21}}{T_{34} T_{56}} \cdot \frac{1}{s^2 + \frac{1}{T_{55}} s + \frac{\alpha_{21}}{T_{34} T_{56}}} \tag{7}$$

where $T_{34} = C_3/G_4$, $T_{56} = C_5/G_6$, $T_{55} = C_5/G_5$ and $\alpha_{21} = G_2/G_1$. Eqn. (7) represents a low-pass function the pole-zero configuration of which is plotted in Fig. 5a (PZC1). The corresponding network is shown in Fig. 6a.

The time constants $T_{\mu\nu}$ and the conductance ratios $\alpha_{\mu\nu}$ in the transfer functions given in Fig. 6 are defined as

$$T_{\mu\nu} = C_\mu / G_\nu \,, \quad \alpha_{\mu\nu} = G_\mu / G_\nu \,. \tag{8}$$

Additional conditions are made to get the desired pole-zero configuration for some networks in Fig. 6, e. g. the condition $G_3 G_5 = G_4 G_6$ for the network in Fig. 6d. These conditions are included in Fig. 6 and in Table 1.

It may be noted that the circuits in Fig. 6i and 6k have imaginary zeros without any additional condition. Thus, the real part of the zeros will be zero if all passive elements have finite non-zero values. Furthermore, the real part of the zeros will remain zero in case the element values deviate from their nominal values.

4. Conclusions

Four equivalent resonators with two operational amplifiers and six passive elements have been given from which general filter structures can be derived. By specifying the admittances in the general structure as capacitances or conductances several filter networks have been given for realizing the most commonly used second-order transfer functions.

References

[1] *Sallen, R. P.; Key, E. L.:* A practical method of designing RC-active filters. Instn. Radio Engrs. Trans. CT—2 (1955) pp. 74 to 85.

[2] *Saraga, W.:* Sensitivity of second-order Sallen-Key RC-active filters. Electron. Letters 3 (1967) pp. 442 to 444.

[3] *Geffe, P. R.:* A Q-invariant active resonator. Proc. IEEE 57 (1969) p. 1442.

[4] *Soderstrand, M. A.; Mitra, S. K.:* Extremely low sensitivity active RC filter. Proc. IEEE 57 (1969) pp. 2175 to 2176.

[5] *Deliyannis, T.:* A low-pass filter with extremely low sensitivity. Proc. IEEE 58 (1970) pp. 1366 to 1367.

[6] *Holt, A. G. J.; Lee, M. R.:* General rule for the compensation of the sensitivity of a network to its active devices. Electron. Letters 5 (1969) pp. 324 to 325.

[7] *Huelsman, L. P.:* Active filters: lumped, distributed, integrated, digital, and parametric. New York: McGraw-Hill, 1970, pp. 31 to 40.

[8] *Bedri, Y.; Deliyannis, T.:* Active RC bandpass and high-pass network. Electron. Letters 5 (1969) pp. 331 to 332.

[9] *Fliege, N.:* Complementary transformation of feedback systems. IEEE Trans. CT—20 (1973) pp. 137 to 139.

[10] *Riordan, R. H. S.:* Simulated inductors using differential amplifiers. Electron. Letters 3 (1967) pp. 50 to 51.

[11] *Antoniou, A.:* Gyrator using operational amplifiers. Electron. Letters 3 (1967) pp. 350 to 352.

[12] *Antoniou, A.:* New gyrator circuits obtained by using nullors. Electron. Letters 4 (1968) pp. 87 to 89.

[13] *Antoniou, A.:* Stability properties of some gyrator circuits. Electron. Letters 4 (1968) pp. 510 to 512.

(Manuscript received: 19th January 1973)

A Versatile Active *RC* Building Block with Inherent Compensation for the Finite Bandwidth of the Amplifier

DAG ÅKERBERG AND KARE MOSSBERG

Abstract—An earlier paper by the authors describes an easily trimmed universal building block for active *RC* filters which possesses the valuable characteristic that, with suitable design, the Q-value can be made approximately independent of the gain-bandwidth product of the operational amplifiers. This makes the filter usable for high frequencies, while at the same time the dependence of the Q-value on temperature variations in the operational amplifiers is drastically reduced. Design formulas are presented, as well as comparative measurements which verify the theory. The building block is shown to have excellent characteristics both as a universal second-order building block and as a standard block for active ladder synthesis of bandpass filters.

Manuscript received September 5, 1972; revised April 13, 1973. This work was supported in part by the Swedish Board for Technical Development and in part by the L. M. Ericsson Telephone Company.

D. Åkerberg (M'72) is with Svenska Radio AB, Stockholm, Sweden.
K. Mossberg is with the Department of Telecommunication Theory, Royal Institute of Technology, Stockholm, Sweden.

INTRODUCTION

ACTIVE *RC* building blocks of the type "three operational amplifiers in a ring" are widely known, from many points of view, as the best second-order networks available. Such building blocks are made commercially by several firms. Compared with other networks [2], these networks have low sensitivity to variations in active and passive components. The

Reprinted from *IEEE Trans. Circuits Syst.*, vol. CAS-21, pp. 75-78, Jan. 1974.

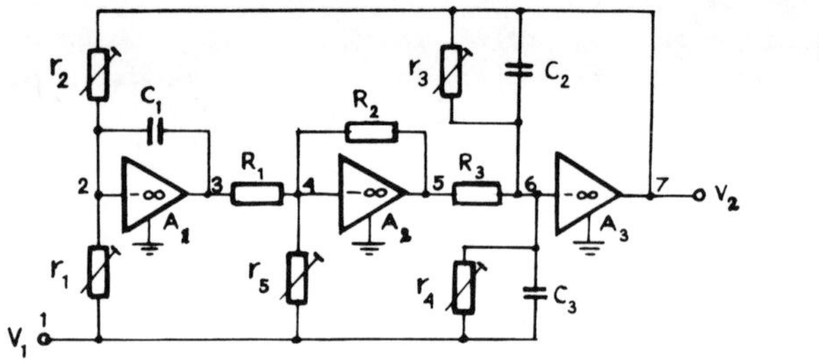

Fig. 1. Building block realizing a general biquadratic transfer function.

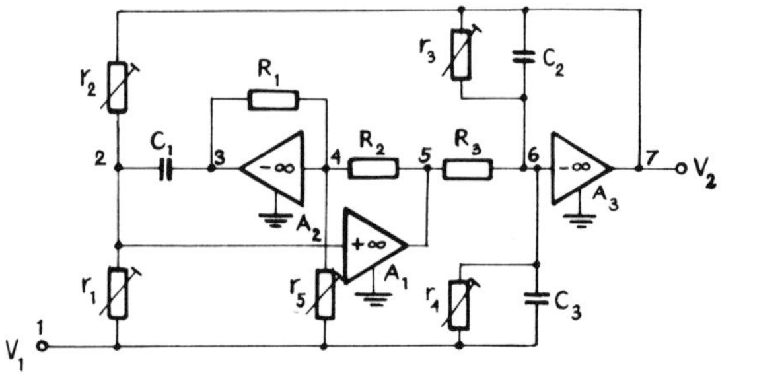

Fig. 2. Modified amplifier connection of Fig. 1 giving the same transfer function, but with improved stability.

best known of them is probably that reported by Kerwin *et al.* [3]. The variant earlier described by the authors [1],[1] and shown in Figs. 1 and 2, has, in addition, the valuable characteristic that frequency and attenuation—ω and σ in (4) *both* for poles and zeros—can be trimmed independently with four resistances, one for each parameter.

In practice, as is known, the amplifiers have a finite frequency-dependent gain

$$A \simeq -\frac{1}{A_0^{-1} + \dfrac{s}{\omega_1}} \tag{1}$$

where A_0 is the dc gain and ω_1 is the gain-bandwidth product in radians per second.

The frequency dependence of the amplifiers limits the usability of the filters at high frequencies, as the Q-value for the type of network under consideration becomes frequency dependent [2] according to the formula

$$Q_p = Q_{0p} \frac{1}{1 - k\, Q_{0p} \dfrac{\omega_{0p}}{\omega_1}} \tag{2}$$

where ω_{0p} is the pole frequency of the filter and Q_{0p} is the Q-value for ideal operational amplifiers; k is typically 4.

The Q-sensitivity with respect to the gain-bandwidth product ω_1 will be

$$S_{\omega_1}^{Q_p} = -\, kQ_{0p} \frac{\omega_{0p}}{\omega_1}. \tag{3}$$

Tarmy and Ghausi [4], using three operational amplifiers with dual outputs and inputs, have designed a network for which $k = 0$. Arbitrarily situated zeros, however, are not realized so simply with this network. However, $k = 0$ can also be obtained with our network by modification of the connection of the amplifiers in the manner earlier indicated [1], and is described as follows.

We present, accordingly, a unique building block for active filters, consisting of three single-ended operational amplifiers, which realizes arbitrarily situated poles and zeros. The frequency and attenuation of the poles and zeros can be trimmed independently with a resistance for each parameter. The Q-value is nearly independent of the limited bandwidth of the amplifiers and of variations in the bandwidth caused by temperature changes.

Calculation of Sensitivity and Design Formulas for Compensation of the Effect of the Limited Bandwidth of the Amplifiers

Fig. 1 shows the universal building block with traditional amplifier connections. Fig. 2 shows the network with modified amplifier connections achieved through nullator-norator equivalents, as described in [1].

The ideal transfer function, i.e., with infinite gain of the amplifiers, will be the same for the two networks:

$$\frac{V_2}{V_1} = K \frac{s^2 + 2\sigma_0 s + \omega_0^2}{s^2 + 2\sigma_p s + \omega_p^2} =$$

$$-\frac{C_3}{C_2} \cdot \frac{s^2 + s\left(\dfrac{1}{r_4} - \dfrac{R_2}{r_5 R_3}\right)\dfrac{1}{C_3} + \dfrac{R_2}{r_1 R_1 R_3 C_1 C_3}}{s^2 + s\dfrac{1}{r_3 C_2} + \dfrac{R_2}{r_2 R_1 R_3 C_1 C_2}} \tag{4}$$

$$Q_p = \frac{\omega_p}{2\sigma_p}. \tag{5}$$

If the transfer function for frequency-dependent gains A_1, A_2, and A_3 is calculated according to (1), a fifth-order polynomial is obtained in s in the denominator. In [2] it is shown that, for high Q-values, the denominator can be approximated with good accuracy to a second-order polynomial by disregarding the s^5 and s^4 terms and substituting for s^3 the term

$$s^3 = -\, s\omega_{0p}^2. \tag{6}$$

For pole frequencies above 1 kHz, the gains for typical amplifiers can be approximated by

$$A_1 = -\frac{\omega_{11}}{s},\quad A_2 = -\frac{\omega_{12}}{s},\quad A_3 = -\frac{\omega_{13}}{s}. \tag{7}$$

Then, both for Figs. 1 and 2, the real pole frequency ω_p (ω_{0p} is the ideal value) for the typical values $R_1 = R_2 = R_3 = r_1 = r_2$, r_3 and $r_4 >> R_1$, and $C_1 = C_2$ will be

$$\omega_p \simeq \omega_{0p} \left(1 + \frac{2\omega_{0p}}{\omega_{11}} + \frac{\omega_{0p}}{\omega_{13}}\right)^{-1/2} \tag{8}$$

which gives the sensitivity

$$S_{\omega_1}^{\omega_p} \simeq \frac{3}{2} \frac{\omega_{0p}}{\omega_1}, \qquad \text{if } \omega_{11} = \omega_{13} = \omega_1. \tag{9}$$

The real Q-value for the network in Fig. 1 (Q_{0P} is the ideal value) will be

[1] Amplifier A_1 in [1, fig. 2] is turned the wrong way.

$$Q_p \simeq Q_{0p} \cdot \left[1 - Q_{0p}\omega_{0p}\left(\frac{1 + R_2/R_1 + R_2/r_5}{\omega_{12}} + \frac{1}{\omega_{11}} + \frac{1 + C_3/C_2}{\omega_{13}}\right)\right]^{-1}. \tag{10}$$

With $\omega_{11} = \omega_{12} = \omega_{13} = \omega_1$ we obtain

$$Q_p \simeq Q_{0p}\left[1 - kQ_{0p}\frac{\omega_{0p}}{\omega_1}\right]^{-1} \tag{2}$$

where

$$k = 3 + \frac{R_2}{R_1} + \frac{R_2}{r_5} + \frac{C_3}{C_2} > 3. \tag{11}$$

With $r_5 = \infty$, $C_3 = 0$, and $R_2 = R_1$, we get $k = 4$, which gives a typical sensitivity for conventional triple operational-amplifier networks:

$$S_{\omega_1}^{Q_p} = -4Q_{0p}\frac{\omega_{0p}}{\omega_1}. \tag{12}$$

As appears from (11), the sensitivity may be considerably greater in the event of unsuitable design.

For the network in Fig. 2,

$$Q_p \simeq Q_{0p}\left[1 + Q_{0p}\omega_{0p} \cdot \left(\frac{1 + R_1/R_2 + R_1/r_5}{\omega_{12}} - \frac{R_2/R_1}{\omega_{11}} - \frac{1 + C_3/C_2}{\omega_{13}}\right)\right]^{-1}. \tag{13}$$

The interesting point about (13) is that the terms in this case have different signs, so that—with suitable design of the *RC* components—the effects on Q_p of the various operational amplifiers can compensate for one another. This also results in a temperature compensation, as one may reckon that the gain-bandwidth products of the three operational amplifiers (monolithic) have practically the same temperature coefficient. With $\omega_{11} = \omega_{12} = \omega_{13} = \omega_1$, we get, according to (2),

$$k = R_2/R_1 + C_3/C_2 - R_1/R_2 - R_1/r_5. \tag{14}$$

Compensation is obtained for $k = 0$, in which case $S_{\omega_1}^{Q_p} = 0$.

The ordinary case without zeros, i.e., with $r_4 = r_5 = \infty$ and $C_3 = 0$, gives

$$k = 0, \quad \text{for } R_1 = R_2. \tag{15}$$

Another common case is zeros on the $j\omega$-axis, i.e., $r_4 = r_5 = \infty$, which gives

$$k = 0, \quad \text{for } \frac{C_3}{C_2} = \frac{R_1}{R_2} - \frac{R_2}{R_1}. \tag{16}$$

A suitable value of R_1/R_2 thus gives zero sensitivity. However, to obtain a signal level as uniform as possible in the network, R_1 must be equal to R_2, which conflicts with minimization of the sensitivity according to (16). If, however, a resistor R_{in} is placed between the amplifier input—node 4—and earth in accordance with Fig. 3, then ($r_5 = \infty$)

$$k = \frac{C_3}{C_2} + \frac{R_2}{R_1} - \frac{R_1}{R_2} - \frac{R_1}{R_{in}} \tag{17}$$

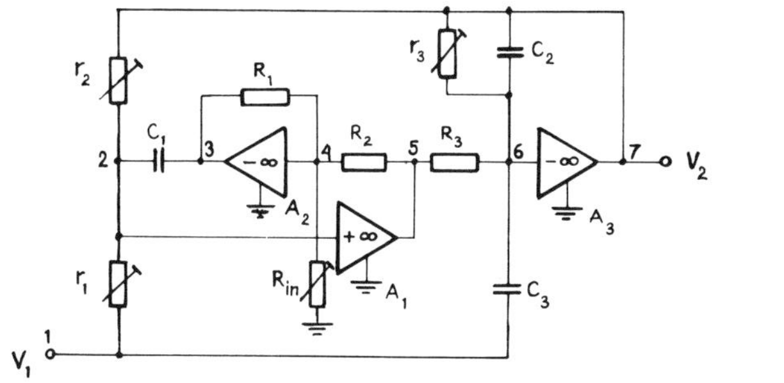

Fig. 3. Compensation by R_{in} providing equal peak voltages at nodes 3, 5, and 7.

and

$$k = 0, \quad \text{for } R_1 = R_2 \text{ and } \frac{C_3}{C_2} = \frac{R_1}{R_{in}}. \tag{18}$$

If one follows the given formulas for the design of $k = 0$, one gets for the network with the modified amplifier connection a Q-value with considerably smaller frequency dependence than for the ordinary network (Fig. 1), even if the amplifiers do not have exactly the same bandwidth.

If one wishes to compensate more exactly for the differences in bandwidth of the amplifiers, the network can be trimmed with R_{in} up to the limit for self-oscillation with r_3 disconnected.

Comment

The network in Fig. 1 can be compensated for by connecting a small capacitor c across R_1 or R_3, as proposed by Thomas [5]. If $R_1 = R_2 = R_3 = r_2 = R$, $C_1 = C_2 = C$ (suitable standard design), and $r_5 = r_4 = \infty$, then, for $\omega_{11} = \omega_{12} = \omega_{13} = \omega_1$,

$$k = 4 + \frac{C_3}{C} - \frac{c\omega_1}{C\omega_{0p}} \tag{19}$$

and

$$k = 0, \quad \text{for } c = C\left(4 + \frac{C_3}{C}\right)\frac{\omega_{0p}}{\omega_1}. \tag{20}$$

As c and ω_1 differ in their temperature dependence, however, this compensation provides no temperature compensation, which is provided by the network in Fig. 2.

Results of Measurements

The results shown below are some of those obtained from more extensive measurements and calculations reported in [6].

Fig. 4 shows the drastic difference in temperature stability between the networks in Figs. 1 and 2. Both networks are designed for $Q_{0p} = 50$, $f_{0p} = \omega_{0p}/2\pi = 10$ kHz, and with a zero on the $j\omega$-axis at 7.5 kHz, which gives $R_1 = R_2 = R_3 = r_2 = 10$ kΩ, $r_1 = 18$ kΩ, $r_3 = 500$ kΩ, $C_1 = C_2 = C_3 = 1.5$ nF, and $r_4 = r_5 = \infty$. The amplifiers used are the μA 741 with $f_1 = \omega_1/2\pi \simeq 500$ kHz. The ordinary network (Fig. 1), without compensation according to (20), would be unconditionally unstable according to (2) and (11) for pole frequencies:

$$f_{0p} > \frac{f_1}{5Q_{0p}} = 2 \text{ kHz}. \tag{21}$$

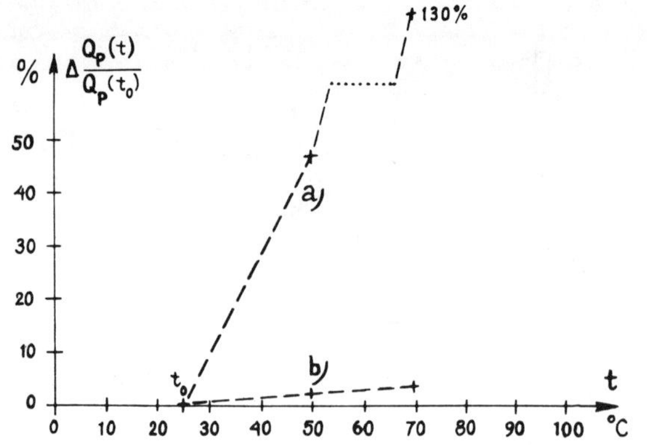

Fig. 4. Temperature dependence of networks with a pole pair at 10 kHz, $Q_p(t_0) = 50$, and a zero pair on the $j\omega$-axis at 7.5 kHz. a) Network according to Fig. 1 and compensated with a capacitor c according to (20). b) Network and compensation according to Fig. 3 and (18).

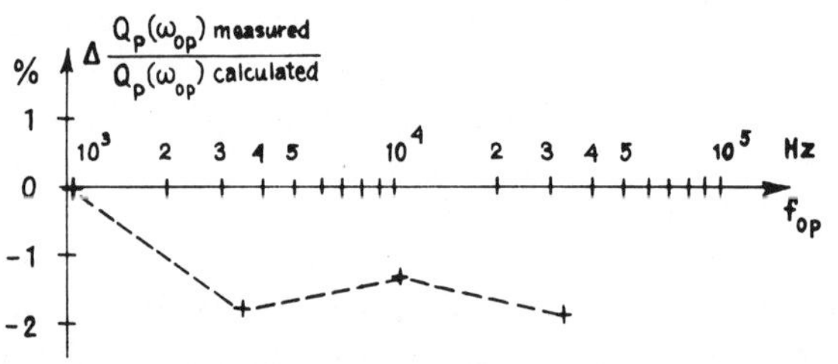

Fig. 5. Change in Q-value in percent when $f_{op} (= \omega_{op}/2\pi)$ is changed from 1 to 30 kHz by changing the capacitors.

The network has therefore been compensated for according to (20) for f_{op} = 10 kHz, which gives

$$c \simeq 150 \text{ pF}.$$

The network with modified amplifier connection has been compensated for as shown in Fig. 3 [(18)], which ideally would give

$$R_{in} \simeq 10 \text{ k}\Omega.$$

Measurement of the bandwidth of the operational amplifiers, however, showed that the choice should be

$$R_{in} = 7.9 \text{ k}\Omega.$$

Fig. 5 shows how good the compensation can be for the network in Fig. 3. On variation of the pole frequency (replacement of C_1, C_2, and C_3) from 1 to 30 kHz, Q_p varied by less than 2 percent.

Comment

The network in Fig. 2 requires stricter compensation of the operational amplifiers A_1 and A_2 than the network in Fig. 1 to be stable from the high-frequency aspect. This is because A_1 and A_2 in Fig. 2 form part of a closed loop without any of the amplifiers being fed back with a capacitor which short-circuits at high frequencies (cf., Fig. 1). Practical tests show that most makes of the amplifier type 741, e.g., Fairchild, SG, and SGS, are sufficiently compensated for the network in Fig. 2. The Texas Instruments amplifier functions very well in Fig. 1, but in Fig. 2 it produced throughout a high-frequency oscillation of about 1 MHz on node 3 and 5. The high-frequency oscillation is limited in amplitude by the slew rate of the amplifiers, and does not appear to disturb the function of the filter to any great extent. The high-frequency oscillation is generally not observable on the output–node 7–as it is heavily damped by C_2.

At high Q-values and high pole frequencies, the networks often have a stable and an unstable mode. In their unstable mode the filter self-oscillates with maximum amplitude at a frequency close to the pole frequency, probably due to nonlinear effects in the amplifiers. In this case, the network in Fig. 2 performs better than that in Fig. 1. Self-oscillation may arise when the main voltage is switched on, or in the case of too high an input signal to the filter. By connecting in Fig. 2 a couple of silicon diodes in parallel, but with opposite polarity, across the input of amplifier A_2 and/or A_1, a temporarily too high input signal can be prevented without affecting the normal function of the filter. Practical trials show that the unstable mode is entirely eliminated in the network in Fig. 2 with μA 741 amplifiers at, for example, $Q = 50$, $f_p = 10$ kHz, in which a couple of silicon diodes have been connected across the input of A_2. For further details on this point, see [2] and [6].

The usual method of realizing active filters is the cascade connection of second-order networks. It is known, however [7], [8], that no cascade connection can compete with resistively matched double-terminated LC networks with regard to the sensitivity to variations in passive components. This applies especially to higher order bandpass filters. Solutions for active RC bandpass filters must therefore be sought among structures which simulate currents and voltages in a resistively terminated LC network [9]. We have developed a couple of such methods using the network of Fig. 2 as a building block with excellent performance [7].

Acknowledgment

The authors wish to thank Prof. L. H. Zetterberg for his encouragement and support in this work, and K. Engström for his practical design and measurements.

References

[1] D. Åkerberg and K. Mossberg, "Low sensitive easily trimmed standard building block for active RC-filters," *Electron. Lett.*, vol. 5, pp. 528–529, 1969.

[2] D. Åkerberg, "Comparison of methods for active RC synthesis. Part III–Measurements and extended theory," Telecommunication Theory, Royal Inst. Technol., Stockholm, Sweden, Tech. Rep. 31, Sept. 1970.

[3] W. J. Kerwin, L. P. Huelsman, and R. W. Newcomb, "State-variable synthesis for insensitive integrated circuit transfer function," *IEEE J. Solid-State Circuits*, vol. SC-2, pp. 87–92, Sept. 1967.

[4] R. Tarmy and M. S. Ghausi, "Very high Q insensitive active RC networks," *IEEE Trans. Circuit Theory*, vol. CT-17, pp. 358–367, Aug. 1970.

[5] L. C. Thomas, "The Biquad: Part I–Some practical design considerations," *IEEE Trans. Circuit Theory*, vol. CT-18, pp. 350–357, May 1971.

[6] K. Engström, "DIG-nät med kommersiella operationsförstärkare," Telecommunication Theory, Royal Inst. Technol., Stockholm, Sweden, Tech. Rep. 42, Jan. 1971.

[7] D. Åkerberg and K. Mossberg, "Comparison of methods for active RC synthesis. Part IV–High order networks," Telecommunication Theory, Royal Inst. Technol., Stockholm, Sweden, Tech. Rep. 47.

[8] H. J. Orchard, "Inductorless filters," *Electron. Lett.*, vol. 2, pp. 224–225, June 1966.

[9] F. E. J. Girling and E. F. Good, "The leapfrog or active ladder synthesis," *Wireless World*, pp. 341–345, July 1970.

Active Adjustable Loss and Delay Equalizers

PAUL E. FLEISCHER, MEMBER, IEEE

***Abstract*—The design of active amplitude and delay equalizers to be used in private line data systems is considered. These equalizers provide sufficient flexibility of adjustment so that one type of amplitude equalizer and one type of delay equalizer are sufficient for all applications. They replace families of passive equalizers which permitted relatively limited or no adjustment.**

I. INTRODUCTION

THE PAST five years have witnessed a tremendous growth in the development and use of active filters. This activity has been spurred primarily by the availability of good, inexpensive IC operational amplifiers. Active *RC* filters in the audio band have many advantages when compared to their passive counterparts. Elimination of the inductors permits the realization of higher, more stable *Q*'s, and also leads in many cases to filters which are cheaper, smaller, more easily manufactured, and more readily tuned.

The literature reflects the extensive activity in the field and, to date, a very large number of different circuit topologies have been proposed for realizing different classes of transfer functions. Most of these are designed to accentuate one or several desirable properties of active filters. In particular, the ease and convenience of tuning a circuit are often considered important benefits. The easy tunability can be directly traced to two properties of active *RC* filters. First, the fact that transfer function parameters such as resonant frequency and *Q* are adjustable by varying resistors only and, second, because operational amplifiers provide buffering, many circuits permit a single resistor to control a single circuit parameter.

The purpose of this paper is to show that this adjustability, when fully exploited, can lead to highly desirable equalizer topologies. This represents an important new application for active filters since equalizers are extensively used for reducing the amplitude and/or delay distortions

Manuscript received March 12, 1974. This paper was presented at the 1974 IEEE International Symposium on Circuits and Systems, San Francisco, Calif., April 22–25, 1974.

The author is with Bell Laboratories, Holmdel, N. J. 07733.

Reprinted from *IEEE Trans. Circuits Syst.*, vol. CAS-21, pp. 527-531, July 1974.

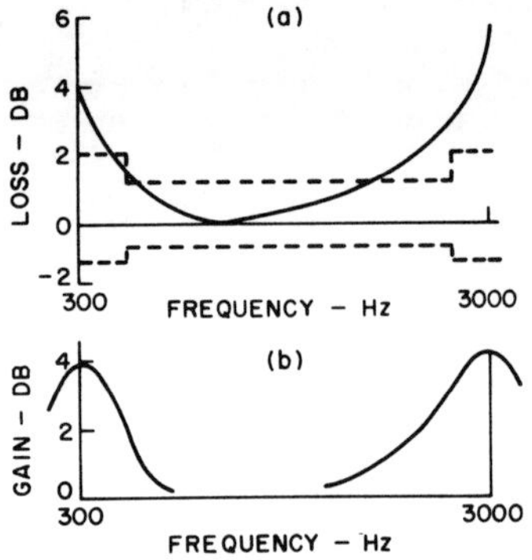

Fig. 1. Amplitude equalization. (a) Typical facility characteristics and requirements. (b) Bump equalizers.

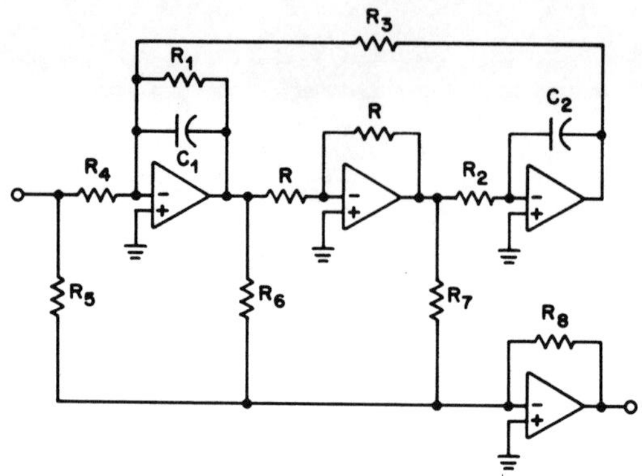

Fig. 2. Basic equalizer topology.

of various transmission systems. The equalizers to be described here[1] were developed in response to a need to equalize private line data circuits to fall within stringent limits. This function was previously fulfilled by using families of passive loss and delay equalizers. While passive equalizers performed quite well in most applications, their use, nevertheless, had several shortcomings. First, due to the limited adjustability available from the passive shaping networks, it is difficult to equalize to tight tolerances and may require a large number of equalizers. Second, a complete stock of a large number of different equalizers has to be maintained. Finally, the passive networks, while providing frequency shaping, also introduce a flat loss which requires the use of an amplifier to recover the original signal level. The active realizations overcome all these shortcomings. As will be seen, a single amplitude equalizer and a single delay equalizer provide enough flexibility and adjustability to act as general-purpose equalizers for all applications.

II. AMPLITUDE EQUALIZER

A large portion of data transmission takes place over the voice frequency band, i.e., from 300 Hz to 3 kHz. In order to improve data transmission, the amplitude and envelope delay distortion in this band must be controlled more stringently than for ordinary voice transmission. It is the function of equalizers to smooth out these amplitude and delay variations. The amplitude response of a typical voice frequency channel is shown in Fig. 1(a). Superimposed is shown a typical set of transmission requirements which must be satisfied through equalization. While these "conditioning requirements" vary, depending on the application and the number of links between the terminals their shape is always similar to the one shown.

One suitable equalizer shape which readily suggests itself is a gain bump in the frequency domain. As shown in Fig. 1(b), two properly selected gain bumps, one at low frequencies and one at high frequencies, does produce the desired overall equalization. Extensive simulations using a large number of randomly selected transmission facility characteristics have verified that bump equalizers are entirely satisfactory provided that a fairly fine adjustment of the bump height and bump width is available. It was also found that occasionally there was a need for bumps in the middle of the band to "mop up" residual errors left after the major portion of the equalization was completed. For this reason, it was highly desirable to provide a bump center frequency which covered the whole audio range, not just its extremes. Also, some cases were found where it was necessary to provide a modest loss bump rather than the more usual gain bump.

To tie these ideas down, the following family of equalizer transfer functions was found to be adequate:

$$T(s) = \frac{s^2 + hbs + \omega_0^2}{s^2 + bs + \omega_0^2} \tag{1}$$

where the resonant angular frequency ω_0, the bandwidth b, and the bump height h must each be adjustable.

To realize this biquadratic transfer function, the Biquad [1], [2] was chosen. This topology, in contrast to most others, permits the independent adjustment of the three required parameters through resistive tuning. A simplified circuit which served as the starting point for realizing $T(s)$ is shown in Fig. 2. The transfer function of this circuit is given by

$$T(s) = -\frac{R_8}{R_5} \cdot \frac{s^2 + [1 + (R_1/R_4)(R_5/R_7 - R_5/R_6)](1/R_1C_1)s + 1/R_2R_3C_1C_2}{s^2 + (1/R_1C_1)s + 1/R_2R_3C_1C_2}. \tag{2}$$

Note that the resonant frequency is adjustable by means of R_2 or R_3, while the height is adjustable from loss through gain by either R_7 or R_6. The bandwidth is nominally adjustable by R_1; however, in order to prevent the bandwidth adjustment from affecting the height, it is necessary to have R_4 track R_1. The manner in which this is realized will be discussed shortly. Because no flat gain or loss was desired, R_8 was set equal to R_5. Also, to simplify the height adjustment, R_6 was set equal to R_5. With these simplifications and assuming that R_4 tracks R_1, the desired transfer

[1] The loss and delay equalizers discussed here are the subject of U. S. Patents 3 715 679 and 3 715 680 issued on February 6, 1973.

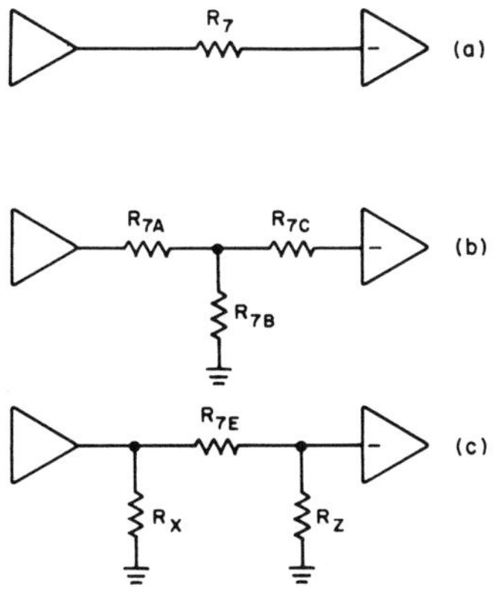

Fig. 3. Transformation of height-controlling resistor. (a) Original circuit. (b) T network replacement. (c) π network equivalent.

function is obtained:

$$T(s) = -\frac{s^2 + (R_5/R_7)(1/R_1C_1)s + 1/R_2R_3C_1C_2}{s^2 + (1/R_1C_1)s + 1/R_2R_3C_1C_2}. \quad (3)$$

Although this solves the design problem in principle, many modifications were necessary to produce a design suitable for the desired physical realization. The principal physical requirement was that the circuit contain no potentiometers so that the three adjustable parameters are easily and accurately settable. The circuit was therefore realized using thin-film resistors on a ceramic substrate which is packaged in a plastic box. Specially designed slide switches make contact with rhodium-plated contacts on the substrate and selectively short out the thin-film resistors. In effect, this produces an adjustable two-terminal potentiometer which has discrete settings available.

Similar discrete variable resistors are used to adjust all three parameters in the circuit. However, because of the large number of discrete settings desired for each parameter, it is impractical to use only one resistor to adjust that parameter. Thus, to adjust the resonant frequency, both R_2 and R_3 are adjustable. By providing 6 discrete settings for each of these resistors, 36 discrete resonant frequencies are obtained.

It was not so straightforward to provide the 36 discrete settings for the height adjustment. Although R_6 and R_7 both control height, the adjustment pattern they provide is not sufficiently uniform. For this reason, R_6 was fixed as stated before. The resistor R_7, on the other hand, was replaced by a T network of resistors as shown in the circuit diagram of Fig. 3(b). The effect of introducing this T network is readily assessed by performing the T to π transformation shown in Fig. 3(c). This shows that the T network is equivalent to the original single resistor except for the addition of the "superfluous" resistors R_x and R_z. The resistor R_x is an extra load on the operational amplifier and is negligible unless it becomes too small. R_z being between the inverting terminal and ground is also negligible, although if it becomes small, it has several adverse effects on the performance of the operational amplifier. First, it lowers the effective gain of the operational amplifier, and second, it enhances the noise and offset voltages produced by that op amp. However, by careful design, R_x and R_z can both be maintained sufficiently large to prevent these effects from becoming significant. With these precautions then, we have produced an equivalent R_7 which is controllable by more than one resistor. The effective resistance is readily shown to be

$$R_{7E} = \frac{R_{7A}R_{7B} + R_{7B}R_{7C} + R_{7C}R_{7A}}{R_{7B}}. \quad (4)$$

In order to best satisfy the diverse requirements on R_x and R_z, it was decided to fix R_{7A} and to provide R_{7B} and R_{7C} each with 6 discrete settings. In this way, 36 discrete height settings are also made available. The advantage of the T network is that it permits the height settings in decibels to be almost uniformly distributed. To see how this is achieved, consider the following form of (4):

$$R_{7E} = R_{7A} + R_{7C}\left(1 + \frac{R_{7A}}{R_{7B}}\right) \quad (5)$$

and let the 6 discrete values of R_{7B} and R_{7C}, respectively, be given by

$$R_{7B}^{(i)} = \frac{R_{7A}}{K_1(1+\delta)^i - 1}, \qquad i = 0,1,\cdots,5 \quad (6)$$

and

$$R_{7C}^{(j)} = K_2(1+\delta)^{6j}, \qquad j = 0,1,\cdots,5. \quad (7)$$

Then, as a function of the switch settings i and j, R_{7E} becomes

$$R_{7E} = R_{7A} + K_1K_2(1+\delta)^{6j+i}, \qquad i = 0,1,\cdots,5; \quad j = 0,1,\cdots,5. \quad (8)$$

Note that except for the fixed portion represented by R_{7A}, the equivalent resistance has an exponential characteristic. Thus, the height which is given by R_5/R_7 is nearly exponential in character so that, when expressed in decibels, one obtains a nearly constant increment between settings.

To obtain 36 discrete settings of bandwidth, the same kind of T network is used to replace R_1. The one additional complication is the necessity of having R_4 track R_1. This problem is neatly solved by connecting R_4 to the junction of the three resistors forming the T network corresponding to R_1, as shown in Fig. 4(a). This produces a resistive star topology which may be replaced [3], for purposes of analysis, by an equivalent circuit consisting of resistors connected between every pair of terminal nodes, as shown in Fig. 4(b). In this case, four parasitic resistors are generated: R_X, R_Y, and R_W represent loads on the operational amplifier shown as well as the driving source (assumed to be another op amp), while R_Z loads the inverting input of the op amp. The conditions under which the presence of these resistors is negligible have already been discussed.

Consider now the other two equivalent resistors, R_{1E} and R_{4E}. It may readily be shown that

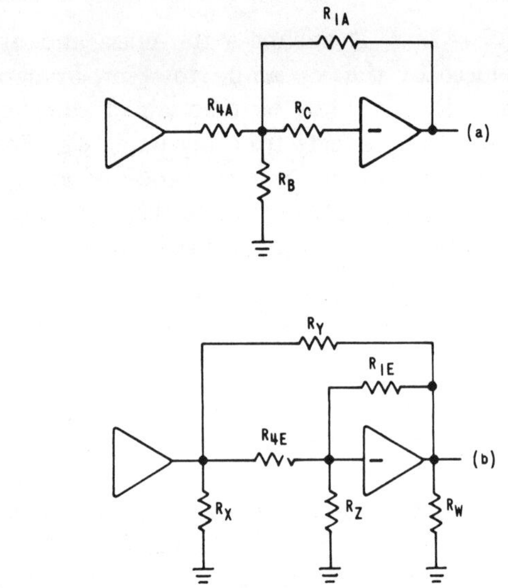

Fig. 4. Transformation of bandwidth-controlling resistor. (a) Star circuit. (b) Mesh equivalent.

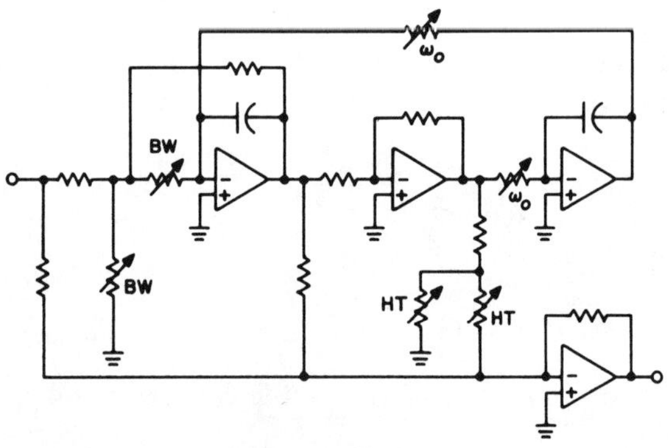

Fig. 5. Amplitude equalizer circuit showing adjustable resistors.

$$R_{1E} = R_{1A} + R_C + \frac{R_{1A}R_C}{R_B} + \frac{R_{1A}R_C}{R_{4A}} \tag{9}$$

and

$$R_{4E} = R_{4A} + R_C + \frac{R_{4A}R_C}{R_B} + \frac{R_{4A}R_C}{R_{1A}}. \tag{10}$$

Thus, when $R_{4A} = R_{1A}$, the equivalent resistors corresponding to R_4 and R_1 track:

$$R_{1E} = R_{4E} = R_{1A} + R_C\left(2 + \frac{R_{1A}}{R_B}\right). \tag{11}$$

Furthermore, by comparing (11) to (5), it is clear that by proper design of the adjustable resistors R_B and R_C, the equivalent resistance, and therefore the bandwidth, can be made to increase in approximately constant percent increments. Note by the way, that the connections of C_1 and R_3 are not being changed as a result of these transformations.

The resulting amplitude equalizer circuit diagram is shown in Fig. 5 where each of the 6 variable resistors has 6 discrete settings. This permits ω_0, BW, and height each to assume 36 independent discrete values, resulting in more than 45 000 different shapes being available from a single equalizer. Because of the very large number of settings provided, a computer program has been prepared [4] to aid in the use of these equalizers. The program requires as input a description of the facility to be equalized and the requirements on equalization. The program will then compute the number of equalizers needed and their optimal settings.

A picture of the amplitude equalizer is shown in Fig. 6.

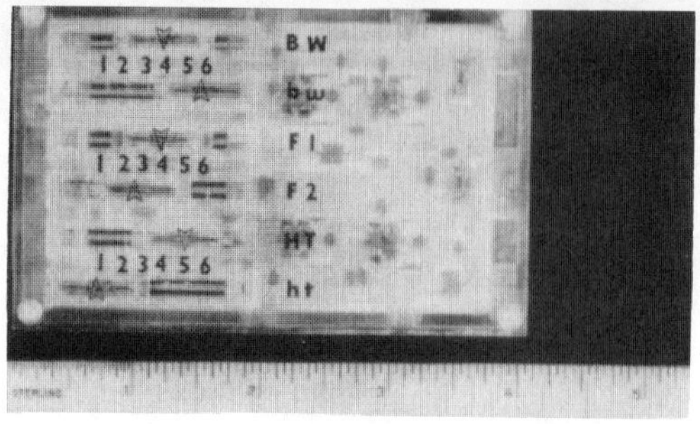

Fig. 6. Amplitude equalizer.

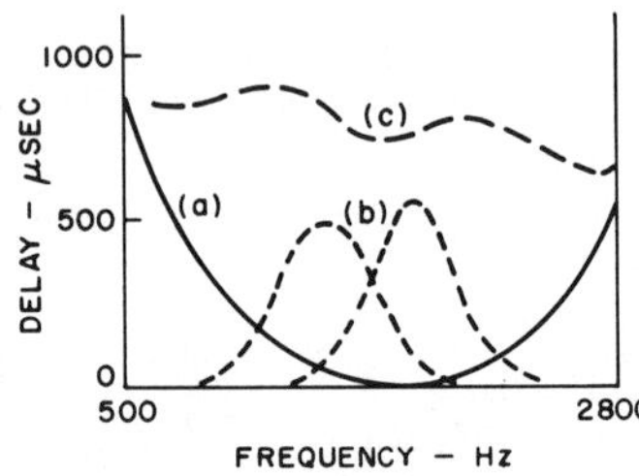

Fig. 7. Delay shapes. (a) Facility delay. (b) Equalizer delay. (c) Residual delay.

III. DELAY EQUALIZER

Most data transmission systems operating in the voice frequency band also require delay equalization. The delay distortion of a typical transmission facility is shown in Fig. 7. Although the requirements for the equalized facility vary, depending on the application, they always establish limits on the peak-to-peak envelope delay distortion permitted in various frequency bands. For example, a typical requirement is the so-called $C2$ conditioning which requires envelope delay distortion of less than 500 μs from 1000 to 2600 Hz, less than 1500 μs from 600 to 2600 Hz, and less than 3000 μs from 500 to 2800 Hz.

In contrast to amplitude equalization where loss is equalized by the addition of gain, delay equalization is achieved by the addition of delay bumps to the facility delay so as to flatten out the overall delay. This is pictorially demonstrated in Fig. 7.

The basic transfer function for delay equalization has traditionally been the biquadratic all-pass function

$$T(s) = \frac{s^2 - bs + \omega_0^2}{s^2 + bs + \omega_0^2}. \tag{12}$$

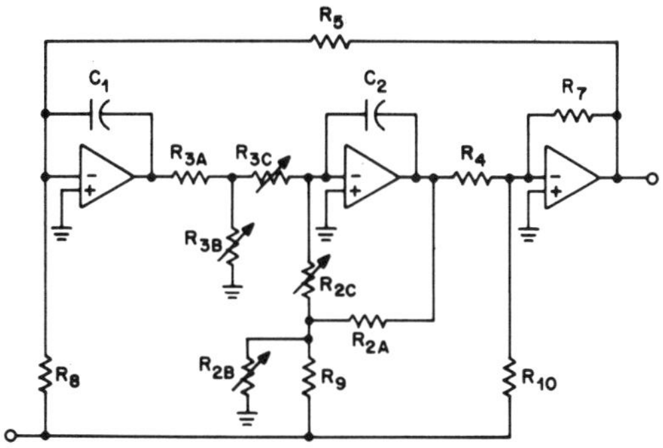

Fig. 8. Delay equalizer circuit showing adjustable resistors.

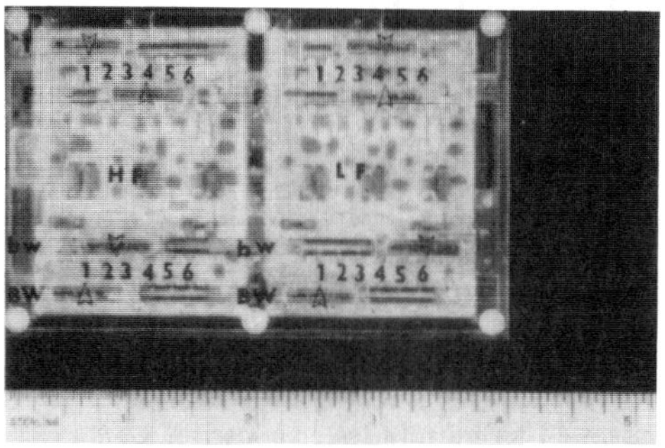

Fig. 9. Delay equalizer.

The delay corresponding to this transfer function is

$$D(j\omega) = \frac{2}{b} \frac{1 + (\omega_0/\omega)^2}{1 + (\omega_0/b)^2(\omega/\omega_0 - \omega_0/\omega)^2}. \quad (13)$$

This function has the general bump shape depicted in Fig. 7 with a maximum approximately at $\omega = \omega_0$ and a width controlled by the bandwidth b. Note that the delay equalizer has only two adjustable parameters.

A straightforward realization of this transfer function could be obtained using the amplitude equalizer topology (Fig. 2) by properly choosing fixed height control resistors R_6 and R_7. Alternative circuits for delay equalization which also use four operational amplifiers and employ capacitive switching have been reported in the literature [5]. A more economical realization which has only three operational amplifiers and retains the resistive switching property is available by using the "feed-forward" topology [6]. Such a circuit is shown in Fig. 8. It is readily shown that the transfer function of this circuit is given by

$$T(s) = -\frac{s^2 - (1/R_2C_2)s + (2/R_3R_5C_1C_2)}{s^2 + (1/R_2C_2)s + (2/R_3R_5C_1C_2)} \quad (14)$$

when $R_7 = R_{10} = 2R_4$, $R_5 = R_8$, and $R_2 = R_9$. In order to obtain the desired adjustments of the parameters, the resistor R_3 has been converted to a T network to provide 36 discrete settings of ω_0. Also, the bandwidth adjustment is provided by the two variable "R_2" resistors. As in the case of the amplitude equalizer, the required R_2 to R_9 tracking is obtained by the star connection shown.

In the physical realization, two delay sections are included in each equalizer box, one covering the lower half of the audio band, while the other covers the upper half. In each section, 36 discrete resonant frequencies and 36 discrete bandwidths are made available. A complete delay equalizer thus has eight 6-position switches and provides more than 1 600 000 different discrete settings. The aforementioned program may be used for designing the optimal settings of the delay equalizer also.

A picture of the delay equalizer is shown in Fig. 9.

IV. SUMMARY

It has been shown that the flexibility and adjustability of active *RC* filters make them particularly suitable for equalizer applications. The amplitude and delay equalizers presented here, which have been released for manufacture, were designed to perform the functions of sizeable families of passive equalizers which were either fixed or had limited adjustability. Besides a significant reduction of the "inventory" problem, these equalizers, because of their fine adjustment capability, are able to equalize to much tighter requirements.

ACKNOWLEDGMENT

The author would like to acknowledge the helpful interactions and discussions with many of his colleagues in the course of this project. In particular, thanks are due to L. C. Thomas, C. F. Simone, K. W. Sussman, and F. E. Stehlik. D. Padnes is responsible for the excellent mechanical design of these equalizers.

REFERENCES

[1] J. Tow, "A step-by-step active filter design," *IEEE Spectrum*, vol. 6, pp. 64–68, Dec. 1967.

[2] L. C. Thomas, "The Biquad: Part I—Some practical considerations" and "The Biquad: Part II—A multipurpose active filtering system," *IEEE Trans. Circuit Theory*, vol. CT-18, pp. 350–357 and pp. 358–361, May 1971.

[3] L. Versfeld, "Remarks on star mesh transformation of electrical networks," *Electron. Lett.*, vol. 6, Sept. 17, 1970.

[4] D. B. Snyder, unpublished work.

[5] A. G. J. Holt and J. P. Gray, "Tunable active allpass sections," *Electron. Lett.*, vol. 4, Jan. 26, 1968.

[6] P. E. Fleischer and J. Tow, "Design formulas for biquad active filters using three operational amplifiers," *Proc. IEEE* (Lett.), vol. 61, pp. 662–663, May 1973.

Minimally Active *RC* Variable Equalizers

FRANC BRGLEZ, MEMBER, IEEE

Abstract—Novel variable equalizer structures[1] based on the concept of inductorless simulation of classical Bode variable equalizers are discussed. It is shown that one can remove the need for negative resistance (in a full range equalizer) by switching between negative and positive feedback, thus saving an operational amplifier. A number of typical equalizer responses with a variable range of ± 10 dB are realized with *RC* components and a single operational amplifier. The measured responses are shown to agree closely with the computer-simulated nominal network responses. In a particular example, a ± 10 dB variable bump equalizer is realized with only two operational amplifiers in contrast to four required in a recent realization.

I. Introduction

RECENTLY [1], [2], new network structures have been proposed which realize transmission properties of the classical Bode equalizer [3], [4] without inductors. The general transfer function of these structures has the form

$$T(s) = \frac{1 - yH(s)}{1 + yH(s)} \tag{1}$$

where

$$|H(\omega)| \leq 1 \qquad |y| \leq 1.$$

A network example from [1], [2] that realizes (1) is shown in Fig. 1(a). This realization is based on the original unsymmetrical lattice network in [1], [2]. Realizing transmission properties of an unsymmetrical lattice in unbalanced form requires at least one controlled source, hence the need for the operational amplifier 1 in Fig. 1(a). The function of the second amplifier in Fig. 1(a) is to create a negative resistance that produces the inverse of (1) and thus extends the variable range of the equalizer to its full value.

This paper shows that it is possible to remove the need for negative resistance and the associated operational amplifier by incorporating a simple switch that switches between an all-negative and a combination of negative and positive feedback structure. Two such networks are shown in Figs. 1(b) and 2(a), and their respective nominal transfer functions are

$$T(s) = -\frac{1 - yH(s)}{1 + yH(s)}, \text{ SW connects to 1}$$

$$T(s) = -\frac{1 + yH(s)}{1 - yH(s)}, \text{ SW connects to 2} \tag{2}$$

where

$$y = R_v/R_0 \qquad 0 \leq R_v \leq R_0$$

$$H(s) = 1/(1 + Z_1(s)/R_0), \qquad \text{(Fig. 1(b))}$$

or

$$y = G_v/G_0 \qquad 0 \leq G_v \leq G_0$$

$$H(s) = 1/(1 + Y_2(s)/G_0), \qquad \text{(Fig. 2(a))}.$$

Note that by removing the operational amplifier that forms negative resistance in Fig. 1(a), one arrives at the single amplifier structure shown in Fig. 1(b) with switch SW connected to position 1. Nominally, sensitivity performance of both networks in Fig. 1(a) and (b) is identical, however in practice the new network in Fig. 1(b) is preferable.

II. Stability Considerations

Networks in Figs. 1(b) and 2(a) are not the only ones that have the transfer properties of (2). It can be verified that interchanging R_v with $(Z_1(s) + R_0)$ in Fig. 1(b) or G_v with $(Y_2(s) + G_0)$ in Fig. 2(a) will also result in (2), however both networks would be unstable with SW connected to position 2. This phenomena has been noted also in the laboratory and can be explained by observing the respective input impedances.

The input impedance of the network in Fig. 1(b) is given below (SW connecting to 2):

$$Z_{i1} = R_a + Z_{a-a'} = R_a\left(1 - \frac{R_v}{R_0} \cdot \frac{1}{1 + Z_1/R_0}\right),$$

$$\mathrm{Re}\,[Z_{i1}(\omega)] \geq 0 \text{ for } R_v \leq R_0 \tag{3}$$

and for the network in Fig. 2(a) (SW connecting to 2)

$$Z_{i2} = R_a + Z_{b-b'} = R_a\left(1 - \frac{G_v}{G_0} \cdot \frac{1}{1 + Y_2/G_0}\right),$$

$$\mathrm{Re}\,[Z_{i2}(\omega)] \geq 0 \text{ for } G_v \leq G_0. \tag{4}$$

Interchanging impedances in Fig. 1(b) results in the impedance below (SW connecting to 2):

$$Z_{i3} = R_a\left(1 - \frac{R_0}{R_v}(1 + Z_1/R_0)\right),$$

$$\mathrm{Re}\,[Z_{i3}(\omega)] \leq 0 \text{ for } R_v \leq R_0 \tag{5}$$

and in Fig. 2(a) (SW connecting to 2)

$$Z_{i4} = R_a\left(1 - \frac{G_0}{G_v}(1 + Y_2/G_0)\right),$$

$$\mathrm{Re}\,[Z_{i4}(\omega)] \leq 0 \text{ for } G_v \leq G_0. \tag{6}$$

Manuscript received December 9, 1974.

The author is with Transmission Network Development, Bell-Northern Research, Ottawa, Ont., Canada.

[1] Canadian and U.S. patent application pending.

Reprinted from *IEEE Trans. Circuits Syst.*, vol. CAS-22, pp. 688-691, Aug. 1975.

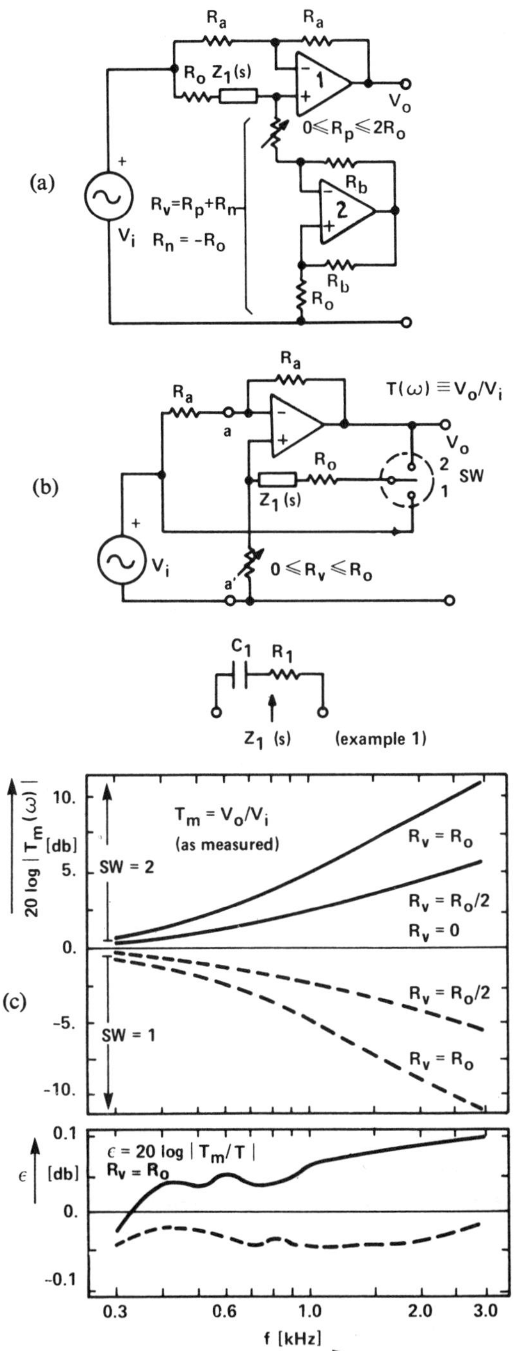

Fig. 1. (a) Two-amplifier, single impedance variable equalizer from [1], [2]. (b) Single-amplifier, single impedance variable equalizer with a switch. (c) Performance evaluation responses of Example 1. (MOT 1458 operational amplifier is used in all examples.)

The expressions in (3) and (4) show that the equalizers in Figs. 1(b) and 2(a) are short-circuit stable for the complete variable range of R_v or G_v, respectively, while networks with solutions in (5) and (6) are not.

III. Example 1

This example is based on the network in Fig. 1(b) and, for the sake of illustration, employs the simplest *RC* impedance to equalize the upper bandedge frequency response over a range of ± 10 dB. The measured responses are shown in Fig. 1(c). The *RC* impedance used is the same as in the two-operational amplifier circuit in Fig. 1(a) and described in [1], [2]. Comparison of error curves in Fig. 1(c) with the ones shown in [1], [2] shows that the performance of both networks (in Fig. 1(a) and (b)) is very close to its nominal response $T(\omega)$ given by (2). As in [1], [2], the response varies very predictably with variations of R_v.

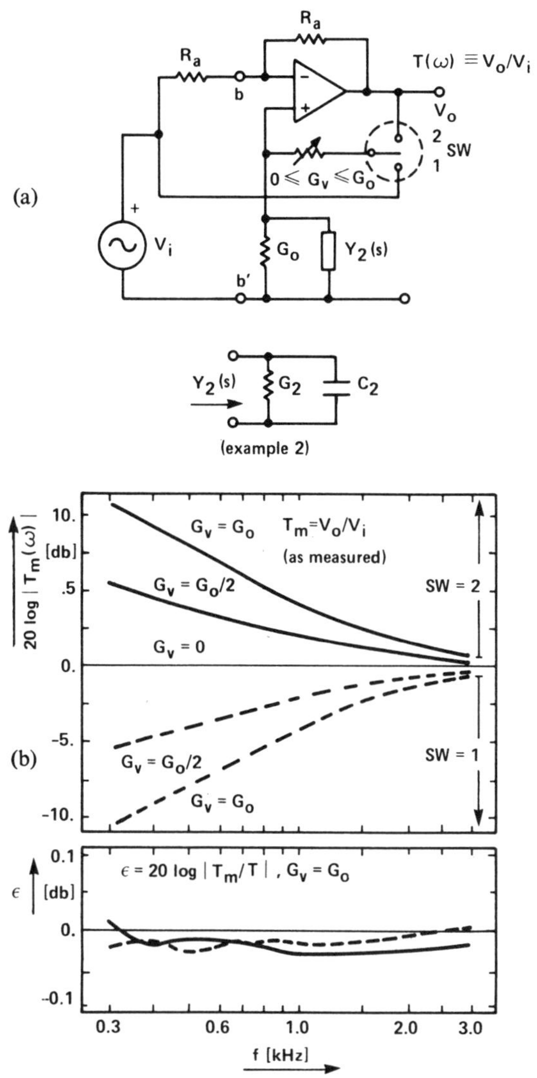

Fig. 2. (a) Single-amplifier, single admittance variable equalizer with a switch. (b) Performance evaluation responses of Example 2.

IV. Example 2

This example is based on the network in Fig. 2(a) and the simple *RC* admittance is used to equalize the lower bandedge frequency response over a range of ± 10 dB. The error curves in Fig. 2(b) show that the measured responses (obtained without tuning) are in close agreement with the nominal response $T(\omega)$ in (2) at both extreme settings ($G_v = G_0$, SW = 1; $G_v = G_0$, SW = 2). Comparison of measured versus calculated responses shows equally good agreement for $G_v < G_0$.

V. Example 3

In this example we use the structure in Fig. 2(a) to realize a very simple yet wide-range resistively tunable

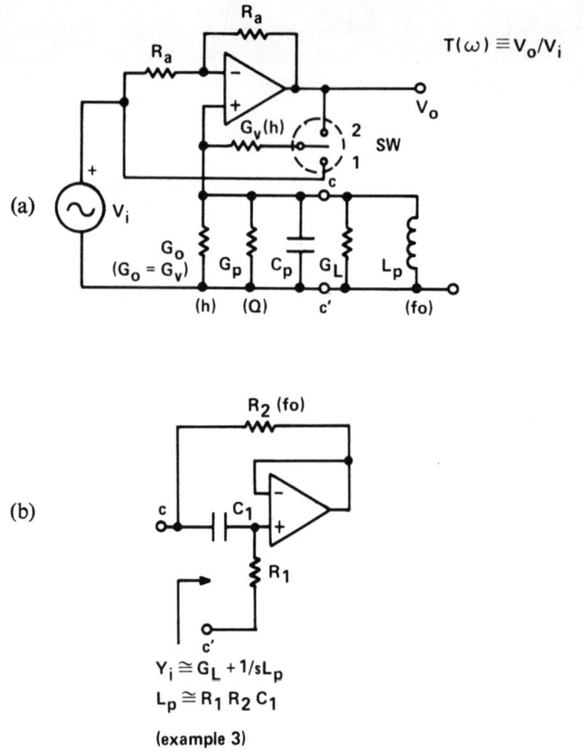

Fig. 3. (a) Variable bump equalizer based on Fig. 2(a). (b) A simple single-amplifier inductance simulation network used in Example 3.

bump equalizer. The network under consideration is shown in Fig. 3(a) where the representation of the lossy inductance $G_L + 1/sL_p$ needs to be considered only as symbolic and for purpose of analysis. In practice, the lossy inductance can be replaced by several single or two-amplifier circuits [5]–[7], depending on the required value of the inductor Q. The transfer function of the network in Fig. 3(a) is

$$T(s) = -\frac{s^2 + (h\omega_0/Q)s + \omega_0^2}{s^2 + (\omega_0/Q)s + \omega_0^2}, \qquad SW = 2$$

and

$$T'(s) = 1/T(s), \qquad SW = 1 \tag{7}$$

where

$$\omega_0^2 = \left(\frac{1}{C_p}\right)\cdot\frac{1}{L_p} = f_1(R_{1x}R_{2x}) \text{ (in active realizations)}$$

$$Q = \frac{\omega_0 C_p}{G_L + G_P + G_0 - G_v} = \left.\frac{(\omega_0 C_p)}{G_L + G_P}\right|_{G_0=G_v} = f_2(G_P)$$

$$h = \frac{G_L + G_P + G_0 + G_v}{G_L + G_P + G_0 - G_v} = 1 + \left.\frac{2G_v}{G_L + G_P}\right|_{G_0=G_v}$$

$$= f_3(G_v).$$

It is apparent from (7) that by providing tracking between G_0 and G_v the network in Fig. 3 can be tuned by sequentially adjusting L_p (a resistor in active realizations) for center frequency $f_0 = \omega_0/2\pi$, G_p for the Q of the bump frequency response and G_v for the bump height h at f_0. It should be noted from (4) that instability problem does not arise for $G_0 = G_v$.

An experimental bump equalizer has been constructed and a relatively low Q single amplifier [5] realization is used to simulate the lossy inductance as shown in Fig. 3(b). The bump equalizer specifications are

$$0.3 \text{ kHz} \leq f_0 \leq 3 \text{ kHz}$$

$$Q \leq 5$$

$$-10 \text{ dB} \leq 20 \log (h) \leq +10 \text{ dB}. \tag{8}$$

Three settings of f_0 have been used for measurements: 0.3, 1.0, and 3.0 kHz with $Q = 5$ and 20 log $(h) = \pm 10$ dB at all settings; the respective measured responses are shown in Fig. 4. Observe the close agreement between the measured and the nominal responses predicted from (7); only slight tuning for f_0 (by means of the resistor R_2 in Fig. 3(b)) was required to reduce the initial error of about ± 0.2 dB to less than ± 0.1 dB as shown in Fig. 4.

The variation of the bump height in (7) can be achieved by varying G_v alone while keeping G_0 constant or by varying G_v and $G_0 = G_v$ simultaneously. In the first case the network functions as the usual (Bode type) variable equalizer with associated minimal distortion discussed in [1], [2]; in the second case tracking between G_0 and G_v ensures the constancy of Q despite the bump height variations. A set of sample responses for ± 5 dB bump height at 1 kHz is shown in Fig. 4, and one can notice the difference between the nonconstant (dotted-line) and constant Q response. In general, network and systems specifications will determine whether the bump height shall be varied by means of G_v with $G_0 = G_v$ or, more simply, by means of G_v alone.

Potentially, the structure in Fig. 3 is capable of a very wide variable range; Q is limited by the $Q_L = 1/(\omega_0 L_p G_L)$ of the simulated inductance and the bump height h is limited by the saturation and/or instability of the amplifier

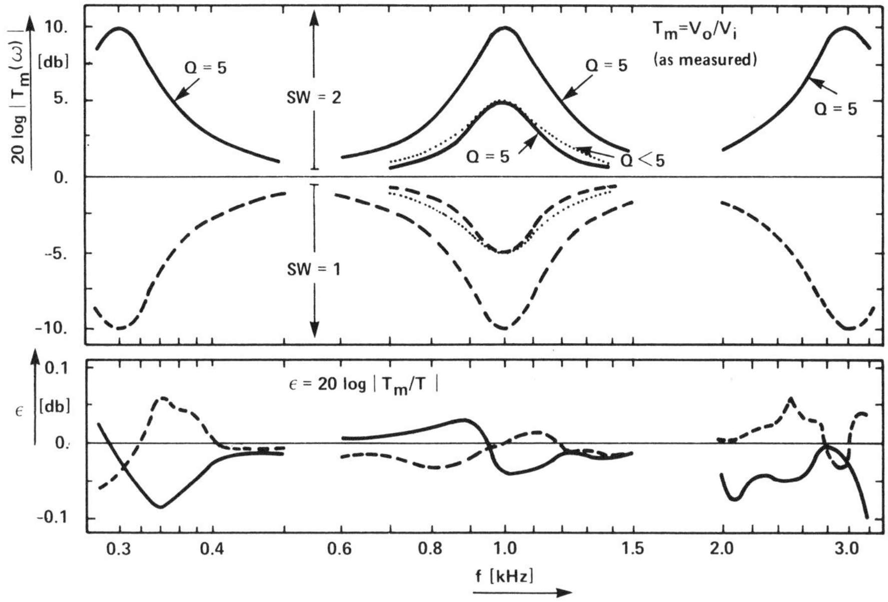

Fig. 4. Performance evaluation responses of the bump equalizer in Fig. 3 (see text).

since the amount of positive feedback increases with increasing G_v. The experimental circuit has been measured at 1 kHz with a maximum Q of 15.2 and bump height ranging from +40 dB to −40 dB.

VI. Conclusions

Very simple yet wide-range inductorless variable equalizer networks have been presented along with the experimental data that is consistently in close agreement with computer-simulated responses. A wide-range variable bump equalizer that is derived from a general structure is shown to contain only two operational amplifiers and is thus simpler and more economical than a recently published one [8] which requires four operational amplifiers. It should be noted that the realization of the bump equalizer in this paper is based on a direct *RC* simulation of a simple tuned circuit.

Acknowledgment

The author wishes to thank G. Stephen for building and testing the three prototype equalizers.

References

[1] F. Brglez, "Inductorless variable equalizers using all-pass network structures," in *Proc. 1974 IEEE Int. Symp. Circuits and Systems*, San Francisco, Calif., pp. 448–452, Apr. 22–25, 1974.
[2] ——, "Inductorless variable equalizers," *IEEE Trans. Circuits Syst.*, vol. CAS-22, pp. 415–419, May 1975.
[3] H. W. Bode, "Attenuation equalizer," U.S. Patent 2 096 027, October 19, 1937.
[4] ——, "Variable equalizers," *Bell Syst. Tech. J.*, vol. 17, pp. 229–244, Apr. 1938.
[5] D. F. Berndt and S. C. Dutta Roy, "Inductor simulation using a single unity gain amplifier," *IEEE J. Solid-State Circuits*, vol. SC-4, pp. 161–162, June 1969.
[6] J. M. Rollett, "Economical *RC* active lossy ladder filters," *Electron. Lett.*, vol. 9, pp. 70–72, Feb. 1973.
[7] L. T. Bruton, "Nonideal performance of two amplifier positive-impedance converters," *IEEE Trans. Circuit Theory*, vol. CT-17, pp. 541–549, Nov. 1970.
[8] P. E. Fleischer, "Active adjustable loss and delay equalizers," *IEEE Trans. Circuits Syst.*, vol. CAS-21, pp. 527–531, July 1974.

Part III
Design of High-Order Filters

Three main methods have been proposed for the design of active filters of order $N \geq 4$: cascade, multiple feedback and ladder simulations. Each method has its advantages and disadvantages, as will be pointed out in connection with the representative papers included in this chapter.

For cascade realization, a high-order transfer function H(s) is factored into a product of second-order sections $T_i(s)$ such that $H(s) = \pi_i T_i(s)$. If H(s) is of odd order, one section is of first or third order. Which zeros and poles of H(s) are to be placed into a particular $T_i(s)$ and how the sections should best be ordered in the cascade, is a question affecting sensitivity and dynamic range (compare papers 1-7 and 1-8). With these questions being answered, each $T_i(s)$ is realized by any suitable circuit, as discussed in the previous group of papers, and all circuits are connected in tandem. Many practical problems related to this approach, such as sensitivity, tolerances, tuning and manufacturability, are discussed in *paper 3-1* by Friedenson, Daniels, Dow and McDonald who use sections related to those in paper 2-3 for the design of fifth-order channel bank filters. The paper should give the reader a good flavor for the design process that results in practical cascade filters. For some of the programs useful in this process, refer to paper 4-1.

Cascade design has a number of advantages, notably modularity, a simple design process and ease of tuning; its main drawback is the relatively large passband sensitivity to component variations which renders the performance of cascade filters of high order, say $N > 6$, very unsatisfactory. Lower sensitivities can be obtained by coupling the sections $T_i(s)$ in different multiple feedback (MF) topologies, an approach which retains the modularity of cascade filters, but complicates tuning and, in general, requires computer-aided design and optimization routines. The procedure is discussed in detail in the review *paper 3-2* by Laker, Schaumann and Ghausi. Synthesis and optimization methods for all currently advocated MF filters are described, and computed and experimental results are given which verify that the sensitivity performance is improved and approaches the low values realized by lossless ladder filters. Among MF filters, the leapfrog (LF) topology comes closest to simulating passive LC ladder filters, including their very low passband sensitivities. The close relationship between LF and LC filters is explored in *paper 3-3* by Dubois and Neirynck. The Applicability of the LF concept is generalized, finite transmission zeros are included and several heretofore open questions are answered.

The already mentioned low sensitivity of LC filters gave rise to a successful search for methods of simulating LC ladders by active networks, in the hope of inheriting the former's low sensitivities. An additional advantage of the resulting structures which can be subdivided into component simulation and operational simulation approaches, is that the active network can be obtained directly from an LC prototype. Bruton in *paper 3-4* discusses in detail the sensitivity motivation for ladder simulations and gives an outline of both approaches. The use of the general impedance converters (GICs), mentioned already in connection with paper 2-5, is emphasized both for the direct simulation of inductors and for transformations of the ladder, leading to the use of the "frequency-dependent negative resistor" (FNDR). Many practical details of this approach are discussed in paper 4-5. An operational simulation, specifically, the signal-flow-graph (SFG) implementation, is presented in *paper 3-5* by Martin and Sedra, along with a number of points of practical interest for the designer. The reader may have noticed that the SFG and LF approaches are closely related; this is illustrated further by Dimopoulos and Constantinides in *paper 3-6* where a linear transformation is introduced which leads from the original passive filter to the RC-active configuration. Depending on the choice of transformation, SFG, LF or the so-called 'wave active filters' [1] are obtained as special cases. The generality of the transformation allows variants of the above topologies to be derived which may have theoretical or practical advantages.

It should be emphasized that, in general, MF realizations and ladder simulations have sensitivity advantages over cascade filters only in the passband. In the stopband, in particular if finite transmission zeros are to be realized, cascade circuits may be no worse or even show better performance than MF or simulated ladder filters (see e.g. the data in paper 3-3!). The question of which approach may result in the lowest stopband sensitivity (see also papers 3-2, 3-3, [2] and [3] for possible methods) is addressed by Gadenz in *paper 3-7* by heuristically comparing several existing and novel design procedures.

With the increasing emphasis on ladder simulations, considerable effort has gone into the development of high-quality circuits for active implementation of FDNRs and inductors (gyrators). One of the best two-amplifier circuits for this purpose is Antoniou's GIC (compare papers 2-5 and 3-4) whose performance and imperfections are discussed by Antoniou and Naidu in *paper 3-8*. A simple compensation for amplifier-induced deviations is introduced that results in a considerable increase in the GIC's useful bandwidth. The device's performance is illustrated in the design of a channel bank filter. A simpler circuit for inductance simulation, using only one amplifier, is discussed in *paper 3-9* by Orchard and Willson. Although the quality of the realized inductor is not as good as that obtained via a two-amplifier simulation, the circuit uses fewer elements and consumes less power. A pair of circuits for realizing both FDNR's and inductors are given in *paper 3-10* by Cheng and Lin. The reader should also refer to paper 5-2 for a different approach to component simulation that results in active ladders suitable for high frequencies.

The last reprint in this group, *paper 3-11* by Bruton and Pederson, discusses an elliptic lowpass FDNR ladder filter with an interesting variation that has many applications in communication systems. The filter is made electronically tunable or adaptive by locking it to an external control signal. For other methods to achieve similar results, see [4-7]. It is pointed out here that the same principle can be used to electronically tune and stabilize integrated filters and thereby overcome some of the component tolerance limitations encountered during IC processing (see papers 5-1 and 5-4).

[1] H. Wupper and K. Meerkotter, "New active filter synthesis based on scattering parameters", IEEE Trans. on Circuits and Systems, vol. CAS-22, pp. 594-602, July 1975.

[2] J. Tow, "Some results on two-section generalized FLF active filters", IEEE Trans. on Circuits and Systems, vol. CAS-25, pp. 181-184, April 1978.

[3] P. R. Padukone, M. S. Ghausi and J. Mulawka, "Sensitivity minimization in generalized follow-the-leader feedback active filters", Proc. IEEE Int. Symp. on Circuits and Systems, pp. 801-806, April 1980.

[4] T. A. Hamilton, "Stable digitally programmable active filters", IEEE J. on Solid-State Circuits, Vol. SC-9, No. 1, pp. 27-29, February 1974.

[5] J. J. Hill, "Digital control of active filter characteristics", Int. J. of Electronics, vol. 41, No. 4, pp. 405-510, April 1976.

[6] R. W. Harris and H. T. Lee, "Digitally controlled conductance tunable active filters", IEEE J. on Solid-State Circuits, vol. SC-10, No. 2, pp. 182-185, June 1975.

[7] G. Gocal and C. S. Lindquist, "Analysis and design of tracking filters", IEEE Trans. on Circuits and Systems, vol. CAS-27, No. 1, pp. 45-50, January 1980.

Paper 3-1

RC Active Filters for the D3 Channel Bank

By R. A. FRIEDENSON, R. W. DANIELS, R. J. DOW, and P. H. McDONALD

(Manuscript received August 1, 1974)

The development of the voice-frequency active filters for the D3 channel bank is described. These filters are the first single-substrate RC active filters using thin-film tantalum RC and silicon integrated-circuit technology to be produced on a large scale by Western Electric. To create complete confidence in both the design and the new technology, an extensive model building and testing program was undertaken. In addition, continuous interaction with manufacturing engineers resulted in a design that facilitated the introduction of this new technology in a large-scale manufacturing environment.

Significant advances were made in reducing the complexity of tuning active filters. In fact, the tuning and testing procedure has been adapted for use with a totally automated computer-controlled test set. Furthermore, a Monte Carlo statistical simulation of the manufacturing process of the filters was developed. This model includes tolerances of the manufactured components, test set errors in measuring gain and component values, resistor adjustment accuracies, and temperature and aging behavior of the components. This computer program has been an invaluable tool in determining the requirements for tuning, testing, and optimizing the final design for minimum manufacturing cost.

I. INTRODUCTION

Resistance-capacitance active filters are a relatively new addition to the family of frequency selective networks. Although active filters have been in existence for 20 years, they have not been widely used because passive filters have been less expensive. However, the concurrent development of silicon integrated circuits, tantalum thin-film technology, and automated computer-controlled test sets have made RC active filters a practical alternative to passive filters at voice frequencies.

This paper deals with the design and development of the voice-frequency RC active filters used in the D3 channel bank.[1,2] These filters are realized in thin-film tantalum technology with beam-leaded silicon integrated-circuit operational amplifiers on one ceramic substrate.[3]

New analysis[4,5] and modeling techniques, more efficient optimization algorithms,[6] the concept of statistical design,[7] and more flexible machine aids to physical design gave the network designers powerful tools with which to attack the design process.

II. DEVELOPMENT PROCESS

The development process involves a number of interrelated steps. Neglecting any of the steps can result in a design that either does not meet requirements or is unmanufacturable at a reasonable cost. Traditionally, the first step is a careful analysis of the filter requirements. This will determine the order of the filter and, possibly, the technology needed for the physical realization (e.g., LC, RC active, or mechanical technology). Next, the initial design is undertaken. This step requires accurate models for the components and general-purpose analysis and synthesis routines. Unfortunately, when initial designs using new technologies are breadboarded, requirements are very rarely met. In addition, the physical realization is either too large and/or the network is too costly. Thus, we must optimize the network to meet not only electrical requirements but also size, environmental, and cost restrictions. Next, sensitivity studies must be undertaken to determine both the viability of the design and whether or not manufacturing adjustments (tuning) are necessary. Since the designer cannot afford to wait 10 or 20 years to see if his design meets performance objectives in the field, a statistical simulation of the manufacturing process and field performance is useful. This simulation can analyze the effects of component and adjustment tolerances, simulate the tuning procedure and temperature and aging effects, and statistically optimize the design to give both the greatest manufacturing yield and a prediction of the end-of-life performance.

In the design of filters, early attention to the physical realization and tuning and testing for manufacture is important. Lack of attention to these details can result in an unmanufacturable design and/or a design that has excessive cost. In addition, when new technology is introduced, the designer must closely follow the initial manufacture of his design to aid the manufacturer in overcoming initial production hurdles.

III. FILTER ENVIRONMENT AND REQUIREMENTS

The D1 and D3 channel banks[1] are the terminals for the T1 PCM repeatered line.[8,9] The T1 PCM line carries 24 telephone conversations over two cable pairs. A simplified sketch of the voice frequency end of the D3 channel bank is shown in Fig. 1. A voice-frequency analog signal from the switching equipment passes through a hybrid trans-

Reprinted with permission from *Bell Syst. Tech. J.*, vol. 54, pp. 507-529, March 1975.

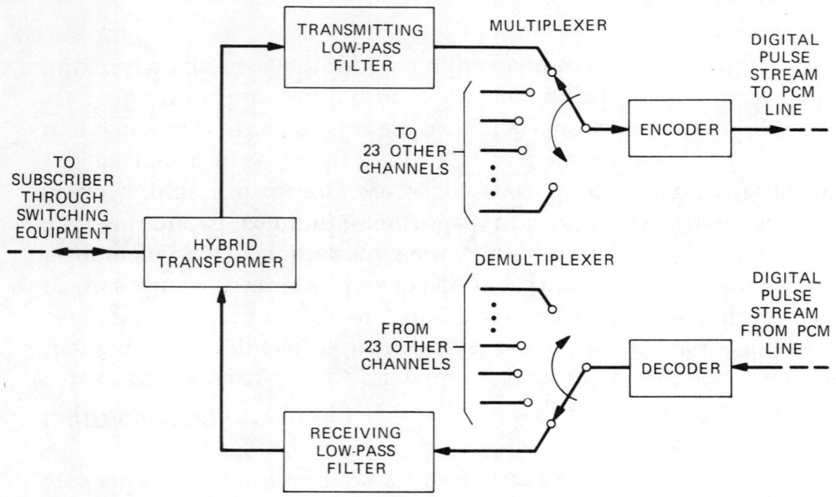

Fig. 1—Simplified model of PCM channel bank.

former. It is then amplified and bandlimited by the transmitting low-pass filter. Subsequently, it is converted into a pulse-amplitude modulated (PAM) signal by a JFET (junction field-effect transistor) switch operating at an 8-kHz rate. The PAM signals from the 24 channels are sequentially encoded into a binary bit stream (1.544 megabits/s) and sent out as PCM signals over a line. On the receiving end, the incoming PCM signals are sequentially decoded and converted to PAM signals. The analog signal for each individual channel is then recovered from the PAM signal by passing it through an interpolating receiving low-pass filter. The analog signal then passes through the hybrid transformer to the switching equipment and subsequently to the subscriber.

The transmitting band-limiting filter must present a good impedance to the hybrid transformer and negate the effect of the switched load impedance. In addition, it must pass frequencies between 200 and 3200 Hz with less than ±0.09-dB ripple and suppress frequencies above 5 kHz by at least 30 dB. At half the sampling rate of 8 kHz (or 4 kHz), there must be at least 15 dB of suppression. An RC active filter is ideal for this voice-frequency application, since the operational amplifier output will be impervious to the time-varying load, and the passband ripple performance is not degraded by inductor losses. The above requirements can be met with a fifth-order Cauer-Chebyshev filter. To obtain a low-cost, highly reliable filter conducive to high-volume manufacture, thin-film tantalum technology[3] was chosen for the physical realization.

One method of realizing a multiple-order active filter is cascading noninteracting lower-order sections. This realization consists of two stages. The first stage is a modification of a low-pass notch section (Fig. 2) developed by J. J. Friend.[10] It is a differential-input single operational-amplifier section. The second stage is a twin-T notch section[11] that has been modified by the addition of an RC network (Fig. 2) to include the pole on the negative real axis. Their respective transfer functions are

$$T_1(s) = \frac{s^2 + \omega_{z1}^2}{s^2 + \dfrac{\omega_{p1}}{Q_{p1}} s + \omega_{p1}^2} \tag{1}$$

and

$$T_2(s) = \frac{s^2 + \omega_{z2}^2}{\left(s^2 + \dfrac{\omega_{p2}}{Q_{p2}} s + \omega_{p2}^2\right)(s + \alpha)}. \tag{2}$$

The nominal performance of each section and the overall design is given in Fig. 3.[12]

The smoothing filter on the receiving side of the channel presents the designer with a difficult problem. It has a time-varying, rather than a time-invariant, generator impedance. This is caused by the JFET switch, which operates at an 8-kHz rate (125 μs) and is "on" for 3.5 μs. The switch working in conjunction with the input impedance of the network contributes an additional frequency-dependent gain characteristic to the filtering function of the network. A detailed description of the design of this filter is given in the next section.

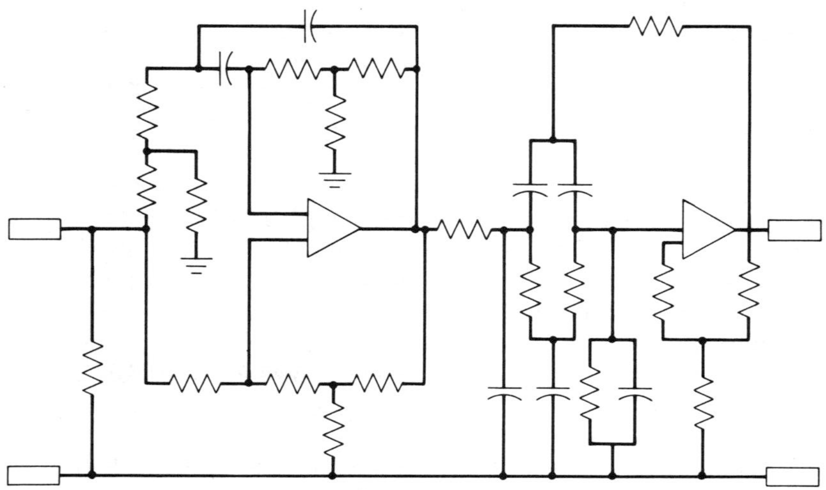

Fig. 2—Configuration of transmitting filter.

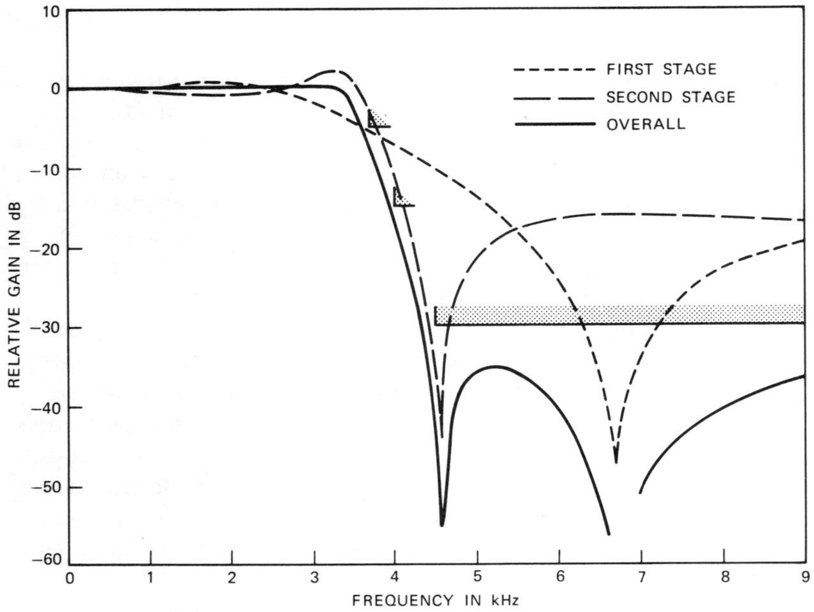

Fig. 3—Nominal performance of transmitting filter.

The design objectives of the receiving filter are as follows:*

(*i*) Passband ripple (200 Hz–3200 Hz) $\leqq \pm 0.16$ dB.
(*ii*) At 3300 Hz: $0.16 \text{ dB} \geqq \text{gain} \geqq -0.60$ dB.
(*iii*) At 3400 Hz: $0 \text{ dB} \geqq \text{gain} \geqq -1.20$ dB.
(*iv*) At 4000 Hz: gain < -15 dB.
(*v*) Above 5000 Hz: gain < -30 dB.
(*vi*) Gain at 1000 Hz: 4.75 dB $\pm$ 0.02 dB.

In addition, the filter must absorb the following manufacturing and environmental variations:

(*i*) Switch "on time": 3.1 μs to 3.7 μs.
(*ii*) Switch "on resistance": 50 Ω to 200 Ω.
(*iii*) Temperature: 0°C to 60°C.
(*iv*) Aging: 20-year life.

IV. INITIAL DESIGN OF THE RECEIVING FILTER

The configuration and nominal performance of the receiving filter are given in Figs. 4 and 5, respectively.[13,5] If the switch were not present, the requirements in Section III could be met by a fifth-order

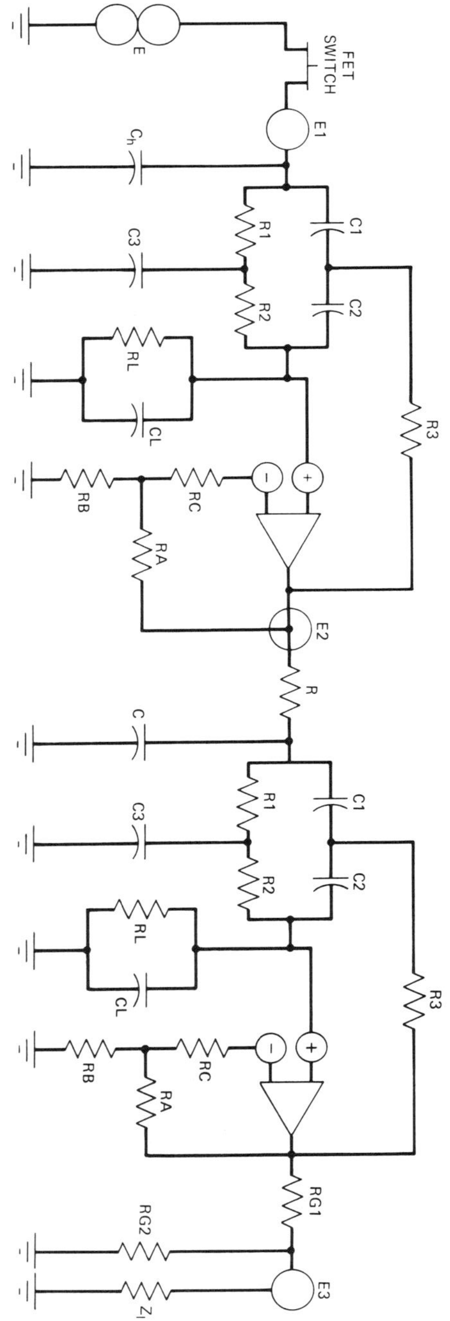

Fig. 4—Configuration of receiving filter.

* All gains and losses are relative to 1000 Hz.

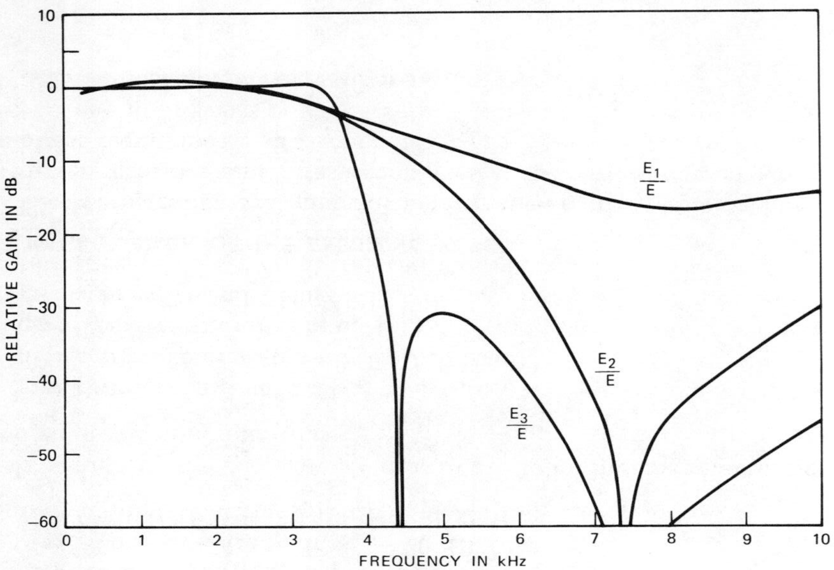

Fig. 5—Nominal performance of receiving filter.

Cauer-Chebyshev transfer function of the form

$$T(s)=\frac{V_{\text{out}}}{V_{\text{in}}}=K\frac{(s^2+\omega_{z1}^2)(s^2+\omega_{z2}^2)}{[s^2+(\omega_{p1}/Q_{p1})s+\omega_{p1}^2][s^2+(\omega_{p2}/Q_{p2})s+\omega_{p2}^2](s+\alpha)}. \quad (3)$$

The first pole-zero pair can be realized with a second-order tuned twin-T.[14] The real pole at $s = -\alpha$ could be realized with a first-order section, although an operational amplifer can be saved by combining this pole with the second pole-zero pair to form a third-order twin-T section.

To maintain the gain level and reduce the sensitivity to switch "on time" and "on resistance" variations, the input stage must provide some "holding" function.[13] An ideal sample-and-hold network would accomplish this, but it would require an additional operational amplifier. Therefore, a holding capacitor was added to the input stage. When the switch is on (for a nominal 3.5 μs), the capacitor C_h is charged. During the off time (121.5 μs), the capacitor will slowly discharge through the network if the input impedance is at a high level (>100 kΩ). The switch working in conjunction with the high input impedance of the first stage acts as a lossy sample-and-hold network with a $(\sin x)/x$-type frequency-dependent gain characteristic. This gain characteristic has minima at multiples of the 8-kHz switch rate (E_1/E in Fig. 5).

To design this filter, we could first assume that no switch or holding capacitor is present. An initial network would then be developed, to which the switch-and-holding capacitor would be added. Because of the roll-off introduced by the lossy sample and hold, further optimization would be required. Using a state-variable switched-filter analysis program in conjunction with a general-purpose optimization program,[4,5] the desired frequency characteristic is obtained (Fig. 6).

The program output gives the element values of the first stage of the filter (including the holding capacitor) and the transfer function coefficients of the second stage of the filter. To realize the second stage, the element values must be described as a function of the transfer function coefficients. The transfer function for a tuned, twin-T, second-order section with a preceding RC stage can be written as

$$T(s) = K\frac{s^2+\omega_z^2}{[s^2+(\omega_p/q_p)s+\omega_p^2](s+\alpha)} \quad (4)$$

or

$$T(s) = K\frac{s^2+\omega_z^2}{s^3+A_1's^2+A_2's+A_3'}. \quad (5)$$

If we define

$$\begin{aligned}\beta &= 1+(RA/RB),\\ c_1 &= (2CL/C1),\\ c_2 &= (CL/C),\\ r_1 &= (2R1/RL),\\ r_2 &= R/RL,\end{aligned} \quad (6)$$

and assume the twin-T is symmetrical and tuned, the coefficients of (5) can be written as

$$K = \frac{\beta}{RC(1+c_1+c_2)} \quad (7)$$

$$\omega_z^2 = \frac{1}{(R1\,C1)^2} \quad (8)$$

$$A_1' = \frac{1}{(1+c_1+c_2)}\left[\frac{1+c_1}{RC}+\frac{2}{R1\,C1}\left(2-\beta+\frac{r_1+c_1}{2}\right)+\frac{2}{R1\,C}\left(2-\beta+c_1+\frac{r_1}{4}\right)\right]=\frac{\omega_p}{q_p}+\alpha \quad (9)$$

$$A_2' = \frac{1}{(1+c_1+c_2)}\left[\frac{1+r_1+c_2}{(R1\,C1)^2}+\frac{2}{RC\,R1\,R2}\left(2-\beta+\frac{r_1+c_1}{2}\right)+\frac{2}{R1^2\,C1\,C}(2-\beta+r_1)\right]=\omega_p^2+\frac{\omega_p}{q_p}\alpha \quad (10)$$

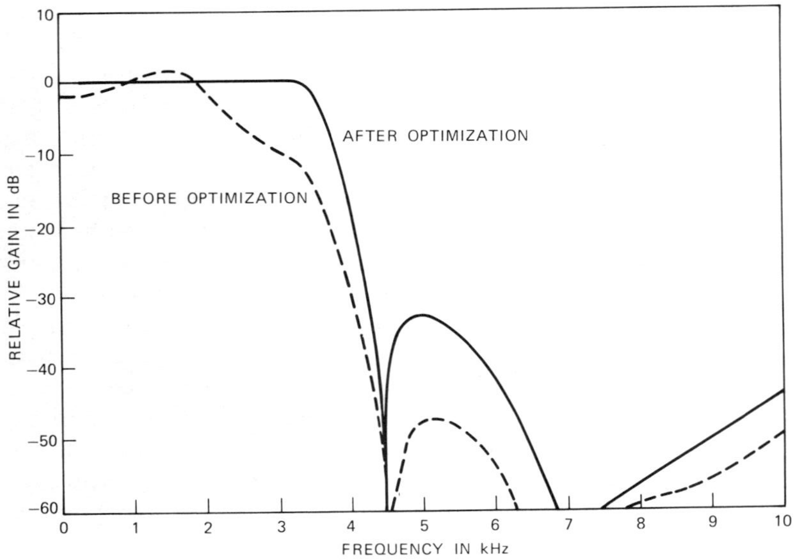

Fig. 6—Receiving filter before and after optimization.

and

$$A_3' = \frac{1}{RC}\left(\frac{1 + r_1 + r_2}{1 + c_1 + c_2}\right)\frac{1}{(R1\ C1)^2} = \omega_p^2\alpha. \tag{11}$$

Since there are more parameters (elements) than constraints (coefficients), the designer can optimize both the absolute values of the elements and the element value ratios, in addition to matching the required coefficients. An interactive computer program[15] was used to obtain the element values of the third-order twin-T.

V. SENSITIVITY CONSIDERATIONS

An important concern in the design of filters is the degradation of the frequency response because of environmental and element value changes. Some portions of this degradation can be controlled by specifying components with tight tolerances, or by carefully tuning the filter. Others can be controlled by matching the temperature coefficients of the components used. However, there are some uncontrollable changes, e.g., aging, adjustment tolerances, and measurement tolerances. The sensitivity of the filter design to these changes will determine whether a particular technology can be used in the realization, the amount of tuning required, the margin requirements of the design, the tolerance of the components, and, subsequently, the cost of the filter.

Many approaches can be taken for analyzing the sensitivity of the filter to element changes. One is the degradation of the frequency response caused by all elements taken one at a time. This method is very useful in determining which elements require tight tolerances and whether tuning or tracking is required. Another is the variation of the frequency response under true manufacturing and environmental variations. This would involve all components varying according to some statistical description.

As an illustration of the first type of sensitivity, Fig. 7 shows the degradation of the frequency response of the receiving filter when each capacitor is increased in value by 1 percent. Although the response degrades significantly with these capacitor variations, the degradations tend to cancel if the capacitors vary in the same direction (track). In fact, as can be seen from eqs. (5) through (11), if the second-stage capacitors track perfectly, i.e., all capacitors increase by 1 percent, the only effect on the frequency response of that stage will be a shift in frequency.

Also, from eqs. (5) through (11), it can be shown that if, in addition, the resistors vary in the opposite direction from the capacitors, then even this degradation is cancelled. Another point evident from Fig. 7 is that even ±1-percent capacitor variations would cause the response to miss the frequency requirements. Since thin-film capacitors can only be manufactured to ±5-percent tolerances, it is evident that initial tuning of the network is necessary.

Fortunately, thin-film resistors and capacitors have opposite temperature and aging characteristics. Thus, once tuned, the frequency response is quite stable. However, since the tracking is not perfect, further analysis must be undertaken.[16] This is explained in Section VII.

VI. TUNING

Many elegant tuning procedures have been developed for hybrid integrated circuits.[17–19] Most rely on gain and/or phase measurements at a number of frequencies (one for each transfer function coefficient) and adjustment of element values in some algorithmic fashion. Usually, some form of descent algorithm, depending upon component sensitivity, with a number of iterations is used to solve this multiparameter optimization problem. Thin-film networks contain critical constraints. The capacitors cannot be adjusted and the resistor values can only be increased either by anodization[20] or laser trimming.[21]

For the above network, nine constraints must be satisfied. They are the zero frequencies of each stage, the frequency of the real pole of the second stage, the frequencies and q's of the two complex poles, and the 1-kHz flat gain. In addition, the manufacturing deviations of the switch "on resistance" and the holding capacitor must be compen-

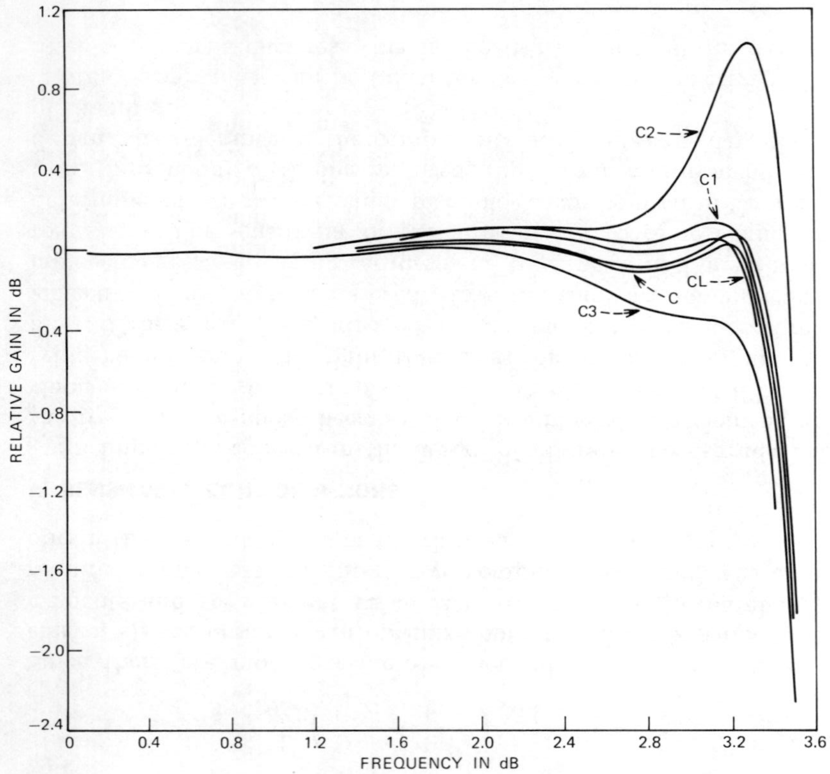

Fig. 7—Sensitivity of frequency response to capacitor changes of +1 percent.

sated for. Consultations with manufacturing engineers indicated that most iterative approaches that involved measurements of the performance at a number of frequencies were impractical from a throughput viewpoint, i.e., they took too much time.

From the manufacturing viewpoint, two points are critical: (*i*) the adjustment sequence must be fast and (*ii*) the adjusted network must meet the given frequency requirements. To accomplish both these aims, the tuning procedure shown in Fig. 8 was developed. It consists of two parts. The first is parametric; that is, it depends only upon component measurements and component adjustments. The second is functional; the gain is measured at *one* frequency for each stage, and an adjustment is made to bring the gain at that frequency to its nominal value. Finally, the voltage divider network at the output is used to trim the absolute through gain at 1 kHz to a predetermined level.

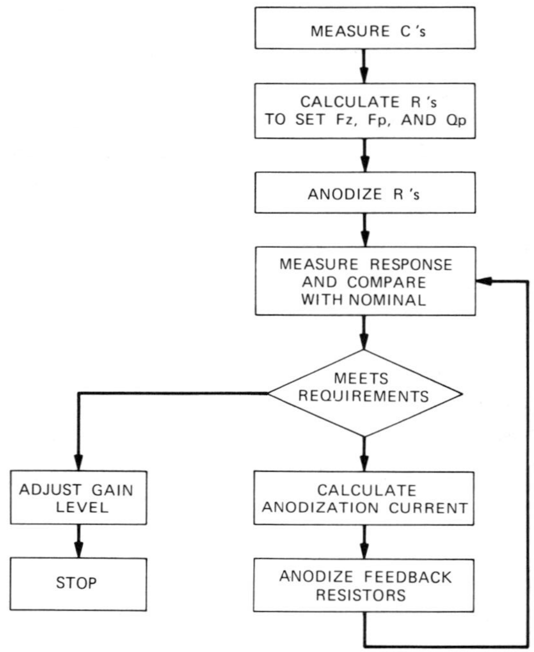

Fig. 8—Tuning procedure.

To tune the first stage of the filter, the switch-and-holding capacitor are initially neglected. The remaining part is a second-order twin-T notch filter whose transfer function is given in the appendix. Since the capacitors $C1$, $C2$, $C3$, and CL cannot be adjusted, their values are measured. To obtain a null at the prescribed frequency, eqs. (22), (25), and (27) dictate that

$$\frac{1}{(C1 + C2)R3} = \frac{R1 + R2}{R1\ R2\ C3}, \tag{12}$$

$$\omega_z^2 = \frac{1}{R1\ R2\ C3\ CS}. \tag{13}$$

Since the capacitor values are known, the constraints given by (12) and (13) can be satisfied if we pick

$$R1 = R2 = \frac{1}{\omega_z (CS\ C3)^{\frac{1}{2}}} \tag{14}$$

and

$$R3 = \frac{R1}{2}\frac{C3}{C1 + C2}. \tag{15}$$

To set the pole frequency, eqs. (22) and (32) can be used. Thus,

$$\omega_p^2 = \frac{1 + (2R1/RL)}{1 + (CL/CS)}\, \omega_z^2. \tag{16}$$

Since all the variables but RL of (16) are known, RL can be used to set ω_p. Rearranging (16),

$$RL = \frac{2R1}{[1 + (CL/CS)](\omega_p/\omega_z)^2 - 1}. \tag{17}$$

At this point, we have calculated the resistor values necessary to set the null, null frequency, and pole frequency of (28). To set the pole q, eq. (30) can be used. It is repeated here for convenience.

$$a_1 = \frac{\omega_p}{q_p} = \left[\frac{1-\beta}{R3\,C1} + \frac{1}{R2\,CS} + \frac{2CL}{R1\,C3\,CS} + \frac{1}{RL\,CS}\right] \Big/ \left(1 + \frac{CL}{CS}\right). \tag{18}$$

At this point, all the capacitors have been measured and $R1$, $R2$, $R3$, and RL have been adjusted. Thus, the only additional adjustment that can be made is on β, which is controlled by negative feedback resistors RA and RB. In fact,

$$\beta = 1 + \frac{RA}{RB}. \tag{19}$$

Solving (18) for β, we have

$$\beta = 1 + R3\,C1 \left\{\frac{1}{CS}\left[\frac{1}{R2} + \frac{2CL}{R1\,C3} + \frac{1}{RL}\right] - \frac{\omega_p}{q_p}\left(1 + \frac{CL}{CS}\right)\right\}. \tag{20}$$

To adjust β to coincide with that calculated by (20), RA and RB are adjusted.

Because we have neglected the holding capacitor and switch "on resistance," it is necessary to functionally adjust this stage. This adjustment is done at 2100 Hz, where the desired nominal gain with respect to 1000 Hz is known. Through sensitivity studies, it was found that the gain deviation from nominal at 2100 Hz was linearly dependent upon the ratio of the β resistors, RA and RB. Hence, either RA or RB is adjusted to complete the tuning of the first stage.

Since there is interaction between the real pole, the complex pole, and the transmission zero, the tuning of the second section is more complicated. With $C1$, $C2$, and $C3$ (Fig. 5) measured and the zero frequency ω_z given, the twin-T resistors $R1$, $R2$, and $R3$ can be determined. With C and CL measured and α, ω_p, and q_p given, eqs. (9) through (11) are a nonlinear set of equations in terms of β, R, and RL. Using the nominal values of these parameters as starting guesses, the new values which set α, ω_p, and q_p to their design values are easily found.

After anodizing the resistors R, $R1$, $R2$, $R3$, RL, RA, and RB to their calculated values, a functional adjustment is performed at 3300 Hz where the nominal gain value is known. This adjustment mops up the previous capacitor measurement and resistor adjustment errors and operational-amplifier variations. As in the first stage, the resistor ratio of the negative feedback network is touched up in a functional adjustment.

Finally, the gain at 1000 Hz must be adjusted to within ±0.02 dB. To increase the output level, $RG2$ is increased. To decrease the level, $RG1$ is increased.

VII. TOLERANCE ANALYSIS

The network designer's work would be cut significantly if inexpensive components could be manufactured to their nominal values and have no variations with time and temperature. However, manufacturing processes are such that the designer must consider variations in the component values and characteristics from sample to sample. He could do a worst-case analysis in which he would specify tolerances that would ensure that, under all component combinations, the final manufactured network would meet specifications. In general, this would result in specifying tighter tolerances than needed and necessarily increase the cost of the network.

To get a realistic estimate of the performance of the filter as it left the manufacturing facility and a good prediction of its field performance, a statistical simulation of the manufacturing process and environmental behavior was developed. Included in this model were manufactured element variations, measurement errors, adjustment errors, temperature and aging characteristics of the components, and switch timing variations. The variables were the tolerances and distributions associated with the above errors, the adjustment procedure, and the adjustment frequencies. The figure of merit was the highest possible end-of-life yield at the lowest possible cost. Figure 9 is a flowchart for this simulation.

The simulation proceeds in the following manner. First, a set of random numbers is generated for a network. Next, the manufactured capacitor and JFET "on resistance" values are calculated. For the thin-film capacitors, the absolute tolerance and the tracking tolerances are included. The switch "on resistance" has a nominal value of 125 ohms and can vary between 50 and 200 ohms. The next step is a simulation of the tuning procedure, where the capacitors are first measured, the resistor values calculated, and the resistors adjusted.

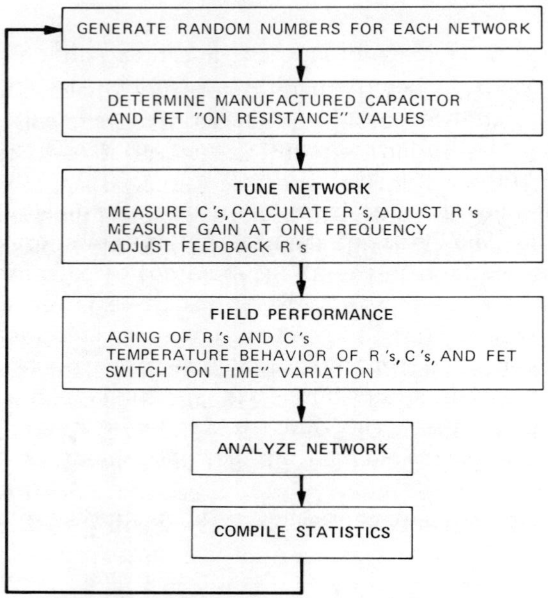

Fig. 9—Flow chart of tolerance analysis program.

For this parametric adjustment, the errors in capacitor measurement and the anodization errors for the resistors are included. During the functional touch-up adjustment, gain measurement errors are simulated. To evaluate the field performance, we use the temperature and aging characteristics of the components.

Finally the network is analyzed for the final element values that are determined by adding the temperature and aging deviations to the manufactured element values. Then the network performance is compared with the specifications at a number of frequencies, and statistics are compiled. If the requirements are not met at any one frequency, the network has failed. Next, a new network is picked, and we repeat the process.

The statistical performance of this filter is shown in Figs. 10 and 11. Only the minimum and maximum deviations at each frequency for 1000 sample networks are shown. At each frequency, there is a normal distribution of the networks' performance between the minimum and maximum deviations. At room temperature, 25°C, 94.5 percent of the samples fell within the design objectives, while at 60°C, 89.3 percent fell within. To ensure that the mathematical modeling was correct, these results were compared with the measured performance of the first 1000 filters manufactured. All those filters fell within the given statistical bands. This type of extensive simulation is justified when we have a well-characterized technology, automated adjustment and testing, and high-volume manufacturing.

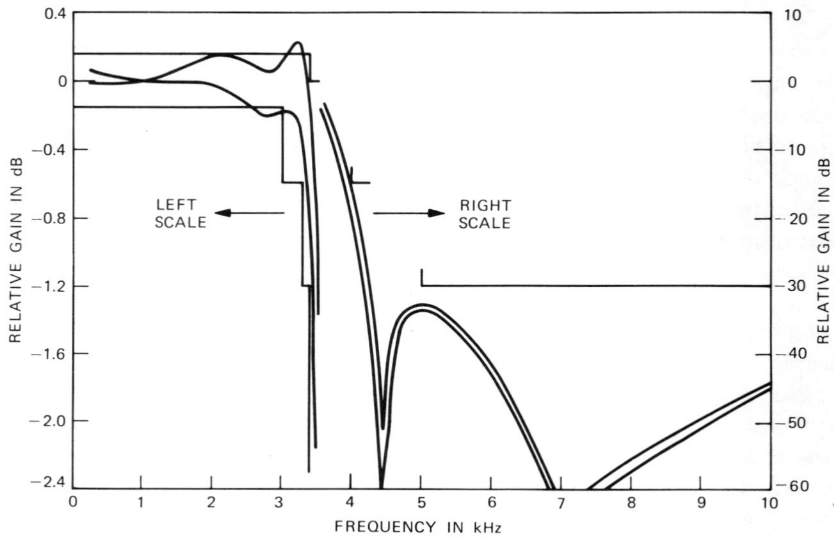

Fig. 10—Overall statistical performance of filter.

VIII. PRODUCTION EXPERIENCE

As previously mentioned, the tolerance analysis program has aided in obtaining a practical tuning procedure. If temperature and aging variations are ignored, the program predicts the filter yield at the end of the manufacturing process. The assigned component tolerances have a significant effect on manufacturing costs; thus, a trade-off takes place between manufacturing cost and filter yield. Ideally, we would like to find the set of component tolerances that minimizes manufacturing cost. This problem has been attacked by Karafin[22] and Pinel and Roberts.[23] However, their work has been restricted to networks where: (*i*) the performance criteria (i.e., gain or loss constraints at given frequencies) is met 100 percent of the time; (*ii*) tunable elements are banned, and (*iii*) there is no correlation between the elements.

In the case of the D3 filters, it is known that the 100-percent performance criteria restriction will not produce a minimum cost network. In other words, if the yield at final test drops from 100 to 96.7 percent by increasing the tolerance on the resistors during parametric adjustment, the cost drops by more than 3.3 percent. Thus far, cost minimization has been attacked in a heuristic manner. Figure 12 shows the

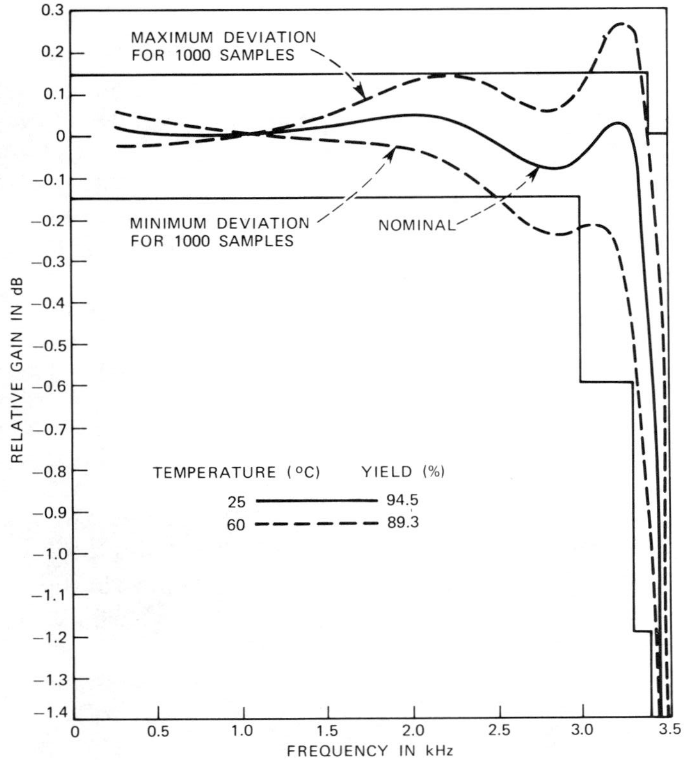

Fig. 11—Statistical performance of passband at temperature extreme.

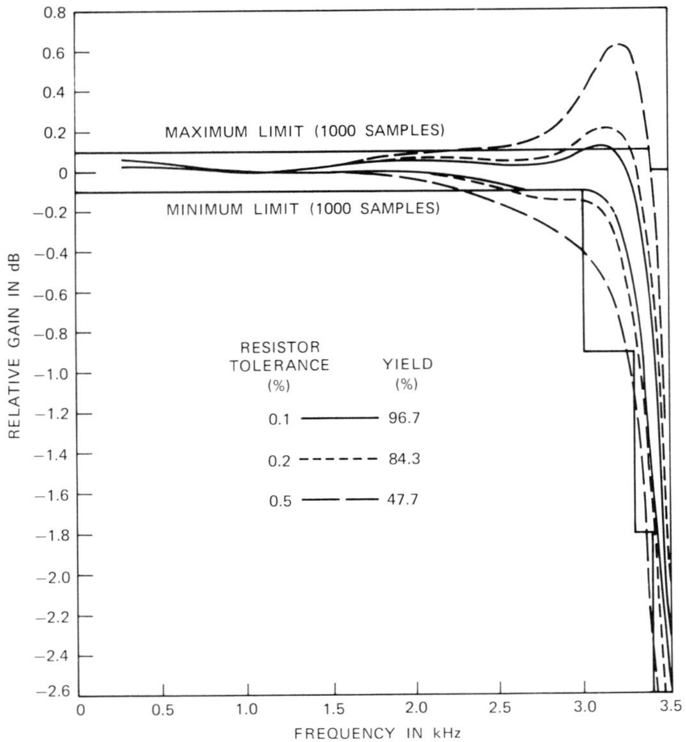

Fig. 12—Statistical performance of transmitting filter with different resistor tolerances.

theoretical manufacturing yields for the transmit filter at 0.1, 0.2, and 0.5-percent resistor tolerances. Current production experience indicates that widening the tolerance on certain resistors to 0.3 percent results in a lower cost, even though the overall yield decreases.

One might ask, Why not remove all tolerance restrictions on the parametric adjustments and accept any filters as long as they are functionally adjusted to meet the frequency requirements at a finite set of test points? The answer is quite simple. If all resistors were adjusted at the parametric step to, say, 0.5 percent tolerance, then only 50 percent of the filters could meet requirements after functional adjustment. Since the silicon is bonded after parametric adjustment, a much higher final yield is needed to justify the additional investment. In addition, as the tolerance is relaxed at the parametric adjustment step, more frequency tests are required at final test.

The tolerance analysis program also helped to provide a temporary solution to a capacitor tracking problem. If the ratio of certain capacitors is not within specified limits, then some resistors may have to be anodized excessive amounts—excessive because the resistors will then age poorly. However, the tracking requirement on all sets of capacitors need not be the same; thus, in a particular case, with the tolerance program, it was possible to demonstrate that the tracking requirement on a pair of capacitors could be relaxed and therefore solve a temporary production problem.

A tolerance analysis program can be used only if we have good estimates for the various tolerances that affect the manufacturing process. This was possible for the parameters that influence the frequency performance, but not possible for the dc gain requirement. The 1-kHz frequency gain is required to be adjusted to an accuracy of ± 0.02 dB. This stringent requirement was chosen so that no adjustments would be necessary when D3 channel banks were installed in the field. The no-adjustment philosophy saves on installation and maintenance cost, but it does require critical tuning for the D3 filters.[24]

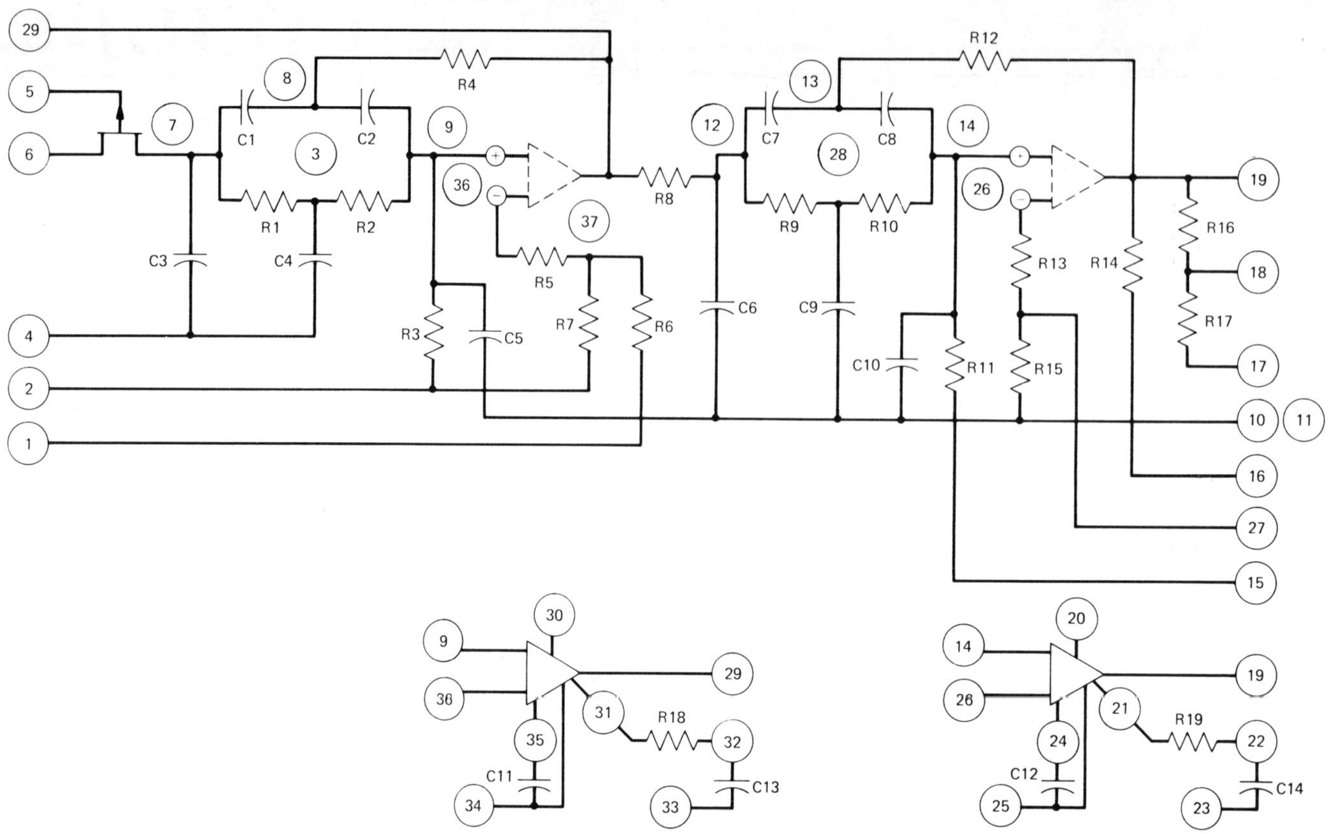

Fig. 14—Electrical layout of HIC realization.

IX. PHYSICAL REALIZATION

In this case, the physical realization uses single-substrate hybrid-integrated circuit (HIC) technology.[3] The resistors are tantalum nitride thin film, and the capacitors have a base electrode of β-tantalum, a dielectric of tantalum pentoxide, and a counter electrode of nichrome paladium gold. The conductor paths are also nichrome paladium gold, while the operational amplifiers and JFET switch are beam-leaded silicon integrated circuits. They are all placed on a 33-by-20 mm glazed ceramic substrate with a 34-terminal lead frame (Fig. 13).

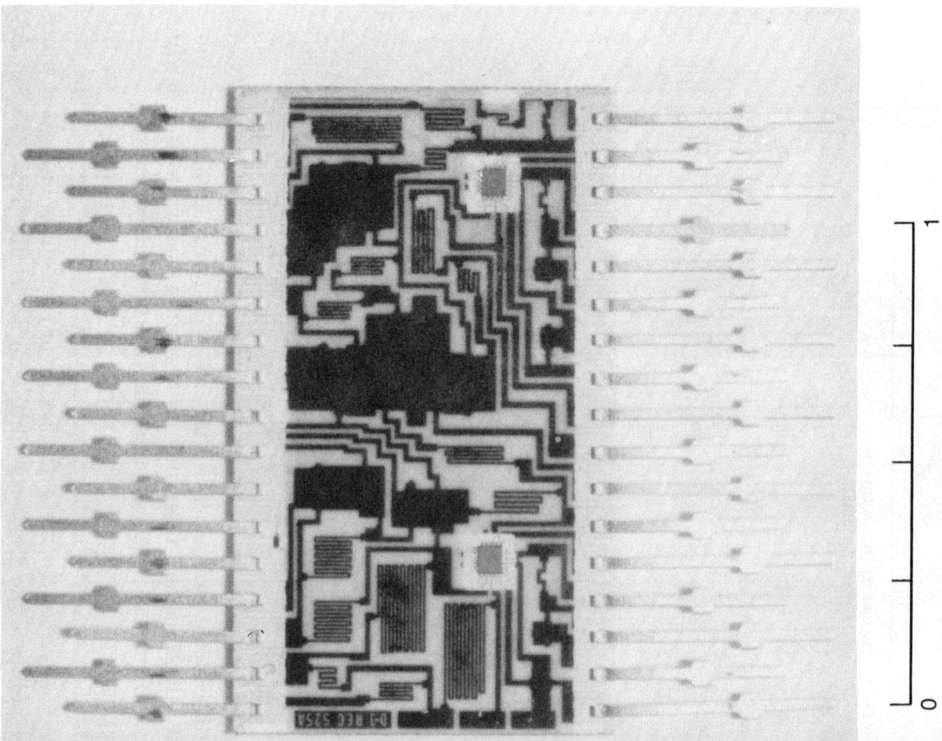

Fig. 13—Hybrid integrated circuit realization of filter.

To minimize both the tuning time and the complexity of the test fixtures, the substrate layout was constrained. Thus, all measurements are made from the sides and all adjustments from the top of the substrates. To incorporate the tuning procedure of Section VI, the layout must provide the capability of measuring all element values. Thus, all network nodes must be brought to the edge of the substrate.

Traditionally, layouts follow the flow of a network. Thus, for a network of this complexity, crossovers are generally required, as well as break points, to measure the component values. Crossover and break closures are normally processed following filter tuning. To eliminate breaks and crossovers and separate the measurement and adjustment functions, a unique circuit layout was developed. These aims were accomplished by judiciously separating common grounds and incorporating break points. These redundant points are brought to the substrate edge (Fig. 14) and are subsequently connected when the substrate is inserted into a printed wiring board. Three terminal capacitor measurements and capacitor electrode sharing were necessary to make these measurements. To realize this layout with its subsequent 11 mask levels required considerable reliance on computer-aided graphics.[25,26]

X. ACKNOWLEDGMENTS

The authors are grateful to W. Thelen for his work on the initial design, R. C. MacLean for his unique circuit layout, J. Novak and M. Fiore for building, tuning, and testing many models of the filters, C. F. Kurth, M. L. Liou, and F. J. Witt for many stimulating discussions, and numerous other colleagues who helped to bring this project to fruition.

APPENDIX

Transfer Function of Second-Order Twin-T Section

A.1 Untuned case

The transfer function of the untuned, unsymmetrical second-order twin-T is

$$T(s) = K\frac{(s+\tau_2)s^2 + (s+\tau_1)\omega_z^2}{(s+\tau_1)D(s) + \delta_\tau E(s)}, \tag{21}$$

with the following definitions:

$$\begin{aligned} CP &= C1 + C2, & CS &= (C1\,C2)/CP,\\ RS &= R1 + R2, & RP &= (R1\,R2)/RS,\\ \tau_1 &= 1/(R3\,CP), & \tau_2 &= 1/(RP\,C3),\\ \delta_\tau &= \tau_2 - \tau_1, & \beta &= 1 + (RA/RB),\\ c_1 &= CL/CS, & r_1 &= RS/RL. \end{aligned} \tag{22}$$

The polynomials $D(s)$ and $E(s)$ of (21) are

$$D(s) = s^2 + \left[\frac{1-\beta}{R3\,C1} + \frac{1}{R2\,CS} + c_1\tau_2 + \frac{1}{RL\,CS}\right]\frac{s}{(1+c_1)} + \left(\frac{1+r_1}{1+c_1}\right)\omega_z^2, \tag{23}$$

$$E(s) = \left[\frac{1}{1+c_1}s + \frac{(1-\beta)}{R3\,C1(1+c_1)}\right]s, \tag{24}$$

and

$$\omega_z^2 = \frac{1}{R1\,R2\,C3\,CS}, \tag{25}$$

$$K = \frac{\beta}{1+c_1}. \tag{26}$$

A.2 Tuned case

When the twin-T is tuned,

$$\tau_1 = \tau_2, \qquad \delta_\tau = 0. \tag{27}$$

Thus, (21) reduces to

$$\begin{aligned} T(s) &= K\frac{s^2+\omega_z^2}{s^2 + a_1 s + a_2}\\ &= K\frac{s^2+\omega_z^2}{s^2 + (\omega_p/q_p)s + \omega_p^2}. \end{aligned} \tag{28}$$

If, in addition, a symmetric twin-T is picked, i.e.,

$$\begin{aligned} R1 &= R2 = 2R3\\ C1 &= C2 = C3/2\\ CL &= (C1\,c_1)/2\\ RL &= (2R1)/r_1, \end{aligned} \tag{29}$$

then

$$a_1 = \frac{2}{R1\,C1}\left(2 - \beta + \frac{r_1 + c_1}{2}\right)\frac{1}{1+c_1}, \tag{30}$$

$$a_2 = \left(\frac{1+r_1}{1+c_1}\right)\omega_z^2, \tag{31}$$

and

$$\omega_p = \left(\frac{1+r_1}{1+c_1}\right)^{\frac{1}{2}}\omega_z, \tag{32}$$

$$q_p = \frac{[(1+r_1)(1+c_1)]^{\frac{1}{2}}}{2[2 - \beta + (r_1+c_1)/2]}. \tag{33}$$

REFERENCES

1. W. B. Gaunt and J. B. Evans, Jr., "The D3 Channel Bank," Bell Laboratories Record, *50,* No. 7 (August 1972), pp. 229–233.
2. R. A. Friedenson, "Active Filters Make it Small in the D3 Channel Bank," Bell Laboratories Record, *51,* No. 4 (April 1973), pp. 104–111.
3. W. H. Orr et al., "Integrated Tantalum Film RC Circuits," Proc. 20th Elec. Components Conf.. May 1970, pp. 602–612.
4. C. Pottle, "State-Space Techniques for General Active Network Analysis," in *System Analysis by Digital Computer,* F. F. Kuo and J. F. Kaiser, eds., New York: John Wiley, 1966, Ch. 3.
5. M. L. Liou, "Exact Analysis of Linear Circuits Containing Periodically Operated Switches with Applications," IEEE Trans. Circuit Theory, *CT-19,* No. 2 (March 1972), pp. 146–154.
6. P. E. Fleischer, "Optimization Techniques in System Design," in *System Analysis by Digital Computer,* F. F. Kuo and J. F. Kaiser, eds., New York: John Wiley, 1966, Ch. 6.
7. A. C. Dickieson, J. Chernak, et al., "Statistical Circuit Design," B.S.T.J., *50,* No. 4 (April 1971), pp. 1099–1310.
8. D. F. Hoth, "The T1 Carrier System," Bell Laboratories Record, *40,* No. 10 (November 1962), pp. 358–363.
9. K. E. Fultz and D. R. Penick, "The T1 Carrier System," B.S.T.J., *44,* No. 7 (September 1965), pp. 1405–1452.
10. J. J. Friend, "A Single Op-Amp Biquadratic Filter Section," Proc. 1970 IEEE International Symposium on Circuit Theory, Atlanta, Georgia, December 1970.
11. G. S. Moschytz, "FEN Filter Design Using Tantalum and Silicon Integrated Circuits," Proc. IEEE, *58,* April 1970, pp. 550–556.
12. R. A. Friedenson, "Computer-aided Design . . . A Network Designer's Viewpoint," IEEE Computer-Aided Design Short Course, *72SC-02,* March, April 1972, pp. 18–42.
13. M. L. Liou, P. H. McDonald, and W. Thelen, "Computer Optimization of Active and Passive Switched Low-Pass Filters for PCM Systems." Proc. 20th Electronic Components Conf., May 1970, pp. 301–306.
14. G. S. Moschytz and W. Thelen, "Design of Hybrid Integrated Circuit Building Blocks," IEEE J. Solid-State Circuits, *SC-5,* No. 2 (June 1970), pp. 99–107.
15. R. L. Adams, private communication.
16. D. Hilberman, "An Approach to the Sensitivity and Statistical Variability of Biquadratic Filters," IEEE Trans. Circuit Theory, *CT-20,* No. 4 (July 1973), pp. 382–390.
17. G. S. Moschytz, "Two-step Precision Tuning of Twin-T Notch Filter," Proc. IEEE, *54,* No. 5 (May 1966), pp. 811–812.
18. G. S. Moschytz, "A General Approach to Twin-T Design and Its Application to Hybrid Integrated Linear Active Networks," B.S.T.J., *49,* No. 6 (July-August 1970), pp. 1105–1149.
19. W. H. Orr, "The Computer Design and Precision Tuning of Thin Film Filters," Proc. 3rd Annu. Seminar on Integrated Circuits, New York Chapter IEEE Basic Sciences Division, February 1966, pp. 45–56.
20. D. A. McLean, M. Schwarz, and E. D. Tidd, "Tantalum-film Technology," Proc. IEEE, *52,* No. 12 (December 1964), pp. 1450–1462.
21. M. J. Cohen, B. A. Unger, and J. F. Milkosky, "Laser Machining of Thin Films and Integrated Circuits," B.S.T.J., *47,* No. 3 (March 1968), pp. 305–405.
22. B. J. Karafin, "The Optimum Assignment of Component Tolerances for Electrical Networks," B.S.T.J., *50,* No. 4 (April 1971), pp. 1225–1242.
23. J. F. Pinel and K. A. Roberts, "Tolerance Assignment in Linear Networks Using Nonlinear Programming," IEEE Trans. Circuit Theory, *CT-19,* No. 5 (September 1972), pp. 475–479.
24. J. Dupcak and R. H. DeGroot, "The Manufacture of Thin-Film Active Filters," The Western Electric Engineer, *18,* No. 3 (July 1974), pp. 18–25.
25. Fowler, "XYMASK," Bell Laboratories Record, *47,* No. 7 (June-July 1969), pp. 204–209.
26. J. J. Degan, "Focus on Efficiency: Computer Graphics for HIC Deisgn," Bell Laboratories Record, *52,* No. 9 (October 1974), pp. 286–292.

Multiple-Loop Feedback Topologies for the Design of Low-Sensitivity Active Filters

KENNETH R. LAKER, MEMBER, IEEE, ROLF SCHAUMANN, MEMBER, IEEE, AND MOHAMMED S. GHAUSI, FELLOW, IEEE

Invited Paper

Abstract—**Improved sensitivity performance of active filters in multiloop feedback (coupled) topologies is well known. Of the numerous design procedures available in the literature, in this paper those methods are reviewed which are based on first- or second-order building blocks in a feedback configuration. A statistical sensitivity measure is given which results in a realistic performance measure for the filters and which can be used to compare circuits and to optimize their design to achieve minimum sensitivity. Design procedures are provided for each topology, which result in realizations having minimum sensitivity and maximum dynamic range. Also, design tables for several examples are given, and the performance of some filters is demonstrated experimentally.**

I. Introduction

OVER THE past decade the realization of active RC networks with voltage transfer functions of the form

$$H(s)=\frac{N(s)}{D(s)}=\frac{b_m s^m+b_{m-1}s^{m-1}+\cdots+b_1 s+b_0}{s^n+a_{n-1}s^{n-1}+\cdots+a_1 s+a_0} \quad (1)$$

where $m \leqslant n$ has been the subject of numerous technical articles [1]–[5]. From this large body of literature one can categorize the methods for realizing (1) into several approaches, namely, direct, inductance simulation, cascaded biquad blocks, frequency dependent negative resistance, and multiple-loop feedback (MF) [1]–[45].

With the direct approach [1]–[5] $H(s)$ is realized using a general active network configuration comprised of one or more active elements embedded in a passive RC network. The sensitivities of direct realizations to element tolerances are so large that this method is only of theoretical interest. With the inductance-simulation approach [6]–[8], the inductances are replaced by capacitor-loaded gyrators, and the classical LC filter synthesis methods carry over directly. Floating inductors may be realized by gyrator-floatation circuits [9], semifloating gyrators [10], or by cascading two-grounded gyrators. Wave filter and signal flow graph approaches to simulate ladder structures have also been used [74]. These ladder-derived analogs, depending on the quality of the active elements, generally retain much of the good sensitivity behavior of their passive counterparts. This class of filters tends to be nonmodular[1] which is a disadvantage in some cases.

Manuscript received April 7, 1978. This work was supported in part by the National Science Foundation under Grants ENG 76-11218 and ENG 78-07247, and in part by the Computer Center of the University of Minnesota.

K. R. Laker is with Bell Telephone Laboratories, Holmdel, NJ 07733.
R. Schaumann is with the Department of Electrical Engineering, University of Minnesota, Minneapolis, MN 55455.
M. S. Ghausi is with Oakland University, School of Engineering, Rochester, MI 48063.

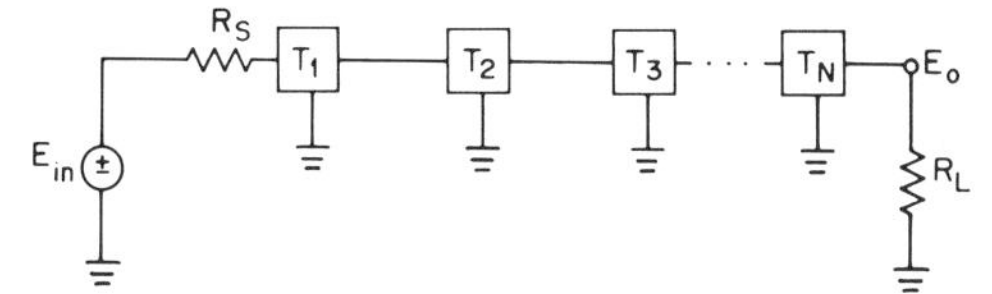

Fig. 1. Cascade of N biquad sections.

Because of the simplicity of the design and the use of modular biquad circuits, the cascade approach has become the workhorse of active filter design. In this method (1) is factored into a product of N subnetwork transfer functions such that

$$H(s)=\prod_{i=1}^{N} T_i(s), \qquad \text{where } N \leqslant n. \quad (2)$$

For n even, $N=n/2$, and all the T_i's are typically biquadratic in form, i.e.,

$$T_i(s)=\frac{N_i(s)}{D_i(s)}=\frac{as^2+bs+c}{s^2+ds+e}$$
$$=\frac{k_i\left(s^2+\omega_{zi}/Q_{zi}s+\omega_{z_i}^2\right)}{s^2+\omega_{0i}/Q_i s+\omega_{0i}^2}. \quad (3)$$

When n is odd, one of the $T_i(s)$ must be either a first- or third-order section. Many noninteracting, or cascadable, active RC [50]–[55] and active R [56]–[58] circuits, capable of realizing (3) and its special cases, are discussed in the literature. The circuits presented in [50]–[58], [87] possess, in our opinion, unique characteristics and/or wide popularity. To synthesize $H(s)$, the N subnetworks, realizing transfer functions in the form of (3), are connected in tandem, as shown schematically in Fig. 1. The

[1]By modular, we mean that a high-order filter can be dissected into a reasonable number of common manufacturable subnetworks. For obvious reasons, it is highly desirable that these subnetworks be sufficiently versatile to satisfy a wide spectrum of requirements. An excellent example is an active RC biquad circuit.

Reprinted from *IEEE Trans. Circuits Syst.*, vol. CAS-26, pp. 1-21, Jan. 1979.

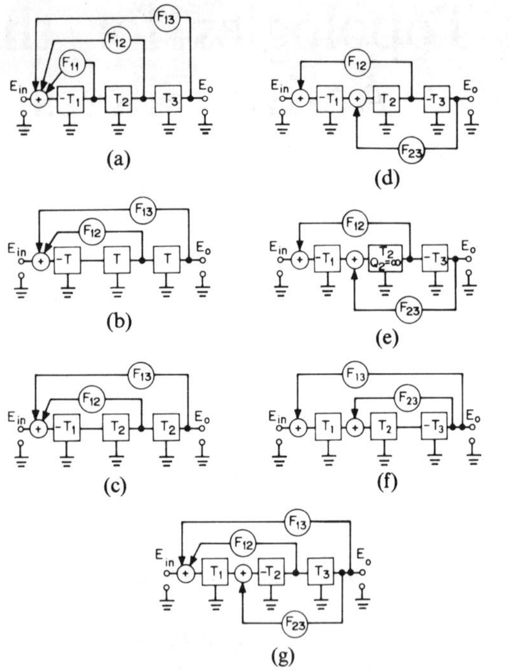

Fig. 2. Multiple-loop feedback topologies and design philosophies for all-pole bandpass realization (shown for sixth-order as an example). (a) FLF; (b) PRB; (c) SCF; (d) MLF; (e) LF, CB; (f) IFLF; (g) MSF.

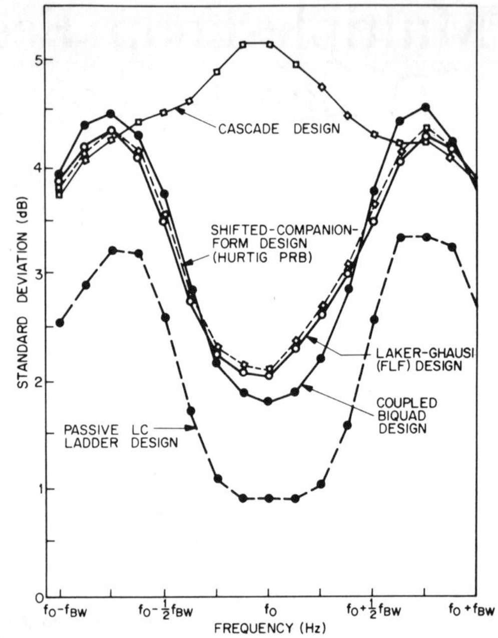

Fig. 3. Simulated variation (Monte Carlo analysis) of three-section Butterworth bandpass filters of different topologies. (Assumed passive component tolerance: 1.732 percent.)

approach is completely modular, with any one of several circuits capable of realizing (3). In addition, since the blocks $T_i(s)$ are assumed to be noninteracting, the synthesis and tuning of each T_i can be performed independently, providing for easy design. The only drawback to this method is that the resulting designs are inherently more sensitive to component variations than ladder-derived analogs and multiple-loop feedback topologies. More specifically, one can show [20], [23], [59] on a statistical basis that variations in T_i tend to be additive and produce a larger variation in H. Nevertheless, the cascade method has been by far the most widely used approach and still retains considerable popularity.

Another approach to the realization of (1) is the introduction of an active device whose driving-point impedance equals $Z(s)=K/s^2$ where K is a constant (i.e., $Z(s)$ is a frequency-dependent negative resistance (FDNR) [11]). In this method an impedance scaling factor K/s is used in the classical RLC filter design. The scaling, which does not affect $E_o(s)/E_{in}(s)$, transforms the elements R to C, L to R, and C to FDNR, and hence makes possible an inductorless realization. The sensitivity behavior of these circuits also depends on the quality of the active elements and is similar to inductance-simulation methods. Note again that the circuit does not possess modularity.

MF topologies [12]–[45] arose from a search for realizations of transfer functions of order $n>3$ which retain the convenient and practical modular implementation of the cascade approach but provide superior sensitivity performance.

To realize an MF filter, such as one of those shown schematically in Fig. 2 cascaded biquads, with T_i's of the form of (3), are coupled together through an external resistive feedback (and perhaps feedforward) network. Fig. 2 displays all the feedback topologies and design philosophies which have appeared in the literature, namely follow-the-leader-feedback (FLF) [19], [22], primary-resonator-block (PRB) [17], [18], shifted-companion-form (SCF) [25], [28], modified-leap-frog (MLF) [24], [30], leap-frog (LF) [12]–[14], [21], coupled-biquad (CB) [15], [16], inverse FLF (IFLF) [32], [43], and the minimum-sensitivity-feedback (MSF) [41] topologies. Although only the configurations for the sixth-order filter design are shown, all but the MSF [41] can be generalized to arbitrary order by simply adding the appropriate number of biquad sections and repeating the feedback patterns shown in Fig. 2. Observe that for the fourth-order design, with two biquad sections, all topologies but the FLF[2] [19], [22] reduce to the same graph.

The improved sensitivity performance of MF circuits over that of cascade design (not surprising to those familiar with feedback systems) is demonstrated by the Monte Carlo simulations shown [28] in Fig. 3 for a sixth-order Butterworth bandpass filter. These curves represent the standard deviation (in decibels) of gain variation $|\Delta|H||$ as a function of frequency. The active devices are considered ideal, and all passive elements (R's and C's) are assumed to be random variables, uniformly distributed about their nominal with a tolerance of ± 1.732

[2]The first feedback loop F_{11}, historically has been part of the FLF topology [19]. This loop has little effect on filter sensitivity; as a result it has been omitted in other topologies. Nevertheless, this loop does provide added design and tuning flexibility. When appropriate, this loop can be used as a means [41] to scale the section quality factors. It is noted that self-loops F_{ii} can be placed around any section within the MF topologies in Fig. 2.

percent (equivalent to a standard deviation $\sigma = 1$ percent). Note the comparison of the MF realizations CB, PRB, FLF, and SCF with the cascade design at the upper limit, and the resistor terminated passive LC-ladder design at the lower limit. The much reduced [12]–[45] passband sensitivities, as compared to cascade designs, are typical of MF bandpass (BP) filters. Note that at the band edge the CB realization has poor sensitivity. The factor of two or so difference in sensitivity between the active MF designs and the passive LC ladder is primarily due to the fact that active RC resonance requires at least four elements and LC resonance requires only two, and thus more components contribute to the variance in active filters. For practical active filter realizations, this component count disadvantage exists regardless of the synthesis approach.

In this paper, the synthesis methods for MF active filters with minimized sensitivity and maximized dynamic range will be reviewed. The solution of the design equations together with the optimization of sensitivity and dynamic range requires in general a digital computer program (see the Appendix) but for simple cases, say $n \leq 6$, a programmable hand-held calculator is adequate. Simple synthesis methods for realizing suboptimal filters and a number of design tables will be presented. The performance of the different MF design techniques will be compared, and their salient features will be pointed out.

II. General Remarks About MF Filter Synthesis

In the succeeding sections the design of specific MF filters will be discussed in detail. In order to make this treatment efficient and to avoid unnecessary repetition, this section will be used to introduce and define the concepts which will be commonly applied in all cases.

Fig. 2 shows the various MF topologies discussed in the literature. Note that among the six configurations in Figs. 2(a)–(f), there are really only three distinct topologies: PRB and SCF are special cases of FLF, and LF(CB) is a special case of MLF. In addition, all the networks in Figs. 2(a)–(f) are seen to be special cases of MSF. Since MSF filters are always derived from a full graph [41] containing all possible feedback loops (including the self-loops F_{ii}, if desired), the generality of MSF shown for $n=6$ in Fig. 2 extends to arbitrary n. The total of seven networks reflects differences only in design philosophy associated with some topologies. Accordingly, only the FLF, MLF, IFLF, and MSF topologies will be discussed in detail in the following, and the special procedures leading to the remaining configurations will be pointed out briefly.

Most of the reported work in MF filters has been devoted to realizing narrow band, symmetrical BP responses which are derivable from a known lowpass (LP) prototype via an LP-to-BP transformation [46]–[49]. This interest in narrow-band BP responses is primarily due to the need to reduce the intrinsically large sensitivities of cascade realizations which increase with the filter quality factor Q. For symmetrical BP responses, the mathematics associated with the MF filter synthesis simplify significantly but in most cases still require computer or calculator aids. It is emphasized, however, that MF filters are not restricted [21], [25], [31], [37], [39], [40] to BP requirements. For example, MF techniques have been shown to be effective in reducing the sensitivities of band reject filters [25], [39].

If the desired BP transfer function $H(s)$ of the form (1) is symmetrical (n even), then by means of the LP-to-BP transformation [46]–[49] the filter can be designed directly from the LP prototype transfer function. The LP prototype transfer function takes the form

$$H_{\mathrm{LP}}(p) = \frac{N_{\mathrm{LP}}(p)}{D_{\mathrm{LP}}(p)} = \frac{c_M p^M + c_{M-1} p^{M-1} + \cdots + c_1 p + c_0}{p^N + d_{N-1} p^{N-1} + \cdots + d_1 p + d_0} \tag{4}$$

where $M < N = n/2$ and p is the LP frequency variable normalized such that the 3-dB bandwidth of H_{LP} equals a given value γ. Note that in general, the tabulated LP prototypes [46]–[49] have nonunity normalized 3-dB bandwidths except for the Butterworth case, where $\gamma = 1$. If $H(s)$ is an all-pole BP function, as is the case for our later examples and comparisons, the LP prototype becomes

$$H_{\mathrm{LP}}(p) = \frac{N_{\mathrm{LP}}(p)}{D_{\mathrm{LP}}(p)} = \frac{H_{MN} d_0}{p^N + d_{N-1} p^{N-1} + \cdots + d_1 p + d_0}. \tag{5}$$

Note that the normalized BP transfer function (1) is obtained from (4) via the LP-to-BP transformations [47] $p = \gamma \hat{Q}(s^2+1)/s$, $\hat{Q} = \omega_0/\Delta\omega_{3\,\mathrm{dB}}$ denotes the overall BP filter quality factor, s is normalized such that the BP center frequency ω_0 equals unity, and $\Delta\omega_{3\,\mathrm{dB}}$ is the 3-dB bandwidth.

When the MF networks in Fig. 2 are used to realize the all-pole BP functions (5), the biquad T_i's are also BP functions as they would be for a cascade design, i.e.,

$$T_i(s) = \frac{N_i(s)}{D_i(s)} = \frac{H_{Bi} s/Q_i}{s^2 + s/Q_i + 1} = H_{Bi} t_i(s). \tag{6}$$

In contrast to cascade designs, each T_i is synchronously tuned to the same center frequency ω_0. One can show that stagger tuning the sections, even slightly, will degrade at least the passband sensitivities [42].

To realize the finite transmission zeros implied in (1) and (4), there are three methods [15], [21], [22], [31], [37], [39], [40] as shown in Fig. 4. In principle, any one of the zero forming methods in Fig. 4 can be used with any one of the MF networks in Fig. 2 to realize a general symmetrical BP response. With the first method [21], [31], [37], [39], the finite transmission zeros are obtained as in cascade fashion with general biquad T_i's of the form of

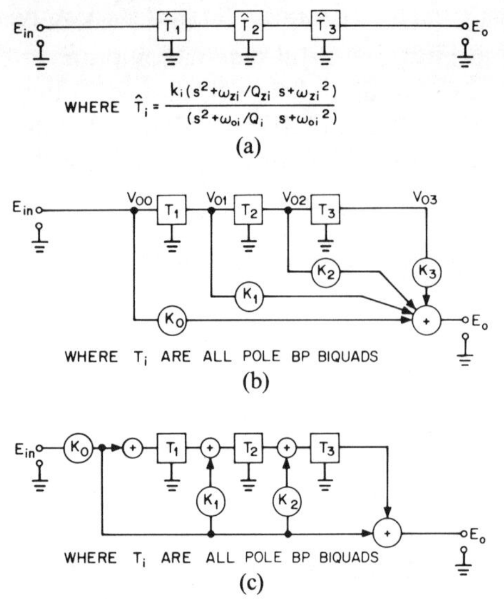

Fig. 4. Methods for realizing finite transmission zeros in multiple feedback active filter topologies. For simplicity and generality, the feedback paths are not shown. (a) LF (Szentirmai), GFLF—finite transmission zero biquad sections. (b) FLF, SCF, CB—feed forward summation. (c) CB, SCF—feed forward.

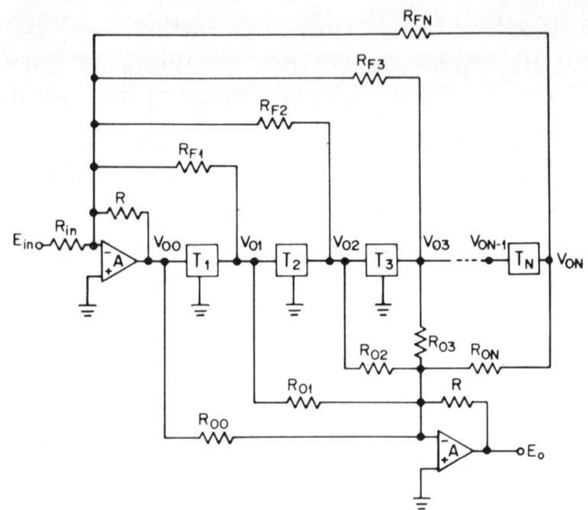

Fig. 5. General $2N$th-order FLF topology.

(3). Forming zeros in this manner, however, does severely complicate the synthesis equations, necessitating a very careful design without which the sensitivity performance can be quite poor [84]. Therefore, in order to get good sensitivity performance, a complex synthesis procedure involving sensitivity minimization is required. This method, however, is particularly useful when the desired response is nonsymmetric, i.e., it is not derivable from a LP prototype. The two remaining methods use feedforward to realize the desired transmission zeros. The technique in Fig. 4(b) is to tap, scale, and sum the input voltage and each of the section output voltages. In Fig. 4(c) the technique is to feed scaled replicas of the input signal forward to each of the section inputs. When the desired BP response is symmetrical, all-pole synchronously tuned BP biquad sections are typically used [22], [28]. It is noted that when $m<n$ in (1), the use of method (c) can result in the elimination [28] of the output summing amplifier. One can combine method (a) with (b) or (c) to obtain a more general synthesis technique with several new design freedoms. These freedoms can perhaps be exploited to improve the stopband sensitivity performance. To our knowledge, no attempt has been made to optimize the sensitivity behavior of such a general multiple-loop feedback–feedforward structure.

The feedforward techniques have been found to work well for moderate-order filters where $n \leqslant 8$. However, when the techniques are applied to filters of higher order, the reliance on subtraction to achieve transmission nulls tends to limit their stopband performance [39], and also the synthesis algorithm can result in impractical element values. Nevertheless, our experience has been that the feedforward techniques are quite adequate for a wide variety of requirements, particularly where stopband sensitivities are not critical. An example, demonstrating experimentally the performance of this design, is given in Section VI.

The summations in Fig. 4 are implemented by means of resistive or, if needed, opamp summers as in Fig. 5. Depending on the particular topology of the sections T_i, the interstage summations in Fig. 4(c) can frequently be incorporated into the input stages of the section T_i. In any event, the weighting factors K_i, as well as the feedback factors F_{ij} in Fig. 2, are always expressed as resistor ratios.

The statistical multiparameter sensitivity measures, to be used for the comparison and optimization, were introduced in [59]–[63]. These measures, which have been proven [61] to agree quite favorably with the Monte Carlo analysis but which are much less time consuming, have been shown to be valuable analytical tools [22], [24], [35], [41] for both the evaluation and design of MF filters. One has the option of considering either magnitude response (gain, α) sensitivity, phase (β) sensitivity, or transfer function (H) sensitivity. The measures can readily be expressed for any MF network, comprised of BP sections, in the following general form:

$$S_\alpha(x) = E\int_{\omega_1}^{\omega_2} \left|\frac{\Delta\alpha}{\alpha'}\right|^2 d\omega \simeq \int_{\omega_1}^{\omega_2} \sum_{i=1}^{N} S_{\alpha i}\, d\omega = \int_{\omega_1}^{\omega_2} I_\alpha\, d\omega \tag{7a}$$

$$S_\beta(x) = E\int_{\omega_1}^{\omega_2} |\Delta\beta|^2 d\omega \simeq \int_{\omega_1}^{\omega_2} \sum_{i=1}^{N} S_{\beta i}\, d\omega = \int_{\omega_1}^{\omega_2} I_\beta\, d\omega \tag{7b}$$

$$S_H(x) = E\int_{\omega_1}^{\omega_2} \left|\frac{\Delta H}{H'}\right|^2 d\omega \simeq \int_{\omega_1}^{\omega_2} \sum_{i=1}^{N} S_{Ti}\, d\omega = \int_{\omega_1}^{\omega_2} I_H\, d\omega \tag{7c}$$

where $S_H(x) = S_\alpha(x) + S_\beta(x)$ and $N(j\omega)/D(j\omega) = \alpha(\omega)\exp[j\beta(\omega)]$ from (1). In (7), E denotes the expected value, ω_1 and ω_2 define a designer-specified frequency band, $\Delta H \equiv H' - H$, $H' = H(j\omega, x + \Delta x)$, $\Delta\alpha = \alpha' - \alpha$, $\Delta\beta = \beta' - \beta$, x is a vector containing the nominal values of the network elements, and the vector Δx denotes the multiparameter random element variations. The quantities S_{Ti}, $S_{\alpha i}$, and $S_{\beta i}$ reflect the contribution of the ith section to the overall sensitivity measures. Let us briefly remark on the physical meaning attached to these measures. The integrands $I_{\alpha,\beta,H}$

$= E\{|\cdot|^2\}$ approximate the statistical second moments, as a function of frequency, for the complex random functions $|\Delta\alpha/\alpha'|$, $|\Delta\beta|$, and $|\Delta H/H'|$, respectively. When the component nominal and mean values are identical, it can be shown [60] that $I_{\alpha,\beta,H}$ reduce to the variance (i.e., standard deviation squared) of these random functions. Plotting $I_{\alpha,\beta,H}$ versus frequency produces curves similar to those shown in Fig. 3. The functional or integral sensitivity measures $S_{\alpha,\beta,H}$ then denote "average sensitivity" values for $I_{\alpha,\beta,H}$ over the specified frequency band $\omega_1 \leqslant \omega \leqslant \omega_2$. The functional sensitivity measures are particularly applicable to minimizing sensitivity with nonlinear programming techniques [22], [23].

Much of the practical value attached to these sensitivity measures is due to their ability to serve as figures of merit for quickly screening or comparing active filter designs or objective functions for sensitivity optimization. Consequently, one is encouraged to apply judicious approximations to simplify (7). Usually, the following five simplifying assumptions are valid in narrow-band ($\hat{Q} \geqslant 10$) BP design: 1) only sensitivities within the passband and transition bands are of interest, 2) statistical correlation between circuit elements, imposed either by the manufacturing process or by tuning [64], exists only within each section, 3) the sensitivities of the section denominator predominate over those of the section numerator, 4) variations in the section center frequencies are the predominant effects which determine the section denominator sensitivities ($\boldsymbol{S}_{\boldsymbol{x}}^{D_i}$), and 5) in active RC designs, variations in the passive components predominate over the active element variations. These assumptions are usually valid at voice frequencies, for many of the low-sensitivity active RC biquads [51]–[55] in use today. Applying assumptions 1) through 3) to the computation of $S_{\alpha i}$, $S_{\beta i}$, and S_{Ti} to (7) results in the following approximate expressions.

$$S_{Ti} = \left[\frac{\mathfrak{L}_i}{D}\boldsymbol{S}_{\boldsymbol{x}}^{D_i}\right]^{t*}(K)_i\left[\frac{\mathfrak{L}_i}{D}\boldsymbol{S}_{\boldsymbol{x}}^{D_i}\right] \tag{8a}$$

$$S_{\alpha i} = \left[\mathrm{Re}\left\{\frac{\mathfrak{L}_i}{D}\boldsymbol{S}_{\boldsymbol{x}}^{D_i}\right\}\right]^{t}(K)_i\left[\mathrm{Re}\left\{\frac{\mathfrak{L}_i}{D}\boldsymbol{S}_{\boldsymbol{x}}^{D_i}\right\}\right] \tag{8b}$$

$$S_{\beta i} = \left[\mathrm{Im}\left\{\frac{\mathfrak{L}_i}{D}\boldsymbol{S}_{\boldsymbol{x}}^{D_i}\right\}\right](K)_i\left[\mathrm{Im}\left\{\frac{\mathfrak{L}_i}{D}\boldsymbol{S}_{\boldsymbol{x}}^{D_i}\right\}\right] \tag{8c}$$

where superscripts t and * denote, respectively, the matrix transpose and complex conjugate operations. The quantities

$$\mathfrak{L}_i = D_i\frac{\partial D}{\partial D_i} \tag{9}$$

are complex frequency-dependent scale factors which depend on the feedback topology and are readily determined from the denominator $D(s)$ of the transfer function $H(s)$. Matrix $(K)_i$ denotes the covariance matrix [22], [24], [62] which contains the component statistics of section i).

Further, applying assumptions 4) and 5) to (8) yields the following expressions for S_{Ti}, $S_{\alpha i}$, and $S_{\beta i}$ in terms of the passive element statistics (correlation coefficients ρ_{rs} and standard deviations σ_{x_r}).

$$S_{Ti} \simeq 4\left|\frac{\mathfrak{L}_i}{DD_i}\right|^2\left\{\sum_{r=1}^{l_i}(S_{x_r}^{\omega_{0i}})^2\sigma_{x_r}^2 + \sum_{r=1}^{l_i-1}\sum_{s=r+1}^{l_i}(S_{x_r}^{\omega_{0i}})(S_{x_s}^{\omega_{0i}})\rho_{rs}\sigma_{x_r}\sigma_{x_s}\right\} \tag{10a}$$

$$S_{\alpha i} \simeq 4\left(\mathrm{Re}\frac{\mathfrak{L}_i}{DD_i}\right)^2\left\{\sum_{r=1}^{l_i}(S_{x_r}^{\omega_{0i}})^2\sigma_{x_r}^2 + \sum_{r=1}^{l_i-1}\sum_{s=r+1}^{l_i}(S_{x_r}^{\omega_{0i}})(S_{x_s}^{\omega_{0i}})\rho_{rs}\sigma_{x_r}\sigma_{x_s}\right\} \tag{10b}$$

$$S_{\beta i} \simeq 4\left(\mathrm{Im}\frac{\mathfrak{L}_i}{DD_i}\right)^2\left\{\sum_{r=1}^{l_i}(S_{x_r}^{\omega_{0i}})^2\sigma_{x_r}^2 + \sum_{r=1}^{l_i-1}\sum_{s=r+1}^{l_i}(S_{x_r}^{\omega_{0i}})(S_{x_s}^{\omega_{0i}})\rho_{rs}\sigma_{x_r}\sigma_{x_s}\right\}. \tag{10c}$$

The total number of circuit elements l_i, determining the ith section center frequency ω_{0i} ($\omega_{0i} = 1$ in (10)) is typically 4 or 6 for most useful active RC biquads. The reader should not be discouraged by the apparent complexity of (10) since in most practical cases [50]–[55] $|S_{x_r}^{\omega_{0i}}|$ are independent of x_r (typically $|S_{x_r}^{\omega_{0i}}| = 1/2$). If we further assume that all passive elements possess the same standard deviation σ (i.e., $\sigma_{x_i} = \sigma$ for all R's and C's), then (10a)–(10c) reduce to the following simple forms

$$S_{Ti} \simeq 4K\left|\frac{\mathfrak{L}_i}{DD_i}\right|^2\sigma^2 \tag{11a}$$

$$S_{\alpha i} \simeq 4K\left(\mathrm{Re}\frac{\mathfrak{L}_i}{DD_i}\right)^2\sigma^2 \tag{11b}$$

$$S_{\beta i} \simeq 4K\left(\mathrm{Im}\frac{\mathfrak{L}_i}{DD_i}\right)^2\sigma^2 \tag{11c}$$

where K depends on the biquad topology and the correlation coefficients ρ_{rs}. For example, when the elements are uncorrelated ($\rho_{rs} = 0$ for all r and s), $K = 1$ for STAR [55] and Tarmy-Ghausi (TG) [50] biquads, and $K = 1.5$ for the state-variable-type [51]–[53] biquads.

The measures (7) may also be used to evaluate response variations due to temperature drifts [63], where element changes are highly correlated not only within but also between all sections T_i in the MF filter.

In order to obtain reliable results from the above expressions, consistent [61] with those obtained by Monte Carlo methods, attention must be paid to the choice [60]–[62] of functions on which the sensitivity measures are based, namely $\Delta H/H'$ and $\Delta\alpha/\alpha'$, and, as discussed in [60], on the convergence properties of the Taylor series of the integrands in (7).

As suggested earlier, the measures can be used in either integral ($S_{\alpha,\beta,H}(\boldsymbol{x})$) or in integrand $I_{\alpha,\beta,H} = \sum_{i=1}^{N} S_{\alpha i,\beta i,Ti}$

form. In integrand form, one may plot sensitivity as a function of frequency to obtain curves similar to those in Fig. 3 determined by Monte Carlo analysis. When used in integral form, $S_{\alpha,\beta,H}(x)$ provide convenient functionals for sensitivity minimization and design screening. It is interesting to note that minimizing S_α, S_β, and S_H individually, usually results in the same optimum design [41], [70].

As shown later, MF design equations are obtained by equating the transfer function E_o/E_{in} of one of the circuits in Fig. 2, where T_i is given in (6), to a prescribed BP characteristic and comparing the coefficients. This process yields equations for the unknown sectional quality factors Q_i, the gain constant H_{Bi}, and the feedback factors F_{ij}. All of these parameters are not determined by the design equations, and thus the free parameters can be used in optimization.

It is customary in MF design to use $F_{ij}=F_{ij}(Q_i,H_{Bi})$ for satisfying the synthesis equations so that Q_i and H_{Bi} remain, within certain limits, free variables. Typically, sensitivity is minimized by adjusting in an iterative manner [22], [23] the section Q_i values subject to constraints, such as topological realizability conditions and practical fabrication requirements. For example, $F_{ij}(Q_i,H_{Bi})$ must clearly be real and finite, and $Q_i>0$; also, one may specify that $Q_i \leqslant Q_M$, where Q_M is some maximum realizable Q value, obtained, e.g., by considerations of element value spread. Mathematically, sensitivity is optimized by minimizing any one of the sensitivity measures in (7) as a function of the N variables Q_i, $i=1,\cdots,N$; the constraints are handled in terms of an appropriately defined penalty function [22], [23]. The degree to which this minimization can improve sensitivity performance over a good educated "guess" is very much topology dependent [35]. However, it should be noted that a poor MF design can be more sensitive than the cascade design we are seeking to improve [35]. Thus it is usually good practice to perform a complete sensitivity analysis or Monte Carlo simulation before a design is finalized.

Once the Q_i values have been determined, the remaining free parameters, i.e., the gain constants H_{Bi} in (6), are adjusted to maximize signal swing. Just as in cascade designs [65]–[67], the objective of this step is to maximize the undistorted peak-to-peak signal handling capability of the filter which of course determines the upper limit of the dynamic range. This procedure is particularly important for operation at higher frequencies (e.g., for μ741 $f_0>10$ kHz) where signal level limits are determined by opamp slewing [48]. In order to insure the full signal handling capability of the filter, it is necessary to control the voltage maxima at the output of all operational amplifiers such that for increasing signal levels no opamp behaves nonlinearly sooner than any other one. Clearly, this implies that the gains H_{Bi} should be chosen such that the voltage maxima at all critical nodes within the filter are equal [34], [38], [42]–[44], [75]. Typically, these critical nodes may be assumed to be the outputs of the sections T_i and of the summers.[3] In this case the requirement can be put into the form[4]

$$\max_{0\leqslant\omega<\infty} |H_i(j\omega)| = \max_{0\leqslant\omega<\infty} |H(j\omega)|, \qquad \text{for all } i \leqslant N-1 \tag{12}$$

where $H_i(j\omega)$, $i=1,\cdots,N$ are the voltage transfer functions (V_{oi}/E_{in}) from the input to the output of the ith section and where $H(j\omega)$ is the specified transfer function for the total filter. With the definitions

$$H(j\omega)=H_N(j\omega)\equiv H_{MN}h_N(j\omega) \tag{13a}$$

and

$$M_N \equiv \max_{0\leqslant\omega\leqslant\infty} |h_N(j\omega)| \tag{13b}$$

where $H_{MN}=|H_N(j\omega_0)|$ is the desired midband gain of the total filter, Equation (12) can be rewritten as[5]

$$\max_{0\leqslant\omega\leqslant\infty} |H_i(j\omega)| = H_{MN}M_N, \qquad i\leqslant N-1. \tag{14}$$

For all-pole BP filters, the functions $H_i(s)$ for any of the MF topologies depend on the section gains H_{Bi} defined in (6), so that via (12) those values of H_{Bi} can be determined which result in a uniform signal level throughout the MF topology. Observe that this procedure is independent of and does not affect the already completed sensitivity minimization. The specific equations and design tables for a number of MF all-pole BP filters are given in the subsequent sections.

Our comments thus far regarding dynamic range have been primarily directed towards all-pole BP MF filter design. When finite transmission zeros are realized via feedforward summation (Figs 4(b) and (c)), nothing in the design process changes in principle, except that the output of the feedforward summer must also be included in satisfying (12) or (14). When transmission zeros are realized in cascade fashion (Fig. 4(a)) using general biquad sections of the form of (3), the process of sensitivity and dynamic range optimization becomes considerably more complex. For this class of MF filters there is the additional problem of pole-zero pairing which affects to some extent both sensitivity and dynamic range. A further com-

[3]The assumption that the output voltage of a section T_i is the critical voltage is certainly true for single amplifier biquads (SAB's) where the section output corresponds to the amplifier output. In multiamplifier biquads it is assumed that the largest signal occurs at the section output (see, for example, circuits in [58]). If this assumption is not true for a particular biquad topology, then a simple scaling factor can be incorporated into (12) to account for this fact.

[4]Computational effort will be reduced if (12) is evaluated for the equivalent prototype LP functions, i.e., before using the LP-to-BP transformation. This is, of course, possible only for symmetrical functions $H(s)$.

[5]Satisfaction of (12) or (14) maximizes signal swing, i.e., the upper limit imposed on dynamic range. The lower limit, the noise floor, can always be improved by reducing the noise sources within the MF filter, notably by decreasing the impedance level and the noise contributions of the amplifiers [77]–[79]. The degree to which MF design parameters such as Q_i, can be used to reduce noise (with a corresponding trade-off in sensitivity performance) has not yet been investigated.

plication is that the feedback factors and pole-zero pairing are interdependent, thus a change in pole-zero pairing requires a complete redesign of the filter. Pole-zero pairing in cascade active filters has been given considerable treatment in the literature [65]–[67], but at this writing little is known about its effects in MF filters.

The concepts introduced here will be applied to specific MF topologies in the next sections. Synthesis equations will be derived which allow the designer to determine, for a given transfer function $H(s)$ or $H_{LP}(p)$, the feedback and feedforward factors F_{ij} and K_i in Figs. 2 and 4, and the design parameters (usually just Q_i and H_{Bi}) for the sections $T_i(s)$. Once the biquad design parameters are determined, it is assumed that the reader is able to obtain the appropriate realizations [1]–[5], [50]–[59]. It is noted again that any cascadable biquad can be used to realize $T_i(s)$.

For clarity of presentation, a unified approach is used in the sequel for all topologies. The approach may not be the most efficient nor the most elegant one for every case; however, its execution will result in a well performing filter. When appropriate, alternative synthesis methods are also recommended to the reader. Some illustrative examples are provided in sufficient detail for the reader to verify synthesis equations, and if desired, to build or simulate the resulting filters. Each example will be optimized according to sensitivity and dynamic range as per (7) and (12), respectively.

III. FLF-Type Filters

A. The FLF Synthesis

The FLF-type circuits are generally represented by the circuit schematic shown in Fig. 5. To remove some interaction and simplify tuning, the feedback and feedforward coefficients F_{1j} and K_i are implemented by single-input opamp summers. Straightforward analysis of Fig. 5 yields

$$H_N(s)=\frac{V_{ON}}{E_{in}}=\frac{-\alpha\prod_{j=1}^{N}T_j(s)}{1+\sum_{k=1}^{N}F_{1k}\prod_{j=1}^{k}T_j(s)} \tag{15}$$

$$H_i(s)=V_{oi}/E_{in}=H_N(s)\cdot\prod_{j=i+1}^{N}T_j^{-1}(s),\quad \text{for } i=0,1,\cdots,N-1 \tag{16}$$

and

$$H(s)=\frac{E_o}{E_{in}}=-\sum_{j=0}^{N}K_jH_j(s)=\alpha\frac{K_0+\sum_{j=1}^{N}K_j\prod_{k=1}^{j}T_k(s)}{1+\sum_{j=1}^{N}F_{1j}\prod_{k=1}^{j}T_k(s)} \tag{17a}$$

where

$$\alpha=R/R_{in}\quad F_{1j}=R/R_{fj}\quad\text{and}\quad K_j=R/R_{oj}. \tag{17b}$$

In some situations, where T_i with both positive and negative gains are not available, the single-input summers can be replaced with differential input summers [51]. For all-pole BP designs the output summing amplifier is not needed and $H(s)=-H_N(s)$ in (15).

The FLF synthesis equations are obtained by inserting T_i from h6) into (17), performing a BP-to-LP transformation,[6] and comparing coefficients between (17a) and (4). This procedure can be readily written in matrix form [22] as follows:

$$\boldsymbol{Lt}=\boldsymbol{\hat{d}} \tag{18a}$$

$$\boldsymbol{Lk}=\boldsymbol{\hat{c}} \tag{18b}$$

and

$$k_0=c_N \tag{18c}$$

where L is an $N\times N$ lower triangular matrix whose nonzero elements contain the unknown parameters q_i

$$q_i\equiv\gamma\hat{Q}/Q_i \tag{19}$$

i.e., the inverse of the relative sectional quality factors. q_i is introduced for convenience of notation. $\boldsymbol{t}$ is an $N\times 1$ vector of "pseudoloop gains" with elements

$$t_{1j}\equiv F_{1j}\prod_{k=1}^{j}H_{Bk}q_k,\qquad j=1,\cdots,N \tag{20a}$$

and the $N\times 1$ vector k contains the terms

$$k_j=K_j\prod_{k=1}^{j}H_{Bk}q_k,\qquad j=1,\cdots,N \tag{20b}$$

and $k_0=K_0$. The $N\times 1$ vectors $\boldsymbol{\hat{d}}$ and $\boldsymbol{\hat{c}}$ depend, respectively, on the denominator and numerator coefficients of (4) and on q_i. To illustrate the structure of (18), it is written for $N=3$, i.e., sixth-order BP functions, in (21):

$$\begin{bmatrix}1&0&0\\q_2+q_3&1&0\\q_2q_3&q_3&1\end{bmatrix}\begin{bmatrix}t_{11}\\t_{12}\\t_{13}\end{bmatrix}=\begin{bmatrix}d_2-(q_1+q_2+q_3)\\d_1-(q_1q_2+q_1q_3+q_2q_3)\\d_0-q_1q_2q_3\end{bmatrix} \tag{21a}$$

$$\begin{bmatrix}1&0&0\\q_2+q_3&1&0\\q_2q_3&q_3&1\end{bmatrix}\begin{bmatrix}k_1\\k_2\\k_3\end{bmatrix}=\begin{bmatrix}c_2-(q_1+q_2+q_3)\\c_1-(q_1q_2+q_1q_3+q_2q_3)\\c_0-q_1q_2q_3\end{bmatrix} \tag{21b}$$

and

$$k_0=K_0=c_3.$$

[6]For asymmetrical BP functions, where BP-to-LP transformation cannot be performed, one compares coefficients betwen (17) and (1). In principle, the procedure does not change. Performing the BP-to-LP transformation whenever possible reduces the degree of the functions and thereby simplifies the algebra significantly.

The lower triangularity of matrix L in (18) assures that for any LP prototype and any selection of q_i values, there exist finite t_{1j}'s and k_j's. This property is particularly important when iterative optimization methods [68] are used to optimize the filter's behavior.

The arbitrariness of the sections' Q_i values is used to select the parameters q_i such that the sensitivity measures in (7) or (10) are minimized. To evaluate the quantities in (8), (10), or (11) one must choose a section topology (to calculate $S_x^{D_i}$) and determine the quantities $\mathfrak{L}_i$ of (9) which for FLF design can readily be written in the following recursive form [22]:

$$\mathfrak{L}_1 = \prod_{j=1}^{N} D_j \tag{22a}$$

and

$$\mathfrak{L}_i = \mathfrak{L}_{i-1} + t_{1,i-1}\left(\frac{j\omega}{\gamma\hat{Q}}\right)^{i-1} \prod_{j=i}^{N} D_j, \qquad \text{for } 2 \leqslant i \leqslant N. \tag{22b}$$

Once the q_i values are determined, the parameters t_{1j} and k_j can be calculated from (18). It is clear from (20) that the resistor ratios F_{1j} and K_j in (17b) can still be obtained for an arbitrary set of gains H_{Bk} which are chosen to satisfy (12) for a maximized dynamic range.

Using (6) and (16) and defining

$$M_i = \max_{\omega} |h_N(j\omega) \prod_{j=i+1}^{N} t_j^{-1}(j\omega)|, \qquad i=0,\cdots,N-1 \tag{23}$$

which can readily be evaluated by computer, (14) results in

$$M_i H_{MN} \prod_{j=i+1}^{N} H_{Bj}^{-1} = M_N H_{MN} \tag{24}$$

or [34]

$$H_{Bi} = M_{i-1}/M_i, \qquad i=1,\cdots,N. \tag{25a}$$

Relatively accurate approximate values, which permit the computer evaluation of (23) to be bypassed are [75]

$$H_{Bi} \approx (1+q_i^{-2})^{1/2}. \tag{25b}$$

H_{B1} is determined from (25) only if the input opamp summer in Fig. 5 is used; if a resistive summer is employed instead, H_{B1} is arbitrary. Finally, the parameter α in (17b) sets the overall gain H_{MN} of the function in (15) as [34]

$$-H_{MN} = (\alpha/d_0) \prod_{j=1}^{N} H_{Bj} q_j. \tag{26}$$

When (17a) is used, $\alpha = 1$ since all the coefficients c_i in (4) are already realized via (18b). If this gain overdrives the output opamp, α must be reduced appropriately.

In summary, for a given LP prototype, element statistics (σ_{xi}, ρ_{rs}) and frequency band $\omega_1 \leqslant \omega \leqslant \omega_2$ for the sensitivity minimization, the design procedure for minimum sensitivity FLF BP filters can be stated as follows:

1) Adjust the Q_i values and compute t_{1j} to minimize either $S_\alpha(x)$ or $S_\beta(x)$ or $S_H(x)$ over the frequency band $\omega_1 \leqslant \omega \leqslant \omega_2$; note that t_{1j}'s must be computed from (18a) with each iteration of the Q_i's to compute sensitivity i.e., $\mathfrak{L}_i$ in (22b).

2) From the optimum Q_i's, compute the feedforward k_i's according to (18b) and (18c).

3) Adjust gain constants H_{Bi} to equalize the voltage maxima according to (23)–(25).

4) From the Q_i's and H_{Bi}'s, compute the feedback (F_{1j}'s) and feedforward (K_j's) factors according to (20a) and (20b), respectively, and α according to (26).

5) Determine the resistor values R_{in}, R_{fj}, and R_{oj} (for all j) from the computed α, F_{1j}'s, and K_j's according to $R_{in} = R/\alpha$, $R_{fj} = R/F_{1j}$, and $R_{oj} = R/K_j$ where R determines the summer impedance level and is usually set to some convenient value such as 10 kΩ.

6) Compute the biquad section element values to realize the BP T_i's specified by Q_i, H_{Bi}, and ω_0; note how the biquad sections are designed independently, as they would be in cascade synthesis. Examples illustrating the above procedure are given in Section VI.

B. The SCF and the PRB Synthesis

As noted previously, SCF and PRB are special cases of FLF. The SCF differs from FLF design essentially by eliminating the self-loop around block T_1, i.e., $F_{11} = 0$ and the assignment of identical sections to T_2 through T_N. This implies (see (18a) and (21a)) that $q_1 = d_{N-1} - \Sigma_{i=2}^{N} q_i$, and $q_2 = \cdots = q_N = q$. The remaining procedure is the same as in Section III-A

In PRB design all sections T_i are assumed to be identical, thus $Q_i = Q$ and $H_{Bi} = H_B$, $i=1,\cdots,N$, also $F_{11}=0$. Using identical sections provides three manufacturing advantages, namely, 1) increases the volume serviced by each biquad, 2) reduces the engineering time required to design the filter, and 3) minimizes the maximum section Q_i. With these simplifications the PRB synthesis equations are readily obtained from the FLF case, i.e.,

$$Q = \frac{N\gamma\hat{Q}}{d_{N-1}} \qquad \text{(implying } t_{11}=0) \tag{27}$$

$$t_{1i} = \hat{d}_{N-i} - \sum_{k=1}^{i-1} L_{ik} t_{1k} \tag{28a}$$

$$k_i = \hat{c}_{N-i} - \sum_{k=1}^{i-1} L_{ik} t_{1k} \tag{28b}$$

and

$$k_0 = c_N \tag{28c}$$

where, (recall that $q = \gamma\hat{Q}/Q$),

$$\hat{d}_{N-i} = d_{n-i} - \frac{N!}{(i)!(N-i)!}(q)^i \tag{29a}$$

$$\hat{c}_{N-i}=c_{N-i}-\frac{c_N N!}{(i)!(N-i)!}(q)^i \tag{29b}$$

and

$$L_{ik}=\frac{(N-k)!}{(i-k)!(N-i)!}(q)^{i-k}. \tag{29c}$$

We note that PRB design, due to the simple and systematic nature of (28) and (29), can be readily performed with even modest capacity programmable calculators. Due to the broad unimodal sensitivity minimum [35] which typifies FLF designs, sensitivity is only modestly sacrificed [23] in PRB circuits where $Q_i=Q$ is chosen *a priori* rather than through sensitivity minimization.

Strictly speaking, dynamic range cannot be maximized because of the requirement that all second-order sections in the PRB circuit be the same so that arbitrary voltage scaling via H_{Bi} is not available. However, some suboptimum ought to be found because dynamic range can suffer significantly [34] for an improper selection of H_{Bi} due to the wide range of voltage maxima in FLF(PRB) designs. A very good choice is the approximate solution indicated in (25b), i.e.,

$$H_{Bi}\equiv H_B=\sqrt{1+q^{-2}}=\sqrt{1+[Q_i/(\gamma\hat{Q})]^2} \tag{30}$$

which is valid for symmetrical BP responses with $T(s)$ given in (6).

In summary, the PRB synthesis procedure can be stated as follows:

1) Compute Q from (27).

2) Compute t_{1j} and k_j according to (28).

3) Adjust H_{Bi} according to (30) to achieve approximate equalization of the voltage maxima.

4) From the Q and H_{Bi} values, compute the F_{1j} and the K_j using (20), i.e., $F_{1j}=t_{1j}(H_Bq)^{-j}$ and $K_j=k_j(H_Bq)^{-j}$.

5) Compute R_{fj} and R_{oj} from the F_{1j} and K_j according to (17b); α determines the overall gain as in (26).

6) Compute the biquad element values for each section to realize the T_i's.

C. Generalized FLF(GFLF) Synthesis

The GFLF [31], [37], [40] method uses the FLF topology in Fig. 2(a) with general biquad T_i's of the form given in (3). Its transfer function can be readily obtained in terms of the individual section transfer functions $T_i=N_i/D_i$. For an nth-order design from (15), the transfer function is given by

$$H_N(s)=\frac{N(s)}{D(s)}=\frac{-\alpha\prod_{j=1}^{N}N_j(s)}{\prod_{j=1}^{N}D_j(s)+\sum_{k=1}^{N-1}F_{1k}\prod_{j=1}^{k}N_j(s)\prod_{i=k+1}^{N}D_i(s)+F_{1N}\prod_{j=1}^{N}N_j(s)}. \tag{31a}$$

For example, for the sixth-order design of Fig. 2(a) with $(F_{11}=0)$, (31a) becomes

$$H_N(s)=\frac{N}{D}=-\frac{\alpha N_1N_2N_3}{D_1D_2D_3+F_{12}N_1N_2D_3+F_{13}N_1N_2N_3}. \tag{31b}$$

From (31) it is seen that the GFLF transmission zeros are derived from the zeros of T_i in cascade fashion as shown in Fig. 4(a). Unfortunately, however, biquad transmission zeros enter into the denominator D of transfer function H_N via the external feedback. As mentioned earlier, this condition complicates the nature of the feedback signals and places extra burden on the stability and ideality of the section transmission zeros. Nevertheless, GFLF designs offer the ability to realize nonsymmetrical transfer functions and an improved stopband attenuation relative to all-pole filters of the same degree. Also, reduced stopband sensitivities in high-order filters may be obtained. A further by-product [40] is that GFLF filters can often be designed with sections of lower Q than cascade circuits, a feature which can be used to reduce noise and element value spreads within the biquad sections.

One procedure [31], [37], [40] for synthesizing GFLF filters is to successively reduce the desired transfer function by degree 2 with the extraction of a biquadratic section. The extraction process begins with section T_N and terminates when all feedback gains F_{1j} and sections T_i have been determined. Exercising complete freedom in selecting F_{1j}'s can result in biquad sections with right-half-plane poles. To insure all T_i's to have left-half-plane poles, certain restrictions [31] must be placed on the values of F_{1j}.

To begin the synthesis process, assume $F_{1j}=0$ for $j=1,\cdots,N-1$, yielding from (31)

$$H_N=\frac{\alpha N^{(1)}}{D^{(1)}}=\frac{\alpha N^{(1)}}{\tilde{D}^{(1)}+F_{1N}N^{(1)}} \tag{32}$$

where

$$N^{(1)}(s)=\prod_{j=1}^{N}N_j(s) \tag{33a}$$

$$\tilde{D}^{(1)}(s)=D^{(1)}(s)-F_{1N}N^{(1)}(s). \tag{33b}$$

Note that the superscript (1) denotes the first stage of the design and $N^{(1)}$, $D^{(1)}$ are the specified transfer function numerator and denominator, respectively. From (33) the design procedure is as follows:

1) Select a value of F_{1N} and compute the resulting $\tilde{D}^{(1)}$ according to (33b).

2) Factor $\tilde{D}^{(1)}$ into its N second-order factors, $D_i(s)$.

3) Select a D_j from the $\tilde{D}^{(1)}$ factors in (33b).

4) Pair D_j with any factor of $N^{(1)}(s)$ to yield $T_N = N_k/D_j$.

5) Compute $D^{(2)} = \tilde{D}^{(1)}/D_j$ and $N^{(2)} = \tilde{N}^{(1)}/N_k$ and form the relation $\tilde{D}^{(2)} = D^{(2)} - F_{1,N-1}N^{(2)}$, and repeat the process.

Note that at the end of the ith cycle one obtains two transfer functions, namely, a biquad T_{N-i+1} and another, probably more complex, transfer function $H^{(i)} = (N^{(i+1)})/(D^{(i+1)})$ whose degree is two less than that of the preceeding $H^{(i-1)}$. Also $\tilde{D}^{(i)} = D^{(i)} - F_{1,N-i+1}N^{(i)}$ as i increases from 1–N. The i design steps provide the following synthesis information 1) the i feedback factors $F_{1,N-i+1}$ to F_{11} and 2) the biquad section transfer functions T_{N-i+1} to T_1. The synthesis process illustrates the considerable amount of freedom one has in selecting F_{1j}, D_j, and the (N_i, D_i) pairings. These selections should be made with the following criteria in mind: 1) realizability of D_i's and hence of $T_i(s)$, 2) sensitivity, and 3) dynamic range. For good dynamic range the voltage maxima must be equalized as per (12). The procedure should closely parallel that outlined for the FLF designs with the final result being the values of the gains K_i in (3). Pole zero pairings affect both sensitivity and dynamic range. The relationship between these interdependent factors is complicated and unknown at this time.

D. IFLF Synthesis

Although FLF [22] and IFLF [32], [43] circuits are topologically different, an IFLF design can be readily obtained from an FLF design through a simple transformation, which completely preserves the FLF sensitivity behavior [43].

An IFLF filter, using N sections and resistor summers, is shown schematically in Fig. 6. This circuit is described by

$$H_N(s) = \frac{E_o}{E_{in}} = \frac{-\alpha \prod_{j=1}^{N} T_j(s)}{1 + \sum_{k=1}^{N} F_{kN} \prod_{j=k}^{N} T_j(s)} \tag{34a}$$

and

$$H_i(s) = \frac{V_{oi}}{E_{in}}$$

$$= -H_N(s) \frac{1 + \sum_{j=i+1}^{N} F_{jN} \prod_{k=j}^{N} T_k(s)}{\prod_{j=i+1}^{N} (G_j/G_{Tj}) T_j(s)}, \qquad i = 1, \cdots, N-1 \tag{34b}$$

where (with $G_{Ti} = G_i + G_{fi} + G_{i,in}$)

$$\alpha = \prod_{i=1}^{N} \frac{G_i}{G_{Ti}}; \quad F_{iN} = \frac{G_{fi}}{G_i} \prod_{j=i}^{N} \frac{G_j}{G_{Tj}}, \qquad i = 1, \cdots, N. \tag{35}$$

For all-pole filters, $T_i(s)$ is given in (6).

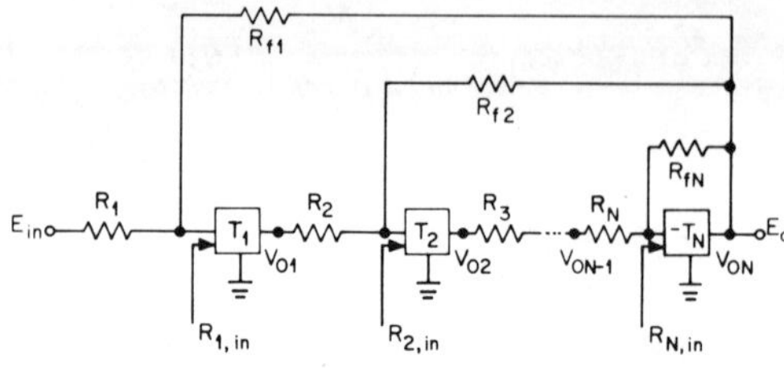

Fig. 6. General 2Nth-order IFLF topology.

A comparison of (34a) and (15) shows that a minimum sensitivity IFLF filter can be obtained directly from a similarly optimized FLF filter by setting

$$T_j^{(\mathrm{IFLF})}(s) = T_{N+1-j}^{(\mathrm{FLF})}(s) \tag{36}$$

with the optimum values of quality factors and loop gains given by

$$Q_j^{(\mathrm{IFLF})} = Q_{N+1-j}^{(\mathrm{FLF})}, \qquad t_{jN}^{(\mathrm{IFLF})} = t_{1,N+1-j}^{(\mathrm{FLF})} \tag{37}$$

where t_{jN} is defined as

$$t_{jN} = F_{jN} \prod_{k=j}^{N} H_{Bk} q_k, \qquad j = 1, \cdots, N \tag{38}$$

and q_K is defined in (19). From (38) it is clear that the gains H_{Bj} remain free parameters which can be chosen to maximize dynamic range according to (12). Recalling (6), (38), and (34b), and defining

$$M_i = \max_{\omega} \left| \frac{h_N(j\omega)\left[1 + \sum_{j=i+1}^{N} t_{jN} \prod_{k=j}^{N} q_k^{-1} t_k(j\omega)\right]}{\prod_{j=i+1}^{N} t_j(j\omega)} \right|, \quad \text{for } i < N \tag{39}$$

which can be readily evaluated by computer, results with (14) in the gain constants

$$H_{Bi} = (G_{Ti}/G_i) M_{i-1}/M_i, \qquad i = 2, \cdots, N \tag{40a}$$

which equalize the signal levels throughout the filter. From (34a), (35), (38), and (40a) one determines

$$H_{B1} = \frac{H_{MN} M_N}{M_1} \frac{G_{T1}}{G_1} |D(j\omega_0)| \tag{40b}$$

which sets the desired gain H_{MN}. $D(s)$ is the denominator of (34a).

One interesting feature of IFLF circuits is that the range of the internal signals varies only moderately, as compared [43] to FLF, so that dynamic range optimization is not quite as critical (see Fig. 10(c)). On the other hand, the design of FLF filters is somewhat easier than that of equivalent IFLF circuits, even if the transformations (36) and (37) are used, because of the dependence of

H_{Bi} and F_{iN} on the input admittances $G_{i,in}$ of the blocks $T_i(s)$. Notice in particular that the synthesis becomes very complicated if these admittances are frequency dependent.

IV. LF Type Filters

A. LF and Coupled Biquad (CB) Synthesis

The LF topology shown in Fig. 2(e) is an active *RC* simulation of doubly terminated passive *LC* ladder networks. In contrast to FDNR [11] and gyrator [6]–[9], [69] active ladders, which are impedance analogs of passive ladder networks, LF active ladders are voltage analogs. One result of this analogy, is that all but the first and last biquad sections have infinite pole $Q(s)$, emulating the lossless internal structure of the passive ladder. A second, and more significant, result is that the low passband magnitude-response sensitivities associated with properly matched passive ladders are preserved [6]. In fact, when properly terminated, the magnitude-response sensitivities for passive ladder filters achieve minima at frequencies of maximum power transfer, and experience indicates that this property is transferred to active ladders [12]–[16], [21], [62]. Thus, for LF filters which achieve maximum power transfer at distinct frequencies over a band about the center frequency, the passband magnitude response sensitivities are indeed very low. Butterworth and Chebyshev filters fall within this class. The low passband sensitivity for Butterworth LF designs was illustrated previously in Fig. 3.

Not all filters possess flat or equiripple passband nor is one always solely interested in passband magnitude-response sensitivities. When requirements demand, transition and stopband magnitude-response sensitivities, or phase or transfer function sensitivities, can be of predominant interest. For these situations, a more general *RLC* ladder [23], or, analogously, an MLF active ladder [30], provides the freedom to vary the internal Q_i's to optimize the filter to meet such requirements. To emphasize the analogy between the passive and active ladder structures, sixth-order BP doubly terminated *LC*, *RLC*, active LF and active MLF circuits are shown together in Fig. 7.

A brief remark regarding the distinction between the LF [12]–[14], [21] and CB [15], [16] design philosophies is in order. Both LF and CB use the LF topology in Fig. 2(e) for symmetrical all-pole BP design. For more general realizations, the two design philosophies are quite different. LF designs [21] are synthesized in much the same manner as one might synthesize a passive ladder, and the T_i's are general biquads of the form of (3). Thus the general LF, like the GFLF, derives its finite transmission zeros in cascade manner from the biquads T_i's. Furthermore both symmetrical and nonsymmetrical transfer functions can in principle be realized with general LF circuits. The CB, like the FLF, uses BP T_i's and derives its finite zeros using either one of the feedforward techniques illustrated in Figs. 4(b) and (c). As a result, only all-pole CB's are truly passive ladder derived.

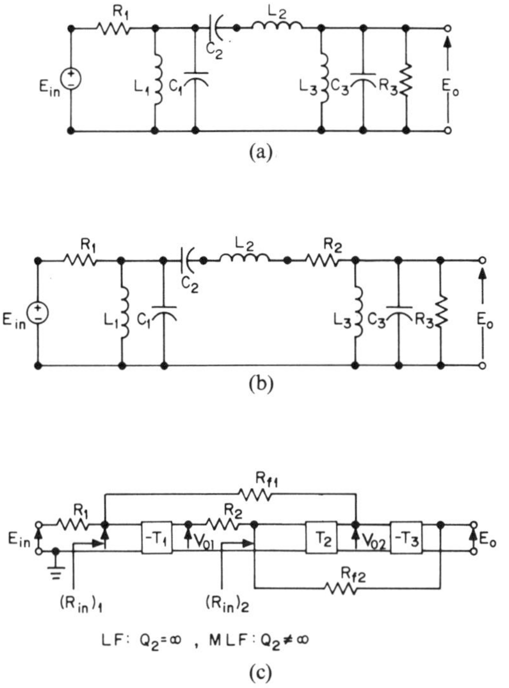

Fig. 7. Doubly terminated *LC* and *RLC* ladders and the resulting LF(MLF) active simulations, shown for sixth-order. (a) Double terminated *LC* ladder. (b) Double terminated *RLC* ladder. (c) MLF(LF) active ladder.

The transfer function for the general LF circuit, i.e., the configuration in Figs. 2(e) or 7(c) extended to *N* sections, is best derived using Mason's formula, i.e.,

$$H_N(s)=\frac{E_o}{E_{in}}=\frac{\alpha\prod_{i=1}^{N}T_i(s)}{\Delta_N(s)} \tag{41}$$

where $\Delta_N(s)$ is the determinant of the feedback configuration and α is given by a ratio of resistors. For example, for $N=4$,

$$\Delta_4(s)=1+F_{12}T_1T_2+F_{23}T_2T_3+F_{34}T_3T_4 + F_{12}F_{34}T_1T_2T_3T_4 \tag{42}$$

where

$$(G_{Ti}=G_i+G_{i,in},\qquad i=1,2,3)$$

$$F_{12}=\frac{G_{f1}G_2}{G_{T1}G_{T2}}\quad F_{23}=\frac{G_{f2}G_3}{G_{T2}G_{T3}}\quad F_{34}=\frac{G_{f3}}{G_{T3}} \tag{43a}$$

and

$$\alpha=G_1G_2G_3/(G_{T1}G_{T2}G_{T3}). \tag{43b}$$

Similarly, the transfer function from input E_{in} to the output of section *i* can be expressed as

$$H_i(s)=\frac{V_{oi}}{E_{in}}=\frac{\Delta_{Ni}(s)H_N(s)}{\prod_{j=i+1}^{N}(G_j/G_{Tj})T_j(s)},\qquad i<N \tag{44}$$

where $H_N(s)$ given in (41) and $\Delta_{Ni}(s)$ is the partial determinant of the feedback configuration with all feedback loops touching the forward path from E_{in} to V_{oi} removed.

For example, for $N=4$

$$\Delta_{41}=1+F_{23}T_2T_3+F_{34}T_3T_4$$
$$\Delta_{42}=1+F_{34}T_3T_4$$
$$\Delta_{43}=1. \quad (45)$$

The case $N=3$ is obtained from (41)–(45) by setting $F_{34}=0$, i.e., $R_3=G_{f3}=0$ and $T_4=1$.

For symmetrical all-pole LF BP designs, all sections T_i are synchronously tuned to ω_0, and $Q_i=\infty$, $i\neq 1$, N. When $F_{ij}=+1$, the LF structure is directly equivalent to a passive LC ladder. To achieve $F_{ij}=1$ requires replacing the resistive summers in Fig. 7(c) with opamp summers; however, in many active sections the desired summation function is naturally achieved at the input of the section, so that additional opamps are not necessary.

Using (44) with (41) in the feedforward topology of Fig. 4(b) results in the transfer function of a $2N$th-order CB design, i.e., with $g_j=G_j/G_{Tj}$,

$$H(s)=\frac{E_o}{E_{in}}=K_0+\sum_{i=1}^{N}K_iH_i(s)$$
$$=\frac{1}{\Delta_N(s)}\left[K_0\Delta_N(s)+\sum_{i=1}^{N}K_i\Delta_{Ni}(s)\prod_{j=1}^{i}g_jT_j(s)\right] \quad (46)$$

which, of course, is useful only for symmetrical BP transfer functions $H(s)$.

In principle, one can derive synthesis equations for LF and CB design by equating, respectively, the coefficients of (41) and (46) with those of (1). The equations derived in this manner become nonlinear for $N>3$ and thus are difficult to solve and time-consuming in the sensitivity optimization [30]. More efficient and elegant methods, which also yield more insight, for designing LF and CB circuits can be obtained. One method is to derive [21], [72], [74] a procedure which parallels well-known passive synthesis methods and involves taking advantage of the voltage equivalence between passive and LF ladders. A "pseudoladder" network is formed by assigning to the series impedances and shunt admittances biquad voltage transfer functions T_i. With this equivalence the LF ladder can be synthesized in a manner similar to passive ladders. Details can be found in references [21], [72], [74]. Also, a computer program [73] has recently become available for implementing this method. Another procedure [16] used for CB, symmetrical BP design, is to transform a passive LP prototype ladder into a BP all-pole CB design using state space techniques. This process yields only the all-pole function, i.e., the denominator of (46). To obtain the desired transmission zeros, the feedforward gains can be derived in a manner similar to (18).

The statistical multiparameter sensitivity measures in (7)–(11) can be used to screen LF and CB designs. For sixth-order LF and CB designs, e.g., where T_i are BP sections, the topology-dependent $\mathfrak{L}_i$ in (9) can be written as

$$\mathfrak{L}_1=D_1D_2D_3+\left(\frac{j\omega}{\gamma\hat{Q}}\right)^2 t_{23}D_1 \quad (47a)$$

$$\mathfrak{L}_2=D_1D_2D_3 \quad (47b)$$

and

$$\mathfrak{L}_3=D_1D_2D_3+\left(\frac{j\omega}{\gamma\hat{Q}}\right)^2 t_{12}D_3 \quad (47c)$$

where

$$t_{ij}=F_{ij}(\gamma\hat{Q})^2\prod_{k=i}^{j}a_k. \quad (48)$$

For finite values of Q_k, $a_k\equiv H_{Bk}/Q_k$ and for $Q_k\equiv\infty$, $a_k\equiv H_{Bk}$. The Nth order case is quite complicated, and no general expressions are available for $\mathfrak{L}_i$. Perhaps closed-form equations can be obtained using continuants [6]. It is noted that CB and LF all-pole symmetrical realizations do have one degree of freedom for sensitivity minimization, namely, either Q_1 or Q_N. Some improvements in transition band and stopband sensitivities can be obtained [30] in this manner.

No matter how the LF and CB circuits are derived, the T_i's must be properly scaled according to (12) to achieve good dynamic range. The procedure for equalizing the signal maxima is identical to the one used for MLF filters and will be discussed in the next section.

B. MLF Synthesis

An MLF active ladder [30] as noted earlier, is the analogue of a doubly terminated RLC ladder as illustrated in Fig. 7. Due to the internal resistors in the passive circuit, all blocks $T_i(s)$ in an MLF filter, in contrast to LF design, have finite pole Q's. These additional degrees of freedom are used to minimize the sensitivity (7)–(11). MLF has an additional advantage as compared to LF in that parasitic losses, due to the finite gain bandwidth (GB) of the opamps and capacitor dissipation factors, can be taken into account by predistorting the ideal, finite Q_i values. If a high-frequency active R realization is desired, MLF is particularly advantageous [38], since infinite Q active R biquads give rise to nonlinear oscillations. Equations (41)–(48), derived for LF design, apply similarly to the MLF topology, with the understanding that the blocks $T_i(s)$ have now finite Q_i values. Unfortunately, however, the insight and systematic procedures available for LC ladder synthesis do not exist for RLC synthesis. Therefore, for MLF filters, a procedure for symmetrical BP design has been used which parallels that outlined earlier for FLF synthesis. The method [30], [38], based on (41) and (46), is to select the free Q_i values to minimize the desired sensitivity measure (7). It has been found that for some designs both passband and stopband sensitivities can be improved [23], [30], [62] by an appropriate choice of Q_i. A sensitivity comparison is given in Section VI. Once the optimal values of Q_i are known, the synthesis equations, obtained by equating (41) and (1), are solved for the loop gains t_{ij} defined in (48). As in the FLF case, the gain constants H_{Bi} remain free parameters which are chosen to satisfy (12).

With (6) and (44) in mind, define

$$M_i = \max_{\omega} \left| \frac{h_N(j\omega)\Delta_{Ni}(j\omega)}{\prod_{j=i+1}^{N} t_j(j\omega)} \right|, \qquad i<N \tag{49}$$

which with (14) yields the gain constants [38]

$$H_{Bi} = (G_{Ti}/G_i)M_{i-1}/M_i, \qquad i=2,\cdots,N \tag{50a}$$

which equalize the signal level in the MLF structure. From (41) and (50a), one determines

$$H_{B1} = H_{MN}\frac{M_N}{M_1}\frac{G_{T1}}{G_1}|D(j\omega_0)| \tag{50b}$$

which sets the desired midband gain H_{MN}. $D(s)$ is the denominator of (41).

As pointed out before, dynamic range maximation of LF and CB filters proceeds in exactly the same fashion; care must be taken only to avoid possible difficulties due to the constraint $Q_i = \infty$, $i \neq 1, N$.

The (M)LF design procedure follows essentially the same sequence of steps 1–6 as the FLF method summarized at the end of Section III-A, only the equations cited there must be replaced by the corresponding ones in the present Section IV. The essential difference is that, in contrast to the linear synthesis equations (18a) for FLF, the corresponding equations for (M)LF become nonlinear for $N \geqslant 4$. As a result, the MLF procedure is complicated by a number of factors which do not appear to have been solved as yet: the coefficients of the nonlinear MLF synthesis equations depend on the set q_i provided by the sensitivity minimization process. The q_i set may fall into a region in q space for which the equations have no real solutions t_{ij}, i.e., a design is impossible. Or it may happen that, for a given set q_i and t_{ij}, the sensitivity "does not exist" (becomes infinite [35]) which, of course, physically is not meaningful. Satisfactory answers to these problems require further investigations.

The synthesis procedure [30], [38] for minimum sensitivity, maximum dynamic range (M)LF all-pole BP design has been written into a general MF design computer program [76]; the LF situation is handled by simply constraining Q_i, $i \neq 1, N$, to be a "very large" number ($Q_i = 10^5$). In Section VI, this program is used to establish design tables and a sensitivity comparison for several Butterworth and Chebyshev filters.

V. The MSF Topology

MSF [41], [70] represents the most general form of MF design; in principle, all the aforementioned MF topologies are special cases of MSF. One can thus visualize MSF as not only a distinct MF topology but as a general formalism for MF synthesis.

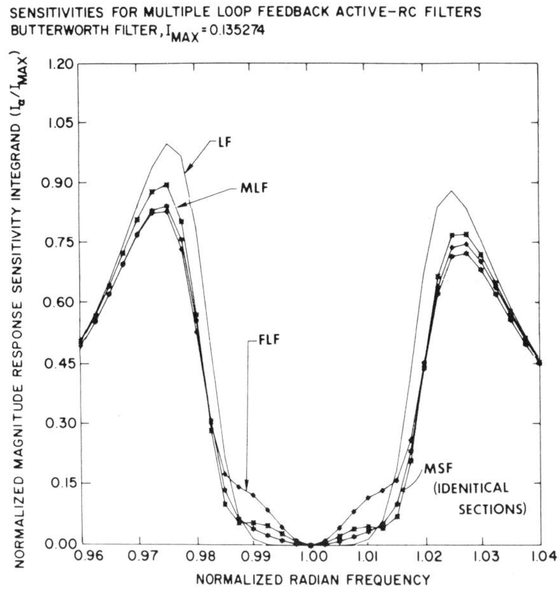

Fig. 8. Comparison of the normalized sensitivity integrands I_α as functions of frequency for different multiple feedback active RC Butterworth bandpass filters.

The purpose of the MSF development was to determine the feedback configuration resulting in the lowest possible sensitivity. In particular, it was to be investigated whether good stopband and passband sensitivities could be achieved simultaneously in the same filter. This objective was based on the following observations:

1) Optimized LF and MLF active filters yield the lowest sensitivities in the passband.

2) Optimized FLF active filters generally yield sensitivities lower than (M)LF designs at the band edges and in the stopband.

These observations are illustrated by the magnitude-response sensitivity spectra in Fig. 8 for sixth-order LF, MLF, and FLF BP designs with $\hat{Q}=25$. For these curves, the T_i's were assumed to be realized with TG biquads [50], with R's and C's normally distributed, correlated random variables with standard deviations of 0.2, and correlation coefficients $\rho_{R_iR_j} = \rho_{C_iC_j} = 0.8$ and $\rho_{R_jC_j} = 0$. In normalized form, however, these curves are essentially independent [70] of the biquad realization as long as the assumptions stated in Section II for the evaluation of (8) are valid, which is usually the case for most practical active RC biquads [50]–[55], [70].

With the full freedom provided by MSF one can achieve, to some degree, improvements in both passband and stopband sensitivities over those other MF topologies [70]. Using the FLF topology as a starting structure, one can outline a systematic process for adding new feedback loops in a manner which minimizes sensitivity. For example, for sixth- and eighth-order filters, this method results in the MSF topologies in Fig. 9, where the non-FLF loops are shown as dashed lines. FLF is a convenient starting structure in that its dependent variables (t) are readily determined from (18a). As noted previously, solutions to

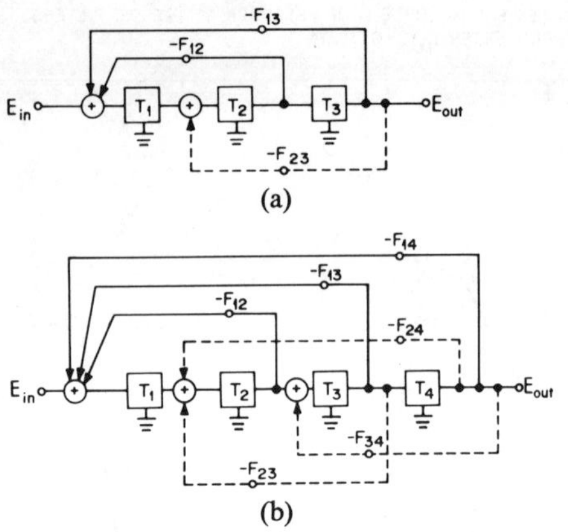

Fig. 9. MSF topologies for (a) sixth-order and (b) eighth-order bandpass realizations.

(18) always exist, and the addition of one or more feedback loops as independent variables does not compromise this important property.

As illustrated in Fig. 9, MSF configurations do not possess a repetitive feedback pattern; also, their complexity increases rapidly as order increases. To date, synthesis equations and sensitivity measures for all-pole sixth- and eighth-order BP designs have been derived [41], [70]. Although it is possible to generalize this process by using any one of the zero-forming methods in Fig. 4, such general MSF synthesis procedures are not yet available. For symmetrical all-pole BP designs, the MSF synthesis equations can be written in a form similar to that in (18), namely,

$$Lt=\hat{d}-g. \tag{51}$$

When L, t, and $\hat{d}$ are the FLF parameters and g is an $N\times 1$ vector which takes into account the "pseudoloop gains" of the non-FLF loops, e.g., for $N=3$ shown in Figs. 2(g) and 9(a),

$$g_1=0 \quad g_2=t_{23} \quad \text{and} \quad g_3=\frac{\gamma\hat{Q}}{Q_1}t_{23}=q_1t_{23}. \tag{52}$$

Corresponding expressions for eighth-order ($N=4$) design can be found in [70]. MSF sensitivities are readily computed using (7)–(11) where, for $N=3$,

$$Z_1=D_1D_2D_3+\left(\frac{j\omega}{\gamma\hat{Q}}\right)^2 t_{23}D_1 \tag{53a}$$

$$Z_2=D_1D_2D_3 \tag{53b}$$

$$Z_3=D_1D_2D_3+\left(\frac{j\omega}{\gamma\hat{Q}}\right)^2 t_{12}D_3 \tag{53c}$$

with t_{ij} defined as in (20a) and (48), namely,

$$t_{ij}=F_{ij}\prod_{k=i}^{j}H_{Bk}q_k. \tag{54}$$

Note that (19) is used in (54).

The independent parameters of MSF synthesis, the Q_i and the non-FLF t_{ij}'s, are selected to minimize sensitivity as discussed in Section II. The sensitivity improvements achievable are also illustrated in Fig. 8, which shows the sensitivity spectrum for a PRB-type (i.e., identical T_i blocks) MSF design together with those equivalent LF, MLF, and FLF circuits. The improved sensitivity behavior in the passband and at the band edge is evident. The design was achieved [41], [70] by optimizing S_α over the band $0.96\leqslant\omega\leqslant 1.04$, which is twice the filter bandwidth. Full optimization (i.e., nonidentical Q_i's) over the narrower band $0.99\leqslant\omega\leqslant 1.01$ gives different results [70] which interestingly possess a slightly broader sensitivity minimum than the LF design.

The remaining free parameters of MSF design H_{Bi} are selected to equalize the signal level according to (12). No general equations or results are available because the transfer functions to the internal outputs $H_i(s)$ depend on the "non-FLF" loop gains t_{ij} and thus can be calculated only after the MSF topology has been established by means of sensitivity minimization.

The MSF synthesis procedure can be summarized as follows:

1) Choose the Q_i's and the non-FLF t_{ij}'s to minimize either S_α, S_β, or S_H given by (7); with each iteration compute the FLF t_{1j}'s according to (51).

2) After the MSF topology has been established in step 1, determine H_{Bi}'s to equalize the signal maxima, by means of (12); compute F_{ij}'s according to (54).

3) Design independently the BP biquads for the parameters Q_i, H_{Bi}, and ω_0.

VI. Design Examples and Comparative Discussion

The primary goal of any MF design procedure is always to find the values of center frequencies ω_{0i}, pole quality factors Q_i, and gain constants H_{Bi} so that sections $T_i(s)$ of the desired type can be implemented. The feedback factors (or loop gains) are then determined which result in the prescribed transfer function $H(s)$. In this section a few examples will be presented to provide the reader with some information about the numbers involved and the ease or complexity of the calculations. Before discussing the examples, however, a few general observations for MF BP design procedures should be noted.

1) The center frequencies of all sections are the same and equal to the center frequency of the overall filter, i.e., $\omega_{0i}=\omega_0$, $1\leqslant i\leqslant N$. For the optimization, the actual value of ω_0 is not important due to normalization; it only affects the capacitor values in the active RC biquads which realize the $T_i(s)$.

2) For symmetrical BP functions, the actual value of the quality factor $\hat{Q}$ which determines the bandwidth of the total filter, $\omega_{3\text{ dB}}\equiv\omega_0/\hat{Q}$, is not important because only the relative quality factors $q_i=\gamma\hat{Q}/Q_i$ enter the synthesis procedure (comp. (18)–(21), (27)–(29), (38), and (41)–(44) with the LP-to-BP transformation). $\hat{Q}$ affects only the element values in the blocks $T_i(s)$.

3) Any cascadable second-order filter can be used to realize the functions $T_i(s)$. The specific topology chosen enters the optimization algorithm only through the values

TABLE I

DESIGN PARAMETERS AND RELATIVE SENSITIVITIES $S(x)/S^c(x)$ FOR OPTIMIZED BUTTERWORTH AND 0.5-dB RIPPLE CHEBYSHEV FILTERS OF ORDERS 4, 6, AND 8. BOTH PASSBAND AND BROADBAND SENSITIVITY VALUES ARE GIVEN. "——" ENTRIES DENOTE NONAPPLICABILITY, "x" ENTRIES DENOTE QUANTITIES NOT EVALUATED. FOR THE FLF FILTERS, Q_1 WAS DETERMINED SUCH THAT $t_{11}=0$ ACCORDING TO (18A) AND (21A)

	N	FEEDBACK TOPOLOGY	$Q_1/\hat{Q}$	$Q_2/\hat{Q}$	$Q_3/\hat{Q}$	$Q_4/\hat{Q}$	M_0	M_1	M_2	M_3	M_4	t_{12}	t_{13}	t_{23}	t_{14}	t_{34}	$1-\frac{1}{4\hat{Q}} \le \omega \le 1+\frac{1}{4\hat{Q}}$ S_α/S_α^c	$1-\frac{1}{\hat{Q}} \le \omega \le 1+\frac{1}{\hat{Q}}$ S_α/S_α^c
BUTTERWORTH	2	ALL	1.414	1.414	–	–	2.453	1.620	1.000	–	–	0.500	–	–	–	–	0.005	0.567
	3	FLF (a)	1.724	1.786	1.163	–	5.216	2.200	1.218	1.000	–	0.695	0.123	–	–	–	0.050	0.521
	3	LF	1.351	∞	0.794	–	–	0.0794	1.159	1.000	–	0.404	–	0.664	–	–	0.001	0.582
	3	MLF_1	1.351	20.000	0.826	–	–	4.295	1.576	1.000	–	0.447	–	0.560	–	–	9×10^{-4}	×
	3	MLF_2	1.852	2.857	0.901	–	–	4.179	1.596	1.000	–	0.607	–	0.217	–	–	×	0.512
	3	MSF	1.370	1.961	1.316	–	–	×	×	×	–	0.340	0.207	0.345	–	–	×	0.483
	4	FLF (b)	1.715	1.639	1.639	1.235	7.309	4.023	1.840	1.281	1.000	0.870	0.283	–	0.165	–	0.063	0.530
	4	LF, MLF_1'	0.665	∞	∞	0.901	–	0.0050	0.0770	1.125	1.000	0.861	–	0.295	–	0.589	3×10^{-4}	0.594
	4	MLF_2	0.831	1.250	5.000	2.439	–	4.079	1.623	1.232	1.000	0.239	–	0.269	–	0.640	×	0.500
CHEBYSHEV 0.5 dB	2	ALL	1.950	1.950	–	–	4.251	2.384	1.059	–	–	1.008	–	–	–	–	0.005	0.509
	3	FLF (c)	3.506	3.244	2.085	–	26.41	8.420	2.694	1.000	–	1.027	0.074	–	–	–	0.190	0.536
	3	LF, MLF	1.816	∞	1.914	–	–	0.1225	1.784	1.000	–	0.570	–	0.573	–	–	0.032	0.429
	3	MSF	∞	0.965	∞	–	–	×	×	×	–	0.724	0.683	0.759	–	–	0.025	×
	4	FLF (d)	4.091	5.202	3.524	2.664	168.3	37.74	8.223	2.853	1.059	1.190	0.067	–	0.193	–	0.097	0.642
	4	LF, MLF	1.861	∞	∞	1.791	–	0.0080	1.319	1.899	1.059	0.488	–	0.354	–	0.516	0.004	0.361

(a) FOR PRB $Q_1/\hat{Q} = Q_2/\hat{Q} = Q_3/\hat{Q} = 1.500$

(b) FOR PRB $Q_1/\hat{Q} = Q_2/\hat{Q} = Q_3/\hat{Q} = 2.800$

(c) FOR PRB $Q_1/\hat{Q} = Q_2/\hat{Q} = Q_3/\hat{Q} = Q_4/\hat{Q} = 1.531$

(d) FOR PRB $Q_1/\hat{Q} = Q_2/\hat{Q} = Q_3/\hat{Q} = Q_4/\hat{Q} = 3.650$

$S_{x_i}^{\omega_{0i}}$ in (10). The fabrication technology chosen (discrete, integrated-hybrid, or monolithic) will also influence the choice of the blocks, e.g., active RC or active R, and enters through the element statistics ρ and σ in (10).

4) The topology of the total filter (FLF, MLF, MSF, etc.) is reflected in the form of the parameters $\mathfrak{L}_i$ defined in (9). Thus, the sensitivity measure (10) is separated into section-dependent and topology-dependent parts.

5) The gain constraints H_{Bi} which satisfy (12) are independent of the section type and of the desired overall gain H_{MN} (comp. (25), (30), (40a), (50a)). They do not affect any prior sensitivity minimization.

6) PRB-type FLF and IFLF do not require sensitivity optimization and can be readily synthesized using (27)–(29) for FLF, and (27)–(29) with (36)–(38) for IFLF. These topologies will usually [23] yield good, suboptimum sensitivity behavior. For better performance, one can employ varying degrees of sophistication from FLF to MSF to meet tough requirements.

7) As in any other active filter, the performance of MF filters can be improved with postfabrication tuning [64]. Since thin-film capacitors can have tolerances on the order of 5 percent [55], [64], postfabrication tuning and laser trimiming have become somewhat standard practices in the production of precision active filters [55]. Optimum tuning procedures [64] have been developed for tuning cascaded biquads. With some modification, these procedures are also applicable [85] to MF filters.

8) Typically, the most critical section in a cascade of biquads is the highest Q section [86]. That is, variations in the response of the highest Q section will yield a more significant variation in the overall response than equivalent variations in the responses of the lower Q sections. Unfortunately, the highest Q section is also the one which is most difficult to control, statistically, in production. In MF filters, on the other hand, the highest Q sections are not always the most critical [86]. More importantly, the impact of variations in the higher Q sections on the overall response is much reduced in MF filters. The LF and CB topologies, comprised of $N-2$ infinite Q_i sections, are excellent examples of this phenomenon.

The examples chosen in the following are Butterworth and 0.5-dB ripple Chebyshev BP filters of order 4, 6, and 8. The second-order blocks assumed in the examples are TG sections [50], which are the minimum sensitivity active RC biquad blocks. Other biquad blocks for active RC [28], [39], [40] or active R [33], [36], [38] MF designs can also be used. For the realizations, FLF, LF, and MLF topologies are considered so that the differences and similarities of the approaches and results can be compared. The results for sixth-order MSF are also shown. The design parameters presented in Table I have been calculated by means of a FORTRAN computer program (SENSMIN [76]), developed at the University of Minnesota for realizing minimum sensitivity, maximum dynamic range FLF- and MLF-type all-pole BP filters.

SCF, PRB, LF, CB, and IFLF topologies are treated as special cases, as is the sensitivity calculation of cascade structures. SENSMIN is based on the FLF synthesis program outlined in [20] and on much of the theory contained in this paper; it is limited to handling BP filters up to order 14 for MLF and 16 for FLF topologies, but can easily be extended to filters of higher order (see Appendix). The sensitivity data in Table I have been normalized to those of the corresponding cascade realizations, denoted by S_α^c, in order to demonstrate the improvement obtainable with MF synthesis.

The design data and sensitivites in Table I are based on the element statistics $\sigma = 0.2$ percent and $\rho_{rs} = 0$. The sensitivity comparisons shown are, however, more general and apply to realizations employing any active biquad for which the assumptions stated in Section II remain valid. In addition, they will be essentially unaltered by changes in σ and ρ_{rs}, as long as the element variations determined by σ do not become excessively large [60] to invalidate the linear approximation.

The table gives those values of $Q_i/\hat{Q}$ which minimize the sensitivity $S_\alpha(\mathbf{x})$ of (7a). Each design was carried out twice, once to minimize $S_\alpha(\mathbf{x})$ over the passband only, exclusive of the bandedges, i.e., $1-1/(4\hat{Q}) \leqslant \omega \leqslant 1+1/(4\hat{Q})$, and a second time over the broader band $1-1/\hat{Q} \leqslant \omega \leqslant 1+1/\hat{Q}$, which includes the transition regions. The passband sensitivity improvements are quite dramatic, particularly for the ladder derived topologies. Observe that the sensitivity improvement over that of cascade design, when calculated over the broader band, is not as striking, but nevertheless the MF sensitivities are still lower by about a factor 2. This result ought to be expected in view of Fig. 3 since now the band edges with their large sensitivity contributions are included in the evaluation. It should be noted that Q_i and H_{Bi} variations are excluded from the sensitivity measure (comp. Section II); if these variations are included in the analysis, the ratio S_α/S_α^c for passband evaluations will increase somewhat. This increase will, however, not substantially alter the percent improvement $|S_\alpha^c - S_\alpha|/S_\alpha^c$. It is interesting to compare the sensitivity improvements as a function of filter order. For the examples evaluated, this improvement appears to be a function of response shape, MF topology, and frequency band of interest. For example, the sensitivity improvements with FLF Chebyshev designs can either increase or decrease with filter order, depending on the frequency band of interest. Before drawing general conclusions about the trends suggested by the sensitivities in Table I, more careful investigations extended to higher order need to be made.

Since, with the exception of the FLF topology, the actual section gains H_{Bi} can be determined only after the section type and its input conductance G_i are chosen (comp. (40) and (50)), Table I lists only the values M_i, defined in (23) and (49), which are needed for calculating H_{Bi}. M_0, required for the FLF topology with input summer (Fig. 5), is also included.

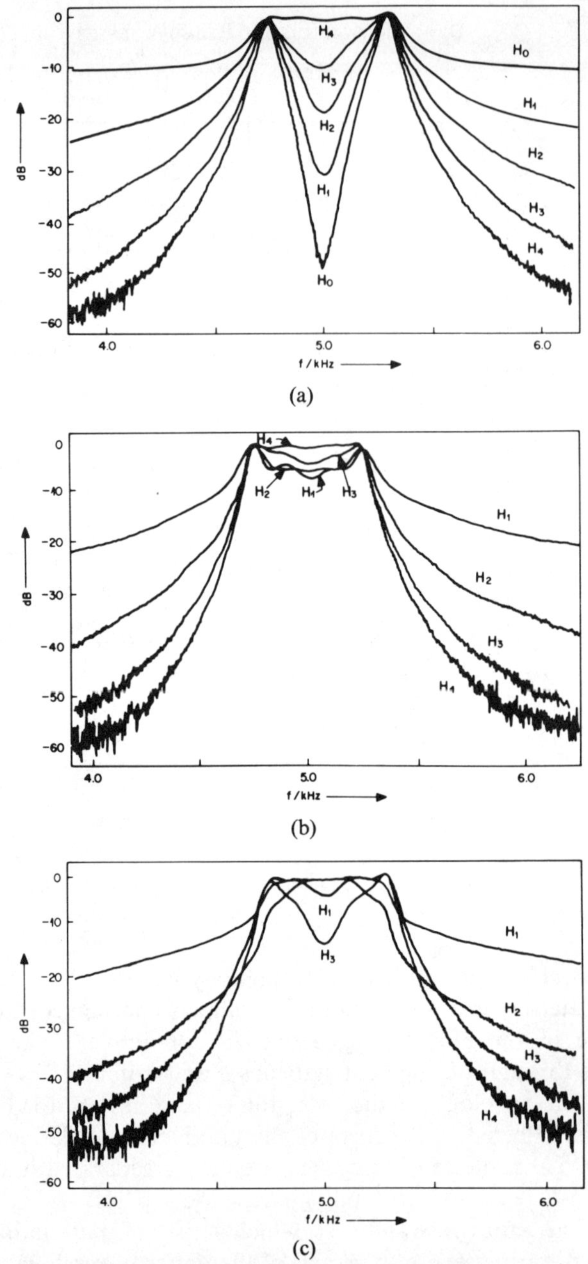

Fig. 10. Experimentally determined output voltages V_{oi}, $i \leqslant 4$, for optimized (a) FLF, (b) MLF, and (c) IFLF realizations of an eighth-order 0.5-dB ripple Chebyshev bandpass filter with $f_0 = 5$ kHz, using Tarmy-Ghausi sections to realize $T_i(s)$.

The equalization of the maximum signal levels is demonstrated by the experimental results shown in Fig. 10 [44], obtained for eighth-order 0.5-dB ripple Chebyshev FLF, MLF, and IFLF BP filters with $f_0 = 5$ kHz and a 0.5-dB bandwidth of 500 Hz, designed using the parameters in Table I. Although the output V_4 is of course the same for Figs. 10(a)–(c), it is noted that the internal voltages V_i, $i<4$ occupy widely differing signal ranges for the three topologies.

A further example is included to demonstrate experimentally the low-sensitivity behavior of a particular fourth-order MF filter. The filter is to realize the following LP prototype transfer function.

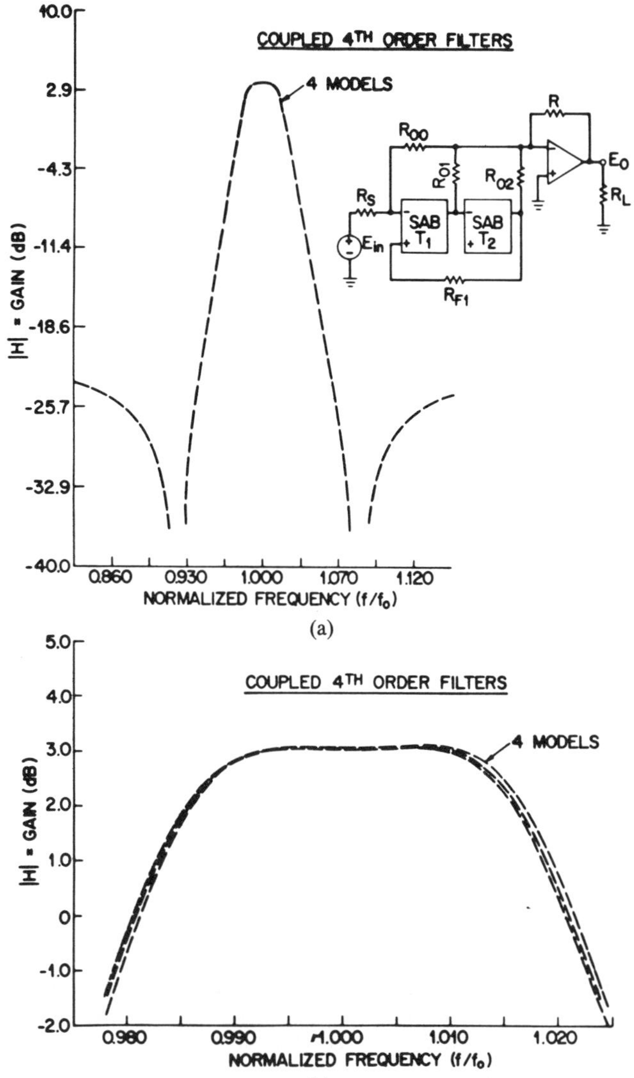

Fig. 11. Overall (a) and passband (c) response measured for four models of the MF topology (b) used to realize (55).

$$H_{LP}(p) = \frac{p^2 + 18.9756}{p^2 + 1.4126p + 1.2772} \tag{55}$$

with $f_0 = 1$ kHz and $\hat{Q} = 25.64$. The desired BP response is shown in Fig. 11(a). One objective of this design was to realize this filter with the minimum number of operational amplifiers, namely three for MF realization. To achieve this goal, the topology shown in Fig. 11(b), which uses feedforward summation to realize the transmission zeros, was selected. Note that to avoid the need for an additional operational amplifier to properly phase the feedback path, the output of the SAB Section 2 is fed directly into the positive terminal of the op-amp in SAB Section 1. A generalization of this approach for higher order designs is briefly discussed in the next section. The SAB BP sections [55] and output summer were realized using STAR [55] technology. The design parameters for each section were: SAB1: $f_0 = 1$ kHz, $Q_i = 37.01$, $H_{Bi} = 2.985$, and SAB2: $f_0 = 1$ kHz, $Q_2 = 37.01$, and $H_{B2} = 1.496$. Note, $Q_1 = Q_2 = 37.01$ is somewhat beyond what is commonly

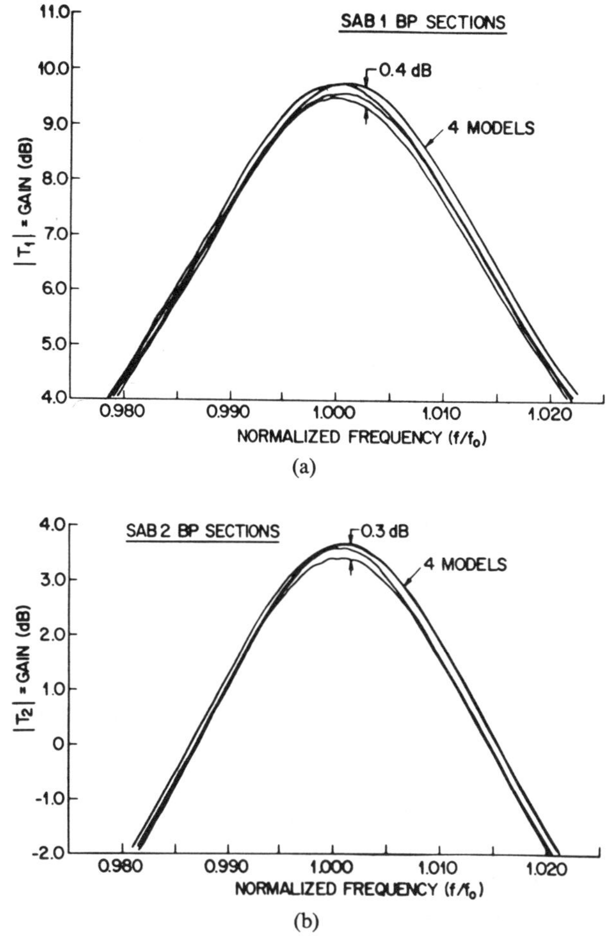

Fig. 12. Experimental responses of four SAB sections used to implement (a) section 1 and (b) section 2 of the filters of Fig. 11(b).

considered to be the upper limit ($Q \leqslant 30$) [40] for precision SAB filters.

Four models of this design were implemented, and the resulting measurements are summarized in Figs. 11 and 12. The filters were tuned by using the measured capacitor values of each STAR section to calculate the desired resistor values via the synthesis equations [55]. The final step is to laser trim the STAR resistors to these values. Laser trimming is not perfect and trim tolerances on the order of 0.2 percent are typical [64]. Fig. 12 shows the resulting responses for the four pairs of STAR sections which comprise the four MF BP filters shown schematically in Fig. 11(b). Note the variations in the sections; the spread about the center frequency is about 0.4 dB (± 0.2 dB) for each section. After the sections are measured, they were randomly combined to realize the four MF filters with responses shown in Figs. 11(a) and (c). Note especially the experimental curves in Fig. 11(c) where the four MF filter responses are plotted to the same scales (1 dB/div and 0.005 Hz/div) as the STAR section responses in Fig. 12. The deviation in the MF responses at and about the center frequency is almost indiscernible. In particular, the response deviations over the band $0.99 \leqslant f/f_0 \leqslant 1.01$ are much smaller than the deviation of any one section. If these sections had been cascaded, the individual section variations at each frequency would sum to yield the overall response variations, which would be expected to be on the average significantly larger than those shown in Fig. 11(c) for the MF realization. Note also the four responses plotted over a wider frequency range in Fig. 11(a). The stopband performance, including the notch areas, are not degraded by the feedforward summation. The experimental results do indeed illustrate that the sensitivity trends in Table I do occur in practice. The extent to which the dramatic improvements cited in Table I will be achieved depends on the design requirements, the quality of the passive and active components and the degree to which the assumptions stated in Section II are valid.

VII. Concluding Remarks

The design of active filters using MF topologies is reviewed in this paper. The synthesis procedures discussed result in realizations with minimum sensitivity and maximum dynamic range; the procedures for each of the MF topologies are summarized in the corresponding sections. Since modular design is emphasized in this paper, only biquad sections in the form of (6) are treated. However, the use of general biquads (3) in MF configurations has been pointed out where appropriate.

In a number of practical applications, it may be desirable to minimize the number of opamps and to get good sensitivity performance. Minimization of opamps arises mainly from the requirement to reduce power consumption and cost. Thus, SAB's are good candidates for the design of MF active filters [50], [80]–[83], in particular, since one attractive feature of MF approaches is that good sensitivity behavior [28] can be achieved with SAB's [55]. With judicious modifications to the topology, CB and IFLF designs can be implemented with a minimum number of opamps and with SAB's of the same topology. The minimum number of opamps for an $2N$-order all-pole design is N. If feedforward summation is used, then $N+1$ opamps are required. SAB–MF topologies are shown in Fig. 13. In lieu of using different [32] SAB topologies to achieve the proper sign, the feedback signal is directed [28] to either the positive or negative terminals of the opamp. The resulting transfer functions are not ideal BP's; however, the error in the approximation can be made negligible.

In addition to fully coupled MF designs one can use partially [39] coupled MF topologies such as that shown in Fig. 13(c). Partially coupled designs do not require sophisticated synthesis procedures and require fewer[7] resistors than fully coupled designs. With the partially coupled topology in Fig. 13(c), the design can be conveniently dissected into the independent synthesis of coupled-biquad pairs. It has been demonstrated that with coupled-biquad pairs one can improve the stopband sensitivities at the expense of higher passband sensitivities as compared to those obtained with some fully coupled structures. Nevertheless the passband sensitivities of coupled-biquad pairs are still lower than those achieved with cascade design.

[7]Fewer resistors need not imply less total resistance.

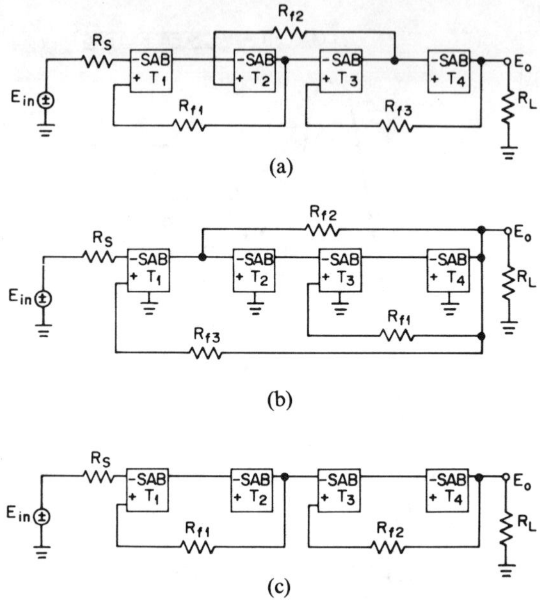

Fig. 13. Minimum opamp topologies using SAB sections (shown for eighth-order bandpass realizations). (a) LF, MLF. (b) IFLF. (c) Mixed coupled cascade (MCC).

Appendix
Brief Description of Program SENSMIN

SENSMIN [76] is a FORTRAN program for batch processing, designed to find an optimized design of transfer functions of the form of (5) with FLF- or MLF-type topologies. The current limits are order $N=9$ for FLF and $N=7$ for MLF topologies; these limits can easily be extended by redefining the appropriate dimension statements and loop counts.

The required input is: 1) degree and coefficients of $H_{LP}(p)$ in (5); 2) type of topology desired (LF, MLF, FLF, Cascade); 3) type of sections desired for the realization of $T_i(s)$ in (6) and their center frequencies ω_{0i}; 4) statistics (ρ_{rs} and σ_i) of the elements used in T_i's; 5) the frequency interval ($\omega_1 \leqslant \omega \leqslant \omega_2$) over which one of the sensitivity measures (7) is to be minimized and calculated; 6) and an initial set of N Q_i values for the second-order sections $T_i(s)$. The output consists of: 1) the set Q_i, $1 \leqslant i \leqslant N$, which minimizes the chosen measure (7) for the FLF- or MLF-type topology; 2) the gains H_{Bi} which equalize the signal level according to (12); 3) the loop gains t_{ij}, defined in (20a) and (48), which satisfy the synthesis equations, such as (18a) for FLF; 4) the minimized value of the sensitivity measures S_α, S_β, and S_H of (7); 5) and a plot of the sensitivity integrands $I_{\alpha,\beta,H}(\omega)$ in $\omega_1 \leqslant \omega \leqslant \omega_2$. The internal outputs $|H_i(j\omega)|$ defined in (12) can be plotted if desired.

SENSMIN is "compartmentalized" into numerous independent sections in order to facilitate additions and changes to the program. Thus separate subroutines have been written to handle: sensitivity minimization, solution

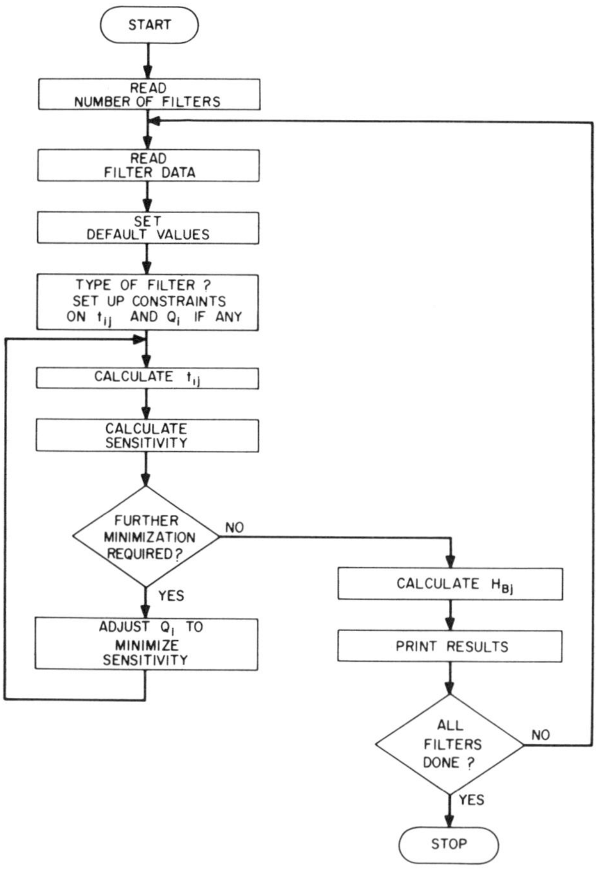

Fig. 14. Simplified flowchart of program SENSMIN.

of the synthesis equations for FLF, and for MLF-type topologies, evaluation of the sensitivity contributions $S_{\alpha i}$, $S_{\beta i}$, or S_{Ti} (see (10)) of the individual sections, the topology-dependent terms $\mathfrak{L}_i$ of (9); and the signal-level maximization of (12). The flow of the program (see flowchart Fig. 14) follows essentially the sequence of steps outlined at the end of Section III-A. For the MLF case with $N \geqslant 4$, however, the present status of the program does not guarantee a solution because of the difficulties alluded to in Section IV-B.

The various special cases of FLF and MLF topologies are handled by placing constraints on the variables. For example, cascade is a special case of FLF with given values of ω_{0i} and Q_i and $t_{1i}=0$; no optimization takes place, but sensitivity is calculated. Similarly, PRB is a special case of FLF with $Q_i = Q$ for all i and no sensitivity minimization, LF is special case of MLF with $Q_i \to \infty$, $i \neq 1, N$, etc.

The second-order sections included are TG, active R, and state-variable biquads. Further sections [53]–[55], [87] can be considered simply by adding small subroutines for section evaluation.

The minimization is handled by a library routine of the University of Minnesota Computer Center, GPM, which permits arbitrary linear constraints to be placed on the independent variables Q_i.

SENSMIN is written in 1100 FORTRAN statements, exclusive of library subroutines. Total CPU time required for optimizing an eighth-order Butterworth bandpass filter in FLF, MLF, LF, and Cascade topologies is 16 s on a CDC Cyber 74 computer.

Acknowledgment

The authors are indebted to the personnel of Department 4391, Bell Telephone Laboratories, Holmdel, NJ, for their support during the preparation of this paper. In particular, we acknowledge E. Y. McBride for the very capable design of the STAR sections in the fourth-order filter, and D. M. Kennedy and E. J. Yurek for the experimental measurements depicted in Figs. 11 and 12. Finally we wish to express our gratitude to J. J. Friend, R. N. Gadenz, and J. Tow for their often sought comments during the evolution of this manuscript.

References

[1] S. K. Mitra, *Active Inductorless Filters.* New York, NY: IEEE Press, 1971.
[2] G. Szentirmai, *Computer-Aided Filter Design.* New York, NY: IEEE Press, 1973.
[3] L. P. Huelsman, *Active RC Filters: Theory and Application.* Stradburg, PA: Dowden, Huthinson and Ross, Inc., 1976.
[4] W. E. Heinlein and W. H. Holmes, *Active Filters for Integrated Circuits.* Oldenbourg, Munich, 1974.
[5] G. S. Moschytz, *Linear Integrated Networks.* New York, NY: Van Nostrand, 1975.
[6] H. J. Orchard, "Inductorless filters," *Electron. Lett.*, vol. 2, pp. 224–225, June 1966.
[7] A. Antoniou, "Realization of gyrators using operational amplifiers and their use in *RC*-active network synthesis," in *Proc. Inst. Elec. Eng.* (London), pp. 1838–1850, Nov. 1969.
[8] R. H. S. Riordan, "Simulated inductors using differential amplifiers," *Electron. Lett.*, pp. 50–51, Feb. 1967.
[9] D. F. Sheahan, "Gyrator-floatation circuit," *Electron. Lett. 3*, pp. 39–40, 1967.
[10] W. H. Holmes, S. Gruetzmann, and W. E. Heinlein, "Direct-coupled gyrators with floating ports," *Electron. Lett. 3*, pp. 46–47, 1967.
[11] L. T. Bruton, "Network transfer functions using the concept of frequency dependent negative resistance," *IEEE Trans. Circuit Theory*, CT-16, pp. 406–408, Aug. 1969.
[12] F. E. J. Girling and E. F. Good, "Active filters, part 12: The leapfrog or active ladder synthesis," *Wireless World*, pp. 341–345, July 1970.
[13] ——, "Active filters, part 13: Applications of active ladder synthesis," *Wireless World*, pp. 445–450, Sept. 1970.
[14] ——, "Active filters, part 14: Bandpass types," *Wireless World*, pp. 505–510, Oct. 1970.
[15] R. L. Adams, "On reduced sensitivity filters," in *Proc. 14th Midwest Symp. on Circuit Theory*, pp. 14.3-1–14.3-8, May 1971.
[16] J. Tow and Y. L. Kuo, "Coupled biquad active filters," in *Proc. IEEE Int. Symp. on Circuit Theory*, pp. 164–167, Apr. 1972.
[17] G. Hurtig, III, "The primary resonator block technique of filter synthesis," in *Int. Filter Symp.*, p. 82 (abstract only), Apr. 1972.
[18] ——, "Filter network having negative feedback loops," U.S. Patent 3 720 881, Mar. 1973.
[19] K. R. Laker and M. S. Ghausi, "A low sensitivity multiloop feedback active *RC* filter," *Tech. Rep. 400-225*, New York University, New York, NY, Oct. 1972. Also, in *Proc. IEEE Int. Symp. on Circuit Theory*, pp. 123–129, Apr. 1973.
[20] K. R. Laker, "On multiparameter sensitivity in active *RC* networks," Ph.D. Dissertation, New York University, New York, NY, May 1973.
[21] G. Szentirmai, "Synthesis of multiple-feedback active filters," *Bell Syst. Tech. J.*, vol. 52, pp. 527–555, Apr. 1973.
[22] K. R. Laker and M. S. Ghausi, "Synthesis of a low sensitivity multi-loop feedback active *RC* filter," *IEEE Trans. Circuits Syst.*, vol. CAS-21, pp. 252–259, Mar. 1974.
[23] ——, "A comparison of active multiple-loop feedback techniques for realizing high-order bandpass filters," *IEEE Trans. Circuits Syst.*, vol. CAS-21, pp. 774–763, Nov. 1974.
[24] K. R. Laker, M. S. Ghausi, and J. J. Kelly, "Minimum sensitivity leap frog active *RC* filters," in *Proc. IEEE Int. Symp. Circuits Syst.*, pp. 201–204, Apr. 1974.
[25] J. Tow, "Design and evaluation of shifted companion form (follow the leader feedback) active filters," in *Proc. IEEE Int. Symp. Circuits Syst.*, pp. 656–660, Apr. 1974.

[26] G. Hurtig, III, "Voltage tunable multipole bandpass active filters," in *Proc. IEEE Int. Symp. Circuits Syst.*, pp. 569–572, Apr. 1974.

[27] L. T. Bruton and A. I. A. Salama, "Frequency limitations of coupled biquadratic active ladder structures," *IEEE J. Solid-State Circuits*, vol. SC-9, pp. 70–72, Apr. 1974.

[28] J. Tow, "Design and evaluation of shifted companion form active filters," *Bell Syst. Tech. J.*, vol. 54, no. 3, pp. 545–568, Mar. 1975.

[29] S. F. Wetencamp and M. C. VanValkenburg, "Another approach to multiloop feedback," in *Proc. IEEE Int. Symp. Circuits Syst.*, pp. 258–260, Apr. 1975.

[30] K. R. Laker, M. S. Ghausi, and J. J. Kelly, "Minimum sensitivity active (leap frog) and passive ladder bandpass filters," *IEEE Trans. Circuits Syst.*, vol. CAS-22, pp. 670–677, Aug. 1975.

[31] R. Biernacki and J. Mulawka, "Synthesis of active filters in the basic FLF configuration," in *Proc. 3rd Int. Symp. on Network Theory* (Split, Yugoslavia), pp. 459–466, 1975.

[32] D. J. Perry, "New multiple feedback active *RC* network," *Electron. Lett.*, vol. 11, no. 16, pp. 364–365, Aug. 1975.

[33] K. R. Laker, R. Schaumann and J. R. Brand, "Multiple loop feedback active *R* filters," in *Proc. IEEE Int. Symp. Circuits Syst.*, pp. 279–282, Apr. 1976.

[34] R. Schaumann, W. A. Kinghorn and K. R. Laker, "Minimizing signal distortion in FLF active filters," *Electron. Lett.*, vol. 12, no. 9, pp. 211–213, Apr. 1976.

[35] K. R. Laker and M. S. Ghausi, "Computer-aided analysis and design of follow-the-leader feedback active *RC* filters," *Int. J. Circuit Theory Appl.*, vol. 4, pp. 177–187, Apr. 1976.

[36] K. R. Laker, R. Schaumann, and J. R. Brand, "A comparison of techniques for realizing high order active *R* filters," in *Proc. 19th Midwest Symp. on Circuits Syst.*, pp. 190–196, Aug. 1976.

[37] D. Dubois and J. Neirynck, "General synthesis methods for FLF active filters," in *Proc. 1976 European Conf. on Circuit Theory Design* (Genoa, Italy), pp. 565–571, Sept. 1976.

[38] R. Schaumann, R. E. Chalstrom, and K. R. Laker, "Maximizing signal swing capabilities of leap frog and modified leap frog active filters," *Proc. 10th Annual Asilomar Conf. on Circuits, Systems, Comput.*, pp. 341–345, Nov. 1976.

[39] R. N. Gadenz, "On low-sensitivity realizations of band elimination active filters," *IEEE Trans. Circuits Syst.*, vol. CAS-24, pp. 175–183, Apr. 1977.

[40] J. Tow, "Some results on generalized follow-the-leader feedback active filters," in *Proc. IEEE Int. Symp. on Circuits Syst.*, pp. 462–465, Apr. 1977.

[41] K. R. Laker and M. S. Ghausi, "Minimum sensitivity multiple loop feedback bandpass active filters," in *Proc. IEEE Int. Symp. Circuits Syst.*, pp. 458–461, Apr. 1977.

[42] R. Schaumann, K. R. Laker, and W. A. Kinghorn, "The maximization of dynamic range of FLF bandpass filters by use of stagger tuning," in *Proc. IEEE Int. Symp. Circuits Syst.*, pp. 450–453, Apr. 1973.

[43] R. Schaumann, R. E. Chalstrom, and K. R. Laker, "Optimization of sensitivity and dynamic range of IFLF active filters," *Electron. Lett.*, vol. 13, no. 12, pp. 362–368, June 1977.

[44] R. Schaumann and R. E. Chalstrom, "The dynamic range of properties of high-order active bandpass filters," in *Proc. IEEE Int. Symp. on Circuits Syst.*, Apr. 1978.

[45] K. R. Laker, "Phase and group delay sensitivities in multiple-loop feedback bandpass active filters," in *Proc. IEEE Int. Symp. on Circuits Syst.*, Apr. 1978.

[46] R. W. Daniels, *Approximation Methods for Electronic Filter Design.* New York, NY: McGraw-Hill, 1974.

[47] M. S. Ghausi, *Principles and Design of Linear Active Circuits.* New York, NY: McGraw-Hill, 1965.

[48] G. Daryanani, *Principles of Active Network Synthesis and Design.* New York, NY: Wiley, 1976.

[49] E. Christian and E. Eisenmann, *Filter Design Tables and Graphs.* New York, NY: Wiley, 1966.

[50] R. Tarmy and M. S. Ghausi, "Very high-Q insensitive active *RC* networks," *IEEE Trans. Circuit Theory*, vol. CT-17, pp. 358–360, 1970.

[51] W. J. Kerwin, L. P. Huelsman, and R. W. Newcomb, "State variable synthesis for insensitive integrated circuit transfer functions," *IEEE J. Solid-State Circuits*, vol. SC-2, pp. 87–92, 1967.

[52] J. Tow, "A step-by-step active filter design," *IEEE Spectrum*, vol. 6, pp. 64–68, Dec. 1969.

[53] L. C. Thomas, "The biquad: Part I—Some practical design considerations; Part II—A multipurpose active filter system," *IEEE Trans. Circuit Theory*, vol. CT-18, pp. 350–362, 1971.

[54] W. B. Mikhael and B. B. Bhattacharyya, "Practical design for insensitive *RC*-active filters," *IEEE Trans. Circuits Syst.*, vol. CAS-22, no. 5, 1975.

[55] J. J. Friend, C. A. Harris, and D. Hilberman, "STAR: An active biquadratic filter section," *IEEE Trans. Circuits Syst.*, vol. CAS-22, 1975.

[56] R. Rao and S. Srinivasan, "Low-sensitivity active filter using the operational amplifier pole," in *Proc. IEEE*, pp. 1713–1714, 1974.

[57] R. Schaumann, "Low-sensitivity, high-frequency, tunable active filter without external capacitors," *IEEE Trans. Circuits Syst.*, vol. CAS-22, pp. 39–44, 1975.

[58] ——, "On the design of active filters using only resistors and voltage amplifiers," *AEU-Electron. Commun.*, vol. 30, pp. 245–252, 1976.

[59] A. L. Rosenblum and M. S. Ghausi, "Multiparameter sensitivity in active *RC* networks," *IEEE Trans. Circuit Theory*, CT-15, pp. 592–599, Nov. 1971.

[60] K. R. Laker and M. S. Ghausi, "Large change sensitivity—A dual pair of approximate statistical sensitivity measures," *J. Franklin Inst.*, vol. 298, pp. 395–412, Dec. 1974.

[61] ——, "Statistical multiparameter sensitivity—A valuable measure for CAD," in *IEEE Int. Symp. on Circuits Syst.*, pp. 333–336, Apr. 1975.

[62] C. Acar, M. S. Ghausi, and K. R. Laker, "Statistical multiparameter sensitivity measures in high Q networks," *J. Franklin Inst.*, vol. 300, no. 4, pp. 281–297, Oct. 1975.

[63] K. R. Laker, R. Schaumann, J. R. Brand, and M. S. Ghausi, "Statistical multiparameter sensitivity measures which include effects due to temperature variations," in *Proc. IEEE Int. Symp. on Circuits Syst.*, pp. 454–457, Apr. 1977.

[64] P. V. Lopresti, "Optimum design of linear tuning algorithms," *IEEE Trans. Circuits Syst.*, vol. CAS-24 pp. 144–151, Mar. 1977.

[65] S. Halfin, "An optimization method for cascaded filters," *Bell Syst. Tech. J.*, vol. 49, no. 2, pp. 185–190, Fig. 1970.

[66] E. Lueder, "A decomposition of a transfer function minimizing distortion and inband losses," *Bell Syst. Tech. J.*, vol. 49, no. 3, pp. 455–469, Mar. 1970.

[67] ——, "Optimization of the dynamic range and the noise of *RC*-active filters by dynamic programming," *Int. J. Circuit Theory Applications*, vol. 3, pp. 365–370, 1975.

[68] D. A. Pierre, *Optimization Theory with Applications.* New York, NY: Wiley, 1969.

[69] H. R. Orchard and D. F. Sheehan, "Inductorless bandpass filters," *IEEE J. of Solid-State Circuits*, vol. SC-5, pp. 108–118, June 1970.

[70] K. R. Laker and M. S. Ghausi, "Design of minimum sensitivity multiple loop feedback bandpass active filters," to be published.

[71] M. S. Ghausi, *Electronic Circuits.* New York, NY: Van Nostrand-Reinhold, 1971, Appendix C.

[72] D. Tuttle, Jr., "A common basis for the Darlington and Szentirmai network synthesis," in *Proc. IEEE*, vol. 62, pp. 1389–1390, Oct. 1974.

[73] G. Szentirmai, "FILSYN: A general purpose filter synthesis program," in *Proc. IEEE*, vol. 65, pp. 1443–1458, Oct. 1977.

[74] D. Dubois and J. J. Neirynck, "Synthesis of a leapfrog configuration equivalent to an *LC*-ladder filter between generalized terminations," *IEEE Trans. Circuits Syst.*, vol. CAS-24, pp. 590–597, Nov. 1977.

[75] R. Schaumann, W. A. Kinghorn, and K. R. Laker, "Optimal design parameters for minimizing sensitivity and distortion in FLF active filters," in *Proc. 19th Midwest Symp. on Circuits Syst.*, pp. 390–395, Aug. 1976.

[76] R. C. Reitan, "SENSMIN—A design program for multiple-loop feedback filters," Masters thesis, University of Minnesota, 1977.

[77] L. T. Bruton, F. N. Trofimenkoff, and D. H. Treleaven, "Noise performance of low sensitivity active filters," *IEEE J. Solid-State Circuits*, vol. SC-8, pp. 85–91, Feb. 1973.

[78] F. N. Trofimenkoff, D. H. Treleaven, and L. T. Bruton, "Noise performance of *RC*-active quadratic filter sections," *IEEE Trans. Circuit Theory*, CT-20, pp. 524–532, Sept. 1973.

[79] W. Guggenbühl and M. Künzli, "Noise performance comparison and optimization of different active filters," in *Proc. 1977 IEEE Int. Symp. on Circuits Syst.*, pp. 336–339, Apr. 1977

[80] T. Deliyannis, "High-Q factor circuit with reduced sensitivity," *Electron. Lett.*, p. 577, Dec. 1968.

[81] T. A. Hamilton and A. S. Sedra, "A single-amplifier biquad active filter," *IEEE Trans. Circuit Theory*, CT-19, pp. 398–403, July 1972.

[82] A. S. Sedra, "Generation and classification of single amplifier filters," *Int. J. of Circuit Theory Appl.*, pp. 1–57, Mar. 1974.

[83] P. E. Fleisher, "Sensitivity minimization in a single amplifier biquad circuit," *IEEE Trans. Circuits Syst.*, vol. CAS-23, pp. 45–55, Jan. 1976.

[84] J. Tow, private communication.

[85] P. V. Lopresti and K. R. Laker, "Optimum tuning of multiple loop feedback active-*RC* filters, to be published.

[86] C. Acar, M. S. Ghausi, and K. R. Laker, "Critical block analysis for the transfer, gain, and phase functions in multiple-loop feedback networks," *Int. J. Circuit Theory Appl.*, vol. 6, no. 1, pp. 89–104, Jan. 1978.

[87] D. Akerberg and K. Mossberg, "A versatile active *RC* building block with inherent compensation for the finite bandwidth of the amplifier," *IEEE Trans. Circuits Syst.*, vol. CAS-21, pp. 75–78, Jan. 1974.

Synthesis of a Leapfrog Configuration Equivalent to an *LC*-Ladder Filter Between Generalized Terminations

DANIEL DUBOIS AND JACQUES NEIRYNCK, SENIOR MEMBER, IEEE

Abstract—**Leapfrog *RC* active filters are known to be insensitive to the value of the elements thanks to their analogy to *LC* filters. It is shown that the class of LF filters amenable to such an analogy can be enlarged by considering nondissipative 2-ports between generalized dissipative terminations. There is no necessity for transmission zeroes at the origin or at infinity, which was the case in the previous approaches. The formulas linking the given transmittance to the *LC* 2-port description are derived and the optimal choice of the remaining degrees of freedom are discussed. The method presented here preserves the analogy between the LF design and its *LC* equivalent since this analogy guarantees the low sensitivity; yet it is totally free from all constraints originating from the *LC* implementation and irrelevant for the LF design.**

I. Introduction

LOSSLESS ladder filters between resistive terminations are known for their low sensitivity to the element variations [3] combined with easy independent tuning properties. Leapfrog *RC* active networks, simulating such lossless ladder filters, are known to enjoy also a high insensitivity [1], [2]. Yet there are two different sets of constraints on the lossless ladder synthesis depending whether the filter is to be actually realized by passive *LC* components or to be simulated by *RC* active biquadratic cells in a leapfrog configuration. In the first case, the synthesis aims at a minimal number of coils and at a circuit realizable with positive elements. In the second case, one wishes a minimal number of biquadratic cells and the constraint of realizability by positive *LC* elements is irrelevant.

Further, one should notice that the approximation problem is also linked to the realization by an *LC* ladder or a leapfrog configuration. In the first case, it is compulsory to locate at least one (in some instances up to three) transmission zero(s) at infinity and (or) at the origin in order to yield shifting capacitances for the extraction of resonant arms; this location of transmission zeros may be far from optimal in the process of approximating the specifications. In the second case, this constraint can be avoided and the choice of attenuation zeros should only be aimed at meeting the selectivity requirements with the transfer function of the lowest possible degree.

Manuscript received August 24, 1976; revised June 21, 1977. This work was supported by the AGEN Foundation, Hasler AG, Berne, Switzerland.

The authors are with the Ecole Polytechnique Fédérale de Lausanne, Department d'Electricite, 16, chemin de Bellerive, CH-1007 Lausanne, Switzerland.

The purpose of this paper is to show that it is possible to get rid of the constraints proper to a classical *LC* realization. The solution amounts to the design of an *LC* 2-port between generalized terminations, in the sense that they include not only resistances but also *FDNC*'s (elements whose admittance is proportional to p^2) [6]. The *FDNC*'s are introduced for the purpose of calculation only, as dummy elements. They do not have to be implemented into the *LF* realization where the general biquadratic sections are used instead. These generalized terminations allow us to realize any voltage transfer ratio by a lossless ladder 2-port, where all ladder arms yield biquadratic sections in the LF simulation and where the location of transmissioh zeros is only linked to the selectivity requirements. The method is summarized in the following steps.

1) A transfer function $N(p)/D(p)$ is given by an approximation procedure meeting the selectivity requirements.

2) $N(p)/D(p)$ is identified with the voltage transfer function of a lossless 2-port between generalized terminations and the scattering matrix of this circuit is determined.

3) The scattering matrix of the same 2-port between ordinary resistive terminations is deduced.

4) The 2-port is synthesized as a ladder involving only third-degree reactances by using a general-purpose filter synthesis program.

5) From the knowledge of the ladder and of the generalized termination arm reactances, one deduces the transfer function of the biquadratic cells.

II. Choice of the Generalized Terminations

The structures of the *LC* equivalent and the *RC* active filter are shown, respectively, in Figs. 1 and 2 for n even. The case of n odd may be treated similarly with a few trivial modifications.

In the *LC* ladder structure, the loop and node equations are, respectively,

$$(V_j - V_{j+1})Y_j = I_{j+1}$$
$$(I_j - I_{j+1})Z_j = V_j. \tag{1}$$

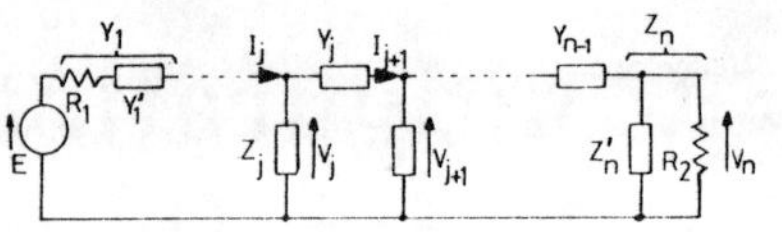

Fig. 1. Structure of the LC 2-port for n even.

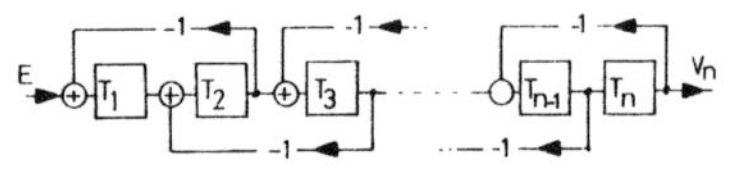

Fig. 2. Structure of the leapfrog configuration.

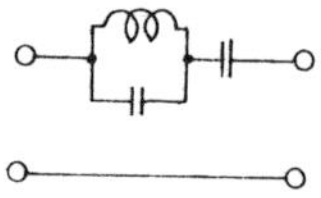

Fig. 3. Structure of a series arm.

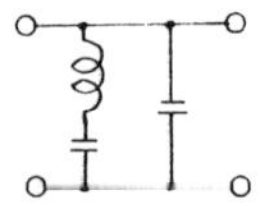

Fig. 4. Structure of a shunt arm.

In the LF structure, these equations are implemented by the block diagram of Fig. 2. The partial transfer functions T_i could be directly identified to the ladder immittances. Yet, since these are reactances, i.e., odd functions, this method will certainly not yield biquadratic transfer functions. Hence, a preliminary impedance transformation is usually introduced before the identification. Equations (1) may be rewritten as

$$(V_j - V_{j+1})\frac{Y_j}{p} = \frac{I_{j+1}}{p}$$

$$\left(\frac{I_j}{p} - \frac{I_{j+1}}{p}\right)pZ_j = V_j. \tag{2}$$

The transfer functions T_i are now related to the ladder immittances by

$$T_i = Y_i/p, \qquad i \text{ odd}$$

$$T_i = pZ_i, \qquad i \text{ even}. \tag{3}$$

By the way, this impedance transformation establishes an analogy between the LF structure and a resistive ladder (including *FDNC*'s) terminated by reactances. The well-known properties of insensitivity formulated by Orchard [3] remain valid [4]. Since this impedance transformation is trivial and occurs only at step 5 of the method, the following discussion is presented in terms of a lossless ladder between resistive terminations.

It is necessary to choose branch impedances which have the right degrees in order to yield biquadratic factors by (3). With the exception of the first and last branches which include resistive terminations, the appropriate reactances are represented in Figs. 3 and 4. Both yield the

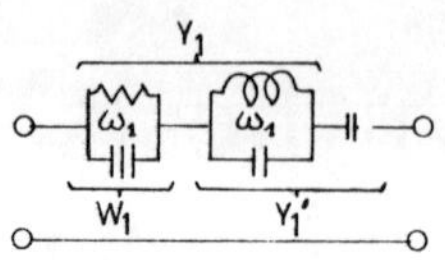

Fig. 5. Structure of the first arm with a generalized termination.

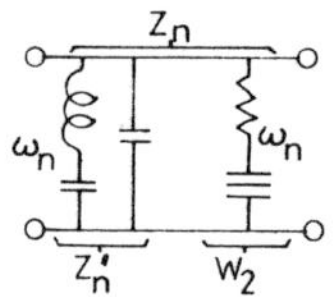

Fig. 6. Structure of the last arm with a generalized termination.

same transfer functions.

$$T_i = h_i \frac{p^2 + \omega_i^2}{p^2 + \omega_{0i}^2}. \tag{4}$$

The first and the last branches cannot be composed, respectively, by placing a resistance in series with the branch of Fig. 3 and another resistance in parallel with the branch of Fig. 4. Indeed, these branches yield T_1 and T_n of degree 3, unsuitable for the LF design.

Furthermore, the occurence of a first-series (last-shunt) arm as in Fig. 3 (4) implies the existence of a reflection pole between ordinary resistive terminations: in other words, the input and output impedances must, respectively, have a pole and a zero at the frequencies ω_1 and ω_n corresponding to transmission zeros. This constitutes a severe constraint on the transfer function which is totally artificial from the standpoint of the LF design.

These obstacles could be circumvented, as in [1], by choosing a first-series arm composed of a series-resonant circuit, and a last-parallel arm composed of a parallel-resonant circuit. Yet, this choice imposes a precise location for the transmission zeros: the 2-port has a double transmission zero both at the origin and at infinity; the first and last arms do not create transmission zeros at frequencies ω_1 and ω_n freely chosen.

Furthermore, a generalization of the LF design has been described in [1] and [2] without restrictions on the location of transmission zeros; however, in this case the relation between the LF design and the LC equivalent has not been defined. It is precisely the purpose of the present paper to show that any LF design with biquadratic building blocks has an LC equivalent. This will allow to answer many of the open questions mentioned by Szentirmai in [1] and [2].

This becomes possible by considering the first and last arms represented, respectively, in Figs. 5 and 6. For instance, the branch of Fig. 5 has an impedance given by

$$1/Y_1' = \frac{(p^2 + \omega_{01}^2)}{h_1 p(p^2 + \omega_1^2)}$$

$$W_1 = \frac{\alpha/h_1}{p^2 + \omega_1^2}$$

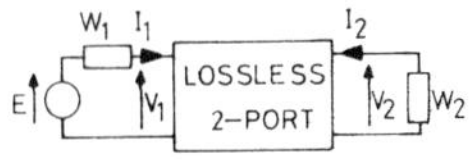

Fig. 7. The lossless filter between generalized terminations.

and

$$Y_1 = \frac{h_1 p(p^2+\omega_1^2)}{p^2+\alpha p+\omega_{01}^2}. \quad (5)$$

In this expression, the pole of W_1 is chosen identically to that of $1/Y_1'$. For the shunt arm of Fig. 6, by duality Z_n has a similar expression. In other words, the generalized terminations are "tuned" to the same frequency as the first (last) resonant circuit realizing a transmittance zero at $\omega_1(\omega_n)$. This particular choice allows to solve the problem mentioned before: on the one hand, to start with a transfer voltage ratio $N(p)/D(p)$ defined only in order to meet the selectivity requirements with the lowest degree possible; on the other hand, to synthesize a 2-port with reflection poles at ω_1 and ω_n and a ladder structure with arms of the right degree.

Hence, the problem at hand is to study the behavior of the circuit represented in Fig. 7, where the generalized terminations are respectively tuned as the lossless arm which is the closest to the port. The generalized terminations have the expression

$$W_1 = \frac{1}{p^2+\omega_1^2}$$

$$W_2 = \frac{p^2+\omega_n^2}{p^2} \quad (6)$$

within a constant factor. They only absorb active power, which in some frequency range is possibly negative.

III. Scattering Matrix Formulation

There is a classical scattering formalism which has been developed especially for the case of lossless 2-ports terminated by ordinary resistances [7]. The purpose of this section is to extend this formalism to the case of generalized terminations.

In the following, one considers the same lossless 2-port first between ordinary resistances R_1 and R_2, secondly between generalized terminations W_1 and W_2, and one establishes the formulas linking both representations. The aim is to deduce from the behavior between W_1 and W_2, interesting for the purpose at hand, the scattering matrix of exactly the same 2-port between R_1 and R_2, since there exist programs to synthesize this last circuit.

The scattering matrix of a lossless 2-port between terminations R_1 and R_2 has the general form

$$s = \frac{1}{g}\begin{bmatrix} h & f \\ f & \mp h_* \end{bmatrix} \quad (7)$$

where f, g, and h are polynomials in the complex variable p that are pairwise prime; g is strictly Hurwitz if the 2-port is passive; $h_* = h(-p)$; $f = \pm f_*$ which amounts to the statement that $f(p)$ is an even or odd polynomial and

$$gg_* = hh_* + ff_*. \quad (8)$$

The input and output terminated impedances of the 2-port are expressed respectively as

$$\zeta_1 = \frac{g+h}{g-h} \qquad \zeta_2 = \frac{g\mp h_*}{g\pm h_*}.$$

Let β_i and α_i be the reflected and incident waves at port i, defined by

$$\alpha_i = \frac{V_i}{\sqrt{R_i}} + \sqrt{R_i}\, I_i$$

$$\beta_i = \frac{V_i}{\sqrt{R_i}} - \sqrt{R_i}\, I_i. \quad (9)$$

One has

$$\beta = s\alpha. \quad (10)$$

The corresponding transfer matrix θ [7] is defined by grouping the waves in an input and output vector, w_1 and w_2.

One has

$$w_1 = \theta w_2 \quad (11)$$

with

$$w_1 = \begin{bmatrix} \beta_1 \\ \alpha_1 \end{bmatrix} = \begin{bmatrix} R_1^{-1/2} & -R_1^{1/2} \\ R_1^{-1/2} & R_1^{1/2} \end{bmatrix}\begin{bmatrix} V_1 \\ I_1 \end{bmatrix} \quad (12)$$

$$w_2 = \begin{bmatrix} \alpha_2 \\ \beta_2 \end{bmatrix} = \begin{bmatrix} R_2^{-1/2} & -R_2^{1/2} \\ R_2^{-1/2} & R_2^{1/2} \end{bmatrix}\begin{bmatrix} V_2 \\ -I_2 \end{bmatrix} \quad (13)$$

and let X_1 and X_2 denote the 2×2 matrices in (12) and (13).

It follows that

$$\begin{bmatrix} V_1 \\ I_1 \end{bmatrix} = X_1^{-1}\theta X_2 \begin{bmatrix} V_2 \\ -I_2 \end{bmatrix}$$

and $X_1^{-1}\theta X_2$ is the chain matrix of the lossless 2-port, i.e., a description which does not involve the terminations R_1 and R_2.

Similarly, let β_i', α_i', s', and θ' be the reflected and incident waves, the scattering and the transfer matrices, when the same 2-port is between W_1 and W_2. The matrices X_1' and X_2' denote matrices similar to X_1 and X_2 with R_1 and R_2 replaced by W_1 and W_2. The input and output wave vectors are w_1' and w_2'.

One has

$$w_1' = \theta' w_2' \quad (14)$$

or

$$\begin{bmatrix} V_1 \\ I_1 \end{bmatrix} = (X_1')^{-1}\theta' X_2' \begin{bmatrix} V_2 \\ -I_2 \end{bmatrix}. \quad (15)$$

The identification between both expressions of the chain matrix of the same 2-port yields

$$X_1^{-1}\theta X_2=(X_1')^{-1}\theta' X_2' \tag{16}$$

or

$$\theta=X_1X_1'^{-1}\theta' X_2'X_2^{-1} \tag{17}$$

$$\theta'=X_1'X_1^{-1}\theta X_2X_2'^{-1}. \tag{18}$$

The matrix θ is known [7] to have the form

$$\theta=\frac{1}{f}\begin{bmatrix} \pm g_* & h \\ \pm h_* & g \end{bmatrix}. \tag{19}$$

Let f', g', and h' denote the functions (not necessarily polynomials) involved in the similar expression of θ'. One has by (18) the relations

$$Q(p)g'=p^2(p^2+\omega_1^2)B(g,h)+p^2D(g,h)$$
$$+(p^2+\omega_1^2)(p^2+\omega_n^2)A(g,h)+(p^2+\omega_n^2)C(g,h) \tag{20}$$

$$Q(p)h'=p^2(p^2+\omega_1^2)B(g,h)-p^2D(g,h)$$
$$+(p^2+\omega_1^2)(p^2+\omega_n^2)A(g,h)-(p^2+\omega_n^2)C(g,h) \tag{21}$$

$$Q(p)f'=4\sqrt{p^2(p^2+\omega_1^2)(p^2+\omega_n^2)}\ f \tag{22}$$

where R_1 and R_2 have been assumed equal to unity for the sake of simplicity and where A, B, C, and D are the linear operators

$$A=g+h\pm(g_*+h_*) \tag{23}$$

$$B=g+h\mp(g_*+h_*) \tag{24}$$

$$C=g-h\mp(g_*-h_*) \tag{25}$$

$$D=g-h\pm(g_*-h_*) \tag{26}$$

and

$$Q(p)=16p(p^2+\omega_1^2)(p^2+\omega_n^2).$$

It follows that g' and h' are still polynomials whereas f' becomes an irrational function. It should be stressed that (18) defines only the ratio h'/f' and h'/g'. Hence, f', g', and h' are defined arbitrarily within a common factor and any other repartition of the irrational factor could yield other formulas. The factor $Q(p)$ has been introduced in order to define f', g', and h' in their irreducible form. Since the 2-port between R_1 and R_2 must have reflection poles, at $\pm\omega_1$ and $\pm\omega_n$, $f, g-h$, and $g\mp h_*$ are, respectively, divisible by $(p^2+\omega_1^2)$ and $(p^2+\omega_n^2)$: it follows that the right-hand members of (20), (21), and (22) are divisible by these factors. Further, the attenuation pole at the origin must be realizable by a series capacitance; hence, the constant terms of g and h are equal and the divisibility by p follows.

By the same argument, the shunt shifting capacitances, appearing in branches as in Fig. 4, may be extracted only if the coefficients of highest degree in g and h are of opposite sign. By denoting g_i the coefficient of p^i in the polynomial g and by assuming $2n+2$ to be the degree of g, then one has

$$g_0=h_0$$
$$g_{2n+2}=-h_{2n+2}. \tag{27}$$

The LC ladder composed exclusively of branches as in Fig. 3 or 4 and terminated by R_1 and R_2 has a single transmission zero at the origin. The polynomial f includes n factors $(p^2+\omega_i^2)$ and one factor p: it is an odd polynomial of degree $2n+1$. The lower sign must be considered in (23) to (26).

By (20) and (21), it follows that g' and h' have degree $2n$ and that their coefficients are related by

$$g_0'=h_0'$$
$$g_{2n}'=h_{2n}'. \tag{28}$$

Since the 2-port is lossless, the relation

$$g'g_*'=h'h_*'+f'f_*' \tag{29}$$

still holds. It can be checked by substitution of (20), (21), and (22) in (29) that it is equivalent to (8).

IV. Computation of f', g', and h'

Let the transfer function $V_2/E=N(p)/D(p)$ without zeros at $\omega=0$, and $\omega=\infty$ be prescribed where N and D are relatively prime polynomials with the same degree, i.e., deg $D=$deg $N=2n$. The roots of N are supposed to be located on the imaginary axis (except at the origin); hence, N is an even polynomial. Among these roots, we choose two pairs to be $\pm\omega_1$ and $\pm\omega_n$ which will be the resonant frequencies, respectively, of the first and last arm; they yield the factors $(p^2+\omega_1^2)$ and $(p^2+\omega_n^2)$. The polynomial D is assumed to be monic and strictly Hurwitz. The general form of N is $K\Pi(p^2+\omega_i^2)$ where K is a scale factor to be determined later on. The roots of N and D are chosen in such a way that the corresponding attenuation $\alpha=20\log|D(j\omega)/N(j\omega)|$ has $2n$ equal attenuation minima. It will be shown in Section VII that this choice gives the best insensitivity in the passband.

The problem is to identify this transfer function given as a starting point with the transfer voltage ratio of the terminated 2-port of Fig. 7.

The transmittance of the 2-port between W_1 and W_2 is

$$s_{12}'=2\sqrt{\frac{W_1}{W_2}}\frac{V_2}{E} \tag{30}$$

which gives

$$\frac{N(p)}{D(p)}=\frac{1}{2}\sqrt{\frac{W_2}{W_1}}\frac{f'}{g'} \tag{31}$$

with, by (6), the following identification

$$g'=D \tag{32}$$

$$f'=\sqrt{\frac{p^2}{(p^2+\omega_1^2)(p^2+\omega_n^2)}}\;2N$$

$$=2\sqrt{p^2(p^2+\omega_1^2)(p^2+\omega_n^2)}\;K\prod_{i\neq 1,n}(p^2+\omega_i^2). \quad (33)$$

The form of g' and f' coincides with the one determined in (20) and (22). By (27), it follows that

$$h'h'_*=DD_*-4K^2p^2(p^2+\omega_1^2)(p^2+\omega_n^2)\prod_{i\neq 1,n}(p^2+\omega_i^2)^2. \quad (34)$$

The right side of this relation is an even polynomial which could be factorized in real polynomials h' and h'_* as long as it has roots of an even multiplicity on the imaginary axis. In order to minimize the sensitivity, the scale factor K is chosen such that there is a double pair of roots on the imaginary axis. It is readily seen that for $K=0$ the roots of $h'h'_*$ coincide with the roots of DD_* which are by assumption not on the imaginary axis, since D is strictly Hurwitz; for K increasing, the roots of $h'h'_*$ shift towards the roots of f', all located on the imaginary axis. It must be noticed that (34) is identical to (24) in [1].

For $p=j\omega$, (34) may be rewritten as

$$h'h'_*=|D|^2-4|N|^2\frac{W_1}{W_2} \quad (35)$$

which shows that $h'h'_*$ may be equal to zero only if W_1/W_2 is positive or if the scale factor K in N tends to infinity.

Starting from $K=0$, one increases K till the right-hand member of (35) is equal to zero, which means that $h'h'_*$ has a double root on the imaginary axis. This is the limiting value $K_{\max}$; any further increase of K yields a splitting of the double root into two simple roots which cannot be shared between the real polynomials h' and h'_*. It will be shown in Section VII that $K_{\max}$ is the optimal value from the standpoint of sensitivity. The computation of $K_{\max}$ is described in Section VI.

The roots of $h'h'_*=|h'|^2$ on the $j\omega$ axis correspond to the maximum value of $|s'_{12}|$ since by (29) $|s'_{11}|=0$ implies $|s'_{12}|=1$. Hence, they occur in the passband and W_1/W_2 must be positive in the passband. By (6), on the $j\omega$ axis

$$\frac{W_1}{W_2}=\frac{-\omega^2}{(\omega_1^2-\omega^2)(\omega_n^2-\omega^2)}.$$

The ratio is positive if and only if ω_1 and ω_n, transmission zeros occuring in the stopbands, are located on either side of the passband. This choice of ω_1 and ω_n is possible for a bandpass filter.

In the case of a low- or high-pass filter, this choice is not possible and the formula (6) must be modified by changing the sign either of W_1 or of W_2. If one defines a parameter κ equal to 1, with ω_1 and ω_n both in the same stopband and equal to zero when they are chosen on either side of the passband, one has

$$W_1=\frac{(-1)^\kappa}{p^2+\omega_1^2} \quad \text{or} \quad W_2=(-1)^\kappa\frac{(p^2+\omega_n^2)}{p^2}. \quad (36)$$

With this modification, (34) becomes

$$h'h'_*=DD_*-(-1)^\kappa 4K^2p^2(p^2+\omega_1^2)\cdot(p^2+\omega_n^2)\prod_{i\neq 1,n}(p^2+\omega_i^2)^2. \quad (37)$$

Once K is determined, (37) gives the roots of $h'h'_*$. The partitioning of the roots may be performed in several ways as long as the relations (28) are satisfied. Since g' is Hurwitz and monic, all its coefficients are positive. A simple (although not necessary) way to satisfy (28) is to specify h' to be Hurwitz.

V. Computation of f, g, and h

From the knowledge of f', g', and h', it is possible to compute f, g, and h.

By (17), one finds

$$p\cdot g=p^2(p^2+\omega_1^2)C(g',h')+p^2A(g',h') + (p^2+\omega_1^2)(p^2+\omega_n^2)D(g',h')+(p^2+\omega_n^2)B(g',h') \quad (38)$$

$$p\cdot h=-p^2(p^2+\omega_1^2)C(g',h')+p^2A(g',h') - (p^2+\omega_1^2)(p^2+\omega_n^2)D(g',h')+(p^2+\omega_n^2)B(g',h') \quad (39)$$

$$p\cdot f=4\sqrt{p^2(p^2+\omega_1^2)(p^2+\omega_n^2)}\;f'. \quad (40)$$

Since f' is an even function by (33), the upper sign must be considered in the operators of (38) and (39). By (28), the right-hand members of (38) to (40) are divisible by p. Similarly, the relations (27) can be checked. By subtracting (39) from (38), it is readily seen that $g-h$ is divisible by $(p^2+\omega_1^2)$; similarly $(g+h_*)$ is divisible by $(p^2+\omega_n^2)$. This insures the existence of the reflection poles which are needed for the extraction of the first and the last arm of the ladder.

To summarize, starting from $N(p)/D(p)$ it is possible to compute the polynomials f, g, and h which define the scattering matrix of the 2-port between unit resistances. These polynomials are introduced as data in a conventional filter synthesis program which computes a ladder of third-degree reactances. By adding the generalized terminations W_1 and W_2 to this lossless 2-port, one realizes a terminated 2-port whose voltage transfer function is exactly $N(p)/D(p)$.

VI. Computation of $K_{\max}$

In an ordinary LC design, the first-order sensitivity [5] of attenuation to the elements is proportional to the reflection coefficient modulus $|hh_*/ff_*|$. By choosing a scale factor such that there is no flat loss, i.e., at least one attenuation zero and as much as $2n$ if the minima of attenuation are equalized in the passband, the sensitivity is minimal; moreover, it vanishes at the roots of hh_* on the $j\omega$ axis [3].

In a LF design, one may also start from a voltage transfer function N/D, given by a standard approximation program or by tables. It has $2n$ equalized maxima of $|N/D|$ in the passband. The problem is to select the scale factor in N such that the sensitivity of $|s'_{12}|$ to the elements is minimal.

In order to have a stable system one must have $|s'_{12}|<1$ on the $j\omega$ axis; in order to minimize sensitivity there should be no flat loss. By (30), one has

$$|s'_{12}|=2K\left|\sqrt{\frac{W_1}{W_2}}\right|\left|\frac{\Pi(p^2+\omega_i^2)}{D(p)}\right|. \tag{41}$$

Since $|\Pi(p^2+\omega_i^2)/D|$ has $2n$ equalized maxima and since $|\sqrt{W_1/W_2}|$ is not a constant, it is impossible to achieve $2n$ equalized maxima for $|s'_{12}|$. The value $K_{\max}$ realizes at least one pair of attenuation zeros for s'_{12}. Its determination amounts to compute the maximum of the right-hand member of (41) and to choose K such that $|s'_{12}|=1$ at the corresponding frequency.

As far as $\sqrt{W_1/W_2}$ is close to a constant within the passband, the other maxima of $|N/D|$ will also provide values of $|s'_{12}|$ close to unity. This will depend on the relative bandwith of the filter and on the location of ω_1 and ω_n: in any case, the optimal choice for ω_1 and ω_n is to select the transmission zeros in the first and last arm as far as possible from the passband.

It will be shown in Section VII that the LF design has the same sensitivity as its LC equivalent and, thanks to the introduction of the generalized terminations, it has been shown that any LF design has an LC equivalent. Yet it must be realized that these generalized terminations prohibit an optimal LC design from the standpoint of sensitivity since it is impossible in general to place the maximum number of attenuation zeros on the $j\omega$ axis. How far one is from this ideal situation depends on the relative bandwith of the filter. In the example treated in Section IX it will be seen that the effect of the generalized terminations is not very important.

VII. Sensitivity

The sensitivity of a function F to a parameter x is denoted by

$$S(F,x)=\frac{x}{F}\frac{\partial F}{\partial x}. \tag{42}$$

Let $T=V_2/E$ denote the transfer function realized by the LF structure; the second-order transfer functions of the building blocks are

$$T_i=\frac{h_i(p^2+\omega_i^2)}{p^2+\frac{\omega_{0i}}{Q_i}p+\omega_{0i}^2} \tag{43}$$

where $Q_i\to\infty$, except for $i=1$ and n. The set of parameters ω_i, Q_i, ω_{0i}, and h_i of the cell i are denoted by q_{ij} and the set of components of the same cell is x_{ik}.

One has

$$S(|T|,x_{ik})=\operatorname{Re}\left|\sum_j S(T,q_{ij})S(q_{ij},x_{ik})\right|$$
$$=\sum_j S(|T|,q_{ij})S(q_{ij},x_{ik}) \tag{44}$$

since $S(q_{ij},x_{ik})$ is real. Formula (44) shows that the sensitivity of an RC active design depends on two factors: the sensitivity of T to the parameters of the building block; the sensitivity of the parameters to the components. This second factor depends on the choice of the biquadratic cell and the limitation of this cell: it will not be discussed further. The first factor depends on the structure. It will be shown that for the LF design it is small in the passband.

By (31) one has

$$S(T,q_{ij})=S(s'_{12},q_{ij})+S\left(\sqrt{W_2/W_1},q_{ij}\right). \tag{45}$$

For $i\neq 1,n$, (45) reduces to

$$S(T,q_{ij})=S(s'_{12},q_{ij}). \tag{46}$$

By [5], (46) yields a low sensitivity in the passband, where $|s'_{12}|\cong 1$, i.e., $|h'h'_*/f'f'_*|\ll 1$.

For $i=1,n$

$$S(T,q_{1j})=S(s'_{12},q_{1j})-\frac{1}{2}S(W_1,q_{1j}) \tag{47}$$

$$S(T,q_{nj})=S(s'_{12},q_{nj})+\frac{1}{2}S(W_2,q_{nj}). \tag{48}$$

Since W_1 and W_2 are parts of the first and the last branches, they can be expressed with respect to the set of parameters q_{1j} and q_{nj}, by taking the even part of the first and last branch, as

$$W_1=\frac{\omega_{01}}{h_1Q_1(p^2+\omega_1^2)}$$

$$W_2=\frac{(p^2+\omega_n^2)h_nQ_n}{\omega_{0n}p^2}.$$

Hence, one has

$$S(W_2,q_{ij})=S(W_1,q_{ij})=\begin{cases}\pm 1, & q_{ij}=\omega_{0i},h_i,Q_i \quad (49)\\ \dfrac{\pm 2\omega_i^2}{p^2+\omega_i^2}, & q_{ij}=\omega_i. \quad (50)\end{cases}$$

Finally, for $s'_{12}=1$ and $i=1,n$

$$S(T,q_{ij})=\begin{cases}\pm\dfrac{1}{2}, & q_{ij}=\omega_{0i},h_i,Q_i \quad (51)\\ \pm\dfrac{\omega_i^2}{p^2+\omega_i^2}, & q_{ij}=\omega_i. \quad (52)\end{cases}$$

In other words, at a zero of attenuation of the transmit-

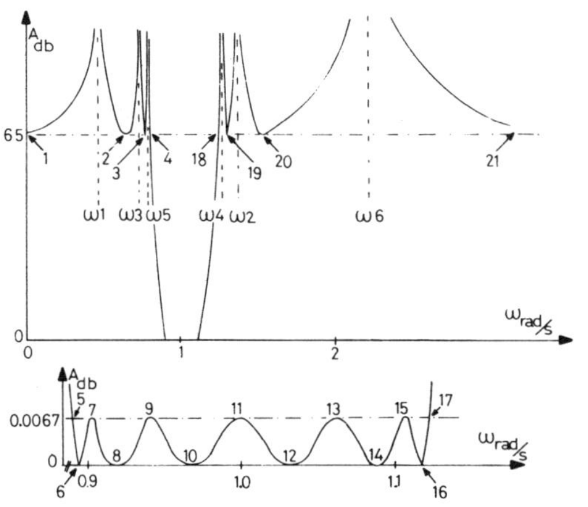

Fig. 8. Computed attenuation of the example filter.

tance s'_{12}, the sensitivity of the voltage transfer function reduces to the sensitivity of the terminations, which in the present case is a function of frequency.

As mentioned by Szentirmai [2], the extraction order of the ω_i has an influence on sensitivity. Since it has now been demonstrated that there is an LC equivalent for all LF design, this rule may be linked to the well-known rules [8] for LC filters.

VIII. Stability and Positiveness of the Components

Starting from a transfer function $T = N/D$ (where D is Hurwitz of degree $2n$), the synthesis procedure yields n second-order blocks in the leapfrog configuration; therefore, the structure is minimal and there is no pole–zero cancellation. Consequently, all the natural frequencies of the system are those of the polynomial D. Since D is supposed to be Hurwitz, the system is stable. It must be stressed that the choice of K_{max} ensures the stability of the equivalent 2-port between W_1 and W_2. On the contrary, (38) does not necessarily yield a polynomial g which is Hurwitz. This does not contradict the property of stability, since g is only an intermediate step in the computation. The 2-port between resistive terminations must not be realized as such.

IX. Example

To illustrate the theory with an example, we have chosen a bandpass filter of the 12th order. The attenuation curve is represented in Fig. 8; it also indicates the angular frequencies ω_j (j: 1 to 21) where the sensitivity has been calculated, i.e., the extrema of the attenuation and the band edges.

The roots of the three polynomials N, D, and h' in (37) are represented in Table I which also indicates the scale factor K^2. This factor has been chosen at its maximum value K^2_{max} for which a double pair of roots in the polynomial $h'h'_*$ occurs on the imaginary axis.

Table II represents the values of the parameters $(h_i, \omega_i, \omega_{0i}, Q_i)$ of each cell yielded by the synthesis.

TABLE I
The Roots of the Three Polynomials N, D, and h' and the Scale Factor K^2

ROOTS OF N	ROOTS OF D
0 ± j 0.464495708285	-5.87597032401E-2 ± j 1.11096098539
0 ± j 0.730926977177	-1.45565231242E-2 ± j 0.874887721576
0 ± j 0.788205241432	-1.90120700615E-2 ± j 1.14268691178
0 ± j 1.26870307352	-8.13964338749E-2 ± j 0.954304681637
0 ± j 1.36812192780	-4.74756777332E-2 ± j 0.897610358265
0 ± j 2.15283077125	-8.87324002634E-2 ± j 1.04031515352

$K^2 = K^2_{max} =$ 2.21294730768E-7

ROOTS OF h′
-1.85457563728E-4 ± j 0.892639257194
-1.77462847895E-2 ± j 0.912141412266
-3.77213725126E-2 ± j 0.963340374638
-4.05835546727E-2 ± j 1.03646683581
-2.13172673715E-2 ± j 1.09590747581
0.0 ± j 1.12027028120

TABLE II
The Parameters of Each Cell i

I	hi	ωi	ωoi	Qi
1	0.210662180	0.464495708	1.031195491	4.895019547
2	0.503366031	1.268703073	1.018270973	∞
3	0.248147671	0.788205241	0.992592552	∞
4	0.343612676	1.368121927	0.997090529	∞
5	0.244487413	0.730926977	0.986043507	∞
6	0.227370958	2.152830771	0.959337350	2.134087961

TABLE III
The Sensitivity Factor ρ Calculated at the Points j of Fig. 8.

j I	1	2	3	4	5	6	7	8	9	10	11	12	13
Leapfrog	2.99	11.8	54.5	185	2.56	2.55	2.91	2.61	2.84	3.10	2.43	2.67	3.09
Cascade	3.46	12.9	57.9	182	40.6	39.2	33.6	26.8	21.7	17.9	16.8	18.1	21.9

j	14	15	16	17	18	19	20	21
Leapfrog	2.87	2.78	2.57	2.63	186	55.1	12.0	3.05
Cascade	26.9	33.8	39.2	40.5	182	57.9	13.0	3.46

The sensitivity criterion is based upon the individual sensitivities of the transfer function to all the cell components x_{ik}. According to the definition of Schoeffler, it is written

$$\rho(\omega) = \sqrt{\sum_{i,k} |S(|T|, x_{ik})|^2}$$

where the sensitivity is defined by (42). The sensitivity $\rho(\omega)$ has been computed in the leapfrog and cascade cases; it is represented in Table III. The type of cells used for the realization is the three-operational amplifier biquad as described by Fleischer and Tow in [9]. The sensitivity analysis has been computed by assuming ideal op-amps.

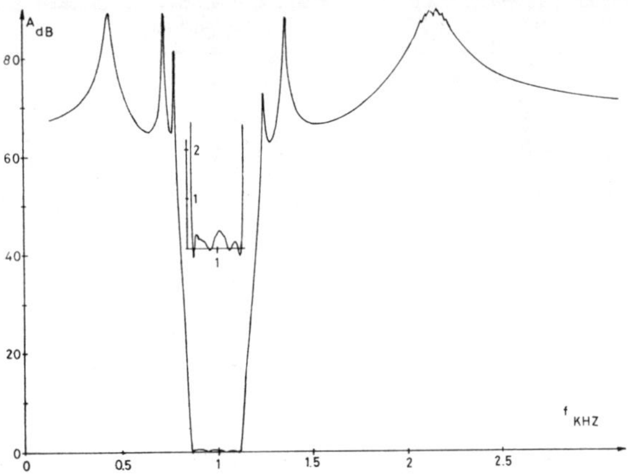

Fig. 9. Measured attenuation of the example filter.

The filter, scaled to 1 kHz center frequency, was implemented with discrete components (tolerances ± 1 percent) by using quad-op-amps Motorola MC 3403 p. Thc computed and measured losses are shown, respectively, in Fig. 8 and Fig. 9.

X. Conclusion

It has been shown that any LF design has an *LC* equivalent even when the *RC* active filter is built in the most economical way. This shows that the low sensitivity of the LF design is linked to the well-known insensitivity of the classical *LC* filters.

The optimal choice of the parameters has been discussed and formulas are given which allow to design the LF filter by using existing programs for lossless filters.

Acknowledgment

The authors would like to thank Dr. M. Hasler for many stimulating discussions, R. Palomera for providing the sensitivity analysis program, and N. C. Bui for the realization of the circuit.

They also wish to thank warmly the reviewers who have paid a considerable amount of attention to their paper and have helped to clarify the authors' own ideas.

References

[1] G. Szentirmai, "Synthesis of multiple feedback active filters," *Bell Syst. Tech. J.*, vol. 52, no. 4, pp. 527–555, Apr. 1973.

[2] ——, "On multiple feedback active filter structures," in *Proc. Seventh Asilomar Conf. on Circuits, Systems and Computers*, pp. 368–377, 1973.

[3] H. J. Orchard, "Inductorless filters," *Electron. Lett.*, vol. 2, pp. 224–225, June 1966.

[4] D. F. Tuttle, "A common basis for the Darlington and Szentirmai network synthesis," *Proc. IEEE*, vol. 62, pp. 1389–1390, Oct. 1974.

[5] G. C. Temes and H. J. Orchard, "First-order sensitivity and worst case analysis of doubly terminated reactance two-ports," *IEEE Trans. Circuit Theory*, vol. CT-20, no. 5, pp. 567–571, Sept. 1973.

[6] L. T. Bruton, "Network transfer functions using the concept of frequency-dependent negative resistance," *IEEE Trans. Circuit Theory*, vol. CT-16, pp. 406–408, Aug. 1969.

[7] V. Belevitch, *Classical Network Theory*. San Francisco, CA: Holden-Day, 1968.

[8] J. Neirynck and Tran Con, "On the extraction order of the attenuation poles in *LC*-filters," in *Proc. 1974 European Conf. on Circuit Theory and Design*, vol. 16, pp. 20–25, July 1974.

[9] P. E. Fleischer and J. Tow, "Design forms for biquad active filters using three operational amplifiers," *Proc. IEEE*, vol. 61, pp. 662–663, May 1973.

Multiple-Amplifier RC-Active Filter Design with Emphasis on GIC Realizations

L. T. BRUTON, MEMBER, IEEE,

Invited Paper

Abstract—**The use of more than one amplifier to realize an RC-active filter is now economically viable because of the relatively low cost of high-performance operational amplifiers. In this review article, multiple-amplifier RC-active synthesis and design techniques are considered. Multiple-amplifier circuits are described that use generalized immittance converters or integrators to realize low-sensitivity high-quality biquadratic circuits and active ladder circuits.**

I. Introduction

THE TOPIC of multiple-amplifier RC-active filter design has become increasingly important in recent years; this is basically because multiple-amplifier RC-active filters may be designed so that the magnitude transfer function is highly insensitive to the tolerance errors associated with the RC elements and to the imperfections of the op amp's. It is now feasible to implement microelectronic multiple-amplifier RC-active filters that meet very stringent specifications on the passband and stopband. Typically, transfer functions of orders greater than ten and with a highest pole-pair Q-factor q_P in excess of 200 have been implemented with passband frequencies as high as 100 kHz. Variations of the gain function with time and with temperature have been shown to be comparable with that obtained using conventional LC filter technology.

The material described in this review concentrates on those multiple-amplifier methods that have led to microelectronic implementations of very high-quality RC-active filters. For all of the circuits discussed here, the tolerance requirements on the RC elements are of the same order of magnitude as is required of the LC elements of a corresponding passive LC filter. Further, the tolerance performance (that is, sensitivity performance) of the filters described in this part of our paper is generally superior to that of the single-amplifier biquadratic filters.

The sensitivity performance of *second-order* RC-active filters is often derived in terms of the perturbations of Q-factors q_P and natural frequency ω_P. This approach is usually satisfactory in the second-order case because we can easily determine the change in the transfer function $T(s)$ that results from specific perturbations of q_P and ω_P. In the case of *high-order* multiple-amplifier structures, the large numbers of poles and zeros make such an approach impractical. Section II of this paper briefly describes the excellent sensitivity performance of resistively terminated LC filters and also the algebraic sensitivity criteria that are particularly useful for determining the sensitivity performance of multiple-amplifier RC-active filters.

Sections III and IV describe the two major techniques that are currently used to implement high-quality RC-active filters by simulating the behavior of resistively terminated LC filters. The direct methods are based on *immittance-simulation techniques* whereby each LC element of the passive filter has a corresponding passive or active element in the resultant RC-active filter; these methods are summarized in Section III. Less direct methods of simulating LC filters are based on signal-flow-graph (SFG) descriptions of the LC filter in which each element is represented by an integrating transmittance; the corresponding RC-active filter employs single-op amp integrators to replace each LC element as described in Section IV. It is stressed that there are design techniques, often referred to as *multiple-feedback* or *coupled-biquadratic* structures [32], [33] that are direct topographical extensions of those described in Section IV which are omitted from this review in the interest of brevity.

Multiple-amplifier second-order circuits may be used in cascade connections to implement high-order transfer functions. Some particularly important two- , three- , and four-amplifier second-order RC-active filters are discussed in Section V. There is a large number of quite different multiple-amplifier second-order circuits that have been proposed and analyzed in the literature; from this extensive body of information, circuits have been selected for inclusion in this review on the basis of their good sensitivity performance, ease of tuning, and the extent to which they are successfully employed in industry as practical microelectronic active filter second-order sections. A tabulated comparison of multiple-amplifier second-order RC-active filters is included in Section V.

Manuscript received September 13, 1976; revised December 5, 1976 and November 23, 1977.

The author is with the Department of Electrical Engineering, University of Calgary, Calgary, Alta., Canada.

Reprinted from *IEEE Trans. Circuits Syst.*, vol. CAS-25, pp. 830-845, Oct. 1978.

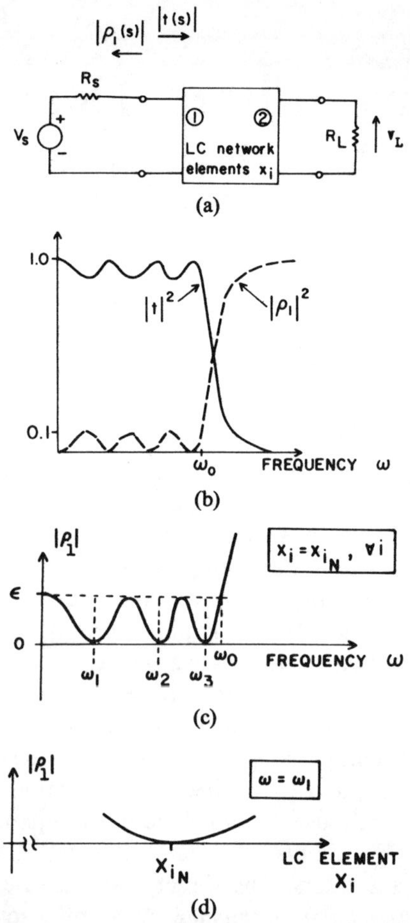

Fig. 1. (a) Resistively terminated lossless LC two-port structure. (b) Typical curves for the transmission and reflection coefficients $|t|$ and $|\rho_1|$. (c) Reflection coefficient versus frequency for exact (nominal) values of LC elements. (d) Reflection coefficient versus an LC element value x_i at frequency ω_1.

II. Review of Sensitivity Theory

A. On the Sensitivity Performance of Resistively Terminated LC Filters

A useful starting point in a discussion of low-sensitivity RC-active filters is the sensitivity performance of passive resistively terminated LC filters. Designers of passive LC filters have generally not encountered the serious sensitivity or tolerance problems that have until recently plagued the designers of RC-active filters. The good *passband* performance of resistively terminated LC filters has been very well explained by Orchard [1] and was often unconsciously assumed by LC filter designers; in terms of the passive structure in Fig. 1(a), the normalized transducer coefficient $t(s)$, and the normalized *input* reflection coefficient $\rho_1(s)$ are related by the well-known Feldtkeller energy-conservation equation which is

$$t(s)t(-s)+\rho_1(s)\rho_1(-s)=1 \tag{1}$$

so that, letting $s=j\omega$ and recognizing that $t(s)$ and $\rho_1(s)$ are functions of the LC elements x_i, we have

$$|t(j\omega,x_i)|^2+|\rho_1(j\omega,x_i)|^2=1. \tag{2}$$

Throughout a low-ripple passband, it is possible to design an LC filter such that $|\rho_1(j\omega)|\ll 1$ or, equivalently, $|t(j\omega)|\approx 1$. Differentiating (2) with respect to x_i and writing $|t|\equiv|t(j\omega,x_i)|$ and $|\rho_1|\equiv|\rho_1(j\omega,x_i)|$ gives

$$\frac{\partial|t|}{\partial x_i}=-\left|\frac{\rho_1}{t}\right|\frac{\partial|\rho_1|}{\partial x_i}, \qquad \forall i$$

so that

$$S_{x_i}^{|T(j\omega)|}=S_{x_i}^{|t|}=\frac{x_i}{|t|}\cdot\frac{\partial|t|}{\partial x_i}=-\frac{x_i}{|t|}\left|\frac{\rho_1}{t}\right|\frac{\partial|\rho_1|}{\partial x_i}, \qquad \forall i. \tag{3}$$

The above expression for the sensitivity of the magnitude function $|T(j\omega)|$ to the LC elements is low-valued *in the passband* for essentially two reasons. (i) The term $[|\rho_1/t|]$ in (3) is much less than unity over the entire passband, as can be seen from the typical equiripple values shown in Fig. 1(b). If the LC filter is designed such that the minimum values of $|\rho_1|$ are zero (as in the diagram), then this term is zero wherever $|\rho_1|=0$. (ii) The term $\partial|\rho_1|/\partial x_i$ in (3) may also be shown to equal zero for all i wherever $|\rho_1|$ is zero (and differentiable). This is briefly explained as follows. In Fig. 1(c), $|\rho_1|$ is zero at ω_1, ω_2, and ω_3 and all LC elements x_i have their exact (or nominal) values which we define as x_{iN}. Now, $|\rho_1|$ is a continuous nonnegative function of each of the LC elements x_i with value zero at ω_1, ω_2, and ω_3. For example, Fig. 1(d) shows the relationship between $|\rho_1|$ and any LC element x_i at the particular frequency ω_1. Clearly, at the frequency ω_1, where $|\rho_1|$ is zero, it also follows that $\partial|\rho_1|/\partial x_i$ is zero. This argument is valid at all such frequencies $\omega_1,\omega_2,\omega_3,\cdots$ where $|\rho_1|$ is zero and is valid for all of the LC elements x_i.

The above argument implies that low-valued passband sensitivities to LC elements are to be expected if $|\rho_1|$ is close to zero throughout the passband. It may be shown that the *upper bound* on the deviation of the magnitude transfer function $|T(j\omega)|\equiv M(\omega)$ is given by [46]

$$\left|\frac{\Delta M(\omega)}{M(\omega)}\right|\leqslant\left|\frac{\epsilon\rho_1(j\omega)}{1-|\rho_1(j\omega)|^2}\right|\omega\tau(\omega), \qquad \frac{\Delta x_i}{x}\leqslant\epsilon \tag{4}$$

where $\tau(\omega)$ is the group delay. This *upper bound* turns out to be quite tight in practice.

Insofar as perturbations of the resistive terminations $R_{s,L}$ are concerned, it may be shown [3] that

$$|S_{R_{s,L}}^{M(\omega)}|\leqslant 1 \tag{5}$$

regardless of the particular transfer function $T(s)$. Thus the LC filter is relatively insensitive to its terminations. In fact, it may also be shown [3] that the major effect of a small perturbation $\Delta R_s|R_s$ or $\Delta R_L|R_L$ is to scale $|M(\omega)|$ by approximately $\frac{1}{2}[\Delta R_{s,L}/R_{s,L}]$.

The active simulations of LC filters that are discussed herein have a nonideal performance that is directly analogous to the effects of incidental dissipation (i.e., finite Q-factors) of the LC elements of the prototype LC filter.

If the Q-factor $Q_l(\omega)$ of the inductance elements and $Q_c(\omega)$ of the capacitance elements are uniform across all L and C elements, respectively, then the resultant first-order deviation of the magnitude function $M(\omega)$ is given by [3]

$$\frac{\Delta M(\omega)}{M(\omega)} \approx \frac{1}{2}\left[\frac{1}{Q_l(\omega)}+\frac{1}{Q_c(\omega)}\right]\omega\tau(\omega) + \frac{1}{2}\left[\frac{1}{Q_l(\omega)}-\frac{1}{Q_c(\omega)}\right]|\rho_1(j\omega)+\rho_2(j\omega)| \quad (6)$$

where $\rho_1(j\omega)$ and $\rho_2(j\omega)$ are the input and output reflection coefficients, respectively. For most practical applications involving high-order high-selectivity transfer functions, the second term containing $\rho_1(j\omega)$ and $\rho_2(j\omega)$ is completely dominated by the first term containing $\tau(\omega)$. This is because $|\rho_{1,2}(j\omega)| \ll 1$ and because $|\tau(\omega)| \gg 1$ over the frequency range of interest. Thus for our purposes,

$$\boxed{\frac{\Delta M(\omega)}{M(\omega)} \approx \frac{1}{2}\left[\frac{1}{Q_l(\omega)}+\frac{1}{Q_c(\omega)}\right]\omega\tau(\omega).} \quad (7)$$

Equations (4) and (7) summarize the sensitivity performance of LC filters.

B. *Useful Algebraic Sensitivity Criteria*

For RC-active networks, the passive RC elements are represented by the parameters $x_1, x_2, x_3, \cdots, x_n$. Let them have uniform tolerance specificiations given by

$$(1-\epsilon)x_{iN} \leqslant x_i \leqslant (1+\epsilon)x_{iN} \quad (8)$$

where x_{iN} are the exact or nominal values of x_i. The *worst case multiparameter sensitivity criterion* WS is defined as

$$WS_x^{M(\omega)} \equiv \sum_{i=1}^{n} |S_{x_i}^{M(\omega)}|. \quad (9)$$

It then follows directly from (8) and (9) that the deviation $|\Delta M(\omega)/M(\omega)|$ is bounded from above by

$$\left|\frac{\Delta M(\omega)}{M(\omega)}\right| \leqslant |\epsilon WS_x^{M(\omega)}|. \quad (10)$$

This expression becomes an equality (implying that worst case performance exists) only under the improbable circumstances that all n elements x_i actually take on their extremum values $x_{iN}(1 \pm \epsilon)$ in such a direction $(\pm\epsilon)$ as to perturb $M(\omega)$ in the same direction. Thus the right side of (10) is a highly pessimistic bound but it must be used if the worst case performance must be known: for example, in a 100-percent yield situation. The computation of $WS_x^{M(\omega)}$ is lengthy because it requires prior computation of all n single-element sensitivities $S_{x_i}^{M(\omega)}$. However, there does exist an invariant lower bound for $WS_{x_i}^{M(\omega)}$ (that is closely approached by all the circuits described herein) and it is worthwhile to compare the performance of particular circuits with this invariant lower bound. It has been shown [2] that for *any* RC-active implementation of the voltage transfer function $T(s)$, the worst case multiparameter sensitivities are necessarily bounded from below by the following expressions:

$$\boxed{\begin{aligned} WS_R^{M(\omega)} &\geqslant \left|\frac{\partial \ln M(\omega)}{\partial \ln \omega}\right| \\ WS_C^{M(\omega)} &\geqslant \left|\frac{\partial \ln M(\omega)}{\partial \ln \omega}\right| \end{aligned}}, \quad \text{invariant lower bounds on worst case sensitivities.} \quad (11)$$

The term on the right side of (11) is simply the logarithmic derivative of the gain function $M(\omega)$ and this is clearly independent of the particular circuit that is chosen to implement $T(s)$. Furthermore, highly selective high-order magnitude functions $M(\omega)$ inherently correspond to large-valued invariant lower bounds in (11) implying a high worst case sensitivity of $M(\omega)$ to perturbations of RC elements in the regions of ω where $M(\omega)$ is highly selective. The worst case passive-sensitivity criteria $WS_{R,C}^{M(\omega)}$ for the high-Q circuits discussed herein are within a factor of 1.5 of the lower bounds in (11) and are, from the worst case viewpoint, therefore roughly equivalent to each other. Consequently, the relative sensitivity behavior of low-sensitivity multiple amplifier circuits is usually based on the dependence of the transfer function on the imperfections of the op amp's. We refer to this dependence of $T(s)$ on op amp imperfections as the *active-sensitivity performance* of the circuit. For all of the circuits described herein, the most important nonideal parameters of the op amp are usually the dc open loop gain A_0 and the gain-bandwidth product B; the nonideal voltage transfer function $A(s)$ of the op amp is usually approximated by the expression

$$A^{-1}(s) \approx A_0^{-1} + \frac{s}{B} \quad (12)$$

where, for an economy op amp, typical numerical values for A_0 and B are

$$A_0 \approx 10^5 \qquad B \approx 2\pi \times 10^6 \text{ rad/s}. \quad (13)$$

It may be shown that, for sufficiently small deviations $\Delta M(\omega)$ of the gain function, the effects of finite A_0 and finite B on $M(\omega)$ are given by

$$\frac{\Delta M(\omega)}{M(\omega)} \approx -S_{A_0}^{M(\omega)}, \quad \text{deviation due to finite } A_0 \quad (14)$$

and

$$\boxed{\frac{\Delta M(\omega)}{M(\omega)} \approx -S_B^{M(\omega)}}, \quad \text{deviation due to finite } B. \quad (15)$$

With the typical numerical values given in (13), it is easily shown that the magnitude of (15) is much larger than that of (14) in the frequency range above about 50 Hz. Therefore, (15) is a most important active-sensitivity criterion by which multiple-amplifier RC-active filters are compared. Of course, there exists a term corresponding to (15) for each op amp. It should be stressed that gainbandwidth products B are far less accurately known and exhibit far larger perturbations as a function of temperature than do the passive RC elements; this usually results in a variation of $M(\omega)$ with temperature that is almost totally attributable to the perturbations of the gainbandwidth products. It may be shown that a small fractional perturbation $\Delta B/B$ causes a deviation $\Delta M(\omega)$ of the gain function $M(\omega)$ given approximately by

$$\boxed{\frac{\Delta M(\omega)}{M(\omega)} \approx S_B^{M(\omega)} \frac{\Delta B}{B}}, \quad \text{deviation due to perturbation of } B. \tag{16}$$

To summarize this review of sensitivity behavior, the worst case sensitivity performance with respect to small perturbations of RC elements is bounded from below by (11), implying highest sensitivity where $M(\omega)$ is highly frequency selective. Most of the circuits to be derived here approach the bounds in (11) and are therefore roughly equivalent from the worst case passive-sensitivity viewpoint. Active-sensitivity performance due to the finiteness of B and the perturbations ΔB of each op amp both depend directly on $S_B^{M(\omega)}$ and the circuits to be discussed are limited in performance by their active-sensitivity behavior.

III. Immittance Simulation of LC Filters

There are two broad classes of RC-active filters that simulate LC filters and thereby attempt to retain the passive insensitivity that applies in the passband as a consequence of (2) and (3). The first class uses direct immittance simulation and the concepts of immittance conversion and/or immittance inversion. This class is the subject of this section. The second class of circuits uses inverting integrators and noninverting integrators to simulate the SFG of the LC filter and is discussed in Section IV.

A. *Immittance Conversion*

Given a general two-port with transmission matrix

$$T = \begin{bmatrix} A(s) & B(s) \\ C(s) & D(s) \end{bmatrix} \tag{17}$$

and a terminating impedance Z_2 at port two, it is easily shown that the driving-point impedance at port one is given by

$$Z_{1d}(s) = \frac{A(s)Z_2(s) + B(s)}{C(s)Z_2(s) + D(s)}. \tag{18}$$

A *generalized-immittance converter* (GIC) is defined as a two-port for which the transmission matrix elements are given by $B(s) = C(s) = 0$ so that

$$Z_{1d}(s) = \left[\frac{A(s)}{D(s)}\right] Z_2(s). \tag{19}$$

The function $[A(s)/D(s)]$ is defined as the *conversion function* $K(s)$ of the GIC and it is straightforward to prove that, since

$$Z_{1d}(s) = K(s)Z_2(s) \tag{20}$$

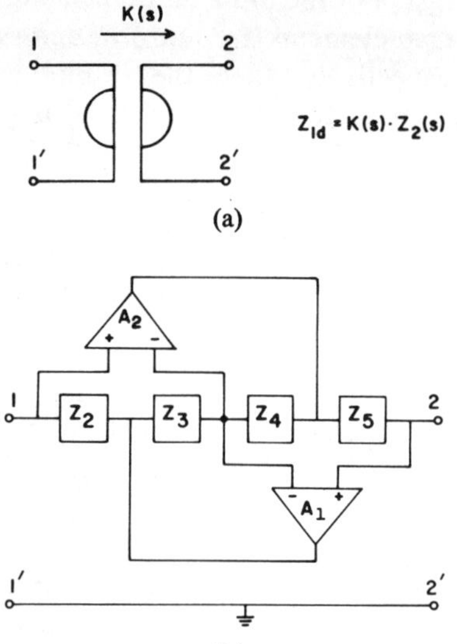

Fig. 2. (a) Symbol for the GIC. (b) A GIC circuit due to Antoniou.

then, if we have a termination Z_1 at port 1, the driving-point impedance at port two is given by

$$Z_{2d}(s) = K^{-1}(s)Z_1(s). \tag{21}$$

Most practical op amp realizations of the GIC [7]–[9] employ at least two op amp's and are constrained, for practical reasons, to have the transmission matrix elements $A(s)$ equal to unity and to have a common grounded terminal at both ports. For conciseness, we will use the symbol in Fig. 2(a) to describe the GIC where it is understood that practical versions correspond to $K(s) = D^{-1}(s)$ and the terminals $1'$ and $2'$ at ground potential. One particularly useful GIC of this type [8], [9] is given in Fig. 2(b) for which it is easily shown that, if $A_1(s) = A_2(s) = \infty$,

$$T = \begin{bmatrix} 1 & 0 \\ 0 & \dfrac{Z_3Z_5}{Z_2Z_4} \end{bmatrix} \tag{22}$$

corresponding to a conversion function $K(s) = Z_2Z_4/Z_3Z_5$.

1) Grounded Inductance Simulation: Selecting either Z_3 or Z_5 in Fig. 2(b) as a capacitor, the remaining branches as resistors, and the termination $Z_2(s) = R_6$ leads to two possible inductance-simulated driving-point impedances at port one given by

$$Z_{1d} = s\left[\frac{C_3R_2R_4R_6}{R_5}\right] \text{ or } Z_{1d} = s\left[\frac{C_5R_2R_4R_6}{R_3}\right] \tag{23}$$

corresponding to effective grounded inductances of

$$L_1 = \frac{C_3R_2R_4R_6}{R_5} \text{ or } L_1 = \frac{C_5R_2R_4R_6}{R_3}. \tag{24}$$

The derivation of the *nonideal* function Z_{1d} may be performed [9]–[12] to determine the fractional magnitude error $\epsilon_l(\omega)$ of Z_{1d} (and of L_1) and also the finite Q-factor $Q_l(\omega)$. The dependence of $Q_l(\omega)$ and $\epsilon_l(\omega)$ on the GBP's B_1 and B_2 of the two op amp's are the major causes of error and the approximate expressions are therefore given in

TABLE I
GIC INDUCTANCE-SIMULATION CIRCUIT

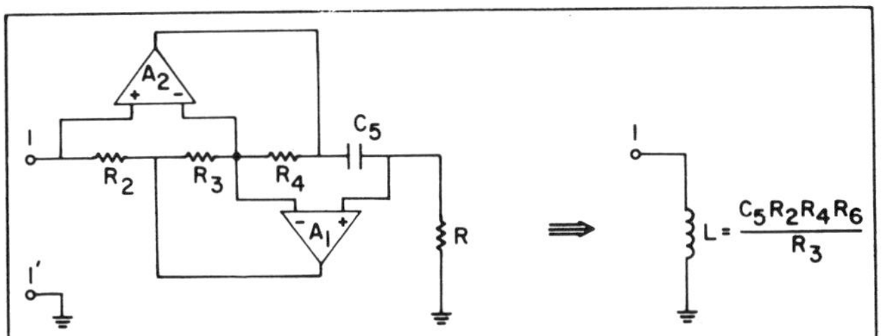

Q FACTOR:

$$Q_\ell^{-1}(\omega) \approx \omega\left[\frac{1}{B_2}\left(\frac{R_4}{R_3} - 1\right) + \frac{1}{B_1}\left(1 - \frac{R_3}{R_4}\right)\right] + \omega^3 \frac{C_5 R_6}{B_1 B_2}$$

ENHANCEMENT CONDITION:

$$R_3 = R_4$$

INDUCTANCE FRACTIONAL MAGNITUDE ERROR:

$$\epsilon_\ell(\omega) \approx \frac{1}{B_1}\left[\frac{R_3}{C_5 R_4 R_6} + \omega^2 C_5 R_6\right] + \frac{1}{B_2}\left[\frac{1}{C_5 R_6} + \omega^2 \frac{C_5 R_4 R_6}{R_3}\right]$$

TABLE II
GIC FDNR-SIMULATION CIRCUIT

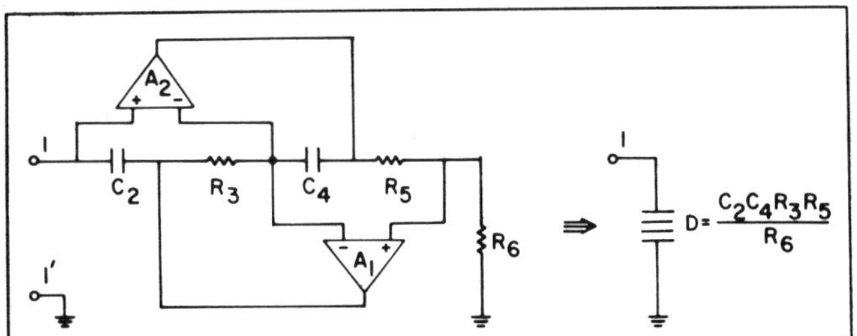

Q FACTOR:

$$Q_d^{-1}(\omega) \approx -\omega\left(\frac{1}{R_5} + \frac{1}{R_6}\right)\left[\frac{R_6}{B_1} - \frac{R_5}{B_2} + \frac{\omega^2 C_4 R_3 R_5}{B_1 B_2}\right]$$

ENHANCEMENT CONDITION:

$$B_1 R_5 = B_2 R_6$$

D-ELEMENT FRACTIONAL MAGNITUDE ERROR:

$$\epsilon_d(\omega) \approx \left(\frac{1}{R_5} + \frac{1}{R_6}\right) \cdot \left[\frac{R_6}{B_2 C_4 R_3} + \frac{\omega^2 C_4 R_3 R_5}{B_1}\right]$$

Table I for the preferred circuit which is the one with the capacitance located in branch 5. It is most important to note that $Q_l^{-1}(\omega)$ may be suppressed to approximately $\omega^3 C_5 R_6 / B_1 B_2$ by simply maintaining the Q-factor enhancement condition $R_3 = R_4$. This technique of *Q-enhancement* is often used in inductance-simulation circuits; in this case, enhancement only requires the accurate matching of two resistors R_3 and R_4, and does not require that the somewhat unpredictable GBP's be matched. Simple second-order *RLC* circuits may be simulated using grounded *RC*-active simulated inductances: for example, the *HP* and *BP* functions in Table V. The outputs *HP*1 and *BP*1 are at the output terminal of op amp A_2 in the GIC of Fig. 2(b) and correspond to a zero output resistance circuit.

2) Grounded Frequency Dependent Negative Resistance (FDNR): If we choose the branches of the GIC in Fig. 2(b) so that Z_2 and Z_4 are capacitors and the remaining branches are resistors, then

$$Z_{1d}(s) = \frac{1}{s^2}\left[\frac{R_6}{C_2 C_4 R_3 R_5}\right] \equiv \frac{1}{s^2 D} \tag{25}$$

where

$$D \equiv C_2 C_4 R_3 R_5 / R_6. \tag{26}$$

Under conditions of sinusoidal steady-state excitation, $Z_{1d}(j\omega) = -1/\omega^2 D$; that is, we have an FDNR element [13]. The symbol and circuit realization of this FDNR D element are given in Table II together with the approximate expressions for Q-factor $Q_d(\omega)$, enhancement and fractional magnitude error $\epsilon_d(\omega)$. The Q-factor $Q_d(\omega)$ of the D element [9] is defined as

$$Q_d(\omega) \equiv -\mathrm{Re}\left[Z_{1d}(j\omega)\right] / \mathrm{Im}\left[Z_{1d}(j\omega)\right]. \tag{27}$$

It is usually possible to enhance the Q-factor $Q_d(\omega)$ by maintaining the enhancement condition $B_1 R_5 = B_2 R_9$ [9] where $B_{1,2}$ are the GBP's of the op amp's. In this case, it is necessary to *employ dual op amp's* so that $B_1 \approx B_2$ in order to maintain $B_1 R_5 \approx B_2 R_6$ and thereby enhance $Q_d(\omega)$. Using economy op amp's, $A_0 \approx 10^5$ and $B_{1,2} \approx 2\pi \times 10^6$ rad/s, and maintaining the enhancement conditions in Tables I and II usually allows $|Q_l(\omega)|$ or $|Q_d(\omega)|$ to be maintained in excess of 1000 over the audio frequency range 200 Hz to 10 kHz. The applications of the FDNR concept are discussed in Section III-C; it should be noted that recent techniques allow *single-amplifier* circuits to be used to realize grounded FDNR elements [48].

3) Impedance Scaling by Embedding Between GIC's: A principal disadvantage of the GIC circuit in Fig. 2(b) is that terminals 1'–2' are grounded so that the simulation of ungrounded immittances is not directly possible. The concept of *GIC embedding* [14] overcomes this problem in many cases: let the network N in Fig. 3(a) be embedded between GIC A and GIC B with conversion functions $K(s)$ and $K^{-1}(s)$. If the transmission matrix T of the embedded network is given by

$$T = \begin{bmatrix} a & b \\ c & d \end{bmatrix} \tag{28}$$

and we employ GIC's with the transmission matrix element $A(s)$ equal to unity, then the transmission matrix T_1

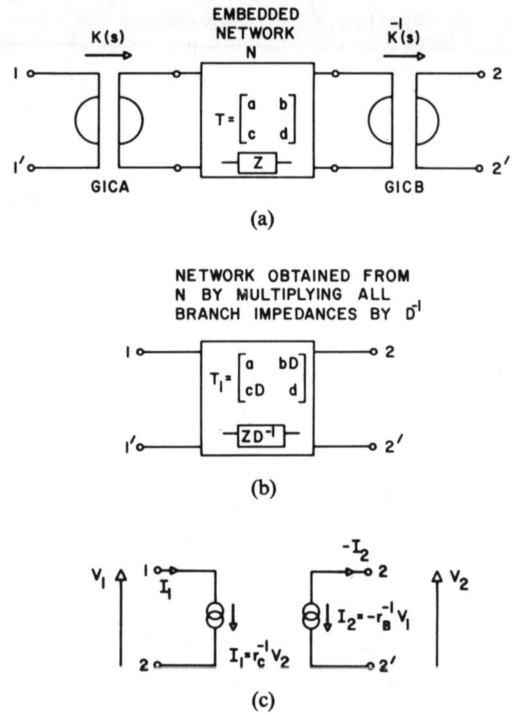

Fig. 3. (a) Network N embedded between two matched GIC's. (b) Equivalent network for Fig. 3(a). (c) Positive immittance inverter.

of the overall network in Fig. 3(a) is given by

$$T_1 = \begin{bmatrix} 1 & 0 \\ 0 & D(s) \end{bmatrix} \begin{bmatrix} a & b \\ c & d \end{bmatrix} \cdot \begin{bmatrix} 1 & 0 \\ 0 & D^{-1}(s) \end{bmatrix} = \begin{bmatrix} a & bD^{-1}(s) \\ cD(s) & d \end{bmatrix}. \quad (29)$$

This transmission matrix T_1 is identical to that which is obtained by impedance scaling every branch impedance Z of the network N by the function $K(s) = D^{-1}(s)$. That is, embedding as shown in Fig. 3(a) is exactly equivalent, insofar as the resultant two-port parameters are concerned, to branch impedance scaling by $K(s)$. The network T_1 in Fig. 3(b), obtained by the scaling $Z \rightarrow K(s) \cdot Z$, is equivalent to the network in Fig. 3(a). Consequently, a two-port subnetwork T of resistors may be made to simulate a subnetwork of inductors by simply embedding the network T between two GIC's with $K(s) \equiv s/\omega_0$, ω_0 constant. Applications of this technique are described in Section III-C.

B. Immittance Inversion

A *generalized-immittance inverter* (GII) is defined as a two-port for which the transmission matrix elements are given by $A(s) = D(s) = 0$ so that, from (18)

$$Z_{1d}(s) = \left[\frac{B(s)}{C(s)} \right] \frac{1}{Z_2(s)}. \quad (30)$$

Successful transistor immittance inverters have been designed and implemented in microelectronic form for the special case where $B(s)$ and $C(s)$ are real, so that

$$B(s) \equiv r_B \qquad C(s) \equiv r_C^{-1} \quad (31)$$

TABLE III
SOME TYPICAL LC FILTER STRUCTURES

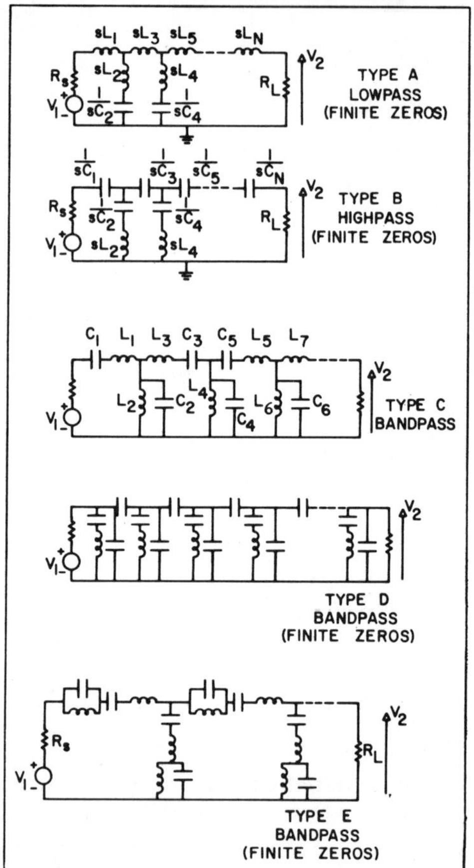

corresponding to $K(s) = r_B r_C$ and, when terminated in Z_2 at port two

$$Z_{1d}(s) = [r_B r_C] \frac{1}{Z_2}. \quad (32)$$

This type of two-port is a positive impedance inverter (PII) or, if $r_B = r_C$, a gyrator. It may be used to realize a simulated inductance by choosing $Z_2 = 1/sC_2$. Implementation of this type of PII is by the parallel connection of two voltage-controlled current sources (VCCS's) as shown in Fig. 3(c). Special-purpose transistorized *differential-input* VCCS's have been implemented [15], [16] that may be employed to realize high-quality floating inductance elements, $L = r_B r_C C_2$, with Q-factors in excess of 1000 over the audio spectrum.

C. Applications of Immittance-Conversion Concepts

The LC filters in Table III are typical of the types of configurations that are commonly encountered in communications circuitry. The type A low-pass circuit is a widely used configuration for obtaining finite stopband zeros, the type B high-pass circuit also achieves finite stopband zeros and the type C bandpass circuit is obtained from the all-pole ($L_2 = L_4 = L_6 \cdots = 0$) low-pass circuit by conventional low-pass to bandpass transformation. The highly selective types D and E bandpass configurations have finite stopband zeros; the type D is char-

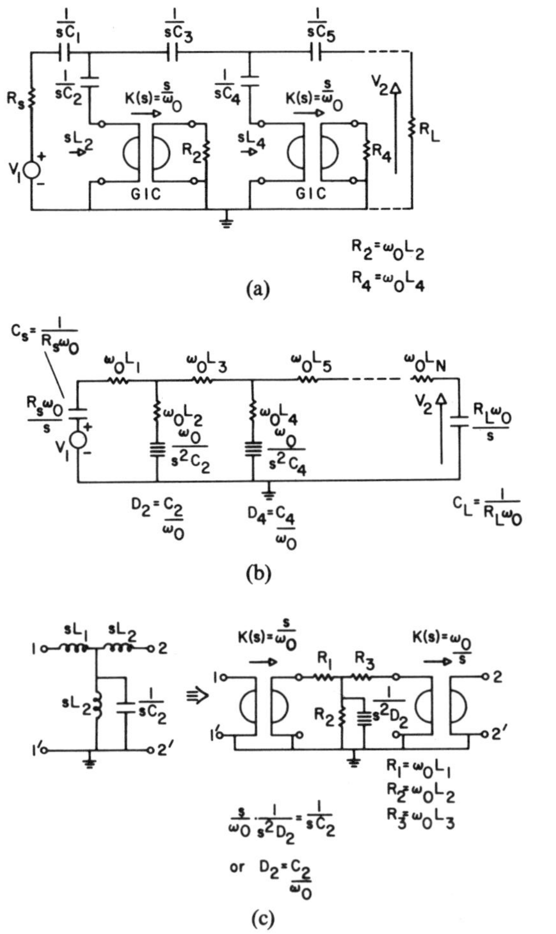

Fig. 4. (a) A high-pass active ladder filter (type B) with transmission zeros using inductance simulation. (b) A low-pass active ladder filter (type A) with transmission zeros using ω_0/s transformation and FDNR elements. (c) Use of embedding techniques to realize a section of the bandpass ladder filter (type C).

acterized by the fact that all inductors are grounded and the type E is obtained by bandpass transformation of the type A circuit. We will use these five LC filters to explain some of the methods that have been employed to realize RC-active simulations of LC filters. It is assumed that equivalued source and load resistances are employed so that $|\rho_1| \ll 1$ throughout the passband, thereby ensuring that $|S_{x_i}^{M(\omega)}| \to 0$ in the passband.

1) Grounded Inductance-Simulation Filters—(Types B or D): The grounded-inductance GIC circuit in Table I may of course be used directly to replace the inductors in the types B and D LC filters. The type B high-pass circuit is shown in Fig. 4(a) where the GIC branches (R_2, C_3, R_4, and R_5 in Fig. 2(a)) are chosen to correspond to $K(s) = s/\omega_0$ where $R_{2,4} = \omega_0 L_{2,4}$. The nonideal nature of the driving-point impedances $sL_{2,4}$, via $Q_l(\omega)$ and $\epsilon_l(\omega)$, determines the nonideal behavior of the filter transfer function. The principal cause of passband error is the finite Q-factor $Q_l(\omega)$; the expression for $Q_l(\omega)$ in Table I implies that typically $|Q_l(\omega)| > 1000$ so that, via (7), it may be estimated that [30]

$$\left|\frac{\Delta M(\omega)}{M(\omega)}\right| \leqslant \frac{1}{2}\left[\frac{1}{1000}\right]\omega\tau(\omega), \quad \text{passband deviation bound due to } Q_l(\omega),\ |Q_l(\omega)| > 1000. \tag{33}$$

The reader may not have any feeling for the typical magnitude of the group delay function $\tau(\omega)$; it may be shown that highly selective "brick-wall" type magnitude functions $M(\omega)$ possess group delay functions $\tau(\omega)$ that have a maximum magnitude of approximately $2q_P/\omega_P$ where q_P is the Q-factor of the highest Q pole pair and where ω_P is close to the edge of the passband. This maximum of $\tau(\omega)$ occurs where $\omega \approx \omega_P$. Therefore, it follows from (33) that

$$\boxed{\left|\frac{\Delta M(\omega)}{M(\omega)}\right| \leqslant \frac{q_P}{1000}}, \quad \text{passband deviation bound due to } Q_l(\omega),\ |Q_l(\omega)| > 1000. \tag{34}$$

Thus if the highest Q pole pair has a Q-factor q_P of 100, then the fractional deviation of the gain function is not greater than 10 percent where, according to (33), this maximum fractional deviation will occur at the band edge where the group delay is a maximum.

The fractional deviation $\Delta M(\omega)/M(\omega)$ due to the inductance magnitude errors $\epsilon_l(\omega)$ given in Table I may be estimated by using (4); it is usually found that throughout the passband, (4) predicts a negligible error due to inductance magnitude errors compared with the error due to the finite inductance Q-factors $Q_l(\omega)$.

The passive-sensitivity performance of simulated-inductance active filters is directly related to that of the sensitivity performance of the LC prototype filter. Thus (3) applies to all the RC elements, thereby ensuring excellent insensitivity of $M(\omega)$ to RC elements throughout the passband if $|\rho_1(j\omega)|$ is approximately zero in this region. It has also been shown [17] that, in the transition region where the term $|\partial \ln M(\omega)/\partial \ln \omega|$ is much greater than unity,

$$WS_{RC}^{M(\omega)} \approx 3\left|\frac{\partial \ln M(\omega)}{\partial \ln \omega}\right|, \quad \text{transition region} \tag{35}$$

which is a factor of approximately 1.5 in excess of the lower bound obtained from (11).

In summary, grounded-inductance simulation techniques, with appropriate enhancement of $Q_l(\omega)$ [8], [12] lead to very high-quality RC-active ladder filters. The active-sensitivity performance is excellent and the worst case passive-sensitivity performance is close to the optimum in the transition region and low-valued in the passband if $|t(j\omega)| \approx 1$. A major disadvantage is that many useful LC filters contain *ungrounded* (floating) inductance elements. The FDNR and GIC-embedding techniques are used in these situations.

2) Grounded-FDNR Filters and ω_0/s Transformation [13]: The type A or type D circuit may be realized by simply impedance scaling of all branches by the function ω_0/s so that we obtain the FDNR network in Fig. 4(b). The dimensionless transfer function $T(s, R_i, C_i, L_i)$ of the LC filter remains unaltered under the ω_0/s transformation and, of equal importance, the sensitivities $S_{R_i}^{T}(s)$, $S_{C_i}^{T(s)}$, and $S_{L_i}^{T(s)}$ also remain unaltered and apply also to the FDNR circuit of Fig. 4(b). Using the GIC FDNR circuit of Table II to realize each *grounded* FDNR element results in a circuit containing N op amp's. The floating inductance subnetwork $L_1, L_2, L_3, \cdots, L_N$ of the type A LC filter transforms to a floating *resistance* subnetwork $\omega_0 L_1, \omega_0 L_2, \cdots, \omega_0 L_N$. It may easily be shown that (34) and (35) also apply to the passband and transition region, respectively, of FDNR filters. Therefore, very high-quality filters may be designed using this technique and have been used in such applications as frequency division multiplexing systems and PCM systems.

The interested reader may wish to consult the references to pursue the design of FDNR bandpass filters [20], the use of FDNR elements of the form $Z_d(j\omega) = -\omega^2 E$ [19] and the implementation of single-amplifier FDNR circuits [48].

3) Ungrounded-Immittance Simulation using Embedding [14], [20], [27], [28]: The type C bandpass LC filter contains both ungrounded L and ungrounded C elements; it is not directly suitable for straightforward grounded-FDNR implementation because capacitors such as C_1, C_3, C_5, etc. would transform into *ungrounded-FDNR* elements that are *not* realizable using the two-amplifier circuit in Table II. Embedding techniques may be used to overcome this difficulty; the subnetwork containing L_1, L_2, L_3, and C_2 in the type C filter may be realized by simply embedding a resistance-FDNR network between two GIC's as shown in Fig. 4(c). The GIC's have conversion functions $K(s)$ of s/ω_0 and ω_0/s so that, by the impedance-scaling argument in Section III-A the three resistors $R_{1,2,3}$ ($\equiv \omega_0 L_{1,2,3}$) are scaled to inductive reactances $sL_{1,2,3}$ and the *grounded FDNR* with the transform impedance ω_0/s^2C_2 is transformed to the required grounded capacitance impedance $1/sC_2$. The complete type C bandpass filter may be realized this way, using embedding circuits to implement L_1, L_2, C_2, L_3, and L_5, L_6, C_6, L_7, etc., and single GIC's to implement L_4, L_8, $L_{12}, \cdots$, etc. Such circuits employ N operational amplifiers and it has been shown that the passband and transition region sensitivities are essentially described by (33) and (35), respectively, when the embedding GIC's are permitted to have imperfectly matched conversion functions due to RC-element errors [27].

There is a variety of other types of immittance-simulating circuits that have been published in the literature; some of these circuits achieve a reduction in op amp's by abandoning the passband insensitivity properties of the LC filter. Such techniques are justifiable if it can be shown that the resultant $\Delta M(\omega)/M(\omega)$ is within acceptable limits.

IV. Signal-Flowgraph (SFG) Simulation of LC Filters Using Integrators [21]–[23]

An SFG may usually be written for a resistively terminated LC filter so that the branches R_s, R_L, $1/sC_i$, and $1/sL_i$ appear as individual transmittances in the

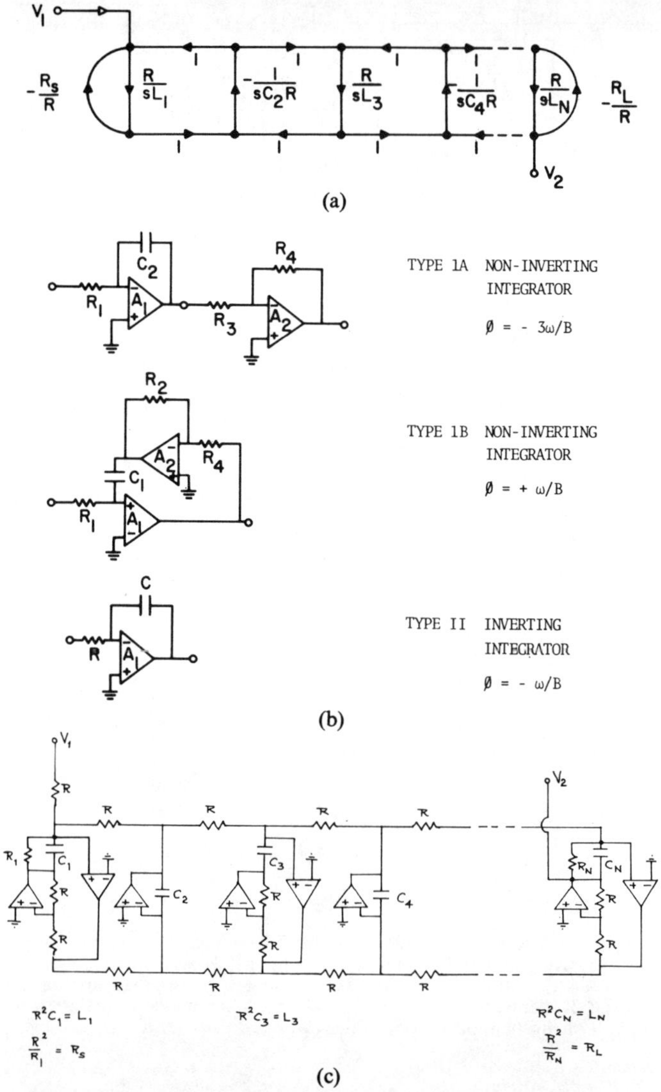

Fig. 5. (a) Leapfrog signal flowgraph of the type A low-pass ladder filter with $L_2 = L_4 = L_6 \cdots = 0$. (b) Single-op amp integrators: types IA, IB, and II. (c) Implementation of the leapfrog low-pass SFG of Fig. 5(a) using types IB and II integrators.

SFG. For example, if $L_2 \equiv L_4 \equiv L_6 \cdots \equiv 0$ in the type A lowpass filter, then the corresponding SFG is that given in Fig. 5(a). The SFG consists of inverting- and noninverting-integrating transmittances of the form $\pm 1/sx_i$. The RC-active circuits of Fig. 5(b) may be used to implement the $\pm 1/sx_i$ integrating voltage transfer function; they are classified here as type IA and IB [24] (noninverting) and type 2 (inverting). The nonideal transfer functions may be written in the form $\pm(1/sx_i)[M_i(s)]$ so that the nonideal phase error ϕ_i is given by $\phi_i \equiv \angle M(j\omega)$. To a first approximation, these phase errors ϕ_i are given by $-3\omega/B$, $+\omega/B$, and $-\omega/B$ for types IA, IB, and II, respectively. It may be shown [25] that a phase error ϕ_i in integrator $\pm 1/sx_i$ of the SFG in Fig. 5(a) is exactly equivalent to a capacitance or inductance x_i in the LC filter with finite Q-factor ϕ_i^{-1}.

Substituting type IB and type II integrators into the SFG of Fig. 5(a) [22], [26] leads directly to the circuit in Fig. 5(c). This is known as a SFG *leapfrog* ladder simulation structure; it has the property that the passive sensitivities to all the RC elements in the *vertically drawn paths* are directly analogous to the passive sensitivities of the LC filter to the LC elements [31]. Thus they are zero wherever $|\rho_1|=0$ and low valued in the passband. Equations (2) and (3) apply to these RC elements. However, the horizontally drawn resistors R do *not* have a one-to-one correspondence with L or C elements.

The active sensitivities of this SFG leapfrog lowpass filter are extremely low valued because the phase error ϕ associated with the noninverting type IB integrator is $+(\omega/B)$ and the phase error ϕ for the inverting type II integrator is exactly opposite and equals $-(\omega/B)$. Thus in terms of the LC prototype and using $Q=\phi^{-1}$, we have $Q_l=B/\omega$ for the inductors and $Q_c=-(B/\omega)$ for the capacitors. Consequently, to a first-order approximation the term $(1/Q_l)+(1/Q_c)$ is zero in (6) resulting in a significant suppression of the dominant group delay term in this equation. The consequence is a much reduced passband error $\Delta M(\omega)/M(\omega)$ than is obtained if type IA integrators are used; in the latter case, $Q_l=-(B/3\omega)$ and $Q_c=-(B/\omega)$ leading, via (7), to $\Delta M/M \approx -2\omega^2\tau(\omega)/B$. Although the Q enhancement that is achieved by selecting type IB–type II integrators depends on equivalued GBP's over all op amp's, the end result is usually a far preferable passband error than is obtained using type IA–type II integrators. With Q enhancement, the active sensitivities of leapfrog RC-active filters are about the same as for the Q-enhanced FDNR filters. Typically, the leapfrog RC-active filter requires at least $3N/2$ op amp's. This is a factor of 1.5 in excess of the number required using grounded inductance, grounded FDNR, or embedding-GIC techniques. A number of designers favor SFG leapfrog methods because it is possible to separately align each two-integrator loop ($\pm(1/sC_i)$ and $\mp(1/sL_i)$) by causing each loop to resonate at $\omega=(L_iC_i)^{-1/2}$ and because it is possible to maximize dynamic range by scaling internal RC element values without altering the transfer function $T(s)$. However, similar alignment operations are possible using GIC techniques [20]. A significant advantage of SFG techniques is that they are well suited to the realization of *electronically adjustable* filters.

The SFG leapfrog methods that have been mentioned partially retain the low passive-sensitivity properties of the LC prototype by retaining a one-to-one correspondence between LC element values and the integrating transmittances of the SFG. However, the basic feedback structure of the flowgraph in Fig. 5(a), suitably generalized by replacing each integrator $\pm(1/sx_i)$ by a general biquadratic function $F_i(s)$, *may be used to synthesize transfer functions without any reference to an existing LC prototype* [32], [33]. Low-sensitivity circuits have been designed and made using such direct synthesis techniques; the sensitivity properties appear to be comparable to those that derive from (3). Current and future studies may reveal further insight concerning the generality of leapfrog structures and their sensitivity performance.

V. Biquadratic Circuits

A. The Biquadratic Transfer Function

There is a large number of multiple-op amp realizations of the biquadratic transfer function

$$T(s)=\frac{a_2s^2+a_1s+a_0}{b_2s^2+b_1s+b_0} \text{ or } K\left[\frac{s^2+\left(\dfrac{\omega_Z}{q_Z}\right)s+\omega_Z^2}{s^2+\left(\dfrac{\omega_P}{q_P}\right)s+\omega_P^2}\right]. \tag{36}$$

A widely adopted approach to the realization of high-order transfer functions is to employ a cascade connection of high-quality multiple-op amp biquadratic sections. If this approach is used, it is usually necessary that the biquadratic section $T(s)$ be able to produce the following types of functions:

$$\begin{aligned} &\text{low-pass (LP):}\ a_1=a_2=0\\ &\text{bandpass (BP):}\ a_0=a_2=0\\ &\text{high-pass (HP):}\ a_0=a_1=0\\ &\text{notch (N):}\ a_1=0. \end{aligned} \tag{37}$$

In addition, the all-pass (AP) function is characterized by $q_P=-q_Z$ and $\omega_P=\omega_Z$ and is also widely used.

Most multiple-op amp high-performance biquadratic RC-active filters, hereafter termed BIQUAD's, are based on the use of integrators or GIC's in conjunction with zero-forming summation networks. The various circuits are usually compared on the basis of the passive sensitivities $S_{RC}^{q_{P,Z}}$ and $S_{RC}^{\omega_{P,Z}}$ and the active sensitivities $S_{B_i}^{q_{P,Z}}$ and $S_{B_i}^{\omega_{P,Z}}$ where B_i are the gainbandwidth products in (12). A comparison of BIQUAD circuits is simplified by means of the following sensitivity considerations.

A. Perturbations of $M(\omega)$ due to Δq_B and $\Delta\omega_P$

We consider here the effect of pole perturbations Δq_P and $\Delta\omega_P$ on the magnitude function $M(\omega)=|T(j\omega)|$. The fractional deviation of $M(\omega)$ is obtained from (36) as

$$\frac{\Delta M(\omega)}{M(\omega)}=S_{q_P}^{M(\omega)}\left[\frac{\Delta q_P}{q_P}\right]+S_{\omega_P}^{M(\omega)}\left[\frac{\Delta\omega_P}{\omega_P}\right]+S_{\omega_Z}^{M(\omega)}\left[\frac{\Delta\omega_Z}{\omega_Z}\right]+S_{q_Z}^{M(\omega)}\left[\frac{\Delta q_Z}{q_Z}\right]+S_K^{M(\omega)}\left[\frac{\Delta K}{K}\right]. \tag{38}$$

The last term in this equation is a constant term that therefore contributes to a flat change in gain of the BIQUAD and may usually be neglected. Consider now the sensitivities $S_{q_P}^{M(\omega)}$ and $S_{\omega_P}^{M(\omega)}$ which, by direct analysis of (36), are given by

$$S_{q_P}^{M(\omega)}=\frac{\partial \ln M(\omega)}{\partial \ln q_P}=\left[\frac{-1}{1+q_P^2\left(\dfrac{\omega}{\omega_P}-\dfrac{\omega_P}{\omega}\right)^2}\right], \quad M(\omega)\equiv|T(j\omega)| \tag{39}$$

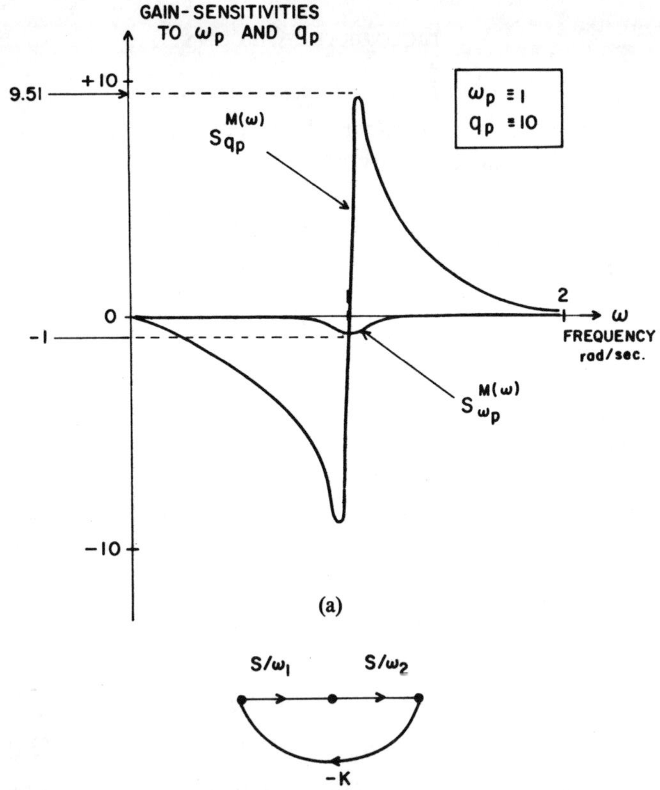

Fig. 6. (a) Sensitivities of gain $M(\omega)$ to q_P and ω_P. (b) Basic TIL flowgraph.

and

$$S_{\omega_P}^{M(\omega)} \equiv \frac{\partial \ln M(\omega)}{\partial \ln \omega_P} = \frac{2q_P^2\left[\frac{\omega_P}{\omega}\left(\frac{\omega}{\omega_P}-\frac{\omega_P}{\omega}\right)\right]-1}{1+q_P^2\left(\frac{\omega}{\omega_P}-\frac{\omega_P}{\omega}\right)^2}. \quad (40)$$

The important observation concerning these two sensitivities is that $|S_{q_P}^{M(\omega)}|$ is much less than $|S_{\omega_P}^{M(\omega)}|$ over the frequency range of interest if $|q_P| \gg 1$. For example, (39) and (40) are shown in Fig. 6(a) for $q_P = 10$ from which it is observed that $S_{q_P}^{M(\omega)}$ is negligible compared with $S_{\omega_P}^{M(\omega)}$. It has been shown that the maxima and minima of $S_{\omega_P}^{M(\omega)}$ are approximately $\pm q_P$ whereas the function $S_{q_P}^{M(\omega)}$ is always bounded between 0 and -1. *Clearly, the* ω_P *sensitivity* $S_{\omega_P}^{M(\omega)}$ *is approximately* $2q_P$ *times as important as the* q_P *sensitivity* $S_{q_P}^{M(\omega)}$. This result was published in 1971 [47] and has significant implications that were not always fully appreciated by designers of single- or multiple-op amp BIQUAD's. For example, the terms $\Delta q_P/q_P$ and $\Delta\omega_P/\omega_P$ that result from a parameter perturbation Δx_i are given by

$$\frac{\Delta q_P}{q_P} = S_{x_i}^{q_P}\left[\frac{\Delta x_i}{x_i}\right] \text{ and } \frac{\Delta \omega_P}{\omega_P} = S_{x_i}^{\omega_P}\left[\frac{\Delta x_i}{x_i}\right] \quad (41)$$

so that, if

$$|S_{x_i}^{q_P}| \ll 2q_P|S_{x_i}^{\omega_P}|, \qquad |q_P| \gg 1 \quad (42)$$

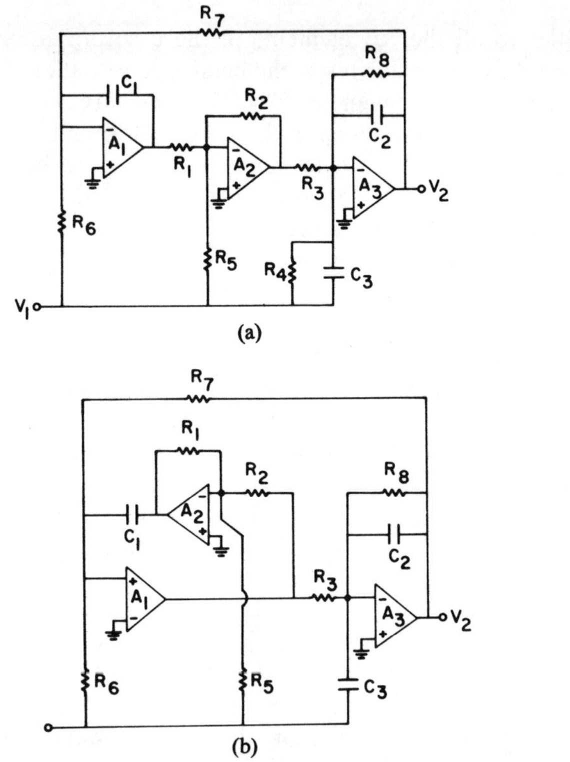

Fig. 7. Biquadratic TIL's. (a) TIL-3A biquadratic circuit. (b) TIL-3B biquadratic circuit.

it is possible to neglect q_P sensitivities $S_{q_P}^{M(\omega)}$ and $S_{x_i}^{q_P}$ and compare and analyze BIQUAD's simply on the basis of ω_P sensitivities $S_{x_i}^{\omega_P}$. Many of the multiple-op amp BIQUAD's to be discussed in the following sections fall into this category where it is generally permissible to neglect variations of q_P in that they are, by virtue of (42), negligible.

C. Two Integrator-Loop (TIL) Biquadratic Circuits

The basic building block from which the TIL circuits are realized [21] is represented by the signal flowgraph in Fig. 6(b). It consists of a single loop with two inverting-integrator transmittances and an inversion transmittance. (This flowgraph will be recognized as a subnetwork of the general ladder flowgraph in Fig. 5(a)). The various TIL circuits differ in the detailed *RC*-active implementations of these transmittances in the methods by which signals are injected into or removed from the TIL and by the use of various auxiliary summing networks to realized specific numerators of (36).

A particularly versatile version [35] of the TIL is shown in Fig. 7(a); it has the biquadratic transfer function

$$T(s) = -\frac{C_3}{C_2}\left[\frac{s^2+s\left[\frac{1}{R_4}-\frac{R_2}{R_5R_3}\right]+\frac{R_2}{R_6R_1R_3C_1C_3}}{s^2+s\left[\frac{1}{R_8C_2}\right]+\frac{R_2}{R_7R_1R_3C_1C_2}}\right] \quad (43)$$

TABLE IV
A Comparative Summary of Multiple-Amplifier Biquadratic/Quadratic Circuits

BIQUAD' TYPE	ACTIVE SENSITIVITIES $S_B^{\omega_P}$	$S_B^{Q_P}$	NUMBER OF OPAMPs LP	HP	BP	N	NO. OF CAPACITORS	REMARKS
TIL-3A Fig. (a)	$\frac{3\omega_P}{2B}$	$\frac{-4Q_P\omega_P}{B}$	3	3	3	3	3	Require type (i) Q-enhancement. Non-constant input impedance.
TIL-3B Fig. (b)	$\frac{3\omega_P}{2B}$	$\ll \frac{Q_P\omega_P}{B}$	3	3	3	3	3	Q-enhanced; type (ii). Non-constant input impedance.
TIL-4	$\frac{-\omega_P}{B}$	$\frac{-4Q_P\omega_P}{B}$	3	4	3	4	2	Requires type (i) Q-enhancement. Constant input impedance. Uses 4 OPAMPs for HP and N functions.
GIC-3	$\frac{-\omega_P}{2B}$	$\ll \frac{Q_P\omega_P}{B}$	3	3	3	3	2	Q-enhanced; type (ii) Low $S_B^{\omega_P}$, only two capacitors, constant input impedance.
GIC-2LP -2BP -2HP -2N	$\frac{-2\omega_P}{B}$	$\ll \frac{Q_P\omega_P}{B}$	2	2	2	2	3	Q-enhanced; type (ii) or (iii). Only 2 OPAMPs. Non-constant input impedance.

from which it follows that, by appropriate selection of the RC elements, it is possible to realize LP, BP, HP, N, and AP functions as described by (37). This circuit is a particularly useful BIQUAD that is representative of a large number of three- or four-amplifier circuits that are based on the TIL concept. In this case, *three* amplifiers are used and we classify the circuit as type TIL-3A; it is basically a TIL with four elements (R_4, R_5, R_6, and C_3) involved in feedforward summation to realize the numerator of (43). Note that three capacitors are required and that the integrators are types IA and II. Consequently, the integrators contribute a phase error of approximately $-3\omega_P/B$ and $-\omega_P/B$, respectively (assuming equal GBP's) and, as shown in Table IV, the resultant q_P sensitivity is given by [24], [36]

$$S_B^{q_P} \approx -k\frac{\omega_P}{B}, \qquad k=4q_P \qquad \text{type TIL-3A.} \tag{44}$$

As might be expected, the type IB noninverting integrator may be used instead of the type IA noninverting integrator [24]. By careful design, the *positive* phase error $+\omega/B$ may be arranged to *approximately* cancel the *negative* phase error $-\omega/B$ of the type 2 inverting integrator. The resultant circuit is given as TIL-3B [24] in Fig. 7(b) for which it may be shown that the abovementioned cancellation of phase errors around the TIL loop results in

$$\boxed{S_{PB}^{q}=k\frac{\omega_P}{B}}, \qquad |k|\ll q_P \qquad \text{type TIL-3B.} \tag{45}$$

Both TIL-3A and TIL-3B may be shown [24] to have

$$\boxed{S_B^{\omega_P} \approx \frac{3\omega_P}{2B}}, \quad B_1 \equiv B_2 \equiv B \quad \text{types TIL-3A and TIL-3B.} \tag{46}$$

An important observation regarding the active-sensitivity performance now follows, that is, from (45) and (46),

$$\boxed{|S_B^{q_P}| \ll 2q_P|S_B^{\omega_P}|} \tag{47}$$

implying that $S_B^{q_P}$ may be neglected insofar as the resultant effect on $\Delta M(\omega)$ is concerned. Equation (46) alone is then sufficient to estimate $\Delta M(\omega)$; thus the active-sensitivity performance is given by

$$\frac{\Delta M(\omega)}{M(\omega)} \approx -S_{\omega_P}^{M(\omega)} S_B^{\omega_P} = -S_B^{\omega_P}\left[\frac{\partial \ln M(\omega)}{\partial \ln \omega_P}\right] \tag{48}$$

which is simply the multiple of (40) and (46). It is easily shown from (40) that we may write $S_{\omega_P}^{M(\omega)}$ as $[-\partial \ln M(\omega)/\partial \ln \omega]$ so that, from (48) and (46),

$$\boxed{\left|\frac{\Delta M(\omega)}{M(\omega)}\right| \approx \frac{3\omega_P}{2B}\left|\frac{\partial \ln M(\omega)}{\partial \ln \omega}\right|}, \quad \text{deviation due to finite gainbandwidths B TIL-3B.} \tag{49}$$

The reader may inspect Table IV, which summarizes the performance of five types of second-order circuits, and note that inequality (47) is valid for all the circuits except TIL-3A and TIL-4 so that an expression of similar form to (49) applies to all of these circuits. The appropriate $S_B^{\omega_P}$ term must be substituted in (49) in place of the $3\omega_P/2B$ term. For the TIL-3B circuit, it follows by inspection of (49) that $\Delta M(\omega)/M(\omega)$ due to finite gainbandwidth products has extremum values of $\pm 3\omega_P q_P/2B$ occurring at the frequencies $\omega_P(1\pm(\omega_P/2q_P))$.

The *worst case passive-sensitivity* performance is now estimated for the case where the zeros of $T(s)$ in (43) are at zero or infinity corresponding to the LP, BP, and HP functions in (37). It then follows from (43) that for the elements C_1, C_2, R_1, R_2, R_3, R_7, and R_8, we have

$$|S_{R_1}^{\omega_P}|=|S_{R_2}^{\omega_P}|=|S_{R_3}^{\omega_P}|=|S_{R_7}^{\omega_P}|=|S_{C_1}^{\omega_P}|=|S_{C_2}^{\omega_P}|=\frac{1}{2} \quad (50)$$

and

$$|S_{R_8}^{q_P}|=1 \qquad |S_{C_2}^{q_P}|=\frac{1}{2}$$

so that for $q_P \gg 1$ it follows that inequality (42) is valid. We may therefore neglect the Q sensitivities $S_{R_8}^{q_P}$ and $S_{C_2}^{q_P}$ in (50); thus

$$WS_{RC}^{M(\omega)} \approx |S_{\omega_P}^{M(\omega)}|\, WS_{RC}^{\omega_P}. \quad (51)$$

Using (51) and (50), together with the fact that $|S_{\omega_P}^{M(\omega)}|=|S_{\omega}^{M(\omega)}|=|\partial \ln M(\omega)/\partial \ln \omega|$ to give

$$WS_{RC}^{M(\omega)} \approx 3\left|\frac{\partial \ln M(\omega)}{\partial \ln \omega}\right| \quad (52)$$

corresponding to a worst case fractional deviation given by

$$\left|\frac{\Delta M(\omega)}{M(\omega)}\right|_{\text{worstcase}} \approx 3\left|\epsilon\frac{\partial \ln M(\omega)}{\partial \ln \omega}\right|, \quad \text{worst case passive-sensitivity performance TIL-3A/B} \quad (53)$$

where ϵ is the upper tolerance error of the RC elements. In a practical circuit implementation, it is often true that the tolerance specifications on the RC elements are much tighter than ω_P/B; that is, $2|\epsilon| \ll \omega_P/B$, and we may therefore conclude that the nonideal effect on $M(\omega)$ of the finite gainbandwidths B are far more serious than that of the RC element tolerances.

1) RC Temperature Coefficients: In order to minimize the effect of variations of RC elements with temperature, it is customary to use resistances and capacitances with temperature coefficients of the forms

$$\epsilon_R=\epsilon_{OR}\pm\epsilon_r \qquad \text{ppm}/^\circ\text{C}$$
$$\epsilon_C=\epsilon_{OC}\pm\epsilon_C \qquad \text{ppm}/^\circ\text{C}$$

where the *mean* or nominal coefficients ϵ_{OR}, ϵ_{OC} are chosen so that $\epsilon_{OR}\approx-\epsilon_{OC}$. Then, a similar analysis to that leading to (53) gives an upper bound on the temperature dependence of $M(\omega)$ as

$$\left|\frac{\Delta M(\omega)}{M(\omega)}\right| \leqslant \left|\frac{\partial \ln M(\omega)}{\partial \ln \omega}\right| |\epsilon_{RO}-\epsilon_{CO}+2\epsilon_r+\epsilon_c| \ \text{ppm}/^\circ\text{C}. \quad (54)$$

Thus it is clearly advantageous to minimize the deviation $\Delta M(\omega)$ by selecting $\epsilon_{RO}\approx-\epsilon_{CO}$ and ϵ_r and ϵ_c to be as small as possible. Typical practical figures are $\epsilon_{RO}=-\epsilon_{CO}=50$ ppm/°C and $\epsilon_r=\epsilon_c=20$ ppm/°C.

2) Q-Enhancement: The technique of reducing $|k|$ in (44) from $4q_P$ to much less than q_P by using phase cancellation is not unique to the TIL-3B circuit; the method is described as *Q-enhancement* and has been widely used by BIQUAD circuit designers. There are basically three approaches to the cancellation of a spurious loop phase shift of the form $\phi=k\omega_P/B$.

Type 1): This is done by external capacitive trimming across a resistance element; this method is temperature sensitive and, since B is not known precisely, involves a functional adjustment.

Type 2): This is done by arranging for inherent cancellation of GBP terms; this involves the introduction of a term $-k\omega_P/B_2$ to cancel ϕ and depends on GBP matching. Dual or quad-op amp's are necessary to achieve good cancellation because matched GBP's are necessary.

Type 3): This is done by producing cancellation of phase shifts as in 2) but in a way that does not require matched GBP's but only matched resistances.

Early techniques of Q-enhancement employ method 1) and more recently methods 2) and 3) have been used. Method 3) is preferable because it is easier to match resistances than to match GBP's.

There is a variety of four-amplifier TIL biquadratic circuits, [36]–[41], most of which have the feedforward summation network at the output side of the TIL. Some of these circuits use type 2) Q-enhancement [41] to reduce $|k|$ to a value much less than unity, but most of them have used type IA noninverting integrators resulting in $|k|$ that is of the order q_P. A typical four-amplifier TIL circuit with output summation (amplifier no. 4) is shown in Fig. 8 [36], [39] and classified here as TIL-4A. It uses a type IA noninverting integrator and has a $|k|$ term equal to $-4q_P$ [36] as shown in Table IV, so that the $S_{q_P}^{M(\omega)}$ and $S_B^{q_P}$ terms *cannot* be neglected. Of course, one can easily reduce $|k|$ and achieve Q-enhancement by simply using a type IB noninverting integrator in Fig. 8. The TIL-4A circuit requires four amplifiers for HP, N, and AP functions, whereas only three amplifiers are required for TIL-3 circuits; furthermore, TIL-3 circuits allow *independent* adjustment of ω_P, ω_Z, q_P, and q_Z via resistors R_4, R_5, R_6, R_7, and R_8. The TIL-4 circuits have the advantages that they require only two capacitors and that they have a constant input impedance.

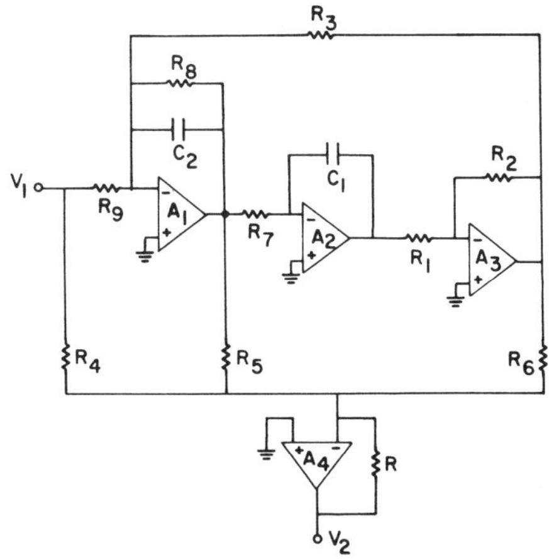

Fig. 8. TIL-4A biquadratic circuit.

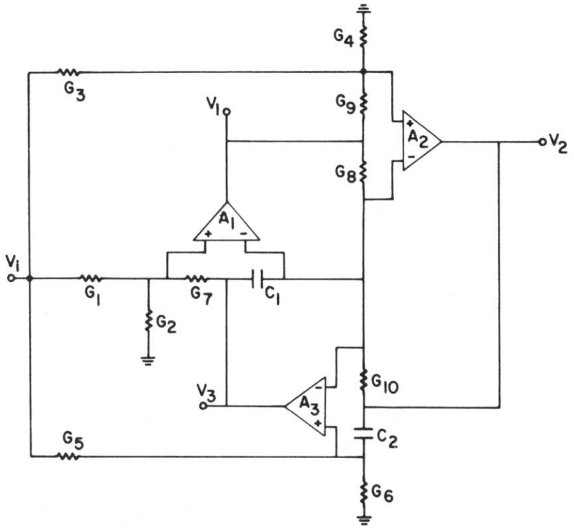

Fig. 9. GIC-3 biquadratic circuit.

D. GIC Biquadratic Circuits

The GIC may be modified by the addition of an op amp A_2 and conductances G_1, G_3, G_5, G_4, G_8, and G_9, to give the general *three*-amplifier biquadratic circuit [42] that is shown in Fig. 9 and classified here as type GIC-3. The exact expressions for the transfer functions V_1/V_i, V_2/V_i, and V_3/V_i are rather complicated; however, it is quite straightforward to show that LP and BP functions are available at V_3 by setting $G_1 = G_3 = 0$ and $G_1 = G_5 = 0$, respectively, and also that the HP and N functions are available at V_1 by selecting $G_3 = G_5 = 0$ and $G_3 = 0$, respectively. The *passive* sensitivities for $|q_{P,Z}| \gg 1$ are similar to the abovementioned TIL circuits with (53) applying also to this circuit. This circuit achieves type 3) Q-enhancement so that, for $B_1 = B_2 = B_3 \equiv B$, the active sensitivities have been shown [42] to be given by

$$S_B^{q_P} \approx \frac{\omega_P}{2B} \quad \text{and} \quad S_B^{\omega_P} \approx -\frac{\omega_P}{2B} \qquad \text{GIC-3.} \tag{55}$$

This circuit behaves very well by comparison with the TIL circuits. It requires only three op amp's and two capacitors. Significantly, the term $S_B^{\omega_P}$ is smaller in magnitude than for the TIL circuits and it is this term that is a limiting factor in the supression of active sensitivities (since $|k| \ll q_P$).

TABLE V
HIGH-PASS, LOW-PASS, AND BANDPASS QUADRATIC GIC CIRCUITS

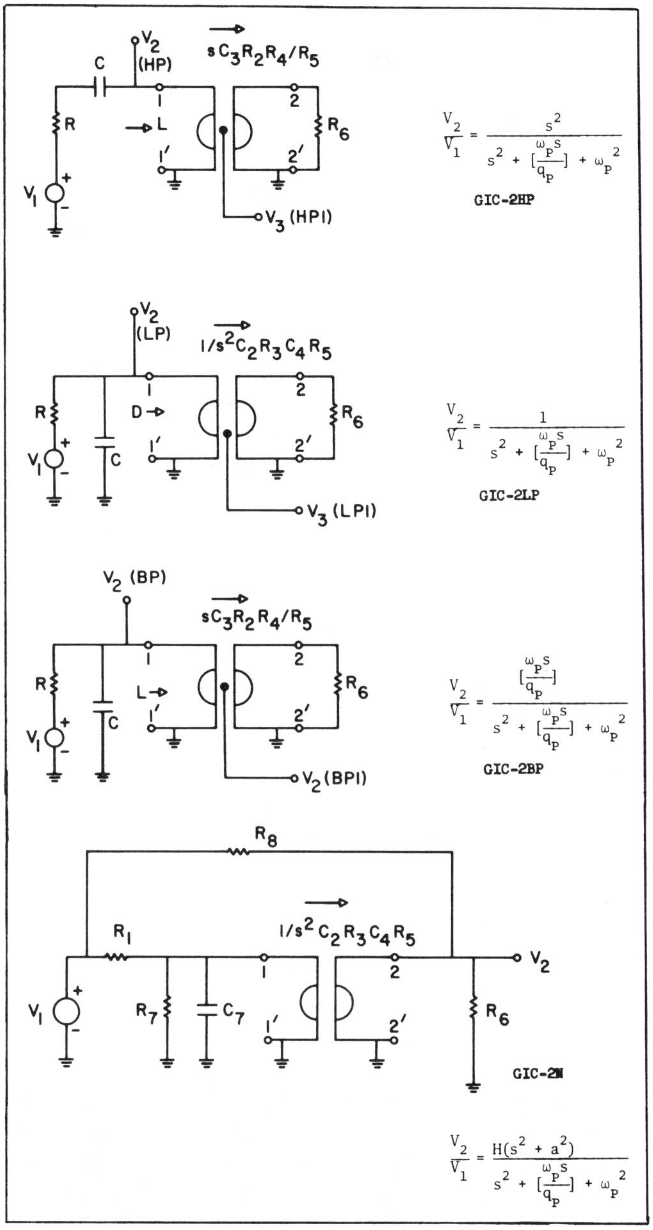

It should perhaps be noted that all of the BIQUAD's that have so far been mentioned achieve Q-enhancement by requiring that GBP's be matched. However, by glancing at Table I, the reader will note that we may realize a simulated inductance with a cancellation of phase terms by simply selecting $R_3 = R_4$. That is, the circuit possesses a type 3) Q-enhancement property that allows $|k| \ll q_P$ by resistance matching instead of GBP matching. An inductance-simulation GIC circuit may be employed in several of the two-amplifier GIC biquadratic circuits shown in Table V. For example, we may use a simple *RCL* series circuit to realize the HP function of circuit GIC-2HP in Table V where the GIC circuit is that given in Fig. 2(b). The output V_3 is taken from the GIC amplifier A_2 output terminal if a voltage-source output is required. The type GIC-2BP provides a BP function with voltage source output at V_3 by means of a simple parallel *RLC* circuit. Both GIC-2LP and GIC-2BP may be designed to achieve type 3) Q-enhancement by selecting Z_5 as the capacitance branch in which case the outputs are at V_2 and do not correspond to a voltage source.

Using Z_3 as the capacitance branch, the circuits shown in Table V permit voltage source outputs at V_3, but at the expense of using type 3) enhancement and therefore matched GBP's.

The two-amplifier LP function is achieved by the simple connection of R, C, and FDNR elements shown as the type GIC-2LP circuit in Table V. The Q-enhancement is type 2) if we select $B_1R_5=B_2R_6$ according to Table II. The GIC-2LP circuit may be modified by the addition of two resistors, R_7 and R_8, to realize the notch N function [43] shown as the type GIC-2N circuit in Table V. A voltage-source output is not available from this circuit. The two-amplifier GIC circuits compare very favorably with other types of multiple amplifier BIQUAD's; they have the important advantage that they require fewer amplifiers. The passive sensitivities are essentially the same as the TIL's and GIC-3 and for simple LP and BP functions, they achieve type 3) Q-enhancement (but without a voltage-source output). It should be noted, however, that this is not achieved without penalty because the dominant term $S_B^{\omega_P}$ is in fact marginally larger (see Table IV) than for the other BIQUAD's.

E. Cascaded Biquadratic Sections Compared with Ladder Simulation

There does *not* exist unanimous agreement among circuit theorists and circuit designers concerning the relative merits of cascaded biquadratic structures compared with ladder structures.

It is generally true, however, that properly designed (i.e., Q-enhanced) simulations of equiterminated LC filters using TIL's or GIC's possess better worst case *passband* passive sensitivities $WS_{RC}^{M(\omega)}$ than that of the equivalent properly designed cascaded-biquadratic filter [22]. For example, consider the BIQUAD implementation of an Nth-order transfer function that employs the cascade connection of $N/2$ BIQUAD's so that

$$M(\omega)=\prod_{i=1}^{N/2} M_i(\omega) \tag{56}$$

where $M_i(\omega)=|T_i(j\omega)|$.

Once more, for simplicity, we assume that the zeros of $T_i(s)$ and, therefore, of $T(s)$ are at either the origin or infinity. Then, from (56) and (52),

$$WS_{RC}^{M(\omega)}\approx\frac{3}{2}\sum_{i=1}^{N/2}\left|\frac{\partial \ln M_i(\omega)}{\partial \ln \omega}\right|. \tag{57}$$

This result implies that the worst case sensitivity is determined by the *sum* of the logarithmic gradients $|\partial \ln M_i(\omega)/\partial \ln \omega|$ of the *individual BIQUAD sections*. These individual gradients have peak magnitudes of approximately $2q_{Pi}$, where the q_{Pi} is the Q-factor of the poles of the ith section. Furthermore, these peak magnitudes occur at $\omega_{Pi}(1\pm 1/2q_{Pi})$ and therefore occur *in the passband of* $M(\omega)$. Thus *for the case of (57)*, cascaded-BIQUAD structures are inferior to ladder structures insofar as worst case passband performance is concerned. For the more general case of finite imaginary zeros of $T(s)$, it has recently been shown [49] that it is possible to optimize the pairing of poles and zeros between sections and obtain a passband sensitivity that is comparable to that obtained using ladder structures.

Cascaded multiple amplifier BIQUAD's have the advantage that they may be designed as building blocks with single-ended input op amp's (TIL-3 or TIL-4), thereby ensuring low common-mode voltages and low harmonic distortion. It has also been claimed that TIL-BIQUAD methods lend themselves to functional trimming of q_P, ω_P, q_Z, and ω_Z, whereas ladder circuits do not allow q_P and ω_P to be functionally trimmed. On the other hand, it may be argued that the superior passband sensitivity of ladders removes the need for trimming q_P and ω_P.

VI. Summary

Two broad classes of multiple-amplifier *RC*-active filters have been reviewed; those which directly or indirectly simulate resistively terminated *LC* filters and those which use multiple-amplifier second-order stages in a cascade connection. Of the class of active filters that simulate *LC* filters, immittance-simulation using GIC's and signal flowgraph simulation using inverting integrators and two integrator loops (TIL's) have been emphasized because these particular methods have been shown to be capable of realizing high-quality high-order functions $M(\omega)$; typically, passband accuracies of ± 0.25 dB with orders N in excess of 12 and Q pole pairs in excess of 200 at passband frequencies from 20 Hz to 100 kHz have been achieved.

The excellent passive-sensitivity performance of *LC* filters is retained in the cases of GIC- and TIL-active ladder simulations if the transmission coefficient $|t(j\omega)|$ is close to unity in the passband. The resultant multiple-op amp circuits contain approximately N op amp's, where N is the order of $T(s)$, and the sensitivity of the gain $M(\omega)$ to the *RC* elements is low valued in the passband and close to optimum (in the worst case sense) in regions of ω where $M(\omega)$ is highly selective; that is, where $|\partial \ln M(\omega)/\partial \ln \omega|$ is much greater than unity. The active-sensitivity performance is largely determined by the finiteness and temperature dependence of the gainbandwidth products B of the op amp's. By employing Q-enhancement techniques, it is possible to drastically reduce the active sensitivity; this implies meeting the GIC enhancement conditions in Table I and II or, in the case of TIL designs, using type IB noninverting integrators to cancel the phase shift of the type II inverting integrators.

The multiple op amp biquadratic circuits are approximately equivalent insofar as passive-sensitivity performance is concerned. This is because inequality (47) is satisfied and therefore $S_{R,C}^{\omega_p}$ are the only significant single-element terms and ω_p is a similar function of six RC elements for every biquadratic circuit considered herein. The active-sensitivity performance is improved by using Q-enhancement because it is then possible to satisfy inequality (47) and thereby neglect $S_B^{q_p}$ and only consider $S_B^{\omega_p}$. The $S_B^{\omega_p}$ terms are different for each circuit that is summarized in Table IV; of the Q-enhanced circuits TIL-3B, GIC-3, and GIC-2, the circuit GIC-3 has the lowest $|S_B^{\omega_p}|$ term and therefore has lowest overall sensitivity with respect to gainbandwidth products. The circuit TIL-3B is flexible, easy to align, has good active-sensitivity performance, uses only three op amp's, and its only real drawbacks are the use of three capacitors and the fact that the input impedance is capacitive.

The rapid decrease in the cost of op amp's is increasing the economic viability of multiple-op amp RC-active microelectronic filters. Recent innovations that suppress the active sensitivity, using types 2) and 3) Q-enhancement, are drastically increasing the quality of RC-active biquadratic and active ladder filters. The eventual widespread application of these techniques depends very much on the technical advances and economic trends associated with the most probable alternate techniques of digital discrete filtering and analog (e.g., charge-coupled device) discrete filtering. In these latter techniques, it is clear that many of the problems of insensitivity remain to be solved.

Acknowledgment

The author is especially grateful to G. S. Moschytz for many helpful suggestions and comments.

References

[1] J. Orchard, "Inductorless filters," *Electron. Lett.*, vol. 2, 224, Sept. 1966.

[2] M. L. Blostein, "Some bounds on sensitivity in *RLC* networks," in *Proc. 1st Allerton Conf. Circuit System Theory*, pp. 488–501, 1963.

[3] ——, "Sensitivity analysis of parasitic effects in resistance terminated *LC* two-ports," *IEEE Trans. Circuit Theory*, pp. 21–25, Mar. 1967.

[4] J. D. Schoeffler, "The synthesis of minimum sensitivity networks," *IEEE Trans. Circuit Theory*, vol. CT-11, pp. 270–276, June 1964.

[5] L. T. Bruton, "Sensitivity comparison of high-*Q* second-order active filter synthesis techniques," *IEEE Trans. Circuits Syst.*, pp. 32–38, Jan. 1975.

[6] M. N. S. Swamy, C. Bushnan, and K. Thulasiraman, "Bounds on the sum of element sensitivity magnitudes for network functions," in *Proc. IEEE Int. Symp. on Circuit Theory*, IEEE Cat. No. 72CHO 594-2CT, pp. 21–25, Apr. 1972.

[7] R. H. S. Riordan, "Simulated inductors using differential amplifiers," *Electron. Lett.*, vol. 3, pp. 50–51, Feb. 1967.

[8] A. Antoniou, "Realization of gyrators using operational amplifiers and their use in *RC*-active network synthesis," *Proc. Inst. Elec. Eng.*, vol. 116, pp. 1838–1850, Nov. 1969.

[9] L. T. Bruton, "Nonideal performance of two-amplifier positive-impedance converters," *IEEE Trans. Circuit Theory*, vol. CT-17, pp. 541–549, Nov. 1970.

[10] L. T. Bruton and J. T. Lim, "High-frequency comparison of GIC-simulated inductance circuits," *Int. J. Circuit Theory Appl.*, vol. 2, pp. 401–404, 1974.

[11] L. T. Bruton, "Frequency selectivity using positive impedance converter type networks," *Proc. IEEE*, vol. 56, pp. 1378–1379, Aug. 1968.

[12] L. T. Bruton and K. Ramakrishna, "High frequency limitations of active ladder networks," presented at the Asilomar Conf., Pacific Grove, Monterey, CA, Dec. 1974.

[13] L. T. Bruton, "Network transfer functions using the concept of frequency-dependent negative resistance," *IEEE Trans. Circuit Theory*, vol. CT-16, pp. 406–408, Aug. 1969.

[14] J. Gorski-Popiel, "*RC*-active synthesis using PIC," *Electron. Lett.*, vol. 3, pp. 381–382, 1967.

[15] H. R. Trimmel and W. E. Heinlein, "Fully floating chain-type gyrator circuit using operational transconductance amplifiers," *IEEE Trans. Circuit Theory* (Corresp.), vol. CT-18, pp. 719–721, Nov. 1971.

[16] W. H. Holmes, W. E. Heinlein, and S. Grutzmann, "Sharp cut-off low-pass filters using floating gyrators," *IEEE J. Solid-State Circuits*, vol. SC-4, pp. 38–50, Feb. 1969.

[17] L. T. Bruton and A. B. Haase, "High frequency limitations of inductorless ladder structures," in *Proc. 16th Midwest Symp. on Circuit Theory*, Waterloo, Ont., Canada, vol. 1, session IX, paper 6, pp. 1–13, Apr. 1973.

[18] H. R. Trimmel, "Realization of canonical bandpass filters with frequency-dependent and frequency-independent negative resistances," in *Proc. 1973 IEEE Int. Symp. on Circuit Theory*, pp. 134–137.

[19] A. Antoniou, "Bandpass transformation and realization using frequency-dependent negative-resistance elements," *IEEE Trans. Circuit Theory*, (Corresp.), vol. CT-18, pp. 297–299, Mar. 1971.

[20] L. T. Bruton and D. Treleaven, "Active filter design using generalized impedance converters, *EDN*, pp. 68–75, Feb. 1975.

[21] F. E. J. Girling and E. F. Good, "Active filters 7 and 8: The two integrator loop," *Wireless World*, vol. 76, pp. 117–119, Mar. 1970, and pp. 134–139, Apr. 1970.

[22] ——, "Active filters 12 and 13: The leapfrog or active ladder synthesis; Applications of the active ladder synthesis," *Wireless World*, vol. 76, pp. 341–345, July 1970, and pp. 445–450, Sept. 1970.

[23] P. O. Brackett and A. S. Sedra, "Direct SFG simulation of *LC* ladder networks with applications to active filter design," *IEEE Trans. Circuits Syst.*, vol. CAS-23, pp. 61–67, Feb. 1976.

[24] D. Akerberg and K. Mossberg, "A versatile active *RC* building block with inherent compensation for the finite bandwidth of the amplifier," *IEEE Trans. Circuits Syst.*, vol. CAS-21, pp. 75–78, Jan. 1974.

[25] L. T. Bruton and A. I. A. Salama, "Frequency limitations of coupled-biquadratic active ladder structures," *IEEE J. Solid-State Circuits*, vol. SC-9, pp. 69–71, Apr. 1974.

[26] P. O. Brackett and A. S. Sedra, "Active compensation for high-frequency effects in op-amp circuits with applications to active *RC* filters," *IEEE Trans. Circuits Syst.*, vol. CAS-23, pp. 68–72, Feb. 1976.

[27] L. T. Bruton and A. B. Haase, "Sensitivity of generalized immittance converter-embedded ladder structures," *IEEE Trans. Circuits Syst.*, vol. CAS-21, pp. 245–250, Mar. 1974.

[28] H. J. Orchard and D. F. Sheahan, "Inductorless bandpass filters," *IEEE J. Solid-State Circuits*, vol. SC-5, pp. 108–118, June 1970.

[29] A. G. J. Holt and M. R. Lee, "Sensitivity comparison of active-cascade and inductance-simulation schemes," *Proc. Inst. Elec. Eng.*, vol. 119, pp. 277–282, 1972.

[30] L. T. Bruton, J. T. Lim and J. Valihora, "The feasibility of active filtering in frequency division multiplex systems," in *Proc. Int. Symp. on Circuit and System Theory*, San Francisco, CA, April 1974.

[31] L. T. Bruton, "Topological equivalence of inductorless ladder structures using integrators," *IEEE Trans. Circuit Theory*, vol. CT-20, pp. 434–437, July 1973.

[32] G. Szentirmai, "Multiple-feedback filters," in *Proc. IEEE Int. Symp. Circuit Theory*, pp. 339, Apr. 1972.

[33] K. R. Laker and M. S. Ghausi, "Synthesis of a low-sensitivity multi-loop feedback active *RC* filter," *IEEE Trans. Circuits Syst.*, vol. CAS-21, pp. 252–259, Mar. 1974.

[34] L. T. Bruton, "Sensitivity of narrowband cascaded bandpass filters," *IREE J.*, vol. 43, no. 5, pp. 325–328, May 1973.

[35] D. Akerberg and K. Mossberg, "Low-sensitivity easily trimmed standard building block for active *RC* filters," *Electron. Lett.*, vol. 5, no. 21, pp. 528–529, Oct. 1969.

[36] L. C. Thomas, "The biquad: Part 1—Some practical design considerations," *IEEE Trans. Circuit Theory*, vol. CT-18, pp. 350–357, May 1971.

[37] A. Budak, *Passive- and Active Network Analysis and Synthesis*. Houghton Mifflin, 1974, chs. 10–12.

[38] J. Tow, "Design formulas for active *RC* filters using operational amplifier biquad," *Electronc. Lett.*, pp. 339–341, July 1969.

[39] ——, "A step-by-step active-filter design," *IEEE Spectrum*, vol. 6, pp. 64–68, Dec. 1969.

[40] W. J. Kerwin, L. P. Huelsman, and R. W. Newcomb, "State-variable synthesis for integrated circuit transfer functions," *IEEE J. Solid-State Circuits*, vol. SC-2, pp. 87–92, Sept. 1967.

[41] R. Tarmi and M. S. Ghausi, "Very high-*Q* insensitive active *RC* networks," *IEEE Trans. Circuit Theory*, vol. CT-17, pp. 358–366, Aug. 1970.

[42] W. B. Mikhael and B. B. Bhattacharyya, "A practical design for insensitivity *RC*-active filters," *IEEE Trans. Circuits Syst.*, vol. CAS-22, pp. 407–415, May 1975.

[43] L. T. Bruton, "Biquadratic sections using generalized impedance converters," *IREE J.*, vol. 41, no. 11, pp. 510–512, Nov. 1971.

[44] D. Vaughan-Pope and L. T. Bruton, "Transfer function synthesis using generalized doubly-terminated two-port networks," *IEEE Trans. Circuits Syst.*, to be published.

[45] L. T. Bruton, "Biquadratic sections using generalized impedance converters," *IREE J.*, vol. 41, no. 11, pp. 510–512, Nov. 1971.

[46] G. C. Temes and H. J. Orchard, "First-order sensitivity and worst case analysis of doubly terminated reactance two-ports," *IEEE Trans. Circuit Theory*, vol. CT-20, pp. 567–571, Sept. 1973.

[47] G. S. Moschytz, "A note on pole frequency and Q sensitivity," *IEEE J. Solid-State Circuits*, vol. SC-6, Aug. 1971.

[48] B. Wise, "Active filters using a predistortion technique," *Electron. Lett.*, vol. 4, pp. 25–26, 1968.

Design of Signal-Flow Graph (SFG) Active Filters

KEN MARTIN AND ADEL S. SEDRA, MEMBER, IEEE

Abstract—A systematic procedure is presented for the design of active filters obtained as operational simulations of *LC* ladder prototypes. The method is based on the signal-flow graph (SFG) approach, but allows one to obtain the final circuit without ever considering signal-flow graphs. Practical design problems, namely, scaling for maximum dynamic range, stabilizing the reciprocator amplifiers, predistorting the *LC* prototype to compensate for the parasitic reactances introduced by reciprocators, eliminating the inherent dc offsets, and tuning are solved and illustrated by examples. Complete design and experimental results are given for a twelfth-order bandpass channel bank filter.

I. Introduction

BECAUSE OF their low-sensitivity properties, active simulations of doubly terminated lossless ladder filters (designed on an insertion-loss basis for maximum power transfer in the passband) have become quite popular [1]–[10]. There are basically two distinct approaches to the design of active filters based on *LC* ladder simulations. These are the component-simulation approach and the operational-simulation approach. The first [2]–[5] is based on simulating the ladder inductances and employs various devices such as gyrators, generalized-immittance converters (GIC's) and frequency-dependent negative resistances (FDNR's). The operational simulation approach [6]–[10] is based on simulating the current-voltage relationships existing in the *LC*-ladder prototype. The operational-simulation methods can be further grouped into two categories: those based on the *direct* simulation of the current and voltage variables [6], [7], and those which simulate linear transformations of these variables [8]. A special case of the latter, which has received considerable attention in the digital-filter area, uses the wave variables [9], [10]. This paper is specifically concerned with the direct operational-simulation method.

Brackett and Sedra [7] have given a general procedure to derive the signal-flow graph which operationally simulates *LC* ladders. This procedure is essentially a number of rules stating how one connects together the signal-flow graphs of each element (of the passive filter), so that the active filter is realized. This interconnection has been made possible by the introduction of elements that ideally simulate "a pair of wires." These elements have been termed "port reciprocators." In the special cases where reciprocators are not needed, signal-flow graph (SFG) filters degenerate to the well-known leap-frog filters [6].

Manuscript received March 31, 1977; revised August 30, 1977. This work was supported in part by the National Research Council of Canada under Grant A7394.

K. Martin was with the Department of Electrical Engineering, University of Toronto, Toronto, Ontario M5S 1A4, Canada. He is now with Bell Northern Research, 1E81, P. O. Box 3511 Station C, Ottawa, Ontario K1Y 4H7, Canada.

A. S. Sedra is with the Department of Electrical Engineering, University of Toronto, Toronto, Ontario M5S 1A4, Canada.

The purpose of this paper is to give a simple step-by-step design procedure for SFG filters. The procedure allows one to draw the circuit of the SFG filter without ever having to deal with the signal-flow graphs. Also, the questions of optimum dynamic range, compensating the reciprocators for filter stability, minimizing attenuation deviations due to the reciprocators, eliminating dc offsets, and tuning SFG filters are dealt with in detail.

It should be noted that the primary intention here is to give a practical design procedure. For the theoretical basis underlying the method we refer the reader to [7]. Nevertheless, especially in areas not dealt with before, justifications and brief explanations are offered. The nontheoretician can safely ignore these explanations and just apply the design equations given.

Every step in the design procedure is illustrated by a general example. Finally, a twelfth-order bandpass filter at 20 kHz is used to show the excellent performance available from SFG filters.

II. Circuit Topology

The first step in the design of the SFG filter is to synthesize a passive-ladder network. It should be impedance normalized and have the smallest possible number of inductors so as to minimize the number of op-amps needed. This can be done through the use of filter tables [11] or computer algorithms [12] and has been extensively explained in the past.

Next, each branch of the passive filter is replaced by a circuit that actively simulates its operation.

The circuits used to simulate reactive elements consist only of positive and negative integrators. The well-known Miller integrators are used to simulate the capacitors. They have a quality factor (Q_C) given by

$$Q_C \cong -\frac{\omega_t}{\omega} \tag{1}$$

where ω_t is the gain-bandwidth product of the op-amp [13].

Reprinted from *IEEE Trans. Circuits Syst.*, vol. CAS-25, pp. 185-195, Apr. 1978.

The positive integrators, used to simulate inductors, are the phase-lead integrators [13]. They have a quality factor (Q_L) given by

$$Q_L \cong \frac{\omega_t}{\omega}. \tag{2}$$

Equations (1) and (2) can be substituted into Blostein's formula that relates the deviation in the transfer function to the dissipative elements [14]. If one assumes that identical op-amps are used in both types of integrators then, to a first-order approximation, the deviation of attenuation will be zero in the passband [13]. This is because the quality factors are equal in magnitude but of opposite sign, and thus cancel each other.

The circuit implementations of the SFG's of the reactive elements are shown in Fig. 1(a). It should be noted that, in most cases, the input resistors to the integrators have been omitted. These will be added when the different sections are connected together. The circuit simulations of the resistive terminations are shown in Fig. 1(b), while the circuit realizations of the reciprocators (used to simulate a pair of wires) are shown in Fig. 1(c). It should be noted that the capacitors denoted C_a and C_b, and the resistor denoted r, are included in the L_0 reciprocator for stability reasons, as will be explained subsequently.[1]

The circuit configuration of the SFG filter is now easily drawn by following a simple procedure.

1) Draw the active simulations of all the reactive elements of the passive filter. They should be placed in the same order as the elements of the passive filter.

2) If there exist two shunt-C sections, that are not separated by a series-L section, then an L_0 reciprocator should be placed anywhere between them. Similarly, series-L sections should be separated either by a shunt-C section or a C_0 reciprocator.[1] (See Section III of [7].)

3) If a shunt-C section is closer to one end of the filter, than any series-L section, then a positive-gain resistive section should be placed at that end. If the opposite is true, a negative-gain resistive section should be placed at that end. If there are no shunt-C or series-L sections (i.e., a high-pass filter) then a positive-gain resistive section should be placed at one end and a negative-gain resistive section should be placed at the other end. This ensures that the end loops will always have negative loop gains.

4) Resistors should be connected between the output of each series-L section and the inputs of all the sections on either side of it, up to and including the first shunt-C section or C_0 reciprocator. Similarly, resistors should be connected between the output of each shunt-C section and the inputs of all the sections on either side of it, up to and including the first series-L section or L_0 reciprocator.

[1]In minimum inductance filters the only series inductors are those used to fully remove loss poles at infinity, from the transfer function [12]. Because of the removal procedure, any two of these series inductors are always separated by shunt capacitors and thus C_0 reciprocators are never needed. For this reason the question of compensating C_0 reciprocators was not investigated.

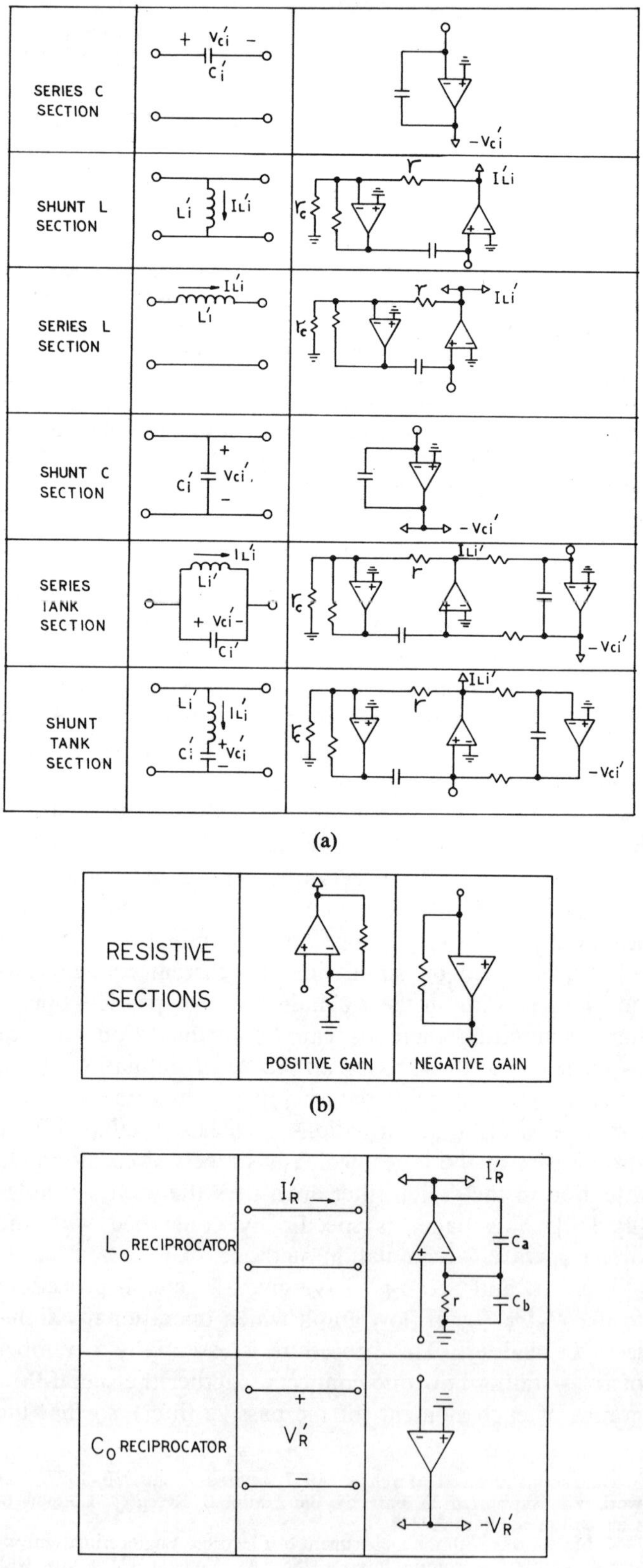

Fig. 1. Circuit implementation. (a) The different sections of LC ladders. (b) The resistive terminations of the ladder. (c) The port reciprocators.

L_0 reciprocators and positive-gain resistive sections should be connected in the same way as series-L sections, while C_0 reciprocators and negative-gain resistive sections should be connected in the same way as shunt-C sections. Note that this procedure implements the SFG interconnection rules given in Section II of [7].

5) The shunt-L, series-C, and tank sections will now have input resistors connected to them from the outputs of other sections. Resistors should then be connected between the inputs of these other sections and the output of the shunt-L, series-C or tank sections, thus causing the filter to consist only of two-amplifier loops.

6) If a resistive section has only one input resistor then it can be eliminated. The integrator, which was connected in a loop with the resistive section, should then have a damping resistor placed in parallel with its capacitor. It is easy to show that in all low-pass and bandpass filters, both resistive sections can be eliminated.

7) The final step is to inject the filter input signal through a resistor connected to the resistive section or damped integrator on the left side of the filter. The filter output should be taken from the output of the resistive section or damped integrator on the right side of the circuit.

III. Element Values

A. Integrator Input Resistors

The time constant of each integrator should be equal to the value of the corresponding reactive component in the LC prototype. As each integrator may have more than one input resistor, the value of these resistors should be equal and determined as follows. It is assumed that the output of the integrator is taken from the output of op-amp $\# i$ and that the capacitor used in the integrator is denoted C_i. An input resistor "coming from" the output of the jth op-amp is denoted R_{ij}. If the integrator is being used to simulate a reactive element of size τ_i, then take

$$k_i C_i R_{ij} = \tau_i, \qquad j = 1, \cdots, n \tag{3}$$

for the n input resistors, where k_i is either equal to 1 for the capacitive simulations or equal to the gain of the invertor used in inductive simulations.

For stable operation of the phase-lead integrator, the gain k_i of its associated invertor should be chosen less than unity. It has been found [15] that a gain value of 0.7 to 0.8 is appropriate. For simplicity we will choose all the invertor gains to be equal. As shown in [15], in order to keep the value of the integrator Q equal to that given by (2), we need to connect a resistor r_c between the negative input of the invertor and ground with the value of r_c given by

$$r_c = r \Big/ \left(\frac{1}{k_i^2} - 1 \right) \tag{4}$$

where r is the magnitude of the invertor input resistor (see Fig. (1a)).

B. Reciprocators

The input resistors to a reciprocator can be arbitrarily chosen as long as they are all equal.

C. Negative Resistive Sections

Since this section simulates the ladder termination but may have more than one "feed-in" resistor, all these input resistors should be chosen equal. If R_i' is the size of the termination being simulated, then each input resistor should be chosen such that the resistive section has a gain K_{R_i} equal to R_i'.

D. Positive Resistive Sections

If R_i' is the size of the resistor being simulated, take $1/R_i' = K_{R_i}/n$ where K_{R_i} is the gain of the positive resistive section, and n is the number of input resistors. The size of the input resistors is arbitrary as long as they are all equal.

E. Damping Resistors

If the jth integrator is being used to simulate both inductor L_i' and resistor R_i' (i.e., it has a damping resistor in parallel with its capacitor C_j) then its damping resistor R_{jj} should be chosen according to the formula

$$R_{jj} C_j = L_i' / R_i' \tag{5}$$

which ensures that the damping time-constant in the active circuit is equal to the damping time-constant associated with the ladder termination. Similarly, if the jth integrator is being used to simulate both capacitor C_i' and resistor R_i', then its damping resistor R_{jj} should be chosen according to the formula

$$R_{jj} C_j = C_i' R_i' \tag{6}$$

where C_j is the capacitor used in the integrator.

If the input resistor to the total filter R_{j0} is connected to a damped integrator with damping resistor R_{jj}, then R_{j0} should be chosen equal to R_{jj}. This results in the active filter having a gain equal to that of its LC prototype.

The above design procedure was applied to the passive filter shown in Fig. 2(a). The resulting SFG filter is shown in Fig. 2(b). The design details and element values of the active realization are listed in Table I.

IV. Scaling for Maximum Dynamic Range

The dynamic range of an active filter (i.e., the ratio of the maximum output voltage, without noticeable distortion, to the noise output, with no input signal) is of considerable importance to the designer. A method for optimizing the dynamic range has been given elsewhere [16].

The procedure consists of scaling the individual gain constants while keeping all loop-gains unchanged. This scaling is done so that the maximum of the absolute value

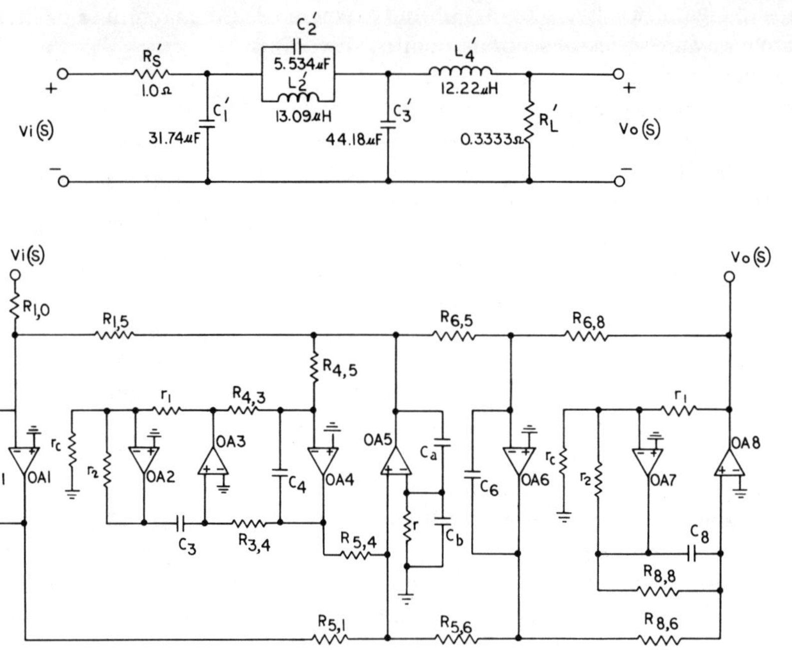

Fig. 2. (a) A fourth-order elliptic low-pass filter. (b) Its corresponding SFG active realization.

TABLE I
DESIGN DETAILS OF THE FILTER IN FIG. 2

$\tau_1 = C_1'$, $\tau_3 = L_2'$, $\tau_4 = C_2'$, $\tau_6 = C_3'$, $\tau_8 = L_4'$
Let r_1 = 3.57 KΩ, r_2 = 2.49 KΩ, ∴ $k_3 = k_8 = 0.6975$, $r_c = r_1/(1/k_3^2 - 1)$ = 3.382 KΩ
Let $C_1 = C_3 = C_4 = C_6 = C_8$ = 2.4 nF, $R_{5,1} = R_{5,4} = R_{5,6}$ = 2 KΩ
$R_{1,5} = \frac{\tau_1}{C_1}$ = 13.23 KΩ , $R_{3,4} = \frac{\tau_3}{k_3 C_3}$ = 7.820 KΩ , $R_{4,3} = R_{4,5} = \frac{\tau_4}{C_4}$ = 2.306 KΩ , $R_{6,5} = R_{6,8} = \frac{\tau_6}{C_6}$ = 18.41 KΩ , $R_{8,6} = \frac{\tau_8}{k_8 C_8}$ = 7.300 KΩ
$R_{1,0} = R_{1,1} = \frac{R_s' C_1'}{C_1}$ = 13.23 KΩ , $R_{8,8} = \frac{L_4'}{R_L' C_8}$ = 15.28 KΩ

TABLE II
THE MAXIMUM MAGNITUDES OF VOLTAGES AND CURRENTS IN THE CIRCUIT OF FIG. 2(a)

OUTPUT OF OP. AMP.	SIMULATED VOLTAGE OR CURRENT	MAXIMUM OF VOLTAGE OR CURRENT
OA1	Voltage across C_1'	M_1 = 0.6992V
OA3	Current through L_2'	M_3 = 1.550A
OA4	Voltage across C_2'	M_4 = 1.262V
OA5	Current through $(L_2' \| C_2')$	M_5 = 1.125A
OA6	Voltage across C_3'	M_6 = 0.6900V
OA8	Current through L_4'	M_8 = 0.8661A

of the voltages, at the output of each op-amp in the active filter, are equal for a constant-amplitude swept-frequency input signal. This ensures that, when the output of the filter is at its maximum then the signal levels everywhere else in the filter are as large as possible.

The first step is to analyze the passive filter for the maximum of the voltages and currents that correspond to the op-amp outputs. The maximum of the output of the ith op-amp is denoted M_i. The integrator whose output is taken at the output of op-amp i, originally has a time constant τ_i that is equal to the size of the corresponding element in the passive filter which is being simulated. After scaling, the ith integrator will have a time constant equal to $k_i R_{ij} C_i$, where k_i is either equal to 1 for all capacitive simulations or equal to the gain of the invertor used in the inductive simulations. The rules for scaling for optimum dynamic range are as follows (for a derivation, see [16]).

a) If op-amp i is used to realize an integrator, take

$$k_i R_{ij} C_i = \tau_i \frac{M_i}{M_j}. \tag{7}$$

b) If op-amp i is used to realize a reciprocator, take

$$R_{ij} = R_{R_i} \frac{M_i}{M_j} \tag{8}$$

where R_{R_i} is arbitrary.

TABLE III
DETAILS OF OPTIMIZING THE CIRCUIT OF FIG. 2(b) TO OBTAIN MAXIMUM DYNAMIC RANGE

$\tau_1 = C_1'$, $\tau_3 = L_2'$, $\tau_4 = C_2'$, $\tau_6 = C_3'$, $\tau_8 = L_4'$	
Let $R_{R_5} = 2\ K\Omega$, $C_1 = C_3 = C_4 = C_6 = C_8 = 2.4\ nF$	
Let $r_1 = 3.57\ K\Omega$, $r_2 = 2.49\ K\Omega$, $\therefore\ k_3 = k_8 = 0.6975$ $r_c = r_1/(1/k_3^2 - 1) = 3.382\ K\Omega$	
$R_{1,5} = \frac{\tau_1 M_1}{C_1 M_5} = 8.219\ K\Omega$	$R_{5,1} = \frac{R_{R5}}{M_1} = 3.218\ K\Omega$
$R_{3,4} = \frac{\tau_3 M_3}{k_3 C_3 M_4} = 9.604\ K\Omega$	$R_{4,3} = \frac{\tau_4 M_4}{C_4 M_3} = 1.877\ K\Omega$
$R_{4,5} = \frac{\tau_4 M_4}{C_4 M_5} = 2.587\ K\Omega$	$R_{5,4} = \frac{R_{R_5} M_5}{M_4} = 1.783\ K\Omega$
$R_{6,5} = \frac{\tau_6 M_5}{C_6 M_5} = 11.29\ K\Omega$	$R_{5,6} = \frac{R_{R_5} M_5}{M_6} = 3.261\ K\Omega$
$R_{6,8} = \frac{\tau_6 M_6}{C_6 M_8} = 14.67\ K\Omega$	$R_{8,6} = \frac{\tau_8 M_8}{k_8 C_8 M_6} = 9.163\ K\Omega$
$R_{1,1} = \frac{R_s' \tau_1}{C_1} = 13.23\ K\Omega$	$R_{8,8} = \frac{\tau_8}{C_8 R_L'} = 15.28\ K\Omega$
$R_{1,0} = \frac{M_1}{M_8} R_{1,1} = 10.68\ K\Omega$ ($R_{1,0}$ can be changed to give filter any desired gain)	

c) If any of the integrators is damped, then the damping time constants should be the same as in the unscaled realization of the filter. This will keep the transfer function unchanged.

d) The filter input resistor can be chosen to give any desired gain. If it is desired to have a gain equal to that of the passive filter, then the filter input resistor denoted R_{i0} should be chosen according to the formula

$$R_{i0} = R_{ii}\frac{M_i}{M_j} \tag{9}$$

where the input signal is being fed into the ith op-amp, which is damped with resistor R_{ii}, and the output of the filter is being taken from the jth op-amp.

Table II lists the maximum voltages and currents, of the passive filter of Fig. 2(a). The design equations and element values for the optimized active filter, shown in Fig. 2(b), are listed in Table III.

V. RECIPROCATORS

A. Compensating the L_0 Reciprocator

Ideally, the L_0 reciprocator has infinite gain [7]. The closest approximation to this, is an open-loop op-amp (see Fig. 3(a)).

The feed-in resistors (denoted R_{ij}, $j = 1, 2, \cdots, n$) are typically connected to the output of sections used to simulate capacitors. Assuming that the op-amp has a transfer function given by

$$A(s) \cong \frac{\omega_t}{s} \tag{10}$$

the transfer function of the reciprocator is given by

$$\frac{V_0}{V_j} = \frac{R_p}{R_{ij}}\frac{\omega_t}{s} \tag{11}$$

where R_p is the parallel equivalent of the feed-in resistors. From (11), it is seen that the open-loop op-amp reciprocator is a positive integrator with a very small time constant $\tau = [R_{ij}/(R_p\omega_t)]$. This gives a phase shift of $-90°$. The capacitive sections in loops with the reciprocator give an ideal phase shift of approximately $-270°$. Thus the total loop phase shift is approximately 360°, and any nonidealities, such as a stray capacitance between the reciprocator-op-amp output and the positive-input terminal, will cause the filter to be unstable.[2] Thus, to ensure stability it is necessary to compensate the reciprocators. The circuit to do this is shown in Fig. 3(b). Using (10) and straightforward analysis, the transfer function $G(s)$ of the

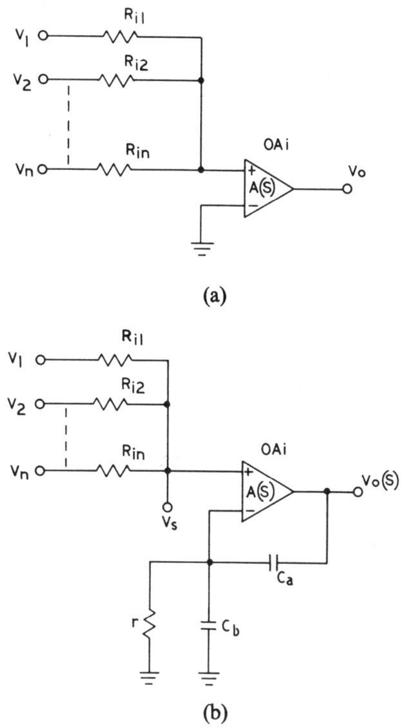

Fig. 3. Circuit implementation of the L_0 reciprocator. (a) Without the stabilization network. (b) With the stabilization network.

[2]It might be wondered why SFG filters are ever stable since they consist of many two-integrator loops. However, the crossover frequency of these loops occurs when the total loop phase shift is much less than 360°, because some of the integrators are damped, whereas the crossover frequency of reciprocator loops is typically at such a high frequency that the damping is negligible.

compensated op-amp is found to be given by

$$G(s)=\frac{V_0}{V_s}=\left(\frac{\omega_t}{s}\right)\frac{1+sr(C_a+C_b)}{(1+\omega_t C_a r)\left[1+s\dfrac{r(C_a+C_b)}{(1+\omega_t C_a r)}\right]}. \tag{12}$$

Thus the compensation network introduces a zero at

$$\omega_1=\frac{1}{r(C_a+C_b)} \tag{13}$$

and a pole at

$$\omega_2=\frac{1+rC_a\omega_t}{r(C_a+C_b)}. \tag{14}$$

At lower frequencies, the compensated op-amp still behaves as a positive integrator with a gain given by

$$\frac{V_0(s)}{V_s(s)}=\frac{\omega_t}{s(1+rC_a\omega_t)}. \tag{15}$$

Thus instead of ideally simulating a pair of wires, at lower frequencies the addition of the reciprocator is equivalent to having a parasitic series inductance between the two shunt capacitors (that are being separated) in the passive prototype.

At higher frequencies (between ω_1 and ω_2) the reciprocator is equivalent to a lossy parasitic inductance. It is this "damping" that is used to ensure stability.

To determine where ω_1 and ω_2 should be placed, it is necessary to evaluate the feedback around the reciprocator at high frequencies. Looking at the passive prototype it is seen that at high frequencies the inductors and resistors of the ladder filter can be ignored. Thus the total voltage across the parasitic inductance is equal to the voltage across the capacitors "seen" by the inductance.

The input to the reciprocator, in the unscaled SFG filter, is the total voltage appearing across the parasitic inductance which the reciprocator is simulating. Thus the feedback around the reciprocator $[H_1(s)]$, at high frequencies, is simply given by

$$H_1(s)=\frac{1}{sC_{\text{eq}}} \tag{16}$$

where C_{eq} is the equivalent capacitance "seen" by the parasitic series inductance. Taking the n input resistors of the reciprocator into account, the feedback around the compensated op-amp, used in the reciprocator, is given by

$$H(s)=\frac{1}{snC_{\text{eq}}} \tag{17}$$

at high frequencies.

For example, in the circuit of Fig. 2(a) the reciprocator is equivalent to a parasitic series inductance located between C_2' and C_3'. The equivalent capacitance seen by this parasitic is given by

$$C_{\text{eq}}=\frac{1}{\dfrac{1}{C_1'}+\dfrac{1}{C_2'}+\dfrac{1}{C_3'}}. \tag{18}$$

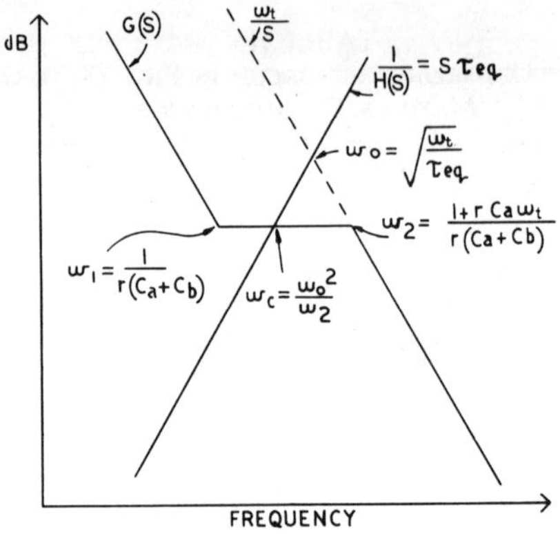

Fig. 4. Bode plots to illustrate the method for reciprocator stabilization.

Taking into account the 3 feed-in resistors, the feedback around the compensated reciprocator op-amp is given by

$$H(s)=\frac{1}{3s}\left(\frac{1}{C_1'}+\frac{1}{C_2'}+\frac{1}{C_3'}\right) \tag{19}$$

at high frequencies.

During the scaling process the n feed-in resistors to the reciprocator were changed so as to optimize the dynamic range. If we denote the scaling constant used for R_{ij} as M_i/M_j (i.e., the original feed-in resistor was multiplied by M_i/M_j to become R_{ij}), then the feedback around the compensated op-amp i in the scaled SFG filter is given by

$$H(s)=\frac{1}{s\tau_{\text{eq}}} \tag{20}$$

where

$$\tau_{\text{eq}}=C_{\text{eq}}\sum_{j=1}^{n}\left(\frac{M_j}{M_i}\right). \tag{21}$$

The Bode plots of $G(s)$ and $1/H(s)$ are shown in Fig. 4. The crossover frequency of the loop (denoted ω_c) occurs at the intersection of the Bode plots of $G(s)$ and $1/H(s)$. We would like to place the zero (ω_1) and the pole (ω_2) such that there is enough phase margin to ensure stability, while maintaining the low-frequency gain of the compensated op-amp as large as possible.

Experimentally it was found that the SFG filter is almost always stable if one chooses

$$\omega_1=0.8\omega_c \tag{22}$$

and

$$\omega_2=2\omega_1. \tag{23}$$

Now, from Fig. 4 it can be seen that the crossover frequency (ω_c) is given by

$$\omega_c=\frac{\omega_t}{\tau_{\text{eq}}\omega_2}. \tag{24}$$

Equations (13), (14), (22), (23), and (24) are easily solved for C_a and C_b to give

$$C_a=\frac{1}{r\omega_t} \tag{25}$$

TABLE IV
DESIGN OF THE COMPENSATION NETWORK FOR THE RECIPROCATOR IN THE FILTER OF FIG. 2

$C_{eq} = \dfrac{1}{\frac{1}{C_1'} + \frac{1}{C_2'} + \frac{1}{C_3'}} = 4.258\ \mu F$
$\tau_{eq} = \dfrac{C_{eq}}{M_5} (M_1 + M_4 + M_6) = 1.003 \times 10^{-5}$
Let $r = 1\ K\Omega$, $\omega_t = 2\pi \times 3 \times 10^6$
$C_a = \dfrac{1}{\omega_t r} = 53\ pF$ $C_b = \dfrac{1}{r} (\sqrt{\dfrac{\tau_{eq}}{0.4\ \omega_t}} - \dfrac{1}{\omega_t}) = 1.1\ nF$

and

$$C_b = \frac{1}{r}\left[\sqrt{\frac{\tau_{eq}}{0.4\omega_t}} - \frac{1}{\omega_t}\right]. \tag{26}$$

The value for r can be chosen to give reasonable component values. If, in the exceptional case, the SFG filter is still unstable after choosing C_a and C_b according to (25) and (26), then C_a should be experimentally increased until stability is achieved.

The design equations and element values for the reciprocator used in the circuit of Fig. 2(b) are given in Table IV. It has been assumed that the op-amps have $\omega_t = 2\pi \times 3 \times 10^6$ rad/s.

B. Minimizing the Attenuation Deviations Due to the L_0 Reciprocator

Using (15) and taking the attenuation due to the feed-in resistors into account, it is seen that the effects of the ith reciprocator can be found by analyzing the passive prototype filter after inserting the parasitic inductance (L_0) into the circuit, where L_0 is given by

$$L_0 = \left(\frac{1}{\omega_t} + rC_a\right)\frac{1}{M_i}\sum_{j=1}^{n} M_j \tag{27}$$

where M_i/M_j is the constant used to scale input resistor R_{ij}. Substituting (25) into (27) obtains

$$L_0 = \frac{2}{\omega_t M_i}\sum_{j=1}^{n} M_j. \tag{28}$$

Using (28), the effects of the reciprocator, on the active-filter transfer function, can be determined to a good approximation before constructing the active filter. This has been experimentally verified.

If the attenuation deviations caused by the reciprocators are found to be excessive, then they can be eliminated at the most critical frequency $\hat{\omega}$ (i.e., where the group delay is greatest, usually close to one of the passband edges). This is done by predistorting either the tank circuit or capacitor (in the passive filter) which the parasitic inductance is in series with, so that at the crucial frequency ($\hat{\omega}$) the total series impedance seen by the rest of the filter will be the ideal one.

TABLE V
PREDISTORTING THE *LC* PROTOTYPE TO COMPENSATE FOR THE PARASITIC INDUCTANCE INTRODUCED BY THE L_0 RECIPROCATOR IN THE CIRCUIT OF FIG. 2(b)

$L_o = \dfrac{2}{\omega_t M_5} (M_1 + M_4 + M_6) = 0.250\ \mu H$
Let $\hat{\omega} = 2\pi \times 10^4$ rad./sec.
$L_2'' = L_2' - L_o (1 - \hat{\omega}^2 C_2' L_2') = 12.91\ \mu H$
$C_2'' = \dfrac{C_2' L_2'}{L_2''} = 5.610\ \mu F$

If a series tank is being predistorted from the original value of C_i' and L_i' to C_i'' and L_i'', then the equations to be satisfied are

$$L_i'' = L_i' - L_0\left(1 - \hat{\omega}^2 C_i' L_i'\right) \tag{29}$$

and

$$C_i'' = \frac{L_i' C_i'}{L_i' - L_0\left(1 - \hat{\omega}^2 C_i' L_i'\right)} = \frac{L_i' C_i'}{L_i''} \tag{30}$$

where L_0 is the parasitic inductance and $\hat{\omega}$ is the critical frequency. This procedure leaves the resonance frequency of the series tank unchanged. Equations (29) and (30) are derived by equating the reactances of the antiresonant circuit formed by L_i' and C_i' to that of an antiresonant circuit, formed by L_i'' and C_i'', in series with the inductance L_0, at the frequency $\hat{\omega}$.

If a series capacitor is being predistorted from its original value of C_i' to C_i'' one should take

$$C_i'' = \frac{C_i'}{1 + \hat{\omega}^2 C_i' L_0}. \tag{31}$$

This equation is obtained by equating the reactance of capacitor C_i' to the reactance of a capacitor C_i'' in series with the inductance L_0, at the frequency $\hat{\omega}$.

For the example of Fig. 2, the design equations for predistorting the series tank are listed in Table V. The critical frequency was the upper passband edge ($\hat{\omega} = 2\pi \times 10^4$). For this example, it was found (using the computer) that the nonideal reciprocator caused the passband ripple to increase from its ideal value of 1.25 dB to 1.42 dB. When the passive filter was predistorted, the passband ripple was restored to 1.26 dB. In the active filter of Fig. 2(b) the predistortion has the effect of changing $R_{4,5}$ to 2.622 kΩ, $R_{4,3}$ to 1.903 kΩ, and $R_{3,4}$ to 9.472 kΩ.

VI. DC OFFSET PROBLEMS

In a passive filter there are often nodes that are capacitively decoupled from ground. If all the capacitors were ideal, the dc voltage at such a node is undefined. (i.e., in

ciprocator). Thus, one need adjust only one of the integrators to obtain the proper total loop gain.

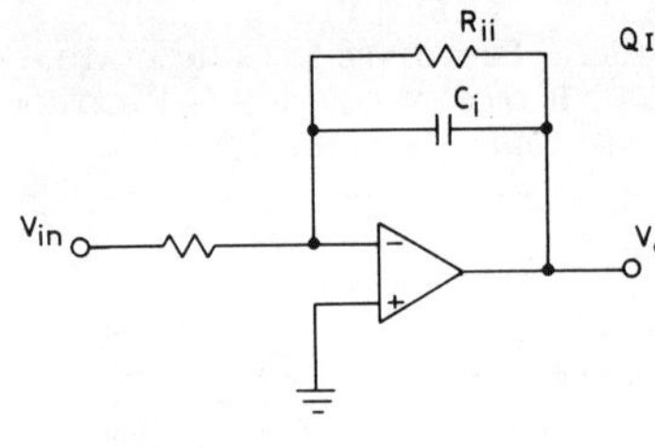

Fig. 5. Damping a capacitive simulation in order to solve the offset problem.

control-system terminology the dc voltage is both uncontrollable and unobservable). Since the capacitors are never perfectly ideal this presents no difficulty (in a passive circuit).

In the SFG realization the integrators simulating the decoupling capacitors also have their output voltages theoretically undefined at dc. Practically, this means that there is very little feedback around these capacitive simulations at dc. Thus the input offset voltages and input bias currents of the op-amps will cause large dc offsets in the capacitive simulations.

An obvious solution is to damp one of the capacitive simulations (see Fig. 5). This gives the integrator a finite quality factor of $\omega C_i R_{ii}$, where R_{ii} is the damping resistor and C_i is the capacitor used in the simulation. To evaluate the effect of the damping resistor, one need only perform an analysis on the passive circuit with the corresponding capacitor (C_j') having a resistor of size $C_i R_{ii}/C_j'$ placed in parallel with it.

The integrator chosen to be damped should never be one simulating a capacitor that produces a finite loss pole. This is because the transfer function of the filter is most sensitive to the quality factor of these capacitors. This constraint is easily met, since series capacitors, without inductors in parallel with them, are only used to produce loss poles at zero (if there are parallel inductors then the node is dc coupled).

In critical situations (filters with high-Q natural modes) one can also connect dc-offset compensating circuits to the damped integrator. This will allow larger damping resistors to be used. Experimentally it was found that the dc offsets could always be eliminated with negligible effect on the transfer function.

It is interesting to note that the number of nodes with dc offsets, in a passive filter, added to the number of reciprocators, needed in an SFG simulation of that filter, is equal to the number of the finite loss poles in the ideal transfer function.

The fourth-order filter used in the example does not have any inherent dc offsets, but the twelfth-order filter described later will be used to illustrate the compensation scheme.

VII. Tuning Procedure

In an SFG filter the normalized transfer function is determined by the loop gains only. Each loop typically consists of two integrators (or an integrator and a reciprocator). Thus, one need adjust only one of the integrators to obtain the proper total loop gain.

The tuning procedure is quite simple. First, one isolates a signal loop and grounds all unused input resistors. For example, to tune the loop consisting of op-amps 2, 3, and 4 in Fig. 2(b) one grounds $R_{4,5}$. Next, the second-order section is adjusted for infinite Q. In the example this is done by adjusting the grounded resistor (r_c), connected to the negative input of 0A2, until the second-order circuit is just marginally stable. The loop should then be damped by placing a large resistor (2.2 MΩ-10 MΩ) in parallel with the capacitor in the integrator to which the input signal is going to be fed ($R_{4,5}$ in our example). The loop can now be tuned to its proper resonant frequency $[\omega_0 = 1/\sqrt{C_2' L_2'}\,]$ by adjusting either $R_{4,3}$ or $R_{3,4}$.

This procedure can be repeated for all of the other loops. When a positive integrator is in more than one loop, then its Q enhancing resistor (r_c) should be adjusted to give the loop with the highest resonant frequency an infinite Q-factor. Loops with finite Q's should be tuned with their damping resistors in the circuit. Their Q-factor can be also adjusted.

To tune loops consisting of a reciprocator and negative integrators, the most accurate method is to replace the reciprocator by a positive integrator, which has the reciprocator input resistors connected to it. The different loops can be tuned so that their resonant frequencies are in the proper ratio.

In practice, tuning every loop is both time-consuming and expensive, if a large quantity of the active filters are to be manufactured. Since the filter-transfer function is far more sensitive to loops simulating finite-loss poles than to any other loops, these are the only ones that should be tuned. However, the input resistors of integrators in other loops should then be chosen slightly smaller than their ideal values to compensate for the finite gain–bandwidth product (ω_t) of the op-amps. The procedure to do this is quite simple. Assume integrator i has nominal input resistor R_{ij} and capacitor C_i. Let the parallel equivalent of all the input resistors be R_i. (This should include any damping resistors.) The new values for the input resistors ($\overline{R_{ij}}$) can be calculated according to the formula

$$\overline{R_{ij}} = R_{ij} \frac{1}{1 + \dfrac{1}{\omega_t R_i C_i}}. \tag{32}$$

Any damping resistor should also be decreased according to (32).

VIII. Design Example—A Twelfth-Order Channel Bank Filter

A twelfth-order bandpass filter was used as a design example to illustrate the excellent performance available from the SFG method. The passband extends from 17 kHz to 19.7 kHz, with a 0.2-dB ripple. It should be mentioned that the highest natural-mode Q-factor is 150, which makes a cascade realization very difficult. In order

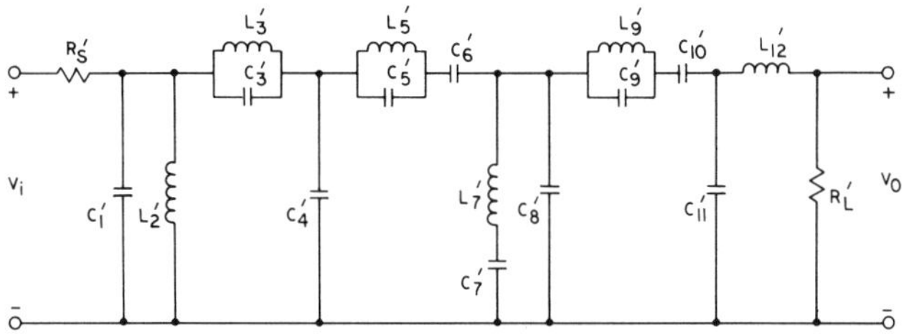

Fig. 6. A passive realization of the twelfth-order bandpass filter. Element values are given in Table VI.

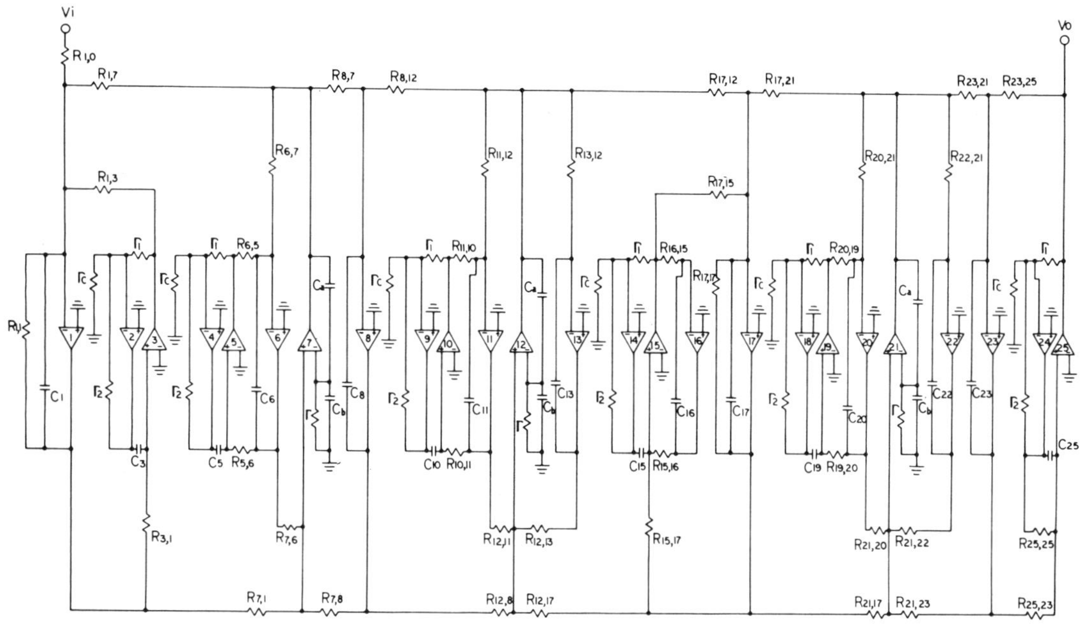

Fig. 7. Active realization of the twelfth-order bandpass filter.

TABLE VI
ELEMENT VALUES OF THE PASSIVE REALIZATION OF THE TWELFTH-ORDER BANDPASS FILTER OF FIG. 6

R'_S = 1.000 Ω	C'_1 = 75.508 μF	L'_2 = 0.94014 μH
L'_3 = 1.4208 μH	C'_3 = 43.423 μF	C'_4 = 4.7578 μF
L'_5 = 9.5764 μH	C'_5 = 6.2375 μF	C'_6 = 1.4520 μF
L'_7 = 37.364 μH	C'_7 = 2.6958 μF	C'_8 = 8.4137 μF
L'_9 = 21.959 μH	C'_9 = 2.3933 μF	C_{10}= 1.0399 μF
C'_{11}= 23.987 μF	L'_{12}= 3.0384 μH	R'_L = 0.037685 Ω

to provide a basis for comparing the SFG method with the component-simulation method, we have chosen the same filter as that discussed in [5].

In order to gain an appreciation of the difficulty involved in realizing this filter one may use the results of [14] and [17] to evaluate the sensitivities of its *LC*-ladder realization. It follows that a 0.1-percent uniform element tolerance results in 0.64-dB attenuation deviation (upper bound) at 19.7 kHz. Also, a uniform element *Q*-factor of 4000 results in 0.73-dB attenuation deviation at 19.7 kHz.

A passive-*LC* realization, obtained using computer-synthesis programs [18], is shown in Fig. 6. The element values are listed in Table VI. The active realization, obtained by following the procedure outlined above, is shown in Fig. 7. The maximum of the op-amp outputs of the unscaled filter are listed in Table VII. The element values for the optimized active realization are listed in Table VIII.

When the passive prototype was analyzed, to determine the effects of the reciprocators, it was found that the passband ripple was 0.52 dB. It was also found, by computer analysis, that a predistorted filter had 0.22-dB ripple. The bandwidth of the Raytheon 4135 op-amps used was assumed to be 3 MHz.

From Fig. 6, it can be seen that the active filter will have a dc offset at the op-amp output which corresponds to the voltage across C'_8. This was eliminated by connecting a 2.2-MΩ resistor ($R_{17,17}$) in parallel with C_{17}, as shown in Fig. (7). This eliminated the dc offset at the output of 0A17 without having any noticeable effect on the passband response.

Each loop of the SFG filter was individually tuned. This ensured that the best possible performance was obtained. The measured stopband response is shown in Fig. 8 superimposed on the ideal one. The measured passband

TABLE VII

THE MAXIMUM MAGNITUDES OF VOLTAGES AND CURRENTS IN THE CIRCUIT OF FIG. 6 TOGETHER WITH THE CORRESPONDING SIMULATING OP-AMP IN THE CIRCUIT OF FIG. 7

Output of Op. Amp.	Simulated Voltage or Current	Maximum of Voltage or Current
OA1	Voltage Across C_1'	M_1 = 0.946V
OA3	Current Through L_2'	M_3 = 8.56A
OA5	Current Through L_3'	M_5 = 15.70A
OA6	Voltage Across C_3'	M_6 = 2.767V
OA7	Current Through $(L_3' \| C_3')$	M_7 = 1.488A
OA8	Voltage Across C_4'	M_8 = 2.037V
OA10	Current Through L_5'	M_{10}= 5.07A
OA11	Voltage Across C_5'	M_{11} = 6.01V
OA12	Current Through $(L_5' \| C_5')$	M_{12}= 0.715A
OA13	Voltage Across C_6'	M_{13}= 4.634V
OA15	Current Through L_7'	M_{15}= 1.968A
OA16	Voltage Across C_7'	M_{16}= 6.85V
OA17	Voltage Across C_8'	M_{17}= 1.860V
OA19	Current Through L_9'	M_{19}= 2.096A
OA20	Voltage Across C_9'	M_{20}= 5.69V
OA21	Current Through $(L_9' \| C_9')$	M_{21}= 0.4919A
OA22	Voltage Across C_{10}'	M_{22}= 4.420V
OA23	Voltage Across C_{11}'	M_{23}= 0.972V
OA25	Current Through L_{12}'	M_{25}= 2.576A

TABLE VIII

COMPONENT VALUES OF THE TWELFTH-ORDER BANDPASS FILTER OF FIG. 7

r = 1.0 KΩ	C_a = 47 pf	C_b = 1.2 nF	
All capacitors used in integrators are 2.2 nF			
r_1 = 3.57 KΩ, r_2 = 2.49 KΩ, r_c = 3.38 KΩ			
$R_{1,3}$ = 3.793 KΩ	$R_{3,1}$ = 5.544 KΩ	$R_{1,7}$ = 21.82 KΩ	$R_{7,1}$ = 15.73 KΩ
$R_{5,6}$ = 5.544 KΩ	$R_{6,5}$ = 3.534 KΩ	$R_{6,7}$ = 37.29 KΩ	$R_{7,6}$ = 5.378 KΩ
$R_{7,8}$ = 7.305 KΩ	$R_{8,7}$ = 2.961 KΩ	$R_{8,12}$ = 6.161 KΩ	$R_{12,8}$ = 8.775 KΩ
$R_{10,11}$ = 5.265 KΩ	$R_{11,10}$ = 3.361 KΩ	$R_{11,12}$ = 23.83 KΩ	$R_{12,11}$ = 2.974 KΩ
$R_{12,13}$ = 3.857 KΩ	$R_{13,12}$ = 4.092 KΩ	$R_{12,17}$ = 9.610 KΩ	$R_{17,12}$ = 9.949 KΩ
$R_{15,16}$ = 6.995 KΩ	$R_{16,15}$ = 4.265 KΩ	$R_{15,17}$ = 25.67 KΩ	$R_{17,15}$ = 3.615 KΩ
$R_{17,21}$ = 14.46 KΩ	$R_{21,17}$ = 10.58 KΩ	$R_{19,20}$ = 5.271 KΩ	$R_{20,19}$ = 2.953 KΩ
$R_{20,21}$ = 12.58 KΩ	$R_{21,20}$ = 3.458 KΩ	$R_{21,22}$ = 4.452 KΩ	$R_{22,21}$ = 4.077 KΩ
$R_{21,23}$ = 20.24 KΩ	$R_{23,21}$ = 21.54 KΩ	$R_{23,25}$ = 4.114 KΩ	$R_{25,23}$ = 5.247 KΩ
$R_{1,0}$ = 5.230 KΩ	$R_{1,1}$ = 34.32 KΩ	$R_{17,17}$ = 2.2 MΩ	$R_{25,25}$ = 36.65 KΩ

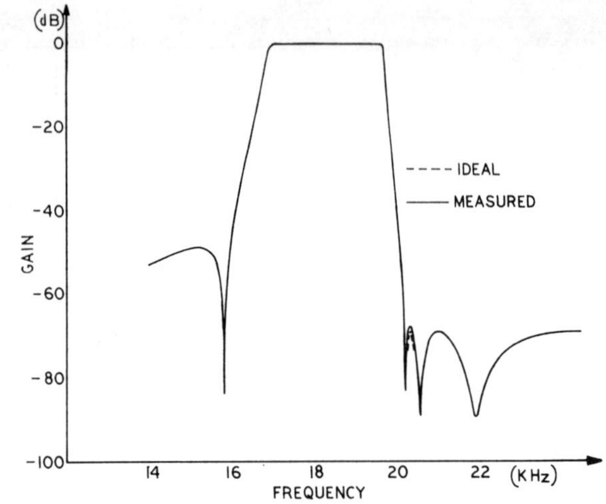

Fig. 8. Measured response of the filter in Fig. 7, superimposed on the ideal response.

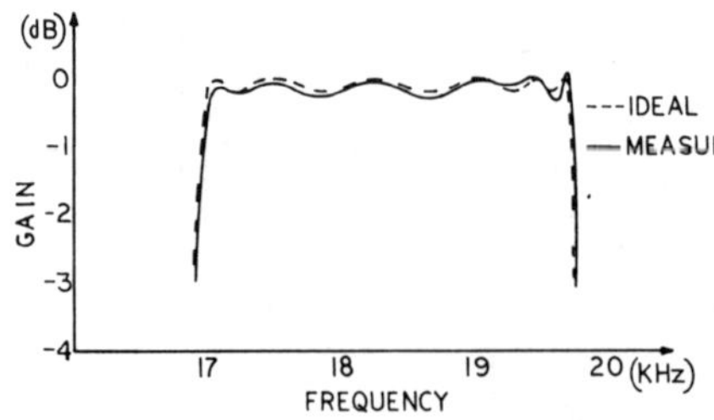

Fig. 9. Measured passband response of the filter of Fig. 7, superimposed on the ideal response.

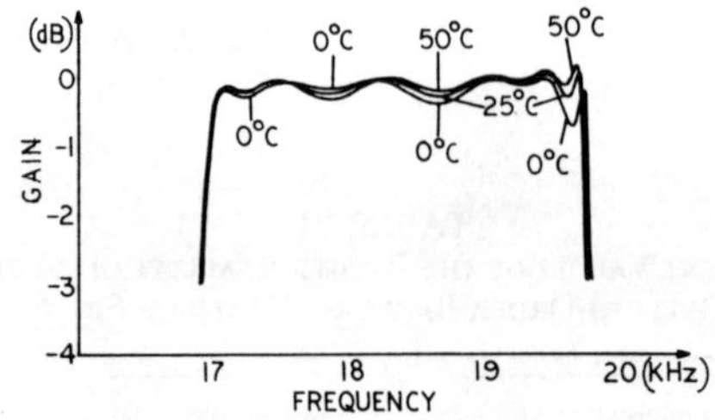

Fig. 10. Passband response measured at different temperatures.

response is shown in Fig. 9 superimposed on the ideal one. The passband response at different ambient temperatures is shown in Fig. 10. Here we should mention that the capacitors used were polystyrene, and the resistors were metal film. The *RC* time constant typically has a temperature coefficient of -100 ppm/°C. The output noise (measured with a 100-Hz bandwidth setting on an HP3580A Spectrum Analyzer) was found to be -81 dB (0 dB corresponds to 1V rms). This gave a dynamic range of 90 dB.

We now offer a brief comparison of the SFG-design method and the method employing GIC's [5].

(1) The SFG filter typically requires a much larger number of op-amps (for our example: 25 versus 12).

(2) Both methods lead to designs having very similar active and passive sensitivities.

(3) The GIC filter is much more sensitive to stray capacitances than the SFG filter. This sensitivity is inhe-

rent because neither of the inputs of any of the op-amps in the GIC filter are connected to ground. From this point of view, one may conclude that the GIC filter would not greatly benefit from better op-amp designs (e.g., wider bandwidth).

(4) The choice of *LC*-ladder prototype is far more critical if the filter is to be designed using the GIC method. In fact, one normally has to apply various network transformations on the *LC* ladder before carrying out a GIC design. Among other things, the *LC* prototype considerably affects the dynamic range of the resulting GIC filter. On the other hand, one can obtain a good SFG design starting from almost any *LC* prototype.

(5) No systematic procedure is available for scaling the GIC filter for maximum dynamic range. Nevertheless, a "well-designed" GIC filter may have a dynamic range as large as, or sometimes even greater than, the SFG filter. This may be due to the smaller number of amplifiers used in the GIC filter.

(6) The SFG design does not require extremely accurate capacitors as the GIC method does. This is due to the fact that one can easily tune every loop in the SFG filter using resistors only.

IX. Conclusion

A systematic design procedure for SFG filters has been presented. It is completely general in that once a passive ladder filter has been designed, then the active simulation can be arrived at by following a simple set of rules. This active filter will have the optimum dynamic range, for the given configuration. Every practical design consideration has been explained in detail and illustrated by the use of a simple example. Finally a twelfth-order channel bank filter has been used to demonstrate the ability of SFG filters to meet stringent design specifications. A brief comparison between the SFG method and that using GIC's has been presented. However, further work is needed in comparing SFG filters with those designed using wave variables.

References

[1] H. J. Orchard, "Inductorless filters," *Electron. Lett.*, vol. 2, pp. 224–225, Sept. 1966.

[2] A. Antontiou, "Realization of gyrators using operational amplifiers and their use in RC-active network synthesis," *Proc. Inst. Elect. Eng.*, vol. 116, pp. 1838–1850, Nov. 1969.

[3] L. T. Bruton and D. Treleavan, "Active filter design using generalized impedance converters," *EDN*, pp. 68–75, Feb. 1975.

[4] J. Valihora, J. T. Lim, and L. T. Bruton, "The feasibility of active filtering in FDM systems," in *Proc. 1974 IEEE Int. Symp. on Circuits Syst.*, pp. 121–125, San Francisco, CA.

[5] K. Martin and A. Sedra, "Optimum design of active filters using Generalized Immittance Converters," in *Proc. 1977 IEEE Int'l Symp. on Circuits Syst.*, pp. 335–338, Phoenix, AZ, Apr. 1977.

[6] F. E. J. Girling and E. F. Good, "Active filters 12: the leap-frog or active ladder synthesis," *Wireless World*, vol. 76, pp. 341–345, July 1970.

[7] P. O. Brackett and A. Sedra, "Direct SFG simulation of LC ladder networks with applications to active filter design," *IEEE Trans. Circuits Syst.*, vol. CAS-23, pp. 61–67, Feb. 1976.

[8] A. G. Constantinides and H. G. Dimopoulos, "Active-RC filters derivable from LC-ladder filters via linear transformations," *Electron. Circuits and Syst.*, vol. 1, no. 1, pp. 17–21, Sept. 1976.

[9] H. Wupper and K. MeerKotter, "New active filter synthesis based on scattering parameters," *IEEE Trans. Circuits Syst.*, vol. CAS-22, pp. 594–602, July 1975.

[10] P. O. Brackett, "Circuits and limitations of wave filters," in *Proc. 1976 IEEE Int'l Symp. Circuits Syst.*, pp. 69–72, Munich, Apr. 1976.

[11] A. I. Zverev, *Handbook of Filter Synthesis*. New York: Wiley, 1967.

[12] A. S. Sedra and P. O. Brackett, *Filter Theory and Design: Active and Passive*, Champaign, Il: Matrix Publishers (in press).

[13] P. O. Brackett and A. Sedra, "Active compensation for high-frequency effects in op-amp circuits with applications to active-RC filters," *IEEE Trans. Circuits Syst.*, vol. CAS-23, pp. 68–73, Feb. 1976.

[14] M. L. Blostein, "Sensitivity analysis of parasitic effects in resistance terminated *LC* filters," *IEEE Trans. Circuit Theory*, vol. CT-14, pp. 21–25, Mar. 1967.

[15] K. Martin and A. Sedra, "On the stability of the phase-lead integrator," *IEEE Trans. Circuits Syst.*, vol. CAS 24, pp. 321–324, June 1977.

[16] K. Martin and A. Sedra, "Designing Leap-Frog and SFG filters with optimum dynamic range," *Proc. IEEE* (Lett.), vol. 65, pp. 1210–1211, Aug. 1977.

[17] G. C. Temes and H. J. Orchard, "First order sensitivity and worst-case analysis of doubly terminated reactance two-ports," *IEEE Trans. Circuit Theory*, vol. CT-20, pp. 567–571, Sept. 1973.

[18] W. M. Snelgrove and A. Sedra, FILTOR 2 Manual. Champaign, Il, Matrix Publishers (in preparation).

Linear Transformation Active Filters

HERCULES G. DIMOPOULOS, MEMBER, IEEE, AND A. G. CONSTANTINIDES, MEMBER, IEEE

***Abstract*—An approach is presented in this paper for the design of active filters from doubly terminated lossless ladder filters. The method relies on the linear transformation of port variables from the V-I domain to a new domain in which the filter is realized actively. The resulting active filters are referred to as linear transformation active (LTA) filters. The transformations are applied on an element-by-element basis thus producing a systematic and effective way of achieving new active structures while others like the leapfrog and the wave active filter structures appear as special cases. Examples are presented in the paper to show the efficiency of the method and the importance of the results.**

I. Introduction

A METHOD is presented in this paper for the design of active inductorless filters from doubly terminated lossless ladder filters [1], [2]. The fundamental principles are based on the linear transformation of port variables of a network from the V-I domain to a new domain in which active realizations are effected. The original ladder lossless elements are transformed one by one to the new domain and are there individually realized actively. Subsequently, their overall connection is achieved by means of a "connecting network" purposely chosen in this paper to contain no extra passive or active components. The resulting filter structures are referred to as linear transformation active (LTA) filters.

Manuscript received March 1, 1977; revised November 17, 1977.
The authors are with the Department of Electrical Engineering, Imperial College of Science and Technology, London SW7 2BT, England.

Due to the general approach of the method, first outlined in a recent publication [3], it is possible to choose the transformations at will. This makes the method very flexible indeed to the extent that the active realizations can be tailored to the needs of the designer. In addition other well known methods such as the leapfrog design [8], the wave active filters [4], [5], [7], and the SFG simulation [9] can be considered as special cases of our approach. Indeed, when these methods are viewed from the LTA point of view they are then amenable to alternative and perhaps preferable interpretations and realizations.

A further concept is introduced in the paper with respect to the frequency dependence of the linear transformation (LT) matrices. This concept appear to be very powerful indeed in its application to active filter realizations as it is capable of producing canonic structures. This is shown through one example.

II. Theoretical Considerations

A. The Analytical Approach

The basic ideas and principles of the port-transformation approach have been outlined in a recent publication [3]. An alternative presentation incorporating many desirable features suitable for the purposes of this paper is adopted as shown in summary form below.

The port variable vectors of a two-port network N_i, $[\,V_{1i} \quad I_{1i}\,]^T$ and $[\,V_{2i} \quad I_{2i}\,]^T$ are linearly transformed by

Reprinted from *IEEE Trans. Circuits Syst.*, vol. CAS-25, pp. 845-852, Oct. 1978.

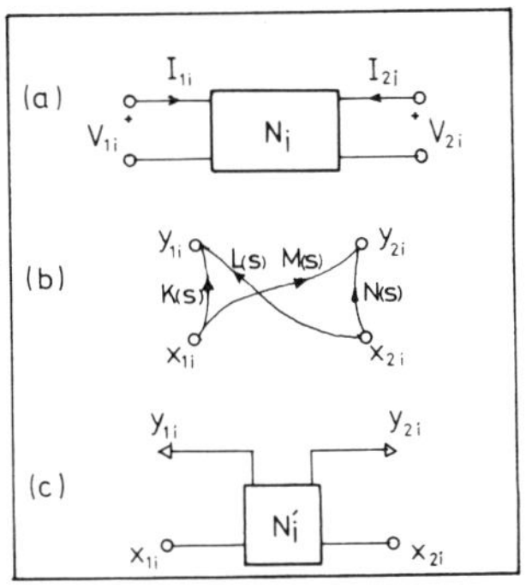

Fig. 1. (a) The lossless passive two-port. (b) The SFG model of the corresponding LT structure. (c) A more suitable model for the corresponding LT structure.

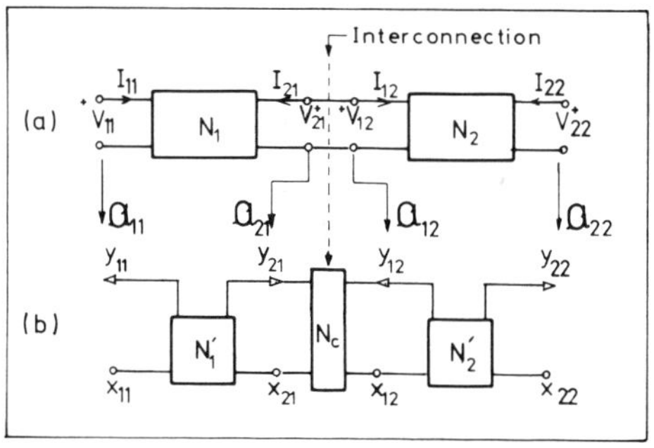

Fig. 2. (a) The cascade interconnection in the V-I domain. (b) The interconnection of the corresponding LT structures.

the two nonsingular transformation matrices $\mathbb{S}_{1i}$ and $\mathbb{S}_{2i}$ as follows:

$$\begin{bmatrix} x_{1i} \\ y_{1i} \end{bmatrix} = \begin{bmatrix} \alpha_{1i} & \beta_{1i} \\ \gamma_{1i} & \delta_{1i} \end{bmatrix} \begin{bmatrix} V_{1i} \\ I_{1i} \end{bmatrix} = \mathbb{S}_{1i} \begin{bmatrix} V_{1i} \\ I_{1i} \end{bmatrix}$$

$$\begin{bmatrix} x_{2i} \\ y_{2i} \end{bmatrix} = \begin{bmatrix} \alpha_{2i} & \beta_{2i} \\ \gamma_{2i} & \delta_{2i} \end{bmatrix} \begin{bmatrix} V_{2i} \\ I_{2i} \end{bmatrix} = \mathbb{S}_{2i} \begin{bmatrix} V_{2i} \\ I_{2i} \end{bmatrix}. \tag{1}$$

In general the modified chain matrix[1] can be used to describe the relationship between the two-port variable vectors in accordance with Fig. 1(a)

$$\begin{bmatrix} V_{1i} \\ I_{1i} \end{bmatrix} = \begin{bmatrix} A_i & B_i \\ C_i & D_i \end{bmatrix} \begin{bmatrix} V_{2i} \\ I_{2i} \end{bmatrix} = \mathcal{T}_i \begin{bmatrix} V_{2i} \\ I_{2i} \end{bmatrix}. \tag{2}$$

From (1) and (2) we have

$$\begin{bmatrix} x_{1i} \\ y_{1i} \end{bmatrix} = \mathbb{S}_{1i} \mathcal{T}_i \mathbb{S}_{2i}^{-1} \begin{bmatrix} x_{2i} \\ y_{2i} \end{bmatrix} = \begin{bmatrix} a_i & b_i \\ c_i & d_i \end{bmatrix} \begin{bmatrix} x_{2i} \\ y_{2i} \end{bmatrix} \tag{3}$$

where

$$a_i = \frac{1}{\Delta_i} [\alpha_{1i}(\delta_{2i} A_i - \gamma_{2i} B_i) + \beta_{1i}(\delta_{2i} C_i - \gamma_{2i} D_i)]$$

$$b_i = \frac{1}{\Delta_i} [\alpha_{1i}(\alpha_{2i} B_i - \beta_{2i} A_i) + \beta_{1i}(\alpha_{2i} D_i - \beta_{2i} C_i)]$$

$$c_i = \frac{1}{\Delta_i} [\gamma_{1i}(\delta_{2i} A_i - \gamma_{2i} B_i) + \delta_{1i}(\delta_{2i} C_i - \gamma_{2i} D_i)]$$

$$d_i = \frac{1}{\Delta_i} [\gamma_{1i}(\alpha_{2i} B_i - \beta_{2i} A_i) + \delta_{1i}(\alpha_{2i} D_i - \beta_{2i} C_i)]$$

$$\Delta_i = \alpha_{2i}\delta_{2i} - \beta_{2i}\gamma_{2i}.$$

In addition, $[y_{1i} \quad y_{2i}]^T$ can be found from the above equation in terms of $[x_{1i} \quad x_{2i}]^T$ as

$$\begin{bmatrix} y_{1i} \\ y_{2i} \end{bmatrix} = \begin{bmatrix} d_i/b_i & -(a_i d_i - b_i c_i)/b_i \\ 1/b_i & -a_i/b_i \end{bmatrix} \begin{bmatrix} x_{1i} \\ x_{2i} \end{bmatrix} = \begin{bmatrix} K(s) & L(s) \\ M(s) & N(s) \end{bmatrix} \begin{bmatrix} x_{1i} \\ x_{2i} \end{bmatrix}. \tag{4}$$

[1]The form of the chain matrix is modified in the sense that B and D appear with sign opposite to the normal. This is a consequence of using $[V_2 \quad I_2]^T$ instead of $[V_2 \quad - \quad {}_2]^T$. This is done to keep the definitions of (1) the same.

Equation (4) can be considered as describing a two-input-two-output system N_i' (referred to as the corresponding LT network or structure) where y_{1i} and y_{2i} are the responses due to the excitations x_{1i} and x_{2i} [11]. The dimensions of x and y are determined by the dimensions of the elements of the transformation matrices, as it can be observed from (1). For the purposes of this work α and γ are dimensionless and β, δ have the dimensions of impedances making x and y have the dimensions of voltages and consequently a, b, c, and d are dimensionless voltage transfer ratios.

The model of the transformed network implied by (4) can be represented using SFG representation as in Fig. 1(b). However, for our purposes the model of Fig. 1(c) is more suitable where terminals on which the stimuli x are impressed (input terminals) are denoted by a small circle while terminals on which the responses y are observed (output terminals) are denoted by a triangle appropriately oriented.

B. Interconnection Constraints

The cascade connection of the two-ports N_1 and N_2 is described in the V-I domain by

$$\begin{bmatrix} V_{12} \\ I_{12} \end{bmatrix} = \begin{bmatrix} 1 & 0 \\ 0 & -1 \end{bmatrix} \begin{bmatrix} V_{21} \\ I_{21} \end{bmatrix} \tag{5}$$

as shown in Fig. 2(a). The corresponding LT structures N_1' and N_2' are shown in Fig. 2(b) where their interconnection to be determined is represented as shown.

In order that the original parameters of N_1' and N_2' remain unaltered as a result of the interconnection, due to loading, we assume that y_{12} and y_{21} are observed across norators [3], [4]. On combining (1) applied to the two adjacent ports of the networks in Fig. 2(a) with (5) we obtain the interconnection constraints for the corresponding LT networks of Fig. 2(b), i.e., we define the "interconnecting network" N_c as follows:

$$\begin{bmatrix} x_{12} \\ y_{12} \end{bmatrix} = \mathbb{S}_{12} \begin{bmatrix} 1 & 0 \\ 0 & -1 \end{bmatrix} \mathbb{S}_{21}^{-1} \begin{bmatrix} x_{21} \\ y_{21} \end{bmatrix}. \tag{6}$$

N_c in fact is seen to produce linear combinations of y_{12} and y_{21} to generate x_{12} and x_{21}.

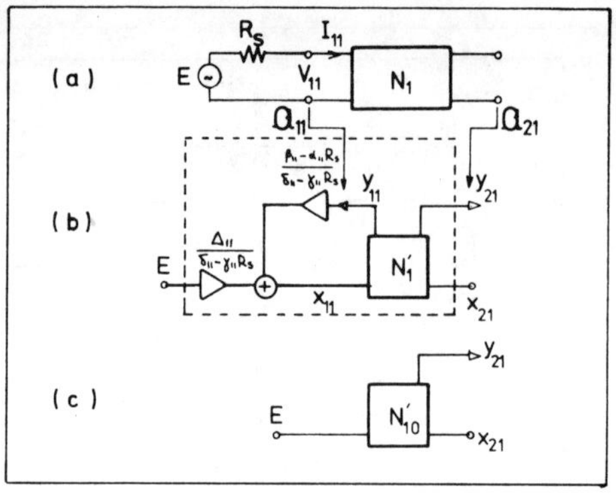

Fig. 3. (a) The resistive input termination of the ladder filter. (b) The corresponding LT structure. (c) The two-input-one-output model.

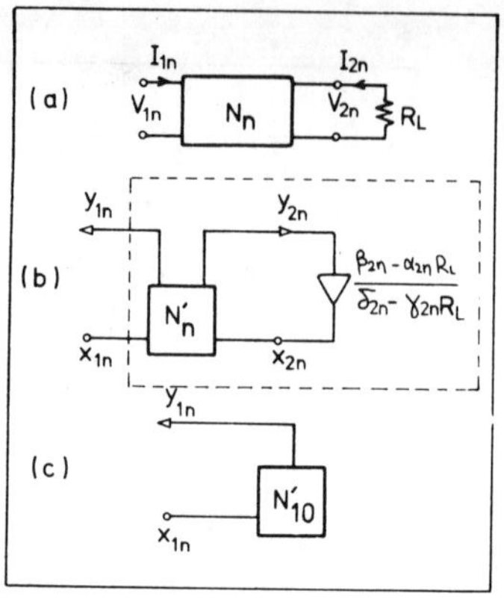

Fig. 4. (a) The resistive output termination of the ladder filter. (b) The corresponding LT structure. (c) The single-input-single-output model.

C. The Termination Requirements

The input termination for a lossless doubly terminated ladder filter, as shown in Fig. 3(a) is described by

$$\begin{bmatrix} 1 & R_s \end{bmatrix} \begin{bmatrix} V_{11} \\ I_{11} \end{bmatrix} = E. \tag{7}$$

N_1 represents the first reactive arm of the lossless ladder. Using $\mathbb{S}_{11}$ and $\mathbb{S}_{21}$ to transform N_1, N_1' can be obtained and (7) yields

$$\begin{bmatrix} 1 & R_s \end{bmatrix} \mathbb{S}_{11}^{-1} \begin{bmatrix} x_{11} \\ y_{11} \end{bmatrix} = E. \tag{8}$$

Equation (8) shows the way in which x_{11} must be generated from E and y_{11} in order to include the resistive input termination (Fig. 3(b)). However, the two-input-one-output network so obtained (Fig. 3(c)) is easily found to be described by

$$\begin{bmatrix} 1 & R_s \end{bmatrix} T_1 \mathbb{S}_{21}^{-1} \begin{bmatrix} x_{21} \\ y_{21} \end{bmatrix} = E. \tag{9}$$

This equation describes the LT structure which corresponds to the first reactive arm N_1 of the ladder terminated by the source resistance R_s. It is worth noticing that only the $\mathbb{S}_{21}$ matrix is involved in (9).

In a similar manner it can be found that if the last reactive arm N_n of the ladder is linearly transformed to the corresponding LT structure N_n' (Fig. 4(a)) using $\mathbb{S}_{1n}$ and $\mathbb{S}_{2n}$, the resistive termination will be described by

$$\begin{bmatrix} 1 & R_L \end{bmatrix} \mathbb{S}_{2n}^{-1} \begin{bmatrix} x_{2n} \\ y_{2n} \end{bmatrix} = 0. \tag{10}$$

This is shown in Fig. 4(b). Again the single-input-single-output corresponding LT structure shown in Fig. 4(b) can be found to be described by

$$\begin{bmatrix} 1 & R_L \end{bmatrix} T_n^{-1} \mathbb{S}_{1n}^{-1} \begin{bmatrix} x_{1n} \\ y_{1n} \end{bmatrix} = 0 \tag{11}$$

where T_n is the modified chain matrix of N_n (Fig. 4(c)). Again only one transformation matrix is involved in the description of the LT structure corresponding to the last reactive arm in conjunction with the terminating resistance R_L. Table I gives (9) and (11) in their explicit form.

TABLE I
THE GENERAL TERMINATION REQUIREMENTS

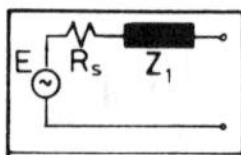	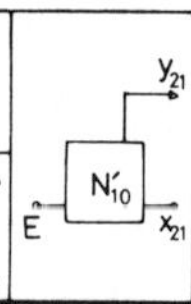	$y_{21}=\frac{\delta_{21}+\gamma_{21}(Z_1+R_S)}{\beta_{21}+\alpha_{21}(Z_1+R_S)}x_{21}-\frac{\alpha_{21}\delta_{21}-\beta_{21}\gamma_{21}}{\beta_{21}+\alpha_{21}(Z_1+R_S)}E$
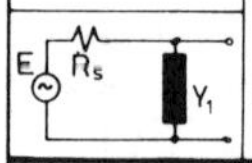		$y_{21}=\frac{\gamma_{21}R_S+\delta_{21}(Y_1R_S+1)}{\alpha_{21}R_S+\beta_{21}(Y_1R_S+1)}x_{21}-\frac{\alpha_{21}\delta_{21}-\beta_{21}\gamma_{21}}{\alpha_{21}R_S+\beta_{21}(Y_1R_S+1)}E$
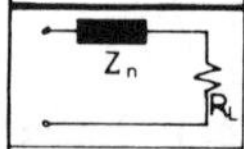	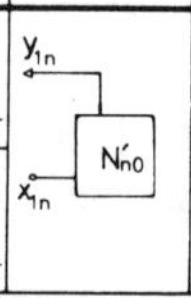	$\alpha_{1n}=0$ or $\gamma_{1n}=0$ for direct output $\quad y_{1n}=\frac{\delta_{1n}+\gamma_{1n}(Z_n+R_L)}{\beta_{1n}+\alpha_{1n}(Z_n+R_L)}x_{1n}$
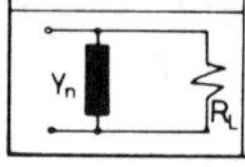		$\beta_{1n}=0$ or $\delta_{1n}=0$ for direct output $\quad y_{1n}=\frac{\gamma_{1n}R_L+\delta_{1n}(Y_nR_L+1)}{\alpha_{1n}R_L+\beta_{1n}(Y_nR_L+1)}x_{1n}$

D. Factors Affecting the Choice of the Transformation Matrices

The procedure to be followed for the purpose of simulating a doubly terminated LC ladder filter using LT can be described as follows:

i) each reactive arm is transformed using two nonsingular transformation matrices as described in Section II-A;
ii) the resistively terminated ladder elements N_1 and N_n are transformed according to Section II-C;
iii) finally the interconnections of the LT structures are made according to (6).

However, the application of the above procedure in its general form is not easy due to the large degree of freedom associated with the choice of the transformation matrices. Therefore, some rules on the choice of the transformation matrices are necessary for the efficient application of the proposed general method.

Starting from the fact that the general interconnection described by (6) requires an active network N_c the complexity of which is evidently dependent on the transformation matrices of the adjacent ports, we impose an external relationship between these transformation matrices so that N_c will be as simple as possible. The relationship

$$\mathbb{S}_{12}\begin{bmatrix} 1 & 0 \\ 0 & -1 \end{bmatrix} = \begin{bmatrix} 0 & 1 \\ 1 & 0 \end{bmatrix} \mathbb{S}_{21} \tag{12}$$

which in practice implies that each of the adjacent transformation matrices is derivable from the other by inter-

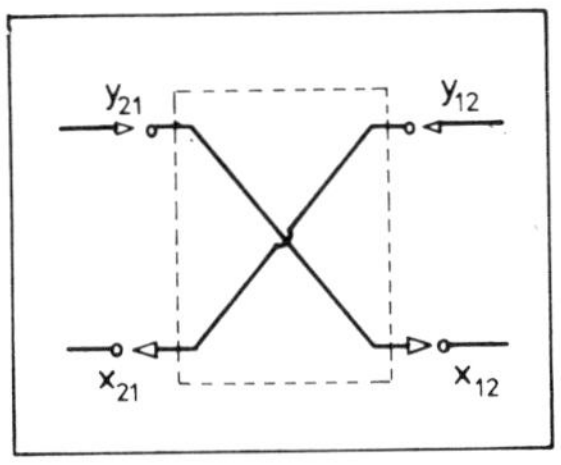

Fig. 5. The cross-cascade interconnection.

changing its rows and changing the sign of the second column, is referred to as the compatibility relationship for cross cascade connection since it reduces (6) to

$$\begin{bmatrix} x_{12} \\ y_{12} \end{bmatrix} = \begin{bmatrix} 0 & 1 \\ 1 & 0 \end{bmatrix} \begin{bmatrix} x_{21} \\ y_{21} \end{bmatrix}. \tag{13}$$

This equation defines the cross-cascade interconnection which is shown in Fig. 5.

Thus if the cross-cascade interconnection of the LT structures is to be used and so reducing the complexity of the overall structure, the adjacent matrices at the interconnections points must satisfy this compatibility relationship.

The complexity of the LT network of a ladder arm depends strongly on the transformation matrices chosen for its transformation. In practice it has been found that if some elements of the transformation matrices are set equal to zero then the active realization of the LT structure may be effectively simpler. The LT network N_i' corresponding to a two-port N_i realizes in fact two linear combinations of the inputs x_{1i} and x_{2i} to generate the outputs y_{1i} and y_{2i} the general form of which is given below

$$\begin{aligned} y_{1i} &= K(s)x_{1i} + L(s)x_{2i} \\ y_{2i} &= M(s)x_{1i} + N(s)x_{2i}. \end{aligned} \tag{14}$$

There are, therefore, in the general case four voltage transfer ratios in (14), K, L, M, and N, to be realized. However since these transfer ratios are function of the elements of the transformation matrices as shown in (4) they can be simplified or, if desired, be made equal to zero [11]. If for example, in the transformation matrices of a series arm Z_i we set $\gamma_{1i} = \gamma_{2i} = 0$ and $\delta_{1i} = \delta_{2i}$ then we have

$$y_{1i} = y_{2i} = M(s)x_{1i} + N(s)x_{2i}$$

where the voltage transfer ratio $M(s)$ can be made equal to the voltage transfer ratio $N(s)$ by a suitable choice of the nonzero transformation matrix elements. Table II gives the zero entries of the transformation matrices and the simplified form of (14).

Finally, when the single-input-single-output model is used for the realization of the last resistively terminated reactive arm, using only one transformation matrix ($\mathbb{S}_{1n}$ in Section II-C) the choice of this matrix is of vital importance for the existence of a convenient output node. The fact that in this case only x_{1n} and y_{1n} are present explicitly whilst the output V_{2n} of the filter (Fig. 4(a)) is related directly to y_{2n}, calls for a proper choice of $\mathbb{S}_{1n}$ in order to avoid extra operational amplifiers to generate the required

TABLE II
THE ZERO ENTRIES OF THE TRANSFORMATION MATRICES AND THE CORRESPONDING SIMPLIFIED EXPRESSION OF (4). THE QUANTITY b_i OF (4) BECOMES EQUAL TO ZERO FOR THE TWO CASES LABELED "INADMISSIBLE"

	Transformation matrices	Transfer functions to be realized
SERIES ARM Z	$\begin{bmatrix} \alpha_1 & \beta_1 \\ 0 & \delta_1 \end{bmatrix} \begin{bmatrix} \alpha_2 & \beta_2 \\ 0 & \delta_2 \end{bmatrix}$	$y_1 = -\frac{\delta_1}{\delta_2} y_2 \qquad y_2 = \frac{-\alpha_2\delta_2}{\alpha_1\beta_2 + \beta_1\alpha_2 + \alpha_1\alpha_2 Z}\left(x_1 - \frac{\alpha_1}{\alpha_2} x_2\right)$
	$\begin{bmatrix} 0 & \beta_1 \\ \gamma_1 & \delta_1 \end{bmatrix} \begin{bmatrix} \alpha_2 & \beta_2 \\ 0 & \delta_2 \end{bmatrix}$	$y_1 = \frac{\gamma_1\alpha_2 Z + \gamma_1\beta_2 + \delta_1\alpha_2}{\beta_1\alpha_2} x_1 + \frac{\gamma_1}{\alpha_2} x_2 \qquad y_2 = -\frac{\delta_2}{\beta_1} x_1$
	$\begin{bmatrix} \alpha_1 & \beta_1 \\ 0 & \delta_1 \end{bmatrix} \begin{bmatrix} 0 & \beta_2 \\ \gamma_2 & \delta_2 \end{bmatrix}$	$y_1 = \frac{-\delta_1}{\beta_2} x_2 \qquad y_2 = \frac{\gamma_2}{\alpha_1} x_1 + \frac{\alpha_1\gamma_2 Z + \alpha_1\delta_2 + \beta_1\gamma_2}{\alpha_1\beta_2} x_2$
	$\begin{bmatrix} 0 & \beta_1 \\ \gamma_1 & \delta_1 \end{bmatrix} \begin{bmatrix} 0 & \beta_2 \\ \gamma_2 & \delta_2 \end{bmatrix}$	INADMISSIBLE
SHUNT ARM Y	$\begin{bmatrix} \alpha_1 & \beta_1 \\ \gamma_1 & 0 \end{bmatrix} \begin{bmatrix} \alpha_2 & \beta_2 \\ \gamma_2 & 0 \end{bmatrix}$	$y_1 = \frac{\gamma_1}{\gamma_2} y_2 \qquad y_2 = \frac{\beta_2\gamma_2}{\alpha_1\beta_2 + \alpha_2\beta_1 + \beta_1\beta_2 Y}\left(x_1 + \frac{\beta_1}{\beta_2} x_2\right)$
	$\begin{bmatrix} \alpha_1 & 0 \\ \gamma_1 & \delta_1 \end{bmatrix} \begin{bmatrix} \alpha_2 & \beta_2 \\ \gamma_2 & 0 \end{bmatrix}$	$y_1 = \frac{\delta_1\beta_2 Y + \delta_1\alpha_2 + \gamma_1\beta_2}{\alpha_1\beta_2} x_1 - \frac{\delta_1}{\beta_2} x_2 \qquad y_2 = \frac{\gamma_2}{\alpha_1} x_1$
	$\begin{bmatrix} \alpha_1 & \beta_1 \\ \gamma_1 & 0 \end{bmatrix} \begin{bmatrix} \alpha_2 & 0 \\ \gamma_2 & \delta_2 \end{bmatrix}$	$y_1 = \frac{\gamma_1}{\alpha_2} x_2 \qquad y_2 = -\frac{\delta_2}{\beta_1} x_1 + \frac{\beta_1\delta_2 Y + \beta_1\gamma_2 + \alpha_1\delta_2}{\beta_1\alpha_2} x_2$
	$\begin{bmatrix} \alpha_1 & 0 \\ \gamma_1 & \delta_1 \end{bmatrix} \begin{bmatrix} \alpha_2 & 0 \\ \gamma_2 & \delta_2 \end{bmatrix}$	INADMISSIBLE

output from the existing x_{1n} and y_{1n}. From the linear transformation of N_n and its modified chain matrix description we have

$$\begin{bmatrix} x_{1n} \\ y_{1n} \end{bmatrix} = \mathbb{S}_{1n} T_n \begin{bmatrix} V_{2n} \\ I_{2n} \end{bmatrix}. \tag{15}$$

The modified chain matrix of any series of shunt arm is of the form

$$T_i = \begin{bmatrix} 1 & B_i \\ C_i & -1 \end{bmatrix}$$

where $B_i = -Z_i$, $C_i = 0$ for any series arm Z_i and $B_i = 0$, $C_i = Y_i$ for any shunt arm Y_i (Note that since $B_i C_i = 0$ for any ladder arm, then det $T_i = -1$ and $T_i = T_i^{-1}$). Thus using $V_{2n} = -R_L I_{2n}$ and the above form of T_i for $i = n$, equation (15) yields

$$\begin{aligned} x_{1n} &= \left[\alpha_{1n}(1 - B_n G_L) + \beta_{1n}(C_n + G_L)\right] V_{2n} \\ y_{1n} &= \left[\gamma_{1n}(1 - B_n G_L) + \delta_{1n}(C_n + G_L)\right] V_{2n}. \end{aligned} \tag{16}$$

Therefore if N_n is a series arm Z_n ($B_n = -Z_n$, $C_n = 0$) in order to have a direct output without extra active elements either α_{1n} or γ_{1n} must be zero and moreover either β_{1n} or δ_{1n} must be equal to R_L, respectively. Such a choice makes the output of the LTA filter either x_{1n} or y_{1n}. Alternatively, if N_n is a shunt arm Y_n ($C_n = Y_n$, $B_n = 0$) it can be found from (16) that either β_{1n} must be zero and $\alpha_{1n} = 1$ thereby making $x_{1n} = V_{2n}$, or $\delta_{1n} = 0$ and $\gamma_{1n} = 1$ and hence $y_{1n} = V_{2n}$.

In both cases an appropriate zero entry is required and there are two alternative choices. From the practical point of view the nature of the last ladder element dictates the appropriate zero entry. For a shunt capacitor for example δ_{1n} must be made zero (with y_{1n} as output of the LTA filter) because by choosing $\beta_{1n}=0$ differentiation operation will be unavoidable in the realization of the corresponding active LT structure. Similarly if N_n is a shunt inductor then β_{1n} must be made zero for similar reasons.

The complexity of an LT structure corresponding to a ladder arm depends of course not only on the transformation matrices but on the impedance of the arm too. It is found that by making some elements in the transformation matrices frequency dependent a reduction in the complexity of the realization can be achieved. This is illustrated in an example below.

III. The Application of the Method

A. Practical Considerations

The application of the LT transformation approach to the simulation of a doubly terminated ladder filter may start from any arm of the ladder. It is advisable, however, to start from the last reactive arm incorporated with the termination, in order to assure the existence of a direct output as already discussed in the last section.

The mathematical manipulation of several matrix equations is effectively simplified by the fact that the modified chain matrix of the series and shunt arms is of the form described above. The basic equation to be employed in the application of the method is (3) from which $[y_{1i}\ \ y_{2i}]^T$ can be determined. The two cases of the series and the shunt arms are shown separately below ((17) and (18)) in their general form. Thus for a series arm impedance Z, we have

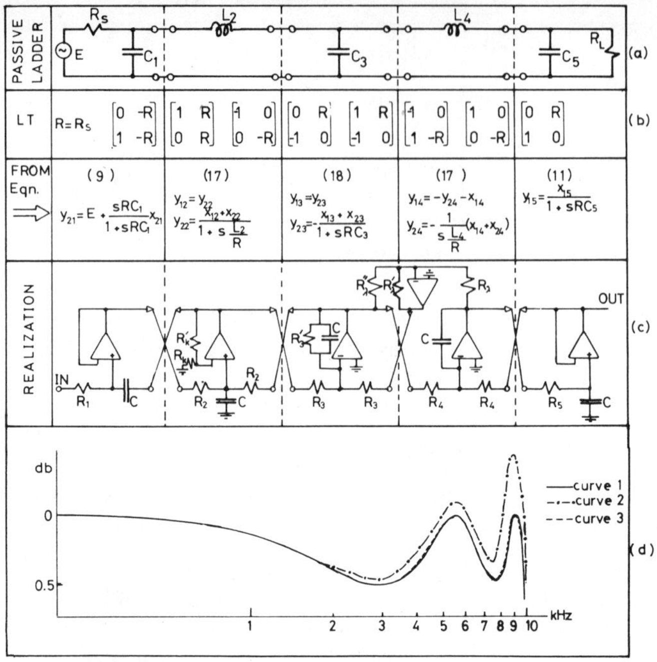

Fig. 6. (a) The fifth-order passive polynomial low-pass filter. (b) The transformation matrices. (c) The LTA filter. (d) The passband response of the LTA filter.

$$\begin{bmatrix} x_1 \\ y_1 \end{bmatrix} = \frac{1}{\Delta_2}\begin{bmatrix} \alpha_1(\delta_2+\gamma_2 Z)+\beta_1\gamma_2 & -\alpha_1(\alpha_2 Z+\beta_2)-\beta_1\alpha_2 \\ \gamma_1(\delta_2+\gamma_2 Z)+\delta_1\gamma_2 & -\gamma_1(\alpha_2 Z+\beta_2)-\delta_1\alpha_2 \end{bmatrix}\begin{bmatrix} x_2 \\ y_2 \end{bmatrix}. \tag{17}$$

Similarly for the shunt-arm admittance Y, (3) becomes

$$\begin{bmatrix} x_1 \\ y_1 \end{bmatrix} = \frac{1}{\Delta_2}\begin{bmatrix} \beta_1(\delta_2 Y+\gamma_2)+\alpha_1\delta_2 & -\beta_1(\beta_2 Y+\alpha_2)-\alpha_1\beta_2 \\ \delta_1(\delta_2 Y+\gamma_2)+\gamma_1\delta_2 & -\delta_1(\beta_2 Y+\alpha_2)-\gamma_1\beta_2 \end{bmatrix}\begin{bmatrix} x_2 \\ y_2 \end{bmatrix}. \tag{18}$$

These equations become even simpler when zero entries in the transformation matrices are taken into consideration in the manner described earlier.

B. Examples

In this section we demonstrate the efficiency and the flexibility of the proposed LTA method through two examples which for comparison reasons have been selected from the existing literature.

The fifth-order low-pass Chebyshev filter (0.5-dB ripple) shown in Fig. 6(a) has been denormalized for $f_0 = 10$ kHz and 1-kΩ impedance level. The application of the LTA method has started from the terminated capacitor C_5 with the transformation matrix shown in Fig. 6(b). According to Section II-D the output will be the node labelled y_{15}. This figure shows the rest of the transformation matrices which have been employed in accordance with Section II-D. The voltage transfer ratios to be realized are also shown in Fig. 6(c) together with their RC-active realizations. It is important to note that from the 6 operational amplifiers used in the overall structure only two have a frequency dependent gain. A simplification is possible in the structure of Fig. 6(c) by disconnecting resistors R_3' and R_λ'' but their presence improves the adjustability. Indeed the frequency response of the constructed filter (curve 2 in Fig. 6(d)) had been easily adjusted to be very close to the ideal by means of R_k and R_λ (curve 3) R_3' and R_λ'' included. This filter may be favorably compared with the corresponding leapfrog structure of Fig. 8(a) or (b) not only with respect to the number of sections having frequency dependent gain, but also with respect to the overall behavior and adjustability.

As a second example, we shall simulate the eighth-order bandpass filter taken from [9]. The passive ladder is shown in Fig. 7(a) and for the transformation matrices shown below that figure the same active structure as in [9] is obtainable. When these transformation matrices are employed it becomes apparent that the "port reciprocator" of [9] is equivalent to a noninverting differentiator. Indeed in the direct SFG simulation method circuits containing a loop of capacitors or a T of inductors will

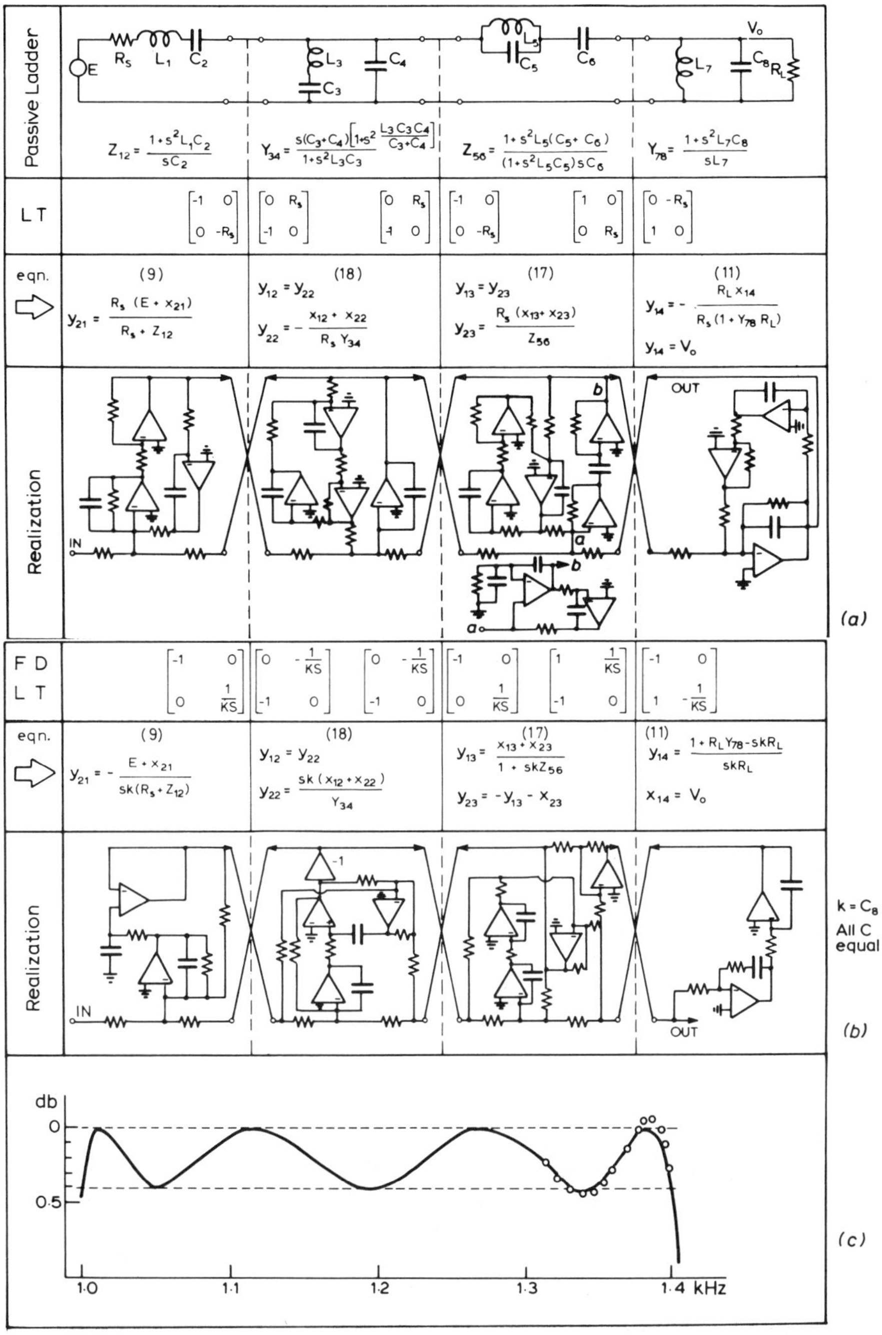

Fig. 7. (a) The LTA simulation of the eighth-order bandpass filter leading to the same active filter as in [9]. (b) A more effective LTA simulation of the same ladder using frequency dependent matrices. (c) The passband response of the LTA filter of (b).

inevidably lead to differentiation in their SFG active realization. In the LTA method using the frequency dependent (FD) transformation matrices shown in Fig. 7(b) the undesirable differentiators are completely avoided from the realization and the number of the operational amplifiers and the capacitors is drastically reduced. Compare for example 15 operational amplifiers and 12 capacitors in [9] and Fig. 7(a) with 12 operational amplifiers and 8 capacitors (for an eighth-order passive original) in the design of Fig. 7(b). This *RC*-active filter has been constructed using polysterene capacitors of 2-percent tolerance, resistors of 5-percent tolerance and three 4136 quad operational amplifier chips. The passband response without any compensation or adjustment is shown in Fig. 7(c).

IV. Special Cases of the LTA Method

It will be observed from the discussion so far that the approach under examination is quite general in nature and it encompasses many other hitherto known methods. These known methods appear as special cases of our procedure contained in this paper and owing to the generality of the LTA approach it is possible sometimes to improve designs based on such known methods. These LTA special cases are examined in the following.

A. The Wave Active Filters [4]–[7]

A particular choice of transformation matrices produces the wave active filters structures as follows. Let $\mathbb{S}_{1i}=\mathbb{S}_{2i}=\mathbb{S}$ for all i's and under the cross-cascade connection the transformation matrix $\mathbb{S}$ becomes self compatible. Then it will have the form

$$\mathbb{S}=\begin{bmatrix}\alpha & \beta\\ \alpha & -\beta\end{bmatrix}=\mathbb{S}_{1i}=\mathbb{S}_{2i}. \tag{19}$$

If the above matrices are used to transform all the arms of the ladder filter with $\alpha=1$ and $\beta=R$ the circuits proposed in [4]–[7] will be obtained. It must be observed, however, that the single-input-single-output model for the last terminated reactive arm employed in Section II-C and its consequent saving of active components cannot be used here since there is no freedom to set an element of the transformation matrices equal to zero to create an output node.

B. The Leapfrog Synthesis Case [8]

The leapfrog structures can also be obtained using the LTA method for a special choice of transformation matrices as follows:

i) For a series-arm transformation $\mathbb{S}_{1i}=\begin{bmatrix}\alpha_{1i} & 0\\ 0 & \delta_{1i}\end{bmatrix}$ $\mathbb{S}_{2i}=\begin{bmatrix}\alpha_{2i} & 0\\ 0 & \delta_{2i}\end{bmatrix}$

ii) For a shunt-arm transformation $\mathbb{S}_{1j}=\begin{bmatrix}0 & \beta_{1j}\\ \gamma_{1j} & 0\end{bmatrix}$ $\mathbb{S}_{2j}=\begin{bmatrix}0 & \beta_{2j}\\ \gamma_{2j} & 0\end{bmatrix}$

(20)

Naturally each right transformation matrix for a series (shunt) arm must be compatible according to (13) to the left transformation matrix of the following shunt (series) arm. Consequently $\alpha_{2i}=\gamma_{1j}$ and $\delta_{2i}=-\beta_{1j}$ ($\beta_{2j}=-\delta_{1i}$ and $\gamma_{2j}=\alpha_{1i}$) for a cross-cascade interconnection between N_i' and N_j'. The last reactive arm N_n of the ladder can be transformed according to (11) since the zero entries in the transformation matrices assure the existence of a direct output node.

The simulation of Fig. 8(a) illustrates how the leapfrog structure for a fifth-order polynomial low-pass filter can be derived using the LTA approach with transformation matrices chosen according to (20). The so derived active leapfrog structure is the same as in [8], [10] while Fig. 8(b) gives an alternative LTA-leapfrog structure.

For the high-pass case the transformation matrices to be used for all integrator active leapfrog structures are as shown below. For both series and shunt arm transformations

$$\text{either } \mathbb{S}_1=\begin{bmatrix}0 & \beta_1\\ \gamma_1 & 0\end{bmatrix} \text{ and } \mathbb{S}_2=\begin{bmatrix}\alpha_2 & 0\\ 0 & \delta_2\end{bmatrix}$$

$$\text{or } \mathbb{S}_1=\begin{bmatrix}\alpha_1 & 0\\ 0 & \delta_1\end{bmatrix} \text{ and } \mathbb{S}_2=\begin{bmatrix}0 & \beta_2\\ \gamma_2 & 0\end{bmatrix}$$

Note that in this case the same transformation matrix forms are for both series and shunt arms.

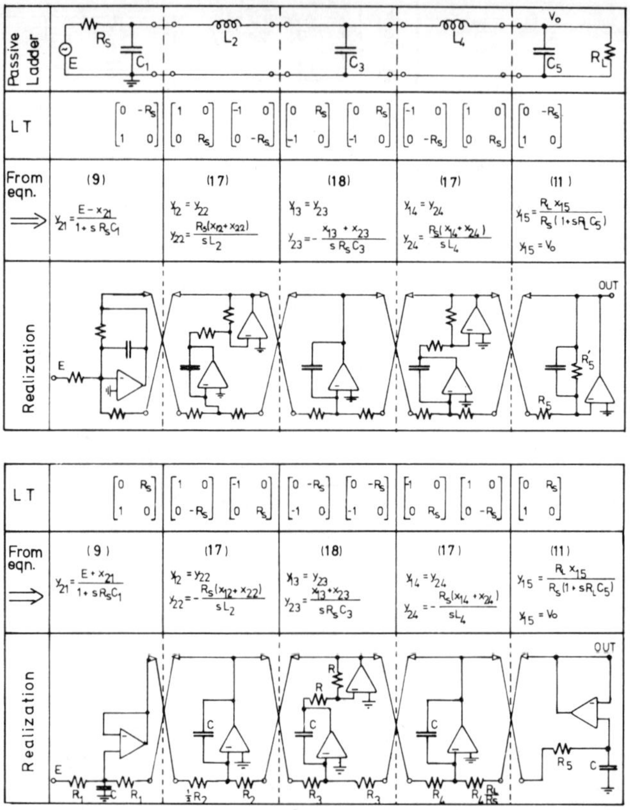

Fig. 8. (a) LTA simulation of a fifth-order polynomial low-pass filter leading to the leapfrog structure. (b) LTA simulation of the same filter leading to a more economical leapfrog structure.

Conclusions

We have shown in this paper that by means of LT of the port-variable vectors of a network it is possible to derive new RC-active filter realizations from LC doubly terminated ladder filters. This can be achieved by constraining the general LTA procedure so as to yield RC-active structures with certain desirable features. This was the case with the self-dual LTA filters derived elsewhere [11], [12]. Moreover, due to the general nature of the LTA approach, many already known LC-ladder simulation techniques, such as the leapfrog synthesis [8], [9] and the wave active filters, may be interpreted as special LTA cases and when seen as such they are amenable to further improvements. Finally it has been shown that frequency-dependent linear transformations (FDLT) are not only possible but in certain cases may yield highly desirable results.

References

[1] H. J. Orchard, "Inductorless filters," *Electron. Lett.*, vol. 2, no. 6, pp. 224–225, 1966.

[2] D. F. Sheahan, "*LC* filter simulation techniques," in *Proc. IEEE Int. Symp. Circuits, Systems and Computers* (Oaxtepec, Mexico), pp. 340–344, 1971.

[3] A. G. Constantinides and H. G. Dimopoulos, "Active *RC* filters derivable from *LC* ladder filters via linear transformations," *IEE J. Electronic Circuits and Syst.*, vol. 1, no. 1, pp. 17–21, Sept. 1976.

[4] A. G. Constantinides and G. Haritantis, "Wave active filters," *Electron. Lett.*, vol. 11, pp. 254–256, 1975.

[5] H. Wupper and K. Meerkötter, "New active filter synthesis based on scattering parameters," *IEEE Trans. Circuits Syst.*, vol. CAS-22, pp. 594–602, 1975.

[6] ——, "New active filter synthesis based on scattering parameters," in *Proc. IEEE Int. Symp. Circuits and Systems*, pp. 254–257, 1975.

[7] I. Haritantis, A. G. Constantinides, and T. Deliyannis, "Wave active filters," *Proc. Inst. Elec. Eng.*, vol. 123, no. 7, pp. 676–682, 1976.

[8] F. E. J. Girling and E. F. Good, "Active filters—A practical approach on feedback amplifier theory," Parts 12, 13, and 14, *Wireless World*, July, Sept., and Oct. 1970.

[9] P. O. Brackett and A. S. Sedra, "Direct SFG simulation of *LC* ladder networks with applications to active filter design," *IEEE Trans. Circuits Syst.*, vol. CAS-23, pp. 61–67, 1976.

[10] ——, "Active conpensation for high frequency effects in op. amp. circuits with applications to active *RC* filters," *IEEE Trans. Circuits Syst.*, vol. CAS-23, pp. 68–72, 1976.

[11] H. G. Dimopoulos, "The design of electronic active filters from ladder networks using linear transformations," Ph.D. dissertation, Univ. London, London, England, Apr. 1978.

[12] H. G. Dimopoulos and A. G. Constantinides, "Self-Dual LTA filters," *IEE J. Electronic Circuits Syst.*, to appear.

Paper 3-7

On Low-Sensitivity Realizations of Band-Elimination Active Filters

RENATO N. GADENZ, MEMBER, IEEE

Abstract—**The results of an empirical search for low-sensitivity band-elimination active filters are presented. The search was oriented toward coupled topologies (leapfrog, coupled biquad, shifted companion form) and realizations that simulate passive structures (FDNR), as opposed to a solution by optimization. An 8th-order band-elimination filter was used as an example.**

A comparison between the above-mentioned realizations and the cascade shows that those topologies, while improving the sensitivity with respect to the cascade in some frequency bands, offer little or no improvement (if not impairment) in the other frequency bands. A sensitivity improvement over the cascade at all frequencies was obtained with a mixed topology, one which uses a cascade of properly chosen coupled sections. Similar results were achieved using a mixed topology in the realization of a 12th-order bandpass filter with critical transition bands.

Manuscript received February 27, 1976; revised November 23, 1976.

The author is with Bell Laboratories, Holmdel, NJ 07733.

I. Introduction

IN RECENT years, *RC*-active filters have replaced the more conventional *LC* passive ones in many applications in which the use of active filters leads to a smaller and cheaper product. Active filters are also suited for integration. Their main drawback is the relatively high sensitivity to variations of both the active and the passive components. Thus an extensive search has been conducted to minimize or at least reduce their sensitivities.

Reprinted from *IEEE Trans. Circuits Syst.*, vol. CAS-24, pp. 175-183, Apr. 1977.

References [1] to [10] illustrate a direct attempt to solve the problem by optimization techniques that modify both the circuit topology and the parameter values until a minimum of sensitivity is reached. As any optimization problem, this approach is complex. On the other hand, if convergence is achieved, an optimum solution (global or local) is obtained.

Many authors have used alternative methods, easier to implement, which usually provide a better insight and yield networks with reduced (although not necessarily minimum) sensitivity. Since a doubly terminated LC network, designed to deliver maximum power from source to load, has good sensitivity properties in the passband [11], a passive realization of the desired transfer function could be obtained first, followed by

a) the substitution of the inductors by equivalent networks [12]–[18], or
b) a transformation into an FDNR (or RCD) structure and the realization of the frequency-dependent negative-resistance element D by equivalent RC-active networks [19]–[22].

Other efforts [23]–[32] have been directed toward coupled structures in which feedback is used in different ways between sections of the filter to improve the sensitivity.

The basic rationale for using a coupled configuration, instead of a simple cascade of biquadratic sections, is again to imitate the passband behavior of the passive configuration; this behavior is strongly dependent upon the poles of the transfer function which are a characteristic of the overall network, not just of one section. In the cascade realization, obtained by a decomposition of the desired transfer function into a product of biquadratic factors, each pair of poles depends solely on the elements of one section.

Most of the research done so far deals with low-sensitivity bandpass filters or considers the passband behavior of the filters. But configurations that are optimum or give minimum sensitivity in the passband do not necessarily have good sensitivity in the stopband. In this work, the problem of low-sensitivity topologies for band-elimination filters is examined. This problem is also important for other types of filters (bandpass, low-pass, high-pass, etc.) in which the design is critical in the stopband(s) and/or the adjacent transition band(s).

One typical example, the 8th-order (four biquadratic sections) band-elimination filter (BEF) of Fig. 1 and Appendix I, was chosen for studying the sensitivity of different BEF realizations. A 12th-order bandpass filter with critical transition bands and stopbands was also considered. Among the realizations, the leapfrog [26], coupled biquad [25], shifted companion form [31], and FDNR or RCD [19], [20] configurations were examined, plus some newly developed ones. The cascade realization was used as the reference for sensitivity comparisons.

Throughout this work, the sensitivities were determined using either Schoeffler's sensitivity measure E [2] or the

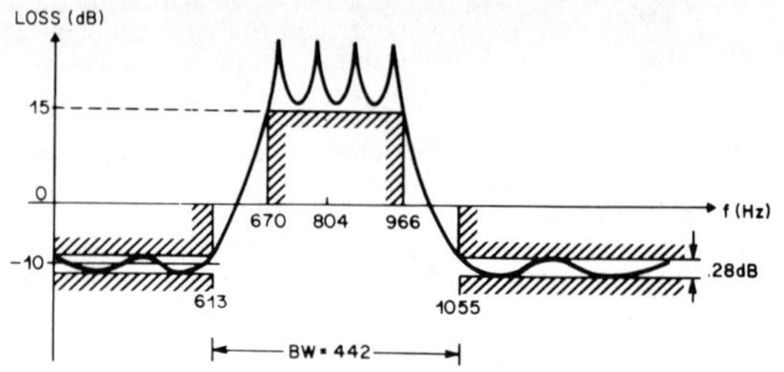

Fig. 1 Eigth-order band-elimination filter.

standard deviation σ of the gain (as obtained from a Monte Carlo analysis) or both. Therefore, at each frequency of interest, the sensitivity of a given topology relative to that of the cascade is given by the numbers E/E_C and/or σ/σ_C.[1] It has been shown [33] that, under certain conditions,[2] $\sqrt{E}$ is proportional to σ. This correspondence between $\sqrt{E}$ and σ has been checked for some realizations and found to be satisfactory.[3]

The investigation has been limited to comparing the sensitivity properties that arise from the topologies themselves, in particular the way in which the biquadratic building blocks of each configuration are interconnected. Therefore, the possibility of improving the behavior of some topologies by tuning or by choosing the least-sensitive available blocks (see Appendix II) has not been incorporated here. This would represent an interesting extension of this work, and it might lead to modified conclusions.[4]

[1]At each frequency of interest, E is defined by [2]

$$E \triangleq \sum_i |S_{x_i}^G|^2$$

where

$$G = 20 \log |T| = 20 \log |V_{out}/V_{in}|$$

$$S_{x_i}^G \triangleq \frac{x_i}{G} \frac{\partial G}{\partial x_i}$$

and x_i is the ith element of the network. The summation is extended over all elements. This definition was adopted because it is the one used by the program SLIC [35], available for the computations. However it would be preferable to define $S_{x_i}^G$ as $x_i \partial G/\partial x_i$. To avoid the problems that the present definition introduces, the same $G(\omega)$ was used for all realizations, so that G does not appear in the ratio of E values corresponding to two configurations. In addition , the exact frequencies for which either $|T|=0$ or $G=0$ were omitted in the computations; frequency points in their neighborhood were used instead, to estimate the sensitivity behavior without incurring in computational errors due to overflow or underflow.

[2]The conditions are: equal and small tolerances for all components, and no correlation between them.

[3]Other sensitivity measures are also found in the literature [16], [26], [34] or can be defined. For example, in [26], the measure is $E\Sigma_j(\text{Re } S_{y_j}^T)^2$ where the y_j are the coefficients of all the individual biquadratic transfer functions used as building blocks in the realization of the overall transfer function T.

[4]All these considerations: which blocks to use, which topology to use (i.e., how should these blocks be interconnected), what kind of tuning is practically possible and to what extent (e.g., the parameters could be tuned to value or to performance; tuning could be used in each block without readjusting the entire network, since this is not straightforward in a coupled topology; etc.) should of course be taken into account in a final product.

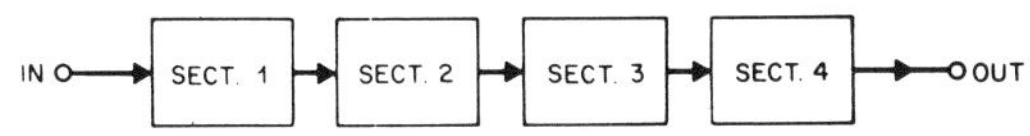

Fig. 2 Cascade realization.

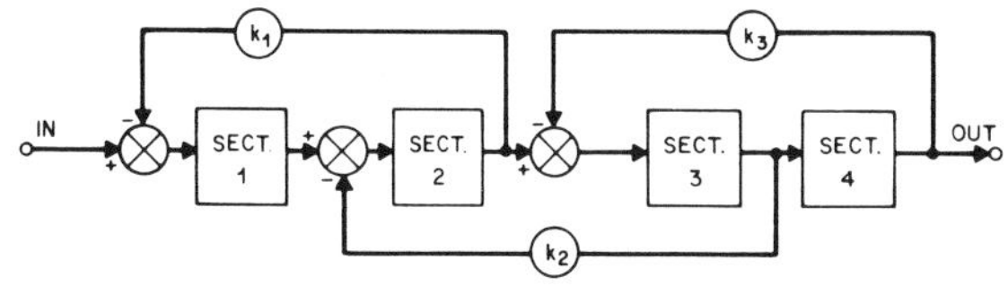

Fig. 3 Leapfrog realization.

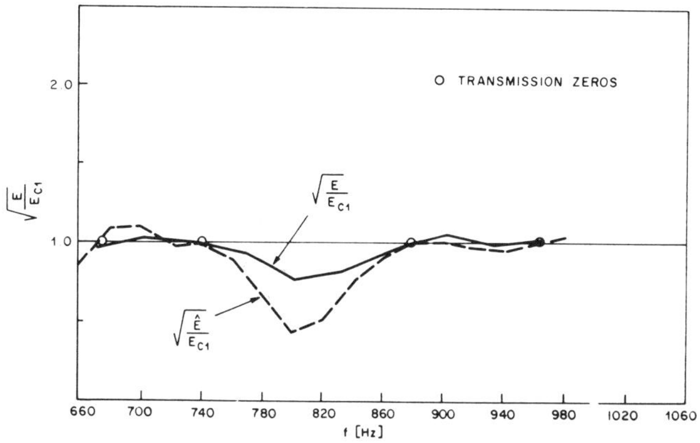

Fig. 4 Sensitivity comparison between leapfrog and cascade realizations.

II. Different Realizations and Their Sensitivities

Several realizations of the 8th-order band-elimination filter of Fig. 1 are presented and their sensitivities are compared.[5] Some interesting conclusions result. Next, some configurations realizing a 12th-order bandpass filter with critical transition bands are considered and similar conclusions are obtained.

A. The Cascade Realization

The cascade arrangement of biquadratic sections is a common way of realizing *RC*-active filters. For the band-elimination filter of Fig. 1 (see also Appendix I), the schematic representation of this configuration is illustrated in Fig. 2.

The poles and zeros of the overall transfer function may be paired in different ways and this affects the sensitivity. For the C1 arrangement shown in Appendix I, the highest Q_p sections are selected first and the zero closest to each pole is put in the same biquad. Another arrangement is possible (C2) in which the lowest Q_p sections are selected first and the zero closest to each pole is put in the same biquad. In the center of the elimination band, C2 was found to be slightly less sensitive than C1, but the converse is true toward the edges of the stopband. The two other arrangements (C3 and C4) considered here are more sensitive.

In what follows, C1 will be preferred, since it is also the optimum for low sensitivity in the passband [25]. This realization will then be used as the reference for comparing the sensitivities of other realizations.

B. The Parallel Realization

An alternative way of realizing the filter is by connecting biquadratic sections in parallel. The sections are of course different from those of the cascade and are obtained by partial fraction expansion of the overall transfer function T.

For the example of Fig. 1, this configuration was found to be much more sensitive than the cascade in the stopband and, therefore, should not be used. The large sensitivity is, at least in part, due to the cancellation of terms at the summation point, required to get the high stopband loss.

C. Leapfrog (LPFG) Realizations

The leapfrog [26] configuration for the band-elimination filter of Fig. 1 is shown in Fig. 3. As in the cascade, there are several possible leapfrog realizations of a given transfer function. They correspond to different pole–zero pairings (i.e., different sequences of the transmission zeros for a given sequence of poles) and different locations of the zeros of an auxiliary polynomial needed for the synthesis [26].

Results obtained for some of these realizations (see, e.g., Fig. 4) indicate the following.

a) Around the transmission zeros, the leapfrog and cascade realizations have similar sensitivities. This can be proven and the proof is based on the fact that each transmission zero is realized independently by each section in either the LPFG or the cascade realizations. Using the notation of Appendix II, near a transmission zero we have

$$\Pi T_i = T \approx \Pi \hat{T}_i.$$

b) The sensitivity of the leapfrog configuration is lower than that of the cascade in some intervals of the stopband,

[5]Due to the low frequencies involved, semi-ideal amplifiers (i.e., having a constant and high, but finite, gain) are assumed; their effect on the overall sensitivity measure E can be neglected. For filters operating at higher frequencies (several kilohertz), the amplifiers are better described by a 2-pole 1-zero model, and their effect on E is no longer negligible. For that case, and in order to preserve $G(\omega)$, the RC elements of the network must be modified to compensate for the amplifier effects. A weighted value of E should then be used to account for the difference in tolerances between the passive and the active components (e.g., if the ratio between the tolerances of active and passive components is 100, a ratio of 10^4 should be used for the corresponding weights). Even then, a sensitivity comparison based on E (first-order sensitivities) may not be accurate enough. In that case, a straight Monte Carlo analysis should be used. This was done, e.g., for the 12th-order channel bank filter whose relative gain sensitivity, as measured by σ/σ_{C1}, is illustrated in Fig. 12.

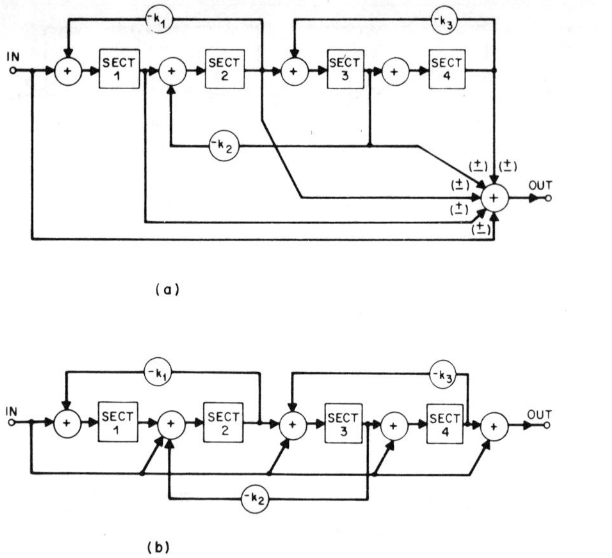

Fig. 5 Coupled biquad realizations. (a) Summation zero-forming. (b) Feedforward zero-forming.

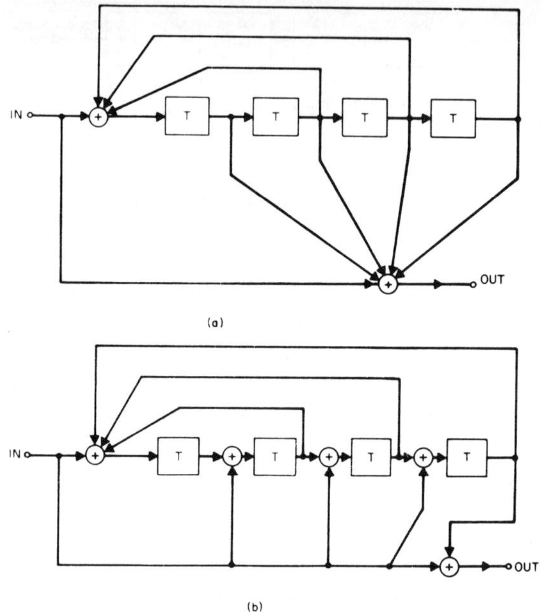

Fig. 6 Shifted companion form realizations. (a) SCF-Σ. (b) SCF-FF.

and higher in some other intervals.

It is also known [26] that the leapfrog realization is less sensitive than the cascade in the passband.

The plot of Fig. 4 also shows a comparison based on Szentirmai's sensitivity measure $\hat{E}$.[3] As expected, the latter is more optimistic than E, predicting a better sensitivity performance in the stopband.

D. Coupled Biquad (CB) Realizations

The state variable coupled biquad realization [25] is another way of coupling biquadratic sections to achieve a given transfer function. The essential difference with the leapfrog configuration is that here the transmission zeros are realized by either combining the outputs of the different sections [Fig. 5(a)] or by multiple input to the different sections [Fig. 5(b)], instead of realizing them independently, one in each section.

The coupled biquad design requires as an initial step the determination of a low-pass (LP) passive circuit that realizes $1/D(s)$ where $D(s)$ is the denominator of the transfer function.[6] Different coupled biquad realizations are obtained by changing the termination ratio and/or the location of the zeros of the reflection coefficient of this LP circuit.

Results obtained for the band-elimination filter of Fig. 1 show the following.

a) The coupled biquad configuration is less sensitive than the cascade design in the passband, but more sensitive in the rejection band. In the latter band, it is even more sensitive than the leapfrog design. The lowest value of $\sqrt{E/E_{C1}}$ obtained around the center of the rejection band was ~ 4.

b) The zeros z_i of the input reflection coefficient must satisfy

$$\sum_i \mathrm{Re}\,(z_i) < 0 \quad \text{and} \quad \left| \sum_i \mathrm{Re}\,(z_i) \right| = \min$$

to reduce the sensitivity in the rejection band.

c) On the effect of the termination ratio, results were inconclusive.

E. Shifted Companion Form (SCF) Realizations

The shifted companion form [27]–[31] is yet another way of realizing a transfer function using coupled sections. For the band-elimination filter of Fig. 1, two arrangements are illustrated in Fig. 6. In Fig. 6(a) the output is a weighted combination of the input and the outputs of all the sections; the configuration is denoted SCF-Σ. In Fig. 6(b) the input is fed forward to all sections and the overall output is a weighted combination of the input and the output of the last section; the arrangement is denoted SCF-FF. These configurations have identical sections T and are obtained from either a low-pass or a high-pass transfer function prototype, using either a low-pass to band-reject (LP to BR) or a low-pass to bandpass (LP to BP) transformation, respectively. In either prototype, each section is of the form $K/(s+a)$ and appropriate feedback and feedforward paths are used.

Experimental results shown in Fig. 7 for SCF configurations designed from LP transfer function prototypes, indicate that the sensitivity around the center f_0 of the rejection band is low, much lower than in the cascade config-

[6] If the filter is designed from a prototype via a frequency transformation, the LP passive circuit realizes $1/D(s)$ where $D(s)$ is the denominator of the prototype transfer function. This prototype transfer function could be either a low-pass or a high-pass function. In the former case, a low-pass to band-reject (LP to BR) transformation is used; in the latter, a low-pass to bandpass (LP to BP) transformation is required.

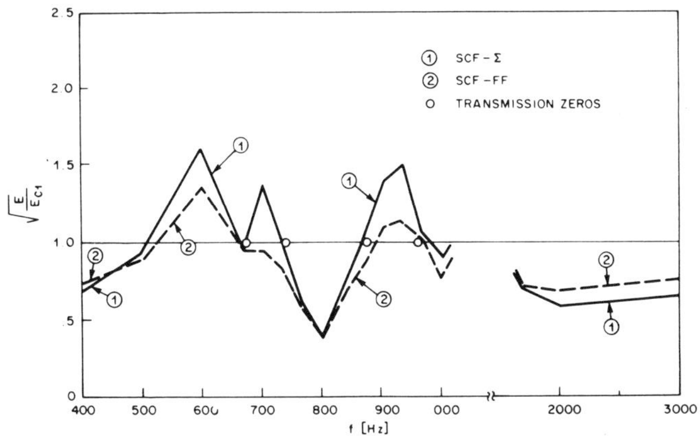

Fig. 7 Sensitivity comparison between shifted companion form and cascade realizations.

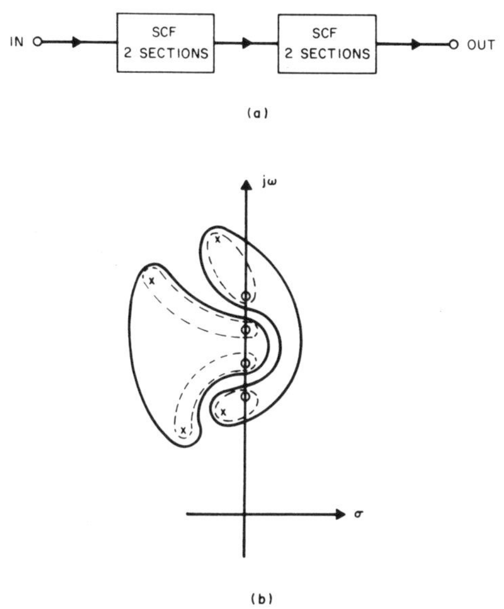

Fig. 8 (a) Cascade of two 2-section SCF. (b) *P-Z* distribution between two biquartics. (Dashed lines indicate *p-z* distribution for C1.)

uration. This was expected, because all sections are identical and have a transmission zero at f_0. (In addition, $\omega_p = \omega_z$.) Then, at f_0, the only significant path to the output is the direct one from the input, through the output adder. This is realized by an amplifier with two resistors which provide the proper weighting, i.e., give the required attenuation. Thus the complete SCF circuit is very much reduced, and the overall sensitivity, which is determined by few significant elements, has to be very low. As we move away from f_0 the sensitivity increases, because each section begins to contribute to the output and the high stopband loss is obtained by a cancellation of these increasing contributions. Toward the edges of the rejection band and in the transition bands, the SCF realization is more sensitive than the cascade. In the passband, the sensitivity of the SCF decreases again. Also note that the SCF-FF arrangement is slightly less sensitive than the SCF-Σ.

If a HP transfer function prototype and the associated LP to BP transformation is used, no transmission zero is created in each section; as a result, the output of each section at and around f_0 is rather high, and all sections contribute significantly to the output. The high loss in this range of frequencies is obtained by a cancellation of large terms, which implies high sensitivities. As a consequence, this realization is very sensitive in the center of the rejection band, even more sensitive than the cascade realization. The argument is reversed in the passband, where the SCF configuration designed from a HP transfer function prototype is less sensitive than the configuration designed from a LP transfer function prototype.

If the SCF-Σ(HP) and SCF-Σ(LP) configurations[7] are compared further we observe that, around f_0, all sections in the former configuration have a high output, thus involving all of them, with all their elements, in the cancellation (of *large* terms). Instead, in the latter configuration, the output of all the sections around f_0 is not only low, but it gets lower and lower as we move away from the input, because of the attenuation in each section; therefore, only the first and perhaps the second sections (thus few elements) count in the final cancellation (of *small* terms). In the passband, the SCF-Σ(HP) configuration is less sensitive than the SCF-Σ(LP) configuration because the output of each section is lower in the former than in the latter. In the SCF-Σ(HP) circuit, the high output (low loss) is provided by the direct path from input to output, through the output adder, i.e., few elements. In the SCF-Σ(LP) circuit, the high output is provided by the contribution of the outputs of all the sections, i.e., many elements.

F. Mixed Configurations

An improved configuration for the same band-elimination filter of Fig. 1, as far as sensitivity in the rejection band is concerned, was found to be a cascade arrangement of two 2-section SCF configurations (Fig. 8(a)). The poles and zeros of each biquartic are shown in Fig. 8(b). (In the same figure, the $p-z$ pairings for C1 are also shown.) Both biquartics were designed from LP transfer function prototypes.

The sensitivity of this realization, as illustrated in Fig. 9, is lower than that of the cascade for the entire rejection band.[8] However, in the transition bands and the edges of the passband, the sensitivity is sharply increased. This high sensitivity in the transition bands was found to be caused by the 2-section SCF realizing the transmission zeros closest to the transition bands. If these two sections are uncoupled, the sensitivity peaks in the transition

[7]SCF-Σ(HP) means a SCF-Σ configuration [see Fig. 6(a)] designed from a HP prototype. Similarly, SCF-Σ(LP) means that a LP prototype has been used.

[8]No significant difference is observed between realizations using either a SCF-Σ or a SCF-FF configuration for each biquartic.

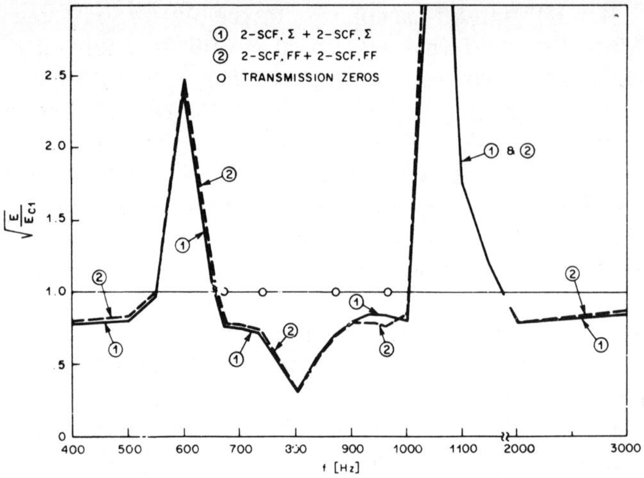

Fig. 9 Sensitivity comparison between (2-SCF)+(2-SCF) realizations (LP prototype) and cascade.

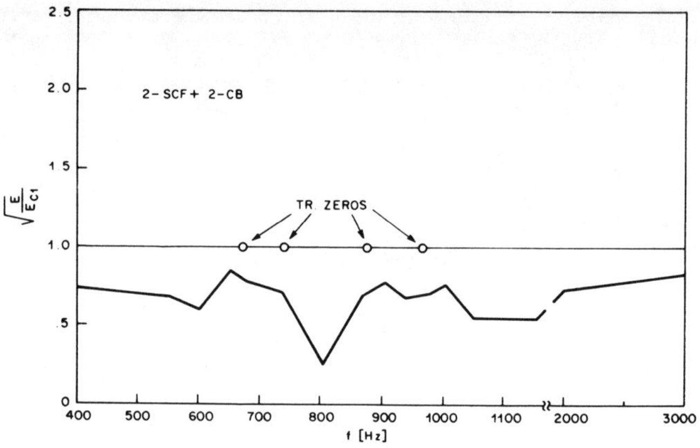

Fig. 11 Sensitivity comparison between mixed coupled-cascade realizationandand cascade.

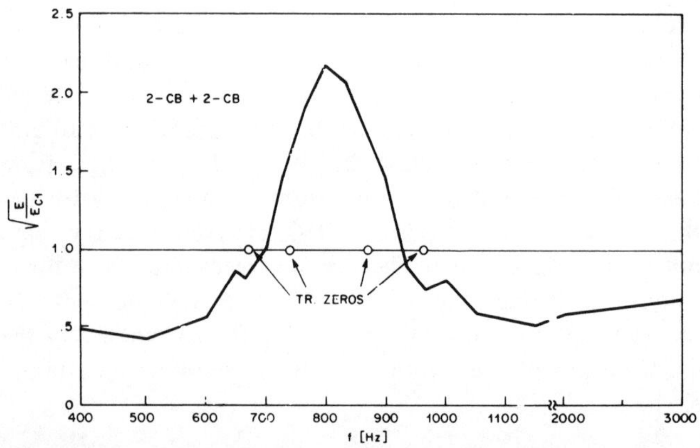

Fig. 10 Sensitivity comparison between (2-CB)+(2-CB) realization (HP prototype) and cascade.

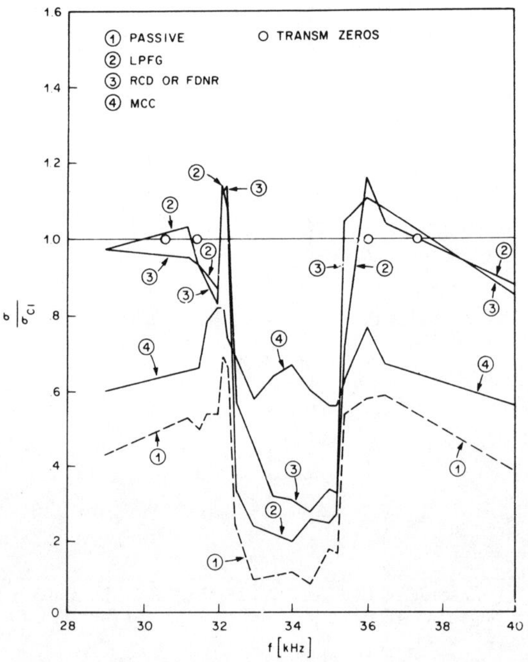

Fig. 12 Sensitivity comparisons between different realizations of 12th-order channel bank filter.

bands are eliminated, without practically affecting the stopband sensitivity, and $\sqrt{E/E_{C1}} \approx 1$ in the transition bands and passbands.

Similar results are obtained with a cascade of two 2-section CB configurations, provided each biquartic is also designed from a LP prototype, via a LP to BR transformation. On the other hand, a cascade of two 2-section CB or SCF configurations in which each biquartic is designed from a HP prototype, using a LP to BP transformation, behaves in just the opposite way, as shown by Fig. 10. The sensitivity is low in the passbands and transition bands, but high in the stopband.

The previous results suggested a new cascade arrangement of a 2-section SCF, designed from an LP transfer function prototype, and a 2-section CB configuration, designed from a HP transfer function prototype. The 2-SCF biquartic realizes the pair of transmission zeros in the center of the rejection band, combined with the low-Q poles, and the 2-CB biquartic realizes the pair of transmission zeros closest to the passbands, combined with the high-Q poles. The plot of Fig. 11 shows that *the sensitivity of this mixed coupled-cascade configuration is lower than that of the 4-section cascade in the rejection band, in the passbands, and in the transition bands*; this is not achieved with a fully coupled structure. However, the sensitivity improvement at all frequencies is achieved by sacrificing some of the sensitivity improvement (over the cascade) in the passband.[9]

[9] The previous comments are also applicable to other symmetrical even-degree band-elimination filters. For unsymmetrical filters, each shifted-companion-form biquartic would have to be realized by using four first-order blocks. More attractive is the idea of redesigning each biquartic so that it can be obtained from a biquadratic (second-order) prototype; then the biquartic can be realized by coupling two biquadratic blocks. For filters with degree n odd or $n/2$ odd, the unpaired first- or second-order section(s) could remain in cascade; other alternatives are possible.

The above conclusions can be extended to other types of filters, e.g., a bandpass filter with critical transition and/or stopbands, when improved sensitivity is required at all frequencies. This is illustrated by Fig. 12, which refers to a 12th-order channel bank filter. (The channel is from 32 to 36 kHz.) The figure shows the standard deviation of the gain for several active realizations of the filter, as obtained from Monte Carlo analyses. The passive realization has also been added as an additional element in the comparison.[10] The mixed coupled-cascade (MCC) configuration, unlike the other active realizations, offers an improved sensitivity over the cascade *at all frequencies*. The figure also shows that among the active configurations the fully coupled ones (like the LPFG or the FDNR [19], [20]) should be preferred if only the passband sensitivity is important.

III. Additional Comments

The experimental data show that coupling is desirable to reduce the passband sensitivity; all fully coupled realizations are less sensitive than a semicoupled one (MCC), which in turn is less sensitive than the uncoupled one (cascade). The essential difference between the coupled, semicoupled, and uncoupled (cascade) realizations is the amount of coupling of the poles. They are fully coupled in the former, partially coupled in the MCC, and uncoupled in the cascade. Little difference is observed between the LPFG and the CB realizations even though the former has uncoupled zeros. On the other hand, significant difference is noted between the LPFG and the cascade, even though both realize the zeros independently (one complex conjugate pair per section). Thus, the common feature of all the realizations with low sensitivity in the passband is their coupling of the poles. This is also true for the passive realization.[11]

[10]As expected [11], the sensitivity of the passive realization is the lowest in the passband. In Fig. 12, this holds true in the other frequency bands as well. For the band-elimination filter of Fig. 1, the LRC realization was also obtained by adding an amplifier to adjust the gain. But the sensitivity near the center f_0 of the rejection band was found to be high, leading to the conjecture that the LRC realization, especially if not tuned, may not always be the optimum in the rejection band. The high sensitivity obtained around f_0 for the filter of Fig. 1 is due, at least in part, to the following. The network was obtained from an LP ladder prototype (coupled inductors were avoided) via a frequency transformation. One of the series inductors and the associated parallel capacitor are negative. These elements can be realized actively, but they make it clear that at f_0 a cancellation of terms occurs to give the required finite loss. No such cancellation exists in the passive realization of the 12th-order bandpass filter whose sensitivity is shown in Fig. 12.

If an FDNR (or RCD) realization is desired, a proper LRC realization must be obtained first. In some cases this step may not be straightforward. In addition, it has to be remembered that the FDNR realization is more sensitive than the LRC because several components are required to realize each D element, especially if this is not grounded. (In the examples mentioned above the LC circuits in the shunt branches were transformed so that all their capacitors, and therefore the corresponding FDNRs, are grounded.) Fig. 12 shows that the FDNR realization, while having a low sensitivity in the passband, offers no significant advantage over the cascade in the other bands.

[11]This condition is not sufficient, however. For example, in the passive realization, maximum (or nearly maximum) transfer of power in the passband is also required.

For low sensitivity in the stopband, a full coupling seems to be not so advantageous for filters of order $n>4$, at least for the topologies considered here. In fact, the highest stopband sensitivities were obtained with a CB structure, in which both poles and zeros (all of them) are coupled. And a lower sensitivity could be achieved by uncoupling the zeros (LPFG and RCD realizations) or uncoupling both poles and zeros (cascade realization). But there is more to it. It is also true that the number of elements involved in the formation of each zero is different in the different realizations, being higher for the CB structure, and this affects the sensitivity. By eliminating part of the couplings in the CB structure, so that isolated 2-CB biquartic sections are obtained, the number of elements affecting each zero is reduced. At the same time the transmission zeros are produced by additional paths around the blocks. Then, in the vicinity of a transmission zero of the overall transfer function, the effect of a change in a section parameter is smaller than if the section itself created that zero, as it occurs in the LPFG or the cascade configurations. This causes a further reduction in sensitivity that goes beyond what can be obtained by fully uncoupling the zeros as in the LPFG or cascade realizations. But there is a limitation. The high loss around the transmission zeros is obtained by a cancellation of terms, and, if large terms are subtracted, the sensitivity will increase again, thus offsetting the reduction achieved. The removed couplings also partially uncouple the poles, and the sensitivity in the passband, while still better than that of the cascade, is higher than that of the fully coupled structure.

IV. Summary

The cascade arrangement of biquadratic sections is a common way of realizing RC-active filters. The addition of coupling between the sections is known to improve the sensitivity in the passband of the filter. Similar improvement can also be obtained by other passive-simulating realizations. Neither improves the sensitivity in the stopband significantly.

Here, an empirical search was carried out to find low-sensitivity configurations for band-elimination filters. Working with an 8th-order BEF as an example, it was found that a mixed coupled-cascade configuration, i.e., a cascade of properly chosen coupled sections, has a reduced sensitivity, with respect to the cascade, at all frequencies. This arrangement represents a compromise that trades part of the sensitivity improvement in the passband by a sensitivity improvement in the other bands.

Similar results were achieved using a mixed topology in the realization of a 12th-order bandpass filter with critical transition bands and stopbands.

An attempt has been made to provide some insight into the sensitivity characteristics of different realizations. Further work is required to put these empirical results and heuristic explanations on a firmer theoretical basis and to determine their limitations.

Appendix I
An Eight-Order Band-Elimination Filter—(Fig. 1)

The transfer function is

$$T(s)=3.111613297\frac{(s^2+36533734.47)}{(s^2+431.4843219s+42498291.47)}$$
$$\times\frac{(s^2+17825823.35)}{(s^2+259.0987887s+15324001.85)}$$
$$\times\frac{(s^2+30153400.12)}{(s^2+3127.921563s+42586732.29)}$$
$$\times\frac{(s^2+21597693.60)}{(s^2+1874.361562s+15292178.15)}.$$

The transmission zeros are at 672, 740, 874, and 962 Hz.

For the cascade configuration, the following four pole–zero arrangements (C_1–C_4) were examined:

			C1	C2	C3	C4
Section No.	Q_p	ω_p	ω_z	ω_z	ω_z	ω_z
1	15.11	6519	6044	5491	4222	4647
2	15.11	3914	4222	4647	5491	6044
3	2.09	6526	5491	6044	4647	4222
4	2.09	3910	4647	4222	6044	5491

where the notation

$$K\frac{s^2+\omega_z^2}{s^2+\dfrac{\omega_p}{Q_p}s+\omega_p^2}$$

has been used for each biquadratic section, and ω_p, ω_z are in radians per second. Considering the sensitivities obtained in both passband and stopband, the C1 arrangement is to be preferred.

Appendix II
Minimizing Overall Sensitivity by Using Least Sensitive Sections in Coupled Configurations

In a cascade realization, each biquadratic section or block is independent, and an improvement in the overall gain sensitivity can be achieved by using the least sensitive topology that is available for each block. In a coupled structure, like the leapfrog realization [26] (Fig. 3), the different sections are no longer independent due to the feedback, and a worse overall gain sensitivity *may* be obtained when sections with lower individual gain sensitivity are used, since the phase sensitivity is also important. This is justified below.

Assuming four biquadratic sections and denoting by T_i the transfer function of section i, the overall transfer function T for the cascade realization is

$$T=T_1T_2T_3T_4.$$

Let the component x_i of section 4 change. Its effect on $|T|$ is given by

$$S_{x_{i4}}^{|T|}=S_{x_{i4}}^{|T_4|}.$$

Therefore, if $S_{x_{i4}}^{|T_4|}$ is reduced, so is $S_{x_{i4}}^{|T|}$.

The overall transfer function for the leapfrog configuration is

$$T=\frac{\hat{T}_1\hat{T}_2\hat{T}_3\hat{T}_4}{1+k_1\hat{T}_1\hat{T}_2+k_2\hat{T}_2\hat{T}_3+k_3\hat{T}_3\hat{T}_4+k_1k_3\hat{T}_1\hat{T}_2\hat{T}_3\hat{T}_4}\triangleq\frac{\Pi\hat{T}_i}{D}.$$

The sensitivity $S_{x_{i4}}^{|T|}$ is obtained from

$$S_{x_{i4}}^{|T|}=aS_{x_{i4}}^{|\hat{T}_4|}-b\nu S_{x_{i4}}^{\nu}$$

where the notation

$$a+jb\triangleq\frac{1+k_1\hat{T}_1\hat{T}_2+k_2\hat{T}_2\hat{T}_3}{D}$$

$$\hat{T}_4=|\hat{T}_4|e^{j\nu}$$

has been used. Clearly, *the sensitivity of* $|T|$ *depends on both the gain and the phase sensitivities of* $\hat{T}_4$. (Also note the dependence on ν.) Therefore, the replacement of one type of section $\hat{T}_4$ by another with reduced $S_{x_{i4}}^{|\hat{T}_4|}$ does not guarantee a reduction in $S_{x_{i4}}^{|T|}$. Similar derivations could be obtained for other sections. Several examples confirmed the above conclusions.

In this work, the same type of section was used to realize the building blocks of the cascade and the coupled configurations.

Acknowledgment

The author is deeply grateful to his colleagues at Bell Laboratories, especially R. C. Drechsler and J. Tow, for their support and many illuminating discussions. The comments and suggestions of Prof. H. J. Orchard and the anonymous reviewers are also greatly appreciated.

References

[1] D. A. Calahan, "Sensitivity minimization in active RC synthesis," *IRE Trans. on Circuit Theory*, vol. CT-9, pp. 38–42, Mar. 1962.

[2] J. D. Schoeffler, "The synthesis of minimum sensitivity networks," *IEEE Trans. on Circuit Theory*, vol. CT-11, pp. 271–276, June 1964.

[3] D. A. Calahan, "Computer generation of equivalent networks," in *IEEE International Convention Record*, Pt. 1, pp. 330–337, 1964.

[4] M. L. Blostein, "Generation of minimum sensitivity networks," *IEEE Trans. on Circuit Theory*, vol. CT-14, pp. 87–88, Mar. 1967.

[5] C. Schmidt and R. Kasper, "On minimum sensitivity networks,"

IEEE Trans. on Circuit Theory, vol. CT-14, pp. 438–440, Dec. 1967.
[6] E. S. Kuh and C. G. Lau, "Sensitivity invariants of continuously equivalent networks," *IEEE Trans. on Circuit Theory*, vol. CT-15, no. 3, pp. 175–177, Sept. 1968.
[7] B. J. Leon and C. F. Yokomoto, "Generation of a class of equivalent networks and its sensitivities," *IEEE Trans. on Circuit Theory*, vol. CT-19, no. 1, pp. 2–8, Jan. 1972.
[8] M. N. S. Swamy, C. Bhushan, and K. Thulasiraman, "Sensitivity invariants for linear time-invariant networks," *IEEE Trans. on Circuit Theory*, vol. CT-20, no. 1, pp. 21–24, Jan. 1973.
[9] B. M. G. Cheetham, "A new theory of continuously equivalent networks," *IEEE Trans. on Circuits and Systems*, vol. CAS-21, no. 1, pp. 17–20, Jan. 1974.
[10] J. Neirynck and P. van Bastelaer, "Sensitivity of equivalent networks," *Electronics Letters*, vol. 10, no. 22, pp. 452–453, Oct. 31, 1974.
[11] H. J. Orchard, "Inductorless filters," *Electronics Letters*, vol. 2, pp. 224–225, June 1966.
[12] R. H. S. Riordan, "Simulated inductors using differential amplifiers," *Electronics Letters*, vol. 3, pp. 50–51, Feb. 1967.
[13] H. J. Orchard and D. F. Sheahan, "Inductorless bandpass filters," *IEEE J. of Solid-State Circuits*, vol. SC-5, no. 3, pp. 108–118, June 1970.
[14] J. Gorski-Popiel, "RC-active synthesis using positive-immittance converters," *Electronics Letters*, vol. 3, pp. 381–382, Aug. 1967.
[15] A. Antoniou, "Realization of gyrators using operational amplifiers and their use in RC-active network synthesis," *Proc. Inst. Elec. Eng. (IEE)*, vol. 116, pp. 1838–1850, Nov. 1969.
[16] L. T. Bruton, "Filter synthesis using generalized impedance converters," NATO Institute of Advanced Signal and Network Theory, Bournemouth, England, pp. 153–159, Sept. 1972.
[17] H. J. Orchard and A. N. Willson, Jr., "New active-gyrator circuit," *Electronics Letters*, vol. 10, no. 13, pp. 261–262, June 27, 1974.
[18] T. Takahashi, S. Hamada, R. Watanabe, and T. Miyata, "A simple method for producing floating semiconductor inductors," *IEEE J. of Solid-State Circuits*, vol. SC-10, no. 1, pp. 75–77, Feb. 1975.
[19] L. T. Bruton, "Network transfer functions using the concept of frequency-dependent negative resistance," *IEEE Trans. on Circuit Theory*, vol. CT-16, no. 3, pp. 406–408, Aug. 1969.
[20] L. T. Bruton and D. Treleaven, "Active filter design using generalized impedance converters," *EDN*, pp. 68–75, Feb. 5, 1973.
[21] F. Molo, "Parallel resonator with a resistance and a frequency-dependent negative resistance realized with a single operational amplifier," *IEEE Trans. on Circuits and Systems*, vol. CAS-21, no. 6, pp. 783–788, Nov. 1974.
[22] C. E. Schmidt and M. S. Lee, "Multipurpose simulation network with a single amplifier," *Electronics Letters*, vol. 11, no. 1, pp. 9–10, Jan. 9, 1975.
[23] F. E. J. Girling and E. F. Good, "Active filters—Part 12: The leapfrog or active-ladder synthesis," *Wireless World*, vol. 76, pp. 341–345, July 1970.
[24] R. L. Adams, "On reduced sensitivity active filters," in *Proc. 14th Midwest Symp. on Circuit Theory*, pp. 14.3-1–14.3-8.
[25] J. Tow and Y. L. Kuo, "Coupled biquad active filters," in *Proc. Int. Symp. on Circuit Theory*, pp. 164–168.
[26] G. Szentirmai, "Synthesis of multiple-feedback active filters," *Bell System Technical Journal*, vol. 52, pp. 527–555, Apr. 1973.
[27] G. Hurtig, III, "The primary resonator block technique of filter synthesis," *International Filter Symposium*, p. 84, and *Feedback*, vol. 1, no. 3, Dec. 1973. Also, U.S. Patent 3,720,881, Mar. 13, 1973.
[28] K. R. Laker and M. S. Ghausi, "A low sensitivity multiloop feedback active *RC* filter," in *Proc. Int. Symp. on Circuit Theory*, pp. 126–129.
[29] K. R. Laker, M. S. Ghausi, and J. J. Kelly, "Minimum sensitivity leapfrog active *RC* filters," in *Proc. Int. Symp. on Circuits and Systems*, pp. 201–204.
[30] K. R. Laker and M. S. Ghausi, "Synthesis of a low-sensitivity multiloop feedback active *RC* filter," *IEEE Trans. on Circuits and Systems*, vol. CAS-21, no. 2, pp. 252–259, Mar. 1974.
[31] J. Tow, "Design and evaluation of shifted companion form (follow the leader feedback) active filters," in *Proc. Int. Symp. on Circuits and Systems*, pp. 656–660, and *Bell System Technical Journal*, vol. 54, no. 3, pp. 545–568, Mar. 1975.
[32] J. Gorski-Popiel, "An active synthesis procedure using multiple feedback paths," in *Proc. 7th Asilomar Conference on Circuits, Systems, and Computers*, pp. 85–90.
[33] H. W. Hanneman and H. N. Linssen, "A relation between sensitivity and variance of network functions," *IEEE Trans. on Circuit Theory*, vol. CT-19, no. 5, pp. 499–502, Sept. 1972.
[34] L. T. Bruton, "Sensitivity comparison of high-Q second-order active filter synthesis techniques," *IEEE Trans. on Circuits and Systems*, vol. CAS-22, no. 1, pp. 32–38, Jan. 1975.
[35] T. E. Idleman, F. S. Jenkins, W. J. McCalla, and D. O. Pederson, "SLIC—A simulator for linear integrated circuits," *IEEE J. of Solid-State Circuits*, vol. SC-6, no. 4, pp. 188–203, Aug. 1971.

Paper 3-8

A Compensation Technique for a Gyrator and Its Use in the Design of a Channel-Bank Filter

ANDREAS ANTONIOU, MEMBER, IEEE, AND KOLLORU SRINIVASULU NAIDU

***Abstract*—A model proposed recently for an operational amplifier gyrator circuit is used to develop a compensation technique whereby the useful bandwidth of the gyrator circuit can be increased. The compensation technique is verified first by computer-aided analysis and then by experiment. It is shown that a significant increase in the useful gyrator bandwidth can be achieved by using an inexpensive capacitor. The compensation technique is then applied in the design of a gyrator-C channel-bank filter. The influence of dc-supply voltage and ambient-temperature variations on the performance of the filter is examined.**

I. Introduction

THE GYRATOR circuit of Fig. 1(a) [1] gives stable inductances and high Q-factors [2] and also, it is easy to fabricate and adjust [3]. Furthermore, it has proven a versatile circuit in that it can be used as a generalized-immittance converter [2], [4]–[6] or as a frequency-dependent negative resistance [7]–[9]. For these reasons, the circuit of Fig. 1(a) has been used for the design of a digitone detector which is currently being manufactured [3]. Also, it is presently considered as a promising circuit for the design of channel-bank filters [10].

Gyrator circuits such as that of Fig. 1(a) can be readily analyzed by using one of the many available circuit analysis programs. However, it is often difficult to derive general conclusions about a circuit or to optimize a particular circuit on the basis of a number of computer runs. To overcome this problem, gyrators and other devices are often represented by suitable simple circuit models.

A circuit model for the gyrator circuit of Fig. 1(a), proposed recently [11], illustrates the influence of amplifier imperfections on the performance of the gyrator circuit. It shows that the finite gain-bandwidth product of the amplifiers gives rise to a parasitic inductance and a parasitic capacitance. The parasitic inductance causes no problems

Manuscript received October 1, 1973; revised June 3, 1974.
The authors are with the Department of Electrical Engineering, Concordia University, Montreal, Quebec, H3G 1M8.

Reprinted from *IEEE Trans. Circuits Syst.*, vol. CAS-22, pp. 316-323, Apr. 1975.

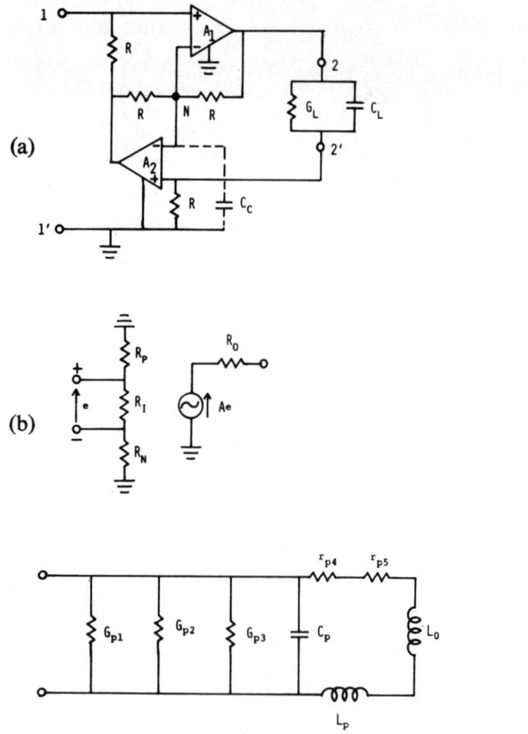

Fig. 1. (a) Gyrator circuit terminated with a lossy capacitor. (b) Model for operational amplifier. (c) Model for capacitively terminated gyrator circuit.

since this can be used as part of the required inductance. The parasitic capacitance, however, tends to limit the useful bandwidth of the gyrator circuit. This paper describes a compensation technique whereby the parasitic capacitance can be eliminated so that the useful bandwidth of the gyrator circuit can be increased. Trimmel and Heinlein have also considered the problem of gyrator broadbanding [12]. However, their technique is applicable to gyrator circuits constructed with floating transconductance amplifiers. Their emphasis is to maximize the bandwidth over which the Q-factor remains high. Our emphasis, on the other hand, is to maximize the bandwidth over which the simulated inductance remains constant.

In order to assess the feasibility of the compensation technique, this was applied in the design of a fairly difficult channel-bank filter which is required to operate accurately over the frequency range 0–24 kHz.

II. Modeling Considerations

By representing each operational amplifier in the gyrator circuit of Fig. 1(a) by the model of Fig. 1(b), where

$$A = \frac{A_0\omega_0}{s + \omega_0} \tag{1}$$

and assuming that

$$|A| \gg 1,\ R_I \gg R \quad \text{and} \quad R_P,\ R_N \gg R_I,\ R \gg R_0 \tag{2}$$

an RLC model for the gyrator circuit has been derived [11] which is valid for the frequency range

$$0 \le \omega \ll A_0\omega_0.$$

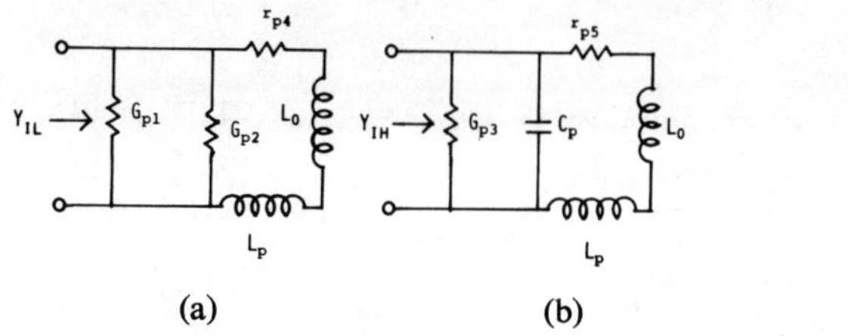

Fig. 2. (a) Simplified low-frequency gyrator model. (b) Simplified high-frequency gyrator model.

This model is illustrated in Fig. 1(c), where

$$G_{p1} = \frac{1}{R_{P1}} + \frac{1}{A_0R}\left(2 + \frac{R_0}{R}\right) + \frac{R}{R_I(2R + 3R_0 + A_0\omega_0R^2C_L)} \tag{3}$$

$$G_{p2} = -\left\{\frac{1}{R_{N1}} + \frac{1}{R_{N2}} + \frac{2R_0}{R(2R + 3R_0 + A_0\omega_0R^2C_L)}\right\} \tag{4}$$

$$G_{p3} = \frac{2\omega^2}{A_0{}^2\omega_0{}^2R}\left[1 + \frac{4R_0}{R} + \frac{1}{(2 + A_0\omega_0RC_L)}\left\{-\frac{R_0}{R} + \frac{R}{R_I}\right\}\right] \tag{5}$$

$$C_p = \frac{1}{A_0\omega_0R}\left(2 + \frac{R_0}{R}\right) \tag{6}$$

$$r_{p4} = \frac{1}{A_0}(2R + 3R_0) + G_LR^2 \tag{7}$$

$$r_{p5} = \frac{2\omega^2}{A_0{}^2\omega_0{}^2}(R + 6R_0) \tag{8}$$

$$L_p = \frac{1}{A_0\omega_0}(2R + 3R_0) \tag{9}$$

$$L_o = R^2C_L. \tag{10}$$

The preceding parameters with the exception of L_o reduce to zero if the amplifiers and the load capacitor are assumed ideal. Therefore, these represent parasitic elements introduced by the imperfections of the amplifiers. The parameter L_o represents the nominal inductance.

The relative magnitudes of the various parasitic impedances tend to change with frequency. At low frequencies, i.e., less than 100 Hz,

$$r_{p4} \gg r_{p5} \qquad G_{p1} \text{ or } |G_{p2}| \gg G_{p3} \qquad \omega C_p \ll \frac{1}{\omega(L_p + L_o)} \tag{11}$$

and hence the gyrator model can be simplified as shown in Fig. 2(a). The low-frequency input admittance can be obtained as

$$Y_{IL} = G_{p1} + G_{p2} + \frac{1}{r_{p4} + j\omega(L_o + L_p)}$$

$$= G_{EL} + \frac{1}{j\omega L_{EL}} \tag{12}$$

where

$$G_{EL} = G_{p1} + G_{p2} + \frac{r_{p4}}{r_{p4}^2 + \omega^2(L_o + L_p)^2} \tag{13}$$

and

$$L_{EL} = \frac{r_{p4}^2 + \omega^2(L_o + L_p)^2}{\omega^2(L_o + L_p)} \tag{14}$$

are the effective input conductance and inductance, respectively. Even for very low frequencies,

$$r_{p4} \ll \omega(L_o + L_p) \tag{15}$$

according to (7) and hence from (14),

$$L_{EL} \simeq L_o + L_p. \tag{16}$$

Equation (9) shows that the parasitic inductance L_p is inversely proportional to the gain-bandwidth product of the amplifiers. Therefore, its tolerance and temperature coefficient can be large. In practice, however, R and C_L can be chosen to give $L_p \ll L_o$, so that the tolerance and temperature coefficient of L_{EL} can be kept small. Consequently, L_p can be used as part of the required inductance. The low-frequency Q-factor is given by

$$Q_{\text{LF}} = \frac{-ImY_{IL}}{ReY_{IL}} = \frac{1}{\omega L_{EL} G_{EL}} \tag{17}$$

and from (13), (14), and (17),

$$Q_{\text{LF}} = \frac{\omega(L_o + L_p)}{(G_{p1} + G_{p2})[r_{p4}^2 + \omega^2(L_o + L_p)^2] + r_{p4}}. \tag{18}$$

Since $G_{p1} > 0$ and $G_{p2} < 0$, Q_{LF} may be positive or negative, depending on the relative magnitudes of the amplifier parameters. In either case, $|Q_{\text{LF}}|$ is large since G_{p1}, G_{p2} and r_{p4} are very small quantities. The conductance G_{p1} can be increased by using a shunt resistance across the input of the gyrator circuit. Similarly, (4) shows that $-G_{p2}$ can be increased by using a shunt resistance between node N and ground in Fig. 1(a) and hence Q_{LF} can be easily increased or reduced as necessary.

For high frequencies, say in excess of 10 kHz,

$$r_{p5} \gg r_{p4}$$

$$G_{p3} \gg G_{p1} \text{ or } |G_{p2}| \tag{19}$$

and hence the model of Fig. 1(c) can now be simplified as shown in Fig. 2(b). If the required inductance appears in parallel with a filter capacitance C_f, the parasitic capacitance C_p can be used as part of C_f. Otherwise, C_p is an undesirable capacitance. The high-frequency input admittance can be obtained from Fig. 2(b) as

$$Y_{IH} = G_{p3} + j\omega C_p + \frac{1}{r_{p5} + j\omega(L_o + L_p)}$$

$$= G_{EH} + \frac{1}{j\omega L_{EH}} \tag{20}$$

where the effective input conductance and inductance are now given by

$$G_{EH} = G_{p3} + \frac{r_{p5}}{r_{p5}^2 + \omega^2(L_o + L_p)^2} \tag{21}$$

and

$$L_{EH} = \frac{(L_o + L_p)^2 + \dfrac{r_{p5}^2}{\omega^2}}{[(L_o + L_p) - C_p\{r_{p5}^2 + \omega^2(L_o + L_p)^2\}]}. \tag{22}$$

Clearly, C_p tends to increase the effective inductance by an amount which increases with frequency. Since the simulated inductance is required to be constant with frequency, C_p imposes a limit on the useful bandwidth of the gyrator circuit. The high-frequency Q-factor is obtained from (21) and (22) as

$$Q_{\text{HF}} = \frac{-ImY_{IH}}{ReY_{IH}} = \frac{1}{\omega L_{EH} G_{EH}}$$

$$= \frac{\omega[(L_o + L_p) - C_p\{r_{p5}^2 + \omega^2(L_o + L_p)^2\}]}{G_{p3}\{r_{p5}^2 + \omega^2(L_o + L_p)^2\} + r_{p5}}$$

$$\simeq \frac{\omega(L_o + L_p)}{G_{p3}\{r_{p5}^2 + \omega^2(L_o + L_p)^2\} + r_{p5}} \tag{23}$$

Since G_{p3} and r_{p5} are positive, Q_{HF} is also positive.

III. Compensation Technique

The useful bandwidth of the gyrator circuit of Fig. 1(a) can be extended if by some means the parasitic capacitance C_p can be eliminated. Equation (4) shows that R_{N1}, which can be considered as an external resistance between node N and ground, gives rise to a negative resistance $-R_{N1}$ across the input terminals of the gyrator circuit. It is expected, therefore, that a positive capacitance C_c between node N and ground will give rise to a negative capacitance in parallel with the positive parasitic capacitance C_p so that C_p may be eliminated. This possibility is now examined.

The use of a compensating capacitance between node N and ground in Fig. 1(a) will modify the model of Fig. 1(c). The changes introduced can be deduced by repeating the analysis reported in [11]. For the sake of brevity, however, only the changes will be indicated. The input resistances R_{N1}, R_{N2}, and R_{P1} have been assumed to be infinite since these are very large in practice, of the order of 500 MΩ. The parameter y_{11} for the gyrator circuit as given by (14) of [11] is now obtained as

$$y_{11} = G_a + G_b + \frac{s}{A_0\omega_0 R}\left(2 + \frac{R_0}{R} - A_0\omega_0 R C_c\right) \tag{24}$$

where

$$G_a = \frac{1}{A_0 R}\left(2 + \frac{R_0}{R}\right) \tag{25}$$

$$G_b = \frac{\omega^2}{A_0\omega_0}\left\{\frac{2}{A_0\omega_0 R} + \frac{8R_0}{A_0\omega_0 R^2} - C_c\left(2 + \frac{3R_0}{R}\right)\right\} \tag{26}$$

The parameters y_{12}, y_{21} and y_{22}, however, remain the same as for the uncompensated gyrator circuit. For a lossy load capacitor in which

$$Y_L = G_L + sC_L \tag{27}$$

the input admittance of the gyrator circuit is given by

$$Y_I = y_{11} - \frac{y_{12}y_{21}}{y_{22} + G_L + sC_L}. \tag{28}$$

By using (2), (24)–(28) and also (15)–(17) of [11], the input admittance can be expressed as

$$Y_I = G_{p1} + G_{p2} + G_{p3} + sC_p + \frac{1}{r_{p4} + r_{p5} + s(L_o + L_p)} \tag{29}$$

where G_{p1}, G_{p2}, r_{p4}, r_{p5}, L_p, and L_o are still given by (3)–(4) and (7)–(10). Now

$$G_{p3} = \frac{\omega^2}{A_0{}^2\omega_0{}^2R}\left\{2 + \frac{8R_0}{R} + \frac{2R}{R_I(2 + A_0\omega_0RC_L)} - \frac{2R_0}{R(2 + A_0\omega_0RC_L)} - A_0\omega_0RC_c\left(2 + \frac{3R_0}{R}\right)\right\} \tag{30}$$

and

$$C_p = \frac{1}{A_0\omega_0R}\left(2 + \frac{R_0}{R} - A_0\omega_0RC_c\right). \tag{31}$$

The new model for the gyrator circuit is still given by Fig. 1(c) except that the values for G_{p3} and C_p are now modified by C_c. As anticipated earlier, C_c reduces C_p, and if

$$C_c = \frac{2R + R_0}{A_0\omega_0R^2}. \tag{32}$$

$C_p = 0$, that is, the parasitic capacitance is completely eliminated. The compensating capacitance C_c has another effect on the performance of the gyrator circuit. It introduces a negative term in G_{p3} which represents Q-factor enhancement, according to (23). For a sufficiently large value of C_c, G_{p3} can actually become negative, and then Q_{HF} can also become negative. Negative Q-factors are sometimes undesirable, since they tend to cause peaking in the filter response.

The modified model has been used for the analysis of a compensated gyrator circuit. An exact computer-aided analysis has also been carried out in order to test a) the validity of the modified model; and b) the validity of the compensation technique. The percentage inductance deviation is defined as

$$\text{ID} = \frac{100(L_E - L_o)}{L_o} \tag{33}$$

where L_E is the effective inductance and $L_o = R^2C_L$ is the nominal inductance. The Q-factor and inductance deviation have been computed for a range of frequencies.

TABLE I

Parameter	ML748C	μA741	Units
A_0	150	200	k
ω_0	18.8	37.6	rad/s
R_0	75	75	Ω
R_I	2	2	MΩ
R_P	500	500	MΩ
R_N	500	500	MΩ

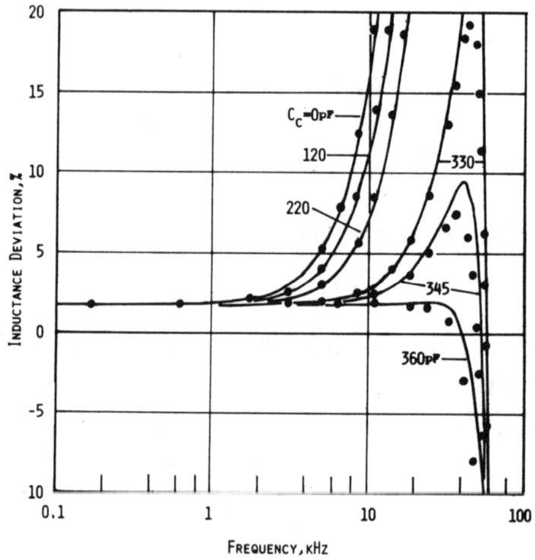

(a)

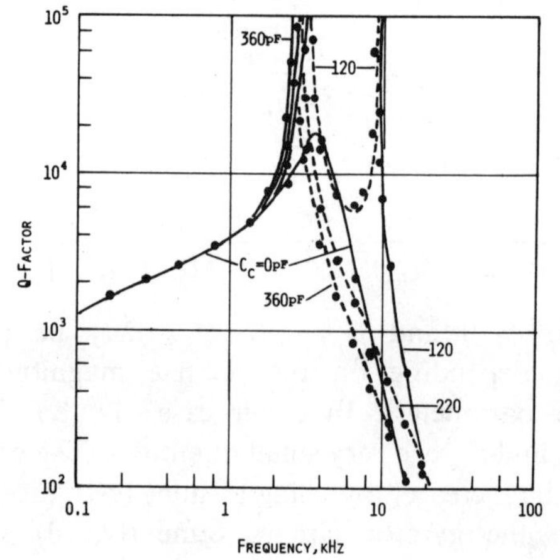

(b)

Fig. 3. Computed frequency response of gyrator circuit. (a) Percentage inductance deviation. (b) Q-factor. Exact analysis ———; model analysis • • • • •. (Dotted lines indicate negative Q-factor.)

The specifications of amplifier type ML748C (Microsystems International Ltd.) given in Table I were used for the analysis. The gyration resistance R and load capacitance C_L were assumed to be 2 kΩ and 22 nF, respectively. The Q-factor for the load capacitor was assumed constant and equal to 3000. The results for various values of C_c in the range 0–360 pF are illustrated in Figs. 3(a) and (b) where close agreement is evident between model-analysis and exact-analysis curves. The validity of the modified model is thus confirmed.

Fig. 3(a) illustrates the influence of C_p on the effective inductance. As may be expected from (22), the high-frequency inductance deviation tends to increase rapidly with frequency and thus the useful bandwidth of the gyrator

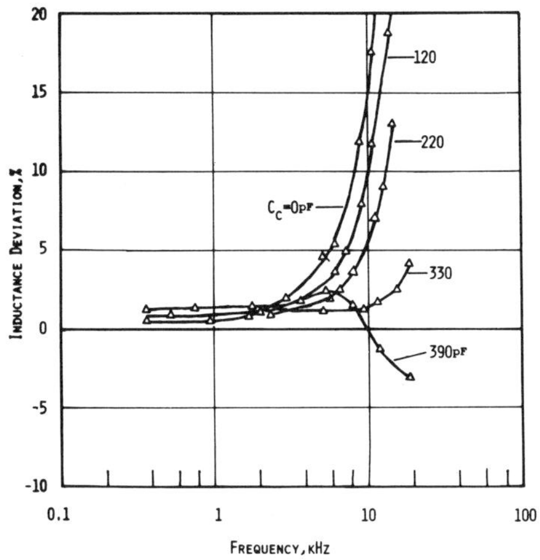

(a)

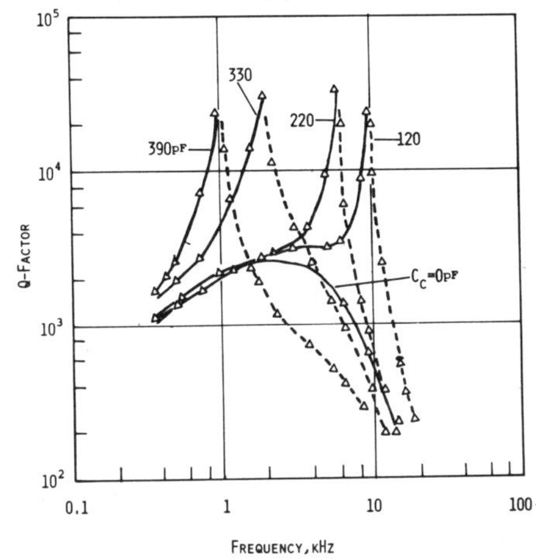

(b)

Fig. 4. Measured frequency response of gyrator circuit. (a) Percentage inductance deviation. (b) Q-factor. (Dotted lines indicate negative Q-factor.)

circuit is limited. As C_c is increased, C_p is reduced according to (31) and, therefore, the high-frequency inductance deviation is reduced. As a consequence, the useful gyrator bandwidth is increased. Fig. 3(b) shows that the low-frequency Q-factor is independent of C_p, as can be expected from (18). The high-frequency Q-factor is positive for small values of C_c, but as C_c is increased, Q-factor enhancement is introduced and eventually the Q-factor becomes negative, as anticipated earlier.

In order to test the feasibility of the proposed compensation technique experimentally, a gyrator circuit was built using ML748C-type amplifiers. The gyration resistance and load capacitance used were 2 kΩ and 22 nF (L_o = 88 mH), respectively, and the external amplifier compensating capacitance was 30 pF. The effective inductance and Q-factor were measured, maintaining a constant voltage of 0.3 V rms across the inductance and using a dc-supply voltage of ±15 V. The measured responses are shown in Figs. 4(a) and 4(b). Some differences are evident between experimental and theoretical responses which are mainly due to the large tolerances on the amplifier parameters. In general, however, the predicted pattern is maintained.

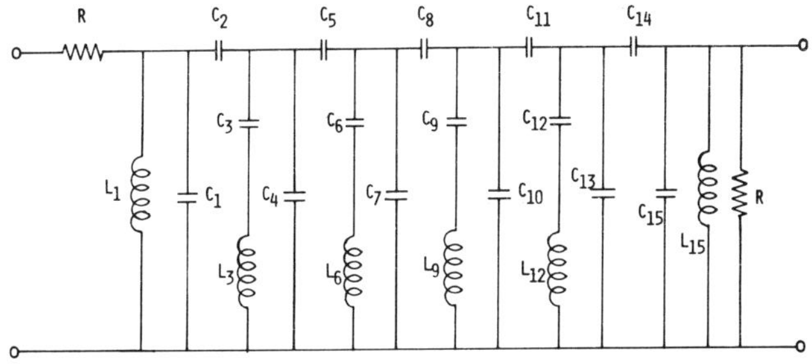

Fig. 5. Prototype channel-bank filter.

TABLE II

n	L_n, mH	C_n, nF
1	6.9417	6.6464
2	—	1.3293
3	50.0	1.6355
4	—	0.7467
5	—	1.7152
6	50.0	1.4033
7	—	1.9090
8	—	3.5273
9	50.0	1.3378
10	—	0.1105
11	—	8.1488
12	50.0	1.2922
13	—	9.0199
14	—	4.5935
15	6.9417	4.2480

The effective inductance with $C_c = 0$ and with $C_c = 330$ pF remains constant to within ±1 percent over the frequency range 0–2 kHz and 0–10.3 kHz, respectively. Therefore, the compensation technique introduces a five-fold increase in the useful bandwidth of the gyrator circuit.

The tolerance and temperature coefficient of C_p, like those of L_p, can be large, since C_p is inversely proportional to the gain-bandwidth product of the amplifiers. Consequently, exact cancellation of C_p is difficult to achieve and maintain in practice. However, even if only 80–90 percent of C_p is cancelled, or even if C_p is overcompensated by 10–20 percent, the resulting increase in the gyrator bandwidth may be adequate for many applications. Precise compensation can be achieved by trimming C_c for each gyrator circuit individually and by choosing the temperature coefficient of C_c to be equal and opposite to that of $A_0\omega_0R$.

IV. Design of a Gyrator-C Channel-Bank Filter

The modeling considerations of Section II and the compensation technique of Section III have been used in the design of a gyrator-C channel-bank filter. The prototype RLC filter employed is described by Valihora and Lim in [10]. The schematic and element values for this filter are given in Fig. 5 and Table II, respectively.

The gyrator-C filter was initially designed on the basis of ideal gyrators so that the influence of gyrator imperfections may be studied. A gyration resistance of 2 kΩ was used throughout. The filter was first analyzed assuming ideal inductors and lossy capacitors (Q_c = 2000). The frequency response obtained is shown as curve A in Figs. 6(a)–(c). The filter was then analyzed with each inductor replaced by the

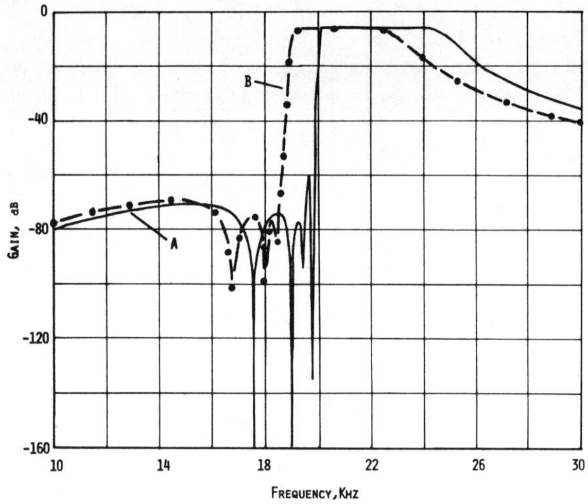

(a)

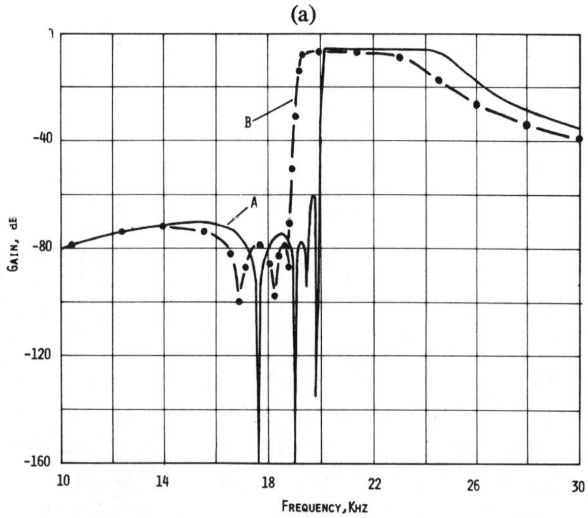

(b)

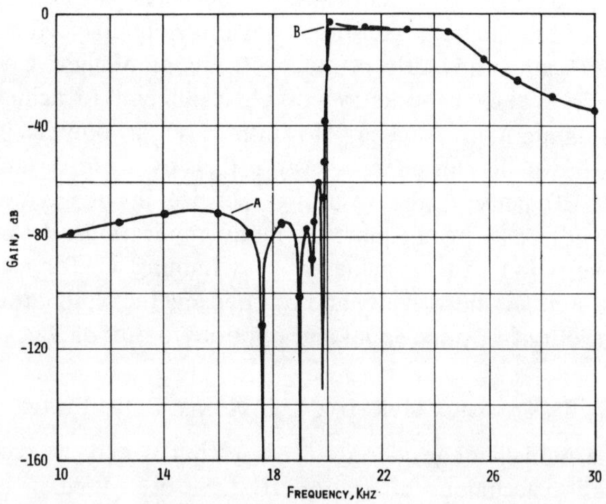

(c)

Fig. 6. Computed frequency responses of gyrator-C filter. Curve A: Filter response with ideal inductors. Curve B: Filter response with non-ideal gyrators, (a) with no corrections, (b) with corrections for L_p, (c) with corrections for L_p and C_p.

gyrator model of Fig. 1(c), assuming μA741 amplifiers (see Table I for specifications) and lossy capacitors (Q_c = 2000). The response obtained, illustrated as curve B in Fig. 6(a), shows that the gyrator imperfections introduce a significant lateral shift in the response of the filter. If no corrections were possible for the gyrator imperfections, amplifier μA741 would clearly be unsuitable for this application because of its relatively low gain-bandwidth product.

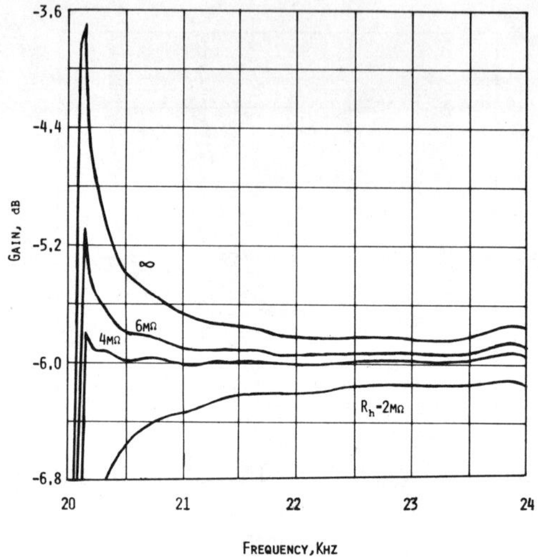

Fig. 7. Computed passband responses for R_h = 2, 4, 6, and ∞ MΩ.

In order to eliminate errors due to L_p, the gyrator-C filter was redesigned using L_p as part of each filter inductance. The response obtained, illustrated as curve B in Fig. 6(b), shows that lateral shift is eliminated for frequencies up to 14 kHz. The compensation technique was then applied to each gyrator circuit so as to eliminate errors due to C_p. The response obtained, illustrated as curve B in Fig. 6(c), shows that lateral shift is completely eliminated. However, peaking is introduced near the cutoff frequency of the filter caused by negative Q-factors. Peaking can be eliminated by using a shunt resistance $R_h \simeq 3.8$ MΩ at the input of each gyrator circuit, as illustrated in Fig. 7.

The compensated gyrator-C filter was constructed using commercial grade μA741 amplifiers, metal-film resistors (tolerance ≤ 1 percent and temperature coefficient + 100 ppm/°C) and polystyrene capacitors (tolerance ≤ 1 percent and temperature coefficient − 100 ± 50 ppm/°C). The filter was then aligned and tested with an input signal level of 0.4 V. The experimental response for an ambient temperature of 20°C and a dc-supply voltage of ±15 V, illustrated as curve B in Figs. 8 and 9, agrees with the theoretical response.

The influence of dc-supply-voltage variations on the performance of the filter is illustrated in Fig. 8. A change in the supply voltage introduces lateral shift in the response of the filter which is mainly due to variations in C_p produced by variations in $A_0\omega_0$. Note that a similar shift would result even if the gyrator circuits were uncompensated. The improvement brought about by the compensation is that any shift in the response occurs about the nominal response of the filter.

The influence of ambient-temperature variations is illustrated in Fig. 9. Again, lateral shift is introduced which, in this case, is due to the combined effects of variations in $A_0\omega_0$ and in the RC products.

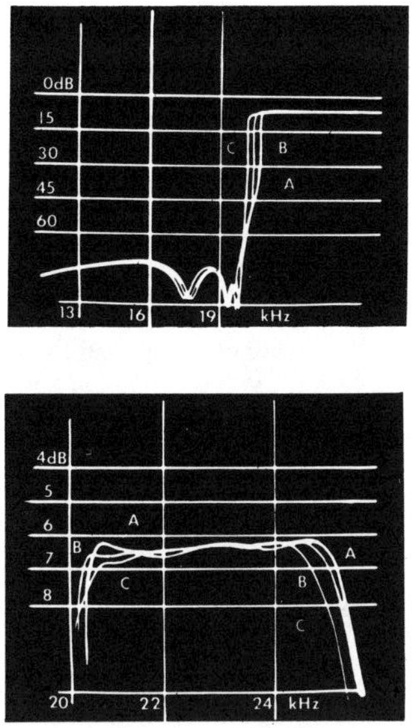

Fig. 8. Experimental frequency responses for various dc-supply voltages. Curve A: ± 11 V. Curve B: ± 15 V. Curve C: ± 17 V.

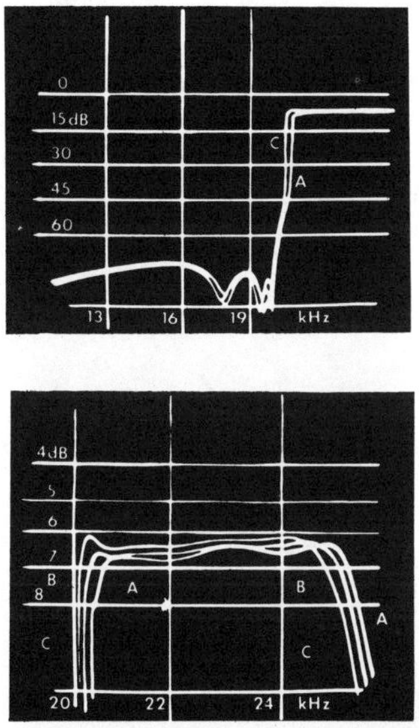

Fig. 9. Experimental frequency responses for various ambient temperatures. Curve A: 0°C. Curve B: 20°C. Curve C: 70°C.

The stability of the filter can be improved by optimizing the gyration resistances as shown in Section V. Further improvements can be brought about by employing filter and gyrator-load capacitors with temperature coefficient equal and opposite to that of the gyrator resistors and by employing compensating capacitors with temperature coefficient equal and opposite to that of $A_0\omega_0R$. Since the temperature coefficient of $A_0\omega_0$ is usually large and negative (about -2000 ppm/°C for the μA471), the temperature coefficient of the compensating capacitors should be large and positive.

V. Optimum Gyration Resistance

The effective input inductance of a compensated gyrator for the frequency range $0 \leq \omega \ll B\ (= A_0\omega_0)$ can be readily obtained as

$$L = \frac{\omega^2(L_o + L_p)^2 + (r_{p4} + r_{p5})^2}{\omega^2[(L_o + L_p) - C_p\{(r_{p4} + r_{p5})^2 + \omega^2(L_o + L_p)^2\}]}$$

by using the modified gyrator model. In practice,

$$\omega^2(L_o + L_p)^2 \gg (r_{p4} + r_{p5})^2$$

and thus

$$L \simeq \frac{(L_o + L_p)}{1 - \omega^2 C_p(L_o + L_p)}. \tag{34}$$

Also, $R \gg R_0$, and hence (9) and (31) give

$$L_p \simeq \frac{2R}{B} \tag{35}$$

$$C_p \simeq \frac{2}{BR} - C_c. \tag{36}$$

In effect, both L_p and C_p are small quantities and thus (34) can be simplified as

$$\begin{aligned} L &\simeq L_o + L_p + \omega^2 C_p(L_o + L_p)^2 \\ &\simeq L_o + L_p + \omega^2 C_p L_o^2. \end{aligned} \tag{37}$$

From (35)–(37), we can now write

$$L = L_o(1 - \omega^2 L_o C_c) + \frac{2}{B}\left(R + \frac{\omega^2 L_o^2}{R}\right)$$

and, consequently

$$\frac{\partial L}{\partial B} = -\frac{2}{B^2}\left(R + \frac{\omega^2 L_o^2}{R}\right).$$

For a given frequency $\omega = \omega_c$, $\partial L/\partial B$ is minimum when

$$R = \omega_c L_o \tag{38}$$

that is, the rate of change in the effective inductance with respect to variations in the gain-bandwidth product can be minimized by choosing the gyration resistance to be equal to the reactance of L_o at ω_c. This value of R is also optimum in the case of an uncompensated gyrator circuit, as was shown by Valihora and Lim in [10]. Since, in practice, variations in L are most objectionable for frequencies in the transition region of a filter, ω_c should be chosen to be the cutoff frequency of the filter.

With $C_p \simeq 0$, (35), (37), and (38) give

$$L \simeq \frac{R}{\omega_c} + \frac{2R}{B}$$

and hence the optimum value of R is obtained as

$$R \simeq \frac{\omega_c L}{1 + \dfrac{2\omega_c}{B}}.$$

The optimum gyration resistances for the filter of Fig. 5 can be calculated as 844 Ω for L_1 and L_{15} and 6080 Ω for $L_3 - L_{12}$.

VI. Conclusions

A compensation technique has been described whereby a gyrator parasitic capacitance introduced by the finite bandwidth of the amplifiers can be eliminated. The technique has been verified first by computer-aided analysis and then by experiment. By using an inexpensive compensating capacitor, a significant increase in the useful bandwidth of the gyrator circuit can be achieved.

The compensation technique has been applied successfully in the design of a difficult channel-bank filter. Despite the relatively low gain-band width product of the amplifiers used, a reasonable frequency response was achieved. DC-supply voltage and ambient-temperature variations tend to introduce lateral shift in the response of the filter. This can, however, be reduced by optimizing the gyration resistances and by choosing the temperature coefficients of the elements employed.

Although the work reported is primarily concerned with type ML748C and μA741 operational amplifiers, the compensation technique is valid for any other operational amplifier which can be represented by a single-pole model. Also the technique can be applied to other types of filters, as long as the inductors are grounded.

Acknowledgment

The authors are grateful to Bell-Northern Research Ltd. and to the National Research Council of Canada for supporting this research. Thanks are due to J. Valihora and J. T. Lim for useful discussions and comments.

References

[1] A. Antoniou, "New gyrator circuits obtained by using nullors," *Electron. Lett.*, vol. 4, pp. 87–88, Mar. 1968.

[2] ——, "Realization of gyrators using operational amplifiers, and their use in *RC*-active network synthesis," vol. 116, pp. 1838–1850, Nov. 1969.

[3] J. Valihora, "Modern technology applied to network implementation," *Proc. Int. Symp. Circuit Theory*, pp. 169–173, Apr. 1972.

[4] L. T. Bruton, "Nonideal performance of two-amplifier positive-impedance converters," *IEEE Trans. Circuit Theory*, vol. CT-17, pp. 541–549, Nov. 1970.

[5] L. T. Bruton and R. T. Pederson, "Tunable *RC*-active filters using periodically switched conductances," *IEEE Trans. Circuit Theory*, vol. CT-20, pp. 294–301, May 1973.

[6] B. B. Bhattacharyya, W. B. Mikhael, and A. Antoniou, "Design of *RC*-active networks using generalized-immittance converters," *J. Franklin Inst.*, vol. 297, pp. 45–58, Jan. 1974.

[7] L. T. Bruton, "Network transfer functions using the concept of frequency-dependent negative-resistance," *IEEE Trans. Circuit Theory*, vol. CT-16, pp. 406–408, Aug. 1969.

[8] A. Antoniou, "Bandpass transformation and realization using frequency-dependent negative-resistance elements," *IEEE Trans. Circuit Theory*, vol. CT-18, pp. 297–299, Mar. 1971.

[9] H. R. Trimmel, "Realization of canonical bandpass filters with frequency-dependent and frequency-independent negative resistances," *Proc. Int. Symp. Circuit Theory*, pp. 134–137, Apr. 1973.

[10] J. Valihora and J. T. Lim, "The feasibility of active filtering in frequency division multiplex systems," *Proc. Int. Symp. Circuit Theory*, pp. 121–125, Apr. 1974.

[11] A. Antoniou and K. S. Naidu, "Modeling of a gyrator circuit," *IEEE Trans. Circuit Theory*, vol. CT-20, pp. 533–540, Sept. 1973.

[12] H. R. Trimmel and H. E. Heinlein, "Fully floating chain-type gyrator circuit using operational transconductance amplifiers," *IEEE Trans. Circuit Theory*, vol. CT-18, pp. 719–721, Nov. 1971.

Paper 3-9

NEW ACTIVE-GYRATOR CIRCUIT

Indexing terms: Active networks, Gyrators

Ideal passive gyrators can be made using two operational amplifiers, and it has been proved that they cannot be made with only one amplifier. However, this latter proof led to the discovery that an ideal active gyrator can be made with only one amplifier. A circuit for doing this is presented here, and its behaviour when used in an inductance-simulating mode is analysed.

The gyrator is now well established as an element in the design of inductorless networks, being used, together with a capacitor, primarily in an inductance-simulating mode, either directly[1, 2] or indirectly.[3] The practical nonreciprocal components that are presently available for making a gyrator are either transistors or, more conveniently, complete transistor amplifiers of which, for this purpose, the operational amplifier is the most commonly used form. Over the past few years, several ideal* gyrator circuits have been described in the literature, all being variants of the original Riordan circuit[4] and using two such amplifiers in conjunction with a few resistors. In contrast to this, no one has described an ideal gyrator circuit using only one amplifier, although a few nonideal 1-amplifier circuits have been published.[5] This led the authors to conjecture, and subsequently to prove,[6] that an ideal gyrator cannot, in fact, be made with resistors† and only one amplifier. In the formulation of this proof, it was expected that the argument would not depend critically on the passivity of the gyrator, and that it would show equally well that an active gyrator having an impedance matrix

$$\begin{bmatrix} 0 & R_a \\ -R_b & 0 \end{bmatrix} \quad R_a > 0, \quad R_b > 0, \quad R_a \neq R_b \quad \ldots \quad (1)$$

also could not be made with only one amplifier. To the authors' surprise, the proof turned out to be valid only if R_a and R_b were equal. This immediately suggested the possibility that a circuit having an impedance matrix of expr. 1 could be provided exactly by a network containing only one amplifier. An extensive search, starting from the point where the proof failed, eventually generated a sequence of such networks, of increasing practical utility, that culminated in the circuit of Fig. 1.

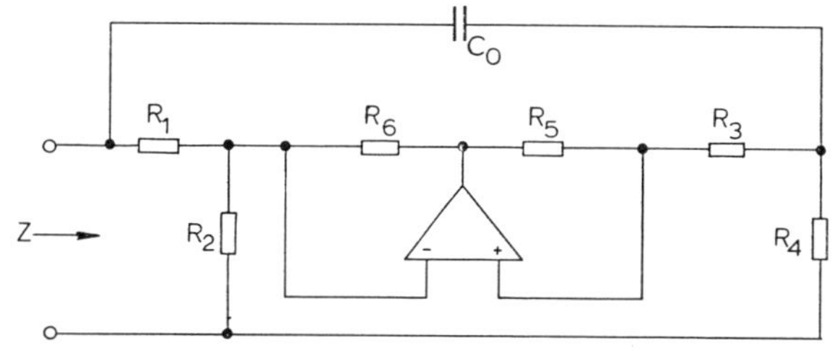

Fig. 1 *New active-gyrator circuit*

A capacitor C_0 is shown terminating one of the ports of the gyrator. The impedance Z seen into the other port then simulates the impedance of an inductor

In those applications of a gyrator where behaviour as a proper 2-port network is required, the condition $R_a = R_b$, which causes it to be passive, is usually essential, and here the new circuit of Fig. 1 would have no place. However, when the gyrator is used in an inductance-simulation role, where it behaves effectively as a 1-port network, the inequality of R_a and R_b is unimportant because the simulated inductance depends only on the product $R_a R_b$. For this reason, rather than to analyse the active gyrator by itself, it is simpler and more directly useful to analyse the circuit in terms of the quality of the simulated inductance presented at the grounded port when the nongrounded, but nevertheless not floating, port is terminated by a capacitor C_0. This capacitor is already included in Fig. 1. A straightforward analysis shows that the impedance Z presented at the grounded port is

$$Z = \frac{(A_1 + \varepsilon A_2) + j\omega C_0 (B_1 + \varepsilon B_2)}{(C_1 + \varepsilon C_2) + j\omega C_0 (D_1 + \varepsilon D_2)} \quad \ldots \quad (2)$$

where

$$A_1 = R_1 G_4 R_5 G_6 - (1 + R_1 G_2)(1 + R_3 G_4)$$

$$B_1 = R_1 R_5 G_6 - (1 + R_1 G_2) R_3$$

$$C_1 = G_4 R_5 G_6 - (1 + R_3 G_4) G_2$$

$$D_1 = A_1 + 1 - R_3 G_2$$

$$A_2 = \{1 + R_1(G_2 + G_6)\}\{1 + G_4(R_3 + R_5)\}$$

$$B_2 = \{1 + R_1(G_2 + G_6)\}(R_3 + R_5)$$

$$C_2 = (G_2 + G_6)\{1 + G_4(R_3 + R_5)\}$$

$$D_2 = A_2 + (G_2 + G_6)(R_3 + R_5)$$

$$\varepsilon = \frac{\text{input voltage to amplifier}}{\text{output voltage from amplifier}} = \alpha + j\omega/\omega_1$$

and $G_i = 1/R_i$.

In the ideal case, when the amplifier gain is infinite, i.e. $\varepsilon = 0$, the impedance Z will be that of an ideal positive inductance if

$$A_1 = D_1 = 0 \qquad \text{and} \qquad B_1/C_1 > 0 \quad \ldots \quad (3)$$

This requires the following two conditions to be satisfied:

(*a*) $R_2 = R_3$

(*b*) $(1 + R_1 G_2)(1 + R_3 G_4) = R_1 G_4 R_5 G_6 \quad \ldots \quad (4)$

and then, with a little rearrangement, one obtains the comparatively simple formula

$$Z = j\omega R_1 R_4 C_0 \left(\frac{1 + R_1 G_2}{1 + R_3 G_4} \right) \quad \ldots \quad (5)$$

One set of resistance values that satisfies eqn. 4 is $R_1 = R_2 = R_3 = R_4 = 2\,\Omega$, $R_5 = 4\,\Omega$ and $R_6 = 1\,\Omega$, which gives $Z = j\omega 4 C_0$. Thus, with ideal components, the circuit can behave as an ideal inductance, although it must be recognised

* The word ideal is used here to imply that, if the practical components behave as ideal elements of circuit theory, i.e. the resistors are perfect and the operational amplifiers have infinite gain, the practical gyrator would also behave as the corresponding ideal circuit element

† In this context, the word resistor refers indistinguishably either to the physical component or to its circuit-theoretic idealisation. A negative resistance is assumed to require for its construction one amplifier and at least one resistor

Reprinted with permission from *Electron. Lett.*, vol. 10, pp. 261-262, 27 June 1974.

that this pleasant state of affairs depends fairly critically on having the correct resistances.‡ Small errors in these resistances will cause the terms A_1 and D_1 to be nonzero, and the inductance simulation then starts to break down. In this respect, the circuit is noticeably inferior to the best 2-amplifier versions, which retain their ideal inductance behaviour with arbitrary positive values for their constituent resistances. Assuming, however, that, with present-day techniques, one can obtain and maintain a match of resistances to within a few tenths of a percent, then, for all but the highest quality of simulated inductances, the new circuit should be satisfactory, and certainly better than any other single-amplifier circuit yet described.

When the resistances are correct, the residual imperfections in the simulated inductance arise almost exclusively from the finite gain and the phase shift of the operational amplifier. A fairly detailed analysis, based primarily on the behaviour at lower frequencies, shows that, for a given value of the parameter $k = (R_5/R_6)^{\frac{1}{2}}$ and for a given amplifier, the resistances that minimise these imperfections are

$$R_1 = R_0/(k-1) \qquad R_2 = R_3 = R_0$$

$$R_4 = (k-1)R_0 \qquad R_5 = kR_0 \qquad R_6 = R_0/k \qquad (6)$$

where R_0 is the geometric mean of the two gyration resistances of the active gyrator.§ Good approximations for the Q factor and the inductance, expressed in terms of a normalised frequency variable $\Omega \triangleq \omega C_0 R_0$ and a parameter

$$\beta = (\omega_1 C_0 R_0)^{-1}$$

which measures, roughly, the ratio of the working frequency to the frequency ω_1 of unity gain in the amplifier, are then

$$Q = \frac{\Omega\{k(k-1)^2 - \beta(k^2+1)^2\Omega^2\}}{\left[\alpha + 2\left(\frac{k+1}{k-1}\right)\beta^2\Omega^2\right]\left[4k^2 + (k^2+1)^2\Omega^2\right]} \qquad (7)$$

and

$$L = \frac{R_0^2 C_0\left\{1 + \frac{4k\beta}{(k-1)^2}\right\}}{1 - \frac{\beta}{k}\left(\frac{k^2+1}{k-1}\right)^2\Omega^2} \qquad (8)$$

If the circuit is designed to operate at sufficiently low frequencies where the amplifier gain is independent of frequency, i.e. so that β can be set to zero, the expressions for Q and L simplify to

$$Q = \frac{\Omega k(k-1)^2}{\alpha\{4k^2 + (k^2+1)^2\Omega^2\}} \qquad (9)$$

and

$$L = R_0^2 C_0 \qquad (10)$$

Maximum Q factor then occurs when Ω has the value

$$\Omega_{max} = \frac{2k}{k^2+1} \qquad (11)$$

and is

$$Q_{max} = \frac{1}{4\alpha}\left(1 - \frac{2k}{k^2+1}\right) = \frac{1}{4\alpha}(1-\Omega_{max}) \qquad (12)$$

Q_{max} thus approaches a limiting value of $1/4\alpha$ as k tends to infinity. A practical limit to k is set by the maximum ratio of resistances, $k^2 (= R_5/R_6)$, which can conveniently be manufactured. Setting $k = 10$ as an upper limit gives, almost exactly, Q_{max} = (amplifier gain)/5, which is not much lower than the corresponding figure of (amplifier gain)/2 obtainable from the best 2-amplifier circuits.[2, 7] In a practical design, one would usually be given both the inductance L and the operating frequency ω at which the highest Q factor is required. For a fixed k, this gives the pair of equations

$$\omega C_0 R_0 = \frac{\omega L}{R_0} = \frac{2k}{k^2+1} \qquad (13)$$

whence

$$R_0 = \omega L\frac{k^2+1}{2k} \quad \text{and} \quad C_0 = \frac{1}{\omega^2 L}\left(\frac{2k}{k^2+1}\right)^2 \qquad (14)$$

The effect of increasing k is thus seen to be equivalent to increasing the mean gyration resistance R_0 and decreasing the required capacitance C_0. This raises the internal impedance level of the circuit, and may, in certain circumstances, also be a factor in deciding the practical upper limit for k. When the operating frequency is increased into the range where the operational amplifier gain is dropping at 20 dB/decade, the terms in eqns. 7 and 8 containing the factor β can no longer be neglected. However, to retain a reasonable behaviour of the Q factor and the inductance, one must limit β to less than about 10^{-2}; in other words, one must have at least 40 dB gain in the operational amplifier at the top of the working frequency range. The main effects of the sloping gain response are that, at higher frequencies, the inductance rises slightly and the Q factor falls off more rapidly past its peak. The behaviour of the Q factor is best appreciated from the curves shown in Fig. 2, which have been

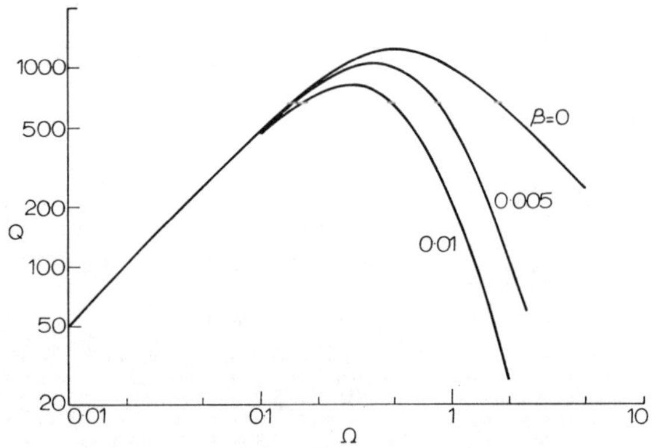

Fig. 2 *Q factor against frequency curves for simulated inductance*

The normalised frequency variable $\Omega \triangleq \omega C_0 R_0$, where R_0 is the mean gyration resistance and β is the ratio of the frequency corresponding to $\Omega = 1$ to the frequency of unity gain in the operational amplifier

computed for $\alpha = 10^{-4}$, corresponding to 80 dB gain at zero frequency, for $\beta = 0$, 0·005 and 0·01 and for the $k (= 2+\sqrt{3})$ that makes $\Omega_{max} = 0{\cdot}5$. In practice, of course, such curves are not likely to be reproduced very accurately, since the performance would be noticeably worsened by the effects of errors in the resistances, and it is unlikely that Q factors much over 100 could be held reliably for long periods.

Acknowledgment: This work was supported in part by the US National Science Foundation under Grant GK 35904X1.

H. J. ORCHARD *7th May 1974*
A. N. WILLSON, JUN.

Electrical Sciences & Engineering Department
School of Engineering & Applied Science
University of California, Los Angeles, Calif. 90024, USA

References

1 ORCHARD, H. J.: 'Inductorless filters', *Electron. Lett.*, 1966, **2**, pp. 224–225
2 ORCHARD, H. J., and SHEAHAN, D. F.: 'Inductorless bandpass filters', *IEEE J. Solid-State Circuits*, 1970, **SC-5**, pp. 108–118
3 HOLMES, W. H.: 'On canonical second-order active *RC* filters', *IEEE Trans.*, 1971, **CT-18**, pp. 721–723
4 RIORDAN, R. H. S.: 'Simulated inductors using differential amplifiers', *Electron. Lett.*, 1967, **3**, pp. 50–51
5 BERNDT, D. F., and DUTTA ROY, S. C.: 'Inductor simulation using a single unity-gain amplifier', *IEEE J. Solid-State Circuits*, 1969, **SC-4**, pp. 161–162
6 WILLSON, A. N., JUN., and ORCHARD, H. J.: 'Realization of ideal gyrators', *IEEE Trans.*, **CT** (to be published)
7 BRUTON, L. T.: 'Nonideal performance of two-amplifier positive-impedance converters', *ibid.*, 1970, **CT-17**, pp. 541–549

‡ The principal diagonal terms in the impedance matrix of the gryator are zero by virtue of a cancellation of terms rather than, as for the 2-amplifier circuits, by a division by an (infinite) amplifier gain

§ When $\varepsilon = 0$, this choice of resistances causes the active gyrator to have an impedance matrix of expr. 1 with the very simple relations $R_a = R_5$ and $R_b = R_6$

SINGLE OPAMP NETWORKS FOR SHUNT ARMS IN LADDER FILTERS

RUDOLPH H. CHENG and JIN TWAN LIM
BELL-NORTHERN RESEARCH, P.O. Box 3511, Station C, Ottawa, Canada

Abstract

This paper describes two networks which are capable of simulating a series-capacitor-inductor branch and series-resistor - FDNR[1] (frequency dependent negative resistance) branch suitable for shunt branches in ladder filters at audio frequencies. Each network is comprised of resistors, 2 loose-tolerance capacitors, and one opamp. Active ladder filter implementations using thick-film hybrid technology with parametric trimming of resistors are discussed. Fifth order elliptic high-pass and low-pass filters with high pole Q's are presented.

1. Introduction

Filters using discrete inductor and capacitor at voice frequencies are heavy and bulky. Today, small low-cost filters are used whenever possible and are found to be comparable to passive filters in performance. The desirable sensitivity properties possessed by active ladder filters realized from their passive LC ladder prototypes are known[2,3]. Simulated inductances [4] and frequency dependent negative resistances (FDNR)[1] produce good quality filters, but two opamps are required for each simulation. To further reduce size, cost, and power consumption of active filters, single opamp arrangements for inductor or FDNR simulations are proposed[5,6,7] but there are disadvantages: Loose tolerance components cannot be used; independent realization of impedance and Q is not possible, and only low Q impedances can be realized without involving wide ranges of resistor values. Orchard and Willson[8] proposed an attractive single opamp gyrator for inductor simulation, having two cancellation requirements in its realization method to ensure the elimination of certain coefficients in the impedance function. To maintain performance, cancellation constraints must be satisfied over temperature and time, as well as at manufacture. Schmidt and Lee [9] proposed a single opamp impedance branch which can simulate a series-resistor-FDNR branch, but, again, the presence of two cancellation constraints in this method makes it less attractive than the one discussed below. In this paper, two single-opamp networks, capable of simulating a series-capacitor-inductor branch and series-resistor-FDNR branch are presented. Each network has only one cancellation requirement, thus imposing less severe stability and tracking requirements on components with temperature and aging. Furthermore, the impedances and the Q's can each be independently trimmed accurately, therefore, achieving both the resonant frequency and the impedance level of the simulated branch.

2. The Active Branch for L-C Simulation

The active branch for simulating a series-capacitor-inductor branch is shown in Figure 1 and its equivalent circuit is shown in Figure 2. Assuming that the amplifier has a dominant response with the gain function $A = A_o/(1 + sT)$ and a gain bandwidth product $B = A_o/T$, usually denoted by f_T, the input impedance Z_{in} of the active branch in Figure 1 is:

$$Z_{in} = R_8 + \frac{s^3\frac{H}{B} + s^2(E+\frac{J}{B}) + s(F+\frac{K}{B}) + G}{s\left\{s^2\frac{N}{B} + s\frac{P}{B} + (M+\frac{Q}{B})\right\}} \quad (1)$$

Where E,F,G,H,J,K,M,N,P, and Q are independent of frequency, w, and are shown in Appendix A.

The finite gain-bandwidth product, B, of opamps introduces strays to the simulated L, C, and R_K, and frequency dependence to C and R_K. However, we can ensure independent trimming of L, C, and R_K, minimize the effect of gain-bandwidth product variation and minimize the dependence on frequency by observing some constraints when designing the active branch. The constraints are formulated in equations (2) to (5).

$$\left|\frac{w^2N}{Q+BM}\right| << 1 \quad (2)$$

$$\left|\frac{wP}{Q+BM}\right| << 1 \quad (3)$$

$$\left|\frac{w^2H}{BF+K}\right| << 1 \quad (4)$$

$$\frac{F}{M} < 0 \quad (5)$$

Where (2) ensures that L is dominantly inductive and C is independent of frequency, (3) ensures that C and R are dominantly capacitive and resistive, respectively; (4) ensures that R is independent of frequency and (5) ensures that R is negative and, therefore, can be cancelled out by a positive R_8. When they are incorporated in the design, the active branch will have an

Reprinted from *IEEE 1977 Int. Symp. on Circuits and Syst.*, Apr. 1977, pp. 313-316.

equivalent circuit, shown in Figure 2, where C, L, R, and R_K are expressed in (6) to (9).

$$C = \frac{MB+Q}{BG} \tag{6}$$

$$L = \frac{EB+J}{Q+BM} \tag{7}$$

$$R = \frac{BF+K}{MB+Q} \tag{8}$$

$$R_K = R_8 + R \tag{9}$$

By expanding (6) to (9) and re-arranging terms, C and L can be obtained and R_K eliminated by trimming R_1, R_4, and R_8, respectively, according to formulas (10), (11), and (12).

$$R_1 = \frac{BC_4g_7 + g_3(g_5+g_7)}{Bg_3\{C(g_5+g_7)-C_4g_7\}} \tag{10}$$

$$R_4 = \frac{\left| g_1\{L[g_2g_3(g_5+g_7) + BC_4g_2g_7] - C_4(g_5+g_7) - g_3(C_4+C_6)\} + LBC_4g_2g_7g_3 - g_2\{C_4(g_5+g_7) + g_3(C_4+C_6)\} - C_6g_3(g_5+g_7) \right|}{BC_6C_4g_3(g_5+g_2) + C_4g_3(g_5+g_7)(g_1+g_2)} \tag{11}$$

$$R_8 = -\frac{BF+K}{MB+Q} \tag{12}$$

where (12) is the only cancellation requirement in the design. For these formulas, B is the typical gain bandwidth product of the particular opamp used and w is usually the resonant frequency of the simulated L-C branch. Both B and w are expressed in radians per second. The four conditions (2) to (5) and the three equations (6), (7), and (12) constitute seven constraints to the active branch, in Figure 1, which has two capacitors and seven resistors. By choosing the seven resistors properly, we can satisfy the seven constraints and have some freedom in choosing capacitors C_4 and C_6. The seven constraints are quite complex and it is necessary to use computer optimization to implement the design of the active branch.

3. The Active Branch for R-FDNR Simulation

The active branch for simulating a series-resistor-FDNR branch is shown in Figure 3 and its equivalent circuit is shown in Figure 4. With the same assumption on the amplifier gain, the input impedance of the active branch in Figure 3 can be represented by equation (13).

$$Z_{in} = R_8 + \frac{1}{s}\left\{\frac{ms^3+ns^2+ps+q}{ts^2+vs+y}\right\} = R_8 + \frac{1}{s}Z' \tag{13}$$

where m, n, p, q, t, v, and y are independent of frequency, w, and are shown in Appendix B. The existence of R_8 is not necessary, but it allows R_1 and R_3 to take on smaller values, thus reducing the spread of resistor values. The design constraints are expressed by (14) and (15), which ensure that the simulated R and D are free from strays and independent of frequency.

$$\left|\frac{wt}{v}\right| \ll 1 \tag{14}$$

$$\left|\frac{y}{wv}\right| \ll 1 \tag{15}$$

The R, D, and C realized are:

$$R = R_8 + \frac{n}{v} \tag{16}$$

$$D = v/q \tag{17}$$

$$C = \frac{v}{p - w^2m} \tag{18}$$

and by rearranging terms, the trimming equations:

$$R_5 = \frac{\left\{D - \frac{C_4g_1}{B}\left(\frac{1}{g_3}+\frac{1}{g_1}\right)\right\}}{\left\{\frac{C_4C_7(g_1+g_3)}{g_3} + \frac{C_4g_1}{B} + \frac{C_7g_1}{B} + \frac{(C_4+C_7)g_3}{B}\right\}} \tag{19}$$

$$R_2 = \frac{\left\{C_4g_3 + C_7g_3 + \frac{g_3g_5}{B}\right\}}{\left\{C_4g_1g_5 - \frac{1}{B}g_3g_5g_1\right\}} \tag{20}$$

$$R_8 = R - \frac{\left\{C_4C_7g_2 + \frac{1}{B}(C_4g_5[g_1+g_2] + C_4g_3[g_1+g_2] + C_7g_3[g_1+g_2])\right\}}{\left\{C_4C_7g_2(g_1+g_3) + \frac{1}{B}(C_4g_5[g_1(g_3+g_2) + g_2g_3] + g_1g_2g_3[C_4+C_7])\right\}} \tag{21}$$

(19), (20), and (21) are obtained.

The formulae, in both cases, are valid for $B \gg w \gg B/Ao$ and the capability of the branches for realizing difficult filters in this frequency range is only limited by the characteristics of the components used.

4. Practical Examples

a) A doubly terminated filter is designed using the active branch in Figure 1, to meet the specification as shown in Figure 5 for the temperature range of 0°C to 75°C. The specification can be satisfied by a 5th order elliptic ladder filter, with the highest pole Q = 12.78 in its transfer function. The corresponding active filter, shown in Figure 7, is realized by thick-film hybrid technology using appliqued NPO chip capacitors and 741-type opamps. The capacitors have a temperature coefficient (TC) of ± 30 ppm/°C, which translates to a temperature tolerance of ± .15% over 0°-75°C. Available thick film resistor inks have standard TC's of ± 100 ppm/°C and ± 50 ppm/°C or ± .5% and ± .25%, respectively, in the same temperature range. The 10-MΩ resistors, RB1 and RB2, provide dc returns for the opamps, they have negligible effect on the frequency response of the filter and only cause a little extra absolute loss for the filter. The filter can be matched to the system impedances by adding a matching pad at the input and a buffer at the output, which also provides the required gain. Resistors R3, R5, R10, R13, R15, and R20 are designed to provide a range of resistance values as shown in Table 1. The magnitudes of the range are dependent on the trimming tolerance of the laser and the capacitor tolerances in the active branch. In manufacturing, all resistors, except R3, R5, R10, R13, R15, and R20, are trimmed to desired values by laser. These resistors and capacitors are then measured, their values are used to calculate R3, R5, R10, R13, R15, and R20, which are then trimmed to their calculated values; this is called parametric trimming.

In the present method, both the capacitor and the inductor are trimmed independently. Therefore, both the pole frequency and the impedance level will be accurate in spite of 10% capacitors used. If capacitors with tolerance larger than 10% were used for the above example, a larger trimming range for the trimming resistors would be required.

b) A doubly terminated filter is designed using the active branch in Figure 3 to meet the specification as shown in Figure 6 for the temperature range of 0^oC to 75^oC. Incidentally, the specification can be satisfied by a 5th order elliptic ladder filter with the highest pole Q = 12.78 in its transfer function. The active filter realization is shown in Figure 8. Resistors R5, R8, R34, R16, R19, and R35 are designed so that they can provide a range of resistances, as shown in Table 2. The properties of the material used and the laser trimming procedure used are similar to those described in the previous example.

5. Conclusion

Two new active branches are discussed and used in the realization of difficult fifth order elliptic high-pass and low-pass filters.

The capability of the branches to realize filters in the frequency range of $B \gg w \gg B/A_o$, where B is the gain-bandwidth product f_T and A_o is the DC gain of the opamp, is only limited by the characteristics of the components used. The Q of the simulated L-C or R-FDNR branch can be controlled adequately by a single resistor and its sensitivity to opamp parameter variations, due to temperature and manufacture can be minimized by proper choice of component values through computer optimization and by proper choice of component characteristics. Furthermore, loose tolerance capacitors can be used and the simulated L and C, or R and FDNR, can both be trimmed independently.

COMP.	VALUE	MFG. TOL.(%)	RANGE OR COMMENT
R1	21.737 K	0.5	
C2	3.0 nf	2.0	
R3	7.66 K	0.5	11K to 3K
R4	20.33 K	0.5	
R5	42.97 K	0.5	55K to 35K
C6	6.8 nf	10.0	
R7	4.901 K	0.5	
C8	6.8 nf	10.0	
R9	3.52 K	0.5	
R10	5.28 K	0.5	6.5K to 4K
R11	42.94 K	0.5	
C12	2.7 nf	2.0	
R13	14.88 K	0.5	20K to 4K
R14	34.95 K	0.5	
R15	32.79 K	0.5	42K to 27K
C16	6.8 nf	10.0	
R17	6.19 K	0.5	
C18	6.8 nf	10.0	
R19	1.02 K	0.5	
R20	9.17 K	0.5	11.5K to 7.5K
R21	20.61 K	0.5	
C22	3.9 nf	2.0	
R23	21.737 K	0.5	
RB1	10.0 M	5.0	External resistors
RB2	10.0 M	5.0	or printed resistors

Table 1.

Component values for filter in Figure 7

COMP.	VALUE	MFG. TOL.(%)	RANGE OR COMMENT
C1	15.0 nf	2.0	
R2	22.99 K	0.5	
R34	1.058 K	0.5	1.09K to 1.03K
R3	4.560 K	0.5	
R4	26.42 K	0.5	
R5	1.954 K	0.5	2.3K to 1.6K
C6	6.8 nf	10.0	
C7	6.8 nf	10.0	
R8	5.654 K	0.5	7.5K to 4K
R13	24.80 K	0.5	
R35	1.208K	0.5	1.29K to 1.13K
R14	22.75 K	0.5	
R15	34.32 K	0.5	
R16	12.96 K	0.5	15K to 11K
C17	6.8 nf	10.0	
C18	6.8 nf	10.0	
R19	9.762 K	0.5	12.5K to 7.5K
R24	17.46 K	0.5	
C25	15.0 nf	2.0	
RB1	10.0 M	5.0	External resistors
RB2	10.0 M	5.0	or printed resistors

Table 2.

Component values for filter in Figure 8

Appendix A

The parameters E,F,G,H,J,K,M,N,P, and Q in equation (1) are shown below:

$E = C_6 C_4 R_4 g_3 \, (g_5+g_2)$

$F = C_6 g_3 g_5 + C_4 g_2 g_3 + C_6 g_2 g_3 + C_4 g_2 g_7 + C_4 R_4 g_2 (g_5+g_7) g_3 - C_4 g_1 g_5$

$G = g_2 (g_5+g_7) \, g_3$

$H = C_4 C_6 \, (g_5+g_7+g_1+g_2+g_3+g_3 R_4 [g_1+g_2+g_5+g_7])$

$J = C_4 (g_5+g_7)(g_1+g_2)(1+R_4 g_3)+g_3 (g_1+g_2)(C_4+C_6) + C_6 g_3 (g_5+g_7)$

$K = g_3 (g_5+g_7)(g_1+g_2)$

$M = C_4 g_2 g_7 (g_1+g_3)$

$N = C_4 C_6 \, (g_2+g_5+g_7)(g_1 g_3 R_4+g_1+g_3)$

$P = C_4 (g_5+g_7)[g_1 (g_3+g_2)+g_3 g_2]+g_1 g_3 \{g_2 (C_4+C_6) + (g_5+g_7)(C_4 R_4 g_2+C_6)\}$

$Q = g_1 g_2 g_3 (g_5+g_7)$

and g_i are the reciprocal of R_i in Figure 1.

Appendix B

The parameters m,n,p,q,t,v, and y of equation (13) are shown below:

$m = \frac{1}{B} C_4 C_7 (g_1+g_2)$

$n = \{C_4 C_7 g_2 + \frac{1}{B}(C_4 g_5 [g_1+g_2] + C_4 g_3 [g_1+g_2] + C_7 g_3 [g_1+g_2]\}$

$p = \{-C_4 g_1 g_5 + C_4 g_2 g_3 + C_7 g_2 g_3 + \frac{1}{B} \, g_3 g_5 (g_1+g_2)\}$

$q = g_2 g_3 g_5$

$t = \frac{1}{B} \, C_4 C_7 [g_1 (g_3+g_2)+g_2 g_3]$

$v = \{C_4 C_7 g_2 (g_1+g_3) + \frac{1}{B} \, (C_4 g_5 [g_1 (g_3+g_2)+g_2 g_3] + C_4 g_1 g_2 g_3 + C_7 g_1 g_2 g_3)\}$

$y = \frac{1}{B} \, g_1 g_2 g_3 g_5$

and g_i are the reciprocal of R_i in Figure 3.

References

1. L.T. Bruton, "Network Transfer Functions Using The Concept of Frequency-dependent Negative Resistance:, IEEE Trans. Circuit Theory, Vol. CT-16, pp. 406-408, Aug.1969.
2. H. J. Orchard, "Inductorless Filters," Electronics Letters, Vol. 2, pp. 224-225, June 1966.
3. G. C. Temes and H. J. Orchard, "First Order Sensitivity and Worse Case Analysis of Doubly Terminated Resistance Two-ports," IEEE Trans. on Circuit Theory, Vol. CT-20, pp. 567-571, Sept. 1973.
4. A. Antoniou, "Realization of Gyrators Using Operational Amplifiers and Their Use in RC-active Network Synthesis," Proc. IEEE, Vol. 116, pp, 1838-1850, Nov. 1969.
5. D. F. Berndt and S.C.D. Roy, "Inductor Simulation Using a Single Unity Gain Amplifier," IEEE J. of Solid-State Circuits, Vol. SC-4, pp. 161-162, June 1969.
6. J. M. Rollett, "Economical RC Active Lossy Ladder Filters," Electronics Letters, Vol.9, pp. 70-72, No. 3, Feb. 8, 1973.
7. J. M. Rollett and D. R. Wise, "Filter Networks," US Patent 3,886,469, May 27,1975.
8. H. J. Orchard and A.N. Willson, Jr., "New Active-gyrator Circuit," Electronics Letters, Vol. 10, pp. 261-262, June 27,1974.
9. C. E. Schmidt and M. S. Lee, "Multipurpose Simulation Network with a Single Amplifier," Electronics Letters, Vol. 9, pp. 69-70, No. 3, Feb. 8, 1973.

Acknowledgement: The authors wish to thank Mr. J. F. Pinel for his encouragement.

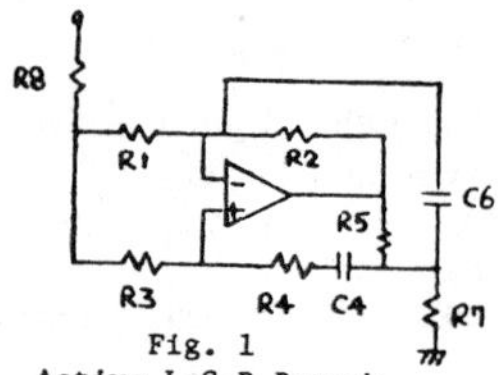

Fig. 1
Active L-C-R Branch

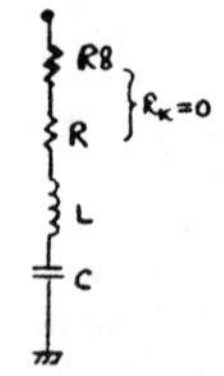

Fig.2 - Equivalent Circuit for Fig. 1.

Fig.3 - Active R-FDNR-C Branch.

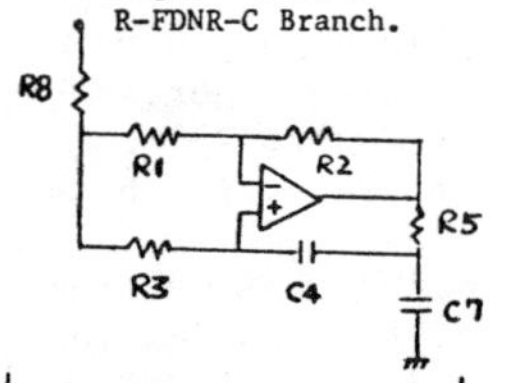

Fig.4 Equivalent circuit of Fig.3.

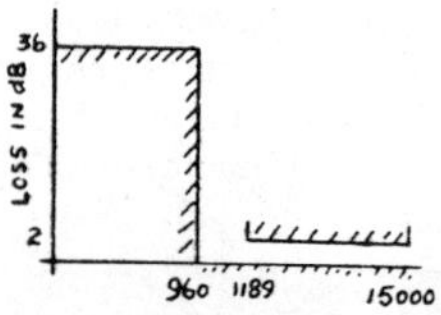

Fig. 5
Spec. for highpass filter

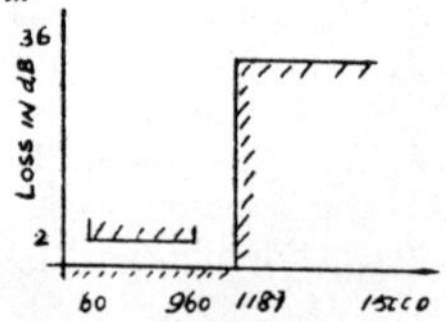

Fig. 6
Spec. for lowpass filter

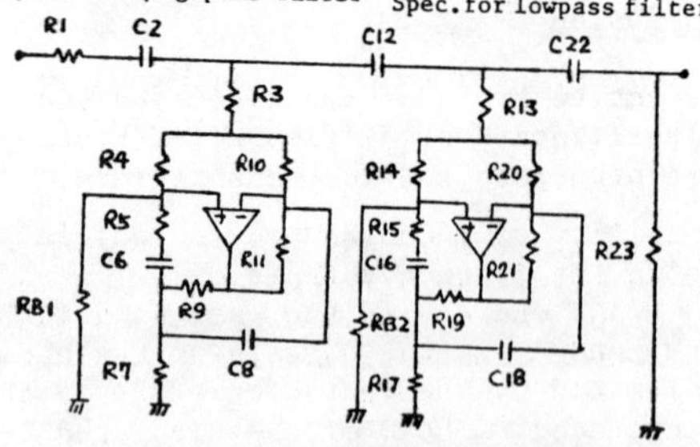
Fig. 7
The active highpass filter

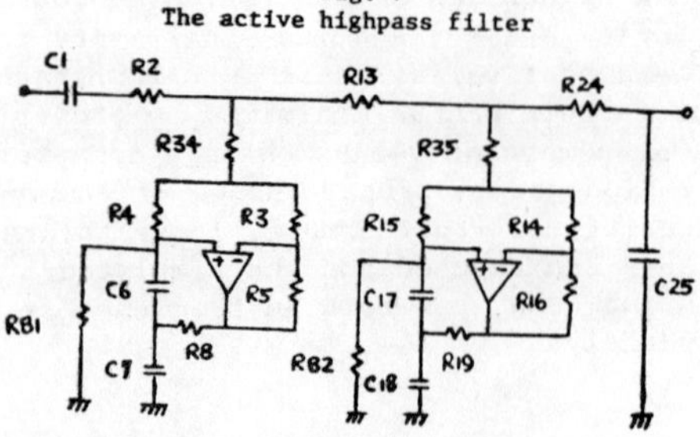
Fig. 8
The active lowpass filter.

A FREQUENCY-LOCKED ADAPTIVE FILTER

L. T. BRUTON AND R. T. PEDERSON
Electrical Engineering Department, The University of Calgary, Calgary, Alberta, Canada

SUMMARY

A frequency-locked RC active ladder filter is described, which is frequency-locked to a sinusoidal control tone. The tunable ladder filter is used as a frequency discriminator by using two of the filter branch voltages to generate a control voltage. The control voltage is then used to tune or frequency-scale the ladder filter transfer function to lock the filter to the frequency of the control tone. A simple linear frequency-signal control system is developed to describe the frequency-locked performance of the adaptive filter. The frequency-locked filter can be designed so that the sinusoidal control tone is rejected at the filter output, making it possible to adapt the filter transfer function during transmission of a message. A practical frequency-locked filter is given in which a frequency-locked third order elliptic lowpass filter is designed. The performance of the practical filter is analysed using a simple linear control system model.

INTRODUCTION

Adaptive filters have found wide use in such areas as the equalization of communication channels[1] and separating signals from noise.[2,3,4] Generally, the filter transfer functions are adapted using a test signal, or are continuously adapting according to some measure of the system performance. Adaptive filters have been constructed using several methods. For example, transversal filters have been realized where the weighting factors of each path adjust to adapt the filter transfer function. Adaptive bandpass filters have been constructed using active filters in which the dominant time constants of the network are adjusted automatically to optimize S/N.[2] The adaptive transversal filter is complex, requires many components, and is therefore costly to construct.

A new adaptive active ladder filter is proposed that has the property that the transfer function is self-scaling in the frequency domain according to the frequency of a sinusoidal control signal. The method employs a voltage tunable structure and depends on the availability, within the topology of the tunable filter, of two voltages that may be used as the input voltages of a simple frequency discriminator. A particularly useful application of this idea is illustrated by using the two element voltages of the shunt arm of an elliptic filter as the input voltages of the frequency discriminator so that it is possible to realize an adaptive elliptic ladder filter that will lock-in such that the frequency of the zero of transmission is identical to the frequency of the control sinusoid. In this case, the control signal cannot appear at the output port of the adaptive filter. A linearized control system model of the adaptive elliptic filter is proposed that allows the stability criterion and lock-in time constants to be determined.

REVIEW OF TUNABLE LADDER FILTER REALIZATIONS

Adaptive ladder filters require a voltage tunable ladder filter in which the transfer function can be frequency scaled. Consider now several recently proposed methods of realizing frequency scaled RC active ladder filters that can be used to realize adaptive filters. First, a filter transfer function can be written in the form

$$T(s) = \frac{\sum_{i=0}^{r} b_i s^i}{\sum_{j=0}^{q} c_j s^j} \tag{1}$$

Received 27 June 1972
Revised 8 September 1972

Reprinted with permission from *Int. J. of Circuit Theory Appl.*, vol. 1, pp. 71-82, Mar. 1973.

which can be frequency scaled by introducing a scaling factor as follows

$$T(s/K) = \frac{\sum_{i=0}^{r} b_i(s/K)^i}{\sum_{j=0}^{q} c_j(s/K)^j} \tag{2}$$

where b_i and c_j are functions of conductances and capacitances, and K is the scaling factor. It has been shown[5] that a set of conductances within the network may be periodically switched at a high switching rate, using the conductance waveform of Figure 1, so that the network essentially responds to the average values of the conductances. The periodically switched conductances may be chosen such that the variations of the average values of the conductances correspond to the frequency-scaling described by equation (2), where the parameter K is *equal* to the average value of the time-varying waveform of Figure 1. Thus

$$\bar{G}_N = KG_N = \frac{K_1}{T} G_N \left[\tau + \frac{K_2}{K_1}(T - \tau) \right] \tag{3}$$

where K_1, K_2, τ and T are defined in Figure 1. The average value $\bar{G}_N$ and therefore K can be varied by changing the pulse width to period ratio τ/T of the switching waveform.

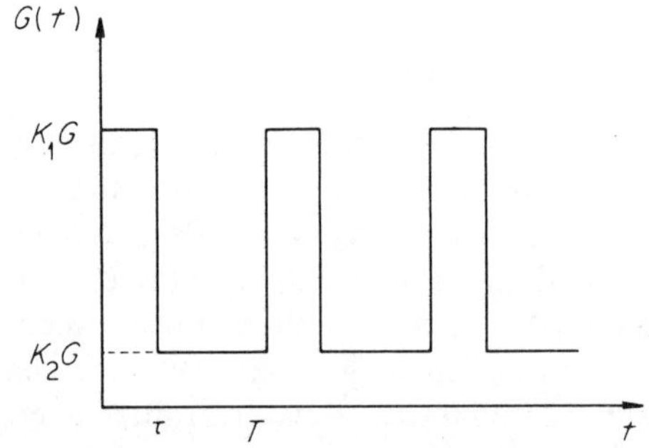

Figure 1. Time variation of switched conductances

This method of realizing electronically tunable filters has been applied to low-sensitivity RC active ladder filters. RC active ladder filters can be realized using several approaches. The inductance simulation approach,[10,11] in which each inductance is simulated by an active network, can be used, or by transforming a LCR ladder network into a topologically similar DCR network[12] having the same voltage transfer function. Since it is difficult to simulate floating inductors, the DCR network approach is briefly reviewed. The DCR network is realized by scaling each impedance in the LCR network by $1/s$, transforming an inductance to a resistance, a resistance to a capacitance, and a capacitance to a Frequency Dependent Negative Resistance (FDNR)[12] having an impedance of the form $1/s^2D_i$. The FDNRs are realized using Generalized Immittance Converters (GICs) which can have a transmission matrix of the form[13]

$$\mathbf{G}(s) = \begin{bmatrix} 1 & 0 \\ 0 & s^2M \end{bmatrix} \tag{4}$$

which when terminated in a resistance has an input impedance corresponding to a FDNR.

The transfer function of a DCR network can be written in the form

$$T_{DCR}(s) = T_{DCR}(1/s^2D_i, 1/sC_i, R_i) \tag{5}$$

which becomes

$$T_{DCR}(s/K) = T_{DCR}(K^2/s^2D_i, K/sC_i, R_i) \tag{6}$$

when frequency-scaled.

To frequency-scale the network, only the FDNRs and capacitances need be frequency-scaled. It has been shown that the input impedance of the GIC can be frequency-scaled using periodically switched conductances within the GICs.[5] Since a variable capacitance is difficult to obtain, a GIC is also used to realize a frequency-scaled capacitance impedance as well as a frequency-scaled FDNR. This makes it possible to frequency-scale or tune RC active ladder filters using periodically switched conductances.

An analogue method of frequency-scaling the transfer junction of a DCR network has also been proposed.[14] Four-quadrant analogue multipliers are used to realize frequency-scaled impedances of the form

$$Z_T(s) = KZ(s) \tag{7}$$

and

$$Z_T(s) = K^2 Z(s) \tag{8}$$

and K is a linear function of a tuning-voltage applied to the analogue multipliers, and $Z(s)$ is a branch of the unscaled network. Equation (7) is defined as a scaled impedance (SI) and equation (8) is defined as a square scaled impedance (SSI). Thus equation (7) is associated with every capacitance in the DCR network, giving

$$Z_T(s/K) = K/sC \tag{9}$$

and equation (8) is associated with every FDNR element, giving

$$Z_T(s/K) = K^2/s^2 D_i \tag{10}$$

so that the voltage transfer function $T(s)$ becomes $T(s/K)$ and therefore frequency-scaling is achieved by voltage control of K via the analogue multiplier.

It is assumed in the following sections that a frequency-scaled (voltage tunable) inductorless ladder filter is available. Either the periodically switched conductance method or the analogue-multiplier method may be employed.

A FREQUENCY-LOCKED LADDER FILTER

A frequency-locked ladder filter exhibits a transfer function of the form $H(s')$, where $s' = As/\omega_L$ and A is a design constant, s is the complex frequency variable, and ω_L is the *frequency* of a control sinusoid. Clearly, frequency-scaling of $H(s')$ occurs by means of the control frequency ω_L. In the case of a frequency-locked elliptic ladder filter, exhibiting a zero of transmission at $s' = j\omega_z$, we may choose $A = \omega_z$ so that the tunable transfer function $H(j\omega')$ is given by $H(j\omega(\omega_z/\omega_L))$. Thus, at the frequency ω_L of the control tone

$$H(j\omega') = H[j\omega(\omega_z/\omega_L)]|_{\omega=\omega_L} = H(j\omega_z) = 0 \tag{11}$$

That is, if $A = \omega_z$, the control tone does not appear at the output port of the filter.

Principle of operation

Consider the voltage-tuned network in Figure 2 having four signal port voltages V_1, V_2, V_3 and V_4, and a tuning voltage V_T, so that

$$H_2(s) = \frac{V_2(s)}{V_1(s)}; \quad H_3(s) = \frac{V_3(s)}{V_1(s)}; \quad \text{and} \quad H_4(s) = \frac{V_4(s)}{V_1(s)} \tag{12}$$

where $V_3(s)$ and $V_4(s)$ are available from branches within the filter. The signals $V_3(s)$ and $V_4(s)$ are chosen such that, for a sinusoidal input $V_1(j\omega)$, the quantity

$$V_5\left(\frac{j\omega}{AV_T}\right) = K_3\left|V_3\left(\frac{j\omega}{AV_T}\right)\right| - K_4\left|V_4\left(\frac{j\omega}{AV_T}\right)\right| \tag{13}$$

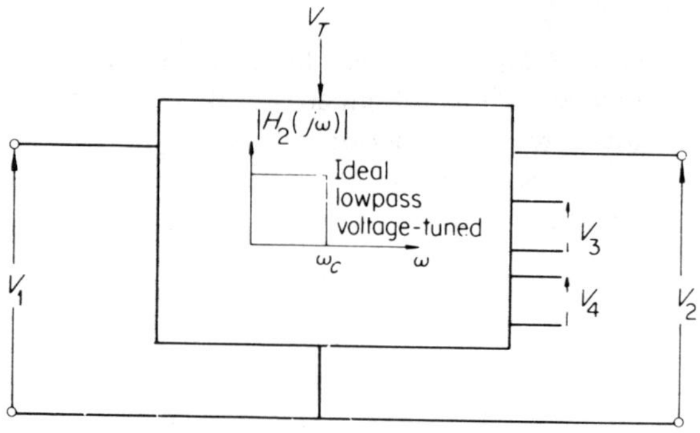

Figure 2. Four-port voltage tunable filter

is monotonic. That is,

$$K_3\left|H_3\left(\frac{j\omega}{AV_T}\right)\right| - K_4\left|H_4\left(\frac{j\omega}{AV_T}\right)\right| \tag{14}$$

is also monotonic. Now, the function in equation (13) is created by employing gain blocks K_3 and K_4 and ideal rectifiers as shown in Figure 3. The function $V_5(j\omega/AV_T)$ is amplified by A_0 in the simple *voltage-signal*

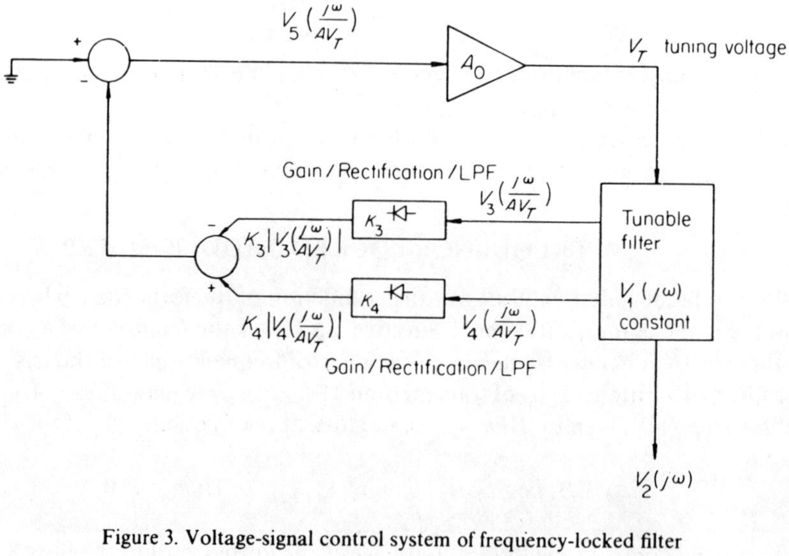

Figure 3. Voltage-signal control system of frequency-locked filter

control system of Figure 3. The voltage $V_5(j\omega/AV_T)$ is given by

$$V_5\left(\frac{j\omega}{AV_T}\right) = \frac{A_0 f[V_T, V_1(j\omega)]}{1 + A_0 f[V_T, V_1(j\omega)]} \tag{15}$$

where

$$f[V_T, V_1(j\omega)] \equiv \frac{V_5(\omega/AV_T)}{V_T}$$

and, for $A_0 f[V_T, V_1(j\omega)] \to \infty$, we have

$$V_5\left(\frac{\omega}{AV_T}, V_1(j\omega)\right) = 0 \tag{16}$$

That is, from equations (13) and (16),

$$\frac{|V_3(j\omega/AV_T)|}{|V_4(j\omega/AV_T)|} = \frac{K_4}{K_3}, \text{ a constant} \tag{17}$$

and since $V_1(j\omega)$ is constant,

$$\frac{|H_3(j\omega/AV_T)|}{|H_4(j\omega/AV_T)|} = \frac{K_4}{K_3} \tag{18}$$

Thus, the steady-state condition is determined by K_4/K_3 and will settle at $|H_3|/|H_4|$ equal to this value. Typical functions for $|H_2(j\omega/AV_T)|$, $|H_3(j\omega/AV_T)|$, $|H_4(j\omega/AV_T)|$ and $|H_5(j\omega/AV_T)| = |V_5(j\omega/AV_T)/V_1(j\omega)|$ are given in Figure 4 from which it is clear that a monotonic $|H_5(j\omega/AV_T)|$ transfer function guarantees only

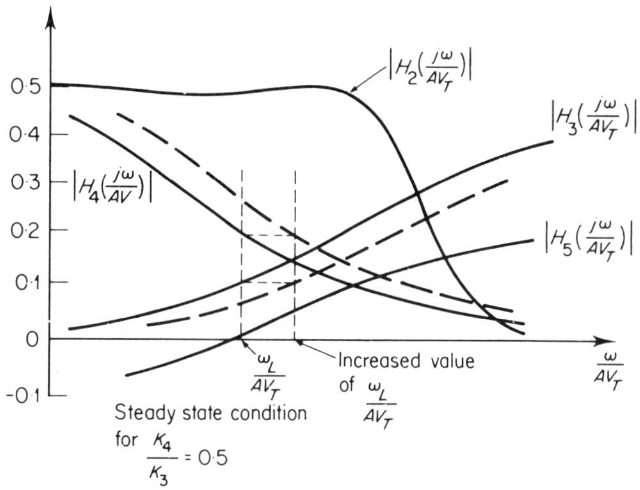

Figure 4. Typical transfer functions of frequency-locked filter

one steady-state condition for the system. We define the locked-frequency ω_L by $H_5(\omega_L/AV_T) = 0$. Now, the four transfer functions in Figure 4 have magnitudes that have a constant ratio, for any particular ω/AV_T, so that we may define

$$K_0 \equiv \left| H_2\left(\frac{\omega_L}{AV_T}\right)\right| \tag{19}$$

where $K_0 = g(K_4/K_3)$. The function $g(K_4/K_3)$ will, of course, depend on the precise filter transfer functions $|H_2|$, $|H_3|$ and $|H_4|$. In terms of the inverse function $|H_2^{-1}|$

$$AV_T = \omega_L |H_2^{-1}(K_0)|^{-1} \tag{20}$$

This result indicates that the scaling parameter AV_T is determined by the initial filter characteristics and a control parameter K_0. By varying K_0 via K_3 and K_4 we can control the frequency scaling parameter; this corresponds to a scaling of the horizontal axis of Figure 4. Note that K_0, via K_3 and K_4, is completely independent of the tunable filter transfer function.

Transient analysis and frequency control via $V_1(j\omega)$

The *voltage*-signal control system of Figure 3 is useful for investigating the steady-state behaviour of the system because it is clear that $V_5(j\omega/AV_T)$ is always zero under steady-state conditions. However, the most useful applications occur when the *input sinusoid $V_1(j\omega_L)$ is variable in frequency ω_L* and K_0 is a fixed parameter of the system. In this case, K_4/K_3 in equation (18) remains constant and the new steady-state condition is once more given by equation (16), and the scaling parameter AV_T must adjust accordingly. Thus, in terms of Figure 4, the input sinusoid frequency ω_L is varied and the transfer function $|H_2(j\omega/AV_T)|$ will be frequency-scaled to maintain $|H_2(j\omega L/AV_T)|$ equal to K_0. The dotted curves in Figure 4 show the effect of increasing ω_L. A most useful way of studying the stability and transient behaviour of this frequency-locked adaptive filter is to develop a simplified model in which *frequency* is the signal. Thus, the reference control frequency to which the steady-state response must settle is defined as ω_L, the final frequency-locked value. The next step is to analyze the filter and, from expressions for $|H_3(j\omega)|$ and $|H_4(j\omega)|$ develop a relationship between $H_5(j\omega)$ and frequency in the form of a Taylor Series Expansion. (We will only use the first term of the expansion.) Thus, by analysis of the particular tunable-filter topology, we derive a relationship between the magnitude $|V_5(j\omega)|$ and the frequency deviation $\omega(t)-\omega_0(t)$, where $\omega_0(t)$ is the slowly varying locked frequency in the control system shown in Figure 5. The error voltage $V_5(j\omega/AV_T)$ is ideally zero at the

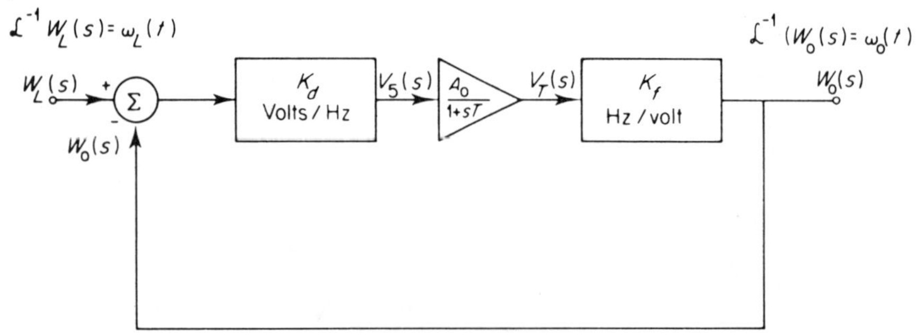

Figure 5. Frequency-signal control system

lock-in frequency ω_L. However, if a finite error is available then we may write down a Taylor Series Expansion for $V_5(j\omega)$ in the form

$$V_5(j\omega) = \frac{dV_5(j\omega)}{d\omega}(\omega-\omega_0) + \frac{1}{2!}\frac{d^2V_5(j\omega)}{d\omega^2}(\omega-\omega_0)^2 + \frac{1}{3!}\frac{d^3V_5(j\omega)}{d\omega^3}(\omega-\omega_0)^3 + \ldots \tag{21}$$

Since the system is assumed to remain in the vicinity of the origin of the V_5 curve (Figure 4), then we may simplify and linearize this expression by assuming sufficiently small $[\omega(t)-\omega_0(t)]$ so that

$$V_5(j\omega) \simeq \frac{dV_5(j\omega)}{d\omega}[\omega(t)-\omega_0(t)] \tag{22}$$

Thus from equation (22), K_d as shown in Figure 6 is defined as

$$K_d = \frac{V_5(j\omega)}{\omega_L-\omega_0} = \frac{dV_5(j\omega)}{d\omega} \tag{23}$$

The error voltage $V_5(j\omega)$, after amplification and filtering, then controls the *actual* lock-frequency ω_0 via a factor K_f that will depend on the filter characteristic. Thus, using Laplace representation $W_L(s)$ and $W_0(s)$ for real quasi-steady-state frequencies $\omega_L(t)$ and $\omega_0(t)$,

$$K_f = \frac{W_0(s)}{V_T} \tag{24}$$

where V_T is the tuning voltage at the filter input and the complete frequency-signal control system is as

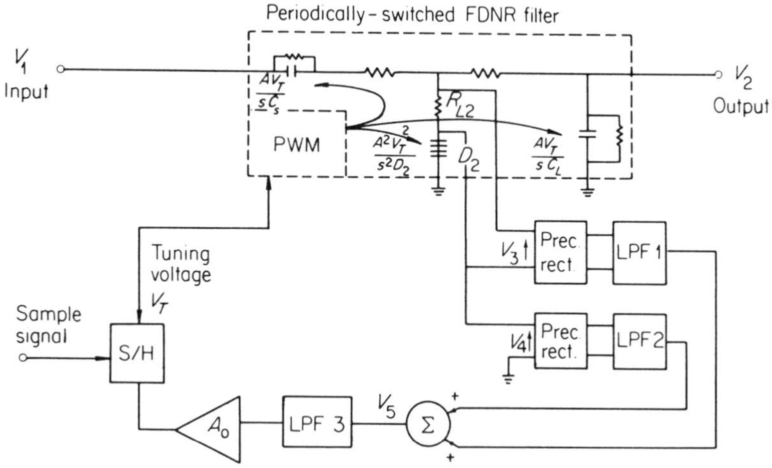

Figure 6. System diagram of a proposed frequency-locked filter

shown in Figure 5. The reference frequency is $\omega_L(t)$ and will normally be constant. Analysis of Figure 5 gives

$$W_0(s) = \frac{1}{1+(sT/A_0K_dK_f)}W_L(s) \tag{25}$$

where the pole at $s = -1/T$ is designed to dominate all other system time constants, including the delay involved in the lock-in frequency $\omega_L(t)$ following the tuning-voltage $V_T(t)$. (This is similar to neglecting the time constant of a voltage controlled oscillator in a conventional phase-locked oscillator system.) Thus the settling-time of the frequency-locked filter may be estimated from equation (25) using the quasi-steady-state representation of frequency $\omega_L(t)$ and $\omega_0(t)$. Thus by inverse Laplace transformation of equation (25),

$$\omega_0(t) = \omega_{in}(t)[1 - e^{-A_0K_fK_d t/T}] \tag{26}$$

Rejection of control signal at filter output

A useful method of rejecting the control signal $V_1(j\omega_L)H_2(j\omega_L)$ at the output port 2 is to design for K_0 equal to zero. Then, for all locked conditions,

$$K_0 = \left| H_2\left(\frac{\omega_L}{KV_T}\right) \right| = 0 \tag{27}$$

and the control signal cannot appear at the filter output.

Message transmission

The frequency-locked filter is, of course, not generally useful unless it can remain locked while performing the usual filtering operation. There are basically two methods of achieving message transmission through the filter from port 1 to port 2.

Sample-hold feedback: intermittent message. The most simple method is to design for K_0 equal to zero, so that the control signal sinusoid $V_1(j\omega_L)$ is absent at port 2, and to incorporate a sample-hold network in the feedback loop of Figure 3 that maintains (holds) V_T at a constant value while a message is being transmitted. This method implies some external control over the adaptive filter in order to produce a message-operated hold network. Of course, in the absence of a message, the sample-hold reverts to the sample mode, and the filter frequency locks in the normal way. This method will be illustrated by an example.

Low-level message transmission: continuous message. The control signal has an input power density spectrum ideally given by the impulse function $N\delta(\omega_1)$ (neglecting $\omega < 0$). Let the message power density spectrum be $|M(j\omega)|^2$. Then the spectra at ports 2, 3 and 4 are

$$|V_2(j\omega)|^2 = |M(j\omega)|^2|H_2(j\omega)|^2; \text{ ideal filtered output} \tag{28}$$

$$|V_3(j\omega)|^2 = |H_3(j\omega)|^2[|M(j\omega)|^2 + N\delta(\omega_L)] \tag{29}$$

$$|V_4(j\omega)|^2 = |H_4(j\omega)|^2[|M(j\omega)|^2 + N\delta(\omega_L)] \tag{30}$$

where it is assumed that K_0 is zero so that $N\delta(\omega_L)$ does not appear in the first equation. Then, ideal frequency-locked operation depends on the adaptive filter not responding significantly to $|M(j\omega)|^2$ in equations (29) and (30). In general, it is sufficient for $|M(j\omega)|^2$ to be much smaller than $N\delta(\omega_L)$ so that the feedback signal $V_5(j\omega)$ in equation (13) does not contain a significant term due to the message. However, in practical terms, this implies that the control sinusoid must be transmitted at a power level that is much greater than that of the message, thereby causing deterioration in dynamic range.

A frequency-locked elliptic ladder filter using periodically switched conductances

A frequency-locked ladder filter of the type described in Figure 3 is required with an elliptic transfer function exhibiting the following specifications:

Passband ripple: 1 dB
Stopband attenuation: 45 dB
Cut-off frequency $\omega_c/2\pi$ frequency locked over the range: 60–260 Hz
Zero of transmission: $3{\cdot}279\,f_c, (f_c = \omega_c/2\pi)$.

It is required that the control sinusoid shall not appear at the output port.

The first step in the design is to realize a voltage tunable FDNR lowpass filter exhibiting the required response characteristic over the above frequency range. This is possible using a periodically switched FDNR realization[5] and a voltage to pulse-width converter (PWM) as shown in the system diagram in Figure 6. The control frequency ω_L must correspond to the zero of transmission of the transfer function $H_2(s)$. Therefore we design for $K_0 = 0$. The voltages corresponding to V_3 and V_4 are chosen as the voltages existing across the shunt resistance D_2 and R_{L2} respectively. The transfer functions $|H_2(j\omega)|$, $|H_3(j\omega)|$ and $|H_4(j\omega)|$ are sketched in Figure 7 where it is seen that the shunt arm containing R_{L2} and D_2 is series resonant

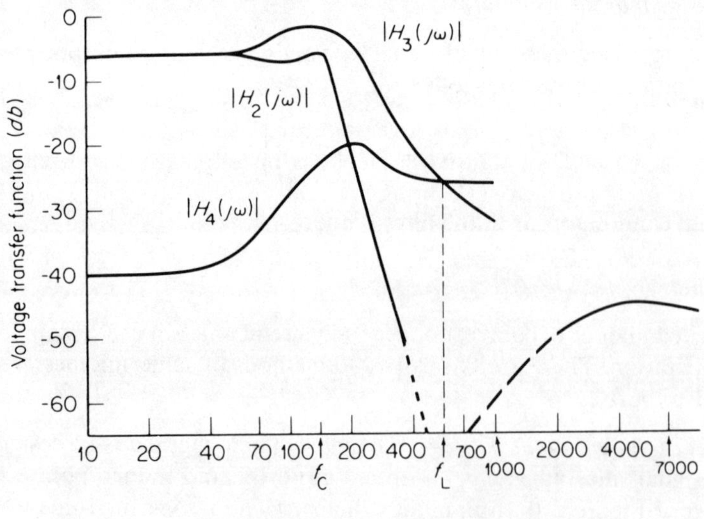

Figure 7. Frequency response of third order elliptic lowpass filter

at the required lock-in frequency. It is then not necessary to determine K_0 as a function $g(K_3/K_4)$ because, for $K_0 = 0$, $K_3 = K_4$.

The next step in the design of the frequency-locked filter is to design the control system such that $\omega_0(t) = \omega_L(t)$. The voltages V_3 and V_4 are precision rectified and lowpass filtered to obtain DC voltages corresponding to $K_3|V_3(j\omega)|$ and $K_4|V_4(j\omega)|$ which are required for the summation described by equation (13). The voltage $V_5(j\omega)$ thus obtained is lowpass filtered and amplified to give the tuning voltage. LPF3 is included in Figure 6 to ensure the system is described by the dominant pole transfer function given in equation (25).

The sample/hold network in Figure 6 has been designed into the control system to realize a sample/hold feedback system for intermittent message transmission. The control for the sample/hold network must be obtained from the input signal $V_1(j\omega)$, in which the system samples and adapts when the control signal is applied and holds when a message is transmitted. The practical version of Figure 6 is shown in Figure 8 with each block in Figure 6 outlined in Figure 8.

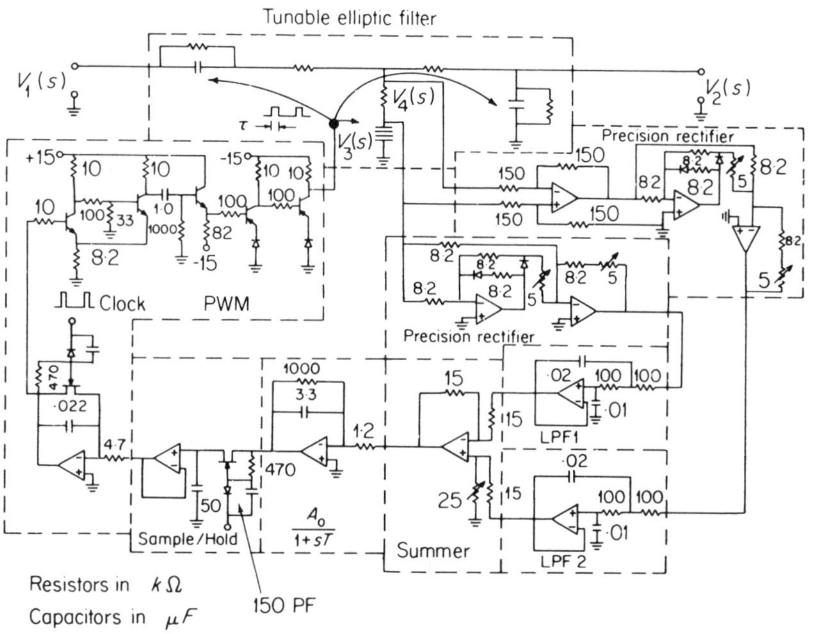

Figure 8. Circuit diagram of a practical frequency-locked filter

To evaluate the performance of the frequency-locked adaptive filter, the system shown in Figure 6 must be related back to the frequency control system in Figure 5. First, consider the factor K_d given by equation (23). This can be evaluated graphically by plotting $V_5(j\omega)$ as a function of the frequency deviation $[\omega_L(t)-\omega_0(t)]$ for each value of $\omega_L(t)$. Since $\omega_0(t)$ varies with the frequency-scaling factor K, $V_5(j\omega)$ is plotted as a function of $[\omega_L(t)-\omega_0(t)]/\omega_0(t)$ as shown in Figure 9. This curve is obtained directly from $|H_3(j\omega)|$ and $|H_4(j\omega)|$ in Figure 7 by evaluating $K_3|H_3(j\omega)|-K_4|H_4(j\omega)|$. Thus, the shape of the curve for our example ($K_3 = K_4$) is identical to the difference between $|H_3(j\omega)|$ and $|H_4(j\omega)|$ in Figure 7. The value of K_d is given by the slope of the curve in the region of interest, in this case $V_5(j\omega) = 0$. Instead of taking the slope of Figure 9 at the origin, improved accuracy can be obtained if the typical instantaneous frequency deviation can be estimated. In this design, it was required that the filter should adapt to step frequencies of the form $\omega_L[1 \pm Q\omega(t)]$, where Q is typically 0·2. Thus the frequency deviation $[\omega_0(t)-\omega_L(t)]$ will typically be in the range $(1 \pm 0{\cdot}2)\omega_L(t)$ representing ± 100 Hz steps on either side of the origin in Figure 9. Average slopes of the curve in Figure 9 for the two regions are

$$\frac{\mathrm{AB}}{\mathrm{OA}} = 0{\cdot}25\ V, \quad 0 < \frac{\omega_L(t)-\omega_0(t)}{\omega_0(t)} < 0{\cdot}2, \tag{31}$$

$$K_d \simeq 0{\cdot}25\left(\frac{2\pi}{\omega_0}\right)$$

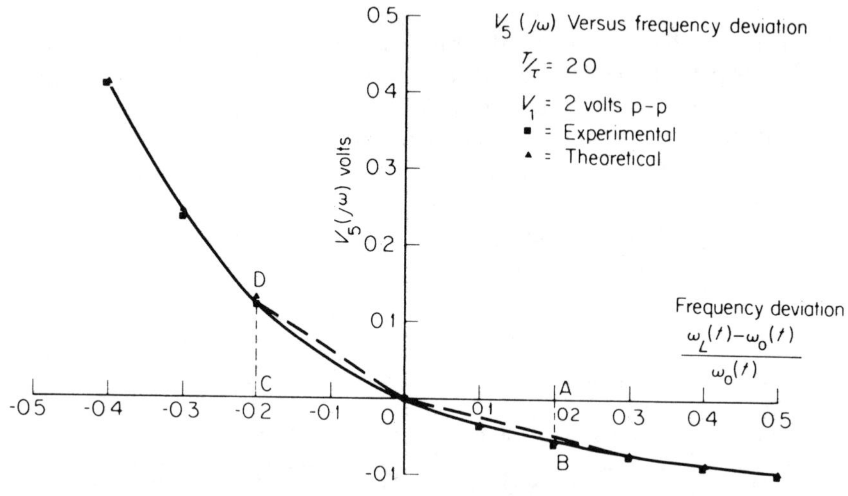

Figure 9. $|V_5(j\omega)|$ as a function of normalized frequency deviation

and

$$\frac{\mathrm{CD}}{\mathrm{CO}} = 0{\cdot}6\ V, \quad -0{\cdot}2 < \frac{\omega_L(t)-\omega_0(t)}{\omega_0(t)} < 0, \tag{32}$$

$$K_d \simeq 0{\cdot}6\left(\frac{2\pi}{\omega_0}\right)$$

This somewhat crude selection of an average slope in the region of interest can now be used to estimate transient performance.

The value of K_f in Figure 5 can be evaluated by calculating ω_0 for each value of V_T. In the system in Figure 8, the PWM is realized using a ramp generator having a slope of V_T/RC and period T, and a level detector which switches state at a threshold voltage V_{th}. For this configuration, the scaling factor K in the transfer function (equation (2)) is related to V_T as follows

$$K = K_1\frac{RCV_{th}}{V_T T} + K_2\left(1-\frac{RCV_{th}}{V_T T}\right) \tag{33}$$

which is obtained by substituting for the pulse width $\tau = RCV_{th}/V_T$ in equation (3). Using equation (33), a plot of $\omega_0(t)$ against V_T can be obtained as shown in Figure 10. This graph is verified by experimental measurement on the PWM and tunable filter as shown in Figure 10. The value of K_f can be calculated from Figure 10 using equation (24), where it is seen that a mid-range value of K_f = 71·6 Hz/volt is calculated.

These values of K_f and K_d can be substituted into equations (25) and (26) to describe the performance of the frequency-locked filter.

A useful method of evaluating the performance of the filter is to determine the step response. In the frequency-locked filter, step changes in frequency $\omega_L/2\pi$ of 450–550 Hz and 550–450 Hz were sequentially applied. Using equation (26), noting that $A_0 = -833$, T = 3·3 s in Figure 8 and denormalizing K_d to $\omega_0/2\pi$ = 500 Hz in equations (31) and (32), we obtain a *theoretical* step response of the form

$$f_0(t) = 100(1-e^{-t/0{\cdot}05})+450,\ f_L = 450 \rightarrow 550\ \text{Hz}$$

$$K_f = 71{\cdot}6\ \text{Hz/volt},\ K_d = 0{\cdot}5\ \text{mV/Hz} \tag{34}$$

$$f_0(t) = 100(e^{-t/0{\cdot}11}-1)+550,\ f_L = 550 \rightarrow 450\ \text{Hz}$$

$$K_f = 71{\cdot}6\ \text{H/volt},\ K_d = 1{\cdot}2\ \text{mV/Hz} \tag{35}$$

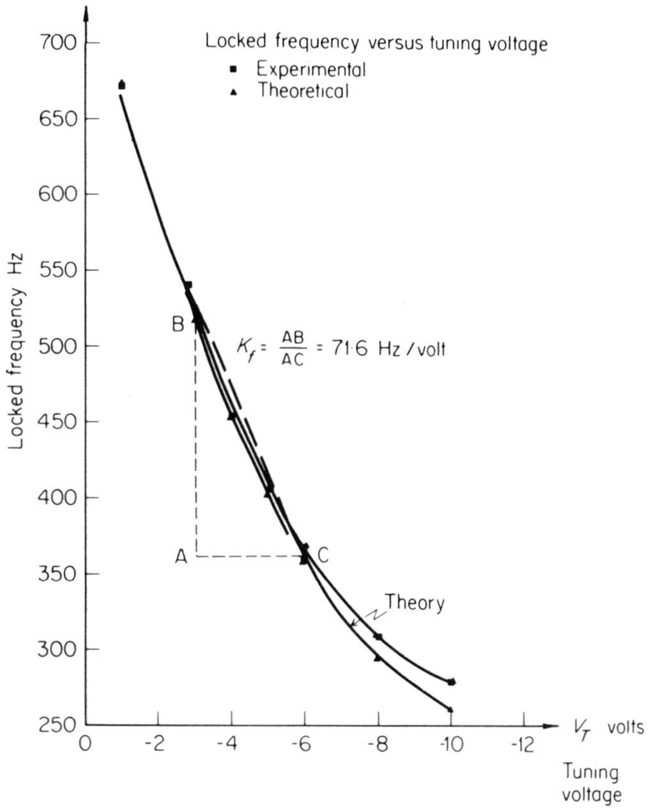

Figure 10. $\omega_0(t)$ as a function of tuning voltage V_T

The *measured* time constants were respectively 0·05 and 0·10 s for steps 450–550 Hz and 550–450 Hz in $\omega_L(t)$. As can be seen, favourable agreement is obtained between the measured and theoretical responses. For small steps in control frequency $\omega_L(t)$, the linear frequency control system describes the frequency-locked transient behaviour of the adaptive filter.

The frequency-locked filter is designed for $K_0 = 0$, measurements of $|H_2(j\omega_L)|$ for the system in Figure 8 over the range 270 Hz $< \omega_L/2\pi <$ 850 Hz give $|H_2(j\omega_L)| < -60$ dB, thereby confirming the virtual elimination of the control sinusoid from the output port.

CONCLUSIONS

It is possible to use a voltage-tuned (frequency-scaled) elliptic ladder filter as part of a feedback system in which the transfer function of the ladder filter is automatically adaptive in the frequency domain. The control signal is the frequency of a control sinusoid and it is possible to design the frequency of the zero of transmission of the filter to coincide with the frequency of the control sinusoid. The shunt series DC branch of an FDNR lowpass elliptic tunable ladder filter is used to achieve frequency discriminator operation.

A simplified linear control system model is proposed for the analysis of the transient performance of a frequency-locked ladder filter and it is shown by example that the response of the system to a small step of the control frequency may be estimated.

The application of this filter for message reception and adaptive filtering is feasible by one of two methods. Either the control sinusoid is transmitted at a much higher signal level than that of a *continuous* message or the operation of the adaptive filter is restricted to *intermittent* message transmission so that a sample-hold network may be employed to hold the lock condition during the periods of message transmission.

In the practical example that has been given, the periodically switched conductance method of realizing a voltage-tunable filter has been employed. Of course, exactly the same results may be obtained by employing the analogue-multiplier method.

REFERENCES

1. H. R. Rudin, 'A continuously adaptive equalizer for general purpose communication channels,' *Bell System Tech. Journ.* **48**, 1865–1884 (1969).
2. Basil Barber, 'A self adjustable bandpass filter picks off weak signals despite noise', *Electronics*, **42**, 104–107 (1969).
3. P. E. A. Cowley, 'Self-adaptive filter', *IBM Tech. Bulletin*, **12**, 2098–2099 (1970).
4. P. C. Majhee and K. P. Rajappan, 'Improved self tuning filter for low frequencies', *Electronic Letters*, **5**, 116–119 (1969).
5. L. T. Bruton and R. T. Pederson, 'Tunable RC-active filters using periodically-switched conductances', *Proc. IEEE Int. Symp. Circuit Theory*, 350–354, April 18–21 (1972).
6. J. A. Kaehler, 'Periodically-switched filter networks—a means of amplifying and varying transfer functions', *IEEE J. Solid State Circuits*, SC-**4**, 225–230 (1969).
7. Y. Sun and I. T. Frisch, 'Resistance multiplication in integrated circuits by means of switching', *IEEE Trans. Circuit Theory*, CT-**15**, 189–192 (1968).
8. K. Hirano and S. Nishimura, 'Active RC filters containing periodically-operated switches', *IEEE Trans. Circuit Theory*, CT-**19**, 253–260 (1972).
9. Ming-Lei Liou, 'Exact analysis of linear circuits containing periodically operated switches with applications', *IEEE Trans. Circuit Theory*, CT-**19**, 146–154 (1972).
10. A. Antoniou, 'Realization of gyrators using operational amplifiers and their use in RC-active network synthesis', *Proc. IEE* **116**, 1838–1850 (1969).
11. H. J. Orchard and O. F. Sheahan, 'Inductorless bandpass filters', *IEEE J. Solid State Circuits*, SC-**5**, 108–118 (1970).
12. L. T. Bruton, 'Network transfer functions using the concept of frequency-dependent negative resistance', *IEEE Trans. Circuit Theory*, CT-**16**, 406–408 (1969).
13. J. Gorski-Popiel, 'RC-active synthesis using positive immittance converters', *Electronics Letters*, **3**, 381–382 (1967).
14. L. T. Bruton, 'Electronically tunable analog active filters', *IEEE Trans. Circuit Theory*, CT-**19**, 299–301 (1972).

Part IV
Fabrication

Although fabrication and design are frequently treated as unrelated topics, the designer can develop better, more reliable and economical active filters by being aware of both the possibilities and the limitations that a particular technology imposes on filter fabrication. This is true even if a filter is built in discrete form, but especially if monolithic or hybrid integrated circuit (IC) technology is chosen. Modern computer-aided design programs often include fabrication steps that the designer should understand; more importantly, good designs often make use of characteristics of the technology, such as temperature tracking, matched passive and/or active devices and high-accuracy processing steps, such as laser trimming. At the same time, the designer is limited by certain aspects of the technology, such as restrictions on power, frequency range, chip or substrate area, and characteristics and size of devices.

Fortunately, the designer does not have to be an expert in fabrication, but some familiarity with the basic properties of available technologies will, as stated before, generally result in better designs. The papers in this chapter are not intended to explain fabrication technology, but instead are selected to give the reader a basic understanding of how the fabrication process affects active filter design. The papers deal exclusively with hybrid IC technology because of its practical importance and predominance in active filter implementation. In addition to the articles here, the reader should consult papers 2-3 and 3-1 for further considerations related to hybrid processing. Design for monolithic technology is treated to some extent in papers 5-3, 5-4 and also in Part VI.

The first two papers do not deal with fabrication directly but with questions related to circuit simulation, tuning and performance evaluation based on a chosen fabrication technology. In *paper 4-1,* Semmelman, Walsh and Daryanani describe two computer-aided design programs in use in the Bell System for simulating circuit performance under the assumption of known statistical variations and aging characteristics of components. The paper illustrates the considerations important for the fabrication of practical filters. A quite general and efficient tuning algorithm is discussed in *paper 4-2* by Lopresti. The algorithm is used to adjust the resistors of a hybrid design, based on measured values of capacitors and resistors, such that the transfer function coefficient errors due to element variations are minimized. In addition to its efficiency, the main advantages of the method are more accurate filter performance because trimming errors are largely compensated, and increased production yield and speed because of relaxed tolerance requirements during fabrication. *Paper 4-3* by Worobey and Rutkiewicz details the processing steps and the materials used for fabricating a thin-film hybrid active filter package (the STAR of paper 2-3). Renz et al. in *paper 4-4* describe design and fabrication of tantalum thin film discrete and distributed RC active filters for telephone applications. Of particular interest are the simplified fabrication and the reduced size made possible by the use of distributed RC structures. The design of miniaturized active filters, using thick film resistors, for applications at higher frequencies, is discussed in *paper 4-5* by Saraga, Haigh and Barker. Because of critical performance requirements and high order, the design is based on active simulations of passive RLC ladders (compare papers 3-3 to 3-5). Design steps, optimization methods and fabrication procedures that result in practical microelectronic filters are presented in detail (see also the companion paper [1]).

[1] W. Saraga, D. Haigh and R. G. Barker, "A Design Philosophy for Microelectronic Active-RC Filters" *Proc. of the IEEE,* vol. 67, No. 1, pp. 24-33, January 1979.

Statistical Circuit Design:

Linear Circuits and Statistical Design

By C. L. SEMMELMAN, E. D. WALSH and G. T. DARYANANI

(Manuscript received November 17, 1970)

The design of linear circuits requires the designer to consider element tolerances, distributions and correlations and how they interact with proposed manufacturing test limits and service conditions such as temperature and aging. Analytical methods can be used in a limited number of cases; simulation methods using large modern computers permit study of more complex design problems. The principal themes of this paper are the designer's needs for computational assistance and the ways in which computer programs may be organized to meet these needs. Two computer programs for this purpose are described. One is a general-purpose analysis program for large networks having any topology. The other is a more specialized program suitable for "biquad" networks which may be used as building blocks to form a variety of filter and equalizer networks.

I. INTRODUCTION

In the Bell System, the origin of interest in the statistical design of linear circuits is lost in history. It could have been no later than the earliest planning for repeaters and filters for carrier systems, as no such system is possible without precise control of frequency and loss characteristics. As the total installed cost of a carrier system is always much greater than the cost of the design effort, use of the most advanced design procedures and techniques known, including statistical methods, has always been justified.

The selection of tolerances for component parts is one of the many topics which is under the control of the designer. He must select values and specify manufacturing tests so as not to increase costs either by paying a premium for unnecessarily precise elements or by rejecting correctly assembled networks in production. In the past, the tools that have been available to the designer to assist in the determination of tolerance limits have ranged from primitive to highly sophisticated and mathematical. In the present era, which is dominated by electronic computers, simulation methods may prove the most valuable of all. The purpose of this paper is to review some of the basic methods of calculating and combining deviation characteristics, pointing out the powers and limitations of these methods, and to describe additional capabilities obtainable from two general purpose computer programs. These enable the designer to determine the result, in production and in service, of specifying a given set of tolerances, distributions, correlations, tuning and testing procedures and service conditions. All the material is confined to linear networks, such as filters, equalizers and feedback amplifiers, which are designed and built to frequency domain specifications.

1.1 *Analytical Methods*

Engineers engaged in designing electrical networks and selecting tolerances for their components have traditionally used the classical design technique which consists of making a first approximation, computing or measuring its performance and refining the approximation. That this procedure worked well and converged rapidly when applied to network design can be seen from the following. The image parameter design process for filters and comparable methods for other networks gave the engineer an understanding of the sensitivity of the various components, i.e., which component has its greatest effect in what part of the frequency spectrum. This led intuitively to both the specification of the tests that would be sufficient to assure accurate manufacture and to the knowledge of which few components need careful scrutiny at any frequency, in order to select appropriate tolerances. A few calculations or laboratory measurements followed by some experimental arithmetic would then produce the necessary tolerance limits. As the element tolerances being used were small, the resulting network deviations were proportional and could be combined by simple addition. Adding the absolute values of the maximum permissible deviations would produce worst-case behavior and if system performance limits were met the job was done.

Mathematicians came to the assistance of the design engineers and devised clever labor-saving methods of determining the deviation effects and of combining them linearly. For example, the rate of change of the driving point impedance of a reciprocal network with respect to a branch impedance is given by

$$\frac{\partial Z}{\partial Z_x} = \left(\frac{I_x}{I}\right)^2$$

and the rate of change of a transfer impedance is

$$\frac{\partial Z_T}{\partial Z_x} = \left(\frac{I_x}{I_2}\right)\left(\frac{I_x'}{I_1'}\right)$$

where the symbols all represent complex quantities and are defined in Fig. 1.* One or two solutions for all the branch currents then permit each individual driving point or transfer impedance deviation to be calculated and scaled to represent the assumed element tolerances. As stated above, these scaled deviations may be added to evaluate the behavior of a network with any one set of element values.

More insight, however, may be obtained by drawing a polygon in the complex plane within which the impedance must lie. This may be done in the following manner: Select a plus or minus sign for the departure of each element from its nominal, as required to make the real component of each scaled deviation positive. Adding the scaled deviations with these signs will give the rightmost point of the polygon. Successive corners are obtained by reversing the signs of the scaled deviations, one at a time, in the sequence of their increasing angles. When all signs have been reversed, the leftmost point is obtained and a second reversal in the same sequence generates the other half of the convex polygon and returns to the first point.

Still more may be learned by superimposing a family of ellipses each of which contains a specified percent of the universe of networks which would be manufactured from elements having the assumed tolerances. This may be done if the distribution of each element value is gaussian and its standard deviation is known.[1,2]

Figure 2 illustrates both the impedance polygon and ellipses for a simple filter where the standard deviation is taken as 40 percent of the tolerance limit assumed for the polygon.

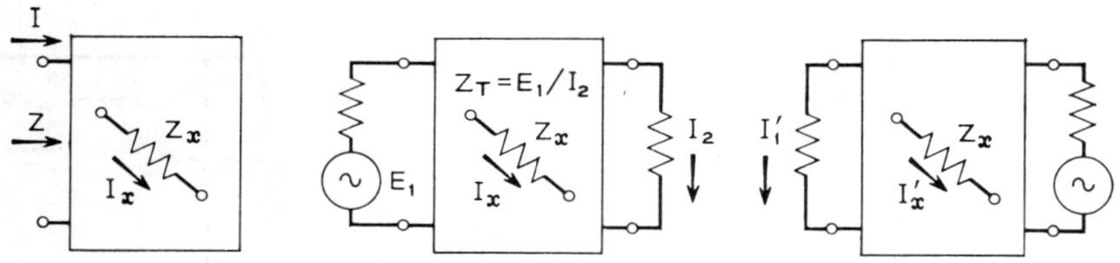

Fig. 1—Definition of symbols for driving point and transfer impedance deviations.

* The origin of these equations cannot be determined. They were in use in 1933 in the design of passive networks. They may be derived in a number of ways, including the use of network theorems and differentiating familiar expressions and identifying terms.

1.2 *Limitations to These Methods*

The first limitations are the facts that large tolerances do not permit linear extrapolation to large deviation shapes and that these do not combine linearly. Other difficulties arise when sophisticated design techniques such as filter synthesis by insertion loss design techniques[3] are used. The designer develops no "feel" for his design and the network performance becomes more sensitive to every element. Further, the elements may not have gaussian distributions. For example, 10 percent components may have a bimodal distribution resulting from a supplier's sorting the product and selling the 5 percent elements elsewhere. It may be impossible to determine whether high or low values of the elements should be used to evaluate worst-case behavior because a large change in one element's value may reverse the sign of the sensitivity for another element. Finally, the worst case may be so bad that system performance requirements are not met and tighter tolerances, factory adjustments, selective assembly, or reduced yield must be considered. Any of these increases the cost of manufacture, complicates the designer's task, and requires more powerful methods of determining the tolerances, adjustments and tests for production.

From the above discussion, it is apparent that only in rare cases will the selection of tolerances be a straightforward process where linear extrapolation and simple addition of deviation shapes will suffice. Frequently, the mathematical and engineering complexities put the designer's task beyond the limits of analytic methods available today.

II. MONTE CARLO TECHNIQUES

"Monte Carlo" computer techniques permit simulating the statistical behavior of random variables and imitating the tuning and testing actions performed in the factory. Such computer programs can take the place of a pilot production run, and furnish comparable information in a much shorter time and at a fraction of the cost.

In these methods the computer is programmed to generate a sequence of numbers which appear to a casual observer to be random, although their distribution and periodicities are known and carefully controlled. By mathematical transformations, these numbers are modified to represent possible values for each network element and to display the appropriate nominal value, tolerance, distribution shape and correlation with other elements. Further explanation of the

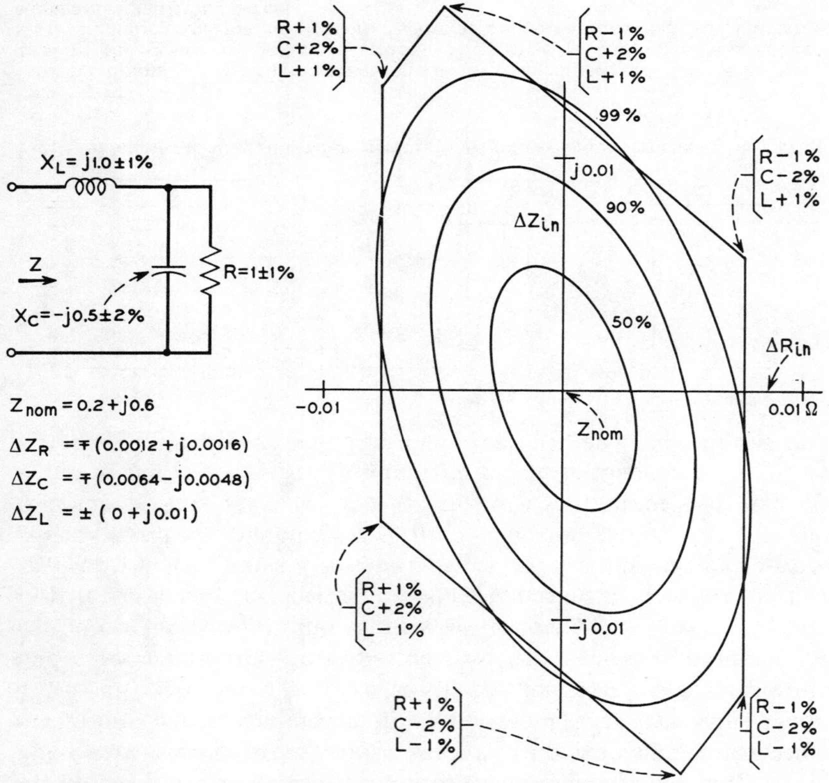

Fig. 2—Simple filter driving point impedance polygon and ellipses.

methods will be found in Appendix A. Many networks composed of such simulated elements are computed and the results compared with proposed test limits. A study of the computer output will enable the designer to determine the suitability of the nominal values, tolerances and tests.

The remainder of this article describes two computer programs, BELTAP and BITAP, which perform this simulation. The programs are addressed to different needs, but within its own area of application each is a general-purpose program.

BELTAP is a program for simulating and analyzing networks such as filters, equalizers, feed-back amplifiers, etc., having many elements and arbitrary topology. It computes insertion loss and return losses at both ends of the network. With it, the designer can obtain information on test yields, the similarity between tests in this respect, element sensitivities, and selected scatter plots and histograms. This information should help him understand the complex interrelationships among element values, tolerances, distributions and tests which are not obvious from the network topology.

BITAP, on the other hand, can deal only with networks of a fixed topology, whose transfer function is restricted to a biquadratic function of frequency [defined in equation (1)]. By varying the element values used and connecting a number of such sections in tandem lowpass, highpass, bandpass filters and various types of equalizers can be constructed. BITAP facilitates specifying correlation among element values, factory tuning procedures, service conditions such as high or low temperature exposure and aging, and performance tests such as gain, phase and delay. If the service conditions are omitted, the tests become factory production tests; if included, end-of-life system performance checks.

III. DESCRIPTION OF BELTAP OPERATION

BELTAP is capable of performing tolerance analysis computations on a vast majority of the transmission networks in production today. The networks may consist of both lumped elements, having specified nominal values, tolerances and distributions, and fixed subnetworks, described by admittance matrices which are obtained either by calculation or measurement. BELTAP builds perturbed circuits by deviating each lumped element by a random amount within the specified tolerance and having a uniform, gaussian or arbitrary distribution shape. If correlation of deviations or tuning is required, special subroutines may be inserted. The perturbed circuits are then analyzed and insertion loss and input and output return losses are computed at each specified frequency. Special subroutines may be prepared to compute other quantities. For each perturbed network the complete set of elements and all measures of circuit performance are recorded on a disk file for later review by a postprocessor. The computed performance values are compared to the circuit requirements, to determine if the network passes or fails, and to calculate the production yields. The data on disk are then reviewed to collect other standard outputs and the listings, scatter plots and histograms requested by the designer. The output will be described more fully below. The variety of outputs available from BELTAP should be of great assistance to a designer who has little insight into the sensitivities of the elements, the interaction of their deviations and the relationship of both to the test limits.

3.1 *Specification of the Network*

A variety of types of information can be used to specify the network to be analyzed. It can contain resistors, inductors, capacitors and controlled current sources. Also, a quality factor, or Q, may be specified for the inductors and capacitors. Parallel and series resonant circuits may be specified by supplying only one of the reactive components and the resonant frequency. All the above items may be assigned tolerance limits and distribution shapes as required. Uniform and gaussian shapes are built-in; other special shapes may be read in at execution time. If correlations or factory tuning operations must be simulated, special subroutines can be written for this purpose.

BELTAP can allow fixed subnetworks to be included in the overall network. These subnetworks are specified by supplying their definite admittance matrices. The matrix data are generally obtained from an analysis of the subnetwork by a general-purpose linear network analysis program but may also be obtained from measurements.

The subnetwork feature allows BELTAP to perform a tolerance analysis on the critical components of a large network in a feasible amount of time. Consider, for example, the wideband amplifier shown in Fig. 3. The complete amplifier contains 80 nodes and 111 components. By blocking the circuit into two four-port subnetworks, as shown, the network to be computed is reduced to 24 nodes and 36 elements. The 36 elements are the critical components in the feedback paths and in sensitive positions in the forward path. The complete circuit contains

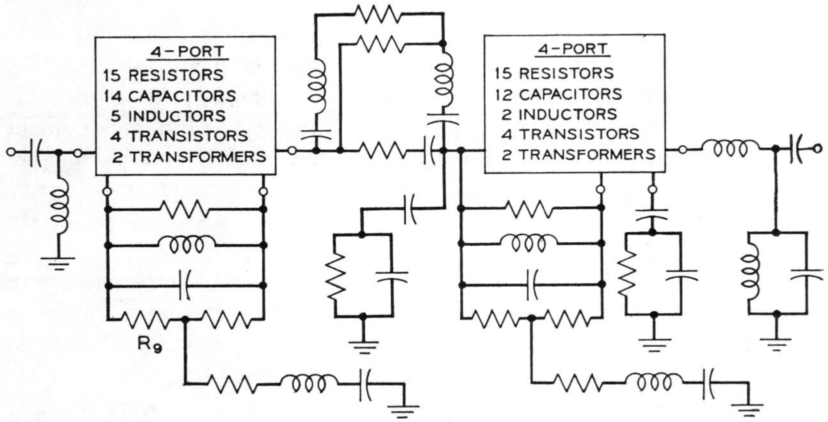

Fig. 3—Network for analysis by BELTAP, reduced to subnetworks.

four-port transformers, two-port transformers, and high-frequency transistors. Accurate lumped element models for these devices, particularly the transformers, do not exist. However, they can be easily incorporated into a BELTAP analysis by supplying their measured Y-matrix data (or, more likely, measured data reduced and transformed to give Y-matrix data).

3.2 BELTAP *Output*

A Monte Carlo analysis can produce such vast quantities of data that the designer, confronted with the complete results, might have difficulty identifying and extracting significant facts. It is the task of a postprocessor program to read the disk files and organize these data into brief comprehensible forms.

The postprocessor has two general classes of output, default output which is always produced and optional output which is produced only as specified. The default output is of a general nature and not of great detail or length. The requested output can be as voluminous as the engineer desires.

3.2.1 *Default Output*

The default output consists of four sections. The first section gives a summary of the test statistics. This includes the percent yield, average value, standard deviation, maximum value and minimum value for each test. It also gives the overall percent yield, combining all tests.

The second section of default output is a pseudo-correlation among pairs of tests. The pseudo-correlation is defined as the fraction of the networks for which the pass-fail result of the two tests agreed. For example, suppose ten networks were produced, the first seven passed test 1, the first nine passed test 2, and the remainder failed in both cases. The pseudo-correlation would then be 0.8 as networks 1 through 7 and 10 gave identical pass-fail results on the two tests. This output can be useful in reducing the number of tests performed without sacrificing reliability of performance. It may also be useful in determining an optimal sequence in which to perform the tests.

The third section of default output tabulates each parameter's name, nominal value, average value and standard deviation. This information is useful as a check on the random distributions and on the adequacy of the sample size. For example, the nominal value and the average value should be the same for all the parameters with a symmetrical density function.

The final section of default output gives the correlation among test values and parameters. An approximation to the sensitivity of each test value to each parameter is also printed.

From these correlations and sensitivity estimates, the engineer is given insight into how individual circuit components affect his circuit's overall performance. More important, they give him a direction to follow in choosing those elements that may require tighter tolerances and those which may have broader tolerances.

3.2.2 *Optional Output*

The optional output has three sections: tabular output, scatter plots and histograms. Tabular output is merely a listing of the random element values and test values for each circuit. Circuits which failed tests are flagged for easy identification. Test results and element values may be intermixed so that the results of a test may be listed in a column adjacent to the values of an element to which the test is very sensitive.

There are three types of scatter plots available: test vs. test, element vs. test, and element vs. element. Figure 4 shows a scatter plot

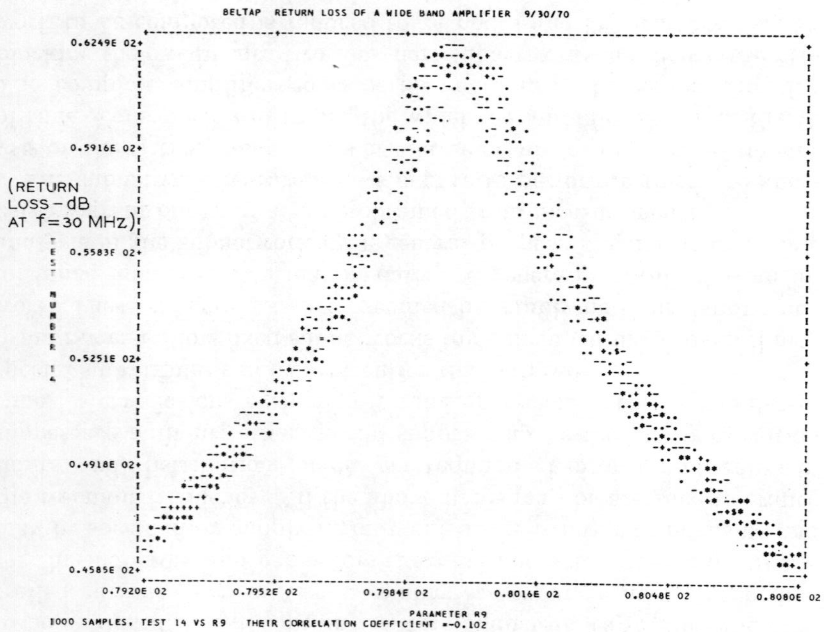

Fig. 4—Scatter plot of return loss vs. an element value.

of a test result vs. an element value. The various symbols printed on a scatter plot indicate the number of points that fell into a particular rectangular cell, the more points the blacker the symbol.

The scatter plots give an engineer more insight into how quantities in his circuit interact. They provide him with a type of pictorial information that is not available from correlation coefficients or sensitivity numbers. In Fig. 4, for example, the nominal value of the element occurs in the center of the plot, and at this point the slope of the test curve is near zero. Hence the partial derivative of the test result with respect to the element value and the correlation coefficient would be small. However, the test result is greatly influenced by the element value.

BELTAP can produce histograms of any number of element values and test values. Histograms of element values can be used as a check on the random number distributions and the density function routine in BELTAP. Histograms of the test values give the approximate shape of the probability density for the test.

3.3 *Summary of* BELTAP *Characteristics*

BELTAP provides a user with considerable flexibility in several respects. The network description language permits any configuration. The numbers of distribution types, branches, nodes and subnetworks are essentially unlimited as dynamic storage methods are used. As a result the size and number of networks computed will probably be restricted by computer running time, cost and numerical precision, rather than by array dimensions. Experience has indicated that when a very large network is encountered, the most sensitive elements can be selected by preliminary computer runs on portions of the entire structure. One or more additional runs can then be made with only these elements varying to obtain final data.

In addition to insertion loss and return loss calculations, which are provided by the program, users may write subroutines to evaluate other measures of network performance. Subroutines for simulating tuning adjustments, for introducing correlation effects, and for computing other measures of the network performance may also be written. The standard and optional outputs are designed to cover a broad range of types of data and methods of presentation. This variety should be helpful in discovering unsuspected interrelationships among tests and between tests and parameter values. For this reason, it is probably desirable to investigate considerably more tests than will actually be used in production.

IV. BITAP

The BITAP program has a very different purpose from that of BELNAP. BITAP can compute the behavior of only one kind of network, the "biquad." This network derives its name from the fact that its transfer function is a biquadratic function of frequency. It is an active network and has a fixed topology. The values of the resistors and capacitors used determine whether the biquad will be a lowpass, highpass, bandpass, loss equalizer or delay equalizer type of network. A number of biquad networks can be connected in tandem without interaction, so their gain and phase characteristics can be combined by simple addition and more complex structures realized. BITAP facilitates the investigation of realistic manufacturing conditions and environmental effects for biquad networks.

4.1 *Area of Application*

A common approach used in designing active filters is to build them with cascadable network sections. After the required transfer function has been determined, it is decomposed into a product of biquadratic functions of the form

$$\frac{V_{\text{out}}}{V_{\text{in}}} = \prod_{i=1}^{k} \frac{m_i s^2 + c_i s + d_i}{n_i s^2 + a_i s + b_i} \tag{1}$$

where m_i, c_i, d_i, n_i, a_i and b_i are real coefficients and $s = j\omega$.This decomposition may be done for any lumped, linear time-invariant network. The synthesis procedure consists of realizing each biquadratic function with an active network, which may be connected in tandem with other sections and will not interact with them. Examples of biquads are the state variable four amplifier biquad,[4] Moschytz's FENs[5] and the Sallen and Key networks.[6]

One of the aims in the realization of biquads is to make them isotopic, that is, to maintain the same network topology for all values of the coefficients m, c, d, n, a and b. If this can be done, the manufacturing process can be standardized, with obvious economic advantages. The state variable four amplifier biquad is completely isotopic; the FEN and Sallen and Key networks need a small number of topologies to realize all forms of the general biquadratic function. A circuit diagram of the four-amplifier biquad is shown in Fig. 5. By selecting the proper element values and making the right connections this network can be made to produce all the various filter functions—lowpass, highpass, bandpass, band-reject, notches and equalizers and delay shapes. The advantages of the realization justify writing the BITAP program.

We are concerned with the analysis of biquad networks under realistic manufacturing conditions and with environmental changes such as temperature, humidity and aging with time. Manufacturing conditions must include correlations, which will be present in thin-film and integrated circuit realizations and tuning adjustments needed to realize high-precision networks. Environmental factors must be taken into account as new manufacturing methods and materials are to be used and extensive experience with them is not available.

4.2 *Description of* BITAP

The basic approach is similar to that described above for BELTAP. For each element a nominal value, a tolerance and a distribution type are specified. The program generates random numbers which are transformed to become deviated element values. These are combined to form the six biquad coefficients, which then permit evaluation of the gain, phase or delay of each biquad at each frequency. These quantities are summed, the total is compared to the network requirements, and performance statistics are accumulated.

A number of additional considerations arise immediately. Because of the variety of types of biquad sections, a section type number is required to designate which set of formulas is to be used to calculate the biquad coefficients. The need to introduce correlation effects re-

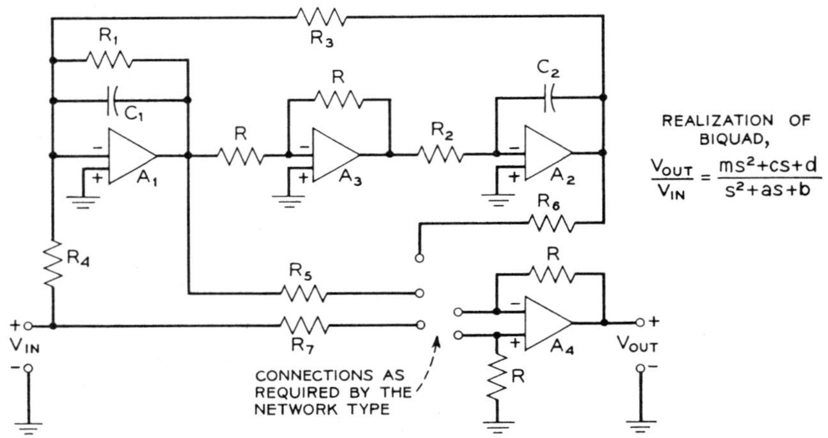

Fig. 5—Four-amplifier biquad circuit.

quires that every element value have an identifying serial number to permit cross-referencing. These serial numbers will also be useful in specifying tuning operations and in introducing environmental effects. Because of the limited number of section types, it has been possible to develop and include quite general methods of producing correlation, tuning, environmental effects and tests. More information on the methods of accomplishing these objectives will be given below. Where the standard methods are inadequate, new subroutines can be written and included in the program.

4.2.1 *Simulation of Manufacturing Conditions*

Due to the production process, parameters such as resistors, capacitors and amplifier gains have an actual value spread about the nominal value. This production tolerance on a parameter such as a resistor is represented by:

$$R_1 = R_0(1 + n_1) \tag{2}$$

where R_0 is the nominal value and n_1 is a random number whose range is the production tolerance on R_0. The distribution of n_1 could be gaussian, uniform, or any of several special shapes. The algorithm used to generate random numbers on the computers is discussed in Appendix A.

In integrated circuits, where all the elements are on the same chip, the values of the resistors track each other to some extent. Consider k resistors whose initial tolerances track. It has been found empirically that their values may be represented by:

$$\begin{aligned} R_1 &= R_{01}(1 + \rho n + (1 - \rho)n_1), \\ R_2 &= R_{02}(1 + \rho n + (1 - \rho)n_2), \\ &\vdots \\ R_k &= R_{0k}(1 + \rho n + (1 - \rho)n_k). \end{aligned} \tag{3}$$

The same random number n is used to perturb all k resistors. Each resistor $R_1, R_2, \cdots R_k$ is also perturbed by a separate random number $n_1, n_2, \cdots n_k$, respectively. ρ is a correlation factor which determines how closely the resistors track. The range of ρ is from 0 to 1. Complete correlation can be simulated by $\rho = 1$, and no correlation by $\rho = 0$. This correlation factor is different from the correlation coefficient used in statistics but has proven useful in the present application. The coding has been arranged so that the serial numbers mentioned earlier effectively refer to both R and n values in equation (3), so that multiple correlations may be specified.

A practical problem in using the above sets of formulas, equations (2) and (3), is that of obtaining data on the distributions of the parameter values in production and on the correlation factor, ρ. The need for accurate statistical information may require that a major effort be made to obtain such data and keep it up-to-date. A companion paper, "Characterization and Modeling for Statistical Design," describes transistor characterization activities.

4.2.2 *Tuning*

The deviations from the nominal performance caused by the initial tolerances of components can be totally or partially corrected by tuning some of the network elements. The elements are tuned during the manufacture of the filter, at room temperature. Two types of tuning steps have been simulated on BITAP. In the first type, an element is adjusted so that a transmission requirement such as loss or phase is met at a given frequency. This type of tuning is used in adjusting for the bandwidth of a biquad section. In the second type of tuning, an element is adjusted for a peak (of gain or loss) at a given frequency. This is used in adjusting the resonant frequency of a pole or a zero. In both these tuning methods the variable element is changed continuously over a range. An additional subroutine could be written to simulate tuning in discrete steps, i.e., using a finite supply of element values close to the nominal, as this is an economical production method of adjusting the network performance. In production, the accuracy with which the element can be tuned depends on the sensitivity of the network to that element and on the precision of the measuring equipment. This is simulated by specifying a given nominal value for the tuning with a tolerance and a distribution.

4.2.3 *Simulation of Field Conditions*

The factors that caused the filter response to deviate from the nominal during manufacture were the initial tolerances of the components and inaccuracy in the tuning step. The environmental conditions that were assumed to exist during manufacture were room temperature and humidity. In the field, any change from these environmental conditions will cause the filter characteristics to change to some extent. The filter response will also change with time due to the chemical processes associated with aging. The temperature, aging and humidity effects in the field are modeled in the following way:

The value of a resistor (R_2) at T°C above room temperature is given by

$$R_2 = R_{02}(1 + \alpha T)$$

where R_{02} is the nominal value. The temperature coefficient (T.C.), α, is

$$\alpha = \alpha_0 + n_2$$

where α_0 is the nominal T.C. and n_2 the randomly distributed part of the T.C. The distribution function describing n_2 is usually gaussian.

The change in the value of a resistor or capacitor (R_3) with time is represented by

$$R_3 = R_{03}(1 + n_3) \tag{4}$$

where R_{03} is the nominal and n_3 simulates the random part.

The change due to humidity is represented by an equation similar to equation (4).

The combined effects of initial tolerance, temperature, aging and humidity is given by

$$R = R_0(1 + n_1)(1 + (\alpha_0 + n_2)T)(1 + n_3)(1 + n_4). \tag{5}$$

The random numbers associated with temperature coefficients (n_2), for components on an integrated circuit, are correlated. The same is true for the aging coefficient (n_3) and the humidity coefficients (n_4). These correlations are simulated just as in equation (3).

4.2.4 *Testing*

The values of the deviated parameters are used in the equations describing the biquad to evaluate the performance of the network. The biquad is analyzed several times using different sets of random numbers to simulate environmental and manufacturing conditions. The performance tests are spelled out in the form of filter function requirements (loss, phase or delay) at various frequencies. Three standard forms have been built into the program. These are:

(*i*) loss (phase, delay) at a frequency f_1 is between two limits, dB_1 and dB_2,
(*ii*) loss at a frequency f_1 is bounded below (or above) by dB_1, and
(*iii*) loss at a frequency f_1 is greater (less) than the measured loss at a reference frequency f_2 plus the amount dB_1.

More than one performance requirement may be specified at a frequency. It is expected that special subroutines will be written for more sophisticated tests such as a resonant frequency, Q, and in-band ripple. It should be noted that if environmental conditions are included, the tests become service condition or end-of-life tests. If they are omitted, the tests may be interpreted as factory production tests.

4.2.5 *Output*

If, for a particular set of deviated elements, the network fails the required specifications, the complete network performance is stored. The output consists of a table which details the failed networks. If a test was failed, the performance function for that test is printed; otherwise, a blank is printed for the test.

The overall yield is computed as the ratio of the number of networks passing all tests to the total number of networks tested. Yields are also calculated for individual tests. The exact form of the output is illustrated by the example in the next section.

4.3 *Example*

The four amplifier biquad circuit of Fig. 5 will be used to illustrate the main features of BITAP. Its transfer function, in terms of the elements, is

$$\frac{V_{out}}{V_{in}} = \frac{+\frac{R}{R_7}\cdot S^2 + \frac{R}{R_7}\cdot\frac{1}{C_1}\left[\frac{1}{R_1} - \frac{R_7}{R_4R_5}\right]\cdot S + \frac{R}{R_7R_2C_1C_2}\left[\frac{1}{R_3} + \frac{R(R_5+R_7)+R_5R_7}{R_4R_5(R+R_6)}\right]}{S^2 + \frac{1}{R_1C_1}S + \frac{1}{R_2R_3C_1C_2}}.$$

The complete network contains two such sections connected in tandem as shown in Fig. 6, which also gives the element values. The resistors have a one percent manufacturing tolerance with a gaussian distribution. The capacitors are one percent elements with a flat distribution. The correlation factors, ρ, are 50 percent for both resistors and capacitors. The pole frequency of the first section is at 5 kHz, and R_3 is tuned to obtain this frequency within ±1 Hz. This ±1 Hz represents the accuracy of tuning and it is assumed to be distributed uniformly around the nominal 5 kHz. The bandwidth of the first section is tuned by R_1 to 100 Hz ±1 Hz. Section 2 is not tuned. These initial tolerances and the tuning steps describe the manufacturing conditions. It is assumed that a BITAP run simulating the above conditions and manufacturing tests has already been made and 100 percent of the network passed. The filter performance is now to be tested at +50°C and at a time 20 years from manufacture. The T.C. of resistors is 130 ±30 parts per million (ρ = 80 percent) and that of capacitors −135 ±10 ppm ($\rho = 50$ percent). The random parts (± 30 ppm and ± 10 ppm) are distributed normally about the nominal. The aging coefficient for the resistors and the capacitors in 20 years is ± 2 percent, uniformly distributed ($\rho = 50$ percent).

The nominal performance and the test requirements to be met at end-of-life are shown in Fig. 7. Figure 8 shows the results of a BITAP analysis of 100 filters and indicates that 82 percent will meet these service condition requirements. Figure 8 indicates both the frequencies where improvements are needed and the magnitude of the changes required. With the biquad type of realization and such data, the designer usually has no difficulty deciding which elements or tuning adjustments are responsible.

In order to improve the performance in the 1000- to 4000-Hz passband, the designer might examine several alternatives. A new nominal design with a smaller ripple or a better centered nominal loss would

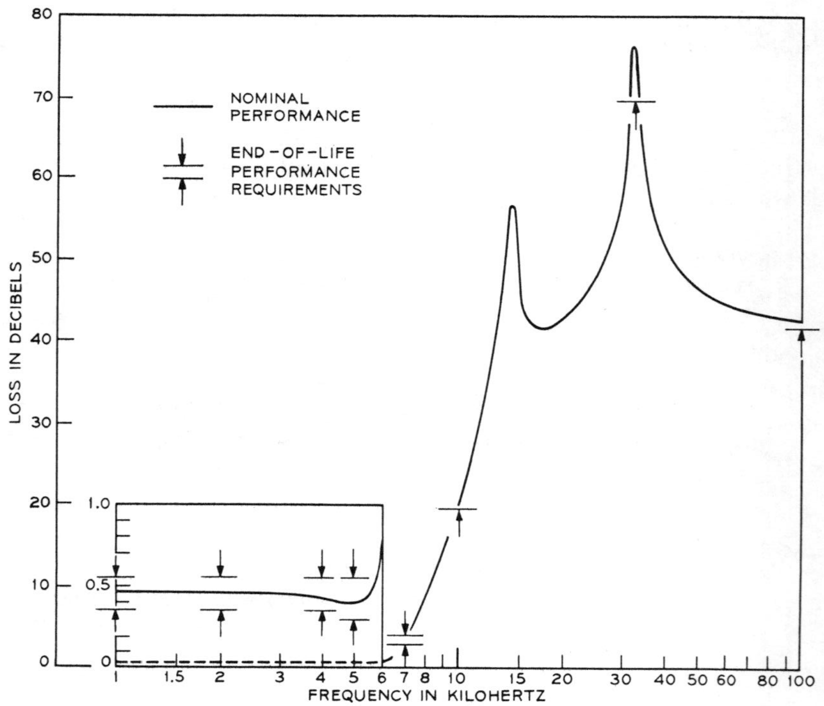

Fig. 7—Nominal performance and end-of-life requirement.

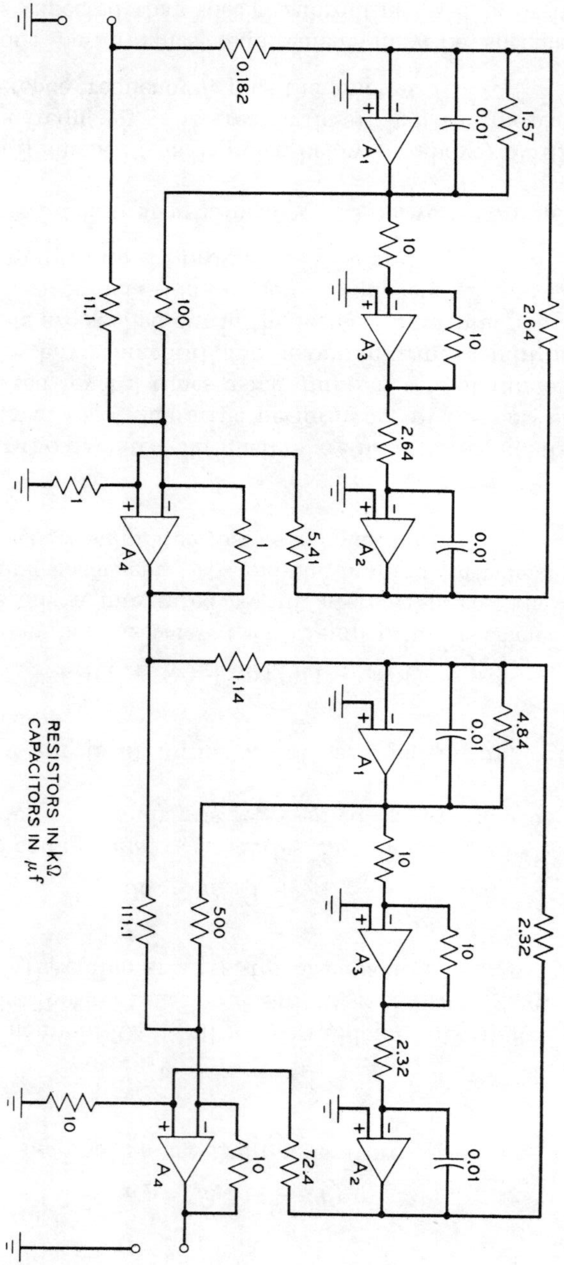

Fig. 6—Lowpass filter network for BITAP analysis.

entail no added cost. Use of closer tolerances for some components or more precise tuning might also provide sufficient improvement, but at some added expense. If none of these works, addition of a third section should be evaluated.

4.4 *Appraisal of* BITAP

BITAP is capable of analyzing up to 20 biquad sections. The number of performance tests is limited to 30. Ten types of biquad sections are currently included in BITAP. The addition of more biquad sections merely involves describing the biquad coefficients (m, c, d, n, a, b) in terms of the elements of the biquad and providing equations for the tuning steps. The rest of the program is identical for all biquads.

BITAP is an exact analysis package and as such it can be used in other steps of the design process. For instance, it could be used to determine whether or not a filter needs to be tuned, and to decide on the number of tuning steps needed. BITAP could also be used in determining the best (cheapest) components that could be used in a circuit while meeting all requirements.[7]

It has been shown how BITAP can be used to evaluate the performance of networks of the biquad family at the time of manufacture and under (end-of-life) field conditions. The information obtained here can be used to decide on the tests to specify during manufacture. The analysis under end-of-life conditions gives a measure of the adequacy of the design under extreme conditions.

V. CONCLUSIONS

We have described some simple analytic methods of calculating and combining deviation shapes and have discussed the areas in which they are useful and the respects in which they are limited. We have also presented two simulation procedures, which rely on the speed and large storage capacity of modern digital computers. Each is a general-purpose program in that it can be used to analyze statistically a large number of networks of an appropriate type. Although the BITAP networks form a subset of those analyzable by BELTAP, the two programs provide entirely different facilities for their users. BELTAP provides complete freedom in the network topology; BITAP uses a functional input form, which reduces computing time. BELTAP facilitates exploration of concealed interactions among component values and test results; in BITAP it is assumed that the few simple network functions allowed are completely understood in this respect. BELTAP provides no built-in cor-

FREQ	1000.00	2000.00	4000.00	5000.00	7000.00	10000.0	32000.0	100000
MEAS	LOSS	LOSS	LOSS	LOSS	LOSS	LOSS	LOSS	LOSS
BELOW	0.5500	0.5500	0.5500	0.5500	4.000	—	—	—
ABOVE	0.3500	0.3500	0.3500	0.3000	3.000	20.40	70.00	42.00
NOM	0.4845	0.4738	0.4468	0.3947	3.391	20.83	76.70	42.93
NET. NO.								
8	0.5569	0.5535	—	—	—	—	—	—
10	—	—	—	—	—	20.21	—	—
13	—	—	—	—	—	20.17	—	—
32	—	—	—	0.2998	—	20.32	—	—
36	—	—	—	—	—	20.39	—	—
45	—	—	—	0.2829	—	—	—	—
46	—	—	—	—	—	20.40	—	—
49	0.5506	—	—	—	—	—	—	—
50	—	—	—	—	—	20.21	—	—
51	—	—	—	—	—	20.16	—	—
57	—	—	0.3375	0.2819	—	—	—	—
60	—	—	—	0.2937	—	—	—	—
67	—	—	0.3364	0.2770	—	—	—	—
73	0.6009	0.6018	0.5891	—	—	—	—	—
76	0.5584	0.5641	0.5606	—	—	—	—	—
78	0.5657	0.5544	—	—	—	—	—	—
85	—	—	—	—	2.961	20.13	—	—
93	0.5620	0.5637	0.5564	—	—	—	—	—
NO. FAIL	6	5	5	5	1	8	0	0

FINAL YIELD = 82.000 PERCENT

Fig. 8—BITAP yield analysis.

relation, tuning, environmental exposure, or phase or delay evaluation but does include return losses; BITAP includes all of these except return losses which are of no interest.

Under certain conditions, the least expensive manufacturing method may require discarding some of the product. The simulation programs should be capable of demonstrating this, if it is true. They may also provide information that is useful in correcting troubles in the manufactured product. Their usefulness will, of course, depend on the availability of accurate statistical information.

Clearly, useful as they are, neither of these programs can be termed completely general purpose. Much work remains to be done in both areas of application. It is considered that the two programs make a very worthwhile contribution to a designer's repertoire and may point the way for still more comprehensive tools in the future.

VI. ACKNOWLEDGMENTS

The authors wish to express their appreciation to F. P. Sansone and Mrs. H. D. Reinecke for their work leading to BITAP.

APPENDIX A

Both BELTAP and BITAP use a congruential random number generator in which the modulus is an integral power of a prime number and the multiplier is a primitive root.[8] This method permits generating a sequence of numbers which has a flat distribution and a period which is made just large enough to produce the number of random numbers required. The modulus, M, is obtained by raising three to a power, n, such that the period, $2 \cdot 3^{n-1}$, exceeds the required value. The multiplier used is $2 + 9 \cdot k$, where k is an integer chosen to make the multiplier approximately equal to the square root of the modulus. It was found empirically that this function of k produces a primitive root, that is, a multiplier having the period given above, which is the longest that can be obtained for the assumed modulus.

A simple example using small numbers will illustrate the process. Suppose five networks are to be calculated and each contains three elements with distributions. Fifteen random numbers will be required. A value of three for n will produce a sequence of $2 \cdot 3^2 = 18$ numbers before repetition begins. The modulus M is $3^3 = 27$ and multipliers of 2, 11 or 20 are possible. Suppose the sequence begins with 1 and a multiplier of 11 is used. The second number is $1 \cdot 11 = 11$. The third is obtained by multiplying the second by 11, dividing by 27 and using the remainder, 13. The fourth is 13 times 11 modulo 27 = 8, etc. The complete sequence is: 1, 11, 13, 8, 7, 23, 10, 2, 22, 26, 16, 14, 19, 20, 4, 17, 25, 5, 1. It should be noted that the nineteenth number, 1, is a repetition of the first and that all numbers less than 27 are present except multiples of three. The first three numbers will be used for the three random elements in the first network, the next three for the second network, etc.

The random numbers are shaped to become element values having the desired tolerance and distribution in the following manner. The first transformation is to put them into a range from -1 to $+1$ by dividing by half the modulus and subtracting 1.

$$n_b = \frac{n_a \cdot 2}{M} - 1. \tag{6}$$

The smallest random number, 1, yields $-0.925925 \cdots$, while the largest gives $+0.925925 \cdots$. With a larger modulus the numbers will approach -1 and $+1$ more closely. If an element has a flat distribution, the random value may be produced as follows:

$$R = R_0(1 + t \cdot n_b/100) \tag{7}$$

where R is the random value, R_0 is the nominal value for that element, and t is its (symmetrical) tolerance in percent. If the distribution is not flat, the values of n_b are transformed as shown in Fig. 9. For each value of n_b the proper line segment is identified and a value of n_c is calculated by interpolation. n_c is then used in place of n_b in equation

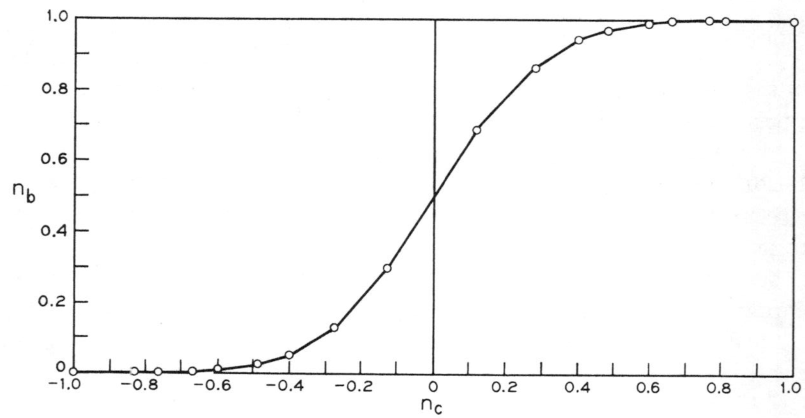

Fig. 9—Cumulative gaussian probability distribution approximated by straight-line segments.

(7) to calculate R. The coordinates for the gaussian distribution are permanently stored in the program. Other distributions can be used by reading in the appropriate sets of breakpoints.

If correlation is required, values of n_b and/or n_c may be combined as follows:

$$n_d = \rho n_x + (1 - \rho) n_y \,. \tag{8}$$

Here ρ is a correlation factor and n_d, n_x and n_y all represent random numbers in the -1 to $+1$ range, associated with selected elements, parameters, field conditions, and/or tuning tolerances. It is to be noted that n_d may be used to evaluate an R, using equation (7), or it may be used as an n_x or n_y in additional equations of the same form as (8). If the latter is done, the result will be partially correlated to three other parameters.

When equation (8) is used to introduce partial correlation and n_x and n_y have average values of zero and are uncorrelated, it can be shown that the variance of n_d is given by

$$\sigma_d^2 = \rho^2 \sigma_x^2 + (1 - \rho)^2 \sigma_y^2 \,. \tag{9}$$

It may also be shown that the tracking coefficient between n_d and n_x is given by

$$r = \sum_1^N n_d n_x / N \sigma_d \sigma_x = \rho \sigma_x / \sigma_d . \tag{10}$$

The tracking between two different distributions n_d formed using the same n_x and different n_y is discussed in the following paper.

REFERENCES

1. Hoyt, R. S., "Probability Theory and Telephone Transmission Engineering," B.S.T.J., *12,* No. 1 (January 1933), pp. 35–75.
2. Boghosian, W. H., unpublished memorandum, May 3, 1938.
3. Darlington, S., "Synthesis of Reactance 4-Poles," J. Math. and Phys., XVIII, (September 1939), pp. 257–353.
4. Tow, J., "A Step-by-Step Active Filter Design," Spectrum, *6,* No. 12 (December 1969), pp. 64–68.
5. Moschytz, G. S., "Active RC Filter Building Blocks using Frequency Emphasizing Networks," IEEE J. Solid State Circuits, *SC-2,* No. 2 (June 1967), pp. 59–62.
6. Sallen, R. P., and Key, E. L., "A Practical Method of Designing RC Active Filters," I.R.E. Trans. Circuit Theory, *CT-2* (March 1965), pp. 74–85.
7. Karafin, B. J., "The Optimum Assignment of Component Tolerances for Electrical Networks," B.S.T.J., this issue, pp. 1225–1242.
8. Hutchison, D. W., "A New Uniform Pseudo-Random Number Generator," N66-32469, Stat. Dept., U. of Calif., Berkeley, April 27, 1965.

Optimum Design of Linear Tuning Algorithms

PHILIP V. LOPRESTI, MEMBER, IEEE

Abstract—A high-performance *RC* active filter realized as a thick or thin-film hybrid integrated circuit presents special fabrication problems. A filter of this type is so sensitive to its parameters that in a large proportion of cases, a circuit so constructed will not meet its performance specifications even if the tightest practicable tolerances are held on the circuit components. As a consequence, the filter network must be tuned after fabrication. It is not feasible to adjust circuit capacitors, hence all tuning must be accomplished by resistor adjustment, and these adjustments can only increase resistance.

In this paper, a generalized tuning technique is described which makes the design of a tuning algorithm, and automated tuning, amenable to a universal computer mechanization. In addition to eliminating the analysis and software preparation required to set up a process-control system, the generalized algorithm improves the yield and speed of operation of the tuning process.

The central idea of this tuning scheme is to estimate transfer function coefficient errors based on capacitor measurements and then to choose resistance values for the circuit in such a way as to minimize these errors. The operation of the algorithm is sequential, with succeeding resistance values chosen to minimize the cumulative error due to preceeding process steps.

I. Introduction

Active *RC* filtering techniques have assumed a central role in low- and voice-frequency filtering applications with the ready availability of inexpensive monolithic operational amplifiers. In this paper, attention will be focused on one important aspect of the manufacture of high-performance active filters realized as hybrid thin- or thick-film circuits; specifically the automation of the filter "tuning" process. Tuning is a required process step since batch fabrication techniques produce typically resistors and capacitors with tolerances in the ±5- percent range, while a high-performance filter may require uncompensated element tolerances of ±0.1 percent. In practice, each filter is adjusted to obtain desired performance characteristics by trimming the circuit resistors through precision anodization of the resistor material, or by selectively cutting the resistor body with a laser beam [1], [2]. Capacitors, as a rule, are not trimmed.

Much theoretical work on the combined problems of design centering, tolerancing and tuning has been carried out [3]–[5], but in practice two rather restricted methods of tuning called *functional* and *deterministic* are usually employed for active filter adjustment.

Functional tuning [6], [7] is carried out by exciting the filter with a sinusoidal input at a critical filter frequency and then adjusting specific resistors until required gain or phase values are obtained. Functional tuning is straightforward and quite efficient when the filter is relatively simple and when the filter design is carefully chosen with the tuning process kept in mind. Generally speaking, however, design requirements and economic considerations lead to filter configurations where the interactions between circuit elements are difficult to analyze and the choices of resistors to trim and their trim order are substantially impossible to make, thus precluding the functional approach. Functional trimming is inherently slow also, since resistors must be adjusted on a one-by-one basis while steady-state measurements are made on the circuit.

The second approach, deterministic tuning, is based upon circuit analysis principles [8]. Since the filter is characterized by the coefficients of its transfer function, and since these coefficients are functions of the circuit resistances and capacitances, transfer function invariance can be obtained by solving these equations for the circuit resistances in terms of the actual circuit capacitances as measured during manufacture. If the circuit resistors are then trimmed to these calculated values with high precision, the manufactured filter should meet its performance specifications.

The above discussion presumes that the filter is accurately modeled by the ideal circuit equations; this is the usual case for thin-film voice-frequency filters, and performance predictions from theoretical models agree very closely with actual filter performance.

In order to be deterministically tuned, a filter is not completely interconnected on its substrate; instead a sufficient number of leads are brought out to the edges of the substrate in order that all passive elements can be measured by an appropriate connection to two pins on the package. Certain pins are connected when the circuit is powered up.

The major drawbacks of deterministic tuning as described above are as follows. 1) In many cases of practical interest, it is a formidable task to derive the required expressions. In fact, they may not even exist. 2) Required resistor trim tolerances are generally tight, often approaching or exceeding the state of the art of the available trimming technology. 3) The "tuning algorithm" (i.e., the set of resistance computation expressions) is usually so involved as to require a sophisticated on-line computer with a large software support system for its implementation.

In this paper, a technique of deterministic tuning will be described which eliminates the above mentioned difficulties associated with the conventional deterministic method. In essence, the procedure is to estimate transfer function coefficient errors after a resistor is adjusted, and

Manuscript received April 28, 1976; revised November 16, 1976.
The author is with the Engineering Research Center, Western Electric Company, Princeton, NJ 08540.

Reprinted from *IEEE Trans. on Circuits Syst.*, vol. CAS-24, pp. 144-151, Mar. 1977.

then calculate a value for the next resistor to be trimmed so that a weighted sum of the squares of these coefficient errors is minimized.

II. Mathematical Formulation

Assume a given filter has been designed with the ideal transfer function given by (1)

$$H(s)=K\frac{s^n+a_{n-1}s^{n-1}+\cdots+a_0}{s^n+b_{n-1}s^{n-1}+\cdots+b_0}. \tag{1}$$

Also, assume that the filter has been designed for the ideal infinite gain operational amplifier case. Then there exist $2n+1$ algebraic expressions of the form

$$K=f_0(\boldsymbol{G},\boldsymbol{C},\boldsymbol{G}_c) \tag{2a}$$

$$a_i=f_{i+1}(\boldsymbol{G},\boldsymbol{C},\boldsymbol{G}_c), \qquad i=0,\cdots,n-1 \tag{2b}$$

$$b_i=f_{n+i+1}(\boldsymbol{G},\boldsymbol{C},\boldsymbol{G}_c), \qquad i=0,\cdots,n-1. \tag{2c}$$

Where $\boldsymbol{G}$ is an m-vector representing the m circuit design conductances

$$\boldsymbol{G}=\begin{bmatrix} G_1\\ \cdot\\ \cdot\\ \cdot\\ G_m\end{bmatrix}. \tag{3}$$

The p-vector $\boldsymbol{C}$ represents the p circuit design capacitances

$$\boldsymbol{C}=\begin{bmatrix} C_1\\ \cdot\\ \cdot\\ \cdot\\ C_p\end{bmatrix} \tag{4}$$

and the p-vector $\boldsymbol{G}_c$ represents the shunt leakage conductances of the circuit capacitors

$$\boldsymbol{G}_c=\begin{bmatrix} G_{c1}\\ \cdot\\ \cdot\\ \cdot\\ G_{cp}\end{bmatrix}. \tag{5}$$

It is assumed that $\boldsymbol{G}_c$ is identically zero in the ideal coefficient equations (2).

For simplicity, let z be the $2n+1$-vector of transfer function coefficients

$$z=\begin{bmatrix} K\\ b_0\\ \cdot\\ \cdot\\ \cdot\\ b_{n-1}\\ a_0\\ \cdot\\ \cdot\\ \cdot\\ a_{n-1}\end{bmatrix} \tag{6}$$

and let $\boldsymbol{f}(\boldsymbol{G},\boldsymbol{C},\boldsymbol{G}_c)$ be the $2n+1$-vector of coefficient equations:

$$\boldsymbol{f}(\boldsymbol{G},\boldsymbol{C},\boldsymbol{G}_c)=\begin{bmatrix} f_0(\boldsymbol{G},\boldsymbol{C},\boldsymbol{G}_c)\\ \cdot\\ \cdot\\ \cdot\\ f_{2_n}(\boldsymbol{G},\boldsymbol{C},\boldsymbol{G}_c)\end{bmatrix}. \tag{7}$$

We then can formulate the conventional deterministic tuning problem as follows.

Given the equations for the ideal system and an actual circuit whose capacitors deviate from the ideal by an amount $\Delta\boldsymbol{C}$ in capacitance and $\Delta\boldsymbol{G}_c$ in leakage conductance, choose $\Delta\boldsymbol{G}$ such that the transfer function deviation Δz is zero

$$\Delta z=\boldsymbol{f}(\boldsymbol{G}+\Delta\boldsymbol{G},\boldsymbol{C}+\Delta\boldsymbol{C},\Delta\boldsymbol{G}_c)-\boldsymbol{f}(\boldsymbol{G},\boldsymbol{C},\boldsymbol{0})=\boldsymbol{0}. \tag{8}$$

This problem is not well-posed: it may have no solution, one solution or many solutions depending upon circuit topology and element values. Even if a solution exists, the algebraic and numerical problems associated with finding this solution may be quite formidable.

Two steps will be taken in order to simplify these algebraic operations and to ensure existence and uniqueness of a solution to the tuning problem. The first is to linearize the coefficient deviation equations and the second is to treat tuning as a minimization problem.

Since element deviations from design values will be relatively small, a differential approximation can be used to estimate transfer function deviations via the expression

$$dz=\sum_{i=1}^{p} C_i\frac{\partial f}{\partial C_i}\frac{dC_i}{C_i}+\sum_{i=1}^{p} C_i\frac{\partial f}{\partial G_{ci}}\frac{dG_{ci}}{C_i}+\sum_{i=1}^{m} G_i\frac{\partial f}{\partial G_i}\frac{dG_i}{G_i} \tag{9}$$

where the partial derivatives are evaluated at the design element values, e.g.,

$$\frac{\partial f}{\partial C_i}=\begin{bmatrix} \frac{\partial f_0}{\partial C_i}(\boldsymbol{G},\boldsymbol{C},\boldsymbol{0})\\ \cdot\\ \cdot\\ \cdot\\ \frac{\partial f_{2n}}{\partial C_i}(\boldsymbol{G},\boldsymbol{C},\boldsymbol{0})\end{bmatrix}. \tag{10}$$

These partial derivatives are quickly and accurately obtained by a method to be described in a following section. For compactness define the quantities x_c, h_k, and u_k as follows:

$$x_c = \sum_{i=1}^{p} C_i \frac{\partial f}{\partial C_i} \frac{dC_i}{C_i} + \sum_{i=1}^{p} C_i \frac{\partial f}{\partial G_{ci}} \frac{dG_{ci}}{C_i} \tag{11a}$$

$$h_k = G_k \frac{\partial f}{\partial G_k} \tag{11b}$$

$$u_k = \frac{dG_k}{G_k}. \tag{11c}$$

Then the differential expression (9) can be rewritten as the recursion relation

$$x_{k+1} = x_k + h_k u_k, \qquad k = 1, \cdots, m \tag{12a}$$

$$x_1 = x_c. \tag{12b}$$

The initial condition, x_c, represents the vector of transfer function coefficient errors due to capacitor deviations from design values. The $2n+1$ "sensitivity" vector h_k converts a deviation in circuit design conductance, u_k, into a corresponding deviation of the transfer function coefficients, and the $2n+1$-vector x_{k+1} represents the cumulative coefficient error due to the capacitor deviations and trim value deviations of the first k resistors. The quantity of primary interest is the final value of transfer function coefficient error, x_{m+1} (the differential dz).

The tuning problem can now be recast as a conventional problem in quadratic optimal control theory. First, reduction in coefficient error is formulated as a quadratic minimization problem, thus avoiding the question of existence of a solution for the algebraic system associated with a given circuit. Then uniqueness is ensured by appending to the quadratic coefficient expression, a weighted sum of the squares of the circuit conductance deviations. A useful by-product is accrued by including resistor deviations in the quantity to be minimized: resistor deviations from design values tend to be small, leading to improved long-term circuit performance.

The solution of this linear optimal tuning problem is the set of conductance value deviations $\{u_k^\circ\}$ which minimizes $I(u_1, \cdots, u_m; x_1)$ where

$$I(u_1, \cdots, u_m; x_1) = x'_{m+1} Q x_{m+1} + \sum_{i=1}^{m} \gamma_i u_i^2 \tag{13}$$

subject to the difference equation constraint

$$x_{k+1} = x_k + h_k u_k \tag{14a}$$

$$x_1 = x_c. \tag{14b}$$

The *terminal weighting matrix*, Q, is symmetric nonnegative definite, i.e.,

$$q_{ij} = q_{ji}, \qquad i,j = 1, \cdots, 2n+1 \tag{15a}$$

$$x'_{m+1} Q x_{m+1} \geqslant 0 \tag{15b}$$

where the superscript apostrophe denotes matrix transpose, and the *control weights*, γ_i, are strictly positive

$$\gamma_i > 0, \qquad i = 1, \cdots, m. \tag{16}$$

The minimizing value of u_k is given by the expression [9], [10]

$$u_k^\circ = F_k x_k \tag{17}$$

where this *feedback control law* is obtained from a recursive solution of the related matrix Riccati difference equation

$$P_{i+1} = P_i - \frac{P_i h_{m-i} h'_{m-i} P_i}{h'_{m-i} P_i h_{m-i} + \gamma_{m-i}}, \qquad i = 0, \cdots, m-1 \tag{18a}$$

with initial condition

$$P_0 = Q. \tag{18b}$$

The *feedback gain* sequence $\{F_k\}$ is obtained from the Riccati equation via the expression

$$F_{m-i} = -\frac{h'_{m-i} P_i}{h'_{m-i} P_i h_{m-i} + \gamma_{m-i}}, \qquad i = 0, \cdots, m-1. \tag{18c}$$

Finally, the minimum value of the performance index, I, is given as

$$I(u_1^\circ, \cdots, u_m^\circ; x_1) = x_1' P_m x_1. \tag{18d}$$

To reiterate and expand upon the preceding ideas it is pointed out that by suitable weighting of the final error term, by a diagonal matrix

$$Q = \begin{bmatrix} \frac{1}{K^2} & & \mathbf{0} \\ & \frac{1}{b_0^2} & \\ & & \ddots \\ \mathbf{0} & & \frac{1}{a_n^2} \end{bmatrix} \tag{19}$$

for example, and by choosing suitably small values for γ_i, e.g.,

$$\gamma_i = 0.0001. \tag{20}$$

One can obtain a sequence of conductance values

$$\hat{G}_i = (1 + u_i^\circ) G_i, \qquad i = 1, \cdots, m \tag{21}$$

which will drive the vector of transfer function coefficient errors essentially to zero if a solution to the conventional deterministic tuning problem exists, or will approach zero error as closely as possible if an identically zero error vector does not exist. There are various tradeoffs which could be made in arriving at acceptable tuning behavior for practical circuits. One obvious approach is to assign a zero weight to the gain error and use all available resistor adjustments to hold the pole-zero pattern of the filter.

Gain deviations can usually be corrected elsewhere in the circuit.

III. Practical Implications

One major advantage of the linearized tuning algorithm described in the preceding section is its simplicity for on-line operation, and the universality of the basic algorithm. The problem data—the coefficient sensitivities and the sequence of gain matrices $\{\boldsymbol{F}_k\}$—change with filter changes, but the basic rule can be coded as a general tuning subroutine requiring only additions and multiplications.

There are two ways in which a filter can be tuned in practice, one will be called *open loop* and the other *closed loop*, although the second method, strictly speaking, does not employ feedback in a control theoretic sense. The open-loop approach is employed when it is necessary to compute all resistance values before any resistors can be trimmed and is carried out simply by measuring circuit capacitances and capacitor leakages (or equivalently, dissipation factors) and then recursively computing $u_1°, \boldsymbol{x}_2, u_2°, \boldsymbol{x}_3, \cdots,$ through $u_m°$. The trim values of circuit resistance, then are

$$\hat{R}_i = \frac{R_i}{(1+u_i°)}, \qquad i=1,\cdots,m \tag{22}$$

where the caret is used to distinguish tuning values from design values of resistance. The open-loop procedure is illustrated in Fig. 1 by a flow chart which can be used to construct the on-line tuning subroutine. Note that in this procedure, estimated values of $\boldsymbol{x}_k$, based upon calculated values of $u_1, \cdots, u_{k-1}$ are used to obtain u_k.

The open-loop tuning approach is a simple and efficient method of arriving at resistance values for computer controlled production tuning, but it does not make maximum use of the great flexibility of the optimization technique.

The recursive solution of the original minimization problem also contains the solution for the sequence of "partial minimizations"

$$I(u_j°, \cdots, u_m°, \boldsymbol{x}_j) = \min_{u_j, \cdots, u_m} \boldsymbol{x}'_{m+1}\boldsymbol{Q}\boldsymbol{x}_{m+1} + \sum_{i=j}^{m} \gamma_i u_i^2 \tag{23a}$$

subject to the constraint

$$\boldsymbol{x}_{k+1} = \boldsymbol{x}_k + \boldsymbol{h}_k u_k, \qquad k=j,\cdots,m \tag{23b}$$

$$\boldsymbol{x}_j = \boldsymbol{x}_a \tag{23c}$$

where $\boldsymbol{x}_a$ is an arbitrary initial condition and j takes any value from 1 to m. In other words, the subsequence $\{\boldsymbol{F}_k | k=j,\cdots,m\}$ chosen from the initial sequence is also the optimal sequence of feedback gains for the partial minimization problem. With a simple modification in tuning procedure, actual trimmed values of resistance can be used to update the computer's estimate of transfer function error after each resistor is trimmed. Succeeding resistance calculations can be based upon this updated value of $\boldsymbol{x}_j$, thus minimizing the cumulative error in transfer function coefficients due to both capacitor errors and trimming errors of those resistors already adjusted (Fig. 2). In high-performance applications this correction effect

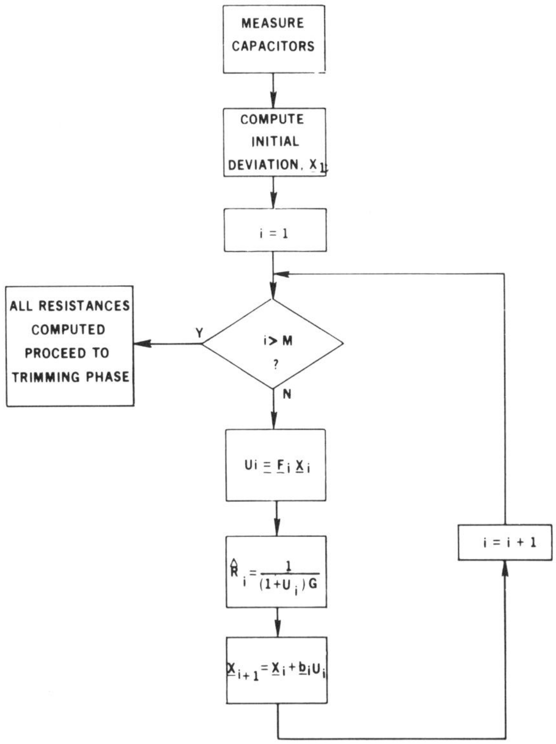

Fig. 1. Open-loop linear optimal tuning.

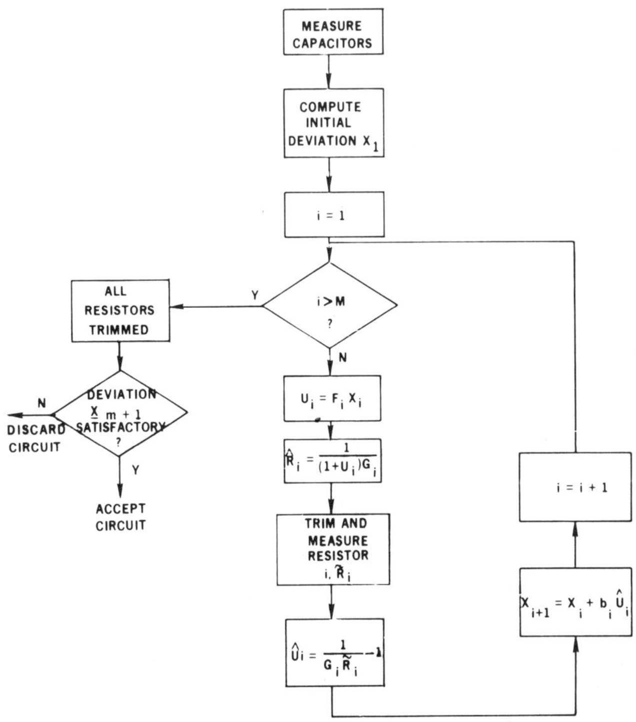

Fig. 2. Closed-loop linear optimal tuning.

can substantially relax the tolerance requirements on the resistor trimming apparatus, thus increasing process yield and throughput. In practice, it is often necessary to use a "mixed" tuning scheme, where several resistor values are calculated in one step. These resistors are then trimmed and measured, with error updating occurring after each group is trimmed. The derived set of feedback gains is also optimal for this configuration.

IV. Partial Derivative Algorithm

The key factor in the applicability of the linear optimal tuning scheme mentioned in the preceding sections is the ability to obtain accurate partial derivatives of the transfer function coefficients with respect to each of the passive circuit elements comprising the circuit. These derivatives can be obtained easily on a computer via the methods described in this section.

A. System Circuit Equations

It is possible to write the circuit equations for an arbitrary RC active filter in a standard form which facilitates derivative computation. The rules are quite simple. For concreteness assume that the active filter contains q nodes in addition to the ground node which will be the reference. Furthermore, assume that the input voltage source is connected between node 1 and ground and the output voltage is taken between node q and ground. Finally, assume one ideal finite gain, differential operational amplifier in the filter with its input nodes labeled p and n and its output node j. An independent set of $q-2$ equations in the node-to-reference voltages can then be obtained by employing Kirchhoff's current law to write "node-voltage" equations at each node except for nodes 1 and j. The two remaining equations are the trivial ones as follows:

for node one

$$e_1 = e_{in} \tag{24a}$$

and for the operational amplifier output node, j,

$$e_p - e_n = \frac{1}{A} e_j. \tag{24b}$$

In the infinite gain approxi:nation, the right-hand side of (24b) is zero.

A set of circuit equations written in this manner takes the form

$$\left[\sum_{i=1}^{p} M_i s C_i + \sum_{i=1}^{p} M_i G_{ci} + \sum_{i=1}^{m} N_i G_i + P\right] e = p_1 e_{in}. \tag{25}$$

Where e is the q-vector of node-to-reference voltages, C_i, G_{ci}, and G_i have been previously defined and the $q \times q$ "connection matrices" M_i and N_i contain ones and minus ones in locations corresponding to the element node labels. The $q \times q$ matrix P contains all system constants and the q-vector p_1 is the unit natural basis vector

$$p_1 = \begin{bmatrix} 1 \\ 0 \\ \cdot \\ \cdot \\ \cdot \\ 0 \end{bmatrix}. \tag{26}$$

B. Partial Derivative Computation

To illustrate the partial differentiation process for a typical element, rewrite the circuit equation (25) as

$$\left[M_i s C_i + N(s)\right] e = p_1 e_{in} \tag{27}$$

where all matrix terms, except the one of interest, have been grouped in the sum matrix $N(s)$.

The system transfer function is, by Cramer's rule,

$$H(s) = \frac{(-1)^{1+q} |M_{i(1q)} s C_i + N_{(1q)}(s)|}{|M_i s C_i + N(s)|} \tag{28a}$$

or

$$H(s) = \frac{\alpha_n s^n + \alpha_{n-1} s^{n-1} + \cdots + \alpha_0}{\beta_n s^n + \beta_{n-1} s^{n-1} + \cdots + \beta_0}. \tag{28b}$$

The enclosed subscripts, $(1q)$, on M_i and N indicate the submatrices obtained by deleting the first row and last column of M and N, respectively. Now it is easy to show that the derivative of the determinant, $|aP + Q|$, with respect to the variable a is

$$\frac{\partial}{\partial a} |aP + Q| = \text{Tr } P[aP + Q]_{\text{adj}}. \tag{29}$$

Where Tr denotes the trace operator and X_{adj} means the adjoint (transpose cofactor matrix) of X. Thus

$$\frac{\partial}{\partial C_i}(\beta_n s^n + \cdots + \beta_0) = s \text{ Tr } M_i[M_i s C_i + N(s)]_{\text{adj}}. \tag{30}$$

To obtain the derivative of the entire set of denominator coefficients, with respect to C_i it is only necessary to combine the elements of the system cofactor matrix as selected by the element "connection matrix." Similar expressions are easily obtained and programmed to obtain partials for each numerator and denominator coefficient with respect to each passive element.

Since the minimum parameter description of the transfer function is desired, the leading numerator and denominator coefficients are factored from the expression (28b) to obtain

$$H(s) = K \frac{s^n + a_{n-1} s^{n-1} + \cdots + a_0}{s^n + b_{n-1} s^{n-1} + \cdots + b_0}. \tag{31}$$

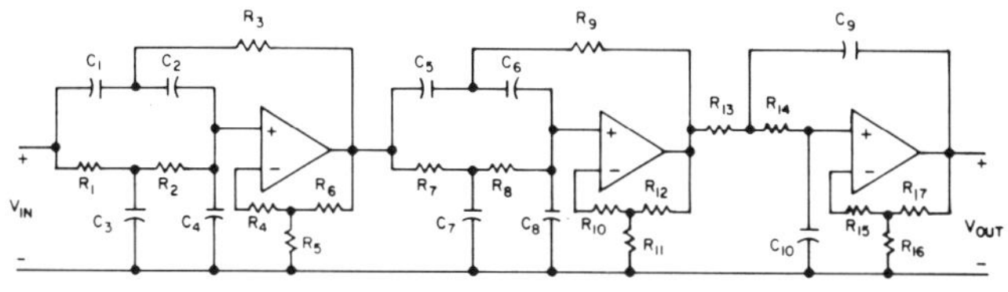

Fig. 3. Six-pole four-zero low-pass elliptic filter.

Where the identifications are made

$$K = \frac{\alpha_n}{\beta_n} \tag{32a}$$

$$a_i = \frac{\alpha_i}{\alpha_n}, \qquad i = 0, \cdots, n-1 \tag{32b}$$

$$b_i = \frac{\beta_i}{\beta_n}, \qquad i = 0, \cdots, n-1 \tag{32c}$$

and the desired partials are obtained from the ratio rule for derivatives, e.g.,

$$\frac{\partial b_j}{\partial C_i} = \frac{1}{\beta_n}\left[\frac{\partial \beta_j}{\partial C_i} - \frac{\beta_j}{\beta_n}\frac{\partial \beta_n}{\partial C_i}\right]. \tag{33}$$

The complete set of operations for all partials is easily programmed and extremely accurate results can be obtained if the network is properly frequency scaled.

V. An Example

In order to evaluate the effects of the tuning technique proposed in the preceding sections, a high-performance filter was chosen as a candidate for tuning and a hypothetical manufacturing situation was simulated. Specifically, a six-pole four-zero elliptic low-pass filter designed by Kerwin was chosen for the test case [11]. The circuit diagram is given in Fig. 3 and the element values are listed in Table I.

Then for this filter, individual tuning algorithms were derived for each of the three stages of the filter under the three conditions given below.

1) The weighting matrix, $\boldsymbol{Q}$, is identically zero and the conductance weights, γ_i, are equal to 0.0001.

2) The weighting matrix is chosen so as to penalize transfer function coefficients in proportion to their effect on the filter response in the 3-kHz region.

The conductance weights are as in case 1 and the resistances are all computed before any resistor is trimmed ("open-loop" tuning).

3) Case 2 weighting is employed and resistances are sequentially computed in reverse schematic order for each stage followed by a simulated resistor trim and subsequent error update ("closed-loop" tuning).

The effect of case 1 weighting is to trim all resistors to their design values regardless of the value of the circuit capacitors; the remaining two cases effectively minimize a weighted sum of the squares of the percent deviations of the transfer function coefficients. In case 2, however, no attempt is made to compensate for resistor deviations from their calculated values, while case 3 employs feedback of the resistance measurements.

The manufacturing process was then simulated via Monte Carlo techniques, with random values chosen from element value and measurement populations as defined in Table II. It is assumed that the first capacitance value is triangularly distributed in a 5-percent band about its design value and subsequent capacitors have a deviation triangularly distributed symmetrically about this initial deviation with limits of ±4 percent within an absolute bound of ±5 percent (4-percent tracking). All other quantities in Table II are also assumed to be symmetrically triangularly distributed between the limits given.

The manufacture of 100 circuits under each of the three conditions was then simulated using the distributions of Table II and the high and low gain limits were tallied across the frequency range from dc to 20 kHz. The results are displayed on the plots in Figs. 4 through 6.

TABLE I
Element Values, Low-Pass Filter

	RESISTANCE - KOHMS		CAPACITANCE - PF
R_1	50	C_1	638
R_2	100	C_2	319
R_3	33.3	C_3	957
R_4	118.1	C_4	278
R_5	50	C_5	810
R_6	88.1	C_6	405
R_7	50	C_7	1215
R_8	100	C_8	230
R_9	33.3	C_9	1000
R_{10}	124.1	C_{10}	1070
R_{11}	50		
R_{12}	53.7		
R_{13}	50		
R_{14}	100		
R_{15}	128.7		
R_{16}	50		
R_{17}	37		

TABLE II
Monte Carlo Simulation Parameters

PARAMETER	RANGE	MEASUREMENT ERROR
CAPACITANCE (C_1)	± 5%	± 0.1%
TRACKING	± 4%	
DISSIPATION FACTOR	001 TO .003	± 40%
RESISTANCE TRIM ERROR	± 0.5%	
	TRIANGULAR DENSITY FUNCTIONS	

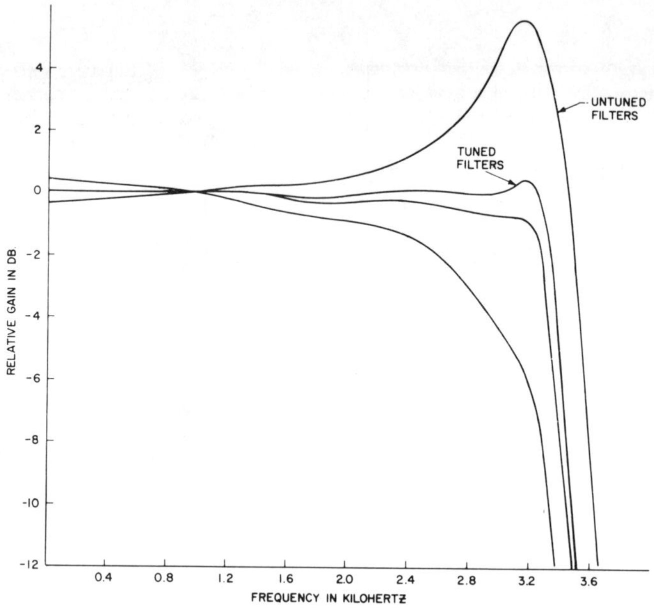

Fig. 4. Passband response limits—no tuning and linear optimal tuning.

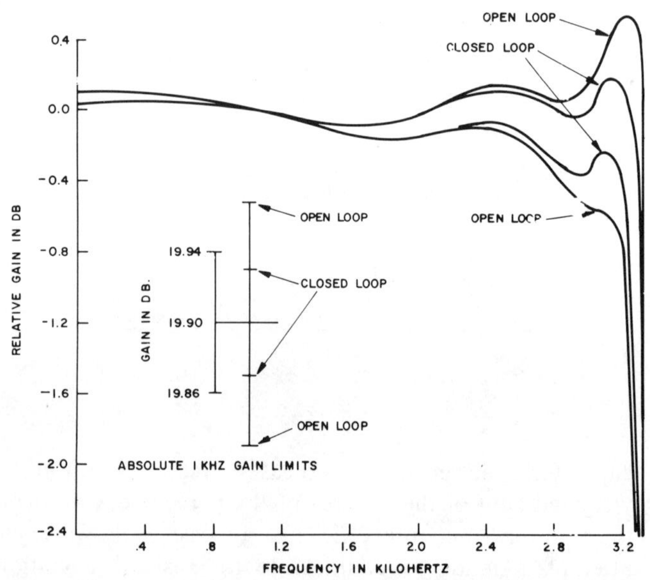

Fig. 5. Passband response limits—open-loop and closed-loop tuning.

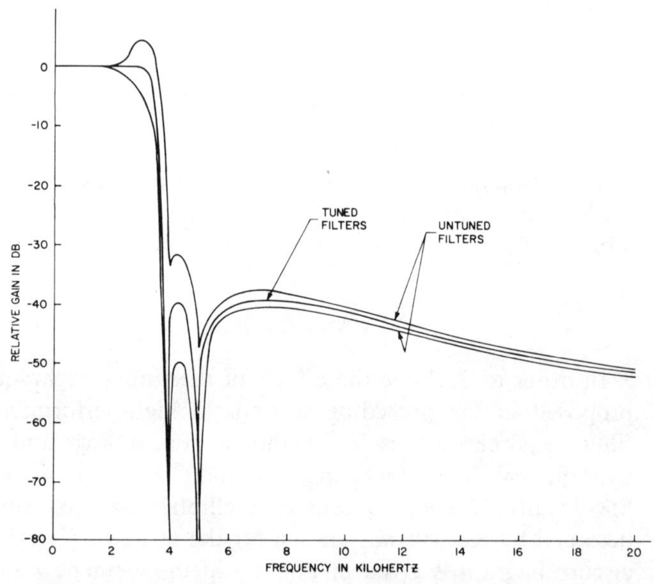

Fig. 6. Stopband response limits—no tuning and linear optimal tuning.

Under these not unrealistic assumptions, the effect of linear optimal tuning is pronounced and the benefits of adding "feedback" are apparent. Practically stated, a filter of this complexity is unmanufacturable without some form of tuning and the addition of resistor updating information gives an additional cost benefit to the process by decreasing the product spread for a given trim tolerance.

VI. Advantages and Conclusions

In this paper, a technique for communications filter tuning is described which has broad applicability and a number of economic benefits in filter design and manufacture. Tuning rule design and on-line software can be standardized resulting in the following:

1) minimal engineering effort required to develop tuning rules for new product;
2) shortened intervals between filter design and full-scale production.
3) better utilization of high expense resistance trimming apparatus through standardized software and unified tuning procedures for a variety of products;
4) improved manufacturing yields and higher throughput through relaxed component tolerance requirements;
5) better product quality through trim error compensation.

The method has been successfully implemented and used in large scale production for a variety of filter products. Extensions of the technique and applications to other types of communications networks are currently under investigation.

References

[1] D. A. McLean, M. Schwartz, and E. D. Tidd, "Tantalum-film technology," *Proc. IEEE*, vol. 52, pp. 1450–1462, Dec. 1964.

[2] M. J. Cohen, B. A. Unger, and J. F. Milkosky, "Laser machining of thin films and integrated circuits," *BSTJ*, vol. 47, pp. 305–405, Mar. 1968.

[3] J. W. Bandler, P. C. Liu, and H. Tromp, "Practical design centering, tolerancing, and tuning," in *Proc. IEEE ICAS Symposium*, 1975.

[4] J. W. Bandler and P. C. Liu, "The tolerance-tuning problem: a nonlinear-programming approach," in *Proc. 12th Allerton Conf.*, 1974.

[5] J. F. Pinel, K. A. Roberts, and K. Singhal, "Tolerance assignment in network design," in *Proc. IEEE ICAS Symposium*, 1975.

[6] G. S. Moschytz, "A general approach to twin-T design and its application to hybrid integrated linear active networks," *BSTJ*, vol. 49, pp. 1105–1149, July 1970.

[7] W. H. Orr, "The computer design and precision tuning of thin film filters," in *Proc. 3rd Ann. Seminar on Integrated Circuits*, New York Chapter IEEE Basic Sciences Div., Feb. 1966, pp. 45–56.

[8] R. A. Friedenson, R. W. Daniels, R. J. Dow, and P. H. McDonald, "RC active filters for the D3 channel bank," *BSTJ*, vol. 54, pp. 507–529, Mar. 1975.

[9] R. E. Kalman and R. W. Koepcke, "Optimal synthesis of linear sampling control systems using generalized performance indexes," *Trans. ASME*, vol. 80, pp. 1820–1838, Nov. 1958.

[10] J. T. Tou, *Optimum Design of Digital Control Systems*. New York: Academic Press, 1963.

[11] L. P. Huelsman, Ed., *Active Filters: Lumped, Integrated, Digital and Parametric*. New York: McGraw-Hill, 1970.

Tantalum Thin-Film *RC* Circuit Technology for a Universal Active Filter

WALTER WOROBEY, SENIOR MEMBER, IEEE, AND JOHN RUTKIEWICZ

Abstract—This paper describes the physical layout, process sequence, and component properties of a universal active filter designated the standard tantalum active resonator (STAR). This active-filter building block was realized as a hybrid integrated circuit (HIC) using tantalum thin-film technology. Several different circuit configurations are possible using only one set of process photomasks and one HIC. This can be accomplished by a wide range of precise resistor values through the use of laser trimming and various interconnecting schemes for the components on the HIC via the printed wiring board.

The STAR circuit is comprised of nine laser-adjustable tantalum thin-film resistors totalling a maximum of 1 MΩ, two 5100-pF tantalum capacitors, and one operational amplifier. The capacitors are formed by anodizing tantalum to 190 V and depositing NiCr-Pd-Au as the adherent counterelectrode. The resistors are made from 300-Ω/□ tantalum oxynitride film which is preaged at 350°C for 1 h. The single-substrate *RC* process is amenable to batch processing and precision tuning, making it attractive for high-level production of precision filters. New thin-film design criteria and process developments permitted this circuit to be fabricated on a 16-pin dual-in-line package ceramic substrate which could be packaged for machine insertion with commercially available insertion equipment.

I. INTRODUCTION

The use of active *RC* filters for analog applications has been rapidly expanding in the electronics industry [1]-[3]. This has been brought about largely by significant improvements in the performance and cost of sophisticated miniature operational amplifiers made with active silicon devices, as well as the reduced size and relatively low cost of *RC* hybrid circuits. Such developments have made the *RC* active filter economically viable and competitive with *LC* designs in terms of size, cost, and performance characteristics. In addition to the process technology advancements, there has been a considerable effort to design universal building blocks for active filters [4]. In general, commercially available universal active filters are attractive to designers because: 1) they can become a readily available item, providing rapid turnaround between application concept and realization; 2) they can reduce the cost of filters since one standard design may be used for many filter applications by simply adding a few external resistors and/or making the appropriate external connections between circuit terminals. This latter feature is especially valuable to customers with low-volume applications who want to take advantage of the low cost offered by high-volume standard codes.

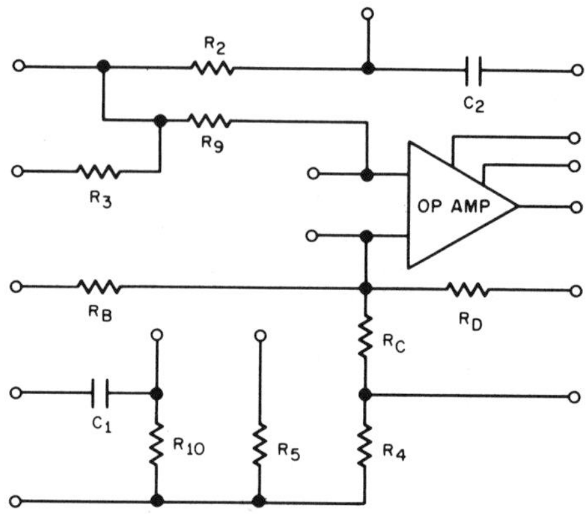

Fig. 1. DIP STAR filter schematic.

An active-filter realization based on a building-block concept using the single amplifier biquad (SAB) circuits was developed by Friend [4] and is shown in Fig. 1. This design was realized as a thin-film hybrid integrated circuit (HIC) and called the standard tantalum active resonator (STAR). Its components were configured so that one physical HIC could be used to realize all the standard biquadratic transfer functions. For example, the STAR building block can be used to realize a wide variety of low-pass notch, high-pass notch, bandpass, or other filter functions [4]. Many commercial universal building blocks for active filters require the user to provide critical external resistors to tune the filter or to make mask changes to change the value and interconnection of the components. The STAR avoids the necessity for external components and different masks by using one HIC for all filters, setting the resistor values by laser trimming rather than by changing masks, and interconnecting the components on the HIC via the printed wiring board (PWB).

The first models of the STAR circuit were made in 1973 using *RC* design rules and processing procedures which were standard for Western Electric manufacture at that time. These models were 2.67 X 2.54 cm in size and were considered too large for some applications which required many STAR circuit sections on the same PWB. In order to reduce the size of the STAR circuit, an extensive development program was carried out to evaluate the effect of a new set of fabrication procedures and circuit-layout rules on reliability and manufacturability. The result was a new STAR circuit one fifth the size

Manuscript received May 15, 1976; revised July 21, 1976. "This paper was presented at the International Society for Hybrid Microelectronics Symposium of 1976 and is reproduced with permission of ISHM." This paper was separately reviewed for this publication and modified and expanded from the original publication.

The authors are with Bell Laboratories, Inc., Allentown, PA 18103.

Reprinted from *IEEE Trans. Parts, Hybrids, Packag.*, vol. PHP-12, pp. 276-282, Dec. 1976.

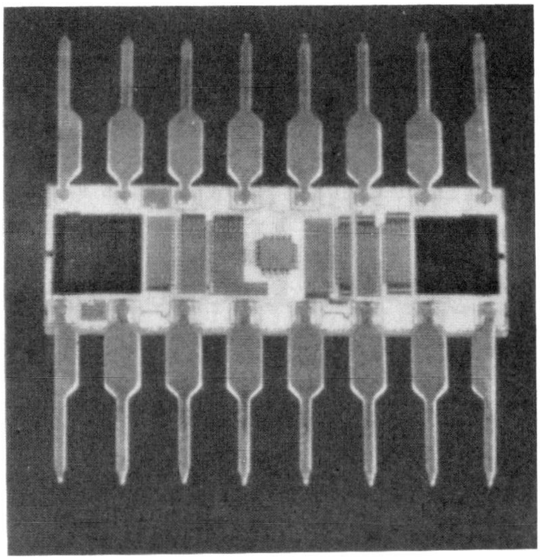

Fig. 2. DIP STAR before encapsulation, showing the lead frame and operational amplifier thermal-compression-bonded to the ceramic substrate.

Fig. 3. Substrate containing 60 DIP STAR circuits after completion of FIC processing prior to circuit separation.

of the original HIC and one which could be packaged in a machine-insertable 16-pin dual-in-line package (DIP).

This paper describes the physical design and processing and component properties of the new DIP STAR circuit which has led to a new standard for the next generation of *RC* integrated circuits.

II. PHYSICAL DESIGN

Fig. 2 shows the new STAR circuit on a 1.93 X 0.66-cm ceramic substrate before encapsulation. It is comprised of nine laser-adjustable resistors, two thin-film capacitors, conductors for interconnections, and an operational amplifier which is thermal-compression-bonded in the center. There are no crossovers in this circuit. Each side of the ceramic substrate has eight external lead-frame bonding pads on 2.54-mm centers. The layout allows each component of the circuit to be individually measured from the external pad locations. For batch fabrication purposes, 60 individual STAR circuits are laid out on a large substrate which is 9.5 X 11.4 cm (see Fig. 3). After the processing is completed, these circuits are separated by laser scribing and snapping them apart. The designs of the various thin-film components and interconnections are described below.

A. Capacitors

The two capacitors required for this circuit are 5100 pF each and are placed on selectively glazed portions of the substrate [5]. The glazed areas are below the capacitors only. The capacitors are formed with a 190-V anodic oxide, which results in a capacitance density of 69 nF/cm^2 when sputtered beta tantalum [6] is used. Each of the circuits on the large substrate is electrically isolated except for the capacitors. The base electrodes of the capacitors are interconnected to allow mass anodization for the capacitor dielectric.

An adherent NiCr-Pd-Au layer is used for the counterelectrode. The capacitors are surrounded on three sides by this

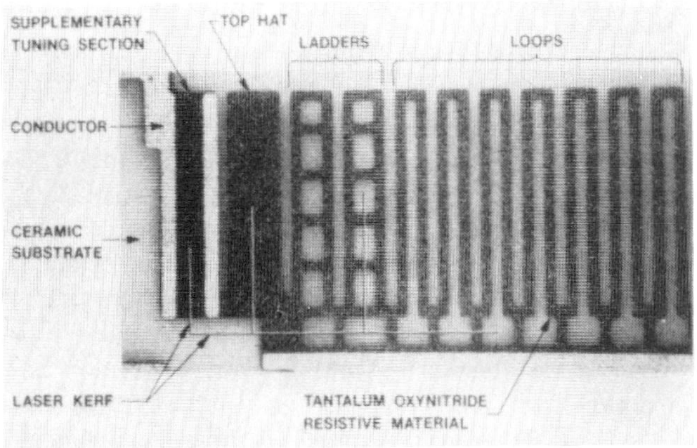

Fig. 4. Resistor pattern for laser trimming.

conductor film, which has electrical contact to the tantalum base electrode in order to reduce the series resistance and thus improve the dissipation factor at higher frequencies.

B. Resistors

The thin-film resistors consist of 300-Ω/□ tantalum oxynitride [7] and are formed on unglazed areas of the substrate. The patterns are composed of loops, ladder sections, top hats, and, in some cases, supplementary tuning sections for fine tuning (see Fig. 4). The linewidths and spaces of the loops and ladders are 50 μm and the top-hat widths are 250 μm. The resistors are terminated with contact areas which consist of successive layers of TiPd and NiCrPdAu on top of the tantalum oxynitride (TaO_xN_y) resistor film. The resistor loops are shorted out by tantalum as shown in Fig. 4. All the resistors are laser-trimmed to value after completion of the *RC* processing.

The resistors can be adjusted over a wide range of values by the laser scribing the tantalum paths between the loop and ladder sections. For example, some resistors can be adjusted between 1000 and 200 000 Ω. Continuous adjustment of the top hats and supplementary tuning sections allows resistors over a few thousand ohms to be adjusted to within 0.05 percent of a specified value. The adjustment range factor of approximately 200 allows the STAR circuit to be tailored to a wide range of precision filter applications without requiring additional photomasks or adding external resistors. These adjustments can be made before or after silicon integrated-ciruit (SIC) attachment prior to packaging. The DIP STAR circuit (see Fig. 2) includes two additional resistor sections between the bonding pads for applications that require higher valued resistors than are attainable with the adjustable loop and ladder sections. The nine adjustable resistors allow a maximum total of 1 MΩ of resistance on each circuit.

C. Conductors and Bonding Pads

Conductor interconnections and SIC bonding pads are composed of NiCr-Pd-Au and are defined during the final pattern delineation step. The minimum conductor width and separation is 75 μm. Lead-frame bonding pads are comprised of β-Ta, TaO_xN_y, and NiCr-Pd-Au. They are located on 2.54-mm centers.

III. CIRCUIT FABRICATION

A. *RC* Processing

Fig. 5 outlines the process sequence used for the *RC* circuit. One screen for selective glazing and six photomasks for capacitor anodization and metal etching are required to produce the *RC* film integrated circuit. This single-substrate *RC* process is similar in many respects to that described by Orr *et al.* [8] but has some differences which will be discussed later.

The top view of Fig. 5 shows the front surface of a completed circuit which has a resistor on the left and a capacitor on the right. Below this top view are cross sections of the circuit at various processing stages.

The substrate is first prepared by forming selectively glazed [5] areas in the capacitor regions and then covering the substrate with approximately 1000 Å of thermally formed Ta_2O_5 to protect the glaze from subsequent hydrofluoric acid etchants. The first step shown in the process sequence of Fig. 5 is the deposition of approximately 3000 Å of tantalum for the capacitors over the entire substrate. The film normally used is beta-tantalum which has a tetragonal structure and has been qualified for high-reliability capacitors [6],[9]. The tantalum film is then etched to define the capacitor base electrode and external lead-frame bonding pads, and to clear areas or windows for the resistors (view B). A matrix of tantalum paths remains on the large substrate to interconnect all the capacitors so that they can be mass anodized from a single electrical contact on the top edge of the large substrate. The capacitor areas are next defined with a photoresist anodization maskant, and the tantalum is anodized to about 125 V to form an initial capacitor dielectric (view C).

A tantalum oxynitride resistor film is sputtered over the substrate (view D) and titanium and palladium are deposited on top (view E) to ensure good ohmic contact with the resistor

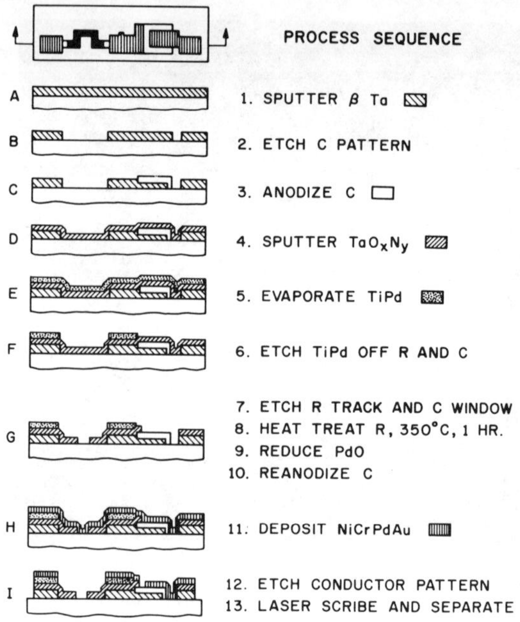

Fig. 5. *RC* process sequence for fabricating the DIP STAR circuit using tantalum thin-film resistors and capacitors.

film. The titanium and palladium films are etched to define contact spots on the capacitor and resistor lead-ins (view F). The resistor pattern is generated in the next photolithographic step (view G). It is during this step (i.e., when the tantalum resistor film is etched off the capacitor area) that the capacitor dielectric serves as an etch stop and prevents etching of the capacitor base electrode.

The resistors are thermally stabilized at 350°C for 1 h and the palladium oxide which is formed during this heat treatment is reduced to metallic palladium by exposing the circuits to forming gas at room temperature for a few minutes [10]. Because the resistor deposition and stabilization steps degrade the capacitor dielectric, it is reformed and brought to the final thickness by anodizing to 190 V. Nichrome, palladium, and gold are evaporated over the entire substrate to provide conductors and a top electrode for the capacitors (view H). In the final patterning step, the NiCr-Pd-Au multilayer is etched to define the interconnection pattern and capacitor counterelectrode (view I). At this point all of the chemical processing is completed. The substrate can now be laser-scribed, the film circuits separated from the parent substrate and then tested. The tantalum bus bars between the capacitor anodes on the parent substrate are opened during the circuit separation step. More complex circuits which could not have capacitor bus bars passing over the edge of the circuit would require an additional etch step to open them.

All the above steps are carried out by batch processing techniques. No mechanical masks are used and no individual substrate handling is required except when transferring substrates from one carrier to another. The large substrate measures 9.5 X 11.4 cm and contains 60 individual DIP STAR circuits (see Fig. 3). This allows the cost of each process step to be distributed over many circuits, thus reducing the overall cost per circuit to a relatively low value.

As mentioned previously, the *RC* process described above

TABLE I

Comparison of Original and New Processes for Fabrication of Thin-Film *RC* Circuits

Process Step	Original Process	New Process
Sputter capacitor film	X	X
Etch capacitor pattern	X	X
Pre-anodize capacitor	X	X
Sputter resistor film	100 Ω/□	300 Ω/□
Deposit TiPd contact film	X	X
Etch contact pattern	X	X
Etch resistor pattern	X	X
Pre-anodize resistors	X	No
Stabilize resistors	250°C, 5 hrs.	350°C, 1 hr.
Reduce PdO	No	X
Reanodize capacitors	230V	190V
Deposit conductor film	X	X
Etch conductor & tantalum	X	No*
Etch conductor film	X	X
Trim anodize resistors	X	No
Separate circuits	X	X
Bond Leads	X	X
Final Resistor Adjustment	Anodic	Laser
Package & Final Test	X	X

*For the DIP STAR circuit the capacitor bus bars are opened when the circuits are separated and therefore no tantalum etching is required at this point. More complex circuits which could not have capacitor bus bars passing over the edge of the circuit would require an etch step at this point to open them.

is similar in many respects to the process described by Orr *et al.* [8] but does have some important differences. These differences allowed the circuit layout to be reduced substantially in size so that the DIP STAR circuit could be made on a 1.93 X 0.66 cm ceramic. The HIC can then be packaged in a machine-insertable 16-pin DIP. The process differences are outlined below.

1) The capacitance density was increased to 69 nF/cm^2 from 57 nF/cm^2 by using 190-V anodization instead of 230 V. This reduction also allowed smaller anodization clearance borders to be used.

2) The resistor stabilization temperature was changed to 350°C for 1 h from the previous 250°C for 5 h. The higher temperature permitted stable resistors to be formed without the need for resistor anodization. This, coupled with the fact that resistors are laser-trimmed instead of anodically trimmed, eliminates all resistor anodization steps and thus the need for any resistor anodization clearance borders. The higher stabilization temperature did require that a PdO reduction step be introduced to eliminate the contact resistance between the conductors and other thin-film components.

3) The resistors were formed from 300-Ω/□ resistor film, in this case tantalum oxynitride, which is greater, by a factor of three, in sheet resistance than the previous standard. This increase allowed the resistor areas to be reduced by the same factor. Placing the resistors on bare ceramic made it possible to reduce the resistor linewidths and space widths to 50 μm without degrading the circuit yield.

4) In addition to the more compact circuit geometry which is possible with this new process, fewer process steps are required than in the original *RC* process described by Orr. For example, only six photomasks are necessary compared to nine in the original process using anodic resistor trimming and 250°C heat treatment. A comparison of the original and new processes is shown in Table I.

B. Assemble, Tune, and Package

After the circuits are separated from the parent substrate, they are tested and examined to eliminate any which are defective. Gold-plated copper lead frames are thermal-compression-bonded to the substrate and the resistors are laser-trimmed to value after the capacitors are measured. The SIC's are usually bonded to the film circuit after the trimming step; however, if functional tuning of frequency selective circuits is desired, the laser trimming of resistors can be done after bonding the SIC's. The circuit is then encapsulated with a conformal coating of RTV silicone rubber [11] and a plastic housing placed over the HIC.

The DIP STAR circuit, before encapsulation, is shown in Fig. 2. It contains one 5100-pF thin-film capacitor on each

Fig. 6. DIP STAR after packaging.

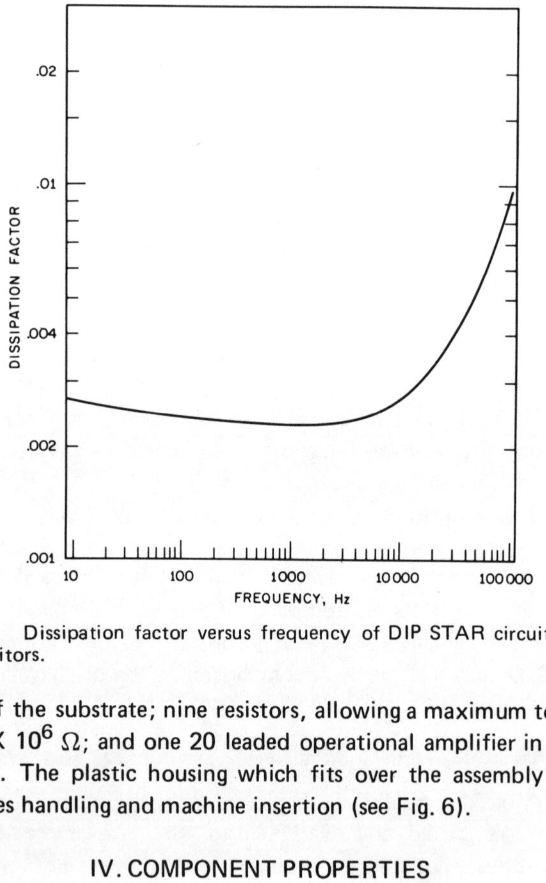

Fig. 7. Dissipation factor versus frequency of DIP STAR circuit capacitors.

end of the substrate; nine resistors, allowing a maximum total of 1 X 10^6 Ω; and one 20 leaded operational amplifier in the center. The plastic housing which fits over the assembly facilitates handling and machine insertion (see Fig. 6).

IV. COMPONENT PROPERTIES

A. Capacitors

Thin-film capacitors for the DIP STAR *RC* circuits are made by anodizing a patterned tantalum film to 190 V in a 0.01 percent aqueous solution of citric acid and forming an adherent counterelectrode of NiCr-Pd-Au on top of the dielectric. The dielectric thickness at this anodization voltage is approximately 3400 Å. Of great importance in fabricating good-quality thin-film capacitors is the base material from which the dielectric is formed. For this work, sputtered beta-tantalum with an added few atomic percent of nitrogen was used [12],[13]. This film has a tetragonal structure and an electrical resistivity of about 200 $\mu\Omega$·cm. Capacitors formed in this manner will typically have the properties shown below. Except for the capcitance density these properties are similar to those for standard *RC* circuits [8].

1) Capacitance density	69 nF/cm^2
2) Dissipation factor (at 1000 Hz and 25°C)	0.002 - 0.003
3) Temperature coefficient of capacitance (at 1000 Hz)	
−40 - 25°C	190 ±25 ppm/°C
25 - 65°C	220 ±25 ppm/°C
4) Moisture sensitivity of capacitance (0 - 85 percent RH)	<+0.02 percent
5) Leakage current (50V, 25°C, 1 min)	<2 A/F
6) Initial capacitance tolerance	±3 percent
7) Capacitor aging (20 years at 65°C, estimate)	−0.4 percent (maximum)
8) DC working voltage at 85°C	
forward bias (Ta positive)	10 V
reverse bias (Ta negative)	0.5 V
9) AC working voltage at 85°C	7 V rms

The dissipation factor as a function of frequency for a 5100-pF capacitor on a DIP STAR circuit is shown in Fig. 7. The increase above 10 kHz is due to the resistance of the tantalum base electrode. The ac losses in the capacitor can be expressed in terms of a series equivalent circuit, and can be given by

$$\tan \delta_T = \tan \delta_i + \omega R_S C_S$$

where $\tan \delta_i$ is the intrinsic dissipation factor due to the dielectric losses, C_S is the equivalent series capacitance, ω is the measurement frequency in radians per second, and R_S is the series resistance due to the capacitor electrodes and interconnections. For a DIP STAR circuit capacitor which is designed and processed as described in this paper, R_S is approximately 2 Ω. This series resistance can be reduced by increasing the tantalum thickness, reducing the base electrode resistivity, or by using a bipolar interdigitated capacitor design [14]. The temperature coefficient of capacitance (TCC) is nonlinear with temperature between −40 and 65°C. The values can be shifted slightly by thermal exposures such as those seen during the counterelectrode evaporation or leadframe bonding. All capacitors are normally screened at 50 V for 1 min. The leakage current is usually one or two orders of magnitude below the specification limit of 2 A/F.

The catastrophic failure characteristics of tantalum thin-film capacitors depend on the direction of the field in the dielectric. Fig. 8 shows the step-stress breakdown voltage at 85°C for forward (i.e., tantalum positive) and reverse bias. The voltage was increased in increments every 0.2 h. It can be seen that the capacitor can withstand higher dc voltages in the forward direction than in the reverse. Acceleration factors are

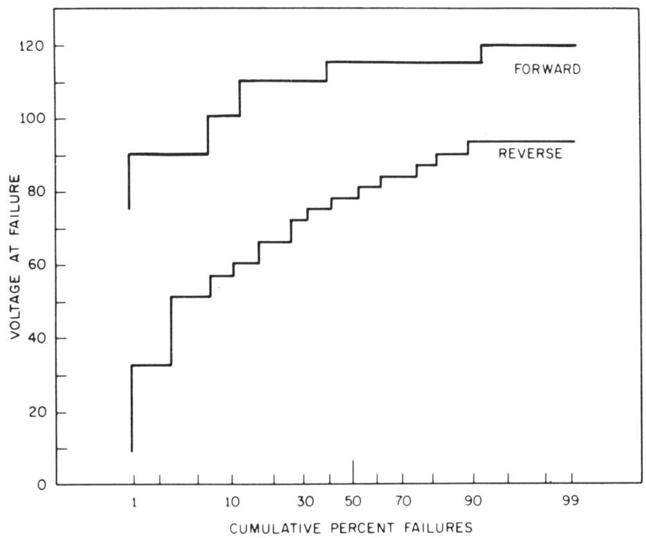

Fig. 8. Two-hour 85°C Step Stress of DIP STAR capacitors.

usually calculated from step-stress and static-stress tests over a limited range of stress levels and extrapolations made to determine the working stress conditions for long periods of time [9]. The dc working voltage of a 10-V forward bias and a 0.5-V reverse bias have been estimated for a 20-year operation at 85°C with a maximum instantaneous failure rate of approximately 100 FIT's (1 FIT = 1 failure in 10^9 component hours). The allowable ac working voltage is ± 10 V (peak-to-peak).

B. Resistors

The thin-film resistors for the DIP STAR circuit described in this paper were formed from tantalum oxynitride films [7] having a sheet resistance of 300 Ω/□ after thermal stabilization at 350°C for 1 h. The resistors are deposited on unglazed high-alumina ceramic substrates with a Ta_2O_5 underlay and have the typical properties shown below after the processing is completed.

1) Sheet resistance	300 Ω/□
2) Resistance density	6 MΩ/cm^2
3) Humidity sensitivity	0
4) Temperature coefficient of resistance (–40 - +65°C)	–200 ±20 ppm/°C
5) Initial tolerance	±0.05 percent
6) Aging (20 years at 60°C)	<+0.1 percent

The temperature coefficient of resistance (TCR) of tantalum resistor film is relatively constant with temperature and can be adjusted by varying the oxygen concentration during the reactive sputtering of tantalum. For precision applications, it is desirable to minimize the temperature coefficient of the *RC* product in order to make the filter functions less temperature sensitive. Therefore, the TCR of the resistor film is adjusted to be equal and opposite to the TCC of the tantalum thin-film capacitor which is in the neighborhood of 200 ppm/°C. Precision matching of the two temperature coefficients over a wide temperature range is not possible due to the nonlinearity of the TCC. However, present technology can provide overall control of the temperature coefficient of the *RC* product to within ±50 ppm/°C over the temperature range from –40 to +65°C, which includes the operating range of most circuits.

The initial tolerance of resistors is determined by the capability of the laser-trimming process and the control of the short-term stability of the resistor's laser-trimmed section. Continuous adjustment of the STAR circuit resistors is provided by the top hat and the supplementary tuning sections (see Fig. 4). The supplementary tuning sections provide approximately one sixth of a square of resistance. The resistance change per unit length of laser cut is very low and permits high precision to be achieved with the present commercial laser trimmers. The initial tolerance of ±0.05 percent for the present design is adequate for STAR filter applications.

Long-term resistor aging is determined by the aging of both the untrimmed and laser-trimmed parts of a resistor [15]. Laser trimming within the current path of a resistor tends to degrade the stability of the resistor film and, therefore, should be kept to a minimum. The loop and ladder sections of the resistor are unaffected by laser trimming because the laser-trimmed edge is not in the vicinity of the current path. Therefore, the resistor design used for the DIP STAR circuit tends to improve trimmed resistor stability since, except for low-valued resistors, a relatively small percentage of the resistor path is made up of a continuously tuned top-hat section. Accelerated aging studies of typical laser-trimmed DIP STAR circuits indicate a 20-year resistance aging of less than 0.1 percent at 60°C. It should be noted that long-term parametric aging of *RC* filters is determined primarily by the aging of the capacitors since they age more than the resistors.

V. SUMMARY AND CONCLUSIONS

The STAR, which is an active biquadatic filter section in thin-film form, has been designed to fit on a 16-pin DIP ceramic substrate. This package is machine-insertable with standard commercially available insertion machines. The processing and design innovations which enabled the five to one reduction in size over the original design have proven to be realizable and valuable for future *RC* circuit designs. The significant changes are: a) 350°C 1-h resistor stabilization which eliminates the need for resistor anodization, b) 190-V anodization for the capacitor dielectric, c) the use of polar capacitors for small reverse-voltage applications, d) 300-Ω/□ resistor film, e) new laser-trimmable resistor designs, and f) tighter design tolerances on components and interconnection patterns. This process is amenable to batch processing and precision tuning which makes it attractive for high-level production of precision circuits. The initial properties as well as the long-term behavior of the tantalum resistors and capacitors are typical of standard *RC* circuits. This new DIP STAR circuit should meet the future needs of circuit designers for a high-precision low-cost active-filter building block in a machine-insertable 16-pin standard DIP package.

ACKNOWLEDGMENT

The prove-in of the film design and process of the DIP STAR required the efforts of many engineers both at Bell

Laboratories and the Western Electric Company. They included: F. N. Fuss, A. R. Adolt, and P. W. Wyatt of Bell Laboratories and J. R. Bitler and M. W. Bodnar of the Western Electric Company.

REFERENCES

[1] J. Tsantes, "Active filters gain acceptance in new designs—Filter report," *Electronic Engineering Times,* pp. 22, Sept. 8, 1975.

[2] W. H. Orr and T. R. Robillard, "Thin film hybrid integrated circuits for communication systems," *IEEE Trans. Parts, Hybrids, Packag.*, vol. PHP-8, June 1972.

[3] H. Kurosawa, A. Iwata, and I. Miyata, "Hybrid integrated *RC* active channel filter," *Rev. Elec. Commun. Lab.,* vol. 23, no's. 5 and 6, May-June 1975.

[4] J. J. Friend, C. A. Harris, and D. Hilberman, "STAR: An active biquadratic filter section," *IEEE Trans. Circuits and Systems,* vol. CAS-22, Feb. 1975.

[5] R. H. Minetti and J. Demchik, "Thin selective glazing," presented at the Meeting of the Electronic Division of the American Ceramic Society, Boston, MA, Sept. 1969.

[6] M. H. Read and C. Altman, "A new structure in tantalum thin films," *Appl. Phys. Lett.,* vol. 7, no. 3, p. 51, 1965.

[7] J. K. Werner and W. Worobey, "Performance of reactively sputtered tantalum oxynitride thin film resistors," *Proc. Electronic Components Conf.,* pp. 362-367, 1972.

[8] W. H. Orr, D. O. Melroy, R. J. Moore, F. P. Pelletier, W. J. Pendergast, and W. H. Yocom, "Integrated tantalum film *RC* circuits,"' in *Proc. Electronic Components Conf.,* pp. 602-612, 1970.

[9] J. S. Fisher, "Failure characteristics of thin film capacitors," in *Proc. Reliability Physics Symp.,* pp. 183-187, 1972.

[10] G. V. Samsonov, *The Oxide Handbook.* New York: IFI/Plenum, 1973, p. 391.

[11] D. Jaffe, "Encapsulation of integrated circuits containing beam leaded devices with a silicone RTV dispersion," in *Proc. Electronic Components Conf.,* pp. 376-381, May 1976.

[12] R. D. Huttemann, J. M. Morabito, and D. Gerstenberg, "Effect of light elements (N,C,O) in tantalum on tantalum film capacitors," *IEEE Trans. Parts, Hybrids, Packag.,* vol. PHP-11, p. 67, Mar. 1975.

[13] H. Y. Kumagai, I. M. Fletcher, K. R. Baurle, J. R. Hightower, and C. A. Steidel, "Properties of thin film capacitors from nitrogen doped beta tantalum films," *Proc. Electronic Components Conf.,* pp. 257-258, May 1973.

[14] R. W. Wyndrum, Jr., P. D. Stark, S. C. Lee, and F. P. Pelletier, "Analysis of bipolar interdigitated thin film capacitors," in *Proc. Electronic Components Conf.,* pp. 350-358, 1969.

[15] G. A. Bulger, "Stability analysis of laser trimmed thin film resistors," *IEEE Trans. Parts, Hybrids, and Packag.*, vol. PHP-11, Sept. 1975.

RC-Active Filters in Single-Layer Tantalum *RC*-Film Technology

HANS-WERNER RENZ, HOLM BAEGER, ERNST LUEDER, HANS-JUERGEN TILCH, AND TRAUGOTT KALLFASS

Abstract—A new composition of reactively sputtered Ta-oxinitride is used as a basic material for thin-film resistors as well as for thin-film capacitors; this single-layer technique enables an economic fabrication process for integrated *RC* circuits with lumped or distributed *RC* components. A bandpass circuit with lumped components for a dialing tone receiver and a voiceband low pass with distributed elements for PCM channel filtering are described; in either case a very small size allowed a complete filter to be packed in regular DIL-packages, and good performance and stability were obtained.

I. Introduction

CONVENTIONAL integrated thin-film *RC* circuits with temperature compensated elements [1], [2] use two different tantalum layers: resistors are usually made from Ta-oxinitride, whereas thin-film capacitors are preferably fabricated by anodizing β-Ta or nitrogen-doped α-Ta [3], [4]. Another approach is based on two different Ta–aluminum alloys [5]. Processing of a Ta double-layer structure, however, requires a large number of fabrication steps. A great simplification of the manufacturing process was achieved by introducing a new composition of Ta-oxinitride (to make resistors as well as capacitors) which has been found recently [6] and which has been called "Ta-*RC*-film". This Ta-*RC*-film will be used for the manufacture of miniaturized circuits with lumped and distributed elements. To minimize the substrate area needed, lumped circuits are mesh-star transformed as will be demonstrated by tone dialing filters [7], whereas in distributed circuits capacitors and resistors are placed on top of one another [8]. Contrary to lumped circuits, distributed *RC* filters required more basic work on circuit design and on the physical tunability problem [9]–[11]. Earlier attempts and extensive references may be found in [12]–[14]. The new methods [9]–[11] can only be briefly reviewed in Section V of this paper, where one circuit example will be presented in detail.

Manuscript received May 5, 1978; revised August 22, 1978. This work has been sponsored by the German Bundesministerium fuer Forschung und Technologie.

H. W. Renz was with the Institut für Netzwerk- und Systemtheorie, Universität Stuttgart, Stuttgart, Federal Republic of Germany. He is now with Standard-Elektrik-Lorenz AG (ITT), Stuttgart, Federal Republic of Germany.

H. Baeger was with the Institut für Netzwerk- und Systemtheorie, Universität Stuttgart, Stuttgart, Federal Republic of Germany. He is now with VDO AG, Schwalbach, Federal Republic of Germany.

E. Lueder, H. J. Tilch, and T. Kallfass are with the Institut für Netzwerk- und Systemtheorie, Universität Stuttgart, Stuttgart, Federal Republic of Germany.

II. "Ta-*RC*-Film" Technology

The new Ta-oxinitride for the manufacturing of integrated thin-film resistors and capacitors ("Ta-*RC*-film") is deposited by reactive sputtering of Ta in a mixture of argon, nitrogen, and oxygen. Contrary to the well-known Ta-oxinitride, the

Reprinted from *Proc. IEEE,* vol. 67, pp. 37-42, Jan. 1979.

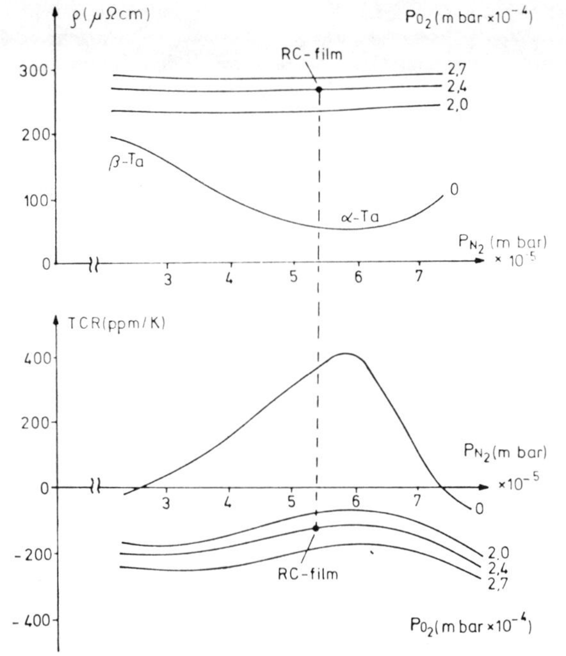

Fig. 1. Resistivity ρ and TCR versus nitrogen pressure with additional oxygen doping (for the RF-diode).

RC-film is composed of 10 at % nitrogen and 30 at % oxygen. Corning Glass 7059 with a Ta_2O_5 etch-stop layer is used as a substrate material. Two different sputtering units have been used alternatively: an *RF*-diode (MRC 8620) and a triode sputtering unit (Balzer's Sputron). Fig. 1 shows the resistivity ρ and the temperature coefficient of the resistance TCR depending upon the partial pressure of nitrogen in the unit prior to sputtering, with the oxygen pressure as a parameter. The *RC*-film as indicated in Fig. 1 is deposited (in the *RF*-diode) at the partial pressures 5.4×10^{-5} mbar nitrogen and 2.4×10^{-4} mbar oxygen. As known from [16], the permittivity of the anodically formed oxide decreases only by 1 percent for each added at % nitrogen. In accordance with [4], we found [6], [15] that an amount of about 10 at % nitrogen considerably improves the temperature stability of normally processed[1] capacitors without decreasing the capacitance density too much. As shown in Fig. 1, the dependence of resistivity and TCR on the nitrogen pressure is rather strong without presence of oxygen, but sufficiently small with the oxygen doping applied for the *RC*-film. This is quite favorable in order to get reproducible values of resistivity and TCR in a production line.

The oxygen pressure is chosen to adjust the TCR to -155 ppm/K in order to compensate the TCC which is found to be +155 ppm/K; a maximum mismatch of ± 10 ppm/K between both temperature coefficients is maintained using the Balzer's Sputron unit. The incorporated oxygen does not degrade the good temperature stability of the capacitors which is mainly due to the nitrogen doping. Hence it is possible to anneal capacitors (complete with top electrodes) at temperatures as high as 250 to 300°C which are applied for the stabilisation of resistors. This feature further simplifies the *RC*-film process which will be detailed in the following section. Finally, the high long-term stability of resistors and capacitors made from the *RC*-film should be emphasized; results of accelerated aging are given in Table I where the most important properties of the Ta-*RC*-film are summarized.

[1] The only difference to the β-Ta process is that the density of the anodization current should not exceed about 0.2 mA/cm².

TABLE I
PROPERTIES OF THE "*TA-RC* LAYER"

Resistivity	270 $\mu\Omega$cm
TCR	-155 ± 5 ppm/K
TCC	$+155 \pm 5$ ppm/K
Relative Permittivity of Anodically Formed Oxide	21
Dielectric Loss Factor (tan δ) at kHz	1.6×10^{-3}
Accelerated Aging After 3300 h at 175°C	$\Delta R/R = +0.4$ percent $\Delta C/C = -0.25$ percent

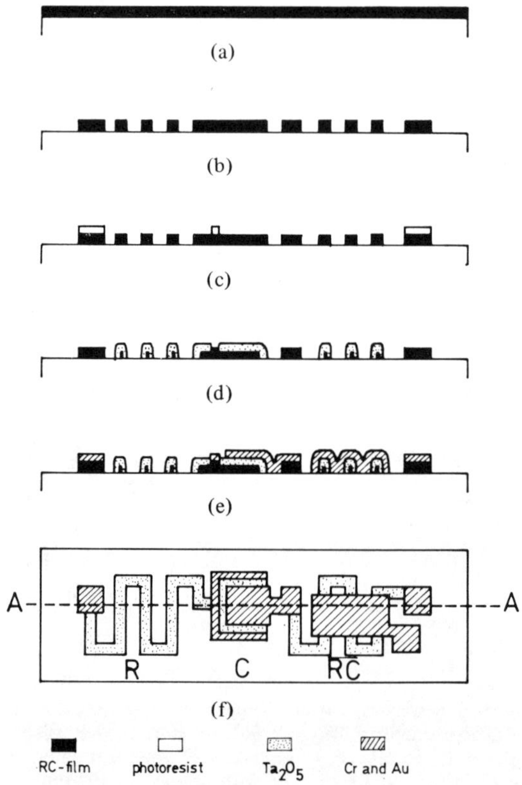

Fig. 2. *RC*-film process.

III. TA-*RC*-FILM PROCESS FOR INTEGRATED *RC* CIRCUITS

The process is illustrated by a circuit given in Fig. 2(f) in top view and in Fig. 2(a)–(e) in cross section along line A-A (in Fig. 2(f): it contains a resistor R, a capacitor C and a distributed *RC*-element $\overline{RC}$ further detailed in Section V. The first step is the reactive sputtering of the *RC*-film (Fig. 2(a), thickness approximately 190 nm, sheet resistance 14Ω/square). The first photomask is applied for etching of meander patterns and bottom electrodes (Fig. 2(b)). The second photomask (Fig. 2(c)) prevents the contact areas from being oxidized during the following anodization step when capacitor electrodes, resistors, and distributed *RC*-elements are anodized simultaneously (Fig. 2(d)). A method to control the *RC*-products of the components to within $\pm$ 10 percent is based on monitoring the *RC*-product during the anodization [9]. This control may be replaced by a tight control of the film parameters and of the anodization voltage (usually 200 V).

For example, an initial RC-film thickness of 190 nm corresponding to 14Ω/square and a 200-V anodization yields a final sheet resistance of about 70Ω/square and a capacitance density of 52 nF/cm^2. After Cr–Au or Ti–Pd–Au is evaporated, the third and last mask is needed to etch the conductor pattern (Fig. 2(e), capacitor top electrodes, interconnections, contact pads). Note that the bottom electrode of a capacitor should exhibit a highly conductive rim, as shown in Fig. 2(e)–(f), to lower the capacitor loss; for a 6-nF capacitor, e.g., this results in loss factors tan δ of 0.18 percent at a frequency of 1 kHz and a 0.5 percent at 15 kHz. Finally, the complete circuits are tempered (e.g. 24 h at 250°C) and further processed as usually (separating, soldering, laser trimming etc.).

IV. Bandpass Filters with Lumped Elements

Selection of signal frequencies in tone dialing receivers can be done by second-order bandpass filters with the transfer function

$$T(s) = \frac{V_{out}}{V_{in}} = K \frac{\omega_p/q_p \cdot s}{s^2 + \omega_p/q_p \cdot s + \omega_p^2}$$

where, e.g., $K = 12$ and $q_p = 33$ may be specified. Out of the well-known eight tone dialing frequencies (from 697 to 1633 Hz), the lowest one, for example, leads to a specified pole frequency $\omega_p = 2\pi f_p = 2\pi \cdot 697$ rad/s. With this specification, usual [17]–[20] single op amp bandpass circuits were found to need a total resistance of about 500 kΩ and a total capacitance of 10 nF, and thus a rather large substrate area. Optimization of these circuits towards lower sensitivity of the filter characteristics with respect to temperature effects (op amp) further increases the element spread and consequently the substrate area needed. Starting out from the well-known [17] circuit in Fig. 3(a) for a single bandpass section, a highly miniaturized circuit was designed as follows. The original circuit was expanded by additional resistors R_d and R_p as shown in Fig. 3(a) (dotted elements), which have little or no influence on the behavior of the circuit, provided the op amp is regarded nearly ideal, with zero input admittance and output impedance, but finite and complex open loop-gain v (one-pole model). The relatively high quality-factor q_p requires a minimization [7] of the gain sensitivity product

$$\text{GSP} = \sqrt{(|v| S_{|v|}^{\omega_p})^2 + \frac{\left(|v| S_{|v|}^{q_p}\right)^2}{(4 q_p^2 - 1)}}$$

(where $S_{|v|}^x = (\partial x / \partial |v| \cdot |v|/x)$ for $x = \omega_p, q_p$) as a measure for the stability of the magnitude of $T(s)$ for $s = j\omega$.

A detailed study [7] shows that the influence of the resistor R_d on GSP can be made negligibly small, whereas R_p has no influence at all. Taking advantage of the additional degrees of freedom introduced by R_p and R_d, a simple mesh–star transformation led to the new circuit in Fig. 3(b) with much smaller resistor values. For a filter with $f_p = 697$ Hz the total resistance is now 64 kΩ compared with 621 kΩ for the circuit in Fig. 3(a). The layout of the circuit, capable of realizing all eight bandpass filters of a tone-dialing receiver, has been made in such a way that the op amp (741) in a miniaturized plastic-encapsulated form is placed directly over the thin-film capacitors without waste of substrate area. The upper part of

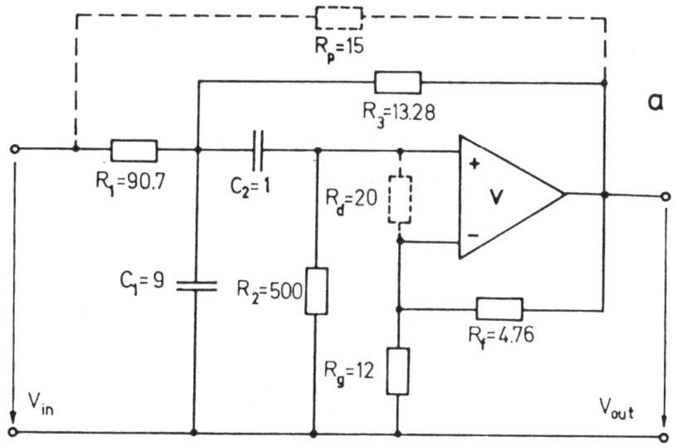

(a)

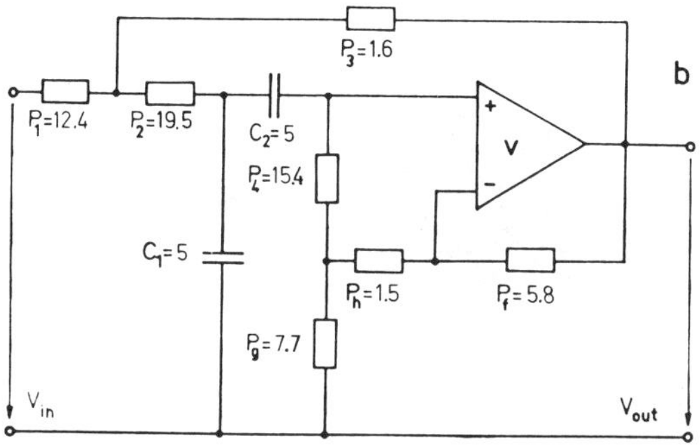

(b)

Fig. 3. Bandpass circuits (R-values in kΩ, C-values in nF).

Fig. 4 shows a photograph of the complete circuit containing two bandpass sections on one 14-pin dual-in-line shaped package. In the lower part of the photograph the thin-film substrate (measuring 8.4 mm × 19.4 mm) is shown, consisting of two symmetrical halves, one for each filter section, where the capacitors are located in the middle, surrounded by the resistors and the contact areas for soldering of the leads of the op amp and the circuit pins. For ease of laser trimming, the resistors P_1–P_4, P_f–P_h of Fig. 3(b) are placed next to the op amps. Functional trimming was done, yielding an accuracy in q_p of ± 2 percent and in f_p of ± 0.5 percent, quite sufficient for the tone-dialing receiver. With 741 op amps, a stability in q_p of 3 percent in the temperature range of 20 to 60°C is achieved, whereas f_p changes by 0.2 percent.

V. Distributed RC-Circuits

A. Circuits and Design Principles

The mathematical description of distributed RC networks generally leads to complicated expressions [14] unless, e.g., the two following constraints are established [9], [10].

1) Confine yourself to the class of circuits consisting merely of uniformly distributed RC-transmission lines and voltage controlled voltage sources with zero input admittance and zero output impedance.

2) Let

$$R_i C_i = n_i^2 \cdot T, \qquad \text{with } n_i = \text{integer} \tag{1}$$

where $i = 1, 2, 3, \cdots$, etc. is the numbering of the RC-lines involved, R_i is the total resistance, and C_i the total capacitance of the line with number i, n_i are free variables, but restricted to integer values, and T is a real parameter involved also in the design procedure.

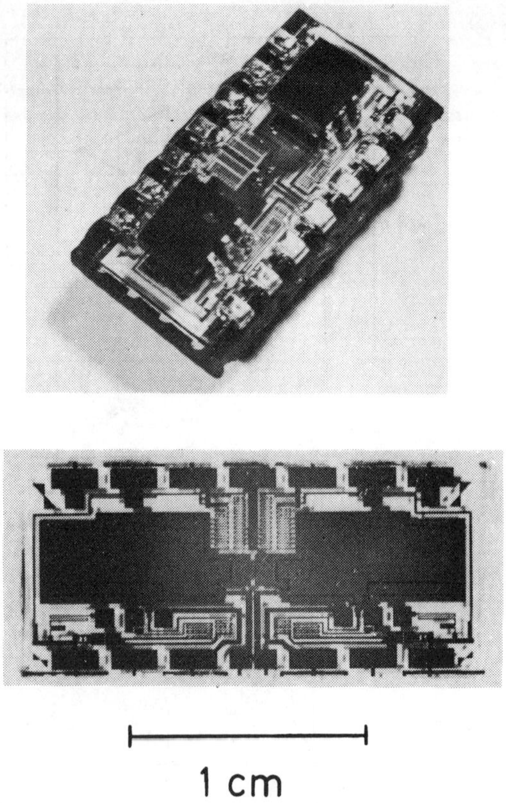

Fig. 4. Photograph of the bandpass circuit.

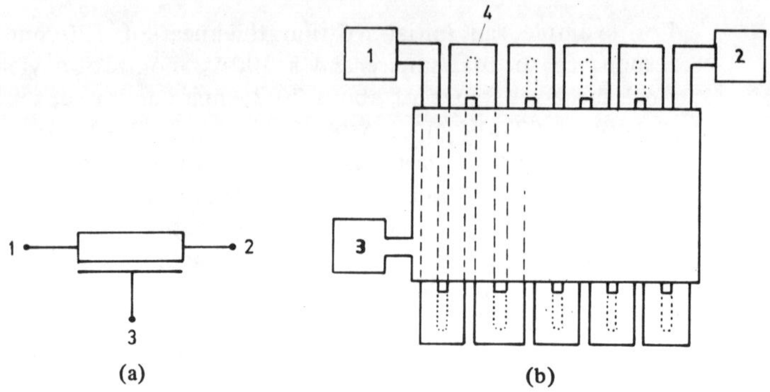

Fig. 5. Distributed RC-element.

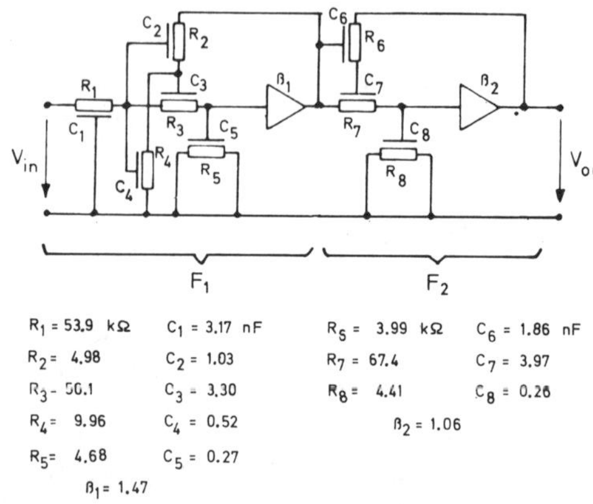

Fig. 6. PCM low pass (R-values in kΩ, C-values in nF).

Since the length of an RC transmission line is known to be proportional to $\sqrt{R_iC_i}$, constraint (1) means that $\sqrt{T}$ is to be interpreted as an "electrical" unit length and n_i as the relative length of the line with number i. Hence, constraint (1) leads to "commensurate" RC-lines but with variable relative lengths n_i. Transfer functions of the considered type of circuits (see, e.g., Fig. 6 which is described later) are known to be rational functions of cosh $\sqrt{pR_iC_i}$; applying (1) and substituting

$$\exp\sqrt{pT} = z \tag{2}$$

yields rational voltage transfer functions in z,

$$F(z) = \frac{V_{\text{out}}}{V_{\text{in}}} = \frac{\sum_{\upsilon=1}^{N} A_\upsilon z^{m_\upsilon}}{\sum_{\upsilon=1}^{N} B_\upsilon z^{m_\upsilon}} \tag{3}$$

with

$$m_\upsilon = \sum_{i=1}^{i_n} g_{\upsilon i} n_i$$

where N, i_n, and $g_{\upsilon i}$ are integer, and A_υ and B_υ are real. In [8], [9], a selection of various useful two-ports is presented to realize transfer functions such as (3) which can be conveniently handled by rational design procedures if the n_i are chosen according to given recommendations. This freedom of choosing the line lengths n_i enables the implementation of design objectives such as arbitrary zeros of transfer functions, low sensitivities, and small circuit size. The design method detailed in [9], [10] is based on dominant poles and zeros of the meromorphic transfer function $F(\exp\sqrt{pT})$ and on the mapping relation (2) between p⁻ and z⁻ plane. To provide highly miniaturized circuits, analytic tools for minimizing the substrate area have been derived.

B. Tunable Distributed RC Elements

The structure shown in Fig. 5(b) (in top view) [9], [11] is used to realize a uniformly distributed RC-transmission line (symbol in Fig. 5(a)). The device consists of a resistive tantalum meander 4 contacted at its two ends 1 and 2 and covered by an oxide layer which is generated by anodization. An evaporated highly conductive top electrode 3 provides the distributed capacitance along the meander path. The vertical sequence of layers corresponds to the RC-film process illustrated in Fig. 2 where a cross-sectional view of a meander-$\overline{RC}$-element was included. It has been shown that this device named "meander-$\overline{RC}$-element" realizes a uniform RC transmission line with an error which can be made negligibly small by applying simple design rules [9], [11]. By setting laser cuts as indicated in Fig. 5(b) by dotted lines, the total resistance can be precisely adjusted in any stage of the production.

C. A PCM Low-Pass Filter

Band-limiting in PCM systems is done by a fifth-order elliptic low-pass filter [21]. The design of a distributed RC circuit approximating such a filter-function has been detailed in [10]. Based on this result, an improved circuit given in Fig. 6 has been fabricated which will be described now. Its meromorphic voltage transfer function $T(p) = V_{\text{out}}/V_{\text{in}}$ was syn-

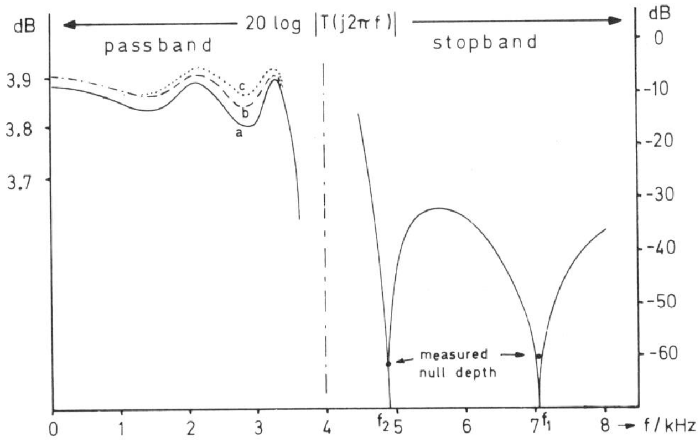

Fig. 7. Magnitude response of the PCM low pass.

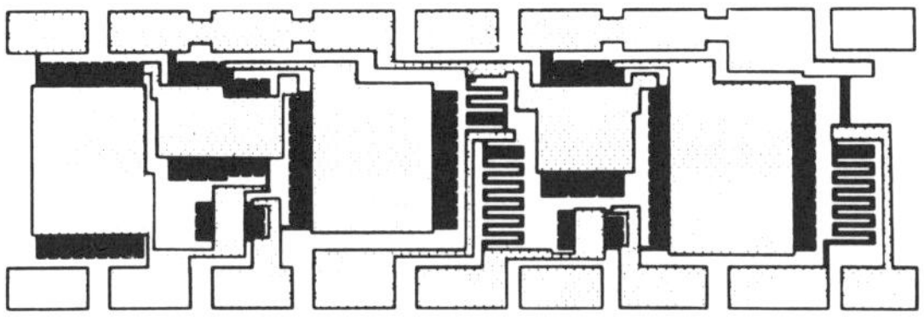

Fig. 8. Layout of the PCM low pass.

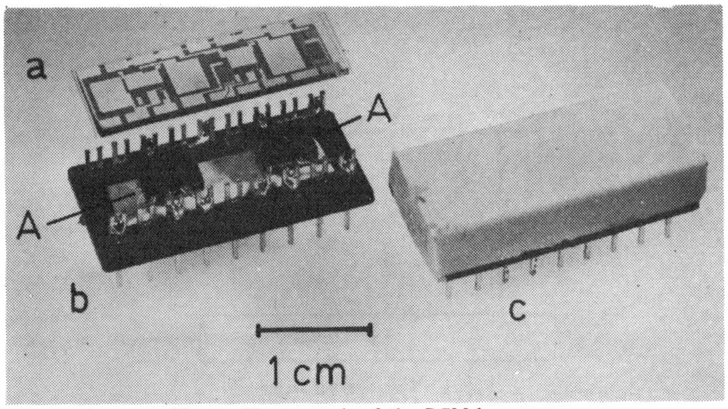

Fig. 9. Photograph of the PCM low pass.

thesized as the product of the two transfer functions $F_1(z_1)$ and $F_2(z_2)$ belonging to the cascaded two-ports (Fig. 6) where $z_1 = \exp\sqrt{pT_1}$ and $z_2 = \exp\sqrt{pT_2}$; The calculated magnitude response $|T(j2\pi f)|$ of the actual design is plotted in Fig. 7, Curve a. The layout of the filter using meander-$\overline{RC}$-elements is given in Fig. 8. The arrangement of the complete filter is explained by Fig. 9: it shows the thin-film substrate (a) which is soldered onto a dual-in-line header (b) where the two op amps (A) are mounted; the encapsulated filter (c) is also shown. The substrate size is 7 × 22 mm; it is achieved with a minimum line width of 100 and 50 μm meander spacings.

D. Filter Tuning and Performance

The frequency notch at f_1 (see Fig. 7) is generated alone by the three meander-$\overline{RC}$-elements $\{R_2, C_2\}$, $\{R_3, C_3\}$, and $\{R_4, C_4\}$. Similarly, the notch at f_2 is due only to the elements $\{R_6, C_6\}$ and $\{R_7, C_7\}$. Functional tuning, e. g., of the notch at f_2 is carried out by increase of R_7 which increases the notch depth, and by another increase of R_7 together with R_6 which shifts the notch frequency f_2. This example intends to demonstrate the ease of functional tuning provided by the rather simple network topology of appropriate distributed RC circuits. R_1 through R_4 are trimmed in a similar simple way [9]; adjustment of R_5 and R_8 was found to be not necessary. Curve b in Fig. 7 indicates a typical response of a tuned PCM low-pass filter using standard 741-operational amplifiers. A temperature dependence is mainly observed in the passband: the maximum deviation observed in the temperature range from 20 to 70°C was as small as 0.02 dB as shown by the corresponding Curve c in Fig. 7. Thus the excellent temperature compensation achieved by the RC-film process is again confirmed. We assume that the as-monolithic structure of miniaturized distributed RC elements with its optimum thermal coupling between R's and C's further improves the temperature stability.

VI. Conclusions

A single-layer technique for the manufacture of thin-film integrated RC circuits based on a new "Ta-RC-film" was described. The main advantages of this technique are:

1) simple processing, small number of steps (only 3 photomasks required) which increases the yield and decreases the costs;
2) high throughput of the sputtering equipment because of only one single sputtering run with the thickness of the sputtered layer being only 190 nm;
3) good thermal stability of resistors and capacitors which allows for annealing and heat treatment up to 300°C.

Lumped RC circuits made in Ta-RC-film technology are preferably operated up to about 20 kHz. At higher frequencies, where the series resistance of the bottom capacitor electrodes causes a considerable increase of the capacitor loss (tan δ), distributed meander-$\overline{RC}$-elements can be employed without difficulties; experimental circuits have been operated up to 5 MHz. Thus the RC-film technology in conjunction with lumped and distributed circuitry covers a wide range of applications. For example, an optimized bandpass with lumped components for tone-dialing receivers and a PCM low pass with distributed elements were described. The circuit miniaturization with distributed RC-elements is due to the fact that resistors and capacitors are located upon each other rather than next to each other as in the case of lumped components; hence, a distributed circuit (if optimized) sometimes saves nearly half of the area of a comparable lumped circuit.

References

[1] W. H. Orr *et al.*, "Integrated tantalum film RC circuits," in *Proc. Electron. Comput. Conf.*, pp. 602–612, 1970.

[2] W. Worobey, "Tantalum thin-film RC circuit technology for a universal active filter," *IEEE Trans. Parts. Hybrid, Packaging*, vol. PHP-12, pp. 276–282, 1976.

[3] W. Anders, "Nitrogen-doped tantalum thin film capacitors with improved temperature stability," *Thin Solid Films*, vol. 27, pp. 135–140, 1975.

[4] O. J. Duff *et al.*, "A high stability RC-circuit using high nitrogen doped tantalum," *in Proc. Electron. Comput. Conf.*, p. 229, 1978.

[5] H. W. Poetzelberger, "Thin film integrated RC-Networks with compensated temperature coefficients of R and C," *Electorcomp. Sci. Technol.*, vol. 4, pp. 139–142, 1977.

[6] H. Baeger, "Herstellung temperaturkompensierter Tantal-Duenn-schichtwiderstaende und Kondensatoren aus einer einzigen Schicht mit definiertem Gaseinbau," Ph.D. dissertation, Univ. Stuttgart, Stuttgart, Germany, 1978.

[7] H. J. Tilch, "Miniaturisierte und kanonische Baueinheiten fuer RC-aktive Filter in Schichttechnik," Ph.D. dissertation, Univ. Stuttgart, Stuttgart, Germany, to be published.

[8] R. W. Berry, P. M. Hall, and M. T. Harris, *Thin Film Technology*. New York: Van Nostrand Reinhold, 1968, ch. 11.4, pp. 533–542.

[9] H. W. Renz, "Entwurf und Herstellung miniaturisierter Filter mit

RC-Leitungen in Duennschichttechnik," Ph.D. dissertation, Univ. Stuttgart, Stuttgart, Germany, 1977.
[10] —, "Filter design with a catalogue of distributed RC-active circuits," to be published.
[11] H. W. Renz and H. Baeger, "Tunable distributed RC-elements in single-layer tantalum thin-film technology," to be published.
[12] J. J. Golembeski, "Distributed RC network tuning," in *Proc. Electron. Comp. Conf.* (Washington, DC), pp. 111–114, 1969.
[13] W. Worobey and M. H. Röttersmann, "Fabrication and performance of adjustable distributed RC networks made with tantalum films, in *Proc. Electron. Comput. Conf.*, (Washington, DC), pp. 152–158, 1968.
[14] M. S. Ghausi and J. J. Kelly, *Introduction to Distributed-Parameter Networks with Application to Integrated Circuits.* New York: Holt, Rinehart, and Winston, 1968.
[15] R. Riekeles, "Tantal-Duennschichtkondensatoren mit verbesserten Eigenschaften," Ph.D. dissertation, Univ. Stuttgart, Stuttgart, Germany, 1978.
[16] D. Gerstenberg, "Properties of anodic films formed on reactively sputtered tantalum," *J. Electrochem. Soc.*, no. 113, pp. 542–547, 1963.
[17] R. P. Sallen and E. L. Key, "A practical method of designing RC active filters," *IRE Trans. Circuit. Theory*, vol. CT-2, no. 3, pp. 74–85, 1955.
[18] T. Deliyannis, "High-Q factor circuit with reduced sensitivity," *Electron Lett.*, vol. 4, no. 26, pp. 577–579, 1968.
[19] A. Antoniou, "Synthesis of active filters with optimum sensitivity," *Radio Electron. Eng.*, pp. 135–147, Sept. 1968.
[20] J. W. Craig, "Practical designs for RC active filters using op amps," Tech. Note 1968–26, M.I.T. Lincoln Lab., 1968.
[21] R. A. Friedenson *et al.*, "RC active filters for the D3 channel bank," *Bell Syst. Tech. J.*, vol. 54, pp. 507–529, 1975.

Microelectronic Active-*RC* Channel Bandpass Filters in the Frequency Range 60–108 kHz for FDM SSB Telephone Systems

WOLJA SARAGA, FELLOW, IEEE, DAVID G. HAIGH, MEMBER, IEEE, AND ROBERT G. BARKER

Abstract—**The paper describes the development of microelectronic active-*RC* 4-kHz-spaced channel bandpass filters (CBF's) for the frequency range 60–108 kHz, suitable for the formation and decomposition of the 12-channel bank (primary group) in *direct-modulation*-type FDM SSB telephone systems. The active filters are designed by simulation (essentially by Gorski–Popiel's method) of low-sensitivity *LC* filters which are the duals of some *LC* CBF's currently used in FDM systems in the U.K. Our circuit development is firmly based on the microelectronic technology adopted by us, in which naked-chip dual- or quad-operational amplifiers, laser-adjustable thick- or thin-film resistors, and NPO ceramic-chip capacitors are used. In order to minimize cost, the circuits are designed to accept, as far as possible, capacitors with preferred (and preferably also equal) nominal values, and a resistor adjustment procedure has been developed which permits wide capacitor manufacturing tolerances of ± 10 percent. This adjustment procedure also provides compensation for the nonideal characteristics of the amplifiers, leading to near-perfect simulation of the nominal dissipationless *LC* prototype filters.**

Microelectronic models of channel 12 CBF (60–64 kHz) and channel 1 CBF (104–108 kHz) have been built (size of each filter: 50 mm$\times$30 mm$\times$5 mm; dc power 400 mW). Channel 12 CBF meets the loss-frequency specification over the required temperature range 10–40°C. For channel 1 CBF the specification is met at ambient temperature (25°C) but not at 10°C or 40°C; this will be remedied when amplifiers with higher f_T's and/or lower f_T-temperature coefficients become available. Neither filter requires a channel equalizer (whereas their *LC* counterparts do), and preliminary nonlinearity, noise, and intermodulation tests had promising results.

I. Introduction

THE BULK OF the world's long-distance telephone traffic is based on frequency division multiplex (FDM) systems, in which each conversation occupies two separate frequency translated single-sideband (SSB) channels (one for each direction). To satisfy the large demand for increased traffic capacity, a continuously growing number of channels is required, and extreme cost minimization of channel equipment, particularly of channel bandpass filters (CBF's) which form its most costly part, is essential. Therefore, both the technology and the economics of CBF's are under almost continuous review [1]. CBF's have to meet very rigorous performance specifications, and the simultaneous demands of high quality and low cost have made the design and manufacture of CBF's, which are highly selective high-precision filters, a challenge to circuit engineers.

The spectacular progress in recent years of silicon integrated-circuit and hybrid technology has led to significant engineering and economic advances in the microelectronic realization of highly selective high-precision filters in general. The purpose of this paper is to report on results obtained by the authors in the development of microelectronic active-*RC* CBF's[1] for use in the formation and in the decomposition of the basic 12-channel bank, the so-called primary group (PG) of FDM SSB telephone systems. The CBF's described in this paper are of the active-*RC* type designed to simulate both the structure and the elements of low-sensitivity *LC* ladder filters, i.e., of *LC* filters which are resistively terminated at both ports so as to provide maximum possible power transfer at the passband loss minima [4].

II. Performance Requirements

The PG is formed from 12 voice-frequency baseband signals as shown in Fig. 1. Each of these signals is band-limited to 3400 Hz[2] by means of a low-pass filter and then modulates one of the 12 carrier signals (with carrier frequencies $f_{c1} = 108$ kHz, $f_{c2} = 104$ kHz, $\cdots$, $f_{c12} = 64$ kHz); each of the double sideband signals generated in this way is reduced to a single (lower) sideband signal by means of a transmit CBF. The decomposition of the PG into individual channels is performed by a corresponding bank of receive CBF's (Fig. 1). The separated SSB signals are retranslated to baseband by 12 modulators followed by low-pass filters.

In view of the very wide margins within which performance specifications for CBF's are chosen, we decided to bypass any arguments about the practical relevance of the performance specification selected by us as a basis for our filter development, by choosing as a design target the performance characteristics of a set of 12 CBF's which

Manuscript received December 27, 1977; revised August 8, 1978. This work was supported by the Procurement Executive, Ministry of Defense, U.K.

W. Saraga and D. G. Haigh are with the Department of Electrical Engineering, Imperial College, London SW7 2BT, U.K.

R. G. Barker is with the Admiralty Underwater Weapons Establishment (AUWE), Portland, Dorset, U.K.

[1] Some of these results have been given briefly in [2], [3].

[2] We are not considering systems with out-of-band signaling facilities, in which frequencies up to about 3850 Hz have to be transmitted.

Reprinted from *IEEE Trans. on Circuits Syst.*, vol. CAS-25, pp. 1022-1031, Dec. 1978.

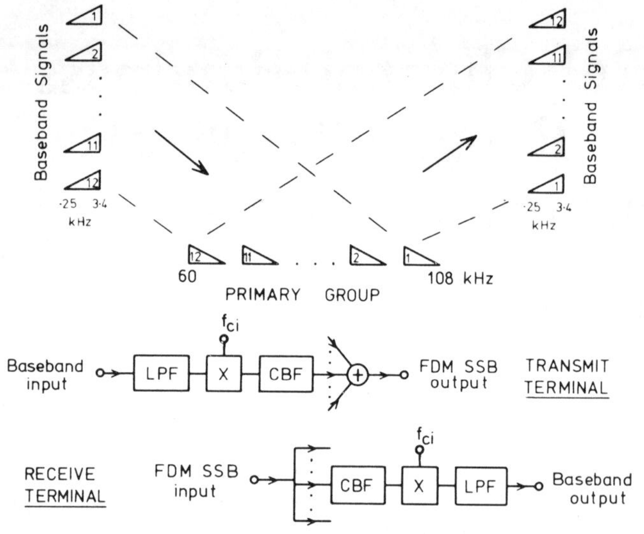

Fig. 1. Formation and decomposition of PG—single-modulation system.

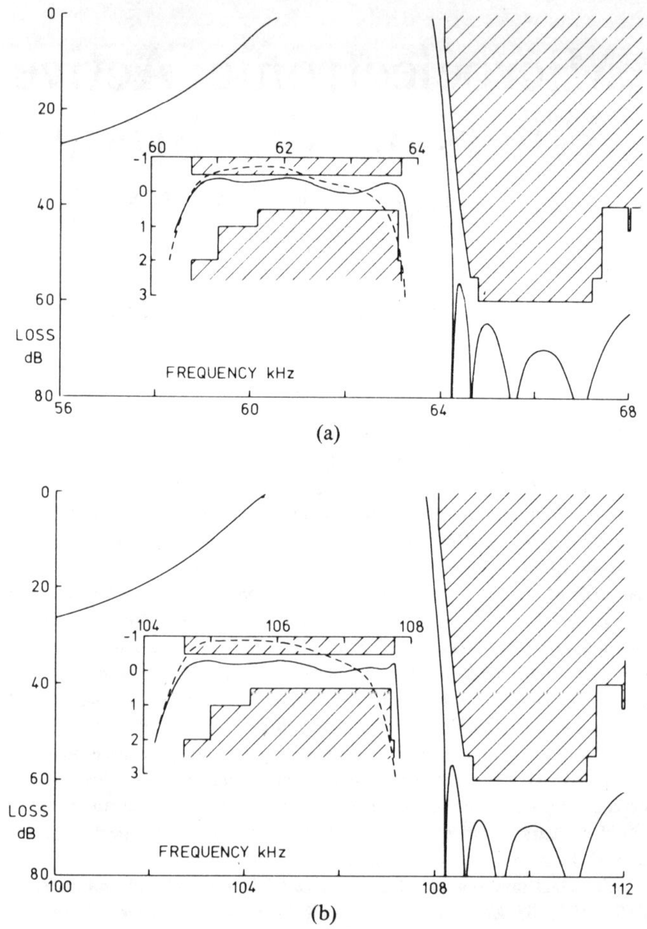

Fig. 2. Computed and measured loss-frequency responses of *LC* CBF's for (a) channel 12 and (b) channel 1. ——— computed responses assuming ideal elements. ---- measured responses (passband only).

form a PG meeting CCITT requirements. In the U.K., such CBF's are of two types: *LC*-crystal filters and *LC* filters. Since we intended to design our active filters by simulation of *LC* filters, we initially chose as prototypes for simulation some *LC* CBF's used in U.K. systems.

The measured passband loss-frequency responses of these *LC* filters for channel 12 (60–64 kHz) and channel 1 (104–108 kHz) can be seen in Fig. 2 together with the loss-frequency specifications.[3] We note that neither filter meets the passband specification; therefore, a loss-frequency equalizer for each channel is required. The main reason for these *LC* filters not meeting the specification is that the tuned-circuit *Q* values of about 800 are not sufficiently high.[4] Since, as we shall show later, active-*RC* simulation can lead to much higher *Q* values than are obtainable in *LC* filters, it is relevant to show in Fig. 2 the computed responses of the *LC* CBF's assuming infinite element *Q* values. As these curves meet the specification, it becomes highly desirable to aim, in our work on active-*RC* CBF's, at obtaining *Q* values which are so high that there is no need for loss-frequency equalizers.

The realization of *LC* CBF's can be made easier by adopting an alternative approach, double modulation, in which the wanted sideband selection is initially performed at lower than PG frequencies. Typically, 4 subgroups, each formed from 3 baseband signals, are assembled in the band from 12 to 24 kHz and are subsequently translated to the PG frequency band as shown in Fig. 3.

The advantages of double-modulation systems, with regard to making the realization of *LC* CBF's easier, apply in general also to active-*RC* CBF's, and, consequently, most published work on active-*RC* CBF's has been aimed at producing filters for double modulation

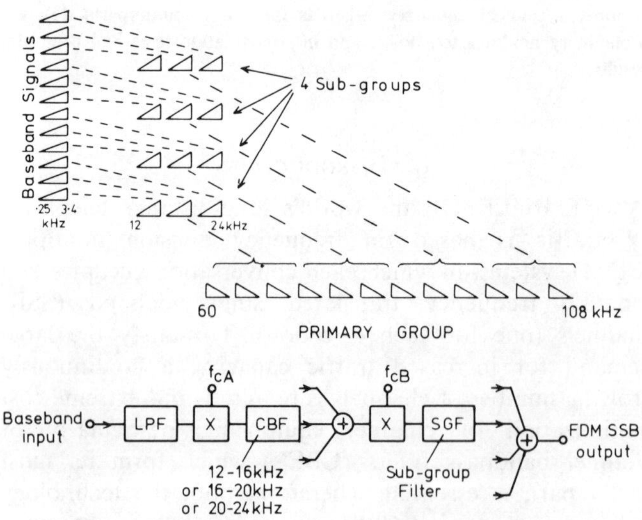

Fig. 3. Formation of PG—double-modulation system.

systems;[5] however, a notable exception is [6], where a 104–108-kHz CBF for single-modulation systems is de-

[3]Note that a high loss is required only in the upper stopband, to suppress the unwanted sideband; the lower stopband loss is supplemented—at baseband frequencies—by the channel low-pass filter.
[4]Inductor *Q* value≈1000; capacitor *Q* value≈4000.

[5]Such a CBF, for the band 20–24 kHz, is described in [5]; a discrete component model met the requirements. For other CBF's for double-modulation systems, see [6]–[11].

scribed (as well as a CBF for double-modulation systems). In practice it would be necessary to weigh the increased system complexity of the double-modulation system against the increased difficulty of realizing the CBF's for a single-modulation system. However, in order to try to extend the scope of our general design capability, we decided to aim at meeting the more severe requirements, in terms of frequency and Q values, of CBF's for single-modulation systems, at the same time fully matching, in order to minimize cost, the circuit design to the particular microelectronic technology adopted.

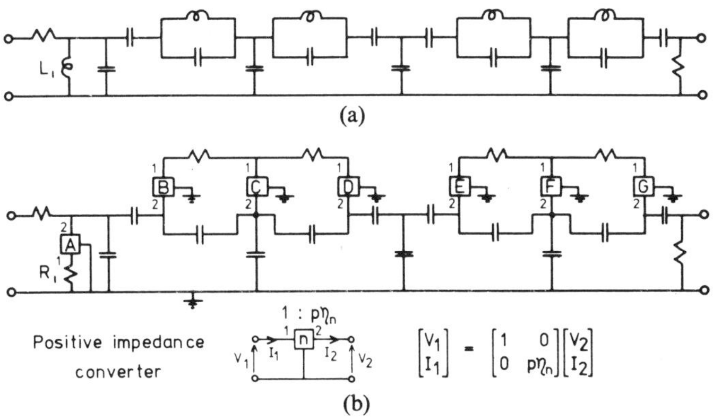

Fig. 4. Simulation of *LC* CBF's.

III. Microelectronic Technology

Due to component tolerances and imperfections, it is not possible to achieve the highly selective high-precision loss-frequency characteristics aimed at without some kind of accurate adjustment. Thick-film or thin-film resistors are ideally suited for this purpose since they can be trimmed by means of a laser (thick-film resistors with qualities and accuracies approaching those of thin-film resistors became available about 4 years ago [12]). As regards amplifiers, we prefer for reasons of cost and size to use commercial integrated-circuit dual- or quad-operational amplifiers in naked-chip form [12] which require, for reasons of thermal dissipation, the use of alumina substrates.

We are not using thin-film capacitors, because SiO_2 capacitors have very low capacitance per unit area and tantalum capacitors suffer from low Q values, but we are using miniature NPO ceramic-chip capacitors which have high Q values and whose capacitance varies very little with temperature. These capacitors are relatively inexpensive if they have nominal values which are manufacturers' preferred values and if they also have wide manufacturing tolerances of ± 10 percent; however, they are rather expensive otherwise.

Previously published active-*RC* CBF's [5]–[11] require considerable numbers of nonpreferred-value narrow-tolerance capacitors, and therefore these CBF's are not economical for microelectronic fabrication using the particular technology which we have adopted. Another disadvantage for us of the published designs is that compensation of the amplifier imperfections involves the provision and adjustment of very-high-value compensating resistors or of very-low-value compensating capacitors, neither of which is desirable in our technology.

The features we are aiming at for our CBF's can now be stated:

1) acceptability of wide capacitor manufacturing tolerances of ± 10 percent, their effect to be compensated by laser trimming of certain resistors;
2) acceptability of capacitors with preferred nominal values;
3) nominal values of all capacitors to be the same, as far as possible; this considerably simplifies manufacture because ceramic chip capacitors do not, in general, have individual markings to denote their value;
4) acceptability of commercial integrated-circuit operational amplifiers (the 104–108-kHz CBF described in [6] uses nonintegrated operational amplifiers);
5) compensation of active and passive component imperfections by laser-trimming of resistors in the basic circuit, rather than by the addition of compensating resistors or capacitors with very high impedances.

IV. Design of Active-*RC* CBF's

A. Simulation of *LC* CBF's

The structure of the *LC* CBF's, shown in Fig. 4(a), is the same for all 12 filters in the group, and identical filters are used at transmit and receive terminals. They are eleventh order filters of the minimum inductor form [13] and have a number of redundant capacitors, introduced in order to reduce the spread of element values.

In principle, the *LC* circuit in Fig. 4(a) can be realized in active-*RC* (inductorless) form using Gorski–Popiel's method [14], and the active-*RC* circuit obtained is shown in Fig. 4(b). The inductor subnetworks of the *LC* filter are replaced by topologically equivalent resistor subnetworks with a common inductance/resistance ratio λ; $(1:p)$-type[6] positive impedance converters (PIC's)[7] with impedance conversion ratios $1:p\eta_n$ (for $n=A,B,\cdots,G$) are inserted in all the links between the resistor subnetworks and the remainder of the network, where the conversion constants η_n have to be chosen so that $\eta_n=\lambda$. However, the circuit in Fig. 4(b) has the following disadvantages:

1) like the *LC* filter in Fig. 4(a), it requires a large number of nonpreferred-value narrow-tolerance capacitors;
2) the sensitivity of the loss of the filter to changes in the passive elements which are used to realize the PIC's is not necessarily low, except in the case of

[6] p is the complex frequency variable.
[7] Often referred to as generalized impedance converters (GIC's).

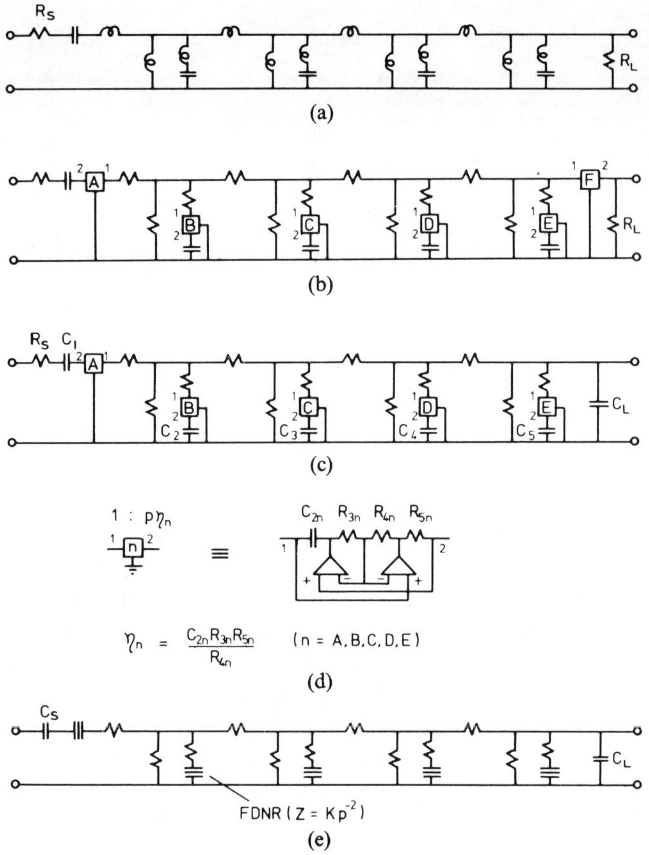

Fig. 5. Simulation of dual of LC filter in Fig. 4(a).

PIC A, for which changes in any of the passive elements are equivalent to a corresponding change of L_1 in the low-sensitivity LC filter in Fig. 4(a) [4]. For PIC's B–G, changes in the passive elements cannot be interpreted as corresponding to changes in the elements of the LC filter;

3) the number of amplifiers is 14 (2 amplifiers per PIC, as we shall show later), which is rather high as regards power requirement, heat generation, and cost;

4) the circuit would be difficult to adjust because there is not a one-to-one correspondence between tuned circuits in the LC filter and PIC's in the active-RC filter.

These disadvantages can be overcome by using, instead of the original LC prototype filter in Fig. 4(a), its dual shown in Fig. 5(a), which has the same loss-frequency characteristic.

B. Simulation of the LC Circuit in Fig. 5(a)

1) The Active-RC Circuit: Application of Gorski–Popiel's method to the LC circuit in Fig. 5(a) leads to the active-RC circuit in Fig. 5(b). By moving the filter output from terminal 2 to terminal 1 of PIC F (the PIC's are of the current converting type in which the voltages at terminals 1 and 2 are identical) and by subsequently replacing the resistively terminated PIC F by an equivalent capacitor C_L, we obtain the circuit in Fig. 5(c) which we have adopted for microelectronic realization.[8] The circuit in Fig. 5(c) is preferable to the circuit in Fig. 5(b) because of the saving of two amplifiers (or at least one amplifier if the replacement of a resistive by a capacitive filter terminating impedance necessitates the use of a separate output buffer amplifier).

For the PIC's, the circuit in Fig. 5(d) was chosen [16], since this particular circuit, when terminated at terminal 1 by a resistor and at terminal 2 by a capacitor so as to simulate an LC tuned circuit, has a Q value which is relatively independent of the gain-bandwidth products of the two amplifiers [17].

2) Acceptability of Capacitors with Equal Preferred Nominal Values and Wide Manufacturing Tolerances:

a) Preferred nominal values: By appropriate design it is possible to give all capacitors in Fig. 5(c) nominal values which are manufacturers' preferred values. This can be done for C_L by appropriately choosing the impedance level of the whole filter, but this leads, in general, to nonpreferred values for the remaining capacitors; it is then necessary to carry out the following design modifications:

i) For C_1 to have a preferred value, the value of R_S is altered to keep R_SC_1 unchanged, and the conversion constant η_A of PIC A is altered to keep η_AC_1 unchanged.

ii) For C_2, C_3, C_4, and C_5 to have preferred values, the associated PIC conversion ratios η_B, η_C, η_D, and η_E are altered to keep the η-C products unchanged.

iii) For the capacitors C_{2n} $(n=A,B,C,D,E)$ within the PIC's to have preferred values, an appropriate choice of the values of resistors within the PIC's is necessary in order to maintain the required η values (see Fig. 5(d)).

b) Equal nominal values: For the 60–64-kHz filter, it was possible to achieve the aim of equal nominal values for all capacitors (using the same design techniques as described for achieving preferred nominal capacitor values): the chosen value is 1 nF. For the 104–108-kHz filter, it was decided to give all capacitors except C_L the same nominal value, 0.47 nF, but to choose $C_L = 1$ nF in order to obtain optimum resistor values.

c) Wide manufacturing tolerances: The acceptability of wide tolerance capacitors depends on the development of resistor adjustment techniques to compensate for the capacitor value variations. In principle, the design techniques described above for obtaining preferred nominal capacitor values at the design stage can be translated into corresponding resistor adjustment procedures to allow for capacitor tolerances. Thus, for C_1, C_2, C_3, C_4, and C_5, tolerance compensation is possible by modifying the PIC

[8]This circuit can also be interpreted (see Fig. 5(e)) as an impedance scaled version (by a factor proportional to p^{-1}) of the LC circuit [15]. The impedances proportional to p^{-2} in Fig. 5(e) are realized in Fig. 5(c) as $(1:p)$-type PIC's terminated at terminal 2 by capacitors, and the capacitor C_S in Fig. 5(e) is realized by R_S (at the terminal 2 side of PIC A).

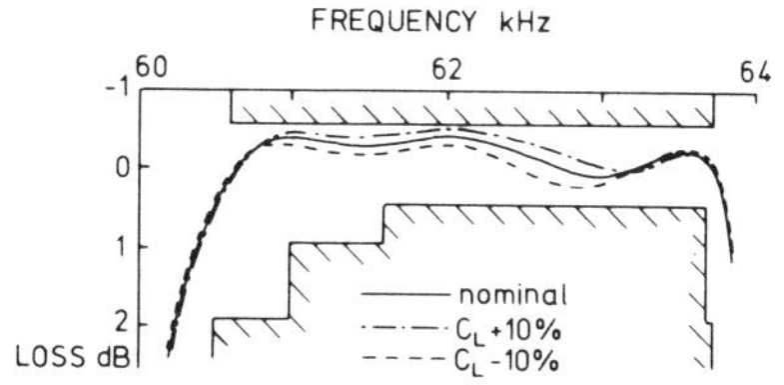

Fig. 6. Computed effect of ±10-percent variations in C_L.

conversion constants η_A, η_B, η_C, η_D, and η_E by means of resistors within the PIC's (in the case of C_1, the resistor R_S also has to be adjusted to maintain the required design value of $R_S C_1$). Since these conversion constants depend also on the capacitors C_{2n} $(n=A,B,C,D,E)$, the adjustment of the PIC resistors compensates at the same time for the tolerances of the C_{2n}. In practice, this compensation is achieved by the functional adjustment procedure which will be described in Section IV-B3.

In order to compensate for C_L variations, it would be necessary to adjust all resistors R which correspond to inductors in the LC filter in Fig. 5(a) so that all products $C_L R$ retain their design values. This would be very inconvenient, and we therefore investigate the effect of ±10 percent variations in the value of C_L. The result is shown, for the 60–64-kHz filter, in Fig. 6. (A variation of about ±0.5 dB in the basic loss, which can be compensated during the channel gain adjustment, is not shown.) The effect on the loss-frequency curve can be considered as just acceptable; alternatively, it may be necessary to impose a slightly narrower tolerance on C_L.

3) Compensation for Active and Passive Component Imperfections by Resistor Adjustment:

a) General: In the microelectronic fabrication procedure which we are using, resistors in the active-RC circuit which correspond to inductors in the LC circuit are adjusted (by means of a laser) to their nominal design values to within ±0.1 percent. Thus the simulation of the inductance values in the LC filter can be regarded as near-perfect. The functional adjustment of PIC resistors, which will be described, leads to perfect tuning of the simulated series tuned circuits to their design resonant frequencies. Therefore, also the simulation of the capacitance values of the LC filter can be regarded as near-perfect.

However, the nonideal parameters of the operational amplifiers have a considerable effect on the *Q value* of a simulated tuned circuit,[9] and the actual values of these nonideal parameters can vary (from sample to sample) from the nominal values; therefore, in order to achieve the required high selectivity, it is necessary to adjust each simulated tuned circuit not only for correct resonant frequency (f_r) but also so that the modulus of the Q value $|Q|$ is above a specified minimum.

b) Adjustment circuit: For adjustment, each PIC in

[9]The most significant amplifier imperfections are finite gain-bandwidth product, common-mode input capacitance, and output resistance. Typical 'unadjusted' tuned-circuit Q values are of the order of 50 at 100 kHz (assuming, in Fig. 5(d), $R_{3n}=R_{4n}$).

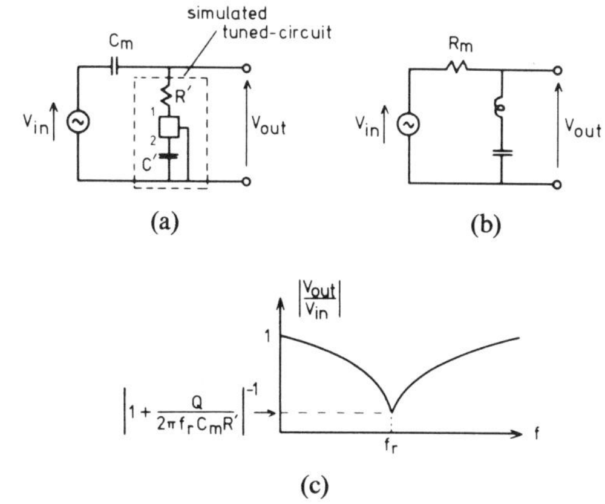

Fig. 7. Adjustment circuit and response.

Fig. 5(c) is associated with the resistor connected to terminal 1 and the capacitor connected to terminal 2 to form a simulated series tuned circuit. The adjustment circuit (Fig. 7(a)) may be regarded as an impedance-scaled (by a factor proportional to p^{-1}) version of the RLC circuit in Fig. 7(b). The voltage gain of the circuit exhibits a 'null', i.e., a minimum, as illustrated in Fig. 7(c). The null frequency is equal to the resonant frequency f_r of the simulated tuned circuit, and the depth of the null is related in a simple way to the Q value (see Fig. 7(c)). This circuit is particularly suitable when the *modulus* of Q is to be above a specified minimum, because the amplitude response in Fig. 7(c) is in practice almost independent of the *sign* of Q.[10]

c) Adjustment strategy: For the PIC circuit adopted (Fig. 5(d)), f_r depends on all its resistors R_3, R_4, and R_5 (i.e., R_{3n}, R_{4n}, and R_{5n});[11] Q, on the other hand, depends mainly on the ratio R_3/R_4 and is relatively independent of R_5 [17]. Therefore (apart from a preliminary f_r adjustment, see below) we adjust Q first by means of R_3 or R_4, and then we adjust f_r by means of R_5. We use R_4 for Q adjustment rather than R_3 because during Q adjustment increase of R_4[12] causes f_r to increase;[11] since f_r decreases during f_r adjustment (by means of R_5), Q adjustment by means of R_4 may allow us to repeat the adjustment cycle if $|Q|$ is not sufficiently high.

A sketch of the variation during adjustment of the Q value of a typical simulated tuned circuit, as a function of its resonant frequency, is shown in Fig. 8. The initial resonant frequency f_{ri} is subject to considerable variations (from one PIC sample to another) due mainly to the wide capacitor tolerances. Therefore we carry out an initial f_r adjustment (by means of R_5) to make f_r equal to its design value f_{ro}. For this to be always possible, it is necessary that

[10]This is so because $|Q|\gg 1$ and because we choose C_m so that $2\pi f_r C_m R' \approx 1$ (see Fig. 7(c)). If required, the sign of Q may be ascertained by observing the phase $\beta=\angle(V_{out}/V_{in})$ at $f=f_r$; for $Q>0$, $\beta=0$ and for $Q<0$, $\beta=\pi$.

[11]Assuming ideal amplifiers, $f_r=(R_4/C_2C'R'R_3R_5)^{1/2}(2\pi)^{-1}$; C' and R' are the PIC terminating elements—see Fig. 7(a). With nonideal amplifiers this equation is still applicable (although only approximately).

[12]By laser-trimming of a resistor, it is only possible to produce an increase in resistance value.

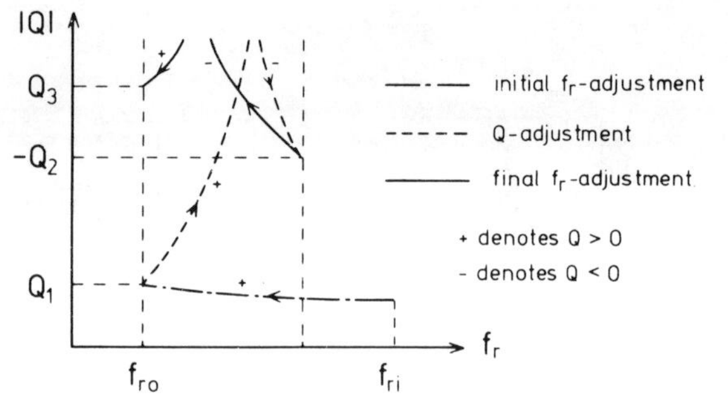

Fig. 8. Typical $|Q|$ vs f_r curves during adjustment of simulated tuned circuit.

the initial (i.e., preadjustment) values of the PIC resistors are such that $f_{ri} \gg f_{ro}$ for all PIC samples. Another constraint on the initial resistor values (particularly on R_3 and R_4) is that the Q value Q_1 obtained after initial f_r adjustment should always be positive (in spite of variations of amplifier imperfections from one PIC sample to another); this is to ensure that Q adjustment by increase of R_4, which is the next adjustment, increases $|Q|$ [17]. Q adjustment is continued until Q has passed through infinity and fallen to a high negative value $-Q_2$ (Fig. 8). Q is adjusted to a negative value because we have found that in this case the effect of final f_r adjustment, which is carried out next, on Q (this is a second order effect) is, at least initially, to increase $|Q|$ (if Q is adjusted to be positive, the effect of final f_r adjustment is to reduce $|Q|$). The optimum value of Q_2 is that value which leads to a statistical distribution of final Q values (Q_3 in Fig. 8) which is centered at infinity.

This adjustment procedure leads to an active-RC filter which corresponds to an LC filter with very high Q values for the tuned circuits at their resonant frequencies. Although in the active-RC filter these Q values are highly frequency dependent, the practical results to be described in Section V show that the Q values remain high enough over a sufficiently wide frequency band for satisfactory CBF realization.

4) Sensitivity:

a) General: The adjustment methods described above lead to a filter loss-frequency response which is near-ideal. However, we are interested in the sensitivity of the loss-frequency response to subsequent component variations due to environmental factors, such as temperature variation and aging.

b) Sensitivity to passive components: It is convenient to consider the passive components in 3 groups:

i) Passive components which appear explicitly in Fig. 5(c), with the exception of R_S and C_L: each of these components corresponds to a reactive element in the LC filter in Fig. 5(a), and the loss of the simulating active-RC filter will have low sensitivities to variations of these components [4].

ii) R_S and C_L: the sensitivity of the loss to C_L has already been discussed in Section IV-B2c, but the variations of C_L due to temperature variations and aging are expected to be smaller by two orders of magnitude than the variations of ± 10 percent considered in Section IV-B2c; therefore, the resulting loss variation is negligible. The loss variation due to changes in R_S is similar to that due to changes in C_L.

iii) The passive components within the PIC's (Fig. 5(d)): assuming initially that the operational amplifiers are ideal, changes in any one of the passive components within a PIC will cause a proportional (or inversely proportional) change in the associated impedance conversion constant η_n. In the case of PIC's B to E, a change in η_n is equivalent to the same change in the PIC terminating capacitance, to which, as already mentioned, the filter loss has a low sensitivity. For PIC A, a change in η is equivalent to the same changes in C_1 and $1/R_S$. The sensitivity to C_1 is low since C_1 corresponds to a reactive element in the LC filter, and the sensitivity to R_S has already been discussed. When the amplifiers are nonideal, variations in the passive components within the PIC's will affect not only the η values but also the Q values of the simulated tuned circuits. However, as mentioned in Section IV-B3, the Q value depends mainly on the resistor ratio R_3/R_4 and therefore is expected to vary very little because R_3 and R_4 have similar values and should vary with temperature and age in a similar way.

c) Sensitivity to amplifier parameters: As mentioned above, the amplifier parameters which are most significant in the design of active-RC filters using PIC's are gain-bandwidth product, common-mode input capacitance, and output resistance; we have shown that at the time of adjustment we can compensate for the nonideal values of these parameters. As far as subsequent variations from these nonideal values are concerned, the most important parameter seems to be the gain-bandwidth product f_T which varies with temperature by typically -0.25 percent/°C.

With regard to the effect of f_T on the Q *value* of a simulated tuned circuit, it has been shown [17] that, for the particular PIC circuit adopted, Q is relatively independent of the f_T's of the amplifiers.

The effect of f_T variations $\Delta(f_{T1}^{-1})$ and $\Delta(f_{T2}^{-1})$ on the *resonant frequency* of a simulated tuned circuit, neglecting second and higher order terms in $\Delta(f_{T1}^{-1})$ and $\Delta(f_{T2}^{-1})$, is

$$\frac{\Delta f_r}{f_r} \approx -\frac{1}{2} f_r \left[\Delta(f_{T1}^{-1}) \sqrt{\frac{R_3}{R_4}} + \Delta(f_{T2}^{-1}) \sqrt{\frac{R_4}{R_3}} \right] \left(M + \frac{1}{M} \right) \tag{1}$$

where $M = (2\pi f_r) C' R_5 \sqrt{R_3/R_4}$ (C' = value of PIC terminating capacitor). Since we are using dual- or quad-operational amplifiers, f_{T1} and f_{T2} will tend to vary with temperature in the same way. With $\Delta(f_{T1}^{-1}) = \Delta(f_{T2}^{-1}) = \Delta(f_T^{-1})$,

$$\frac{\Delta f_r}{f_r} \approx -\frac{1}{2} f_r \Delta(f_T^{-1}) \left[\sqrt{\frac{R_3}{R_4}} + \sqrt{\frac{R_4}{R_3}} \right] \left(M + \frac{1}{M} \right). \tag{2}$$

The effect of f_T variations on f_r is minimized [10] if $R_3 = R_4$ and $M=1$ (i.e., $C'R_5 = 1/(2\pi f_r)$). Then

$$\frac{\Delta f_r}{f_r} \approx -2f_r \Delta(f_T^{-1}). \quad (3)$$

This equation can be used to determine that part of the variation of f_r with temperature which is due to variation of f_T with temperature. This, as we shall show in Section V, is in practice of critical importance.[13]

V. Practical Results

A. Microelectronic Fabrication

A microelectronic hybrid module, suitable for these active-*RC* CBF's, has been developed at the Microelectronics Section of the AUWE. Fig. 9(a) shows a photograph of the module (size: 50 mm×30 mm×5 mm), together with an *LC* CBF currently manufactured and used in the U.K. (shown for comparison). Fig. 9(b) gives a more detailed view of the hybrid module. It consists of a single alumina substrate, on which high-quality thick-film resistor and conductor patterns are deposited. Thick-film resistors of sufficiently high quality (TCR within ±30 ppm/°C values adjustable very accurately by means of laser) became available a few years ago [12].

The capacitors are of the NPO ceramic-chip type with a TCC within ±10 ppm/°C (for $-20°C < T < +70°C$) and typical Q factors of 2000; they are attached to the substrate by reflow soldering. As mentioned in Section IV-B2c, the circuit will accept wide capacitor manufacturing tolerances of ±10 percent throughout, with only a minor effect on the response due to changes in the value of C_L (see Fig. 6). However, for the small number of prototype models which have been manufactured so far, the value of C_1 was selected to be within 2 percent of its design value in order to avoid the need to adjust R_S, and the value of C_L was also chosen to be within 2 percent of its design value in order to obtain the best possible response. Two wide-tolerance 4.7 nF ceramic-chip capacitors (K1200 type) for power supply decoupling are also included in the hybrid circuit.

The amplifiers are Harris HA 2735 dual amplifiers in naked-chip form connected to the circuit by ultrasonically bonded aluminum wires. The HA 2735 amplifier is a "programmable" amplifier and each dual chip requires two resistors, the values of which can be chosen to provide a compromise between the requirements of high gain-bandwidth product and high slew rate, on the one hand, and low dc power consumption, on the other hand. We chose to have the maximum gain-bandwidth product of 10 MHz, and this gave a dc power consumption for the whole filter of 400 mW, with ±15 V power supplies. In order to minimize the loading effect caused by following equipment on the response of the filter, a unity-gain buffer amplifier (an LM310 voltage follower in chip form) was included at the output of the filter.

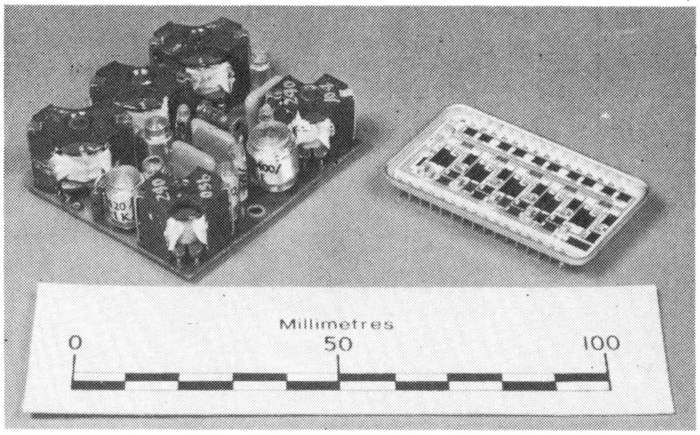

(a)

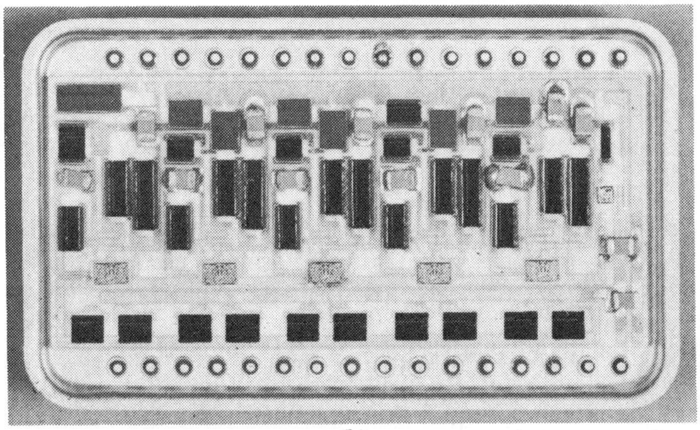

(b)

Fig. 9. (a) Photograph of microelectronic CBF module together with conventional *LC* CBF. (b) Close-up view of microelectronic CBF module.

In order to simplify adjustment of the PIC's, the layout was arranged so that the simulated tuned circuits can be adjusted separately and then connected to form the filter by means of a number of bonded wire connections. Each PIC tuned circuit was adjusted, as described in Section IV-B3, by laser trimming two resistors in the PIC to obtain a $|Q| > 5000$ and a resonant frequency accurate to ±5 Hz.

B. Results for 60–64-kHz Filter

The measured loss-frequency response of the 60–64-kHz microelectronic active-*RC* CBF is shown in Fig. 10. It can be seen that the loss-frequency specification, which is suitable for *LC* CBF's used in U.K. systems, is met over the specified temperature range of 10°C–40°C. The input voltage level used for these measurements was 100 mV_{rms}, and the output voltage level in the passband is similar to the input voltage level.

For a CBF to be usable in an FDM system, it has to satisfy not only a loss-frequency specification, but also requirements on noise, nonlinearity, and intermodulation. Up till now, some measurements of noise, nonlinearity, and intermodulation have been carried out on a discrete component model of the 60–64-kHz filter, using Burr-Brown BB 3505J single amplifiers ($f_T = 12$ MHz). The test

[13]After PIC adjustment as described in Section IV-B3, the condition $R_3 = R_4$ may not be exactly satisfied. However, in practice, the value of R_3/R_4 after adjustment is such that $|\Delta f_r/f_r|$ is close to its minimum.

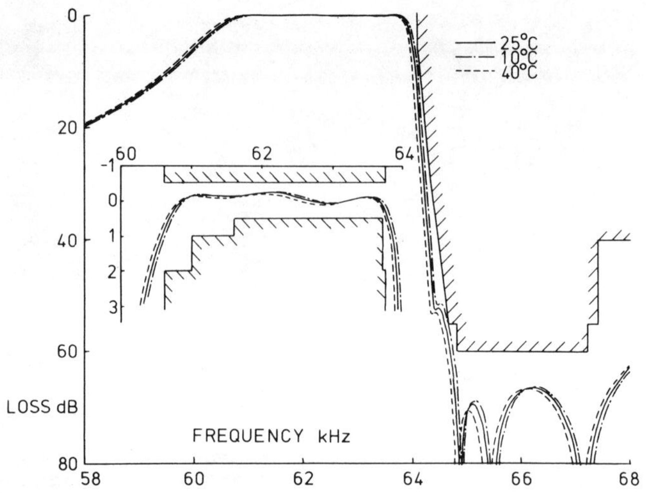

Fig. 10. Measured response of 60–64-kHz microelectronic active-*RC* CBF.

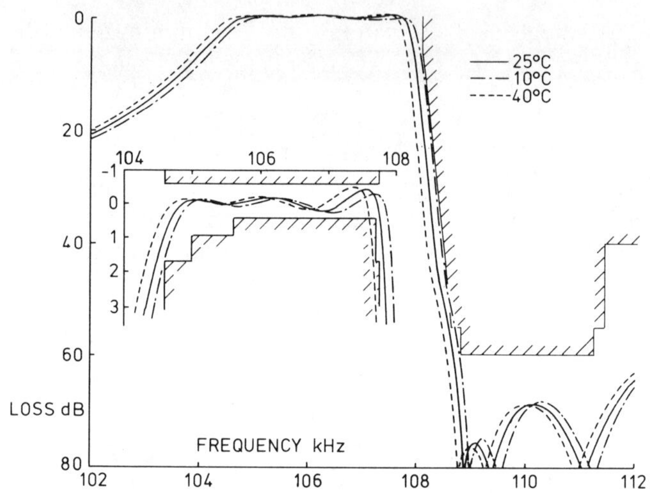

Fig. 11. Measured response of 104–108-kHz microelectronic active-*RC* CBF.

results are given below. The intermodulation tests carried out so far are the same as those normally used for evaluating *LC* CBF's; since active-*RC* CBF's may produce intermodulation products at different frequencies, these test results cannot be regarded as final.

1) Nonlinearity: when the input voltage level was varied between 300 and 10 mV_{rms} (a range of 30 dB), the loss of the filter in the passband varied by a maximum of 0.16 dB. The specified loss variation is $\leqslant 0.15$ dB.

2) Noise: adopting a nominal input voltage level of 150 mV_{rms} (in order to allow a 6-dB overload margin within which the nonlinearity specification is met), the signal-to-noise ratio (SNR) measured over a 3.5-kHz weighted bandwidth centered at 63 kHz was 85.7 dB. Since for *LC* CBF's an SNR$\geqslant$90 dB is specified, this active-*RC* CBF did not meet the SNR requirements. However, apart from the possibility of redesigning the filter circuit for a higher SNR, it may be possible to satisfy the SNR requirement for the complete channel (SNR to be $\geqslant$78 dB) using the present active-*RC* CBF, because the higher signal levels admissible in active-*RC* CBF's compared to those in *LC* CBF's (typically 30 mV_{rms}) may allow a reduction in the noise contribution from other constituent units of the channel.

3) Intermodulation distortion: For a nominal input voltage level of 150 mV_{rms}, comprising two equal-amplitude signals at 59.5 kHz (f_A) and 55.5 kHz (f_B), the level of the intermodulation signal at 63.5 kHz ($2f_A - f_B$) is -77 dB relative to the input voltage. The specified figure is < -70 dB.

C. Results for 104–108 kHz Filter

The measured loss-frequency response of the 104–108-kHz microelectronic active-*RC* CBF is shown in Fig. 11. The input voltage level used for these measurements was 30 mV_{rms}. The following comments on these results are appropriate.

1) At room temperature (25°C), the active-*RC* CBF meets the loss-frequency specification. This is not true for the *LC* CBF (see Fig. 2(b)).

2) Although the loss-frequency response of the active-*RC* CBF at 25°C is acceptable, it has a larger ripple than the design curve (see Fig. 2(b)). With more experience in the adjustment of PIC tuned circuits and possibly also by improving the microelectronic circuit layout, we hope to be able to improve the response.

3) Although the effect of temperature changes on the *shape* of the response is very small, there is a *frequency shift* of +131 Hz at 10°C and of −131 Hz at 40°C which causes the specification to be violated; the corresponding f_r-temperature coefficient is -81 ppm/°C. This frequency shift is due mainly to two causes; variation of amplifier f_T's with temperature and variation of passive components. For our amplifiers, the measured variation of f_T with temperature is -0.25 percent per °C, and with this value (3) gives an f_r temperature coefficient of -54 ppm/°C. This suggests that the f_r temperature coefficient due to passive component changes is -27 ppm/°C, which is consistent with the technology we are using (for high-quality thick-film resistors $|TCR| < 30$ ppm/°C, and for NPO ceramic-chip capacitors $|TCC| < 10$ ppm/°C). The high contribution to the total frequency shift due to changes in the amplifier f_T's is the main reason for the specification not being met at the temperature limits.

The frequency shift due to f_T changes can be reduced in two ways: 1) by using amplifiers with the same relative variation of f_T with temperature but higher nominal f_T values of the order of 20 MHz and 2) by using amplifiers with the same nominal f_T value (10 MHz) but reduced variation of f_T with temperature. It is relevant to mention, in this context, a promising development in integrated-circuit amplifiers, namely the development of an amplifier which has a temperature-compensated f_T [18]. Whereas for existing amplifiers the variation of f_T with temperature is typically -0.25 percent per °C, it is only 0.05 percent per °C for the new amplifier.

VI. Conclusions

The development of microelectronic active-*RC* channel bandpass filters (CBF's) for the frequency range 60–108 kHz (for the formation and decomposition of the primary group in single-modulation-type FDM SSB telephone systems) has been described and discussed. We adopted as our design target a specification for CBF's meeting CCITT requirements. In order to minimize the sensitivity of the active-*RC* CBF's, they were designed by simulation of suitably chosen low-sensitivity *LC* ladder filter prototypes—the duals of *LC* CBF's used in FDM systems in the U.K.

The design of the active-*RC* filters is firmly based on the low-cost microelectronic technology adopted by us. We use hybrid modules consisting of alumina substrates with deposited thick- or thin-film conductor and resistor patterns, to which high-Q NPO-type ceramic-chip capacitors and dual- or quad-operational amplifiers in naked-chip form are attached. The film resistor values can be accurately laser trimmed, and it is this adjustability which forms the basis of our functional adjustment strategy.

Cost minimization, as far as capacitors are concerned, is achieved by steering the design towards the acceptability of capacitors with manufacturer-preferred (and equal) nominal values and by the development of an adjustment procedure which permits very wide manufacturing tolerances (± 10 percent) on the nominal capacitor values. It is found that the same adjustment procedure provides an additional bonus: at and near the adjustment frequencies all imperfections of passive and active components are compensated so that in these frequency regions the ideal nominal dissipationless *LC* prototype filter is simulated nearly perfectly. The reasons for this—perhaps surprising—result are twofold.

1) The positive impedance converters adopted provide resistors both for Q adjustment and for resonant frequency (f_r) adjustment of the simulated tuned circuits; hence we can adjust f_r to within ± 5 Hz and obtain $|Q|$ values>5000.

2) The inductors of the *LC* filters are represented by resistors which can be laser adjusted to ± 0.1 percent; therefore, the accurate adjustment of f_r's corresponds to the adjustment to ∓ 0.1 percent of the capacitors in the *LC* filters. This is a much higher accuracy than is achievable in modern *LC* filters using ferrite-cored inductors and polystyrene or ceramic capacitors, where only the inductors are adjustable and capacitor manufacturing tolerances can be compensated only partly—by inductor adjustment to obtain correct f_r's. This leads to larger L/C ratio deviations, and thus limits acceptable capacitor manufacturing tolerances for an *LC* CBF to typically ± 1 to ± 1.5 percent.

Thus, we obtain active *RC* CBF's which have higher Q values and more accurate (simulated) L and C values, and, therefore, better loss-frequency responses than their practical *LC* counterparts. This is demonstrated most strongly by the fact that, whereas practical *LC* CBF's meet the passband specification only with the help of additional loss-frequency channel equalizers, for the active-*RC* CBF's no additional equalizers would be required.

The lowest frequency CBF, for channel 12 (60–64 kHz), and the highest frequency CBF, for channel 1 (104–108 kHz), have been fabricated in microelectronic form and tested, and since both of them meet their loss-frequency specification, at least at room temperature (25°C), and improve on their passive prototypes, the basic aim of the development described in this paper can be considered as having been achieved: to design and fabricate minimum-cost microelectronic filters, using commercially available amplifiers and wide tolerance, preferred value capacitors, to satisfy highly selective high-precision requirements at frequencies up to about 110 kHz.

As to the possible introduction of the active CBF's into practical FDM systems, one problem remains (assuming that the satisfactory provisional intermodulation results are confirmed by more detailed tests and that the signal-to-noise ratio requirements can be met in the way indicated in Section V-B): when the temperature varies over the specified temperature range 10–40°C, the loss-frequency response of the 104–108-kHz CBF suffers a frequency shift which moves it beyond the specification limits. This is mainly due to the variation with temperature of the gain-bandwidth products f_T of the operational amplifiers used. It is interesting to note that Orchard and Sheahan [6] encountered a similar limitation. However, we hope that this will be overcome by the availability in the near future of amplifiers with higher f_T's and/or lower f_T temperature coefficients. At the same time, it is desirable to use amplifiers with reduced power consumption, since the reduction of dc power requirements and of heat generation are of paramount importance in most channel equipment. We also hope to use amplifiers which do not require programming resistors, because this would lead to a 20-percent reduction in the area of the microelectronic substrate.

Finally, in order to take maximum advantage of the use of microelectronic circuits, it seems desirable to produce a single microelectronic assembly (anticipated size: 50 mm $\times 60$ mm$\times 5$ mm) which incorporates all the constituent units of an FDM SSB channel. The constituent units of such an assembly (apart from the CBF) would be as follows:

1) Channel Low-Pass Filter: This filter could be realized [19] using circuit design techniques and a manufacturing technology similar to those described in this paper. As an example, Fig. 12 shows a hybrid module (size: 27 mm$\times 20$ mm$\times 14$ mm) which contains four fifth-order Cauer low-pass filters. This module was developed at the AUWE Microelectronics Section and uses low power-consumption quad-operational amplifiers; the dc power consumed by each filter is 30 mW, with ± 15 V power supplies.

2) Modulator: This could be an integrated-circuit multiplier chip attached directly to a substrate.

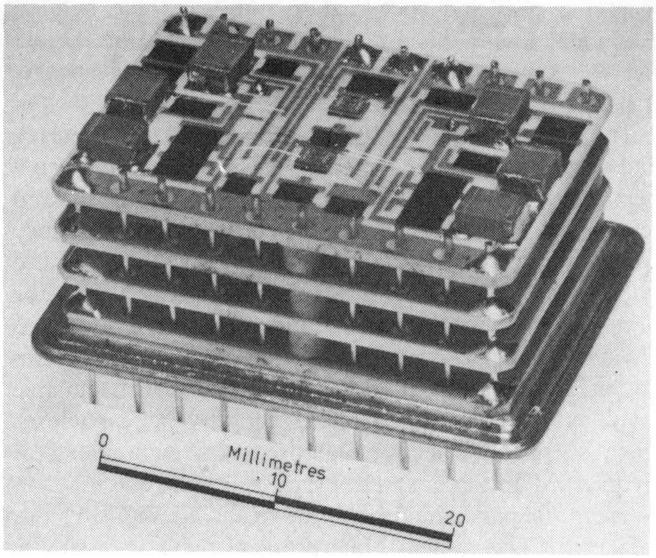

Fig. 12. Photograph of microelectronic module comprising four low-pass filters.

3) Loss-Frequency Equalizer: An equalizer may not be needed due to the high-tuned-circuit Q values obtainable in active CBF's. If still required, the equalizer would be simpler than that required for an *LC* CBF and could easily be realized as a simple active-*RC* circuit.

Acknowledgment

We wish to thank Pye TMC Limited, and in particular John A. Bonnett, for giving us the information about their *LC* CBF's, and also for carrying out some nonlinearity, noise, and intermodulation tests on one of our active-*RC* CBF's.

References

[1] C. F. Kurth, "Analog and digital filtering in multiplex communication systems," *IEEE Trans. Circuit Theory*, vol. CT-20, pp. 408–415, July 1973.

[2] W. Saraga, D. G. Haigh, and R. G. Barker, "Microelectronic active *RC* channel band pass filters for FDM SSB telephone systems," in *Proc. 1976 European Conf. Circuit Theory Design* (part-sponsored by IEEE and IEE) (Genoa, Italy), Sept. 1976, pp. 580–587.

[3] ——, "A design 'philosophy' for microelectronic active RC filters," in *Proc. 1977 IEEE Int. Symp. Circuits Syst.* (Phoenix, AZ), Apr. 1977, pp. 195–198.

[4] H. J. Orchard, "Inductorless filters," *Electron. Lett.*, vol. 2, pp. 224–225, June 1966.

[5] J. Valihora, J. T. Lim, and L. T. Bruton, "The feasibility of active filtering in frequency division multiplex systems," in *Proc. 1974 IEEE Int. Symp. Circuits Syst.* (San Francisco, CA), Apr. 1974, pp. 121–125.

[6] H. J. Orchard and D. F. Sheahan, "Inductorless band-pass filters," *IEEE J. Solid-state Circuits*, vol. SC-5, pp. 108–118, June 1970.

[7] R. Thimm, "Active filters for channel translating equipment," *Elec. Commun.*, vol. 52, pp. 130–136, 1977.

[8] A. Antoniou and K. S. Naidu, "A compensation technique for a gyrator and its use in the design of a channel-bank filter," *IEEE Trans. Circuits Syst.*, vol. CAS-22, pp. 316–323, Apr. 1975.

[9] P. O. Brackett and A. S. Sedra, "Active compensation for high-frequency effects in op-amp circuits with applications to active *RC* filters," *IEEE Trans. Circuits Syst.*, vol. CAS-23, pp. 68–72, Feb. 1976.

[10] K. Martin and A. S. Sedra, "Optimum design of active filters using the generalised immittance converter," *IEEE Trans. Circuits Syst.*, vol. CAS-24, pp. 495–503, Sept. 1977.

[11] K. Yakuwa, S. Okuda, and M. Yanagi, "Development of new channel band-pass filters," in *Proc. 1974 IEEE Int. Symp. Circuits Syst.*, (San Francisco, CA), Apr. 1974, pp. 100–105.

[12] P. Kirby, "Thick film advances simplify complex hybrid module design," *Electron. Eng.*, vol. 48, pp. 35–38, Mar. 1976.

[13] W. Saraga, "Minimum inductor or capacitor filters," *Wire. Eng.*, pp. 163–174 July 1953.

[14] J. Gorski-Popiel, "*RC*-active synthesis using positive impedance converters," *Electron. Lett.*, vol. 3, pp. 381–382, Aug. 1967.

[15] L. T. Bruton, "Network transfer functions using the concept of frequency-dependent negative resistance," *IEEE Trans. Circuit Theory*, vol. CT-16, pp. 406–408, Aug. 1969.

[16] A. Antoniou, "Realisation of gyrators using operational amplifiers, and their use in RC-active network synthesis," *Proc. Inst. Elec. Eng.*, vol. 116, pp. 1838–1850, Nov. 1969.

[17] L. T. Bruton and J. T. Lim, "High-frequency comparison of GIC-simulated inductance circuits," *Int. J. Circuit Theory Appli.*, vol. 2, pp. 401–404, 1974.

[18] D. L. Cave and W. R. Davis, "A quad JFET wide-band operational-amplifier integrated circuit featuring temperature-compensated bandwidth," *IEEE J. Solid-State Circuits*, vol. SC-12, pp. 382–388, Aug. 1977.

[19] R. Jeffers and D. G. Haigh, "Active *RC* low-pass filters for FDM and PCM systems," *Proc. Inst. Elec. Eng.*, vol. 170, pp. 945–953, Sept. 1973.

Part V
Techniques for Monolithic Continuous Analog Active Filters

Until very recently, analog active filters have proven to be "integration resistant", i.e., design techniques and technologies were not available that lend themselves to building integrated, monolithic, active filters. The two major problems are the need for large-valued capacitors, requiring unrealistically large areas on a Si-chip, and the need for accurate resistor and capacitor values to set filter parameters, a requirement not compatible with integrated circuit (IC) processing. Clearly, what is needed are designs that depend on only ratios of components and require smaller element (capacitor) values; in addition, techniques for cancelling the inaccuracies inherent in IC processing must be developed. This chapter contains four papers that describe specific monolithic continuous analog filters and a general design procedure whose practical implementation in IC technology—MOS or bipolar—is currently being investigated in a number of laboratories.

The design techniques in this chapter are aimed primarily at applications at higher frequencies although, as illustrated in paper 5-4, low-frequency implementations are, of course, possible. For alternative excellent low-frequency monolithic designs, the reader should consult Part VI where analog sampled-data switched-capacitor filters are discussed.

Paper 5-1 by Brand and Schaumann reviews the design of so-called "active R" filters. Although initially developed only with a view on extending the frequency range of active filters by including the frequency-dependence of operational amplifiers (compare papers 1-5 and 2-2) directly into the design, these circuits also turned out to be quite suitable for monolithic implementation because all "external" capacitors are avoided (hence: active R filters) by treating the amplifiers as integrators, and because filter parameters are established via ratios of resistors. The only capacitors needed in these filters are "internal" to the amplifiers and used for frequency compensation. (For the parallel and analogous development of "active C" filters, see [1] and [2].) Active R filters of low and high order have been designed and tested for operation from audio frequencies until a few megahertz; the one serious problem with these circuits is the dependence of filter pole-frequency on the amplifiers' gain-bandwidth product GB. Thus, as discussed in this paper, and also in papers 5-3 and 5-4, for most applications, special steps must be taken to stabilize the pole-frequency against drifts.

The same concept is also employed in *paper 5-2* by Soderstrand for the design of high-frequency active ladders. Using procedures similar to those in papers 3-4 to 3-6, he uses only resistors and amplifiers for an operational simulation of voltages and currents, such that the integrating behavior of inductors and capacitors is implemented by means of active R integrators.

The development of a bipolar chip for filtering applications at video frequencies is described by Moulding et al. in *paper 5-3*. The authors base their design on active resonance circuits, implemented via high-frequency gyrators connected to junction capacitors that can be varied (tuned) by changing the junction bias voltages. The chip is a substitute for several discrete filter circuits and requires no post-design tuning or alignment. In *paper 5-4*, Tan and Gray describe the design of a high-order monolithic lowpass filter using the leap-frog approach. As illustrated in papers 3-2 and 5-2, leap-frog filters are designed from LC ladders by simulating inductors and capacitors as active integrators. The long RC time constants required for the low-frequency filter described in this paper are implemented as C/g_m where g_m is the transconductance of an amplifier stage and is of the order of μmhos. Because g_m depends on the bias current, the integrator time constants and, thereby, the filter response can be controlled—tuned—by varying a bias voltage and, in particular, can be locked to an external control signal via a phase-locked-loop approach (compare papers 5-1 and 5-3). Using this method and relying on matching and tracking of components on a chip, the authors succeed not only in compensating for any design tolerances but also in eliminating drifts due to temperature or aging. The circuits meet design specifications without requiring any post design trimming.

[1] R. Schaumann and J. R. Brand, "Design method for monolithic analog filters", Electronics Letters, vol. 14, No. 22, pp. 710-711, October 1978.

[2] R. Schaumann and J. R. Brand, "Integrable analog active filters using amplifiers and ratios of capacitors", Proc. Int. Symp. on Circuits and Systems, pp. 914-917, July 1979.

Active *R* filters: review of theory and practice

J.R. Brand and R. Schaumann

Indexing term: *Active filters*

Abstract: Active filters that derive their frequency response from internal amplifier dynamics, but use no external capacitors in their implementation, are referred to as 'active *R*' filters. Because of their potential advantages in terms of miniaturisation (i.c. fabrication), ease of design and high-frequency performance, these circuits are receiving increasing attention in the technical literature. The paper reviews the published synthesis procedures for 1st-, 2nd- and higher-order active *R* filters; it is shown that the 1st- and 2nd-order circuits developed to date are special cases of a general active *R* topology. The paper discusses nonideal aspects and limitations of active *R* filter performance, which are usually neglected in the literature; in particular, the potentially severe effect of temperature-induced drift in the amplifier's dynamic response is considered, and methods of drift compensation are discussed. Experimental results are presented to verify filter performance and the theoretical derivations.

1 Introduction

Traditionally, the design of active filters was based on embedding into a passive *RC* network one or more amplifiers whose gain was assumed to be large and independent of frequency. In actuality, however, active devices, such as transistors and amplifiers, have complex gains that exhibit lowpass behaviour, i.e. their magnitudes decrease with increasing frequency and they have a nonzero phase. This nonideal behaviour of amplifiers has long been known to cause sensitivity and stability problems,[1-3] in addition to limiting applications of active filters to relatively low frequencies. As a result, several workers in the field have developed criteria for comparing the performance of active filters based on their sensitivity to the gain roll-off of the amplifiers;[4-7] others have, with varying degrees of success, employed special design techniques or compensation methods that attempted to eliminate or minimise errors introduced by the frequency-dependent amplifier gain.[2,8-10]

In an alternative approach to this problem, the internal parasitic reactances of transistors, instead of being considered undesirable, were exploited to implement active inductive impedances, which together with capacitors were used to build active microwave filters.[11-12] Later, recognising that the lowpass characteristic of transistors could be employed to eliminate all external reactances, Capparelli and Liberatore[13-16] and Paphitis and Murata[17] built pure transistor-resistor filters, using a method that was shown by Berman and Newcomb[18] to be a general state-variable synthesis procedure, using transistors as lossy integrators.

Although this method does work in principle for arbitrary transfer functions and provides the means of building high-frequency active filters without the need for reactances, there are a number of difficulties that apparently discouraged further development of this design technique. The most important are:

(i) the extremely poor repeatability of parasitic transistor parameters and their dependence on temperature and bias conditions

(ii) the finite input and output impedances of transistors, which necessitate buffering and complicate any tuning procedure

(iii) the need for d.c. bias circuitry, which will interfere with the a.c. transfer function.

It was soon recognised that Berman and Newcomb's method[18] could still be used and that the problems associated with (ii) and (iii) above could largely be avoided if, instead of on individual transistors, the design would be based on well developed and inexpensive integrated operational amplifiers (o.a.s), whose gain also exhibits a lowpass or integrator characteristic. Thus, in rapid succession, a large number of papers appeared in which o.a.s and resistors were used to implement inductors,[19-20,79] oscillators[21-22,71] and 2nd-[23-44,70] or higher-order[45-49] active filters. (A closely related technique uses c.m.o.s. inverters for the active devices.[50-52]) A few of those circuits also use 'external' lumped capacitors,[21,23,34,39,40,72] but the majority employ only resistors and internally compensated o.a.s, thus being denoted 'active *R*' filters in contrast to the usual active *RC* designs.

In this paper we shall review the active *R* synthesis techniques developed to date and discuss a number of practical aspects of active *R* design that are usually neglected or receive only cursory treatment in the literature. In Section 2 we shall present the basic building block, an integrating summer, and show that all 2nd-order active *R* filters can be derived as special cases from an appropriate interconnection of two of these blocks. After deriving the most versatile 2nd-order sections, the realisation of active *R* filters of higher order, together with a number of practical considerations, is discussed briefly. Nonideal aspects of active *R* filter operation, all traceable to amplifier imperfections, are examined in Section 3; the potentially most severe of these imperfections, that of temperature dependence of the dynamic response, is considered in detail in Section 3.1, as are the methods of compensation necessary for temperature-stabilised filter response. In the concluding Section 4 are presented experimental results that demonstrate the performance of active *R* filters and provide verification of the theoretical relationships developed throughout the paper.

2 Active *R* synthesis techniques

2.1 The integrating summer

The integrating summer, shown schematically in Fig. 1*a*, can be implemented by the circuit in Fig. 1*b*, where A_1 represents an internally compensated o.a. having a gain

Paper T191 E, first received 3rd January and in revised form 21st March 1978

Mr. Brand and Dr. Schaumann are with the Department of Electrical Engineering, University of Minnesota, 123 Church Street S.E., Minneapolis, Mn. 55455, USA

Reprinted with permission from *IEE J. Electron. Circuits Syst.*, vol. 2, pp. 89-101, July 1978.

given by

$$A_1(s) = \frac{GB_1}{s + \sigma_1} e^{-s\tau_1} \tag{1a}$$

GB_1 is the gain-bandwidth product, σ_1 is the 3 dB frequency, and the term $e^{-s\tau_1}$ is included to model the delay τ_1 or the excess phase $-\omega\tau_1$,[43] which is not accounted for by the simpler single-pole model $GB_1/(s + \sigma_1)$ usually found in the literature. Since the low-frequency pole $-\sigma_1$ in integrated o.a.s is usually less than 100 rad/s, and since active R techniques are normally employed only at frequencies above about 20 kHz, eqn. 1a may be simplified to

$$A_1(s) \approx \frac{GB_1}{s} e^{-s\tau_1} \tag{1b}$$

by setting $\sigma_1 = 0$. Routine analysis of the circuit in Fig. 1b yields

$$V_{01} = \frac{c_1}{1/A_1 + c_0} V_i + \frac{c_2}{1/A_1 + c_0} \hat{V}_1 \tag{2}$$

where the following abbreviations have been introduced:

$$c_0 = b_3 - b_7; \quad c_1 = b_5 - b_1; \quad c_2 = b_6 - b_2 \tag{3}$$

$$b_1 = \frac{a_1 + a_{on}a_5}{1 - a_{on}a_{op}}; \; b_2 = \frac{a_2 + a_{on}a_6}{1 - a_{on}a_{op}}; \; b_3 = \frac{a_3 + a_{on}a_7}{1 - a_{on}a_{op}}$$

$$b_5 = \frac{a_5 + a_{op}a_1}{1 - a_{on}a_{op}}; \; b_6 = \frac{a_6 + a_{op}a_2}{1 - a_{on}a_{op}}; \; b_7 = \frac{a_7 + a_{op}a_3}{1 - a_{on}a_{op}} \tag{4}$$

$$a_{on} = G_o/G_{Tn}; \quad a_i = G_i/G_{Tn} \quad (i = 1, 2, 3) \tag{5a}$$

$$a_{op} = G_o/G_{Tp}; \quad a_i = G_i/G_{Tp} \quad (i = 5, 6, 7) \tag{5b}$$

$$G_{Tn} = \sum_{i=0}^{4} G_i; \quad G_{Tp} = G_o + \sum_{i=5}^{8} G_i \tag{6}$$

Assuming $\tau_1 = 0$ and setting $R_2 = R_6 = \infty$, i.e. $c_2 = 0$, the circuit in Fig. 1b according to eqn. 2 is a 1st-order lowpass of form

$$V_{01}/V_i = \frac{GB_1 c_1}{s + c_0 GB_1}, c_0 \geqslant 0 \tag{7}$$

which may be used in its own right or in the design of filters of odd order by setting (including to 0 or ∞) the remaining seven resistors in the circuit such that c_0 and c_1 give the desired pole and the positive or negative gain of the filter. In addition, a number of 2nd-order lowpass or bandpass filters may be derived from this circuit by replacing one of the resistors by a capacitor, (e.g. References 23, 34 and 39).

2.2 2nd-order active R filters

We shall show next that by interconnecting two of the blocks in Fig. 1 a general circuit can be developed from which all the 2nd-order active R filters in the literature can be derived as special cases. To this end, let us define an integrating summer, identical in topology to Fig. 1b, built around an amplifier A_2 with resistors labelled r_i $(i = 0, \ldots, 8)$ and abbreviations α_j, β_j and γ_j analogous to a_i, b_i and c_i in eqns. 3–6. The resulting transfer function for block 2 is, as in eqn. 2,

$$V_{02} = \frac{\gamma_1}{1/A_2 + \gamma_0} V_i + \frac{\gamma_2}{1/A_2 + \gamma_0} \hat{V}_2 \tag{8}$$

The proposed, completely symmetrical, interconnection of the two blocks is shown in Fig. 2. Apart from connecting the input V_i to block 2, it forms a regular state-variable circuit,[53] using two amplifiers (integrators) and 18 resistors. Using eqns. 2 and 8, we readily find from Fig. 2 that

$$T_1 = V_{01}/V_i = (c_1/A_2 + c_1\gamma_0 + c_2\gamma_1)D^{-1} \tag{9a}$$

$$T_2 = V_{02}/V_i = (\gamma_1/A_1 + \gamma_1 c_0 + \gamma_2 c_1)D^{-1} \tag{9b}$$

where the denominator D equals

$$D = (A_1A_2)^{-1} + c_0/A_2 + \gamma_0/A_1 + c_0\gamma_0 - c_2\gamma_2 \tag{10}$$

Assuming for the time being that $\tau_1 = \tau_2 = 0$, it is clear from eqns. 1, 9 and 10 that T_1 and T_2 are 2nd-order filters realising transfer functions having one zero on the real axis, including one with zero at the origin (bandpass) and one with zero at infinity (lowpass). The implementation of complete biquadratic transfer functions will be discussed below.

In order to determine which of the 18 resistors R_i and r_i $(i = 0, \ldots, 8)$ are either redundant or detrimental to the performance, and hence can be eliminated, we note first that no generality is lost by setting $R_0 = r_0 = \infty$, i.e.

$$a_{on} = a_{op} = \alpha_{on} = \alpha_{op} = 0 \tag{11}$$

because, according to eqn. 4, the parameters b_i and β_i are restricted to $0 \leqslant (b_i, \beta_i) \leqslant 1$, as are a_i and α_i. It is noted

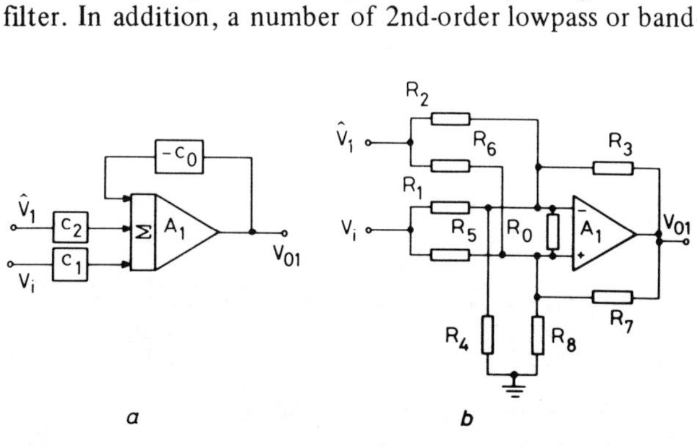

Fig. 1 *Integrating summer*

a Block diagram
b Circuit

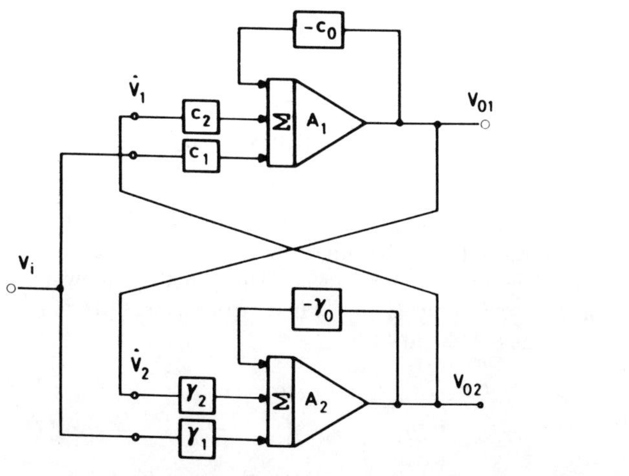

Fig. 2 *Schematic representation of a general 2nd-order active R filter*

also that the following sensitivity argument would result in eqn. 11 as the best choice. As a further important criterion for eliminating resistors, we require that the sensitivity of pole frequency ω_o and pole quality factor Q to all elements be as small as possible. With $\tau_1 = \tau_2 = 0$, we find from eqns. 1*b* and 10 that

$$D(s) = \frac{1}{GB_1 GB_2}[s^2 + s(c_0 GB_1 + \gamma_0 GB_2) + GB_1 GB_2(c_0\gamma_0 - c_2\gamma_2)] \qquad (12)$$

so that

$$\omega_o^2 = GB_1 GB_2(c_0\gamma_0 - c_2\gamma_2) \qquad (13a)$$

and

$$Q = \omega_0/(c_0 GB_1 + \gamma_0 GB_2) \qquad (13b)$$

Using the definition $S_x^F = \frac{x}{F}\frac{dF}{dx}$, routine calculation gives the sensitivities

$$S_{GB_1}^{\omega_o} = S_{GB_2}^{\omega_o} = \tfrac{1}{2}$$

$$S_{GB_1}^{Q} = -S_{GB_2}^{Q} = -\frac{1}{2}\frac{\rho c_0 - \gamma_0}{\rho c_0 + \gamma_0} \qquad (14)$$

where $\rho = GB_1/GB_2$. Furthermore,

$$S_{c_0}^{\omega_o} = S_{\gamma_0}^{\omega_o} = \frac{1}{2}\frac{c_0\gamma_0}{c_0\gamma_0 - c_2\gamma_2} \qquad (15a)$$

$$S_{c_2}^{\omega_o} = S_{\gamma_2}^{\omega_o} = S_{c_2}^{Q} = S_{\gamma_2}^{Q} = -\frac{1}{2}\frac{c_2\gamma_2}{c_0\gamma_0 - c_2\gamma_2} \qquad (15b)$$

$$S_{c_0}^{Q} = S_{c_0}^{\omega_o} - \frac{\rho c_0}{\rho c_0 + \gamma_0}; \quad S_{\gamma_0}^{Q} = S_{\gamma_0}^{\omega_o} - \frac{\gamma_0}{\rho c_0 + \gamma_0} \qquad (15c)$$

These equations show that, to keep sensitivities low, we must choose either $c_0\gamma_0 = 0$ or $c_2\gamma_2 = 0$, thereby avoiding difference effects in addition to redundant elements. The latter case is of no interest, since, by breaking the feedback loop in Fig. 2, it would result in a transfer function having two real simple poles. Assuming also that a zero on the real axis is not of interest, we observe from eqns. 9 and 13 that with $c_0\gamma_0 = 0$ only the following choices exist:

(i) $\gamma_0 = \gamma_1 = 0, c_0 > 0, c_2\gamma_2 < 0$ (16)

so that $T_1 \to$ bandpass, $T_2 \to$ lowpass

(ii) $c_0 = c_1 = 0, \gamma_0 > 0, c_2\gamma_2 < 0$

so that $T_1 \to$ lowpass, $T_2 \to$ bandpass.

Observe also that, from the symmetry of the equations and of the circuit, (ii) gives the same circuit as (i) with blocks 1 and 2 in Fig. 2 interchanged. Thus, in the following, we shall discuss only (i), for which, from eqns. (13) and (16) with eqns. 3, 4 and 11,

$$\omega_o^2 = GB_1 GB_2(a_2 - a_6)(\alpha_6 - \alpha_2) \qquad (17a)$$

$$Q = \omega_o/[GB_1(a_3 - a_7)] \qquad (17b)$$

Q, being established through a difference effect, has sensitivities to a_3 and a_7 that are inversely proportional to $(a_3 - a_7)$. Thus, to keep these sensitivities small, we choose

$$a_7 = 0 \qquad (18)$$

A similar argument for ω_0 in eqn. 17*a* results in the two possible requirements

(I) $a_6 = \alpha_2 = 0$ (19*a*)

(II) $a_2 = \alpha_6 = 0$ (19*b*)

Thus, with eqns. 11, 18 and 19, the general structure in Fig. 2 has been reduced to two cases:

Case I

$a_{on} = a_{op} = a_6 = a_7 = \alpha_{on} = \alpha_{op} = \alpha_2 = 0$, i.e. from eqn. 5, $R_0 = R_6 = R_7 = r_0 = r_2 = \infty$. The circuit realising

$$T_1(s) = \frac{V_{01}}{V_i} = \frac{sGB_1(a_5 - a_1)}{s^2 + sGB_1 a_3 + GB_1 GB_2 a_2 \alpha_6} \qquad (20a)$$

$$T_2(s) = \frac{V_{02}}{V_i} = \frac{GB_1 GB_2(a_5 - a_1)\alpha_6}{s^2 + sGB_1 a_3 + GB_1 GB_2 a_2 \alpha_6} \qquad (20b)$$

is shown in Fig. 3*a*.

Case II

$a_{on} = a_{op} = a_2 = a_7 = \alpha_{on} = \alpha_{op} = \alpha_6 = 0$, i.e. from eqn. 5, $R_0 = R_2 = R_7 = r_0 = r_6 = \infty$. The circuit realising

$$T_1(s) = \frac{V_{01}}{V_i} = \frac{sGB_1(a_5 - a_1)}{s^2 + sGB_1 a_3 + GB_1 GB_2 a_6 \alpha_2} \qquad (21a)$$

$$T_2(s) = \frac{V_{02}}{V_i} = \frac{-GB_1 GB_2(a_5 - a_1)\alpha_2}{s^2 + sGB_1 a_3 + GB_1 GB_2 a_6 \alpha_2} \qquad (21b)$$

is shown in Fig. 3*b*.

In eqns. 20 and 21, because of eqn. 16, we have assumed that

$$\alpha_1 - \alpha_5 = 0, \quad \alpha_3 - \alpha_7 = 0 \qquad (22)$$

The two circuits in Fig. 3 realise, at the output terminals

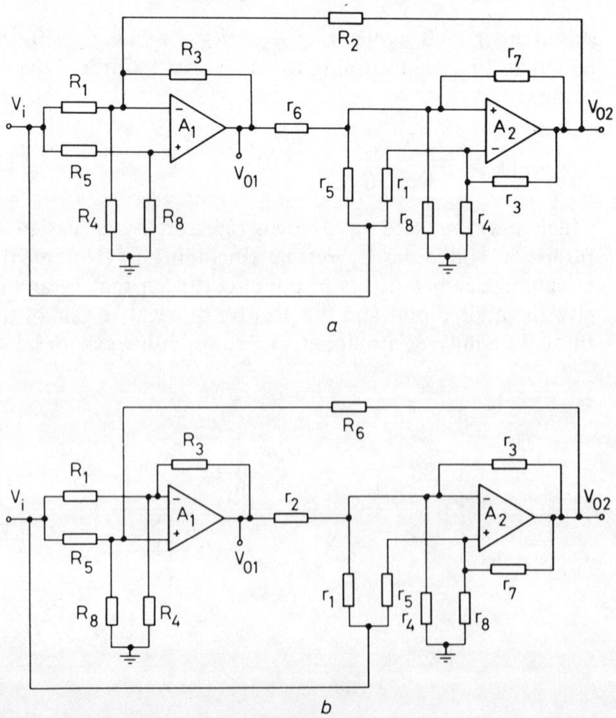

Fig. 3 *General 2nd-order active R filters*
Realising (*a*) eqn. 20 and (*b*) eqn. 21

identified, bandpass and lowpass functions of either positive or negative gain. It should be noted that, in contrast to Fig. 3*a*, there is always a sign inversion between T_1 and T_2 of Fig. 3*b*. In addition, labelling the inverting and noninverting terminals of the amplifier A_i as V_{in} and V_{ip}, respectively, we obtain, for example, from Fig. 3*a*

$$V_{2p}/V_i = \alpha_5 + \alpha_6 T_1 + \alpha_7 T_2 \qquad (23a)$$

and from Fig. 3*b*

$$V_{2n}/V_i = \alpha_1 + \alpha_2 T_1 + \bar{\alpha}_3 T_2 \qquad (23b)$$

where T_1 and T_2 are given in eqns. 20 and 21. Eqn. 23 represents biquadratic transfer functions, which can realise, for example, highpass, general-notch, and allpass filters by a suitable choice of resistor ratios. This approach of obtaining a biquadratic transfer function at the input terminal of an o.a. was taken by Kim *et al.*[38,70] and in a simplified form by Soderstrand.[27,36] In our experience, it suffers from the serious disadvantage that any load at these nodes, especially if it is not purely resistive, tends to severely perturb some or all of the filter parameters. A less load-sensitive circuit for realising a biquadratic transfer function is obtained[33] by feeding the input and the two output signals of Fig. 3 into a resistive summer (Fig. 4), so that a load resistor can be accounted for in R_{s4} and a load capacitor contributes only a simple negative real pole which can be placed sufficiently far away from the dominant poles. Defining $k_i = G_{si}/\sum_{j=1}^{4} G_{sj}$ $(i = 1, 2, 3)$, we obtain from Fig. 4 with eqns. 20 or 21

$$T(s) = V_o/V_i = k_1 + k_2 T_1(s) + k_3 T_2(s) \qquad (24)$$

which, by analogy with eqn. 23, results in a general biquadratic function. If loading is severe, a buffered (o.a.) summer can be used.[27,35,36]

The circuit in Fig. 3*a*, using a noninverting integrator in block 2 and feedback into the inverting terminal of A_1, contains as special cases the circuits reported in References 24–27, 29–32, 35–37 and 42. Observing that eqn. 22 implies that the contributions of V_i and V_{02} to the input signal of amplifier A_2 are zero, it is clear that this condition is most economically satisfied by setting $\alpha_i = 0$ $(i = 1, 3, 5, 7)$, i.e. $g_1 = g_3 = r_4 = g_5 = g_7 = 0$,[31] rather than by carefully balancing resistor ratios.[37,38,41,70] Eliminating also the clearly redundant resistor R_4, i.e. $R_4 = \infty$, we are left with the simple circuit in Fig. 5*a*, described by eqn. 20, which retains enough resistors to realise arbitrarily prescribed values of ω_0, Q and positive $(G_1 = 0)$[31] or negative $(G_5 = R_8 = 0)$[26] gain.

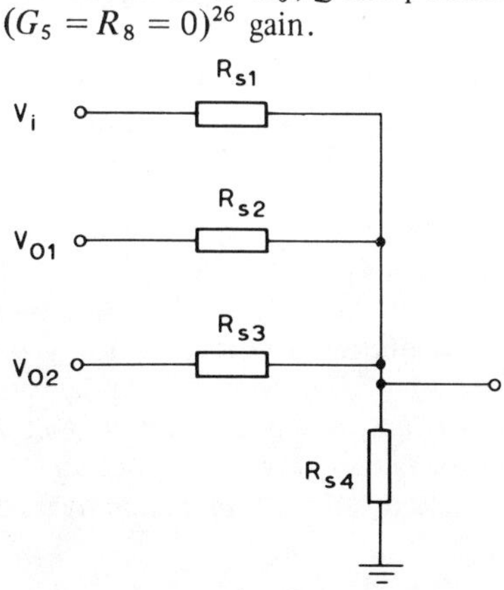

Fig. 4 *Resistive summer*

The dual structure in Fig. 3*b* contains as special cases the circuits discussed in References 33, 38 and 70. The use of analogous considerations as before, i.e. satisfying eqn. 22 by choosing $g_1 = g_3 = g_5 = g_7 = r_8 = 0$, yields the circuit in Fig. 5*b*, described by eqn. 21, which again can be used to realise arbitrary values of ω_0, Q, and positive $(G_1 = G_8 = 0)$ or negative $(G_5 = 0)$[33] bandpass gain. Note that in Fig. 5*b* lowpass and bandpass gains have opposite signs. The latter structure[33] can further be simplified by choosing $R_6 = G_8 = 0$, i.e. $a_6 = 1$, thus saving two resistors and gaining some advantages as far as sensitivity to parasitic elements is concerned.[44,*] The design equations for the circuits in Fig. 5 are summarised in Table 1.

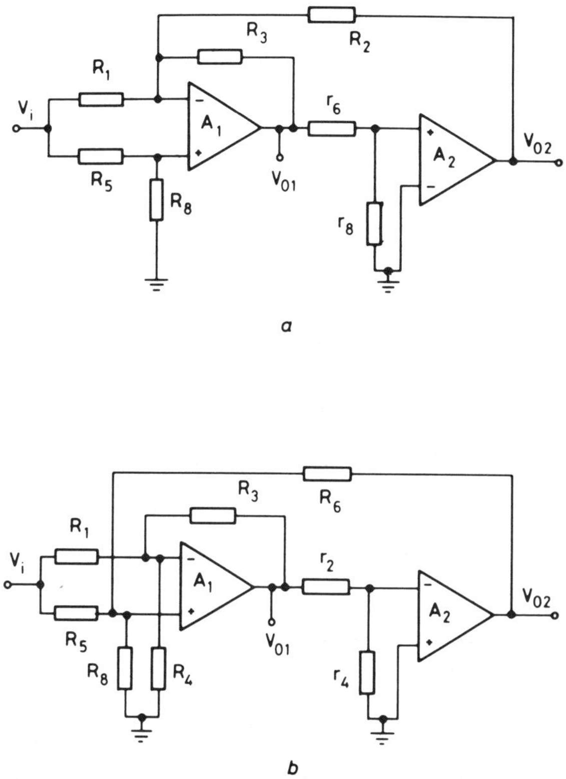

Fig. 5 *Practical 2nd-order active R filters*
Realising (*a*) eqn. 20 and (*b*) eqn. 21

Both configurations in Fig. 5, together with Fig. 4, can be used to implement general biquadratic functions to realise highpass, allpass and notch filters. The describing equations are, from eqns. 20 and 24 for Fig. 5*a*

$$T(s) = k_1 \frac{s^2 + sGB_1[a_3 + R_{s1}G_{s2}(a_5 - a_1)] + GB_1 GB_2 \alpha_6 [a_2 + R_{s1} G_{s3}(a_5 - a_1)]}{s^2 + sGB_1 a_3 + GB_1 GB_2 a_2 \alpha_6} \qquad (25a)$$

*The effects of parasitic capacitances on circuit performance are in general not negligible in filters operating at higher frequencies. Using computer simulation and experimental verification, it has been shown in Reference 44 for the active *R* circuits reported in References 31 and 33 that parasitic capacitors in general will change the pole frequency ω_o only slightly, but that the pole quality factor Q may be drastically altered. The most sensitive locations for parasitic elements are the $r_6 - r_8$ and $r_2 - r_4$ voltage dividers in Figs. 5*a* and 5*b*, respectively. Q is decreased or enhanced, depending on whether the *RC* dividers form a phase-lead or a phase-lag configuration. The effect is similar to that of the amplifier delay discussed in Section 3.2; it shows also that the phase lead introduced by a small capacitor across r_6 and r_2 in Fig. 5 could be used to compensate for the amplifiers' excess phase $-\omega\tau_1$ of eqn. 1*a*. However, the problem is more easily solved by tuning (see Section 3.2).

Table 1: Design equations for circuits of Fig. 5

Parameters	Fig. 5*a*	Fig. 5*b*
$H_B > 0$	$\frac{G_5}{G_5 + G_8}\left(1 + \frac{G_2}{G_3}\right)$	$\frac{G_5}{G_5 + G_6}\left(1 + \frac{G_4}{G_3}\right)$
	$(G_1 = 0)$	$(G_1 = G_8 = 0)$
H_L	$\frac{G_5}{G_5 + G_8}\left(1 + \frac{G_3}{G_2}\right)$	$-\frac{G_5}{G_6}$
$H_B < 0$	$-\frac{G_1}{G_3}$	$-\frac{G_1}{G_3}$
	$(G_5 = R_8 = 0)$	$(G_5 = R_6 = G_8 = 0)$
H_L	$-\frac{G_1}{G_2}$	$\frac{G_1}{G_1 + G_3 + G_4}$
ω_o^2	$GB_1 GB_2 \frac{G_2}{G_1 + G_2 + G_3} \frac{g_6}{g_6 + g_8}$	$GB_1 GB_2 \frac{G_6}{G_5 + G_6 + G_8} \frac{g_2}{g_2 + g_4}$
Q	$\frac{\omega_o}{GB_1}\left(1 + \frac{G_1 + G_2}{G_3}\right)$	$\frac{\omega_o}{GB_1}\left(1 + \frac{G_1 + G_4}{G_3}\right)$

ω_o = pole frequency, Q = quality factor, H_B = midband gain of bandpass function, H_L – d.c. gain of lowpass function

and from eqns. 21 and 24 for Fig. 5*b*

$$T(s) = k_1 \frac{s^2 + sGB_1 [a_3 + R_{s1} G_{s2}(a_5 - a_1)] + GB_1 GB_2 \alpha_2 [a_6 - R_{s1} G_{s3}(a_5 - a_1)]}{s^2 + sGB_1 a_3 + sGB_1 GB_2 \alpha_2 a_6} \tag{25b}$$

It should be noted here that the bandpass topologies reported in References 24–26 and 35, although being special cases of Fig. 3, do not satisfy eqn. 22, and therefore have a small but nonzero d.c. transmission. Also, a number of circuits have been discussed in the literature[24,25,27,35,36] that use fewer resistors than are indicated in Fig. 5. Again, they can be obtained from Fig. 3 by eliminating the appropriate resistors, but suffer from the disadvantage that of the three parameters ω_o, Q and gain, one is a function of the other two and so they cannot be specified independently.

2.3 High-order active R filters

High-order ($N > 2$) transfer functions may be implemented by a direct state-variable synthesis procedure,[18,53] using active *R* integrators (see Section 2.1). However, because of the resulting large sensitivities it is preferable to use one of the numerous techniques available for high-order active *RC* synthesis, such as a cascade or any of the multiple-loop feedback approaches.[54] In these topologies, active *R* 2nd-order blocks (Figs. 3, 5) can be used directly,[46,58] but care must be taken to include buffering in the realisation if the outputs are not taken from a low-impedance node (e.g., V_{2n} in Fig. 3*b*[38,70] or V_o in Fig. 4 (Reference 33), or if the resistive interfaces between sections cannot be combined in a manner allowing sufficiently independent parameter adjustment.

In general, one of the high-order structures is chosen for active *R* synthesis for the same reasons as for active *RC* design; however, with high-frequency filters, parasitic effects will be much more evident, and their existence may influence the choice. It is shown in Section 3.2 that, in addition to perturbations of the filter parameters caused by parasitic capacitors,[44] the signal delay τ_1 of eqn. 1*a* causes significant deviations of some parameters of the basic blocks. Even though functional tuning, or predistortion, can largely eliminate the effects of τ_1 on the blocks' response, there are residual errors (eqn. 34) which affect the transfer characteristic of high-order filters, particularly those synthesised by multiple-loop feedback techniques. Cascaded filters tend to be less affected, and, from this point of view, appear to have an advantage over multifeedback topologies unless suitable phase-compensation techniques can be applied (see Section 3.2).

Because of the o.a.'s slew-rate limitations,[55] it is important in the design of high-order high-frequency active *R* filters to maximise the dynamic range determined by the maximum undistorted output signal and by the output noise level (computed in Section 3.3). Since the noise level is generally alterable only by changing the impedance level and/or the o.a.s, the maximum output signal can be considered as determining the dynamic range, and high-order active *R* filters should always be designed so that the undistorted signal level is maximised. The necessary design procedures, based on equalising the maximum signal level throughout the filter, are available in the literature for cascade[56] and for a number of multiloop feedback[57–60] structures. Recent work indicates that there appears to be no significant difference in the dynamic range properties of any of these topologies.[60]

Using an approach different from the previous ones, active *R* techniques have been applied by Soderstrand[45,47] to active ladder synthesis. In his method, the 'active *R* ladder' is an active implementation of the signal flow graph (s.f.g.) of a double terminated *LC* network, using the basic units shown in Fig. 6[47] as active *R* integrators simulating the passive *L*s and *C*s of the ladder. The s.f.g. procedure has been generalised by Brackett and Sedra for active *RC* filters.[61]

To date, Soderstrand's procedure is complete for a large number of common filter structures; from the reported experimental results,[45,47] the method appears to provide an active R technique potentially capable of the excellent sensitivity and dynamic range-scaling properties known to be available in the active RC analogues[61] of lossless networks. In addition, this method has the advantage of allowing the designer to use readily available tables on LC ladder design;[62] the effects of amplifier parasitics, leading to the realisation of lossy reactances,[47] can also be accounted for by predistorting the filter using these tables. It should be noted, however, that the active R realisation of certain ladders containing capacitive Πs and/or inductive Ts leads to a difficulty in synthesis if the designer attempts to ensure a one-to-one correspondence between filter integrators and ladder reactances; this problem[61] appears to be surmountable in high-frequency active R ladders only by introducing additional amplifiers of exceptionally high GB compared to those used in the basic units of Fig. 6. Finally, whereas it would appear that the tuning problems of such a structure would be nontrivial, it has been noted[49] that interactive computer-aided optimisation of the active R ladders, using accurate amplifier models, can be successfully accomplished with readily available commercial computer programs.

3 Nonideal aspects of active R filter performance

The behaviour of active R filters deviates in a number of ways from the ideal operation described by the equations in Section 2. All of these effects can be traced to o.a. imperfections, and can in our opinion not be significantly affected by improved filter design methods. Nevertheless, this nonideal operation should be clearly understood by the designer so that, as far as possible, corrective measures can be taken.

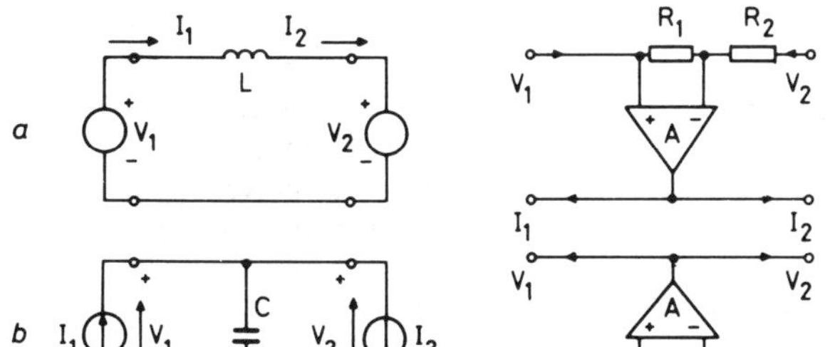

Fig. 6 *Basic active R ladder units for simulating inductors (a, e, f: $R_2/R_1 = LGB - 1$) and capacitors (b, c, d: $R_2/R_1 = CGB - 1$)*
Σ = summer, either resistive or designed with a broadband o.a.

3.1 Effect of temperature drifts

It is clear from dimensional considerations that in active R filters, as shown in eqn. 13, the pole frequency ω_0 must be proportional to $\sqrt{(GB_1 GB_2)}$, whereas the quality factor Q must depend on a ratio of gain-bandwidth products. Similarly, the filter gain must either depend on the ratio GB_1/GB_2 or be independent of GB. In practice, active R filters will be built with dual o.a.s on a common semiconductor die, so that it may safely be assumed that the o.a.s are nominally identical and their parameters track. Thus, assuming $GB_1 = GB_2 = GB$, it is apparent that gain and Q are independent of GB and that $\omega_o = kGB$, where k is a function only of resistor ratios.

GB in commercially available o.a.s is a strong function of temperature T; it is essentially proportional to T^{-1} and varies by as much as 15–20% in the range $0°C \leqslant T \leqslant 70°C$ (e.g. μA741).[63] Clearly, pole-frequency variations of this magnitude are quite unacceptable in a filter, and so compensation methods must be employed. The early procedures proposed in the literature use the temperature coefficient (t.c.) of a capacitor[64] or of a resistor[22,24,25] carefully selected to match the t.c. of GB. These methods work quite well and have the advantage of being very simple. Their drawbacks are that the compensation elements enter the signal path rather than only controlling the d.c. bias points of the o.a.s, and that matching the t.c.s of electrically dissimilar elements is required, implying an expensive selection process.

A superior method uses the dependence of GB on the bias conditions in the o.a. to counteract temperature drifts: a control bias voltage $V(T)$ is applied to the o.a.s in either an open-loop[30] or a closed-loop stabilisation system based on the tracking properties of the GB of similar amplifiers.[42,74,75] A similar scheme[73] applies $V(T)$ to voltage-dependent resistors (j.f.e.t.s) in the filter's signal path to achieve centre-frequency stabilisation. The open-loop method[30] simply generates a $V(T)$ based on a known linear approximation to the behaviour of $GB(T, V)$; it is inferior to the closed-loop approaches[42,73,74,75] in that the latter derive $V(T)$ from the actual monitored (via ω_o) variation of $GB(T, V)$ by using phase-lock[77] techniques, as will be described.

In Reference 42, an external stable oscillator signal, preferably (but not necessarily[42]) obtained from a portion of the system in which the active R filters are embedded, passes through an active R 'reference' section connected as a lowpass filter (in Reference 42, the section of Fig. 5*a* with $R_1 = \infty$ was used; V_{02} is the lowpass output). As shown in Fig. 7*a*, the external-oscillator and reference-filter signals are then brought into a precise phase relationship by the feedback system comprising the phase comparator/integrator and the bias-controlled amplifier GB of the reference filter. Hence ω_o of the reference filter is locked to the oscillator. The controlling bias $V(T)$ is then applied to the o.a.s of the signal-processing filters, resulting in filters with zero steady-state sensitivity of ω_o to step changes in ambient temperature, provided that the only significant deviation in ω_o in all filters is that contributed by GB (note that if the reference filter and all controlled filter sections are identical, this restriction is unnecessary). Using 741-type o.a.s, a 3-filter system with the following nominal ($T = 25°C$) parameters was examined: filter no. 1 and reference filter, $\omega_o = 2\pi$ (63·00 kHz), $Q = 7{\cdot}24$ and $H_B = 0{\cdot}473$; filter no. 2, $\omega_o = 2\pi$ (63·10 kHz), $Q = 32{\cdot}69$, and $H = 2{\cdot}09$; filter no. 3, $\omega_o = 2\pi$ (122·90 kHz), $Q = 44{\cdot}69$, and $H = 1{\cdot}37$. All used the section of Fig. 5*a* with $R_1 = \infty$. The results[42,74] for the pole-frequency deviation for the three filters were virtually identical over the commercial (0° – 70°C) temperature range; without stabilisation, the bounds on $\Delta\omega_o/\%$ were $+6{\cdot}6$ and $-12{\cdot}3$; with stabilisation, these bounds became $+0{\cdot}05$ and $-0{\cdot}71$. Owing to problems with temperature-dependent excess phase [see Section 3.2[42,43,74]], the quality of pole Q (and hence H) stabilisation was somewhat poorer than that of ω_o and varied for the different filters;[74] as noted elsewhere,[42,74] a system for excess phase compensation, operating simultaneously with one for ω_o stabilisation, should be possible using the same closed-loop tracking approach.

In Reference 73, results very similar to those of Reference 42 were obtained on a single filter for ω_o stabilisation with 741 type o.a.s, but with the controlling bias $V(T)$ being applied to voltage-dependent attenuators (j.f.e.t. resistors) in the signal path of the stabilised filters (Fig. 7*b*) instead of to the o.a.s; since these results were taken for filters with lower $\omega_o Q$ products (i.e. $\omega_o = 2\pi$ (50·00 kHz), $Q = 10$) than those of Reference 42 and over a more limited temperature range (30° – 60°C), the problems due to excess phase were not observed.[74]

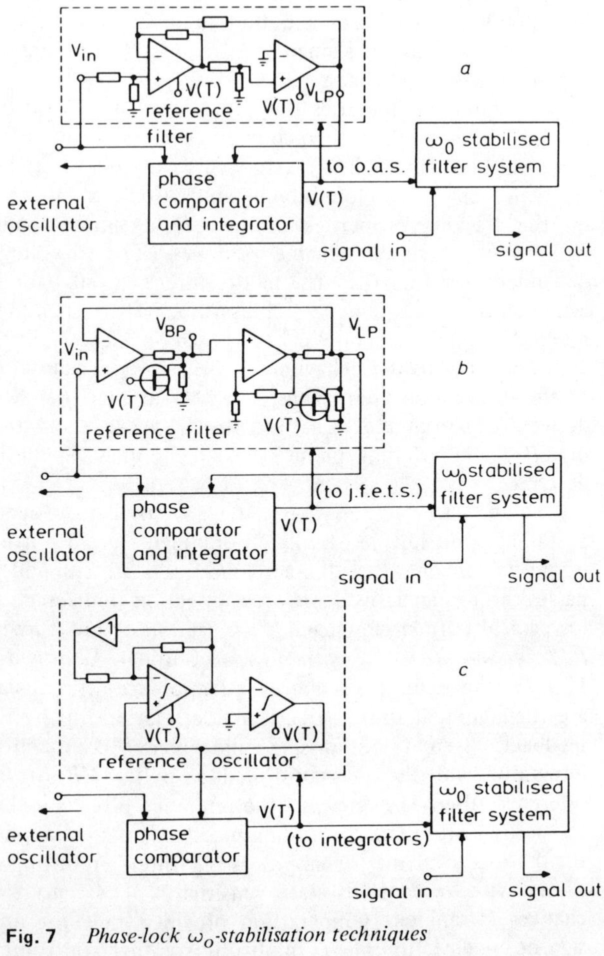

Fig. 7 *Phase-lock ω_o-stabilisation techniques*

a Brand-Schaumann-Skei[42,74]
b Rao-Sethuraman-Neelakantan[73]
c Tan-Gray[75]

In Reference 75, high-order low-frequency (< 10 kHz) filters are designed using monolithic integrators with bias-controllable gains; these integrators, using bipolar j.f.e.t. technology and including on-chip selectable integrating capacitors, can be used at low frequencies in active R circuits and in active ladder circuits.[45,47,49,61] The ability to alter the gain constant from integrator to integrator via the selectable capacitors allows for many realisations without the use of resistors (compare, for example, the active ladders discussed in Section 2.3]. The temperature-stabilisation technique for these integrators (Fig. 7*c*) is directly applicable to any active R circuit and also uses a phase-lock approach; in contrast to the circuits of Figs. 7*a* and 7*b*, however, this method locks the frequency of an oscillator made of two of the bias-controlled integrators to the external oscillator, and thereby generates the control bias for the filter integrators. For a 5th-order 8 kHz Chebÿshev lowpass filter with 0·1 dB passband ripple, it was reported[75] that a 50°C change in ambient temperature resulted in a passband deviation of 0·1 dB and a cutoff frequency variation of 1·5%; it was further reported that the cutoff frequency could be varied by ±2 kHz via external oscillator adjustment, and that a 20% variation in power-supply voltage resulted in less than 0·05 dB passband deviation.

Although the phase-lock ω_o-stabilisation techniques[42,73,74,75] are seemingly rather complicated, and are clearly suited primarily for systems of several filters, they can, as demonstrated in Reference 75, be implemented in inexpensive integrated form. They also have the advantage of stabilising filter performance not only against temperature variations, but also against drifts in power-supply voltages and aging. In addition, electronic tuning is possible as a feature of this method. The discussion of excess phase in Section 3.2 (compare also References 42, 43 and 74) will illustrate the necessity of extending such techniques to active phase compensation of amplifiers/integrators used in high-frequency filters with narrow passbands (or containing high-Q sections).

The most practical and economical solution of temperature-drift problems would be the use of temperature-compensated o.a.s. Such units are beginning to become available, e.g. National's LM 324,[65] but their performance is not quite satisfactory yet because their GB variations of approximately 3·5% in the range $0°C \leqslant T \leqslant 70°C$ are too large for highly stable filter applications.[66]

3.2 *Effects of o.a. excess phase*

The excess phase $-\omega\tau_1$, defined in eqn. 1, causes deviations in the high-frequency performance of any circuit using o.a.s.* Being concerned in particular with 2nd-order active R filters, we shall base the discussion of this problem on

*Note that $e^{s\tau} \approx [1 + s\tau]^{-1}$ for $|s\tau| \ll 1$. Hence the effects of the 'delay' τ may in an equivalent manner also be analysed by considering an amplifier model having a second, distant, pole at $-1/\tau$. In this paper, the delay approach is used because the model in eqn. 1 does correctly represent the magnitude of $A(j\omega)$, in agreement with published o.a. data,[65] whereas a second pole has some effect on $|A(j\omega)|$. Furthermore, this delay approach is preferable because (i) for filters with narrow passbands, the use of a measured delay τ provides a good 1st-order approximation to an arbitrary phase behaviour, and (ii) the delay formulation appears analytically to be more convenient than the use of an additional pole, and it provides the same results.

eqn. $9a$ with eqn. 16, and the practical assumption $A_1 = A_2 \equiv A(s)$, i.e.

$$T_1(s) = \frac{V_{01}}{V_i} \frac{[H_B \omega_o/(GB\,Q)]/A}{A^{-2} + [\omega_o/(GB\,Q)]/A + (\omega_o/GB)^2} \tag{26}$$

where the notation

$$\omega_o/GB = \sqrt{(-c_2\gamma_2)}, \quad \omega_o/(GB\,Q) = c_0 \tag{27a}$$

and

$$H_B\omega_o/(GB\,Q) = c_1 \tag{27b}$$

was introduced to retain normal filter nomenclature, and where $A(s)$ is defined in eqn. $1b$. Eqn. 26, showing clearly the dependence on amplifier gain A, is convenient for analysing the effect of a nonzero delay term τ on filter parameters.

In the A^{-1} plane, the poles of T_1 are located where

$$\frac{1}{A} = \frac{1}{2}\frac{\omega_o}{GB\,Q}\left[-1 \pm j\sqrt{(4Q^2-1)}\right] \tag{28}$$

which for $\tau = 0$, i.e. $A^{-1} = s/GB$, gives the nominal pole locations in the s-plane for the *desired* parameters ω_o and Q. To determine the actual *realised* parameters, ω_R and Q_R, and their dependence on τ, we subsitute eqn. $1b$ with $s = -\sigma + j\omega$ into eqn. 28 and equate the magnitudes, and phases, of each side of the resulting expression. With the definitions $\sqrt{(\sigma^2 + \omega^2)} = \omega_R$ and $\sigma = \omega_R/(2Q_R)$, we obtain, after some manipulation,[43,†]

$$\omega_o = \omega_R \exp[-\omega_R\tau/(2Q_R)] \tag{29a}$$

$$\sqrt{(4Q^2-1)} = \frac{\sqrt{(4Q_R^2-1)} - \tan\{\omega_R\tau\sqrt{[1-(2Q_R)^{-2}]}\}}{1 + \sqrt{(4Q_R^2-1)}\tan\{\omega_R\tau\sqrt{[1-(2Q_R)^{-2}]}\}} \tag{29b}$$

and

$$H_B = H_{BR}2Q\left\{\left[\cosh\left(\frac{\omega_R\tau}{2Q_R}\right) - \frac{\sin\omega_R\tau}{2Q}\right]^2 - \left(1 - \frac{1}{4Q^2}\right)\cos^2\omega_R\tau\right\}^{1/2} \tag{29c}$$

Eqn. 29 directly provides the designer with the predistorted values of ω_o, Q, and H_B, which, for a given τ, lead to the realised parameters ω_R, Q_R, and H_{BR} being as required. A better feeling for these relationships can be gained by assuming $2Q_R \gg \tan\omega_R\tau$ and $4Q_R^2 \gg 1$; for these usually satisfied conditions, eqn. 29 yields

$$\omega_R \approx \omega_o[1 + \omega_R\tau/(2Q_R)] \approx \omega_o \tag{30a}$$

$$Q_R \approx \frac{1}{2}\frac{\sqrt{(4Q^2-1)}}{1 - \sqrt{(4Q^2-1)}\tan\omega_R\tau} \approx \frac{Q}{1 - 2Q\omega_R\tau} \tag{30b}$$

and

$$H_{BR} \approx \frac{H_B}{1 - 2Q\omega_R\tau} \tag{30c}$$

The second approximation in eqns. $30b$ and $30c$ is valid for 'large' Q and 'small' $\omega_R\tau$. These equations show clearly that the pole frequency is approximately unaffected by τ, but that pole quality factor and gain are strongly enhanced. As an example, designing a filter with $Q = 10$ and $\omega_o = 2\pi$ (100 kHz), using μA 741 amplifiers having $\tau \approx 75$ ns at 100 kHz, results in realised values of $\omega_R \approx 2\pi$ (100 kHz) and $Q_R = 168(!)$; if the circuit was meant to have $Q_R = 10$, it should have been designed with $Q = 5{\cdot}17$.

The rapid Q enhancement for even moderate values of τ indicates also that these filters can become unstable, in contrast to eqns. 20 or 21 which seem to promise unconditional stability. Evaluation of the conditions for oscillation exactly, by setting the denominator of eqn. 26 to zero, with eqn. $1b$ and $s = j\omega_{osc}$, yields

$$\omega_{osc} = \omega_o, \quad Q = 1/(2\sin\omega_o\tau) \tag{31}$$

which shows that the maximum designable‡ quality factor Q is $1/(2\sin\omega_o\tau)$ and that the oscillation frequency equals the designed pole frequency ω_o.

We have demonstrated that, by the use of predistortion techniques or resistive tuning, the correct values of ω_o and Q of a 2nd-order block can be obtained in the presence of a delay term τ. Similarly, the gain H_B of a block can be realised correctly. This, however, does not alleviate the problems connected with high-order multiple-feedback filters alluded to in Section 2.3. The difficulties can be understood by evaluating eqn. 26 with eqn. $1b$ for $s = j\omega$, i.e.

$$T_1(j\omega) = \frac{j\omega H_B\omega_o/Q}{(\omega_o^2 - \omega^2)\cos\omega\tau + j[\omega\omega_o/Q - (\omega_o^2 + \omega^2)\sin\omega\tau]} \tag{32}$$

Denoting the function to be realised by $T_{1R}(j\omega)$, i.e.

$$T_{1R}(j\omega) = \frac{j\omega H_{BR}\omega_R/Q_R}{\omega_R^2 - \omega^2 + j\omega\omega_R/Q_R} \tag{33}$$

we can evaluate the magnitude error, $\epsilon = |T_1(j\omega)/T_{1R}(j\omega)| - 1$, and the phase error, $\Delta\phi = \phi_1 - \phi_{1R}$, where ϕ_1 is the angle of $T_1(j\omega)$, as follows:

$$\epsilon = \left\{\frac{(\omega_o/\omega_R)^2(Q_R/Q)^2(H_B/H_{BR})^2[(\omega_R^2 - \omega^2)^2 + (\omega\omega_R/Q_R)^2]}{(\omega_o^2 - \omega^2)^2\cos^2\omega\tau + [\omega\omega_o/Q - (\omega_o^2 + \omega^2)\sin\omega\tau]^2}\right\}^{1/2} - 1 \tag{34a}$$

$$\Delta\phi = \arctan\frac{\omega\omega_R/Q_R}{\omega_R^2 - \omega^2} - \arctan\left[\frac{\omega\omega_o/Q}{\omega_o^2 - \omega^2}\frac{1}{\cos\omega\tau} - \frac{\omega_o^2 + \omega^2}{\omega_o^2 - \omega^2}\tan\omega\tau\right] \tag{34b}$$

In eqns. 32 and 34, ω_o, Q and H_B are determined according to eqn. 29 to approximate the behaviour of eqn. 33 as dictated by the parameters ω_R, Q_R and H_{BR}. Note that, with eqn. 29 or eqn. 30, it follows from eqns. 32 – 34 that $T_1 \equiv T_{1R}$, $\epsilon = 0$ and $\Delta\phi = 0$ if $\tau = 0$.

For example, for $\omega_R = 2\pi(100\text{ kHz})$ and $\tau = 75$ ns, one finds, for frequencies in the range $\omega_R(1 \pm 0{\cdot}1)$ and for Q_R values between 5 and 400, that from eqn. $34a$ $|\epsilon| < 4 \times 10^{-7}$. For these same conditions, $\Delta\phi$ of eqn. $34b$ is represented to within 0·5% by the term $-\omega\tau$, which yields

†Eqn. $29c$ is arrived at by comparing the expressions obtained by evaluating $|T_1(j\omega)|$ of eqn. 26 with eqn. $1b$, once for $\tau \neq 0$, H_B, ω_o and Q, and once for $\tau = 0$, H_{BR}, ω_R and Q_R from eqns. $29a$ and $29b$

‡Note that this does not imply that the realisable quality factor Q_R cannot be larger than $1/(2\sin\omega_o\tau)$

an excess phase of $-2{\cdot}7°$ at ω_R. The magnitude error is clearly insignificant; the phase error, although hardly noticeable in a single 2nd-order section, results in a characteristic slanting of the desired level passband characteristic (Fig. 8) of multiple feedback (m.f.) bandpass filters. As expected, this behaviour is not observed in cascaded filters, where no feedback exists around sections and therefore no phase errors affect the overall magnitude characteristic. It has been shown experimentally[58] that in some cases this slant in the transfer characteristic of m.f. filters can be corrected by tuning, but a systematic procedure has not yet been developed.

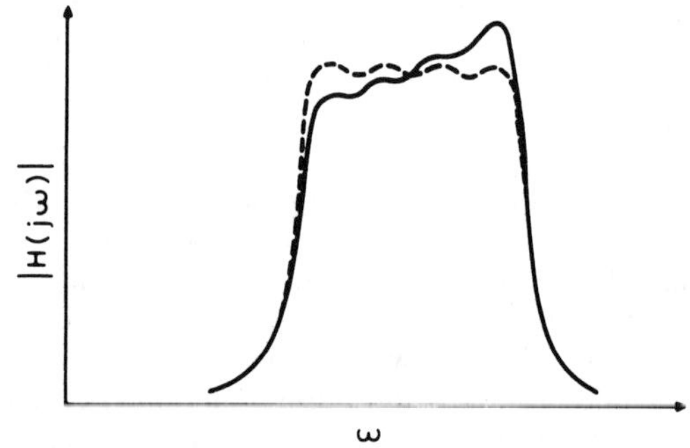

Fig. 8 *Experimentally observed deviation of untuned magnitude characteristic of 8th-order Chebyshev multiple-loop feedback active R filter*

Dashed line: ideal characteristic

In an alternative method for eliminating the transfer function slant of m.f. filters, a leading phase shift ($+2{\cdot}7°$ at ω_R is needed in the above example) can be introduced[76] in 'cascade' with each section by shunting a section's series input resistor (R_1 or R_5 in Fig. 5, assuming that $R_1 \neq 0$ implies $R_5 = 5$ and vice versa) with a small capacitor of, typically, a few picofarads for practical circuits with $\omega_R \geqslant 2\pi(100\,\text{kHz})$. This compensation of active R m.f. circuits is analogous to the one used in Reference 78 on low-frequency lowpass active RC filters. It should be pointed out that these approaches for eliminating the effects of $\tau \neq 0$, if necessary,[||] may be completely satisfactory only for a nearly constant thermal environment, owing to the temperature dependence of τ.[42,73,74] Preliminary efforts at τ stabilisation indicate that techniques similar to those[42,73,75] discussed in Section 3.1 can be used for electronically controlling the excess phase response of the amplifiers. Whereas such an approach has the potential advantage of eliminating both the need for individual section predistortion and that for overall filter retuning or phase-lead compensation, the only practical implementation seems to be through the use of specially designed amplifiers.

3.3 Dynamic range

At high frequencies, active filters tend to have a limited dynamic range, owing to amplifier restrictions and potentially high device noise. In addition to resulting in signal distortion, violating the dynamic range limits of active R filters can produce serious distortion of the filter's transfer characteristic, such as is illustrated in Fig. 10 for a second-order bandpass. It is seen for this filter that even small violations of the linearity constraints will cause an apparent decrease in ω_o, coupled with a marked asymmetry in the 3 dB points. A further increase in signal level will lead to the well known 'jump' phenomenon.[67]

If such distortion occurs in a high-order filter where feedback is used around the sections, the total filter may become unstable. Thus it is very important to obey the restrictions placed on the maximum signal level by a particular filter topology.

Dynamic range will be defined by

$$DR = 20\log_{10}\frac{V_{OM}}{\sqrt{2}E_{ON}}\ \text{dB} \tag{35}$$

where, in the frequency range of interest, V_{OM} is the maximum distortionless output voltage and E_{ON} is total r.m.s. noise voltage measured at the filter output with all inputs grounded.

The maximum distortionless output signal at low frequencies is set by o.a. output voltage and/or current limitations; at high frequencies, the limits are given by the o.a's slew rate, which is generally defined simply as the maximum permissible time rate-of-change of o.a. output voltage. Since the usual operating frequencies for active R filters are such that the total gain through the amplifiers is very low (e.g. an o.a. with $GB = 2\pi\,(1\,\text{MHz})$ has a gain of only 10 at 100 kHz), the behaviour of the nonlinearity of the o.a.'s input differential stage[77] *before* it saturates becomes very important; hence, it should be noted that the slew rate (SR) stated on an o.a.'s data sheet is obtained from pulse measurements, and that relating this value to the maximum slope of a sinusoid $A\sin\omega t$, i.e. $SR = A\omega$,[55] usually leads to grossly optimistic assumptions about the maximum undistorted output signal V_m that an o.a. can process at the frequency ω. Instead, $V_m(\omega)$ should be determined experimentally for a given type of o.a. under sinusoidal operation; usually one finds $V_m(\omega) < SR/\omega$. Once $V_m(\omega)$ is known, it must be observed that *all* o.a.s, i.e. both the lowpass and the bandpass outputs in Figs. 3 and 5, must satisfy the constraint

$$\max\left[V_{in}|T_k(j\omega)|\right] = \max|V_{ok}(j\omega)| \leqslant V_m(\omega)$$

$$\text{for } k = 1,2 \tag{36}$$

The upper signal level at the filter output, $V_{OM}(\omega)$, needed for eqn. 35, is either determined directly from eqn. 36 or, for the more general sections that use a summer, from eqns. 36 and 25.

If the linearity constraints are exceeded, distortion of the output signal and the transfer characteristic must be expected, as pointed out earlier. It is also worth noting that, for many 2nd-order sections, owing to the filtering action of the network even during nonlinear operation, the wave shape of the output signal (or its harmonic content) is not nearly as sensitive an indicator of slew-rate limiting as is the observation of the transfer-function magnitude (frequency response), say, on a continuously swept spectrum analyser.

For the evaluation of the noise voltage E_{ON} in eqn. 35, the resistors and o.a.s are replaced by their respective noise models (Fig. 9), and the contribution of each noise source to the output noise voltage spectral density $e_{ON}(\omega)$ is calculated under the assumption that all souces are uncorrelated. E_{ON} is obtained from

$$E_{ON} = \left[\int_{\Omega_N} e_{ON}^2(\omega)\,d\omega\right]^{1/2} \tag{37}$$

[||]Complete elimination of τ variations may not be required in some high-order multiloop feedback active filters[54] because of their low sensitivity to Q variations[66]

where Ω_N is defined as the noise passband, or the range of frequencies to which the stage following the section is sensitive, and where

$$e_{ON}^2(\omega) = \sum_i e_i^2(\omega)|t_{oi}(j\omega)|^2 + \sum_j i_j^2(\omega)|z_{oj}(j\omega)|^2$$

is defined as the total squared output noise voltage spectral density. The quantities e_i, i_j, t_{oi}, and z_{oj}, respectively, are the ith noise voltage and jth noise current source (Fig. 9), the voltage transfer function to the output from the ith voltage source, and the transfer impedance to the output from the jth current source.

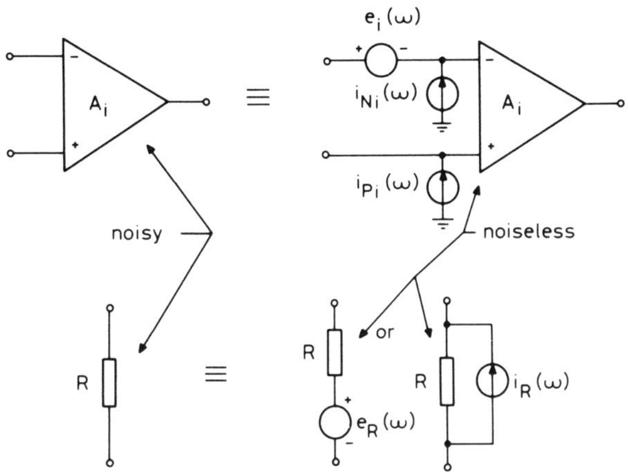

Fig. 9 *Noise models of active R elements*

Using this method of analysis, E_{ON} is readily found[68] for the general 1st-order section with $R_o = \infty$, but including the summer of Fig. 4 (with $V_{02} = 0$),

$$E_{ON} \approx \left[\frac{\pi}{2} k_2^2 \left(\frac{GB}{\omega_{3db}}\right)^2 N_1 + \frac{2}{\pi} kTR_s\Omega_N\right]^{1/2} \qquad (38a)$$

and for the general 2nd-order section, again with the summer

$$E_{ON} \approx \left(\frac{\pi}{2}\omega_o Q\left[\left\{k_2^2\left(\frac{GB}{\omega_o}\right)^2 + \left[k_3(\alpha_6 - \alpha_2) - k_2(\alpha_7 - \alpha_3)\right]^2\left(\frac{GB}{\omega_o}\right)^4\right\}N_1 + \left\{k_3^2\left(\frac{GB}{\omega_o}\right)^2 + \left[k_2(a_6 - a_2) - k_3(a_7 - a_3)\right]^2\left(\frac{GB}{\omega_o}\right)^4\right\}N_2\right] + \frac{2}{\pi} kTR_s\Omega_N\right)^{1/2} \qquad (38b)$$

and where

$$N_i = e_i^2 + i_{Ni}{}^2 R_{Ni}^2 + i_{Pi}^2 R_{Pi}^2 + \frac{2}{\pi} kT(R_{Ni} + R_{Pi})$$

$$(i = 1, 2)$$

$$R_{N1} = 1/\left(\sum_{i=1}^{4} G_i\right) \qquad R_{P1} = 1/\left(\sum_{i=5}^{8} G_i\right)$$

$$R_{N2} = 1/\left(\sum_{i=1}^{4} g_i\right) \qquad R_{P2} = 1/\left(\sum_{i=5}^{8} g_i\right)$$

$$R_s = 1/\left(\sum_{i=1}^{4} G_{si}\right)$$

and where e_i, i_{Ni}, and i_{Pi} are defined in Fig. 9, a_i and α_i in eqn. 5 and k_i is as for eqn. 24. Eqn. 38 is obtained under the assumptions that (*a*) resistors contribute only Johnson (thermal) noise, i.e. $e_R^2 = (2/\pi)kTR$, and (*b*) o.a.s contribute only 'white' noise over the frequencies of interest for most high-frequency active R filters, i.e e_{OA}^2 and i_{OA}^2 are independent of frequency. For situations in which these assumptions are not sufficiently accurate, the reader is referred to the literature[68] for techniques appropriate for evaluating eqn. 37. Eqns. 38 are approximate because of the choice of a finite Ω_N, necessitated by the requirement of a finite noise contribution of R_s. For the cases requiring no summer, ($R_s = 0$, and either $k_2 = 1$ and $k_3 = 0$ or $k_2 = 0$ and $k_3 = 1$), eqn. 38 gives the exact result obtained for $\Omega_N \to \infty$ under the assumption that all noise sources are 'white'. In practice, the interval Ω_N equals $(\omega_H - \omega_L)$ where ω_L is the frequency above which the o.a.'s noise source spectra are sufficiently flat and $\omega_H \geqslant 5(\omega_o$ or $\omega_{3db})$;[68] generally, the summer contribution will be negligible.

With V_{OM} and E_{ON} determined according to eqns. 36 and 38, the dynamic range of a particular section can be evaluated from eqn. 35. An example of d.r. calculation and measurement is given in Section 4.

4 Experimental results

To demonstrate the practical behaviour of an active R filter relative to the preceding discussion, we examine the circuit of Fig. 5*a* with $R_1 = \infty$. A dual o.a. (Motorola MC 1458S[69]) with specified characteristics of SR = 10 V/μs and GB = 1·0 MHz is used; measurements on a group of these units indicated a nominal GB of 1·36 MHz, an input noise voltage spectral density $e^2 = 2{\cdot}5 \times 10^{-16}\,\mathrm{V^2/Hz}$, and an input noise current spectral density $i^2 = 1 \times 10^{-24}\,\mathrm{A^2/Hz}$. For the particular dual unit used in the filter, the measured GB was 1·2 MHz, and the measured τ, at 300 kHz, was found to be 73·7 ns.

The experimental filter parameters to be *realised* were $\omega_R = 2\pi$ (300 kHz), $Q_R = 100$ and $H_{BR} = 1{\cdot}00$; to achieve this with the above value of τ, the necessary *design* parameters according to eqn. 29 are $\omega_o = 2\pi$ (300 kHz), Q = 3·486 and H_B = 0·035. Using the appropriate (Fig. 5*a*, $G_1 = 0$) design equations from Table 1, the following resistor values were chosen: R_7 = 1·0 kΩ (± 1%), R_3 = 12·944 kΩ, $R_5 = r_6$ = 10 kΩ (± 1%), R_8 = 25·1 Ω and r_8 = 722 Ω; using these values, the measured filter parameters were $\omega_R = 2\pi$(296·92 kHz), Q_R = 119·73, and H_{BR} = 0·997. The most significant error, that in Q_R, can be at least partially attributed to the extreme sensitivity of Q_R to Q and τ, as shown by the second expression in eqn. 30*b*. The values of $\omega_R = 2\pi$(299 kHz), Q_R = 96·45 and H_{BR} = 1·00 were attained by adjusting R_3, R_8 and r_8 to R_3 = 12·582 kΩ, R_8 = 31·7 Ω, and r_8 = 736·2 Ω. These values of ω_R, Q_R and H_{BR} are used as the parameters of the ideal curve in Fig. 10, in which the response of this adjusted filter is displayed for various input signal levels.

In calculating dynamic range DR, it is clear from Fig. 10 that the maximum undistorted signal that the filter can process is 70 mV r.m.s.; this is a dramatic reduction from the 23·6 V r.m.s. predicted is the specified SR! (Of course, with an o.a. power supply of ± 15 V, the maximum attainable signal at any frequency is approximately 10 V r.m.s.[69]) The broadband noise is calculated with eqn. 38*b*, using the parameters and resistor values of the adjusted filter along with the group measurements of MC1458S GB and noise spectra; the result of 2·14 mV r.m.s. compares well with the measured value of 2·09 mV r.m.s. (Hewlett Packard 400E a.c. voltmeter with Marconi TK 2374 zero-loss probe, −3 dB bandwidth ≈ 10 MHz, rescaled for noise measurements). In addition, the noise in the filter's −3 dB passband was calculated using eqn. 37 to be 1·52 mV r.m.s. and measured at 1·45 mV r.m.s. (Hewlett Packard 3590A wave analyser with Marconi TK 2374 zero-loss probe; the bandwidth = 3100 Hz, rescaled for noise measurements). Thus, with the definition of eqn. 35, the measured *DR* is 30·5 dB compared to a calculated 30·3 dB.*

It should be noted that the effects of o.a. output impedance (Z_{OUT})† and common-mode rejection ratio (*CMRR*)‡ have largely been ignored in the preceding theoretical and experimental discussions; in particular, at high frequencies where Z_{OUT} may deviate significantly from being resistive and where *CMRR* rapidly decreases (see, for example, the characteristics for μA741[63]), the perturbations introducted into the filter behaviour could be severe.[32] It is the authors' experience that, at the present stage of development of active *R* filters, the most efficient approach to compensating for these effects is that of post-design tuning (assuming that loading precautions have been taken); for example, in the experimental filter above, the o.a.'s specified Z_{OUT} and *CMRR* are 75 Ω and 90 dB, respectively, at low frequencies;[69] still, only minor tuning is required to compensate for errors due to *all* o.a. imperfections at high frequencies. As higher operating frequencies are considered, however, a more rigorous approach to the Z_{OUT} problem will doubtless be necessary.

*Clearly, this dynamic range is limited by the choice of commercial o.a.s that are not optimised for large differential input signal swing

†Consider the case of $Z_{OUT} = R_{OUT}$,[32] with R_{OUT} being much smaller than any resistors connected to o.a. outputs. This is the only practical situation, and it can usually be achieved with little difficulty. The problems are thus due only to external loads attached to the o.a. outputs. It is clear (from, for example, Fig. 5) that a resistive load R_L results in an effective decrease in *GB* of the affected o.a. by a factor $R_L/(R_{OUT} + R_L)$; the resultant parameter deviations are then determined by straightforward sensitivity methods, as in Section 2.2 (see eqn. 14). In the case of a *small* capacitive load C_L such as a short coaxial cable, the filter behaves as if the excess phase (Section 3.2) of the affected o.a. were increased; for example, in 2nd-order sections where excess phase behaviour is described by eqn. 30, the apparent pole quality factor is given by $Q_R \approx Q/(1 - Q\omega_R\tau_T)$, where ω_R and Q are as described in Section 3.2 and $\tau_T \simeq 2\tau + R_{OUT}C_L$, τ being the nominal 'delay' associated with an o.a. For large enough C_L, of course, the filter will oscillate (compare with eqn. 31).

‡Problems of finite *CMRR* arise only in active *R* circuits using differential-input amplifiers in the differential mode; in practice, this also means that in most practical active *R* sections such problems arise in only one amplifier per section (e.g. A_1 in Fig. 5). Consider the circuit in Fig. 5, and let $V_{01} = A_D(V_{1p} - V_{1n}) + A_C(V_{1p} + V_{1n})/2$ and define $CMRR = A_D/A_C$, where A_D is the differential and A_C the common mode gain, and where V_{1p} and V_{1n} are the positive and negative inputs, respectively, of A_1; letting $A_D(s) = A_2(s) = GB/s$, direct analysis yields

$$\frac{V_{01}}{V_i} = \frac{sGB[K_2G_5/(G_5+G_8) - K_1G_1/(G_1+G_2+G_3)]}{s^2 + sK_1GB\,G_3/(G_1+G_2+G_3) + GB^2G_2g_6K_1/[(G_1+G_2+G_3)(g_6+g_8)]}$$

for the section of Fig. 5*a*, and

$$\frac{V_{01}}{V_i} = \frac{sGB[K_2G_5/(G_5+G_6+G_8) - K_1G_1/(G_1+G_3+G_4)]}{s^2 + sK_1GB\,G_3/(G_1+G_3+G_4) + GB^2G_6g_2K_2/[(G_5+G_6+G_8)(g_2+g_4)]}$$

for that of Fig. 5*b*, where

$$K_1 = 1 - 1/(2\,CMRR) \text{ and } K_2 = 1 + 1/(2\,CMRR).$$

Even without considering the phase of A_C, it is clear from these expressions that *CMRR* must be extremely small for a significant shift in circuit parameters to occur. In such cases, the designer can consider those active *R* filters which use only single-input amplifiers.[27,36,50]

5 Summary

This paper has discussed both the theoretical and practical aspects of active *R* filters with the intention of introducing the reader to this recently developed area of analogue filtering; a design example has been presented to illustrate the ease of design and quality of performance of high-frequency filters using this technique. Numerous further experimental results for both low- and high-order active *R* circuits are available in the cited literature.

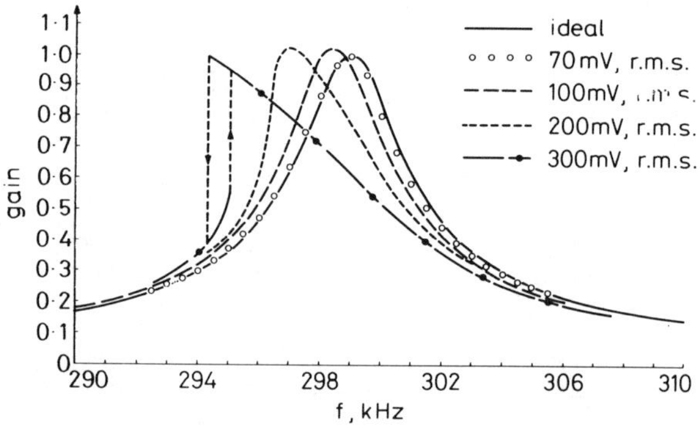

Fig. 10 *Measured magnitude response of 2nd-order bandpass filter subjected to various input signal levels*

Ideal parameters: $\omega_R = 2\pi(299\text{ kHz})$, $Q_R = 96{\cdot}5$, $H = 1{\cdot}00$. For the largest excitation (300 mV r.m.s.) the 'jump' phenomenon is apparent

Throughout, the various limitations on filter performance imposed by amplifier nonidealities have been noted; where possible, compensation techniques have been discussed or demonstrated, notably for the amplifier's excess phase and its temperature-dependent dynamic response, which is probably the single most important practical problem with these filters. It should be emphasised that the economics of the currently most satisfactory compensation schemes ('phase-lock' techniques, Section 3.1) are such that high-order systems are the most suitable choices for active *R* synthesis. However, use of integration techniques and the development of amplifiers with improved temperature-stabilised dynamic response may soon result in low-order high-performance active *R* systems. Based upon the quality of the reported $GB(\omega_o)$ stabilisation techniques (Section 3.1) and the prospect of extending these techniques to $\tau(Q)$ stabilisation through specially-designed amplifiers, it is the authors' opinion that this synthesis method has the potential for evolving into the most suitable form of analogue active filtering for high frequencies (up to about 10 MHz).

In any 2nd-order active *R* filter, the quantities ω_o, Q and H_B, as noted in the beginning of Section 3.1, must always have the same dependence on amplifier parameters, and, therefore, the same limitations. Thus it is the authors' opinion that additional active *R* topologies, which might not be obtainable by the approach presented in Section 2.2, will not have any performance advantages over the circuits discussed in this paper. Rather than searching for new topologies, it appears, therefore, to be more profitable to concentrate further research related to active *R* filters on the development of more suitable amplifiers coupled with economical techniques for compensating amplifier imperfections. From Section 3 it should be clear to the reader that, whereas commercial o.a.s may be satisfactory for some active *R* applications, the perfection of this synthesis method will best be achieved with custom amplifiers as done in Reference 75 for low frequencies.

By realistically portraying both the capabilities and problems of these filters, the authors hope to encourage those interested in high-frequency signal processing with integrated circuits to seriously consider and possibly further develop this unique approach.

6 Acknowledgment

This work was supported in part by US National Science Foundation Grant ENG 7611218 and by a grant from the Computer Center of the University of Minnesota.

7 References

1 FAULKNER, E.A., and GRIMBLEBY, J.B.: 'Active filters and gain-bandwidth product', *Electron. Lett.*, 1970, **6**, (17), pp. 549–550

2 BUDAK, A., and PETRELA, D.M.: 'Frequency limitations of active filters using operational amplifiers', *IEEE Trans.*, 1972, **CT-19**, pp. 322–328

3 SCHWARTZ, E.: 'On the stability of resistance networks with one operational amplifier', *AEÜ*, 1974, **28**, pp. 43–49

4 HAMILTON, T.A., and SEDRA, A.: 'Some new configurations for active filters', *IEEE Trans.*, 1972, **CT-19**, pp. 25–33

5 SEDRA, A.: 'Generation and classification of single-amplifier filters', *Int. J. Circuit Theory & Appl.* 1974, **2**, pp. 51–64

6 SEDRA, A., and ESPINOZA, J.L.: 'Sensitivity and frequency limitations of biquadratic active filters', *IEEE Trans.*, 1975, **CAS-22**, pp. 122–130

7 SRINIVASAGOPALAN, R., and MARTENS, G.O.: 'A comparison of a class of active filters with respect to the operational amplifier gain-bandwidth product', *ibid.*, 1974, **CAS-21**, pp. 377–381

8 MITRA, S.K., BHUMIRATANA, A., and SHANTA PAI, B.H.: 'A two-step procedure for designing active filters using frequency-dependent amplifiers' Proceedings of the 17th Midwest symposium on circuits and systems, 1974

9 SODERSTRAND, M.A., and SZCZUPAK, J.: 'Active *RC* filter design using compensated operational amplifiers', *ibid.*, 1974

10 ALLEN, P.E.: 'Minimising the effect of gain-bandwidth on active filters'. Proceedings of the 8th Asilomar conference on circuits, systems, and computers, 1974, pp. 647–652

11 ADAMS, D.K., and HO, R.Y.C.: 'Active filters for UHF and microwave frequencies', *IEEE Trans.*, 1969, **MTT-17**, pp. 662–670

12 SNYDER, R.V., JUN., and BOZARTH, D.L.: 'Analysis and design of a microwave transistor active filter', *ibid.*, 1970, **MTT-18**, pp. 2–9

13 CAPPARELLI, F., and LIBERATORE, A.: 'Active bandpass network with only resistors as passive elements', *Electron. Lett.*, 1972, **8**, (2), pp. 43–44

14 CAPPARELLI, F., and LIBERATORE, A.: 'An inductorless and capacitorless allpass network', *Trans. Inst. Electron. & Commun. Eng. Japan*, 1972, **55-A**, pp. 304–365

15 CAPPARELLI, F., and LIBERATORE. A.: 'A technique to design active filters without inductors as well as capacitors' Presented at the summer school of circuit theory, Tale, Czechoslovakia, 1971

16 CAPPARELLI, F., and LIBERATORE, A.: 'Sulla possibilità di attuare amplificatori selettivi utilizzando lo sfasamento intrinseco dei transistori', *Alta Freq.*, 1972, **41**, pp. 158–163

17 PAPHITIS, G.A., and MURATA, T.: 'Active *LC*-less bandpass filter using transistor lowpass characteristic' Proceedings of the 10th Allerton conference on circuit and system theory, 1972, pp. 685–691

18 BERMAN, B.D., and NEWCOMB, R.W.: 'Transistor-resistor synthesis of voltage transfer functions', *IEEE Trans.*, 1973, **CT-20**, pp. 591–593

19 ALLEN, P.E., and MEANS, J.A.: 'Inductor simulation derived from an amplifier rolloff characteristic', *ibid.*, 1972, **CT-19**, pp. 395–397

20 SOLIMAN, A., and FAWZY, M.: 'Active *R* simulation of an inductor and a series resonator' Proceedings of the 20th Midwest symposium on circuits and systems, 1977, pp. 348–352

21 RADHAKRISHNA RAO, K., and SRINIVASAN, S.: 'Integrable VCOs using the operational amplifier pole', *AEÜ*, 1974, **28**, pp. 326–328

22 RADHAKRISHNA RAO, K., and SRINIVASAN, S.: 'Voltage controlled oscillators (VCOs) for integrated circuits', *ibid.*, 1975, pp. 374–378

23 RADHAKRISHNA RAO, K., and SRINIVASAN, S.: 'A bandpass filter using the operational amplifier pole', *IEEE J. Solid-State Circuits*, 1973, **SC-8**, pp. 245–246

24 RADHAKRISHNA RAO, K., and SRINIVASAN, S.: 'Low-sensitivity active filters using the operational amplifier pole', *IEEE Trans.*, 1974, **CAS-21**, pp. 260–262

25 RADHAKRISHNA RAO, K., and SRINIVASAN, S.: 'A high-Q temperature insensitive bandpass filter using the operational amplifier pole', *Proc. IEEE*, 1974, **62**, pp. 1713–1714

26 SCHAUMANN, R.: 'Low-sensitivity, high-frequency, tunable active filter without external capacitors' Proceedings of the IEEE International symposium on circuits and systems, 1974, pp. 438–442. Also *IEEE Trans.*, 1975, **CAS-22**, pp. 39–44

27 SODERSTRAND, M.A.: 'Active *R* filters: active filters using only resistors and amplifiers' Proceedings of the 8th Asilomar conference on circuits, sytems and computers, 1974, pp. 675–681

28 SWAMY, M.N.S., and SANKARA RAO, K.: 'A tunable filter using amplifier pole', *Proc. IEEE*, 1975, **63**, pp. 197–199

29 SCHAUMANN, R.: 'A note on state-variable-type active *R* filters'. Proceedings of the 18th Midwest symposium on circuits and systems, 1975, pp. 578–602

30 BRAND, J.R., and SCHAUMANN, R.: 'A temperature-compensated, high-frequency, tunable active-R filter using programmable operational amplifiers', *ibid.*, 1975, pp. 603–607

31 SCHAUMANN, R.: 'On the design of active filters using only resistors and voltage amplifiers', *AEÜ*, 1976, **30**, pp. 245–252

32 BRAND, J.R., and SCHAUMANN, R.: 'The effects of non-zero output resistance and finite common-mode rejection-ratio of operational amplifiers on the performance of active *R* filters'. Proceedings of the 19th Midwest symposium on circuits and systems, 1976, pp. 196–201

33 MITRA, A.K., and AATRE, V.K.: 'Low sensitivity high frequency active *R* filters', *IEEE Trans.*, 1976, **CAS-23**, pp. 670–676

34 MITRA, A.K., and AATRE, V.K.: 'Low sensitivity bandpass filter using the operational-amplifier pole', *Electron. Lett.*, 1976, **12**, (9), pp. 226–227

35 SRINIVASAN, S.: 'Synthesis of transfer functions using the operational amplifier pole', *Int. J. Electron.*, 1976, **40**, pp. 5–13

36 SODERSTRAND, M.A.: 'Design of active *R* filters using only resistors and operational amplifiers', *ibid.*, 1976, **40**, pp. 417–432

37 TALKHAN, E.A., IBRAHIM, M.M., and NOUH, A.S.: 'An integratable tunable active bandpass filter'. Proceedings of the international symposium on circuits and systems, 1977, pp. 340–343

38 KIM, H.K., RA, J.B., and PARK, S.B.: 'An active biquadratic building block without external capacitors', *ibid.*, 1977, pp. 317–320

39 SOLIMAN, A.M., and FAWZY, M.: 'A bandpass filter using the operational amplifier pole', *IEEE J. Solid-State Circuits*, 1977, **SC-12**, pp. 429–430

40 HUERTAS, J.L., ACHA, J.I., and TRUJILLO, M.: 'On the nonlinear behaviour of bandpass filters which use the operational amplifier pole: a comparative study of different configurations'. Proceedings of the 20th Midwest symposium on circuits and systems, 1977, pp. 316–320

41 SOLIMAN, A.M., and FAWZY, M.: 'A bandpass and a highpass active *R* filter', *ibid.*, 1977, pp. 321–323

42 BRAND, J.R., SCHAUMANN, R., and SKEI, E.M.: 'Temperature stabilized active *R* filters', *ibid.*, 1977, pp. 295–300

43 BRAND, J.R., and SCHAUMANN, R.: 'Effects of excess amplifier phase on active *R* filter parameters', *ibid.*, 1977, pp. 301–305

44 CHOI, T.C., and SCHAUMANN, R.: 'The effects of parasitic capacitances in high-frequency active filters', *ibid.*, 1977, pp. 306–310

45 SODERSTRAND, M.A.: 'Lowpass active *R* ladder structures'. Proceedings of the 18th Midwest symposium on circuits and systems, 1975, pp. 594–597

46 LAKER, K.R., SCHAUMANN, R., and BRAND, J.R.: 'Multiple-loop feedback active-*R* filters'. Proceedings of the international symposium on circuits and systems, 1976, pp. 279–282

47 SODERSTRAND, M.A.: 'Design of high-frequency, high-order active-*R* ladders', *ibid.*, 1976, pp. 275–278

48 LAKER, K.R., SCHAUMANN, R., and BRAND, J.R.: 'A comparison of techniques for realizing high-order active *R* filters'. Proceedings of the 19th Midwest symposium on circuits and systems, 1976, pp. 190–195

49 LEE, K.L., and SODERSTRAND, M.A.: 'Computer-aided design of active *R* ladders'. Proceedings of the 20th Midwest symposium on circuit and systems, 1977, pp. 611–616

50 SODERSTRAND, M.A.: 'A simple synthesis technique for active *R* filters'. Proceedings of the IEEE international symposium on circuits and systems, 1975, pp. 261–265

51 NAIMPALLY, S.V.: 'An active *R* filter using CMOS transistor arrays', *Int. J. Electron.*, 1976, **40**, pp. 513–516

52 SODERSTRAND, M.A.: 'An improved CMOS active *R* filter', *Proc. IEEE*, 1977, **65**, pp. 1204–1206

53 NEWCOMB, R.W.: 'Linear multiport synthesis' (McGraw-Hill, New York, 1966)

54 LAKER, K.R., and GHAUSI, M.S.: 'A comparison of active, muliple-loop feedback techniques for realizing high-order bandpass filters', *IEEE Trans.*, 1974, **CAS-21**, pp. 774–783

55 SOLOMON, J.E.: 'The monolithic op amp: a tutorial study', *IEEE J. Solid-State Circuits*, 1974, **SC-9**, pp. 314–332

56 LÜDER, E.: 'Optimization of the dynamic range and the noise distance of *RC*-active filters by dynamic programming', *Int. J. Circuit Theory & Appl.*, 1975, **3**, pp. 365–370

57 SCHAUMANN, R., KINGHORN, W.A., and LAKER, K.R.: 'Minimising signal distortion in f.l.f. active filters', *Electron. Lett.*, 1976, **12**, (9), pp. 211–213

58 SCHAUMANN, R., CHALSTROM, R.E., and LAKER, K.R.: 'Maximizing signal level capabilities of leapfrog and modified leapfrog active filters'. Proceedings of the 10th Asilomar conference on circuits, systems and computers, 1976, pp. 341-345

59 SCHAUMANN, R., CHALSTROM, R.E., and LAKER, K.R.: 'Optimisation of sensitivity and dynamic range of i.f.l.f. active filters', *Electron. Lett.*, 1977, **13**, pp. 367–368

60 SCHAUMANN, R., and CHALSTROM, R.E.: 'The dynamic range properties of high-order active bandpass filters'. Proceedings of the IEEE international symposium on circuits and systems, 1978, pp. 146–150

61 BRACKETT, P.O., and SEDRA, A.S.: 'Direct SFG simulation of *LC* ladder networks with applications to active filter design', *IEEE Trans.*, 1976, **CAS-23**, pp. 61–67

62 ZVEREV, A.I.: 'Handbook of filter synthesis' (Wiley, 1967)

63 'Linear integrated circuits data catalog'. Fairchild Semiconductor, Feb. 1973

64 MEANS, D.R.: 'Compensation of temperature dependence of operational-amplifier gain', *IEEE J. Solid-State Circuits*, 1972, **SC-7**, pp. 507–509

65 'Linear integrated circuits data handbook'. National Semiconductor, Feb. 1975

66 LAKER, K.R., SCHAUMANN, R., BRAND, J.R., and GHAUSI, M.S.: 'Statistical multiparameter sensitivity measures which include effects due to temperature variations'. Proceedings of the IEEE international symposium on circuits and systems, 1977, pp. 454–457

67 TRUXAL, J.G.: 'Automatic feedback control systems synthesis' (McGraw-Hill, New York, 1955), chap. 10

68 TROFIMENKOFF, F.N., TRELEAVEN, D.H., and BRUTON, L.T.: 'Noise performance of RC-active quadratic filter sections', *IEEE Trans.*, 1973, **CT-20**, pp. 524–532

69 'Data sheet on MC 1558S/MC1458S'. Motorola Semiconductors DS9367, 1976

70 KIM, H.K., and RA, J.B.: 'An active biquadratic building block without external capacitors', *IEEE Trans.*, 1977, **CAS-24**, pp. 689–694

71 BHATTACHARYYA, B.B., and NATARAJAN, S.: 'A new continuously tunable sinusoidal oscillator without external capacitors', *Proc. IEEE*, 1977, **65**, pp. 1726–1727

72 NANDI, R.: 'Grounded-capacitor bandpass filter using the operational-amplifier pole', *Electron. Lett.*, 1977, **13**, pp. 713–715

73 RAO, K.R., SETHURAMAN, V., and NEELAKANTAN, P.K.: 'A novel "Follow the master" filter', *Proc. IEEE*, 1977, **65**, pp. 1725–1726

74 BRAND, J.R., and SCHAUMANN, R.: 'Comments on "A novel 'Follow the master' filter"', *ibid.*, 1978, **66**, pp. 590–592

75 TAN, K.S., and GRAY, P.R.: 'High-order monolithic analog filters using bipolar/JFET technology'. Digest of the IEEE international solid-state circuits conference, 1978, pp. 80–81

76 SCHAUMANN, R., and FARRELL, S.: 'The effect of excess phase and its compensation on the design of high-order multiple feedback active filters'. Proceedings of the 21st Midwest symposium on circuits and systems, 1978

77 GRAY, P.R., and MEYER, R.G.: 'Analysis and design of analog integrated circuits' (Wiley, New York, 1977)

78 BRUTON, L.T., and SALAMA, A.I.A.: 'Frequency limitations of coupled-biquadratic active ladder structures', *IEEE J. Solid-State Circuits*, 1974, **SC-9**, pp. 70–72

79 SOLIMAN, A.M., and FAWZY, M.: 'Active-R resonator realization', *Proc. IEEE*, 1978, **66**, pp. 254–255

Active *R* Ladders: High-Frequency High-Order Low-Sensitivity Active *R* Filters Without External Capacitors

MICHAEL A. SODERSTRAND, MEMBER, IEEE

Abstract—**Active *R* filter techniques which make use of the pole of an operational amplifier to produce the state variables without the need for inductors or external capacitors may be employed to simulate all of the elements of both singly and doubly terminated reactance ladders. Although the design techniques are very simple and follow directly from passive synthesis techniques, the resulting filters offer high-order circuits with low sensitivity. However, the primary advantage of these active *R* filters is the ability to design filters which operate in a considerably higher frequency range than obtainable with standard active *RC* designs. While still primarily of research interest due to dynamic range problems and temperature sensitivity problems, active *R* ladders may be adequately compensated to allow many applications in specialized signal processing environments.**

I. Introduction

In recent years, two important concepts have dominated active filter research. One of these has been the development of "active *R*" filters which make use of the parasitic capacitances in the active device to build filters without inductors or capacitors at frequencies higher than possible with active *RC* filters [1]–[4]. Simultaneously, other researchers have investigated the design of low-sensitivity higher order, active filters through simulation of passive reactance ladders [5]–[9]. In this paper, the two fields will be merged by developing "active *R*" realizations of the basic building blocks for simulated passive reactance ladders. Portions of this paper were presented at the 18th Midwest Symposium [10] and the 1976 IEEE International Symposium on Circuits and Systems [11].

II. Basic Building Blocks

Fig. 1 shows the basic passive sections (Darlington A and B sections [12]) and the proposed active *R* simulation. It is important to understand that this is a simulation of the passive circuit where voltages are used to represent both voltages and currents.

For example, the passive circuit of Fig. 1(a) is characterized by the following *ABCD* matrix representation:

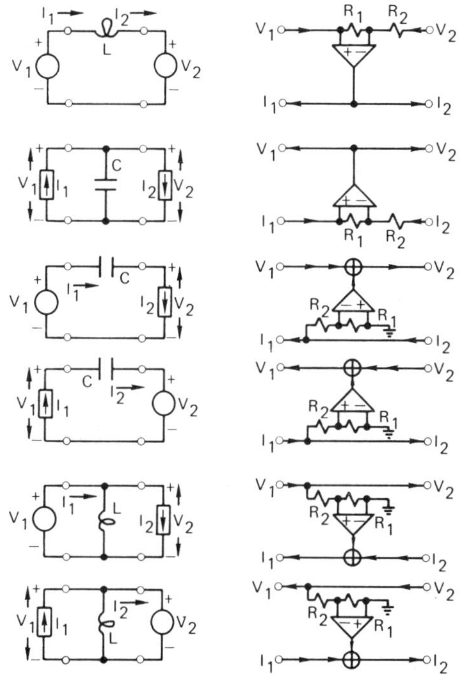

Fig. 1. Basic structures. (a) Series inductor. (b) Shunt capacitor. (c) Series capacitor. (d) Shunt inductor.

$$\begin{pmatrix} V_1 \\ I_1 \end{pmatrix} = \begin{pmatrix} 1 & Ls \\ 0 & 1 \end{pmatrix} \begin{pmatrix} V_2 \\ I_2 \end{pmatrix}. \quad (1)$$

Assuming the gain of the operational amplifier is of the form $A = GB/s$ where GB is the gain-bandwidth product [13], we see that the currents I_1 and I_2 simulated by the *voltage* at the output of the operational amplifier are related to the voltage inputs V_1 and V_2 by

$$\begin{pmatrix} V_1 \\ I_1 \end{pmatrix} = \begin{bmatrix} 1 & \dfrac{(R_1+R_2)s}{R_1 GB} \\ 0 & 1 \end{bmatrix} \begin{pmatrix} V_2 \\ I_2 \end{pmatrix}. \quad (2)$$

For dimensional consistency, one might introduce a normalizing resistance R_0 and replace I_1 and I_2 by $R_0 I$, and $R_0 I_2$. For simplicity, we shall choose $R_0 = 1$. Equations (1) and (2) are identical if we choose

Manuscript received August 9, 1977; revised April 27, 1978. This work was supported in part by Sandia Laboratories for the U.S. Energy Research and Development Agency under Contract AT-(29-1)-789.

The author is with Sandia Laboratories, Livermore, CA, and the University of California, Davis, CA.

Reprinted from *IEEE Trans. on Circuits Syst.*, vol. CAS-25, pp. 1032-1038, Dec. 1978.

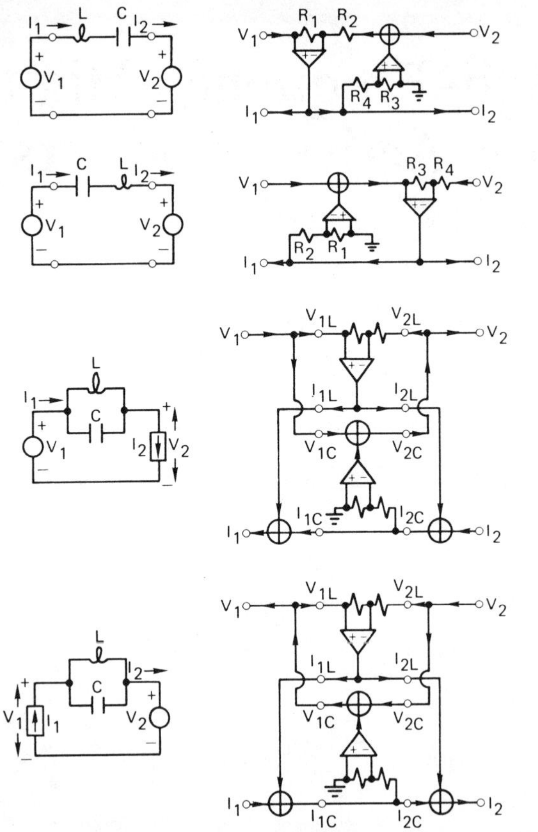

Fig. 2. Combining basic structures. (a) Series resonance first configuration. (b) Series resonance second configuration. (c) Parallel resonance first configuration. (d) Parallel resonance second configuration.

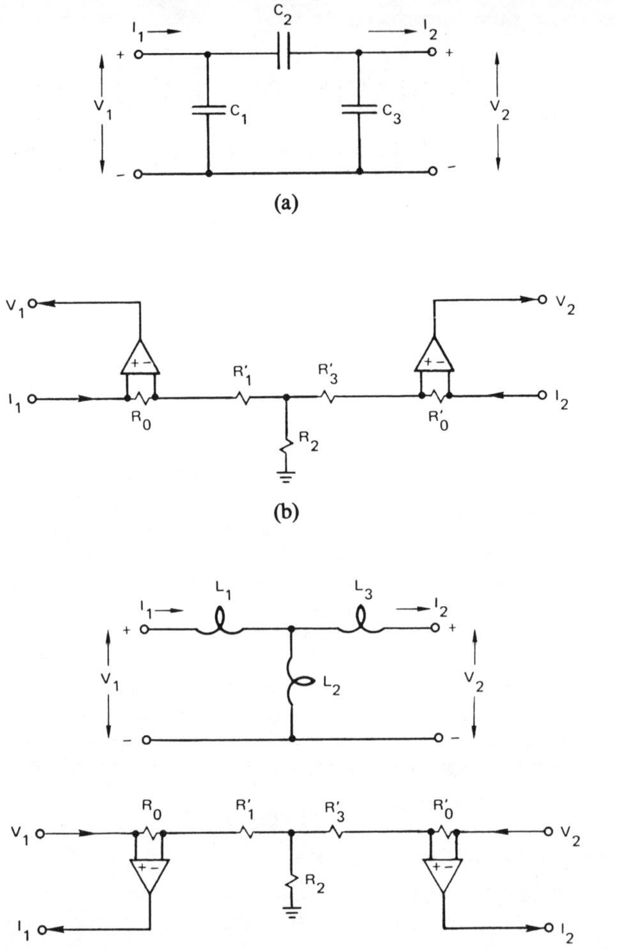

Fig. 3. π of capacitors and T of inductors. (a) π of capacitors. (b) Active R realization of π of capacitors. (c) T of inductors. (d) Active R realization of T of inductors.

$$\frac{R_2}{R_1} = L \cdot GB - 1. \tag{3}$$

For the circuit of Fig. 1(b), we find a similar simulation for a shunt capacitor when

$$\frac{R_2}{R_1} = C \cdot GB - 1. \tag{4}$$

The remaining circuits of Fig. 1 show variations in which either (3) or (4) will result in the appropriate passive simulation.

III. Combining Basic Building Blocks

Two basic concepts govern the combining of basic sections.

1) Active R circuits of Fig. 1 must be combined in such a way that voltage and current directions are properly matched from section to section.

2) Combinations must simulate Kirchoff's current and voltage laws for proper combination.

To illustrate the first point, Fig. 2(a) and 2(b) shows two possible combinations of series inductors and capacitors. Notice that the order of the occurrence of the inductor and capacitor in the passive circuit dictates which section will be used from Fig. 1(c) to simulate a series-tuned circuit. To illustrate the second point, Fig. 2(c) and 2(d) shows how Kirchoff's current and voltage laws govern the combination of inductors and capacitors to simulate a parallel tuned circuit. (Note: currents are summed according to KCL by the summing amplifiers.)

IV. Realization of a π of Capacitors or a T of Inductors

Due to the combinational restrictions mentioned in the last section, some combinations of circuit elements cannot be realized by combining the basic building blocks of Fig. 1. In these cases, the combined circuit elements must be realized directly.

For example, the π of capacitors shown in Fig. 3(a) cannot be realized by combining building blocks from Fig. 1. However, it may be realized directly by the circuit of Fig. 3(b). Note that the π of capacitors has only two state variables (even though it has three capacitors), and thus the active R realization of Fig. 3(b) requires only two operational amplifiers.

Analysis of the π of capacitors of Fig. 3(a) yields the

following V–I relationships:

$$\begin{pmatrix} V_1 \\ V_2 \end{pmatrix} = \frac{C_2}{s(C_1C_2+C_1C_3+C_2C_3)} \begin{bmatrix} \left(1+\frac{C_3}{C_2}\right) & -1 \\ +1 & -\left(1+\frac{C_1}{C_2}\right) \end{bmatrix} \begin{pmatrix} I_1 \\ I_2 \end{pmatrix}. \quad (5)$$

Analysis of the active R simulations of Fig. 3(b) yields the following V–I relationships:

$$\begin{pmatrix} V_1 \\ V_2 \end{pmatrix} = \frac{\frac{R_0'}{R_1R_2}GB}{s\left(\frac{1}{R_1}+\frac{1}{R_2}+\frac{1}{R_3}\right)} \begin{bmatrix} \frac{R_0}{R_0'}\left(1+\frac{R_3}{R_2}\right) & -\frac{R_0}{R_0'} \\ 1 & -\left(1+\frac{R_1}{R_2}\right) \end{bmatrix} \begin{pmatrix} I_1 \\ I_2 \end{pmatrix} \quad (6)$$

where

$$R_1 = R_0 + R_1'$$
$$R_3 = R_0' + R_2'.$$

For (6) to simulate (5), we must have

$$R_0' = R_0$$
$$R_1' = (C_1GB - 1)R_0$$
$$R_2 = C_2GBR_0$$
$$R_3' = (C_3GB - 1)R_0. \quad (7)$$

Similarly, the T of inductors of Fig. 3(c) may be simulated by the active R circuit of Fig. 3(d) with

$$R_0' = R_0$$
$$R_1' = (L_1GB - 1)R_0$$
$$R_2 = L_2GBR_0$$
$$R_3' = (L_3GB - 1)R_0. \quad (8)$$

V. Sources and Terminations

Fig. 4(a) shows simulation of voltage and current sources. Notice that feeding V_1 or I_1 directly is equivalent to a zero impedance voltage source or an infinite impedance current source which results in a singly terminated structure. Fig. 4(b) shows simulations of a load resistance R_L. Notice that for proper simulation, the summing and gain amplifiers in all the figures must be essentially ideal at the frequencies involved, while the operational amplifiers must be adequately modeled by the integrator model GB/s.

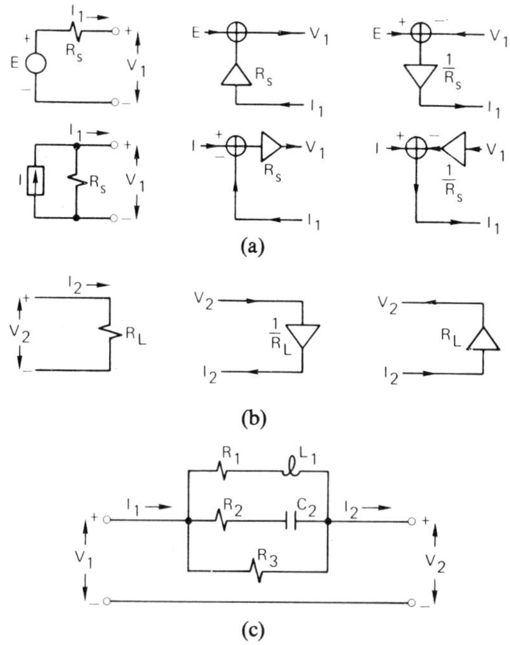

Fig. 4. Active R filter design. (a) Simulation of voltage and current sources. (b) Simulation of terminating impedance. (c) Nonideal terms due to two-pole model of operational amplifier.

VI. Effect of Nonideal Amplifiers

At higher frequencies, the poles in the various summing amplifiers will begin to effect the response of the filter. However, since the gain of the summing stages is at most one, this effect does not occur within the operating range of the active filter. Furthermore, since the gain is less than one, stability of the feedback loops will be guaranteed. However, care must be taken with the source and terminating impedances such that gains of more than one are carefully employed. Generally, 1-Ω terminations (which may be realized by either a buffer or, in most cases, a short circuit) should be used to minimize the effect of amplifier poles and to guarantee stability.

Since the operational amplifier is not really an integrator, there are two effects which are introduced. If the one-pole model of the form $A = GB/(s+\omega_p)$ is employed in the analysis of the circuit of Fig. 1(a), one effect of the operational amplifier can be seen from the $ABCD$ matrix:

$$\begin{pmatrix} V_1 \\ I_2 \end{pmatrix} = \begin{bmatrix} 1 & \frac{(R_1+R_2)s}{R_1GB} + \frac{R_1+R_2}{R_1A_0} \\ 0 & 1 \end{bmatrix} \begin{pmatrix} V_2 \\ I \end{pmatrix} \quad (9)$$

where A_0 is the dc gain, ω_p is the operational amplifier pole, and $GB = A_0\omega_p$. Comparing (9) to (2), we see than the one-pole model predicts a resistance of value $(R_1 + R_2)/R_1A_0$ in series with the inductor of Fig. 1(a). Similarly, the one-pole model would predict resistors in series with any inductors and resistors in parallel with any capacitors in the simulations of the circuits of Figs. 1 and 2. These effects are exactly the same as the inherent resistances in practical inductors and capacitors, and thus can be dealt with by using passive designs which utilize similar nonideal elements. For example, filter tables are available which compensate for this effect [14].

However, if we consider a two-pole model of the form $A=GB\omega_q/(s+\omega_p)(s+\omega_q)$, we obtain the simulated structure of Fig. 4(c) where

$$L_1=\frac{(R_1+R_2)(\omega q-\omega p)}{R_1 GB\omega_q}$$

$$R_1'=\frac{(R_1+R_2)(\omega_q-\omega_p)\omega_p}{R_1 GB\omega_q}$$

$$C_2=\frac{R_1 GB}{(R_1+R_2)\omega_q(\omega_q-\omega_p)}$$

$$R_2'=\frac{(R_1+R_2)(\omega_q-\omega_p)}{R_1 GB}$$

$$R_3'=-R_2. \tag{10}$$

To give some feeling for the relative magnitudes of the above, for a 741-type operational amplifier with $GB=10^6$, $\omega_p=10$, $\omega_q=10^6$, and R_1' normalized to 10 Ω (for reference purposes only), we obtain

$$L_1=1\text{ H}$$

$$R_1'=10\ \Omega$$

$$C_2=1\text{ pF}$$

$$R_2'=1\text{ M}\Omega$$

$$R_3'=-1\text{ M}\Omega. \tag{11}$$

Since $\omega_q \gg \omega_p$, L_1 and R_1' are essentially the inductor and resistor predicted in the one-pole model. The capacitor and resistor branch will come into consideration only at frequencies above ω_q, so the dominant effect of the second pole is the introduction of excess phase shift through the parallel negative resistance R_3'. Since R_3' is quite large in magnitude, it can generally be ignored during design, but it does introduce the possibility of instabilities in the circuit. Fortunately, stability problems occur in practice only, for designs which require critical frequencies near the gain-bandwidth product of the amplifier. In cases where stability is a problem, introduction of small amounts of negative feedback around the individual operational amplifiers can rectify the problem.

VII. Temperature and Signal Level Problems

Active R ladders share with all active R filters the common problems of temperature sensitivity of the amplifier pole and slew rate limitations which adversely effect dynamic range of the filters. For example, a 741-type operational amplifier which is not temperature compensated will drift about 3000 ppm/K, which will result in about a 6-percent shift in frequency response over a 20 K range of temperatures. For most applications, this is not acceptable, and either a better operational amplifier must be used, temperature must be controlled, or temperature compensation must be employed. It is not within the scope of this paper to discuss in detail temperature compensation techniques, but it is appropriate to illustrate how they might be dealt with and refer the reader to appropriate literature.

Basically, there are two schemes of temperature compensation which work fairly well with active R ladders. The first is based on the fact that all of the active R ladder elements control the gain-bandwidth product of the operational amplifier with a resistor divider (for example, R_1 and R_2 in the basic circuits of Fig. 1). By replacing one of these resistors with a temperature-sensitive device which appropriately tracks the temperature shift of the gain-bandwidth product of the operational amplifier, it is possible to greatly improve the temperature stability of the filter. A detailed discussion of this technique, including experimental verification, is given by Rao and Srinisavan [2], [15], [16]. A second technique for temperature compensation of the active R ladder would make use of programmable operational amplifiers with a temperature compensation feedback loop. A description of this technique along with experimental verification has been provided by Brand and Schaumann [17].

We complete this section with a discussion of dynamic range problems. These problems are due to slew rate limitations of the operational amplifier which limit the input voltage levels allowable, particularly for high-frequency operation. The best solution to this problem is a higher slew rate operational amplifier. However, using different frequency compensating schemes, it is possible to improve the slew rate and thus help the problem, at least to some extent [10].

VIII. The Design Procedure

Since the proposed active R building blocks of Figs. 1 and 2 simulate elements of a passive reactance ladder, most of the design techniques for singly or doubly terminated reactance ladders are directly applicable to active R ladder synthesis [12]. The only restrictions imposed by active R realization are those of ordering imposed by voltage and current connection requirements. The easiest design procedure is to simply look up the required passive circuit in a standard table for passive filters [13] and then replace each passive element by its active R simulation. However, since the proposed circuits are simply simulations of passive reactance ladders, any of the many passive synthesis techniques may be used to obtain the passive circuit from which the active R simulation is derived [12]. It should be noted that the active R simulations are usually best derived from the normalized passive circuit presented in the filter tables with L, C, and R about equal to one.

IX. Examples

A. Example 1—Low-Pass Filters

Fig. 5(a) shows the basic configuration obtained for low-pass filters. Notice that 1-Ω termination is obtained

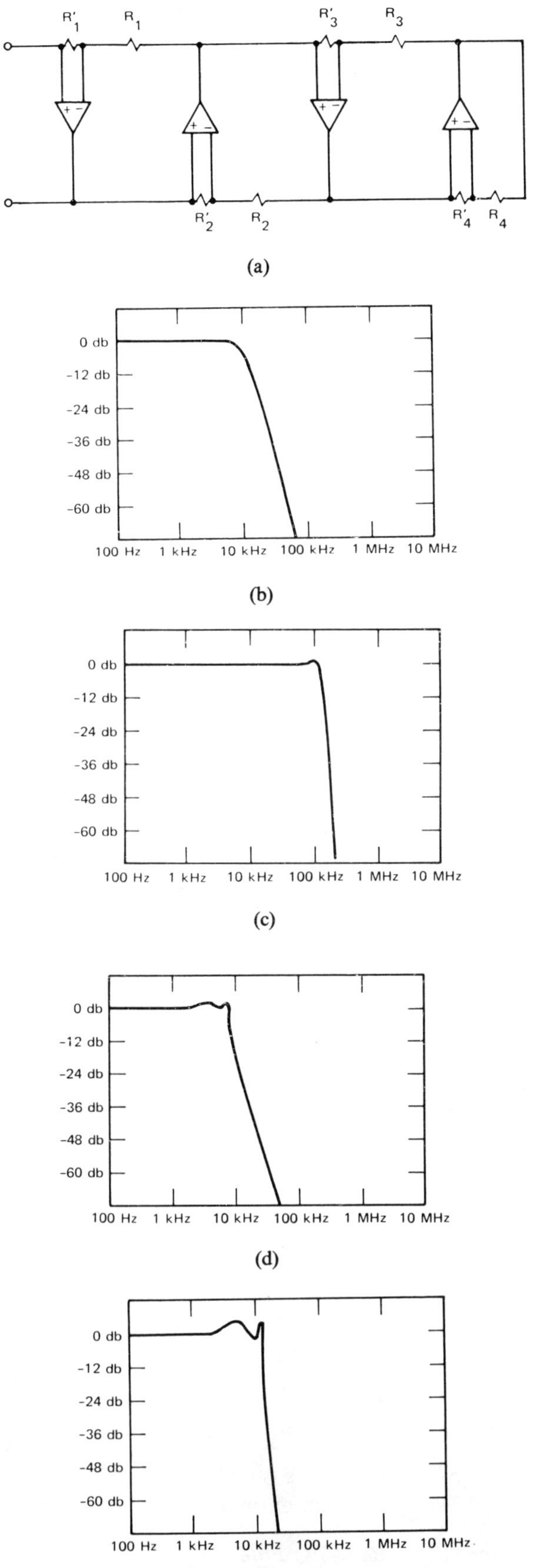

Fig. 5. Low-pass active R filter examples. (a) Low-pass fourth-order ladder. (b) Measured response Butterworth low-pass ($f_0 = 10$ kHz). (c) Measured response Butterworth low-pass ($f_0 = 200$ kHz). (d) Measured response Chebyshev low-pass (1-dB ripple). (e) Measured response Chebyshev low-pass (3-dB ripple).

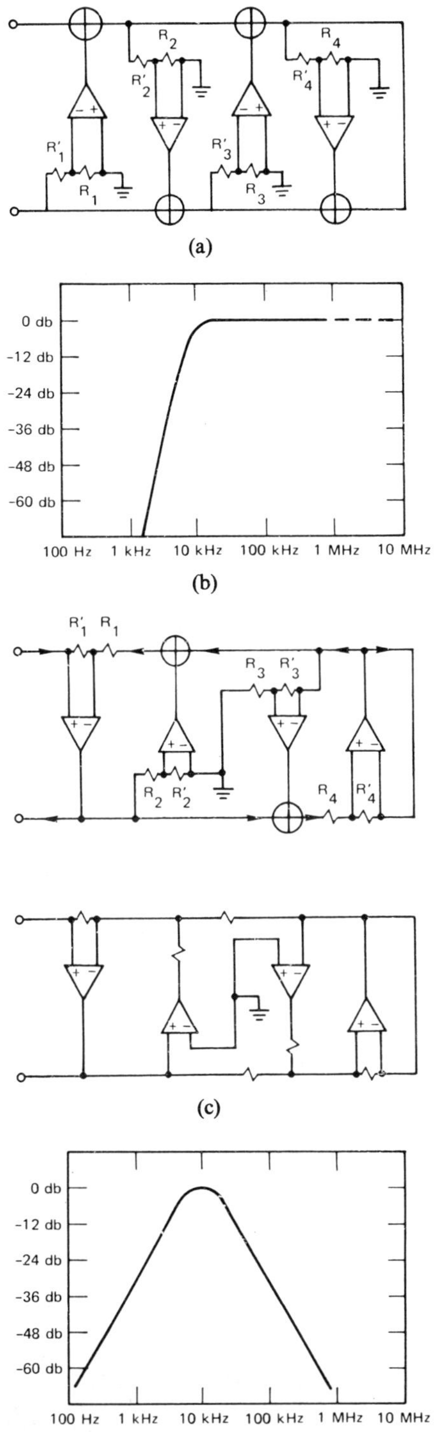

Fig. 6. High-pass and bandpass active R filter examples. (a) High-pass ladder. (b) Measured response Butterworth high-pass filter. (c) Bandpass ladder. (d) Simplified bandpass ladder. (e) Measured response simplified bandpass ladder.

by tying the input to the output at port 2. From the normalized tables, the following element values are obtained.

Fig. 5(b): Butterworth $\omega = 10$ kHz:

$$R_1 = R_2 = R_3 = R_4 = 100\ \Omega$$

$$R_1' = R_2' = 15\ \text{k}\Omega \qquad R_3' = 10\ \text{k}\Omega$$

$$R_4' = 3.83\ \text{k}\Omega.$$

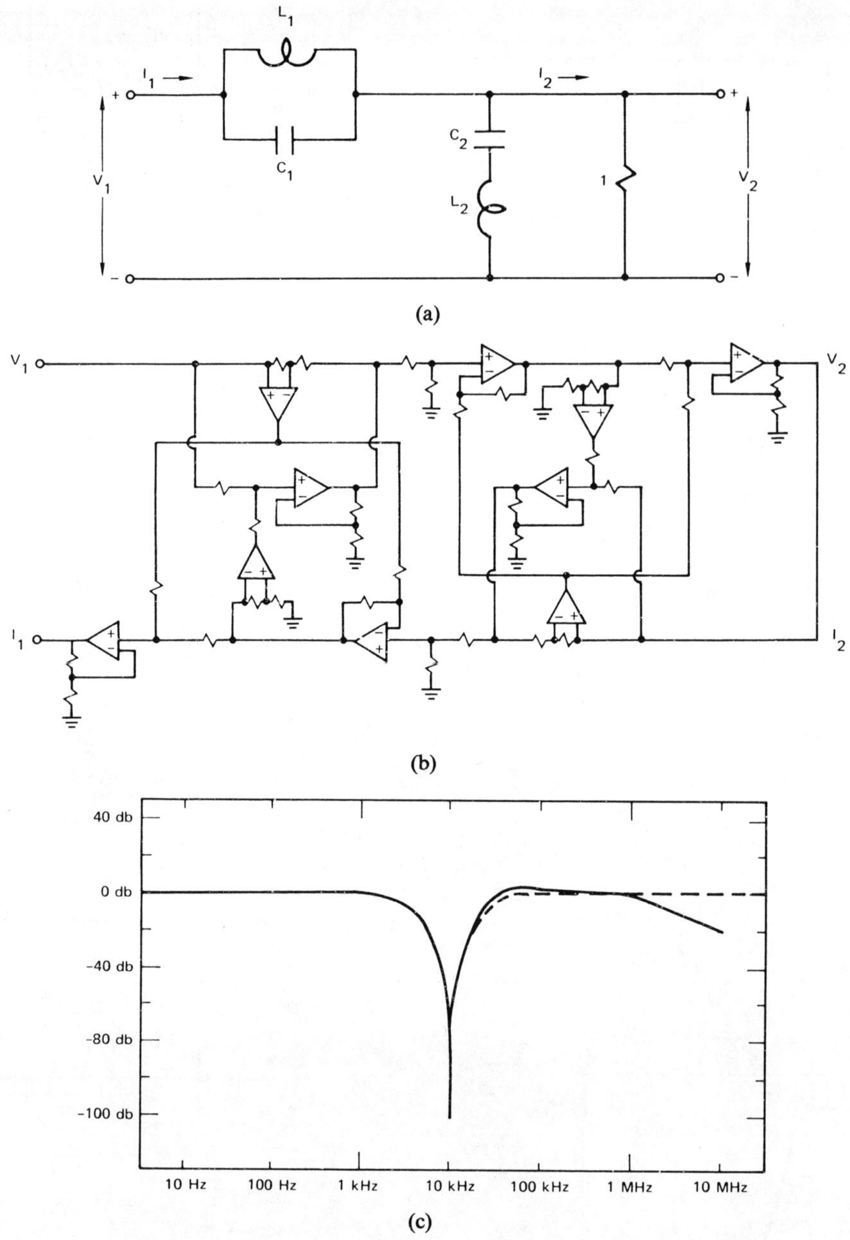

Fig. 7. Band reject active *R* ladders. (a) Passive band reject filter. (b) Fourth-order band reject active *R* ladder. (c) SPICE simulation of band reject active *R* ladder.

Fig. 5(c): Butterworth $\omega = 200$ kHz:

$$R_1 = R_2 = R_3 = R_4 = 100\ \Omega$$

$$R_1' = R_2' = 750\ \Omega \qquad R_3' = 511\ \Omega$$

$$R_4' = 200\ \Omega.$$

Fig. 5(d): Chebyshev $\omega = 10$ kHz, 1-dB ripple:

$$R_1 = R_2 = R_3 = R_4 = 100 \qquad R_1' = 12.7\ \text{k}\Omega$$

$$R_2' = 20\ \text{k}\Omega \qquad R_3' = 10\ \text{k}\Omega \qquad R_4' = 10\ \text{k}\Omega.$$

Fig. 5(e): Chebyshev $\omega = 10$ kHz, 3-dB ripple:

$$R_1 = R_2 = R_3 = R_4 = 400\ \Omega \qquad R_1' = 10\ \text{k}\Omega$$

$$R_2' = 24.3\ \text{k}\Omega \qquad R_3' = 12.1\ \text{k}\Omega \qquad R_4' = 16.2\ \text{k}\Omega.$$

B. Example 2—High-Pass Filters

Fig. 6(a) shows the circuit resulting from a low-pass to high-pass transformation on a Butterworth low-pass filter (fourth-order). The resulting element values are

$$R_1 = R_2 = R_3 = R_4 = 100\ \Omega \qquad R = 1\ \text{k}\Omega$$

$$R_1' = R_2' = 15\ \text{k}\Omega \qquad R_3' = 10\ \text{k}\Omega \qquad R_4' = 3.83\ \text{k}\Omega.$$

Fig. 6(b) shows the measured response for this circuit using μA741 operational amplifiers.

C. Example 3—Bandpass Filters

Fig. 6(c) shows the circuit resulting from a low-pass to bandpass transformation on a second-order (Butterworth) low-pass filter. Fig 6(d) shows a simplification obtained by using resistors to do the summing, and has element values of

$$R_1 = 10\ \Omega \qquad R_2 = 100\ \mathrm{k}\Omega \qquad R_3 = 1.42\ \Omega$$

$$R_4 = 681\ \Omega \qquad R_5 = 100\ \mathrm{k}\Omega \qquad R_6 = 10\ \Omega.$$

Fig. 6(e) shows the measured response, once again using μA741 operational amplifiers.

D. Example 4—Band Reject Filters

Assume a fourth-order band reject filter is required to eliminate 10 kHz. A typical design would start with a second-order Butterworth low-pass prototype filter, and apply a low-pass to band reject transformation to obtain the passive realization of Fig. 7(a). Finally, simulation of the passive elements using the active R building blocks yields the final circuit shown in Fig. 7(b). It should be noted that the summing amplifiers are realized using standard operational amplifier summing techniques. Although this introduces an undesirable pole for each summing amplifier, the low gain of the amplifier minimizes the effect of this pole. Often, the summer poles can be tuned out by appropriate techniques or one can go to an operational amplifier with a larger gain-bandwidth product for the summers.

The band reject section and several others were simulated using the SPICE computer program [18] and a macromodel for the operational amplifier [19]. In Fig. 7(c), the dotted line is the result of the computer analysis of the passive circuit, and the solid line is for the active R circuit. In addition, several of these circuits were built in the laboratory and measurements agreed closely with the computer analysis [20].

References

[1] M. A. Soderstrand, "Active R filters: A new concept in active filter design," *Sandia Lab. Tech. Rep.*, SCL-DR-720066, Oct. 1972.

[2] K. R. Rao and S. Srinisavan, "A bandpass filter using the operational amplifier pole," *IEEE J. Solid-State Circuits*, vol. SC-8, pp. 245–246, June 1973.

[3] R. Schaumann, "Low-sensitivity high-frequency tunable active filter without external capacitors," *IEEE Trans. Circuits Syst.*, vol. CAS-22, pp. 39–44, Jan. 1975.

[4] K. R. Laker, R. Schaumann, and J. R. Brand, "Multiple-loop feedback active R filter," in *Proc. 1976 IEEE Int. Symp. on Circuits and Systems*, Munich, Germany, Apr. 1970.

[5] J. Tow and Y. L. Kuo, "Coupled-biquad active filters," in *Proc. 1972 IEEE Symp. on Circuits and Systems*, Los Angeles, CA, pp. 164–168, Apr. 1972.

[6] G. Szenticmai, "Synthesis of multiple-feedback active filters," *Bell Syst. Tech. J.*, pp. 527–555, Apr. 1973.

[7] P. O. Brackett and A. S. Sedra, "Application of flow graphs to the synthesis of linear systems," in *Proc. 1975 IEEE Int. Symp. on Circuits and Systems*, Boston, MA, pp. 242–249, Apr. 1975.

[8] H. Wupper and K. Meerkotter, "New active filter synthesis based on scattering parameters," *IEEE Trans. Circuits Syst.*, vol. CAS-22, pp. 594–602, July 1975.

[9] W. E. Heinlein and W. H. Homes, *Active Filters for Integrated Circuits*. New York: Springer, 1974, pp. 409–420.

[10] M. A. Soderstrand, "Low pass active R ladder structures," in *Proc. 18th Midwest Symp. on Circuits and Systems*, Montreal, Canada, pp. 594–597, Aug. 1975.

[11] ——, "Design of high frequency, high order active R ladders," in *Proc. 1976 IEEE Int. Symp. on Circuits and Systems*, Munich, Germany, Apr. 1976.

[12] G. C. Temes and S. K. Mitra, *Modern Filter Theory and Design*. New York: Wiley, 1973, pp. 89–98.

[13] M. A. Soderstrand, "Design of active R filters using only resistors and operational amplifiers," *Int. J. Electron.*, vol. 40, no. 5, pp. 417–432, May 1976.

[14] White Electromagnetics, Inc., *Handbook on Electrical Filters*, White Electromagnetics, Inc., 1963.

[15] K. R. Rao and S. Srinivasan, "Low sensitivity active filters using the operational amplifier pole," *IEEE Trans. Circuits Syst.*, vol. CAS-21, Mar. 1974.

[16] ——, "A high Q temperature insensitive bandpass filter using the operational amplifier pole," *Proc. IEEE*, vol. 62, pp. 1713–1714, Dec. 1974.

[17] J. R. Brand and R. Schaumann, "A temperature compensated, high frequency, tunable active R filter using programmable operational amplifiers," in *Proc. 18th Midwest Symp. on Circuits and Systems*, Montreal, P.Q., Canada, pp. 603–607, Aug. 1975.

[18] L. W. Nagel and D. O. Pederson, "Simulation program with integrated circuit emphasis (SPICE)," Electron. Research Lab., Univ. of California, Berkeley, ERL-M382, 1973.

[19] G. R. Boyle, B. M. Cohn, D. O. Pederson, and J. E. Soloman, "Macromodeling of integrated circuit operational amplifiers," *IEEE J. Solid-State Circuits*, vol. SC-9, pp. 353–364, Dec. 1974.

[20] K. L. Lee, "Computer-aided design and analysis of active R ladders," M.S. thesis, Univ. of California, Davis, June 1977.

Paper 5-3

Gyrator Video Filter IC with Automatic Tuning

KENNETH W. MOULDING, JONATHAN R. QUARTLY, PAUL J. RANKIN, ROGER S. THOMPSON, AND GORDON A. WILSON

Abstract—A single-chip gyrator filter for separating the components of the video signal in a TV receiver is described which is suitable for mass production in a standard bipolar process ($f_T \approx 400$ MHz). The 11 mm^2 filter chip operates at frequencies up to 10 MHz, requires no tuning or alignment and has Q-factors which are stable with temperature.

The IC contains an automatic tuning system which tunes the five resonators of the filter by aligning an auxiliary gyrator resonator with the crystal oscillator present in the color decoder of a TV receiver. Problems of matching the frequencies of the individual gyrator resonators are discussed, showing how alignment accuracy of 0.5 percent can be obtained when resistivities and specific capacitances have production spreads of at least 10 percent. Various gyrator circuit configurations are given which minimize the circuit complexity and, hence, the chip area. Computer aided design techniques for the filter using geometrically scaled models and macromodeling are presented and it is shown how a complete simulation of the chip led to a significant improvement in bandstop performance. Finally, the measured responses are presented and the filter performance is discussed in the light of present day requirements. Ideas for larger scale integration of the TV system are prompted by the advent of this IC.

Introduction

Although large sections of the signal path of the color television receiver (Fig. 1) have been integrated, further progress is restricted by the number of components which defy integration, in particular, the filter coils and trimmers. Integration to avoid these components is, therefore, very desirable, especially if it can be achieved using a process which is compatible with the other IC's in the receiver. The integrated filter described in this paper replaces the LC video filter required to separate the components of the video signal and is manufactured in a standard bipolar process. Fewer external connections and the freedom from adjustments and alignment provided by this IC should give greater reliability than its LC counterpart. At the same time, designed-in flexibility allows the IC to interface with many current video demodulator and color decoder arrangements.

Two versions of the filter have been designed, one for use with PAL television system I (sound carrier at 6 MHz) and the other with PAL television system B or G (sound carrier at 5.5 MHz), and the design can easily be adapted for NTSC or SECAM. The description of the filter in this paper relates to the European system G.

Manuscript received May 20, 1980; revised July 24, 1980.

K. W. Moulding, P. J. Rankin, R. S. Thompson, and G. A. Wilson are with Philips Research Laboratories, Redhill, Surrey, England.

J. R. Quartly was with Philips Research Laboratories, Redhill, Surrey, England. He is now with the Welding Institute, Abington, Cambridge, England.

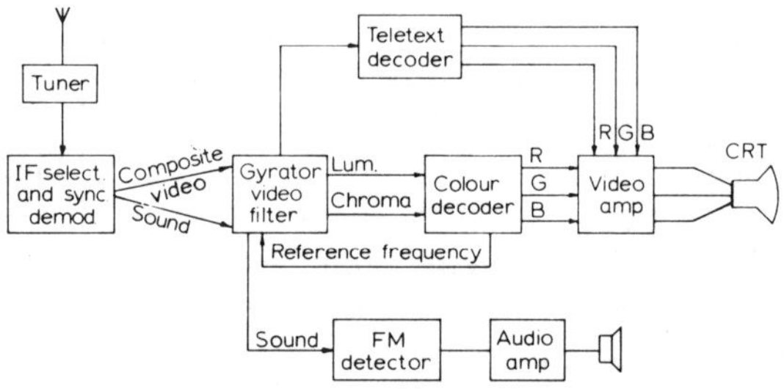

Fig. 1. Television receiver using a gyrator video filter IC.

In the television receiver, the video filter is placed after the synchronous demodulation of the composite signal (Fig. 1) which consists of an 0-5 MHz luminance signal, a 4.43 MHz color subcarrier and a 5.5 MHz FM sound carrier. LC resonators in the filter are simulated with gyrators terminated by junction capacitors, the voltage dependence of the latter being exploited for tuning the filter. Each gyrator employs two voltage controlled current sources, and uses a total of 14 transistors. Using these techniques, resonant frequencies in the range of at least 1-10 MHz can be achieved. One of the main problems encountered in designing such a filter is to achieve accurate resonant frequencies and Q factors which are tolerant of process spreads.

An earlier paper [1] described a basic video filter chip which demonstrated the feasibility of fully integrated selectivity, and also discussed some of the features of the gyrator and its performance limitations with regard to signal handling and Q-factor. The use of gyrator-simulated inductances provides active filters which are least sensitive to component variations and have the widest dynamic range [2], [3]. This paper describes the gyrator filters and new innovations which make the IC very tolerant of process spreads and suitable for mass production, focussing particularly on the automatic tuning circuit, dc biasing and layout design. Computer aided design methods are also discussed as these have featured strongly in the development.

General Description

The separation of the composite signal is performed by five gyrator resonators. First, the input signal is stripped of the intercarrier sound in a two-stage sound trap (Fig. 2) and a low impedance takeoff for a teletext decoder or VCR is provided. The signal is then applied to a single gyrator resonator, with two outputs, to separate luminance and chrominance signals. In the sound channel, a coupled bandpass pair of gyrator resonators provide the necessary selectivity. Further, the chip

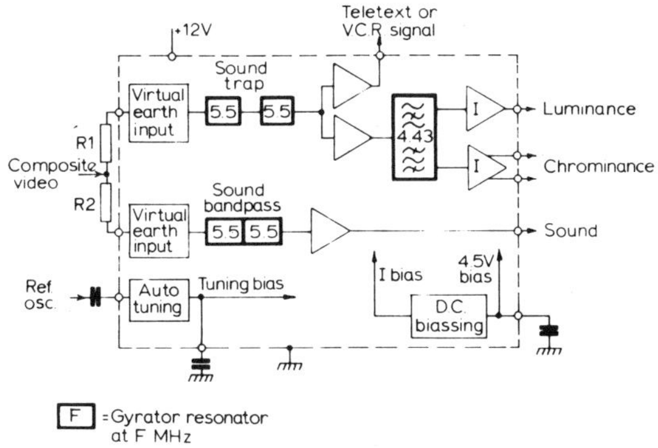

Fig. 2. Schematic of gyrator video filter IC.

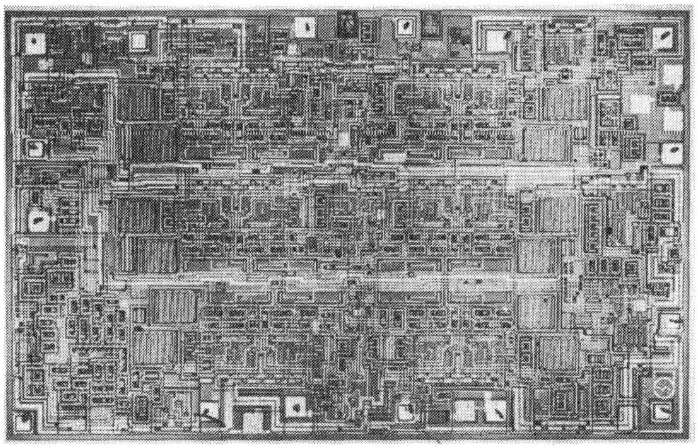

Fig. 3. Gyrator video filter (chip photograph).

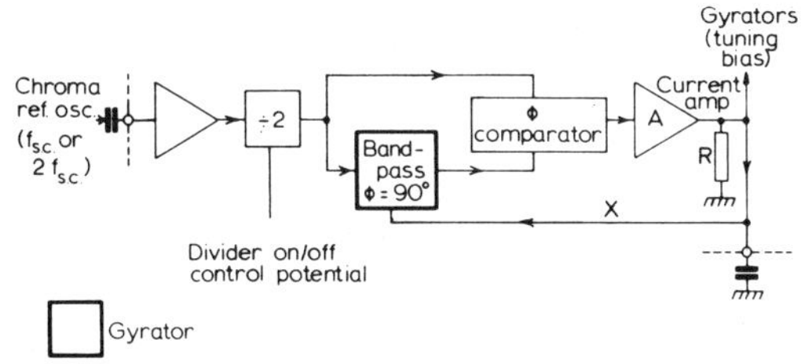

Fig. 4. Automatic tuning circuit.

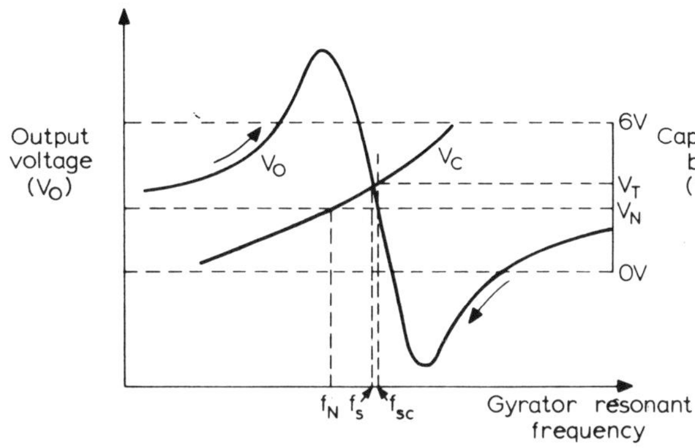

Fig. 5. Open loop characteristics of the tuning circuit (loop broken at X).

contains an automatic tuning system which tunes the five resonators of the filter by aligning a sixth auxiliary gyrator resonator with the crystal oscillator present in the color decoder of the TV receiver. A bandgap voltage reference circuit is used to set voltage and current levels throughout the filter giving immunity to supply voltage variations.

The present filter has in-built flexibility to interface with most color television receiver circuit designs. Interfacing circuits are included to match to the signal levels of adjacent IC's and to isolate the resonators from the effects of external loading. By changing the values of the series resistors R_1 and R_2 at the virtual earth inputs, the filter can accommodate a wide range of video amplitudes from a standard 1 V p-p video, to the 3 V video from a synchronous demodulator IC (e.g., TDA2541). Also, the provision of two inputs allows for separate IF demodulation of video and sound subcarrier signals and permits the individual adjustment of the signal levels applied to the IC. To give optimum signal-to-noise ratio without distortion, the video signal level (black to white) across the port of the first sound trap gyrator internal to the IC is arranged to be ≈120 mV p-p.

The 11 mm^2 chip (Fig. 3), dissipates 650 mW and uses resonating capacitors of about 20 pF. Each gyrator resonator occupies an area of ≈0.7 mm^2, of which ≈0.12 mm^2 are the capacitors. The complete filter IC has 567 components comprising 216 transistors, 339 resistors, and 12 capacitors.

Automatic Tuning

Thanks to the good matching of components on a single chip, all resonators can be tuned with a common capacitor bias voltage. This bias must be adjusted to allow for ambient temperature variations and the relatively large batch-to-batch spreads. Adequate compensation circuits for use with preset adjustment are problematical. An automatic tuning arrangement was therefore chosen [4] in which the filter is kept tuned by locking a reference gyrator to the crystal oscillator in the color decoder (Fig. 4). This oscillator may be either at subcarrier frequency, f_{sc} (4.43 MHz) or at $2f_{sc}$ (8.87 MHz). The reference signal is fed into a limiter amplifier followed by a divider which can be switched on or off as required by the receiver design. The resultant signal is passed through a single 4.43 MHz gyrator bandpass resonator where it suffers a nominal 90° phase shift. This is then compared in phase with a sample of the incoming reference. If the resonant frequency of the gyrator is $<f_{sc}$, the phase shift across the bandpass is $<90°$. This gives a positive-going output voltage at the phase comparator which is amplified and smoothed (Fig. 5). Applying the dc output to the resonating capacitors closes the control loop and tunes the gyrator up towards f_{sc}. When the gyrator resonant frequency is $>f_{sc}$, a negative-going output voltage tunes the gyrator down to f_{sc}.

The residual tuning error δf depends on how far the gyrator is initially off-tune, and on the slope K of the open loop characteristic near f_{sc} in Fig. 5. This figure also shows the typical characteristic of resonant frequency versus capacitor bias. The output voltage swing available to the control loop is restricted to protect the capacitors from forward bias or breakdown. An approximate tuning voltage V_N is first derived by a simple network. With no reference signal present, this voltage, applied to the capacitors, tunes the gyrator to frequency f_N. (V_N also corresponds to the zero-error voltage from the phase comparator when $\phi = 90°$.) When the reference signal f_{sc} is applied, a voltage is generated which is applied to the capacitors and tunes the gyrator to f_s. The tuning frequency error can be shown to be given approximately by $\delta f = (V_N - V_T)/K$ for large loop gains.

The slope of the open loop characteristic K is dependent on a number of factors:

$$K = \frac{d\phi}{df} \cdot \frac{dI}{d\phi} \cdot A \cdot R \ \text{V/Hz}$$

where $d\phi/df$ is the rate of change of phase with frequency of the gyrator bandpass and is proportional to Q; $dI/d\phi$ is the rate of change of output current with phase for the phase comparator; A is the amplifier current gain and R is the load resistor. The present circuit gives a worst-case tuning error of 10 kHz ($\frac{1}{4}$ percent) using a gyrator bandpass with $Q = 25$, and a loop gain of 100.

It is vital that the phase of the two signals arriving at the comparator should differ by exactly 90° when the gyrator is correctly tuned as any deviation from this would cause an additional tuning error. Account is therefore taken during the design of any parasitic phase lags in the two paths.

The useful capacitor bias range is 0.5-6 V, limited at one end by increasing distortion, and at the other by junction breakdown. This can control the capacitance over a range of ±28 percent which is sufficient to accommodate the spread in $R \cdot C$ due to variations in the process and an ambient operating temperature range of 0-60°C.

The temperature coefficient of the filter frequencies with automatic tuning has been measured to be 60 ppm/°C, which is satisfactory.

Filter Designs

The gyrators, with ports 1-1′ and 2-2′, are formed from degenerated long-tailed pair amplifiers (voltage controlled current sources) as shown in Fig. 6. These amplifiers have the advantage that the V_{be}'s of the transistors are balanced and both phases of output are available to form the forward and reverse transconductances, $-g_o$ and g_o. Also, the gyrators so formed are floating. When the gyrator is grounded (that is ac grounded) at terminals 1′ and 2′ and the ports loaded with grounded p-n junction capacitors, tuning bias voltages can be easily applied. Grounded ports also allow indirect input and output points to be used which buffer the resonator from the effects of loading without affecting the operation of the gyrator. A buffered output signal proportional to the port voltage across 1-1′ is provided by a series load resistor R_L connected between the earthed terminal and the corresponding transistor collector. Buffered input is obtained by connecting the source in series with the earthed terminal 2′ and transistor base $B1$, which then injects signal current into the resonator across the gyrator port 1-1′.

The circuit shown in Fig. 7 is a practical gyrator resonator with load capacitors C_1 and C_2. It employs indirect input and output points, and gives a simple bandpass response. The

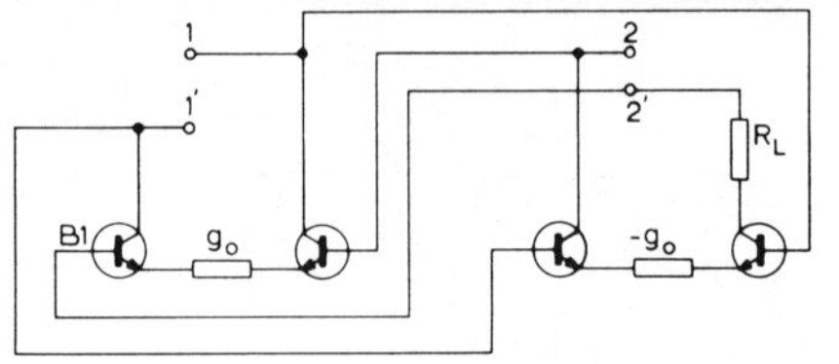

Fig. 6. Gyrator realization using two transconductance amplifiers.

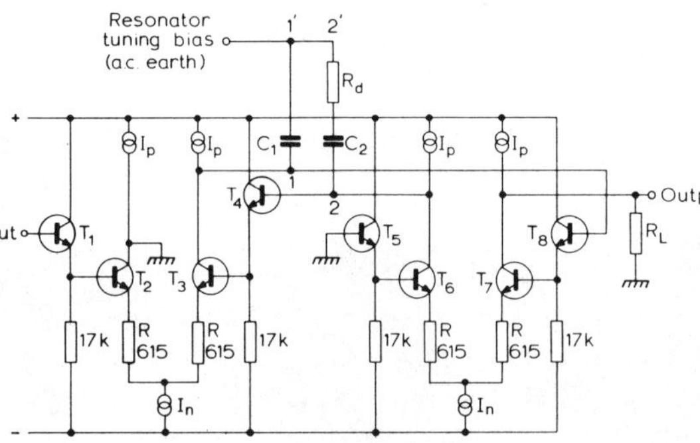

Fig. 7. Gyrator resonator giving a bandpass response.

degeneration resistors R determine the transconductance g_o of the amplifiers. Emitter-follower transistors (T_1, T_4, T_5, and T_8) provide level shifting for dc coupling and, together with the use of dc current sources as collector loads, give a low gyrator port conductance. The current sources in the collectors (I_p) employ lateral p-n-p transistors, while the emitter sources (I_n) are vertical n-p-n devices. A resistor R_d is included in series with one of the terminating capacitors to give the desired Q factor for each resonator. In calculating the value of R_d, the nonideal nature of the gyrator and its terminating capacitors must be modeled. Both gyrator port conductance and capacitor Q are finite and these lower the resonator Q, while phase shift in the amplifiers (due mainly to finite transistor f_T) has the effect of enhancing Q. Port conductance is kept low as indicated above and has only a small effect on Q, but parasitic damping associated with the internal resistance of the terminating capacitors is more significant, especially for the emitter-base capacitance which is used. Nevertheless, the damping arising from these two sources is outweighed by the Q enhancement due to amplifier phase shifts [1] and the resonator Q would be negative. The value chosen for R_d is therefore made up of two contributions: first, the resistance needed to compensate for the amplifier phase shift and give a resonator Q of infinity, and, second, the resistance needed to damp this infinite Q resonator to the Q required for the particular circuit application.

A simplified representation of the symmetric bandpass resonator is shown in Fig. 8(a). Fig. 8(b) shows a single gyrator used to give both trap and asymmetric bandpass functions. For a trap connection, one floating gyrator port is needed to allow the series capacitor to be grounded. The buffered bandpass output shows a positive tilt which arises from the intrinsic high-pass character of this connection. This configuration is used for the chroma filtering at 4.43 MHz (Fig. 2), where it is

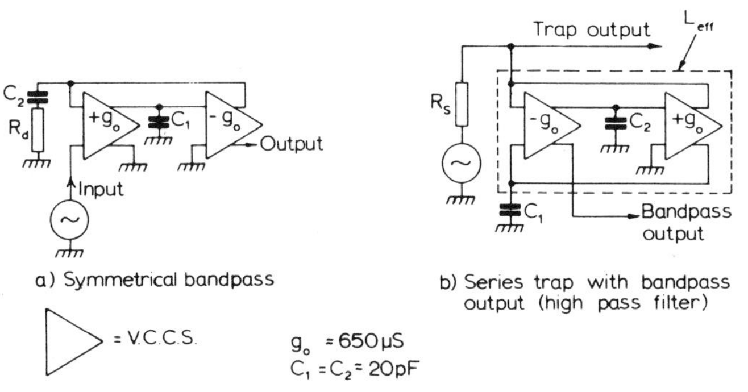

Fig. 8. Gyrator filter arrangements.

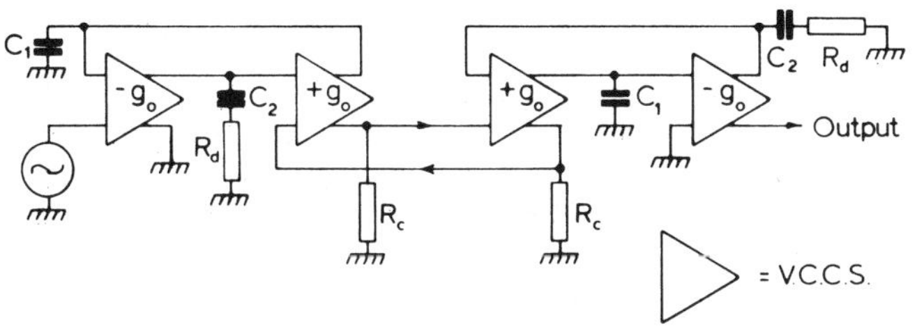

Fig. 9. Actively coupled second-order bandpass.

acceptable to have the same bandwidth of 1.5 MHz for both trap and bandpass responses.

The sound bandpass is a second-order Butterworth filter. Active coupling [6] has been used rather than the more common top-*C* coupling, since the latter would require a floating capacitor. Fig. 9 shows the separate forward and reverse coupling paths, where the coupling coefficient is given by the product $R_c \cdot g_o$. The -3 dB bandwidth of the response is 280 kHz at 5.5 MHz and requires gyrator resonators with Q factors of 27. For the sound trap, the severe specification in the chroma channel demands two stages of trapping in order to guarantee the necessary rejection of >40 dB over the sound bandwidth.

Process Tolerance

Batch-to-batch spreads in the parameters of a standard bipolar process can be up to ±20 percent while across one chip, ±2 percent variation might be found. However, this application demands very close control of frequency characteristics, and, therefore, particular attention has been paid to the effects of processing variations when choosing circuits and their layout.

The resonance of a gyrator resonator is given by $\omega = 1/\sqrt{R_1 R_2 C_1 C_2}$ where R_1 and R_2 are the resistances defining the transconductances and C_1 and C_2 are the load capacitances. In the filter, the sound trap must track the reference gyrator to ±0.5 percent implying component matching of a similar accuracy.

Resistor matching is a function of resistivity, etching, and contact resistance variations. These can be minimized by placing the resistors in close proximity with the same orientation and by using generous dimensions with large contacts.

The capacitors are formed by the parallel connection of the emitter-base and collector-base junctions of a transistor. For best matching of the emitter-base contribution, a single large emitter would be preferred. Unfortunately, the series resistance of such a structure is too high and a striped geometry must be used. As with resistors, close proximity and common orientation are beneficial. The capacitor areas are scaled to obtain the two resonant frequencies required in the filter. This scaling must be applied equally to the two junctions to maintain a constant ratio as the specific capacitances of each vary independently. In the present design, using 60 μm wide resistors and 25 μm wide emitter junction stripes for the capacitors, a 90 percent yield in respect of frequency matching is anticipated.

The chip uses the 12 V ± 10 percent supply normally available in a television receiver. An internally generated 4.5 V supply provides a suitable bias potential for the gyrator amplifiers. This circuit uses a bandgap reference [7] to generate 1.5 V which is then amplified to 4.5 V with low output impedance so that signal currents flowing into it do not give rise to excessive crosstalk between the gyrators. The choice of 4.5 V optimizes the available capacitor bias range while giving sufficient headroom to the n-p-n current sinks in the gyrators.

Extensive use is made of n-p-n and p-n-p current sources and sinks to give immunity to supply rail variations. Special current setting circuits ensure that the absolute values of the operating currents are the same from one gyrator to the next and do not vary too widely with temperature and process spreads. It is impractical to distribute bias *voltages* over the chip to set the gyrator current sources and sinks because of the potential drops down the supply rails. Therefore, to overcome this problem, a "master" current setting circuit is used to produce three equal reference currents from the 4.5 V supply. These *currents* are fed to "slave" circuits close to the gyrators where local bias voltages are generated. Within one gyrator, it is also most important that the p-n-p current sources accurately match the n-p-n current sinks. Offset currents at the ports are then kept low, thus avoiding problems of dc shifts between stages.

CAD Techniques

The design of the filter has relied on extensive computer modeling and simulation. Breadboarding methods for high frequency applications are unsuitable because of their poor representation of the parasitic electrical effects associated with IC components. CAD methods provide for the component tracking and allow experimentation with changes in component dimensions, etc. In designing the filter, only the frequency divider and phase comparator circuits were breadboarded because of their grossly nonlinear mode of operation.

The CAD programs take two files of input data (Fig. 10). The circuit designer manipulates one file by giving the circuit topology and his choices for component dimensions such as resistor widths, length of emitters, number of base contacts, etc. The other file is a process description common to a number of designers; it contains about 110 values of mask clearances, nominal sheet resistances, specific junction capacitance laws, temperature coefficients, etc. The effects of process changes can then be readily tested by altering these values.

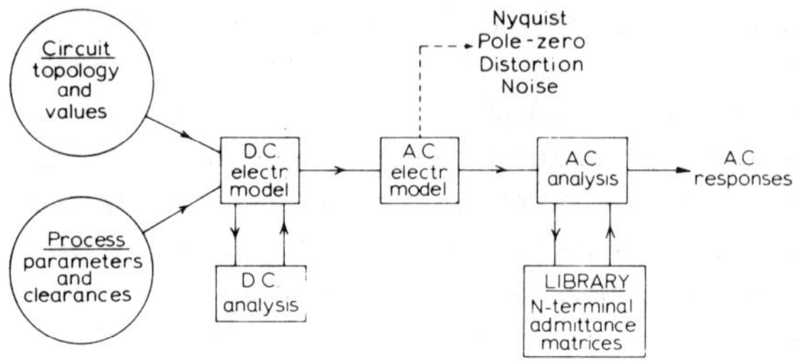

Fig. 10. CAD for modeling and analysis of IC's.

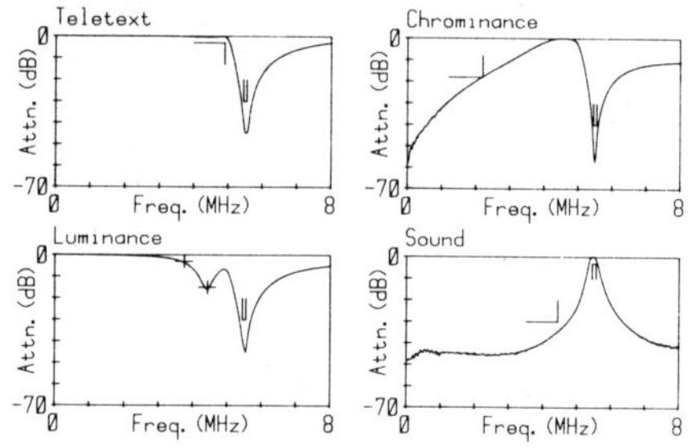

Fig. 11. Measured responses of video filter (with HP 3042A Automatic Network Analyzer).

The two files are combined using geometrically scaled component models, to give a complete circuit representation. These models have evolved in the course of a continuing activity of on-slice dc and HF measurements for the necessary process characterization. Particular attention is paid to accuracy at high frequencies, for example, interconnections are modeled with resistance and stray capacitance.

The filter is first partitioned into circuit blocks which are analyzed, individually. The programs find the dc operating points of the components, then expand the models into a linearized equivalent circuit for interactive ac analysis. Other programs are available for investigating stability via Nyquist plots or pole-zero computation, noise, or nonlinear distortion performances. When a satisfactory design has been reached, the small-signal performance of each circuit block is stored in an admittance matrix library. For instance, a gyrator circuit is stored as a 12-terminal matrix with the gyrator ports, supply and bias rails, and buffered inputs and outputs as terminals of the macromodel.

Although most of the design can be done by considering the circuit blocks separately, it is essential that the complete filter of about 600 components is simulated to study crosstalk effects and to verify the overall responses. This is done efficiently by recalling the macromodels from the admittance matrix library and adding any further components, interconnections, and encapsulation strays. ac analysis on the full circuit, which is equivalent to about 3000 linearized elements can then be rapidly performed.

Crosstalk between the sound channel or tuning circuit and the luminance or chrominance output must be very low for adequate filter rejection performance. This means that any common paths must be carefully considered. Complete simulation of the filter by the methods described above revealed that the common impedances due to bias and supply rails and bonding wire inductances could be a problem. A significant improvement in the bandstop performance was obtained by a rerouting of power rails and an isolation of the resonant currents of individual gyrators.

Performance

Fig. 11 shows the measured responses of a filter for the European TV system *G* with the specification points marked. Samples show very consistent performance and agree closely with design predictions of amplitude and phase response. The tightest specification to be met is the sound trap in the chrominance channel where a minimum attenuation of 40 dB is required over the sound bandwidth of 100 kHz. The trap has been designed to give a tuning tolerance of ±0.5 percent which allows for automatic tuning errors and component mismatch on the chip. The crosstalk performance is very good, thus maintaining deep traps. The signal-to-noise performance is fully adequate, with values better than 80 dB, 70 dB, and 60 dB[1] in the teletext, luminance, and chrominance channels, respectively. This is confirmed by good performance in a television receiver.

Conclusions

This video filter IC is fully compatible with most other television IC's produced in a standard bipolar process and is suitable for mass production. It requires no adjustment and is free of alignment drift. This filter represents an important step towards larger scale integration in television signal processing, as the filtering function could now be combined on one IC with the synchronous demodulator, color decoder, or sound detector.

References

[1] K. W. Moulding, G. A. Wilson, "A fully integrated five-gyrator filter at video frequencies," *IEEE J. Solid-State Circuits*, vol. SC-13, pp. 303–307, June 1978.

[2] A. Fettweis, "Noise performance of capacitor-gyrator filters," *Int. J. Circuit Theory Appl.*, vol. 2, pp. 181–186, June 1974.

[3] J. O. Voorman, "The gyrator as a monolithic circuit in electronic systems," Ph.D. dissertation, Nijmegen, The Netherlands, 1977.

[4] British Patent 1 421 093.

[5] H. T. Van Looij and K. M. Adams, "Phase compensation in electronic gyrator circuits," *Electron. Lett.*, vol. 4, pp. 430–431, Oct. 1968.

[6] British Patent Application 7916112.

[7] R. J. Widlar, "New developments in IC voltage regulators," *IEEE J. Solid-State Circuits*, vol. SC-6, pp. 2–7, Feb. 1971.

[1] Peak-to-peak signal to weighted rms noise ratios measured with filters complying to the CCIR Standard Rec. 451-1.

Fully Integrated Analog Filters Using Bipolar-JFET Technology

KHEN-SANG TAN AND PAUL R. GRAY, SENIOR MEMBER, IEEE

Abstract—A new approach for realizing high-order analog filters which can be fully integrated using a compatible bipolar and ion-implanted JFET process is described. This approach is based on the recognition that what is really needed is a long time constant monolithic integrator which can be effectively realized in a small silicon area. These integrators have been designed, fabricated, and used in a "leapfrog" or active ladder configuration to realize a fifth-order 8 kHz Chebyshev low-pass filter with 0.1 dB passband ripple. No external trimming operations and no anti-aliasing prefilter are required.

Manuscript received May 10, 1978; revised August 3, 1978. This research was supported by the National Science Foundation under Grant ENG75-04986. This paper was presented in part at the International Solid-State Circuits Conference, San Francisco, CA, February 1978.

The authors are with the Department of Electrical Engineering and Computer Sciences, University of California, Berkeley, CA 94720.

I. Introduction

FREQUENCY selective circuits are widely used in communications and control systems. In many cases, these filters must cut off sharply and must have small passband ripple. These tight specifications can only be met with precision high-order filters. Conventional *RC* active filters which use thin film or hybrid technologies often require external trimming [1], [2]. From the standpoint of cost effectiveness, it is desirable to realize a fully integrated filter system which requires small silicon area and no external trimming operations.

Two promising approaches to realizing high-order monolithic filters are the charge-coupled device (CCD) filtering technique, and the switched capacitor metal-oxide-semicon-

Reprinted from *IEEE J. Solid-State Circuits,* vol. SC-12, pp. 814-821, Dec. 1978.

ductor (MOS) filter approach [3], [4]. However, the main disadvantage of these approaches is that they are analog sampled-data systems which require a continuous-time prefilter to minimize aliasing effects.

In this paper, a new approach for realizing precision continuous-time monolithic analog filters is described. This approach is based on the recognition that what is really needed is an integrator with a long time constant which can be effectively realized in a small silicon area. Such an integrator can be used as a building block for realizing different filtering functions. The filter technique described in this paper uses ion-implanted JFET's in a new circuit technique both to achieve the required long integrator time constants and to control them precisely by reference to an external clock frequency.

Since this is a continuous-time analog filter, no anti-aliasing prefilter is required. This approach is demonstrated with a fifth-order 8 kHz Chebyshev low-pass filter with passband ripple of 0.1 dB. Excellent agreement between the experimental and designed values has been obtained.

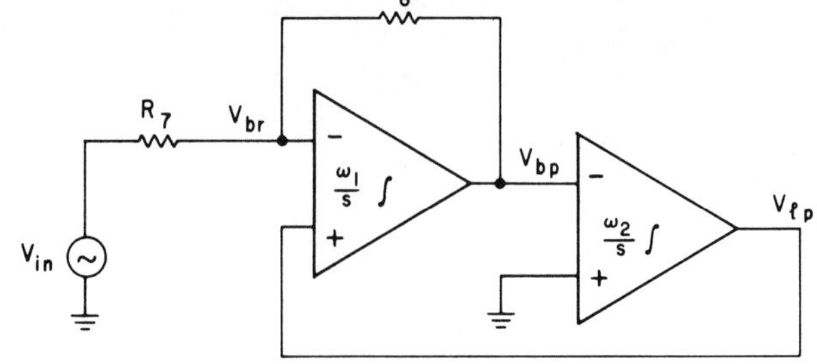

Fig. 1. Second-order filter.

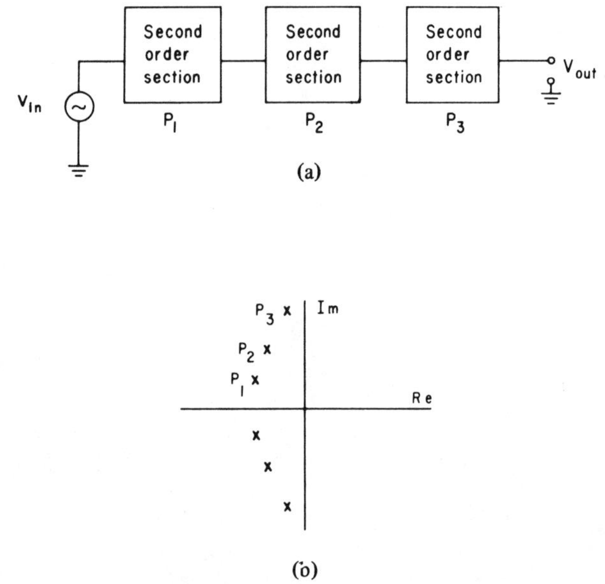

Fig. 2. (a) Sixth-order filter by cascading three second-order sections. (b) Pole locations of the low-pass filter in the s-plane.

II. Integrators as Basic Building Blocks

The key element in a variety of filter configurations is an integrator which can be used to realize different filtering functions. Fig. 1 shows the circuit for a second-order "state variable" filter. It requires two resistors and two differential input integrators. Despite its simplicity, three different transfer functions are available. These transfer functions can be easily derived and are shown as follows:

$$\frac{V_{br}}{V_{\text{in}}} = \left(\frac{R_8}{R_7+R_8}\right)\left(\frac{\omega_1\omega_2+s^2}{s^2+s(\omega_1R_7/(R_7+R_8))+\omega_1\omega_2}\right) \quad (1)$$

$$\frac{V_{bp}}{V_{\text{in}}} = \left(\frac{-R_8}{R_7+R_8}\right)\left(\frac{s\omega_1}{s^2+s(\omega_1R_7/(R_7+R_8))+\omega_1\omega_2}\right) \quad (2)$$

$$\frac{V_{lp}}{V_{\text{in}}} = \left(\frac{R_8}{R_7+R_8}\right)\left(\frac{\omega_1\omega_2}{s^2+s(\omega_1R_7/(R_7+R_8))+\omega_1\omega_2}\right). \quad (3)$$

Examining these expressions carefully, we notice that (1)-(3) correspond to the second-order bandreject, bandpass, and low-pass transfer functions, respectively. For these transfer functions, it is interesting to note that

$$\omega_o = \sqrt{\omega_1\omega_2} \quad (4)$$

$$Q = \left(1+\frac{R_8}{R_7}\right)\sqrt{\frac{\omega_2}{\omega_1}} \quad (5)$$

where ω_o corresponds to the resonance frequency, and Q corresponds to the quality factor of the second-order filter.

These second-order stages can be cascaded to form a higher order filter as shown in Fig. 2(a). In this example, three second-order sections are cascaded to realize a sixth-order filter. Each of the second-order sections realizes a pair of complex poles in the s-plane, as shown in Fig. 2(b). The main advantage of this cascading method is that it is very easy to implement for any realizable transfer function. The main disadvantage is that it typically has a higher sensitivity of network parameters due to component variations as compared with the direct implementation method.

One example of the direct implementation method is the fifth-order doubly terminated low-pass passive ladder network, as shown in Fig. 3(a). From classical circuit theory, it is well known that under the maximum power transfer condition, this network has an extremely low sensitivity due to component variations over the passband region [5].

By standard flowgraph techniques, this passive ladder network can be transformed into the "leapfrog" or active ladder configuration using integrators, as shown in Fig. 3(b) [6], [7]. The transfer functions of these passive and active ladder networks are identical. More importantly, there is a one-to-one correspondence between the reactive element values and the gain constants of the integrators. Because of this one-to-one correspondence, this active ladder network retains the low sensitivity properties of the passive prototype.

For instance, this active ladder network can be used to realize a fifth-order 8 kHz Chebyshev low-pass filter with passband ripple of 0.1 dB. The required unity gain bandwidths of the integrators for this example have values ranging from 4 kHz to 7 kHz, as shown in Table I. To illustrate the sensitivity of this active ladder network, the nominal magnitude response of the fifth-order low-pass filter is plotted across the frequency

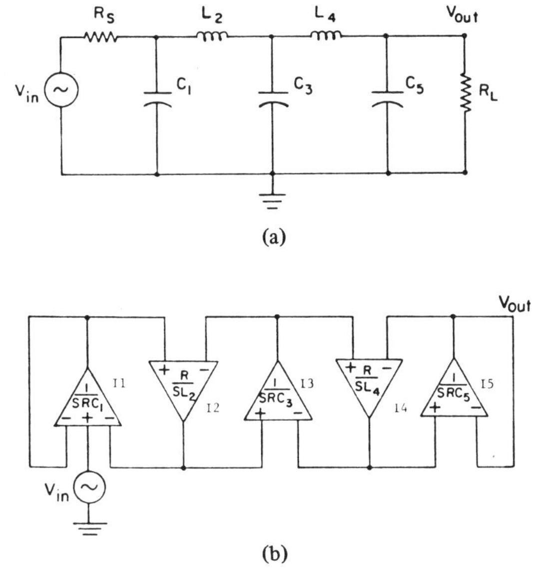

Fig. 3. (a) Doubly terminated *LC* ladder network. (b) Equivalent active ladder network.

TABLE I
THE REQUIRED UNITY-GAIN BANDWIDTHS FOR REALIZING A FIFTH-ORDER 8 kHz CHEBYSHEV LOW-PASS FILTER WITH 0.1 dB PASSBAND RIPPLE

Integrator	Unity-Gain Bandwidth
*I*1	6.9759 kHz
*I*2	5.8343 kHz
*I*3	4.0506 kHz
*I*4	5.8343 kHz
*I*5	6.9759 kHz

band of interest, as shown by the solid line of Fig. 4. The dotted lines indicate the bounds of the worst case deviations of any combination of ±2 percent variations in the gain constants of the integrators. Notice that despite such a wide variation in component values, the maximum deviation is less than 0.1 dB. The extremely low sensitivity of this type of active ladder network enables the monolithic integration of high-order high-precision analog filters without any external trimming.

This direct implementation method is not limited to low-pass realizations. Bandpass, high-pass, and bandreject transfer functions are all realizable using integrators as basic elements [8], [9].

III. MONOLITHIC INTEGRATOR

A. Integrator Requirements

Integrators can be used in a variety of circuit configurations to realize different filtering functions. In this section, the requirements for realizing integrators with time constants in the audio frequency range are described.

A simple differential input integrator can be designed using two resistors, two capacitors, and one operational amplifier, as shown in Fig. 5. The gain constant of this integrator is

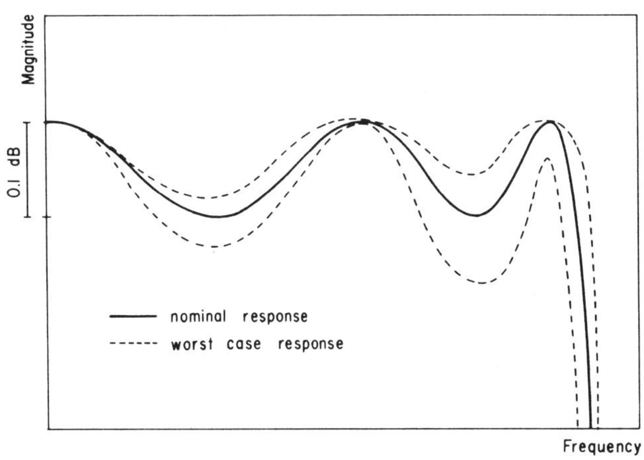

Fig. 4. Worst case deviations for a fifth-order Chebyshev low-pass filter with 0.1 dB nominal passband ripple.

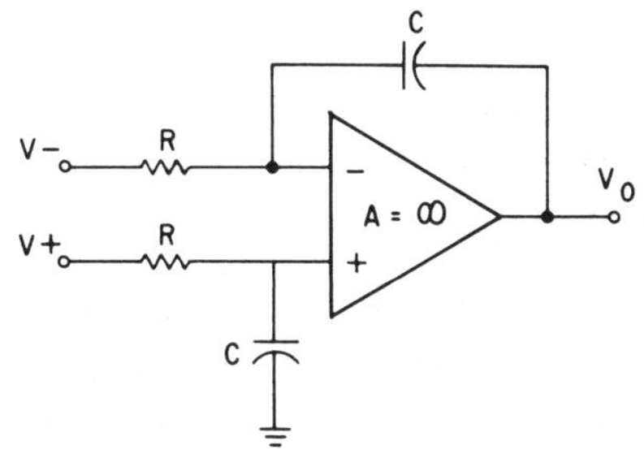

Fig. 5. Simple *RC* differential input integrator.

equal to $1/(RC)$. For low frequency operation, long *RC* time constants are required. For instance, for a frequency of 5 kHz with a monolithic capacitor of 30 pF, a resistor of 1 MΩ is required. This large value resistor is not efficiently realized in conventional monolithic approaches because it requires a large silicon area. Also, the precise value of the *RC* product is difficult to control due to process and temperature variations.

In order to implement integrators monolithically and efficiently, a way to realize integrators with long time constants in small silicon areas is needed. Also, a technique of stabilizing the absolute magnitude of the *RC* product, or equivalently, the gain constant of the integrator is required.

B. Monolithic Integrator

As described earlier, to realize an integrator with a long time constant, a large value of resistance must be realized. The circuit of Fig. 6 effectively realizes a large value resistor by employing a small value of transconductance. In this circuit, the input stage is a transconductance amplifier followed by an inverting gain stage and feedback integrating capacitor. The gain constant of this integrator is precisely defined by the ratio of G_m to C, while the transconductance G_m is given by the following expression:

$$G_m = \frac{2\sqrt{I_d I_{dss}}}{V_p} \tag{6}$$

where I_{dss} is defined as the saturation drain current, I_d is the drain current, and V_p is the pinchoff voltage of the JFET.

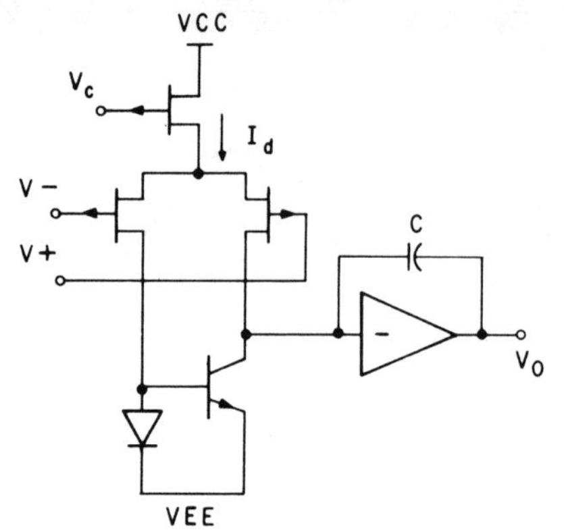

Fig. 6. Basic monolithic differential input integrator.

Again, for a frequency of 5 kHz, using a monolithic capacitor of 30 pF, a transconductance of 1 μmho is required. This small value of G_m can be easily achieved by designing JFET's with small value of Z/L, as well as by operating the input stage with low current. The required chip area for realizing such a small value transconductance (and hence large value of resistance) is small.

By varying the voltage at the controlling node at V_c, the biasing current source I_d can be varied, and by varying I_d, the G_m of the input stage is changed. Hence, by varying V_c, the G_m of the input stage, and thus the gain constant of the integrator can be varied. This feature will be used to control the absolute magnitude of the gain constant of the integrator, as described in the next section.

C. Absolute Bandwidth Control

The absolute magnitude of the gain constant of the integration is difficult to control due to process and temperature variations. This problem can be solved by utilizing the fact that while the absolute values of monolithic component parameters are not well controlled, the matching of one value to another on the same die is quite good. Using this fact, the time constants of the integrators can be locked to an externally supplied precision clock frequency using a phase-locked-loop system, as shown in Fig. 7 [10]. The three main elements of this block diagram are the phase comparator, the voltage-controlled oscillator (VCO), and the voltage-controlled filter (VCF).

The oscillating frequency of the VCO is precisely defined by the gain constants of the two integrators, and is equal to

$$\omega_{osc} = \sqrt{\omega_1 \omega_2} \tag{7}$$

where ω_1, ω_2 are the gain constants. As previously described, the gain constant of the integrator can be controlled through V_c. Hence by controlling V_c, the frequency of oscillation can be controlled. The VCF is designed using the same type of integrators as those of the VCO. The bandwidth of the filter is designed to be exactly proportional to the frequency of the oscillator. By varying V_c, both the frequency of the oscillator and the bandwidth of the filter are varied, and the bandwidth of the filter tracks the frequency of the oscillator. The phase comparator compares the external precision clock frequency f_s to the frequency of the VCO, and generates an error voltage.

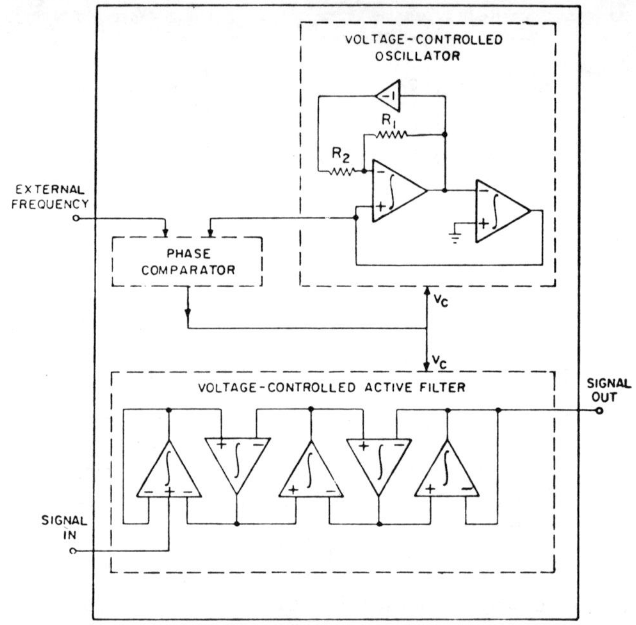

Fig. 7. Block diagram of the filter system.

After passing through a simple low-pass filter, the error voltage V_c is then fed back to the controlling node. In the steady state, V_c forces the gain constants of the integrators to have the desired values, such that the frequency of the oscillator is exactly equal to the external precision clock frequency. Since the bandwidth of the filter is always exactly proportional to the frequency of the oscillator, the bandwidth of the filter is always exactly proportional to the external precision clock frequency, and is independent of process and temperature variations.

The approach described above has several advantages. First, monolithic integrators with long time constants can be realized in a small area. Second, the ratio-matching of the gain constants of integrators depends only on the ratio-matching of the current sources and the ratio-matching of the capacitors. These can be accurately achieved [11], [13]. Finally, the absolute magnitude of the gain constant, and thus the bandwidth of the filter, can be precisely controlled by reference to an external precision frequency, and it is insensitive to process and temperature variations.

IV. Practical Design Considerations

Ideally, an integrator has the $1/f$ frequency response, as shown in Fig. 8(a). The phase is exactly 90° and the gain at dc is infinite. Unfortunately, an actual monolithic integrator deviates from these ideal characteristics. For instance, the nonideal integrator has a finite dc gain. However, this is not an important consideration since for most cases, a dc gain of about 1000 is sufficient to give very accurate frequency responses.

Another nonideal effect is due to the second pole of the monolithic integrator, as shown in Fig. 8(b). In this figure, f_0 corresponds to the gain constant, and f_1 corresponds to the second pole of the nonideal integrator. Because of the presence of f_1, the phase angle at f_0 is not exactly 90°. This

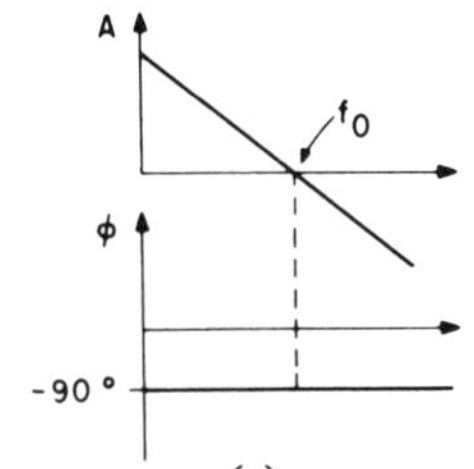

(a)

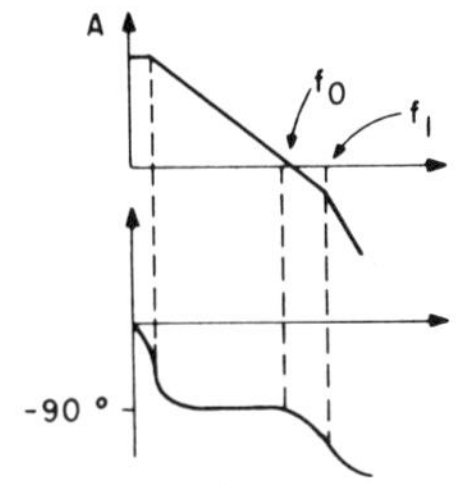

(b)

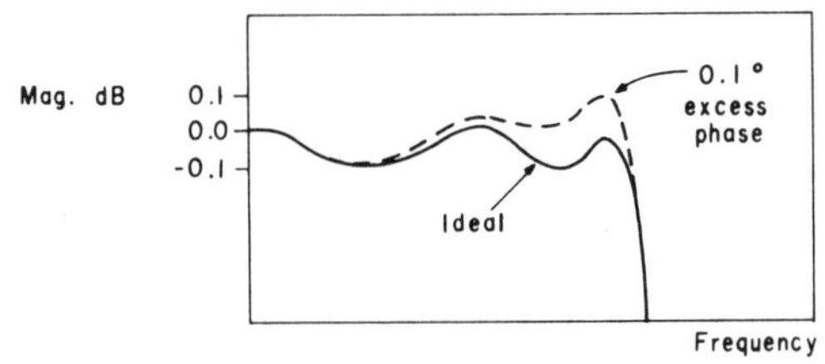

(c)

Fig. 8. (a) Frequency and phase response of an ideal integrator. (b) Frequency and phase response of a nonideal integrator. (c) Effects of excess phase.

excess phase will cause a slight peaking in the magnitude response, as shown in Fig. 8(c). In this case, a 0.1° phase error causes approximately a 0.1 dB peaking in the passband of the magnitude response of a fifth-order Chebyshev low-pass filter. Typically, to achieve a phase error of less than 0.1°, the ratio of f_1 to f_0 has to be greater than 50.

One important practical consideration is the nonlinearity of the monolithic integrator. Fig. 9(a) shows the basic monolithic integrator. A closer look at this circuit discloses that the input transconductance stage is linear for a limited range of input voltage. A plot of output current versus input voltage is shown in Fig. 9(b). Notice that the output current saturates at a maximum input voltage of $^{+}V_p$, where V_p is the pinchoff voltage of the JFET transistors.

To improve the linearity of the transconductance amplifier, the maximum input voltage range can be extended by using the input stage shown in Fig. 10(a). In this circuit, four additional transistors, $Q2$ through $Q5$, are used as source degeneration JFET resistors to increase the linearity of the transconductance amplifier. A plot of output current versus input voltage is shown in Fig. 10(b). Notice that by simply adding four transistors, the maximum input voltage range has been increased by a factor of 3. Actually, we can add as many of the JFET resistors as necessary to improve the linearity of the transconductance amplifier. The only disadvantage is that when the number of transistors increases, the required chip

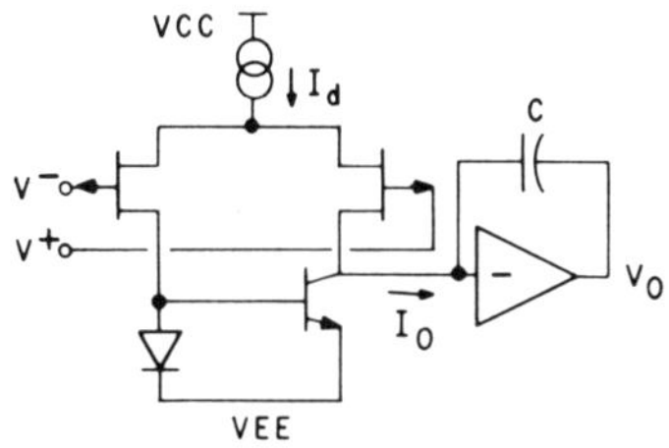

(a)

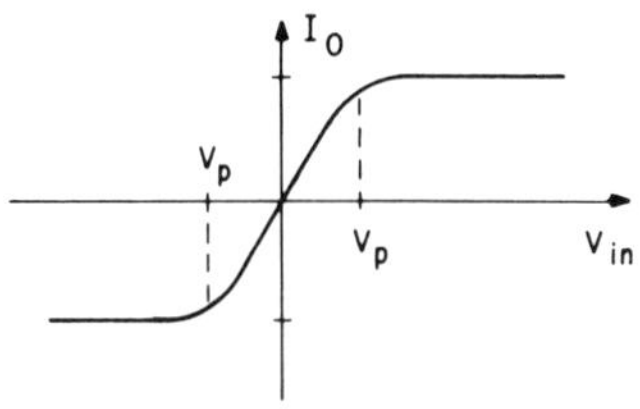

(b)

Fig. 9. (a) Basic monolithic integrator. (b) Plot of I_0 versus V_{in}.

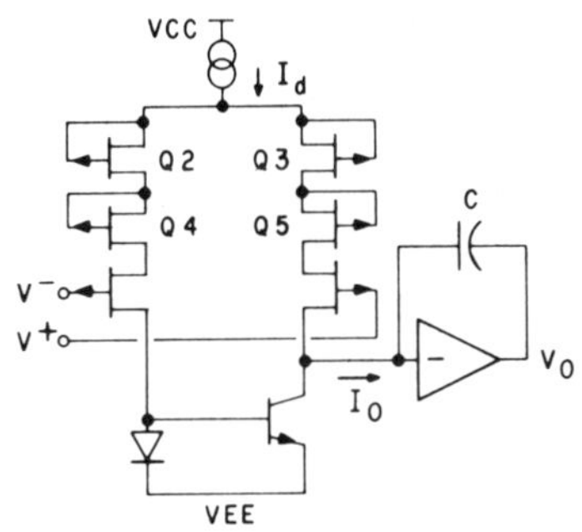

(a)

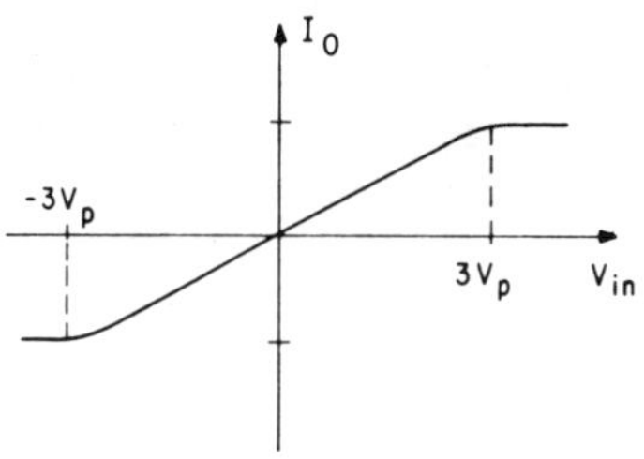

(b)

Fig. 10. (a) Monolithic integrator with a new input stage. (b) Plot of I_0 versus V_{in}.

area also increases. This may impose a practical limit on the maximum number of transistors used.

The final design of the monolithic integrator is shown in Fig. 11. The input transconductance stage consists of transistors $Q1$ through $Q12$. Transistors $Q2$ through $Q5$ are used as source degeneration resistors as mentioned earlier. Transistors $Q8$ through $Q10$ are used as active loads. Transistors $Q11$ and $Q12$ are used as emitter degeneration resistors to improve the noise and offset of the integrator. With no emitter degeneration, the noise and offset of the bipolar devices when referred

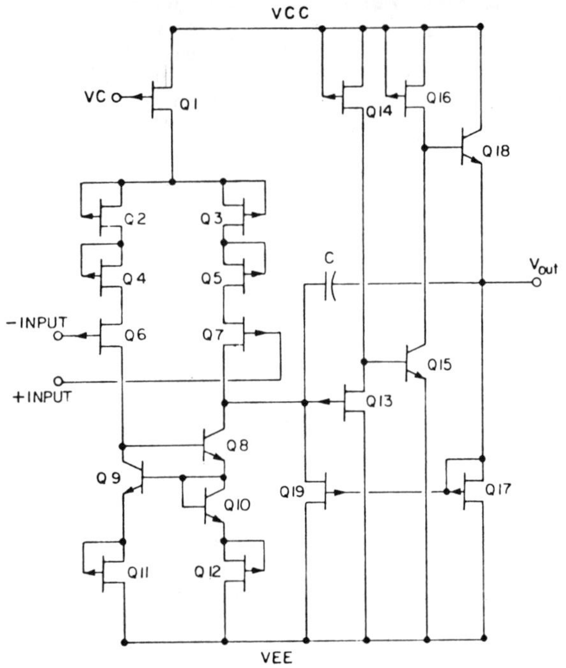

Fig. 11. Schematic of the monolithic integrator.

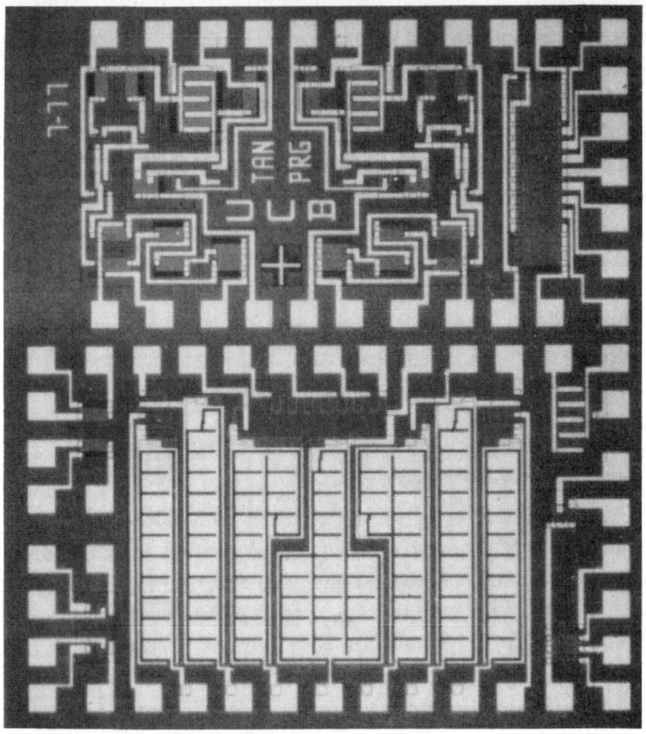

Fig. 12. Photograph of the experimental IC.

to the input will be amplified by the ratio of the bipolar to JFET transconductances. By using JFET transistors as degeneration resistors, we can lower the noise and the offset voltage of the integrator with little increase in silicon area. Transistor $Q1$ is the biasing JFET current source. By controlling the voltage at node V_c, the biasing current, and thus the transconductance of the input stage can be varied. This feature is used to control the bandwidth of the integrator as outlined earlier.

The following stage is the source follower which consists of transistors $Q13$ and $Q14$. This is used to minimize the loading to the input transconductance stage. Transistors $Q15$ and $Q16$ constitute the common emitter inverting gain stage. Transistors $Q18$ and $Q17$ are used as emitter follower output stage to reduce the output impedance to about 10 kΩ.

The output of the integrator is clamped by transistor $Q19$, such that the output common mode range is always smaller than the input common mode range. This prevents the integrator from latching up.

Typically, to realize a given transfer function, integrators of different gain constants are required. These different gain constants are achieved by scaling the ratios of the integrating capacitors while keeping all the input transconductance stages identical. In this way, excellent matching of transconductances can be achieved. Previous work has shown that it is possible to achieve ratio-matching accuracy of a few tenths of one percent for monolithic MOS capacitors [11].

V. Experimental Results

A. Description of the Experimental IC

The photograph of the experimental IC chip is shown in Fig. 12. The overall chip size is 90 × 100 mils. A 10 μm minimum feature size and 5 μm minimum alignment tolerance were used. In this chip, there are seven ratioed MOS capacitors of values ranging from 15 to 30 pF. There are two identical monolithic integrators with each realized in a silicon area of 850 mils^2. A bipolar compatible ion-implanted JFET process has been used [12], [13]. The process is similar to a standard bipolar process, with the addition of two ion-implantation steps. The first implant (boron) is used to create the channel while the second implant (phosphorus) is used to realize the top gate of the JFET transistors. The JFET pinchoff voltage is designed to be 4 V.

TABLE II
Integrator Performance Data

Power supplies	±12 V
DC open-loop gain	10 000
CMRR, PSRR (dc)	70 dB
Nominal unity-gain bandwidth (20 pF)	7 kHz
Input offset voltage (standard deviation)	16 mV
Power dissipation	4 mW
Chip area	850 mil^2

B. Performance of the Integrator

The measured performance of the monolithic integrator agrees very well with the design values. With power supplies of ±12 V, the dc open loop gain is 10 000. Both the power supply rejection ratio (PSRR) and the common mode rejection ratio (CMRR) are 70 dB. The nominal unity gain bandwidth using a 20 pF integrating capacitor is 7 kHz. The slew rate using a 20 pF integrating capacitor is 0.5 V/μs. The total power dissipation for each integrator is only 4 mW. Table II summarizes the performance of the monolithic integrator.

Realization of high-order high-precision filters requires precision ratio-matched integrators. As described earlier, the ratio-matching of integrators depends on the ratio matching of G_m/C. Experimental data on the matching of G_m and C are tabulated and shown in Fig. 13. Based on a sample size of 43, the standard deviation of the matching of MOS capacitors is 0.24 percent for this process, and the standard deviation of G_m based on 94 samples, is found to be 0.75 percent. Since these two parameters are uncorrelated, the standard deviation of the ratio matching of monolithic integrators is $(0.75^2 + 0.24^2)^{1/2} =$ 0.79 percent. This will introduce a worst case deviation of less than 0.05 dB in the passband magnitude response of the fifth-order low-pass filter mentioned earlier.

C. Performance of the Fifth-Order Low-Pass Filter Using Leapfrog Structure

The monolithic integrators described earlier have been used in a "leapfrog" or active ladder configuration to construct a fifth-order 8 kHz Chebyshev low-pass filter with 0.1 dB passband ripple. To realize such a fifth-order low-pass filter, a total of seven integrators are required. Two integrators are used in the VCO and five integrators are used in the filter section. The estimated area to integrate this complete fifth-order low-pass filter using the process previously described would be 10 000 mils^2, and the total power dissipation would be approximately 50 mW.

The measured frequency response of the fifth-order low-pass filter is shown in Fig. 14(a). The measured passband ripple is 0.1 dB to within 0.05 dB, with a cutoff frequency of 8 kHz. By controlling the external frequency, the cutoff frequency of the filter was tunable over a range of 6 kHz to 10 kHz. The measured frequency response over a wider frequency range is shown in Fig. 14(b). Notice that the filter is performing well into the 200 kHz range without any aliasing problem. The stopband rolloff is approximately 100 dB/decade. These measured data were obtained without any external trimming operations, and they agree very well with the expected design values.

The filter is relatively insensitive to power supply variations. A 20 percent variation in the power supply introduced less than a 0.02 dB change in the passband of the magnitude response. A change of 50°C in ambient temperature resulted in a passband deviation of 0.05 dB and a cutoff frequency variation of 1 percent. These results were obtained on a bread-boarded filter system which consists of several packages of monolithic integrators. It is believed that, for a fully monolithic filter system, the sensitivity due to temperature variation will be very much reduced.

The filter has a wide dynamic range of operation. The measured output noise in the frequency band of 100 Hz to 10 kHz is 200 μV rms. At a signal frequency of 1 kHz, the measured total harmonic distortion (THD) for an output voltage of 7.5 V peak-to-peak is only 0.3 percent. When the output voltage is increased to 10 V peak-to-peak, the measured THD is increased to 1 percent. Based on the 1 percent THD output voltage and the 200 μV rms noise level, the dynamic range is 85 dB. Table III summarizes the performance of this fifth-order low-pass filter.

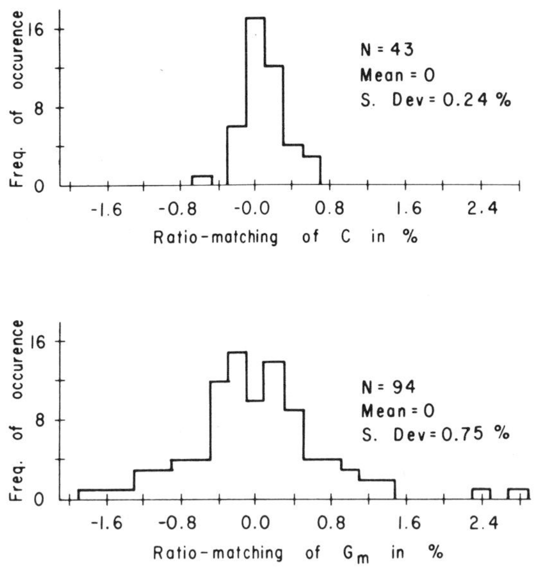

Fig. 13. Distributions of the matching of G_m and C.

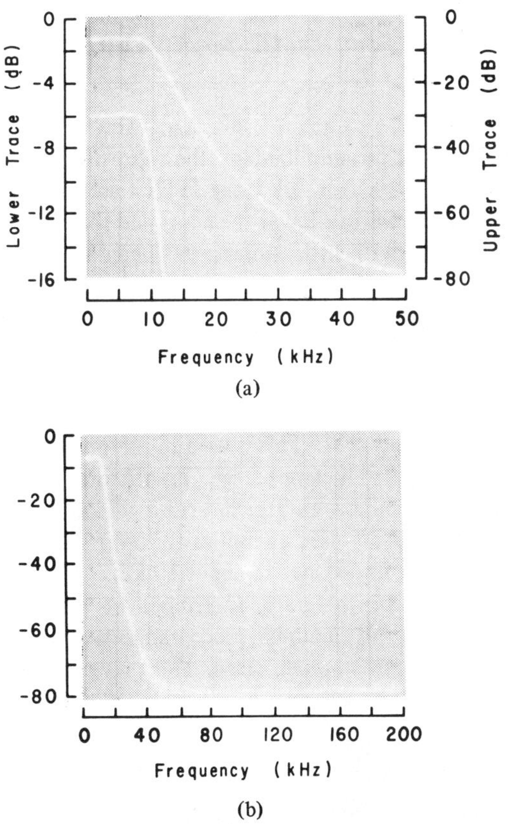

Fig. 14. (a) Measured frequency response of the fifth-order Chebyshev low-pass filter over a narrow frequency range. (b) Measured frequency response of the low-pass filter over a wide frequency range.

TABLE III
FIFTH-ORDER FILTER PERFORMANCE DATA

Ripple bandwidth	8 kHz
Passband ripple	0.1 dB
Output THD (1 kHz, 10 V peak to peak)	1 percent
Output noise (100 Hz–10 kHz)	200 μV rms
Dynamic range	85 dB

VI. Conclusions and Future Developments

The design techniques for realizing fully integrated analog filters using a compatible bipolar and ion-implanted JFET technology have been described. This approach is applicable to a wide variety of filters. Monolithic integrators with long time constants have been designed and fabricated. These integrators have been used in a "leapfrog" or active ladder configuration to realize a fifth-order 8 kHz Chebyshev low-pass filter with 0.1 dB passband ripple. The low sensitivity properties of the active ladder structure enables the realization of this high order filter without any external trimming operations. A dynamic range of 85 dB has been obtained. Since this is a continuous time approach, no anti-aliasing prefilter is required.

Because of this nonsampled-data approach, this technique may be able to achieve a higher frequency of operation. Also, for this active ladder low-pass network, transmission zeros can be easily achieved without any additional integrators. Only two additional capacitors are needed for each pair of transmission zeros [14].

Acknowledgment

The authors want to thank S. K. Lui, who developed the bipolar compatible ion-implanted JFET process used in the experimental devices, and D. J. Allstot for useful discussions and assistance.

References

[1] J. J. Friend *et al.*, "Star: An active biquadratic filter section," *IEEE Trans. Circuits and Systems*, vol. CAS-22, pp. 115–121, Feb. 1975.

[2] R. A. Friedenson *et al.*, "RC active filters for the D3 channel bank," *Bell Syst. Tech. J.*, vol. 54, pp. 507–529, Mar. 1975.

[3] R. D. Baertsch *et al.*, "The design and operation of practical charge transfer transversal filters," *IEEE Trans. Electron Devices*, vol. ED-23, pp. 133–141, Feb. 1976.

[4] D. J. Allstot, R. W. Brodersen, and P. R. Gray, "MOS switched capacitor ladder filters," this issue, pp. 806–814.

[5] H. J. Orchard, "Inductorless filters," *Electron. Lett.*, vol. 2, pp. 224–225, Sept. 1966.

[6] O. Wing, "Ladder network analysis by signal flow graph—Application to analog computer programming," *IRE Trans. Circuit Theory*, vol. CT-3, pp. 289–297, 1956.

[7] W. E. Heinlein and W. H. Holmes, *Active Filters for Integrated Circuits*. Englewood Cliffs, NJ: Prentice-Hall, 1974.

[8] F. E. J. Girling and E. F. Good, "Active filters, Part 12: The leapfrog or active-ladder synthesis," *Wireless World*, pp. 341–345, July 1970.

[9] F. E. J. Girling and E. F. Good, "Active filters, Part 13: Applications of the active-ladder synthesis," *Wireless World*, pp. 445–450, Sept. 1970.

[10] F. M. Gardner, *Phase Lock Techniques*. New York: Wiley, 1966.

[11] J. L. McCreary and P. R. Gray, "All MOS charge-redistribution A-D conversion techniques, Part I," *IEEE J. Solid-State Circuits*, vol. SC-10, pp. 371–379, Dec. 1975.

[12] R. W. Russell and D. D. Culmer, "Ion implanted JFET-bipolar monolithic analog circuits," in *ISSCC Dig. Tech. Papers*, pp. 140–141, 1974.

[13] S. K. Lui *et al.*, "Process related characterization of bipolar compatible ion-implanted JFETs," to be published.

[14] K. S. Tan, "High-order continuous-time monolithic analog filters," Ph.D dissertation, Univ. California, Berkeley, 1978.

Part VI
Techniques for Monolithic Switched-Capacitor Filters

Since their recent rediscovery [1,2], switched-capacitor (SC) filters have captured world-wide attention, both from universities and from industry. This interest is motivated by the proven ability of SC networks to realize precision monolithic analog filters in MOS integrated circuit (IC) technologies. Since the total capacitance required to implement a second-order audio-frequency filter is typically less than 100 pF, many poles and zeros of filtering can be packed onto a small MOS chip. Furthermore, digital circuits, memories and analog circuits can be compatibly fabricated on a single MOS chip, bringing new viability to the concept of single chip systems†.

Switched-capacitor filters are usually recursive, or infinite impulse response (IIR), sampled data filters; therefore, high Q's and flat passbands can be realized efficiently, much like in the active RC realizations treated in earlier Parts. The transfer functions of SC filters are determined by highly stable crystal controlled clock frequencies and ratioed capacitors. This property, coupled with the high degree of correlation that MOS capacitors in close proximity on an IC chip exhibit over wide ranges of temperature [3] yields the mentioned filter precision. In contrast to s-domain transfer functions which describe continuous analog filters, sampled data SC circuits are characterized by z-domain transfer functions. As a direct result of the IIR architecture and the ability to make precision ratios with small capacitors, sampling rates in excess of 100 kHz can be rountinely accommodated. The high sampling rate serves to lessen the burden placed on the continuous antialiasing filter required to band limit the input spectrum to one-half the sampling frequency.

The enthusiasm for SC filters was greatly inspired by the research [1,2] conducted at the University of California at Berkeley. This work represented the first practical marriage of the SC concept with MOS operational amplifiers and with the inherent capacitor ratio precision offered by MOS fabrication. *Paper 6-1* by Broderson, Gray and Hodges presents an excellent review of the work done and the insight into SC filter operation gained by this group.

Many of the SC network techniques used today were explored some ten or more years ago [4-10]. Specifically the theory of resonant transfer circuits described in [5] is employed in *paper 6-3* by Fettweis, the periodically reverse-switched capacitor described in [6,8] is used in circuits given in *paper 6-2* by Temes, Orchard and Jahanbegloo and in paper 6-3, and the toggle-switched capacitor described in [9] is the workhorse in [1,2] and paper 6-1. However, without MOS technology to exploit SC properties, the usefulness of the early results was limited and, as a consequence, this body of literature laid dormant until recently. Because design and implementation of monolithic filters is not addressed in these early papers, they are not reprinted in this chapter. However, the pioneering contributions they represent should not go unrecognized.

Since the design of continuous analog passive RLC and active RC filters is well understood and mature, many of the SC filters appearing in the literature are derived from existing prototype continous analog networks, e.g. papers 6-1, 6-2 and 6-3. The derivation is consummated either by transforming s-domain admittances of a continuous filter schematic into equivalent z-domain SC networks or by transforming s-domain functional blocks, such as integrators, in a network flowgraph into SC equivalents. The SC implementations for the z-domain admittances or flowgraph functions are then inserted directly into the circuit schematic or flowgraph. Since much of the large body of circuits and design tables for continuous analog filters are used directly in these approaches, the design labor is reduced in principle. However, the degree to which the resulting SC filter emulates the performance of the continous analog prototype depends on the type of s-to-z transformation used and on the sampling rate.

Paper 6-5 by Fleischer and Laker presents a somewhat different approach to the design of SC filters and addresses a practical nemesis to the MOS implementation. The approach adopted in this paper is to perform the filter design directly in the z-domain, as one would design a digital filter. That is, first determine the z-domain transfer function which satisfies the frequency domain specifications, and then realize it with the best possible SC network, i.e., one which has good sensitivity properties, is implementable within the allotted chip area, and is tolerant of the unavoidable chip parasitics. In this spirit, a family of SC biquad circuits, providing a general z-domain biquadratic transfer function capability, is developed. Crucial to the successful MOS implementation of these biquads are the switching arrangements which lead to immunity to all parasitic capacitances from the back biased substrate (ac ground) to the top and bottom capacitor plates. This most important innovation was discovered independently by many, including [11-13] and paper 6-5. Unfortunately, all of the SC filters derived by direct transformation of continuous analog prototypes, see e.g. papers 6-1, 6-2, and 6-3, are plagued by at least the top plate parasitics and, in some cases, the bottom plate parasitics as well. Furthermore, the implementations described in papers 6-2 and 6-3 are not realizable with parasitic-immune switching.

† *Many examples of complex single chip analog/digital systems can be found in the digest of papers for the 1979 and 1980 IEEE International Solid State Circuits Conferences.*

Once an SC filter has been designed, it is highly desirable to perform either pencil and paper analysis or computer simulation to verify performance. The final reprint in this section, *paper 6-6* by Liou and Kuo, addresses this task. It provides detailed explanations and guidelines for the formulation of the problem of SC analysis. A rigorous analysis of general SC networks is a most challenging and treacherous undertaking because of the time varying nature of SC circuits which stems from the periodic opening and closing of the switches. Thus in a bi-phase SC network, depending on the relative phasing of switches, different transfer functions can be observed when the output is sampled during either one of the two phases or during both phases. The problem quickly becomes unmanageable with pencil and paper when more than two clock phases are present.

Reference [14] describes the first attempt to provide a rigorous network theoretic framework for the analysis of SC networks. The analysis in these papers was restricted to bi-phase SC circuits, excited by sampled-data signals. The principle contribution is the concept that bi-phase SC filters can be viewed as having two signal paths,one for each switch phase, coupled together, or linked, via time invariant z-domain equivalent circuits. With this concept, familiar linear time invariant circuit theory can be directly applied to the analysis of SC networks. In [15] this concept was used to provide a formalism suitable for pencil and paper analysis, a technique used extensively in paper 6-5. In paper 6-3 the analysis was cast into a state variable framework and generalized to accommodate continuous analog excitations. The formulation is highly suited to efficient computer implementation for frequency-domain simulations. A general time-domain analysis program [16] that obtains frequency-domain results via an FFT operation on the time-domain waveforms has recently appeared in the literature. In addition, the program allows n-phase switching and is sufficiently versatile to permit the study of clock feedthrough, offset drift, noise and nonlinear parasitic effects.

[1] I. A. Young, D. A. Hodges and P. R. Gray, "Analog MOS sampled data recursive filters", IEEE Int. Solid State Circuits Conference, Digest of Technical Papers, pp. 156-157, February 1977.

[2] B. J. Hosticka, R. W. Brodersen and P. R. Gray, "MOS sampled data recursive filters using state variable technques", Proc. IEEE Int. Symp. on Circuits and Systems, pp. 525-529, April 1977.

[3] J. L. McCreary and P. R. Gray, "All-MOS charge redistribution analog-to-digital conversion techniques", IEEE J. Solid State Circuits, vol. SC-10, No. 6, pp. 371-379, December 1975.

[4] W. Poschenrieder, "Frequency filters with periodically operated switches", Proc. NTG Symp. on the Analysis and Synthesis of Networks, Stuttgart, Germany, pp. 220-237, 1966.

[5] A. Fettweis, "Theory of resonant transfer circuits", published in Network and Switching Theory (edited by G. Biorci), Academic Press, pp. 382-446, 1968.

[6] J. A. McKinney and C. A. Haljiak, "The periodically reverse-switched capacitor", IEEE Trans. on Circuit Theory, vol. CT-15, No. 3, pp. 288-290, September 1968.

[7] R. Boite and J. P. V. Thiran, "Synthesis of filters with capacitances, switches and regenerating devices", IEEE Trans. on Circuit Theory, CT-15, No. 4, pp. 447-454, December 1968.

[8] A. Fettweis, "On the theory of periodically reverse-switched capacitor networks", AEU, vol. 24, No. 12, pp. 539-544, December 1970.

[9] D. L. Fried, "Analog sampled-data filters", IEEE J. on Solid State Circuits, vol. SC-7, No. 4, pp. 302-304, August 1972.

[10] H. Weinrichter, "Capacitor/commutator networks", AEU, vol. 26, No. 7/8, pp. 293-305, July/August 1972 (in German).

[11] B. J. White, G. M. Jacobs and G. F. Landsburg, "A monolithic dual-tone multifrequency receiver", IEEE Int. Solid State Circuits Conference Digest of Papers, pp. 36-37, February 1979.

[12] K. Martin and A. S. Sedra, "Strays-insensitive switched capacitor filters based on the bilinear transform", Electronic Letters, vol. 15, No. 13, pp. 365-366, June 1979.

[13] K. Martin, "Improved circuits for the realization of switched capacitor filters", IEEE Trans. on Circuits and Systems, vol. CAS-27, No. 4, pp. 237-244, April 1980.

[14] C. F. Kurth and G. S. Moschytz, "Nodal analysis of switched capacitor networks", and "Two-port analysis of switched capacitor networks using four-port equivalent circuits", IEEE Trans. on Circuits and Systems, vol. CAS-26, No.'s 2 and 3, pp. 93-105 and pp. 166-179, February & March 1979.

[15] K. R. Laker, "Equivalent circuits for the analysis and synthesis of switched capacitor networks", Bell System Tech. Journal, vol. 58, No. 3, pp. 729-769, March 1979.

[16] H. J. DeMan, J. Rebaey, G. Arnoui and J. Vandewalle, "Practical implementation of a general computer aided design technique for switched capacitor circuits", IEEE J. Solid State Circuits, vol. SC-15, No. 2, pp. 190-200, April 1980.

MOS Switched-Capacitor Filters

ROBERT W. BRODERSEN, MEMBER, IEEE, PAUL R. GRAY, SENIOR MEMBER, IEEE AND DAVID A. HODGES, FELLOW, IEEE

Invited Paper

Abstract—In the past several years, much progress has been made in bringing the economies of integrated-circuit technology to bear on the realization of voiceband frequency selective filters. This paper will review one approach to this problem, the use of switched-capacitor techniques. The paper emphasizes the practical aspects of switched-capacitor filter design under the constraints imposed by MOS integrated-circuit technology. The basic operation of switched-capacitor filters is reviewed, followed by a discussion of the properties of the various circuit building blocks in MOS technology. Finally, a summary of several filter organizations which appear to be well suited to switched-capacitor implementation is presented.

I. INTRODUCTION

FREQUENCY selective analog filtering has long evaded the efforts of technologists and designers to develop filter realizations which can be completely integrated on a silicon chip. The development of low-cost high-performance monolithic operational amplifiers led to the utilization of active-*RC* filtering techniques to replace passive *RLC* filters in the 1960's. Classical *RC*-active filters have proven difficult to realize in completely monolithic form because stable precise *RC* products cannot be achieved in standard monolithic processes. As a result, the preferred technological approach to voice-frequency analog filtering until very recently has been discrete component or hybrid integrated-circuit realizations of various types of active filters. Recent developments in monolithic circuit techniques offer the possibility of fully integrating the voice-frequency filtering function, with resultant savings in manufacturing cost. One of these techniques, the switched-capacitor approach, is the subject of this paper.

This technique makes use of the unique properties of metal-oxide–semiconductor (MOS) integrated-circuit technology. This technology has become widely used in the realization of digital LSI circuits because of its superior logic density, which arises from the fact that, in contrast to bipolar, the MOS transistor is self-isolating. While the density advantage of MOS has led to its wide utilization for digital circuits, a second-key advantage, which has only recently been recognized, has led to the increasing use of MOS technology to perform analog signal processing. In contrast to bipolar technology, MOS integrated circuits offer the ability to store charge on a node over a period of many milliseconds and to sense the value of the charge continuously and nondestructively. The former results from the high impedance of the MOS transistor in the off state, and the latter results from the essentially infinite input impedance of the MOS transistor in the active mode of operation. This inherent analog memory capability does not exist in bipolar technology because the recombination current associated with the flow of minority carriers in a bipolar transistor rapidly discharges the small capacitance on which the voltage information is stored.

Manuscript received June 17, 1978; revised September 21, 1978. The research was supported by DARPA and monitored by NRL under Contract N000173-77-C-0238.

The authors are with the Department of Electrical Engineering and Computer Sciences and the Electronics Research Laboratory, University of California, Berkeley, CA 94720.

The charge storage feature of MOS technology was first utilized in dynamic random-access memory design and dynamic logic. In both instances, information storage on small nodal capacitances in the circuit allows the elimination of active or pullup transistors which are otherwise required to preserve the information content of the circuit. This results in greater circuit density. While the information stored on nodal capacitance is essentially analog in that the node voltage has a continuous range of values, in digital applications the information is usually interpreted as being only one of two logic levels; these binary levels are typically refreshed (restored) every two milliseconds.

The charge storage feature of MOS technology was first used to perform analog signal processing with the advent of the bucket brigade shift register. In this device, a time-varying analog signal is converted to a series of charge packets, proportional to the value of the input signal at the sample instants. These charge packets are passed serially through a chain of MOS capacitors, the motion of the charge being brought about by the action of appropriately clocked MOS devices. In analog filtering applications of such devices, the charge at each capacitor site is sensed nondestructively by MOS transistors. These sensed voltages can, for example, be added in a weighted summation to produce a single output, providing realization of sampled-data transversal filter [1].

The charge-coupled device (CCD), which was developed subsequent to the bucket brigade, provides a similar function in analog-signal processing applications, but with the integration of the storage capacitance with the MOS switch structure and MOS sensing device in such a way that transversal filters and other analog signal processing devices can be fabricated in a smaller silicon area and achieve higher performance. These devices provide an extremely effective technological tool for attacking filtering problems in which transversal filter structures are required, for example as matched and adaptive filters [2].

However, the utilization of CCD and bucket-brigade filters in large-quantity low-cost general filtering applications has been hindered by several factors. First, until very recently, the peripheral circuitry required to convert the input signal into charge packets and to recover a low-impedance voltage signal at the output had to be realized off-chip since this circuitry was not thought to be easily realizable on the same chip. Second, the transversal filter itself has certain limitations as a general-purpose analog filtering and signal process-

Reprinted from *Proc. IEEE,* vol. 67, pp. 61-75, Jan. 1979.

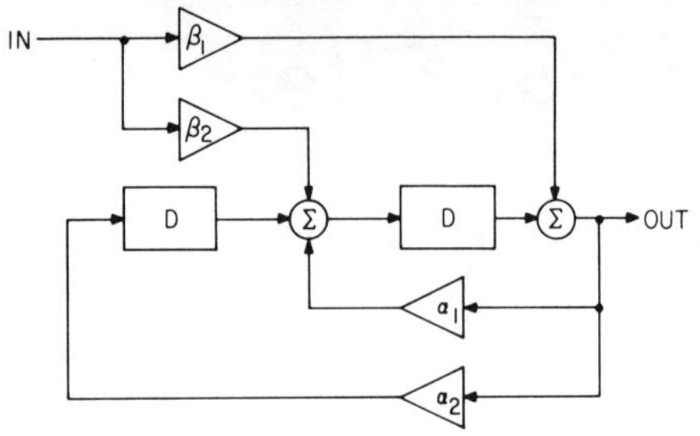

Fig. 1. A second-order direct form discrete time filter.

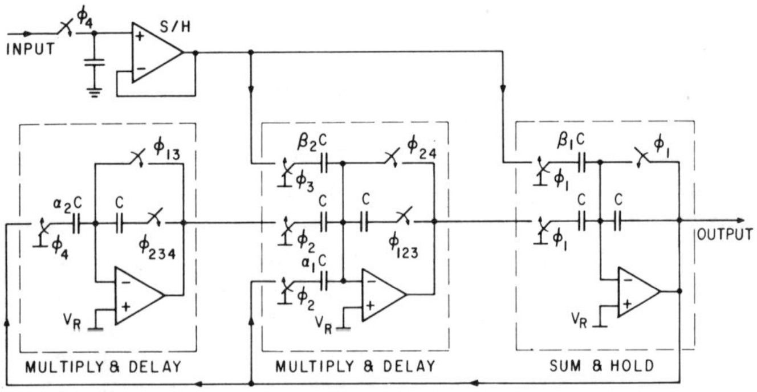

Fig. 2. Analog sampled data realization of the filter shown in Fig. 1 using MOS switches, capacitors, and operational amplifiers.

ing technique. Among these is the fact that since the signal is time sampled at the input to the CCD, a continuous-time antialiasing filter is required to bandlimit the input signal to a frequency below the sampling rate. Since the sampling rate in CCD's must be relatively low (to minimize filter length and conserve silicon area) a relatively complex anti-aliasing filter is required. Also, since CCD filters are nonrecursive, for transfer functions with a narrow passband which have long impulse responses, a very long CCD filter utilizing large silicon area is required.

In the last two years, the concept of analog charge storage and processing in MOS technology has been applied in new ways and has overcome some of the difficulties inherent in the transversal filter structures. First, in order to obtain highly selective filters with long impulse responses in a small silicon area, one is led to adopt a recursive filter structure. The classical second-order recursive filter section, shown in Fig. 1, requires two delay elements (shown as a D) and four multiplication elements with factors which can be greater than unity. While a passive charge-coupled structure can realize the delays, an active amplifier of some type must be used to provide gain. This second-order recursive section can perhaps be most efficiently realized by simply using two capacitive delay elements together with several amplifiers as shown in Fig. 2 [3], [4]. This structure has the advantage that infinitely long impulse responses can be obtained, but it suffers from the disadvantage that as one attempts to make the sampling rate higher to ease the anti-aliasing requirements, the sensitivity of the filter response to variation in the parameters α_1 and α_2 becomes larger. Since in an analog recursive filter like the one shown in Fig. 2, these coefficients are realized using ratioed passive components, a severe practical limitation is imposed on the range of sample rates and the selectivity of responses obtainable from analog realizations of these filters.

The recursive section of Fig. 2 consists of analog switches, operational amplifiers, and capacitors. These same elements can be rearranged to provide many alternate filtering configurations with sensitivity properties more closely resembling those of continuous-time filters in the limit of high sampling rates. These filters have come to be called switched-capacitor filters, and are distinguished by the fact that they utilize a capacitor and MOS switches to simulate the circuit behavior of a resistor, as shown in Fig. 3 [5]-[7].

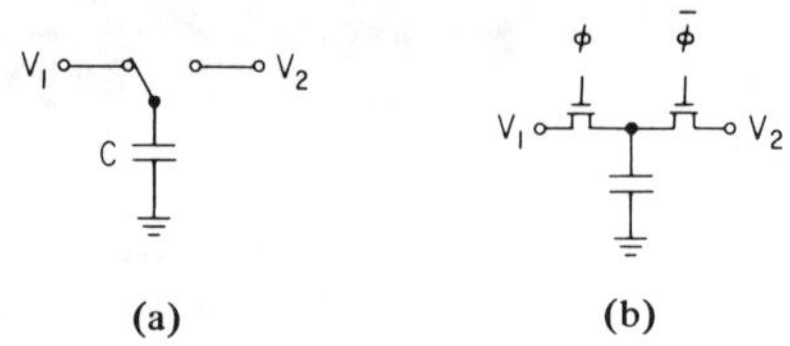

Fig. 3. (a) A switched-capacitor which in many cases can simulate the function of a resistor. (b) An MOS implementation of the circuit shown in part (a).

The operation of this "resistor" is as follows: the switch is initially in the left-hand position so that the capacitor C is charged to the voltage V_1. The switch is then thrown to the right and the capacitor is discharged to the voltage V_2. The amount of charge which flows into (or from) V_2 is thus $Q = C(V_2 - V_1)$. If the switch is thrown back and forth at a clock rate f_c, then the average current flow i from V_1 into V_2 will be $C(V_2 - V_1)f_c$. Thus the size of an equivalent resistor which would give the same average current as this circuit is

$$R = 1/Cf_c. \tag{1}$$

If the switched time is much larger than the signal frequencies of interest then the time sampling of the signal which occurs in this circuit can be ignored in a first-order analysis and the switched capacitor can often be considered as a direct replacement for a conventional resistor. If, however, the switch rate and signal frequencies are of the same order then sampled data techniques are required for analysis and, as for any sampled data system, the input signal should be bandlimited as dictated by the sampling theorem.

Note that we have assumed that V_1 and V_2 are voltage sources, and that the values of V_1 and V_2 are not affected by the switch closures. While this is in effect true in the filters to be described in this paper, which are based on switched-capacitor integrators, it is not true in more general circuits containing capacitors and switches. In these cases simple substitution of switched capacitors for resistors can give results different from the continuous case.

The switched-capacitor resistors require very little silicon area to implement large resistance values. In fact the silicon area *decreases* as the required value of resistance increases. To implement audio-frequency filters a resistance on the order of 10 MΩ is needed if a monolithic capacitor of a reasonable size is to be used (~10 pF). This value of resistance is easily achieved by switching a 1-pF capacitor at 100-kHz rate, requiring approximately a silicon area of 0.01 mm^2 (to be compared to a total chip area of 10–20 mm^2). If a 10-MΩ resistor were implemented by using a polysilicon line or diffusion the area required would be at least 100 times larger.

By using this switched-capacitor resistor in conjunction with other capacitors and op amps it is possible to realize many of the circuit configurations used in conventional RC-active filters. As in the case of the recursive filter of Fig. 1 and the charge-transfer device (CTD) filters, the signal in switched-capacitor filters is sampled in time so that an antialiasing filter is required. However, in contrast to the circuit of Fig. 2, as the

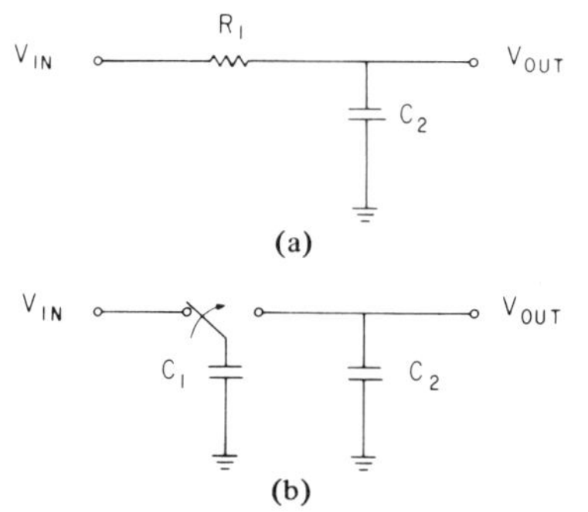

Fig. 4. (a) A single-pole RC continuous low-pass filter. (b) Single-pole low-pass filter utilizing a switched-capacitor resistor.

sampling rate is increased the sensitivities generally decrease to those of the continuous-time RC-active counterpart. Also the penalty in silicon area for a high sample rate is much less in the switched-capacitor filters in comparison to the CTD transversal filters.

Viewed from a conventional RC-active filter standpoint, these filters have the property, which will be demonstrated below, that all the RC time constants of the filter are determined by the clock frequency which is used to drive the switches and by capacitor ratios. Thus the problem of controlling RC products to a high degree of accuracy, which is characteristic of RC-active filter realizations, is reduced for these filters to the problem of maintaining accurate capacitor *ratios* and supplying an accurate external clock frequency. This property makes the filters much more practical for monolithic realization.

A conventional single-pole RC-low-pass filter is shown in Fig. 4(a), and a switched-capacitor implementation of this filter is shown in Fig. 4(b). In spite of the simplicity of this filter it demonstrates some of the most important advantages of the use of switched-capacitors. The 3-dB bandwidth of the conventional RC filter is

$$\omega_{3\mathrm{dB}}^{RC} = \frac{1}{R_1 C_2}. \tag{2}$$

The 3-dB width of the switched-capacitor filter can be found approximately by substituting the effective resistance of the switched capacitor C_1 into (1) to obtain

$$\omega_{3\mathrm{dB}}^{SC} \approx f_c \left(\frac{C_1}{C_2}\right), \qquad f_c >> \omega_{3\mathrm{dB}}^{SC}. \tag{3}$$

(The constraint that $f_c >> \omega_{3\mathrm{dB}}^{SC}$ is imposed in order that the effect of time sampling and charge sharing can be ignored.) Since the bandwidth of the switched-capacitor filter is proportional to a ratio of capacitor values, it can be accurately defined with a high degree of stability.

From (3) it can be seen that $\omega_{3\mathrm{dB}}^{SC}$ is proportional to the clock frequency; thus programmability can be achieved by simply varying the clock rate (a property which is common to all discrete time filters).

If one were restricted to the use of passive components and switches as in the case of Fig. 4(b), a very limited class of filter responses would be possible. It is possible, however, to design MOS operational amplifiers that can be used with switches and capacitors so as to extend greatly the capability of the filters. In the next section the limitations of the MOS technology and its relation to switched-capacitor filter circuits will be described.

II. Properties of MOS Switches, Capacitors, and Operational Amplifiers

The properties of MOS technology and the switched-capacitor resistors place significant limitations on the range of realizable filter configurations. In this section these limitations are explored.

A. MOS Capacitors

The MOS transistor itself is essentially a nonlinear capacitor in which the charge induced on one plate by the applied voltage forms the conducting channel of the transistor. The dielectric of this capacitor is silicon dioxide, which is one of the most stable dielectrics known. However, the MOS transistor itself is not particularly useful as a capacitor because of its inherently nonlinear behavior. Within an MOS integrated circuit, capacitors which have a sufficiently low-voltage coefficient to be useful as precision passive components are generally made in one of several ways:

1) Metal-Oxide-Crystalline Silicon Capacitors: This structure is formed by utilizing a heavily doped (low-resistance) region in the silicon substrate as the capacitor back plate, and the interconnect metallization (usually aluminum) as the top plate. This particular capacitor type is best suited to metal-gate CMOS and MOS processes which do not use self-alignment procedures and can thus realize the capacitors directly without process modifications.

2) Polysilicon-Oxide-Polysilicon Capacitors: In this capacitor structure, the two plates of the capacitor are formed of two separate layers of polysilicon (deposited polycrystalline silicon). This capacitor structure is best suited to silicon gate NMOS and CMOS processes having two polysilicon layers, which can realize these capacitance structures without extra process steps.

3) Other Capacitance Structures: By including extra masking steps, any combination of silicon substrate, polysilicon layer, and metallization layer can be used as the plates of the capacitors. However, such structures are only needed in silicon gate processes which have only one layer of polysilicon.

These capacitance structures tend to have common characteristics. From the standpoint of switched-capacitor filters, the most important of these characteristics are:

a) Ratio accuracy: A key aspect of the performance of any frequency-selective filter is the accuracy and reproducibility of the frequency response. For switched-capacitor filters this requires a certain level of accuracy in the ratios of capacitors. Integrated MOS capacitors have a value which is determined by the dielectric constant, the thickness of the dielectric, and the area of the capacitor. Assuming that the dielectric constant and thickness do not vary, the ratio of two capacitors made within the same integrated circuit will depend only on their area ratio. This is primarily determined by the geometrical shape of the capacitors defined by the photolithographic mask used to make the integrated circuit.

Errors in the geometrically defined ratio occur because the actual definition of the capacitor is done optically, which results in errors in the capacitor edge location due to the finite wavelength of the light used and the control of the chemical or plasma etching of the metallization and, in addition, because the thickness of the dielectric can vary with distance across the integrated circuit. These effects can be alleviated with careful layout of the components and the errors generally get smaller as the capacitor dimensions are made larger. Generally speaking, the achievable ratio accuracies range from

1-2 percent for small capacitor geometries ($\sim$400 μm^2) to on the order of 0.1 percent for capacitor geometries which approach the limit of economical size (40 000 μm^2) [8]. This also implies that as the capacitance ratio gets larger, the achievable accuracy in a given area decreases since the smaller of the two capacitors must be decreasing.

b) Voltage coefficient and temperature coefficient: MOS capacitors made with heavily doped silicon plates display voltage coefficients in the range of 10 to 100 ppm/V. Temperature coefficients are generally in the range of from 20 to 50 ppm/°C and are of course much lower for the value of a ratio [8]. These variations are low enough to be insignificant in almost all applications.

c) Parasitic capacitances: In both polysilicon–polysilicon capacitors and in metal–silicon capacitors, a sizable parasitic capacitance exists from the bottom plate of the capacitor to the substrate. In the case of capacitors made with two polysilicon plates, this is the capacitance of the SiO_2 layer under the first layer of poly. In the case of metal–silicon capacitors, it is the capacitance of the p-n junction surrounding the heavily doped region. Typically, this capacitance has a value of from one-fifth to one-twentieth of the MOS capacitor itself, depending on the technology. Also, because the top plate of the MOS capacitor must be connected to other circuitry, a small capacitance will exist from the top plate to the substrate due to the interconnections. This capacitance can range from 0.01 to 0.001 of the desired MOS capacitance, depending on the capacitor size, layout, and technology. These parasitic capacitances are unavoidable, and the design of switched-capacitor filters must be done in such a way that they do not degrade the performance of the filter.

B. MOS Switches

The second principal component in switched-capacitor filters is the MOS switch. This device behaves like a resistor in the on-state and in the off-state like an open circuit. The important parameters of the MOS transistor from the standpoint of switched-capacitor filters are the on-resistance, off-leakage currents, and the parasitic capacitances. Typical NMOS silicon gate technology is capable of producing switch devices with a channel length of 5 μm and, assuming a length/width ratio of unity, an on-resistance of 5 kΩ (with a gate drive voltage of 5 V with respect to the source). This device would display a leakage current from the source and drain to the substrate of on the order of 10^{-14} A at 70°C. The parasitic capacitance from source and drain to substrate would be about 0.020 pF each, and the overlap capacitance from drain to gate and source to gate would be about 0.005 pF. The charge induced in the channel when the gate potential is five volts more positive than the source and drain is approximately 0.03 pC. The values of capacitances used as passive components in switched-capacitor filters typically range from 1 to 20 pF. Thus the effect of the various parasitic capacitances on the behavior of the filter must be carefully considered, as well as the flow of the charge stored in the channel of the transistor.

C. MOS Operational Amplifiers

Until very recently, virtually all commercially manufactured operational amplifiers were fabricated utilizing either bipolar technology or a mixture of bipolar and MOS technology. Recently, the trend toward higher levels of integration on MOS-LSI chips has led to the need for all-MOS operational amplifiers to be included on such chips. Considerable recent work has been carried out directed towards the realization of CMOS and NMOS operational amplifiers. Typically achieved levels of performance are a gain of from 60 to 80 dB, common-mode and power supply rejection ratios of 60 dB, unity-gain bandwidth of 2 MHz, and power dissipation of 5-15 mW [9]. These devices typically occupy a die area on the order of 0.2 mm^2, so that the inclusion of several tens of them on a single chip is feasible.

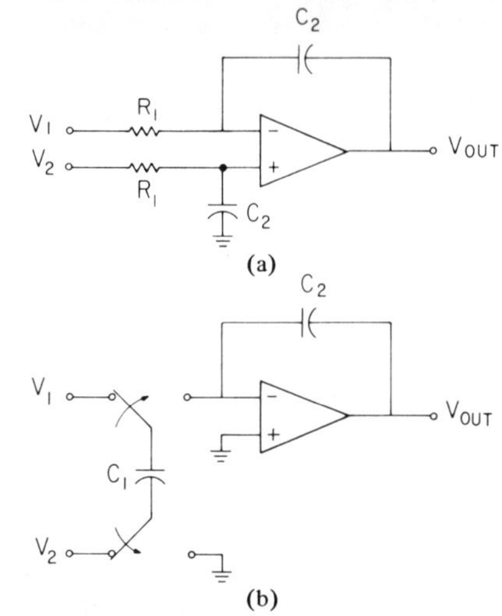

Fig. 5. (a) A conventional differential *RC* integrator and (b) a switched-capacitor differential integrator.

D. Sampled-Data Differential Integrators and Summers

Often in filter design the need arises for a differential integrator, which integrates the difference between two analog voltages. The traditional approach to this function is shown in Fig. 5(a), and requires two matched capacitors and two matched resistors. The input–output relation for this circuit is

$$V_{out} = \frac{-(1/R_1C_2)}{j\omega}(V_1 - V_2). \tag{4}$$

The differential integrator function is quite amenable to realization in switched-capacitor form as shown in Fig. 5(b). Here, the basic switched-capacitor concept has been used to replace a resistor (Fig. 3); but, in addition, the capability of the capacitor to perform a level shift function has been utilized. The result is a particularly efficient realization of the integrator which for frequencies much less than the clock rate, realizes the following input–output relation:

$$V_{out} = \frac{-(f_cC_1/C_2)}{j\omega}(V_1 - V_2). \tag{5}$$

Switched-capacitor differential integrators are subject to several important deviations from ideal behavior. These result from the parasitic capacitances from each plate of the capacitor to ground, and from finite bandwidth and gain in the operational amplifier. The effect of the parasitic capacitances on the circuit are illustrated in Fig. 6. For the integrating capacitor C_2, the parasitic capacitance C_{SUB2} from the op amp output to ground has no effect on circuit operation since it is driven from the op amp output voltage source (unless it is too large, when it can cause stability problems in the op amp). Assuming that an amplifier of sufficiently high gain is used, the parasitic capacitance C_{P2} from the inverting input of

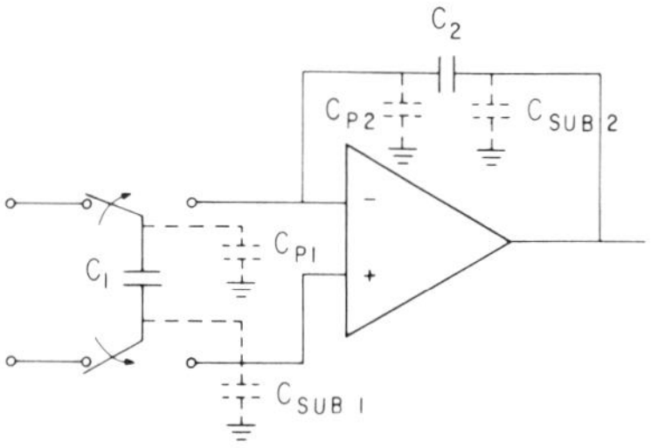

Fig. 6. A switched-capacitor differential integrator showing the parasitic capacitances associated with the switches and capacitors.

the operational amplifier to ground does not affect circuit operation since the steady-state voltage does not change at this input. Further, the parasitic capacitance, C_{SUB1}, from the bottom plate of the sampling capacitor C_1 has no effect on circuit operation because it is always connected to either an input voltage or to ground. Therefore, the only parasitic capacitance of any significance is that from the top plate of the sampling capacitor to ground, C_{p1}. The charge placed on this capacitor when the input sampling switches are on is transferred into the integrating capacitor. This results in a transfer function from input to output of

$$V_{\text{out}} = \frac{-f_c}{j\omega}\left[\frac{C_1}{C_2}(V_1 - V_2) + \frac{C_{P1}}{C_2} V_1\right] \tag{6}$$

which can be rearranged to give

$$V_{\text{out}} = \frac{-f_c}{j\omega}\left[\left(\frac{C_1 + C_{p1}/2}{C_2}\right) V_{dm} + V_{cm}\left(\frac{C_{p1}}{C_2}\right)\right] \tag{7}$$

where $V_{dm} = V_1 - V_2$ and $V_{cm} = (V_1 + V_2)/2$. Thus the effect of C_p causes the integrator to have a degraded common-mode rejection ratio (CMRR) and to change slightly the differential mode gain constant. By making C_1 and C_2 large compared to C_{p1}, the CMRR can be increased to any desired level. Practical circuits generally have a CMRR of 40-60 dB.

If the filter circuit only requires a noninverting integrator (i.e., $V_1 = 0$, $V_2 = V_{\text{in}}$), then from (6a) it can be seen that the integrator is unaffected by the parasitic capacitance. Therefore, it is possible to use a very small value for C_1 and/or C_2 to obtain very large ratios (>200) without consuming large amounts of silicon area.

The frequency response of the switched capacitor integrator shown in Fig. 5(b) deviates from that of the continuous one in Fig. 5(a) in a very important way. The sampling process inherent in the switching introduces an excess phase shift at frequencies approaching the clock frequency in the switched capacitor version. A straightforward analysis (over a complete clock cycle) of the switched-capacitor integrator gives a frequency response:

$$H_1(\omega) \;=\; -\left[\frac{(C_1 f_c/C_2)}{j\omega}\right]\frac{\omega T_c}{2\sin(\omega T_C/2)}\exp(-j\omega T_0/2). \tag{8}$$

The first term in brackets is the response of an ideal continuous integrator, and the rest of the expression is the deviation from that response caused by the sampling process. The most important aspect of this deviation is the delay term. The resulting excess phase shift will distort the frequency response of the complete filter in a way which is similar to the effects of excess phase shift in operational amplifiers in conventional RC-active filters. Typically, this distortion takes the form of Q-enhancement, in which the response of the filter shows some undesired peaking.

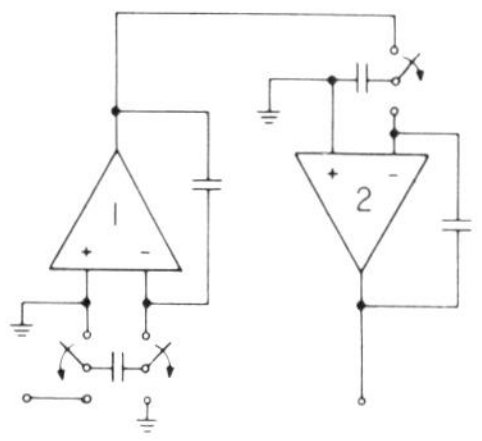

Fig. 7. A two-integrator loop with proper switch phasing to minimize phase errors.

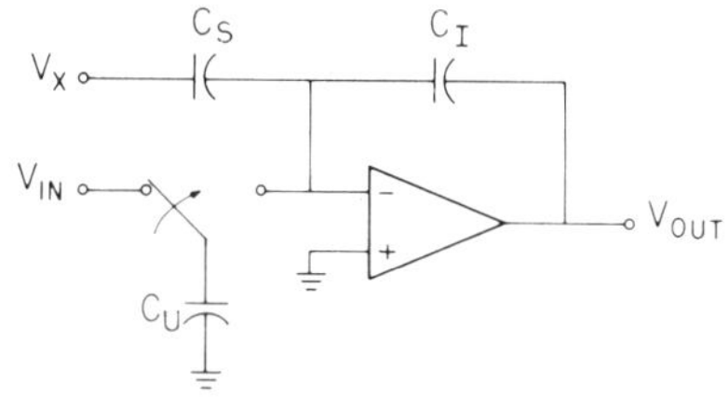

Fig. 8. A switched-capacitor integrator/summer.

The problem of excess phase due to sampled data effects can be attacked in three ways. The filter can be predistorted so that the response is the desired one when the excess phase is present. This approach has the disadvantages of increasing component sensitivity and of considerably complicating the design of filters requiring transmission zeros in the response. A second approach is to use a more complex integrator which inherently has less excess phase shift. Several such configurations have recently been proposed [11]. Perhaps the simplest approach is one based on earlier work by Bruton [10] on digital integrators, in which it was shown that the excess phase in the simple digital integrator could be precisely removed by decreasing the delay time through the integrator block by precisely one-half clock cycle. In ladder filters, in which one integrator always samples the output of another integrator, one can alternatively remove a full unit of delay from every other integrator. Fortunately, this can very easily be accomplished in switched-capacitor filters by simply reversing the phase of alternate clocks in the ladder filters. This clocking technique has been termed lossless digital integrator (LDI) clocking after Bruton. The correct phasing for the LDI clocking scheme is illustrated in Fig. 7 [11].

In addition to integrating a signal it is often necessary to sum one signal with the integrated value of other inputs. The circuit in Fig. 8 performs this function with only one op amp. The output of this circuit is

$$V_{\text{out}} = -\left(\frac{f_c C_u}{C_I}\right)\frac{V_{\text{in}}}{j\omega} - \frac{C_s}{C_I}V_x \tag{9}$$

where it is assumed that the clock rate is high enough that the sampled-data effects are not important.

The effect of finite op amp open-loop gain is to cause the integrator to display a pole at some finite low frequency rather than at zero frequency. Typically, this pole is a factor of 1000 or more below the unity-gain frequency of the integrator (as determined by the op amp open loop gain) so that this behavior does not usually have an important effect on filter performance.

There are two primary sources of noise in switched-capacitor integrators. The first is due to the thermal noise in the MOS transistor switches. When a MOS transistor switch is in the ON state connecting a capacitor C, to a voltage source, the thermal noise power in the resistive channel of the switch ($4kTR$) appears in series with the voltage source. The noise source is bandlimited by the RC circuit formed by the on-resistance of the switch R and the sampling capacitor. This single-pole low-pass circuit has a noise bandwidth of $1/4RC$. The bandlimited noise power is, therefore,

$$V_{\text{rms}}^2 = (4kTR)\left(\frac{1}{4RC}\right) = \frac{kT}{C}. \qquad (10(a))$$

When the switch is turned off, this noise is then sampled and held on the sampling capacitor.

In the switched-capacitor integrator, the noise contributed by each of the two switches must be considered. The analysis outlined above applies directly to the noise contributed by the switch between the input and the sampling capacitor. The analysis of the noise contributed by the switch between the sampling capacitor and the operational amplifier input is more complex because it involves the frequency response of the operational amplifier itself and the timing of the switches sampling the amplifiers output. However, it appears that the noise contributed by this device is similar in magnitude to that contributed by the first switch.

More quantitative noise analysis can be made on the passive low-pass circuit of Fig. 4, which gives insight into the noise to be expected from the use of switched-capacitors in place of resistors. Since the circuit of Fig. 4(a) is the same type of circuit which was analyzed in arriving at (10(a)) (band-limited resistor noise) the output noise of this simple RC circuit is

$$V_{\text{rms}}^2 = \frac{kT}{C_2}. \qquad (10(b))$$

To calculate the noise of the switched-capacitor circuit of Fig. 4(b) it is necessary to take into account the noise of the two switches, the capacitive division of this noise and the two separate sampling operations which occur each clock cycle. This calculation yields an output noise for the switched-capacitor circuit of

$$V_{\text{rms}}^2 = \left[\frac{kT}{C_2}\right]\left\{\frac{(C_1/C_2)(2 + C_1/C_2)}{(1 + C_1/C_2)^2}\right\}. \qquad (10(c))$$

Since the factor in curly brackets is always less than one, by comparison with (10(b)) it can be seen that the switched-capacitor circuit always has less noise than its RC equivalent. In fact if the ratio C_1/C_2 is small, the noise of the switched-capacitor circuit is substantially lower (e.g., if $C_1/C_2 = 0.2$, the factor is 0.17).

The second important noise source is that of the operational amplifiers. Since the most attractive feature of these filters is that they can be completely integrated in NMOS technology, one must carefully consider the effects of the relatively noisy NMOS operational amplifiers (compared to bipolar amplifiers) which must be used. These amplifiers inherently display a relatively high value of $1/f$ noise because of the surface behavior in the channel of the MOS transistors. This low-frequency noise falls in the passband of voice-frequency filters. For typical NMOS–LSI processes, this noise restricts the dynamic range of these filters to the range of 90–100 dB or less.

A second important noise source related to the operational amplifier is the broad-band noise of the amplifiers, which is aliased into the passband by the sampling process. This can be an important source of noise unless the operational amplifier is designed to have a low level of noise output at frequencies beyond its own unity-gain frequency.

E. Practical Considerations for MOS Implementation

The techniques have been shown for implementing all the basic building blocks of conventional RC-active filters using MOS technology. There are, however, some constraints on the type of active filters which can be used that are imposed by some additional practical aspects of the MOS circuit elements, if high-performance filters are required.

1) Switched Capacitor Resistors Cannot Close an Op Amp Feedback Path: Since the switched-capacitor resistor does not provide a continuous time path, it cannot be used to provide the continuous time feedback necessary to stabilize an op amp; as required for example in negative immittance converter filter circuits. However, switched-capacitor resistors can be used in conjunction with capacitors to shape the frequency dependence of the feedback as long as a continuous time path exists.

This constraint complicates the design of circuits which require a precise value of closed loop gain (such as that often required in voltage controlled-voltage source filters) since capacitors and switched capacitors must be connected in parallel in order to insure a continuous feedback path for the op amp as well as stability against charge accumulation. Alternatively, conventional diffused or ion-implanted resistors could also be used, but they result in increased power dissipation.

2) No Floating Nodes: All capacitive plates are subject to charge accumulation from a variety of parasitic sources such as leakage currents, electromagnetic radiation, and transient power-up conditions. In order to insure stability of the circuit there must be a path either directly or through switched-capacitor resistors from every node in the circuit to a voltage source. For example, the top plate of capacitor C_2 in Fig. 4(b) is stabilized against parasitic charge build up because it is indirectly connected to the input voltage source V_{in} through the switched-capacitor resistor C_1.

3) At Least One Plate of Every Capacitor Must Be Connected to a Voltage Source or Switched Between Voltage Sources: The nonlinear parasitic capacitance between the lower plate of any capacitor and the substrate must either be connected to a voltage source or switched between voltage sources. It will then be charged and discharged, but will not affect the filter response. This, therefore, rules out capacitive voltage dividers which are composed of three or more series capacitors or circuits which sequentially switch both ends of a capacitor into an op amp.

4) The Noninverting Op Amp Input Should Be Kept at a Constant Voltage: If the positive input of the op amp is connected to a signal voltage, then the filter response is sensitive to all the parasitic capacitances due to switches, bus lines and substrate that are connected to the inverting input. In addition, increased common-mode performance is required of the op amp.

III. Switched-Capacitor Filter Organizations

In spite of the constraints imposed by the MOS technology there still remains a large variety of possible active filter organizations which can be used for monolithic implementations.

One class of filters which are particularly well suited to switched capacitor techniques are those which use op amps as integrators in the same way they would be used in an analog computer realization [16]. This class of filters is closely related to such commonly used circuits as the infinite-gain multiple-feedback, state variable, and leapfrog (or active ladder) filters. An important advantage of these types of filters is that they can be organized so that there is a close correspondence between these circuits and passive LC ladder networks. By exploiting this correspondence the extensive tables available for the LC networks can be used to considerably decrease the filter design effort. Also by simulating passive LC circuits the active filter retains the low sensitivity of the frequency response to errors in the element values which is present in the passive network [15]. In particular a doubly terminated LC ladder which has been designed for maximum power transfer in the passband has very low sensitivity of the passband response to first-order changes in the values of the L and C elements. For the switched-capacitor implementation this implies an extremely low sensitivity of the filter response to the accuracy of the capacitor ratios.

A second-order low-pass filter will be discussed in detail to demonstrate the design technique as well as the practical considerations. The excitation to higher order low-pass, bandpass, and high-pass filters is then relatively straightforward and will be discussed briefly.

A. *Second-Order Low-Pass Filter Example*

In this section a switched capacitor simulation of a two-pole singly terminated LC ladder (Fig. 9(a)) will be derived and then related to state variable and infinite gain-multiple feedback filters. A single rather than double termination is used because for this low-order filter only a small decrease in sensitivity is obtained with the addition of a second termination which does not justify the extra circuitry (for higher order filters the sensitivity decrease with two terminations is much more dramatic).

The values of the L and C values for the passive prototype filter can be found from standard design tables for a normalized low-pass filter; i.e., $R_T = 1\ \Omega$ and a cutoff of $\omega_{co} = 1$ rad/s. An abbreviated version of these tables is given in Table I which gives the values of L and C for a few of the standard filter responses. The relationships between the table values $\mathcal{L}$ and $\mathcal{C}$ and LC values for arbitrary values of the cutoff frequency ω_{co} and termination resistance R_T are

$$L = \frac{\mathcal{L} R_T}{\omega_{co}} \tag{11(a)}$$

$$C = \frac{\mathcal{C}}{R_T \omega_{co}}. \tag{11(b)}$$

In Fig. 9(a) the node voltages and loop currents are defined for the passive circuit. The equations which describe this network can be written so that they only contain integrations:

$$V_1 = V_{in} - R_T i_1 \tag{12(a)}$$

$$i_1 = \frac{1}{sL}(V_1 - V_2) \tag{12(b)}$$

$$V_2 = \frac{1}{sC} i_1 \tag{12(c)}$$

$$V_{out} = V_2. \tag{12(d)}$$

Since the op amps which will be used to implement this circuit are voltage-controlled voltage sources it is necessary to represent the current i_1 by a voltage V_1'. This can be accomplished by multiplying i_1 by a scaling resistance R_S so that $V_1' = i_1 R_S$. The resulting all-voltage equations which maintain the proper relationships between the voltage and current nodes are

$$V_1 = V_{in} - \frac{R_T}{R_S} V_1' \tag{13(a)}$$

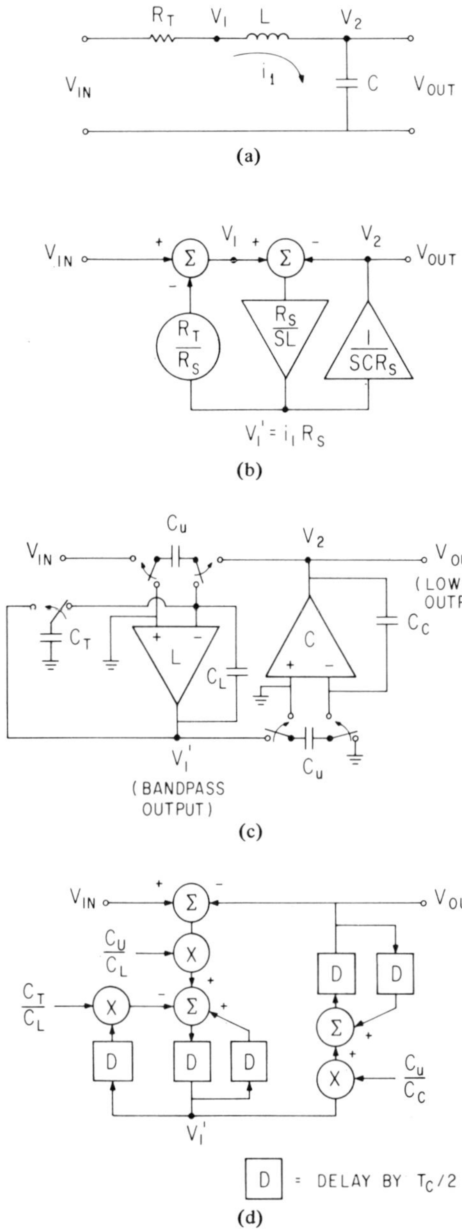

Fig. 9. (a) Two-pole singly terminated LC ladder filter. (b) Analog simulation of the LC ladder filter. (c) Switched-capacitor realization of ladder simulator. (d) Discrete time model for switched-capacitor circuit of (c).

TABLE I
L AND C VALUES FOR NORMALIZED LOW-PASS FILTERS
($R_1 = 1\ \Omega\ \omega_{co} = 1$ rad/s)

	Butterworth	Chebyshev (0.1-dB Passband Ripple)	Bessel
$\mathcal{C}$	1.4142 F	1.3911 F	1.3617 F
$\mathcal{L}$	0.7071 H	0.8191 H	0.4539 H

$$V_1' = \frac{R_S}{sL}(V_1 - V_2) \quad (13(b))$$

$$V_2 = \frac{1}{sR_S C} V_1' \quad (13(c))$$

$$V_{out} = V_2. \quad (13(d))$$

Fig. 9(b) is a schematic representation of the circuit which represents these equations in which the bandwidths of the two integrators are (R_S/L) and $(1/R_S C)$. This two-integrator loop structure is more commonly known as a second-order state variable network. This network is very versatile since in addition to the low-pass response which is available at V_2, a bandpass response is simultaneously present at the V_1' node and a high pass at V_1.

The switched-capacitor implementation of the network of Fig. 9(b) is obtained by replacing the conventional integrators with their switched-capacitor equivalents shown in Fig. 5 which yields the network in Fig. 9(c). The termination path R_T/R_S is implemented by the capacitor C_T, which is switched between the input and output of the first integrator. The capacitor ratios in Fig. 9(c) can be determined in terms of the normalized L and C values given in Table I, $\mathcal{L}$, and $\mathcal{C}$ by combining (1), (11), and (13),

$$\frac{C_T}{C_L} = \left(\frac{\omega_{co}}{f_c}\right)\frac{1}{\mathcal{L}} \quad (14(a))$$

$$\frac{C_L}{C_u} = \mathcal{L}\left(\frac{R_T}{R_S}\right)\left(\frac{f_c}{\omega_{co}}\right) \quad (14(b))$$

$$\frac{C_C}{C_u} = \mathcal{C}\left(\frac{R_S}{R_T}\right)\left(\frac{f_c}{\omega_{co}}\right) \quad (14(c))$$

where C_u is a unit capacitor which other capacitors are ratioed against, f_c is the clock rate of the switches, and ω_{co} is the desired cutoff frequency (in rad/s) of the filter.

The value of the ratio R_S/R_T is a free variable which can be used to maximize the filter dynamic range as well as to adjust the capacitor ratios C_L/C_u and C_C/C_u in order to minimize the silicon circuit area required for implementation. From (14) it is seen that the minimum ratios are achieved when $C_L/C_u = C_C/C_u$ which implies that

$$\frac{R_T}{R_S} = \left(\frac{\mathcal{C}}{\mathcal{L}}\right)^{1/2} \quad (15)$$

The maximum dynamic range of the filter is achieved when all the op amp outputs have the same peak voltage. If the gain from the filter input to the output of an internal op amp is greater than the gain to the output of the filter, that internal op amp will saturate before the output amplifier saturates. On the other hand, if the gain to the interior op amp is low, then this implies that there will be gain from this node to the filter output and thus any noise generated in the internal op amp will be amplified.

For the filter in Fig. 9 the peak amplitude at V_1 and V_{out} occurs at a frequency $\omega_0 = 1/\sqrt{LC}$ which yields a gain to the internal op amp output, V_1', of

$$\frac{V_1'(\omega_0)}{V_{in}(\omega_0)} = \frac{R_S}{R_T} \quad (16(a))$$

and to the filter output V_2 of

$$\frac{V_2(\omega_0)}{V_{in}(\omega_0)} = \left(\frac{\mathcal{L}}{\mathcal{C}}\right)^{1/2} \quad (16(b))$$

From (16(a)) and (16(b)) it is seen that if R_S/R_T is set by (15) then the dynamic range is maximized since all outputs have the same peak gain (as well as simultaneously resulting in the minimum size for the capacitors). The optimization of the design for dynamic range and minimum circuit area is particularly simple for this low-order filter since simple analytic expressions can be obtained. For higher order filters the same general considerations for the R_S/R_T ratio hold, but the optimum values are more dependent on the particular frequency response.

Up to this point in the analysis it has been assumed that the clock rate has been sufficiently high that the magnitude errors in the integrators due to the time delay through the switched capacitors have had a negligible effect on the frequency response of the filters. In order to check the validity of this assumption or if it is desirable to operate the clock at a low rate (a small multiple of the cutoff frequency), a more exact analysis is desirable. The exact frequency response can be found by analyzing the equivalent discrete time circuit for the filter. In Fig. 9(d) such an equivalent circuit is shown for the low-pass filter of Fig. 9(c). The delay blocks (squares with a D) represent a half of a clock cycle delay ($T_c/2 = 1/2f_c$) which is obtained by the switch phasing shown in Fig. 7. The remaining elements are multipliers and summers as in conventional digital filters. The extra delay associated with the termination (the path which contains the C_T/C_L multiplier) results in a phase shift which has a relatively large effect on the frequency response of filters which have both a low $Q(\sim 1)$ and a low ratio of the clock rate to the filter cutoff frequency.

The z-transform transfer function of the discrete time network of Fig. 9(d) is given by

$$H(z) = \frac{C_u^2/C_L C_c}{z^2 - z(2 - C_T/C_L - C_u^2/C_c C_L) + (1 - C_T/C_L)}. \quad (17)$$

The exact frequency response $H(\omega)$ of the filter can be determined by evaluating $H(z)$ along the unit circle defined by $z = \exp(j\omega T_c)$; i.e., $H(\omega) = H(z)_{z=\exp(j\omega T_c)}$. The deviation of this discrete-time-switched capacitor transfer function from the response obtained with a continuous time prototype RLC filter response can be reduced by a modification of the capacitor ratios. The modified capacitor ratios ($(C_T/C_c)_M$, $(C_u/C_c)_M$ and $(C_u/C_C)_M$) can be chosen by matching the Q and center frequency ω_0, of the two transfer functions where ω_0 and Q are defined by the continuous time transfer function as follows:

$$H_{L.P.}(\omega) = \frac{\omega_0^2}{\omega_0^2 - \omega^2 + j\omega\,\omega_0/Q}. \quad (18)$$

The values of the modified ratios in terms of the ratios calculated by using (14) and (15) (C_T/C_L, C_u/C_L, and C_u/C_c) are

$$\left(\frac{C_T}{C_c}\right)_M = \left[1 - \exp\left(\frac{-C_T}{C_c}\right)\right] \quad (19)$$

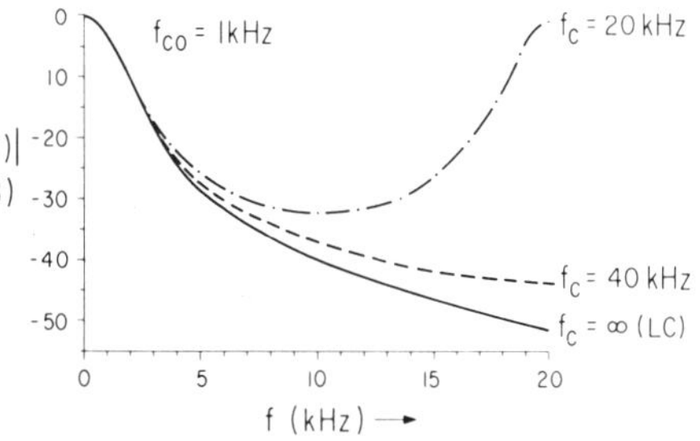

Fig. 10. Magnitude response of example filter of Fig. 9 with various clock frequencies.

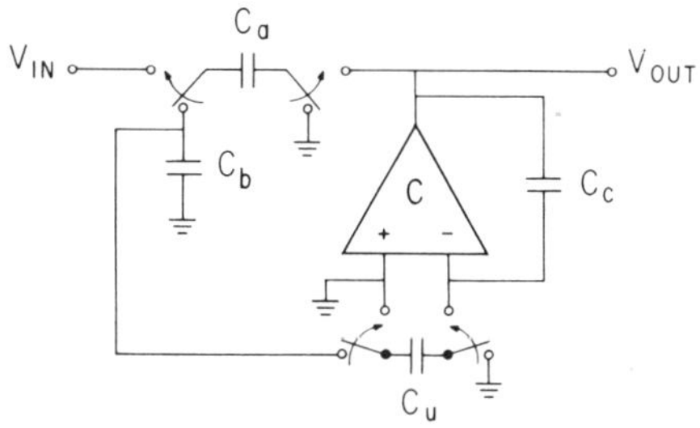

Fig. 11. Switched-capacitor implementation of an infinite gain-multiple feedback low-pass filter.

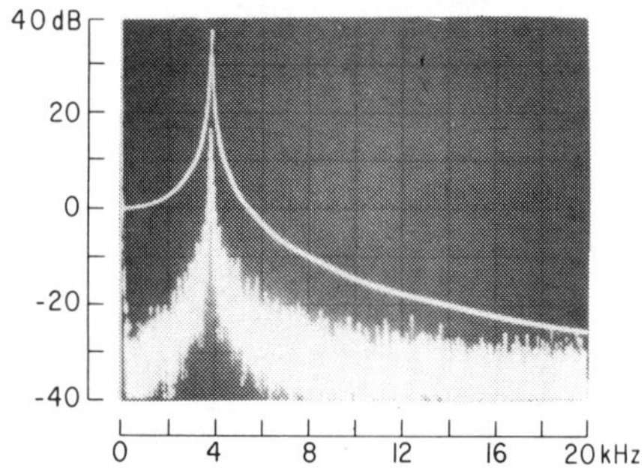

Fig. 12. Frequency response and output noise spectrum of high-Q version of experimental filter of Fig. 11. The noise spectrum was taken with a bandwidth of 30 Hz and amplified by 30 dB.

$$\left(\frac{C_u}{C_L}\right)_M^2 = \left(\frac{C_u}{C_u}\right)_M^2 = 1 - 2\exp\left(\frac{-C_T}{2C_c}\right) \cdot \cos\left\{\left(\frac{C_u}{C_c}\right)^2 - \frac{1}{4}\left(\frac{C_T}{C_c}\right)^2\right\}^{1/2} + \exp\left(\frac{-C_T}{C_c}\right). \tag{20}$$

In Fig. 10 a plot of the magnitude of the switched capacitor transfer function with modified ratios for a Butterworth (Q = 0.707) filter for clock rates 20 and 40 times the cutoff frequency of 1 kHz is compared to the continuous time LC filter response (or for $f_c = \infty$). Note that the responses are very close at frequencies $f \ll f_c$ but deviate as f approaches f_c. The aliased response of the switched capacitor filter which is clocked at 20 kHz is clearly evident in the region near f = 20 kHz.

A simplification of the circuit shown in Fig. 9(c) can be made by recognizing that the integrator labeled L which has the negative feedback through capacitor C_T is in effect a lossy integrator which can be replaced by a switched-capacitor resistor and a capacitor to ground. The resultant circuit with

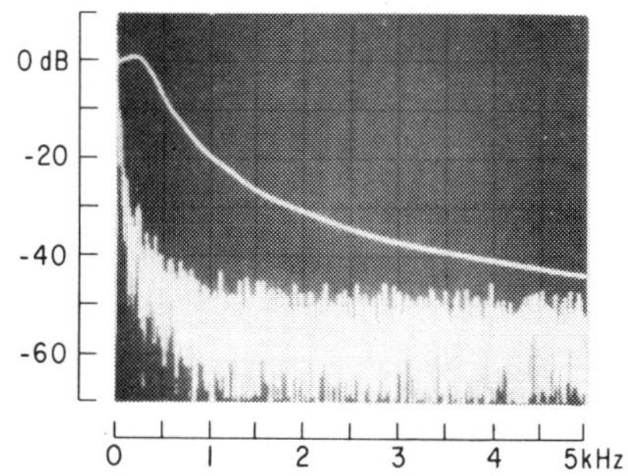

Fig. 13. Frequency response and output noise spectrum of low-Q version of the filter of Fig. 12. The noise spectrum was taken with a bandwidth of 30 Hz and was amplified by 30 dB.

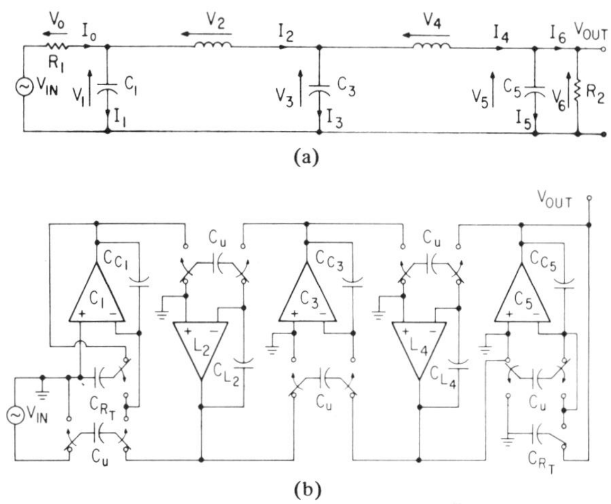

Fig. 14. (a) Doubly terminated RLC fifth-order all-pole low-pass ladder filter. (b) Switched-capacitor version of fifth-order all-pole low-pass filter.

resistors in place of switched-capacitors is known as an infinite-gain multiple-feedback network [13]. Several filters of this form (shown in Fig. 11) were fabricated by Hostica *et al.* [14] using an n-channel metal gate MOS process. The filters which were fabricated included responses which had a high Q(Q = 73) and low Q(Q = 1). The two filters required about 2800 mm^2.

In the top trace in Fig. 12 the frequency response of the high-Q filter is shown. The reference level of 0 dB at dc corresponds to 0-dB insertion loss through the filter. The peak of the response is a factor of Q times larger (Q = 73 yields 37 dB). Also shown in Fig. 12(a) is the noise spectral density (measured with a 30-Hz bandwidth) which has been amplified by 30 dB. As expected because of the 37-dB gain of the filter at the center frequency the noise also shows a peak at f_0.

In Fig. 13 the low-Q filter response is shown as well as the noise of the filter (also after an amplification of 30 dB with a bandwidth of 30 Hz). Since this filter has very little gain in the passband the noise is not amplified; however, the $1/f$ noise of the input MOSFET's of the amplifier can be seen. In Table II the measured and calculated filter performance is shown for a small sample of these filters.

B. High-Order Low-Pass

It is possible to implement higher order (>2) low-pass filters by cascading the second-order sections which were just de-

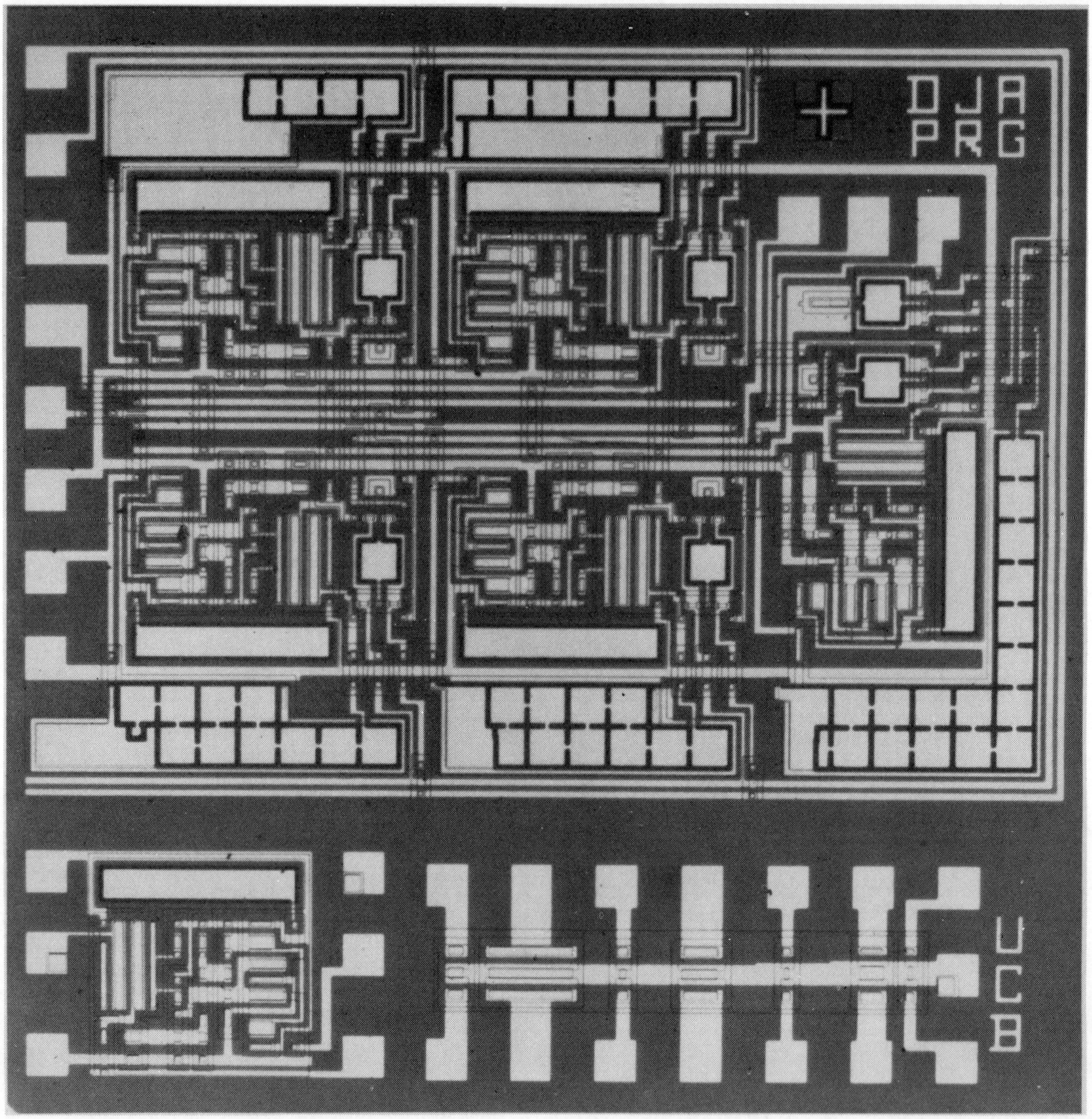

Fig. 15. Die photo of experimental fifth-order all-pole low-pass switched-capacitor filter. Bottom edge of chip contains test devices.

TABLE II
MEASURED PERFORMANCE PARAMETERS OF A SAMPLE OF TWO-POLE FILTERS

		Filter I	Filter II
Clock rate f_c		102.4 kHz	16.0 kHz
Calculated:	center frequency f_0	3.75 kHz	264.0 Hz
	selectivity Q	73.14	0.99
Measured:	center frequency mean	3.715 kHz	287.0 Hz
	center frequency standard derivation	13 Hz	1.0 Hz
	selectivity mean	71.2	0.97
	selectivity standard deviation	2.2	0.004
	output wide-band noise (rms)	0.85 mV	160 μV

scribed. However, the frequency response will then exhibit relatively high sensitivity to component variations. A better method is to simulate a passive-LC low-pass ladder which has the desired order [15]. A five-pole network of this form is shown in Fig. 14(a). Note that resistor terminations are used at each end of the ladder since this results in significant reductions in sensitivity for these high-order filters.

Following a procedure similar to that used to obtain Fig. 9(b) and (c), a switched capacitor simulation of the passive prototype five-pole filter shown in Fig. 14(a) can be obtained and is shown in Fig. 14(b). This network is a switched-capacitor implementation of what is known as the "leapfrog" or "active-ladder" active-filter circuit [16]. Using the switch

TABLE III
MEASURED PERFORMANCE PARAMETERS OF A FIVE-POLE LOW-PASS FILTER

Ripple bandwidth (f_c = 128 kHz)	3400 Hz
Total passband ripple	±0.1 dB
Output THD (3 $V_{p\text{-}p}$)	0.5 percent
Power dissipation	30 mW
Chip area (fifth order)	68 × 98 mils
Noise (300–3 kHz)	180 μV_{RMS}
Dynamic range (1-percent THD)	80 dB

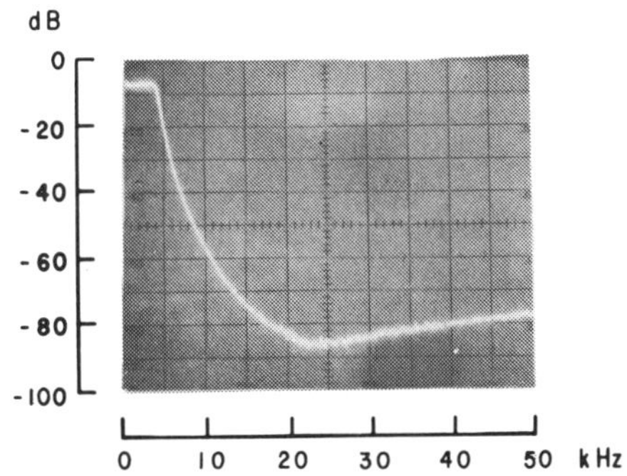

Fig. 16. Frequency response of experimental low-pass fifth-order filter.

phasing described in Fig. 7, the capacitor ratios are given by

$$\frac{C_{R_1}}{C_u} = \frac{R_S}{R_T} \tag{21(a)}$$

$$\frac{C_{C_i}}{C_u} = \mathcal{C}_i \left(\frac{R_S}{R_T}\right)\left(\frac{f_c}{\omega_{co}}\right), \quad i = 1, 3, 5 \tag{21(b)}$$

$$\frac{C_{L_i}}{C_u} = \mathcal{L}_i \left(\frac{R_T}{R_S}\right)\left(\frac{f_c}{\omega_{co}}\right), \quad i = 2, 4 \tag{21(c)}$$

where $\mathcal{C}_i$'s and $\mathcal{L}_i$'s are the normalized values corresponding to a cutoff frequency of 1 rad/s and a 1-Ω termination resistance R_T obtained from design tables. The input and output terminations C_{R_T} were set equal to minimize sensitivity, but if extremely low sensitivity is not required it is sometimes possible to reduce capacitor ratios by not setting them equal. The values of R_S and R_T are free variables as in the second-order case; and, as discussed in the previous section, their ratio should be chosen to minimize the capacitor ratios and maximize filter dynamic range (see the discussion leading to (15) and (16)). The modification of the capacitor ratios which was found to be necessary with low-Q second-order filters (equations (19) and (20)) due to the phase shift in the terminations has been observed not to be necessary for the higher order doubly terminated filters even for precise passband frequency characteristics as long as the clock rate is at least 10–20 times the cutoff frequency of the filter. A fifth-order Chebyshev filter was fabricated by Allstot *et al.* [17] which was designed for a ±0.1-dB passband ripple and a cutoff frequency of 3400 Hz when clocked at a rate of 128 kHz. The die which required 6300 mm^2 is shown in Fig. 15 and the experimental frequency response is shown in Fig. 16. The measured performance is summarized in Table III.

C. The Addition of Transmission Zeros to Filter Response

The addition of finite transmission zeros to a low-pass ladder filter response has great importance in many filter applications. The zero addition is easily accomplished on the LC low-pass prototype by adding a shunting capacitor across the series arm of the ladder network such as C_2 in Fig. 17(a). Imaginary axis zero locations are the resonant frequencies of the L-C tank circuit, i.e., $\omega_0 = [C_2 L_2]^{-1/2}$. The active-filter implementation of this noncanonical network is not as straightforward as the simple low-pass case. The usual approach to simulation of the circuit is suitable only for continuous-time active-RC implementations as it contains voltage attenuators (multiplications separate from op amp integrators) [16]. This is not desirable in a switched-capacitor implementation since additional op amps would be required.

In order to design a switched capacitor network with zeroes which does not require any additional operational amplifiers, it is useful to examine in detail the operations that are performed by the feedthrough capacitors added to the low-pass ladder structure. Referring to Fig. 17(a), a three-pole two-zero RLC filter is shown with voltages and currents defined. Using Kirchhoff's current law at nodes A and B, the following equations are derived to explain the function of C_2 (this is shown symbolically in Fig. 17(b)):

$$V_1 = \frac{(I_0 - I_2)}{s(C_1 + C_2)} + V_3\left(\frac{C_2}{C_1 + C_2}\right) \tag{22}$$

$$V_3 = \frac{(I_2 - I_4)}{s(C_2 + C_3)} + V_1\left(\frac{C_2}{C_2 + C_3}\right). \tag{23}$$

Thus C_2 has been identified as an element that feeds some of the voltage V_3 to node V_1 and vice versa as well as to modify the integrator time constants. The integrator/summer circuit shown in Fig. 8 can be used to perform this function if minor modifications are made to the circuit to take into account the fact that all summations involve a sign inversion [11]. The complete switched-capacitor circuit is shown in Fig. 18. As seen in this figure the method described above for obtaining transmission zeros only requires two switches and four capacitors over the all-pole circuit.

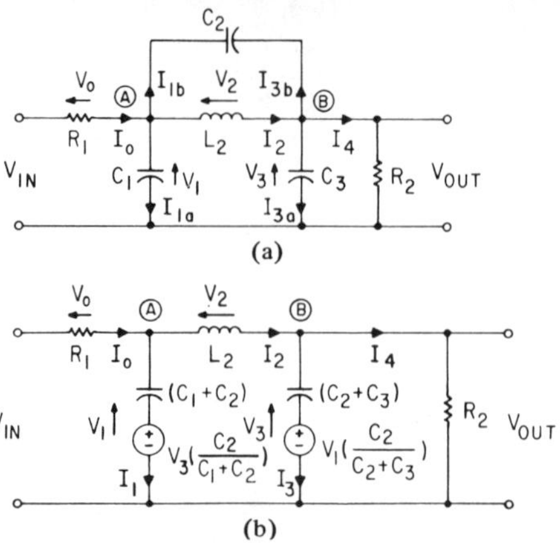

Fig. 17. (a) An *RLC* third-order elliptic low-pass filter. (b) Its equivalent circuit.

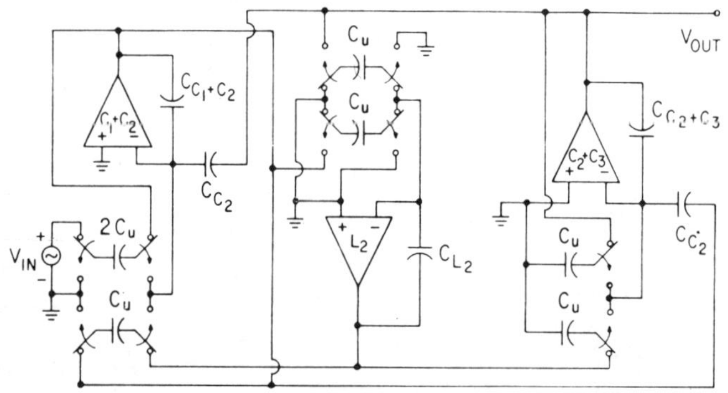

Fig. 18. Switched-capacitor version of third-order elliptic low-pass filter.

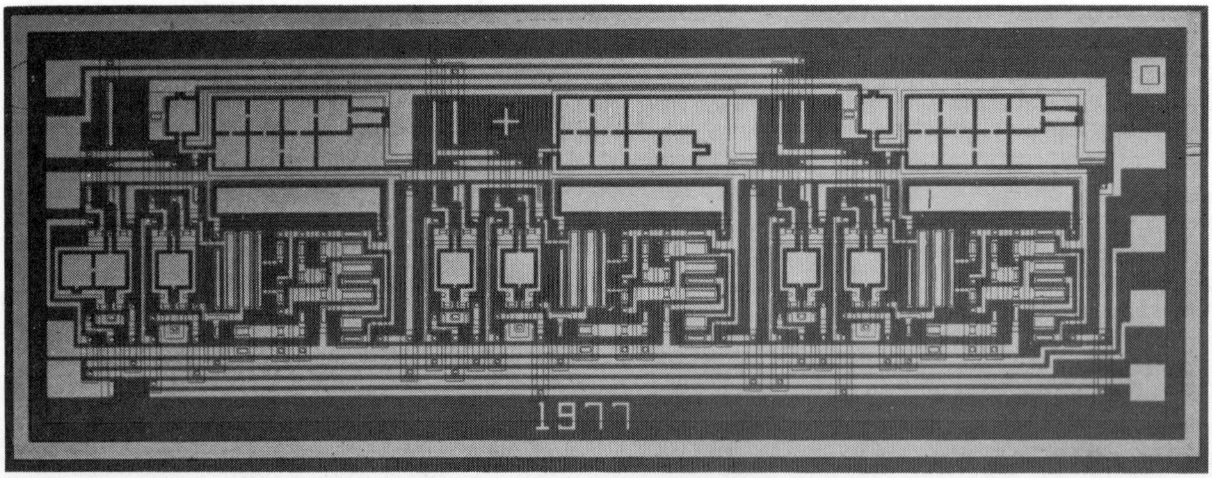

Fig. 19. Die photo of experimental elliptic third-order low-pass filter.

In the filter of Fig. 18 note that the input capacitor is twice tne usual unit size ($2C_u$). This doubles the overall gain of the filter and avoids the 6-dB insertion loss usually associated with doubly terminated filters.

A monolithic implementation of this version of an elliptic filter has been integrated [17] and is shown in Fig. 19 using standard MOS technology. The filter was designed for an elliptic lowpass response with 0.1-dB passband ripple, a cutoff frequency of 3400 Hz, and a transmission zero at 8.8 kHz when clocked at 128 kHz. Again, the measured performance shown in Fig. 20 agrees very closely with the design goals as summarized in Table IV.

D. Bandpass Filters

The technique of simulating passive *RLC* ladders which was used to implement the low-pass filters in the previous section is directly applicable to the design of bandpass filters. An important advantage of these filters is that because of their low sensitivity to component values it is possible to attain relatively high values of Q (up to ~100) with a high degree of precision.

The bandpass design is based on the use of the standard low-pass to bandpass transformation [12] which involves replacing each capacitor and inductor in a normalized low-pass circuit

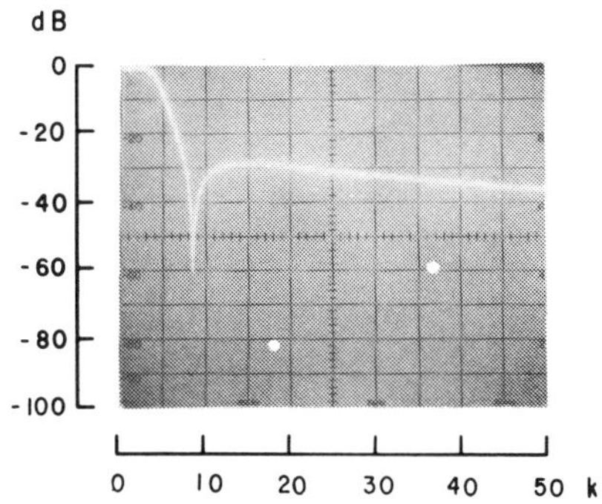

Fig. 20. Frequency response of third-order experimental elliptic low-pass filter.

(e.g., see Table I) with a parallel LC (tank) circuit and series LC circuit respectively. The values of the inductor and capacitor in the tank circuit which replaces a capacitor are

$$C_{\mathcal{C}} = \frac{\mathcal{C}Q}{\omega_0 R_T} \tag{24(a)}$$

$$L_{\mathcal{C}} = \frac{R_T}{\mathcal{C}Q\omega_0} \tag{24(b)}$$

where ω_0 is the center frequency of the desired bandpass filter, Q is the selectivity ($Q = \omega_0/\omega_{BW}$), ω_{BW} is the 3-dB bandwidth, and R_T the termination resistance.

The LC values of the series circuit which replace an inductor $\mathcal{L}$, are

$$L_{\mathcal{L}} = \frac{R_T \mathcal{L} Q}{\omega_0} \tag{25(a)}$$

$$C_{\mathcal{L}} = \frac{1}{\mathcal{L}Q\omega_0 R_T}. \tag{25(b)}$$

The passive prototype circuit is then transformed into a switched capacitor circuit following a method similar to that indicated in Fig. 9. An example of the complete procedure is outlined in Fig. 21 which leads to a second-order bandpass filter. In Fig. 21(a) a normalized low-pass filter ($\omega_{co} = 1$ rad/s) is shown which is transformed using the low-pass to bandpass transformation into the circuit shown in 21(b). An all-integrator representation of the circuit is given in 21(c) and the switched-capacitor implementation follows in 21(d). The values of the capacitor ratios in Fig. 21(d) are found from (1), (2), (4), and Fig. 21(c).

$$\frac{C_T}{C_C} = \left(\frac{\omega_0}{f_c}\right)\left(\frac{1}{Q}\right) \tag{26(a)}$$

$$\frac{C_C}{C_u} = \left(\frac{R_S}{R_T}\right)\left(\frac{f_c}{\omega_0}\right) Q \tag{26(b)}$$

$$\frac{C_L}{C_u} = \left(\frac{R_T}{R_S}\right)\left(\frac{f_c}{\omega_0}\right)\frac{1}{Q}. \tag{26(c)}$$

since $\mathcal{C} = 1$ F, it has not been included in the above expressions. It can be seen from these equations that if the Q is large then the capacitor ratios can also become very large. The size of the ratios can be minimized, however, by requiring that

$$\frac{R_T}{R_S} = Q. \tag{27}$$

TABLE IV
MEASURED PERFORMANCE OF A THIRD-ORDER ELLIPTIC LOW-PASS SWITCHED-CAPACITOR FILTER

Clock frequency	128 kHz
Ripple bandwidth	3400 Hz
Total passband ripple	0.1 dB
Minimum stopband rejection	30.4 dB
RMS output voltage (1-percent THD)	2.6 V
RMS noise (300–3 kHz)	80 μV
Dynamic range	90 dB
Power dissipation	18 mW
Filter die area	4400 mil^2

This condition also matches the peak gain from the input to the internal node V_2' with the gain to the output (which is unity) and, therefore, maximizes the dynamic range. The effect of the sampled data time delays (especially the delay through the termination path) on the transfer function of a second-order bandpass filter is the same as already discussed for a low-pass filter. Therefore, the equations which describe the modification of the capacitor ratios (which retain the same Q and center frequency for the switched capacitor bandpass filter as the prototype LC filter) are the same as those given in (19) and (20). For this to be true, however, the R_T/R_S ratio must be set by the condition given (27) and the switches phased as shown in Fig. 7.

Even after minimizing the capacitor ratios, the ratios can be very large for high-Q filters. It is possible, however, to decrease the size of the ratios by increasing the sensitivity of the filter. This is accomplished by using the switch phasing for the integrators which was described earlier that gives a *full* clock cycle of delay and thus yields an integrator which has an additional phase shift of $\exp(-j\omega T_c/2)$ (see (8)). This precisely defined amount of extra phase shift results in a pole movement which increases the Q of the filter (and also increases the sensitivity).

As in the case of the low-pass filter shown in Fig. 11, single op amp bandpass filters can be implemented which make use of the equivalence between a lossy integrator and an RC circuit. These single op amp structures are essentially switched capacitor versions of the infinite-gain multiple-feedback bandpass circuits of active-filter theory [13].

The implementation of higher order bandpass filters is a straightforward extension of the procedure outlined in Fig. 21. As in the case of the high-order low-pass filters the modifications to the capacitor ratios which compensate for sampled data effects in the second-order filters (equations (19) and (20)) are usually not necessary in the high-order bandpass circuits.

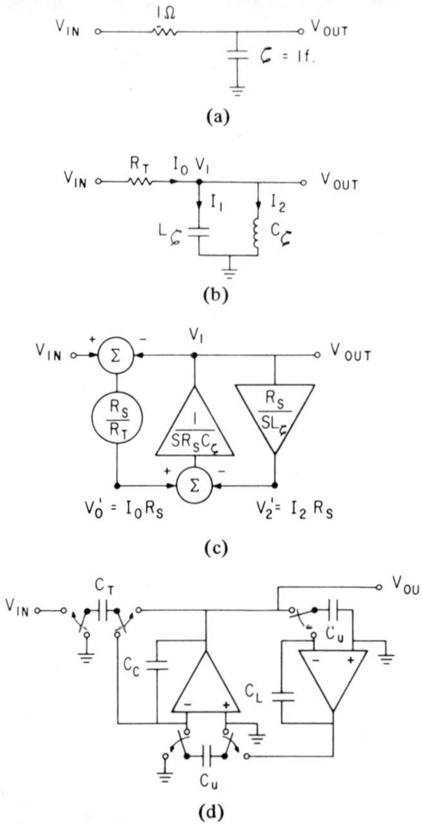

Fig. 21. (a) Normalized single-pole low-pass filter. (b) Bandpass transformed low-pass filter. (c) Integrator realization of bandpass filter. (d) Switched-capacitor implementation of bandpass filter.

E. Other Filter Organizations

The implementation of high-pass and band-reject filters can also be performed by use of the appropriate transformation on a low-pass LC circuit and then developing the switched capacitor realization by the use of the switched-capacitor integrators.

Of course, as mentioned earlier there are many other possible switched-capacitor filters that are based on active-filter circuits that are consistent with the limitations of the MOS implementation.

There are also switched-capacitor filter organizations which are not based on standard active filters [18]-[20]. One example is the filter which was shown in Fig. 2. Another is the N-path circuit which uses modulation techniques to translate a very-narrow-low-pass characteristic up to a center frequency which is controlled by the clock rate of the switches [20]. In this way extremely narrow bandpass and band elimination filters can be realized and in fact Q's in excess of 1000 are readily achievable.

IV. Conclusions

Design considerations for MOS switched-capacitor filters have been described. The major advantage of this approach is that a broad range of monolithic MOS filters can be realized which are insensitive to component variations. MOS filters have been designed and tested, and the results confirm the theoretical predictions.

Commercial exploitation of the switched-capacitor filtering technique is proceeding at a rapid pace. The first areas for application of the technique has been in special-purpose frequency selective filters for telecommunications applications, and in telephone applications in particular. It appears likely that the technique will also be used in monolithic programmable filters for general-purpose applications.

References

[1] F. L. J. Sangster, "The bucket brigade delay line, a shift register for analogue signals," *Philips Tech. Rev.*, vol. 31, pp. 97-110, 1969.

[2] D. D. Buss, D. R. Collins, W. H. Bailey, and C. R. Reeves, "Transversal filtering using charge transfer devices," *IEEE J. Solid-State Circuits*, vol. SC-8, pp. 138-146, Apr. 1973.

[3] W. Kuntz, "A new sample and hold device and its application to the realization of digital filters," *Proc. IEEE*, pp. 2092-2093, Nov. 1968.

[4] I. A. Young, D. A. Hodges, and P. R. Gray, "Analog MOS sampled-data recursive filters," in *Int. Solid-State Circuits Conf. Dig. of Tech. Papers*, pp. 156-157, 1977.

[5] D. L. Fried, "Analog sampled-data filters," *IEEE J. Solid-State Circuits* (Corresp.), SC-7, pp. 302-304, Aug. 1972.

[6] B. J. Hosticka, R. W. Brodersen, and P. R. Gray, "MOS sampled data recursive filters using state variable techniques," in *Proc. Int. Symp. Circuits and Syst.*, Phoenix, AZ, pp. 525-529, Apr. 1977.

[7] J. T. Caves *et al.*, "Sampled analog filtering using switched capacitors as resistor equivalents," *IEEE J. Solid-State Circuits*, vol. 6, SC-12, pp. 592-599, Dec. 1977.

[8] J. L. McCreary and P. R. Gray, "All-MOS charge redistribution analog-to-digital conversion techniques—Part I," *IEEE J. Solid-State Circuits*, vol. SC-10, pp. 371-379, Dec. 1975.

[9] D. Senderowicz, D. A. Hodges, and P. R. Gray, "A high performance NMOS operational amplifier," *IEEE J. Solid-State Circuits*, Dec. 1978.

[10] L. T. Bruton, "Low sensitivity digital ladder filters," *IEEE Trans. Circuits and Syst.*, vol. CAS-22, pp. 168-176, Mar. 1975.

[11] G. M. Jacobs, D. J. Allstot, R. W. Brodersen, and P. R. Gray, "Design techniques for MOS switched capacitor ladder filters," *IEEE Trans. Circuits and Syst.*, Dec. 1978. See also *Proc. Int. Symp. Circuits and Syst.*, New York, May 1978.

[12] E. A. Guillemin, *Synthesis of Passive Networks.* New York: Wiley, 1957.

[13] J. G. Graeme, G. E. Tobey, and L. P. Huelsman, *Operational Amplifier Design and Application.* New York: McGraw-Hill, 1971.

[14] B. J. Hosticka, R. W. Brodersen, and P. R. Gray, "MOS sampled data recursive filters using switched capacitor integrators," *IEEE J. Solid-State Circuits*, vol. SC-12, pp. 600–608, Dec. 1977.

[15] H. J. Orchard, "Inductorless filters," *Electron. Lett.*, vol. 2, pp. 224–225, June 1966.

[16] F. E. J. Girling and E. F. Good, "The leadfrog or active ladder synthesis," *Wireless World*, pp. 341–345, July 1970.

[17] D. J. Allstot, R. W. Brodersen, and P. R. Gray, "MOS switched capacitor ladder filters," *IEEE J. Solid-State Circuits*, Dec. 1978. See also *Int. Solid-State Circuits Conf. Tech. Dig.*, San Francisco, CA, pp. 82–83, Feb. 1978.

[18] A. Fettweis, "Steady state analysis of circuits containing a periodically operated switch," *IRE Trans. Circuit Theory*, vol. CT-6, pp. 252–260, Sept. 1959.

[19] J. McKinney and C. Halijak, "The periodically reverse-switched capacitor," *IEEE Trans. Circuit Theory*, vol. CT-15, pp. 288–290, 1968.

[20] L. E. Franks and I. W. Sandberg, "An alternative approach to the realization of network transfer functions: the *N*-path filter," *Bell Syst. Tech. J.*, pp. 1321–1350, Sept. 1960.

[21] P. R. Gray, D. Senderowicz, H. Ohara, and B. W. Warren, "A single-chip NMOS dual channel filter for PCM telephony applications," in *Dig. Tech. Papers, Int. Solid-State Circuits Conf.*, Philadelphia, PA., Feb., 1979.*

Switched-Capacitor Filter Design Using the Bilinear z-Transform

GABOR C. TEMES, FELLOW, IEEE, H. J. ORCHARD, FELLOW, IEEE, AND MASOOD JAHANBEGLOO

***Abstract*—A general synthesis technique is described for obtaining a switched-capacitor (SC) filter from an active *RC* prototype. The response of the SC filter is related to that of its prototype by the bilinear *z*-transform. This allows the filtering and sensitivity properties of the prototype to be preserved for the SC circuit. It also enables the designer to realize such economical and low-sensitivity circuits as the FDNR and gyrator filters. Finally, it permits the processing of signals at much higher frequencies than do previously known circuit design techniques.**

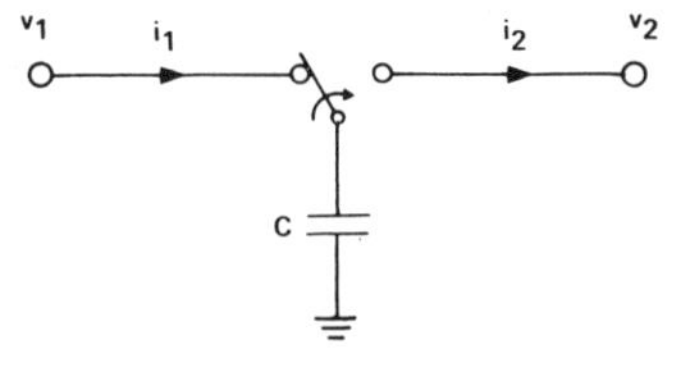

(a)

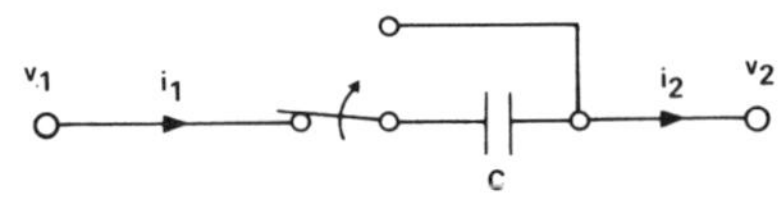

(b)

Fig. 1. Switched-capacitor equivalents of a resistor. (a) Shunt capacitor. (b) Series capacitor.

I. INTRODUCTION

THE RECENT interest [1]–[10] in the design of switched-capacitor (SC) filters is motivated by the fact that such filters can be fully integrated using standard MOS processing technology. This is due to the recent availability of suitable MOS operational amplifiers, as well as high-quality MOS capacitors and switches. As can readily be shown, the voltage gain of an SC filter depends only on the *ratios* of capacitance values; these ratios can be obtained and maintained to an accuracy much better than 1 percent over a wide range of temperatures and of signal amplitudes. Finally, the physical size of all components can be made sufficiently small, so that a high-quality filter can be laid out on a single semiconductor chip.

In many existing designs, the switched capacitor is grounded and switched between two nodes [1], as shown in Fig. 1(a). It can easily be shown that the *average* current i_2 delivered by the capacitor to node 2 is $(C/T_c)[v_1 - v_2]$ where $T_c = 1/f_c$ is the switching (clock) period. Thus in this *restricted* sense, the circuit of Fig. 1(a) is equivalent to a conductance C/T_c connected between nodes 1 and 2. Note, however, that unless v_1 and v_2 do not change with time, $i_1(t) \neq i_2(t)$ and hence the circuit does not act as a two-terminal network. An alternative circuit, which operates by charging and discharging C, and for which $i_1 = i_2$ always holds, is shown in Fig. 1(b) [4].

Using the circuits of Fig. 1 as basic building blocks, it was possible to design quadratic filter sections [2]–[4]. By ingenious modifications of the switching arrangement, it was also feasible to realize leapfrog ("active ladder") circuits [5], [6].

Manuscript received June 14, 1978; revised July 20, 1978. This work was supported in part by the National Science Foundation under Grant ENG 76-81622.

The authors are with the Department of Electrical Sciences and Engineering, University of California, Los Angeles, CA 90024.

It was, however, soon found that the equivalence between a resistor and either of the circuits of Fig. 1 is valid only for signal frequencies satisfying $f \ll f_c$. In the case of integrators, the error made by using the equivalence introduced an extra phase shift, which affected adversely the operation of the active-ladder filter. Hence, Jacobs *et al.* [6] used a modification of the clock phasing to obtain two-integrator loops which were free of such phase error. The technique is, however, not applicable to all active filter circuits.

To eliminate the phase error in a single integrator stage, a slightly more complicated SC circuit was introduced [7]. This circuit has a discrete-time transfer function which is related to that of a continuous-signal integrator through the bilinear z-transform. Similar circuits have also been found for sum and difference integrators, lossy integrators, etc. Subsequently, Dr. R. Gregorian of American Microsystems, Inc., and two of the present authors independently discovered a simplified SC arrangement which can implement the bilinear z-transform [8]–[10] in any filter configuration.

The purpose of this paper is, first, to give a general theory for SC filter design based on an element-by-element replacement of all resistors in a passive or active *RC* filter; next, to point out the analogy between methods of numerical integration and of SC filter design; and, finally, to introduce a design technique for SC filters based on the bilinear z-transform. The latter is illustrated by the design

Reprinted from *IEEE Trans. on Circuits Syst.*, vol. CAS-25, pp. 1039-1044, Dec. 1978.

of a simple FDNR SC filter, constructed from commercially available integrated circuits.

II. The Effects of Resistor Replacement on the Transfer Function

Consider the active-RC filter of Fig. 2(a). It will be used as a model from which an active-SC circuit is derived by replacing each branch resistor R_k by a two-terminal circuit containing a single capacitor C_k and several switches (Fig. 2(b)). The charge carried by the kth resistive branch prior to time t is

$$q_k(t)=\int_{-\infty}^{t} i_k(\tau)\,d\tau = G_k\int_{-\infty}^{t} v_k(\tau)\,d\tau. \tag{1}$$

The Laplace transforms of $q_k(t)$ and $v_k(t)$ are, therefore, related by

$$Q_k(s)=\frac{G_k}{s}V_k(s). \tag{2}$$

Assuming that *all* switches in the circuit of Fig. 2(b) operate every T seconds, the relation between the samples of the charge q_k and the branch voltage v_k is in the form of the first-order difference equation[1]

$$a_1q_k(nT)+a_2q_k(nT-T)=C_k[b_1v_k(nT)+b_2v_k(nT-T)]. \tag{3}$$

In (3), the a_i and b_i depend only on the arrangement of the switches. If the same switching arrangement is used in every branch, then these coefficients are independent of the branch index k. Choosing the ratio C_k/G_k also the same for all branches, and taking the z-transform of (3) gives

$$Q_k(z)=C_k\frac{b_1z+b_2}{a_1z+a_2}V_k(z)=\frac{G_k}{F(z)}V_k(z). \tag{4}$$

Here, $F(z)$ is a bilinear function of z, independent of k.

Consider next the circuit equations of the two filters in Fig. 2. The Kirchhoff laws will give for both circuits the same relations

$$\begin{aligned} \boldsymbol{Aq}&=0 \qquad \text{(KCL)}\\ \boldsymbol{v}&=\boldsymbol{A}^T\boldsymbol{e} \qquad \text{(KVL)}\end{aligned} \tag{5}$$

since their configurations are identical. In (5), $\boldsymbol{A}$ is the branch-to-node incidence matrix, $\boldsymbol{q}$ is the vector of the branch charges $q_k(t)$, $\boldsymbol{v}$ of the branch voltages $v_k(t)$, and $\boldsymbol{e}$ of the node voltages $e_m(t)$. Using Laplace transformation in the equation of the active-RC circuit, and z-transformation in the (sampled data) equations of the active-SC filter, (5) gives

$$\begin{aligned} \boldsymbol{AQ}&=\boldsymbol{0}\\ \boldsymbol{V}&=\boldsymbol{A}^T\boldsymbol{E}.\end{aligned} \tag{6}$$

Next, the branch relations will be considered. We will assume ideal memoryless active elements (such as ideal

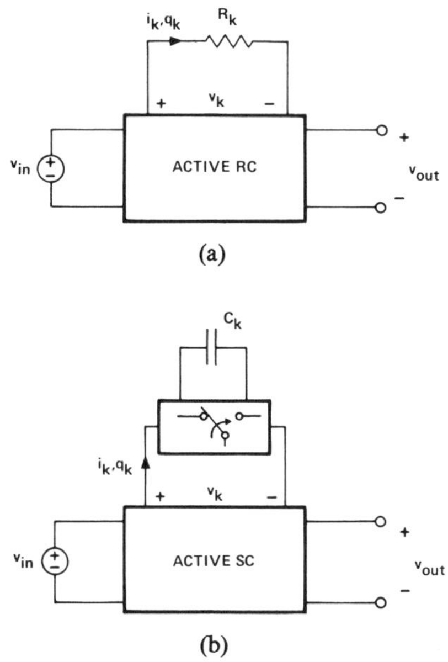

Fig. 2. Equivalent filter circuits. (a) Active-RC filter. (b) Active-SC filter.

operational amplifiers), with zero or infinite branch impedances. The capacitive branches of the active-RC circuit have relations of the form

$$\boldsymbol{Q}=\boldsymbol{CV} \tag{7}$$

which will remain the same for the corresponding branches of the active-SC filter. Finally, the branch relations for the resistive and switched branches are given by (2) and (4), respectively.

In conclusion, the equations relating $\boldsymbol{Q}(s)$, $\boldsymbol{V}(s)$, and $\boldsymbol{E}(s)$ for the RC circuit are (6), (7), and (2); those relating $\boldsymbol{Q}(z)$, $\boldsymbol{V}(z)$, and $\boldsymbol{E}(z)$ for the SC circuit are (6), (7), and (4). A comparison of these relations indicates clearly that all transformed voltages and charges $E_m(z)$, $V_k(z)$, $Q_k(z)$ of the SC circuit can be obtained directly from the corresponding quantities $E_m(s)$, $V_k(s)$, $Q_k(s)$ of the RC filter, simply by replacing s by $F(z)$. Thus all network functions (voltage gain, input impedance, etc.) are also easily derived for the SC filter from the corresponding functions of the RC network, using the same manipulation. This observation forms the basis of the design technique described in the next section.

It is well known from the theory of IIR digital filters that the transformation

$$s\to\frac{2}{T}\frac{z-1}{z+1} \tag{8}$$

provides an efficient tool for deriving useful sampled-data filter functions from continuous-time ones. Hence, in the next sections we will restrict the discussion to SC filters obtained by using a switching scheme which realizes $F(z)=(2/T)(z-1)/(z+1)$ in each switched branch.

A time-domain criterion for the resistor replacement process is discussed in [10]. Interestingly, it is found there that all existing schemes correspond to using some simple techniques of numerical integration in applying this criter-

[1] The form of (3) is determined by the assumption that all switches change positions every T seconds. Hence, the memory of the circuit extends only to T seconds, i.e., only one sample period.

ion. It is also shown why most previously used design techniques led to imperfect filter responses and/or high sensitivities to element variations, unless the operation was restricted to frequencies much smaller than the clock rate.

III. SC Filter Design Using the Bilinear z Transform

Choosing, as explained above

$$F(z)=\frac{2}{T}\frac{z-1}{z+1} \tag{9}$$

(4) shows that $C_k b_1 = C_k b_2 = G_k T/2$ and $a_1 = -a_2 = 1$ should be used. Then, (3) gives the required difference equation

$$q_k(nT)-q_k(nT-T)=\frac{G_k T}{2}\left[v_k(nT)+v_k(nT-T)\right]. \tag{10}$$

Thus the charge passing through the kth branch during one sampling interval T must be proportional to the sum of $v_k(nT)$ and $v_k(nT-T)$. This can be achieved by using the circuit of Fig. 3(a) [8]–[11]. Assuming that the arms of the switches make only momentary contact at sample times,[2] the operation is as follows. At time $(n-1)T$, C_k is charged to $C_k v_k(nT-T)$. At time nT, it is discharged *and* recharged to $C_k v_k(nT)$. Hence, the total charge passing through the branch during the $(n-1)T<t\leqslant nT$ interval is

$$q_k(nT)-q_k(nT-T)=C_k\left[v_k(nT)+v_k(nT-T)\right]. \tag{11}$$

Comparison with (10) shows that if one chooses C_k as $G_k T/2$, (10) is satisfied and hence the bilinear mapping is obtained. Clearly, here $T=T_c/2=1/2f_c$, since all switches change position *twice* every T_c seconds.

If the switch establishes contact at all times, except (momentarily) at $t=nT$, $n=0,\pm1,\pm2,\cdots$, the situation becomes more involved.

As Fig. 3(b) illustrates,[3] at times $t_n = nT-\tau$ $(\tau \ll T)$ the relation

$$\begin{aligned} q_k(t_n)-q_k(t_{n-1}) &= C_k\left[v_k(nT-T)-v_k(t_{n-1})+2v_k(nT-T)\right.\\ &\quad \left.+v_k(t_n)-v_k(nT-T)\right]\cong C_k\left[v_k(t_n)+v_k(t_{n-1})\right] \end{aligned} \tag{12}$$

is satisfied, in agreement with (10). However, the relation between the *continuous-time* waveforms is, from Fig. 3(b)

$$q_k(t)=C_k v_k(t)+2C_k\sum_{n=-\infty}^{[t/T]} v_k(nT)u(t-nT). \tag{13}$$

In (13), the first term represents a charge proportional to v_k, passing through the branch due to the physical presence of C_k between the nodes at all times, not only at the sampling instances t_n. This has an undesirable leakage

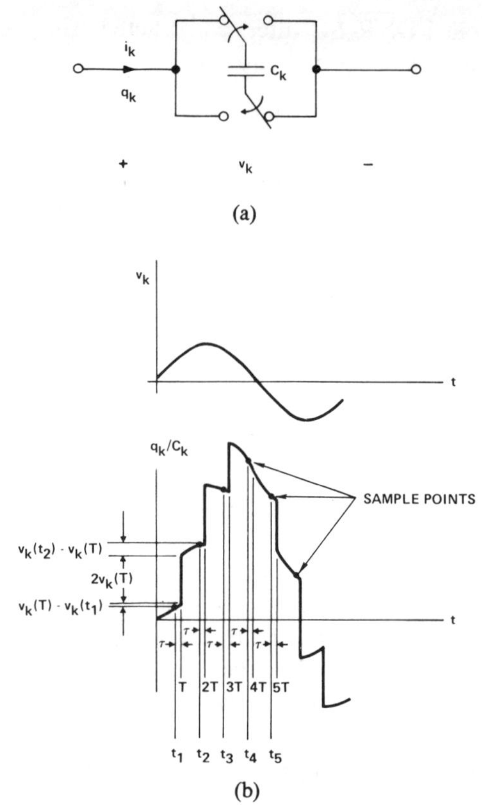

Fig. 3. SC equivalent circuit of a resistor which corresponds to the bilinear z-transform. (a) Circuit. (b) Branch voltage and charge waveforms.

effect, which can prevent, e.g., the attainment of high stopband losses and low distortion using the SC filter. Fortunately, it can easily be eliminated by cascading a sample-and-hold circuit with the SC filter. The resulting filter response then equals the bilinear transform of the RC-filter characteristic, combined with the spectrum of a zero-order hold. Specifically, if the transfer function of the active RC filter is $H(\Omega)$, then that of the derived active-SC circuit will be

$$\hat{H}(\omega)=H\left(\frac{2}{T}\tan\frac{\omega T}{2}\right)\frac{\sin \omega T/2}{\omega T/2}. \tag{14}$$

Therefore, the basehand segment of $\hat{H}(\omega)$ contained in the $-\omega_c \leqslant \omega \leqslant \omega_c$ region is replicated with a period $2\omega_c$. The replicas are attenuated by the weighting function of the zero-order hold.

In conclusion, the design of the active-SC filter can be performed in the following steps:

1) The specified passband and stopband limit frequencies $\omega_{p_1},\omega_{p_2},\cdots,\omega_{s_1},\omega_{s_2},\cdots$ of the SC filter are translated into the corresponding values $\Omega_{p_1},\Omega_{p_2},\cdots,\Omega_{s_1},\Omega_{s_2},\cdots$ of the "model" active-RC filter, using the relation $\Omega=(2/T)$ tan $(\omega T/2)$ which follows from the bilinear z-transformation.

2) The active-RC filter is designed from the transformed band-limit frequencies.

3) Each resistor R_k in the active-RC circuit is replaced by a switched capacitor of value $C_k = 1/(4f_c R_k)$, as shown in Fig. 3(a).

[2]This idealization assumes that the time constants for charge and discharge of C_k are zero; it simplifies the analysis because it eliminates the direct transmission of the input signal through the filter during the periods between switch operations.

[3]For convenience, a continuous $v_k(t)$ is assumed in Fig. 3(b). In fact, all waveforms are discontinuous.

4) A sample-and-hold circuit is added at the input, and (usually) a simple smoothing filter at the output of the SC filter. The sample-and-hold circuit should operate at twice the clock rate.

IV. Design Example

Since the design techniques described in the preceding section are independent of the filter configuration, they can be applied to any of the innumerable active-*RC* filter types, including FDNR and gyrator filters. The design of a simple FDNR-SC filter will be described next, as an illustration. The circuit is derived from a third-degree elliptic filter (Fig. 4(a)) with the following specifications:

$$\text{passband: } 0 \leqslant \Omega \leqslant 1, \quad \rho = 10 \text{ percent}, \; \alpha_p = 0.44 \text{ dB}$$
$$\text{stopband: } 2.79 \leqslant \Omega \leqslant \infty, \quad \alpha_s = 30 \text{ dB}.$$

Fig. 4(b) shows the FDNR realization of this filter for a passband limit of 2.5 kHz. The impedance level was chosen to make construction convenient with *discrete* components; a different level would, of course, be needed for realization with an MOS integrated circuit. The two resistors *R* across the source and load capacitors are the components always needed in low-pass FDNR filters to define the response at zero frequency; the value of *R* is made as high as practicable, typically 0.5 MΩ. The corresponding SC circuit, with a clock frequency of 50 kHz, is shown in Fig. 4(c). All resistors, including those marked *R*, are replaced by switched capacitors.

Both the *RC* and SC filters were built using inexpensive commercially available integrated circuits. The FET switches used were contained in CD 4053 BE units; the operational amplifiers used were LF 356 N types. A sample/hold circuit preceded the SC filter. No attempt was made to trim the elements to their exact values. The measured spectrum of the SC filter is shown in Fig. 5(a); the details of the passband in Fig. 5(b). The responses of the *RC* and SC circuits (corresponding to the frequency warping) are illustrated in Fig. 5(c); the lower trace is the response of the SC circuit.

Next, the SC filter was changed by reducing the values of all unswitched (10 nF) capacitors to 1 nF. This increases the passband limit to 21.224 kHz. The corresponding measured frequency response is shown in Fig. 5(d). Finally, Fig. 5(e) illustrates the output waveform of the modified filter for a 3-kHz input signal.

V. Conclusions

A general design technique was presented for obtaining active switched-capacitor filters from active-*RC* prototypes. The advantages of the process are the following:

1) It now permits the use of any of the many available active-*RC* configurations, including such useful ones as FDNR and gyrator filters, which were not utilized by earlier design processes.

2) It is based on the bilinear *z*-transform. Hence, it does not cause any inherent distortion and/or higher

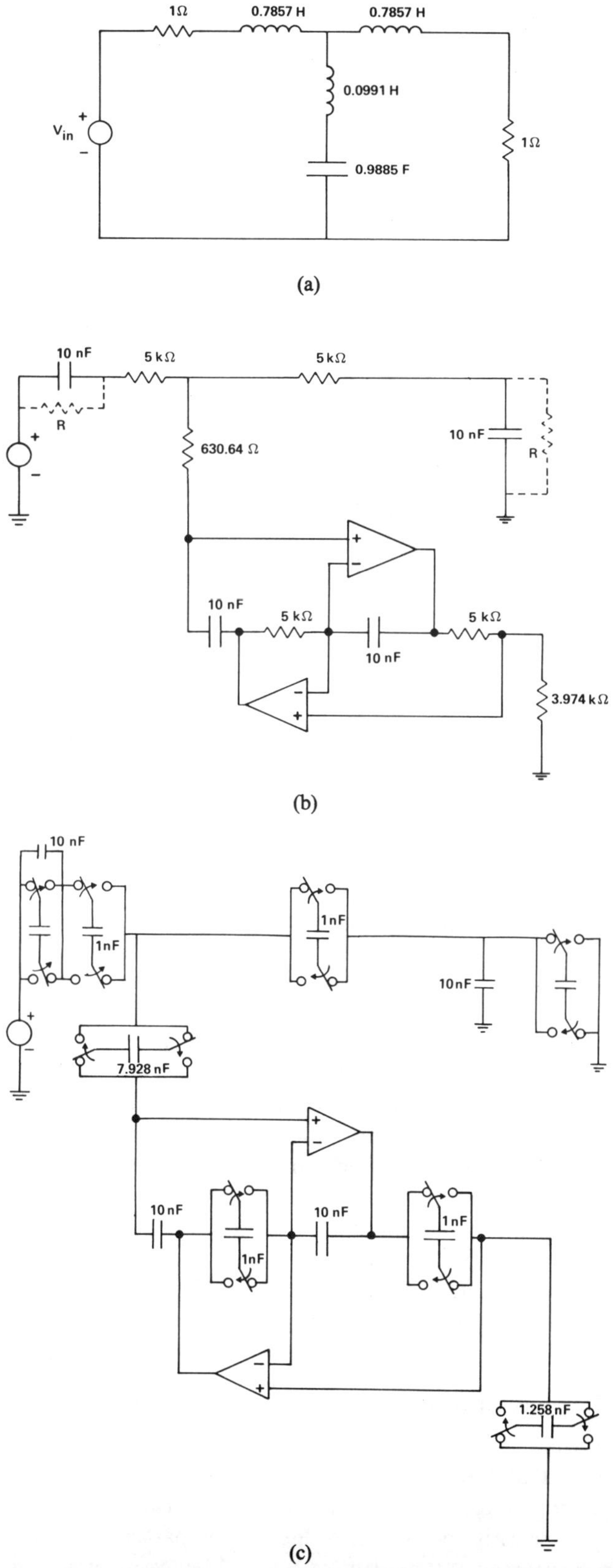

Fig. 4. Elliptic filter circuits. (a) LCR prototype. (b) Analog-FDNR equivalent circuit. (c) SC realization.

sensitivity to element variations due to imperfect modeling of the analog prototype.

3) Because of the perfect mapping of the analog and sampled-data frequency axes into each other, as well as

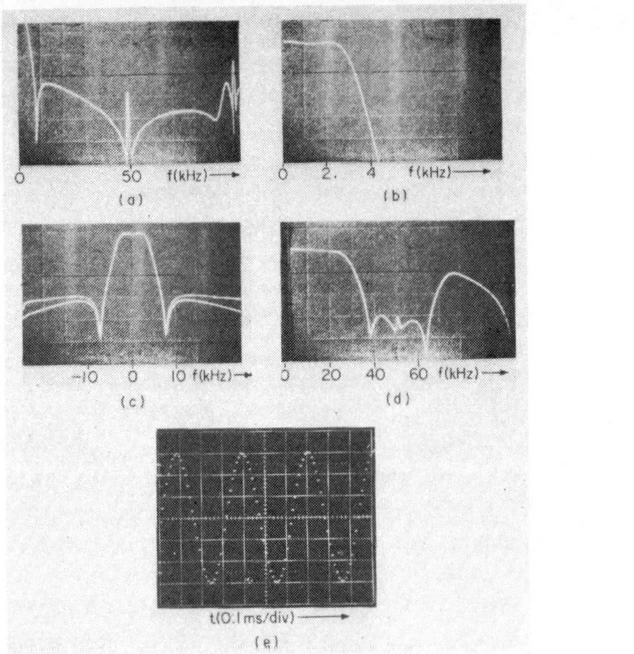

Fig. 5. Measured responses of the filter of Fig. 4(c). (a) Gain response (horizontal scale 10 kHz/div., vertical 10 dB/div.). (b) Expanded passband response (horizontal scale 1 kHz/div., vertical 2 dB/div.). (c) Comparison of *RC* and SC filter gain responses. Top trace-RC circuit; bottom trace-SC circuit (horizontal scale 5 kHz/div., vertical 10 dB/div.). (d) Gain response of the rescaled SC filter (horizontal scale 10 kHz/div., vertical 10 dB/div.). (e) Output waveform for 3-kHz sinewave input voltage (horizontal scale 0.1 ms/div., vertical 0.2 V/div.).

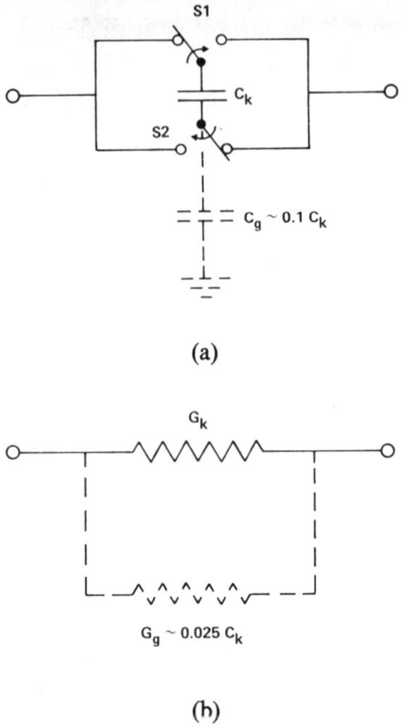

Fig. 6. Parasitic capacitance to ground. (a) Circuit. (b) Equivalent resistive branch.

the doubled effective sampling rate, the useful frequency range of the filter can extend to higher frequencies for a given clock rate than for earlier designs.

A potential disadvantage of the proposed technique is that it requires the realization of floating capacitors. This can complicate the manufacturing process of an MOS integrated circuit, and also introduces sensitivity to parasitic capacitances to ground. Also, if a gyrator or FDNR realization is used, then operational amplifiers with ungrounded input terminals are needed; this requires a reasonably good common-mode rejection performance from the amplifiers, and increases the sensitivity to parasitics. A brief discussion of the effects of parasitic capacitance to ground is given in the Appendix.

Appendix

The Effect of Parasitic Capacitance to Ground

For the usual MOS construction of a floating capacitor, such as that needed in the circuit of Fig. 3(a), one or both capacitor plates have also an appreciable capacitance C_g to ground (Fig. 6(a)). The value of this parasitic capacitance can reach 10 percent of C_k, and hence its effect on the circuit performance should be considered.[4] A comparison of Fig. 6(a) with Fig. 1(a) reveals that C_g, together with switch $S2$, forms the equivalent circuit of a conductance G_g in parallel with the main conductance G_k (Fig. 6(b)). Furthermore, a consideration of the average currents transported by C_k and C_g reveals that if $C_g = 0.1C_k$, then $G_g \cong 0.025G_k$. This 2.5 percent increase in all conductances can often be ignored. If the sensitivities of the response to variations in the element values make this impossible, then various steps can be taken:

1) The realized values of the C_k can be selected a few percent lower than the nominal values to account for some of the effects of the C_g.

2) The processing of the integrated circuit can possibly be modified so as to lower the value of C_g. Decrease by a factor of 10 appears to be feasible.[5]

3) For some important circuit configurations, such as those used in leapfrog filters, small compensating capacitors can be added, which exactly cancel the effects of the parasitic capacitance.

Acknowledgment

The authors are grateful to the (unknown) referee for useful comments and for bringing to their attention the earlier description of the circuit of Fig. 3(a) in [11].

References

[1] D. L. Fried, "Analog sample-data filters," *IEEE J. Solid-State Circuits*, vol. SC-7, pp. 302–304, Aug. 1972.

[2] I. A. Young, P. R. Gray, and D. A. Hodges, "Analog NMOS sampled data recursive filters," in *Proc. Int. Solid-State Circuits Conf.* (Philadelphia, PA), pp. 156–157, Feb. 1977.

[3] B. J. Hosticka, R. W. Brodersen, and P. R. Gray, "MOS sampled data recursive filters using state variable techniques," *IEEE J. Solid-State Circuits*, vol. SC-12, pp. 600–608, Dec. 1977.

[4] J. T. Caves, M. A. Copeland, and C. F. Rahim, "Sampled analog filtering using switched capacitors as resistor equivalents," *IEEE J. Solid-State Circuits*, vol. SC-12, pp. 592–599, Dec. 1977.

[5] D. J. Allstot, R. W. Brodersen, and P. R. Gray, "Fully-integrated high-order NMOS sampled-data ladder filter," in *Proc. Int. Solid-State Circuits Conf.* (San Francisco, CA), pp. 82–83, Feb. 1978.

[6] G. M. Jacobs, D. J. Allstot, R. W. Brodersen, and P. R. Gray, "Design considerations for MOS switched capacitor ladder filters,"

[4]Prof. R. W. Brodersen, private communication.

[5]Prof. R. W. Bower, private communication.

in *Proc. Int. Symp. on Circuits and Systems* (New York, NY), pp. 324–329, May 1978.
[7] G. C. Temes and I. A. Young, "An improved switched-capacitor integrator," *Electron. Lett.*, vol. 14, no. 9, pp. 287–288, Apr. 27, 1978.
[8] G. C. Temes, "The derivation of switched-capacitor filters from active-RC prototypes," *Electron. Lett.*, vol. 14, no. 12, pp. 361–362, June 8, 1978.
[9] R. Gregorian, private communication, May 5, 1978.
[10] G. C. Temes, "Digital-filter design techniques for the synthesis of switched-capacitor active circuits," in *Proc. 1978 Internat. Conf. on Digital Signal Processing* (Florence, Italy), Aug. 1978.
[11] J. A. McKinney and C. A. Halijak, "The periodically reverse-switched capacitor," *IEEE Trans. Circuit Theory*, vol. CT-15, no. 3, pp. 288–290, 1968.

Paper 6-3

Basic Principles of Switched-Capacitor Filters Using Voltage Inverter Switches

by Alfred Fettweis *

A wide class of low-sensitivity switched-capacitor filters simulating LC ladder structures can be built by using capacitances, ordinary switches, and voltage inverter switches, the latter being quite simply realizable by electronic means. The circuits thus obtained are closely related to known filters based on the resonant-transfer principle, but make use of technological means now available and require relatively few critical electronic devices such as operational amplifiers. The theory to be presented is self-contained and does not impose restrictions on the operating rate (other than those resulting from the Nyquist theorem) nor on the location of the attenuation poles.

Grundprinzipien von Schalter-Kondensator-Filtern mit Spannungsumkehrschaltern

Eine umfangreiche Klasse von Schalter-Kondensator-Filtern niedriger Empfindlichkeit kann in Nachbildung von LC-Abzweigschaltungen unter Verwendung von Kapazitäten, einfachen Schaltern und Spannungsumkehrschaltern gebaut werden, wobei die letztgenannten vergleichsweise einfach elektronisch realisiert werden können. Die so erhaltenen Schaltungen sind eng mit bereits bekannten Resonanztransferfiltern verwandt, machen jedoch von jetzt verfügbaren technologischen Möglichkeiten Gebrauch und benötigen nur vergleichsweise wenige kritische elektronische Bausteine wie Operationsverstärker. Die dargelegte Theorie ist in sich abgeschlossen und bedingt keinerlei Einschränkungen bezüglich der Betriebsrate (außer denjenigen, die sich aus dem Nyquisttheorem ergeben) oder der Lage der Dämpfungspole.

1. Introduction

Stimulated by progress in MOS technology, considerable interest has recently arisen in the possibility of realizing filters by means of switched capacitors [1]–[7]. The major advantages to be gained this way are the facts that only capacitances, operational amplifiers, and switches are needed, that nearly perfect switches can easily be built, and especially that all resonant frequencies are determined exclusively by capacitance ratios. It had, however, apparently first been overlooked that possibilities for realizing filters by switched capacitors had already been described in 1966 [8]–[17]. In particular, the decisive property concerning the resonant frequencies had explicitly been pointed out to hold for these older filters ([10], last paragraph of Section 6.2.3, and [14], 1st comment of Section 6.3).

The filters described in 1966 were actually based on the resonant-transfer principle as well as on its generalization and equivalent electronic realizations. Correspondingly, the theory had been derived from the more general theory of resonant-transfer circuits and was thus not easily accessible. Furthermore, the practical realization was handicapped by the imperfections of the technology then available, although some interesting experimental results had later been published [11], [17].

Recent investigations have brought to light that these resonant-transfer filters can easily be modified to make them suitable for the technological means now available. The essential circuit elements which we need are what we shall call *voltage inverter switches*. A voltage inverter switch can be built quite easily by means of an operational amplifier and ordinary switches. The purpose of this paper is to present the fundamental principles of these switched-capacitor filters based on voltage inverter switches. The theory will be developed without reference to resonant-transfer circuits.

The method to be presented offers easy possibilities of simulating lossless ladder networks between resistive terminations. The resulting filters thus enjoy the excellent sensitivity properties of these structures. The design theory does not involve any assumption concerning the sampling (operating) rate, i.e. this rate may be as low as is compatible with the Nyquist theorem. Transmission zeros (attenuation poles) may be placed at arbitrary locations in the stopband(s). These properties combined with the fact that only capacitance ratios are of importance, that the number of critical electronic devices (e.g. operational amplifiers) is substantially lower than for other approaches, and that the overall circuit arrangement is relatively simple make the filters presented in this paper very suited for integrated circuit technology.

In Sections 2 and 4 will be discussed the basic components by means of which a large number of LC ladder filters can be simulated. The realization of such filters will be discussed in Section 5. A variety of further circuits, many of which appear to be of particular importance, will be presented in a subsequent paper [18].

2. Grounded Voltage Inverter Switches

We consider the $(n+1)$-terminal network of Fig. 1 which we assume to be exclusively composed of

* Prof. Dr. A. Fettweis. Lehrstuhl für Nachrichtentechnik, Ruhr-Universität, Postfach 10 21 48, D-4630 Bochum.

Reprinted with permission from *Arch Elek Ubertragunge,* vol. 33, pp. 13-19, Jan. 1979.

capacitances. Its terminals will be numbered $\nu = 0, 1, \ldots, n$, the reference terminal being $\nu = 0$. Assume that at a given time instant charges are present on the capacitances; the resulting values of the v_ν will be designated by $v_{\nu b}$ (b: before). Into each

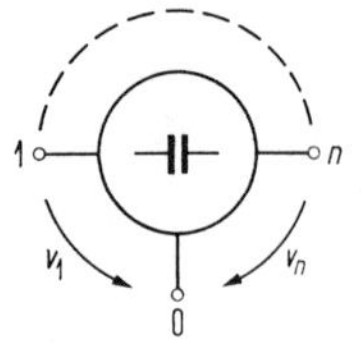

Fig. 1. Symbolic representation of an $(n+1)$-terminal capacitance network.

terminal ν, $\nu = 1, \ldots, n$, we then inject a charge q_ν, respectively; the resulting new values of the v_ν will be designated by $v_{\nu a}$ (a: after). We are interested, in particular in achieving

$$v_{\nu a} = -v_{\nu b}, \quad \nu = 1, \ldots, n. \qquad (1)$$

Devices by which this can be achieved will be called *voltage inverter switches.*

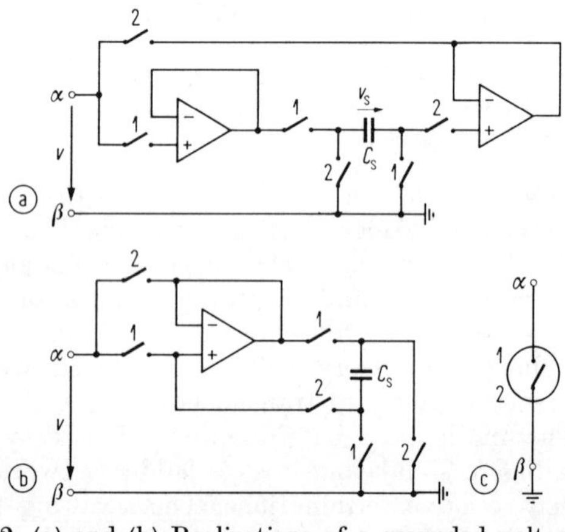

Fig. 2. (a) and (b) Realizations of a grounded voltage inverter switch using two and one operational amplifiers (connected as voltage followers), respectively,

(c) symbolic representation of a (grounded) voltage inverter switch.

An immediate simple means for realizing a grounded voltage inverter switch is shown in Fig. 2a. It comprises two operational amplifiers connected as voltage followers. During a first time interval, the switches numbered 1 are closed while those numbered 2 are open. The voltage v_S across the auxiliary storage capacitance C_S then becomes equal to the voltage of terminal α with respect to ground (terminal β). After the switches 1 are open again, the switches 2 will be closed. Consequently, the voltage then appearing across α and β will be the negative of the one originally present across these terminals.

It is essential, of course, that there is no overlap between the time intervals during which the switches 1 and 2, respectively, are closed, i.e. that the switches 1 are open before the switches 2 are being closed and vice versa. It thus follows in particular that the two operational amplifiers in the circuit of Fig. 2a are never active simultaneously. Hence, the voltage inversion effect can also be achieved by the circuit of Fig. 2b which comprises only one operational amplifier and where the switches operate in the same way as discussed before.

The arrangements of Fig. 2a and b are not the only examples of circuits for realizing a grounded voltage inverter switch [18]. It is not necessary, however, to represent each time all the details, and we therefore henceforth represent a voltage inverter switch by the symbol given in Fig. 2c. (The ground connection being shown explicitly and the numbers 1 and 2 having the same meaning as those in Fig. 2a and b.) We will say that it is *operative* if it is carrying out the process of voltage inversion. The *operating period*, i.e. the period during which it is operative, thus extends from the instant when the switches 1 start being closed until the switches 2 are finally fully open again. An $(n+1)$-terminal capacitance network with simultaneously operating grounded voltage inverter switches is shown, for $n = 4$, in Fig. 3.

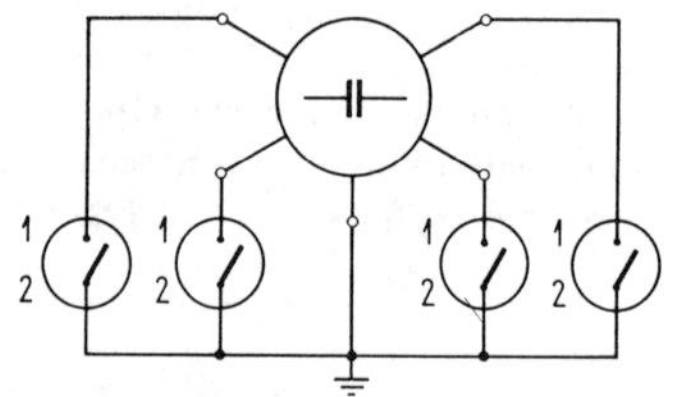

Fig. 3. An $(n+1)$-terminal capacitance network (drawn for $n = 4$) with simultaneously operating grounded voltage inverter switches.

Readers familiar with resonant-transfer circuits will recognize that a voltage inverter switch produces the same overall effect as a resonant-transfer switch. Thus, the circuits of Fig. 2a and b can be considered to be electronic realizations of a grounded resonant-transfer switch.

3. Fundamental Circuit Properties

By proper interpretation, the circuits with which we are dealing can be considered to be of the sampled-data type. An essential point in our theory will be to derive for these appropriate *equivalent circuits* of well-known type. In order to achieve this, a suitable equivalent frequency must first be adopted.

Let F be the *operating rate* of the circuit under consideration, T the corresponding operating period, i.e.

$$F = 1/T, \quad \Omega = 2\pi F = 2\pi/T, \qquad (2)$$

and p the actual complex frequency of the signal. A suitable *equivalent complex frequency* is then the quantity ψ defined by

$$\psi = \tanh(pT/2) = (z-1)/(z+1), \quad z = e^{pT} \qquad (3)$$

which had also formerly been adopted e.g. for resonant-transfer circuits and wave digital filters [8] to [17], [19].

The advantages of the transformation (2) are well-known. Let us recall simply that we have the correspondences

$$\begin{aligned} \operatorname{Re}(p) > 0 &\leftrightarrow \operatorname{Re}(\psi) > 0\,, \\ \operatorname{Re}(p) = 0 &\leftrightarrow \operatorname{Re}(\psi) = 0\,, \\ \operatorname{Re}(p) < 0 &\leftrightarrow \operatorname{Re}(\psi) < 0\,, \end{aligned} \tag{4}$$

that for $p = \mathrm{j}\omega$ we have

$$\psi = \mathrm{j}\varphi\,, \quad \varphi = \tan(\omega T/2)\,, \tag{5}$$

and that the Nyquist interval $0 \leqq \omega \leqq \Omega/2$ (where Ω is defined by eq. (2)) is transformed into the full range $0 \leqq \varphi \leqq \infty$. An important conclusion to be drawn from (4) is that stability in the ψ-domain implies stability in the p-domain.

We are dealing hereafter with *pulse currents,* i.e. with currents $i = i(t)$ consisting of sequences of pulses. The individual pulses are assumed to occur during periodically repetitive intervals, but nothing will be assumed concerning the pulse width except that the pulses must be non-overlapping; in particular, the pulse width does nowhere have to be short compared to T.

To the n-th pulse, we may make correspond a time instant t_n located inside of the pulse width and given by

$$t_n = t_0 + nT\,, \quad n = \cdots, -1, 0, 1, 2, \ldots, \tag{6}$$

t_0 being a constant (Fig. 4). The total charge carried by the n-th pulse will be designated by $q(t_n)$.

Note that for the needs of the present paper it would be sufficient to choose $t_0 = 0$. In view of the theory to be presented later [18], however, it is important to realize that the results of Section 4 are not affected if $t_0 \neq 0$.

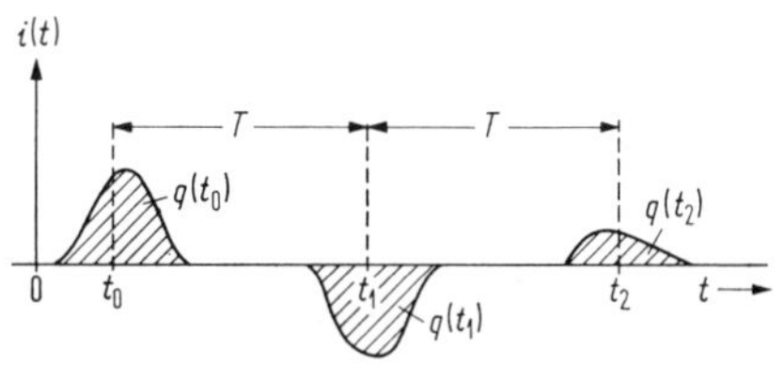

Fig. 4. Current $i(t)$ composed of a sequence of individual nonoverlapping pulses of charges $q(t_n)$ and unspecified width.

4. Basic Elements

We consider a capacitance C into which flows a pulse current $i = i(t)$ of the form discussed in Section 3 (Fig. 5a). Let $v = v(t)$ be the voltage across C.

We are interested in particular in the voltages v_b and v_a which we define to be the values of v just before a pulse begins and just after it ends. We refer these voltages to the instant t_n which corresponds to the pulse under consideration. Consequently, we write more precisely, $v_\mathrm{b} = v_\mathrm{b}(t_n)$ and $v_\mathrm{a} = v_\mathrm{a}(t_n)$, respectively. The equalities

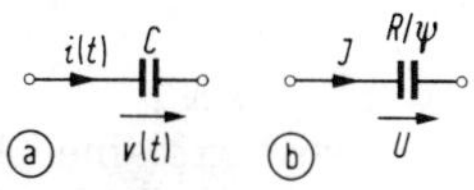

Fig. 5. (a) A capacitance C,
(b) equivalent circuit corresponding to a capacitance of impedance R/ψ, $R = T/2C$.

$$v_\mathrm{a}(t_n) - v_\mathrm{b}(t_n) = q(t_n)/C \tag{7}$$

and

$$v_\mathrm{b}(t_{n+1}) = v_\mathrm{a}(t_n) \tag{8}$$

obviously hold. We furthermore define a voltage $u = u(t_n)$ by

$$u(t_n) = [v_\mathrm{a}(t_n) + v_\mathrm{b}(t_n)]/2\,. \tag{9}$$

In order to express the difference eqs. (7) to (9) under steady-state conditions, we may write

$$v_\mathrm{a}(t_n) = V_\mathrm{a}\exp(pt_n)\,, \quad v_\mathrm{b}(t_n) = V_\mathrm{b}\exp(pt_n)\,, \tag{10}$$

$$u(t_n) = U\exp(pt_n)\,, \qquad q(t_n) = Q\exp(pt_n) \tag{11}$$

where V_a, V_b, U and Q are complex constants. Using eq. (3) as well as eqs. (6) to (9) and eliminating V_a and V_b, we obtain finally

$$U = JR/\psi \tag{12}$$

where J and R are defined by

$$J = Q/T\,, \quad R = T/2C\,. \tag{13}$$

Thus, if we interprete U as an *equivalent voltage* and J as an *equivalent current,* eq. (12) can be interpreted as the voltage-current relationship of a capacitance of impedance R/ψ, i.e. eq. (12) corresponds to the *equivalent circuit of a capacitance* (Fig. 5b). Note that we also have

$$U = (V_\mathrm{a} + V_\mathrm{b})/2\,, \quad RJ = (V_\mathrm{a} - V_\mathrm{b})/2\,, \tag{14}$$

$$V_\mathrm{a} = U + RJ\,, \qquad V_\mathrm{b} = U - RJ \tag{15}$$

and that R has been called *step resistance* in the theory of resonant-transfer circuits [9], [10], [15].

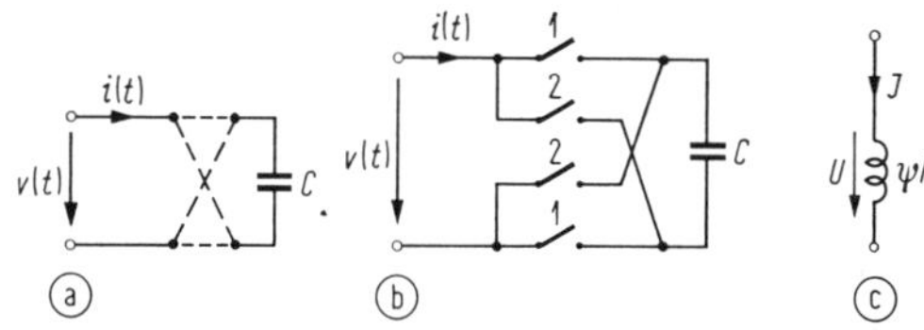

Fig. 6. (a) A capacitance C with a terminal inverter switch,
(b) a realization of (a) using ordinary switches,
(c) equivalent circuit corresponding to an inductance of impedance ψR, $R = T/2C$.

Let us now modify the circuit of Fig. 5a by adding a *terminal inverter switch* (Fig. 6a) which reverses the terminals of C between each two consecutive pulses of $i(t)$. Such an inverter is easily realized by means of four ordinary switches as shown in Fig. 6b. It is then sufficient that e.g. the switches numbered 1 are closed and those numbered 2 are open while the n-th pulse is arriving if n is odd, and the other way around if n is even. The presence of the terminal inverter switch does not alter eq. (7) but replaces eq. (8) by

$$v_\mathrm{b}(t_{n+1}) = -\, v_\mathrm{a}(t_n)\,.$$

Using then again eqs. (9) to (11) as well as eqs. (3), (6), and (13), we obtain

$$U = J\psi R \tag{16}$$

to which corresponds the equivalent circuit of an inductance of impedance R (Fig. 6c).

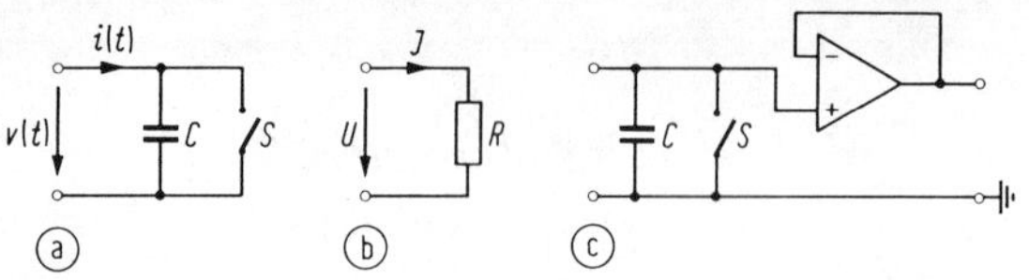

Fig. 7. (a) A capacitance C with an auxiliary switch S which discharges C between any two consecutive pulses of $i(t)$,
(b) equivalent circuit of (a), with $R = T/2\,C$,
(c) circuit for the realization of a resistance, complemented by an output voltage follower.

The circuit of Fig. 7a comprises a capacitance C into which flows a current $i(t)$ as discussed in Section 3, and an auxiliary switch S which is open while a pulse of $i(t)$ is arriving but which discharges C during each interval separating two consecutive pulses. Eq. (7) thus remains valid but eq. (8) has to be replaced by

$$v_{\mathrm{b}}(t_n) = 0\,.$$

Using then again eqs. (9) to (11) as well as eq. (13), we obtain

$$U = J\,R \tag{17}$$

to which corresponds the *equivalent circuit of a resistance* R (Fig. 7b). If such a resistance is to be used as termination of a filter, it has of course to be followed e.g. by a voltage follower as shown in Fig. 7c.

The circuit of Fig. 8a is derived from the one of Fig. 7a by inserting an ideal voltage source $e_0(t)$ in series with S. This auxiliary switch is again open while a pulse of $i(t)$ is arriving but is closed at least during part of the interval separating two consecutive pulses. Let $t_n - T_0$, where T_0 is a constant, be the instant at which S is being opened before the pulse referring to t_n is arriving. We then have

$$v_{\mathrm{b}}(t_n) = e_0(t_n - T_0)$$

which is thus the equation replacing eq. (8), while eq. (7) remains valid. Using then again eqs. (9) to (11) as well as eq. (13) and $e_0(t_n) = E_0 \exp(p t_n)$, we obtain

$$U = E + J\,R\,, \quad E = E_0 \exp(-p\,T_0) \tag{18}$$

to which corresponds the *equivalent circuit* of a *source* of *voltage* E in series with a *resistance* R (Fig. 8b). The difference between E and E_0 is irrelevant since it corresponds simply to a uniform delay T_0. In order to obtain a voltage source with substantially zero internal impedance, it will be necessary in practice, of course, to make use of e.g. a voltage follower as shown in Fig. 8c.

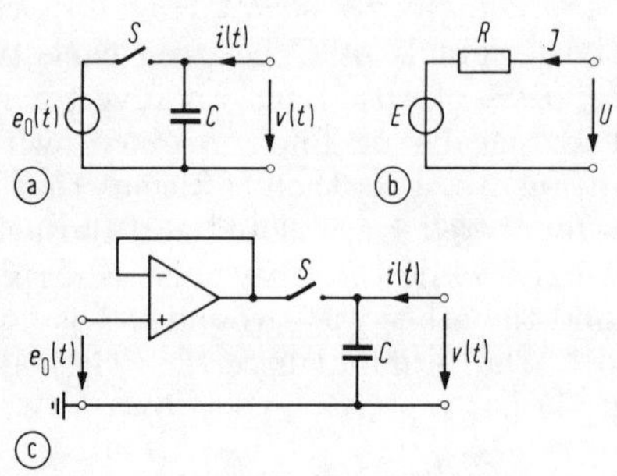

Fig. 8. (a) A capacitance C with an ideal voltage source $e_0(t)$ and an auxiliary switch S connecting C to $e_0(t)$ between any two consecutive pulses of $i(t)$,
(b) equivalent circuit of (a), with $R = T/2\,C$,
(c) practical realization of the circuit in (a).

Observe that eqs. (14) and (15) remain valid for all four circuit arrangements which we have discussed. The equivalent circuits at which we have arrived can also be used for computing the average power transmitted, P. Assume indeed that $p = \mathrm{j}\,\omega$ and let us write

$$U = |U|\,\mathrm{e}^{\mathrm{j}\alpha}\,, \quad Q = T\,|J|\,\mathrm{e}^{\mathrm{j}\beta}\,.$$

By taking the real part of the corresponding complex quantities, we thus have

$$\begin{aligned} u(t_n) &= |U| \cos(\omega t_n + \alpha)\,, \\ q(t_n) &= T\,|J| \cos(\omega t_n + \beta)\,. \end{aligned} \tag{19}$$

On the other hand, the energy trasmitted by one pulse is

$$[v_{\mathrm{a}}^2(t_n) - v_{\mathrm{b}}^2(t_n)]\,C/2 = u(t_n)\,q(t_n) \tag{20}$$

where use has been made of eqs. (7) and (9), i.e. expressions valid in all cases. Comparing eqs. (19) and (20) and using eq. (6), it is now easily seen by taking the time average of eq. (20) that

$$P = \mathrm{Re}(U J^*/2) = \mathrm{Re}(U^* J/2)\,, \tag{21}$$

provided that ω is not a multiple of $\Omega/2$ (cf. eq. (2)).

Since eq. (21) is exactly as for conventional circuits, it is clear that all other classical results derived from such an expression are also valid. As examples, the average power absorbed by a resistance R (Fig. 7) is given by $|J|^2 R/2$, and the maximum power which can be extracted from a source according to Fig. 8a is $|E|^2/8\,R = |E_0|^2/8\,R$.

5. Realization of Ladder Filters whose Series Branches are Composed only of Elements Connected in Series

5.1. High-pass filters

We now attempt to build circuits by using the building blocks described in Section 4. The main principle which we have to adhere to is that currents between capacitances should flow only under control of voltage inverter switches. The validity of this principle is justified by the success to which it leads. The essential conclusion which we derive from it is that any loop containing an inductance, a resistance, or a source may be closed only via a voltage inverter switch. Otherwise, currents not controlled by such switches would flow under the influence of the switching operations implied by Figs. 6 to 8. Another conclusion is that the need for all voltage inverter switches to be grounded restricts us to structures in which all loops of the aforementioned type have one node at ground. The most important structures of this class are ladder circuits whose series arms are composed only of elements connected in series.

The method is easiest explained by means of an example. We use for this the high-pass filter of

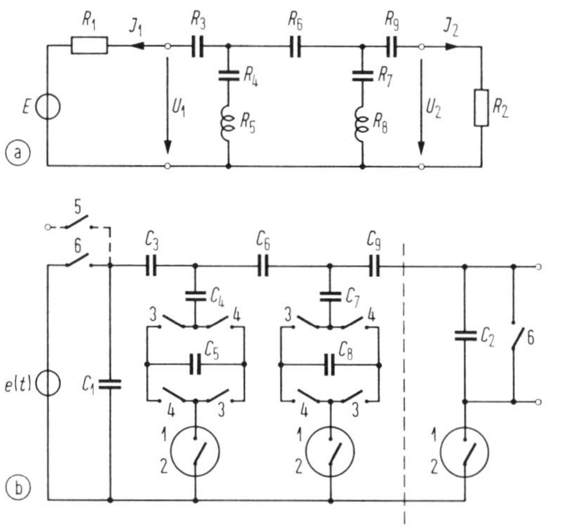

Fig. 9. (a) A 5th degree high-pass filter in the ψ-domain, (b) a corresponding switched-capacitor structure, although without grounded output terminal.

Fig. 9a. Since this filter is to be designed in the ψ-domain, its elements (whether capacitances, inductances, or resistances) can all be characterized by their respective resistive value, as has been done in the figure. Fig. 9b is obtained by replacing the elements in Fig. 9a by their respective realization described in Section 4 and by observing the aforementioned rule concerning the need for including voltage inverter switches. Note that there are three other acceptable possibilities of observing this rule, i.e. by moving anyone of the three present voltage inverter switches over into the shunt branch containing C_1. The values of the C_i are assumed to be related to the R_i by

$$R_i = T/2\,C_i\,, \quad i = 1, \ldots, 9\,.$$

Let us now analyze the way of operation of the circuit of Fig. 9b. We observe first that the output has no terminal on ground. This point will first be ignored and we will discuss later on how it can be corrected. (The dotted line to the right has been drawn in view of this discussion.) The numbers 1 and 2 marked in Fig. 9b refer to the switches actually present in the realizations of the voltage inverter switches (e.g. according to Fig. 2). In the first line of Fig. 10 is marked over a time scale a possible choice for the intervals during which these switches have to be closed, respectively. Corresponding choices for the intervals during which the switches 3 and 4 are closed are shown in the second line, and for the switches 6 they are shown in the fourth

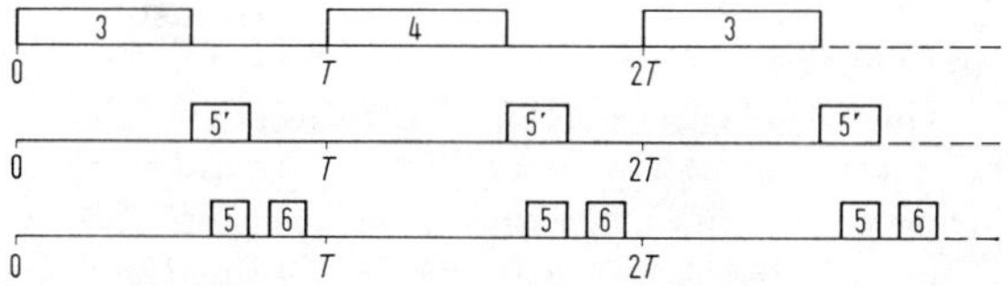

Fig. 10. Intervals during which the switches in the circuits of Figs. 9b, 11, and 12 are closed and during which the correct output signal is available.

line. The third line indicates during which intervals (marked 5′) the correct output signal is available. The duration of these various intervals is, of course, only shown very schematically and could actually be varied in a quite wide range. In particular, the duration of the intervals in the third and fourth line can be quite small, i.e. the duration of e.g. an interval marked 3 or 4 can be quite close to T. The whole mode of operation of the switches repeats itself periodically with a period equal to $2\,T$.

In Fig. 9b, the operation of the voltage inverter switches will cause pulse currents due to which charges q_i are carried through the respective capacitances C_i. Assuming the orientations to be properly chosen, we can write for each node in the circuit an equation of the form

$$\sum q_i = 0 \tag{22}$$

where the sum has to be extended to all those i whose C_i is incident with that node. Let us designate furthermore by $v_{i\mathrm{b}}$ and $v_{i\mathrm{a}}$ the voltages across C_i just before the transfer of q_i begins and just after it ends, respectively, and let $u_i = (v_{i\mathrm{a}} + v_{i\mathrm{b}})/2$. Assuming again appropriately chosen orientations, we can then write for each loop in the circuit,

$$\sum u_i = 0 \tag{23}$$

where the sum has to be extended over all those i whose C_i is located in the loop. While the result (22) is immediate, eq. (23) needs some justification.

Indeed, the sum of the voltages before the charge transfer begins is zero if we extend this sum not only over the voltages across the capacitances in the loop but also over those across the voltage inverter switches. The same is true for the sum of the voltages after the transfer ends. If we then add up these two sums, the contributions of the voltage inverter switches drop out in view of eq. (1), thus leading to eq. (23).

Let us now for instance consider steady-state conditions. In this case, we can define for each q_i a corresponding J_i and for each u_i a corresponding U_i, in analogy with what we have done in Section 4. In this case, eq. (22) gives immediately rise to a corresponding Kirchhoff node equation

$$\sum J_i = 0\,, \tag{24}$$

and eq. (23) similarly to a corresponding Kirchhoff loop equation

$$\sum U_i = 0\,. \tag{25}$$

On the other hand, taking into account the operation of the switches marked 3, 4, and 6, the equations derived in Section 4 are also fulfilled in the way required by Fig. 9a. Consequently, the circuit of Fig. 9b gives rise to precisely the equations describing the circuit of Fig. 9a. This latter is thus an equivalent representation of the former. Note that eq. (22), and thus eq. (24), are most immediately verified for the nodes other than the ground node, although they can also be checked directly for the latter; this, however, is not necessary since the Kirchhoff current equations hold for all nodes as soon as they are known to hold for all except one.

Let us now examine the grounding of the output port. We observe first that this grounding is only required while the output signal is being extracted. We must make sure, however, that the grounding does not close a loop which according to Fig. 9b is interrupted. A possible solution which takes these various requirements into account is obtained if we replace that part of the circuit in Fig. 9b which is

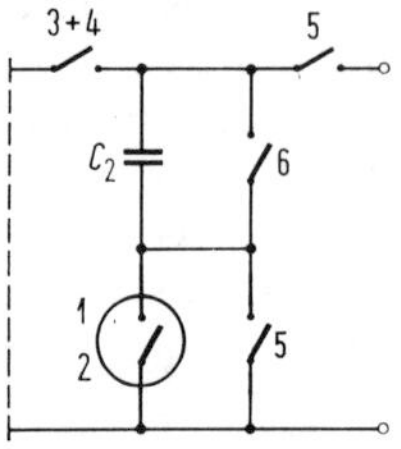

Fig. 11. Arrangement by which the part to the right of the dotted line in Fig. 9b has to be replaced in order to obtain a grounded output port.

to the right of the dotted line by the arrangement of Fig. 11. The switch marked $3+4$ should be closed during all the intervals shown in the second line of Fig. 10, i.e. during those marked 3 as well as those marked 4. The switches marked 5 may close only after the end of the intervals marked 2, as shown in the fourth line of Fig. 10 (compared to the first line). The correct output signal is available during these intervals marked 5.

The output voltage thus obtained is

$$V_{2a} = U_2 + R_2 J_2 = 2\,U_2 .$$

Hence, the ratio V_{2a}/E differs only by the constant factor $\sqrt{C_2/C_1}$ (usually even equal to unity) from the transmittance

$$S_{21} = 2\,(\sqrt{R_1/R_2})\,U_2/E$$

and thus has the same good distortion insensitivity as S_{21}. Another interesting voltage is V_{1a} which can be obtained as a further output signal by connecting to C_1 an additional switch, to be closed during the intervals 5, as shown by a dotted line in Fig. 9a. In view of the equations obtained for a source, we have

$$V_{1a}/E = (U_1 + R_1 J_1)/(U_1 - R_1 J_1) =$$
$$= (Z_1 - R_1)/(Z_1 + R_1) = S_{11}$$

where $Z_1 = -\,U_1/J_1$ is the input impedance of the filter. Hence, the reflectance S_{11} is just as easily available as S_{21}.

The method thus described for obtaining a switched-capacitor high-pass filter is valid, obviously, for any order n of the filter. The number of voltage inverter switches, and thus the number of operational amplifiers which are required is $(n+1)/2$.

5.2. Low-pass and band-pass filters

We will illustrate the realization of *low-pass* filters again by means of an example of degree five. A suitable structure is shown in Fig. 12a. We can derive from it a switched-capacitor structure exactly as we have done in Figs. 9 and 11, except that there will be now five terminal inverter switches (one for each inductance) while only two were needed in Fig. 9b. (This is the reason why we had

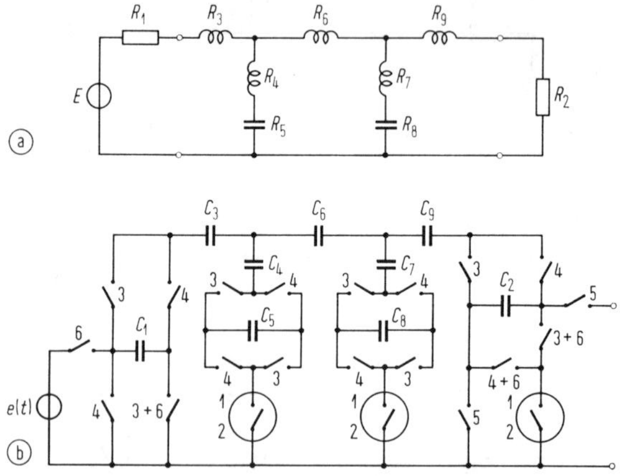

Fig. 12. (a) A 5th degree low-pass filter in the ψ-domain, (b) a corresponding switched-capacitor structure obtained from Figs. 9 and 11 by including terminal inverter switches at both the input and the output.

started out with the high-pass case.) If n is the degree of the filter, the number of voltage inverter switches required is the same as in the high-pass case, i.e. $(n+1)/2$.

One can reduce the total number of terminal inverter switches by making use of the following principle: Let $X(z)$ and $Y(z)$ be the z-transforms of the input and output signal, respectively, and let $H(\psi)$ be the transfer function. We have $Y(z) = H(\psi)\,X(z)$ and thus, making use of eq. (3), $Y(-z) = H(1/\psi)\,X(-z)$. Replacing z by $-z$ in a signal, however, amounts simply to changing the sign of every second sample, while replacing ψ by $1/\psi$ amounts to replacing e.g. a low-pass by a high-pass filter. Hence, realizing the low-pass filter of Fig. 12a is equivalent to realizing the high-pass filter of Fig. 9a if we connect a terminal inverter switch at both the input and the output port. The resulting circuit, with grounded output port (cf. Fig. 11), is shown in Fig. 12b, the numbers next to the switches referring to the time intervals shown in Fig. 10. The circuit of Fig. 12b involves fewer switches than what would have been required in case of a direct realization of Fig. 12a. The saving achievable this way increases with the degree of the filter.

Ladder *band-pass* filters can also be designed according to structures whose series branches involve only elements connected in series. For a filter of degree 6, some typical examples (drawn without their terminations) are given in Fig. 13a—d, all of which can be translated into switched-capacitor filters by using the same principles as those explained so far. The structure of Fig. 13a contains two redundant reactive elements; these can usually be eliminated by Norton transformations, but the way in which this can be achieved without any element becoming negative depends then mainly on the location of the attenuation poles (Fig. 13b—d). Other suitable structures could be obtained from certain band-pass filters with either minimum number of inductances or minimum number of capacitances. The transformation $\psi \to 1/\psi$ can again lead to

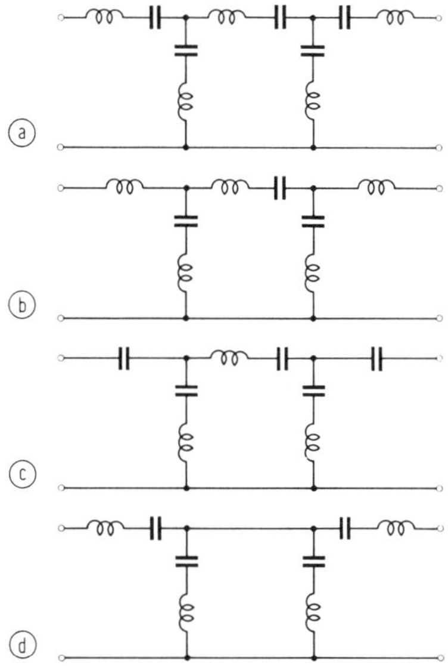

Fig. 13. Four structures of 6th degree band-pass filters which can be realized as switched-capacitor filters with grounded voltage inverter switches.

savings if it implies a sufficient decrease in the number of inductances. If n is the degree of the filter, the number of voltage inverter switches required is $n/2$ for the structures of Fig. 13a—d, but higher for structures with minimum number of either inductances or capacitances.

6. Concluding Remarks

We have seen how a wide class of LC ladder filters can be built by using capacitances, ordinary switches, and voltage inverter switches. A voltage inverter switch in turn can be built quite simply by electronic means, e.g. by using an operational amplifier connected as voltage follower, an auxiliary storage capacitance, and ordinary switches.

The resulting filters are fully described by the dimensionless frequency parameter ψ given by eq. (3) and by a certain number of resistive parameters R_i having the form $R_i = T/2C_i$, C_i being the capacitance realizing the element of value R_i. Since transfer functions are dimensionless rational functions in ψ, their coefficients can only depend on ratios of the R_i, thus only on ratios of the C_i, as desired.

In all circuits, we have easily available not only the transmittance S_{21} but also the reflectance S_{11}. It is known that for lossless twoports we have

$$|S_{21}|^2 + |S_{11}|^2 = 1\,,$$

i.e. S_{21} and S_{11} have complementary properties. For both S_{21} and S_{11}, the known argument for achieving low passband sensitivity does hold. For ladder structures, however, only S_{21} offers good stopband sensitivity.

Acknowledgement

The author gratefully acknowledges the many fruitful discussions which he has had on the subject of this paper with D. Herbst, B. Höfflinger, K. A. Owenier, K. Schumacher, R. Schweer, and H. Wupper.

(Received May 13th, 1978.)

References

[1] Fried, D. L., Analog sample-data filters. IEEE J. Solid-State Circuits SC-**7** [1972], 302—304.

[2] Young, I. A., Hodges, D. A., and Gray, P. R., Analog NMOS sampled-data recursive filter. Digest of Tech. Papers, IEEE Int. Solid-State Circuits Conf., Philadelphia, Feb. 1977, pp. 156—157.

[3] Hodges, D. A., Gray, P. R., and Brodersen, R. W., Potential of MOS technologies for analog integrated circuits. Digest of Tech. Papers, European Solid-State Circuits Conf., Ulm (W. Germany), Sept. 1977, pp. 43—47.

[4] Caves, J. T. et al., Sampled analog filtering using switched capacitors as resistive equivalents. IEEE J. Solid-State Circuits SC-**12** [1977], 592—599.

[5] Hosticka, B. J., Brodersen, R. W., and Gray, P. R., MOS sampled-data recursive filters using switched capacitor integrators. IEEE J. Solid-State Circuits SC-**12** [1977], 600—608.

[6] Allstot, D. J., Brodersen, R. W., and Gray, P. R., Fully-integrated high-order NMOS sampled-data ladder filters. Digest of Tech. Papers, IEEE Int. Solid-State Circuits Conf., Philadelphia, Feb. 1978, pp. 82, 83, and 268.

[7] Jacobs, G. M. et al., Design considerations for MOS switched capacitor filters (to be published).

[8] Poschenrieder, W., Frequenzfilterung durch Netzwerke mit periodisch gesteuerten Schaltern. Proc. NTG-Symp. "Analyse und Synthese von Netzwerken", Stuttgart 1966, pp. 220—237.

[9] Fettweis, A., Theory of resonant-transfer circuits. Lecture notes, NATO Advanced Study Inst. on Network and Switching Theory, Sept. 1966; also in "Network and Switching Theory" (ed. by G. Biorci), pp. 382—446, Academic Press, New York 1968.

[10] Fettweis, A. Theory of resonant-transfer circuits and their application to the realization of frequency-selective networks. Proc. URE Summer School on Circuit Theory, 68 pages, Prague, July 1968.

[11] Boite, R. and Thiran, J. V. P., Synthesis of filters with capacitances, switches, and regenerating devices. Transact. IEEE CT-**15** [1968], 447—454.

[12] Fettweis, A., Input and output networks with constant input pulse impedances for resonant-transfer circuits. Transact. IEEE CT-**16** [1969], 562—564.

[13] Fettweis, A., Resonant-transfer filters. Proc. Third Asilomar Conf. on Circuits and Systems, p. 309, Pacific Grove, Cal., Dec. 1969.

[14] Fettweis, A., Realization of general network functions using the resonant-transfer principle. Proc. Fourth Asilomar Conf. on Circuits and Systems, pp. 663—666 and 671, Pacific Grove, Cal., Nov. 1970; also AEÜ **25** [1971], 295—303.

[15] Fettweis, A., Some properties of pulse impedances and pulse impedance matrices. AEÜ **24** [1970], 506—512.

[16] Boite, R., Discrete-time systems with interwoven transfers. Proc. IEE **118** [1971], 312—316.

[17] Weinrichter, H., Kondensator-Kommutator-Netzwerke. AEÜ **26** [1972], 293—305.

[18] Fettweis, A., Switched-capacitor filters using voltage inverter switches: further design principles. AEÜ **33** [1979], in print.

[19] Fettweis, A., Digital filter structures related to classical filter networks. AEÜ **25** [1971], 79—89.

Switched-Capacitor Filters Using Voltage Inverter Switches: Further Design Principles

by Alfred Fettweis *

The basic principles for designing switched-capacitor filters using voltage inverter switches have previously been described. The catalog of available elements is now extended to include in particular unit elements as well as related reciprocal and nonreciprocal one-port and multiport elements. This way, a variety of further types of circuits can be designed, and the number of voltage inverter switches required per filter can be reduced. It is shown how grounded voltage inverter switches can be built by making use of inverse recharging.

Schalter-Kondensator-Filter mit Spannungsumkehrschaltern: weitere Entwurfsprinzipien

Die Grundprinzipien für den Entwurf von Schalter-Kondensator-Filtern sind in einer früheren Arbeit beschrieben worden. Der Katalog der verfügbaren Elemente wird jetzt erweitert, und zwar insbesondere um Einheitselemente sowie verwandte reziproke und nichtreziproke Ein- und Mehrtorelemente. Hierdurch steht eine Vielfalt weiterer Schaltungstypen zur Verfügung, so daß unter anderem die Anzahl der pro Filter erforderlichen Spannungsumkehrschalter reduziert werden kann. Es wird gezeigt, wie geerdete Spannungsumkehrschalter nach dem Prinzip der inversen Wiederaufladung realisiert werden können.

1. Introduction

Switched-capacitor filters are receiving rapidly increasing attention. An interesting class of such filters are those using grounded voltage inverter switches. These filters are known to be closely related to certain so-called resonant-transfer filters, but make use of the technological means now available. The simplest such filters are designed in order to simulate LC ladder filters of low-pass, high-pass, or band-pass type. The corresponding theory has been presented, in self-contained form, in a previous paper [1].

In the present paper, this previous theory is extended in various directions. First of all, unit elements are introduced as circuit elements and it is shown that the number of voltage inverter switches, and thus the number of critical electronic devices (e.g. operational amplifiers), can this way be further reduced. Secondly, an alternative method of realizing a grounded voltage inverter switch is described, using what we shall call the principle of inverse recharging. This principle requires usually that somewhat faster operational amplifiers be employed, but the influence of the parasitic capacitances between the input terminals of the operational amplifiers and ground can now essentially be eliminated.

Thirdly, the catalog of available elements will further be augmented by generalizing the unit element in various respects. This way, it becomes possible in particular to realize nonreciprocal devices.

In the present paper, we shall freely make use of results already described in the previous paper [1]. In particular, notations already used there will not be redefined. For more details concerning the bibliography we also refer to [1].

Prof. Dr. A. Fettweis, Lehrstuhl für Nachrichtentechnik, Ruhr-Universität, Postfach 10 21 48, D-4630 Bochum.

2. Circuits Involving Unit Elements

2.1. *Realization of unit elements and quasi-reciprocal lines* (QUARL's)

We consider a capacitance C into which are fed two distinct pulse currents $i_1 = i_1(t)$ and $i_2 = i_2(t)$ (Fig. 1a). We assume that the pulses of i_1 are non-overlapping with those of i_2, as shown in Fig. 2. To the pulses of i_1 and i_2, we make correspond, similar to eq. (6) of [1], time instants t_{1n} and t_{2n} for which we can write

$$t_{kn} = t_{k0} + nT, \quad (1)$$
$$n = \cdots, -1, 0, 1, 2, \ldots, \quad k = 1, 2,$$

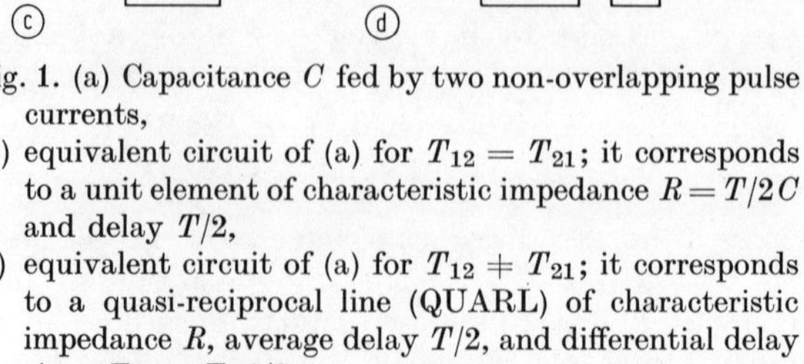

Fig. 1. (a) Capacitance C fed by two non-overlapping pulse currents,
(b) equivalent circuit of (a) for $T_{12} = T_{21}$; it corresponds to a unit element of characteristic impedance $R = T/2C$ and delay $T/2$,
(c) equivalent circuit of (a) for $T_{12} \neq T_{21}$; it corresponds to a quasi-reciprocal line (QUARL) of characteristic impedance R, average delay $T/2$, and differential delay $\Delta = (T_{12} - T_{21})/2$,
(d) decomposition of the QUARL (c) into a cascade of a unit element and a zero-QUARL.

Reprinted with permission from *Arch Elek Ubertragunge,* vol. 33, pp. 107-114, Mar. 1979.

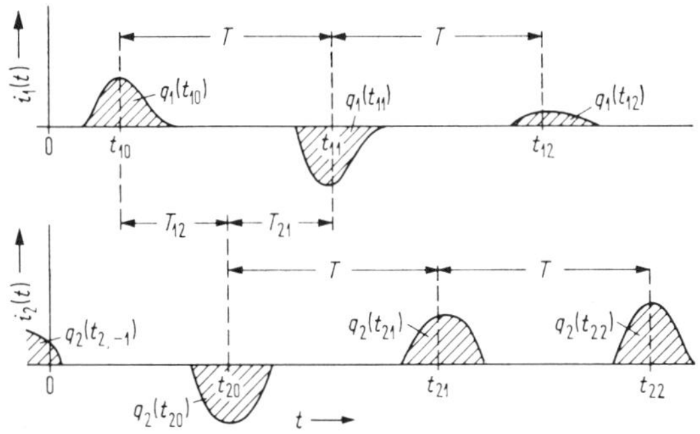

Fig. 2. Representation of two non-overlapping pulse currents $i_1(t)$ and $i_2(t)$.

t_{10} and t_{20} being constants such that

$$t_{10} < t_{20} < t_{10} + T\,.$$

In particular, the quantities T_{12} and T_{21} defined by

$$T_{12} = t_{20} - t_{10}\,, \quad T_{12} + T_{21} = T \tag{2}$$

are positive constants. The charges carried by the n-th pulses of i_1 and i_2 will be designated by $q_1(t_{1n})$ and $q_2(t_{2n})$, respectively.

Let $v = v(t)$ be the voltage across C. We designate by $v_{k\mathrm{b}}(t_{kn})$ and $v_{k\mathrm{a}}(t_{kn})$ the values of v just before and just after the arrival of the n-th pulse of i_k, with $k = 1, 2$. We have

$$v_{k\mathrm{a}}(t_{kn}) - v_{k\mathrm{b}}(t_{kn}) = q_k(t_{kn})/C\,, \quad k = 1, 2, \tag{3}$$

$$v_{1\mathrm{a}}(t_{1n}) = v_{2\mathrm{b}}(t_{2n})\,, \quad v_{2\mathrm{a}}(t_{2n}) = v_{1\mathrm{b}}(t_{1,\,n+1}) \tag{4}$$

and we define

$$u_k(t_{kn}) = [v_{k\mathrm{a}}(t_{kn}) + v_{k\mathrm{b}}(t_{kn})]/2\,, \quad k = 1, 2\,. \tag{5}$$

In order to express these difference equations under steady-state conditions, we write

$$v_{k\mathrm{a}}(t_{kn}) = V_{k\mathrm{a}} \exp(p\,t_{kn})\,, \quad v_{k\mathrm{b}}(t_{kn}) = V_{k\mathrm{b}} \exp(p\,t_{kn})\,, \tag{6}$$

$$u_k(t_{kn}) = U_k \exp(p\,t_{kn})\,, \quad q_k(t_{kn}) = Q_k \exp(p\,t_{kn})\,, \tag{7}$$

where $V_{k\mathrm{a}}$, $V_{k\mathrm{b}}$, U_k, and Q_k are complex constants and $k = 1, 2$. Using eqs. (1) to (7), eliminating the $V_{k\mathrm{a}}$ and $V_{k\mathrm{b}}$, and defining

$$J_k = Q_k/T\,, \quad R = T/2C\,, \quad \Delta = (T_{12} - T_{21})/2\,,$$

we obtain expressions between the U_k and J_k which we can interprete to define a two-port of chain matrix

$$\boldsymbol{K} = \boldsymbol{K}_0\, \mathrm{e}^{p\Delta}\,, \quad \boldsymbol{K}_0 = \begin{pmatrix} \cosh(p\,T/2) & R\sinh(p\,T/2) \\ (1/R)\sinh(p\,T/2) & \cosh(p\,T/2) \end{pmatrix}. \tag{8}$$

In most cases, we have $T_{12} = T_{21} = T/2$, i.e. $\Delta = 0$; $\boldsymbol{K}$ then coincides with $\boldsymbol{K}_0$, which is the chain matrix of a *unit element* of *characteristic impedance* R and *delay* $T/2$ (Fig. 1b). For $\Delta \neq 0$, $\boldsymbol{K}$ corresponds to a *quasi-reciprocal line*, or QUARL (Fig. 1c). Clearly, $\boldsymbol{K}$ and $\boldsymbol{K}_0$ differ only by the factor $\mathrm{e}^{p\Delta}$ involving the *differential delay* Δ. Thus, a QUARL behaves largely like a unit element whose delay is equal to the *average delay* of the QUARL. In particular, if a QUARL appears in a cascade connection, the factor $\mathrm{e}^{p\Delta}$ does not affect the input impedance and its only effect upon the transfer properties is to cause the delay in the direction $1 \to 2$ to exceed the one in the direction $2 \to 1$ by the constant 2Δ [2], [3], [4]. Thus, even if $\Delta \neq 0$, we may usually still simply ignore the factor $\mathrm{e}^{p\Delta}$. On the other hand, we can always separate this factor by decomposing a QUARL into a cascade of a unit element and a *zero*-QUARL (Fig. 1d), the latter being a QUARL with zero average delay, i.e. with $\boldsymbol{K}_0$ in eq. (8), replaced by the unit matrix. For each port of a unit element or QUARL we can write equations corresponding to eqs. (14), (15), and (21) of [1].

2.2. *Filters using inductances, capacitances, and/or unit elements*

The simplest circuit is obtained by connecting unit elements in cascade, an example of degree 5 being shown in Fig. 3a. A corresponding switched capacitor structure is shown in Fig. 3b. A possible choice for the intervals during which the switches in Fig. 3b are to be closed is given in Fig. 3c, the overall period of operation being T. Observe that the voltage inverter switches now have to be operative twice during one period T. The reflected signal (thus the reflectance S_{11} as discussed in [1]) is available through the switch marked 7. As in [1], we assume throughout this paper that the R_i and

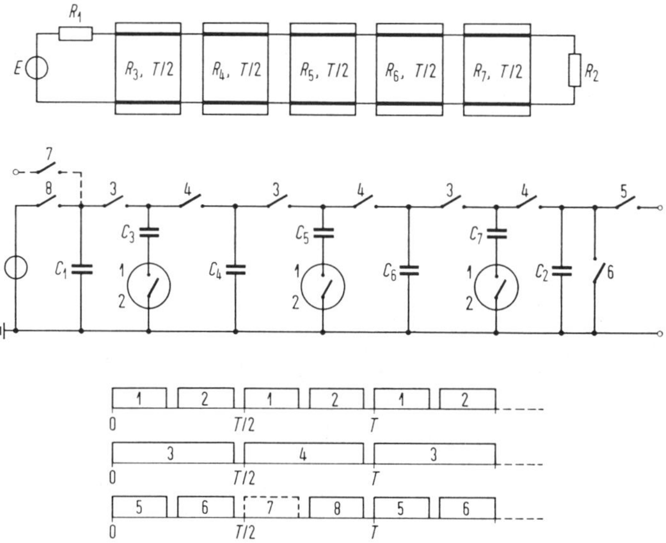

Fig. 3. (a) A 5th degree filter composed of unit elements in cascade,
(b) a corresponding switched-capacitor structure,
(c) intervals during which the switches are to be closed.

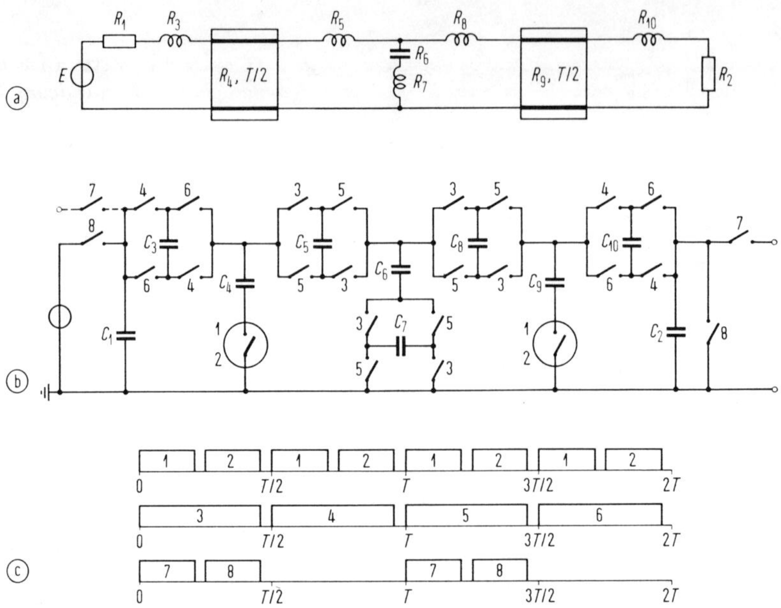

Fig. 4. (a) A 7th degree low-pass filter,
(b) a corresponding switched-capacitor structure,
(c) intervals during which the switches are to be closed.

C_i are related by

$$R_i = T/2C_i\,.$$

In Fig. 4 is shown an example of a low-pass filter of degree 7 involving inductances, capacitances, and unit elements (Fig. 4a), together with a corresponding switched-capacitor structure (Fig. 4b) and a possible choice for the intervals during which the switches are to be closed (Fig. 4c). The overall period of operation of the switches is now equal to $2T$. The present solution requires only two voltage inverter switches, i.e. the same number as for an LC low-pass filter of degree 3. The filtering performance, however, is better, and by proper design even the filtering capability of the unit elements can be exploited. On the other hand, the voltage inverter switches have to operate at a faster rate than for the circuits of e.g. Figs. 9 and 12 in [1]. The reflected signal (thus S_{11}) is available through the switch marked 7 connected to C_1.

The circuit of Fig. 4a can be generalized to any degree $n = 4\nu + 3$ where $\nu = 1, 2, 3, \ldots$ For $n = 11$, the corresponding circuit is shown in Fig. 5a. The number of voltage inverter switches which are required is $(n+1)/4$. A similar band-pass filter of degree $n = 4\nu$, requiring $n/4$ voltage inverter switches, is shown in Fig. 5b for $n = 12$.

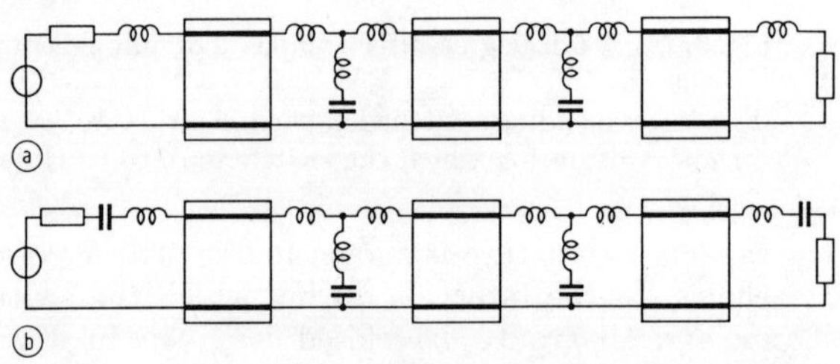

Fig. 5. (a) Extension of the circuit of Fig. 4a to degree $n = 11$, the corresponding switched-capacitor filter requiring $(n+1)/4 = 3$ voltage inverter switches,
(b) a similar band-pass filter of degree $n = 12$, requiring $n/4 = 3$ voltage inverter switches.

By using distinct locations for the intervals during which the switches in the series arms of Fig. 3b are closed one can achieve that the various voltage inverter switches do not have to be operative simultaneously. If we then introduce further auxiliary switches it even becomes possible to time share a single voltage inverter switch among the various places where such a switch is required. The degree n of the filter which can be realized this way by means of a single voltage inverter switch is limited only by the speed at which the operational amplifiers can operate. The same principle can be applied to filters of the type of Figs. 4 and 5, two operational amplifiers being then required independently of the degree, except of course for the limitation just pointed out.

3. Realization of Circuits Using Inverse Recharging

3.1. The principle of inverse recharging

We consider again the $(n+1)$-terminal capacitance network of Fig. 1 in [1], reproduced here for convenience as Fig. 6. As we know, charges q_ν

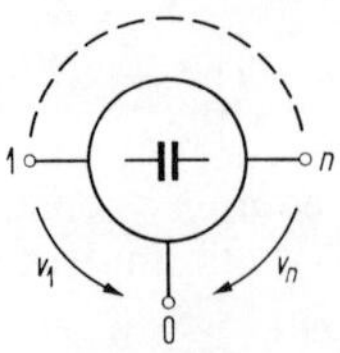

Fig. 6. Symbolic representation of an $(n+1)$-terminal capacitance network.

are assumed to be injected into the respective terminals $\nu = 1$ to n such that the voltages v_ν are changed from $v_{\nu\mathrm{b}}$ to $v_{\nu\mathrm{a}}$. We want to achieve the equalities

$$v_{\nu\mathrm{a}} = -v_{\nu\mathrm{b}}\,, \quad \nu = 1, \ldots, n\,. \tag{9}$$

Let $\boldsymbol{v}_\mathrm{a}$, $\boldsymbol{v}_\mathrm{b}$, and $\boldsymbol{q}$ be the vectors composed of the $v_{\nu\mathrm{a}}$, $v_{\nu\mathrm{b}}$, and q_ν respectively; we have, for any $\boldsymbol{q}$ and the resulting $\boldsymbol{v}_\mathrm{a}$,

$$q = C(v_a - v_b) \quad (10)$$

where C is the capacitance matrix of the $(n+1)$-terminal capacitance network (Fig. 6). If eq. (9) is fulfilled, eq. (10) becomes

$$q = -2 C v_b . \quad (11)$$

On the other hand, let q_0 be that value of q for which v_a becomes zero; we obtain from eq. (10),

$$q_0 = - C v_b . \quad (12)$$

Comparing eqs. (11) and (12), we see that $q = 2 q_0$. Thus, the charges q_ν which produce the voltage reversal are precisely twice those which reduce the voltages to zero. (Observe in this respect that the relation between $v_a - v_b$ and q is unique in the sense that for the networks with which we are dealing C always exists and is nonsingular.)

Consequently, the voltage inversion (9) can be achieved in two steps, i.e. by first connecting to ground all terminals of the capacitance network and determining the charges thus flowing through the short-circuits, and by then shifting these same charges through once again in the same direction as the first time. We shall refer to this method as to the principle of *inverse recharging*. This principle was first used by Schlichte for producing by electronic means the overall effect of a simple resonant-transfer switch [5], [6]. It has subsequently been generalized [2], [7], and experimental results on filters built this way have been published [8], [9]. New methods for implementing the principle of inverse recharging will be presented hereafter, and it is hoped that the shortcomings of the older methods can be avoided this way.

It is true that conceptually very simple means for implementing a voltage inverter switch had already been described in [1]. In the circuits of Fig. 2a and b of [1], however, the parasitic capacitances unavoidably present between the input terminals of the operational amplifiers and ground should be small compared as well to the capacitances of the capacitance network to which the voltage inverter switch is connected as to the auxiliary storage capacitance C_s. This may impose restrictions for highly miniaturized realizations. We will therefore examine realizations in which the input terminals of the operational amplifiers are, at least virtually, on ground so that the parasitic capacitances are short-circuited.

3.2. *Grounded voltage inverter switches using inverse recharging*

In Fig. 7 is shown how the principle of inverse recharging can be applied to one of the terminals (terminal n) of the capacitance network of Fig. 6. In particular, during a first interval the operational amplifier and the auxiliary storage capacitance C_s are connected to the capacitance network as shown in Fig. 7a. The terminal n is then virtually put to ground, but the charge flowing as a result of this is accumulated in C_s. If during a second interval the connections are established as shown

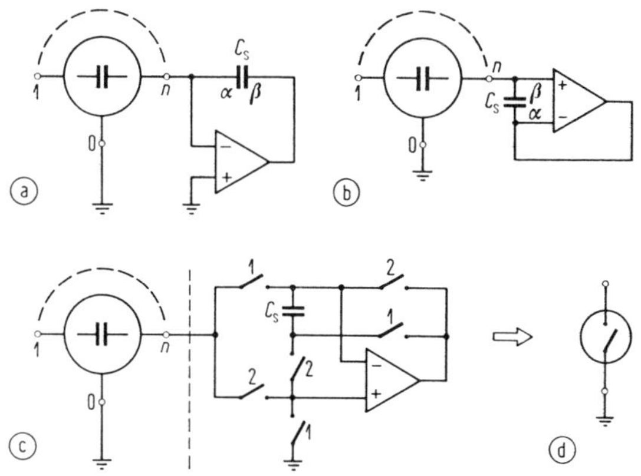

Fig. 7. A possible arrangement for realizing a voltage inverter switch according to the principle of inverse recharging.
(a) and (b) The two consecutive connections to be realized,
(c) realization of the connections (a) and (b) by introducing switches 1 and 2, respectively,
(d) equivalent representation of the arrangement to the right of the dotted line in (c).

in Fig. 7b, C_s will discharge; the resulting charge flow is the same as before and (due to the way in which the terminals α and β of C_s are connected) crosses the capacitance network again in the same direction as before. An arrangement realizing these connections by means of six switches of which those marked 1 are closed during the first of the aforementioned intervals and those marked 2 during the second is shown in Fig. 7c.

The overall effect of the arrangement to the right of the dotted line in Fig. 7c, clearly, corresponds fully to a grounded voltage inverter switch. It may thus be represented by the known symbol, as has been done in Fig. 7d, and it may be used in exactly the same way as the voltage inverter switches of Fig. 2 of [1] (e.g. as shown in Fig. 3 of [1]). Nevertheless, the internal way of operation is entirely different than for the latter; in particular, the changes in the capacitance network occur only during the interval 2 if a voltage inverter switch according to Fig. 2a or b of [1] is used, but during both intervals if the voltage inverter switch is realized according to Fig. 7.

In the solution of Fig. 7, the input terminals of the operational amplifier are on ground potential during the interval 1, but they move away from that potential during the interval 2. This deficiency can be overcome by replacing the arrangement of Fig. 7b by the one of either Fig. 8a or b, i.e. by disconnecting from ground the terminal 0 of the capacitance network during the interval 2. Introducing switches (to be closed again during consecutive intervals 1 and 2, respectively), the connections according to Fig. 7a and 8a are realized by the arrangement of Fig. 8c, and those according to Figs. 7a and 8b by the arrangement of Fig. 8d (where the role of the switch in dotted form will be explained later). In both Fig. 8c and d, that part of the circuit which is to the right

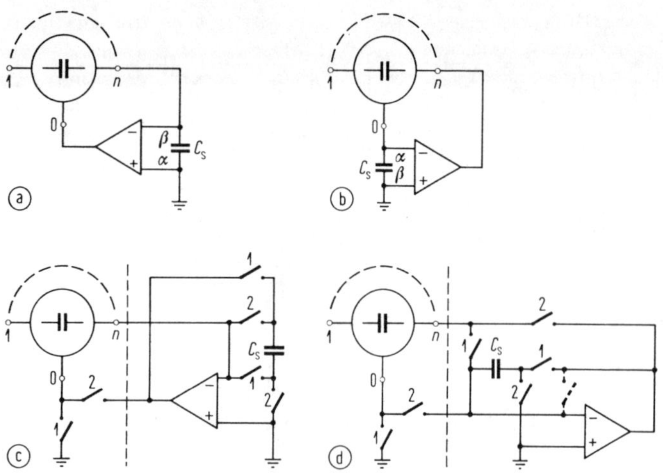

Fig. 8. (a) and (b) Two possible alternatives to the arrangement of Fig. 7b,
(c) and (d) the two corresponding realizations replacing Fig. 7c, respectively; under certain conditions, only (d) is acceptable.

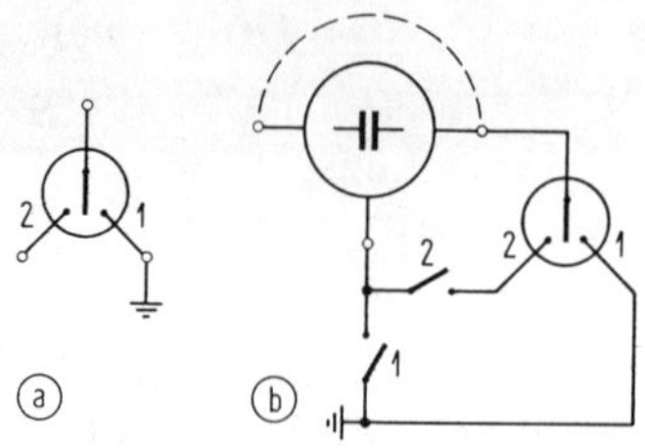

Fig. 9. (a) An inverse recharging device representing the circuit arrangement to the right of the dotted line in either Fig. 8c or d,
(b) representation of the overall arrangements of Fig. 8c and d, using the symbol (a).

of the dotted line will be called an *inverse recharging device*; we represent it by the three-pole symbol of Fig. 9a. Together with the further switches shown in Fig. 8c and d, it fulfills the function of a voltage inverter switch. The numbers next to its lower two terminals indicate that during the intervals 1 and 2 the charge flow is via the terminals marked 1 and 2, respectively, the first of these two terminals being necessarily grounded. Using the symbol of Fig. 9a, the overall arrangements of Fig. 8c and d can be drawn as shown in Fig. 9b. The reason why the two switches to the left in Fig. 9b are kept separately will become clear later on.

The solution of Fig. 8 has indeed the property that the input terminals of the operational amplifier are at ground potential during both relevant time intervals. If we try to apply this solution without further thought, however, we discover that this is possible only if the capacitance network is composed of a single loop. This is due to the fact that during the second interval, only one of the terminals of the capacitance network may be kept at ground potential since the voltages at the other terminals must be allowed to become equal to minus the values which they had initially. This dilemma cannot be resolved by connecting all those terminals of the three-pole voltage inverter switches which are virtually at ground, simultaneously to the common terminal 0 of the capacitance network; it can indeed be checked that the proper flow of the charges cannot be controlled in this case.

Observe, however, that if there are several capacitance loops, only the interval 1 must be coincident for all the loops, i.e. only the original discharge must take place simultaneously for all the loops. Once the charges which have thus flown through the various loops have been determined, the inverse recharging can be carried out during distinct and even nonoverlapping intervals. For the 5-pole capacitance network of Fig. 3 in [1], the corresponding arrangement is shown in Fig. 10a, and possible locations for the various time intervals thus involved are shown in Fig. 10b. Of the realizations of Fig. 8 for the inverse recharging device, however, only the one of Fig. 8d can be used without further modification. Indeed, in the one of Fig. 8c, terminal n of the capacitance network is permanently connected to the negative terminal of the operational amplifier, thus also during those intervals during which other loops are being recharged.

It should now also be clear at least why it was better not to include in the symbol for an inverse recharging device the left most switch in the arrangement of Fig. 9b. Indeed, this switch does not have to be provided individually for each one of these devices, as is clearly seen in Fig. 10a. On the other hand, it may be appropriate to include now an additional switch as shown in dotted form in Fig. 8d. This additional switch allows us to apply a strong feedback during those intervals in which all the other switches of the same inverse recharging device are open, i.e. to prevent undesirable charges from accumulating, during these intervals, on the parasitic capacitances of the operational amplifier.

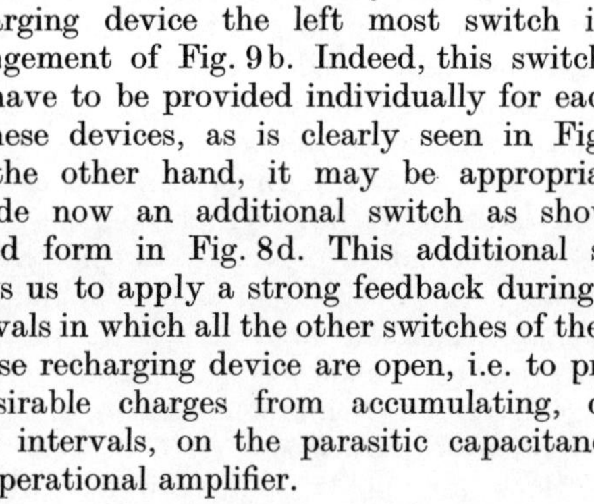

Fig. 10. (a) A 5-pole capacitance network with inverse recharging devices,
(b) schematic representation of the consecutive time intervals during which the various operations take place.

As can be concluded from Fig. 10b, the price which we have to pay for being able to eliminate the effect of the parasitic capacitances is that more individual intervals (five in case of Fig. 10) are needed to produce the required voltage inversion, i.e. that faster operational amplifiers are needed to achieve the same sampling rate. As will be seen, however, this disadvantage can often be kept rather small. In many practical situations, indeed, at most two loops are simultaneously involved, and it may sometimes even be possible to overcome the disadvantage almost entirely by using some form of time sharing.

3.3. Realization of filters by means of voltage inverter switches based on inverse recharging

We illustrate the method by means of the low-pass filter of Fig. 4a. A corresponding switched-capacitor structure using two inverse recharging devices is shown in Fig. 11a, and a possible choice for the intervals during which the switches are to be closed is given in Fig. 11b. The reflected signal (thus S_{11}) is available through the switch marked 3 connected to C_1.

One sampling period T now comprises five separate operational steps of the inverse recharging devices, while only four such steps were required according to Fig. 4c. Hence, for a given sampling rate $F=1/T$, the time available per operational step becomes reduced by a factor 0.8. This factor has to be replaced by the factor 2/3 for the circuits of Fig. 5 as well as for higher order generalizations thereof. The reduction, however, would be larger if filter structures according to [1] were used, and this enhances the interest in filters comprising unit elements. From Fig. 11a, it should now fully be clear why the two switches in the lower left of Fig. 9b have not been included in the symbol of Fig. 9a.

If the inverse recharging devices are fast enough, one might consider time sharing e.g. their operational amplifiers among several filters the auxiliary storage capacitances must be provided individually. It is true that this may be a cause of crosstalk, but this might not be too serious if the time sharing is applied to the filters forming a send and receive pair in a communications channel. By making use of time sharing, it becomes possible to overcome, at least to a large extent, the disadvantage resulting from the reduction of the time available for the individual operational steps.

In order to illustrate this, let us compare the filters of Figs. 4 and 11. As a rough estimate, let us assume that the power consumed by an operational amplifier is equal to a constant K divided by the time available for one operational step. From the first line in Fig. 4c we thus conclude, neglecting the short guard spaces between consecutive intervals, that the power consumed by one voltage inverter switch is $K/(T/4)=4K/T$. The total power consumed by two filters of the type of Fig. 4b is thus $2\times 2\times 4K/T=16\,K/T$.

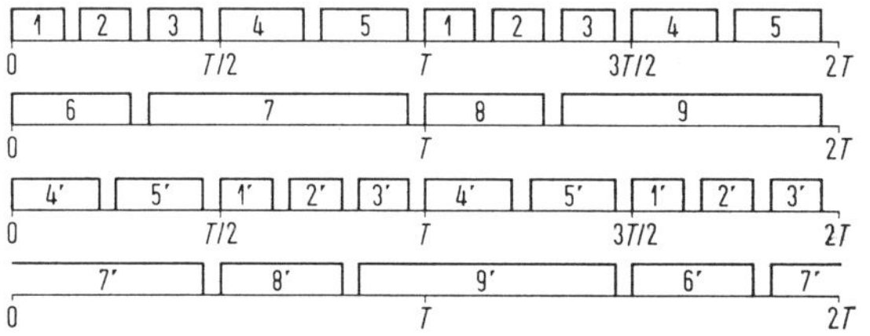

Fig. 12. First two lines: Alternative arrangement for the various time intervals of Fig. 11b,
last two lines: Corresponding arrangement for a second filter according to Fig. 11a and chosen in such a way that the operational amplifiers can be time shared among the two filters.

For the filter of Fig. 11a, the arrangement of the various time intervals given in Fig. 11b may be replaced by the one shown in the first two lines of Fig. 12. During the intervals 1, 2, and 3, two operational amplifiers are operative simultaneously, the time available per operational step being then $T/6$ if we neglect again the short guard spaces between consecutive intervals. During each one of the intervals 4 and 5, however, only one operational amplifier is operative at a time, the time available per operational step being now $T/4$. For a second

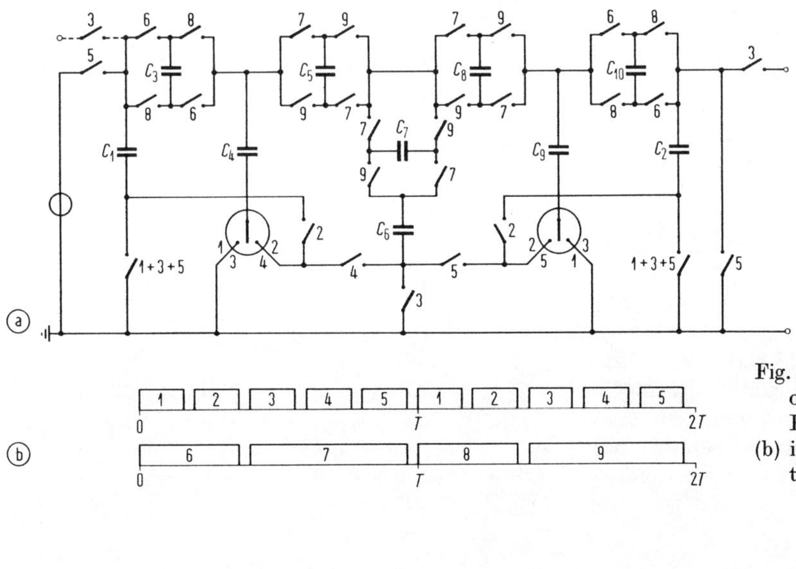

Fig. 11. (a) A switched-capacitor realization of the 7th degree low-pass filter of Fig. 4a, using inverse recharging,
(b) intervals during which the switches are to be closed.

filter according to Fig. 11a, the various time intervals may be arranged as shown in the last two lines of Fig. 12, with dashes added to the various numbers in order to avoid confusion. By time sharing the operational amplifiers among the two filters, clearly, two operational amplifiers of the former type are required altogether, and one of the latter. Hence, the total power consumed is $2 \times K/(T/6) + 1 \times K/(T/4) = 16\,K/T$, which is exactly the same figure as before. The same conclusion can be arrived at also for other filters if the time sharing is properly arranged.

4. Use of Further Circuit Elements

The considerations of Subsection 2.1 can be extended to

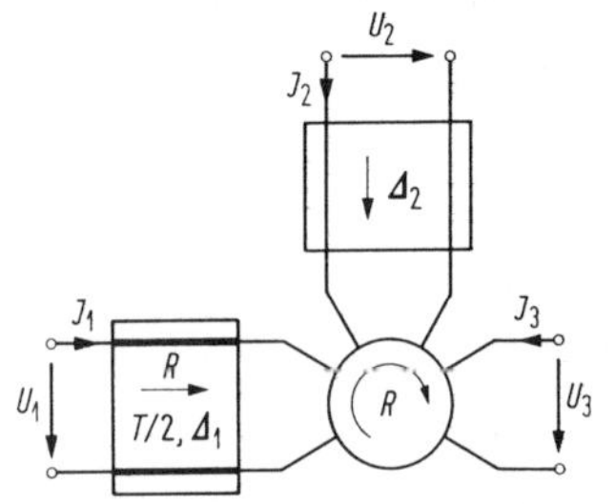

Fig. 13. Equivalent circuit of a capacitance fed by three distinct non-overlapping pulse currents i_1, i_2, i_3.

any number of distinct non-overlapping pulse currents i_k. For $k = 1, 2, 3$, for instance, we simply have to replace eq. (2) by

$$T_{12} = t_{20} - t_{10}\,,$$
$$T_{23} = t_{30} - t_{20}\,,$$
$$T_{12} + T_{23} + T_{31} = T\,,$$

and we may assume $T_{12} > 0$, $T_{23} > 0$, $T_{31} > 0$. The resulting equivalent circuit is shown in Fig. 13. It comprises a three-port circulator, a QUARL, and a zero-QUARL, with $\Delta_1 = (T/2) - T_{31}$, $\Delta_2 = T_{23}$, $R = T/2C$. For $k = 1, 2, 3, 4$, we would arrive at a representation comprising a four-port circulator [4].

Devices of this type can be used e.g. to realize all-pass circuits. An example comprising a section of order 1 and one of order 2 is shown in Fig. 14a (ignoring the irrelevant zero-QUARL's), and the corresponding switched-capacitor circuit in Fig. 14b. It requires two voltage inverter switches. A possible choice for the intervals during which the various switches are to be closed is given in Fig. 14c.

Another possibility of realizing a nonreciprocal element is to reverse the terminals of C in Fig. 1a e.g. between each pulse of i_2 and the next following pulse of i_1. This can be achieved by means of an appropriately operated terminal inverter switch, thus as shown in Fig. 15a. In the first line of Fig. 16 are represented schematically the pulses of i_1 and i_2, and in the second line are shown the intervals during which the switches marked 1 and 2 of Fig. 15a are to be closed, respectively. The only modification to be applied to the analysis of Subsection 2.1 is then that the second equation (4) has to be replaced by

$$v_{2\mathrm{a}}(t_{2n}) = -v_{1\mathrm{b}}(t_{1,\,n+1})\,.$$

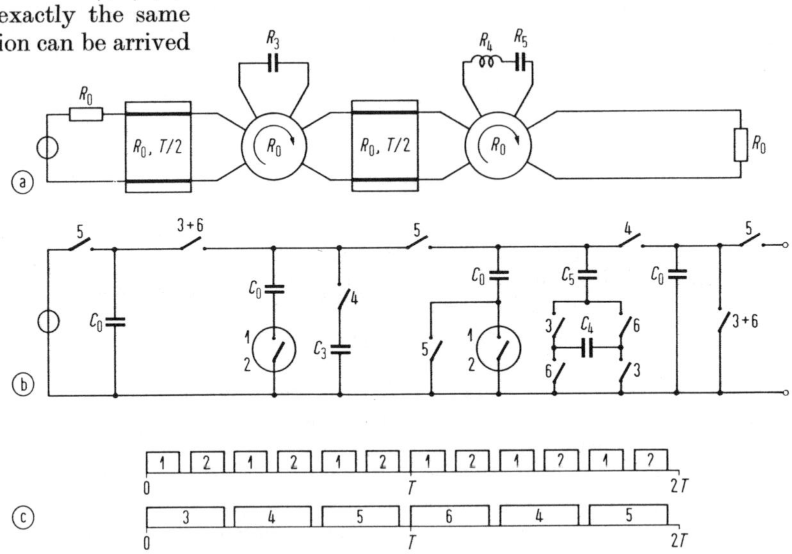

Fig. 14. (a) An all-pass circuit,
(b) a corresponding switched-capacitor structure,
(c) intervals during which the switches in (b) are to be closed.

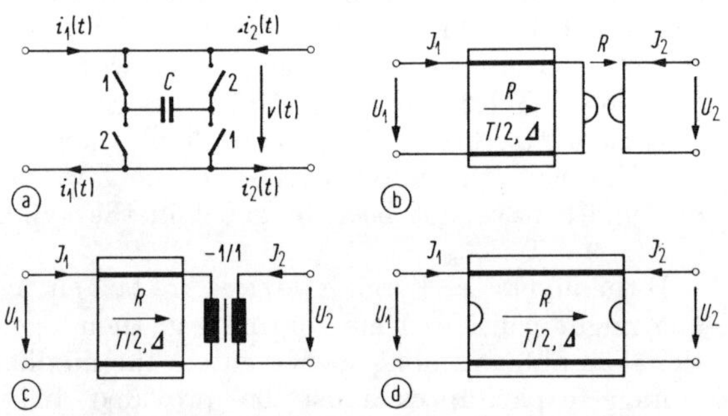

Fig. 15. (a) Circuit arrangement obtained from the one of Fig. 1a by connecting C through four switches connected as terminal inverter switch,
(b) and (c) equivalent circuits of (a) obtained if the switches operate according to the second and fourth line in Fig. 16, respectively,
(d) symbolic representation of a gyro-QUARL, i.e. of the arrangement (b).

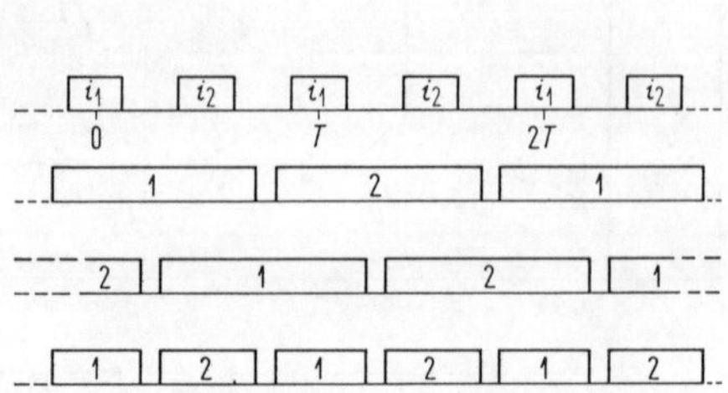

Fig. 16. Schematic representation of the current pulses in Fig. 15a and three distinct ways of operation of the switches marked 1 and 2, respectively.

This in turn causes $\boldsymbol{K}_0$ in eq. (8) to be replaced by

$$\boldsymbol{K}_0 = \begin{pmatrix} \sinh(pT/2) & R\cosh(pT/2) \\ (1/R)\cosh(pT/2) & \sinh(pT/2) \end{pmatrix},$$

i.e. the chain matrix given by eq. (8) is now multiplied by

$$\begin{pmatrix} 0 & R \\ 1/R & 0 \end{pmatrix}.$$

Consequently, the equivalent circuit replacing the one of Fig. 1c is as shown in Fig. 15b. If the switches in Fig. 15a are closed according to the third line in Fig. 16, the arrow in the gyrator of Fig. 15b has to be reversed. Finally, if the switches in Fig. 15a operate according to the fourth line in Fig. 16, the equivalent circuit of Fig. 15c is obtained [4]. It is convenient to call the arrangement of Fig. 15b a *gyro*-QUARL and to represent it by the symbol of Fig. 15d. For $\Delta = 0$, a gyro-QUARL reduces to a *unit gyroelement*, in which case the letter Δ may be omitted in the corresponding symbol.

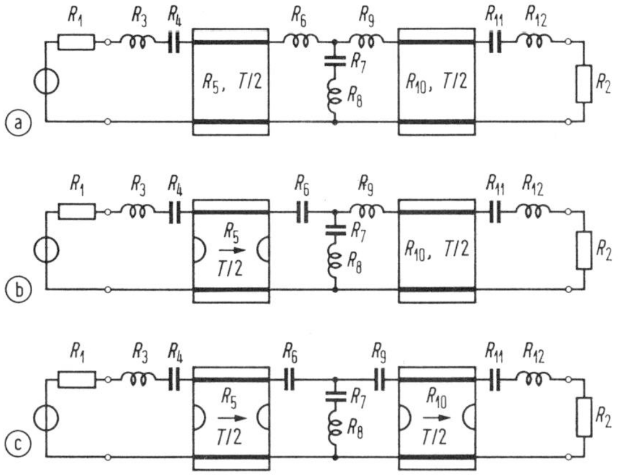

Fig. 17. Eigth order band-pass filters having
(a) a simple pole at zero and a triple pole at infinity,
(b) a double pole at zero and a double pole at infinity, and
(c) a triple pole at zero and a simple pole at infinity.

The circuit of Fig. 15a can again be generalized to any number of non-overlapping pulse currents. It can be shown that if, for a given k, the polarity of C is changed between each pulse i_k and the next following current pulse entering C, a gyrator has to be connected to port k of the equivalent circuit in the same way as to port 2 in Fig. 15b. It is easily checked, using simple equivalences, that this rule holds for all three cases discussed in relation with Fig. 15a. We can apply it, of course, also to the circuit of Fig. 13 or to any circulator arrangement with more than three ports. The various circuit elements thus obtained obviously increase the flexibility for designing circuits.

As an example of the applicability of these concepts, the three eighth order band-pass filters of Fig. 17 may be compared. The first of these has a simple pole at zero and a triple pole at infinity, the second one a double pole at zero as well as at infinity, and the third one a triple pole at zero and a simple pole at infinity. Nevertheless, all three require only two voltage inverter switches. For optimum design of structures such as those of Fig. 17 the interested reader is referred e.g. to [10].

Acknowledgements

The author is grateful to D. Herbst, B. Höfflinger, K.-A. Owenier, K. Schumacher, R. Schweer and H. Wupper for various discussions on the subject of this paper, and to W. Jobs and K.-A. Owenier for carrying out numerical computations.

(Received June 28th, 1978.)

References

[1] Fettweis, A., Basic principles of switched-capacitor filters using voltage inverter switches. AEÜ **33** [1979], 13–19.

[2] Fettweis, A., Theory of resonant-transfer circuits and their application to the realization of frequency-selective networks. Proc. URE Summer School on Circuit Theory, 68 pages, Prague, July 1968.

[3] Fettweis, A., Realization of general network functions using the resonant-transfer principle. Proc. Fourth Asilomar Conf. on Circuits and Systems, pp. 663–666, and 671, Pacific Grove, Cal., Nov. 1970; also AEÜ **25** [1971], 295–303.

[4] Fettweis, A., Some properties of pulse impedances and pulse impedance matrices. AEÜ **24** [1970], 506–512.

[5] Schlichte, M., Schaltungsanordnung zur impulsweisen Energieübertragung, insbesondere für Zeitmultiplex-Vermittlungssysteme. Deutsche Patentschrift (German Patent) 1 227 079, Dec. 20th, 1963.

[6] Poschenrieder, W., Frequenzfilterung durch Netzwerke mit periodisch gesteuerten Schaltern. Proc. NTG-Symp. "Analyse und Synthese von Netzwerken", Stuttgart 1966, pp. 220–237.

[7] Fettweis, A., Theory of resonant-transfer circuits. Lecture notes, NATO Advances Study Inst. on Network and Switching Theory, Sept. 1966; also in "Network and Switching Theory" (ed. by G. Biorci), pp. 382–446, Academic Press, New York 1968.

[8] Boite, R. and Thiran, J. V. P., Synthesis of filters with capacitances, switches, and regenerating devices. Transact. IEEE CT-**15** [1968], 447–454.

[9] Weinrichter, H., Kondensator-Kommutator-Netzwerke. AEÜ **26** [1972], 293–305.

[10] Owenier, K.-A., Optimierung von Filtern, insbesondere von Wellendigitalfiltern mit verringerter Zahl an Multiplizierern. Thesis, Ruhr-Universität Bochum, 1977.

Paper 6-5

A Family of Active Switched Capacitor Biquad Building Blocks

By P. E. FLEISCHER and K. R. LAKER

(Manuscript received May 10, 1979)

Two closely related, low-sensitivity, active switched capacitor filter topologies are presented. Each of these circuits comprises two operational amplifiers and at most nine capacitors. The topologies have been carefully constructed so that they are immune to the various parasitic capacitances normally present in switched capacitor networks. One filter topology is capable of realizing any stable biquadratic z-domain transfer function, while the second one is only slightly less than fully general. Most commonly used transfer functions can be realized with either topology and will require only seven capacitors. The choice between the two topologies will generally be made on the basis of total capacitance required, dynamic range behavior, and sensitivity. A complete set of synthesis equations is given for both circuits which cover both the general and all the important special cases of the biquadratic transfer function. Finally, several examples are given which illustrate the synthesis procedures and the versatility of the filter topologies.

I. INTRODUCTION

Active RC building blocks which realize a biquadratic transfer function[1,2] have played an increasingly important role in supplying filters for the Bell System. Thus, at the present time, STAR building blocks enjoy a very large volume of production and there is a constant effort directed toward the reduction of their costs.[3,4]

Perhaps the most promising recent development in filtering is the emergence of active switched capacitor filters.[5–8] These filters are fully MOS realizable and therefore enjoy the cost advantages of LSI integrated circuit realizations. Further, they can be expected to share in the future cost reductions due to VLSI.

Since biquadratic building blocks have played such a dominant role in the active RC field, they are also expected to be of great importance

Reprinted with permission from *Bell Syst. Tech. J.*, vol. 58, pp. 2235-2269, Dec. 1979.

for active switched capacitor (SC) filter realizations. Many of these basic biquadratic blocks can then be realized on a single chip to implement higher order filter functions.

The purpose of this paper is to describe a pair of similar active SC biquadratic filter topologies[9] capable of realizing any stable biquadratic z-domain transfer function. In arriving at these networks, we have been particularly careful to eliminate the effects of parasitic capacitances. Such capacitances arise mainly from two sources.

First is a sizable (approximately 10 percent) parasitic capacitance from the bottom plates of the capacitors to the epitaxial layer (ac ground). By using two operational amplifiers, it is possible to arrange matters so that the parasitic capacitors are connected either to a voltage source (operational amplifier output) or ground, thereby eliminating any effects due to them. Second are parasitic capacitances from the MOS switches to ground via the power supplies. By using only certain switching arrangements,[10] the parasitics again can be forced to appear at harmless locations.

After a description of the general biquad circuits, a full set of synthesis relations will be developed for them. The paper concludes with several examples that demonstrate the flexibility and general usefulness of these circuits.

II. GENERAL CIRCUIT TOPOLOGY

The transfer function of an active-RC biquad is biquadratic in the Laplace transform variable s:

$$\frac{V_{out}}{V_{in}} = \frac{cs^2 + es + d}{s^2 + as + b}. \tag{1}$$

In contrast, the active-SC biquad voltage transfer function is biquadratic in the z-transform[11] variable z, where $z = e^{sT_s}$ and T_s is the sampling interval, viz.,

$$\frac{V_{out}}{V_{in}}(z) = \frac{N(z)}{D(z)} = \frac{\gamma + \epsilon z^{-1} + \delta z^{-2}}{1 + \alpha z^{-1} + \beta z^{-2}}. \tag{2}$$

This is the general biquadratic transfer function we seek to realize. Of course, by suitable choice of the numerator coefficients, the different generic transfer functions, such as LP, BP, and HP, can be obtained.

All the filter topologies to be considered in this paper are special cases of the general active-SC biquad shown in Fig. 1a. This circuit bears a close resemblance to the three amplifier biquad;[2] however, because of the inversion[7] inherent in capacitor A, the inverter of the active-RC biquad is not needed. In effect, the circuit consists of two integrators, the first stage being inverting while the second stage is noninverting. Damping is provided by the capacitor E and the switched

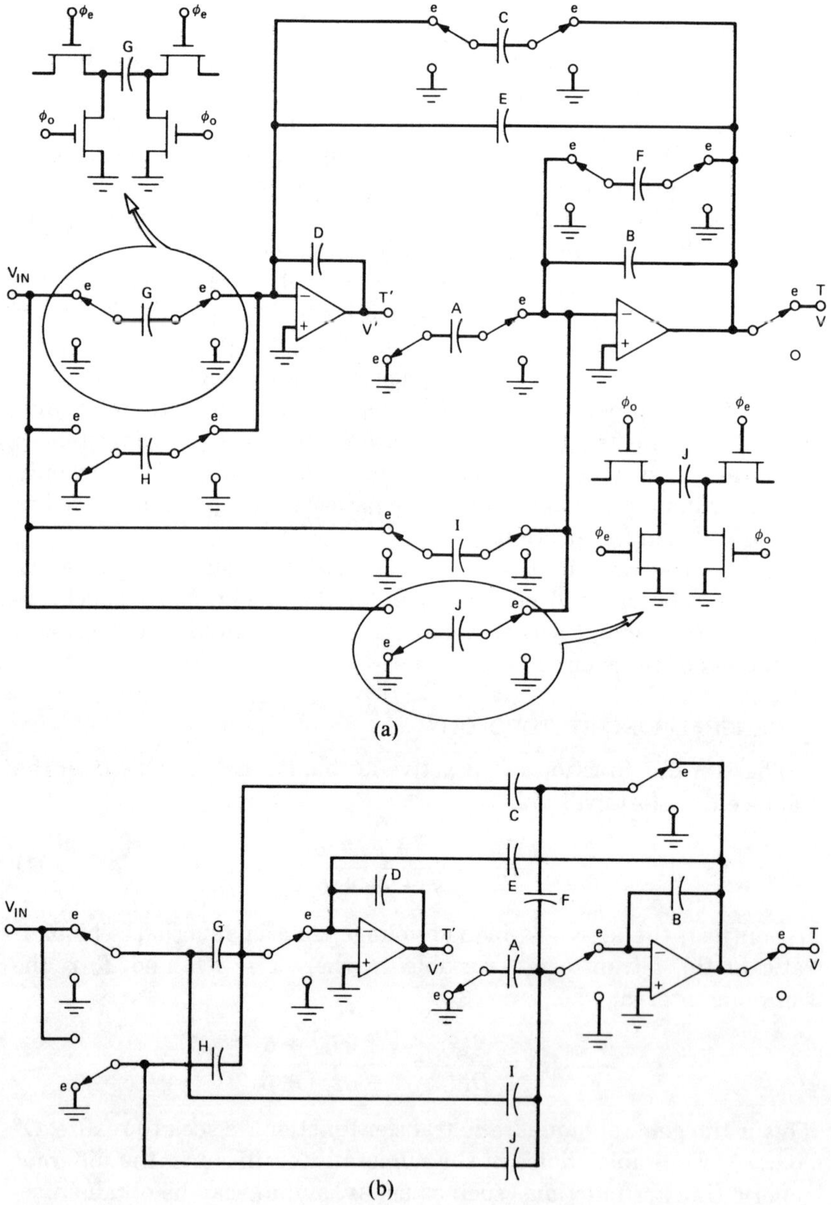

Fig. 1—(a) General active sc topology. (b) General active sc topology (minimum switch configuration).

capacitor *F*. In any particular application, only one of these will be present, leaving a total of nine capacitors, but for analysis purposes it is convenient to handle the two cases together.

The transmission zeros are realized via the multiple feed-forward paths consisting of switched capacitors G, H, I, and J. It is seen later that, typically, no more than two of these capacitors are needed to realize the useful biquadratic transfer functions. Thus, most often it will be found that only seven capacitors are needed. Although the circuit schematic shown in Fig. la facilitates understanding the circuit, a more efficient implementation can be obtained by allowing similarly switched capacitors to share a common switch. Rearranging the circuit schematic in this way results in the minimum switch configuration shown in Fig. 1b. One can readily verify that the electrical behaviors of these circuits are identical.

To minimize[8,10] the deleterious effects of switch parasitics, we have avoided the use of toggle-switched capacitors[7] in which one end of the capacitor is permanently connected to ground. Furthermore, it is noted that the SC elements shown in Figs. 2a and 2b are equivalent in function. However, by discharging the capacitor C through ground, as per Fig. 2b (and the insert in Fig. 1), the parasitic switch capacitances are also discharged. For ideal operational amplifiers, the parasitic switch capacitances have absolutely no effect on the operation of the biquad. It is expected that the effect will only be negligibly enhanced[12] by the nonideal character of practical operational amplifiers.

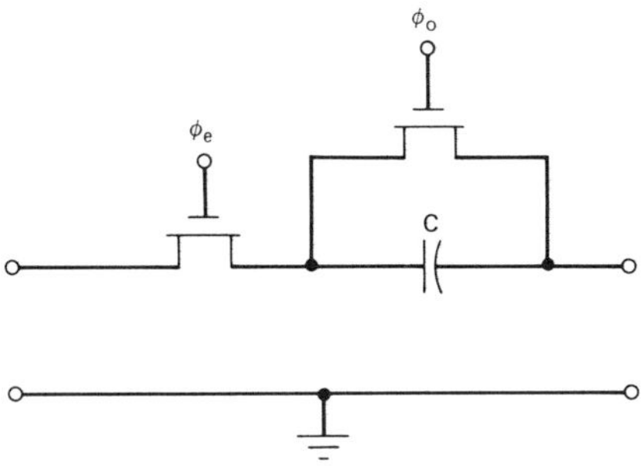

(a)

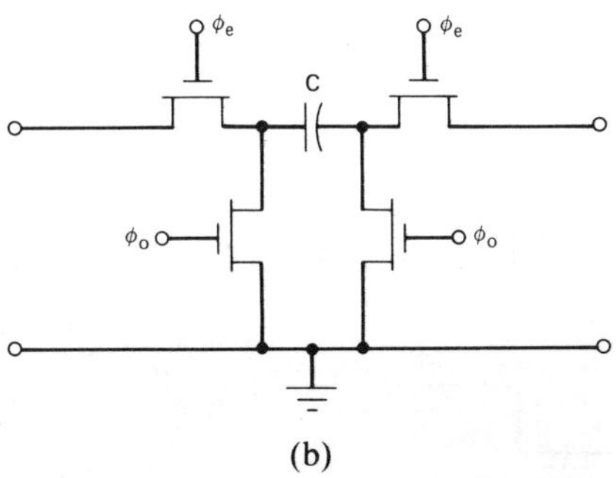

(b)

Fig. 2—Discharge type switched capacitors. (a) Typical realization. (b) Parasitic free realization.

The close analogy of SC filters to active-RC biquads has already been mentioned. In particular, note that the circuit with $E = 0$ is very similar in spirit to the three-operational amplifier multiple input biquad[13] except for the absence of the inverter, which is not needed for the SC filter. However, active-SC biquads offer even further versatility which to this point has not been exploited. Because of the inability to trim capacitors and the relative cost factors, practical active-RC biquads are constructed to be canonic in capacitors, namely, two. This constraint is unnecessarily placed on active-SC topologies when they are derived from an active-RC topology via a resistor-to-switched-capacitor replacement.[5-7,14] As we show, one can achieve interesting and beneficial results when this constraint is removed.

2.1 The transfer functions

Before deriving the transfer functions which characterize the general active-SC network of Fig. 1, it is necessary to consider the timing of the switches. Note that the schematics for some typical SC elements are shown as inserts in Fig. 1.

For simplicity in analysis, we assume the clocks ϕ_e and ϕ_o to be nonoverlapping with 50-percent duty cycle, as in Fig. 3. It is noted that switches clocked by ϕ_e close instantaneously at the $2k\tau$ (even) time instants and those clocked by ϕ_o close instantaneously at the $(2k + 1)\tau$ (odd) time instants. We also assume that the input signal is sampled and held over a full clock period, 2τ, as shown in Fig. 3. In fact, the switch phasings of the SC biquad have been chosen to operate with this kind of input. Under these conditions, we have

$$V_{\text{in}}^{o}(z) = z^{-1/2}V_{\text{in}}^{e}(z). \tag{3}$$

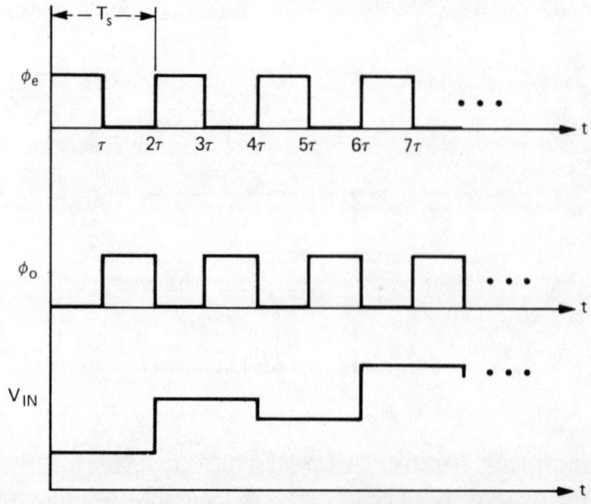

Fig. 3—Clock and input waveforms.

It follows then[16] that the output voltages appearing at the operational amplifier outputs are also held for the full clock period and change only at the even sampling instants:

$$V^{o}_{\text{out}}(z) = z^{-1/2} V^{e}_{\text{out}}(z). \tag{4}$$

This fact can also be surmised by inspection of Fig. 1a. Appendix A shows that the full-cycle S/H assumption can be relaxed. For the present, however, since both the input and the outputs are fully held, the even transfer functions provide all the information we need. For simplicity, therefore, the "even" superscript will be omitted from here on.

Let us now derive the voltage transfer functions T' and T for the networks of Fig. 1. As noted previously, these transfer functions are most conveniently expressed in the z-domain where $z = e^{sT_s}$ and $T_s = 2\tau$ represents a full clock period. Any SC network containing biphased switches can be conveniently transformed into a z-domain equivalent circuit.[15,16] With this equivalent circuit, one can then apply all the network tools available to continuous, linear, time-invariant networks. When the input signal is held over the full clock period, one can readily obtain[16] the equivalent circuit given in Fig. 4.

The desired transfer functions are then derived using straightforward nodal analysis:

$$T \triangleq \frac{V}{V_{\text{in}}}$$

$$= \frac{-Az^{-1}(G - Hz^{-1}) - D(1 - z^{-1})(I - Jz^{-1})}{Az^{-1}(C + E - Ez^{-1}) + D(1 - z^{-1})(F + B - Bz^{-1})} \tag{5a}$$

$$= -\frac{DI + (AG - DI - DJ)z^{-1} + (DJ - AH)z^{-2}}{D(F + B) + (AC + AE - DF - 2DB)z^{-1} + (DB - AE)z^{-2}}, \tag{5b}$$

and

$$T' \triangleq \frac{V'}{V_{\text{in}}}$$

$$= \frac{(I - Jz^{-1})(C + E - Ez^{-1}) - (G - Hz^{-1})(F + B - Bz^{-1})}{Az^{-1}(C + E - Ez^{-1}) + D(1 - z^{-1})(F + B - Bz^{-1})} \tag{6a}$$

$$= \frac{(IC + IE - GF - GB) + (FH + BH + BG - JC - JE - IE)z^{-1} + (EJ - BH)z^{-2}}{D(F + B) + (AC + AE - DF - 2DB)z^{-1} + (DB - AE)z^{-2}}. \tag{6b}$$

Before undertaking further analysis, some extraneous degrees of freedom will be eliminated. First, we arbitrarily set $A = B$. It may be shown that the net effect of this choice is to remove our ability to

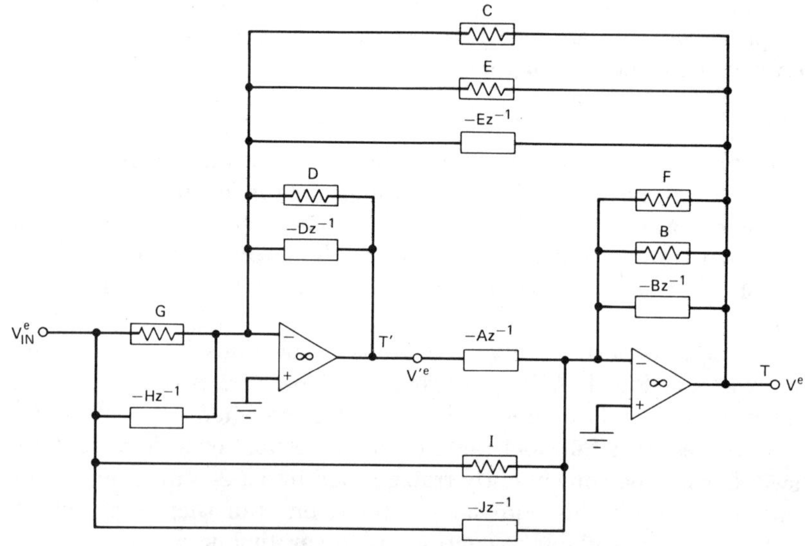

Fig. 4—z-domain equivalent circuit for the biquad in Fig. 1.

control the gain constants associated with T and T' simultaneously. Later we will show how, through scaling techniques, this degree of freedom can be restored to the circuit. Second, it is clear by examining the circuit of Fig. 1 that there is an arbitrary impedance scaling associated with each of the two stages. Thus, the two groups of capacitors (C, D, E, G, H) and (A, B, F, I, J) may each be arbitrarily and independently scaled without changing the transfer functions. Accordingly, we arbitrarily choose $B = 1$ and $D = 1$. This choice will also be ultimately overridden by minimum capacitor realization considerations.

In view of the above, we have

$$A = B = D = 1. \tag{7}$$

Substituting these into (5) and (6) yields the following simplified transfer functions:

$$T = -\frac{I + (G - I - J)z^{-1} + (J - H)z^{-2}}{(F + 1) + (C + E - F - 2)z^{-1} + (1 - E)z^{-2}} \tag{8}$$

and

$$T' = \frac{\begin{matrix}(IC + IE - FG - G) \\ + (FH + H + G - JC - JE - IE)z^{-1} + (EJ - H)z^{-2}\end{matrix}}{(F + 1) + (C + E - F - 2)z^{-1} + (1 - E)z^{-2}}. \tag{9}$$

Let us first examine the salient features of the transfer function T. Note that its poles are determined by C, E, and F, and its zeros by G,

H, I, and J. The fact that the poles and zeros are independently adjustable may prove to be useful in some filter or equalizer applications. It is also clear that the three numerator coefficients are independently adjustable thus permitting arbitrary zeros to be realized. The fact that any stable poles are also realizable will be demonstrated later.

Regarding the transfer function T', we first observe the obvious fact that its poles are identical to those of T. We note, however, that its zeros are formed in a more complicated fashion and they do not have the aforementioned independence property. Nevertheless, there are cases where T' provides a more economical realization of a given transfer function than T.

The zero-forming pairs (I, J) and (G, H) have some interesting alternate realizations. Using the techniques of Ref. 16, it may be shown that the pair I and J, for example, may be realized as shown in Fig. 5. Thus, one capacitor is eliminated not only when $I = 0$ or $J = 0$ but also when $I = J$. The equivalence of a pair of switched capacitors to one unswitched capacitor is quite fascinating in view of the switched-capacitor—resistance equivalence so commonly assumed in dealing with SC networks.

When $I = J$, not only is one capacitor eliminated, but sensitivity is usually improved since I and J now "track" perfectly. Even when $I \neq J$, the transformations may be useful to reduce sensitivity or to lower the total capacitance needed for a given case. For example, if $I = 13$ pf and $J = 12$ pf, transformation (a) can be applied, yielding new element values $I - J = 1$ pf and $J = 12$ pf. The transformation is obviously reversible; thus, the converse transformation can be applied to element pairs C,E and B,F. It should be noted that in arriving at these transformations we have assumed, as in Fig. 1, that terminals 1 and 2 are connected to a full clock period S/H voltage source and to a virtual ground, respectively. In Appendix A, we show that the full cycle S/H assumption can be conditionally relaxed.

Note that the SC elements in Fig. 5 all possess a z-domain admittance of the form $I - Jz^{-1}$. If the inverting switched capacitor J were replaced with a noninverting switched capacitor,[16] a z-domain admittance of the form $I + Jz^{-1}$ would be obtained. At times, it might be convenient from a synthesis point of view to have $I + Jz^{-1}$ admittances; however, as mentioned earlier, noninverting toggle-switched capacitors do introduce parasitics;[8,10] therefore, we avoid using this element. In spite of this omission, the circuit is capable of realizing all stable biquadratic transfer functions.

2.2 The E- and F-circuits

One final simplification we can make to the general biquad in Fig. 1 involves the elements E and F. As already mentioned, E and F are

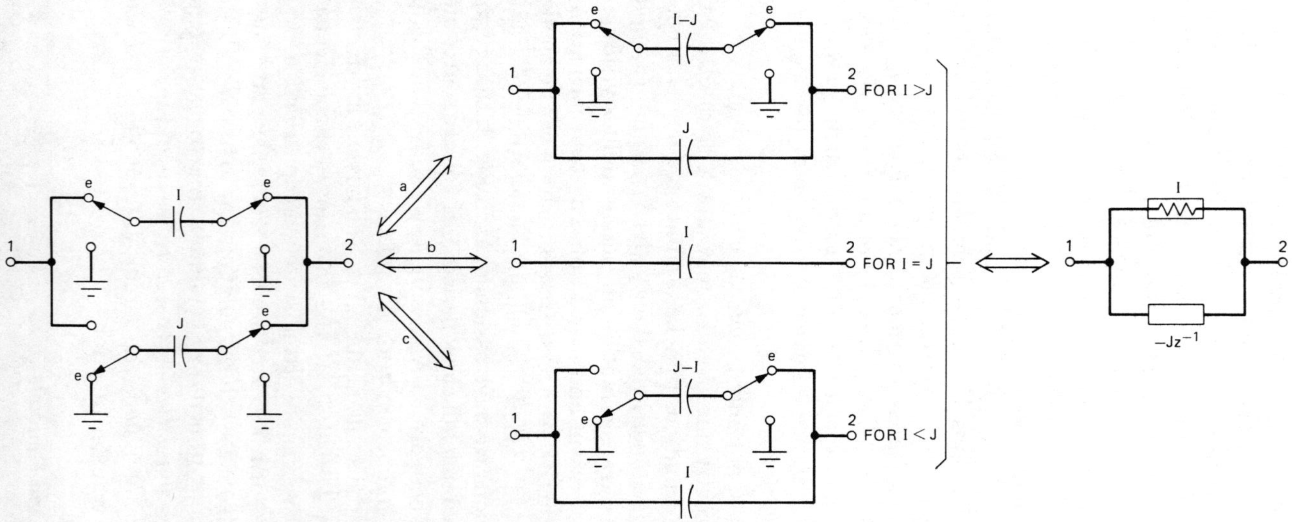

Fig. 5—SC element transformations (ports 1 and 2 are assumed to be buffered).

redundant elements in the sense that they both provide damping. Consequently, they need not both be present in the same circuit. It is, therefore, convenient to define an "E-circuit" in which $E \neq 0$ and $F = 0$ and an "F-circuit" in which $F \neq 0$ and $E = 0$.

The transfer functions for these two circuits are

$$T'_E = \frac{I + (G - I - J)z^{-1} + (J - H)z^{-2}}{1 + (C + E - 2)z^{-1} + (1 - E)z^{-2}} \tag{10a}$$

$$T'_E = \frac{(IC + IE - G) + (H + G - JC - JE - IE)z^{-1} + (EJ - H)z^{-2}}{1 + (C + E - 2)z^{-1} + (1 - E)z^{-2}} \tag{10b}$$

and

$$T_F = -\frac{\hat{I} + (\hat{G} - \hat{I} - \hat{J})z^{-1} + (\hat{J} - \hat{H})z^{-2}}{(\hat{F} + 1) + (\hat{C} - \hat{F} - 2)z^{-1} + z^{-2}} \tag{11a}$$

$$T'_F = -\frac{(\hat{G}\hat{F} + \hat{G} - \hat{I}\hat{C}) + (\hat{J}\hat{C} - \hat{F}\hat{H} - \hat{H} - \hat{G})z^{-1} + \hat{H}z^{-2}}{(\hat{F} + 1) + (\hat{C} - \hat{F} - 2)z^{-1} + z^{-2}}. \tag{11b}$$

The "hats" are placed on the F-circuit elements in order to distinguish them from the E-circuit elements.

Let us briefly examine these transfer functions. Note that the numerators of T_E and T_F are identical, while the numerators of T'_E and T'_F are quite different. Thus for a given design in which the desired output is V, the T' as well as the unscaled dynamic range of V' may be quite different for the two networks. The analogous situation is obtained if the desired output is V'. These differences will ultimately affect the final scaled capacitor values and the total capacitance required to realize the circuit. Significant sensitivity differences between the four possible realizations of a given transfer function may also exist. These points will be illustrated via examples in Section IV.

2.4 Sensitivities

The sensitivities for the E- and F-circuits are at least comparable to any active-RC biquad. One can arrive at this conclusion by examining the Q and ω_o relations and associated sensitivities for a moderate-to-high-Q resonant response.

For any pair of complex conjugate poles in the z-domain, one can write the denominator as:

$$D(z) = 1 + \alpha z^{-1} + \beta z^{-2} \tag{12a}$$

$$= (1 - re^{j\theta}z^{-1})(1 - re^{-j\theta}z^{-1})$$

$$= 1 - 2r\cos\theta z^{-1} + r^2 z^{-2}. \tag{12b}$$

By analogy to the continuous case, the following equations involving the resonant frequency ω_o and Q can be written:

$$\theta \approx 2\pi \frac{\omega_o}{\omega_s} = \omega_o T_s \tag{13}$$

and (14)

$$\frac{1}{2Q} \approx \frac{1-r}{\theta} = \frac{1-r}{\omega_o T_s},$$

which implies

$$r \approx 1 - \frac{\omega_o T_s}{2Q}. \tag{15}$$

Therefore, it follows from (12) through (15) that

$$\alpha \approx -2\left(1 - \frac{\omega_o T_s}{2Q}\right)\cos \omega_o T_s \tag{16a}$$

$$\beta \approx \left(1 - \frac{\omega_o T_s}{2Q}\right)^2. \tag{16b}$$

Whenever $\omega_o T_s \ll 1$, i.e., the sampling rate is high and $Q \gg 1$, the above expressions may be further approximated:

$$\alpha \approx -2\left(1 - \frac{\omega_o T_s}{2Q}\right)\left(1 - \frac{\omega_o^2 T_s^2}{2}\right)$$

$$\approx -2 + \frac{\omega_o T_s}{Q} + \omega_o^2 T_s^2 \tag{17a}$$

and

$$\beta \approx 1 - \frac{\omega_o T_s}{Q}. \tag{17b}$$

Consider first the E-circuit. After suitable manipulations, the denominator of (5) becomes (with $F = 0$)

$$D_E(z) = 1 + \left(-2 + \frac{AC}{DB} + \frac{AE}{DB}\right)z^{-1} + \left(1 - \frac{AE}{DB}\right)z^{-2}. \tag{18}$$

Comparing this with (12a) and (17) immediately yields

$$\frac{\omega_o T_s}{Q} \approx \frac{AE}{DB} \tag{19a}$$

and

$$\omega_o^2 T_s^2 \approx \frac{AC}{DB}. \tag{19b}$$

Therefore,

$$\omega_o T_s \approx \left(\frac{AC}{DB}\right)^{1/2} \tag{20a}$$

and

$$Q \approx \frac{1}{E}\left(\frac{DBC}{A}\right)^{1/2}. \tag{20b}$$

Similarly, it may be shown that, for the F-circuit,

$$\omega_o T_s \approx \left(\frac{\hat{A}\hat{C}}{\hat{D}\hat{B} + \hat{D}\hat{F}}\right)^{1/2} \tag{21a}$$

and

$$Q \approx \left[\frac{\hat{A}\hat{C}}{\hat{D}\hat{F}}\left(1 + \frac{\hat{B}}{\hat{F}}\right)\right]^{1/2}. \tag{21b}$$

From (20) and (21), it is seen that ω_o and Q are controlled by the ratios of four or five capacitors. Furthermore, it is clear that

$$|S_x^{\omega_o}| \leq \tfrac{1}{2} \quad \text{and} \quad |S_x^Q| \leq 1, \tag{22}$$

where x denotes any capacitor in the E- or F-circuits. This situation compares favorably to the active-RC case,[1,2,17] where a minimum of four passive elements, namely, two RC products, determines ω_o. Since, in practice, ratios of capacitors can be more tightly controlled than individual resistors and capacitors, the active-SC realization can be expected to be superior to the active-RC case with respect to initial (untuned) response as well as temperature and aging variations.*

Even though, in practice, the ω_o variation is usually the most significant contributor to the overall variation in the response,[17] we note that the Q sensitivity of the active-SC circuit is also low.

The overall circuit sensitivity is also affected, of course, by the contribution of the numerator. This aspect of the problem is not amenable to a general analysis and has to be handled on a case-by-case basis. It has been our experience that T designs provide lower sensitivity realizations than T' designs. It can be seen from (5) and (6) that the numerator of T is simpler than the numerator of T'. Although one might expect the pole-zero tracking afforded by the T' designs to yield lower sensitivities for some applications, we find that, because of the cancellations which occur in the coefficients, larger sensitivities are often incurred.

---

* Since T_s is normally derived from a very stable clock, it is assumed to be invariant.

III. SYNTHESIS

The synthesis of the biquad begins with the identification of the desired transfer function. This determination ultimately depends upon the frequency domain specifications which can then be transformed in some manner to a z-domain transfer function. The individual biquad transfer functions are then obtained by factoring this higher order z-domain transfer function.

Once the desired biquadratic transfer function is identified, the unscaled capacitor values are determined from (10) or (11) for the E- or F-circuit, respectively. Once this basic design is obtained, the final step consists of scaling the capacitors to adjust the dynamic range at the output of the other operational amplifier. It is then convenient to rescale the admittances in each of the two stages to obtain a minimum capacitance value of 1 unit in each stage. The actual minimum value of capacitance which can be realized depends on the technology, the desired precision of the transfer function, and the estimated effects of parasitics. A minimum capacitance of 1 pf will be used in this paper.

3.1 z-domain biquadratic transfer functions

Like digital filters, switched-capacitor filters are most conveniently synthesized from a z-domain transfer function. Several methods are available[11] for obtaining a z-domain transfer function from frequency domain specifications. Perhaps the most useful of these is based on the bilinear transformation[11] which can be shown to preserve "maximally flat" or "equal ripple" properties. This method starts with a suitably prewarped analog transfer function in the s-domain. This analog transfer function is then converted to a z-domain transfer function using the bilinear transformation,

$$s = \frac{2}{T_s} \frac{1 - z^{-1}}{1 + z^{-1}}, \tag{23}$$

where, it is recalled, T_s denotes the full clock period. The application of the bilinear transform to an analog biquadratic transfer function yields a z-domain transfer function which is also biquadratic. Of consequence to the ultimate form of low-pass and bandpass transfer functions is the fact that analog zeros at $s = \infty$ map into finite z-domain zeros at $z = -1$. Other transformations,[8] such as

$$s = \frac{1}{T_s} \frac{1 - z^{-1}}{z^{-1}}, \tag{24}$$

do not have this mapping property. Using (24), the mapping of the poles is also somewhat different than that obtained via the bilinear transformation.

3.2 Pole placement

At this point, it is appropriate to consider the stability and realizability of the proposed circuits. It is, of course, desirable to be able to realize all stable pole positions. Stability for a biquad can be conveniently expressed[18] in the α, β parameter space as the area within the shaded triangle shown in Fig. 6. The upper parabolic area of the triangle represents the α, β values for stable, complex poles. The remainder of the upper triangular area, where $\beta > 0$, corresponds to real pole pairs which lie pairwise to the left or right of $z = 0$, while the lower triangular area, where $\beta < 0$, corresponds to real poles which lie on alternate sides of $z = 0$. Clearly, the upper portion of the triangle, i.e., $\beta > 0$, represents most of the useful pole locations for frequency selective filters.

Consider first the E-circuit realizability properties. Comparing (2) and (10) yields

$$\alpha = E + C - 2 \tag{25a}$$

$$\beta = 1 - E \tag{25b}$$

and, therefore,

$$\alpha + \beta = C - 1. \tag{25c}$$

Since $E \geq 0$ and $C \geq 0$, we immediately have from (25a) and (25c):

$$\alpha \geq -2 \tag{26a}$$

$$\beta \leq 1 \tag{26b}$$

$$\alpha + \beta \geq -1. \tag{26c}$$

Thus, the α, β values realizable with the E-circuit are confined within the wedge-like area shown in Fig. 7a. This area includes all the stable

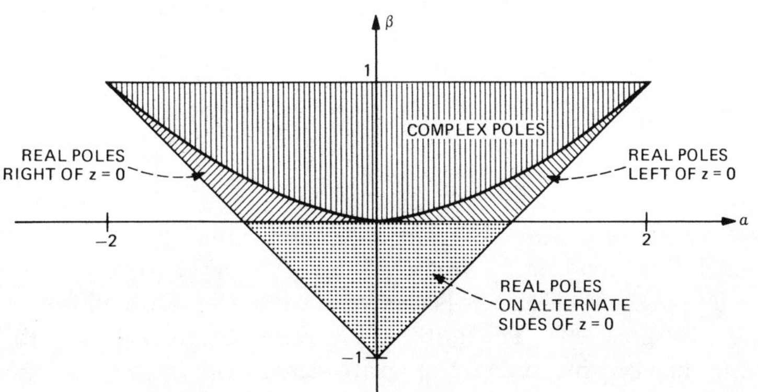

Fig. 6—Triangle of stable pole positions for $D(z) = 1 + \alpha z^{-1} + \beta z^{-2}$.

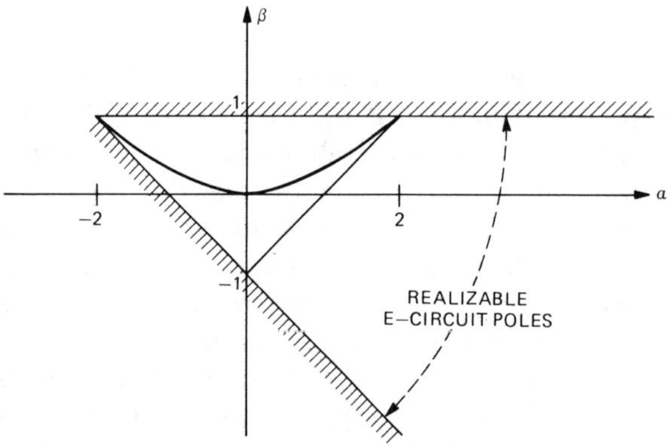

(a)

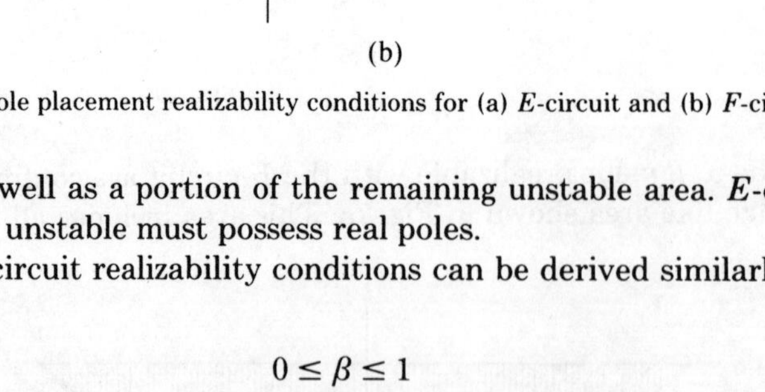

(b)

Fig. 7—Pole placement realizability conditions for (a) E-circuit and (b) F-circuit.

region as well as a portion of the remaining unstable area. E-circuits which are unstable must possess real poles.

The F-circuit realizability conditions can be derived similarly from (11):

$$0 \leq \beta \leq 1 \tag{27a}$$

and

$$\alpha + \beta \geq -1. \tag{27b}$$

These equations define the semiwedge-like area shown in Fig. 7b. Although this area is more restricted than that representing realizable E-circuit poles, the F-circuit is seen to be able to realize all the useful stable pole positions. The only stable case not realizable with the F-circuit has two real poles of opposite signs. Also, unstable F-circuits must have real poles of the same sign.

Now that we have established the realizability conditions for the E-

and F-circuits, let us state the pole placement synthesis equations in terms of the z-domain transfer function coefficients α and β. For the E-circuit, the synthesis equations can be stated as

$$E = 1 - \beta \tag{28a}$$

and

$$C = 1 + \beta + \alpha. \tag{28b}$$

Similarly, for the F-circuit we have

$$\hat{F} = \frac{1 - \beta}{\beta} \tag{29a}$$

and

$$\hat{C} = \frac{1 + \alpha + \beta}{\beta}. \tag{29b}$$

Equations (28) and (29) yield nonnegative values for E, C and $\hat{F}$, $\hat{C}$ within the realizability areas sketched in Figs. 7a and 7b, respectively. Also note that $\hat{F}$ and $\hat{C}$ for the F-circuit can be obtained from values calculated for E and C for the E-circuit using the simple relations

$$\hat{F} = \frac{E}{1 - E} \tag{30a}$$

and

$$\hat{C} = \frac{C}{1 - E}. \tag{30b}$$

As noted previously, it is often convenient to derive the z-domain transfer function from the prewarped analog requirements via the bilinear transformation given by (23). Once an appropriate prewarped s-domain transfer function has been determined, one can then derive simple synthesis relations for the pole-determining capacitors in terms of its coefficients. Let the denominator of the prewarped s-domain transfer function be

$$D(s) = s^2 + as + b. \tag{31}$$

Then, substituting for s the bilinear transformation of (23) and equating the coefficients of the resulting z-domain quadratic polynomial with those of the denominator of T_E (or T'_E) yields

$$E = \frac{aT_s}{1 + \dfrac{aT_s}{2} + b\dfrac{T_s^2}{4}} \tag{32a}$$

and

$$C = \frac{bT_s^2}{1 + \dfrac{aT_s}{2} + b\dfrac{T_s^2}{4}} . \tag{32b}$$

Similar expressions for the F-circuit may be written

$$\hat{F} = \frac{aT_s}{1 - \dfrac{aT_s}{2} + \dfrac{bT_s^2}{4}} \tag{33a}$$

and

$$\hat{C} = \frac{bT_s^2}{1 - \dfrac{aT_s}{2} + \dfrac{bT_s^2}{4}} . \tag{33b}$$

In summary, when the desired z-domain biquadratic transfer function is known, capacitor values E, C or $\hat{F}$, $\hat{C}$ are readily evaluated according to (28) or (29), respectively. Alternatively, when an appropriately prewarped transfer function is known, the capacitor values E, C or $\hat{F}$, $\hat{C}$ can be evaluated in terms of the coefficients of the prewarped analog function and the sampling period T_s according to (32) or (33), respectively.

3.3 Zero placement

Before deriving the synthesis relations for the zeros, it is instructive to list the z-domain transfer functions for the well-known generic forms; namely, low-pass (LP), high-pass (HP), bandpass (BP), low-pass notch (LPN), high-pass notch (HPN), and all-pass (AP).[19] The numerators, with reference to (2), for these generic forms are listed in Table I. The LP and BP functions are particularly interesting in that there are several different forms which can be used. These forms are referred to in Table I as LP_{ij} and BP_{ij}, where i denotes the number of $1 + z^{-1}$ factors and j the number of z^{-1} factors. As already noted, the zeros at $z = -1$ arise only when the bilinear transform is used. These transfer functions, of course, exhibit steeper cutoff in the vicinity of half the sampling rate, but they may not afford the most economical realization. As a rule of thumb, the additional cutoff will become less and less important as the sampling frequency increases with respect to the pole-zero locations, i.e., as $\omega_o T_s \rightarrow$ small.

The number of z^{-1}, i.e., delay, terms will usually be immaterial. In these cases, economy of realization or perhaps sensitivity considerations might indicate the best choice. Note, however, that if there is additional feedback around any biquad block, the delay term becomes critical.

Table I—Generic biquadratic transfer functions

Generic Form	Numerator $N(z)$
LP 20 (bilinear transform)	$K(1 + z^{-1})^2$
LP 11	$Kz^{-1}(1 + z^{-1})$
LP 10	$K(1 + z^{-1})$
LP 02	Kz^{-2}
LP 01	Kz^{-1}
LP 00	K
BP 10 (bilinear transform)	$K(1 - z^{-1})(1 + z^{-1})$
BP 01	$Kz^{-1}(1 - z^{-1})$
BP 00	$K(1 - z^{-1})$
HP	$K(1 - z^{-1})^2$
LPN	$K(1 + \epsilon z^{-1} + z^{-2}), \epsilon > \alpha/\sqrt{\beta}, \beta > 0$
HPN	$K(1 + \epsilon z^{-1} + z^{-2}), \epsilon < \alpha/\sqrt{\beta}, \beta > 0$
AP	$K(\beta + \alpha z^{-1} + z^{-2})$
General	$\gamma + \epsilon z^{-1} + \delta z^{-2}$

As already noted in connection with (10) and (11), T_E and T_F have identical numerators except for the $1/(1 + \hat{F})$ gain constant term. For convenience, the numerator of T_E is repeated below:

$$N(z) = -I + (G - I - J)z^{-1} + (J - H)z^{-2}. \tag{34}$$

It is evident that there are enough degrees of freedom here to choose the three coefficients independently and thus realize arbitrary zero locations; the fact that the leading coefficient is nonpositive is a trivial constraint. In Table II a complete set of design equations is given for the special generic transfer functions of Table I as well as for the general case. For each of the cases, a "simple" solution is also listed. These simple solutions, which are not unique, lead to fewer number of capacitors by setting as many of the capacitors G, H, I, and J as possible to zero, or by having $G = H$ or $I = J$ which, according to Fig. 5, also eliminates a capacitor as well as some switches.

It should be noted that the F-circuit capacitors $\hat{G}$, $\hat{H}$, $\hat{I}$, and $\hat{J}$ are related to G, H, I, and J by

$$\hat{x} = (1 + \hat{F})x \quad \text{where} \quad x = G, H, I, J. \tag{35}$$

For this reason a separate table need not be given for the synthesis of the zeros of T_F.

Similar zero-placement synthesis conditions are listed for transfer functions T'_E and T'_F in Tables III and IV, respectively. It is noted that these synthesis equations require prior knowledge of either E, C or $\hat{F}$, $\hat{C}$, in contrast to the T_E or T_F case.

This completes the material on zero-placement. It will be recognized that, for any given transfer function, four alternative realizations exist, i.e., T_E, T'_E, T_F, and T'_F. In the case of LP or BP designs, additional degrees of freedom exist, as there may be a choice of transfer functions (see Table I). At this point, we cannot state any general rule for

Table II—Zero placement formulas for T_E and T_F

Filter Type	Design Equations	Simple Solution
LP 20	$I = \|K\|$ $G - I - J = 2\|K\|$ $J - H = \|K\|$	$I = J = \|K\|$ $G = 4\|K\|, \quad H = 0$
LP 11	$I = 0$ $G - I - J = \pm\|K\|$ $J - H = \pm\|K\|$	$I = 0, \quad J = \|K\|$ $G = 2\|K\|, \quad H = 0$
LP 10	$I = \|K\|$ $G - I - J = \|K\|$ $J - H = 0$	$I = \|K\|, \quad J = 0$ $G = 2\|K\|, \quad H = 0$
LP 02	$I = 0$ $G - I - J = 0$ $J - H = \pm\|K\|$	$I = J = 0$ $G = 0, \quad H = \|K\|$
LP 01	$I = 0$ $G - I - J = \pm\|K\|$ $J - H = 0$	$I = J = 0$ $G = \|K\|, \quad H = 0$
LP 00	$I = \|K\|$ $G - I - J = 0$ $J - H = 0$	$I = \|K\|, \quad J = 0$ $G = \|K\|, \quad H = 0$
BP 10	$I = \|K\|$ $G - I - J = 0$ $J - H = -\|K\|$	$I = \|K\|, \quad J = 0$ $G = H = \|K\|$
BP 01	$I = 0$ $G - I - J = \pm\|K\|$ $J - H = \mp\|K\|$	$I = 0, \quad J = \|K\|$ $G = H = 0$
BP 00	$I = \|K\|$ $G - I - J = -\|K\|$ $J - H = 0$	$I = \|K\|, \quad J = 0$ $G = H = 0$
HP	$I = \|K\|$ $G - I - J = -2\|K\|$ $J - H = \|K\|$	$I = J = \|K\|$ $G = H = 0$
HPN and LPN	$I = \|K\|$ $G - I - J = \|K\|\epsilon$ $J - H = \|K\|$	$I = J = \|K\|$ $G = \|K\|\{2 + \epsilon\}, \quad H = 0$
AP ($\beta > 0$)	$I = \|K\|\beta$ $G - I - J = \|K\|\alpha$ $J - H = \|K\|$	$I = \|K\|\beta, \quad J = \|K\|$ $G = \|K\|(1 + \beta + \alpha) = \|K\|C$ $H = 0$
General ($\gamma > 0$)	$I = \gamma$ $G - I - J = \epsilon$ $J - H = \delta$	$I = \gamma$ $J = \delta + x$ $G = \gamma + \delta + \epsilon + x$ $H = x \geq 0$

Note: $\hat{G} = G(1 + \hat{F})$, $\hat{H} = H(1 + \hat{F})$, $\hat{I} = I(1 + \hat{F})$, and $\hat{J} = J(1 + \hat{F})$.

Table III—Zero placement formulas for T_E^L

Filter Type	Design Equations	Simple Solution
LP 20	$IC + IE - G = \pm\|K\|$ $H + G - JC - JE - IE = \pm 2\|K\|$ $EJ - H = \pm\|K\|$	$I = \frac{\|K\|(4E + C)}{EC}, \quad J = \frac{\|K\|}{E}$ $G = \frac{\|K\|(2E + C)^2}{EC}, \quad H = 0$
LP 11	$IC + IE - G = 0$ $H + G - JC - JE - IE = \pm\|K\|$ $EJ - H = \pm\|K\|$	$I = \frac{\|K\|(2E + C)}{EC}, \quad J = \frac{\|K\|}{E}$ $G = \frac{\|K\|(E + C)(2E + C)}{EC}, \quad H = 0$
LP 10	$IC + IE - G = \pm\|K\|$ $H + G - JC - JE - IE = \pm\|K\|$ $EJ - H = 0$	$I = \frac{2\|K\|}{C}, \quad J = 0$ $G = \frac{\|K\|(E + C)^2}{EC}, \quad H = 0$
LP 02	$IC + IE - G = 0$ $H + G - JC - JE - IE = 0$ $EJ - H = \pm\|K\|$	$I = \frac{\|K\|(E + C)}{EC}, \quad J = \frac{\|K\|}{E}$ $G = \frac{\|K\|(E + C)^2}{EC}, \quad H = 0$
LP 01	$IC + IE - G = 0$ $H + G - JC - JE - IE = \pm\|K\|$ $EJ - H = 0$	$I = \frac{\|K\|}{C}, \quad J = 0$ $G = \frac{\|K\|(E + C)}{C}, \quad H = 0$
LP 00	$IC + IE - G = \pm\|K\|$ $H + G - JC - JE - IE = 0$ $EJ - H = 0$	$I = \frac{\|K\|}{C}, \quad J = 0$ $G = \frac{\|K\|E}{C}, \quad H = 0$
BP 10	$IC + IE - G = \pm\|K\|$ $H + G - JC - JE - IE = 0$ $EJ - H = \mp\|K\|$	$I = J = \frac{\|K\|}{E}$ $G = \frac{\|K\|(2E + C)}{E}, \quad H = 0$
BP 01	$IC + IE - G = 0$ $H + G - JC - JE - IE = \pm\|K\|$ $EJ - H = \mp\|K\|$	$I = J = 0$ $G = 0, \quad H = \|K\|$
BP 00	$IC + IE - G = \pm\|K\|$ $H + G - JC - JE - IE = \mp\|K\|$ $EJ - H = 0$	$I = 0, \quad J = 0$ $G = \|K\|, \quad H = 0$
HP	$IC + IE - G = \pm\|K\|$ $H + G - JC - JE - IE = \mp 2\|K\|$ $EJ - H = \pm\|K\|$	$I = J = 0$ $G = H = \|K\|$
HPN and LPN	$IC + IE - G = \pm\|K\|$ $H + G - JC - JE - IE = \pm\|K\|\epsilon$ $EJ - H = \pm\|K\|$	See general solution below
AP	$IC + IE - G = \pm\|K\|\beta$ $H + G - JC - JE - IE = \pm\|K\|\alpha$ $EJ - H = \pm\|K\|$	See general solution below

Table III—*Continued*

Filter Type	Design Equations	Simple Solution
General	$IC + IE - G = \gamma$	$I = \frac{\gamma + \delta + \epsilon}{C} + \frac{\delta}{E}$, $J = \frac{\delta}{E}$
$\delta > 0$	$H + G - JC - JE - IE = \epsilon$	$G = I(C + E) - \gamma$ $H = 0$
	$EJ - H = \delta$	

selecting the optimum transfer function from these four functions. We can show for specific designs that considerable differences in total capacitance and sensitivity can be obtained for equivalent T_E, T_F, T'_E, and T'_F designs.

Hopefully, as we gain more experience in active-SC design, some insights will be acquired to shorten the design procedure. In the meanwhile, enough alternatives have to be tried until a satisfactory solution is obtained.

3.4 Capacitor value scaling

The synthesis equations given in the previous subsections result in unscaled capacitor values. To complete the synthesis, some scaling is required. The first order of business is to adjust the voltage level at the "secondary" output. If this voltage is too high, overloads will result, while if it is too low, unnecessary noise penalties may be taken.

Although the voltage levels may be obtained using analysis techniques,[16] the simplest procedure is to simulate the unscaled circuit[20] on a program such as CAPECOD.[21] This also serves as a confirmation of the correctness of the design.

To adjust the voltage level V', i.e., the flat gain of T', without affecting T, only the capacitors A and D need be scaled. More precisely, if it is desired to modify the gain constant associated with V' according to

$$T' \rightarrow \mu T', \tag{36}$$

then it is only necessary to scale A and D as

$$(A, D) \rightarrow \left(\frac{1}{\mu} A, \frac{1}{\mu} D\right). \tag{37}$$

The gain constant associated with T remains invariant under this scaling. The correctness of this procedure follows directly from simple signal flow graph concepts. Note that A and D are the only two capacitors connected to the operational amplifier output node.

In a similar fashion, it can be shown that, if the flat gain associated with V is to be modified, i.e.,

Table IV—Zero placement formulas for T'_F

Filter Type	Design Equations	Simple Solution
LP 20	$\hat{G}\hat{F} + \hat{G} - \hat{I}\hat{C} = \lvert K\rvert(1 + \hat{F})$ $\hat{J}\hat{C} - \hat{F}\hat{H} - \hat{H} - \hat{G} = 2\lvert K\rvert(1 + \hat{F})$ $\hat{H} = \lvert K\rvert(1 + \hat{F})$	$\hat{I} = 0, \quad \hat{J} = \dfrac{\lvert K\rvert(2 + \hat{F})^2}{\hat{C}}$ $\hat{G} = \lvert K\rvert, \quad \hat{H} = \lvert K\rvert(1 + \hat{F})$
LP 11	$\hat{G}\hat{F} + \hat{G} - \hat{I}\hat{C} = 0$ $\hat{J}\hat{C} - \hat{F}\hat{H} - \hat{H} - \hat{G} = \lvert K\rvert(1 + \hat{F})$ $\hat{H} = \lvert K\rvert(1 + \hat{F})$	$\hat{I} = 0, \quad \hat{J} = \dfrac{\lvert K\rvert(1 + \hat{F})(2 + \hat{F})}{\hat{C}}$ $\hat{G} = 0, \quad \hat{H} = \lvert K\rvert(1 + \hat{F})$
LP 10	$\hat{G}\hat{F} + \hat{G} - \hat{I}\hat{C} = \pm\lvert K\rvert(1 + \hat{F})$ $\hat{J}\hat{C} - \hat{F}\hat{H} - \hat{H} - \hat{G} = \pm\lvert K\rvert(1 + \hat{F})$ $\hat{H} = 0$	$\hat{I} = 0, \quad \hat{J} = \dfrac{\lvert K\rvert(2 + \hat{F})}{\hat{C}}$ $\hat{G} = \lvert K\rvert, \quad \hat{H} = 0$
LP 02	$\hat{G}\hat{F} + \hat{G} - \hat{I}\hat{C} = 0$ $\hat{J}\hat{C} - \hat{F}\hat{H} - \hat{H} - \hat{G} = 0$ $\hat{H} = \lvert K\rvert(1 + \hat{F})$	$\hat{I} = 0, \quad \hat{J} = \dfrac{\lvert K\rvert(1 + \hat{F})^2}{\hat{C}}$ $\hat{G} = 0, \quad \hat{H} = \lvert K\rvert(1 + \hat{F})$
LP 01	$\hat{G}\hat{F} + \hat{G} - \hat{I}\hat{C} = 0$ $\hat{J}\hat{C} - \hat{F}\hat{H} - \hat{H} - \hat{G} = \pm K(1 + \hat{F})$ $\hat{H} = 0$	$\hat{I} = 0, \quad \hat{J} = \dfrac{\lvert K\rvert(1 + \hat{F})}{\hat{C}}$ $\hat{G} = 0, \quad \hat{H} = 0$
LP 00	$\hat{G}\hat{F} + \hat{G} - \hat{I}\hat{C} = \pm\lvert K\rvert(1 + \hat{F})$ $\hat{J}\hat{C} - \hat{F}\hat{H} - \hat{H} - \hat{G} = 0$ $\hat{H} = 0$	$\hat{I} = \dfrac{\lvert K\rvert(1 + \hat{F})}{\hat{C}}, \quad \hat{J} = 0$ $\hat{G} = 0, \quad \hat{H} = 0$
BP 10	$\hat{G}\hat{F} + \hat{G} - \hat{I}\hat{C} = -\lvert K\rvert(1 + \hat{F})$ $\hat{J}\hat{C} - \hat{F}\hat{H} - \hat{H} - \hat{G} = 0$ $\hat{H} = \lvert K\rvert(1 + \hat{F})$	$\hat{I} = \dfrac{\lvert K\rvert(1 + \hat{F})}{\hat{C}}, \quad J = \dfrac{\lvert K\rvert(1 + \hat{F})^2}{\hat{C}}$ $\hat{G} = 0, \quad \hat{H} = \lvert K\rvert(1 + \hat{F})$
BP 01	$\hat{G}\hat{F} + \hat{G} - \hat{I}\hat{C} = 0$ $\hat{J}\hat{C} - \hat{F}\hat{H} - \hat{H} - \hat{G} = -\lvert K\rvert(1 + \hat{F})$ $\hat{H} = \lvert K\rvert(1 + \hat{F})$	$\hat{I} = 0, \quad \hat{J} = \dfrac{\lvert K\rvert\hat{F}(1 + \hat{F})}{\hat{C}}$ $\hat{G} = 0, \quad \hat{H} = \lvert K\rvert(1 + \hat{F})$
BP 00	$\hat{G}\hat{F} + \hat{G} - \hat{I}\hat{C} = \pm\lvert K\rvert(1 + \hat{F})$ $\hat{J}\hat{C} - \hat{F}\hat{H} - \hat{H} - \hat{G} = \mp\lvert K\rvert(1 + \hat{F})$ $\hat{H} = 0$	$\hat{I} = \hat{J} = \dfrac{\lvert K\rvert(1 + \hat{F})}{\hat{C}}$ $\hat{G} = \hat{H} = 0$
HP	$\hat{G}\hat{F} + \hat{G} - \hat{I}\hat{C} = \lvert K\rvert(1 + \hat{F})$ $\hat{J}\hat{C} - \hat{F}\hat{H} - \hat{H} - \hat{G} = -2\lvert K\rvert(1 + \hat{F})$ $\hat{H} = \lvert K\rvert(1 + \hat{F})$	$\hat{I} = 0, \quad \hat{J} = \dfrac{\lvert K\rvert\hat{F}^2}{\hat{C}}$ $\hat{G} = \lvert K\rvert, \quad \hat{H} = \lvert K\rvert(1 + \hat{F})$
HPN and LPN	$\hat{G}\hat{F} + \hat{G} - \hat{I}\hat{C} = \lvert K\rvert(1 + \hat{F})$ $\hat{J}\hat{C} - \hat{F}\hat{H} - \hat{H} - \hat{G} = -\lvert K\rvert\epsilon(1 + \hat{F})$ $\hat{H} = \lvert K\rvert(1 + \hat{F})$	See general solution below
AP	$\hat{G}\hat{F} + \hat{G} - \hat{I}\hat{C} = \lvert K\rvert\beta(1 + \hat{F})$ $\hat{J}\hat{C} - \hat{F}\hat{H} - \hat{H} - \hat{G} = \lvert K\rvert\alpha(1 + \hat{F})$ $\hat{H} = \lvert K\rvert(1 + \hat{F})$	See general solution below
General	$\hat{G}\hat{F} + \hat{G} - \hat{I}\hat{C} = \gamma(1 + \hat{F})$	$I = x \geq 0$

Table IV—*Continued*

Filter Type	Design Equations	Simple Solution
$\delta > 0$	$\hat{J}\hat{C} - \hat{F}\hat{H} - \hat{H} - \hat{G} = \epsilon(1 + \hat{F})$	$\hat{J} = \dfrac{\delta(1 + \hat{F})^2 + \epsilon(1 + \hat{F}) + \gamma}{\hat{C}} + \dfrac{x}{1 + \hat{F}}$
	$\hat{H} = \delta(1 + \hat{F})$	$\hat{G} = \gamma + \dfrac{\hat{C}}{1 + \hat{F}} x$
		$\hat{H} = \delta(1 + \hat{F})$

$$T \to \nu T, \tag{38}$$

the following capacitors must be scaled:

$$(B, C, E, F) \to \left(\frac{1}{\nu} B, \frac{1}{\nu} C, \frac{1}{\nu} E, \frac{1}{\nu} F\right) \tag{39}$$

Once satisfactory gain levels have been obtained at both outputs, it is convenient to scale the admittances associated with each stage so that the minimum capacitance value in the circuit becomes unity. This makes it easier to observe the maximum capacitance ratios required to realize a given circuit and also serves to "standardize" different designs so that the total capacitance required can be readily observed. The two groups of capacitors which are to be scaled together are listed below.

$$\text{Group 1: } (C, D, E, G, H).$$

$$\text{Group 2: } (A, B, F, I, J).$$

Note that capacitors in each group are distinguished by the fact that they are all incident on the same input node of one of the operational amplifiers.

This completes the design process for synthesizing practical SC-biquad networks. In the next section, a detailed example is given to demonstrate each step of the design.

IV. DESIGN EXAMPLES

In this section, some illustrative examples will be given. The first example is a low-pass notch network whose design is followed through, step by step, to illustrate the design procedure. The second example is a bandpass. For this case, eight different designs are displayed to demonstrate the versatility of the active-SC topologies and to provide some insight into the relative merits of different realizations.

4.1 Low-pass notch circuit

The transfer function to be realized will be based on the s-domain transfer function shown below:

$$T(s) = \frac{0.891975s^2 + (1.140926 \times 10^8)}{s^2 + 356.047s + (1.140926 \times 10^8)}. \tag{40}$$

This transfer function provides a notch frequency of f_z = 1800 Hz, a peak corresponding to a quality factor Q_p = 30 at f_p = 1700 Hz and 0 dB dc gain. The assigned sampling frequency is 128 kHz, i.e., T_s = 7.8125 μs.

The z-domain transfer function is conveniently obtained via the bilinear transformation shown in (23). Because the band-edge frequency of 1700 Hz is much less than the sampling rate, it is not necessary to prewarp the $T(s)$ given in (40). Applying the bilinear transformation to (40) yields, after some algebraic manipulations:

$$T(z) = 0.89093 \frac{1 - 1.99220z^{-1} + z^{-2}}{1 - 1.99029z^{-1} + 0.99723z^{-2}}. \tag{41}$$

Note that in obtaining this transfer function a high degree of numerical precision is required. However, this does not result in high sensitivities, since the capacitor ratios define only the departures from -2 and $+1$ in the above terms.

Only the T_E and T_F realizations of the above circuit will be given here, as these circuits are more economical in the number of capacitors required for realization. The synthesis itself is straightforward. The capacitors C, E or $\hat{C}$, $\hat{F}$ are determined from (28) or (29), respectively, while the capacitors G, H, I, J, or $\hat{G}$, $\hat{H}$, $\hat{I}$, $\hat{J}$ are obtained from the "simple solution" entry in Table II. Finally, of course, A, B, D or $\hat{A}$, $\hat{B}$, $\hat{D}$ are set equal to unity according to (7). The resulting unscaled capacitor values are given in the appropriate columns of Table V. Note that, in this table and all succeeding ones, the hats are omitted from the F-circuit capacitors for notational convenience. Also note that since $I = J$ these two switched capacitors are replaced by the unswitched capacitor K, ($K = I = J$), in accordance with Fig. 5.

At this point, the unscaled E- and F-circuits were simulated via CAPECOD.[20,21] These results confirmed that the T_E and T_F were both correct. In particular, the maximum gain in both these realizations was approximately 10.56 dB. However, the maximum gains for T'_E and T'_F were very low. It was decided to increase these gains also to a maximum of 10.56 dB. In this way, the first stage is no more susceptible to overloads than the second stage. Specifically, it was found that

$$T'_{E\,\mathrm{MAX}} \approx -11.05 \text{ dB}, \qquad T'_{F\,\mathrm{MAX}} \approx -10.96 \text{ dB}. \tag{42}$$

Therefore, in accordance with (36),

$$\mu = 12.0365, \qquad \hat{\mu} = 11.9124. \tag{43}$$

Using these factors to rescale A, D and $\hat{A}$, $\hat{D}$, respectively, as given in (37), yields the "dynamic range adjusted" capacitor values shown in Table V. Finally, the capacitors associated with each operational amplifier stage are separately rescaled so that the minimum capacitance value becomes 1 pF. These "final" values are also shown in Table V.

In comparing the "final" realizations, we note that the F-circuit requires roughly 12 times the total capacitance of the E-circuit, in spite of the fact that the initial values were almost identical. Thus, alternative designs must be carried to completion before they can be meaningfully compared. It should be noted that other practical examples exist where the F-circuit designs are dramatically more efficient than the corresponding E-circuit designs.

As a final step, Monte Carlo simulations on the two circuits were carried out, assuming each capacitor to have a flat, independent ±1 percent tolerance. It should be pointed out that independent 1-percent capacitor tolerances represent a pessimistic estimate in view of today's technology. Since capacitor deviations, whether they are due to manufacturing tolerances, temperature variations, or aging, are highly correlated, the capacitor ratios are recognized to be achievable[6,7] with considerably better precision. These results, given in Table V, show both circuits to be excellent, with the E-circuit being slightly superior. Note that σ_1 is the absolute standard deviation at 1 Hz, while σ_{1700} is the standard deviation of the relative gain at 1700 Hz with respect to 1 Hz.

Table V—Low-pass notch realization

Capacitor (pf)	*E*-Circuit Initial	*E*-Circuit Dynamic Range Adjusted	*E*-Circuit Final	*F*-Circuit Initial	*F*-Circuit Dynamic Range Adjusted	*F*-Circuit Final
A	1.0000	0.08308	1.0000	1.0000	0.08395	30.1895
B	1.0000	1.0000	12.0365	1.0000	1.0000	359.629
C	0.00694	0.00694	2.5035	0.00696	0.00696	1.0000
D	1.0000	0.08308	29.9613	1.0000	0.08395	12.0591
E	0.00277	0.00277	1.0000	—	—	—
F	—	—	—	0.00278	0.00278	1.0000
G	0.00694	0.00694	2.5035	0.00696	0.00696	1.0000
H	—	—	—	—	—	—
I	—	—	—	—	—	—
J	—	—	—	—	—	—
$K(I = J)$	0.89093	0.89093	10.7238	0.89340	0.89340	321.293
ΣC (pF)	—	—	59.7	—	—	726.1
σ_1 (dB)	—	—	0.068	—	—	0.068
σ_{1700} (dB)	—	—	1.233	—	—	1.271

4.2 High Q bandpass circuits

This example demonstrates the versatility of the topology and examines the various tradeoffs this versatility provides. As noted in the previous sections and highlighted in Tables I through IV, we have the following degrees of freedom in realizing a bandpass active-SC biquad.

(*i*) The transfer function; namely, BP 00, BP 01, or BP 10, as shown in Table I.

(*ii*) The circuit realization; namely, the E- or F-circuits.

(*iii*) The output port; namely, T_E or T'_E for the E-circuit and T_F or T'_F for the F-circuit.

Using the "simple" solutions given in Tables II through IV, these freedoms yield 12 different design possibilities for a bandpass biquad realization. In selecting a design from these 12 possibilities, we adopt the following criterion: The circuit must meet all frequency domain requirements within acceptable tolerances. The circuit that satisfies this criterion, while requiring an estimated minimum silicon area for its realization, is in our view the best design. The primary factor that determines the required silicon area is total capacitance. Of secondary importance are the number of capacitors and the number of switches, i.e., an unswitched capacitor consumes less area than a switched capacitor of the same capacitance value. Let us now consider the various design possibilities in view of these considerations.

The transfer function to be realized will be based on the following s-domain transfer function:

$$T(s) = \frac{2027.9s}{s^2 + 641.28s + (1.0528 \times 10^8)}, \tag{44}$$

which possesses a center frequency $f_o = 1633$ Hz, a quality factor $Q = 16$, and a peak gain of 10 dB at f_o.

For the active-SC design, the assumed sampling frequency is 8 kHz, i.e., $T_s = 125\ \mu$s. One method for obtaining the z-domain transfer function is the application of the bilinear transform given by (23) to the prewarped s-domain transfer function. The prewarped s-domain function is obtained by adjusting the desired 3-dB frequencies f_l and f_h according to the relation[11]

$$\tilde{f}_{l,h} = \frac{1}{\pi T_s} \tan\,(\pi f_{l,h} T_s), \tag{45}$$

where $\tilde{f}_{l,h}$ denotes the prewarped 3-dB frequencies. Upon calculating $\tilde{f}_l$ and $\tilde{f}_h$, the following prewarped transfer function is determined:

$$\tilde{T}(\tilde{s}) = \frac{3159.2\tilde{s}}{\tilde{s}^2 + 999.0289\tilde{s} + (1.4285 \times 10^8)}. \tag{46}$$

Substituting for the prewarped complex frequency variable $\bar{s}$ in (46), the bilinear transform (23) yields the following BP 10 z-domain transfer function:

$$T(z) = \frac{0.1219(1 - z^{-1})(1 + z^{-1})}{1 - 0.5455z^{-1} + 0.9229z^{-2}}. \tag{47}$$

In accordance with the synthesis procedures given in Section III and in Tables II through IV, the four possible "simple" BP 10 designs were evaluated. The total capacitance required for each of these realizations is listed in Table VI. The capacitance values were scaled in the same manner as described in the previous example. As noted in Section II, switched capacitors, G and H, when equal, can be replaced by a single unswitched capacitor. All four designs were simulated on CAPECOD[20,21] to verify the response and to determine their statistical behavior. The Monte Carlo simulations were carried out at the 1633-Hz peak frequency, assuming each capacitor to have a flat, independent ±1 percent tolerance. Note that σ_{1633} is the standard deviation of the absolute gain at 1633 Hz. The results of these simulations are listed in Table VI.

Comparing the four designs, it is apparent that the F-circuit, using either T_F or T'_F, requires less capacitance and is less sensitive than either of the E-circuit designs. The 0.25-dB standard deviation for the F designs indicates the good stability of these circuits. Although the T'_F design consumes slightly less capacitance (~3 pF) than the T_F design, it requires one more capacitor and four additional switches. Depending on the layout, the additional capacitor, switches, and connections can more than offset the total capacitance advantage. With this reasoning, we are inclined to recommend the T_F design.

As noted earlier, one of our degrees of freedom is the choice of the BP transfer function. Exercising this freedom, as we shall soon demonstrate, can significantly impact the character of the designed circuit. Alternative realizations can be obtained by altering the z-domain transfer function in (47) to achieve a BP 00 function. An appropriate BP 00 function is obtained from (47) by removing the zero at one-half the sampling rate and adjusting gain K from 0.1219 to 0.1953 to preserve the desired 10-dB peak gain. The desired BP 00 transfer function is then

Table VI—Comparison of BP 10 and BP 00 biquad realizations

Case		E-Circuit (T_E)	F-Circuit (T_F)	E-Circuit (T'_E)	F-Circuit T'_F)
(1) BP 10, 10-dB	ΣC (pF)	55.0	51.2	75.1	48.3
gain at 1633 Hz	σ_{1633} (dB)	0.2738	0.2524	0.9932	0.2569
(2) BP 00, 10-dB	ΣC (pF)	46.7	32.7	39.6	32.8
gain at 1633 Hz	σ_{1633} (dB)	0.2852	0.2554	0.2833	0.2570

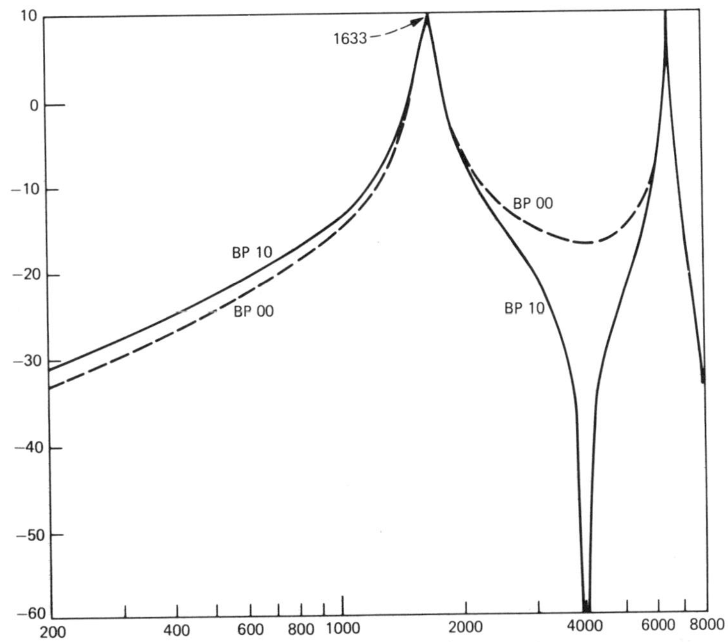

Fig. 8.—Frequency responses for the BP 10 and BP 00 designs with $Q = 16$ and $f_0 = 1633$ Hz.

$$T(z) = \frac{0.1953(1 - z^{-1})}{1 - 0.5455z^{-1} + 0.9229z^{-2}}. \tag{48}$$

Before carrying out the circuit designs, let us examine the frequency domain behavior of (48). The frequency responses for both the BP 10 and BP 00 transfer functions are plotted in Fig. 8. The BP 00 response is seen to be equivalent to the BP 10 response except for frequencies near 4000 Hz, where (47) possesses a zero. In any event, the BP 00 response was considered adequate. The appropriately scaled E- and F-circuit designs are listed in Table VI. In contrast to the first example, the F-circuit designs are seen to require less total capacitance and to be less sensitive than the E-circuit designs. More important, however, is the comparison between the BP 10 and the BP 00 designs. The BP 00 F-circuit designs are seen to require about 20 pF less total capacitance while sacrificing nothing. As far as the F-circuit BP 00 designs are concerned, our inclination is to recommend the T'_F design which requires four fewer switches than the T_F design.

V. CONCLUSIONS

Two closely related two-amplifier, active-SC filter topologies have been presented. These circuits have been constructed so that they are immune to parasitic capacitances normally present in SC networks.

The first topology, the E-circuit, has been shown to permit the realization of arbitrary stable z-domain biquadratic transfer functions at either of its two outputs. The second topology, the F-circuit, has been shown to be only slightly less general in that only certain unimportant pole pairs (real poles of opposite signs) are not realizable. A complete set of synthesis equations is presented for both of these circuits. Since every desired biquadratic transfer function has at least four alternative realizations, the designer can choose among these to best satisfy his economic and sensitivity requirements. If the "simple solutions" given in the tables do not satisfy his requirements, many other possible realizations also exist, especially if an LP or BP is being designed.

Several examples have been given to demonstrate the design process and to highlight the many degrees of freedom these biquad topologies provide.

APPENDIX

In the text, we have assumed that the input signal is sampled and held for the full clock period. While this assumption simplifies the analysis, it is by no means necessary. Thus, consider the more general case where the clock period is still T_s but the desired input signal is sampled and held only for the interval τ_e, ($\tau_e < T_s$). The subscript "e" here is meant to imply the even phase of the clock period. The odd phase of the clock period is referred to as τ_o ($\tau_o = T_s - \tau_e$). The input during this phase is assumed to be "undesirable." These concepts are also shown in Fig. 9.

In certain special cases, the circuits of Fig. 1 will continue to perform correctly even with this less restricted class of inputs. This happens whenever $H = 0$ and $J = 0$. This is readily confirmed by observing that the input voltage during τ_o is coupled into the circuit only via the two capacitors H and J. When these are both absent, the input during the odd clock phase is simply not "seen" by the circuit.

In general, however, the circuits of Fig. 1 must be modified by reversing the switch phasings of the switched capacitors A, H, and I. The resulting active-SC circuits are shown in Fig. 10. Note that the topology is so arranged that only the input during the even phase is coupled into the circuit. Thus, the input during the odd phase is again immaterial. One slight constraint on the operation of this circuit is that now the "correct" output is also only obtained during the even phase. Thus, if a fully held output signal is desired, the circuits of Fig. 10 will have to be followed by a suitable sample-and-hold circuit.

The proof of the above statement is most conveniently obtained by using the equivalent circuit techniques given in Ref. 16. For convenience in analysis, the duty cycle is assumed to be 50 percent, i.e., $\tau_e =$

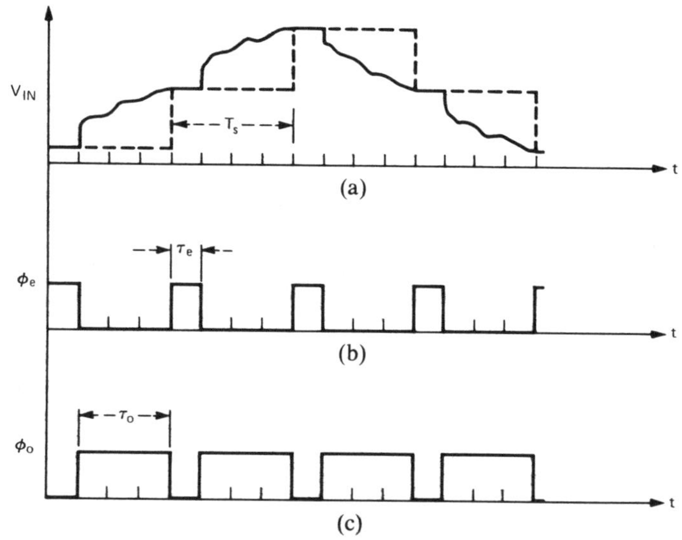

Fig. 9—Waveforms when input is not fully held.

$\tau_o = \frac{1}{2}T_s$; however, this does not detract from the generality of the results.

The equivalent circuit given in Fig. 11 can be readily constructed by substituting the equivalent circuits[16] for each SC element and operational amplifier in the circuit schematic given in Fig. 10. Due to the new switch phasings, V_{in}^{o} does not enter the filter. Writing nodal charge equations at the four virtual ground nodes of this circuit yields the following system of equations:

$$GV_{in}^{e} + DV'^{e} - Dz^{-1/2}V'^{o} + (C + E)V^{e} - Ez^{-1/2}V^{o} = 0, \quad (49)$$

$$-Hz^{-1/2}V_{in}^{e} + DV'^{o} - Dz^{-1/2}V'^{e} + EV^{o} - Ez^{-1/2}V^{e} = 0, \quad (50)$$

$$IV_{in}^{e} + (F + B)V^{e} - Bz^{-1/2}V^{o} = 0, \quad (51)$$

and

$$Jz^{-1/2}V_{in}^{e} - Az^{-1/2}V'^{e} + BV^{o} - Bz^{-1/2}V^{e} = 0. \quad (52)$$

Algebraically eliminating V^{o} and V'^{o} from these equations results in the following pair of equations:

$$(I - Jz^{-1})V_{in}^{e} + (F + B - Bz^{-1})V^{e} - Az^{-1}V'^{e} = 0 \quad (53)$$

and

$$(G - Hz^{-1})V_{in}^{e} + D(1 - z^{-1})V'^{e} + (C + E - Ez^{-1})V^{e} = 0. \quad (54)$$

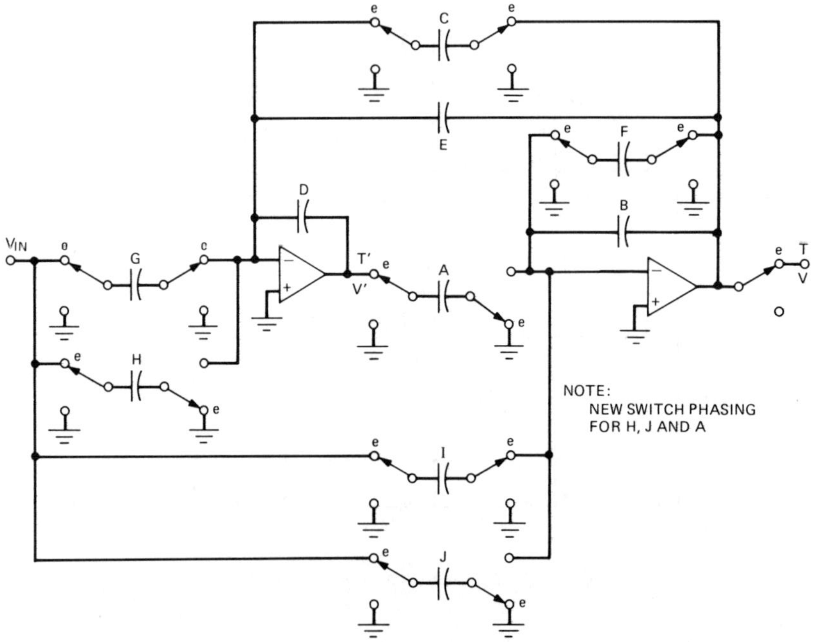

(a)

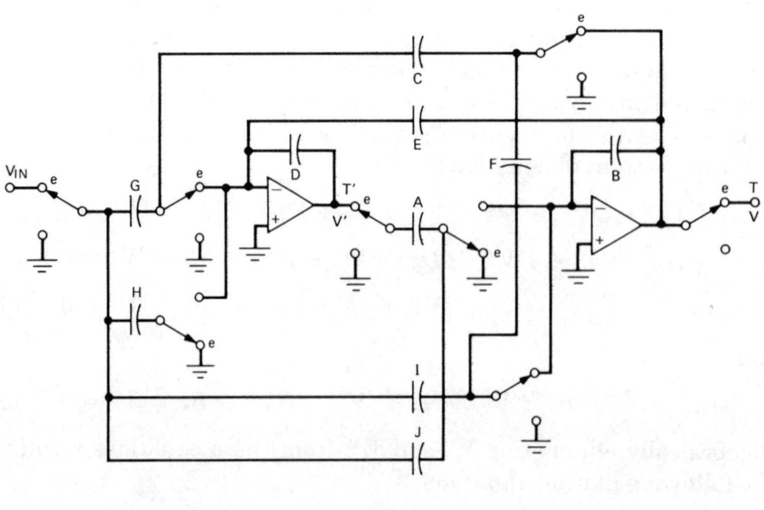

(b)

Fig. 10—(a) General biquad topology for input signals which are not held constant over the full clock period. (b) General biquad topology for input signals which are not held constant over the full clock period (minimum switch configuration).

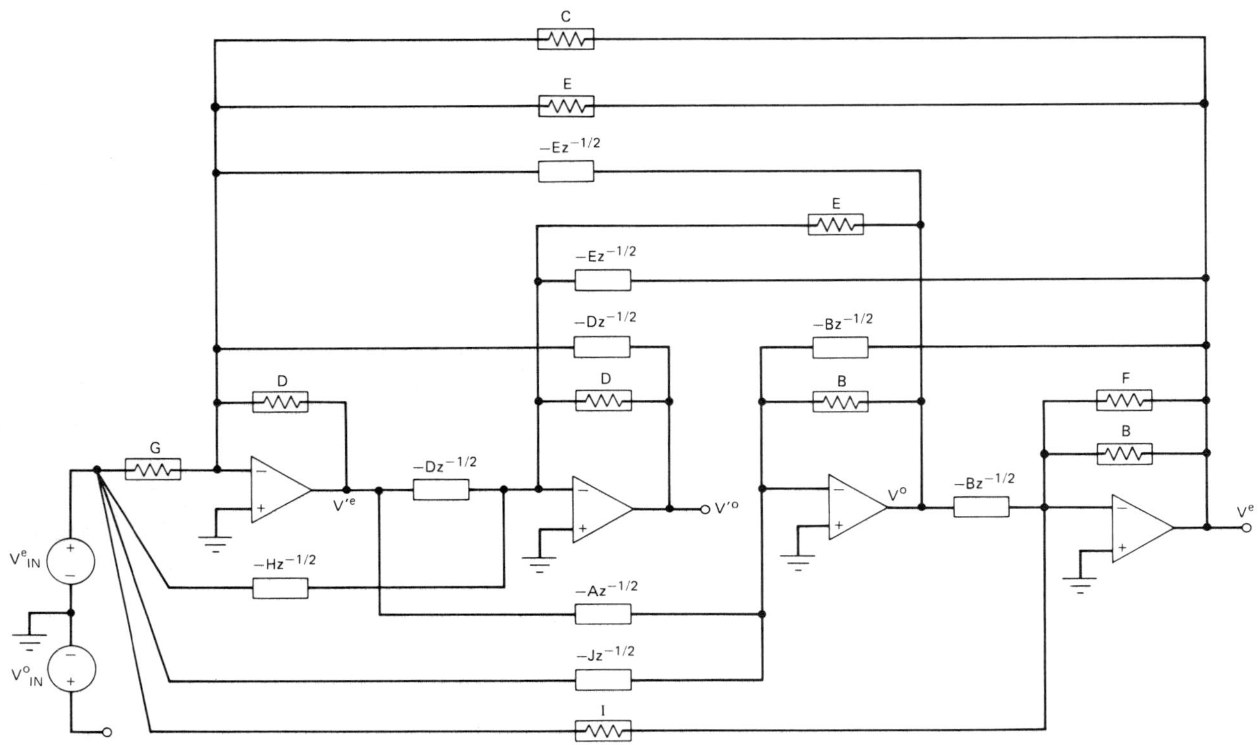

Fig. 11—z-domain equivalent circuit for the active SC biquad in Fig. 10.

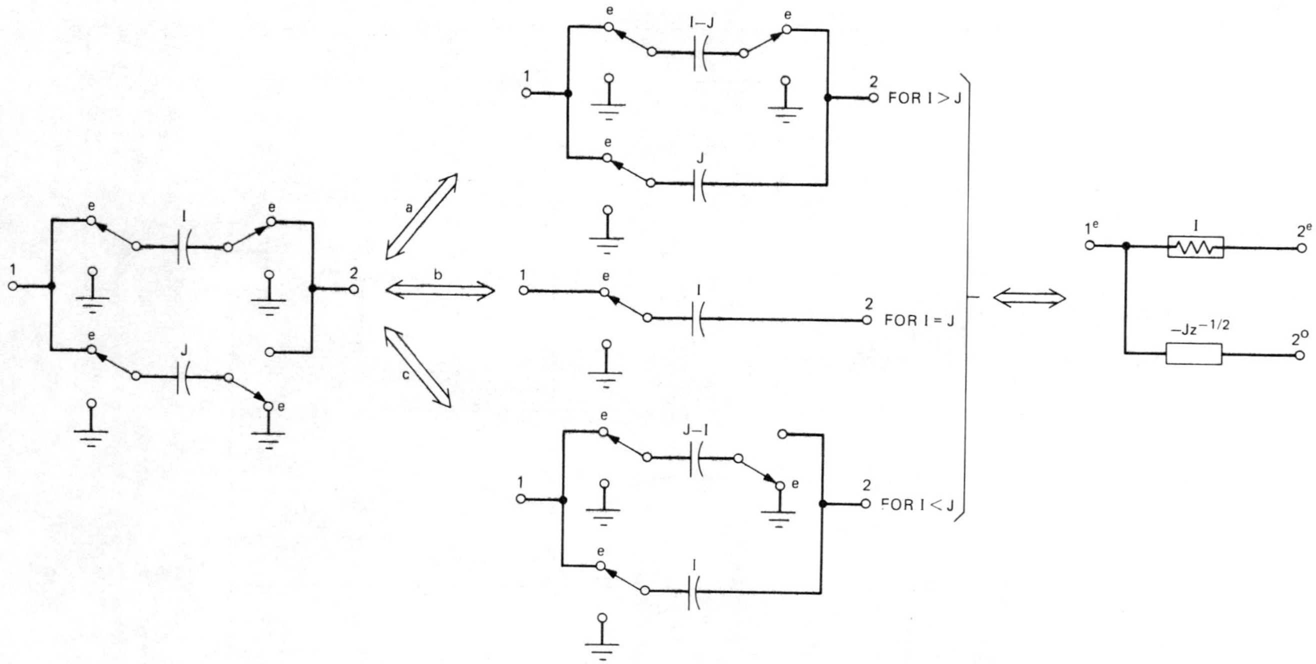

Fig. 12—sc element transformations for the biquad in Fig. 10 (ports 1 and 2 are assumed to be buffered).

Equations (53) and (54) can be readily verified to be the nodal equations which characterize the equivalent circuit given in Fig. 4. Thus, during the even phase the transfer functions will again be those given in (5) and (6), thus proving our contention. Note carefully, however, that, during the odd phase, the transfer functions which relate V_{in}^{e} to V^{o} and V'^{o} are quite different from those which characterize the even phase operation. In fact, we can express the two odd outputs (V^{o} and V'^{o}) as functions of the even outputs (V^{e} and V'^{e}) and the even input (V_{in}^{e}); i.e.,

$$V^{o} = z^{1/2}\left[\left(1 + \frac{F}{B}\right)V^{e} + \frac{I}{B}V_{\text{in}}^{e}\right] \tag{55}$$

and

$$V'^{o} = z^{1/2}\left[z^{-1}V'^{e} - \frac{E}{D}\left(1 + \frac{F}{B} - z^{-1}\right)V^{e} + \left(\frac{EI}{DB} - \frac{H}{D}z^{-1}\right)V_{\text{in}}^{e}\right]. \tag{56}$$

It is noted that, for $F = I = 0$, $V^{e} = z^{-1/2}V^{o}$; thus, V is held for the full clock period. On the other hand, when $E = H = 0$, V' is held for the full clock period. Thus, in some special cases, at least, a fully held output can be obtained. Finally, SC-element equivalences similar to those given in Fig. 5 can be used to reduce sensitivity and/or total capacitance. These element equivalences and their common z-domain equivalent circuit are given in Fig. 12. As in Fig. 5, these equivalences are based on the assumption that port 1 is voltage-driven and port 2 is connected to an operational-amplifier virtual ground.

REFERENCES

1. J. J. Friend, C. A. Harris, and D. Hilberman, "STAR: An Active Biquadratic Filter Section," IEEE Trans. Circuits and Systems, *CAS-22* (February 1975), pp. 115–121.
2. L. C. Thomas, "The Biquad: Part I—Some Practical Considerations," and "The Biquad: Part II—A Multi-Purpose Active Filtering System," IEEE Trans. Circuit Theory, *CT-18* (May 1971), pp. 350–357 and pp. 358–361.
3. W. Worobey and J. Rutkiewicz, "Tantalum Thin-Film RC Circuit Technology for a Universal Active Filter," IEEE Trans. Parts, Hybrids, and Packaging, *PHP-12* (December 1976), pp. 276–282.
4. P. E. Fleischer and J. J. Friend, "Active Filters at Bell Laboratories," presented at the 1977 IEEE International Symposium on Circuits and Systems, April 1977.
5. J. T. Caves et al., "Sampled Analog Filtering Using Switched Capacitors as Resistor Equivalents," IEEE J. Solid State Circuits, *SC-12*, No. 6 (December 1977), pp. 592–599.
6. B. J. Hosticka, R. W. Brodersen, and P. R. Gray, "MOS Sampled Data Recursive Filters Using Switched Capacitor Integrators," IEEE J. Solid State Circuits, *SC-12*, No. 6 (December 1977), pp. 600–608.
7. G. M. Jacobs, "Practical Design Considerations for MOS Switched Capacitor Ladder Filters," unpublished work.

8. B. J. White, G. M. Jacobs, and G. F. Landsburg, "A Monolithic Dual-Tone Multi-frequency Receiver," Digest of the 1979 ISSCC (February 1979), pp. 36–37.
9. P. E. Fleischer and K. R. Laker, internal memorandum, November 8, 1978.
10. P. E. Fleischer, Laboratory Notebook, May 26, 1978.
11. L. R. Rabiner and B. Gold, *Theory and Application of Digital Signal Processing*, Englewood Cliffs, N.J.: Prentice-Hall, 1975, Ch. 4.
12. A. J. Vera, unpublished work.
13. P. E. Fleischer and J. Tow, "Design Formulas for Biquad Active Filters Using Three Operational Amplifiers," Proc. IEEE, *61,* No. 5 (May 1973), pp. 662–663.
14. G. C. Temes, "The Derivation of Switched Capacitor Filters from Active-RC Prototypes," Electron. Lett., *14,* No. 12 (June 8, 1978), pp. 361–362.
15. C. F. Kurth and G. S. Moschytz, "Nodal Analysis of Switched Capacitor Networks," IEEE Trans. Circuits and Systems, *CAS-26* (February 1979), pp. 93–104, and "Two-Port Analysis of Switched Capacitor Networks Using Four-Port Equivalent Circuits," IEEE Trans. Circuits and Systems, *CAS-26* (March 1979).
16. K. R. Laker, "Equivalent Circuits for the Analysis and Synthesis of Switched Capacitor Networks," B.S.T.J., *58* (March 1979), pp. 727–767.
17. P. E. Fleischer, "Sensitivity Minimization in a Single Amplifier Biquad Circuit," IEEE Trans. Circuits and Systems, *CAS-23,* No. 1 (January 1976), pp. 45–55.
18. H. W. Schüssler, *Digitale Systeme Zur Signalverarbeitung*, Berlin: Springer-Verlag, 1973, p. 38.
19. S. A. Tretter, *Introduction to Discrete-Time Signal Processing*, New York: John Wiley, 1976, pp. 219–222.
20. P. E. Fleischer, "Computer Analysis of Switched Capacitor Networks in the Frequency Domain," unpublished work.
21. R. M. M. Chen et al., "L5 System: Role of Computing and Precision Measurements," B.S.T.J., *53,* No. 10 (December 1974), pp. 2249–2267.

Exact Analysis of Switched Capacitor Circuits with Arbitrary Inputs

MING L. LIOU, FELLOW, IEEE, AND YEN-LONG KUO, MEMBER, IEEE

***Abstract*—In this paper a general formulation procedure and an exact analysis in the frequency domain for switched capacitor circuits with arbitrary inputs including cisoidal, sample-and-hold, and noise, are presented. No topological and duty cycle constraints have been imposed on the circuits. Transfer functions are derived explicitly in terms of circuit matrices. Various special cases of practical interest are discussed in detail. Due to the simple form of the solutions much insight has been gained in understanding the working of switched capacitor circuits. The results are numerically attractive and can be readily programmed for computer-aided design of large size circuits. A simple two-capacitor circuit has been used throughout the text for illustration under various conditions.**

I. Introduction

Due to recent advances in MOS technology, switched capacitor circuits have received a great deal of attention from circuit designers [1]–[4]. Many analysis techniques have been proposed. They all considered the case when the input is sampled, except Tsividis' paper [5] where he obtained the result for a simple two-capacitor circuit with a cisoidal input. Various topological and input constraints have also been imposed upon circuits they have considered. On the other hand, general analysis techniques on *RLC* linear circuits containing periodically operated switches have been described in the literature [6], [7]. Unfortunately those results cannot be directly applied to a switched capacitor circuit in the sense that the normal form of state equation, as described in [6] and [7], does not exist for switched capacitor circuits.

In this paper, a new and general formulation procedure for switched capacitor circuits will be described. An exact and explicit frequency domain solution will be subsequently derived for an arbitrary input. Various special input cases such as cisoidal, sample-and-hold, and noise will be discussed in detail. Some implications due to topological constraints and some design insights will also be discussed when appropriate. Last, but not the least important, considerations for program implementation will be highlighted.

II. Equation Formulation of Switched Capacitor Circuits

In this section, the equation formulation of switched capacitor circuits will be discussed. Two equivalent formulations, which differ only in the output equations, are developed via the following theorem.

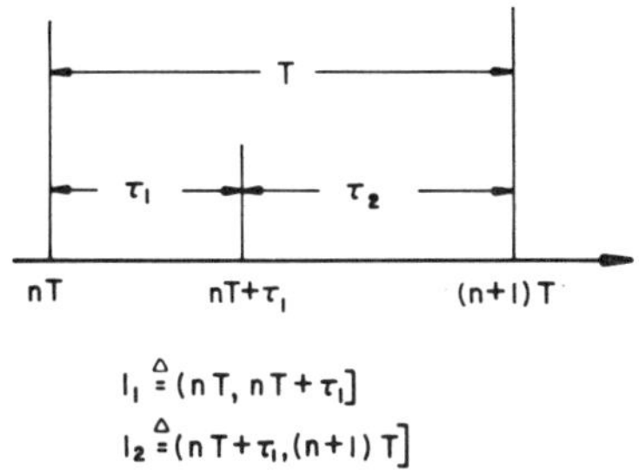

Fig. 1. Definition of switching subintervals.

Theorem

Consider a linear circuit containing capacitors, ideal switches, independent voltage and current sources, voltage-controlled voltage sources (VCVS), and current-controlled current sources (CCCS). Let T be the switching period and consider the time interval $nT<t\leqslant(n+1)T$. For simplicity, assume that each period is divided into two subintervals I_1 and I_2 corresponding to two switching positions as shown in Fig. 1.[1] Then the circuit equations in terms of capacitor voltage vectors, $\boldsymbol{x}_1(t)$ and $\boldsymbol{x}_2(t)$, can be described by

$$\boldsymbol{x}_1(t)=F_1\boldsymbol{x}_2(nT^-)+G_1\boldsymbol{u}(t), \qquad t\in I_1 \tag{1a}$$

$$\boldsymbol{x}_2(t)=F_2\boldsymbol{x}_1(nT+\tau_1^-)+G_2\boldsymbol{u}(t), \qquad t\in I_2 \tag{1b}$$

where the dimension of $\boldsymbol{x}_1$ and $\boldsymbol{x}_2$ is equal to the number of capacitors, $\boldsymbol{u}(t)$ consists of independent voltage sources and the integrals of independent current sources, F_1, F_2, G_1, and G_2 are real constant matrices of appropriate dimensions, and the superscript "$-$" indicates the time just before switching instants. Denoting $\boldsymbol{y}_1(t)$ and $\boldsymbol{y}_2(t)$ as the output voltage vectors in I_1 and I_2, respectively, the output equations can be described by

$$\boldsymbol{y}_k(t)=C_k\boldsymbol{x}_k(t)+D_k\boldsymbol{u}(t), \qquad t\in I_k, \quad k=1,2 \tag{2}$$

or equivalently by

$$\boldsymbol{y}_1(t)=\hat{C}_1\boldsymbol{x}_2(nT^-)+\hat{D}_1\boldsymbol{u}(t), \qquad t\in I_1 \tag{3a}$$

$$\boldsymbol{y}_2(t)=\hat{C}_2\boldsymbol{x}_1(nT+\tau_1^-)+\hat{D}_2\boldsymbol{u}(t), \qquad t\in I_2 \tag{3b}$$

where the C_k's, D_k's, $\hat{C}_k$'s, and $\hat{D}_k$'s are real constant matrices of appropriate dimensions.

Remarks: Before we give the proof of the above theorem, some useful observations about (1)–(3) are noted.

Manuscript received June 12, 1978; revised August 28, 1978.
The authors are with Bell Telephone Laboratories, North Andover, MA 01845.

[1] The results of this paper can be easily extended to cases when there are more than two subintervals in each switching period.

Reprinted from *IEEE Trans. Circuits Syst.*, vol. CAS-26, pp. 213-223, Apr. 1979.

1) Equation (1) is the "state" equation of the circuit, which does not involve the time derivatives of the state vectors. The state vector in a subinterval depends linearly on the input source vector in the same subinterval and on the final value of the state vector in the preceeding interval. Therefore, (1) also includes the switching transformations at $t=nT^+$ and $t=nT+\tau_1^+$, respectively, (the superscript "+" indicates the time immediately after switching instants).

2) The output formulation of (2) is in the standard form for the general switched networks [6]. The C_k's and D_k's for simple switched capacitor circuits can usually be determined by inspection. However, they are generally not unique. On the other hand, the output formulation of (3) can be shown to be unique if the circuit is completely solvable. As a matter of fact, as we shall see in Section VIII, it is much easier to compute $\hat{C}_k$ and $\hat{D}_k$ matrices when implemented on computers. Of course, if both (1) and (2) are known, (3) follows immediately by direct substitution with

$$\hat{C}_k = C_k F_k \qquad \hat{D}_k = C_k G_k + D_k, \qquad k=1,2. \tag{4}$$

3) As can be easily seen in the proof of the theorem, the equation forms of (1)–(3) will not be altered if both "charge"-controlled voltage sources and voltage-controlled "charge" sources are allowed.

Proof: Consider each branch "k" of the circuit as being characterized by a voltage-"charge" relation $[v_k(t), \hat{I}_k(t)]$, where

$$\hat{I}_k(t) \triangleq \int_{\tau^-}^{t} i_k(t')\,dt'$$

is the time integral of the branch current i_k from an initial time immediately before the last switching instant τ^-. With this characterization, the KVL and KCL (now it should be properly termed as Kirchhoff's charge law) still hold, i.e.,

$$\sum_{k \in \text{loop}} v_k(t) = 0 \quad \text{and} \quad \sum_{k \in \text{cut set}} \hat{I}_k(t) = 0.$$

For a capacitor, its branch characteristic is now given by

$$\hat{I}_c(t) \triangleq \int_{\tau^-}^{t} i_c(t')\,dt' = \int_{\tau^-}^{t} c\frac{dv_c}{dt'}\,dt'$$
$$= c\left[v_c(t) - v_c(\tau^-)\right] \tag{5}$$

which is represented by an equivalent composite "resistive"[2] element with "conductance" equal to c as shown in Fig. 2. For an ideal switch, when the switch is thrown from closed to open, the time integral of the switch current is

$$\hat{I}_{sw}(t) = \int_{\tau^-}^{t} i_{sw}(t')\,dt' = 0, \qquad \text{for } t \in \text{"open" subinterval.}$$

When the switch is thrown from open to closed, $\hat{I}_{sw}(t)$ is

[2]Here, "resistive" is used to mean that the "charge" $\hat{I}_c(t)$ is related to the voltage $[v_c(t) - v_c(\tau^-)]$ by a constant multiplier.

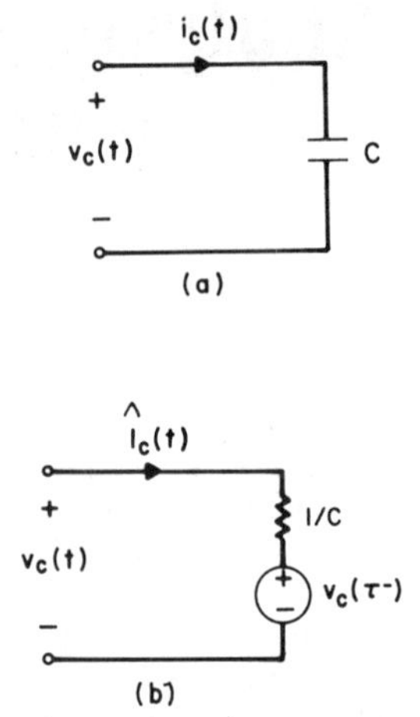

Fig. 2. (a) A capacitor. (b) "Resistive" equivalent.

finite and the voltage across the switch is

$$v_{sw}(t) = 0, \qquad \text{for } t \in \text{"closed" subinterval.}$$

For both the independent and controlled sources (only VCVS's and CCCS's are allowed), the new branch characteristics and the controlling relations remain the same with each branch current replaced by its corresponding time integral. The independent current source $J_s(t)$ is now represented by a "charge" source

$$\hat{Q}_s(t) \triangleq \int_{\tau^-}^{t} J_s(t')\,dt'$$

whereas the independent voltage source $E_s(t)$ is the same as before.

From the above, it is seen that the original switched capacitor circuit is transformed into a "resistive" circuit in the voltage- "charge" domain with the same circuit topology. The independent sources now consist of $E_s(t)$'s, $\hat{Q}_s(t)$'s, and the constant initial capacitor voltages $v_c(\tau^-)$'s. If the linear circuit is completely solvable, the capacitor voltages $v_c(t)$'s for t in each subinterval I_k, $k=1$, 2, are linear functions of these sources. Specifically, let $\boldsymbol{x}(t)$ be the column vector of all M capacitor voltages, i.e.,

$$\boldsymbol{x}(t) = \text{col}\left[v_{c_1}(t), v_{c_2}(t), \cdots, v_{c_M}(t)\right]$$

and denote $\boldsymbol{x}_k(t)$ as $\boldsymbol{x}(t)$ in I_k, $k=1$, 2, and let the input vector $\boldsymbol{u}(t)$ consist of all independent voltage sources $E_s(t)$'s and charge sources $\hat{Q}_s(t)$'s, then we have

$$\boldsymbol{x}_1(t) = F_1\boldsymbol{x}_2(nT^-) + G_1\boldsymbol{u}(t), \qquad t \in I_1$$
$$\boldsymbol{x}_2(t) = F_2\boldsymbol{x}_1(nT+\tau_1^-) + G_2\boldsymbol{u}(t), \qquad t \in I_2.$$

This proves (1).

Similarly, the output voltage vector $\boldsymbol{y}(t)$ is also a linear function of the initial capacitor voltage vector and the input vector $\boldsymbol{u}(t)$, i.e.,

$$\boldsymbol{y}_1(t) = \hat{C}_1\boldsymbol{x}_2(nT^-) + \hat{D}_1\boldsymbol{u}(t), \qquad t \in I_1$$
$$\boldsymbol{y}_2(t) = \hat{C}_2\boldsymbol{x}_1(nT+\tau_1^-) + \hat{D}_2\boldsymbol{u}(t), \qquad t \in I_2$$

which proves (3). Since the circuit is "resistive" and $\boldsymbol{x}(t)$ and $\boldsymbol{u}(t)$ completely determine any voltage in the circuit, (2) then results. Q.E.D.

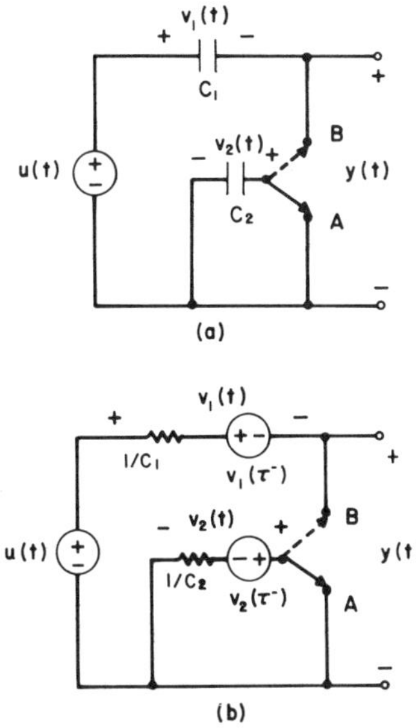

Fig. 3. A two-capacitor high pass circuit. Switch at position A for $t\in I_1 \triangleq (nT, nT+\tau]$. Switch at position B for $t\in I_2 \triangleq (nT+\tau_1, (n+1)T]$.

Example 1: As an example, consider the switched capacitor circuit of Fig. 3(a) and its "resistive" equivalent Fig. 3(b). From Fig. 3(b), for $t\in I_1$ (switch at position A), we readily have

$$\begin{bmatrix} v_1(t) \\ v_2(t) \end{bmatrix} = \begin{bmatrix} 1 & 0 \\ 0 & 0 \end{bmatrix} \begin{bmatrix} v_1(nT^-) \\ v_2(nT^-) \end{bmatrix} + \begin{bmatrix} 0 \\ 0 \end{bmatrix} u(t)$$

and

$$y(t) = [-1 \quad 0] \begin{bmatrix} v_1(t) \\ v_2(t) \end{bmatrix} + u(t)$$

$$\text{or} \quad = [-1 \quad -1] \begin{bmatrix} v_1(t) \\ v_2(t) \end{bmatrix} + u(t).$$

Similarly for $t\in I_2$ (switch at position B), by solving the corresponding resistive network of Fig. 3(b) for $v_1(t)$ and $v_2(t)$ we have

$$\begin{bmatrix} v_1(t) \\ v_2(t) \end{bmatrix} = \begin{bmatrix} 1-\beta & -\beta \\ \beta-1 & \beta \end{bmatrix} \begin{bmatrix} v_1(nT+\tau_1^-) \\ v_2(nT+\tau_1^-) \end{bmatrix} + \begin{bmatrix} \beta \\ 1-\beta \end{bmatrix} u(t)$$

and

$$y(t) = [-1 \quad 0] \begin{bmatrix} v_1(t) \\ v_2(t) \end{bmatrix} + u(t)$$

$$\text{or} \quad = [0 \quad 1] \begin{bmatrix} v_1(t) \\ v_2(t) \end{bmatrix}$$

where

$$\beta = \frac{c_2}{c_1+c_2}.$$

From the above simple example we see that the "state" (capacitor voltage) equations may be dependent to each other (i.e., F_1 or F_2 is singular) and the C_k and D_k matrices may not be unique, even though the solution is unique. However, $\hat{C}_1=[-1 \quad 0]$, $\hat{D}_1=1$, $\hat{C}_2=[\beta-1 \quad \beta]$, and $\hat{D}_2=1-\beta$ are unique.

III. Exact Solution in the Frequency Domain

In this section the exact solution in the frequency domain of a switched capacitor circuit due to an arbitrary input will be derived. Cases for various specific inputs will be discussed subsequently.

We first define the following window functions:

$$w_\tau(t-t_i) = \begin{cases} 1, & t_i < t \leq t_i+\tau \\ 0, & \text{elsewhere} \end{cases}$$

$$w_1(t) = \sum_{n=-\infty}^{\infty} w_{\tau_1}(t-nT)$$

$$w_2(t) = \sum_{n=-\infty}^{\infty} w_{\tau_2}(t-nT-\tau_1)$$

then from (1a) and (1b) we have

$$\boldsymbol{x}_1(t) = F_1 \sum_{n=-\infty}^{\infty} \boldsymbol{x}_2(nT^-) w_{\tau_1}(t-nT) + G_1 \boldsymbol{u}(t) w_1(t) \tag{6a}$$

$$\boldsymbol{x}_2(t) = F_2 \sum_{n=-\infty}^{\infty} \boldsymbol{x}_1(nT+\tau_1^-) w_{\tau_2}(t-nT-\tau_1) + G_2 \boldsymbol{u}(t) w_2(t) \tag{6b}$$

for all t.

The Fourier transforms of (6a) and (6b) are, respectively,

$$\boldsymbol{X}_1(\omega) = F_1 \hat{\boldsymbol{X}}_2(\omega) \frac{1-e^{-j\omega\tau_1}}{j\omega} + G_1 \boldsymbol{U}_1(\omega) \tag{7a}$$

$$\boldsymbol{X}_2(\omega) = F_2 \hat{\boldsymbol{X}}_1(\omega) \frac{1-e^{-j\omega\tau_2}}{j\omega} + G_2 \boldsymbol{U}_2(\omega) \tag{7b}$$

where

$$\hat{\boldsymbol{X}}_1(\omega) = \sum_n \boldsymbol{x}_1(nT+\tau_1^-) e^{-jn\omega T} e^{-j\omega\tau_1} \tag{7c}$$

$$\hat{\boldsymbol{X}}_2(\omega) = \sum_n \boldsymbol{x}_2(nT^-) e^{-jn\omega T} \tag{7d}$$

$$\boldsymbol{U}_1(\omega) = \sum_n \theta_{1,n} \boldsymbol{U}(\omega - n\omega_s) \tag{7e}$$

$$\boldsymbol{U}_2(\omega) = \sum_n \theta_{2,n} \boldsymbol{U}(\omega - n\omega_s) \tag{7f}$$

and where

$$\omega_s = \frac{2\pi}{T}$$

$$\theta_{1,n} = \frac{1-e^{-jn\omega_s\tau_1}}{jn\omega_s T}, \qquad n\neq 0;\ \theta_{1,0} = \frac{\tau_1}{T} \tag{7g}$$

$$\theta_{2,n} = \frac{e^{-jn\omega_s\tau_1} - e^{-jn\omega_s T}}{jn\omega_s T}, \qquad n\neq 0;\ \theta_{2,0} = \frac{\tau_2}{T}. \tag{7h}$$

The $\hat{\boldsymbol{X}}_k(\omega)$'s in the above equations may be solved from the difference equations derived from (1a) and (1b). From (1a) we let $t=nT+\tau_1^-$

$$\boldsymbol{x}_1(nT+\tau_1^-) = F_1 \boldsymbol{x}_2(nT^-) + G_1 \boldsymbol{u}(nT+\tau_1^-) \tag{8a}$$

and from (1b) we let $t=nT+T^-$

$$\boldsymbol{x}_2(nT+T^-) = F_2 \boldsymbol{x}_1(nT+\tau_1^-) + G_2 \boldsymbol{u}(nT+T^-). \tag{8b}$$

Increasing the index n in (8a) by 1 and then substituting (8b) into (8a) yields

$$\boldsymbol{x}_1(nT+T+\tau_1^-)=F_1F_2\boldsymbol{x}_1(nT+\tau_1^-)+F_1G_2\boldsymbol{u}(nT+T^-)+G_1\boldsymbol{u}(nT+T+\tau_1^-).$$

Substituting (8a) into (8b) gives

$$\boldsymbol{x}_2(nT+T^-)=F_2F_1\boldsymbol{x}_2(nT^-)+F_2G_1\boldsymbol{u}(nT+\tau_1^-)+G_2\boldsymbol{u}(nT+T^-).$$

Applying the z transform to the above equations we obtain

$$z\tilde{X}_1(z)=F_1F_2\tilde{X}_1(z)+F_1G_2z\tilde{U}_2(z)+G_1z\tilde{U}_1(z) \quad (9a)$$

$$z\tilde{X}_2(z)=F_2F_1\tilde{X}_2(z)+F_2G_1\tilde{U}_1(z)+G_2z\tilde{U}_2(z) \quad (9b)$$

where

$$\tilde{X}_1(z)\triangleq\sum_n \boldsymbol{x}_1(nT+\tau_1^-)z^{-n} \quad (9c)$$

$$\tilde{X}_2(z)\triangleq\sum_n \boldsymbol{x}_2(nT^-)z^{-n} \quad (9d)$$

$$\tilde{U}_1(z)\triangleq\sum_n \boldsymbol{u}(nT+\tau_1^-)z^{-n} \quad (9e)$$

$$\tilde{U}_2(z)\triangleq\sum_n \boldsymbol{u}(nT^-)z^{-n}. \quad (9f)$$

Thus

$$\tilde{X}_1(z)=\Phi_2(z)\left[F_1G_2z\tilde{U}_2(z)+G_1z\tilde{U}_1(z)\right] \quad (10a)$$

$$\tilde{X}_2(z)=\Phi_1(z)\left[F_2G_1\tilde{U}_1(z)+G_2z\tilde{U}_2(z)\right] \quad (10b)$$

where

$$\Phi_1(z)=(zI-F_2F_1)^{-1} \quad (10c)$$

$$\Phi_2(z)=(zI-F_1F_2)^{-1} \quad (10d)$$

and I is the identity matrix.[3]

$\Phi_1(z)$ and $\Phi_2(z)$ may be viewed as the characteristic matrices of a switched capacitor circuit and they are related by the following equations:

$$\Phi_1(z)F_2=F_2\Phi_2(z) \quad (11a)$$

$$\Phi_2(z)F_1=F_1\Phi_1(z). \quad (11b)$$

Let $\boldsymbol{u}(t)$ be continuous at $t=nT^-$ and $t=nT+\tau_1^-$, then with the relationship $z=e^{j\omega T}$ and using Poisson's formula the $\tilde{U}_k(z)$'s in (9e) and (9f) can be expressed as

$$\tilde{U}_1(e^{j\omega T})=\sum_n \boldsymbol{u}(nT+\tau_1^-)e^{-jn\omega T}=\frac{1}{T}\sum_n U(\omega-n\omega_s)e^{j(\omega-n\omega_s)\tau_1} \quad (12a)$$

$$\tilde{U}_2(e^{j\omega T})=\sum_n \boldsymbol{u}(nT^-)e^{-jn\omega T}=\frac{1}{T}\sum_n U(\omega-n\omega_s). \quad (12b)$$

By comparing (9c) and (9d) with (7c) and (7d) we also have

$$\hat{X}_1(\omega)=\tilde{X}_1(e^{j\omega T})e^{-j\omega\tau_1} \quad (13a)$$

$$\hat{X}_2(\omega)=\tilde{X}_2(e^{j\omega T}). \quad (13b)$$

Substituting (12a) and (12b) into (10a) and (10b) and then into (13a) and (13b) we obtain

$$\hat{X}_1(\omega)=\frac{1}{T}\sum_n P_{2,n}(\omega)U(\omega-n\omega_s) \quad (14a)$$

$$\hat{X}_2(\omega)=\frac{1}{T}\sum_n P_{1,n}(\omega)U(\omega-n\omega_s) \quad (14b)$$

where

$$P_{1,n}(\omega)=\Phi_1(e^{j\omega T})\left(F_2G_1e^{j(\omega-n\omega_s)\tau_1}+G_2e^{j\omega T}\right) \quad (14c)$$

$$P_{2,n}(\omega)=\Phi_2(e^{j\omega T})\left(F_1G_2e^{j\omega\tau_2}+G_1e^{j\omega\tau_2}e^{j(\omega-n\omega_s)\tau_1}\right). \quad (14d)$$

Note that $\Phi_1(e^{j\omega T})$ and $\Phi_2(e^{j\omega T})$ are periodic functions of ω with period ω_s.

From the output equation (2) we let

$$y_k(t)=C_k\boldsymbol{x}_k(t)+D_k\boldsymbol{u}(t)w_k(t), \qquad k=1,2 \text{ for all } t$$

where the $\boldsymbol{x}_k(t)$'s are defined as (6a) and (6b). The Fourier transforms of the above equations are

$$Y_k(\omega)=C_kX_k(\omega)+D_kU_k(\omega), \qquad k=1,2. \quad (15a)$$

Substituting the results obtained in (14a) and (14b) into (7a) and (7b) and then into (15a) together with (7e) and (7f) we have

$$Y_k(\omega)=\sum_n T_{k,n}(\omega)U(\omega-n\omega_s), \qquad k=1,2 \quad (15b)$$

where

$$T_{k,n}(\omega)=\frac{1-e^{-j\omega\tau_k}}{j\omega T}C_kF_kP_{k,n}(\omega)+\theta_{k,n}(C_kG_k+D_k), \qquad k=1,2. \quad (15c)$$

If we use the notations introduced in the second form of output equations (3a) and (3b), (15c) becomes

$$T_{k,n}(\omega)=\frac{1-e^{-j\omega\tau_k}}{j\omega T}\hat{C}_kP_{k,n}(\omega)+\theta_{k,n}\hat{D}_k, \qquad k=1,2. \quad (16)$$

The total output is

$$Y(\omega)=Y_1(\omega)+Y_2(\omega)=\sum_n T_n(\omega)U(\omega-n\omega_s) \quad (17a)$$

where

$$T_n(\omega)=T_{1,n}(\omega)+T_{2,n}(\omega). \quad (17b)$$

The above expression gives the Fourier transform of the output due to an arbitrary input whose Fourier transform is $U(\omega)$. The $T_n(\omega)$ is the transfer function which relates the shifted input spectrum at $n\omega_s$ to the output. It is important to point out that $T_n(\omega)$ in general consists of a constant component due to feedthrough and a component whose magnitude of the envelope is inversely proportional to the frequency.

[3]As a result of the definition of $\tilde{U}_1(z)$ and $\tilde{U}_2(z)$ given in (9e) and (9f), (10a) and (10b) do not appear in symmetrical form.

We now consider a special case of (17a) and (17b) when the input is a cisoidal function. Let

$$\boldsymbol{u}(t)=e^{j\omega_0 t}\hat{\boldsymbol{u}} \tag{18}$$

where $\hat{\boldsymbol{u}}$ is a complex vector. Then

$$U(\omega)=2\pi\delta(\omega-\omega_0)\hat{\boldsymbol{u}}. \tag{19}$$

Thus the transfer function in (17a) should be evaluated at $\omega=\omega_0+n\omega_s$, i.e.,

$$T_n(\omega)=T_n(\omega_0+n\omega_s). \tag{20}$$

The transfer function $T_n(\omega_0+n\omega_s)$ in (20) can also be derived from [6] by setting $A_k=0$ and $B_k=j\omega_0 G_k$, $k=1,2$ (F_3 and G_3 in [6] are equivalent to F_1 and G_1 in this paper). The justification for making the above statement, though not obvious, can be seen by transforming the equation formulation of switched capacitor circuits to the following state equation formulation:

$$\dot{x}_k(t)=G_k\dot{\boldsymbol{u}}(t), \qquad k=1,2.$$

Since an arbitrary input can be viewed as a continuum of cisoidal components in the frequency domain, using the superposition principle, the result for a cisoidal input can also be extended to an arbitrary input.

Example 2: Using the same example given previously, we have

$$F_1=\begin{bmatrix}1 & 0\\ 0 & 0\end{bmatrix} \qquad G_1=\begin{bmatrix}0\\ 0\end{bmatrix}$$

$$C_1=\begin{bmatrix}-1 & 0\end{bmatrix} \qquad D_1=1$$

$$F_2=\begin{bmatrix}1-\beta & -\beta\\ \beta-1 & \beta\end{bmatrix} \qquad G_2=\begin{bmatrix}\beta\\ 1-\beta\end{bmatrix}$$

$$C_2=\begin{bmatrix}-1 & 0\end{bmatrix} \qquad D_2=1.$$

From (10c) and (10d) we have

$$\Phi_1(z)=\frac{1}{z(z-1+\beta)}\begin{bmatrix}z & 0\\ \beta-1 & z-1+\beta\end{bmatrix}$$

$$\Phi_2(z)=\frac{1}{z(z-1+\beta)}\begin{bmatrix}z & -\beta\\ 0 & z-1+\beta\end{bmatrix}.$$

Substituting the above results with $z=e^{j\omega T}$ into (14c) and (14d) yields

$$P_{1,n}(\omega)=\frac{1}{e^{j\omega T}-1+\beta}\begin{bmatrix}\beta e^{j\omega T}\\ (1-\beta)(e^{j\omega T}-1)\end{bmatrix}$$

$$P_{2,n}(\omega)=\frac{1}{e^{j\omega T}-1+\beta}\begin{bmatrix}\beta e^{j\omega\tau_2}\\ 0\end{bmatrix}.$$

From (16) and (17b) with $\omega=\omega_0+n\omega_s$ (for a cisoidal input) we obtain

$$T_n(\omega_0+n\omega_s)=T_{1,n}(\omega_0+n\omega_s)+T_{2,n}(\omega_0+n\omega_s)$$

$$=\frac{\beta\left[\beta\left(e^{j(\omega_0+n\omega_s)\tau_2}-1\right)+1-e^{j\omega_0 T}\right]}{j(\omega_0+n\omega_s)T\left(e^{j\omega 0T}-1+\beta\right)}$$

$$+\frac{\beta\left(1-e^{-jn\omega_s\tau_1}\right)}{jn\omega_s T}, \qquad n\neq 0 \tag{21a}$$

and

$$T_0(\omega_0)=\frac{\beta\left[\beta\left(e^{j\omega_0\tau_2}-1\right)+1-e^{j\omega_0 T}\right]}{j\omega_0 T\left(e^{j\omega_0 T}-1+\beta\right)}+\left(1-\frac{\beta\tau_2}{T}\right). \tag{21b}$$

The above results agree with those obtained by Tsividis [5].

IV. Fifty-Percent Duty Cycle

When a switched capacitor circuit is operated at 50-percent duty cycle more interesting results can be derived from the general expressions obtained earlier. With $\tau_1=\tau_2=T/2$, from (17b) and (16) we have

$$T_n(\omega)=\frac{1-e^{-j\omega T/2}}{j\omega T}\left[\hat{C}_1 P_{1,n}(\omega)+\hat{C}_2 P_{2,n}(\omega)\right]+\frac{1-(-1)^n}{jn2\pi}(\hat{D}_1-\hat{D}_2), \qquad n\neq 0 \tag{22a}$$

$$T_0(\omega)=\frac{1-e^{-j\omega T/2}}{j\omega T}\left[\hat{C}_1 P_{1,0}(\omega)+\hat{C}_2 P_{2,0}(\omega)\right]+\frac{1}{2}(\hat{D}_1+\hat{D}_2) \tag{22b}$$

where

$$P_{1,n}(\omega)=\Phi_1(e^{j\omega T})\left[(-1)^n F_2 G_1 e^{j\omega T/2}+G_2 e^{j\omega T}\right] \tag{22c}$$

$$P_{2,n}(\omega)=\Phi_2(e^{j\omega T})\left[F_1 G_2 e^{j\omega T/2}+(-1)^n G_1 e^{j\omega T}\right]. \tag{22d}$$

Notice that the $P_{k,n}(\omega)$'s given above are rational functions of $e^{j\omega T/2}$. If there is no feedthrough from the input to the output, i.e., $\hat{D}_1=\hat{D}_2=0$, (22a) and (22b) reduce to

$$T_n(\omega)=\frac{1-e^{-j\omega T/2}}{j\omega T}\hat{T}_n(z) \tag{23a}$$

where

$$z=e^{j\omega T}$$

$$\hat{T}_n(z)=\text{a rational function of } z^{1/2}$$

$$=\hat{C}_1\Phi_1(z)\left[(-1)^n F_2 G_1 z^{1/2}+G_2 z\right]+\hat{C}_2\Phi_2(z)\left[F_1 G_2 z^{1/2}+(-1)^n G_1 z\right]. \tag{23b}$$

The factor $(1-e^{-j\omega T/2})/j\omega T$ in (23a) is a zeroth-order hold function of magnitude $1/T$ and duration $T/2$. In many practical applications this factor is nearly a constant of $1/2$ when the switching frequency is much higher than the signal frequency. Consequently the behavior of a switched capacitor circuit can be conveniently analyzed in the rational $z^{1/2}$ domain.

If in addition the input is sampled only once over a switching period T, say $G_1=0$, the $\hat{T}_n(z)$ in (23b) after some matrix manipulation, becomes

$$\hat{T}_n(z)=(C_1+z^{-1/2}C_2F_2)\Phi_2(z)F_1G_2 z \tag{24a}$$

or when $G_2=0$

$$\hat{T}_n(z)=(-1)^n(z^{-1/2}C_1F_1+C_2)\Phi_1(z)F_2G_1 z. \tag{24b}$$

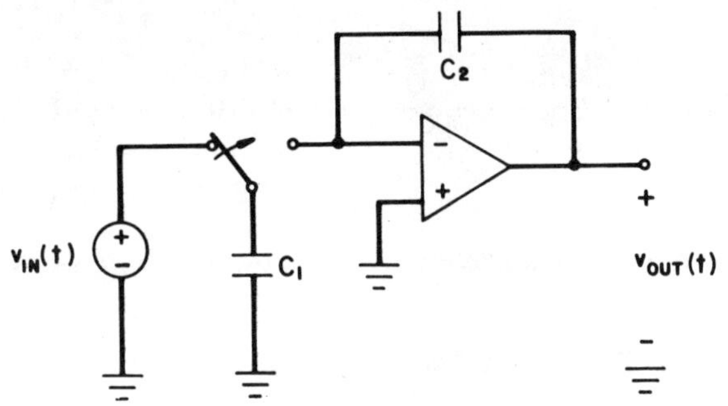

Fig. 4. A switched capacitor integrator.

It is interesting to note that the $\hat{T}_n(z)$ in (24a) is identical for all n and the $\hat{T}_n(z)$ in (24b) differs only in sign for different n.

The $\hat{T}_n(z)$ in (24a) or (24b) is a rational function of $z^{1/2}$. In order to eliminate the dependence on $z^{1/2}$ we impose the topological constraint $C_1 = C_2F_2$ in (24a) or $C_1F_1 = C_2$ in (24b). Under these constraints the factor $(1+z^{-1/2})$ in (24a) or (24b) can be combined with the zeroth-order hold function $(1-e^{-j\omega T/2})/j\omega T$ when we substitute (24a) or (24b) into (23a). Thus we have

$$T_n(\omega) = \frac{1-e^{-j\omega T}}{j\omega T} C_1\Phi_2(z)F_1G_2z \tag{25a}$$

or

$$T_n(\omega) = (-1)^n \frac{1-e^{-j\omega T}}{j\omega T} C_2\Phi_1(z)F_2G_1z. \tag{25b}$$

The factor $(1-e^{-j\omega T})/j\omega T$ in the above two equations is a zeroth-order hold function of magnitude $1/T$ and duration T. It reduces to unity when the switching frequency is much higher than the signal frequency. Under this condition the transfer function $T_n(\omega)$ can be approximated and conveniently analyzed in the rational z domain.

It appears that the topological constraints for (25a) and (25b) to be valid are restrictive. After a careful examination we find that all the circuits discussed by Hosticka *et al.* [2] fall into this category. It is also interesting to note that the exact transfer function for a switched capacitor integrator shown in Fig. 4 is (using (25a))

$$T_n(\omega) = \frac{1-e^{-j\omega T}}{j\omega T} \frac{(c_1/c_2)z}{1-z} = \frac{-c_1/c_2}{j\omega T}, \quad \text{for all } n.$$

Thus if we make $c_1 = T/r_1$ [2] the switched capacitor integrator gives exactly the same frequency response as the analog integrator. Hosticka *et al.* have analyzed this circuit with two stages of approximations which turn out to cancel each other.

V. Sample-and-Hold Input

In many practical situations, in particular when the applied signal is already sampled and held constant during each subinterval it is desirable to analyze switched capacitor circuits through the use of z-domain techniques without any of the topological constraints mentioned in the previous section. In the following we shall derive some useful expressions using some of the results previously obtained in the frequency domain.

Since all the functions are piece-wise constant we can solve the problem initially at discrete points as long as they are sampled in a consistent manner. This has already been done from (8) to (11). If we apply the same sampling process to the output equation (2) or (3a) and (3b) we obtain

$$\tilde{Y}_1(z) \triangleq \sum_n y(nT+\tau_1^-)z^{-n} = C_1\tilde{X}_1(z) + D_1\tilde{U}_1(z) \tag{26a}$$

and

$$\tilde{Y}_2(z) \triangleq \sum_n y(nT^-)z^{-n} = C_2\tilde{X}_2(z) + D_2\tilde{U}_2(z) \tag{26b}$$

or

$$\tilde{Y}_1(z) = \hat{C}_1\tilde{X}_2(z) + \hat{D}_1\tilde{U}_1(z)$$

and

$$\tilde{Y}_2(z) = \hat{C}_2\tilde{X}_1(z)z^{-1} + D_2\tilde{U}_2(z).$$

Substituting (10a) and (10b) into the above equations we obtain

$$\tilde{Y}_1(z) = H_{11}(z)\tilde{U}_1(z) + H_{12}(z)\tilde{U}_2(z) \tag{27a}$$

$$\tilde{Y}_2(z) = H_{21}(z)\tilde{U}_1(z) + H_{22}(z)\tilde{U}_2(z) \tag{27b}$$

where

$$H_{11}(z) = C_1\Phi_2(z)G_1z + D_1 = \hat{C}_1\Phi_1(z)F_2G_1 + \hat{D}_1 \tag{27c}$$

$$H_{12}(z) = C_1\Phi_2(z)F_1G_2z = \hat{C}_1\Phi_1(z)G_2z \tag{27d}$$

$$H_{21}(z) = C_2\Phi_1(z)F_2G_1 = \hat{C}_2\Phi_2(z)G_1 \tag{27e}$$

$$H_{22}(z) = C_2\Phi_1(z)G_2z + D_2 = \hat{C}_2\Phi_2(z)F_1G_2 + \hat{D}_2. \tag{27f}$$

The $H_{ij}(z)$'s in the above expressions are the discrete transfer functions from the $\tilde{U}_k(z)$'s to the $\tilde{Y}_k(z)$'s.

Since the overall output is a piecewise constant function its Fourier transform can be shown as

$$Y(\omega) = \frac{1-e^{-j\omega\tau_1}}{j\omega}\tilde{Y}_1(z) + \frac{1-e^{-j\omega\tau_2}}{j\omega}e^{j\omega\tau_2}\tilde{Y}_2(z) \tag{28}$$

where we have used the facts that $y(nT^+) = y(nT+\tau_1^-)$, $y(nT+\tau_1^+) = y(nT+T^-)$, and $z = e^{j\omega T}$. Substituting (27a) and (27b) into (28) and using the relation given by (12a) and (12b) we obtain

$$Y(\omega) = \sum_n \tilde{H}_n(\omega)U(\omega - n\omega_s) \tag{29a}$$

where

$$\tilde{H}_n(\omega) = \left[\frac{1-e^{-j\omega\tau_1}}{j\omega T} H_{11}(z) + \frac{1-e^{-j\omega\tau_2}}{j\omega T} e^{j\omega\tau_2} H_{21}(z)\right] e^{j(\omega - n\omega_s)\tau_1} + \left[\frac{1-e^{-j\omega\tau_1}}{j\omega T} H_{12}(z) + \frac{1-e^{-j\omega\tau_2}}{j\omega T} e^{j\omega\tau_2} H_{22}(z)\right]. \quad (29b)$$

In the following we shall consider some useful results under certain special conditions.

A. Fifty-Percent Duty Cycle

At 50-percent duty cycle we have $\tau_1 = \tau_2 = T/2$. The overall transfer function can be derived from (29b) as

$$\tilde{H}_n(\omega) = \frac{1-e^{-j\omega T/2}}{j\omega T} \hat{H}_n(z) \quad (30a)$$

where

$$\hat{H}_n(z) = \text{a rational function of } z^{1/2} = (-1)^n \left[z^{1/2} H_{11}(z) + z H_{21}(z)\right] + \left[H_{12}(z) + z^{1/2} H_{22}(z)\right]. \quad (30b)$$

The transfer function in (30a) is for a sample-and-hold input and it allows feedthrough. On the other hand, the transfer function in (23a) is for an arbitrary input but with no feedthrough. If we allow no feedthrough in (30a), then the $\tilde{H}_n(\omega)$ in (30a) is identical to the $T_n(\omega)$ in (23a), although they have different kinds of inputs. This interesting result, valid also for an arbitrary duty cycle, will be discussed more fully in the following section.

Under certain topological constraints the $\hat{H}_n(z)$ in (30b) or its magnitude may be made independent of n. For example, if we let $G_1 = D_1 = 0$ then

$$\hat{H}_n(z) = H_{12}(z) + z^{1/2} H_{22}(z). \quad (31a)$$

If, in addition, we let $\hat{D}_2 = 0$ then the above $\hat{H}_n(z)$ becomes identical to the $\hat{T}_n(z)$ in (24a).

Similarly, if we let $G_2 = D_2 = 0$ then

$$\hat{H}_n(z) = (-1)^n \left[z^{1/2} H_{11}(z) + z H_{21}(z)\right]. \quad (31b)$$

If, in addition, we let $\hat{D}_1 = 0$ then the above $\hat{H}_n(z)$ becomes identical to the $\hat{T}_n(z)$ in (24b).

VI. Relationship Between Solutions Due to Arbitrary and Sample-and-Hold Inputs

The transfer function $T_n(\omega)$ defined in (17b) can also be expressed in terms of results obtained in the previous section when the input is a sample-and-hold function. Using (16), (14c), and (14d), we can easily show that

$$T_n(\omega) = \tilde{H}_n(\omega) + \hat{D}_1\left[\theta_{1,n} - \frac{1-e^{-j\omega\tau_1}}{j\omega T} e^{j(\omega - n\omega_s)\tau_1}\right] + \hat{D}_2\left[\theta_{2,n} - \frac{1-e^{-j\omega\tau_2}}{j\omega T} e^{j\omega\tau_2}\right] \quad (32)$$

where $\tilde{H}_n(\omega)$ is defined as in (29b). Equation (32) implies that the transfer function $T_n(\omega)$ for an arbitrary input can be obtained through the use of transfer function where the input is sampled and held constant in each interval. The last two terms in (32) correspond to the feedthrough of a residue function formed by subtracting the sample-and-hold component from an arbitrary input. They vanish when either the input is a sample-and-hold function or there is no feedthrough in the circuit or both. This observation also explains the fact that when there is no feedthrough in a circuit the transfer function for an arbitrary input is the same as for a sample-and-hold input. A block diagram representation of the relation between the input and output is shown in Fig. 5.

VII. Noise Analysis

From (17a), the time-domain solution of the switched capacitor circuit is readily given by

$$y(t) \triangleq y_1(t) + y_2(t), \qquad \text{for all } t$$

$$= \sum_{n=-\infty}^{\infty} \int_{-\infty}^{\infty} \tilde{T}_n(t-\tau) u(\tau) e^{jn\omega_s\tau}\, d\tau \quad (33a)$$

$$= \int_{-\infty}^{\infty} h(t,\tau) u(\tau)\, d\tau \quad (33b)$$

where $\tilde{T}_n(t)$ is the inverse Fourier transform of $T_n(\omega)$ and $h(t,\tau)$ is the impulse response to an input impulse applied at $t=\tau$, i.e.,

$$h(t,\tau) = \sum_{n=-\infty}^{\infty} \tilde{T}_n(t-\tau) e^{jn\omega_s\tau}. \quad (34)$$

If the input $u(t)$ is a stationary stochastic process, then the output $y(t)$ is generally a nonstationary process. However, $y(t)$ is cyclostationary (in the wide sense) [8], that is

$$E\{y(t+T)\} = E\{y(t)\}$$

and

$$E\{y(t_1+T) y^*(t_2+T)\} = E\{y(t_1) y^*(t_2)\}$$

where $E\{\ \}$ denotes mathematical expectation and $*$ denotes transposition and conjugation.

Let the mean (vector) and the spectral density (matrix) of the stationary input be m_u and $S_u(\omega)$, respectively. The mean (vector) of the output, denoted by $m_y(t)$, is given by (from (33a))

$$m_y(t) = E\{y(t)\} = \sum_{n=-\infty}^{\infty} T_n(n\omega_s) e^{jn\omega_s t} m_u. \quad (35a)$$

The autocorrelation (matrix) of the output, denoted by $R_y(t+\tau, t)$ [8], [9], is given by

$$R_y(t+\tau, t) = E\{y(t+\tau) y^*(t)\} = \sum_{n=-\infty}^{\infty} \sum_{m=-\infty}^{\infty} \int_{-\infty}^{\infty} \int_{-\infty}^{\infty} \tilde{T}_n(t+\tau-\xi_1) \cdot R_u(\xi_1-\xi_2) \tilde{T}_m^*(t-\xi_2) \cdot e^{j(n\omega_s\xi_1 - m\omega_s\xi_2)}\, d\xi_1\, d\xi_2 \quad (35b)$$

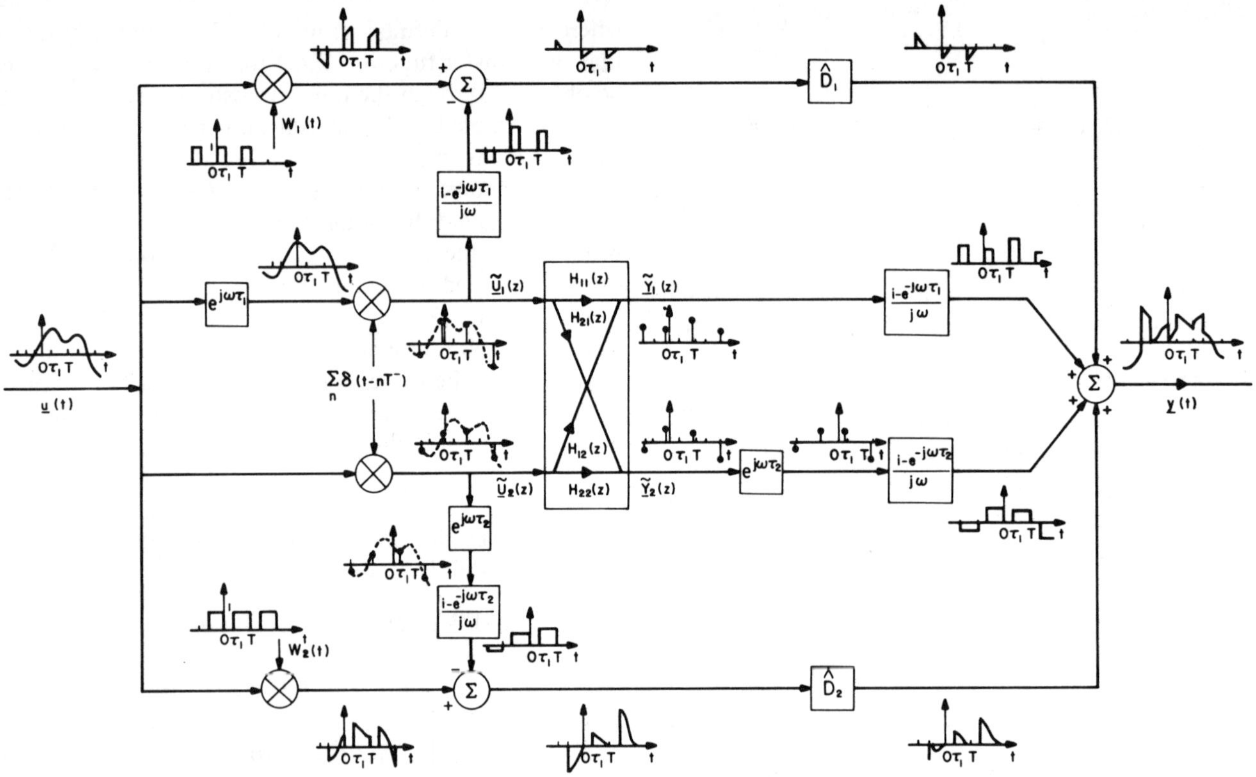

Fig. 5. A block diagram representation of exact solution.

where $R_u(\xi_1-\xi_2)=R_u(\xi)$ $(\xi \triangleq \xi_1-\xi_2)$ is the autocorrelation (matrix) of the stationary input $\boldsymbol{u}(t)$. Observe that both $\boldsymbol{m}_y(t)$ and $R_y(t+\tau,t)$ are periodic with respect to t with period $T=2\pi/\omega_s$. Denote their time averages by $\tilde{\boldsymbol{m}}_y$ and $\tilde{R}_y(\tau)$, respectively. Then

$$\tilde{\boldsymbol{m}}_y = T_0(0)\boldsymbol{m}_u \tag{36a}$$

and

$$\tilde{R}_y(\tau)=\sum_{n=-\infty}^{\infty}\int_{-\infty}^{\infty}\int_{-\infty}^{\infty}\tilde{T}_n(\tau_1)R_u(\tau-\tau_1+\tau_2)\cdot e^{jn\omega_s(\tau-\tau_1+\tau_2)}\tilde{T}_n^*(\tau_2)\,d\tau_1\,d\tau_2. \tag{36b}$$

The Fourier transform of $\tilde{R}_y(\tau)$, i.e., the average output spectral density (matrix) $\tilde{S}_y(\omega)$, is given by

$$\tilde{S}_y(\omega)=\sum_{n=-\infty}^{\infty}T_n(\omega)S_u(\omega-n\omega_s)T_n^*(\omega). \tag{37}$$

Example 3: As an example, consider the switched capacitor integrator of Fig. 6. We assume that the operational amplifier's input noise $\overline{v_n^2}$ is given by

$$\overline{v_n^2}=\overline{v_{n0}^2}\left(1+\frac{\omega_B}{|\omega|}\right)$$

where $\overline{v_{n0}^2}$ is the equivalent thermal noise and ω_B is the $1/f$ noise (flicker noise) corner frequency. The switch is assumed to be ideal. To compute the noise spectrum using (37), the transfer functions $T_n(\omega)$ from the noise input v_n to the output v_{out} must be determined. For 50-percent duty cycle and using (22a) and (22b) we have

$$T_n(\omega)=(-1)^n\frac{c_1}{c_2}\frac{1}{j\omega T}+\begin{cases}1+\dfrac{1}{2}\dfrac{c_1}{c_2}, & \text{for } n=0\\[2mm] \dfrac{1}{jn\pi}\dfrac{c_1}{c_2}, & \text{for } n=\text{odd}\\[2mm] 0, & \text{for } n=\text{even},\quad n\neq 0\end{cases}$$

Now, the input noise spectrum is given by $S_0(1+\omega_B/|\omega|)$ with $S_0=4kT\,R_{\text{eq}}$, where R_{eq} = equivalent thermal resistance. The average output noise spectrum is then given by (see Fig. 7)

$$\begin{aligned}\tilde{S}_y(\omega)&=\sum_{n=-\infty}^{\infty}|T_n(\omega)|^2S_0\left(1+\frac{\omega_B}{|\omega-n\omega_s|}\right)\\ &=S_0\left\{\left[\left(1+\frac{1}{2}\frac{c_1}{c_2}\right)^2+\left(\frac{c_1}{c_2}\frac{1}{\omega T}\right)^2\right]\left(1+\frac{\omega_B}{|\omega|}\right)\right.\\ &\quad+\left[\left(\frac{c_1}{c_2}\right)^2\left(\frac{1}{\omega T}+\frac{1}{\pi}\right)^2\right]\left(1+\frac{\omega_B}{|\omega+\omega_s|}\right)\\ &\quad\left.+\left[\left(\frac{c_1}{c_2}\right)^2\left(\frac{1}{\omega T}-\frac{1}{\pi}\right)^2\right]\left(1+\frac{\omega_B}{|\omega-\omega_s|}\right)\right\}\\ &\quad+\text{higher sideband contributions.}\end{aligned}$$

The above result clearly indicates that $\tilde{S}_y(\omega)$ is unbounded in magnitude. In reality, the operational amplifiers characteristic is band-limited, i.e., all $|T_n(\omega)|\to 0$ for $|\omega|\to\infty$ and $S_u=S_0(1+\omega_B/|\omega|)$ is valid only for $\omega_l<|\omega|<\omega_u$, where ω_l and ω_u are some lower and upper bound frequencies. As a result of the band-limiting nature of

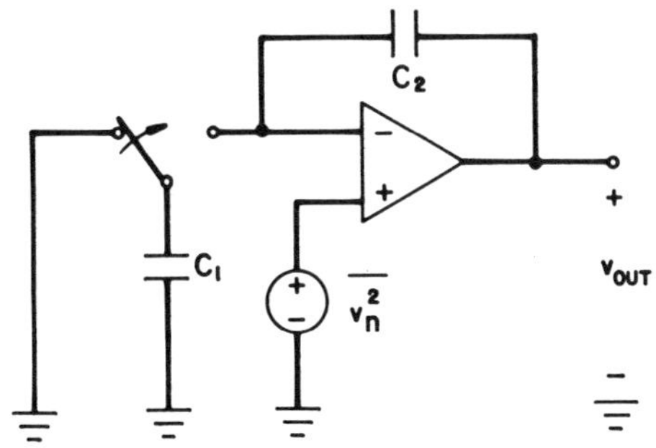

Fig. 6. A noise model of a switched capacitor integrator.

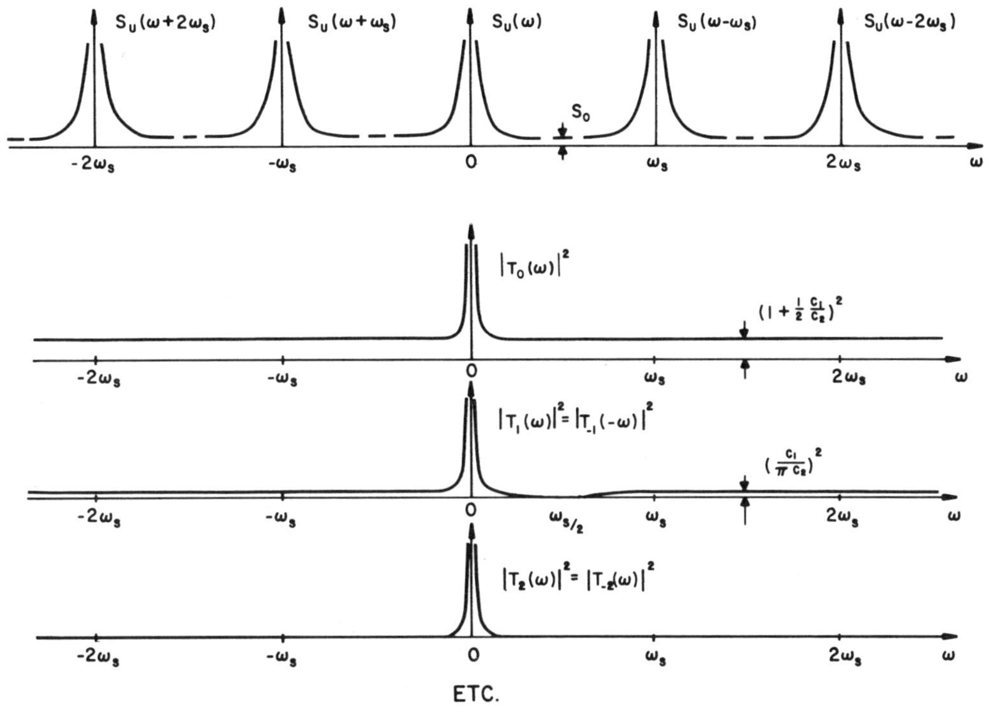

Fig. 7. Noise spectra of the circuit in Fig. 6.

$|T_n(\omega)|$ and S_u, only a finite number of terms in (37) is needed for the determination of $\tilde{S}_y(\omega)$.

VIII. Consideration for Program Implementation

As discussed in Section II, the circuit to be solved for the determination of F_k, G_k, $\hat{C}_k$, and $\hat{D}_k$ matrices is linear and "resistive." There exist several efficient analysis methods for resistive circuits. In this section we shall briefly describe the modified nodal analysis method for possible program implementation.

The classical nodal method is perhaps the simplest and easiest to implement on computers for solving linear, resistive circuits containing independent current sources and voltage-controlled current sources (VCCS) as the only sources. For the "resistive" circuit under consideration, one must convert the independent voltage sources into their Norton equivalent current sources, and VCVS's and CCCS's into equivalent VCCS's before the nodal method can be applied. This conversion usually involves additional programming effort and it turns out to be not as straightforward as it appears for the computation of F_k, G_k, $\hat{C}_k$, and $\hat{D}_k$ matrices. To overcome this shortcoming, the modified nodal method as proposed in [10] shall be employed.

Let the circuit be connected with $(n+1)$ nodes and b composite branches as shown in Fig. 8. Each composite branch is made up of a two-terminal element b_k, a voltage source $\mathcal{E}_k(t)$ and a current source $\mathcal{Q}_k(t)$. For a "resistive" branch (corresponding to a capacitor in the original circuit), $\hat{I}_k(t)=c_k\hat{v}_k(t)$ $(=c_k(v_{c_k}(t)-v_{c_k}(\tau^-)))$, and $\mathcal{E}_k(t)=v_{c_k}(\tau^-)$. For an independent voltage source branch, $\hat{v}_k(t)=0$ and $\mathcal{E}_k(t)=E_{s_k}(t)$. For a voltage-controlled voltage source, $\hat{v}_k(t)=0$ and $\mathcal{E}_k(t)$ is some linear combination of $v_k(t)$'s and independent voltage sources $E_{s_k}(t)$'s. Indepen-

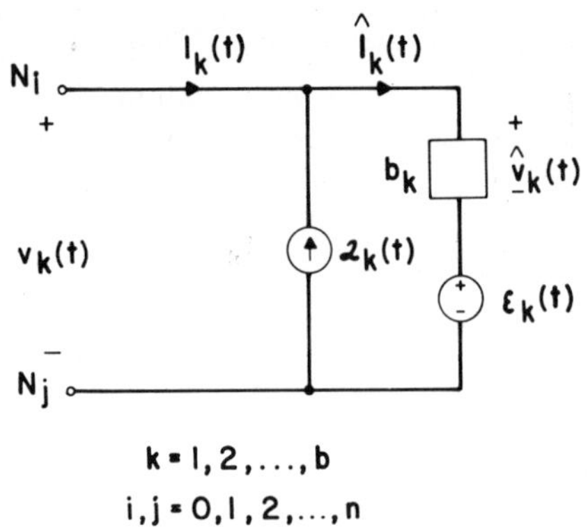

Fig. 8. A "resistive" equivalent composite branch of a switched capacitor circuit.

dent "current" sources $\hat{Q}_{s_k}(t)$'s may be combined to form the composite current source $\mathcal{Q}_k(t)$. For a current controlled current source whose controlling current is flowing through a nonzero "resistor," it is assumed that the conversion to the equivalent voltage-controlled current source has already been made.

Define the voltage vectors

$$\boldsymbol{v} \triangleq \begin{bmatrix} v_1 \\ \cdot \\ \cdot \\ \cdot \\ v_b \end{bmatrix} \qquad \hat{\boldsymbol{v}} \triangleq \begin{bmatrix} \hat{v}_1 \\ \cdot \\ \cdot \\ \cdot \\ \hat{v}_b \end{bmatrix} \qquad \boldsymbol{\mathcal{E}} \triangleq \begin{bmatrix} \mathcal{E}_1 \\ \cdot \\ \cdot \\ \cdot \\ \mathcal{E}_b \end{bmatrix} \tag{38}$$

and the current vectors

$$\boldsymbol{I} \triangleq \begin{bmatrix} I_1 \\ \cdot \\ \cdot \\ \cdot \\ I_b \end{bmatrix} \qquad \hat{\boldsymbol{I}} \triangleq \begin{bmatrix} \hat{I}_1 \\ \cdot \\ \cdot \\ \cdot \\ \hat{I}_b \end{bmatrix} \qquad \boldsymbol{\mathcal{Q}} \triangleq \begin{bmatrix} \mathcal{Q}_1 \\ \cdot \\ \cdot \\ \cdot \\ \mathcal{Q}_b \end{bmatrix}. \tag{39}$$

Let $\hat{\boldsymbol{I}}$ be partitioned into two subvectors,

$$\hat{\boldsymbol{I}} = \begin{bmatrix} \hat{\boldsymbol{I}}_a \\ \hat{\boldsymbol{I}}_b \end{bmatrix}$$

where $\hat{\boldsymbol{I}}_b$ corresponds to the current vector associated with the currents in the voltage sources, including all the independent and controlled sources, and $\hat{\boldsymbol{I}}_a$ corresponds to the remaining currents. Let node 0 be the datum node, and denote $\boldsymbol{v}_N \triangleq \text{col}\,[v_{N_1}, v_{N_2}, \cdots, v_{N_n}]$ as the node-to-datum voltage vector and A as the $n \times b$ reduced incidence matrix obtained from the complete incidence matrix by deleting the row corresponding to the datum node. Clearly, $\boldsymbol{v} = A^t\boldsymbol{v}_N$. We partition all the voltage vectors in (38) and A in the same manner as in $\hat{\boldsymbol{I}}$, i.e.,

$$\boldsymbol{v} = \begin{bmatrix} \boldsymbol{v}_a \\ \boldsymbol{v}_b \end{bmatrix} \qquad \hat{\boldsymbol{v}} = \begin{bmatrix} \hat{\boldsymbol{v}}_a \\ \hat{\boldsymbol{v}}_b \end{bmatrix}$$

$$\boldsymbol{\mathcal{E}} = \begin{bmatrix} \boldsymbol{\mathcal{E}}_a \\ \boldsymbol{\mathcal{E}}_b \end{bmatrix} \quad \text{and} \quad A = [A_a, A_b].$$

The KCL equation for the circuit is given by

$$A\boldsymbol{I} = A(\hat{\boldsymbol{I}} - \boldsymbol{\mathcal{Q}}) = 0. \tag{40}$$

With the above described partitioning, (40) can be rewritten as

$$[A_a, A_b]\begin{bmatrix} \hat{\boldsymbol{I}}_a \\ \hat{\boldsymbol{I}}_b \end{bmatrix} = A\boldsymbol{\mathcal{Q}} \tag{41}$$

where $\hat{\boldsymbol{I}}_a$ is related to $\hat{\boldsymbol{v}}_a$ by the branch-admittance matrix Y_a, which always exists as the result of the partitioning procedure, i.e.,

$$\hat{\boldsymbol{I}}_a = Y_a\hat{\boldsymbol{v}}_a. \tag{42}$$

On the other hand, $\hat{\boldsymbol{v}}_a$ can be expressed as

$$\hat{\boldsymbol{v}}_a = \boldsymbol{v}_a - \boldsymbol{\mathcal{E}}_a = A_a^t\boldsymbol{v}_N - \boldsymbol{\mathcal{E}}_a. \tag{43}$$

Substituting (42) and (43) into (41) we obtain

$$[A_aY_aA_a^t, A_b]\begin{bmatrix} \boldsymbol{v}_N \\ \hat{\boldsymbol{I}}_b \end{bmatrix} = A_aY_a\boldsymbol{\mathcal{E}}_a + A\boldsymbol{\mathcal{Q}}. \tag{44}$$

Equation (44) is a set of n equations in n node-to-datum voltages ($\boldsymbol{v}_N$) and the currents in all the independent and controlled voltage sources ($\hat{\boldsymbol{I}}_b$). Denote the number of these voltage sources by b_1. The branch relations of these b_1 elements are given by[4]

$$K\begin{bmatrix} \boldsymbol{v}_a \\ \boldsymbol{v}_b \end{bmatrix} = H\boldsymbol{E}_s \quad \text{or} \quad KA^t\boldsymbol{v}_N = H\boldsymbol{E}_s \tag{45}$$

where $\boldsymbol{E}_s(t)$ = independent voltage source vector of dimension $(1 \times l) \triangleq \text{col}\,[E_{s_1}(t), \cdots, E_{s_l}(t)]$, K and H are real constant matrices of dimensions $(b_1 \times b)$ and $(b_1 \times l)$, respectively.

Let $\hat{\boldsymbol{Q}}_s(t)$ be the independent "current" source vector of dimension $(1 \times m)$, i.e., $\hat{\boldsymbol{Q}}_s(t) = \text{col}\,[\hat{Q}_{s_1}(t), \cdots, \hat{Q}_{s_m}(t)]$ and $\boldsymbol{v}_c(\tau^-)$ be the initial capacitor voltage vector of dimension $(1 \times M)$, i.e., $\boldsymbol{v}_c(\tau^-) = \text{col}\,[v_{c_1}(\tau^-), \cdots, v_{c_M}(\tau^-)]$. Without loss of generality, we assume that the "resistive" branches (corresponding to the M capacitors) are numbered first, then,[5]

$$\boldsymbol{\mathcal{E}}_a = \begin{bmatrix} I_M \\ 0 \end{bmatrix}\boldsymbol{v}_c(\tau^-) \tag{46}$$

and

$$A_aY_a\boldsymbol{\mathcal{E}}_a = A_a\tilde{Y}_a\boldsymbol{v}_c(\tau^-) \tag{47}$$

where $\tilde{Y}_a$ is obtained from Y_a by retaining only the first M columns. Also, since $\boldsymbol{\mathcal{Q}}$ is a linear combination of $\hat{Q}_{s_k}(t)$'s, we can express

$$A\boldsymbol{\mathcal{Q}} = \tilde{A}\hat{\boldsymbol{Q}}_s(t) \tag{48}$$

where $\tilde{A}$ is a real matrix of dimension $(n \times m)$, which is a sort of reduced incidence matrix involving only the independent current source branches.

[4]Note that all the controlled sources are voltage controlled.

[5]I_M is the unit matrix of dimension M.

Combining (44) and (45), and using (47) and (48) we obtain the $(n+b)$ modified nodal equations

$$\begin{bmatrix} A_a Y_a A_a^t & A_b \\ KA^t & 0 \end{bmatrix} \begin{bmatrix} v_N \\ \hat{I}_b \end{bmatrix} = \begin{bmatrix} A_a \tilde{Y}_a \\ 0 \end{bmatrix} v_c(\tau^-) + \begin{bmatrix} 0 & \tilde{A} \\ H & 0 \end{bmatrix} \begin{bmatrix} E_s(t) \\ \hat{Q}_s(t) \end{bmatrix}. \qquad (49)$$

In (49), the submatrix $A_a Y_a A_a^t$ is the node-admittance matrix excluding the contributions due to all the voltage sources. Equation (49) can be solved to obtain

$$v_N(t) = \tilde{F} v_c(\tau^-) + \tilde{G} u(t), \qquad u(t) \triangleq \begin{bmatrix} E_s(t) \\ \hat{Q}_s(t) \end{bmatrix}.$$

Then,

$$\begin{aligned} v_c(t) &= \text{linear combination of } v_{N_1}, \ldots, v_{N_n} \\ &= A_c^t v_N = F v_c(\tau^-) + G u(t) \qquad (50) \end{aligned}$$

where A_c is obtained from A by deleting the columns (the last $b-M$ columns) which do not correspond to the M capacitor branches, and $F = A_c^t \tilde{F}$, $G = A_c^t \tilde{G}$.

The output voltage vector $y(t)$ is obtained in a similar manner, i.e.,

$$\begin{aligned} y(t) &= \text{linear combination of } v_{N_1}, \cdots, v_{N_n} \\ &= A_0 v_N = \hat{C} v_c(\tau^-) + \hat{D} u(t) \qquad (51) \end{aligned}$$

where A_0 is the connection matrix relating v_N to y. Since A_0 is unique, the matrices $\hat{C}$ and $\hat{D}$ are also unique.

IX. Conclusions

In this paper a general formulation procedure and an exact analysis in the frequency domain for switched capacitor circuits with arbitrary inputs have been presented. No topological constraint has been imposed on the circuits and no approximation has been used throughout the analysis. Explicit solutions have been derived for sample-and-hold as well as continuous inputs including noise. Due to the simple form of the solutions much insight has been gained in understanding the working of a switched capacitor circuit. The results are numerically attractive and can be readily programmed for computer-aided design of large size circuits.

Acknowledgment

The authors wish to thank their colleagues R. L. Adams, P. E. Fleischer, D. L. Fraser, C. F. Kurth, and A. J. Vera for numerous stimulating discussions on this extremely exciting topic.

References

[1] D. L. Fried, "Analog sampled-data filters," *IEEE J. Solid-State Circuits*, vol. SC-7, pp. 302–304, Aug. 1972.

[2] B. J. Hostika, R. W. Broderson, and P. R. Gray, "MOS sampled data recursive filters using switched capacitor integrators," *IEEE J. of Solid-State Circuits*, vol. SC-12, pp. 600–608, Dec. 1977.

[3] J. T. Caves *et al.*, "Sampled analog filtering using switched capacitors as resistor equivalents," *IEEE J. Solid-State Circuits*, vol. SC-12, pp. 592–599, Dec. 1977.

[4] C. F. Kurth and G. S. Moschytz, "Two-port analysis of cascaded switched-capacitor networks," presented at *1978 IEEE Int. Symp. Circuits Syst.*, New York, May 1978.

[5] Y. P. Tsividis, "Analytical and experimental evaluation of a switched-capacitor filter and remarks on the resistor/switched-capacitor correspondence," *IEEE Trans. Circuits Syst.*, vol. CAS-26, pp. 140–144, Feb. 1979.

[6] M. L. Liou, "Exact analysis of linear circuits containing periodically operated switches with applications," *IEEE Trans. Circuit Theory*, vol. CT-19 pp. 146–154, Mar. 1972.

[7] T. Ström and S. Signell, "Analysis of periodically switched linear circuits," *IEEE Trans. Circuits Syst.*, vol. CAS-24, pp. 531–541, Oct. 1977

[8] A. Papoulis, *Probability, Random Variables and Stochastic Processes*. New York: McGraw-Hill, 1965, ch. 12.

[9] L. E. Franks, *Signal Theory*. Englewood Cliffs, N.J.: Prentice-Hall, 1969, pp. 204–214.

[10] C. W. Ho, A. E. Ruehli, and P. A. Brennan, "The modified nodal approach to network analysis," *IEEE Trans. Circuits Syst.*, vol. CAS-22, pp. 504–509, June 1975.

Bibliography

BOOKS

0001 RC-ACTIVE CIRCUITS; THEORY AND DESIGN.
BRUTON LT
PRENTICE HALL, 1980

0002 ACTIVE FILTERS FOR INTEGRATED CIRCUITS; FUNDAMENTALS AND DESIGN METHODS.
BUDAK A
HOUGHTON MIFFLIN, 1974

0003 APPROXIMATION METHODS FOR ELECTRONIC FILTER DESIGN, WITH APPLICATIONS TO PASSIVE, ACTIVE AND DIGITAL NETWORKS.
DANIELS RW
MCGRAW HILL, 1974

0004 PRINCIPLES OF ACTIVE NETWORK SYNTHESIS AND DESIGN.
DARYANANI G
WILEY, 1976

0005 MANUAL OF ACTIVE FILTER DESIGN.
HILBURN JL + JOHNSON DE
MCGRAW HILL, 1973

0006 ACTIVE FILTERS FOR INTEGRATED CIRCUITS; FUNDAMENTALS AND DESIGN METHODS.
HOLMES WE + HEINLEIN WH
SPRINGER-VERLAG, 1974

0007 ACTIVE RC FILTERS; THEORY AND APPLICATION.
HUELSMAN LP
HUTCHINSON AND ROSS, 1976

0008 INTRODUCTION TO THE THEORY AND DESIGN OF ACTIVE FILTERS.
HUELSMAN LP + ALLEN PE
MCGRAW HILL, 1980

0009 RAPID PRACTICAL DESIGNS OF ACTIVE FILTERS.
JOHNSON DE + HILBURN JL
WILEY, 1975

0010 A HANDBOOK OF ACTIVE FILTERS.
JOHNSON DE + JOHNSON JR + MOORE HP
PRENTICE HALL 1980

0011 ANALOG AND DIGITAL FILTERS; DESIGN AND REALIZATION.
LAM HYF
PRENTICE HALL, 1979

0012 ACTIVE FILTER COOKBOOK.
LANCASTER DE
SAMS, 1975

0013 ACTIVE NETWORK DESIGN.
LINDQUIST CS
STEWARD AND SONS, 1977

0014 ACTIVE INDUCTORLESS FILTERS
MITRA SK
IEEE PRESS, 1971

0015 LINEAR INTEGRATED NETWORKS; FUNDAMENTALS.
MOSCHYTZ GS
VAN NOSTRAND REINHOLD, 1976

0016 LINEAR INTEGRATED NETWORKS; DESIGN.
MOSCHYTZ GS
VAN NOSTRAND REINHOLD, 1975

0017 FILTER THEORY AND DESIGN; ACTIVE AND PASSIVE.
SEDRA AS + BRACKETT PO
MATRIX PUBLISHERS, 1977

0018 INTRODUCTION TO CIRCUIT SYNTHESIS AND DESIGN.
TEMES GC + LAPATRA JW
MCGRAW HILL, 1977

0019 MODERN FILTER THEORY AND DESIGN.
TEMES GC + MITRA SK
WILEY, 1973

0020 INTRODUCTION TO OPERATIONAL AMPLIFIER THEORY AND APPLICATIONS.
WAIT JV + HUELSMAN LP + KORN GA
MCGRAW HILL, 1975

0021 ACTIVE FILTER DESIGN.
WILLIAMS AB
ARTECH HOUSE, 1975

DESIGN CONSIDERATIONS

0022 HUM FILTER USING ELECTRONICALLY SIMULATED INDUCTOR.
ELEKTOR 4(7-8): 22 (JULY-AUG 1978)

0023 COMPENSATING CIRCUIT FOR AMPLIFIER ELEMENT.
UK PAT 1404645, 3 SEPT 1975, P

0024 CRITICAL ELEMENT ANALYSIS FOR THE TRANSFER, GAIN AND PHASE FUNCTIONS IN LINEAR ACTIVE FILTERS.
ACAR C
INT J CIRCUIT THEORY APPL 7(2): 201-8 (APRIL 1979)

0025 CRITICAL ELEMENT ANALYSIS FOR THE TRANSFER, GAIN AND PHASE FUNCTIONS IN LINEAR ACTIVE FILTERS.
ACAR C
P139-40 OF PROC OF THE 1978 IEEE INT SYMP ON CIRCUITS AND SYSTEMS, IEEE, NEW YORK, USA 17-19 MAY 1978, IEEE, 1978

0026 STATISTICAL TOLERANCE ANALYSIS AND DESIGN OF A LARGE INTEGRATED CIRCUIT.
ADBY PR
P9-10 OF 1ST EUROPEAN SOLID STATE CIRCUITS CONF-ESSCIRC (EXTENDED ABSTRACTS ONLY) IEE, ASSOC ELLETROTECNICA AND ELLETRONICA ITALIANA, ET AL, CANTERBURY, KENT, 2-5 SEPT 1975, IEE, 1975

0027 EVALUATING THE FREQUENCY LIMITATIONS OF OPERATIONAL AMPLIFIER RC NETWORKS - THE DOMINANT-POLE TECHNIQUE AND ITS APPLICATION TO INDUCTANCE SIMULATION.
AHMED MT + DUTTA ROY SC
J INST ELECTRON TELECOMMUN ENG 22(11): 703-8 (NOV 1976)

0028 A MODEL FOR SLEW-INDUCED DISTORTION IN SINGLE-AMPLIFIER ACTIVE FILTERS.
ALLEN PE
IEEE TRANS CIRCUITS SYST CAS-25(8): 565-72 (AUG 1978)

0029 INCREASED FREQUENCY CAPABILITY FOR A GIVEN ACTIVE FILTER CONFIGURATION.
ALLEN PE
IEEE TRANS INSTRUM MEAS IM-25(1): 18-27 (MARCH 1976)

0030 MINIMIZING THE EFFECTS OF GAIN-BANDWIDTH ON ACTIVE FILTERS.
ALLEN PE
P647-52 OF 8TH ASILOMAR CONF ON CIRCUITS, SYSTEMS AND COMPUTERS NAVAL POSTGRADUATE SCHOOL, UNIV SANTA CLARA, IEEE, ET AL, PACIFIC GROVE, CALIF, USA, 3-5 DEC 1974, PARKER, SR(ED) WESTERN PERIODICALS, 1975

0031 DESIGN OF ACTIVE FILTERS WITH MINIMUM S(GB)(PI).
ALLEN PE
P63-72 OF 17TH MIDWEST SYMP ON CIRCUITS AND SYSTEMS, LAWRENCE, KAN, USA, 16-17 SEPT 1974, WESTERN PERIODICALS 1974

0032 PRACTICAL CONSIDERATIONS OF ACTIVE FILTER DESIGN.
ALLEN PE
J AUDIO ENG SOC 22(10): 770-82 (DEC 1974)

0033 Q-SENSITIVITY OF CERTAIN RC ACTIVE FILTERS TO GAIN-BANDWIDTH PRODUCT.
ANANDA MOHAN PV
ELECTRON LETT 15(3): 95-7 (1 FEB 1979)

0034 INFLUENCE OF AMBIENT-TEMPERATURE VARIATIONS IN TWO GYRATOR CIRCUITS.
ANTONIOU A
PROC INST ELECTR ENG 123(6): 490-4 (JUNE 1976)

0035 COMMENT ON - GAIN-SENSITIVITY PRODUCT-A FIGURE OF MERIT FOR HYBRID-INTEGRATED FILTERS USING SINGLE OPERATIONAL AMPLIFIERS.
ARBEL AF
IEEE J SOLID-STATE CIRCUITS SC-7(3): 273 (JUNE 1972)

0036 NOISE AND SENSITIVITY OPTIMIZATION OF A SINGLE-AMPLIFIER BIQUAD.
BACHLER HJ + GUGGENBUHL W
IEEE TRANS CIRCUITS SYST CAS-26(1): 30-6 (JAN 1979)

0036 NOISE AND SENSITIVITY PERFORMANCE OF COMPLEMENTARY AND DUAL SECOND ORDER ACTIVE FILTER BUILDING BLOCKS.
BACHLER HJ + GUGGENBUHL W
P256-60 OF CIRCUIT THEORY AND DESIGN, LAUSANNE, SWITZERLAND, 4-8 SEPT 1978, MOSCHYTZ, GS, NIERYNCK, J(ED), GEORGI 1978

0037 STATISTICAL MULTIPARAMETRIC SENSITIVITY MEASURING METHODS COMPARISON.
BARBINI P + LIBERATORE A
ALTA FREQ 46(7): 319-20 (JULY 1977) (IN ITALIAN)

0038 PROBLEMS CONCERNING EXPENSE OPTIMIZATION AND MULTIPARAMETER SENSITIVITY OF ACTIVE RC FILTERS.
BENCSATH P + KUN L
HIRADASTECHNIKA 28(7): 210-19 (JULY 1977) (IN HUNGARIAN)

0039 FILTER DESIGN BY MEANS OF OPTIMIZATION TECHNIQUES.
BENSASSON A + FABRE J
ONDE ELEC 52(3): 147-58 (MARCH 1972) (IN FRENCH)

0040 TEMPERATURE STABILITY OF THE Q-FACTOR OF AN ACTIVE BANDPASS RC FILTER WITH A NULLING CIRCUIT IN THE FEEDBACK LOOP.
BERENSHTEIN BSH + VOROB'EV AM
IZV VUZ RADIOELEKTRON 18(8): 59-63 (AUG 1975) (IN RUSSIAN)

0041 SENSITIVITY AND NOISE OF SECOND ORDER ACTIVE RC-FILTERS. II. FILTERS WITH TWO AND THREE OPERATIONAL AMPLIFIERS.
BERGMANN K + EYSEN S
FREQUENZ 31(6): 168-75 (JUNE 1977) (IN GERMAN)

0042 SENSITIVITY AND NOISE OF SECOND ORDER ACTIVE RC-FILTERS. I. FUNDAMENTALS AND SINGLE-AMPLIFIER FILTERS.
BERGMANN K + EYSEN S
FREQUENZ 31(5): 156-9 (MAY 1977) (IN GERMAN)

0043 NOISE MINIMIZATION OF SECOND-ORDER ACTIVE RC-FILTERS WITH A SINGLE OPERATIONAL AMPLIFIER.
BERGMANN K + EYSEN S
ARCH ELEKTRON UEBERTRAGUNGSTECH 29(10): 421-8 (OCT 1975) (IN GERMAN)

0044 ANALYSIS OF WORST-CASE GAIN FOR NTH ORDER TRANSFER FUNCTIONS REALISED BY CASCADED 2ND-ORDER NETWORKS.
BERKOVICS IE + MOSCHYTZ GS
ELECTRON LETT 8(8): 199-200 (20 APRIL 1972)

0045 A PHASE DISTORTION FREE 50 HZ INTERFERENCE FILTER.
BETTS RP
ELECTRON ENG 51(625): 17, (JULY 1979)

0046 W CLASS OF MULTIPLE-REAL-ROOT EQUAL-RIPPLE (M.U.R.R.O.E.R.) POLYNOMIALS FOR THE DESIGN OF ACTIVE FILTERS BY CASCADING G 3RD-ORDER BLOCKS.
BIEY M + PREMOLI A
IEE J ELECTRON CIRCUITS SYST 3(2): 53-7 (MARCH 1979)

0047 HARMONIC DISTORTION IN A CLASS OF LINEAR ACTIVE FILTER NETWORKS.
BILLAM PJ
J AUDIO ENG SOC 26(6): 426-9 (JUNE 1978)

0048 COMPUTER AIDED SYNTHESIS OF THE REACTIVE-ACTIVE FILTERS.
BINELLI P
ALTA FREQ 43(7): 429-35 (JULY 1974)

0049 AN ALGORITHM FOR COMPUTING THE MAXIMUM PERMISSIBLE TOLERANCES FOR THE COMPONENTS OF AN ACTIVE FILTER.
BIZOUARN R + GIMENES C
REV TECH THOMSON-CSF 6(1): 137-43 (MARCH 1974) (IN FRENCH)

0050 SOME APPLICATIONS OF THE OPTGE OPTIMISATION PROGRAM TO THE DESIGN OF ACTIVE AND PASSIVE FILTERS.
BIZOUARN R + GIMENES C
REV TECH THOMSON-CSF 6(1): 123-36 (MARCH 1974) (IN FRENCH)

0051 SOME EXPERIMENTS WITH ACTIVE LOW-PASS FILTERS. II. RELATIONSHIP BETWEEN AC RESPONSE AND TRANSIENT RESPONSE.
BOGLE AG
AEU 7(1): 26-9 (1974)

0052 COMPARISON OF THE NOISE BEHAVIOUR OF REALIZATIONS OF SOME BAND-PASS FILTER CIRCUITS.
BOHME R + SCHREIBER H
WISS Z TECH UNIV DRESDEN 26(2): 355-60 (1977) (IN GERMAN)

0053 MODERN FDM PILOT FILTERS.
BOROWSKI W + WOLLMERSHAUSER P
ELECTR COMMUN 53(2): 148-52 (1978)

0054 AMPLIFIER NONLINEARITIES IN THE MULTIPLE-NEGATIVE-FEEDBACK BANDPASS FILTER.
BOWRON P + MOHAMED MA
INT J CIRCUIT THEORY APPL 6(2): 121-34 (APRIL 1978)

0055 INTERACTION BETWEEN PARAMETERS OF OPERATIONAL AMPLIFIERS AND ONE CLASS OF ACTIVE RC FILTERS.
BOYANOV ID + DIOCHINOVA RI
IZV VMEI 'LENIN' 34(11): 329-36 (1976) (IN BULGARIAN)

0056 SENSITIVITY OF ACTIVE FILTERS USING MONTE CARLO TECHNIQUES.
BOYCE AH
P13/3PP OF COLLOQUIUM DIGEST ON SENSITIVITY THEORY AND TOLERANCE DESIGN OF CIRCUITS, TEE, LONDON, ENGLAND 29 JAN 1973, IEE, 1973

0057 ACTIVE COMPENSATION FOR HIGH-FREQUENCY EFFECTS IN OP-AMP CIRCUITS WITH APPLICATIONS TO ACTIVE RC FILTERS.
BRACKETT PO + SEDRA AS
IEEE TRANS CIRCUITS SYST CAS-23(2): 68-72 (FEB 1976)

0058 HIGH-FREQUENCY LIMITATIONS OF RC-ACTIVE FILTERS CONTAINING SIMULATED-L AND FDNR ELEMENTS.
BRUTON LT + HAASE AB
INT J CIRCUIT THEORY APPL 2(2): 187-94 (JUNE 1974)

0059 A FREQUENCY-LOCKED ADAPTIVE FILTER.
BRUTON LT + PEDERSON RT
INT J CIRCUIT THEORY APPL 1(1): 71-82 (MARCH 1973)

0060 HIGH-FREQUENCY LIMITATIONS OF ACTIVE LADDER NETWORKS.
BRUTON LT + RAMAKRISHNA K
IEEE TRANS CIRCUITS SYST CAS-22(8): 704-8 (AUG 1975)

0061 HIGH-FREQUENCY LIMITATIONS OF ACTIVE LADDER NETWORKS.
BRUTON LT + RAMAKRISHNA K
P642-6 OF 8TH ASILOMAR CONF ON CIRCUITS, SYSTEMS AND COMPUTERS NAVAL POSTGRADUATE SCHOOL, UNIV SANTA CLARA, IEEE, ET AL, PACIFIC GROVE, CALIF, USA, 3-5 DEC 1974, PARKER, SR(ED) WESTERN PERIODICALS, 1975

0062 ELECTRICAL NOISE IN LOW-PASS FDNR FILTERS.
BRUTON LT + TRELEAVEN DH
IEEE TRANS CIRCUIT THEORY CT20(2): 154-8 (MARCH 1973)

0063 NOISE PERFORMANCE OF LOW-SENSITIVITY ACTIVE FILTERS.
BRUTON LT + TROFIMENKOFF FN + TRELEAVEN DH
IEEE J SOLID-STATE CIRCUITS SC 8(1): 85-91 (FEB 1973)

0064 FREQUENCY LIMITATIONS OF ACTIVE FILTERS USING OPERATIONAL AMPLIFIERS.
BUDAK A + PETRELA DM
IEEE TRANS CIRCUIT THEORY CT19(4): 322-8 (JULY 1972)

0065 PRACTICAL DESIGN CONSIDERATIONS FOR A VARIABLE CENTER FREQUENCY, CONSTANT BANDWIDTH, AND CONSTANT PEAK-VALUE ACTIVE FILTER.
BUDAK A + ZELLER ER
IEEE J SOLID-STATE CIRCUITS SC-7(4): 308-11 (AUG 1972)

0066 A QUAD JFET WIDE-BAND OPERATIONAL-AMPLIFIER INTEGRATED CIRCUIT FEATURING TEMPERATURE-COMPENSATED BANDWIDTH.
CAVE DL + DAVIS WR
IEEE J SOLID-STATE CIRCUITS SC-12(4): 382-8 (AUG 1977)

0067 ANALYSIS OF THE INPUT IMPEDANCE OF AN ACTIVE BAND-PASS FILTER.
CULEN K + HANG DQ
ELEKTROTECH CAS 29(9): 712-13 (1978) (IN SLOVAK)

0068 ACTIVE FILTER WITH CONTROLLED PHASE SHIFT.
CULTER DA
ELECTRON ENGINEERING 47(571): 18-19 (SEPT 1975)

0069 ACTIVE HYBRID FILTERS - EVOLUTION AND MARKETS.
CUNTZ G
ELECTRON MICROELECTRON IND NO 219: 15-21 (15 APRIL 1976) (IN FRENCH)

0070 FIRST ORDER PRECORRECTION FOR EXACT SYNTHESIS OF SINGLE AMPLIFIER BIQUADS (SABS) USING REAL AMPLIFIERS.
DATAR RB + DHILLA IF + DHABHAR TJ
P145-9 OF CIRCUIT THEORY AND DESIGN, LAUSANNE, SWITZERLAND, 4-8 SEPT 1978, MOSCHYTZ, GS, NIERYNCK, J(ED), GEORGI 1978

0071 ACTIVE FILTERS - SOME SIMPLE DESIGN METHODS FOR BIOMEDICAL WORKERS.
DERRETT CJ + TAVNER PJ
MED BIOL ENG 13(6): 883-8 (NOV 1975)

0072 SOME FURTHER SENSITIVITY THEOREMS IN ACTIVE EQUIVALENT-NETWORK THEORY.
DUNNING-DAVIES J + FAULKNER FD + SEWELL JI
IEE J ELECTRON CIRCUITS SYST 2(6): 193-8 (NOV 1978)

0073 SENSITIVITY COMPARISON OF LOW-PASS RC STRUCTURES.
DUNNING-DAVIES J + STEPHENSON FW
PROC IEEE 65(3): 488-9 (MARCH 1977)

0074 A HIGH-FREQUENCY-OPTIMIZED RC-ACTIVE FILTER BASED ON THE WIEN BRIDGE.
FAULKNER EA + GRIMBLEBY JB
RADIO ELECTRON ENG 45(9): 462-4 (SEPT 1975)

0075 EFFECT OF AMPLIFIER GAIN-BANDWIDTH PRODUCT ON THE PERFORMANCE OF ACTIVE FILTERS.
FAULKNER EA + GRIMBLEBY JB
RADIO ELECTRON ENG 43(9): 547-52 (SEPT 1973)

0076 DYNAMIC RANGE, LINEARITY AND GAIN-BANDWIDTH PRODUCT IN ACTIVE FILTERS.
FAULKNER EA + GRIMBLEBY JB
P4/1 OF COLLOQUIUM ON ELECTRONIC CIRCUIT ASPECTS OF ACTIVE FILTERS, IEE, LONDON, ENGLAND, 13 OCT 1972, IEE, 1972

0077 NOISE PERFORMANCE OF CAPACITOR-GYRATOR FILTERS.
FETTWEIS A
INT J CIRCUIT THEORY APPL 2(2): 181-6 (JUNE 1974)

0078 SENSITIVITY ASSESSMENT OF PARASITIC EFFECTS IN SECOND-ORDER ACTIVE-FILTER CONFIGURATIONS.
FIDLER JK
IEE J ELECTRON CIRCUITS SYST 2(6): 181-5 (NOV 1978)

0079 SLOPE-NORMALISED SENSITIVITY - A NEW SENSITIVITY MEASURE.
FIDLER JK + NIGHTINGALE C
ELECTRON LETT 15(2): 54-6 (18 JAN 1979)

0080 SENSITIVITY MINIMIZATION IN A SINGLE AMPLIFIER BIQUAD CIRCUIT.
FLEISCHER PE
P202-5 OF 1975 IEEE INT SYMP ON CIRCUITS AND SYSTEMS, IEEE, NEWTON, MASS, USA, 21-23 APRIL 1975, IEEE 1975

0081 SENSITIVITY MINIMIZATION IN A SINGLE AMPLIFIER BIQUAD CIRCUIT.
FLEISCHER PE
IEEE TRANS CIRCUITS SYST CAS-23(1): 45-55 (JAN 1976)

0082 A SIMPLE METHOD FOR TUNING ACTIVE FILTERS.
FLIEGE N
INT J CIRCUIT THEORY APPL 7(2): 270-2 (APRIL 1979)

0083 BROADBANDING TECHNIQUES FOR RC-ACTIVE FILTERS WITH TWO OPERATIONAL AMPLIFIERS.
FLIEGE N
NACHRICHTENTECH Z 31(4): 289-92 (APRIL 1978)

0084 ANALYSIS AND SYNTHESIS OF RC-ACTIVE FILTERS WITH NON-IDEAL AMPLIFIERS.
FLIEGE N
NACHRICHTENTECH Z 25(6): 281-7 (JUNE 1972) (IN ENGLISH)

0085 JUMP PHENOMENON IN ACTIVE RC NETWORKS.
FORSEN A + KRISTIANSSON L
P331-5 OF 1974 EUROPEAN CONF ON CIRCUIT THEORY AND DESIGN, IEE, IEEE, INST PHYS, IERE, LONDON, ENGLAND 23-26 JULY 1974, IEE, 1974

0086 SENSITIVITY CONSIDERATIONS OF FEEDBACK-TYPE ACTIVE RC NETWORKS.
FUJII N
ELECTRON LETT 12(12): 313-15 (10 JUNE 1976)

0087 SYNTHESIS OF ACTIVE RC FILTERS WITH PRESCRIBED POLE SENSITIVITY USING AN AMPLIFIER WITH ONE-POLE ROLL-OFF CHARACTERISTIC.
FUJII N
ELECTRON COMMUN JAP 57(5): 58-66 (MAY 1974)

0088 MINIMUM POWER OF FEEDBACK RC ACTIVE NETWORKS.
FUJII N
VOL54NO6

0089 BREADBOARD ACTIVE FILTERS AT A COMPUTER.
GECKLE WA
ELECTRON DES 21(13): 104-10 (21 JUNE 1973)

0090 HOW TO BUILD HIGH-QUALITY FILTERS OUT OF LOW-QUALITY PARTS.
GEFFE PR
ELECTRONICS 49(23): 111-13 (11 NOV 1976)

0091 ACTIVE FILTERS WITH ZERO AMPLIFIER SENSITIVITY.
GEIGER RI + BUDAK A
IEEE TRANS CIRCUITS SYST CAS-26(4): 277-88 (APRIL 1979)

0092 ZERO SENSITIVITY ACTIVE FILTERS EMPLOYING FINITE GAIN VOLTAGE AMPLIFIERS.
GEIGER RL
P1064-8 OF PROC OF THE 1978 IEEE INT SYMP ON CIRCUITS AND SYSTEMS, IEEE, NEW YORK, USA 17-19 MAY 1978, IEEE, 1978

0093 TRANSIENTS IN SWITCHED ACTIVE RC FILTERS.
GEISSLER E
FERNMELDETECHNIK 17(3): 106-10 (1977) (IN GERMAN)

0094 NOISE CHARACTERISTICS OF ACTIVE RC-CIRCUITS EMPLOYING INTEGRATED OPERATIONAL AMPLIFIERS.
GELL' PP + LUNDIN VZ
POLUPROVODN PRIB TEKH ELEKTROSVYAZI NO 16: 12-23 (1975) (IN RUSSIAN)

0095 STABILITY AND FREQUENCY LIMITS OF RC ACTIVE FILTERS.
GENSEL J
NACHRICHTENTECH ELEKTRON 25(9): 334-6 (1975) (IN GERMAN)

0096 A HIGHLY SELECTIVE BAND PASS FILTER WITH NOISE REJECTION.
GILISSEN M
ELECTRON APPL IND NO 262S: 41-3 (15 JAN 1979) (IN FRENCH)

0097 SYNTHESIS AND CALCULATION OF SENSITIVITIES OF ACTIVE FILTERS WITH OPERATIONAL AMPLIFIERS AND RC ELEMENTS AND DESIGN CONSTRAINTS.
GIMENES C
REV TECH THOMSON-CSF 4(4): 799-831 (DEC 1972) (IN FRENCH)

0098 AN INVESTIGATION INTO NETWORK TOLERANCES USING INTERVAL ANALYSIS.
GLESNER M
NTG-FACHBER 51: 23-30 (1975) (IN GERMAN)

0099 STATISTICAL TOLERANCE INVESTIGATIONS ON ELECTRICAL NETWORKS BY THE GENERALIZED QUANTILE ARITHMETIC.
GLESNER M + HAUBRICHS K + LINSLER HJ
ARCH ELEKTRON UEBERTRAGUNGSTECH 32(1): 1-7 (JAN 1978) (IN GERMAN)

0100 STATISTICAL TOLERANCE INVESTIGATIONS ON ELECTRICAL NETWORKS BY THE GENERALIZED QUANTILE ARITHMETIC.
GLESNER M + HAUBRICHS K + LINSLER HJ
P212-15 OF PROC OF THE 1976 IEEE INT SYMP ON CIRCUITS AND SYSTEMS, IEEE, MUNICH, GERMANY 27-29 APRIL 1976, IEEE, 1976

0101 A THEORY OF EQUIVALENT ACTIVE NETWORKS.
GRANT LG + SEWELL JI
P497-500 OF PROC OF THE 1976 IEEE INT SYMP ON CIRCUITS AND SYSTEMS, IEEE, MUNICH, GERMANY 27-29 APRIL 1976, IEEE, 1976

0102 A COMPARATIVE ANALYSIS OF ACTIVE AND PASSIVE ELECTRICAL FILTERS FOR PULSED OPTOELECTRONIC SYSTEMS.
GRISHKO MI + ZHURAVLEV VK
SOV J OPT TECHNOL 39(9): 530-2 (SEPT 1972)

0103 ACTIVE FILTER DESIGN FOR MINIMAL NOISE AND GAIN SENSITIVITY.
GUGGENBUHL W + BACHLER H
P328-31 OF 1977 IEEE INT SYMP ON CIRCUITS AND SYSTEMS PROC, IEEE, PHOENIX, ARIZ, USA 25-27 APRIL 1975, IEEE, 1977

0104 NOISE PERFORMANCE COMPARISON AND OPTIMIZATION OF DIFFERENT ACTIVE FILTERS.
GUGGENBUHL W + KUNZLI M
P336-9 OF PROC OF THE 1976 IEEE INT SYMP ON CIRCUITS AND SYSTEMS, IEEE, MUNICH, GERMANY 27-29 APRIL 1976, IEEE, 1976

0105 NOISE OPTIMISATION IN RC-ACTIVE FILTER SECTIONS.
HAASE AB + BRUTON LT
IEE J ELECTRON CIRCUITS SYST 1(4): 117-24 (JULY 1977)

0106 PRINCIPLES AND PRACTICE OF LINEAR AMPLIFIERS. XXXIV.
HARMS G
ELEKTROTECHNIK 59(4): 25-6 (18 FEB 1977) (IN GERMAN)

0107 A STUDY ON DESIGN OF SIGNALING EQUIPMENT USING ACTIVE FILTER TECHNOLOGY.
HASEGAWA K + OHNO M + UEKI I + YASUDA K
P129-34 OF PROC OF THE 1978 IEEE INT SYMP ON CIRCUITS AND SYSTEMS, IEEE, NEW YORK, USA 17-19 MAY 1978, IEEE, 1978

0108 OPTIMUM DESIGN OF FILTERS BY THE AID OF LARGE CHANGE SENSITIVITIES.
HASHIMOTO T
P194-7 OF 1975 IEEE INT SYMP ON CIRCUITS AND SYSTEMS, IEEE, NEWTON, MASS, USA, 21-23 APRIL 1975, IEEE 1975

0109 COMPARISON OF EXPECTED YIELDS OF ACTIVE FILTERS FOR THE PCM CHANNEL BANK.
HASHIMOTO T + TOYAMA N + KUSUNOKI H + NAGATA K + MINATO Y
BULL KYUSHU INST TECHNOL (SCI NO 37: 87-94 (SEPT 1978) (IN JAPANESE)

0110 NOISE PERFORMANCE LIMITATIONS OF SINGLE AMPLIFIER RC ACTIVE FILTERS.
HASLETT JW
IEEE TRANS CIRCUITS SYST CAS-22(9): 743-7 (SEPT 1975)

0111 NOISE PERFORMANCE OF THE NEW NORTON OP AMPS.
HASLETT JW
IEEE TRANS ELECTRON DEVICES ED-21(9): 571-7 (SEPT 1974)

0112 SYNTHESIS OF ACTIVE RC BAND-PASS FILTER BY RC TRANSFORMATION.
HAYAHARA E + YAMADA S
TRANS INST ELECTRON COMMUN ENG JPN SECT E E60(10): 557-8 (OCT 1977)

0113 AN APPROACH TO THE SENSITIVITY AND STATISTICAL VARIABILITY OF BIQUADRATIC FILTERS.
HILBERMAN D
IEEE TRANS CIRCUIT THEORY CT-20(4): 382-90 (JULY 1973)

0114 IMPULSE RESPONSE OF RECURSIVE FILTERS.
HIRATA Y
ACUSTICA 42(4): 263-4 (JUNE 1979)

0115 FUNCTIONAL CALIBRATION OF MICROELECTRONIC FILTERS.
HIRSL J
SDELOVACI TECH 26(11): 427-8 (NOV 1979) (IN CZECH)

0116 AMPLIFIERS AND ACTIVE FILTERS.
HOLT AGJ
P1/1 OF COLLOQUIUM ON ELECTRONIC CIRCUIT ASPECTS OF ACTIVE FILTERS, IEE, LONDON, ENGLAND, 13 OCT 1972, IEE, 1972

0117 FEASIBILITY STUDY OF ACTIVE FILTER STRUCTURES FOR HYBRID THICK FILM IMPLEMENTATION.
HORN P + LEUENBERGER R
P165-9 OF CIRCUIT THEORY AND DESIGN, LAUSANNE, SWITZERLAND, 4-8 SEPT 1978, MOSCHYTZ, GS, NIERYNCK, J(ED), GEORGI 1978

0118 TAKING THE NONIDEALITY OF THE PASSIVE ELEMENTS OF AN ACTIVE TUNABLE BAND-PASS FILTER INTO ACCOUNT.
INSHAKOV YUM
IZV VUZ ENERG NO 7: 122-4 (JULY 1975) (IN RUSSIAN)

0119 A NOTE ON THE NOISE OF 2ND-ORDER RC ACTIVE FILTER REALIZED BY SINGLE AMPLIFIER.
ISHIBASHI Y
TRANS INST ELECTRON COMMUN ENG JPN SECT E E61(12): 981 (DEC 1978)

0120 A CONSIDERATION ON THE SENSITIVITY OF A 2ND-ORDER RC ACTIVE FILTER.
ISHIBASHI Y
TRANS INST ELECTRON COMMUN ENG JPN SECT E E61(11): 920-1 (NOV 1978)

0121 SENSITIVITY OF A 2ND-ORDER RC-ACTIVE FILTER REALIZED BY SINGLE AMPLIFIER.
ISHIBASHI Y
TRANS INST ELECTRON COMMUN ENG JPN SECT E E61(5): 380-1 (MAY 1978)

0122 DESIGN OF HYBRID INTEGRATED RC ACTIVE FILTER.
IWATA A + YOSHIMURA H + YANAGAWA H
REV ELECTR COMMUN LAB 26(3-4): 565-77 (MARCH-APRIL 1978)

0123 POWER CONSUMPTION IN FDM CHANNEL BANK RC ACTIVE FILTERS.
JACOVITTI G + MARTINELLI G
ALTA FREQ 45(2): 124-9 (FEB 1976)

0124 COMPUTER-AIDED DESIGN OF ACTIVE FILTERS.
JUGS A
RCA ENG 19(4): 33-5 (DEC1973- JAN 1974)

0125 IC OP AMPS HAVE EVOLVED FROM GENERAL-PURPOSE DIFFERENTIAL-INPUT AMPLIFIERS INTO MANY SPECIALIZED TYPES.
JUNG WG
ELECTRON DES 26(1): 94-101 (4 JAN 1978)

0126 ULTRA-LOW FREQUENCY ACTIVE FILTER USING SWITCHING-MODE OPERATIONS.
KAKU S + FUJII A + OKUMURA Y + MIYAUCHI M
TRANS INST ELECTRON COMMUN ENG JPN SECT E E61(6): 486-7 (JUNE 1978)

0127 NONINVERTING INTEGRATORS WITH IMPROVED HIGH-FREQUENCY RESPONSE.
KELLNER W
ELECTRON LETT 15(9): 257-8 (26 APRIL 1979)

0128 PRE-DISTORTION OF ACTIVE RC FILTERS.
KERWIN WJ + ARAMRATTANA M + HAMILTON DJ
P37-40 OF CONF RECORD OF THE TWELFTH ASILOMAR CONF ON CIRCUITS, SYSTEMS AND COMPUTERS NAVAL POSTGRADUATE SCHOOL MONTEREY, IEEE, UNIV SANTA CLARA, PACIFIC GROVE, CA, USA, 6-8 NOV 1978, IEEE, 1978

0129 SENSITIVITIES OF THE PHASE AND GROUP DELAY CHARACTERISTIC OF A TRANSMISSION NETWORK.
KIDA T
P359-62 OF 7TH ASILOMAR CONF ON CIRCUITS, SYSTEMS AND COMPUTERS NAVAL POSTGRADUATE SCHOOL, UNIV SANTA CLARA, ET AL, PACIFIC GROVE, CALIF, USA, 27-29 NOV 1973, WESTERN PERIODICALS 1974

0130 ACTIVE FILTERS.
KIMEC JJ + TARR LA
GTE AUTOM ELECTR TECH J 15(2): 96-104 (APRIL 1976)

0131 AN AUTOMATIC OVER-RANGE CORRECTING CIRCUIT.
KORBA LW + COUSIN AJ
MED BIOL ENG COMPUT 17(1): 133-4 (JAN 1979)

0132 DESIGNING ACTIVE FILTERS WITH OPERATIONAL AMPLIFIERS.
KUCAROV SI
ELEKTRO PROM-ST PRIBOROSTR NO 3: 84-7 (1977) (IN BULGARIAN)

0133 SENSITIVITY MINIMIZATION IN A RC SELECTIVE AMPLIFIER.
KUNIEDA T + FUKUI A
TRANS INST ELECTRON COMMUN ENG JPN SECT E E60(9): 489 (SEPT 1977)

0134 NOISE PERFORMANCE OF ACTIVE RC-FILTERS.
KUNZLI M + GUGGENBUHL W
NACHRICHTENTECH Z 28(10): 368-73 (OCT 1975)

0135 PHASE AND GROUP DELAY SENSITIVITIES IN MULTIPLE LOOP FEEDBACK BANDPASS ACTIVE FILTERS.
LAKER KR
P1050-5 OF PROC OF THE 1978 IEEE INT SYMP ON CIRCUITS AND SYSTEMS, IEEE, NEW YORK, USA 17-19 MAY 1978, IEEE, 1978

0136 STATISTICAL MULTIPARAMETER SENSITIVITY-A VALUABLE MEASURE FOR CAD.
LAKER KR + GHAUSI MS
P333-6 OF 1975 IEEE INT SYMP ON CIRCUITS AND SYSTEMS, IEEE, NEWTON, MASS, USA, 21-23 APRIL 1975, IEEE 1975

0137 NOISE CHARACTERISTICS OF CASCADED 2ND-ORDER ACTIVE FILTERS.
LEE MR + MECROW S
ELECTRON LETT 12(12): 301-3 (10 JUNE 1976)

0138 CHOICE OF TOLERANCES OF THE PASSIVE COMPONENTS OF RC BAND FILTERS.
LER AM + SHVETSKII BI
OTBOR PEREDACHA INF NO 39: 62-6 (1974) (IN RUSSIAN)

0139 'H' OR 'BOOTSTRAP' LF CIRCUIT FILTER.
LINSLEY HOOD JL
ELECTRON ENGINEERING 48(581): 55-8 (JULY 1976)

0140 EVALUATION OF THE Q-FACTOR OF ACTIVE FILTERS USING INTEGRATORS.
LOVEIKO VN
RADIO ENG ELECTRON PHYS 19(4): 137-9 (APRIL 1974)

0141 OPTIMIZATION OF THE DYNAMIC RANGE AND THE NOISE DISTANCE OF RC-ACTIVE FILTERS BY DYNAMIC PROGRAMMING.
LUEDER E
INT J CIRCUIT THEORY APPL 3(4): 365-70 (DEC 1975)

0142 EFFECT OF FREQUENCY LIMITATIONS OF OPERATIONAL AMPLIFIERS IN ACTIVE RC FILTERS.
LUNDIN VZ
IZV VUZ RADIOELEKTRON 18(8): 126-9 (AUG 1975) (IN RUSSIAN)

0143 A STUDY OF THE EFFECTS OF A COMPREHENSIVE AMPLIFIER MODEL ON ACTIVE FILTER PERFORMANCE.
MACK RJ + FIDLER JK
INT J ELECTRON 40(3): 277-88 (MARCH 1976)

0144 SENSITIVITY AND NOISE COMPARISON OF RC ACTIVE FILTERS.
MARTINELLI G + SALERNO M
P196-214 OF NATO ADVANCED STUDY INSTITUTE ON NETWORK AND SIGNAL THEORY, NATO, IEE, IERE, IEEE, BOURNEMOUTH, HANTS, ENGLAND SEPT 1972, SKWIRZYNSKI, JK, SCANLAN, JO(ED), PETER PEREGRINUS, 1973

0145 ACTIVE FILTERS GET MORE OF THE ACTION.
MATTERA L
ELECTRONICS 45(13): 104-9 (19 JUNE 1972)

0146 FILTER SPECIFICATION AND SELECTION.
MEEK WE
ELECTRON NO 98: 28,30,32 (17 JUNE 1976)

0147 ECONOMICAL LINE FILTERING FOR BUSINESS COMMUNICATION SYSTEMS.
MIKHAEL WB + KEYS LK
IEEE J SOLID-STATE CIRCUITS SC-9(4): 196-9 (AUG 1974)

0148 SENSITIVITY FUNCTIONS OF ACTIVE RC FILTERS.
MILIC L
AUTOMATIKA 17(5-6): 197-9 (1976) (IN CROATIAN)

0149 A TWO-STEP PROCEDURE FOR DESIGNING ACTIVE FILTERS USING FREQUENCY DEPENDENT AMPLIFIERS.
MITRA SK + BHUMIRATANA A + SHANTA PAI BA
P363-7 OF 7TH ASILOMAR CONF ON CIRCUITS, SYSTEMS AND COMPUTERS, NAVAL POSTGRADUATE SCHOOL, UNIV SANTA CLARA, ET AL PACIFIC GROVE, CALIF, USA, 27-29 NOV 1973, WESTERN PERIODICALS, 1974

0150 M.P.U.-CONTROLLED FILTER.
MOORE D
NEW ELECTRON 9(4): 14 (24 FEB 1976)

0151 MORPHOLOGICAL METHODS IN ELECTRICAL ENGINEERING WITH EXAMPLES FROM ACTIVE NETWORK THEORY. II.
MOSCHYTZ GS
BULL ASSOC SUISSE ELECTR 68(8): 401-6 (16 APRIL 1977) (IN GERMAN)

0152 SENSITIVITY PROBLEM IN ACTIVE FILTERS.
MOSCHYTZ GS
SCI ELECTR 21(4): 81-105 (1975)

0153 REDUCING NONIDEAL OP-AMP EFFECTS IN ACTIVE FILTERS BY MINIMIZING THE GAIN-SENSITIVITY PRODUCT (GSP).
MOSCHYTZ GS + HORN P
IEEE TRANS CIRCUITS SYST CAS-24(8): 437-45 (AUG 1977)

0154 GAIN-SENSITIVITY AS A UNIVERSAL MEASURE FOR ACTIVE FILTER POLE VARIATIONS DUE TO NONIDEAL AMPLIFIERS.
MOSCHYTZ GS + HORN P
P122-5 OF PROC OF THE 1976 IEEE INT SYMP ON CIRCUITS AND SYSTEMS, IEEE, MUNICH, GERMANY 27-29 APRIL 1976, IEEE, 1976

0155 DESIGN AND TOLERANCE ANALYSIS OF A LOW FREQUENCY ACTIVE FILTER.
MYOUNG SAM KO + KEUN HO YOO
J KOREAN INST ELECTR ENG 24(3): 316-29 (MAY 1975)

0156 SENSITIVITY ANALYSIS OF FOURTH ORDER FILTERS.
NAIMPALLY SV + MAHAPATRA S
J INST ELECTRON TELECOMMUN ENG 21(12): 664-5 (DEC 1975)

0157 DESIGN AND SOME APPLICATIONS OF EXTENDED BANDWIDTH FINITE GRAIN AMPLIFIERS.
NATARAJAN S + BHATTACHARYYA BB
J FRANKLIN INST 305(6): 321-41 (JUNE 1978)

0158 DESIGN AND SOME APPLICATIONS OF EXTENDED BANDWIDTH FINITE GAIN AMPLIFIERS.
NATARAJAN S + BHATTACHARYYA BB
P700-1 OF PROC OF THE 1978 IEEE INT SYMP ON CIRCUITS AND SYSTEMS, IEEE, NEW YORK, USA 17-19 MAY 1978, IEEE, 1978

0159 OPTIMIZATION OF RC ACTIVE FILTERS FOR EXTENDED BANDWIDTH OPERATION.
NATARAJAN S + BHATTACHARYYA BB
PROC IEEE 66(2): 260-1 (FEB 1978)

0160 A CONTRIBUTION TO THE SENSITIVITY ANALYSIS OF ACTIVE RC NETWORKS.
NEJEPINSKA J + SLAMA J
ACTA POLYTECH III NO 3: 99-108 (1973)

0161 OPTIMIZING THE SYNTHESIS OF SELECTIVE CIRCUITS AND FUNCTIONAL BLOCKS.
NOVAK M
SLABOPROUDY OBZ 37(2): 70-2 (FEB 1976) (IN CZECH)

0162 OPTIMIZATION OF DYNAMIC RATIO BY ACTIVE RC FILTERS.
NOVAK M
P635-9 OF PROC OF THE 1974 IEEE INT SYMP ON CIRCUITS AND SYSTEMS, IEEE, SAN FRANCISCO, CALIF USA, 22-25 APRIL 1974, IEEE, 1974

0163 OPTIMIZATION OF THE DYNAMIC RATIO AND SENSITIVITY BY THE NONCASCADE ACTIVE RC FILTER STRUCTURES.
NOVAK M
P319-24 OF 1974 EUROPEAN CONF ON CIRCUIT THEORY AND DESIGN, IEE, IEEE, INST PHYS, IERE, LONDON, ENGLAND 23-26 JULY 1974, IEE, 1974

0164 OPTIMIZATION OF THE NOISE FACTOR AND DYNAMIC RANGE IN CHAINLIKE ACTIVE RC FILTERS.
NOVAK M
NACHRICHTENTECHNIK 22(10): 342-3 (OCT 1972) (IN GERMAN)

0165 INFLUENCE OF NON-LINEARITY IN ACTIVE FILTERS.
NUYEN WC + REGTIEN PPL
INT J ELECTRON 41(3): 241-7 (SEPT 1976)

0166 A LARGE AND SMALL-CHANGE SENSITIVITY IN BIQUADRATIC ACTIVE FILTERS WITH ONE OP-AMP.
OGRODZKI JT + STYBLINSKI MA
P266-70 OF CIRCUIT THEORY AND DESIGN, LAUSANNE, SWITZERLAND, 4-8 SEPT 1978, MOSCHYTZ, GS, NIERYNCK, J(ED), GEORGI 1978

0167 STABILITY OF OPTIMUM ACTIVE FILTERS.
PERSIANOV AYU
IZV VUZ RADIOELEKTRON 17(9): 104-7 (SEPT 1974) (IN RUSSIAN)

0168 A METHOD OF FREQUENCY CORRECTION OF HIGH Q ACTIVE FILTERS BASED ON EQUIVALENT TRANSFORMATIONS.
PODOL'SKII LZ
RADIO ENG ELECTRON PHYS 22(8): 142-5 (AUG 1977)

0169 A FREQUENCY TRANSFORMATION CHART FOR RC-ACTIVE BAND-PASS FILTERS.
PREMOLI A
ALTA FREQ 41(6): 468-9 (JUNE 1972) (IN ENGLISH)

0170 NOISE MINIMIZATION IN RC-ACTIVE FILTERS USING GENERALIZED IMPEDANCE CONVERTERS.
RAMAKRISHNA K + BRUTON LT
INT J CIRCUIT THEORY APPL 6(2): 135-45 (APRIL 1978)

0171 ERRORS DUE TO CMRR IN A CLASS OF ACTIVE FILTERS.
RANJIT SINGH + VENKATESWARAN S
J INST ELECTRON TELECOMMUN ENG 22(6): 340-2 (JUNE 1976)

0172 A CONTRIBUTION TO THE STUDY OF ACTIVE RC-FILTER QUALITY.
RELJIN B
TEHNIKA 33(2): 263-8 (1978) (IN CROATIAN)

0173 DIMENSIONING OF ACTIVE BAND STOP AND BAND PASS RC FILTERS.
ROHMANN O
ELEKTRONIK 11(4): 22-8 (APRIL 1973) (IN DANISH)

0174 EVALUATION CRITERIA OF RC ACTIVE FILTERS.
ROSZKIEWICZ J + BORYS A + WOJTYNA R
ROZPR ELECTROTECH 23(4): 851-61 (1977) (IN POLISH)

0175 DESIGN AND REALIZATION OF RC-ACTIVE FILTERS.
RUPP W
P340-3 OF PROC OF THE 1976 IEEE INT SYMP ON CIRCUITS AND SYSTEMS, IEEE, MUNICH, GERMANY 27-29 APRIL 1976, IEEE, 1976

0176 RC ACTIVE FILTER WITH ZERO Q-SENSITIVITY.
SAHA SK
PROC INST ELECTR ENG 126(2): 147-8 (FEB 1979)

0177 GET HIGH-Q IN ACTIVE BANDPASS FILTERS WITH A QUADRATURE MODULATION SCHEME.
SANSONE J
ELECTRON DES 26(23): 124-7 (8 NOV 1978)

0178 MINIMIZING SIGNAL DISTORTION IN F.L.F. ACTIVE FILTERS.
SCHAUMANN R + KINGHORN WA + LAKER KR
ELECTRON LETT 12(9): 211-13 (29 APRIL 1976)

0179 INFLUENCE OF CAPACITOR LOSSES ON Q OF ACTIVE FILTERS.
SCHLOTZHAUER K + SINGHAL K
PROC IEEE 61(11): 1662-3 (NOV 1973)

0180 GENERAL FREQUENCY LIMITATIONS OF ACTIVE RC FILTERS USING IC OPERATIONAL AMPLIFIERS.
SCULTETY L
PCT-28/247-56 OF PROC OF THE 5TH COLLOQUIUM ON MICROWAVE COMMUNICATION VOLII CIRCUIT THEORY AND COMPUTER AIDED DESIGN (PREPRINTS), HUNGARIAN ACAD SCI, SCI SOC TELECOMMUNICATION RES INST TELECOMMUNICATION, BUDAPEST HUNGARY, 24-30 JUNE 1974, BOGNAR, G(ED), AKAD KIADO, 1974

0181 SENSITIVITY AND FREQUENCY LIMITATIONS OF BIQUADRATIC ACTIVE FILTERS.
SEDRA AS + ESPINOZA JL
IEEE TRANS CIRCUITS SYST CAS-22(2): 122-30 (FEB 1975)

0182 SENSITIVITY AND FREQUENCY LIMITATIONS OF BIQUADRATIC ACTIVE FILTERS.
SEDRA AS + ESPINOZA JL
P645-50 OF PROC OF THE 1974 IEEE INT SYMP ON CIRCUITS AND SYSTEMS, IEEE, SAN FRANCISCO, CALIF USA, 22-25 APRIL 1974, IEEE, 1974

0183 IMPROVEMENT OF THE SIGNAL-TO-NOISE RATIO OF ACTIVE RC LOW-PASS FILTERS USING THE EQUAL RIPPLE APPROXIMATION.
SEIDEL V
NACHRICHTENTECH ELEKTRON 28(7): 276-8 (1978) (IN GERMAN)

0184 ASPECTS FOR DESIGNING, REALIZING AND COMPARING VARIANTS OF LOW-NOISE AND LOW-DISTORTION RC FILTERS.
SEIDEL V
NACHRICHTENTECH ELEKTRON 27(10): 428-32 (1977) (IN GERMAN)

0185 NOISE CALCULATION OF ACTIVE RC FILTERS BY THE INVERSION PRINCIPLE.
SEIDEL V
NACHRICHTENTECH ELEKTRON 21(1): 18-21 (1975) (IN GERMAN)

0186 EFFECTIVE BAND WIDTHS OF FILTERS.
SENFLUK J
SDELOVACI TECH 24(3): 109 (MARCH 1976) (IN CZECH)

0187 PARTIAL CORRECTION OF POLE-PERTURBATIONS IN SINGLE-POLE-COMPENSATED SINGLE-AMPLIFIER INDUCTORLESS FILTERS.
SHAW GW
INT J ELECTRON 46(4): 427-34 (APRIL 1979)

0188 OPTIMUM VARIABILITY DESIGN AND COMPARATIVE EVALUATION OF THIN-FILM RC ACTIVE FILTERS.
SHENOI BA
IEEE TRANS CIRCUITS SYST CAS-21(2): 263-7 (MARCH 1974)

0189 A NEW STATISTICAL VARIABILITY CRITERION FOR OPTIMAL DESIGN AND COMPARATIVE EVALUATION OF RC-ACTIVE FILTERS.
SHENOI BA
P122-5 OF INT SYMP ON CIRCUIT THEORY 1973, IEEE, TORONTO, CANADA, 9-11 APRIL 1973, IEEE, 1973

0190 WHICH FILTERS.
SHEPARD I
ELECTRON NO 64: 2731-2,47 (21 NOV 1974)

0191 DESIGN TABLES FOR RC ACTIVE CAUER FILTERS.
SLUYTER RJ
J APPL SCI ENG SECT A 3(1): 1-14 (APRIL 1978)

0192 OPTIMIZATION OF DYNAMIC RANGE IN CASCADE ACTIVE FILTERS.
SNELGROVE WM + SEDRA AS
P151-5 OF PROC OF THE 1978 IEEE INT SYMP ON CIRCUITS AND SYSTEMS, IEEE, NEW YORK, USA 17-19 MAY 1978, IEEE, 1978

0193 A SIMPLE FILTER DESIGN PROCEDURE USING COMPENSATED OPERATIONAL AMPLIFIERS.
SODERSTRAND MA + SZCZUPAK J
INT J ELECTRON 46(1): 19-32 (JAN 1979)

0194 ACTIVE RC FILTER DESIGN USING COMPENSATED OPERATIONAL AMPLIFIERS.
SODERSTRAND MA + SZCZUPAK J
P73-81 OF 17TH MIDWEST SYMP ON CIRCUITS AND SYSTEMS, LAWRENCE, KAN, USA, 16-17 SEPT 1974, WESTERN PERIODICALS 1974

0195 OP-AMP. INTEGRATORS WITH INFINITE Q-FACTOR.
SOLIMAN AM + ISMAIL M
FREQUENZ 32(11): 331-4 (NOV 1978)

0196 A UNIVERSAL COMPENSATION SCHEME FOR ACTIVE FILTERS.
SRINIVASAN S
INT J ELECTRON 42(2): 141-51 (FEB 1977)

0197 TOLERANCE MINDED DESIGN OF ACTIVE RC BAND PASS FILTER.
SRIVASTAVA MC + DUTTA ROY SC + SATSANGI PS
J INST TELECOMMUN ENG 18(2): 86-8 (FEB 1972)

0198 RC ACTIVE FILTERS FOR MICROELECTRONIC REALIZATION.
STEPHENSON FW
MICROELECTRONICS 6(3): 7-10 (MARCH 1975)

0199 TRANSITIONAL MURROMAF-MURROER FILTERS.
STOJANOVIC VS + PAVLOVIC VD
ELECTRON LETT 15(12): 356-7 (7 JUNE 1979)

0200 TIME-DOMAIN SYNTHESIS OF TRANSFER FUNCTIONS FOR LOW SENSITIVITY ACTIVE FILTERS.
STOJANOVICH VS + RAKOVICH BD
INT J ELECTRON 45(4): 385-400 (OCT 1978)

0201 DERIVATION OF PROBABILITY CHARACTERISTICS OF THE PROCESS AT THE OUTPUT OF A SECOND-ORDER FILTER.
TARASOV GP + POKROVSKIY AV
RADIO ENG ELECTRON PHYS 23(4): 59 (APRIL 1978)

0202 APPLICATION OF SECOND ORDER SENSITIVITY TO ANALYSIS OF PARAMETER CHANGES IN ACTIVE FILTERS.
TATARA I
ROZPR ELECTROTECH 25(1): 129-44 (1979) (IN POLISH)

0203 NOISE-REDUCING FILTER SWITCHES TIME COSNTANTS.
THOMAS MV
ELECTRONICS 50(8): 109 (14 APRIL 1977)

0204 PREDISTORTION OF ACTIVE FILTERS.
THOMAS RE + RAY DJ
PI4/9PP OF PROC OF THE 15TH MIDWEST SYMP ON CIRCUIT THEORY VOLI, UNIV MISSOURI-ROLLA, ROLLA, MO, USA 4-5 MAY 1972, UNIV MISSOURI-ROLLA, 1972

0205 HIGH FREQUENCY NONLINEARITIES IN ACTIVE FILTERS.
THOMAS RE + RAY DJ
P199-201 OF PROC OF THE NATIONAL ELECTRONICS CONF VOL XXVII, NAT ELECTRONICS CONF INC ET AL CHICAGO, ILL, USA, 9-11 OCT 1972, NAT ELECTRONICS CONF INC, 1972

0206 QUASI-LINEAR RANGE OF OPERATION OF A GYRATOR-CAPACITOR FILTER.
TOMASZEWSKI Z + BIALKO M
P235-7 OF 1975 IEEE INT SYMP ON CIRCUITS AND SYSTEMS, IEEE, NEWTON, MASS, USA, 21-23 APRIL 1975, IEEE 1975

0207 COMPUTER AS A TOOL IN ACTIVE FILTER DESIGN.
TOW J
P489 OF PROC OF THE 1974 IEEE INT SYMP ON CIRCUITS AND SYSTEMS, IEEE, SAN FRANCISCO, CALIF, USA 22-25 APRIL 1974, IEEE, 1974

0208 NOISE IN POSITIVE GAIN SALLEN-KEY RC-ACTIVE FILTERS.
TROFIMENKOFF FN + SMALLWOOD RE + BRUTON LT
P325-30 OF 1974 EUROPEAN CONF ON CIRCUIT THEORY AND DESIGN, IEE, IEEE, INST PHYS, IERE, LONDON, ENGLAND 23-26 JULY 1974, IEE, 1974

0209 NOISE PERFORMANCE OF RC-ACTIVE QUADRATIC FILTER SECTIONS.
TROFIMENKOFF FN + TRELEAVEN DH + BRUTON LT
IEEE TRANS CIRCUIT THEORY CT-20(5): 524-32 (SEPT 1973)

0210 NOISE ANALYSIS OF A FOURTH ORDER MAXIMALLY-FLAT BANDPASS FILTER.
TROFIMENKOFF FN + BRUTON LT + SMALLWOOD RE
P223-7 OF PROC OF THE NATIONAL ELECTRONICS CONF VOL XXVII, NAT ELECTRONICS CONF INC ET AL CHICAGO, ILL, USA, 9-11 OCT 1972, NAT ELECTRONICS CONF INC, 1972

0211 HYBRID ACTIVE RC FILTERS.
UDVARHELYI G
HIRADASTECHNIKA 27(6): 161-74 (JUNE 1976)

0212 SENSITIVITY BOUNDS AND EXPLICIT POLYNOMIAL DECOMPOSITION FOR BIQUADRATIC SINGLE-AMPLIFIER ACTIVE FILTERS.
VAN BOKHOVEN WMG
P118-21 OF PROC OF THE 1976 IEEE INT SYMP ON CIRCUITS AND SYSTEMS, IEEE, MUNICH, GERMANY 27-29 APRIL 1976, IEEE, 1976

0213 NEED AN ACTIVE FILTER TRY THESE DESIGN AIDS.
VAN DALSEN D
EDN 23(20): 105-10 (5 NOV 1978)

0214 POLE SENSITIVITY AND SELECTION OF CIRCUIT SOLUTIONS OF LOW-Q-FACTOR CHEBYSHEV ACTIVE FILTERS.
VLADIMIROV VL + MALYUTA BD
TELECOMMUN RADIO ENG PART 2 31(1): 88-92 (JAN 1977)

0215 GET AROUND OP AMP LIMITATIONS IN ACTIVE FILTER DESIGN.
VOLPE GT
ELECTRON DES 24(19): 104-7 (13 SEPT 1976)

0216 FEEDBACK-CONCEPTUAL OR PHYSICAL.
WALDHAUER FD
P8-12 OF INT SYMP ON CIRCUIT THEORY 1973, IEEE, TORONTO, CANADA, 9-11 APRIL 1973, IEEE, 1973

0217 NOISE ANALYSIS OF ACTIVE RC-FILTERS.
WEINRICHTER H + NOSSEK JA
P344-7 OF PROC OF THE 1976 IEEE INT SYMP ON CIRCUITS AND SYSTEMS, IEEE, MUNICH, GERMANY 27-29 APRIL 1976, IEEE, 1976

0218 NOISE PERFORMANCE OF INTEGRATED ACTIVE RC FILTERS.
WEINRICHTER H
RADIO ELEKTRON SCHAU 52(5): 20-2 (1976) (IN GERMAN)

0219 A USEFUL SENSITIVITY MEASURE FOR SECOND-ORDER RC ACTIVE FILTER DESIGN.
WEYTEN L
IEEE TRANS CIRCUITS SYST CAS-23(8): 506-8 (AUG 1976)

0220 VARIATION OF THE POLES OF A SECOND ORDER RC ACTIVE FILTER DUE TO A FINITE TAN DELTA OF THE FILTER CAPACITORS.
WEYTEN L
ARCH ELEKTRON UEBERTRAGUNGSTECH 28(3): 140-1 (MARCH 1974)

0221 PRECISION NETWORKS FOR COMMUNICATIONS SYSTEMS.
WITT FJ
P159-63 OF IEEE PROC OF THE INT SYMP ON CIRCUIT THEORY, IEEE, NORTH HOLLYWOOD, CALIF, USA 18-21 APR 1972, IEEE, 1972

0222 ELECTRONIC CIRCUITS FOR AUTOMATION. VI. FILTERS.
WOLF R + FIEDLER K
MSR (MESS STEUERN 21(6): 345-6 (JUNE 1978) (IN GERMAN)

0223 A HIGH-FREQUENCY ACTIVE CORRECTOR FILTER.
WOLFF A
ELECTRON APPL IND NO 262S: 39-40 (15 JAN 1979) (IN FRENCH)

0224 USE SLEW-RATE FILTERING TO DISCRIMINATE AGAINST NOISE SPIKES.
WRIGHT MJ
ELECTRON DES 24(19): 110-12 (13 SEPT 1976)

0225 EXTENDING THE FREQUENCY RANGE OF SOME RC ACTIVE FILTERS.
WUPPER H
P324-7 OF 1977 IEEE INT SYMP ON CIRCUITS AND SYSTEMS PROC, IEEE, PHOENIX, ARIZ, USA 25-27 APRIL 1975, IEEE, 1977

0226 SOME ASPECTS OF BAND-PASS FILTER DESIGN PERTAINING TO AMPLIFIER IMPERFECTIONS AND MASS PRODUCTION.
ZEHEB E
INT J CIRCUIT THEORY APPL 3(3): 249-60 (SEPT 1975)

0227 PRACTICAL CONSIDERATIONS IN THE DESIGN OF ACTIVE FILTERS WITH PARTICULAR REFERENCE TO MICROPOWER REQUIREMENTS.
ZEHEB E + MERON P
INT J ELECTRON 36(6): 811-24 (JUNE 1974)

0228 ANALYSIS AND SYNTHESIS OF SECOND ORDER ACTIVE FILTERS.
ZOGG U
ELEKTRONIKER 17(7): 10-16 (1978) (IN GERMAN)

0229 NOISE PERFORMANCE OF ACTIVE RC FILTERS REALIZED BY THE CASCADE APPROACH.
ZURADA J
INT J ELECTRON 42(4): 409-12 (APRIL 1977)

0230 NOISE AND DYNAMICS RANGE OF ACTIVE RC FILTERS USING OPERATIONAL AMPLIFIER.
ZURADA J
ROZPR ELECTROTECH 20(4): 691-712 (1974) (IN POLISH)

0231 SENSITIVITY ANALYSIS AND DESIGN METHOD FOR ACTIVE FILTERS USING VCVS/OP-AMPS WITH FINITE GAIN-BANDWIDTH PRODUCT.
ZURADA J
P115-20 OF 1974 EUROPEAN CONF ON CIRCUIT THEORY AND DESIGN, IEE, IEEE, INST PHYS, IERE, LONDON, ENGLAND 23-26 JULY 1974, IEE, 1974

0232 NOISE AND DYNAMIC RANGE OF ACTIVE FILTERS WITH OPERATIONAL AMPLIFIERS.
ZURADA J + BIALKO M
IEEE TRANS CIRCUITS SYST CAS-22(10): 805-9 (OCT 1975)

0233 NOISE AND DYNAMIC RANGE OF ACTIVE FILTERS WITH OPERATIONAL AMPLIFIERS.
ZURADA J + BIALKO M
P677-81 OF PROC OF THE 1974 IEEE INT SYMP ON CIRCUITS AND SYSTEMS, IEEE, SAN FRANCISCO, CALIF USA, 22-25 APRIL 1974, IEEE, 1974

SECOND ORDER FILTER SECTIONS

0234 AN ACTIVE FILTER FOR 20 HZ TO 20 KHZ.
REV ESP ELECTRON 25(184): 48 (JULY 1978) (IN SPANISH)

0235 AMPLIFIER ACTIVE FILTER.
ELECTRON IND 1(1): 39 (SEPT 1975)

0236 AN OPTIMAL DESIGN OF RC ACTIVE FILTERS USING GROUNDED CAPACITORS.
ABOUGABAL MS + BHATTACHARYYA BB + SWAMY MNS
INT J CIRCUIT THEORY APPL 6(1): 31-40 (JAN 1978)

0237 EASILY INTEGRABLE VERY-LOW-SENSITIVITY ACTIVE RC FILTER.
ABOUGABAL MS + BHATTACHARYYA BB
ELECTRON LETT 8(12): 303-04 (15 JUNE 1972)

0238 SENSITIVITY OF BAND-PASS ACTIVE FILTERS USING ONE OR TWO AMPLIFIERS.
AE T + NAKANISHI K
INT J ELECTRON 41(6): 589-606 (DEC 1976)

0239 ACTIVE FILTERS USING TWO AMPLIFIERS.
AE T + TERUHIRA M + YOSHIDA N
BULL FAC ENG HIROSHIMA UNIV 27(1): 53-61 (OCT 1978) (IN JAPANESE)

0240 A VERSATILE ACTIVE RC BUILDING BLOCK WITH INHERENT COMPENSATION FOR THE FINITE BANDWIDTH OF THE AMPLIFIER.
AKERBERG D + MOSSBERG K
IEEE TRANS CIRCUITS SYST CAS-21(1): 75-8 (JAN 1974)

0241 AN ACTIVE RC BRIDGED-T NOTCH FILTER.
ANANDA MOHAN PV
PROC IEEE 65(8): 1208-9 (AUG 1977)

0242 ACTIVE RC REALISATION OF A 3RD-ORDER BANDPASS BUTTERWORTH CHARACTERISTIC USING EQUAL-VALUED PASSIVE ELEMENTS.
ANDAY F
ELECTRON LETT 11(10): 228 (15 MAY 1975)

0243 COMMENTS ON - FLEXIBLE ACTIVE FILTER DESIGN USING THREE OPERATIONAL AMPLIFIERS.
ANDAY F
PROC IEEE 63(3): 542 (MARCH 1975)

0244 COMMENT ON ACTIVE RC REALISATION OF A 3RD-ORDER LOWPASS BUTTERWORTH CHARACTERISTIC.
ANDAY F
ELECTRON LETT 212 (17 MAY 1973)

0245 ACTIVE FILTER REALIZATION OF BIQUADRATIC TRANSFER FUNCTION.
ARBEL A + ROGOVSKI Z
INT J ELECTRON 36(4): 527-35 (APRIL 1974)

0246 A UNIVERSAL SECOND DEGREE TWO PORT NETWORK USING A SINGLE OPERATIONAL AMPLIFIER.
ATIYA FS + SOLIMAN AM + SAADAWI TN
ALTA FREQ 47(7): 574-9 (JULY 1978)

0247 ACTIVE RC NONMINIMUM PHASE NETWORK USING THE DVCCS/DVCVS.
ATIYA FS + SOLIMAN AM + SAADAWI TN
PROC IEEE 65(11): 1606-7 (NOV 1977)

0248 ACTIVE RC BANDPASS AND LOWPASS FILTERS USING THE DVCCS/DVCVS.
ATIYA FS + SOLIMAN AM + SAADAWI TN
ELECTRON LETT 12(14): 360-1 (8 JULY 1976)

0249 LOW COST ACTIVE LOW PASS AND BAND PASS NETWORK.
ATTIKIOUZEL J + LINGGARD R
INT J CIRCUIT THEORY APPL 2(4): 397-400 (DEC 1974)

0250 RING OF TWO ACTIVE NETWORK.
ATTIKIOUZEL J + LINGGARD R
ELECTRON NO 65: 41-2,44 (5 DEC 1974)

0251 REALISATION OF HIGHLY SELECTIVE FILTERS WITH RC TYPE CIRCUITS AND OPERATIONAL AMPLIFIERS.
BADEA M + PUSCASU S
ELECTROTEH ELECTRON AUTOM AUTOM ELECTRON 19(4): 174-8 (DEC 1975) (IN RUMANIAN)

0252 MORE SELECTIVITY FOR CW SIGNALS (RC FILTER).
BAHR D + WEESE K
ELEKTRON INT NO 2: 47-8 (1978) (IN GERMAN)

0253 REALIZATION OF ACTIVE BANDPASS FILTERS.
BAHRI KK + SETTY CSGK
INDIAN J PURE APPL PHYS 13(8): 533-6 (AUG 1975)

0254 ACTIVE FILTER HAS STABLE NOTCH, AND RESPONSE CAN BE REGULATED.
BAINTER JR
ELECTRONICS 48(20): 115-17 (2 OCT 1975)

0255 TWO-OPERATION-MODE FILTERING DEVICE.
BALLATORE D
IBM TECH DISCLOSURE BULL 20(7): 2718-19 (DEC 1977)

0256 OPTIMALLY GAIN INSENSITIVE ACTIVE RC-FILTERS OF SECOND ORDER WITH A SINGLE OPERATIONAL AMPLIFIER.
BERGMANN K
ARCH ELEKTRON UEBERTRAGUNGSTECH 28(7-8): 288-96 (JULY-AUG 1974) (IN GERMAN)

0257 SECOND ORDER ACTIVE RC FILTERS WITH TWO OPERATIONAL AMPLIFIERS AND VERY LOW SENSITIVITY.
BERGMANN K + EYSEN S
FREQUENZ 30(2): 32-7 (FEB 1976) (IN GERMAN)

0258 CLASSICAL AND Q-SENSITIVITY OF SECOND ORDER ACTIVE RC FILTERS.
BERGMANN K + EYSEN S
NACHRICHTENTECH Z 28(5): 170-1 (MAY 1975)

0259 OPTIMALLY GAIN INSENSITIVE ACTIVE RC-FILTERS OF SECOND ORDER WITH A FREQUENCY DEPENDENT OPERATIONAL AMPLIFIER.
BERGMANN K + STRASSBURGER-NORING E
ARCH ELEKTRON UEBERTRAGUNGSTECH 30(4): 172-6 (APRIL 1976) (IN GERMAN)

0260 A NOTE ON BACH'S LOW-PASS ACTIVE RC FILTER.
BHATTACHARYYA BB + DAS SK + SWAMY MNS
PROC IEEE 64(4): 562-4 (APRIL 1976)

0261 DESIGN OF RC-ACTIVE NETWORKS BY USING GENERALIZED-IMMITTANCE CONVERTERS.
BHATTACHARYYA BB + MIKHAEL WB + ANTONIOU A
P290-3 OF INT SYMP ON CIRCUIT THEORY 1973, IEEE, TORONTO, CANADA, 9-11 APRIL 1973, IEEE, 1973

0262 AN INTEGRABLE ACTIVE RC FILTER.
BHATTACHARYYA BB + ABOUGABAL MS + SWAMY MNS
P256-9 OF INT SYMP ON CIRCUIT THEORY 1973, IEEE, TORONTO, CANADA, 9-11 APRIL 1973, IEEE, 1973

0263 ACTIVE RC FILTERS HAVING ZERO Q-SENSITIVITY.
BIALKO M
P196-200 OF PROC OF THE 1974 IEEE INT SYMP ON CIRCUITS AND SYSTEMS, IEEE, SAN FRANCISCO, CALIF USA, 22-25 APRIL 1974, IEEE, 1974

0264 SYSTEMS OF ACTIVE RC FILTERS WITH LITTLE SENSIVITY ON ELEMENTS VARIATIONS.
BIALKO M + MULAWKA J
ELEKTRONIKA 14(8): 348-51 (1973) (IN POLISH)

0265 ZERO Q-SENSITIVITY ACTIVE RC CIRCUIT SYNTHESIS.
BIALKO M + SIENKO W
IEEE TRANS CIRCUITS SYST CAS-21(2): 239-44 (MARCH 1974)

0266 SOME INDUCTORLESS FILTERS.
BLOM D + HANNEMAN HW + VOORMAN JO
PHILIPS TECH REV 33(10): 294-308 (1973)

0267 SINGLE AMPLIFIER FUNCTIONALLY TUNABLE LOW PASS NOTCH FILTER.
BOCTOR SA
P265-9 OF 1975 IEEE INT SYMP ON CIRCUITS AND SYSTEMS, IEEE, NEWTON, MASS, USA, 21-23 APRIL 1975, IEEE 1975

0268 SINGLE AMPLIFIER FUNCTIONALLY TUNABLE LOW-PASS-NOTCH FILTER.
BOCTOR SA
IEEE TRANS CIRCUITS SYST CAS-22(11): 875-81 (NOV 1975)

0269 A NOVEL SECOND-ORDER CANONICAL RC-ACTIVE REALIZATION OF HIGH-PASS NOTCH FILTER.
BOCTOR SA
IEEE TRANS CIRCUITS SYST CAS-22(5): 397-404 (MAY 1975)

0270 DESIGN OF A THIRD ORDER SINGLE AMPLIFIER FILTER.
BOCTOR SA
P666-71 OF PROC OF THE 1974 IEEE INT SYMP ON CIRCUITS AND SYSTEMS, IEEE, SAN FRANCISCO, CALIF USA, 22-25 APRIL 1974, IEEE, 1974

0271 A NOVEL SECOND ORDER CANONICAL RC-ACTIVE REALIZATION OF HIGH PASS NOTCH FILTER.
BOCTOR SA
P432-7 OF PROC OF THE 1974 IEEE INT SYMP ON CIRCUITS AND SYSTEMS, IEEE, SAN FRANCISCO, CALIF USA, 22-25 APRIL 1974, IEEE, 1974

0272 SYNTHESIS OF RC BANDPASS FILTERS.
BOCTOR SA + SEDRA AS
IEEE TRANS CIRCUIT THEORY CT19(4): 374-6 (JULY 1972)

0273 HIGH-Q BAND-PASS SECTION OF AN ACTIVE RC FILTER.
BODNAR LV + ROZHANKOVSKII RV
OTBOR PEREDACHA INF NO 41: 55-7 (1974) (IN RUSSIAN)

0274 REALIZING THE MAXIMUM Q IN AN ACTIVE RC FILTER WITH ONE AMPLIFIER.
BODNAR LV + ROZHANKOVSKII RV
OTBOR PEREDACHA INF NO 37: 51-7 (1973) (IN RUSSIAN)

0275 5-KHZ BANDSTOP WITH ACTIVE TWIN-T.
BOSCH G
ELEKTRON INT NO 7: 233 (1976) (IN GERMAN)

0276 GENERALIZED FEEDBACK ACTIVE FILTERS.
BOWRON P + STEPHENSON FW
P321-2 OF PROC OF SOUTHEASTCON 1979, IEEE, ROANOKE, VA, USA, 1-4 APRIL 1979, IEEE, 1979

0277 FILTER EXHIBITS NO PHASE SHIFT WITH THE FUNDAMENTAL.
BOYD V
ELECTRON ENGINEERING 49(588): 23 (FEB 1977)

0278 LOW SENSITIVE THIRD ORDER ACTIVE FILTER USING THE DVCCS/DVCVS SOURCES.
BRALKO M + ZURADA J
ARCH ELEKTROTECH 21(4): 857-69 (1972) (IN POLISH)

0279 ACTIVE RESONATORS SAVE STEPS IN DESIGNING ACTIVE FILTERS.
BRANDT R
ELECTRONICS 45(9): 106-10 (24 APRIL 1972)

0280 FAST SYNTHESIS OF MODULAR ACTIVE FILTERS.
BRAUN M
MITT AGEN NO 26: 35-8 (NOV 1978) (IN FRENCH)

0281 'THE BIQUAD' AND CURRENT DIFFERENCING OPERATIONAL AMPLIFIERS.
BRODIE JH
MONITOR 37(10): 296-300 (OCT 1976)

0282 A NOTCH FILTER EMPLOYING CURRENT DIFFERENCING OPERATIONAL AMPLIFIER.
BRODIE JH
INT J ELECTRON 41(5): 501-8 (NOV 1976)

0283 A LOW-PASS BIQUAD DERIVED FILTER REALIZATION.
BRODIE JH
IEEE J SOLID-STATE CIRCUITS SC-11(4): 552-5 (AUG 1976)

0284 NEW MULTIFUNCTION BIQUADRATIC FILTER CIRCUIT WITH INHERENTLY STABLE Q-FACTOR.
BROWN L + SEDRA AS
ELECTRON LETT 13(24): 719-21 (24 NOV 1977)

0285 GAIN-TUNED ACTIVE FILTERS WITH CONSTANT PERCENT BANDWIDTH.
BRUMINHENT S + SU KL
IEEE TRANS CIRCUITS SYST CAS-22(7): 587-94 (JULY 1975)

0286 GAIN-TUNED ACTIVE FILTERS.
BRUMINHENT S + SU KL
P443-7 OF PROC OF THE 1974 IEEE INT SYMP ON CIRCUITS AND SYSTEMS, IEEE, SAN FRANCISCO, CALIF USA, 22-25 APRIL 1974, IEEE, 1974

0287 SENSITIVITY COMPARISON OF HIGH-Q SECOND-ORDER ACTIVE FILTER SYNTHESIS TECHNIQUES.
BRUTON LT
IEEE TRANS CIRCUITS SYST CAS-22(1): 32-8 (JAN 1975)

0288 SIMPLE BANDPASS FILTERS.
BURHANS RW
J AUDIO ENG SOC 21(4): 275-7 (MAY 1973)

0289 FLEXIBLE ACTIVE FILTER DESIGN USING THREE-OPERATIONAL AMPLIFIERS.
CAHIT R + CAHIT I
PROC IEEE 62(6): 851-2 (JUNE 1974)

0290 A HIGH-Q ACTIVE FILTER WITH IMPROVED FREQUENCY RESPONSE.
CHAPOVSKII MZ + LOVEIKO VN + PODOL'SKII LZ
IZV VUZ RADIOELEKTRON 18(9): 105-7 (SEPT 1975) (IN RUSSIAN)

0291 UTILITY OF THE ALL-PASS FILTER.
COMER DJ
IEEE TRANS INSTRUM MEAS IM-28(2): 164-7 (JUNE 1979)

0292 COMMENTS ON 'A LOW-PASS BIQUAD DERIVED FILTER REALIZATION'.
CROSKEY C + BRODIE JH
IEEE J SOLID-STATE CIRCUITS SC-12(3): 329-30 (JUNE 1977)

0293 A CLASS OF HIGH-Q HIGH-FREQUENCY TWO AMPLIFIER ACTIVE RC NETWORKS.
DALPADADO RNG
IEEE TRANS CIRCUITS SYST CAS-25(12): 1099-101 (DEC 1978)

0294 A VOLTAGE-DRIVEN RESONANT FILTER OF 2ND ORDER.
DAMM E
ELEKTRONIK 24(11): 91 (NOV 1975) (IN GERMAN)

0295 ACTIVE IMPEDANCE MATCHING NETWORKS FOR REPEATERS.
DARYANANI G
P40-3 OF 1975 IEEE INT SYMP ON CIRCUITS AND SYSTEMS, IEEE, NEWTON, MASS, USA, 21-23 APRIL 1975, IEEE 1975

0296 ACTIVE FILTERS-WHAT'S AVAILABLE IN THE INDUSTRY.
DARYANANI G
P398-401 OF INT SYMP ON CIRCUIT THEORY 1973, IEEE, TORONTO, CANADA, 9-11 APRIL 1973, IEEE, 1973

0297 VARIABLE BANDWIDTH NOTCH FILTER WITH FLOATING NIC.
DAS AK + GANGULY G
INT J ELECTRON 37(4): 573-6 (OCT 1974)

0298 USE OF AN OTA AND AN OPERATIONAL AMPLIFIER IN ACTIVE RC FILTER SYNTHESIS.
DELIYANNIS T
P140-4 OF CIRCUIT THEORY AND DESIGN, LAUSANNE, SWITZERLAND, 4-8 SEPT 1978, MOSCHYTZ, GS, NIERYNCK, J(ED), GEORGI 1978

0299 SELECTIVITY ENHANCEMENT OF CERTAIN LOW-SENSITIVITY RC ACTIVE NETWORKS.
DELIYANNIS T + BEDRI Y
INT J ELECTRON 34(4): 513-26 (APRIL 1973)

0300 AN APPROACH FOR GENERATION OF SECOND ORDER RC-ACTIVE FILTERS.
DUNN WR
P292-5 OF PROC OF THE FIFTH ASILOMAR CONF ON CIRCUITS AND SYSTEMS UNIV SANTA CLARA, NAVAL POSTGRADUATE SCHOOL, ETAL, PACIFIC GROVE, CALIF, USA, 8-10 NOV 1971, PARKER, SR(ED) WESTERN PERIODICALS CO, 1972

0301 ANALYTICAL REALISATION OF EVC-ARC NETWORKS.
DUNNING-DAVIES J + STEPHENSON FW
ELECTRON LETT 10(7): 103-4 (4 APRIL 1974)

0302 A VERSATILE SECOND ORDER ACTIVE FILTER.
DUTTA ROY SC
ELECTRON ENGINEERING 46(552): 64-6,68 (FEB 1974)

0303 SECOND ORDER ACTIVE RC BAND-PASS FILTERS USING OPERATIONAL AMPLIFIER IN LOW GAIN CONNECTION.
DUTTA ROY SC + AHMED MT
J INST ELECTRON TELECOMMUN ENG 21(5): 248-52 (MAY 1975)

0304 ACTIVE RC REALISATION OF A 3RD-ORDER LOWPASS BUTTERWORTH CHARACTERISTIC.
DUTTA ROY SC + MALIK KK
ELECTRON LETT 8(26): 630-1 (28 DEC 1972)

0305 ACTIVE NOTCH FILTERS FOR ELIMINATING NOISE.
ELLERN F + HAZONY D
ELECTRON ENG 44 NO, : 53565-7 (SEPT 1972)

0306 DIGITALLY CONTROLLED ACTIVE FILTER MODULES.
ENGSTROM K + HEDEVARN A + MOSSBERG K
P523-7 OF CIRCUIT THEORY AND DESIGN, LAUSANNE, SWITZERLAND, 4-8 SEPT 1978, MOSCHYTZ, GS, NIERYNCK, J(ED), GEORGI 1978

0307 USE OF INTEGRATORS IN ACTIVE FILTERS.
ESCUDER AG
MUNDO ELECTRON NO 52: 77-82 (MAY 1976) (IN SPANISH)

0308 A GROUP OF SECOND ORDER ACTIVE RC FILTERS WITH A SINGLE OPERATIONAL AMPLIFIER AND MINIMAL SENSITIVITY.
EYSEN ST
NACHRICHTENTECH Z 27(5): 171-5 (MAY 1974)

0309 ACTIVE FILTERS INCORPORATING THE VOLTAGE-TO-CURRENT TRANSACTOR.
FIDLER JK + MACK RJ + NOAKES PD
MICROELECTRONICS 8 NO : 319-22 (MARCH 1977)

0310 MORE ACTIVE RC FILTERS USING THE DVCCS/DVCVS.
FILANOVSKY IM + STROMSMOE KA
ELECTRON LETT 15(15): 466-7 (19 JULY 1979)

0311 DESIGN FORMULAS FOR BIQUAD ACTIVE FILTERS USING THREE OPERATIONAL AMPLIFIERS.
FLEISCHER PE + TOW J
PROC IEEE 61(5): 662-3 (MAY 1973)

0312 ELLIPTIC-FUNCTION FILTERS IN ACTIVE CASCADING TECHNIQUE.
FLIEGE N
NACHRICHTENTECH Z 28(4): 128-32 (APRIL 1975)

0313 A NEW CLASS OF SECOND-ORDER RC-ACTIVE FILTERS WITH TWO OPERATIONAL AMPLIFIERS.
FLIEGE N
NACHRICHTENTECH Z 26(4): 279-82 (APRIL 1973) (IN ENGLISH)

0314 SALLEN AND KEY-TYPE RC-ACTIVE FILTERS WITH COMPLEX TRANSMISSION ZEROS.
FLIEGE N
INT J CIRCUIT THEORY APPL 1(1): 99-103 (MARCH 1973)

0315 RC ACTIVE FILTERS WITH PRESCRIBED COMPLEX POLE SENSITIVITY USING A DIFFERENCE DECOMPOSITION OF THE DENOMINATOR POLYNOMIAL.
FLIEGE N
NACHRICHTENTECH Z 25(11): 535-6 (NOV 1972) (IN ENGLISH)

0316 USE TRANSCONDUCTANCE AMPLIFIERS TO MAKE PROGRAMMABLE ACTIVE FILTERS.
FRANCO S
ELECTRON DES 24(19): 98-101 (13 SEPT 1976)

0317 STAR - AN ACTIVE BIQUADRATIC FILTER SECTION.
FRIEND JJ + HARRIS CA + HILBERMAN D
IEEE TRANS CIRCUITS SYST CAS-22(2): 115-21 (FEB 1975)

0318 STAR - AN ACTIVE BIQUADRATIC FILTER SECTION.
FRIEND JJ + HARRIS CA + HILBERMAN D
P640-4 OF PROC OF THE 1974 IEEE INT SYMP ON CIRCUITS AND SYSTEMS, IEEE, SAN FRANCISCO, CALIF USA, 22-25 APRIL 1974, IEEE, 1974

0319 POLE ZERO POSITION DATA AND THE COEFFICIENTS OF TRANSFER FACTORS WITH MINIMAL Q FACTOR POLES (C(M)-APPROXIMATION).
FRITZSCHE G + JENTSCHEL H-J
NACHRICHTENTECH ELEKTRON 27(5): 208-10 (1977) (IN GERMAN)

0320 DESIGNING AND REALISING ACTIVE RC FILTERS WITH SIMPLE POSITIVE FEEDBACK.
FROMMHOLD W
RADIO FERNSEHEN ELEKTRON 23(24): 793-4,799-80 (1974) (IN GERMAN)

0321 TWO TYPES OF INSENSITIVE BIQUADRATIC ACTIVE RC FILTERS.
FURUKAWA K
TRANS INST ELECTRON COMMUN ENG JPN SECT E E60(11): 629-30 (NOV 1977)

0322 FEEDFORWARD IN SYNTHESISED LOSSY INDUCTOR REJECTION FILTERS YIELDS RC ALL PASS WITH COMPLEX POLES.
GANGULY US
P104-7 OF 17TH MIDWEST SYMP ON CIRCUITS AND SYSTEMS, LAWRENCE, KAN, USA, 16-17 SEPT 1974, WESTERN PERIODICALS 1974

0323 A UNIVERSAL ACTIVE FILTER MODULE.
GAUTHIER N
TOUTE ELECTRON NO 410: 73-8 (MAY 1976) (IN FRENCH)

0324 ACTIVE FILTERS.
GAZIN J-F
TOUTE ELECTRON NO 376: 77-81 (APRIL 1973) (IN FRENCH)

0325 ACTIVE FILTERS.
GAZIN J-F
TOUTE ELECTRON NO 373: 41-6 (JAN 1973) (IN FRENCH)

0326 ACTIVE FILTERS.
GAZIN J-F
TOUTE ELECTRON NO 369: 53-9 (AUG-SEPT 1972) (IN FRENCH)

0327 EXACT SYNTHESIS WITH REAL AMPLIFIERS.
GEFFE PR
IEEE TRANS CIRCUITS SYST CAS-21(3): 369-76 (MAY 1974)

0328 APPLICATIONS OF CURRENT CONVEYOR.
GOPAL K
J INST ENG (INDIA) ELECTRON TELECOMMUN ENG DIV 56(ET-1): 12-13 (AUG 1975)

0329 ACTIVE RC REALIZATION OF A THIRD-ORDER LOW-PASS BUTTERWORTH CHARACTERISTIC USING EQUAL-VALUED PASSIVE ELEMENTS.
GOPAL K
PROC IEEE 62(1): 137 (JAN 1974)

0330 AN ACTIVE RLC BUILDING BLOCK FOR CASCADE SYNTHESIS.
GORDON H + ZOBRIST GW
PX1/10PP OF PROC OF THE 15TH MIDWEST SYMP ON CIRCUIT THEORY VOL II, UNIV MISSOURI-ROLLA, ROLLA, MO USA, 4-5 MAY 1972, UNIV MISSOURI-ROLLA, 1972

0331 RC-ACTIVE BIQUADRATIC FILTER SECTIONS WITH TWO OPERATIONAL AMPLIFIERS.
GOTZ G + NOSSEK JA
FREQUENZ 32(5): 140-5 (MAY 1978)

0332 SIMPLE ELECTRONICALLY TUNABLE 2ND-ORDER ACTIVE FILTER.
GRIMBLEBY JB
ELECTRON LETT 13(9): 258-60 (28 APRIL 1977)

0333 AN ECONOMICAL BAND-PASS FILTER USING THE OPERATIONAL AMPLIFIER POLE.
GUMUSSOY M + HOLT AGJ
INT J ELECTRON 43(5): 473-9 (NOV 1977)

0334 AN ACTIVE RC-FILTER BASED ON AN INTEGRATED MICROCIRCUIT.
GUSEINOV MSH + RAGIMOV MI + KAFAROV YUM
INSTRUM EXP TECH 18(4): PT1 1160-1 (JULY-AUG 1975)

0335 ZERO S(K)(Q) ACTIVE FILTER USING ACTIVE ELEMENT HAVING GAIN LESS THAN 1.
GUZINSKI A
BULL ACAD POL SCI SER SCI TECH 22(11): 89-94 (1974)

0336 COMPENSATION OF ACTIVE FILTER SECTIONS WITH A SINGLE OPERATIONAL AMPLIFIER.
GYULA S
HIRADASTECHNIKA 25(8): 234-9 (AUG 1974)

0337 STABLE, DIGITALLY PROGRAMMABLE ACTIVE FILTERS.
HAMILTON TA
IEEE J SOLID-STATE CIRCUITS SC-9(1): 27-9 (FEB 1974)

0338 A NOVEL APPLICATION OF A GYRATOR TYPE CIRCUIT.
HAMILTON TA + SEDRA AS
P343-8 OF PROC OF THE FIFTH ASILOMAR CONF ON CIRCUITS AND SYSTEMS UNIV SANTA CLARA, NAVAL POSTGRADUATE SCHOOL, ETAL, PACIFIC GROVE, CALIF, USA, 8-10 NOV 1971, PARKER, SR(ED) WESTERN PERIODICALS CO, 1972

0339 A SINGLE AMPLIFIER BIQUAD ACTIVE FILTER.
HAMILTON TA + SEDRA AS
P355-9 OF IEEE PROC OF THE INT SYMP ON CIRCUIT THEORY, IEEE, NORTH HOLLYWOOD, CALIF, USA 18-21 APR 1972, IEEE, 1972

0340 A SINGLE-AMPLIFIER BIQUAD ACTIVE FILTER.
HAMILTON TA + SEDRA AS
IEEE TRANS CIRCUIT THEORY CT19(4): 398-403 (JULY 1972)

0341 SOME NEW CONFIGURATIONS FOR ACTIVE FILTERS.
HAMILTON TA + SEDRA AS
IEEE TRANS CIRCUIT THEORY (USA) VOL CT-19(1): 25-33 (JAN 1972)

0342 PRINCIPLES AND PRACTICE OF LINEAR AMPLIFIERS (39).
HARMS G
ELEKTROTECHNIK 59(18): 35-6 (23 SEPT 1977) (IN GERMAN)

0343 DIGITALLY CONTROLLED, CONDUCTANCE TUNABLE ACTIVE FILTERS.
HARRIS RW + LEE HT
IEEE J SOLID-STATE CIRCUITS SC10(3): 182-5 (JUNE 1975)

0344 VARIABLE FILTER CHARACTERISTICS WITH AN OP. AMP.
HAUSSMANN H + KUO MCY
ELECTRON ENG 44(528): 41-8,50,2,4, 6 (FEB 1972)

0345 A SELF-TUNING FILTER.
HEINICKE HC + VOGLER G
NACHRICHTENTECH ELEKTRON 27(12): 522-5 (1977) (IN GERMAN)

0346 IMPROVED GYRATOR-RC FILTER STAGES DERIVED FROM THE REACTANCE TRANSISTOR.
HERBERT-LOWE HB + HOLMES WH
PROC INST RADIO ELECTRON ENG AUST 35(11): 470-7 (DEC 1973)

0347 TWO-INTEGRATOR-LOOP ACTIVE RC BIQUADS.
HESZBERGER A + SIMONYI E
P261-5 OF CIRCUIT THEORY AND DESIGN, LAUSANNE, SWITZERLAND, 4-8 SEPT 1978, MOSCHYTZ, GS, NIERYNCK, J(ED), GEORGI 1978

0348 GAIN-SENSITIVITY VECTORS IN MULTIAMPLIFIER ACTIVE RC BIQUADS.
HESZBERGER A + SIMONYI E
P1069-73 OF PROC OF THE 1978 IEEE INT SYMP ON CIRCUITS AND SYSTEMS, IEEE, NEW YORK, USA 17-19 MAY 1978, IEEE, 1978

0349 COMMENTS ON - PRACTICAL DESIGN FOR INSENSITIVE RC-ACTIVE FILTERS.
HESZBERGER A + SIMONYI E
IEEE TRANS CIRCUITS SYST CAS-23(5): 326-7 (MAY 1976)

0350 DESIGN CONSIDERATIONS OF HYBRID INTEGRATED ACTIVE RC FILTERS.
HESZBERGER A + SIMONYI E
PCT-14/131-42 OF PROC OF THE 5TH COLLOQUIUM ON MICROWAVE COMMUNICATION VOLII CIRCUIT THEORY AND COMPUTER AIDED DESIGN (PREPRINTS), HUNGARIAN ACAD SCI, SCI SOC TELECOMMUNICATION RES INST TELECOMMUNICATION, BUDAPEST HUNGARY, 24-30 JUNE 1974, BOGNAR, G(ED), AKAD KIADO, 1974

0351 DIGITAL CONTROL OF ACTIVE FILTER CHARACTERISTICS.
HILL JJ
INT J ELECTRON 41(4): 405-10 (OCT 1976)

0352 ACTIVE FILTER CIRCUIT IS DIGITALLY CONTROLLED.
HILL JJ
ELECTRON ENGINEERING 48(577): 17, 19 (MARCH 1976)

0353 ACTIVE RC ALL-PASS FILTERS CONTAINING PERIODICALLY OPERATED SWITCHES.
HIRANO K + NISHIMURA S + HITANI M
IEEE TRANS CIRCUITS SYST CAS-22(9): 764-7 (SEPT 1975)

0354 A MICROELECTRONIC ACTIVE BAND-PASS FILTER.
HIRSL J
TESLA ELECTRON 10(4): 108-14 (DEC 1977)

0355 MICROELECTRONIC ACTIVE BANDPASSES.
HIRSL J
SDELOVACI TECH 25(6): 203-5 (JUNE 1977) (IN CZECH)

0356 DESIGN OF A SALLEN-KEY TYPE ACTIVE RC FILTER.
HO CF
HONG KONG ENG 5(7): 45-7 (JULY 1977)

0357 COMMENT ON - A HIGH-Q TEMPERATURE INSENSITIVE BANDPASS FILTER USING THE OPERATIONAL AMPLIFIER POLE.
HOANG LE-HUY + MERCIER O
PROC IEEE 63(12): 1725 (DEC 1975)

0358 ACTIVE RC SINGLE-OPAMP DESIGN OF DRIVING-POINT IMPEDANCES.
HORN P + MOSCHYTZ GS
IEEE TRANS CIRCUITS SYST CAS-26(1): 22-30 (JAN 1979)

0359 VCO CONTROLLED BY ONE VARIABLE RESISTOR.
HRIBSEK M + NEWCOMB RW
IEEE TRANS CIRCUITS SYST CAS-23(3): 166-9 (MARCH 1976)

0360 BAND-PASS FILTER OF CONTROLLED CENTRE FREQUENCY AND QUALITY.
HUTER U
ELEKTRONIK 26(5): 80 (MAY 1977) (IN GERMAN)

0361 ACTIVE BAND-PASS FILTER USING 3-PORT AMPLIFIER.
IKEDA K + HONDA S
TRANS INST ELECTRON COMMUN ENG JPN SECT E E60(3): 150 (MARCH 1977)

0362 A CONSTITUTION OF THE ACTIVE BAND-PASS FILTER.
IKEDA K + TOMITA Y + HONDA S
J FAC ENG IBARAKI UNIV NO 26: 131-6 (1978) (IN JAPANESE)

0363 CANONICAL ACTIVE FILTER DESIGN.
INIGO RM
PX4/9PP OF PROC OF THE 15TH MIDWEST SYMP ON CIRCUIT THEORY VOL II, UNIV MISSOURI-ROLLA, ROLLA, MO USA, 4-5 MAY 1972, UNIV MISSOURI-ROLLA, 1972

0364 ACTIVE RC-FILTER WITH VARIABLE BAND-PASS.
INSHAKOV YUM
IZV VUZ ENERG NO 7: 30-4 (JULY 1972) (IN RUSSIAN)

0365 AN INSENSITIVE 2ND-ORDER RC ACTIVE FILTER AND ITS APPLICATION TO BANDPASS CIRCUIT.
ISHIBASHI Y
TRANS INST ELECTRON COMMUN ENG JPN SECT E E61(10): 841 (OCT 1978)

0366 A REALIZATION OF INSENSITIVE RC-ACTIVE FILTERS.
ISHIBASHI Y
TRANS INST ELECTRON COMMUN ENG JPN SECT E E60(8): 418-19 (AUG 1977)

0367 ACTIVE RC FILTERS USING INVERTING AND NONINVERTING VOLTAGE AMPLIFIERS.
ISHIKAWA H
ELECTRON COMMUN JAP 54(2): 40-6 (FEB 1972)

0368 EQUALISER MODULE.
JAHN G
FUNKSCHAU 50(26): 1345-9 (15 DEC 1978) (IN GERMAN)

0369 NULLATORS AND NORATORS SIMPLIFY DESIGN OF RC ACTIVE FILTERS.
JAIN PC + RAMAKRISHNAN KV
J INST ELECTRON TELECOMMUN ENG 20(12): 566-8 (DEC 1974)

0370 A STUDY OF A NEW ACTIVE FILTER SECTION.
JEFFERS R + HAIGH DG
P5/1-4 OF COLLOQUIUM ON ELECTRONIC CIRCUIT ASPECTS OF ACTIVE FILTERS, IEE, LONDON, ENGLAND, 13 OCT 1972, IEE, 1972

0371 ACTIVE FILTER HAS SEPARATE BAND AND FREQUENCY CONTROLS.
JENKINS J
ELECTRONICS 45(19): 110 (11 SEPT 1972)

0372 ACTIVE RC-FILTERS.
JENSEN F
ELEKTRONIK 10(5): 21-2,27-8 (13 MAY 1972) (IN DANISH)

0373 LOW-DISTORTION OSCILLATOR USES STATE-VARIABLE FILTER.
JUNG WG
ELECTRONICS 49(3): 90-1 (5 FEB 1976)

0374 PRACTICAL CONSIDERATIONS OF SINGLE-AMPLIFIER SECOND-ORDER ALL-PASS NETWORKS WITH COMPLEX POLES AND ZEROS.
KANEMITU I + YANAGISAWA T
P658-62 OF 8TH ASILOMAR CONF ON CIRCUITS, SYSTEMS AND COMPUTERS NAVAL POSTGRADUATE SCHOOL, UNIV SANTA CLARA, IEEE, ET AL, PACIFIC GROVE, CALIF, USA, 3-5 DEC 1974, PARKER, SR(ED) WESTERN PERIODICALS, 1975

0375 SYNCHRONOUS FILTER.
KAPTSOV AV + ZAKHAROV VA + IVANOV VV
INSTRUM EXP TECH 20(4): PT2 1085-7 (JULY-AUG 1977)

0376 REALIZATION OF THE BIQUADRATIC TRANSFER FUNCTION USING A SALLEN-KEY TYPE FILTER WITH A GENERAL TWIN-T NETWORK.
KAWAJI S
TRANS SOC INSTRUM CONTROL ENG 10(4): 430-8 (AUG 1974) (IN JAPANESE)

0377 A SENSITIVITY COMPARISON OF SINGLE AND MULTIPLE AMPLIFIER EQUAL CAPACITOR ACTIVE RC STRUCTURES.
KERWIN WJ + AIKENS RS + GROSS WH
P58-62 OF 6TH ASILOMAR CONF ON CIRCUITS AND SYSTEMS, NAVAL POSTGRADUATE SCHOOL, UNIV SANTA CLARA, PACIFIC GROVE CALIF, USA, 15-17 NOV 1972, PARKER, SR(ED), WESTERN PERIODICALS, 1973

0378 AN INTRODUCTION TO ACTIVE FILTERS.
KESNER D
CQ RADIO AMAT J 31(4): 32-4,66 (APRIL 1975)

0379 EFFECT OF COMPONENT-TOLERANCE ON THE TRANSFER PROPERTIES OF ACTIVE WIEN-ROBINSON FILTERS.
KESSLER A + REITZ E
INT ELEKTRON RUNDSCH 29(8): 171-3 (AUG 1975) (IN GERMAN)

0380 NEW ACTIVE FILTER USING THE AUGMENTED INTEGRATOR.
KIM CHUNG DUK + KWAK HOON SEONG
J KOREA INST ELECTRON ENG 15(4): 20-5 (SEPT 1978) (IN KOREAN)

0381 ACTIVE RC BANDPASS FILTER WITH THE INDEPENDENT TUNING AND BANDWIDTH CONTROLS.
KIN SOO JOONG + CHUNG SHIN IL
J KOREAN INST ELECTRON ENG 12(6): 9-13 (FEB 1976) (IN KOREAN)

0382 TIME RESPONSE OF SECOND ORDER BAND PASS FILTER.
KOH BYUNG JOON + KIM JAE KONG
J KOREAN INST ELECTRON ENG 8(5): 221-7 (NOV 1971) (IN KOREAN)

0383 A FAST DESIGN OF ACTIVE LOW-PASS AND HIGH-PASS FILTERS.
KONECNY I
SDELOVACI TECH 25(6): 215-18 (JUNE 1977) (IN CZECH)

0384 BANDPASS/LOWPASS FILTER WITH TWO OP AMPS.
KRAUS K
ELECTRON ENGINEERING 47(573): 25 (NOV 1975)

0385 NARROW BAND ACTIVE FILTER NETWORK WITH OPERATIONAL AMPLIFIER.
KRAUS K
INT ELEKTRON RUNDSCH 28(3): 49-53 (MARCH 1974) (IN GERMAN)

0386 LOW-SENSITIVITY RC SECTION WITH A BIQUADRATIC CHARACTERISTIC.
KRUTCHINSKIY SG
RADIO ENG ELECTRON PHYS 20(5): 144-5 (MAY 1975)

0387 REALIZABILITY CONDITIONS OF VARIABLE-Q, ACTIVE BANDPASS FILTERS USING SEMICONDUCTOR TETRODES AND PENTODES.
KUTOVOY IV + SLEPOV LI
SOV AUTOM CONTROL 11(2): 61-4 (MARCH-APRIL 1978)

0388 GENERAL-PURPOSE GYRATOR FILTER.
KUTSAROV SI
IZV VUZ RADIOELEKTRON 19(5): 92-5 (MAY 1976) (IN RUSSIAN)

0389 UNIVERSAL ACTIVE FILTER (USING GYRATOR).
KUTSAROV SI
ELEKTRO PROM-ST PRIBOROSTR 11(8): 290-2 (SEPT 1976) (IN BULGARIAN)

0390 CERTAIN CIRCUITRIES OF ACTIVE FILTERS WITH OPERATION AMPLIFIERS.
KUTSAROV SI + UZUNOV IS
ELEKTRO PROM-ST PRIBOROSTR 10(6): 252-5 (JUNE 1975) (IN BULGARIAN)

0391 NEW ACTIVE RC FILTERS WITH INHERENT COMPENSATION FOR THE FINITE BANDWIDTH OF OPERATIONAL AMPLIFIERS.
LE QUI THE + YANAGISAWA T
TRANS INST ELECTRON COMMUN ENG JPN SECT E E61(1): 21-2 (JAN 1978)

0392 TUNABLE NOTCH FILTER SUPPRESSES HUM.
LEFFERSON P
ELECTRONICS 49(18): 100 (2 SEPT 1976)

0393 DESIGN OF A TUNABLE NOTCH FILTER WITH VARIABLE Q.
LETELLIER JP
NAVAL RES LAB, WASHINGTON, DC, USA, MARCH 1972, 26P AD-740744

0394 STABLE HIGH Q ACTIVE NETWORKS.
LEVI A + ZEHEB E
PC1-1/1-2 OF 9TH CONVENTION OF ELECTRICAL AND ELECTRONIC ENGINEERS IN ISRAEL, IEEE, TEL AVIV, ISRAEL, 23-24 APRIL 1975 IEEE, 1975 (IN HEBREW)

0395 A SIMPLE TRACKING FILTER.
LINDQUIST CS + GOCAL G
P24-8 OF CONF RECORD OF THE TWELFTH ASILOMAR CONF ON CIRCUITS, SYSTEMS AND COMPUTERS NAVAL POSTGRADUATE SCHOOL MONTEREY, IEEE, UNIV SANTA CLARA, PACIFIC GROVE, CA, USA, 6-8 NOV 1978, IEEE, 1978

0396 AN INERTIAL ELEMENT FOR ACTIVE FILTERS BASED ON OPERATIONAL AMPLIFIERS.
LOVEIKO VN
IZV VUZ RADIOELEKTRON 21(3): 112-13 (MARCH 1978) (IN RUSSIAN)

0397 SYNTHESIS OF PHASE-SHIFT CIRCUITS USING ACTIVE RC-NETWORKS OF THE MINIMUM-PHASE TYPE.
LOVEIKO VN + CHAPOVSKII MZ
TELECOMMUN RADIO ENG PT 1 28(6): 75-7 (1974)

0398 A FAMILY OF RC ACTIVE FILTER SECTIONS.
LOVEIKO VN + PODOL'SKII LZ + CHAPOVSKII MZ
RADIO ENG ELECTRON PHYS 19(12): 149-50 (DEC 1974)

0399 TWO-AMPLIFIER BIQUADS.
LUDER E
FREQUENZ 29(5): 140-2 (MAY 1975)

0400 TWO NEW METHODS FOR THE SYNTHESIS OF RC-ACTIVE PHASE UNITS WITH OPERATIONAL AMPLIFIER.
MARINOV JP + DONEVSKI BD
ELEKTRO PROM-ST PRIBOROSTR 10(3): 96-9 (MARCH 1975) (IN BULGARIAN)

0401 FREQUENCY RANGE OF SELECTIVE RC-CIRCUITS USING INTEGRATED OPERATIONAL AMPLIFIERS.
MASLENNIKOV VV + SIROTKIN AP
POLUPROVODN PRIB TEKH ELEKTROSVYAZI NO 16: 69-81 (1975) (IN RUSSIAN)

0402 DEGENERACY OF SALLEN-KEY TYPE ACTIVE RC FILTERS.
MATSUMOTO T + YAMAMOTO T
TRANS INST ELECTRON COMMUN ENG JPN SECT E E60(6): 308 (JUNE 1977)

0403 A PRACTICAL DESIGN FOR INSENSITIVE RC-ACTIVE FILTERS.
MIKHAEL WB + BHATTACHARYYA BB
IEEE TRANS CIRCUITS SYST CAS-22(5): 407-15 (MAY 1975)

0404 A PRACTICAL DESIGN FOR INSENSITIVE RC-ACTIVE FILTERS.
MIKHAEL WB + BHATTACHARYYA BB
P205-9 OF PROC OF THE 1974 IEEE INT SYMP ON CIRCUITS AND SYSTEMS, IEEE, SAN FRANCISCO, CALIF USA, 22-25 APRIL 1974, IEEE, 1974

0405 A PRACTICAL DESIGN FOR INSENSITIVE RC-ACTIVE FILTERS.
MIKHAEL WB + BHATTACHARYYA BB
P152-3 OF INT ELECTRICAL, ELECTRONICS CONF AND EXPOSITION DIGEST, IEEE, TORONTO, CANADA, 1-3 OCT 1973 IEEE, 1973

0406 ECONOMICAL, LOW SENSITIVITY AND FUNCTIONALLY TUNABLE ACTIVE LOW-PASS FILTERS.
MIKHAEL WB + KEYS LK
P217-22 OF 1974 EUROPEAN CONF ON CIRCUIT THEORY AND DESIGN, IEE, IEEE, INST PHYS, IERE, LONDON, ENGLAND 23-26 JULY 1974, IEE, 1974

0407 A NOTE ON FREQUENCY AND Q LIMITATIONS OF ACTIVE FILTERS.
MITRA AK + AATRE VK
IEEE TRANS CIRCUITS SYST CAS-24(4): 215-18 (APRIL 1977)

0408 LOW-SENSITIVITY BANDPASS FILTERS USING THE OPERATIONAL-AMPLIFIER POLE.
MITRA AK + AATRE VK
ELECTRON LETT 12(9): 226-7 (29 APRIL 1976)

0409 PRECISION TUNING OF RC PARALLEL-T NOTCH FILTER.
MOHAN PVA + UDUPA AH + VENKATARATHINAM BS
PROC IEEE 63(12): 1730 (DEC 1975)

0410 A UNIVERSAL LOW-Q ACTIVE-FILTER BUILDING BLOCK SUITABLE FOR HYBRID-INTEGRATED CIRCUIT IMPLEMENTATION.
MOSCHYTZ GS
IEEE TRANS CIRCUIT THEORY CT-20(1): 37-47 (JAN 1973)

0411 HIGH-Q FACTOR INSENSITIVE ACTIVE RC NETWORK, SIMILAR TO THE TARMY-GHAUSI CIRCUIT BUT USING SINGLE-ENDED OPERATIONAL AMPLIFIERS.
MOSCHYTZ GS
ELECTRON LETT 8(18): 458-9 (7 SEPT 1972)

0412 OPTIMIZING TWO COMMONLY USED ACTIVE FILTER BUILDING BLOCKS USING THE COMPLEMENTARY TRANSFORMATION.
MOSCHYTZ GS + HORN P
P332-5 OF 1977 IEEE INT SYMP ON CIRCUITS AND SYSTEMS PROC, IEEE, PHOENIX, ARIZ, USA 25-27 APRIL 1975, IEEE, 1977

0413 OPTIMISING TWO COMMONLY USED ACTIVE-FILTER BUILDING BLOCKS USING THE COMPLEMENTARY TRANSFORMATION.
MOSCHYTZ GS + HORN P
IEE J ELECTRON CIRCUITS SYST 1(4): 125-32 (JULY 1977)

0414 A MULTIPLE-FEEDBACK UNIVERSAL ACTIVE FILTER.
MOUSTAKAS E + CHAN SP
INT J ELECTRON 41(1): 1-24 (JULY 1976)

0415 SENSITIVITY CONSIDERATIONS IN A MULTIPLE-FEEDBACK UNIVERSAL ACTIVE FILTER.
MOUSTAKAS E + CHAN SP
IEEE TRANS CIRCUITS SYST CAS-24(12): 695-703 (DEC 1977)

0416 A CLASS OF INSENSITIVE ACTIVE FILTERS COMPRISING THREE OPERATIONAL AMPLIFIERS.
MULAWKA J
INT J ELECTRON 43(5): 465-72 (NOV 1977)

0417 MODIFICATIONS OF TARMY-GHAUSI ACTIVE FILTER.
MULAWKA J + BIALKO M
ELECTRON LETT 10(18): 380-1 (5 SEPT 1974)

0418 DESIGN OF SECTIONS OF ACTIVE FILTERS REALIZING BISQUARE TRANSFER FUNCTIONS ON THE GROUND OF PITCHFORK CONFIGURATION OF OPERATIONAL AMPLIFIER.
MULAWKA J + BIALKO M
ARCH ELEKTROTECH 21(4): 819-35 (1972) (IN POLISH)

0419 A NEW TWO-AMPLIFIER-BIQUAD FOR HIGH POLE-Q REALIZATION.
MUTSCHLER J
P135-9 OF CIRCUIT THEORY AND DESIGN, LAUSANNE, SWITZERLAND, 4-8 SEPT 1978, MOSCHYTZ, GS, NIERYNCK, J(ED), GEORGI 1978

0420 ACTIVE RC FILTERS WITH ZERO Q-SENSITIVITY.
NAIMPALLY SV
PROC INST ELECTR ENG 124(9): 754-5 (SEPT 1977)

0421 A HIGH-Q LOW-SENSITIVITY TWIN-T NOTCH FILTER.
NAIMPALLY SV
J INST ELECTRON TELECOMMUN ENG 23(1): 1-4 (JAN 1977)

0422 DESIGN OF INSENSITIVE ACTIVE CIRCUITS FOR SYSTEM TRANSFER FUNCTION REALISATION.
NAIMPALLY SV
P67-9 OF PROC OF THE 8TH ANNUAL SOUTHEASTERN SYMP ON SYSTEM THEORY, UNIV TENNESSEE, KNOXVILLE, TENN, USA 26-27 APRIL 1976, IEEE, 1976

0423 A NEW ACTIVE RC CIRCUIT REALIZATION OF A THIRD-ORDER LOW-PASS BUTTERWORTH CHARACTERISTIC USING GROUNDED CAPACITORS AND EQUAL-VALUED PASSIVE ELEMENTS.
NAIMPALLY SV + MAHAPATRA S
PROC IEEE 63(12): 1732-3 (DEC 1975)

0424 REALIZATION OF ALLPASS OR NOTCH FILTERS - A SIMULATED INDUCTOR APPROACH.
NANDI R
IEE-IERE PROC INDIA 15(6): 254-8 (NOV-DEC 1977)

0425 GROUNDED-CAPACITOR BANDPASS FILTER USING THE OPERATIONAL-AMPLIFIER POLE.
NANDI R
ELECTRON LETT 13(24): 713-5 (24 NOV 1977)

0426 FEN CIRCUITS.
NEMESSZEGHY G
HIRADASTECHNIKA 28(3): 65-70 (MARCH 1977) (IN HUNGARIAN)

0427 THREE ACTIVE BANDPASS FILTERS.
NEWMAN HG
PX2/10PP OF PROC OF THE 15TH MIDWEST SYMP ON CIRCUIT THEORY VOL II, UNIV MISSOURI-ROLLA, ROLLA, MO USA, 4-5 MAY 1972, UNIV MISSOURI-ROLLA, 1972

0428 ACTIVE NOTCH FILTERS.
NEZER Y
WIRELESS WORLD 81(1475): 307-11 (JULY 1975)

0429 AN ACTIVE FILTER WITH BIQUADRATIC TRANSFER-FUNCTION AND LOW SENSITIVITIES.
NOSSEK JA
NACHRICHTENTECH Z 25(9): 415 (SEPT 1972) (IN ENGLISH)

0430 NEW HIGH-QUALITY SECOND-ORDER ACTIVE FILTERS USING ONLY TWO OPERATIONAL AMPLIFIERS.
NOSSEK JA + GOTZ G
P446-9 OF PROC OF THE 1976 IEEE INT SYMP ON CIRCUITS AND SYSTEMS, IEEE, MUNICH, GERMANY 27-29 APRIL 1976, IEEE, 1976

0431 ACTIVE RC FILTER USING VCCSS.
OKINE M
TRANS INST ELECTRON COMMUN ENG JPN SECT E E61(7): 536-7 (JULY 1978)

0432 FORMULA FOR DETERMINING THE STEEPNESS OF THE RESONANCE CURVE OF SECOND-ORDER BAND FILTERS.
OROZOV BL
ELEKTRO PROM-ST PRIBOROSTR 13(11): 414-16 (1978) (IN BULGARIAN)

0433 ALL PASS NOTCH FILTER.
ORR T
ELECTRON TODAY INT 7(5): 70 (MAY 1978)

0434 VOLTAGE/CURRENT CONTROLLED FILTER.
ORR T
WIRELESS WORLD 82(1491): 63 (NOV 1976)

0435 AUTOMATIC NOTCH FREQUENCY CONTROL FOR SERVO STABILISATION.
PALMER RS
IBM TECH DISCLOSURE BULL 19(10): 3860-2 (MARCH 1977)

0436 OSCILLATING NOTCH FILTER.
PALMER RS
IBM TECH DISCLOSURE BULL 19(10): 3858-9 (MARCH 1977)

0437 GENERATION OF ACTIVE RC-CIRCUITS BY THE NULLOR EQUIVALENCY CONCEPT.
PALOMERA-GARCIA R
P53-7 OF CIRCUIT THEORY AND DESIGN, LAUSANNE, SWITZERLAND, 4-8 SEPT 1978, MOSCHYTZ, GS, NIERYNCK, J(ED), GEORGI, 1978

0438 A NOTE ON THE SYNTHESIS OF ACTIVE RC FILTERS.
PATRANABIS D
J INST ELECTRON TELECOMMUN ENG 20(7): 365-7 (JULY 1974)

0439 A SIMPLE ACTIVE RC NOTCH FILTER.
PATRANABIS D
J INST TELECOMMUN ENG 18(6): 278-80 (JUNE 1972)

0440 ACTIVE-RC ALL-PASS FILTER.
PATRANABIS D + TRIPATHI MP
J INST ELECTRON TELECOMMUN ENG 24(1): 4-5 (JAN 1978)

0441 GENERALIZATION OF ACTIVE RC ALL-PASS SCHEMES.
PATRANABIS D + ROY SB + TRIPATHI MP
PROC IEEE 66(3): 354-6 (MARCH 1978)

0442 ADDITIONAL GENERALISATION SCHEME OF ACTIVE RC ALLPASS FILTERS.
PATRANABIS D + TRIPATHI MP
ELECTRON LETT 13(6): 167-8 (17 MARCH 1977)

0443 ALL-PASS WITH AN OPERATIONAL AMPLIFIER AND AN RC LATTICE.
PATRANABIS D + TRIPATHI MP
INT J ELECTRON 39(6): 647-51 (DEC 1975)

0444 ACTIVE FILTERS WITH LOW Q-SENSITIVITY.
PLESHKO AD
IZV VUZ RADIOELEKTRON 19(12): 59-65 (DEC 1976) (IN RUSSIAN)

0445 ACTIVE RC SYNTHESIS OF A NARROW BAND TRANSFER FUNCTION USING YANAGISAWA SCHEME.
PRASAD NK + KHANIJO MK
INT J ELECTRON 36(1): 101-12 (1974)

0446 MODIFIED SCHUSSLER FILTERS WITH LOW SENSITIVITY TO ELEMENT CHANGES.
RAKOVICH BD + STOJANOVICH VS
ELECTRON LETT 12(3): 73-4 (5 FEB 1976)

0447 A NEW DESIGN PROCEDURE FOR INDUCTORLESS FILTERS.
RATHORE TS
J INST ELECTRON TELECOMMUN ENG 22(12): 765-6 (DEC 1976)

0448 SYNTHESIS OF TRANSFER FUNCTION EMPLOYING THE HOLT-AND-GRAY CONFIGURATION.
RATHORE TS
ELECTRON LETT 12(19): 488-9 (16 SEPT 1976)

0449 ACTIVE REALIZATION OF RC TRANSFER FUNCTIONS WITH ARBITRARY GAIN CONSTANT.
RATHORE TS + DASGUPTA SM
IEEE TRANS CIRCUITS SYST CAS-22(10): 835-6 (OCT 1975)

0450 AN ACTIVE FILTER USING EQUAL VALUED EARTHED CAPACITORS.
RATHORE TS + DASGUPTA SM
INT J ELECTRON 38(2): 283-6 (FEB 1975)

0451 ACTIVE FILTER CROSSOVER NETWORKS.
READ DC
WIRELESS WORLD 79(1458): 574-6 (DEC 1973)

0452 AN ACTIVE-RC FILTER FOR HIGH-Q AND HIGH-FREQUENCY WITH ZERO-Q AND ZERO-FREQUENCY-SENSITIVITY TO AMPLIFIER GAIN-BANDWIDTH PRODUCT.
REDDY MA
PROC IEEE 65(5): 814-15 (MAY 1977)

0453 AN INSENSITIVE ACTIVE-RC FILTER FOR HIGH Q AND HIGH FREQUENCIES.
REDDY MA
IEEE TRANS CIRCUITS SYST CAS-23(7): 429-33 (JULY 1976)

0454 OPERATIONAL-AMPLIFIER CIRCUITS WITH VARIABLE PHASE SHIFT AND THEIR APPLICATION TO HIGH-Q ACTIVE RC-FILTERS AND RC-OSCILLATORS.
REDDY MA
IEEE TRANS CIRCUITS SYST CAS-23(6): 384-9 (JUNE 1976)

0455 A HIGH-Q, LOW SENSITIVITY ACTIVE RC FILTER. II. THE EFFECT OF OPERATIONAL-AMPLIFIER NON-IDEALITIES AND EXPERIMENTAL RESULTS.
REDDY MA
NACHRICHTENTECH Z 26(4): 180-3 (APRIL 1973) (IN ENGLISH)

0456 A HIGH-Q LOW SENSITIVITY ACTIVE-RC FILTER. I. INITIAL CONSIDERATIONS-THE NETWORK IN ITS IDEAL FORM.
REDDY MA
NACHRICHTENTECH Z 26(3): 129-32 (MARCH 1973) (IN ENGLISH)

0457 HIGH-Q, LOW SENSITIVITY RC AMPLIFIER-FILTERS.
REDDY MA
Z ELEKTR INF- ENERGIETECH 3(1): 41-8 (1973)

0458 SENSITIVITY OPTIMIZATION TO OPERATIONAL-AMPLIFIER GAIN IN THE KERWIN-HUELSMAN-NEWCOMB HIGH-Q ACTIVE RC FILTER-SECTION.
REDDY MA
ARCH ELEKTRON UBERTRAGUNGSTECH 26(9): 394-6 (SEPT 1972) (IN ENGLISH)

0459 THIRD-ORDER SECTION FOR ACTIVE RC FILTERS.
REINBOT AE + BROVCHENKO EN + KAZ'MIRCHUK DS
IZV VUZ RADIOELEKTRON 15(11): 1368-71 (1972) (IN RUSSIAN)

0460 A MODIFIED MOSCHYTZ HIGH-Q ACTIVE FILTER.
REVANKAR GN + SHANKAR K + DATAR RB
INT J ELECTRON 42(2): 117-20 (FEB 1977)

0461 REVIEW OF UNIVERSAL ACTIVE FILTERS CIRCUITS.
ROGULSKI A
ELEKTRONIKA 20(3): 125-9 (1979) (IN POLISH)

0462 DESIGNING WITH UNIVERSAL ACTIVE FILTERS. II. THE STEP BY STEP DESIGN OF FILTERS.
ROVIRA AA
TEC REGUL MANDO AUTOM 12(85): 21-5 (FEB 1979) (IN SPANISH)

0463 DESIGNS WITH UNIVERSAL ACTIVE FILTERS.
ROVIRA AA
TEC REGUL MANDO AUTOM 11(83): 27-34 (NOV 1978) (IN SPANISH)

0464 DESIGNING A LOW-FREQUENCY ACTIVE NOTCH FILTER.
ROWE NB
ELECTRON ENG 44(530): 43-5 (APRIL 1972)

0465 ACHIEVING A MAXIMUM Q-FACTOR IN AN ACTIVE RC FILTER SECTION WITH TWO AMPLIFIERS.
ROZHANKOVSKII RV
OTBOR PEREDACHA INF NO 41: 57-60 (1974) (IN RUSSIAN)

0466 A STATE-VARIABLE APPROACH FOR GENERATING CANONIC BIQUADRATIC ACTIVE FILTER FUNCTIONS.
RUSSELL HT + CHAN SP
P91-8 OF 7TH ASILOMAR CONF ON CIRCUITS, SYSTEMS AND COMPUTERS, NAVAL POSTGRADUATE SCHOOL, UNIV SANTA CLARA, ET AL PACIFIC GROVE, CALIF, USA, 27-29 NOV 1973, WESTERN PERIODICALS, 1974

0467 REALIZATION OF HIGH-Q INSENSITIVE RC ACTIVE FILTERS USING ALL-PASS NETWORKS.
SAHA SK
INT J ELECTRON 39(1): 81-92 (JULY 1975)

0468 SYNTHESIS OF RC ACTIVE FILTERS WITH ZERO ACTIVE Q-SENSITIVITY.
SAHA SK + SEN JK
J INST ELECTRON TELECOMMUN ENG 21(12): 649-52 (DEC 1975)

0469 AN ACTIVE BANDPASS FILTER WITH SIMPLE MIDDLE FREQUENCY EQUALIZATION.
SCHMIDT D
MESS ZAEHLEN REGIST NO 1: 10-11 (SEPT 1973) (IN GERMAN)

0470 ACTIVE BAND-PASS FILTER WITH OPERATIONAL AMPLIFIER.
SCHNEIDER G
ELEKTRONIKER 12(11): EL12-19 (NOV 1973) (IN GERMAN)

0471 GENERATION AND CLASSIFICATION OF SINGLE AMPLIFIER FILTERS.
SEDRA AS
INT J CIRCUIT THEORY APPL 2(1): 51-67 (MARCH 1974)

0472 A CLASS OF STABLE ACTIVE FILTERS USING UNITY-GAIN VOLTAGE FOLLOWERS.
SEDRA AS
IEEE J SOLID-STATE CIRCUITS SC-7(4): 311-15 (AUG 1972)

0473 A REFINED CLASSIFICATION OF SINGLE AMPLIFIER FILTERS.
SEDRA AS + BROWN L
INT J CIRCUIT THEORY APPL 7(1): 127-37 (JAN 1979)

0474 A REFINED CLASSIFICATION OF SINGLE AMPLIFIER FILTERS.
SEDRA AS + BROWN L
P850-4 OF PROC OF THE 1978 IEEE INT SYMP ON CIRCUITS AND SYSTEMS, IEEE, NEW YORK, USA 17-19 MAY 1978, IEEE, 1978

0475 PREFERRED ACTIVE FILTER FORMS FOR BANDPASS FUNCTIONS.
SEELEY RM
IEEE J SOLID-STATE CIRCUITS SC-7(4): 304-6 (AUG 1972)

0476 ACTIVE RC FILTER MODULES WITH THREE CAPACITORS FOR THE REALIZATION OF CAUER FILTERS.
SEIDEL V
NACHRICHTENTECH ELEKTRON 26(10): 367-71 (1976) (IN GERMAN)

0477 FILTER MODULES FOR REALIZING BASIC ELEMENT CHAINS IN ACTIVE RC TECHNIQUE.
SEIDEL V
NACHRICHTENTECH ELEKTRON 26(11): 416-21 (1976) (IN GERMAN)

0478 ABSOLUTE STABLE ACTIVE RC-FILTER BUILDING BLOCKS USING TWO CAPACITORS AND A DIFFERENTIAL AMPLIFIER HAVE IMAGINARY TRANSFER ZEROS.
SEIDEL V
NACHRICHTENTECH ELEKTRON 26(9): 332-5 (1976) (IN GERMAN)

0479 LOW-PASS PARTIAL FILTERS IN ACTIVE RC TECHNIQUE.
SEIDEL V
NACHRICHTENTECH ELEKTRON 24(11): 418-21 (1974) (IN GERMAN)

0480 ACTIVE RC-PARTIAL FILTER WITH THREE FREQUENCY-DETERMINING CAPACITIES.
SEIDEL V
NACHRICHTENTECH ELEKTRON 23(1): 10-13 (1973) (IN GERMAN)

0481 AN ACTIVE RC FILTER EXHIBITING SELECTIVE, ALL-PASS AND NOTCH CHARACTERISTICS.
SEN PC + PATRANABIS D
INT J ELECTRON 33(5): 583-91 (NOV 1972)

0482 PARAMETER CONTROL OF VARIABLE ACTIVE INDUCTORLESS FILTERS.
SHAW GW
INT J ELECTRON 41(2): 199-207 (AUG 1976)

0483 STABLE OSCILLATOR.
SHILLING M
NEW ELECTRON 9(5): 18 (9 MARCH 1976)

0484 OPTIMUM REALIZATIONS OF SINGLE-AMPLIFIER NETWORKS WITH ZERO POLE FREQUENCY SENSITIVITY AND WITH MINIMIZED POLE-Q DEVIATION.
SINGH NP + RAMACHANDRAN V + BHATTACHARYYA BB
INT J CIRCUIT THEORY APPL 7(1): 9-19 (JAN 1979)

0485 REALIZATIONS OF SINGLE AMPLIFIER NETWORKS WITH SPECIFIED POLE-Q OR POLE-FREQUENCY SENSITIVITY.
SINGH NP + RAMACHANDRAN V + BHATTACHARYYA BB
P674 OF 8TH ASILOMAR CONF ON CIRCUITS, SYSTEMS AND COMPUTERS NAVAL POSTGRADUATE SCHOOL, UNIV SANTA CLARA, IEEE, ET AL, PACIFIC GROVE, CALIF, USA, 3-5 DEC 1974, PARKER, SR(ED) WESTERN PERIODICALS, 1975

0486 ALTERNATE STATE SPACE REALIZATION OF BIQUADRATIC VOLTAGE TRANSFER FUNCTION USING GROUNDED CAPACITORS.
SINGH R + VENKATESWARAN S
J INST ELECTRON TELECOMMUN ENG 21(12): 634-9 (DEC 1975)

0487 FREQUENCY-CONTROLLED GATE MAKES HIGH-Q FILTER.
SIVERTSON NA
ELECTRONICS 50(8): 106 (14 APRIL 1977)

0488 ANALOG FILTER CAN BE PROGRAMMED DIGITALLY.
SMITHLINE LM
ELECTRONICS 47(20): 107 (3 OCT 1974)

0489 BIQUADS USING ONLY TWO OPERATIONAL AMPLIFIERS.
SODERSTRAND MA
P651-5 OF PROC OF THE 1974 IEEE INT SYMP ON CIRCUITS AND SYSTEMS, IEEE, SAN FRANCISCO, CALIF USA, 22-25 APRIL 1974, IEEE, 1974

0490 A SYSTEMATIC METHOD OF GENERATING BIQUADRATIC FILTER SECTIONS.
SODERSTRAND MA
INT J ELECTRON 36(6): 825-39 (JUNE 1974)

0491 DESIGN OF ACTIVE RC FILTERS WITH ZERO GAIN-SENSITIVITY PRODUCT.
SODERSTRAND MA + MITRA SK
IEEE TRANS CIRCUIT THEORY CT-20(4): 441-5 (JULY 1973)

0492 DESIGN OF ACTIVE FILTERS WITH ZERO PASSIVE Q-SENSITIVITY.
SODERSTRAND MA + MITRA SK
IEEE TRANS CIRCUIT THEORY CT20(3): 289-94 (MAY 1973)

0493 ACTIVE RC FILTERS WITH ZERO GAIN SENSITIVITY PRODUCT.
SODERSTRAND MA + MITRA SK
P340-4 OF IEEE PROC OF THE INT SYMP ON CIRCUIT THEORY, IEEE, NORTH HOLLYWOOD, CALIF, USA 18-21 APR 1972, IEEE, 1972

0494 NEW CONVERSION METHODS FOR REALIZING NONMINIMUM PHASE TRANSFER FUNCTIONS.
SOLIMAN AA + AWAD SS
PROC IEEE 66(5): 588-90 (MAY 1978)

0495 ACTIVE COMPENSATED BANDPASS FILTER WITH REDUCED SENSITIVITY TO OP AMP GAIN BANDWIDTH PRODUCT.
SOLIMAN AM
FREQUENZ 33(5): 149-51 (MAY 1979)

0496 A MODIFIED CANONIC ACTIVE-RC BANDPASS FILTER WITH REDUCED SENSITIVITY TO AMPLIFIER GAIN-BANDWIDTH PRODUCT.
SOLIMAN AM
PROC IEEE 67(2): 325-6 (FEB 1979)

0497 A NEW SINGLE OPERATIONAL AMPLIFIER ACTIVE RC BANDPASS NETWORK WITH REDUCED SENSITIVITY TO AMPLIFIER GAIN-BANDWIDTH PRODUCT.
SOLIMAN AM
INT J CIRCUIT THEORY APPL 6(4): 321-6 (OCT 1978)

0498 TWO NOVEL ACTIVE RC CANONIC BANDPASS NETWORKS USING THE CURRENT CONVEYOR.
SOLIMAN AM
INT J ELECTRON 42(1): 49-54 (JAN 1977)

0499 CONVERSION OF A BAND-PASS RESONATOR TO AN ALL-PASS OR A NOTCH FILTER.
SOLIMAN AM
INT J ELECTRON 38(4): 559-62 (APRIL 1975)

0500 ACTIVE RC HIGH SELECTIVITY NOTCH FILTER.
SOLIMAN AM
INT J ELECTRON 37(4): 565-7 (OCT 1974)

0501 ACTIVE RC LOW-PASS FILTER SUITABLE FOR INTEGRATION.
SOLIMAN AM
INT J ELECTRON 36(6): 799-803 (JUNE 1974)

0502 ANOTHER REALIZATION OF AN ALL-PASS OR A NOTCH FILTER USING A CURRENT CONVEYOR.
SOLIMAN AM
INT J ELECTRON 35(1): 135-6 (JULY 1973)

0503 A LOW-SENSITIVITY ACTIVE RC LOW-PASS FILTER.
SOLIMAN AM
IEEE TRANS AUDIO ELECTROACOUST AU-21(4): 372-4 (AUG 1973)

0504 NEW ACTIVE RC CONFIGURATION FOR REALISING A MEDIUM-SELECTIVITY NOTCH FILTER.
SOLIMAN AM
ELECTRON LETT 8(21): 522-4 (19 OCT 1972)

0505 A BANDPASS FILTER USING THE OPERATIONAL AMPLIFIER POLE.
SOLIMAN AM + FAWZY M
IEEE J SOLID-STATE CIRCUITS SC-12(4): 429-30 (AUG 1977)

0506 PHASE CORRECTION IN TWO-INTEGRATOR LOOP FILTERS USING A SINGLE COMPENSATING RESISTOR.
SOLIMAN AM + ISMAIL M
ELECTRON LETT 14(12): 375-6 (8 JUNE 1978)

0507 HYBRID ACTIVE RC FILTERS.
SONKOLY A
HIRADSTECH IPARI KUT INTEZ KOEZL 17(2): 44-64 (1977) (IN HUNGARIAN)

0508 A TRACKING BIQUAD ACTIVE FILTER.
SPARKES RG + SEDRA AS
P585-9 OF PROC OF THE 1974 IEEE INT SYMP ON CIRCUITS AND SYSTEMS, IEEE, SAN FRANCISCO, CALIF USA, 22-25 APRIL 1974, IEEE, 1974

0509 PROGRAMMABLE ACTIVE FILTERS (SPECTRUM ANALYZER APPLICATION).
SPARKES RG + SEDRA AS
IEEE J SOLID-STATE CIRCUITS SC 8(1): 93-5 (FEB 1973)

0510 A COMPARISON OF A CLASS OF ACTIVE FILTERS WITH RESPECT TO THE OPERATIONAL-AMPLIFIER GAIN-BANDWIDTH PRODUCT.
SRINIVASAGOPALAN R + MARTENS GO
IEEE TRANS CIRCUITS SYST CAS-21(3): 377-81 (MAY 1974)

0511 A GENERAL CLASS OF HIGH-Q ACTIVE RESONATORS.
SRINIVASAGOPALAN R + MARTENS GO
P298-300 OF INT SYMP ON CIRCUIT THEORY 1973, IEEE, TORONTO, CANADA, 9-11 APRIL 1973, IEEE, 1973

0512 SINGLE RESISTANCE CONTROLLED ACTIVE RC BAND FILTERS.
SRZEDNICKI J
ELEKTRONIKA 17(7-8): 293-4 (1976) (IN POLISH)

0513 VOLTAGE-TUNABLE ACTIVE FILTER FEATURES LOW, HIGH AND BANDPASS MODES.
STEINER NA
ELECTRON DES 22(25): 96,98 (6 DEC 1974)

0514 EQUIVALENCE OF THIRD-ORDER LOW-PASS STRUCTURES.
STEPHENSON FW + DUNNING-DAVIES J
PROC IEEE 63(2): 319-20 (FEB 1975)

0515 REALIZATION OF HIGH-PASS EQUAL-VALUED CAPACITOR FILTERS.
STEPHENSON FW
INT J ELECTRON 35(6): 847-55 (DEC 1973)

0516 EQUAL CAPACITOR SALLEN AND KEY FILTERS.
STEPHENSON FW
INT J ELECTRON 33(5): 539-44 (NOV 1972)

0517 ACTIVE FILTERS ACHIEVE PROMINENCE.
SUZUKI K
JEE NO 132: 58-62 (DEC 1977)

0518 USE OF FREQUENCY BAND FILTERS AND SUPPRESSION FILTERS IN ALL-PASS DESIGN.
SVIEZENY B
ELEKTROTECH CAS 29(8): 628-34 (1978) (IN SLOVAK)

0519 AN ELECTRONICALLY TUNABLE RC ACTIVE FILTER.
SWAMY MNS + SANKARA RAO K
INT J CIRCUIT THEORY APPL 3(2): 123-33 (JUNE 1975)

0520 A TUNABLE FILTER USING AMPLIFIER POLE.
SWAMY MNS + SANKARA RAO K
PROC IEEE 63(1): 197-9 (JAN 1975)

0521 ANALOG-COMPUTER-TYPE ACTIVE FILTER.
TAFLOVE A
ELECTRON AUST 37(4): 77 (JULY 1975)

0522 A RADIO-FREQUENCY TUNABLE ACTIVE FILTER.
TAFLOVE A + BRODWIN ME
IEEE J SOLID-STATE CIRCUITS SC-9(2): 73-6 (APRIL 1974)

0523 A VERSATILE TUNABLE FILTER USING OPERATIONAL AMPLIFIERS.
TAYLOR PL
P11/1-2 OF COLLOQUIUM ON ELECTRONIC CIRCUIT ASPECTS OF ACTIVE FILTERS, IEE, LONDON, ENGLAND, 13 OCT 1972, IEE, 1972

0524 A LOW-PASS ACTIVE RC FILTER WITH VOLTAGE-CONTROLLED CUT-OFF.
TAYLOR PL
P9/1-3 OF COLLOQUIUM ON ELECTRONIC CIRCUIT ASPECTS OF ACTIVE FILTERS, IEE, LONDON, ENGLAND, 13 OCT 1972, IEE, 1972

0525 CURRENT-CONTROLLED BANDPASS FILTER CAN BE BUILT WITH ONLY ONE IC.
THIBODEAUX E
ELECTRON DES 24(11): 96,98 (24 MAY 1976)

0526 SECOND ORDER ACTIVE RC-FILTERS WITH ADJUSTABLE PARAMETERS.
THIELMANN H
NACHRICHTENTECH Z 27(7): 266-71 (JULY 1974) (IN GERMAN)

0527 VARIABLE SLOPE LOW-PASS FILTER.
TIDEY RJ
WIRELESS WORLD 83(1504): 78 (DEC 1977)

0528 FORBIDDEN REGIONS FOR THE POLES OF A 3RD-ORDER EQUAL-VALUED-RESISTOR ACTIVE LOWPASS FILTER.
TIRTOPRODJO S + VAN MEER A
PROC INST ELEC ENG 119(9): 1281-2 (SEPT 1972)

0529 A CONSIDERATION OF LOW SENSITIVITY ACTIVE RC BAND-PASS FILTER.
TOMITA Y + MICHIYAMA S + IKEDA K + HONDA S
J FAC ENG IBARAKI UNIV NO 24: 181-4 (1976) (IN JAPANESE)

0530 SOME OBSERVATIONS ON SINGLE-ELEMENT CONTROL SINE-WAVE OSCILLATORS.
TRIPATHI MP + ROY SB + PATRANABIS D
INT J ELECTRON 43(5): 513-20 (NOV 1977)

0531 SYNTHESIS OF ALL-PASS FILTERS USING OPERATIONAL AMPLIFIERS AND RC-NETWORKS.
TROSTER R
FREQUENZ 28(3): 78-81 (MARCH 1974) (IN GERMAN)

0532 DIGITALLY CONTROLLED ACTIVE FILTERS.
UKEILEY RL
P573-7 OF PROC OF THE 1974 IEEE INT SYMP ON CIRCUITS AND SYSTEMS, IEEE, SAN FRANCISCO, CALIF USA, 22-25 APRIL 1974, IEEE, 1974

0533 SYNTHESIS OF ACTIVE RC-FILTERS BY A SINGLE DIFFERENTIAL-INPUT OPERATIONAL AMPLIFIER.
USUNOV IS + KOUTZAROV SI
INT J ELECTRON 40(6): 601-10 (JUNE 1976)

0534 INTEGRATED ACTIVE FILTERS. I.
VIGUERIE-NOEL B
TOUTE ELECTRON NO 385: 36-8 (FEB 1974) (IN FRENCH)

0535 ACTIVE FILTERS FOR INTERMEDIATE FREQUENCIES USING LOW-GAIN TRANSDUCERS.
VLACH J + VISWANATHAN TR + SINGHAL K
IEEE TRANS CIRCUITS SYST CAS-21(1): 79-86 (JAN 1974)

0536 ACTIVE RC NETWORK REALIZATION OF STABLE HIGH-O TRANSFER FUNCTIONS.
VOGEL P
MITT AGEN NO 16: 23-38 (DEC 1973)

0537 WIDE-BAND ACTIVE VARIABLE TUNING CIRCUITS.
WATANABE A + MORI S
IEEE J SOLID-STATE CIRCUITS SC-12(4): 389-94 (AUG 1977)

0538 Q- AND OMEGA(0)-SENSITIVITIES IN POSITIVE GAIN SECOND-ORDER RC ACTIVE FILTERS.
WEYTEN L
PROC IEEE 61: 1462-3 (NOV 1972)

0539 DIGITAL-TO-ANALOG CONVERTER CONTROLS ACTIVE FILTER.
WHITMORE J
ELECTRONICS 48(2): 86 (23 JAN 1975)

0540 TUNEABLE MICROELECTRONIC ACTIVE BAND-PASS FILTER.
WHITTEN JR
USA PAT 3643173, 15 FEB 1972, P

0541 SELECTIVE RC FILTERS.
WIEDERHOLD M + SITTIG K
RADIO FERNSEHEN ELEKTRON 22(24): 811-3 (DEC 1973) (IN GERMAN)

0542 RC FILTERS.
WIEDERHOLD M + SITTIG K
RADIO FERNSEHEN ELEKTRON 22(20): 677-80 (OCT 1973) (IN GERMAN)

0543 A 'SIMPLIFIED' NOTCH-FILTER DESIGN.
WILLIAMS RE
ELECTRON DES 24(23): 68-9 (8 NOV 1976)

0544 AN RC-ACTIVE ALL-PASS SECTION WITH REDUCED MAGNITUDE RESPONSE ERROR.
WILSON G
IEEE TRANS CIRCUITS SYST CAS-26(2): 144-8 (FEB 1979)

0545 SINGLE-AMPLIFIER ASYMMETRIC NOTCH FILTERS.
WILSON G
INT J ELECTRON 43(4): 405-13 (OCT 1977)

0546 LOSSY-INTEGRATOR BIQUADRATIC RC-ACTIVE FILTERS.
WILSON G
IEEE TRANS CIRCUITS SYST CAS-24(8): 455-9 (AUG 1977)

0547 DESIGN AND GAIN BANDWIDTH DEPENDENCE OF A CLASS OF SINGLE AMPLIFIER SECOND-ORDER ALL-PASS RC ACTIVE NETWORKS.
WILSON G
IEEE TRANS CIRCUITS SYST CAS-24(8): 446-54 (AUG 1977)

0548 A PARALLEL-T BASED RC ACTIVE NOTCH FILTER.
WILSON G
PROC IEEE 65(4): 580-1 (APRIL 1977)

0549 AN EXTREMELY LOW CAPACITANCE VERY LOW FREQUENCY RC ACTIVE NETWORK.
WILSON G
PROC IEEE 64(11): 1626-7 (NOV 1976)

0550 CANONIC UNITY-GAIN COMPLEX-POLE ALL-PASS/SYMMETRICAL-NOTCH RC ACTIVE NETWORKS.
WILSON G
INT J ELECTRON 41(1): 41-8 (JULY 1976)

0551 NEW COMPLEX-POLE SINGLE-AMPLIFIER ALLPASS FILTER WITH REDUCED ACTIVE-ELEMENT SENSITIVITIES.
WILSON G
ELECTRON LETT 12(13): 330-1 (24 JUNE 1976)

0552 PRACTICAL MINIMUM-CAPACITANCE VERY-LOW-FREQUENCY RC ACTIVE NETWORK.
WILSON G
ELECTRON LETT 11(18): 443-4 (4 SEPT 1975)

0553 HIGH-Q RC ACTIVE BAND-PASS FILTERS WITH LOW SENSITIVITY.
WILSON G + BEDRI Y
INT J ELECTRON 39(2): 201-7 (AUG 1975)

0554 OPTIMIZE SECOND-ORDER ACTIVE FILTERS.
ZICKO CP
ELECTRON DES 20(4): 70-3 (17 FEB 1972)

0555 A FAMILY OF ACTIVE FILTERS WITH SIMPLE MEANS OF SETTING-UP.
ZINNIKER R
BULL ASS SUISSE ELEC 63(5): 223-6 (4 MARCH 1972) (IN FRENCH)

0556 ANALYSIS AND SYNTHESIS OF ACTIVE FILTERS OF THE SECOND ORDER.
ZOGG U
ELEKTRONIKER 17(6): EL1-5 (1978) (IN GERMAN)

0557 DESIGN OF AN EASILY ADJUSTABLE ACTIVE RC FILTER.
ZSCHUNKE W
P205-10 OF 1974 EUROPEAN CONF ON CIRCUIT THEORY AND DESIGN, IEE, IEEE, INST PHYS, IERE, LONDON, ENGLAND 23-26 JULY 1974, IEE, 1974

0558 DESIGN OF AN ACTIVE RC FILTER WITH TRANSFER FUNCTION NULLS.
ZYUKOVA ED + TEPLYUK IN
TELECOMMUN RADIO ENG PT 2 30(4): 92-7 (APRIL 1975)

DESIGN OF HIGH ORDER FILTERS

0559 SELECTIVE BANDPASS FILTER.
ELEKTOR 4(7-8): 30-1 (JULY-AUG 1978)

0560 18 DB PER OCTAVE HIGH/LOWPASS FILTER.
ELEKTOR 4(7-8): 18 (JULY-AUG 1978)

0561 APPLICATIONS OF INTEGRATED CIRCUIT GYRATORS. II.
ELEKTRONIK 27(12): 125-6 (2 NOV 1978) (IN GERMAN)

0562 APPLICATION OF INTEGRATED CIRCUIT GYRATOR DEVICES.
ELEKTRONIK 27(10): 89-90 (OCT 1978) (IN GERMAN)

0563 TCA580 INTEGRATED GYRATOR. II. DESIGN AND APPLICATIONS.
ELECTRON APPL BULL 34(4): 148-61 (AUG 1977)

0564 COMPUTER-AIDED DESIGN OF HIGH ORDER ACTIVE FILTERS IN INTEGRATED CIRCUITS.
ACAR C + GHAUSI MS
P134 OF COMPUTER AIDED DESIGN OF ELECTRONIC AND MICROWAVE CIRCUITS AND SYSTEMS, UNIV HULL, IEEE, HULL, ENGLAND 12-14 JULY 1977, UNIV HULL, 1977

0565 COUPLED BAND-ELIMINATION ACTIVE FILTERS.
ADAMS RL
P845-9 OF PROC OF THE 1978 IEEE INT SYMP ON CIRCUITS AND SYSTEMS, IEEE, NEW YORK, USA 17-19 MAY 1978, IEEE, 1978

0566 ACTIVE RC FILTERS WITH OPTIMUM POLE SENSITIVITY.
AGARWAL SC + MANCHANDA KK
J INST ENG (INDIA) ELECTRON TELECOMMUN ENG DIV 53PT ET 4128-33 MARCH 1973

0567 LINEAR TRANSFORMATIONS FOR THE DESIGN OF DIGITAL AND ACTIVE FILTERS.
ALI AM
P347-50 OF ICASSP 79 1979 IEEE INT CONF ON ACOUSTICS, SPEECH AND SIGNAL PROCESSING, IEEE, WASHINGTON DC, USA, 2-4 APRIL 1979, IEEE, 1979

0568 ACTIVE RC FILTERS OF THE TYPE D-3.4, D-6 AND K-140.
ALIIEV VI + VOROBIENKO PP + ZELYAKH EV + IVANITSKII AM + MIKHAYLOV VI
TELECOMMUN RADIO ENG PT 1 28(8): 42-5 (AUG 1974)

0569 INDUCTANCE SIMULATION AND FILTER DESIGN USING A SINGLE-POLE AMPLIFIER APPROXIMATION.
ALLEN J + GUMUSSOY M + HOLT AGJ
ELECTRON LETT 14(19): 629-31 (14 SEPT 1978)

0570 TABLES SHORTEN DESIGN TIME FOR ACTIVE FILTERS.
AL-NASSER F
ELECTRONICS 45(22): 113-18 (23 OCT 1972)

0571 A COMPENSATION TECHNIQUE FOR A GYRATOR AND ITS USE IN THE DESIGN OF A CHANNEL-BANK FILTER.
ANTONIOU A + NAIDU KS
IEEE TRANS CIRCUITS SYST CAS-22(4): 316-23 (APRIL 1975)

0572 COMPOSITE ACTIVE FILTER - A COMBINED WAVE-ACTIVE AND GYRATOR-C FILTER.
ATIYA FS
IEEE TRANS CIRCUITS SYST CAS-25(8): 573-9 (AUG 1978)

0573 ACTIVE HIGH PASS AND LOW PASS FILTERS USING COMMERCIALLY AVAILABLE RESISTORS AND CAPACITORS.
BERTHOLD W + LOFFLER U
ELEKTRONIK 25(9): 73-4 (SEPT 1976) (IN GERMAN)

0574 DESIGN OF RC-ACTIVE NETWORKS USING GENERALIZED-IMMITTANCE CONVERTERS.
BHATTACHARYYA BB + MIKHAEL WB + ANTONIOU A
J FRANKLIN INST 297(1): 45-58 (JAN 1974)

0575 PROPERTIES AND REALIZATION METHODS OF RC ACTIVE FILTERS.
BIALKO M + MULAWKA J
ELEKTRONIKA 14(6): 261-5 (1973) (IN POLISH)

0576 DESIGN OF THIRD ORDER SELECTIVE RC FILTERS REALIZING HIGH Q-FACTORS WITH SMALL Q-SENSITIVITIES.
BIALKO M + WALCZAK M
ARCH ELEKTROTECH 21(4): 837-55 (1972) (IN POLISH)

0577 SYNTHESIS OF MULTILOOP ACTIVE FILTERS WITH IDENTICAL BIQUADRATIC BLOCKS.
BIERNACKI R + STYBLINSKA B
ACTA POLYTECH SER III NO 7: 45-56 (1977) (IN CZECH)

0578 GENERAL SYNTHESIS OF ACTIVE FILTERS WITH IDENTICAL BIQUADRATIC SECTIONS.
BIERNACKI RM + STYBLINSKA B
P63-7 OF CIRCUIT THEORY AND DESIGN, LAUSANNE, SWITZERLAND, 4-8 SEPT 1978, MOSCHYTZ, GS, NIERYNCK, J(ED), GEORGI, 1978

0579 PROPERTIES OF TRANSFER FUNCTIONS OF THE MULTILOOP FEEDBACK ACTIVE FILTERS WITH IDENTICAL BIQUADRATIC SECTIONS.
BIERNACKI RM + STYBLINSKA B
P871-5 OF PROC OF THE 1978 IEEE INT SYMP ON CIRCUITS AND SYSTEMS, IEEE, NEW YORK, USA 17-19 MAY 1978, IEEE, 1978

0580 LOWPASS RATIONAL TRANSFER FUNCTIONS FOR THE DESIGN OF RC-ACTIVE FILTERS BY CASCADING 3RD-ORDER BLOCKS.
BIEY M + PREMOLI A
P881-5 OF PROC OF THE 1978 IEEE INT SYMP ON CIRCUITS AND SYSTEMS, IEEE, NEW YORK, USA 17-19 MAY 1978, IEEE, 1978

0581 DESIGN OF LOW-PASS MAXIMALLY FLAT RC-ACTIVE FILTERS WITH MULTIPLE REAL POLE - THE MURROMAF POLYNOMIALS.
BIEY M + PREMOLI A
IEEE TRANS CIRCUITS SYST CAS-25(4): 196-200 (APRIL 1978)

0582 A NEW CLASS OF MULTIPLE-CRITICAL-POLE EQUAL-RIPPLE RATIONAL (MCPER) FUNCTIONS FOR THE DESIGN OF C ASCADED RC-ACTIVE FILTERS WITH LOW Q-FACTORS.
BIEY M + PREMOLI A
IEEE TRANS CIRCUITS SYST CAS-25(1): 19-27 (JAN 1978)

0583 DESIGN OF LOWPASS EQUAL-RIPPLE RC-ACTIVE FILTERS BY CASCADING THIRD-ORDER BLOCKS.
BIEY M + PREMOLI A
P135-9 OF COMPUTER AIDED DESIGN OF ELECTRONIC AND MICROWAVE CIRCUITS AND SYSTEMS, UNIV HULL, IEEE, HULL, ENGLAND 12-14 JULY 1977, UNIV HULL, 1977

0584 HIGHLY SELECTIVE RC-ACTIVE FILTERS USING MUCROER POLYNOMIALS.
BIEY M + PREMOLI A
IEE J ELECTRON CIRCUITS SYST 1(4): 155-8 (JULY 1977)

0585 HIGH-Q LOW-COST RC ACTIVE FILTERS - THE RIMF (REAL INTEGRATOR MULTIPLE-FEEDBACK) SECTION.
BIEY M + PREMOLI A
IEEE TRANS CIRCUITS SYST CAS-24(1): 35-41 (JAN 1977)

0586 EQUAL-RIPPLE RATIONAL FUNCTIONS WITH MULTIPLE CRITICAL POLE PAIR.
BIEY M + PREMOLI A
ELECTRON LETT 12(10): 263-4 (13 MAY 1976)

0587 DESIGN OF ACTIVE RC FILTERS USING CASCADE TECHNIQUES.
BOHME R
NACHRICHTENTECH ELEKTRON 27(5): 211-13 (1977) (IN GERMAN)

0588 ACTIVE FOURTH-ORDER BAND PASS (FILTER) WITH ONE AMPLIFIER.
BOHME R + GEISSLER E + REICHE V
NACHRICHTENTECH ELEKTRON 24(5): 177-8,183-5 (1974)

0589 ACTIVE TONE CONTROL IN TANTALUM THIN FILM TECHNOLOGY.
BOSSELMANN W
ELEKTRON INT NO 10: 350-3 (1975) (IN GERMAN)

0590 ACTIVE FILTER DESIGN USING IC GYRATORS.
BOWLES BA + NELSON TU
ELECTRON ENG 49(597): 53-6 (OCT 1977)

0591 CIRCUITS AND LIMITATIONS OF WAVE FILTERS.
BRACKETT PO
P69-72 OF PROC OF THE 1976 IEEE INT SYMP ON CIRCUITS AND SYSTEMS, IEEE, MUNICH, GERMANY 27-29 APRIL 1976, IEEE, 1976

0592 APPLICATION OF SIGNAL FLOW GRAPHS TO THE SYNTHESIS OF LINEAR SYSTEMS. II. DESIGN OF ACTIVE LADDER FILTERS.
BRACKETT PO + SEDRA AS
P246-9 OF 1975 IEEE INT SYMP ON CIRCUITS AND SYSTEMS, IEEE, NEWTON, MASS, USA, 21-23 APRIL 1975, IEEE 1975

0593 DIRECT SFG SIMULATION OF LC LADDER NETWORKS WITH APPLICATIONS TO ACTIVE FILTER DESIGN.
BRACKETT PO + SEDRA AS
IEEE TRANS CIRCUITS SYST CAS-23(2): 61-7 (FEB 1976)

0594 AN 8-TH ORDER FILTER WITH TWO ACTIVE SECTIONS.
BROVCHENKO EN + REINBOT AE
TELECOMMUN RADIO ENG PT 2 28(8): 125-6 (AUG 1973)

0595 A LOW-PASS FILTER WITH A QUASIFLAT APPROXIMATION OF THE FREQUENCY FUNCTION.
BROVCHENKO EN + RYENBOT AE
TELECOMM RADIO ENG PART 1 31(1): 53-5 (JAN 1977)

0596 MULTIPLE-AMPLIFIER RC-ACTIVE FILTER DESIGN WITH EMPHASIS ON GIC REALIZATIONS.
BRUTON LT
IEEE TRANS CIRCUITS SYST CAS-25(10): 830-45 (OCT 1978)

0597 A TRANSISTOR REALIZATION OF THE GENERALIZED IMPEDANCE CONVERTOR.
BRUTON LT
RADIO ELECTRON ENG 42(3): 133-6 (MARCH 1972)

0598 ELECTRONICALLY TUNABLE ANALOGUE ACTIVE FILTERS.
BRUTON LT
IEEE TRANS CIRCUIT THEORY CT-19(3): 299-301 (MAY 1972)

0599 SENSITIVITY OF GENERALIZED IMMITTANCE CONVERTER EMBEDDED LADDER STRUCTURES.
BRUTON LT + HAASE AB
P427-31 OF PROC OF THE 1974 IEEE INT SYMP ON CIRCUITS AND SYSTEMS, IEEE, SAN FRANCISCO, CALIF USA, 22-25 APRIL 1974, IEEE, 1974

0600 SENSITIVITY OF GENERALIZED IMMITTANCE CONVERTER-EMBEDDED LADDER STRUCTURES.
BRUTON LT + HAASE AB
IEEE TRANS CIRCUITS SYST CAS-21(2): 245-50 (MARCH 1974)

0601 TUNABLE RC-ACTIVE FILTERS USING PERIODICALLY SWITCHED CONDUCTANCES.
BRUTON LT + PEDERSON RT
IEEE TRANS CIRCUIT THEORY CT20(3): 294-301 (MAY 1973)

0602 TUNABLE RC-ACTIVE FILTERS USING PERIODICALLY SWITCHED CONDUCTANCES.
BRUTON LT + PEDERSON RT
P350-4 OF IEEE PROC OF THE INT SYMP ON CIRCUIT THEORY, IEEE, NORTH HOLLYWOOD, CALIF, USA 18-21 APR 1972, IEEE, 1972

0603 ELECTRONICALLY TUNABLE TIME-MULTIPLEXED ACTIVE FILTERS.
BRUTON LT + PEDERSON RT
IEEE J SOLID-STATE CIRCUITS SC-7(3): 266-7 (JUNE 1972)

0604 TIME-MULTIPLEXED ACTIVE FILTERS.
BRUTON LT + PEDERSON RT
IEEE J SOLID-STATE CIRCUITS SC-7(3): 259-65 (JUNE 1972)

0605 ELECTRONICALLY PROGRAMMABLE INDUCTORLESS LADDER FILTERS.
BRUTON LT + TARCHUK B
IEEE TRANS INSTRUM MEAS IM-23(1): 35-40 (MARCH 1974)

0606 LOW-PASS GYRATOR FILTERS.
BUI NC
MITT AGEN NO 19: 65-73 (NOV 1975) (IN FRENCH)

0607 ANTONIOU'S METHOD FOR BANDPASS FILTERS WITH FDNR AND FDNC ELEMENTS.
BUI NC + HASLER M + HUYNH LT
IEEE TRANS CIRCUITS SYST CAS-25(3): 169-72 (MARCH 1978)

0608 METHOD OF DESIGN OF ACTIVE LOW-PASS AND HIGH-PASS FILTERS OF HIGH SELECTIVITY.
CACIOTTA M
ALTA FREQ 43(8): 499-508 (AUG 1974) (IN ITALIAN)

0609 BANDPASS STRUCTURES HAVING EQUAL-VALUED CAPACITORS.
CARROLL WR + STEPHENSON FW
INT J ELECTRON 39(5): 583-8 (NOV 1975)

0610 SYNTHESIS OF INTEGRABLE HIGH-Q FILTERS USING GROUNDED UNITY-GAIN AMPLIFIERS.
CHENG EHC + STROMSMOE KA
INT J ELECTRON 33(6): 665-76 (DEC 1972)

0611 SINGLE OPAMP NETWORKS FOR SHUNT ARMS IN LADDER FILTERS.
CHENG RH + LIM JT
P313-16 OF 1977 IEEE INT SYMP ON CIRCUITS AND SYSTEMS PROC, IEEE, PHOENIX, ARIZ, USA 25-27 APRIL 1975, IEEE, 1977

0612 OPEN-CIRCUIT VOLTAGE TRANSFER FUNCTION SYNTHESIS USING THE GENERALIZED POSITIVE IMPEDANCE CONVERTER.
COBB DR + SU KL
P345-9 OF IEEE PROC OF THE INT SYMP ON CIRCUIT THEORY, IEEE, NORTH HOLLYWOOD, CALIF, USA 18-21 APR 1972, IEEE, 1972

0613 ACTIVE RC FILTERS DERIVABLE FROM LC LADDER FILTERS VIA LINEAR TRANSFORMATIONS.
CONSTANTINIDES AG + DIMOPOULOS HG
IEE J ELECTRON CIRCUITS SYST 1(1): 17-21 (SEPT 1976)

0614 WAVE ACTIVE FILTERS.
CONSTANTINIDES AG + HARITANTIS I
ELECTRON LETT 11(12): 254-6 (12 JUNE 1975)

0615 CASCADE TECHNIQUE OF REALISATION OF VOLTAGE TRANSFER FUNCTIONS OF HIGHER ORDER BAND PASS FILTERS.
CULEN K
ELEKTROTECH CAS 28(10): 764-6 (1977) (IN SLOVAK)

0616 TUNABLE ACTIVE FILTER HAS SWITCHABLE RESPONSE.
CUSHING P
ELECTRONICS 46(1): 104 (4 JAN 1973)

0617 A SINGLE OP AMP ACTIVE FILTER FOR WIDELY SEPARATED POLE-ZERO PAIRS.
DANIELS RW
P270-2 OF 1975 IEEE INT SYMP ON CIRCUITS AND SYSTEMS, IEEE, NEWTON, MASS, USA, 21-23 APRIL 1975, IEEE 1975

0618 REALIZATION OF RC ACTIVE FILTERS USING ONE AMPLIFIER.
DAS SK + DAS G
J INST TELECOMMUN ENG 19(5): 227-30 (MAY 1973)

0619 GROUNDED INDUCTOR SIMULATION USING A NON-INVERTING VOLTAGE-CONTROLLED VOLTAGE SOURCE.
DAS SK + DAS G
J INST TELECOMMUN ENG 18(11): 501-3 (NOV 1972)

0620 AN ACTIVE AUDIO BANDPASS FILTER.
DAVIES N
RADIO COMMUN 53(1): 30-1 (JAN 1977)

0621 REACTIVE-ACTIVE FILTERS.
DE JULIO U + GIRETTI A + MARTINELLI G
IEEE TRANS CIRCUIT THEORY CT20(2): 113-19 (MARCH 1973)

0622 APPLY TOPOLOGICAL GRAPHS.
DEFALCO JA
ELECTRON DES 23(9): 84-7 (26 APRIL 1975)

0623 ACTIVE RC-FILTERS FOR VERY LOW FREQUENCIES USING INTEGRATED MICROCIRCUITS.
DEMCHENKO AN + IGNATENKO VN
IZV VUZ RADIOELEKTRON 18(9): 100-2 (SEPT 1975) (IN RUSSIAN)

0624 LINEAR TRANSFORMATION ACTIVE FILTERS.
DIMOPOULOS HG + CONSTANTINIDES AG
IEEE TRANS CIRCUITS SYST CAS-25(10): 845-52 (OCT 1978)

0625 SELF DUAL L.T.A. FILTERS.
DIMOPOULOS HG + CONSTANTINIDES AG
IEE J ELECTRON CIRCUITS SYST 2(5): 137-42 (SEPT 1978)

0626 ACTIVE FILTERS-GYRATORS AND NIC CIRCUITS.
DIVJAK A + PETERNELJ J
ELEKTROTEH VESTN 39(1-3): 33-6 (JAN 1972) (IN SLOVENE)

0627 SYNTHESIS OF A LEAPFROG CONFIGURATION EQUIVALENT TO AN LC-LADDER FILTER BETWEEN GENERALIZED TERMINATIONS.
DUBOIS D + NEIRYNCK J
IEEE TRANS CIRCUITS SYST CAS-24(11): 590-7 (NOV 1977)

0628 SYNTHESIS OF BANDPASS LF-FILTERS BETWEEN GENERALIZED RESISTIVE TERMINATIONS.
DUBOIS D + NEIRYNCK J
P348-50 OF 1977 IEEE INT SYMP ON CIRCUITS AND SYSTEMS PROC, IEEE, PHOENIX, ARIZ, USA 25-27 APRIL 1975, IEEE, 1977

0629 APPLICATION OF CAPACITANCE MINIMISATION TO LOWPASS ACTIVE RC STRUCTURES.
DUNNING-DAVIES J + STEPHENSON FW
ELECTRON LETT 11(9): 198-9 (1 MAY 1975)

0630 SOME RECENT CONTRIBUTIONS TO INDUCTOR SIMULATION AND APPLICATIONS TO LOW SENSITIVITY BIQUAD DESIGN.
DUTTA ROY SC
J INST ELECTRON TELECOMMUN ENG 22(7): 419-25 (JULY 1976)

0631 ACTIVE RC FILTERS USING A SINGLE DIFFERENTIAL INPUT OPERATIONAL AMPLIFIER.
DUTTA ROY SC
J INST TELECOMMUN ENG 18(6): 273-5 (JUNE 1972)

0632 EQUAL-VALUED GROUNDED-CAPACITOR ACTIVE RC REALISATION OF ALL-POLE TRANSFER FUNCTIONS.
DUTTA ROY SC + PRASAD SC + BHARGAVA DK
ELECTRON LETT 13(19): 566-7 (15 SEPT 1977)

0633 ACTIVE FILTER UNIT.
EDWARDS D
ELECTRON AUST 39(11): 50-1,53-4 (FEB 1978)

0634 D3 PCM CHANNEL BANK.
EVANS JB + GAUNT WB
P7D/1-5 OF INT CONF ON COMMUNICATIONS, IEEE, MINNEAPOLIS, MINN, USA, 17-19 JUNE 1974, IEEE, 1974

0635 STRUCTURE INDEPENDENT SYNTHESIS OF ANALOG FILTERS WITH DIGITALLY TUNABLE PARAMETERS.
FERLING HD
P518-22 OF CIRCUIT THEORY AND DESIGN, LAUSANNE, SWITZERLAND, 4-8 SEPT 1978, MOSCHYTZ, GS, NIERYNCK, J(ED), GEORGI 1978

0636 INTERCHANGING OF THE SECOND-ORDER NETWORK WITH HIGH Q-FACTOR POLES BY LOW Q-FACTOR STAGES.
FILANOVSKY IM + STROMSMOE KA
INT J CIRCUIT THEORY APPL 7(2): 151-70 (APRIL 1979)

0637 INTERCHANGEABLE ACTIVE FILTERS.
FLIEGE N
BAUELEM ELEKTROTECH 12(6): 30,32,34,36 (JUNE 1977) (IN GERMAN)

0638 PROPERTIES AND DESIGN OF ACTIVE FILTERS.
FLIEGE N
NACHRICHTENTECH Z 26(9): 423-34 (SEPT 1973) (IN GERMAN)

0639 A NEW RC-ACTIVE REALIZATION OF CAUER-TYPE FILTERS.
FLIEGE N
PCT-12/111-20 OF PROC OF THE 5TH COLLOQUIUM ON MICROWAVE COMMUNICATION VOLII CIRCUIT THEORY AND COMPUTER AIDED DESIGN (PREPRINTS), HUNGARIAN ACAD SCI, SCI SOC TELECOMMUNICATION RES INST TELECOMMUNICATION, BUDAPEST HUNGARY, 24-30 JUNE 1974, BOGNAR, G(ED), AKAD KIADO, 1974

0640 RC ACTIVE FILTERS FOR THE D3 CHANNEL BANK.
FRIEDENSON RA + DANIELS RW + DOW RJ + MCDONALD PH
BELL SYST TECH J 54(3): 507-29 (MARCH 1975)

0641 ACTIVE FILTERS MAKE IT SMALL IN THE D3 CHANNEL BANK.
FRIEDENSON RA
BELL LAB REC 51(4): 104-11 (APRIL 1973)

0642 LOW ACCURACY APPROXIMATION OF NETWORKS WITH GIVEN PARAMETERS.
FRITZSCHE G
NACHRICHTENTECH ELEKTRON 25(9): 332-4 (1975) (IN GERMAN)

0643 POLE DATA AND COEFFICIENTS OF TRANSMISSION FACTORS WITH MINIMAL POLE QUALITY (T(M) APPROXIMATION).
FRITZSCHE G + JENTSCHEL H-J
NACHRICHTENTECH ELEKTRON 25(6): 225-32 (1975) (IN GERMAN)

0644 A NEW PROPOSAL FOR THE UNIFORM APPROXIMATION IN SELECTIVE CIRCUITS WITH MINIMUM POLE QUALITY.
FRITZSCHE G + JENTSCHEL H-J
NACHRICHTENTECH ELEKTRON 24(6): 202-5 (1974) (IN GERMAN)

0645 LOW-SENSITIVITY REALIZATIONS OF BAND-ELIMINATION ACTIVE FILTERS.
GADENZ RN
IEEE TRANS CIRCUITS SYST CAS-24(4): 175-83 (APRIL 1977)

0646 LOW-SENSITIVITY REALIZATIONS OF BAND-ELIMINATION ACTIVE FILTERS.
GADENZ RN
P78-61 OF PROC OF THE 1976 IEEE INT SYMP ON CIRCUITS AND SYSTEMS, IEEE, MUNICH, GERMANY 27-29 APRIL 1976, IEEE, 1976

0647 REALISATION OF ACTIVE ANALOGUE SELECTIVE SYSTEMS.
GEISSLER E
NACHRICHTENTECH ELEKTRON 23(9): 345-7 (1973) (IN GERMAN)

0648 RC FILTERS OF A HIGHER DEGREE WITH MULTIPLE FEEDBACK.
GEISSLER E + PAULSEN D
NACHRICHTENTECH ELEKTRON 23(2): 68-72 (1973) (IN GERMAN)

0649 DESIGN OF RC-ACTIVE FOLLOW-THE-LEADER FEEDBACK FILTERS.
GENSEL J
INT J CIRCUIT THEORY APPL 5(4): 361-5 (OCT 1977)

0650 'STATE' FILTER-AN ACTIVE RC-STRUCTURE OF OPTIMUM SENSITIVITY.
GENSEL J
FERNMELDETECHNIK 12(3): 141-2 (1972) (IN GERMAN)

0651 AN ACTIVE RC FILTER FOR PCM SYSTEMS.
GENSEL J + MEISSNER R + SEIDEL V
INT J CIRCUIT THEORY APPL 1(4): 315-22 (DEC 1973)

0652 VALUE OF COPYING (IN ACTIVE FILTER DESIGN).
GIRLING FEJ
P2/1-4 OF COLLOQUIUM ON ELECTRONIC CIRCUIT ASPECTS OF ACTIVE FILTERS, IEE, LONDON, ENGLAND, 13 OCT 1972, IEE, 1972

0653 DESIGN ASPECTS AND PERFORMANCE LIMITATIONS OF AMPLIFIER ISOLATED ACTIVE R-C FILTERS.
GLASFORD GM
P44-8 OF CONF RECORD OF THE TWELFTH ASILOMAR CONF ON CIRCUITS, SYSTEMS AND COMPUTERS NAVAL POSTGRADUATE SCHOOL MONTEREY, IEEE, UNIV SANTA CLARA, PACIFIC GROVE, CA, USA, 6-8 NOV 1978, IEEE, 1978

0654 NEW TECHNIQUES FOR ACTIVE FILTER DESIGN INCORPORATING MODERN OPERATIONAL AMPLIFIER TECHNOLOGIES.
GLASFORD GM
P63-9 OF ELEVENTH ANNUAL ASILOMAR CONF ON CIRCUITS SYSTEMS AND COMPUTERS UNIV SANTA CLARA, NAVAL POSTGRADUATE SCHOOL, IEEE, PACIFIC GROVE, CA, USA, 7-9 NOV 1977, IEEE, 1978

0655 A COMPARISON OF THE TWO FILTER DESIGNS.
GOOD BF
P3/1-3 OF COLLOQUIUM ON ELECTRONIC CIRCUIT ASPECTS OF ACTIVE FILTERS, IEE, LONDON, ENGLAND, 13 OCT 1972, IEE, 1972

0656 ART OF DESIGNING ACTIVE FILTERS WITHOUT DIFFICULTY.
GRALL J
ELECTRON APPL IND NO 246: 21-6 (15 JAN 1978) (IN FRENCH)

0657 ACTIVE RC LOW PASS FILTERS FOR FDM AND PCM SYSTEMS.
HAIGH DG + JEFFERS R
P6/1-4 OF COLLOQUIUM ON ELECTRONIC CIRCUIT ASPECTS OF ACTIVE FILTERS, IEE, LONDON, ENGLAND, 13 OCT 1972, IEE, 1972

0658 DESIGN OF AN AUDIO-FREQUENCY ACTIVE RC BAND-PASS FILTER FOR A SPECIFIC ENGINEERING REQUIREMENT.
HAIGH DG + JEFFERS R
RADIO ELECTRON ENG 42(8): 373-80 (AUG 1972)

0659 PRACTICAL LC FILTER SIMULATION TECHNIQUES FOR THE DESIGN OF ACTIVE RC BAND-PASS FILTERS.
HAIGH DG + SARAGA W
P336-41 OF 1974 EUROPEAN CONF ON CIRCUIT THEORY AND DESIGN, IEE, IEEE, INST PHYS, IERE, LONDON, ENGLAND 23-26 JULY 1974, IEE, 1974

0660 SIGNAL FLOW FILTERS.
HARITANTIS I
P68-72 OF CIRCUIT THEORY AND DESIGN, LAUSANNE, SWITZERLAND, 4-8 SEPT 1978, MOSCHYTZ, GS, NIERYNCK, J(ED), GEORGI 1978

0661 WAVE ACTIVE FILTERS.
HARITANTIS I + CONSTANTINIDES AG + DELIYANNIS T
PROC INST ELECTR ENG 123(7): 676-82 (JULY 1976)

0662 OPTIMIZATION OF AN ACTIVE FILTER FOR THE PCM CHANNEL BANK.
HASHIMOTO T + AZIKI T + AKIYAMA H + MORINAGA S + KIKUCHIHARA H
TRANS INST ELECTRON COMMUN ENG JPN SECT E E60(2): 80 (FEB 1977)

0663 AN ACTIVE RC FILTER SYNTHESIS BY IMPEDANCE TRANSFORMATION.
HAYAHARA E
TRANS INST ELECTRON COMMUN ENG JPN SECT E E62(1): 25-6 (JAN 1979)

0664 LADDER SYNTHESIS OF RESISTOR ROTATOR NETWORKS.
HAYAHARA E
BULL NAGOYA INST TECHNOL 24: 225-30 (1972) (IN JAPANESE)

0665 SYNTHESIS OF WIDEBAND EQUALIZER-NETWORKS BY ACTIVE RC ELEMENTS.
HENNICKE R
FERNMELDETECHNIK 13(3): 129-32,37 (1973) (IN GERMAN)

0666 POST-SAMPLING RECONSTITUTION FILTER FOR VF SYSTEMS.
HICK B
ELECTRON ENG 51(625): 57-8 (JULY 1979)

0667 INPUT AND GROUND AS COMPLEMENTS IN ACTIVE FILTERS.
HILBERMAN D
P138-41 OF INT SYMP ON CIRCUIT THEORY 1973, IEEE, TORONTO, CANADA, 9-11 APRIL 1973, IEEE, 1973

0668 INPUT AND GROUND AS COMPLEMENTS IN ACTIVE FILTERS.
HILBERMAN D
IEEE TRANS CIRCUIT THEORY CT-20(5): 540-7 (SEPT 1973)

0669 ACTIVE RC FILTERS CONTAINING PERIODICALLY OPERATED SWITCHES.
HIRANO K + NISHIMURA S
IEEE TRANS CIRCUIT THEORY CT-19(3): 253-60 (MAY 1972)

0670 REALISATION OF ACTIVE FILTERS USING GYRATORS.
HO CF
MONITOR 39(8-9): 134-44 (SEPT-OCT 1978)

0671 A SIMPLE METHOD OF DESIGNING NTH ORDER ACTIVE RC FILTERS.
HO CF
HONG KONG ENG 6(10): 21-3 (OCT 1978)

0672 COMPUTER AIDED DESIGN OF ACTIVE RC FILTERS.
HRUBY J
SLABOPROUDY OBZ 37(1): 25-30 (JAN 1976) (IN CZECH)

0673 DESIGN OF ACTIVE LOW-PASS AND HIGH-PASS RC FILTERS WITH ONE OPERATIONAL AMPLIFIER.
HRUBY J
SDELOVACI TECH 23(12): 443-8 (DEC 1975) (IN CZECH)

0674 APPROXIMATION AND REALIZATION OF ACTIVE CASCADE FILTERS WITH A MINIMUM POLE QUALITY.
HRUBY J + KOMINEK Z + NOVAK M + FRITZSCHE G + JENTSCHEL H-J
NACHRICHTENTECH ELEKTRON 25(5): 197-200 (1975) (IN GERMAN)

0675 VOLTAGE TUNABLE MULTIPOLE BANDPASS ACTIVE FILTERS.
HURTIG G
P569-72 OF PROC OF THE 1974 IEEE INT SYMP ON CIRCUITS AND SYSTEMS, IEEE, SAN FRANCISCO, CALIF USA, 22-25 APRIL 1974, IEEE, 1974

0676 ACTIVE FILTERS.
HUYNH LT + HASLER M + DUBOIS D
BULL ASSOC SUISSE ELECTR 69(12): 642-8 (24 JUNE 1978) (IN FRENCH)

0677 REALIZATION OF BANDPASS FILTER WITH FLOATING FREQUENCY-DEPENDENT-NEGATIV CONDUCTANCES USING OPERATIONAL TRANSCONDUCTANCE AMPLIFIERS.
HUYNH LT + KNOB A + ZUFFEREY J-L
P155-9 OF CIRCUIT THEORY AND DESIGN, LAUSANNE, SWITZERLAND, 4-8 SEPT 1978, MOSCHYTZ, GS, NIERYNCK, J(ED), GEORGI 1978

0678 A REALIZATION OF A 3RD-ORDER LOW-PASS NOTCH ACTIVE FILTER.
ISHIBASHI Y
TRANS INST ELECTRON COMMUN ENG JPN SECT E E61(8): 650 (AUG 1978)

0679 4TH-ORDER BANDPASS RC ACTIVE FILTER DERIVED FROM BIQUAD CIRCUIT.
ISHIBASHI Y
TRANS INST ELECTRON COMMUN ENG JPN SECT E E61(7): 549 (JULY 1978)

0680 A REALIZATION OF AN RC ACTIVE FILTER HAVING 6TH-ORDER BANDPASS CHARACTERISTICS.
ISHIBASHI Y
TRANS INST ELECTRON COMMUN ENG JPN SECT E E60(11): 649 (NOV 1977)

0681 A REALIZATION OF A 3RD-ORDER NOTCH ACTIVE FILTER.
ISHIBASHI Y
TRANS INST ELECTRON COMMUN ENG JPN SECT E E60(11): 648 (NOV 1977)

0682 DESIGN OF HYBRID INTEGRATED RC ACTIVE BPF FOR MULTI-FREQUENCY SIGNALING.
IWATA A + YOSHIMURA H + YANAGAWA H
TRANS INST ELECTRON COMMUN ENG JPN SECT E E60(6): 326-7 (JUNE 1977)

0683 ACTIVE RC LOWPASS FILTERS FOR F.D.M. AND P.C.M. SYSTEMS.
JEFFERS R + HAIGH DG
PROC INST ELECTR ENG 120(9): 945-53 (SEPT 1973)

0684 A STATE-SPACE APPROACH TO ACTIVE FILTER DESIGN.
JESS J
IEEE TRANS CIRCUIT THEORY CT-20(5): 586-9 (SEPT 1973)

0685 ACTIVE FILTERS.
KACMAZ M
ELEK MUHENDISLIGI 16(181): 21-37 (1972) (IN TURKISH)

0686 SYNTHESIS OF ACTIVE RC FILTERS.
KHRISTICH VV + KUFLEVSKIY EI
TELECOMMUN RADIO ENG PT 1 26(5): 47-8 (MAY 1972)

0687 ALGORITHM FOR AUTOMATIC DESIGN OF BANDPASS CHEBYSHEV FILTERS.
KISILEWICZ J
ROZPR ELECTROTECH 21(4): 821-42 (1975) (IN POLISH)

0688 A NARROWBAND ACTIVE RC FILTER.
KOTOV AT + KIRSANOV AA
IZV VUZ RADIOELEKTRON 21(7): 73-6 (JULY 1978) (IN RUSSIAN)

0689 SIMPLIFIED DESIGN OF ACTIVE R.C. CIRCUITS WITH PRESCRIBED ATTENUATION CHARACTERISTICS.
KOWALSKI H-J
RADIO FERNSEHEN ELEKTRON 21(8): 267-9 (2 APRIL 1972) (IN GERMAN)

0690 F.D.N.R. ACTIVE FILTER TERMINATION.
KRAICER O
ELECTRON ENG 50(601): 28-9 (JAN 1978)

0691 METHOD FOR CALCULATING THE TRANSMISSION FUNCTION OF A FILTER BY MEANS OF CASCADED FRACTIONS.
KRAUS K
NACHR ELEKTRON 33(6): 196-8 (JUNE 1979) (IN GERMAN)

0692 CAUER-FILTERS WITH GYRATORS TCA 580.
KRAUS K
NACHR ELEKTRON 32(12): 396-400 (DEC 1978) (IN GERMAN)

0693 FILTER CIRCUITS USING THE TCA 580 GYRATOR. II.
KRAUS K
NACHR ELEKTRON 32(7): 225-6 (JULY 1978) (IN GERMAN)

0694 FILTER CIRCUITS USING THE TCA 580 GYRATOR. I.
KRAUS K
NACHR ELEKTRON 32(6): 188-91 (JUNE 1978) (IN GERMAN)

0695 INVESTIGATING A GYRATORY ACTIVE FILTER.
KUCAROV SI
ELEKTRO PROM-ST PRIBOROSTR 10(2): 58-60 (FEB 1975)

0696 SYNTHESIS OF LADDER NETWORKS WITH RESISTANCES AT BOTH ENDS BY THE RECURRENT-CONTINUANT METHOD.
KUMAR JHA D + CHARAN PRASAD S
RADIO ELECTRON ENG 46(8-9): 421-8 (AUG-SEPT 1976)

0697 DESIGN OF RC ACTIVE CHANNEL FILTER.
KUROSAWA H + IWATA A + YAMANE M
ELECTR COMMUN LAB TECH J 23(1): 17-35 (1974) (IN JAPANESE)

0698 FILTER APPLICATIONS IN COMMUNICATIONS AND ELECTRONICS INDUSTRY.
KURTH CF
P33-42 OF PROC OF THE 28TH ANNUAL FREQUENCY CONTROL SYMP 1974, US ARMY, ATLANTIC CITY, NJ, USA 29-31 MAY 1974, ELECTRONIC INDUSTRIES ASSOC, 1974

0699 MINIMUM SENSITIVITY MULTIPLE LOOP FEEDBACK BANDPASS ACTIVE FILTERS.
LAKER KR + GHAUSI MS
P458-61 OF 1977 IEEE INT SYMP ON CIRCUITS AND SYSTEMS PROC, IEEE, PHOENIX, ARIZ, USA 25-27 APRIL 1975, IEEE, 1977

0700 COMPUTER-AIDED ANALYSIS AND DESIGN OF FOLLOW-THE-LEADER FEEDBACK ACTIVE-RC NETWORKS.
LAKER KR + GHAUSI MS
INT J CIRCUIT THEORY APPL 4(2): 177-84 (APRIL 1976)

0701 MINIMUM SENSITIVITY ACTIVE (LEAPFROG) AND PASSIVE LADDER BANDPASS FILTERS.
LAKER KR + GHAUSI MS + KELLY JJ
IEEE TRANS CIRCUITS SYST CAS-22(8): 670-7 (AUG 1975)

0702 MINIMUM SENSITIVITY LEAPFROG ACTIVE RC FILTERS.
LAKER KR + GHAUSI MS + KELLY JJ
P201-4 OF PROC OF THE 1974 IEEE INT SYMP ON CIRCUITS AND SYSTEMS, IEEE, SAN FRANCISCO, CALIF USA, 22-25 APRIL 1974, IEEE, 1974

0703 SYNTHESIS OF A LOW-SENSITIVITY MULTILOOP FEEDBACK ACTIVE RC FILTER.
LAKER KR + GHAUSI MS
IEEE TRANS CIRCUITS SYST CAS-21(2): 252-9 (MARCH 1974)

0704 A LOW SENSITIVITY MULTILOOP FEEDBACK ACTIVE RC FILTER.
LAKER KR + GHAUSI MS
P126-9 OF INT SYMP ON CIRCUIT THEORY 1973, IEEE, TORONTO, CANADA, 9-11 APRIL 1973, IEEE, 1973

0705 MULTIPLE-LOOP FEEDBACK TOPOLOGIES FOR THE DESIGN OF LOW-SENSITIVITY ACTIVE FILTERS.
LAKER KR + SCHAUMANN R + GHAUSI MS
IEEE TRANS CIRCUITS SYST CAS-26(1): 1-21 (JAN 1979)

0706 SOME NEW LOSSLESS FLOATING INDUCTANCE CIRCUITS.
LE QUI THE + YANAGISAWA T
PROC IEEE 65(7): 1071-2 (JULY 1976)

0707 DESCRIBES HOW TO DESIGN SIMPLE ACTIVE FILTERS. ACTIVE FILTERS SIMPLIFIED.
LEE MR
ELECTRON NO 15: 33-4 (9 NOV 1972)

0708 ACTIVE RC-FILTERS BASED ON UNITS WITH INDEPENDENT ADJUSTMENT OF THE POSITION OF THE POLES AND ZEROES OF THE TRANSMISSION CHARACTERISTIC.
LER AM
OTBOR PEREDACHA INF NO 31: 91-4 (1972) (IN RUSSIAN)

0709 ANALYSIS AND DESIGN OF TUNABLE HIGH ORDER ACTIVE FILTERS.
LING XIE-TING
FUDAN J NO 3: 78-87 (SEPT 1978) (IN CHINESE)

0710 APPLICATION OF ACTIVE GYRATOR DEVICES TO FILTER DESIGN.
LOYEZ P + SCHWARTZEL J
CABLES TRANSM 28(2): 104-31 (APRIL 1974) (IN FRENCH)

0711 THIRD-ORDER POLE-ZERO ASSIGNMENT MAXIMIZING THE DYNAMIC RANGE OF RC-ACTIVE FILTERS.
LUDER E
ARCH ELEKTRON UEBERTRAGUNGSTECH 26(11): 487-92 (NOV 1972) (IN GERMAN)

0712 RIGHT GYRATOR TRIMS THE FAT OFF ACTIVE FILTERS.
LYNCH TH
ELECTRONICS 50(15): 115-19 (21 JULY 1977)

0713 SENSITIVITY EVALUATION AND COMPARISON FOR HIGH-ORDER ACTIVE FILTERS.
MACKAY R + SEDRA AS
INT J CIRCUIT THEORY APPL 4(4): 345-56 (OCT 1976)

0714 DESIGN OF SIGNAL-FLOW GRAPH (SFG) ACTIVE FILTERS.
MARTIN K + SEDRA AD
IEEE TRANS CIRCUITS SYST CAS-25(4): 185-95 (APRIL 1978)

0715 A COMPARISON OF HIGH-ORDER ACTIVE FILTERS.
MARTIN K + SEDRA AS
P260-1 OF INT ELECTRICAL, ELECTRONICS CONF AND EXPOSITION, IEEE, TORONTO, CANADA, 26-28 SEPT 1977 IEEE, 1977

0716 OPTIMUM DESIGN OF ACTIVE FILTERS USING THE GENERALIZED IMMITTANCE COVERTER.
MARTIN K + SEDRA AS
IEEE TRANS CIRCUITS SYST CAS-24(9): 495-503 (SEPT 1977)

0717 DESIGNING LEAP-FROG AND SFG FILTERS WITH OPTIMUM DYNAMIC RANGE.
MARTIN K + SEDRA AS
PROC IEEE 65(8): 1210-11 (AUG 1977)

0718 OPTIMUM DESIGN OF ACTIVE FILTERS USING THE GENERALIZED IMMITTANCE CONVERTER.
MARTIN K + SEDRA AS
P336-9 OF 1977 IEEE INT SYMP ON CIRCUITS AND SYSTEMS PROC, IEEE, PHOENIX, ARIZ, USA 25-27 APRIL 1975, IEEE, 1977

0719 MODIFIED CHEBYSHEV FILTERING FUNCTIONS WITH LOW Q-FACTOR.
MASSAD K + YARLAGADDA R
IEEE TRANS CIRCUITS SYST CAS-23(5): 312-7 (MAY 1976)

0720 COMPUTER-ORIENTED DESIGN OF A CLASS OF SIMULATED INDUCTANCE NETWORKS.
MASSARA RE
P28-31 OF COLLOQUIUM ON ELECTRONIC FILTERS, IEEE, LONDON, ENGLAND, 20 JUNE 1978, IEE, 1978

0721 DESIGN CONSIDERATIONS FOR LOWPASS AND HIGHPASS ALL-POLE F.D.N.R. NETWORKS.
MASSARA RE
ELECTRON LETT 14(7): 225-6 (30 MARCH 1978)

0722 DIRECT-COUPLED-GYRATOR WITH FLOATING PORTS.
MATACZ J
ELEKTRONIKA 14(7): 309-11 (1973)

0723 ONE TYPE OF A SYNTHETIC INDUCTOR AND ITS APPLICATION TO THE CONSTRUCTION OF FILTERS.
MIKULA J
SDELOVACI TECH 26(1): 21-2 (JAN 1978) (IN SLOVAK)

0724 FUNCTIONAL BAND-PASS FILTER USING EQUIVALENT GYRATOR.
MIURA M + YONEDA S + KASAI T
TRANS INST ELECTRON COMMUN ENG JPN SECT E E61(7): 556-7 (JULY 1978)

0725 ANALYSIS AND DESIGN OF 'STATE' FILTERS USING INTEGRATORS WITH REAL AMPLIFICATION PROPERTIES.
MOCKEL W
NACHRICHTENTECH ELEKTRON 27(5): 214-17 (1977) (IN GERMAN)

0726 PARALLEL RESONATOR WITH A RESISTANCE AND A FREQUENCY-DEPENDENT NEGATIVE RESISTANCE REALIZED WITH A SINGLE OPERATIONAL AMPLIFIER.
MOLO F
IEEE TRANS CIRCUITS SYST CAS-21(6): 783-8 (NOV 1974)

0727 PARALLEL RESONATOR WITH A RESISTANCE AND A FREQUENCY DEPENDENT NEGATIVE RESISTANCE REALISED WITH A SINGLE OPERATIONAL AMPLIFIER.
MOLO F
P130-3 OF INT SYMP ON CIRCUIT THEORY 1973, IEEE, TORONTO, CANADA, 9-11 APRIL 1973, IEEE, 1973

0728 ACTIVE ELLIPTIC FILTERS.
MOORE HP + JOHNSON DE + JOHNSON JR
P335-6 OF PROC OF THE 1976 IEEE SOUTHEASTCON REGION 3 CONF ON ENGINEERING IN A CHANGING ECONOMY, IEEE CLEMSON, SC, USA, 5-7 APRIL 1976, IEEE, 1976

0729 A BIDIRECTIONAL TONE SIGNALLING SYSTEM USING ACTIVE FILTERS WITH DIGITAL CONTROL.
MOOREHEAD TJ + LAWTON JS
P396-400 OF PROC OF THE 1974 IEEE INT SYMP ON CIRCUITS AND SYSTEMS, IEEE, SAN FRANCISCO, CALIF USA, 22-25 APRIL 1974, IEEE, 1974

0730 A PARAMETRIC 'NIC'.
MOOS P
SDELOVACI TECH 20(4): 139 (1972) (IN CZECH)

0731 A LOW-COST PARALLEL IMPLEMENTATION FOR ACTIVE FILTERS.
MORAN PL
IEE J ELECTRON CIRCUITS SYST 2(1): 21-5 (JAN 1978)

0732 ACTIVE FILTER USING NORTON OP-AMP GYRATORS.
MORRIS JE
ELECTRON ENG 50(608): 17, 19 (JUNE 1978)

0733 GYRATOR C-FILTERS WITH MUTUAL COUPLING OF NON-ADJACENT NETWORK ELEMENTS.
MULLER Z
TESLA ELECTRON 11(3): 73-9 (SEPT 1978)

0734 MUTATOR SIMULATED FLOATING INDUCTORS.
MURATA T + RIKOSKI A
INT J ELECTRON 39(2): 229-32 (AUG 1975)

0735 FLOATING INDUCTANCE BY COMPLEMENTARY DIFFERENTIAL AMPLIFIERS.
NAKAO S + MURAKAMI T + USAMI K + YAMANAKA S
REP UNIV ELECTRO-COMMUN 28(1): 1-9 (AUG 1977) (IN JAPANESE)

0736 EQUAL-VALUED EARTHED-CAPACITOR REALISATION OF A THIRD-ORDER LOWPASS BUTTERWORTH CHARACTERISTIC USING CURRENT CONVEYORS.
NANDI R
ELECTRON LETT 14(22): 699-700 (26 OCT 1978)

0737 ACTIVE INDUCTANCES USING CURRENT CONVEYORS AND THEIR APPLICATION IN A SIMPLE BANDPASS FILTER REALISATION.
NANDI R
ELECTRON LETT 14(12): 373-5 (8 JUNE 1978)

0738 PSEUDOPASSIVITY AND INSENSITIVITY OF DIGITAL FILTERS.
NEIRMYNCK J + VINCKENBOSCH C
P1079-82 OF PROC OF THE 1978 IEEE INT SYMP ON CIRCUITS AND SYSTEMS, IEEE, NEW YORK, USA 17-19 MAY 1978, IEEE, 1978

0739 LOSSLESS FILTER SENSITIVITY PROBLEM.
NEIRYNCK J
J INST ELECTRON TELECOMMUN ENG 23(2): 72-5 (FEB 1977)

0740 SIMULATION OF LC FOUR-POLES WITH ACTIVE NETWORKS.
NEMESSZEGHY G
HIRADASTECHNIKA 26(8): 246-53 (AUG 1975) (IN HUNGARIAN)

0741 AN ACTIVE POWER BANDPASS FILTER.
NIENHAUS HA + BOWERS JC + BROOKS JL
P307-16 OF PECC '78 RECORD, IEEE, ET AL, SYRACUSE, NY, USA, 13-15 JUNE 1978, IEEE, 1978

0742 AN EFFICIENT METHOD FOR OBTAINING ELEMENT VALUES OF LADDER FILTERS WHOSE TRANSFER FUNCTIONS POSSESS COMPLEX ZEROS AND WHOSE REACTIVE COMPONENTS HAVE FREELY ASSIGNABLE LOSSES.
NIGHTINGALE C
INT J CIRCUIT THEORY APPL 6(1): 49-55 (JAN 1978)

0743 EXACT SYNTHESIS OF ACTIVE LOWPASS FREQUENCY-DEPENDENT NEGATIVE-RESISTANC FILTERS.
NIGHTINGALE C + ROLLETT JM
ELECTRON LETT 10(3): 34-5 (7 FEB 1974)

0744 MULTIPLE FEEDBACK ACTIVE RC FILTERS WITH COMPLEX FREQUENCY TRANSMISSION ZEROS.
NISHIKAWA K + TAKEBE T
TRANS INST ELECTRON COMMUN ENG JPN SECT E 59(9): 11-12 (SEPT 1976)

0745 ACTIVE RC REALIZATION OF BILATERAL TRANSMISSION FILTERS.
NISHIKAWA K + HWANG J + TAKEBE T
MEM FAC TECHNOL KANAZAWA UNIV 8(1): 113-14 (MARCH 1974) (IN JAPANESE)

0746 ANALOG ACTIVE FILTERS - SOME DESIGN CONSIDERATIONS AND TECHNOLOGICAL ASPECTS.
NOSSEK JA
P208-17 OF CIRCUIT THEORY AND DESIGN, LAUSANNE, SWITZERLAND, 4-8 SEPT 1978, MOSCHYTZ, GS, NIERYNCK, J(ED), GEORGI 1978

0747 SYNTHESIS OF NON-CASCADE ACTIVE FILTERS.
NOVAK M
NACHRICHTENTECH ELEKTRON 25(9): 337-9 (1975) (IN GERMAN)

0748 NON-CASCADE SYNTHESIS OF RC ACTIVE FILTER NETWORKS.
NOVAK M
SLABOPROUDY OBZ 35(11): 506-10 (NOV 1974) (IN CZECH)

0749 OPTIMUM ARRANGEMENT OF CASCADED ACTIVE ELECTRIC WAVE FILTERS.
NOVAK M
SLABOPROUDY OBZ 35(2): 53-7 (FEB 1974) (IN CZECH)

0750 OPTIMIZED DESIGN OF CASCADED ACTIVE RC BANDPASS FILTERS.
NOVAK M + HRUBY J
INT J CIRCUIT THEORY APPL 3(2): 113-21 (JUNE 1975)

0751 A STATE-VARIABLE REALIZATION OF ACTIVE RC FILTERS WITHOUT INTEGRATORS.
OGUCHI M + TAKAHASHI T + HAMADA N + TAKAHASHI S
TRANS INST ELECTRON COMMUN ENG JPN SECT E E60(11): 647 (NOV 1977)

0752 ACTIVE RC SYNTHESIS OF TRANSFER FUNCTIONS FOR LC FILTER.
OKINE M
TRANS INST ELECTRON COMMUN ENG JPN SECT E E59(10): 15-16 (OCT 1976)

0753 A REALIZATION OF THE RING GYRATOR AND ITS APPLICATIONS.
OKINE M + NAKAMURA M
ELECTRON COMMUN JPN 58(8): 27-36 (AUG 1975)

0754 A REALIZATION OF THE MULTIPORT GYRATOR AND ITS APPLICATION TO TRANSMISSION NETWORKS.
OKINE M + NAKAMURA M + ISHIKAWA H
ELECTRON COMMUN JAP 57(6): 42-9 (JUNE 1974)

0755 SPEEDACTIVE MULTIPOLE FILTER DESIGN WITH A FLEXIBLE COMPUTER PROGRAM THAT CALCULATES THE COMPONENT VALUES FOR OPTIMUM PERFORMANCE.
OLSEN DE
ELECTRON DES 20(24): 142-7 (23 NOV 1972)

0756 SIMPLE MUSICAL CHIME GENERATOR.
ORR T
ELECTRON TODAY INT 7(4): 29 (APRIL 1978)

0757 SYNTHESIS OF ACTIVE RC FILTERS.
OSTAPENKO AG
TELECOMMUN RADIO ENG PT 2 31(3): 120-4 (MARCH 1976)

0758 REALIZATION OF AN ACTIVE FILTER WITH A GIVEN TRANSFER FUNCTION OF THE N-TH ORDER USING A SIGNAL FLOW GRAPH.
OSTAPENKO AG
RADIO ENG ELECTRON PHYS 20(1): 136-8 (JAN 1975)

0759 CASCADE REALISATION OF HIGHER-ORDER TRANSFER FUNCTIONS WITH THE AID OF SIGNAL GRAPHS.
OSTAPENKO AG
IZV VUZ RADIOELEKTRON 18(11): 91-2 (NOV 1975) (IN RUSSSIAN)

0760 ACTIVE BANDPASS FILTER WITH A MINIMUM NUMBER OF CAPACITORS USING FREQUENCY-DEPENDENT NEGATIVE-RESISTANCES AND A GYRATOR.
PANZER K
P38-40 OF PROC OF THE 1976 IEEE INT SYMP ON CIRCUITS AND SYSTEMS, IEEE, MUNICH, GERMANY 27-29 APRIL 1976, IEEE, 1976

0761 ACTIVE BANDPASS FILTER WITH A MINIMUM NUMBER OF CAPACITANCES USING IMPEDANCE CONVERTERS.
PANZER K
NACHRICHTENTECH Z 27(10): 379-82 (OCT 1974) (IN GERMAN)

0762 ACTIVE BANDPASS FILTERS USING FREQUENCY-DEPENDENT NEGATIVE-RESISTANCE ELEMENTS.
PANZER K
NACHRICHTENTECH Z 25(10): 455-6 (OCT 1972) (IN GERMAN)

0763 INDUCTOR REALIZATION WITH RESISTIVE FEEDBACK IN RC FILTERS.
PATRANABIS D
INT J ELECTRON 38(3): 367-74 (MARCH 1975)

0764 LOW PASS ACTIVE FILTERS OF ORDER 5.
PELAT A
TOUTE ELECTRON NO 434: 61-3 (JULY 1978) (IN FRENCH)

0765 PRINCIPLES OF CALCULATION FOR LOW PASS SECOND ORDER FILTERS.
PELAT A
TOUTE ELECTRON NO 404: 60-2 (NOV 1975) (IN FRENCH)

0766 IMPROVEMENT OF DYNAMIC RANGE AND REDUCTION OF LIMITING EFFECTS IN MULTIPLE FEEDBACK FILTERS.
PERRY DJ
P36-40 OF COLLOQUIUM ON ELECTRONIC FILTERS, IEEE, LONDON, ENGLAND, 20 JUNE 1978, IEE, 1978

0767 A MINIMAL GROUNDED CAPACITOR CASCADE SYNTHESIS FOR INTEGRATED CIRCUITS.
PIEKARSKI MS
P239-44 OF 1974 EUROPEAN CONF ON CIRCUIT THEORY AND DESIGN, IEE, IEEE, INST PHYS, IERE, LONDON, ENGLAND 23-26 JULY 1974, IEE, 1974

0768 NETWORKS FOR DIGITAL TRANSMISSION - APPLICATIONS AND TRENDS.
PINEL JF
P183-90 OF CIRCUIT THEORY AND DESIGN, LAUSANNE, SWITZERLAND, 4-8 SEPT 1978, MOSCHYTZ, GS, NIERYNCK, J(ED), GEORGI 1978

0769 ACTIVE FILTER DESIGNS FOR A TONE RECEIVER.
PINEL JF
P342-7 OF 1974 EUROPEAN CONF ON CIRCUIT THEORY AND DESIGN, IEE, IEEE, INST PHYS, IERE, LONDON, ENGLAND 23-26 JULY 1974, IEE, 1974

0770 SYNTHESIS OF ACTIVE FILTERS.
POKORNY L
RADIO ELEKTRON SCHAU 52(3): 30-3 (1976) (IN GERMAN)

0771 COMPUTER AIDED DESIGN OF ACTIVE FILTERS.
POPOVIC M
P1-303/1-3 OF INFORMATICA 78 XIII YUGOSLAV INT SYMP ON INFORMATION PROCESSING, BLED, YUGOSLAVIA 2-7 OCT 1978, RAJKOVIC, V(ED), INFORMATIKA, 1978 (IN CROATIAN)

0772 A THIRD-ORDER ACTIVE-FILTER REALIZATION HAVING A COMPLEX POLE WITH A LOW-Q SENSITIVITY WITH RESPECT TO THE GAIN-BANDWIDTH PRODUCT OF OPERATIONAL AMPLIFIER.
PRASAD NK + MAY EJP
PROC IEEE 67(2): 320-2 (FEB 1979)

0773 MULTIPLE-POLE APPROXIMATION IN THE DESIGN OF RC-ACTIVE FILTERS.
PREMOLI A
P876-80 OF PROC OF THE 1978 IEEE INT SYMP ON CIRCUITS AND SYSTEMS, IEEE, NEW YORK, USA 17-19 MAY 1978, IEEE, 1978

0774 TABLES AND GRAPHS OF THE MUCROER POLYNOMIALS FOR THE REALIZATION OF EQUAL-RIPPLE RC-ACTIVE FILTERS WITH LOW SENSITIVITY.
PREMOLI A
NOTE RECENS NOT 24 SUPPL (5-6): 399-518 (SEPT-DEC 1975) (IN ENGLISH)

0775 DESIGN OF MAXIMALLY-FLAT RC-ACTIVE FILTERS WITH LOW Q-FACTORS - TABLES AND GRAPHS OF THE MUCROMAF POLYNOMIALS.
PREMOLI A
ALTA FREQ 44(11): 684-9 (NOV 1975)

0776 A NEW CLASS OF EQUAL-RIPPLE FILTERING FUNCTIONS WITH LOW Q-FACTORS - THE MUCROER POLYNOMIALS.
PREMOLI A
IEEE TRANS CIRCUITS SYST CAS-21(5): 609-13 (SEPT 1974)

0777 MUCROMAF POLYNOMIALS - AN APPROACH TO THE MAXIMALLY FLAT APPROXIMATION OF RC ACTIVE FILTERS WITH LOW SENSITIVITY.
PREMOLI A
IEEE TRANS CIRCUIT THEORY CT-20(1): 77-80 (JAN 1973)

0778 INDUCTORLESS ACTIVE RC FILTERS.
RADHAKRISHNA G
J INST ELECTRON TELECOMMUN ENG 22(8): 507-9 (AUG 1976)

0779 SYNTHESIS OF ACTIVE RC BAND-PASS FILTERS.
RADULESCU T
POSTA TELECOMUNICATII 2(6): 291-4 (JUNE 1972) (IN RUMANIAN)

0780 AN INTERACTIVE COMPUTER PACKAGE FOR ACTIVE FILTER DESIGN.
RAETH PG + BARKSDALE WJ
P351-4 OF PROC OF SOUTHEASTCON '78 REGION 3 CONF, IEEE, ATLANTA, GA, USA, 10-12 APRIL 1978, IEEE, 1978

0781 LOW SENSITIVITY ACTIVE FILTERS WITH PREDETERMINED GROUP DELAY AND CHEBYSHEV STOPBAND ATTENUATION.
RAKOVICH BD + STOJANOVICH VS
GLAS CCCIV ACAD SERBE SCI ARTS CL SCI TECH NO 12: 65-86 (1978) (IN RUSSIAN)

0782 LOW SENSITIVITY ACTIVE FILTERS WITH PREDETERMINED GROUP DELAY AND CHEBYSHEV STOPBAND ATTENUATION.
RAKOVICH BD + STOJANOVICH VS
BULL ACAD SERBE SCI ARTS CL SCI TECH 58(12): 51-5 (1977)

0783 FREQUENCY AND TIME DOMAIN CHARACTERISTICS OF A CLASS OF FILTER FUNCTIONS WITH MULTIPLE POLES AND PRESCRIBED DELAY DISTORTION.
RAKOVICH BD + STOJANOVICH VS
PUBL ELEKTROTEH FAK SER ELEKTRON TELEKOMUN AUTOM NO 109-113: 19-30 (1976)

0784 HIGHPASS AND BANDPASS FILTERS USING SINGLE-AMPLIFIER SIMULATED INDUCTORS.
RAMSEY WT
P855-9 OF PROC OF THE 1978 IEEE INT SYMP ON CIRCUITS AND SYSTEMS, IEEE, NEW YORK, USA 17-19 MAY 1978, IEEE, 1978

0785 HIGHPASS AND BANDPASS FILTERS USING A NEW SINGLE-AMPLIFIER SIMULATED INDUCTOR.
RAMSEY WT
IEE J ELECTRON CIRCUITS SYST 2(3): 79-85 (MAY 1978)

0786 NOVEL GIC SUITABLE FOR RC ACTIVE-FILTER APPLICATIONS.
RAO MKN + HASLETT JW + BRUTON LT
ELECTRON LETT 15(15): 462-4 (19 JULY 1979)

0787 ACTIVE FILTER HAVING IDEALLY ZERO SENSITIVITY.
REINECK KM
NACHRICHTENTECH Z 26(9): 420-2 (SEPT 1973) (IN ENGLISH)

0788 ACTIVE REALIZATION OF BAND-PASS FILTER FOR CHANNEL TRANSLATING EQUIPMENT.
REWAL MK
J INST ELECTRON TELECOMMUN ENG 22(8): 503-6 (AUG 1976)

0789 DESIGN OF AN ACTIVE FILTER FOR CHANNEL TRANSLATING EQUIPMENT.
REWAL MK
J INST ELECTRON TELECOMMUN ENG 21(8): 409-12 (AUG 1975)

0790 TRIAMPLIFIER, A NEW CIRCUIT ELEMENT, AND ITS USE IN ACTIVE FILTERS.
RIORDAN RH
IEEE J SOLID-STATE CIRCUITS SC-10(6): 505-8 (DEC 1975)

0791 N-PATH FILTER TECHNIQUES.
ROBERTS J
ELECTRON NO 5: 18-19 (1 JUNE 1972)

0792 ECONOMICAL RC ACTIVE LOSSY LADDER FILTERS.
ROLLETT JM
ELECTRON LETT 9(3): 70-2 (8 FEB 1973)

0793 ECONOMICAL INSENSITIVE ACTIVE FILTERS.
ROLLETT JM
P7/1-4 OF COLLOQUIUM ON ELECTRONIC CIRCUIT ASPECTS OF ACTIVE FILTERS, IEE, LONDON, ENGLAND, 13 OCT 1972, IEE, 1972

0794 AFILT-A COMPUTER PROGRAM FOR THE SYNTHESIS OF ACTIVE FILTERS.
ROSTE T
ELEKTRO 86(7): 9-11 (5 APRIL 1973) (IN NORWEGIAN)

0795 ORTHOGONAL FILTER, AN ACTIVE UNIVERSAL FILTER.
ROTTER J + SCHMIDT W
ELEKTRONIK 26(3): 69-74 (MARCH 1977) (IN GERMAN)

0796 AN ORTHONORMAL SYSTEM OF FUNCTIONS DERIVED FROM SIMPLE RC-FUNCTIONS FOR IMPLEMENTING FILTERS WITH EQUAL-VALUED CAPACITORS.
RUPPRECHT W + SCHMIDT W
P142-6 OF IEEE PROC OF THE INT SYMP ON CIRCUIT THEORY, IEEE, NORTH HOLLYWOOD, CALIF, USA 18-21 APR 1972, IEEE, 1972

0797 RC-ACTIVE FILTERS USING EQUAL-VALUED CAPACITORS.
RUPPRECHT W + SCHMIDT W
NACHRICHTENTECH Z 25(4): 169-73 (APRIL 1972)

0798 SYNTHESIS OF ACTIVE RC NTH-ORDER TRANSFER FUNCTIONS VIA A MINIMAL-GAIN POLE AND ZERO SCHEME.
RUSSELL HT + CHAN SP
IEEE TRANS CIRCUITS SYST CAS-23(10): 609-27 (OCT 1976)

0799 GENERATION OF N-TH ORDER ACTIVE FILTER TRANSFER FUNCTIONS USING CANONIC QUADRATIC NETWORKS.
RUSSELL HT + CHAN SP
P672-6 OF PROC OF THE 1974 IEEE INT SYMP ON CIRCUITS AND SYSTEMS, IEEE, SAN FRANCISCO, CALIF USA, 22-25 APRIL 1974, IEEE, 1974

0800 SYNTHESIS OF RC ACTIVE FILTERS USING FINITE GAIN AMPLIFIERS.
SAHA SK
INT J ELECTRON 38(2): 241-5 (FEB 1975)

0801 ACTIVE FILTERS FOR COMMUNICATION.
SARAGA W
P218-27 OF CIRCUIT THEORY AND DESIGN, LAUSANNE, SWITZERLAND, 4-8 SEPT 1978, MOSCHYTZ, GS, NIERYNCK, J(ED), GEORGI 1978

0802 A CONTRIBUTION TO THE DESIGN OF LOW-SENSITIVITY ACTIVE RC NETWORKS.
SARAGA W
P191-5 OF PROC OF THE 1974 IEEE INT SYMP ON CIRCUITS AND SYSTEMS, IEEE, SAN FRANCISCO, CALIF USA, 22-25 APRIL 1974, IEEE, 1974

0803 LOWPASS-BANDPASS TRANSFORMATION FOR ACTIVE RC-FILTERS.
SARAGA W
ELECTRON LETT 8(23): 572-4 (16 NOV 1972)

0804 DESIGN OF CASCADED ACTIVE RC BAND-PASS FILTERS BY MEANS OF LOW-PASS TO BAND-PASS TRANSFORMATIONS.
SARAGA W
P8/1-4 OF COLLOQUIUM ON ELECTRONIC CIRCUIT ASPECTS OF ACTIVE FILTERS, IEE, LONDON, ENGLAND, 13 OCT 1972, IEE, 1972

0805 A DESIGN PHILOSOPHY FOR MICROELECTRONIC ACTIVE-RC FILTERS.
SARAGA W + HAIGH D + BARKER RG
PROC IEEE 67(1): 24-33 (JAN 1979)

0806 MICROELECTRONIC ACTIVE-RC CHANNEL BANDPASS FILTERS IN THE FREQUENCY RANGE 60-108 KHZ FOR FDM SSB TELEPHONE SYSTEMS.
SARAGA W + HAIGH DG + BARKER RG
IEEE TRANS CIRCUITS SYST CAS-25(12): 1022-31 (DEC 1978)

0807 A 'DESIGN PHILOSOPHY' FOR MICROELECTRONIC ACTIVE RC FILTERS.
SARAGA W + HAIGH DG + BARKER RG
P295-8 OF 1977 IEEE INT SYMP ON CIRCUITS AND SYSTEMS PROC, IEEE, PHOENIX, ARIZ, USA 25-27 APRIL 1975, IEEE, 1977

0808 DYNAMIC RANGE PROPERTIES OF HIGH-ORDER ACTIVE BANDPASS FILTERS.
SCHAUMANN R + CHALSTROM RE
P146-50 OF PROC OF THE 1978 IEEE INT SYMP ON CIRCUITS AND SYSTEMS, IEEE, NEW YORK, USA 17-19 MAY 1978, IEEE, 1978

0809 MAXIMIZATION OF DYNAMIC RANGE OF FLF BANDPASS FILTERS BY USE OF STAGGER TUNING.
SCHAUMANN R + LAKER KR + KINGHORN WA
P450-3 OF 1977 IEEE INT SYMP ON CIRCUITS AND SYSTEMS PROC, IEEE, PHOENIX, ARIZ, USA 25-27 APRIL 1975, IEEE, 1977

0810 OPTIMISATION OF SENSITIVITY AND DYNAMIC RANGE OF IFLF ACTIVE FILTERS.
SCHAUMANN R + CHASTROM RE + LAKER KR
ELECTRON LETT 13(12): 367-8 (9 JUNE 1977)

0811 MULTIPURPOSE SIMULATION NETWORK WITH A SINGLE AMPLIFIER.
SCHMIDT CE + MAN SHEK LEE
ELECTRON LETT 11(1): 9-10 (9 JAN 1975)

0812 FILTER COMPONENTS FOR ACTIVE RC FILTERS WITH MULTIPLE FEEDBACK.
SEIDEL V
NACHRICHTENTECH ELEKTRON 27(5): 202-5 (1977) (IN GERMAN)

0813 DETAILED DESIGN OF AN AC-TELEGRAPHY BANDPASS WITH OPERATIONAL AMPLIFIERS AND DOUBLE-T RC-NETWORKS.
SEIDL A
NACHR ELEKTRON 31(1): 1-4 (JAN 1977) (IN GERMAN)

0814 ACTIVE CIRCULATOR SYSTEMS.
SEWELL JI
INT J ELECTRON 33(6): 677-9 (DEC 1972)

0815 SYNTHESIS OF FOUR-TERMINAL RC ACTIVE FILTERS.
SHUKLA RS + MISHRA BN
J INST TELECOMMUN ENG 18(3): 152-7 (MARCH 1972)

0816 PROPERTIES OF ANALOGUE-COMPUTER-TYPE ACTIVE RC FILTERS.
SIENKO W + KRUGER KE
ARCH ELEKTROTECH 26(4): 663-78 (1977) (IN POLISH)

0817 SOME PROPERTIES OF THE ACTIVE RC WAVE FILTER.
SIENKO W + KRUGER KE
NACHRICHTENTECH ELEKTRON 25(5): 192-7 (1975) (IN GERMAN)

0818 A CANONICAL REALIZATION OF ACTIVE RC FILTERS.
SINGHAL K + WING O
INT J ELECTRON 37(1): 83-96 (JULY 1974)

0819 WAVE ACTIVE FILTERS BASED ON UNIT-ELEMENT PROTOTYPES.
SIU-MING NG + KU WH
ELECTRON LETT 11(24): 578-80 (27 NOV 1975)

0820 LOW-FREQUENCY FILTERS FOR MEASURING TRANSDUCERS.
SOKOLOVA NA + TSYVINSKII VG
IZV VUZ RADIOELEKTRON 19(11): 77-82 (NOV 1976) (IN RUSSIAN)

0821 CANONICAL HIGH-SELECTIVITY PARALLEL RESONATOR, USING A SINGLE OPERATIONAL AMPLIFIER, AND ITS APPLICATIONS IN FILTERS.
SOLIMAN AM + AWAD SS
IEE J ELECTRON CIRCUITS SYST 1(4): 145-8 (JULY 1977)

0822 AN APPROACH TO ACTIVE FILTER DESIGN PROBLEMS USING GYRATORS - A FRESH LOOK.
SRINIVASAN S
INT J ELECTRON 34(3): 353-8 (MARCH 1973)

0823 OPTIMIZED RC-NIC BANDPASS STRUCTURES.
STEPHENSON FW + DUNNING-DAVIES J
INT J ELECTRON 37(3): 289-93 (SEPT 1974)

0824 HIGH-PASS REALIZATION OF EQUAL-VALUED CAPACITOR ACTIVE RC NETWORKS.
STEPHENSON FW
PROC IEEE 60(7): 922 (JULY 1972)

0825 DESIGN OF ACTIVE BAND-PASS FILTERS.
STEVANKA D
SDELOVACI TECH 23(12): 456-9 (DEC 1975) (IN SLOVAK)

0826 NETWORK SYNTHESIS BY ADMITTANCE MATRIX EXPANSION.
STEVENSON JK
P5-9 OF COLLOQUIUM ON ELECTRONIC FILTERS, IEEE, LONDON, ENGLAND, 20 JUNE 1978, IEE, 1978

0827 TABULATION OF CHARACTERISTICS AND ACTIVE NETWORK DESIGN VALUES FOR CAUER ELLIPTIC FUNCTION FILTERS OF ORDER 2 TO 9.
STEVENSON JK
RADIO ELECTRON ENG 44(11): 597-604 (NOV 1974)

0828 A LESS COMPONENT-SENSITIVE ACTIVE FILTER FOR A CHANNEL UNIT OF A TIME DIVISION SWITCHING SYSTEM.
SUZUKI T + INOUE H
P135-8 OF PROC OF THE 1978 IEEE INT SYMP ON CIRCUITS AND SYSTEMS, IEEE, NEW YORK, USA 17-19 MAY 1978, IEEE, 1978

0829 ACTIVE FILTERS USING OPERATIONAL AMPLIFIERS WITH A UNIT GAIN.
SVOBODA P
SDELOVACI TECH 22(10): 371-4 (OCT 1974) (IN CZECH)

0830 FILSYN-A GENERAL PURPOSE FILTER SYNTHESIS PROGRAM.
SZENTIRMAI G
PROC IEEE 65(10): 1443-58 (OCT 1977)

0831 MULTIPLE-FEEDBACK ACTIVE FILTER STRUCTURES.
SZENTIRMAI G
P368-77 OF 7TH ASILOMAR CONF ON CIRCUITS, SYSTEMS AND COMPUTERS NAVAL POSTGRADUATE SCHOOL, UNIV SANTA CLARA, ET AL, PACIFIC GROVE, CALIF, USA, 27-29 NOV 1973, WESTERN PERIODICALS 1974

0832 SYNTHESIS OF MULTIPLE-FEEDBACK ACTIVE FILTERS.
SZENTIRMAI G
BELL SYST TECH J 52(4): 527-55 (APRIL 1973)

0833 MULTIPLE-FEEDBACK FILTERS.
SZENTIRMAI G
P339 OF IEEE PROC OF THE INT SYMP ON CIRCUIT THEORY, IEEE, NORTH HOLLYWOOD, CALIF, USA 18-21 APR 1972, IEEE, 1972

0834 SYNTHESIS OF AN UNGROUNDED INDUCTANCE USING OPERATIONAL AMPLIFIERS.
TAE WEON RHEE
J KOREAN INST ELECTRON ENG 11(4): 164-9 (AUG 1974)

0835 INTER-ORDER RESPONSE CHARACTERISTICS FOR SIMPLIFIED ACTIVE FILTERS.
THIELE AN
PROC INST RADIO ELECTRON ENG AUST 35(3): 57-60 (MARCH 1974)

0836 ACTIVE FILTERS FOR CHANNEL TRANSLATING EQUIPMENT.
THIMM R
ELECTR COMMUN 52(2): 130-6 (1977)

0837 REALIZATION OF ACTIVE BANDPASS FILTERS USING NONIDEAL IMPEDANCE CONVERTORS.
THIMM R
P73-7 OF PROC OF THE 1976 IEEE INT SYMP ON CIRCUITS AND SYSTEMS, IEEE, MUNICH, GERMANY 27-29 APRIL 1976, IEEE, 1976

0838 ACTIVE RC-FILTER DESIGN USING NONIDEAL IMPEDANCE CONVERTORS.
THIMM R
ARCH ELEKTROTECH 58(5): 245-58 (NOV 1976) (IN GERMAN)

0839 SYNTHESIS OF RC ACTIVE FILTERS HAVING ZEROS ON THE REAL FREQUENCY AXIS.
TOMLINSON GH
INT J ELECTRON 37(2): 199-207 (AUG 1974)

0840 MINIMUM SENSITIVITY IN A CLASS OF ACTIVE RC FILTERS.
TOMLINSON GH
INT J ELECTRON 33(3): 261-71 (SEPT 1972)

0841 SOME RESULTS ON TWO-SECTION GENERALIZED FLF ACTIVE FILTERS.
TOW J
IEEE TRANS CIRCUITS SYST CAS-25(4): 181-4 (APRIL 1978)

0842 SOME RESULTS ON GENERALIZED FOLLOW-THE-LEADER-FEEDBACK ACTIVE FILTERS.
TOW J
P462-5 OF 1977 IEEE INT SYMP ON CIRCUITS AND SYSTEMS PROC, IEEE, PHOENIX, ARIZ, USA 25-27 APRIL 1975, IEEE, 1977

0843 DESIGN AND EVALUATION OF SHIFTED-COMPANION-FORM ACTIVE FILTERS.
TOW J
BELL SYST TECH J 54(3): 545-68 (MARCH 1975)

0844 DESIGN AND EVALUATION OF SHIFTED COMPANION FORM (FOLLOW THE LEADER FEEDBACK) ACTIVE FILTERS.
TOW J
P656-60 OF PROC OF THE 1974 IEEE INT SYMP ON CIRCUITS AND SYSTEMS, IEEE, SAN FRANCISCO, CALIF USA, 22-25 APRIL 1974, IEEE, 1974

0845 COUPLED-BIQUAD ACTIVE FILTERS.
TOW J + KUO YL
P164-8 OF IEEE PROC OF THE INT SYMP ON CIRCUIT THEORY, IEEE, NORTH HOLLYWOOD, CALIF, USA 18-21 APR 1972, IEEE, 1972

0846 SENSITIVITY COMPARISON OF THREE TYPES OF ACTIVE RC CIRCUITS.
TURPIN RW + SARAGA W
P114-17 OF PROC OF THE 1976 IEEE INT SYMP ON CIRCUITS AND SYSTEMS, IEEE, MUNICH, GERMANY 27-29 APRIL 1976, IEEE, 1976

0847 A REALIZATION OF A LOW PASS FILTER USING A NEW INDUCTANCE CIRCUIT.
UENO F + INOUE T + KUMAZAWA M
TRANS INST ELECTRON COMMUN ENG JPN SECT E E61(9): 761 (SEPT 1978)

0848 MULTILOOP FEEDBACK SYNTHESIS METHOD OF BP ACTIVE FILTERS.
URBAS A
ACTA POLYTECH SER III NO 7: 7-16 (1977) (IN CZECH)

0849 VARIABLE FREQUENCY AND SELECTIVITY NARROW BANDPASS ACTIVE FILTER.
USAMI K + KANAI F + YAMANAKA S
REP UNIV ELECTRO-COMMUN 22(1): 11-15 (AUG 1971) (IN JAPANESE)

0850 FEASIBILITY OF ACTIVE FILTERING IN FREQUENCY DIVISION MULTIPLEX SYSTEMS.
VALIHORA J + LIM JT + BRUTON LT
P121-5 OF PROC OF THE 1974 IEEE INT SYMP ON CIRCUITS AND SYSTEMS, IEEE, SAN FRANCISCO, CALIF USA, 22-25 APRIL 1974, IEEE, 1974

0851 METHODS AND ASPECTS OF ACTIVE-RC FILTER SYNTHESIS.
VAN BOKHOVEN WMG
EINDHOVEN UNIV TECHNOL, NETHERLANDS, DEC 1970, 68P TH 71-E-27

0852 CHARACTERISTIC PROPERTIES OF SECOND-ORDER ACTIVE FILTER UNITS.
VARKONYI T
HIRADASTECHNIKA 28(11): 332-40 (NOV 1977) (IN HUNGARIAN)

0853 DESIGN OF A HYBRID INTEGRATED CIRCUIT ELLIPTIC ACTIVE RC FILTER.
VOLPE GT + FREEMAN LM
IEEE J SOLID-STATE CIRCUITS SC-9(2): 76-9 (APRIL 1974)

0854 A TUNED FILTER WITH A GYRATOR.
VORONENKOV VYU + ZAKHAROV VK + NAZIN GP
IZV VUZ PRIBOROSTR 20(6): 96-101 (1977) (IN RUSSIAN)

0855 A SYNTHESIS OF VARIABLE ACTIVE RC FILTERS CONTAINING PERIODICALLY OPERATED SWITCHES.
WATANABE H + ONDA N
TRANS INST ELECTRON COMMUN ENG JPN SECT E E61(11): 916-17 (NOV 1978)

0856 ANOTHER APPROACH TO MULTILOOP FEEDBACK.
WETENKAMP SF + VAN VALKENBURG ME
P258-60 OF 1975 IEEE INT SYMP ON CIRCUITS AND SYSTEMS, IEEE, NEWTON, MASS, USA, 21-23 APRIL 1975, IEEE 1975

0857 COMMENTS ON SENSITIVITY COMPARISON OF HIGH-Q SECOND-ORDER ACTIVE FILTER SYNTHESIS TECHNIQUES (AND REPLY).
WEYTEN L + BRUTON LT
IEEE TRANS CIRCUITS SYST CAS-26(1): 77 (JAN 1979)

0858 ELECTRONIC CIRCUITS IN AUTOMATION TECHNOLOGY. V. FILTERS AND FILTER CIRCUITS.
WOLF R + FIEDLER K
MSR (MESS STEUERN 21(5): 277-8 (MAY 1978) (IN GERMAN)

0859 ELECTRONIC CIRCUITS FOR AUTOMATION. IV. FILTERS.
WOLF R + FIEDLER K
MSR (MESS STEUERN 21(4): 215-16 (APRIL 1978) (IN GERMAN)

0860 SCATTERING PARAMETER ACTIVE FILTERS WITH REDUCED NUMBER OF ACTIVE ELEMENTS.
WUPPER H
P65-8 OF PROC OF THE 1976 IEEE INT SYMP ON CIRCUITS AND SYSTEMS, IEEE, MUNICH, GERMANY 27-29 APRIL 1976, IEEE, 1976

0861 SCATTERING PARAMETER ACTIVE FILTERS WITH REDUCED NUMBER OF ACTIVE ELEMENTS.
WUPPER H
IEEE TRANS CIRCUITS SYST CAS-23(5): 318-22 (MAY 1976)

0862 NEW CONSTRUCTION METHOD FOR COIL-FREE ACTIVE FILTERS.
WUPPER H
NACHRICHTENTECH ELEKTRON 25(9): 339-42 (1975) (IN GERMAN)

0863 NEW ACTIVE FILTER SYNTHESIS BASED ON SCATTERING PARAMETERS.
WUPPER H + MEERKOTTER K
P254-7 OF 1975 IEEE INT SYMP ON CIRCUITS AND SYSTEMS, IEEE, NEWTON, MASS, USA, 21-23 APRIL 1975, IEEE 1975

0864 NEW ACTIVE FILTER SYNTHESIS BASED ON SCATTERING PARAMETERS.
WUPPER H + MEERKOTTER K
IEEE TRANS CIRCUITS SYST CAS-22(7): 594-602 (JULY 1975)

0865 INDUCTORLESS FILTER APPLICATIONS IN TRANSMISSION SYSTEMS.
YAKUWA K + KOJIMA T + YANAGI M + SHIRAI K
P191-7 OF CIRCUIT THEORY AND DESIGN, LAUSANNE, SWITZERLAND, 4-8 SEPT 1978, MOSCHYTZ, GS, NIERYNCK, J(ED), GEORGI 1978

0866 A METHOD OF APPROXIMATING THE TRANSFER FUNCTION FOR RC ACTIVE-PASSIVE NETWORKS.
YANAGISAWA T + FUJII N
ELECTRON COMMUN JAP 56(5): 23-30 (MAY 1973)

0867 LOW-SENSITIVITY ACTIVE AND DIGITAL FILTERS-NODE-VOLTAGE SIMULATION OF LC LADDER STRUCTURES.
YOSHIHIRO M + NISHIHARA A + YANAGISAWA T
TRANS INST ELECTRON COMMUN ENG JPN SECT E E60(7): 360-1 (JULY 1977)

0868 LOW-SENSITIVITY ACTIVE AND DIGITAL FILTERS BASED ON THE NODE-VOLTAGE SIMULATION OF LC LADDER STRUCTURES.
YOSHIHIRO M + NISHIHARA A + YANAGISAWA T
P446-9 OF 1977 IEEE INT SYMP ON CIRCUITS AND SYSTEMS PROC, IEEE, PHOENIX, ARIZ, USA 25-27 APRIL 1975, IEEE, 1977

FABRICATION

0869 RESEARCH INSTITUTE FOR TELECOMMUNICATION INDUSTRY (THICK AND THIN FILM CIRCUITS).
HIRADASTECHNIKA 27(12): 357-8 (DEC 1976) (IN HUNGARIAN)

0870 ACTIVE FILTERS.
PR INST TELE- RADIOTECH 17(4): 43-7 (1973) (IN POLISH)

0871 HYBRID CIRCUIT OVERTAKING FEATURES.
BESAMAT M
P173-80 OF INT CONF ON MANUFACTURING AND PACKAGING TECHNIQUES FOR HYBRID CIRCUITS ELECTRONIC INDUSTRIES ASSOC FRANCE, ET AL, PARIS, FRANCE, 7-8 APRIL 1976, ELECTRONIC INDUSTRIES ASSOC FRANCE, 1976 (IN FRENCH)

0872 ACTIVE TONE CONTROL DEVICE IN TANTALUM THINFILM TECHNOLOGY.
BOSSELMANN W
COMPONENTS REP 10(4): 104-7 (OCT 1975)

0873 ACTIVE RC FILTERS IN THE HYBRID THIN TANTALUM FILM TECHNIQUE.
BOSSELMANN W
ELEKTRON INT NO 9: 313-16 (1975) (IN GERMAN)

0874 DESIGN AND CONSTRUCTION OF ACTIVE RC FILTERS (INTEGRATED CIRCUITS).
BOSSELMANN W
ELEKTRON IND 5(12): 271-3 (DEC 1974) (IN GERMAN)

0875 ACTIVE RC-FILTERS IN TANTALUM THIN-FILM HYBRID TECHNIQUE.
BOSSELMANN W
COMPONENTS REP 9(3): 77-80 (JULY 1974)

0876 ACTIVE RC FILTERS REALIZED BY THICK-FILM HYBRID TECHNOLOGY.
BOSSELMANN W
SIEMENS ELECTRON COMPONENTS BULL 8(2): 25-6 (MAY 1973)

0877 AN AUDIO FREQUENCY RANGE FILTER BANK USING HYBRID MICROELECTRONICS TECHNIQUES.
COLCLASER RA
P388-92 OF 7TH ASILOMAR CONF ON CIRCUITS, SYSTEMS AND COMPUTERS NAVAL POSTGRADUATE SCHOOL, UNIV SANTA CLARA, ET AL, PACIFIC GROVE, CALIF, USA, 27-29 NOV 1973, WESTERN PERIODICALS 1974

0878 HYBRID INTEGRATION OF A MULTISTAGE ACTIVE BANDPASS FILTER/AMPLIFIER.
COOK KB + KERNS DV + NAGLE HT + SLAGH TD + RUWE VW
IEEE TRANS PARTS, HYBRIDS PACKAG (USA) PHP-12(4): 336-44 (DEC 1976)

0879 COMPUTER-AIDED THERMAL ANALYSIS OF A HYBRID MULTISTAGE ACTIVE BANDPASS FILTER/AMPLIFIER.
COOK KD + KERNS DV + NAGLE HT + SLAGH TD + RUWE VW
IEEE TRANS PARTS, HYBRIDS PACKAG PHP-12(4): 344-50 (DEC 1976)

0880 AN EIGHT SECTION MICROCIRCUIT FILTER MODULE FOR DETECTION OF SUBSCRIBER DIALLING FREQUENCIES.
COWPLAND MCJ + BOOTHROYD AR
P158-9 OF INT ELECTRICAL, ELECTRONICS CONF AND EXPOSITION DIGEST, IEEE, TORONTO, CANADA, 1-3 OCT 1973 IEEE, 1973

0881 A LOW COST THICK-FILM ACTIVE FILTER FOR CONSUMER APPLICATION.
DELL'ORTO G + RAVASCHIO G
P260-3 OF PROC OF THE 28TH ELECTRONIC COMPONENTS CONF, IEEE, ANAHEIM, CA, USA, 24-26 APRIL 1978, IEEE, 1978

0882 MANUFACTURE OF THIN-FILM ACTIVE FILTERS.
DUPCAK J + DEGROOT RH
WESTERN ELECTRIC ENG 18(3): 18-25 (JULY 1974)

0883 REALISE YOUR RC FILTERS IN HYBRID THIN FILM TECHNOLOGY.
DUPONT M
ELECTRON MICROELECTRON IND NO 210: 47-50 (15 OCT 1975) (IN FRENCH)

0884 HIGH STABILITY MATCHED ACTIVE FILTERS FOR SONAR PROCESSING.
ESTEP GJ
P121-9 OF 24TH ELECTRONIC COMPONENTS CONF, ELECTRONIC INDUSTRY ASSOC, IEEE, WASHINGTON, DC, USA, 13-15 MAY 1974 IEEE, 1974

0885 THIN FILM RESISTOR NETWORKS IN HYBRID CIRCUITS.
GROTH L
SOLID STATE TECHNOL 20(3): 45-9 (MARCH 1977)

0886 NEW TENDENCIES IN THE PRODUCTION OF HYBRID INTEGRATED CIRCUITS.
HIRSL J
SDELOVACI TECH 25(9): 335-6 (SEPT 1977) (IN CZECH)

0887 A NEW METHOD OF ADJUSTING ELECTRICAL NETWORKS.
HUBER B
NTG-FACHBER 51: 31-8 (1975) (IN GERMAN)

0888 DEVELOPMENT OF MULTI-FREQUENCY RECEIVERS WITH HYBRID IC TECHNOLOGY.
IKEDA H + ARIYOSHI H
ELECTR COMMUN LAB TECH J 26(6): 1525-40 (1977) (IN JAPANESE)

0889 LASER FUNCTIONAL TRIMMING FOR HYBRID INTEGRATED RC ACTIVE FILTERS.
IWATA A + NAKAYAMA S + SAITO Y + TANSHO K
TRANS INST ELECTRON COMMUN ENG JPN SECT E E60(11): 667-8 (NOV 1977)

0890 HYBRID INTEGRATED RC ACTIVE FILTER DESIGN.
IWATA A + YOSHIMURA H + YANAGAWA H
ELECTR COMMUN LAB TECH J 26(8): 2445-62 (1977) (IN JAPANESE)

0891 HYBRID MICROELECTRONICS IN FINLAND.
JARVINEN ETK
SAEHKOE 51(7-8): 221-3 (JULY-AUG 1978)

0892 A NEW METHOD FOR THE MANUFACTURE OF RC-THIN FILM CIRCUITS USING MULTILAYER STRUCTURES.
KAISER B
NTG-FACHBER NO 60: 207-12 (1977) (IN GERMAN)

0893 TEMPERATURE COMPENSATED RC-ACTIVE NETWORKS IN THICK FILM TECHNOLOGY.
KALLFASS T
FREQUENZ 29(5): 147-51 (MAY 1975)

0894 METHOD OF ADJUSTMENT IN THE MASS PRODUCTION OF BIQUADRATIC ACTIVE RC BASIC SECTIONS.
KOVACS O
HIRADASTECHNIKA 27(10): 314-18 (OCT 1976) (IN HUNGARIAN)

0895 THIN FILM NETWORKS IN FREQUENCY AND TIME DOMAIN.
KULLA K
NACHRICHTENTECH ELEKTRON 25(9): 328-9336 (1975) (IN GERMAN)

0896 HYBRID INTEGRATED RC ACTIVE CHANNEL FILTER.
KUROSAWA H + IWATA A + MIYATA I
REV ELECTR COMMUN LAB 23(5-6): 411-28 (MAY-JUNE 1975)

0897 LASER TRIMMING ACTIVE FILTERS.
LEJEUNE B + GITTOES J
ELECTRON EQUIP NEWS 12-13 (MAY 1978)

0898 OPTIMUM DESIGN OF LINEAR TUNING ALGORITHM.
LOPRESTI PV
IEEE TRANS CIRCUITS SYST CAS-24(3): 144-51 (MARCH 1977)

0899 MEASURE-PREDICT TUNING OF HYBRID THIN-FILM FILTERS.
LUEDER E + MALEK G
IEEE TRANS CIRCUITS SYST CAS-23(7): 461-6 (JULY 1971)

0900 ACTIVE FILTERS ACCOMPLISHED BY MEANS OF THIN STRUCTURE TECHNOLOGY.
MANOLESCU AM + MANOLESCU A + BURILEANU C
ELECTROTEH ELECTRON AUTOM AUTOM ELECTRON 22(3): 111-21 (SEPT 1978) (IN RUMANIAN)

0901 ACTIVE RC FILTER MADE IN THICK FILM TECHNIQUE.
MIJAJLOVIC M
TEHNIKA 30(3): 572-5 (1975) (IN CROATIAN)

0902 STANDARDIZABLE LOW-PASS VOICE BAND HYBRID CIRCUIT FILTERS FOR PULSE CODE MODULATION (PCM) COMMUNICATION SYSTEMS.
MIKHAEL WB + KEYS LK
IEEE J SOLID-STATE CIRCUITS SC-10(4): 229-36 (AUG 1975)

0903 SOME SYSTEM CONSIDERATIONS OF THICK FILMS.
MORAN PL
ELECTRON IND 1(4): 19-21 (DEC 1975)

0904 FUNCTIONAL ADJUSTMENT OF DLA NETWORKS.
MORAN PL + ISMAIL NABN + MAJITHIA PK
ELECTROCOMPON SCI TECHNOL 5(3): 193-5 (OCT 1978)

0905 FUNCTIONAL AND DETERMINISTIC TUNING OF HYBRID INTEGRATED ACTIVE FILTERS.
MOSCHYTZ GS
ELECTROCOMPON SCI TECHNOL 5(2): 79-89 (JUNE 1978)

0906 A NEW RESISTIVE MATERIAL FOR HIGH STABILITY RC NETWORKS. APPLICATION TO THE PRODUCTION OF ACTIVE FILTERS.
MOULIN M
P251-7 OF COLLOQUE INT SUR LES MATERIAUX POUR LES COMPOSANTS ELECTRONIQUES (INT SYMP ON MATERIALS FOR ELECTRONIC COMPONENTS), PARIS, FRANCE, 2-4 APRIL 1975, COMM D'ORGANISATION DU COLLOQUE DE PARIS, 1975 (IN FRENCH)

0907 AUTOMATIC LASER TRIMMING FOR RC ACTIVE FILTERS.
NAKAYAMA S + IWATA A + SAITO Y + TANSHO K
ELECTR COMMUN LAB TECH J 26(9): 2677-89 (1977) (IN JAPANESE)

0908 A THICK FILM ACTIVE FILTER FOR PCM COMMUNICATION SYSTEMS.
NIIZAKI N + NAKAMURA K + TSUBOKAWA I
P183-9 OF PROC OF THE 28TH ELECTRONIC COMPONENTS CONF, IEEE, ANAHEIM, CA, USA, 24-26 APRIL 1978, IEEE, 1978

0909 REALIZATION OF RC ACTIVE FILTERS IN TAAL-SIO(2) THIN FILM TECHNOLOGY.
NOSSEK JA + GOTZ G
NTG-FACHBER NO 60: 93-8 (1977) (IN GERMAN)

0910 GENERAL-PURPOSE HYBRID ACTIVE FILTER.
ODESS L + BEN-GAL J
P76PP OF 8TH CONVENTION OF THE ELECTRICAL AND ELECTRONICS ENGINEERS IN ISRAEL, IEEE, TEL AVIV, ISRAEL 30 APRIL - 3 MAY 1973, ELECTRICAL AND ELECTRONICS ENGRS IN ISRAEL, 1973

0911 TANTALUM THIN FILM APPLICATIONS-A NEW APPROACH FOR CAPACITORS.
PERINATI A
ELECTROCOMPON SCI TECHNOL 4(2): 69-73 (OCT 1977)

0912 A SWITCHABLE ACTIVE FILTER HYBRID.
PROBSTEIN M
IEEE TRANS PARTS, HYBRIDS PACKAG PHP-12(13): 201-6 (SEPT 1976)

0913 A SWITCHABLE ACTIVE FILTER HYBRID.
PROBSTEIN M
P195-8 OF 26TH ELECTRONIC COMPONENTS CONF, IEEE, ELECTRONIC INDUSTRIES ASSOC, SAN FRANCISCO, CALIF, USA 26-28 APRIL 1976, IEEE, 1976

0914 CIRCUITS OF THE HYBRID TECHNIQUE.
RACUROW B + SPIEGEL P + STORCH K + GEIDEL KD
RADIO FERNSEHEN ELEKTRON 27(7): 432-4 (1978) (IN GERMAN)

0915 HIGH-Q HIGH-FREQUENCY RC-ACTIVE FILTERS USING DISTRIBUTED THIN FILM ELEMENTS.
RENZ HW
P170-4 OF CIRCUIT THEORY AND DESIGN, LAUSANNE, SWITZERLAND, 4-8 SEPT 1978, MOSCHYTZ, GS, NIERYNCK, J(ED), GEORGI 1978

0916 SYNTHESIS AND THIN FILM IMPLEMENTATION OF DISTRIBUTED RC ACTIVE FILTERS.
RENZ HW
P224-7 OF PROC OF THE 1976 IEEE INT SYMP ON CIRCUITS AND SYSTEMS, IEEE, MUNICH, GERMANY 27-29 APRIL 1976, IEEE, 1976

0917 RC-ACTIVE FILTERS IN SINGLE-LAYER TANTALUM RC-FILM TECHNOLOGY.
RENZ HW + BAEGER H + LUEDER E + TILCH HJ + KALLFASS T
PROC IEEE 67(1): 37-42 (JAN 1979)

0918 ACTIVE FILTERS AND THE INCIDENCE OF NEW TECHNOLOGIES.
RIEZENMAN MJ
REV TELEGR ELECTRON 67(789): 923-7 (OCT 1978) (IN SPANISH)

0919 SPECIAL REPORT - ACTIVE FILTERS RIDE THE CREST OF NEW TECHNOLOGY.
RIEZENMAN MJ
ELECTRONICS 50(11): 119-24 (26 MAY 1977)

0920 PARAMETERS OF ACTIVE FILTERS.
ROGULSKI A
ELEKTRONIKA 18(11): 439-44 (1977) (IN POLISH)

0921 DESIGNING AND REALISING ACTIVE THICK-FILM RC FILTERS.
RUPP W
NTG-FACHBER 51: 85-9 (1975) (IN GERMAN)

0922 DETERMINISTIC AND FUNCTIONAL TUNING OF THICK FILM ACTIVE RC-FILTERS USING A MINICOMPUTER CONTROLLED LASER TRIMMING SYSTEM.
RUPP W + SCHMIDT E + ULBRICH W
NTG-FACHBER NO 60: 110-14 (1977) (IN GERMAN)

0923 A SET OF THICK FILM R-C ACTIVE FILTERS FOR PCM APPLICATION.
SEN DK
P304-8 OF 1977 IEEE INT SYMP ON CIRCUITS AND SYSTEMS PROC, IEEE, PHOENIX, ARIZ, USA 25-27 APRIL 1975, IEEE, 1977

0924 DESIGN AND PROCESS CONSIDERATIONS FOR HYBRID ACTIVE FILTER APPLICATIONS.
SHIPP RE
P309-12 OF 1977 IEEE INT SYMP ON CIRCUITS AND SYSTEMS PROC, IEEE, PHOENIX, ARIZ, USA 25-27 APRIL 1975, IEEE, 1977

0925 MINICOMPUTER-A PRODUCTION LINE TECHNIQUE FOR TUNING ACTIVE FILTERS.
SHOCKLEY TD + MORRIS CF
P6B-5/1-4 OF PROC OF 1975 IEEE SOUTHEASTCON REGION 3 CONF ON ELECTRICITY AND EXPANDING TECHNOLOGY, IEEE CHARLOTTE, NC, USA, 6-9 APRIL 1975, IEEE, 1975

0926 COMPUTERIZED DESIGN AND TUNING OF ACTIVE FILTERS.
SHOCKLEY TD + MORRIS CF
IEEE TRANS CIRCUIT THEORY CT-20(4): 438-41 (JULY 1973)

0927 VERSATILE ACTIVE RC FILTERS IN HYBRID CIRCUIT REALISATION.
SONKOLY A
HIRADASTECHNIKA 28(8): 225-31 (AUG 1977) (IN HUNGARIAN)

0928 MINIATURIZED THICK-FILM RC-ACTIVE FILTERS FOR PCM APPLICATION.
SUZUKI K + KATSUTA Y
PROC IEEE 67(1): 34-7 (JAN 1979)

0929 HYBRID ACTIVE RC FILTERS.
UDVARHELYI G
HIRADASTECHNIKA 24(7): 215-18 (JULY 1973) (IN HUNGARIAN)

0930 MULTI-FREQUENCY RECEIVER WITH HYBRID IC TECHNOLOGY.
USUDA S + IWATA A + YANAGAWA H
ELECTR COMMUN LAB TECH J 26(6): 1571-602 (1977) (IN JAPANESE)

0931 THICK FILM HYBRID CIRCUITS FOR TELECOMMUNICATIONS EQUIPMENT.
WAER RL
P307-13 OF 27TH ELECTRONIC COMPONENTS CONF, ELECTRONIC INDUSTRIES ASSOC, IEEE, ARLINGTON, VA, USA 16-18 MAY 1977, IEEE, 1977

0932 TANTALUM THIN-FILM RC CIRCUIT TECHNOLOGY FOR A UNIVERSAL ACTIVE FILTER.
WOROBEY W + RUTKIEWICZ J
IEEE TRANS PARTS, HYBRIDS PACKAG PHP-12(4): 276-82 (DEC 1976)

0933 HYBRID INTEGRATED PUSHBUTTON RECEIVER.
YANAGAWA H + USUDA S + IWATA A + KUROSAWA H
ELECTR COMMUN LAB TECH J 26(6): 1541-70 (1977) (IN JAPANESE)

0934 OPTIMIZING ELEMENT VALUES IN HYBRID RC ACTIVE FILTERS.
ZIMMER CR + MYERS G
P333-5 OF 27TH ELECTRONIC COMPONENTS CONF, ELECTRONIC INDUSTRIES ASSOC, IEEE, ARLINGTON, VA, USA, 16-18 MAY 1977 IEEE, 1977

TECHNIQUES FOR MONOLITHIC ACTIVE FILTERS: CONTINUOUS ANALOG

0935 DESIGN OF HIGH-ORDER ACTIVE FILTERS IN INTEGRATED CIRCUITS.
ACAR C + GHAUSI MS
IEEE J SOLID-STATE CIRCUITS SC-13(2): 273-6 (APRIL 1978)

0936 REALIZATION OF AN ACTIVE R BIQUADRATIC CIRCUIT.
AHMED MT + SIDDIQI MA
P49-54 OF CONF RECORD OF THE TWELFTH ASILOMAR CONF ON CIRCUITS, SYSTEMS AND COMPUTERS NAVAL POSTGRADUATE SCHOOL MONTEREY, IEEE, UNIV SANTA CLARA, PACIFIC GROVE, CA, USA, 6-8 NOV 1978, IEEE, 1978

0937 COMMENTS ON 'A NOVEL ''FOLLOW THE MASTER'' FILTER'.
BRAND JR + SCHAUMANN R
PROC IEEE 66(5): 590-2 (MAY 1978)

0938 ACTIVE R FILTERS - REVIEW OF THEORY AND PRACTICE.
BRAND JR + SCHAUMANN R
IEE J ELECTRON CIRCUITS SYST 2(4): 89-101 (JULY 1978)

0939 ACTIVE BANDPASS NETWORK WITH ONLY RESISTORS AS PASSIVE ELEMENTS.
CAPPARELLI F + LIBERATORE A
ELECTRON LETT 8(2): 43-4 (27 JAN 1972)

0940 R-ACTIVE FILTERS. PARALLEL SYNTHESIS.
DI ROSA P + MARTINELLI G
NOTE RECENS NOT 26(1-2): 5-22 (JAN-APRIL 1977) (IN ITALIAN)

0941 SIMULATION OF INDUCTORS AND CAPACITORS USING OPERATIONAL-AMPLIFIER COMPENSATION POLE - A CAUTION.
GILBERT B
ELECTRON LETT 14(25): 832 (7 DEC 1978)

0942 REALISATION OF ACTIVE-R FILTERS USING AMPLIFIER POLE.
HO CF + CHIU PL
PROC INST ELECTR ENG 123(5): 406-10 (MAY 1976)

0943 FREQUENCY LIMITATION OF ACTIVE-R FILTERS USING OPERATIONAL AMPLIFIERS.
KAPUSTIAN V + BHATTACHARYYA BB + SWAMY MNS
P150-4 OF CIRCUIT THEORY AND DESIGN, LAUSANNE, SWITZERLAND, 4-8 SEPT 1978, MOSCHYTZ, GS, NIERYNCK, J(ED), GEORGI 1978

0944 HIGH-ORDER MONOLITHIC ANALOG FILTERS USING BIPOLAR/JFET TECHNOLOGY.
KHEN-SANG TAN + GRAY PR
P80-1 OF 1978 IEEE INT SOLID-STATE CIRCUITS CONF (DIGEST OF TECHNICAL PAPERS), IEEE, SAN FRANCISCO, CA USA, 15-17 FEB 1978, IEEE, 1978

0945 AN ACTIVE BIQUADRATIC BUILDING BLOCK WITHOUT EXTERNAL CAPACITORS.
KIM HK + RA JB
IEEE TRANS CIRCUITS SYST CAS-24(12): 689-94 (DEC 1977)

0946 AN ACTIVE BIQUADRATIC BUILDING BLOCK WITHOUT EXTERNAL CAPACITORS.
KIM HK + RA JB + PARK SB
P317-20 OF 1977 IEEE INT SYMP ON CIRCUITS AND SYSTEMS PROC, IEEE, PHOENIX, ARIZ, USA 25-27 APRIL 1975, IEEE, 1977

0947 MULTIPLE-LOOP FEEDBACK ACTIVE R FILTERS.
LAKER KR + SCHAUMANN R + BRAND JR
P279-82 OF PROC OF THE 1976 IEEE INT SYMP ON CIRCUITS AND SYSTEMS, IEEE, MUNICH, GERMANY 27-29 APRIL 1976, IEEE, 1976

0948 ACTIVE R FILTER USING OPERATIONAL AMPLIFIER POLE.
LI MK + LI CW
ELECTRON ENG 50(602): 34 (FEB 1978)

0949 LOW SENSITIVITY HIGH-FREQUENCY ACTIVE R FILTERS.
MITRA AK + AATRE VK
IEEE TRANS CIRCUITS SYST CAS-23(11): 670-6 (NOV 1976)

0950 A UNIVERSAL ACTIVE-R FILTER.
MOKHTAR A + MOKHTAR M
FREQUENZ 31(4): 107-10 (APRIL 1977)

0951 A FULLY INTEGRATED FIVE-GYRATOR FILTER AT VIDEO FREQUENCIES.
MOULDING KW + WILSON GA
IEEE J SOLID-STATE CIRCUITS SC-13(3): 303-7 (JUNE 1978)

0952 AN ACTIVE R FILTER USING CMOS TRANSISTOR ARRAYS.
NAIMPALLY SV
INT J ELECTRON 40(5): 513-16 (MAY 1976)

0953 COMMENT ON MULTIFUNCTION ACTIVE R FILTER WITH TWO OPERATIONAL AMPLIFIERS.
NANDI R
ELECTRON LETT 15(9): 261-2 (26 APRIL 1979)

0954 A NOVEL 'FOLLOW THE MASTER' FILTER.
RADHAKRISHNA RAO K + SETHURAMAN V + NEELAKANTAN PK
PROC IEEE 65(12): 1725-6 (DEC 1977)

0955 A HIGH-Q TEMPERATURE INSENSITIVE BANDPASS FILTER USING THE OPERATIONAL AMPLIFIER POLE.
RADHAKRISHNA RAO K + SRINIVASAN S
PROC IEEE 62(12): 1713-14 (DEC 1974)

0956 LOW-SENSITIVITY ACTIVE FILTERS USING THE OPERATIONAL AMPLIFIER POLE.
RADHAKRISHNA RAO K + SRINIVASAN S
IEEE TRANS CIRCUITS SYST CAS-21(2): 260-2 (MARCH 1974)

0957 COMMENT ON MULTIFUNCTION ACTIVE R FILTER WITH TWO OPERATIONAL AMPLIFIERS.
SANCHEZ-SINENCIO E
ELECTRON LETT 15(9): 262 (26 APRIL 1979)

0958 DESIGN OF ACTIVE FILTERS USING ONLY RESISTORS AND VOLTAGE AMPLIFIERS.
SCHAUMANN R
ARCH ELEKTRON UEBERTRAGUNGSTECH 30(6): 245-52 (JUNE 1976)

0959 LOW-SENSITIVITY HIGH-FREQUENCY TUNABLE ACTIVE FILTER WITHOUT EXTERNAL CAPACITORS.
SCHAUMANN R
IEEE TRANS CIRCUITS SYST CAS-22(1): 39-44 (JAN 1975)

0960 A LOW-SENSITIVITY, HIGH-FREQUENCY, TUNABLE ACTIVE FILTER WITHOUT EXTERNAL CAPACITORS.
SCHAUMANN R
P438-42 OF PROC OF THE 1974 IEEE INT SYMP ON CIRCUITS AND SYSTEMS, IEEE, SAN FRANCISCO, CALIF USA, 22-25 APRIL 1974, IEEE, 1974

0961 DESIGN METHOD FOR MONOLITHIC ANALOGUE FILTERS.
SCHAUMANN R + BRAND JR
ELECTRON LETT 14(22): 710-11 (26 OCT 1978)

0962 ACTIVE R LADDERS - HIGH-FREQUENCY HIGH-ORDER LOW-SENSITIVITY ACTIVE R FILTERS WITHOUT EXTERNAL CAPACITORS.
SODERSTRAND MA
IEEE TRANS CIRCUITS SYST CAS-25(12): 1032-8 (DEC 1978)

0963 AN IMPROVED CMOS ACTIVE-R FILTER.
SODERSTRAND MA
PROC IEEE 65(8): 1204-6 (AUG 1977)

0964 DESIGN OF HIGH-FREQUENCY HIGH-ORDER ACTIVE R LADDERS.
SODERSTRAND MA
P275-8 OF PROC OF THE 1976 IEEE INT SYMP ON CIRCUITS AND SYSTEMS, IEEE, MUNICH, GERMANY 27-29 APRIL 1976, IEEE, 1976

0965 A SIMPLE SYNTHESIS TECHNIQUE FOR ACTIVE R FILTERS.
SODERSTRAND MA
P261-4 OF 1975 IEEE INT SYMP ON CIRCUITS AND SYSTEMS, IEEE, NEWTON, MASS, USA, 21-23 APRIL 1975, IEEE 1975

0966 DESIGN OF ACTIVE R FILTERS USING ONLY RESISTORS AND OPERATIONAL AMPLIFIERS.
SODERSTRAND MA
INT J ELECTRON 40(5): 417-32 (MAY 1976)

0967 ACTIVE R FILTERS - ACTIVE FILTERS USING ONLY RESISTORS AND AMPLIFIERS.
SODERSTRAND MA
P675-81 OF 8TH ASILOMAR CONF ON CIRCUITS, SYSTEMS AND COMPUTERS NAVAL POSTGRADUATE SCHOOL, UNIV SANTA CLARA, IEEE, ET AL, PACIFIC GROVE, CALIF, USA, 3-5 DEC 1974, PARKER, SR(ED) WESTERN PERIODICALS, 1975

0968 A NEW ACTIVE R BANDPASS FILTER.
SOLIMAN AM + FAWZY M
J FRANKLIN INST 306(2): 159-63 (AUG 1978)

0969 A UNIVERSAL ACTIVE R FILTER.
SOLIMAN AM + FAWZY M
ELECTRON ENG 49(594): 49-50 (JULY 1977)

0970 AN INTEGRATABLE TUNABLE ACTIVE BP FILTER.
TALKHAN EA + IBRAHIM MM + NOUTH AS
P340-3 OF 1977 IEEE INT SYMP ON CIRCUITS AND SYSTEMS PROC, IEEE, PHOENIX, ARIZ, USA 25-27 APRIL 1975, IEEE, 1977

0971 BOOSTING OP AMPS - ROLL-OFF EXTENDS RANGE OF ACTIVE FILTERS.
VAN DER GREER CAJ
ELECTRONICS 51(16): 120-1 (3 AUG 1978)

0972 MULTIFUNCTION ACTIVE R FILTER WITH TWO OPERATIONAL AMPLIFIERS.
VENKATESWARAN S
ELECTRON LETT 14(23): 741-2 (9 NOV 1978)

0973 A REALIZATION OF AN ACTIVE R FILTER.
YOSHIDA N + KUMAZAWA M
TRANS INST ELECTRON COMMUN ENG JPN SECT E E62(5): 339 (MAY 1979)

TECHNIQUES FOR MONOLITHIC ACTIVE FILTERS: SWITCHED CAPACITOR

0974 CONSIDERATIONS FOR THE MOS IMPLEMENTATIONS OF SWITCHED CAPACITOR FILTERS.
ALLSTOT DJ + BRODERSEN RW + GRAY PR
P684-8 OF CONF RECORD OF THE TWELFTH ASILOMAR CONF ON CIRCUITS, SYSTEMS AND COMPUTERS NAVAL POSTGRADUATE SCHOOL MONTEREY, IEEE, UNIV SANTA CLARA, PACIFIC GROVE, CA, USA, 6-8 NOV 1978, IEEE, 1978

0975 AN ELECTRICALLY PROGRAMMABLE ANALOG NMOS SECOND-ORDER FILTER (SPEECH SYNTHESIS APPLICATIONS).
ALLSTOT DJ + BRODERSEN RW + GRAY PR
P76-7 OF 1979 IEEE INT SOLID-STATE CIRCUITS CONF (DIGEST OF TECHNICAL PAPERS), IEEE, PENNSYLVANIA, PA USA, 14-16 FEB 1979, IEEE, 1979

0976 FULLY-INTEGRATED HIGH-ORDER NMOS SAMPLED-DATA LADDER FILTERS.
ALLSTOT DJ + BRODERSEN RW + GRAY PR
P82-3 OF 1978 IEEE INT SOLID-STATE CIRCUITS CONF (DIGEST OF TECHNICAL PAPERS), IEEE, SAN FRANCISCO, CA USA, 15-17 FEB 1978, IEEE, 1978

0977 MOS SWITCHED-CAPACITOR FILTERS.
BRODERSEN RW + GRAY PR + HODGES DA
PROC IEEE 67(1): 61-75 (JAN 1979)

0978 SAMPLED ANALOG FILTERING USING SWITCHED CAPACITORS AS RESISTOR EQUIVALENTS.
CAVES JT + ROSENBAUM SD + COPELAND MA + RAHIM CF
IEEE J SOLID-STATE CIRCUITS SC-12(6): 592-9 (DEC 1977)

0979 INTEGRATION OF FILTERS IN LINEAR MOS CIRCUITS.
CHRISTIANSEN P + GEBHARDT PJ + KOHLBACHER G
WISS BER AEG-TELEFUNKEN 52(1-2): 131-8 (1979) (IN GERMAN)

0980 DYNAMIC AMPLIFIER FOR MOS TECHNOLOGY.
COPELAND MA + RABAEY JM
ELECTRON LETT 15(10): 301-2 (10 MAY 1979)

0981 BASIC PRINCIPLES OF SWITCHED-CAPACITOR FILTERS USING VOLTAGE INVERTER SWITCHES.
FETTWEIS A
ARCH ELEKTRON UEBERTRAGUNGSTECH 33(1): 13-19 (JAN 1979)

0982 SWITCHED-CAPACITOR FILTERS USING VOLTAGE INVERTER SWITCHES - FURTHER DESIGN PRINCIPLES.
FETTWEIS A
ARCH ELEKTRON UEBERTRAGUNGSTECH 33(3): 107-14 (MARCH 1979)

0983 ANALOG SAMPLE-DATA FILTERS.
FRIED DL
IEEE J SOLID-STATE CIRCUITS SC-7(4): 302-4 (AUG 1972)

0984 CHARGE-ROUTING NETWORKS.
GERSHO A + GOPINATH B
IEEE TRANS CIRCUITS SYST CAS-26(2): 81-92 (FEB 1979)

0985 A SINGLE-CHIP NMOS DUAL-CHANNEL FILTER FOR PCM TELEPHONY APPLICATIONS.
GRAY PR + SENDEROWICZ D + O HARA H + WARREN BM
P26-7 OF 1979 IEEE INT SOLID-STATE CIRCUITS CONF (DIGEST OF TECHNICAL PAPERS), IEEE, PENNSYLVANIA, PA USA, 14-16 FEB 1979, IEEE, 1979

0986 A SWITCHED-CAPACITOR HIGH-PASS FILTER.
GREGORIAN R + NICHOLSON WE
P669-73 OF CONF RECORD OF THE TWELFTH ASILOMAR CONF ON CIRCUITS, SYSTEMS AND COMPUTERS NAVAL POSTGRADUATE SCHOOL MONTEREY, IEEE, UNIV SANTA CLARA, PACIFIC GROVE, CA, USA, 6-8 NOV 1978, IEEE, 1978

0987 CMOS SWITCHED-CAPACITOR FILTER FOR A TWO-CHIP PCM VOICE CODER.
GREGORIAN R + HAQUE YA + MAO R + BLASCO R + NICHOLSON WE
P28-9 OF 1979 IEEE INT SOLID-STATE CIRCUITS CONF (DIGEST OF TECHNICAL PAPERS), IEEE, PENNSYLVANIA, PA USA, 14-16 FEB 1979, IEEE, 1979

0988 MOS SWITCHED-CAPACITOR FILTERS.
HERBST D + HOEFFLINGER B + SCHUMACHER K + SCHWEER R + FETTWEIS A + OWENIER K + PANDEL J
P74-5 OF 1979 IEEE INT SOLID-STATE CIRCUITS CONF (DIGEST OF TECHNICAL PAPERS), IEEE, PENNSYLVANIA, PA USA, 14-16 FEB 1979, IEEE, 1979

0989 MOS SAMPLED DATA RECURSIVE FILTERS USING SWITCHED CAPACITOR INTEGRATORS.
HOSTICKA BJ + BRODERSEN RW + GRAY PR
IEEE J SOLID-STATE CIRCUITS SC-12(6): 600-8 (DEC 1977)

0990 MOS SAMPLED DATA RECURSIVE FILTERS USING STATE VARIABLE TECHNIQUES.
HOSTICKA BJ + BRODERSEN RW + GRAY PR
P525-9 OF 1977 IEEE INT SYMP ON CIRCUITS AND SYSTEMS PROC, IEEE, PHOENIX, ARIZ, USA 25-27 APRIL 1975, IEEE, 1977

0991 DESIGN TECHNIQUES FOR MOS SWITCHED CAPACITOR LADDER FILTERS.
JACOBS GM + ALLSTOT DJ + BRODERSEN RW + GRAY PR
IEEE TRANS CIRCUITS SYST CAS-25(12): 1014-21 (DEC 1978)

0992 DESIGN CONSIDERATIONS FOR MOS SWITCHED CAPACITOR LADDER FILTERS.
JACOBS GM + ALLSTOT DJ + BRODERSEN RW + GRAY PR
P324-9 OF PROC OF THE 1978 IEEE INT SYMP ON CIRCUITS AND SYSTEMS, IEEE, NEW YORK, USA 17-19 MAY 1978, IEEE, 1978

0993 NEW REALIZATIONS FOR SWITCHED CAPACITOR FILTERS.
JENKINS WK + TRICK TN + EL-MASRY E
P694-8 OF CONF RECORD OF THE TWELFTH ASILOMAR CONF ON CIRCUITS, SYSTEMS AND COMPUTERS NAVAL POSTGRADUATE SCHOOL MONTEREY, IEEE, UNIV SANTA CLARA, PACIFIC GROVE, CA, USA, 6-8 NOV 1978, IEEE, 1978

0994 EQUIVALENT CIRCUITS FOR THE ANALYSIS AND SYNTHESIS OF SWITCHED CAPACITOR NETWORKS.
LAKER KR
BELL SYST TECH J 58(3): 729-69 (MARCH 1979)

0995 INTEGRATED FILTERS IN VLSI-TECHNOLOGY.
LUDER E
ARCH ELEKTRON UEBERTRAGUNGSTECH 32(10): 381-9 (OCT 1978) (IN GERMAN)

0996 SWITCHED-CAPACITOR-FILTERS - A COMPARISON AND SOME EXPERIMENTAL RESULTS.
NOSSEK JA
FREQUENZ 33(7-8): 219-20 (JULY-AUG 1979)

0997 SPECTRAL ANALYSIS OF SWITCHED-CAPACITOR FILTERS DESIGNED USING THE BILINEAR Z-TRANSFORM.
ORCHARD HJ + TEMES GC
P674-8 OF CONF RECORD OF THE TWELFTH ASILOMAR CONF ON CIRCUITS, SYSTEMS AND COMPUTERS NAVAL POSTGRADUATE SCHOOL MONTEREY, IEEE, UNIV SANTA CLARA, PACIFIC GROVE, CA, USA, 6-8 NOV 1978, IEEE, 1978

0998 ALTERNATIVE APPROACH TO THE DESIGN OF SWITCHED-CAPACITOR FILTERS.
PAIN BG
ELECTRON LETT 15(14): 438-9 (5 JULY 1979)

0999 A PARAMETRIC FILTER WITH CAPACITOR SWITCHING AND THE POSSIBILITY OF SIGNAL SPECTRUM TRANSFER.
TARBAYEV SI
TELECOMMUN RADIO ENG PART 2 33(5): 78-83 (MAY 1978)

1000 DERIVATION OF SWITCHED-CAPACITOR FILTERS FROM ACTIVE-RC PROTOTYPES.
TEMES GC
ELECTRON LETT 14(12): 361-2 (8 JUNE 1978)

1001 SWITCHED-CAPACITOR FILTER DESIGN USING THE BILINEAR Z-TRANSFORM.
TEMES GC + ORCHARD HJ + JAHANBEGLOO M
IEEE TRANS CIRCUITS SYST CAS-25(12): 1039-44 (DEC 1978)

1002 AN IMPROVED SWITCHED-CAPACITOR INTEGRATOR.
TEMES GC + YOUNG IA
ELECTRON LETT 14(9): 287-8 (27 APRIL 1978)

1003 ANALYTICAL AND EXPERIMENTAL EVALUATION OF A SWITCHED-CAPACITOR FILTER AND REMARKS ON THE RESISTOR/SWITCHED-CAPACITOR CORRESPONDENCE.
TSIVIDIS YP
IEEE TRANS CIRCUITS SYST CAS-26(2): 140-4 (FEB 1979)

1004 APPLICATION OF SWITCHED-CAPACITOR RESISTORS IN RC OSCILLATORS.
VISWANATHAN TR + SINGHAL K + METZKER G
ELECTRON LETT 14(20): 659-60 (28 SEPT 1978)

1005 ANALOG NMOS SAMPLED-DATA RECURSIVE FILTER.
YOUNG IA + HODGES DA + GRAY PR
P156-7 OF 1977 INT SOLID-STATE CIRCUITS CONF (DIGEST OF TECHNICAL PAPERS), IEEE, PHILADELPHIA, PA USA, 16-18 FEB 1977, LEWIS WINNER, 1977

1006 SWITCHED-CAPACITOR FILTER SECTIONS IMPLEMENTING THE BILINEAR TRANSFORMATION.
YOUNG IA + SIMONYI E + TEMES GC
P689-92 OF CONF RECORD OF THE TWELFTH ASILOMAR CONF ON CIRCUITS, SYSTEMS AND COMPUTERS NAVAL POSTGRADUATE SCHOOL MONTEREY, IEEE, UNIV SANTA CLARA, PACIFIC GROVE, CA, USA, 6-8 NOV 1978, IEEE, 1978

MISCELLANEOUS

1007 ELECTRONIC GONG.
ELEKTOR 4(7-8): 9 (JULY-AUG 1978)

1008 FORMANT - THE ELEKTOR MUSIC SYNTHESISER. VI.
ELEKTOR 3(12): 1227-33 (DEC 1977)

1009 ACTIVE LOUDSPEAKER-CROSSOVER FILTERS. II.
ELEKTOR 3(6): 37-45 (JUNE 1977)

1010 ACTIVE LOUDSPEAKER-CROSSOVER FILTERS. I.
ELEKTOR 3(5): 546-51 (MAY 1977)

1011 SELF TUNING FILTER CIRCUIT FOR RADIO COMMUNICATION.
UK PAT 1379658, 8 JAN 1975, P

1012 DISTORTION METER.
ELEKTOR 74-5 (JULY-AUG 1977)

1013 SYMMETRICAL CONSTANT-VOLTAGE CROSS-OVER FILTER.
ELEKTOR 42 (JULY-AUG 1977)

1014 PREEMPHASIS AND DEEMPHASIS NETWORKS USING ACTIVE RC-FILTERS.
BABUROV EF + MARIGODOV VK
TELECOMMUN RADIO ENG PART 2 32(6): 138-40 (JUNE 1977)

1015 A HIGH-GAIN NARROW-BAND AMPLIFIER USING ACTIVE FILTERS (IONOSPHERE MEASUREMENT EQUIPMENT).
BAHRI KK + SETTY CSGK
J INST ELECTRON TELECOMMUN ENG 23(1): 29-31 (JAN 1977)

1016 MONOLITHIC RC-STRUCTURES WITH DISTRIBUTED PARAMETERS IN ACTIVE FILTERS AND OSCILLATORS.
BALAZ I + PASTOREK A
ELEKTROTECH CAS 27(2): 105-20 (1976) (IN SLOVAK)

1017 REALISATION AND PROPERTIES OF ACTIVE RC FILTERS WITH DISTRIBUTED PARAMETERS.
BALAZ I + PAVHIK S
ELEKTROTECH CAS 28(10): 748-53 (1977) (IN SLOVAK)

1018 DIGITAL MEASUREMENT OF RESPIRATION FREQUENCY FROM RESPIRATORY FLUCTUATIONS OF BLOOD PRESSURE.
BARTHELMES H + EICHMEIER J
BIOMED TECH 18(4): 148-51 (AUG 1973) (IN GERMAN)

1019 NEW TOUCH CALLING DTMF RECEIVER USING CUSTOM INTEGRATED CIRCUITS.
BEEMAN RH
GTE AUTOM ELECTR TECH J 14(8): 364-9 (OCT 1975)

1020 AN N-PATH FILTER.
BEGAKIS N + GOODE GB
MONITOR 38(7): 125-6 (JULY 1977)

1021 A LINEAR VOLTAGE-TUNABLE DISTRIBUTED NULL DEVICE.
BENZ HF + MATTAUCH RJ
IEEE J SOLID-STATE CIRCUITS SC-7(6): 499-503 (DEC 1972)

1022 NO.2 ESS DESIGN FOR OPERATION IN HIGH 60-HZ INDUCTION ENVIRONMENTS.
BEUSCHER HJ + RICHARDS LW
P281/1-5 OF 1976 NATIONAL TELECOMMUNICATIONS CONF, IEEE, ET AL, DALLAS, TEX, USA, 29 NOV - 1 DEC 1976, IEEE 1976

1023 ACTIVE FILTER WITH DISTRIBUTED RC LINE HAVING ZERO Q-SENSITIVITY.
BIALKO M + GUZINSKI A
IEEE TRANS CIRCUITS SYST CAS-21(1): 87-90 (JAN 1974)

1024 ACTIVE FILTER WITH DISTRIBUTED RC LINE HAVING ZERO Q-SENSITIVITY.
BIALKO M + GUZINSKI A
P341-3 OF INT SYMP ON CIRCUIT THEORY 1973, IEEE, TORONTO, CANADA, 9-11 APRIL 1973, IEEE, 1973

1025 BKCHKH-1 BLOCK FOR COMPENSATING FREQUENCY RESPONSES.
BLOKHIN AV + KOLODIN MI + REBRIN YUI + ROMANOV VA + SHABASHOV VP + GORBUNTSOV AF
INSTRUM EXP TECH 19(3): PT2 929 (MAY-JUNE 1976)

1026 SOME EXPERIMENTS WITH ACTIVE LOW-PASS FILTERS. I. DESIGN AND PERFORMANCE.
BOGLE AG
AEU 6(4): 28-30 (1973)

1027 TWIN-T AND ANALOGOUS DISTRIBUTED-LUMPED-ACTIVE NOTCH FILTERS.
BOWRON P + TAPINOS C
ARCH ELEKTRON UEBERTRAGUNGSTECH 33(5): 193-9 (MAY 1979)

1028 APPLICATION OF CONTROLLABLE FILTER NETWORKS TO SPEECH SYNTHESIS IN AUTOMATIC SPEECH PRODUCTION SYSTEMS.
BRAUN HJ
FERNMELDE-ING 29(12): 1-31 (15 DEC 1975) (IN GERMAN)

1029 MULTILOOP FEEDBACK DISTRIBUTED LUMPED ACTIVE LOW-PASS FILTER.
BRODIE JH + MARISO S
MICROELECTRONICS 5(4): 13-19 (1974)

1030 ACTIVE ONE-THIRD OCTAVE MICROELECTRONIC FILTER USING RC-STRUCTURES.
BULBANYUK AF + VOLKOVA ED + KOZLOV MV + MASYURENKO YUA + NIZHENSKII AD
INSTRUM EXP TECH 20(2): PT1 464-5 (MARCH-APRIL 1977)

1031 COMPARISON BETWEEN ACTIVE LUMPED AND ACTIVE DISTRIBUTED RC FILTERS.
BUNTIG W
NACHRICHTENTECH ELEKTRON 27(3): 100-1 (1977) (IN GERMAN)

1032 VOLTAGE-CONTROLLED UNIVERSAL FILTER.
BURKHARDT H
FUNKSCHAU 50(6): 261-3 (10 MARCH 1978) (IN GERMAN)

1033 SOME APPLICATIONS OF CCD FIXED TAP WEIGHT TRANSVERSAL FILTERS.
BURT DJ + MCCABE APH + SCAIFE TJ
P11/1-2 OF COLLOQUIUM ON MODERN SIGNAL PROCESSING TECHNOLOGY FOR SONAR, LONDON, ENGLAND, 16 NOV 1978, IEE, 1978

1034 E OF STATE VARIABLE ACTIVE FILTERS AS DIFFERENTIATORS, RATE METERS, AND END-POINT LOCATORS (CHEMICAL ANALYSIS APPLIC.). .
CALLICOTT RH + CARR PW
ANAL CHEM 46(12): 1840-2 (OCT 1974)

1035 NEW INSTRUMENTATION AMPLIFIERS AND FUNCTION MODULES FOR INDUSTRIAL CONTROL SYSTEMS - THEORY AND APPLICATIONS.
CATE T + BROWN L
P816/1-12 OF ADVANCES IN INSTRUMENTATION VOL30, ISA, MILWAUKEE, WIS, USA, 6-9 OCT 1975, ISA, 1975

1036 REALIZATION OF MULTI-CHANNEL ACTIVE FILTERS BY USING OPERATIONAL AMPLIFIERS.
CHUNG DUK KIM
J KOREAN INST ELECTR ENG 24(4): 404-6 (JULY 1975) (IN KOREAN)

1037 AN ANALOGUE INVERTING DIFFERENTIATOR.
COMANESCU V
INT J ELECTRON 42(1): 101-3 (JAN 1977)

1038 ACTIVE FILTER IMPROVES TRACKING AND CAPTURE RANGES OF PLL.
CONNELLY JA
ELECTRON DES 23(7): 128-130 (1 APRIL 1975)

1039 SIMPLE LOW SENSITIVITY ACTIVE RC NETWORKS FOR QUADRATURE SIGNAL GENERATION.
DAOUD E + MIKHAEL WB + SWAMY MNS
P254-6 OF PROC OF THE 1978 IEEE INT SYMP ON CIRCUITS AND SYSTEMS, IEEE, NEW YORK, USA 17-19 MAY 1978, IEEE, 1978

1040 SIMPLE CONNECTIONS OF TWIN CAPACITORS.
DOSTAL T + SEDLACEK J
SLABOPROUDY OBZ 39(9): 417-21 (SEPT 1978) (IN CZECH)

1041 INTEGRATING ACTIVE LOW-FREQUENCY FILTER BASED ON INTEGRATED-CIRCUIT OPERATIONAL AMPLIFIERS.
D'YAKONOV VP + KALINENKO AG + MAKSIMCHUK AA
INSTRUM EXP TECH 20(3): PT2 774-6 (MAY-JUNE 1977)

1042 ACTIVE DISTRIBUTED RC LOW-PASS FILTERS.
DUTTA ROY SC + SAH RP
MICROELECTRON REL 11(3): 293-5 (JULY 1972)

1043 MICROELECTRONIC ACTIVE FILTERS EMPLOYING DISTRIBUTED RC STRUCTURES.
EFIMOV IE + KOCHARYAN AG
TELECOMMUN RADIO ENG PT 1 29(5): 28-32 (MAY 1975)

1044 OPTIMAL LOW SENSITIVITY DISTRIBUTED ACTIVE FILTER.
EL-DIWANY M + ABOUGABAL MS + GIGUERE JC + BHATTACHARYYA BB
P151-4 OF 1975 IEEE INT SYMP ON CIRCUITS AND SYSTEMS, IEEE, NEWTON, MASS, USA, 21-23 APRIL 1975, IEEE 1975

1045 WIEN BRIDGE AND OP AMP SELECT NOTCH FILTER'S BANDWIDTH.
FELLOT D
ELECTRONICS 51(25): 124 (7 DEC 1978)

1046 SURFACE ACOUSTIC WAVE REGENERATIVE ACTIVE RESONATOR.
FIELD ME + CHEN CL
P909-12 OF 1977 ULTRASONICS SYMP PROC, IEEE, PHOENIX, AZ, USA, 26-28 OCT 1978, IEEE, 1977

1047 AN EIGHT-CHANNEL COMPUTER-CONTROLLED SCANNING ELECTROMYOGRAPH.
FRIDLUND AJ + FOWLER SC
BEHAV RES METHODS INSTRUM 10(5): 652-62 (OCT 1978)

1048 ACTIVE TRIMMING FOR HILBERT FILTER STRUCTURE.
FROMENT JC
IBM TECH DISCLOSURE BULL 21(1): 217-18 (JUNE 1978)

1049 VOTRAX REAL TIME HARDWARE FOR PHONEME SYNTHESIS OF SPEECH.
GAGNON RT
P175-8 OF PROC OF THE 1978 IEEE INT CONF ON ACOUSTICS, SPEECH AND SIGNAL PROCESSING, IEEE, TULSA OK, USA, 10-12 APRIL 1978, IEEE, 1978

1050 A HARMONIC FILTER.
GARIBYAN MS + GARUNTS FG
MEAS TECH 17(1): 131-4 (JAN 1974)

1051 A NEW DISTRIBUTED RC-ELEMENT FOR USE IN ACTIVE FILTERS HAVING ZERO Q-SENSITIVITY.
GITS HJ
ARCH ELEKTRON UEBERTRAGUNGSTECH 30(6): 232-7 (JUNE 1976)

1052 AN ACTIVE CIRCUIT WITH NEW FUNCTIONAL DEPENDENCIES OF THE INPUT NEGATIVE IMPEDANCE ON FREQUENCY.
GORDIENKO KP + ANTOCH NV
IZV VUZ RADIOELEKTRON 22(1): 117-18 (JAN 1979) (IN RUSSIAN)

1053 EQUIVALENCE OF ACTIVE-FILTER STRUCTURES.
GRABEL A
P47-52 OF CIRCUIT THEORY AND DESIGN, LAUSANNE, SWITZERLAND, 4-8 SEPT 1978, MOSCHYTZ, GS, NIERYNCK, J(ED), GEORGI 1978

1054 TWIN-TEE FILTER REJECTS MORE THAN 70 DB WITH CAPACITANCE-MULTIPLIER CIRCUIT.
GRAEME J
ELECTRON DES 26(2): 104 (18 JAN 1978)

1055 REALIZATION OF TRANSFER FUNCTIONS USING A SPECIAL DISTRIBUTED RC NETWORK STRUCTURE.
GRUNDNER HR
P232-5 OF PROC OF THE 1976 IEEE INT SYMP ON CIRCUITS AND SYSTEMS, IEEE, MUNICH, GERMANY 27-29 APRIL 1976, IEEE, 1976

1056 SINGLE-SECTION LUMPED MODEL FOR DISTRIBUTED RC NOTCH FILTERS.
GUPTA ML
PROC IEEE 65(12): 1723-5 (DEC 1977)

1057 AN ACTIVE FILTER WITH A DISTRIBUTED RGC LINE.
GUZINSKI A
INT J ELECTRON 40(4): 409-13 (APRIL 1976)

1058 IMPROVING THE SELECTIVITY OF A REBUILT HRO.
HALL T
RADIO COMMUN 53(2): 113-14 (FEB 1977)

1059 SOME PSYCHOACOUSTICAL EXPERIMENTS WITH ALL-PASS NETWORKS.
HARTMANN WM
AM J PHYS 47(1): 29-34 (JAN 1979)

1060 ASSESSMENT OF RC ACTIVE NETWORKS AND SOME NEW RESULTS FOR CIRCUITS APPROXIMATING IMPEDANCE K(P)(N).
HOLT AGJ
P175-89 OF NATO ADVANCED STUDY INSTITUTE ON NETWORK AND SIGNAL THEORY, NATO, IEE, IERE, IEEE, BOURNEMOUTH, HANTS, ENGLAND SEPT 1972, SKWIRZYNSKI, JK, SCANLAN, JO(ED), PETER PEREGRINUS, 1973

1061 NONDOMINANT POLES OF A DISTRIBUTED-ACTIVE LOW-PASS NETWORK.
HUELSMAN LP + TONG RM
PROC IEEE 62(3): 403-4 (MARCH 1974)

1062 MEASUREMENT OF PRESSURE AND FLOW IN THE MICROCIRCULATION - APPLICATION TO PHYSIOLOGICAL AND CLINICAL PROBLEMS.
INTAGLIETTA M
MED PROG TECHNOL 4(1-2): 55-60 (1976)

1063 ACTIVE PIEZOELECTRIC RC FILTERS.
IVANITSKII AM + MIKHAILOV VI
IZV VUZ RADIOELEKTRON 20(12): 21-6 (DEC 1977) (IN RUSSIAN)

1064 A CRYSTAL BANDPASS (FILTER) WITH VARIABLE BANDWIDTH FOR TELECOMMUNICATIONS MEASURING INSTRUMENTS.
JIRU V + SMID V
SLABOPROUDY OBZ 39(4): 173-7 (APRIL 1978) (IN CZECH)

1065 STABILITY OF A DISTRIBUTED-LUMPED-ACTIVE RC NETWORK.
KAPUSTYAN VI
TELECOMMUN RADIO ENG PT 1 28(2): 51-2 (FEB 1974)

1066 DESIGN OF OSCILLATORS AND ACTIVE BANDPASS FILTERS UTILIZING THE DOUBLE-RESISTIVE URC.
KAVANAUGH MF + BOURQUIN JJ
P412-16 OF 7TH ASILOMAR CONF ON CIRCUITS, SYSTEMS AND COMPUTERS NAVAL POSTGRADUATE SCHOOL, UNIV SANTA CLARA, ET AL, PACIFIC GROVE, CALIF, USA, 27-29 NOV 1973, WESTERN PERIODICALS 1974

1067 APPROXIMATE ANALYSIS OF ACTIVE DISTRIBUTED RC NETWORKS.
KLYUKIN VI + VINNIK AYA + SYNOROV VF
IZV VUZ RADIOELEKTRON 17(2): 60-5 (FEB 1974) (IN RUSSIAN)

1068 MAINS REJECTION TRACKING FILTER.
KNOTT KJ + UNSWORTH L
WIRELESS WORLD 80(1466): 375-9 (OCT 1974)

1069 ACTIVE FILTERS IN D.C. SUPPLY SYSTEMS.
KOSICKI M
PRZEGL ELEKTROTECH 50(11): 485-9 (NOV 1974) (IN POLISH)

1070 AN ON-LINE INTEGRATOR FOR ALPHA QUANTIFICATION.
KRAUSMAN DT + LENOX JR
BEHAV RES METHODS INSTRUM 6(3): 317-20 (MAY 1974)

1071 ELECTRONICS FOR THE LONGITUDINAL ACTIVE DAMPING SYSTEM FOR THE CERN PS BOOSTER.
KRIEGBAUM B + PEDERSEN F
IEEE TRANS NUCL SCI NS-24(3): 1695-7 (JUNE 1977)

1072 SYNTHESIS OF ACTIVE FILTERS WITH DISTRIBUTED PARAMETERS.
KVASIL J + PSENICKA B
Z ELEKTR INF- ENERGIETECH 7(5): 493-500 (1977) (IN GERMAN)

1073 SYNTHESIS OF ACTIVE FILTERS WITH DISTRIBUTED PARAMETERS.
KVASIL J + PSENICKA B
NACHRICHTENTECH ELEKTRON 27(10): 425-7 (1977) (IN GERMAN)

1074 CO-CHANNEL INTERFERENCE FILTER.
LATHAM AJ
TELEVISION 28(2): 79 (DEC 1977)

1075 ACTIVE LOW-PASS FILTER WITH A DISTRIBUTED NULL RC CIRCUIT.
LOGINOV BN
RADIO ENG ELECTRON PHYS 17(1): 74-9 (JAN 1972)

1076 SYNTHESIS OF LOW-PASS LINEAR PHASE ACTIVE CIRCUITS USING IDENTICAL UNIFORMLY OR EXPONENTIALLY DISTRIBUTED RC NETWORKS.
MAHDI H
P503-8 OF 6TH ASILOMAR CONF ON CIRCUITS AND SYSTEMS, NAVAL POSTGRADUATE SCHOOL, UNIV SANTA CLARA, PACIFIC GROVE CALIF, USA, 15-17 NOV 1972, PARKER, SR(ED), WESTERN PERIODICALS, 1973

1077 PROGRAMMABLE FILTER HELPS FAULTY HEARING.
MANGOLD S + RISSLER-AKESSON G
ELTEK AKTUELL ELEKTRON A 20(15): 64-6 (OCT 1977) (IN SWEDISH)

1078 AN ELECTRONIC MODEL OF THE BASILAR MEMBRANE.
MANOLI SH
P52 OF 29TH ANNUAL CONF ON ENGINEERING IN MEDICINE AND BIOLOGY (PAPERS IN SUMMARY FORM ONLY RECEIVED), BOSTON MASS, USA, 6-10 NOV 1976, ALLIANCE FOR ENGNG IN MEDICINE AND BIOLOGY, 1976

1079 BANDPASS ACTIVE FILTERS FOR FSK SIGNALS DISCRIMINATION.
MARTI JA + SANTOS S
COMUNICACIONES NO 18: 13-26 (1976) (IN SPANISH)

1080 AN ACTIVE RC DELAY LINE FOR MATCHED FILTER CORRELATORS.
MASKARA SL
IEEE TRANS COMMUN COM25(8): 856-61 (AUG 1977)

1081 INDUCTORLESS FILTER DESIGN USING ACTIVE ELEMENTS AND PIEZOELECTRIC RESONATORS.
MEANS DR + GHAUSI MS
IEEE TRANS CIRCUIT THEORY CT-19(3): 247-53 (MAY 1972)

1082 ACTIVE LOW-PASS FILTERS FOR N-PATH FILTERS.
MEISZNER R
NACHRICHTENTECH ELEKTRON 24(11): 422-4 (1974) (IN GERMAN)

1083 CIRCUIT SYNTHESIS USING MC GRAPHS.
MIKULA J
SLABOPROUDY OBZ 39(9): 398-402 (SEPT 1978) (IN SLOVAK)

1084 ACTIVE FILTERS FOR AC HARMONIC SUPPRESSION.
MOHAN N + PETERSON HA + LONG WF + DREIFUERST GR + VITHAYATHIL JJ
PA77 026-8/1-7 OF IEEE 1977 POWER ENGINEERING WINTER MEET, IEEE, NEW YORK, USA, 30 JAN - 4 FEB 1977, IEEE, 1977

1085 ACTIVE LOWPASS AND BANDPASS FILTERS USING RC DISTRIBUTED LINES.
NEELAKANTAN MN
J INST ELECTRON TELECOMMUN ENG 21(8): 395-8 (AUG 1975)

1086 DESIGN OF ACTIVE FILTERS USING DOUBLE LAYER RC DISTRIBUTED LINES.
NEELAKANTAN MN
INT J CIRCUIT THEORY APPL 3(3): 307-10 (SEPT 1975)

1087 DIRECT SYNTHESIS OF THE GAUSSIAN FILTER FOR NUCLEAR PULSE AMPLIFIERS.
OHKAWA S + YOSHIZAWA M + HUSIMI K
NUCL INSTRUM METHODS 138(1): 85-92 (1 OCT 1976)

1088 ACTIVE VS. PASSIVE DC FILTERS HOW AND WHEN TO USE EACH.
OLSON FJ
TELEPH ENG MANAGE 80(18): 54-5 (15 SEPT 1976)

1089 A PROGRAM-CONTROLLED NOISE FILTER.
ORBAN R
J AUDIO ENG SOC 22(1): 2-9 (JAN-FEB 1974)

1090 SIMPLE SPEECH FILTER.
ORR T
ELECTRON TODAY INT 7(5): 71 (MAY 1978)

1091 AN ACTIVE DISTRIBUTED RC NOTCH FILTER.
PAL K + AHMED S + KUMAR S + SINGH R
MICROELECTRONICS 9(1): 22-4 (SEPT 1978)

1092 A DIGITAL SUB-BAND CODER FOR SPEECH COMMUNICATION AT 9.6 KB/S.
PRESTI AJ + JOHNSTON JD + CROCHIERE RE + GOODMAN DJ
P193-7 OF 1977 INT CONF ON COMMUNICATIONS, IEEE, CHICAGO, ILL, USA, 12-15 JUNE 1977, IEEE, 1977

1093 AUTOMATICALLY TRACKING ANALOGUE FILTERS.
REGTIEN PPL
PB5-6/2PP OF EUROPEAN CONF ON ELECTROTECHNICS EUROCON '74 DIGEST (EXTENDED ABSTRACTS ONLY) CONVENTION OF NAT SOC ELECTRICAL ENGRS WESTERN EUROPE, IEEE, AMSTERDAM, NETHERLANDS, 22-26 APRIL 1974 ROYAL INSTN ENGRS, NETHERLANDS, 1974

1094 HIGH-Q ACTIVE CRYSTAL BAND-PASS FILTER.
REVANKAR GN + BAPAT VB
STUD J INST ELECTRON TELECOMMUN ENG 17(4): 197-8 (OCT 1976)

1095 CONSTRUCTION OF EQUIVALENT CIRCUITS WITH OPERATIONAL AMPLIFIERS.
ROZENKO VP
IZV VUZ RADIOELEKTRON 20(3): 123-5 (MARCH 1977) (IN RUSSIAN)

1096 BAND-STOP, NARROWBAND, LIGHT- AND BIAS VOLTAGE-CONTROLLED RC MICROFILTER WITH DISTRIBUTED PARAMETERS.
RUDNEV VV + NIFONTOV NG
RADIO ENG ELECTRON PHYS 17(7): 1178-80 (JULY 1972)

1097 AN ACTIVE LOW-PASS FILTER WITH APPLICATION OF DISTRIBUTED RC LINE ELEMENTS.
SAKAMOTO T + SUEZAKI T
VOL54NO1

1098 A SIMPLE FORM OF THE DISTRIBUTED RC NOTCH FILTER.
SALAWU RI + OSIBO OO
INT J ELECTR ENG EDUC 15(4): 341-4 (OCT 1978)

1099 TEACHING A RC ACTIVE FILTER COURSE WITH THE AID OF A DIGITAL COMPUTER.
SANCHEZ-SINENCIO E + RAJASEKARAN PK
INT J ELECTR ENG EDUC 14(4): 325-32 (OCT 1977)

1100 A GENERAL ACTIVE ALL-PASS NETWORK USING LUMPED OR LUMPED-DISTRIBUTED ELEMENTS.
SCHAUMANN R
IEEE TRANS CIRCUIT THEORY CT-20(5): 589-91 (SEPT 1973)

1101 REALIZATION OF ARBITRARY VOLTAGE TRANSFER FUNCTIONS WITH ACTIVE LUMPED-DISTRIBUTED NETWORKS.
SCHAUMANN R
ARCH ELEKTRON UEBERTRAGUNGSTECH 27(3): 139-44 (MARCH 1973) (IN ENGLISH)

1102 FOLLOW-UP REJECTOR FILTER FOR THE POWER-LINE FREQUENCY AND ITS HARMONICS.
SHCHEKOTOV AYU + GOLYAVIN AM
INSTRUM EXP TECH 21(4): PT2 1034-7 (JULY-AUG 1978)

1103 AN ACTIVE FILTER FOR USE IN AN A.C. CARRIER CONTROL SYSTEM.
SIMPSON RJ + GLEDHILL CS
RADIO ELECTRON ENG 43(8): 471-4 (AUG 1973)

1104 SIMPLE DESIGN PROCEDURE FOR ACTIVE LOWPASS FILTERS USING EXPONENTIAL RC LINES.
SWAMY MNS + WALSH J
PROC INST ELECTR ENG 120(1): 41-4 (JAN 1973)

1105 AN OPTIMIZATION TECHNIQUE FOR THE DESIGN OF ACTIVE DISTRIBUTED RC FILTERS.
TAM KY + FAHMY MM
P956-7 OF PROC OF THE 1978 IEEE INT SYMP ON CIRCUITS AND SYSTEMS, IEEE, NEW YORK, USA 17-19 MAY 1978, IEEE, 1978

1106 CONSTRUCTION OF ACTIVE RC-FILTERS USING OPERATIONAL AMPLIFIERS.
TEPLYUK IN
TELECOMMUN RADIO ENG PT 1 27(9): 52-5 (SEPT 1973)

1107 LOW-FREQUENCY FILTERS USING ANALOGUE INTEGRATED-CIRCUIT MULTIPLIERS.
THOMAS JG
MED BIOL ENG 10(6): 791-2 (NOV 1972)

1108 EQUI-RIPPLE ACTIVE DISTRIBUTED RC LOW-PASS FILTERS.
TOMLINSON GH + UNSWORTH L
INT J ELECTRON 45(4): 411-16 (OCT 1978)

1109 DESIGN OF AN ACTIVE DISTRIBUTED RC LOWPASS NETWORK.
TOMLINSON GH + UNSWORTH L
ELECTRON LETT 13(19): 575-6 (15 SEPT 1977)

1110 AUDIO FILTERS AS AN AID TO RECEPTION-WITH SPECIAL REFERENCE TO THE DATONG FREQUENCY AGILE AUDIO FILTER MODEL FL1.
TONG DA
RADIO COMMUN 54(2): 114-18 (FEB 1978)

1111 INTEGRATING FILTER FOR THRESHOLD DETECTORS.
TRNKA JT
IBM TECH DISCLOSURE BULL 20(8): 3254 (JAN 1978)

1112 AN ALGORITHM FOR THE CASCADE SYNTHESIS OF A MULTIVARIABLE ACTIVE FILTER.
VANDEWALLE J
P283-6 OF PROC OF THE 1976 IEEE INT SYMP ON CIRCUITS AND SYSTEMS, IEEE, MUNICH, GERMANY 27-29 APRIL 1976, IEEE, 1976

1113 NEW TYPES OF FILTER USED IN TOUCH TONE RECEIVERS.
VOLLMER N
TELETEKNIK (ENGL 18(1): 11-21 (1974)

1114 ACTIVE DISTRIBUTED RC-FILTER DESIGN.
WALSH J + SWAMY MNS
J FRANKLIN INST 295(5): 395-409 (MAY 1973)

1115 DESIGN OF ACTIVE DISTRIBUTED RC-FILTERS USING EXPONENTIAL LINES.
WALSH J + SWAMY MNS
PXIV5/11PP OF PROC OF THE 15TH MIDWEST SYMP ON CIRCUIT THEORY VOL II, UNIV MISSOURI-ROLLA, ROLLA, MO USA, 4-5 MAY 1972, UNIV MISSOURI-ROLLA, 1972

1116 DESIGN OF ACTIVE DISTRIBUTED RC-FILTERS USING EXPONENTIAL LINES.
WALSH J + SWAMY MNS
RADIO ELECTRON ENG 42(9): 409-15 (SEPT 1972)

1117 DESIGN AND APPLICATION OF ACTIVE COMPENSATION CIRCUITS FOR SERVO CONTROL SYSTEMS.
WILSON DR + BURL M
RADIO ELECTRON ENG 43(6): 379-83 (JUNE 1973)

1118 ACTIVE REJECTION FILTERS WITH PIEZOELECTRIC RESONATORS.
ZELYAKH EV + NOVIKOV AA
TELECOMM RADIO ENG PART 1 31(1): 45-50 (JAN 1977)

1119 AN ACTIVE-FILTER DEVICE FOR EXTREME SELECTIVE MEASUREMENTS AT LOW FREQUENCIES.
ZINKERNAGEL J
IEEE TRANS INSTRUM MEAS IM-26(4): 304-9 (DEC 1977)

1120 PIEZOELECTRIC TUNING FORK FILTERS.
ZMUDZKI B
ELEKTRONIKA 19(12): 525-6 (1978) (IN POLISH)

Author Index

Subject Index

V

Z

Editors' Biographies

Rolf Schaumann (M'70-SM'79) was born in Nurnberg, Germany, on July 29, 1941. He received the degree of Diplom Ingenieur from the University of Stuttgart, Germany, in 1967 and the Ph.D degree from the University of Minnesota in 1970, both in electrical engineering.

His industrial experience includes work for Standard Elektrik Lorenz and Telefunken in Germany and for the Standard Telecommunication Laboratories in England. Since 1970 he has been on the staff of the Department of Electrical Engineering of the University of Minnesota where he currently holds an appointment as Associate Professor. His teaching and research interests are in linear systems, electronics, integrated circuits and active filters, an area in which he has published numerous technical papers.

Dr. Schaumann is a member of the IEEE Circuits and Systems Society, Eta Kappa Nu, and the New York Academy of Sciences.

Michael A. Soderstrand (S'70-M'72) was born in Berkeley, CA, on July 31, 1946. He received the B.S.E.E., M.S., and Ph.D. degrees from the University of California, Davis, CA, in 1968, 1969, and 1972, respectively.

He worked at various radio and television stations until 1968, when he joined Sandia Laboratories where he was a Member of the Technical Staff in the Electronics Development Division. Also, since 1972 he has held concurrently an appointment as Adjunct Assistant Professor in the Electrical Engineering Department at the University of California, Davis, CA, and is currently an Associate Professor. He is author of the Engineering Mathematics Review Course (Sandia Laboratories, 1975), which has been taught as part of the N.S.F. sponsored Women in Engineering program administered through the University of California Extension, Davis, CA. He has also authored numerous technical articles on active filters, digital signal processing, and microprocessor applications to signal processing. He was one of the early researchers to study active filters designed without reactive elements, and coined the name "active-R filters." He holds the basic U.S. patent on active-R filters (U.S. Patent 3955150, 1976), and has authored numerous technical papers on the subject. He is Editor of the IEEE Circuits and Systems Magazine.

Dr. Soderstrand is a member of Sigma XI, the National Society of Professional Engineers (NSPE), and the IEEE Circuits and Systems, and Acoustics Speech and Signal Processing Societies. He has served as Secretary and Vice President of the local chapter of the IEEE Circuits and Systems Society, and is completing a two-year term as Vice President of the California Society of Professional Engineers. He also was in charge of the Engineers' Week activities for the San Francisco Bay Area Engineering Council in 1978, and President of the Livermore chapter of NSPE.

Kenneth R. Laker (S'70-M'73-SM'78) was born in Newburg, New York on September 10, 1946. He received the B.E. degree in electrical engineering from Manhattan College, New York, New York, in 1968, and the M.S. and Ph.D. degrees in electrical engineering from New York University, New York, New York, in 1970 and 1973, respectively.

From 1973 to 1977, he served as an Air Force Officer with the Microwave Physics Laboratory, Air Force Cambridge Research Laboratories, Hanscom AFB, MA. This organization later became part of RADC. Since June, 1977, he has been a member of the technical staff of Bell Laboratories, Holmdel, New Jersey. His work has been in the areas of active networks and surface acoustic wave (SAW) devices. Currently he is engaged in the development of switched capacitor filters for various Bell System applications. He has authored and coauthored more than 50 papers on these topics, collectively. In addition, he has been awarded three patents in the area of SAW devices and has two patents pending.

Dr. Laker is a member of Eta Kappa Nu, Sigma Xi, and IEEE. In the past he has served the IEEE Circuits and Systems Society as Cochairman of the Optical, Microwave and Acoustical Circuits Technical Committee and as Publicity Committee Chairman for the 1978 ISCAS. Presently he serves as the Society Secretary/Treasurer and as a member of the Administrative Committee.